Programs for Digital Signal Processing

Automatic Speech & Speaker Recognition, *Edited by N. R. Dixon and T. B. Martin*
Speech Analysis, *Edited by R. W. Schafer and J. D. Markel*
The Engineer in Transition to Management, *I. Gray*
Multidimensional Systems: Theory & Applications, *Edited by N. K. Bose*
Analog Integrated Circuits, *Edited by A. B. Grebene*
Integrated-Circuit Operational Amplifiers, *Edited by R. G. Meyer*
Modern Spectrum Analysis, *Edited by D. G. Childers*
Digital Image Processing for Remote Sensing, *Edited by R. Bernstein*
Reflector Antennas, *Edited by A. W. Love*
Phase-Locked Loops & Their Application, *Edited by W. C. Lindsey and M. K. Simon*
Digital Signal Computers and Processors, *Edited by A. C. Salazar*
Systems Engineering: Methodology and Applications, *Edited by A. P. Sage*
Modern Crystal and Mechanical Filters, *Edited by D. F. Sheahan and R. A. Johnson*
Electrical Noise: Fundamentals and Sources, *Edited by M. S. Gupta*
Computer Methods in Image Analysis, *Edited by J. K. Aggarwal, R. O. Duda, and A. Rosenfeld*
Microprocessors: Fundamentals and Applications, *Edited by W. C. Lin*
Machine Recognition of Patterns, *Edited by A. K. Agrawala*
Turning Points in American Electrical History, *Edited by J. E. Brittain*
Charge-Coupled Devices: Technology and Applications, *Edited by R. Melen and D. Buss*
Spread Spectrum Techniques, *Edited by R. C. Dixon*
Electronic Switching: Central Office Systems of the World, *Edited by A. E. Joel, Jr.*
Electromagnetic Horn Antennas, *Edited by A. W. Love*
Waveform Quantization and Coding, *Edited by N. S. Jayant*
Communication Satellite Systems: An Overview of the Technology, *Edited by R. G. Gould and Y. F. Lum*
Literature Survey of Communication Satellite Systems and Technology, *Edited by J. H. W. Unger*
Solar Cells, *Edited by C. E. Backus*
Computer Networking, *Edited by R. P. Blanc and I. W. Cotton*
Communications Channels: Characterization and Behavior, *Edited by B. Goldberg*
Large-Scale Networks: Theory and Design, *Edited by F. T. Boesch*
Optical Fiber Technology, *Edited by D. Gloge*
Selected Papers in Digital Signal Processing, II, *Edited by the Digital Signal Processing Committee*
A Guide for Better Technical Presentations, *Edited by R. M. Woelfle*
Career Management: A Guide to Combating Obsolescence, *Edited by H. G. Kaufman*
Energy and Man: Technical and Social Aspects of Energy, *Edited by M. G. Morgan*
Magnetic Bubble Technology: Integrated-Circuit Magnetics for Digital Storage and Processing, *Edited by H. Chang*
Frequency Synthesis: Techniques and Applications, *Edited by J. Gorski-Popiel*
Literature in Digital Processing: Author and Permuted Title Index (Revised and Expanded Edition), *Edited by H. D. Helms, J. F. Kaiser, and L. R. Rabiner*
Data Communications via Fading Channels, *Edited by K. Brayer*
Nonlinear Networks: Theory and Analysis, *Edited by A. N. Willson, Jr.*
Computer Communications, *Edited by P. E. Green, Jr. and R. W. Lucky*
Stability of Large Electric Power Systems, *Edited by R. T. Byerly and E. W. Kimbark*
Automatic Test Equipment: Hardware, Software, and Management, *Edited by F. Liguori*
Key Papers in the Development of Coding Theory, *Edited by E. R. Berkekamp*
Technology and Social Institutions, *Edited by K. Chen*
Key Papers in the Development of Information Theory, *Edited by D. Slepian*
Computer-Aided Filter Design, *Edited by G. Szentirmai*
Laser Devices and Applications, *Edited by I. P. Kaminow and A. E. Siegman*
Integrated Optics, *Edited by D. Marcuse*
Laser Theory, *Edited by F. S. Barnes*
Digital Signal Processing, *Edited by L. R. Rabiner and C. M. Rader*
Minicomputers: Hardware, Software, and Applications, *Edited by J. D. Schoeffler and R. H. Temple*
Semiconductor Memories, *Edited by D. A. Hodges*
Power Semiconductor Applications, Volume II: Equipment and Systems, *Edited by J. D. Harnden, Jr. and F. B. Golden*
Power Semiconductor Applications, Volume I: General Considerations, *Edited by J. D. Harnden, Jr. and F. B. Golden*
A Practical Guide to Minicomputer Applications, *Edited by F. F. Coury*
Active Inductorless Filters, *Edited by S. K. Mitra*
Clearing the Air: The Impact of the Clean Air Act on Technology, *Edited by J. C. Redmond, J. C. Cook, and A. A. J. Hoffman*

Programs for Digital Signal Processing

Edited by the

Digital Signal Processing Committee
IEEE Acoustics, Speech, and Signal Processing Society

Committee members:

Clifford J. Weinstein (Chairman), M.I.T. Lincoln Laboratory
James W. Cooley, IBM T. J. Watson Research Center
Ronald E. Crochiere, Bell Laboratories
Marie T. Dolan, Bell Laboratories
Joseph R. Fisher, Signal Processing Systems, Inc.
Howard D. Helms, American Telephone and Telegraph Company
Leland B. Jackson, University of Rhode Island
James F. Kaiser, Bell Laboratories
James H. McClellan, Massachusetts Institute of Technology
Carol A. McGonegal, Bell Laboratories
Russell M. Mersereau, Georgia Institute of Technology
Alan V. Oppenheim, Massachusetts Institute of Technology
Lawrence R. Rabiner, Bell Laboratories
Charles M. Rader, M.I.T. Lincoln Laboratory
Ronald W. Schafer, Georgia Institute of Technology
Harvey F. Silverman, IBM T. J. Watson Research Center
Kenneth Steiglitz, Princeton University
Jose M. Tribolet, Instituto Superior Tecnico
John W. Woods, Rensselaer Polytechnic Institute

A volume published under the sponsorship of the
IEEE Acoustics, Speech, and Signal Processing Society.

The Institute of Electrical and Electronics Engineers, Inc. New York

PRINTED IN THE UNITED STATES OF AMERICA

IEEE International Standard Book Numbers: Clothbound: 0-87942-127-4
Paperbound: 0-87942-128-2

Library of Congress Catalog Card Number 79-89028

Sole Worldwide Distributor (Exclusive of the IEEE):

JOHN WILEY & SONS, INC.
605 Third Ave.
New York, NY 10016

Wiley Order Numbers: Clothbound: 0-471-05962-5
Paperbound: 0-471-05961-7

Contents

Preface

During the past fifteen years, digital signal processing has been an extremely active and dynamic field. Advances in integrated circuit technology and in processor architecture have greatly enlarged the scope of the technical areas to which digital signal processing techniques can be applied. Research in fundamental signal processing techniques and algorithms has led to dramatic improvements in the efficiency of signal processing systems.

An important facet of the progress in digital signal processing has been the development of algorithms and their embodiment in computer programs, both for the execution of processing operations on signals and for the design of signal processing filters and systems. The purpose of this book is to make widely available, in a directly usable form, a comprehensive set of computer programs useful for digital signal processing. In addition, the book serves as an outlet for excellent programming effort and enables authors of programs to receive credit for their work.

The programs have been carefully selected to cover a broad spectrum of digital signal processing applications and design techniques. The programs are categorized into eight chapters, and separate summaries (authored by chapter editors who were also asked to make final checks on the documentation and code) are provided with each chapter. The first chapter focuses on the Discrete Fourier Transform (DFT) and presents a variety of Fast Fourier Transform (FFT) and related algorithms. Chapter 2 includes algorithms for the periodogram and correlation methods of power spectrum estimation and for coherence and cross spectrum estimation. A program for high speed, FFT-based convolution is presented in Chapter 3. Chapter 4 presents several algorithms related to the linear prediction techniques of signal processing, including the autocorrelation, covariance, and lattice methods. The design and synthesis of Finite Impulse Response (FIR) digital filters are the subjects of Chapter 5. Algorithms for optimal (minimax), windowed, and maximally-flat filter designs, and a design program which incorporates finite-word-length effects, are included. Chapter 6 presents a comprehensive set of programs dealing with the design and synthesis of Infinite Impulse Response (IIR) digital filters. The first program in that chapter includes most of the classical filter design techniques as well as consideration of finite word length issues such as pairing and ordering in a single but modular package. The other programs in that chapter are specialized either to the finite word length design problem or to filter design based on iterative optimization. Chapter 7 deals with cepstral and homomorphic algorithms, with specific attention to the difficult problem of phase calculation in a homomorphic system. Finally, Chapter 8 presents several programs for interpolation and decimation, the fundamental operations necessary for changing sampling rates in a digital system. Included are multistage implementations and sampling rate conversion by rational ratios.

The book is the culmination of a project undertaken in early 1976 by the Digital Signal Processing Committee of the IEEE Acoustics, Speech, and Signal Processing Society. In contrast to previous books in the IEEE Press reprint series, this book consists largely of material which has not been published before. In addition, a unique and complex set of issues regarding the publication of computer programs had to be dealt with. These issues included: (1) clarity and uniformity of documentation and program presentation; (2) independent review and verification of all programs; (3) portability and machine-independence of code; and (4) availability of computer-readable source programs. The project has therefore been a very ambitious and time-consuming one for the committee, for the program contributors, and for the reviewers.

In order to assess the feasibility of the project, the committee distributed to the digital signal processing community in early 1976 a description of the proposed project and a request for descriptions of candidate programs. The distribution of this request was carried out both through direct mailing and through handouts at the 1976 International Conference on Acoustics, Speech, and Signal Processing. In a short time, enough very promising candidate programs had been identified to convince the committee that the book was not only feasible, but potentially a landmark contribution to the digital signal processing field. In addition, a reasonable organization of the material into the eight sections now included in the book began to emerge.

Because of the large effort to be required of program contributors and of reviewers, it was deemed desirable to pre-screen program candidates based on the offered descriptions, and invite submission of only those programs judged most likely to fit our rather stringent requirements. In cases where two programs were offered to accomplish the same functions, a choice was generally made based on which of the programs was closer to publishable form in terms of documentation, portability, verification through wide previous use, or other similar criteria.

After this pre-screening, invitations were sent to prospective authors to submit full code and documentation for their programs. It was required of the authors that all programs meet the following conditions:

(1) complete user documentation must be included;

(2) the program must be in FORTRAN and conform to ANSI FORTRAN standards;

(3) the code should be fully commented and debugged, and as clear and understandable as possible;

(4) the code should be transportable and machine-independent;

(5) comprehensive test examples with corresponding input and output should be supplied.

Of course, in addition to printouts, a computer-readable source of all code had to be provided by the authors. This led to the rather onerous task of collecting the submitted tapes and cards, reading them into a central filing system, and distributing them to reviewers in one of several possible formats.

The committee viewed it to be an essential part of the project that computer-readable source material for the programs be made available to readers of the book. The collection of all programs into a central file was clearly necessary to achieve this. In addition, with the cooperation of Reed Crone of IEEE Press and Jack Fraum of the IEEE Service Center, we were able to arrange for IEEE to offer the service of reproducing and distributing, upon request and at a nominal cost, a tape of all programs in the book. For information on how to order such a tape, contact the IEEE Press, 345 East 47th St., New York, NY 10017. It is worth emphasizing that this is a new service which IEEE has not offered in the past, and which can perhaps be utilized again in other projects.

As authors submitted programs, the review process was able to proceed. Each reviewer was asked to act from points of view (1) of a highly critical program user, and (2) of an editorial reviewer of journal-quality technical material. The reviewer was asked to read the program description, compile the program, and attempt to run the test examples described as well as other test examples considered appropriate. The review process was an open one, with reviewers frequently consulting directly with authors on points of difficulty. In most cases it was found that the material submitted by authors was excellent and well-debugged, and the review process went quite smoothly. However, in some cases serious difficulties, particularly in the area of program portability, were uncovered and corrected in the review process. At the end of the book, we have identified the reviewer(s) as well as the computer(s) and operating system(s) on which each program was tested. In addition to giving credit for the efforts of the reviewers, this serves to identify for readers at least one successful user of each program besides the author(s).

Program portability was the objective of major importance in this project. One successful review was not considered sufficient to insure that this goal was met; therefore an extensive parallel effort in program standardization was undertaken. The standardization procedures are described in detail in the standards section.

Much care has been taken to insure the accuracy, clarity, and uniformity of the printed material in the book. All listings of program code have been printed directly from the centrally-collected computer-readable source material, with particular attention to the selection of a readable type font. This was considered necessary to avoid either direct photoreproduction of the assorted type fonts produced by authors or the inevitable errors which would be caused by attempting to transcribe and typeset the listings. The authors' manuscripts documenting their programs were typed into a computer system, after which final editing and correcting were carried out. All the printed material in the book was then produced by computer-driven phototypesetters.

Although this has been a project of the full Digital Signal Processing Committee as listed on the title page, a number of outstanding individual contributions deserve special mention here. The idea for this project was first presented to the committee by Ronald Crochiere in early 1976, and the project was initiated under the committee chairmanship of Alan Oppenheim. Crochiere was responsible for the initial solicitation and collection of candidate programs from the digital signal processing community. Clifford Weinstein took over as committee chairman in September 1976 and since that time served as general coordinator for the program book project.

The task of collecting submitted tapes and cards from the authors, reading them into a central file, and distributing them for review in specified formats, was initially carried out by Howard Helms and later taken over by Marie Dolan. Helms was also instrumental in arranging for IEEE reproduction of the program tape for users. To insure the standardization and portability of the programs, a standardization subcommittee including James Kaiser, Dolan, Carol McGonegal, Lawrence Rabiner, and Jose Tribolet was made responsible for detailed specification and execution of the standards imposed on the programs in this book. The typing of author manuscripts into a computer file was carried out by Penny Blaine and Carmela Patuto. Art editing and page layout were skillfully handled by Madeline Wilson. Kaiser coordinated the handling of manuscripts and arranged for the phototypesetting of the book. Much of this phototypesetting was done by Yourdon, Inc., New York City. The subcommittee of Kaiser, Dolan, McGonegal, and Rabiner took charge of tracking and carrying out the many details required to bring the book into final form.

Special thanks are due to Marie Dolan and Carol McGonegal, who joined the committee for the express purpose of participating in the program book project and who put forth enormous and very productive effort in seeing it through to completion. We would like to acknowledge the excellent cooperation and assistance provided to this project by IEEE Press Managing editor Reed Crone, and the specific efforts of Jack Fraum in arranging for IEEE reproduction of the program tape. Finally, the Committee must thank the program authors, reviewers, and chapter editors for the large amounts of time and effort they contributed to this project.

Standards

In putting together this collection of programs on digital signal processing, it was immediately recognized that users would be trying to run the different programs on a wide variety of computing machines and under several types of operating systems. For this reason FORTRAN was chosen as the principal language and a great deal of attention was given to the problem of program portability. The goal was for each program to compile and execute, according to its defined performance criteria, on a wide range of computers. Adopted was the PORT Mathematical Subroutine Library approach toward portability of Fox, Hall, and Schryer [1]. That is, all programs "had to be portable in the IFIP sense*, with the exception that the three machine-constant defining functions had to be particularized to the host computer once, at installation". The techniques used in the PORT library and in the programs in this book to make them easily portable were: (1) the programs are written in a subset of ANSI FORTRAN, and (2) the target environment is specified in terms of machine-dependent parameters. In addition to portability, program readability was given consideration in this collection of programs. The goal was to enhance the readability of the program code by standardizing the format of the programs. Following is a discussion of the techniques used to make the programs portable and enhance their readability.

Language

The programs were restricted to the particular portable subset of ANSI FORTRAN known as PFORT (Portable FORTRAN) [3]. The PFORT verifier program [3] checks that individual program units conform to the American National Standard FORTRAN [4]; it also checked that the inter-program-unit communication through the use of COMMON and argument lists is consistent with the standard. During the review each program was passed through the PFORT verifier to guarantee its adherence to this language requirement.

Machine-Dependent Quantities

All machine-dependent quantities (I/O device codes, machine constants, etc.) are specified by the I1MACH, R1MACH and D1MACH FORTRAN function subprograms used in the PORT [1,5] library. Table 1 is a list of all the machine-dependent quantities which are specified by these routines. They are delivered by means of a single integer argument indicating the particular quantity desired. Complete listings of I1MACH, R1MACH and D1MACH are given in the Appendix. Any program in this book which is not truly portable, that is, which uses machine-dependent quantities not given in Table 1, is however still transportable in the IFIP sense. Additional machine-dependent quantities have been kept to a minimum; instructions for changing their values are documented in the program code.

To move the programs to a new environment, only the DATA statements in the I1MACH, R1MACH and D1MACH subprograms need to be changed. Values are provided in these routines for a number of different computing systems. They include the Burroughs 1700 system, the Burroughs 5700/6700/7700 systems, the CDC 6000/7000 series, the CRAY 1, the Data General Eclipse S/200, the Harris Slash 6 and Slash 7, the Honeywell 6000 series, the IBM 360/370 series, the Xerox SIGMA 5/7/9, the SEL Systems 85/86, the DEC PDP 10 (KA and KI processors), the DEC PDP 11, the UNIVAC 1100 series, and the VAX-11.

To aid in adding values for a target environment not provided in these subprograms, a description of the integer and floating-point variables follows. This description appears in the PORT Mathematical Subroutine Library report [1].

---

* The IFIP Working Group (on Numerical Software) (WG2.5) has proposed, in a working draft, the following definitions [2]
Portable. A program is portable over a given range of machines and compilers if, *without any alteration*, it can compile and run to satisfy specified performance criteria on that range.
Transportable. In transferring a program between members of a given range of machines and compilers, some changes may be necessary to the base version before it satisfies specified performance criteria on each of the machines and compilers. The program is transportable if (1) the changes lend themselves to mechanical implementation by a processor, and (2) the changes, ideally, are limited in number, extent, and complexity.

Integer variables: Let the values for integer variables be written in the s-digit, base-a form:

$$\pm(x_{s-1}a^{s-1} + x_{s-2}a^{s-2} + \cdots + x_1 a + x_0)$$

where $0 \leqslant x_i < a$ for $i = 0,...s-1$. Then specify the base a, the maximum number of digits s, and the largest integer a^s-1. Although the quantity a^s-1 can easily be computed from s and the base a, it is provided because a naive evaluation of the formula would cause overflow on most machines and hence result in an incorrect value. (Storage of integers as magnitude and sign or in a complement notation is not specified since PORT subprograms must be independent of the storage mode.)

Floating-point variables: If floating-point numbers are written in the t-digit, base-b form:

$$\pm b^e \left(\frac{x_1}{b} + \frac{x_2}{b^2} + \cdots + \frac{x_t}{b^t} \right)$$

where $0 \leqslant x_i < b$ for $i=1,...,t$, $0 < x_1$, and $e_{\min} \leqslant e \leqslant e_{\max}$, then for a particular machine, choose values for the parameters t, $e_{\min}$, and $e_{\max}$ such that all numbers expressible in this form are representable by the hardware and usable from FORTRAN. Note that the formula is symmetrical under negation but not reciprocation. On some machines a small portion of the range of permissible numbers may be excluded. Also, for 2's complement machines, care must be taken in assigning the values.

For each machine one must specify for real (single-precision) floating-point numbers, the base, b, the number t of base-b digits, the minimum exponent $e_{\min}$ and the maximum exponent $e_{\max}$. For double-precision numbers, b remains the same, but t, $e_{\min}$, and $e_{\max}$ are replaced by T, $E_{\min}$, and $E_{\max}$.

The 16 parameters discussed above are all integers and are obtained by invoking the function I1MACH with the appropriate argument. The floating-point single-precision and double-precision quantities provided by the functions R1MACH and D1MACH can be derived from the given integer quantities, but are provided for efficiency and convenience.

The single-precision floating-point quantities provided in R1MACH are the smallest positive magnitude, $b^{e_{\min}-1}$, the largest magnitude, $b^{e_{\max}}(1-b^{-t})$, the smallest relative spacing between values, b^{-t}, the largest relative spacing between values, $b^{(1-t)}$, and the logarithm of the base b, $\log_{10} b$. The relative spacing is $|(y-x)/x|$, when x and y are successive floating-point numbers. Equivalent values for the double-precision floating-point quantities are provided by D1MACH, with $e_{\min}$, $e_{\max}$, and t replaced by $E_{\min}$, $E_{\max}$, and T.

Portable Random Number Generator

In addition to the machine-dependent quantities, the PORT library was the source of the portable random number generator, UNI. UNI, written by Alan Gross, is a uniform generator which will produce the same sequence of values, to the accuracy of the computer, on any computer with 16 or more bits. UNI is implemented as a FORTRAN function subprogram and returns a single real random variate from the uniform [0,1) distribution. A complete listing of UNI is given in the Appendix. There are three local variables (FIRST, CSEED in UNI and TSEED in the routine R1UNIF) which must retain their most recently assigned values. In order to adhere to this requirement on a Data General Computer, the STATIC mode must be set when UNI is compiled.

Program Readability

In order to enhance the readability of the programs, two steps were taken. First, the authors were asked to precede each program unit with a specific comment structure. This structure was designed to allow the reader to easily identify a program unit. Second, the programs (except for 1.7, 1.8, and 5.1) were passed through the POLISH program [6]. POLISH was written at the University of Colorado at Boulder and is intended to improve and standardize the format of FORTRAN programs. It does this by renumbering all statement labels, by inserting CONTINUE statements to force every DO to end on a unique CONTINUE statement, and by systematically spacing and indenting each statement in the program.

References

1. P. A. Fox, A. D. Hall, and N. L. Schryer, "The PORT Mathematical Subroutine Library", *ACM Transactions on Mathematical Software,* Vol. 4, No. 2, pp. 104-126, June 1978.

2. B. Ford and B. T. Smith, "Transportable mathematical software: A substitute for portable mathematical software", Position paper, IFIP Working Group 2.5 (on Numerical Software), 1975.

3. B. G. Ryder, "The PFORT Verifier", *Software-Practice and Experience,* Vol. 4, No. 4, pp. 359-377, Oct.-Dec. 1974.

4. American National Standard X3.9-1966 (ISO 1539-1972), FORTRAN, American National Standards Institute (ANSI), New York, 1966.

5. P. A. Fox, A. D. Hall, and N. L. Schryer, "ALGORITHM 528, Framework for a Portable Library [Z]", *ACM Transactions on Mathematical Software,* Vol. 4, No. 2, pp. 177-188, June 1978.

6. J. Dorrenbacher, D. Paddock, D. Wisneski, and L. D. Fosdick, "POLISH, A FORTRAN Program to Edit FORTRAN Programs", Department of Computer Science, Technical Report No. CU-CS-050-74, University of Colorado, Boulder, Colorado, July 1974.

Table 1

machine-dependent quantity	function (argument)
standard input unit	I1MACH (1)
standard output unit	I1MACH (2)
standard punch unit	I1MACH (3)
standard error message unit	I1MACH (4)
number of bits per integer storage unit	I1MACH (5)
number of characters per integer storage unit	I1MACH (6)
base of an integer, a	I1MACH (7)
number of base a digits, s	I1MACH (8)
largest integer, a^s-1	I1MACH (9)
base of a real number, b	I1MACH (10)
number of single-precision real base b digits, t	I1MACH (11)
smallest single-precision real exponent, $e_{\min}$	I1MACH (12)
largest single-precision real exponent, $e_{\max}$	I1MACH (13)
number of double-precision base b digits, T	I1MACH (14)
smallest double-precision exponent, $E_{\min}$	I1MACH (15)
largest double-precision exponent, $E_{\max}$	I1MACH (16)
smallest positive real magnitude, $b^{e_{\min}-1}$	R1MACH (1)
largest positive magnitude, $b^{e_{\max}}(1-b^{-t})$	R1MACH (2)
smallest relative spacing between values, b^{-t}	R1MACH (3)
largest relative spacing between values, $b^{(1-t)}$	R1MACH (4)
$\log_{10}b$	R1MACH (5)
smallest positive double-precision magnitude, $b^{E_{\min}-1}$	D1MACH (1)
largest positive double-precision magnitude, $b^{E_{\max}}(1-b^{-T})$	D1MACH (2)
smallest relative spacing between double-precision values, b^{-T}	D1MACH (3)
largest relative spacing between double-precision values, $b^{(1-T)}$	D1MACH (4)
double-precision, $\log_{10}b$	D1MACH (5)

Appendix

```
C
C-----------------------------------------------------------------------
C  FUNCTION:  I1MACH
C  THIS ROUTINE IS FROM THE PORT MATHEMATICAL SUBROUTINE LIBRARY
C  IT IS DESCRIBED IN THE BELL LABORATORIES COMPUTING SCIENCE
C  TECHNICAL REPORT #47 BY P.A. FOX, A.D. HALL AND N.L. SCHRYER
C-----------------------------------------------------------------------
C
      INTEGER FUNCTION I1MACH(I)
C
C  I/O UNIT NUMBERS.
C
C    I1MACH( 1) = THE STANDARD INPUT UNIT.
C
C    I1MACH( 2) = THE STANDARD OUTPUT UNIT.
C
C    I1MACH( 3) = THE STANDARD PUNCH UNIT.
C
C    I1MACH( 4) = THE STANDARD ERROR MESSAGE UNIT.
C
C  WORDS.
C
C    I1MACH( 5) = THE NUMBER OF BITS PER INTEGER STORAGE UNIT.
C
C    I1MACH( 6) = THE NUMBER OF CHARACTERS PER INTEGER STORAGE UNIT.
C
C  INTEGERS.
C
C    ASSUME INTEGERS ARE REPRESENTED IN THE S-DIGIT, BASE-A FORM
C
C               SIGN ( X(S-1)*A**(S-1) + ... + X(1)*A + X(0) )
C
C               WHERE 0 .LE. X(I) .LT. A FOR I=0,...,S-1.
C
C    I1MACH( 7) = A, THE BASE.
C
C    I1MACH( 8) = S, THE NUMBER OF BASE-A DIGITS.
C
C    I1MACH( 9) = A**S - 1, THE LARGEST MAGNITUDE.
C
C  FLOATING-POINT NUMBERS.
C
C    ASSUME FLOATING-POINT NUMBERS ARE REPRESENTED IN THE T-DIGIT,
C    BASE-B FORM
C
C               SIGN (B**E)*( (X(1)/B) + ... + (X(T)/B**T) )
C
C               WHERE 0 .LE. X(I) .LT. B FOR I=1,...,T,
C               0 .LT. X(1), AND EMIN .LE. E .LE. EMAX.
C
C    I1MACH(10) = B, THE BASE.
C
C  SINGLE-PRECISION
C
C    I1MACH(11) = T, THE NUMBER OF BASE-B DIGITS.
C
C    I1MACH(12) = EMIN, THE SMALLEST EXPONENT E.
C
C    I1MACH(13) = EMAX, THE LARGEST EXPONENT E.
C
C  DOUBLE-PRECISION
C
C    I1MACH(14) = T, THE NUMBER OF BASE-B DIGITS.
C
C    I1MACH(15) = EMIN, THE SMALLEST EXPONENT E.
C
C    I1MACH(16) = EMAX, THE LARGEST EXPONENT E.
C
C  TO ALTER THIS FUNCTION FOR A PARTICULAR ENVIRONMENT,
C  THE DESIRED SET OF DATA STATEMENTS SHOULD BE ACTIVATED BY
C  REMOVING THE C FROM COLUMN 1.  ALSO, THE VALUES OF
C  I1MACH(1) - I1MACH(4) SHOULD BE CHECKED FOR CONSISTENCY
C  WITH THE LOCAL OPERATING SYSTEM.
C
      INTEGER IMACH(16),OUTPUT
C
      EQUIVALENCE (IMACH(4),OUTPUT)
C
C     MACHINE CONSTANTS FOR THE BURROUGHS 1700 SYSTEM.
C
C     DATA IMACH( 1) /    7 /
C     DATA IMACH( 2) /    2 /
C     DATA IMACH( 3) /    2 /
C     DATA IMACH( 4) /    2 /
C     DATA IMACH( 5) /   36 /
C     DATA IMACH( 6) /    4 /
C     DATA IMACH( 7) /    2 /
C     DATA IMACH( 8) /   33 /
C     DATA IMACH( 9) / Z1FFFFFFFF /
C     DATA IMACH(10) /    2 /
C     DATA IMACH(11) /   24 /
C     DATA IMACH(12) / -256 /
C     DATA IMACH(13) /  255 /
C     DATA IMACH(14) /   60 /
C     DATA IMACH(15) / -256 /
C     DATA IMACH(16) /  255 /
C
C     MACHINE CONSTANTS FOR THE BURROUGHS 5700 SYSTEM.
C
C     DATA IMACH( 1) /   5 /
C     DATA IMACH( 2) /   6 /
C     DATA IMACH( 3) /   7 /
C     DATA IMACH( 4) /   6 /
C     DATA IMACH( 5) /  48 /
C     DATA IMACH( 6) /   6 /
C     DATA IMACH( 7) /   2 /
C     DATA IMACH( 8) /  39 /
C     DATA IMACH( 9) / O0007777777777777 /
C     DATA IMACH(10) /   8 /
C     DATA IMACH(11) /  13 /
C     DATA IMACH(12) / -50 /
C     DATA IMACH(13) /  76 /
C     DATA IMACH(14) /  26 /
C     DATA IMACH(15) / -50 /
C     DATA IMACH(16) /  76 /
C
C     MACHINE CONSTANTS FOR THE BURROUGHS 6700/7700 SYSTEMS.
C
C     DATA IMACH( 1) /   5 /
C     DATA IMACH( 2) /   6 /
C     DATA IMACH( 3) /   7 /
C     DATA IMACH( 4) /   6 /
C     DATA IMACH( 5) /  48 /
C     DATA IMACH( 6) /   6 /
C     DATA IMACH( 7) /   2 /
C     DATA IMACH( 8) /  39 /
C     DATA IMACH( 9) / O0007777777777777 /
C     DATA IMACH(10) /   8 /
C     DATA IMACH(11) /  13 /
C     DATA IMACH(12) / -50 /
```

```
C     DATA IMACH(13) /  76 /
C     DATA IMACH(14) /  26 /
C     DATA IMACH(15) / -32754 /
C     DATA IMACH(16) /  32780 /
C
C     MACHINE CONSTANTS FOR THE CDC 6000/7000 SERIES.
C
C     DATA IMACH( 1) /    5 /
C     DATA IMACH( 2) /    6 /
C     DATA IMACH( 3) /    7 /
C     DATA IMACH( 4) /    6 /
C     DATA IMACH( 5) /   60 /
C     DATA IMACH( 6) /   10 /
C     DATA IMACH( 7) /    2 /
C     DATA IMACH( 8) /   48 /
C     DATA IMACH( 9) / 00007777777777777777B /
C     DATA IMACH(10) /    2 /
C     DATA IMACH(11) /   48 /
C     DATA IMACH(12) / -974 /
C     DATA IMACH(13) / 1070 /
C     DATA IMACH(14) /   96 /
C     DATA IMACH(15) / -927 /
C     DATA IMACH(16) / 1070 /
C
C     MACHINE CONSTANTS FOR THE CRAY 1
C
C     DATA IMACH( 1) /   100 /
C     DATA IMACH( 2) /   101 /
C     DATA IMACH( 3) /   102 /
C     DATA IMACH( 4) /   101 /
C     DATA IMACH( 5) /    64 /
C     DATA IMACH( 6) /     8 /
C     DATA IMACH( 7) /     2 /
C     DATA IMACH( 8) /    63 /
C     DATA IMACH( 9) /  777777777777777777777B /
C     DATA IMACH(10) /     2 /
C     DATA IMACH(11) /    47 /
C     DATA IMACH(12) / -8192 /
C     DATA IMACH(13) /  8190 /
C     DATA IMACH(14) /    95 /
C     DATA IMACH(15) / -8192 /
C     DATA IMACH(16) /  8190 /
C
C     MACHINE CONSTANTS FOR THE DATA GENERAL ECLIPSE S/200
C
C     DATA IMACH( 1) /   11 /
C     DATA IMACH( 2) /   12 /
C     DATA IMACH( 3) /    8 /
C     DATA IMACH( 4) /   10 /
C     DATA IMACH( 5) /   16 /
C     DATA IMACH( 6) /    2 /
C     DATA IMACH( 7) /    2 /
C     DATA IMACH( 8) /   15 /
C     DATA IMACH( 9) /32767 /
C     DATA IMACH(10) /   16 /
C     DATA IMACH(11) /    6 /
C     DATA IMACH(12) /  -64 /
C     DATA IMACH(13) /   63 /
C     DATA IMACH(14) /   14 /
C     DATA IMACH(15) /  -64 /
C     DATA IMACH(16) /   63 /
C
C     MACHINE CONSTANTS FOR THE HARRIS SLASH 6 AND SLASH 7
C
C     DATA IMACH( 1) /       5 /
C     DATA IMACH( 2) /       6 /
C     DATA IMACH( 3) /       0 /
C     DATA IMACH( 4) /       6 /
C     DATA IMACH( 5) /      24 /
C     DATA IMACH( 6) /       3 /
C     DATA IMACH( 7) /       2 /
C     DATA IMACH( 8) /      23 /
C     DATA IMACH( 9) / 8388607 /
C     DATA IMACH(10) /       2 /
C     DATA IMACH(11) /      23 /
C     DATA IMACH(12) /    -127 /
C     DATA IMACH(13) /     127 /
C     DATA IMACH(14) /      38 /
C     DATA IMACH(15) /    -127 /
C     DATA IMACH(16) /     127 /
C
C     MACHINE CONSTANTS FOR THE HONEYWELL 600/6000 SERIES.
C
C     DATA IMACH( 1) /    5 /
C     DATA IMACH( 2) /    6 /
C     DATA IMACH( 3) /   43 /
C     DATA IMACH( 4) /    6 /
C     DATA IMACH( 5) /   36 /
C     DATA IMACH( 6) /    6 /
C     DATA IMACH( 7) /    2 /
C     DATA IMACH( 8) /   35 /
C     DATA IMACH( 9) / O377777777777 /
C     DATA IMACH(10) /    2 /
C     DATA IMACH(11) /   27 /
C     DATA IMACH(12) / -127 /
C     DATA IMACH(13) /  127 /
C     DATA IMACH(14) /   63 /
C     DATA IMACH(15) / -127 /
C     DATA IMACH(16) /  127 /
C
C     MACHINE CONSTANTS FOR THE IBM 360/370 SERIES,
C     THE XEROX SIGMA 5/7/9 AND THE SEL SYSTEMS 85/86.
C
C     DATA IMACH( 1) /   5 /
C     DATA IMACH( 2) /   6 /
C     DATA IMACH( 3) /   7 /
C     DATA IMACH( 4) /   6 /
C     DATA IMACH( 5) /  32 /
C     DATA IMACH( 6) /   4 /
C     DATA IMACH( 7) /   2 /
C     DATA IMACH( 8) /  31 /
C     DATA IMACH( 9) / Z7FFFFFFF /
C     DATA IMACH(10) /  16 /
C     DATA IMACH(11) /   6 /
C     DATA IMACH(12) / -64 /
C     DATA IMACH(13) /  63 /
C     DATA IMACH(14) /  14 /
C     DATA IMACH(15) / -64 /
C     DATA IMACH(16) /  63 /
C
C     MACHINE CONSTANTS FOR THE PDP-10 (KA PROCESSOR).
C
C     DATA IMACH( 1) /    5 /
C     DATA IMACH( 2) /    6 /
C     DATA IMACH( 3) /    5 /
C     DATA IMACH( 4) /    6 /
C     DATA IMACH( 5) /   36 /
C     DATA IMACH( 6) /    5 /
C     DATA IMACH( 7) /    2 /
C     DATA IMACH( 8) /   35 /
```

```
C     DATA IMACH( 9) / "377777777777 /
C     DATA IMACH(10) /    2 /
C     DATA IMACH(11) /   27 /
C     DATA IMACH(12) / -128 /
C     DATA IMACH(13) /  127 /
C     DATA IMACH(14) /   54 /
C     DATA IMACH(15) / -101 /
C     DATA IMACH(16) /  127 /
C
C     MACHINE CONSTANTS FOR THE PDP-10 (KI PROCESSOR).
C
C     DATA IMACH( 1) /    5 /
C     DATA IMACH( 2) /    6 /
C     DATA IMACH( 3) /    5 /
C     DATA IMACH( 4) /    6 /
C     DATA IMACH( 5) /   36 /
C     DATA IMACH( 6) /    5 /
C     DATA IMACH( 7) /    2 /
C     DATA IMACH( 8) /   35 /
C     DATA IMACH( 9) / "377777777777 /
C     DATA IMACH(10) /    2 /
C     DATA IMACH(11) /   27 /
C     DATA IMACH(12) / -128 /
C     DATA IMACH(13) /  127 /
C     DATA IMACH(14) /   62 /
C     DATA IMACH(15) / -128 /
C     DATA IMACH(16) /  127 /
C
C     MACHINE CONSTANTS FOR PDP-11 FORTRAN SUPPORTING
C     32-BIT INTEGER ARITHMETIC.
C
C     DATA IMACH( 1) /    5 /
C     DATA IMACH( 2) /    6 /
C     DATA IMACH( 3) /    5 /
C     DATA IMACH( 4) /    6 /
C     DATA IMACH( 5) /   32 /
C     DATA IMACH( 6) /    4 /
C     DATA IMACH( 7) /    2 /
C     DATA IMACH( 8) /   31 /
C     DATA IMACH( 9) / 2147483647 /
C     DATA IMACH(10) /    2 /
C     DATA IMACH(11) /   24 /
C     DATA IMACH(12) / -127 /
C     DATA IMACH(13) /  127 /
C     DATA IMACH(14) /   56 /
C     DATA IMACH(15) / -127 /
C     DATA IMACH(16) /  127 /
C
C     MACHINE CONSTANTS FOR PDP-11 FORTRAN SUPPORTING
C     16-BIT INTEGER ARITHMETIC.
C
C     DATA IMACH( 1) /    5 /
C     DATA IMACH( 2) /    6 /
C     DATA IMACH( 3) /    5 /
C     DATA IMACH( 4) /    6 /
C     DATA IMACH( 5) /   16 /
C     DATA IMACH( 6) /    2 /
C     DATA IMACH( 7) /    2 /
C     DATA IMACH( 8) /   15 /
C     DATA IMACH( 9) / 32767 /
C     DATA IMACH(10) /    2 /
C     DATA IMACH(11) /   24 /
C     DATA IMACH(12) / -127 /
C     DATA IMACH(13) /  127 /
C     DATA IMACH(14) /   56 /
C     DATA IMACH(15) / -127 /
C     DATA IMACH(16) /  127 /
C
C     MACHINE CONSTANTS FOR THE UNIVAC 1100 SERIES.
C
C     NOTE THAT THE PUNCH UNIT, I1MACH(3), HAS BEEN SET TO 7
C     WHICH IS APPROPRIATE FOR THE UNIVAC-FOR SYSTEM.
C     IF YOU HAVE THE UNIVAC-FTN SYSTEM, SET IT TO 1.
C
C     DATA IMACH( 1) /    5 /
C     DATA IMACH( 2) /    6 /
C     DATA IMACH( 3) /    7 /
C     DATA IMACH( 4) /    6 /
C     DATA IMACH( 5) /   36 /
C     DATA IMACH( 6) /    6 /
C     DATA IMACH( 7) /    2 /
C     DATA IMACH( 8) /   35 /
C     DATA IMACH( 9) / O377777777777 /
C     DATA IMACH(10) /    2 /
C     DATA IMACH(11) /   27 /
C     DATA IMACH(12) / -128 /
C     DATA IMACH(13) /  127 /
C     DATA IMACH(14) /   60 /
C     DATA IMACH(15) /-1024 /
C     DATA IMACH(16) / 1023 /
C
C     MACHINE CONSTANTS FOR THE VAX-11 WITH
C     FORTRAN IV-PLUS COMPILER
C
C     DATA IMACH( 1) /    5 /
C     DATA IMACH( 2) /    6 /
C     DATA IMACH( 3) /    5 /
C     DATA IMACH( 4) /    6 /
C     DATA IMACH( 5) /   32 /
C     DATA IMACH( 6) /    4 /
C     DATA IMACH( 7) /    2 /
C     DATA IMACH( 8) /   31 /
C     DATA IMACH( 9) / 2147483647 /
C     DATA IMACH(10) /    2 /
C     DATA IMACH(11) /   24 /
C     DATA IMACH(12) / -127 /
C     DATA IMACH(13) /  127 /
C     DATA IMACH(14) /   56 /
C     DATA IMACH(15) / -127 /
C     DATA IMACH(16) /  127 /
C
      IF (I .LT. 1  .OR.  I .GT. 16) GO TO 10
C
      I1MACH=IMACH(I)
      RETURN
C
 10   WRITE(OUTPUT,9000)
 9000 FORMAT(39H1ERROR    1 IN I1MACH - I OUT OF BOUNDS)
C
      STOP
C
      END
```

```
C
C-----------------------------------------------------------------------
C  FUNCTION:  R1MACH
C  THIS ROUTINE IS FROM THE PORT MATHEMATICAL SUBROUTINE LIBRARY
C  IT IS DESCRIBED IN THE BELL LABORATORIES COMPUTING SCIENCE
C  TECHNICAL REPORT #47 BY P.A. FOX, A.D. HALL AND N.L. SCHRYER
C  A MODIFICATION TO THE "I OUT OF BOUNDS" ERROR MESSAGE
C  HAS BEEN MADE BY C. A. MCGONEGAL - APRIL, 1978
C-----------------------------------------------------------------------
C
      REAL FUNCTION R1MACH(I)
C
C  SINGLE-PRECISION MACHINE CONSTANTS
C
C  R1MACH(1) = B**(EMIN-1), THE SMALLEST POSITIVE MAGNITUDE.
C
C  R1MACH(2) = B**EMAX*(1 - B**(-T)), THE LARGEST MAGNITUDE.
C
C  R1MACH(3) = B**(-T), THE SMALLEST RELATIVE SPACING.
C
C  R1MACH(4) = B**(1-T), THE LARGEST RELATIVE SPACING.
C
C  R1MACH(5) = LOG10(B)
C
C  TO ALTER THIS FUNCTION FOR A PARTICULAR ENVIRONMENT,
C  THE DESIRED SET OF DATA STATEMENTS SHOULD BE ACTIVATED BY
C  REMOVING THE C FROM COLUMN 1.
C
C  WHERE POSSIBLE, OCTAL OR HEXADECIMAL CONSTANTS HAVE BEEN USED
C  TO SPECIFY THE CONSTANTS EXACTLY WHICH HAS IN SOME CASES
C  REQUIRED THE USE OF EQUIVALENT INTEGER ARRAYS.
C
      INTEGER SMALL(2)
      INTEGER LARGE(2)
      INTEGER RIGHT(2)
      INTEGER DIVER(2)
      INTEGER LOG10(2)
C
      REAL RMACH(5)
C
      EQUIVALENCE (RMACH(1),SMALL(1))
      EQUIVALENCE (RMACH(2),LARGE(1))
      EQUIVALENCE (RMACH(3),RIGHT(1))
      EQUIVALENCE (RMACH(4),DIVER(1))
      EQUIVALENCE (RMACH(5),LOG10(1))
C
C     MACHINE CONSTANTS FOR THE BURROUGHS 1700 SYSTEM.
C
C     DATA RMACH(1) / Z400800000 /
C     DATA RMACH(2) / Z5FFFFFFFF /
C     DATA RMACH(3) / Z4E9800000 /
C     DATA RMACH(4) / Z4EA800000 /
C     DATA RMACH(5) / Z500E730E8 /
C
C     MACHINE CONSTANTS FOR THE BURROUGHS 5700/6700/7700 SYSTEMS.
C
C     DATA RMACH(1) / O1771000000000000 /
C     DATA RMACH(2) / O0777777777777777 /
C     DATA RMACH(3) / O1311000000000000 /
C     DATA RMACH(4) / O1301000000000000 /
C     DATA RMACH(5) / O1157163034761675 /
C
C     MACHINE CONSTANTS FOR THE CDC 6000/7000 SERIES.
C
C     DATA RMACH(1) / 00014000000000000000B /
C     DATA RMACH(2) / 37767777777777777777B /
C     DATA RMACH(3) / 16404000000000000000B /
C     DATA RMACH(4) / 16414000000000000000B /
C     DATA RMACH(5) / 17164642023241175720B /
C
C     MACHINE CONSTANTS FOR THE CRAY 1
C
C     DATA RMACH(1) / 200004000000000000000B /
C     DATA RMACH(2) / 577767777777777777776B /
C     DATA RMACH(3) / 377224000000000000000B /
C     DATA RMACH(4) / 377234000000000000000B /
C     DATA RMACH(5) / 377774642023241175720B /
C
C     MACHINE CONSTANTS FOR THE DATA GENERAL ECLIPSE S/200
C
C     NOTE - IT MAY BE APPROPRIATE TO INCLUDE THE FOLLOWING CARD -
C     STATIC RMACH(5)
C
C     DATA SMALL/20K,0/,LARGE/77777K,177777K/
C     DATA RIGHT/35420K,0/,DIVER/36020K,0/
C     DATA LOG10/40423K,42023K/
C
C     MACHINE CONSTANTS FOR THE HARRIS SLASH 6 AND SLASH 7
C
C     DATA SMALL(1),SMALL(2) / '20000000, '00000201 /
C     DATA LARGE(1),LARGE(2) / '37777777, '00000177 /
C     DATA RIGHT(1),RIGHT(2) / '20000000, '00000352 /
C     DATA DIVER(1),DIVER(2) / '20000000, '00000353 /
C     DATA LOG10(1),LOG10(2) / '23210115, '00000377 /
C
C     MACHINE CONSTANTS FOR THE HONEYWELL 600/6000 SERIES.
C
C     DATA RMACH(1) / O402400000000 /
C     DATA RMACH(2) / O376777777777 /
C     DATA RMACH(3) / O714400000000 /
C     DATA RMACH(4) / O716400000000 /
C     DATA RMACH(5) / O776464202324 /
C
C     MACHINE CONSTANTS FOR THE IBM 360/370 SERIES,
C     THE XEROX SIGMA 5/7/9 AND THE SEL SYSTEMS 85/86.
C
C     DATA RMACH(1) / Z00100000 /
C     DATA RMACH(2) / Z7FFFFFFF /
C     DATA RMACH(3) / Z3B100000 /
C     DATA RMACH(4) / Z3C100000 /
C     DATA RMACH(5) / Z41134413 /
C
C     MACHINE CONSTANTS FOR THE PDP-10 (KA OR KI PROCESSOR).
C
C     DATA RMACH(1) / "000400000000 /
C     DATA RMACH(2) / "377777777777 /
C     DATA RMACH(3) / "146400000000 /
C     DATA RMACH(4) / "147400000000 /
C     DATA RMACH(5) / "177464202324 /
C
C     MACHINE CONSTANTS FOR PDP-11 FORTRAN'S SUPPORTING
C     32-BIT INTEGERS (EXPRESSED IN INTEGER AND OCTAL).
C
C     DATA SMALL(1) /    8388608 /
C     DATA LARGE(1) / 2147483647 /
C     DATA RIGHT(1) /  880803840 /
C     DATA DIVER(1) /  889192448 /
C     DATA LOG10(1) / 1067065499 /
C
C     DATA RMACH(1) / O00040000000 /
```

```
C     DATA RMACH(2) / O17777777777 /
C     DATA RMACH(3) / O06440000000 /
C     DATA RMACH(4) / O06500000000 /
C     DATA RMACH(5) / O07746420233 /
C
C     MACHINE CONSTANTS FOR PDP-11 FORTRAN'S SUPPORTING
C     16-BIT INTEGERS  (EXPRESSED IN INTEGER AND OCTAL).
C
C     DATA SMALL(1),SMALL(2) /   128,      0 /
C     DATA LARGE(1),LARGE(2) / 32767,     -1 /
C     DATA RIGHT(1),RIGHT(2) / 13440,      0 /
C     DATA DIVER(1),DIVER(2) / 13568,      0 /
C     DATA LOG10(1),LOG10(2) / 16282,   8347 /
C
C     DATA SMALL(1),SMALL(2) / O000200, O000000 /
C     DATA LARGE(1),LARGE(2) / O077777, O177777 /
C     DATA RIGHT(1),RIGHT(2) / O032200, O000000 /
C     DATA DIVER(1),DIVER(2) / O032400, O000000 /
C     DATA LOG10(1),LOG10(2) / O037632, O020233 /
C
C     MACHINE CONSTANTS FOR THE UNIVAC 1100 SERIES.
C
C     DATA RMACH(1) / O000400000000 /
C     DATA RMACH(2) / O377777777777 /
C     DATA RMACH(3) / O146400000000 /
C     DATA RMACH(4) / O147400000000 /
C     DATA RMACH(5) / O177464202324 /
C
C     MACHINE CONSTANTS FOR THE VAX-11 WITH
C     FORTRAN IV-PLUS COMPILER
C
C     DATA RMACH(1) / Z00000080 /
C     DATA RMACH(2) / ZFFFF7FFF /
C     DATA RMACH(3) / Z00003480 /
C     DATA RMACH(4) / Z00003500 /
C     DATA RMACH(5) / Z209B3F9A /
C
      IF (I .LT. 1  .OR.  I .GT. 5) GOTO 100
C
      R1MACH = RMACH(I)
      RETURN
C
 100  IWUNIT = I1MACH(4)
      WRITE(IWUNIT, 99)
  99  FORMAT(24HR1MACH - I OUT OF BOUNDS)
      STOP
      END
```

```
C
C-----------------------------------------------------------------------
C  FUNCTION:  D1MACH
C  THIS ROUTINE IS FROM THE PORT MATHEMATICAL SUBROUTINE LIBRARY
C  IT IS DESCRIBED IN THE BELL LABORATORIES COMPUTING SCIENCE
C  TECHNICAL REPORT #47 BY P.A. FOX, A.D. HALL AND N.L. SCHRYER
C  A MODIFICATION TO THE "I OUT OF BOUNDS" ERROR MESSAGE
C  HAS BEEN MADE BY C. A. MCGONEGAL - APRIL, 1978
C-----------------------------------------------------------------------
C
      DOUBLE PRECISION FUNCTION D1MACH(I)
C
C  DOUBLE-PRECISION MACHINE CONSTANTS
C
C  D1MACH( 1) = B**(EMIN-1), THE SMALLEST POSITIVE MAGNITUDE.
C
C  D1MACH( 2) = B**EMAX*(1 - B**(-T)), THE LARGEST MAGNITUDE.
C
C  D1MACH( 3) = B**(-T), THE SMALLEST RELATIVE SPACING.
C
C  D1MACH( 4) = B**(1-T), THE LARGEST RELATIVE SPACING.
C
C  D1MACH( 5) = LOG10(B)
C
C  TO ALTER THIS FUNCTION FOR A PARTICULAR ENVIRONMENT,
C  THE DESIRED SET OF DATA STATEMENTS SHOULD BE ACTIVATED BY
C  REMOVING THE C FROM COLUMN 1.
C
C  WHERE POSSIBLE, OCTAL OR HEXADECIMAL CONSTANTS HAVE BEEN USED
C  TO SPECIFY THE CONSTANTS EXACTLY WHICH HAS IN SOME CASES
C  REQUIRED THE USE OF EQUIVALENT INTEGER ARRAYS.
C
      INTEGER SMALL(4)
      INTEGER LARGE(4)
      INTEGER RIGHT(4)
      INTEGER DIVER(4)
      INTEGER LOG10(4)
C
      DOUBLE PRECISION DMACH(5)
C
      EQUIVALENCE (DMACH(1),SMALL(1))
      EQUIVALENCE (DMACH(2),LARGE(1))
      EQUIVALENCE (DMACH(3),RIGHT(1))
      EQUIVALENCE (DMACH(4),DIVER(1))
      EQUIVALENCE (DMACH(5),LOG10(1))
C
C     MACHINE CONSTANTS FOR THE BURROUGHS 1700 SYSTEM.
C
C     DATA SMALL(1) / ZC00800000 /
C     DATA SMALL(2) / Z000000000 /
C
C     DATA LARGE(1) / ZDFFFFFFFF /
C     DATA LARGE(2) / ZFFFFFFFFF /
C
C     DATA RIGHT(1) / ZCC5800000 /
C     DATA RIGHT(2) / Z000000000 /
C
C     DATA DIVER(1) / ZCC6800000 /
C     DATA DIVER(2) / Z000000000 /
C
C     DATA LOG10(1) / ZD00E730E7 /
C     DATA LOG10(2) / ZC77800DC0 /
C
C     MACHINE CONSTANTS FOR THE BURROUGHS 5700 SYSTEM.
C
C     DATA SMALL(1) / O1771000000000000 /
C     DATA SMALL(2) / O0000000000000000 /
C
C     DATA LARGE(1) / O0777777777777777 /
C     DATA LARGE(2) / O0007777777777777 /
C
C     DATA RIGHT(1) / O1461000000000000 /
C     DATA RIGHT(2) / O0000000000000000 /
C
C     DATA DIVER(1) / O1451000000000000 /
C     DATA DIVER(2) / O0000000000000000 /
C
C     DATA LOG10(1) / O1157163034761674 /
C     DATA LOG10(2) / O0006677466732724 /
C
C     MACHINE CONSTANTS FOR THE BURROUGHS 6700/7700 SYSTEMS.
C
C     DATA SMALL(1) / O1771000000000000 /
C     DATA SMALL(2) / O7770000000000000 /
C
C     DATA LARGE(1) / O0777777777777777 /
C     DATA LARGE(2) / O7777777777777777 /
C
C     DATA RIGHT(1) / O1461000000000000 /
C     DATA RIGHT(2) / O0000000000000000 /
C
C     DATA DIVER(1) / O1451000000000000 /
C     DATA DIVER(2) / O0000000000000000 /
C
C     DATA LOG10(1) / O1157163034761674 /
C     DATA LOG10(2) / O0006677466732724 /
C
C     MACHINE CONSTANTS FOR THE CDC 6000/7000 SERIES.
C
C     DATA SMALL(1) / 00604000000000000000B /
C     DATA SMALL(2) / 00000000000000000000B /
C
C     DATA LARGE(1) / 37767777777777777777B /
C     DATA LARGE(2) / 37167777777777777777B /
C
C     DATA RIGHT(1) / 15604000000000000000B /
C     DATA RIGHT(2) / 15000000000000000000B /
C
C     DATA DIVER(1) / 15614000000000000000B /
C     DATA DIVER(2) / 15010000000000000000B /
C
C     DATA LOG10(1) / 17164642023241175717B /
C     DATA LOG10(2) / 16367571421742254654B /
C
C     MACHINE CONSTANTS FOR THE CRAY 1
C
C     DATA SMALL(1) / 200004000000000000000B /
C     DATA SMALL(2) / 000000000000000000000B /
C
C     DATA LARGE(1) / 577767777777777777777B /
C     DATA LARGE(2) / 000007777777777777776B /
C
C     DATA RIGHT(1) / 376424000000000000000B /
C     DATA RIGHT(2) / 000000000000000000000B /
C
C     DATA DIVER(1) / 376434000000000000000B /
C     DATA DIVER(2) / 000000000000000000000B /
C
C     DATA LOG10(1) / 377774642023241175717B /
C     DATA LOG10(2) / 000007571421742254654B /
```

```
C
C     MACHINE CONSTANTS FOR THE DATA GENERAL ECLIPSE S/200
C
C     NOTE - IT MAY BE APPROPRIATE TO INCLUDE THE FOLLOWING CARD -
C     STATIC DMACH(5)
C
C     DATA SMALL/20K,3*0/,LARGE/77777K,3*177777K/
C     DATA RIGHT/31420K,3*0/,DIVER/32020K,3*0/
C     DATA LOG10/40423K,42023K,50237K,74776K/
C
C     MACHINE CONSTANTS FOR THE HARRIS SLASH 6 AND SLASH 7
C
C     DATA SMALL(1),SMALL(2) / '20000000, '00000201 /
C     DATA LARGE(1),LARGE(2) / '37777777, '37777577 /
C     DATA RIGHT(1),RIGHT(2) / '20000000, '00000333 /
C     DATA DIVER(1),DIVER(2) / '20000000, '00000334 /
C     DATA LOG10(1),LOG10(2) / '23210115, '10237777 /
C
C     MACHINE CONSTANTS FOR THE HONEYWELL 600/6000 SERIES.
C
C     DATA SMALL(1),SMALL(2) / O402400000000, O000000000000 /
C     DATA LARGE(1),LARGE(2) / O376777777777, O777777777777 /
C     DATA RIGHT(1),RIGHT(2) / O604400000000, O000000000000 /
C     DATA DIVER(1),DIVER(2) / O606400000000, O000000000000 /
C     DATA LOG10(1),LOG10(2) / O776464202324, O117571775714 /
C
C     MACHINE CONSTANTS FOR THE IBM 360/370 SERIES,
C     THE XEROX SIGMA 5/7/9 AND THE SEL SYSTEMS 85/86.
C
C     DATA SMALL(1),SMALL(2) / Z00100000, Z00000000 /
C     DATA LARGE(1),LARGE(2) / Z7FFFFFFF, ZFFFFFFFF /
C     DATA RIGHT(1),RIGHT(2) / Z33100000, Z00000000 /
C     DATA DIVER(1),DIVER(2) / Z34100000, Z00000000 /
C     DATA LOG10(1),LOG10(2) / Z41134413, Z509F79FF /
C
C     MACHINE CONSTANTS FOR THE PDP-10 (KA PROCESSOR).
C
C     DATA SMALL(1),SMALL(2) / "033400000000, "000000000000 /
C     DATA LARGE(1),LARGE(2) / "377777777777, "344777777777 /
C     DATA RIGHT(1),RIGHT(2) / "113400000000, "000000000000 /
C     DATA DIVER(1),DIVER(2) / "114400000000, "000000000000 /
C     DATA LOG10(1),LOG10(2) / "177464202324, "144117571776 /
C
C     MACHINE CONSTANTS FOR THE PDP-10 (KI PROCESSOR).
C
C     DATA SMALL(1),SMALL(2) / "000400000000, "000000000000 /
C     DATA LARGE(1),LARGE(2) / "377777777777, "377777777777 /
C     DATA RIGHT(1),RIGHT(2) / "103400000000, "000000000000 /
C     DATA DIVER(1),DIVER(2) / "104400000000, "000000000000 /
C     DATA LOG10(1),LOG10(2) / "177464202324, "476747767461 /
C
C     MACHINE CONSTANTS FOR PDP-11 FORTRAN'S SUPPORTING
C     32-BIT INTEGERS (EXPRESSED IN INTEGER AND OCTAL).
C
C     DATA SMALL(1),SMALL(2) /    8388608,           0 /
C     DATA LARGE(1),LARGE(2) / 2147483647,          -1 /
C     DATA RIGHT(1),RIGHT(2) /  612368384,           0 /
C     DATA DIVER(1),DIVER(2) /  620756992,           0 /
C     DATA LOG10(1),LOG10(2) / 1067065498, -2063872008 /
C
C     DATA SMALL(1),SMALL(2) / O00040000000, O00000000000 /
C     DATA LARGE(1),LARGE(2) / O17777777777, O37777777777 /
C     DATA RIGHT(1),RIGHT(2) / O04440000000, O00000000000 /
C     DATA DIVER(1),DIVER(2) / O04500000000, O00000000000 /
C     DATA LOG10(1),LOG10(2) / O07746420232, O20476747770 /
C
C     MACHINE CONSTANTS FOR PDP-11 FORTRAN'S SUPPORTING
C     16-BIT INTEGERS (EXPRESSED IN INTEGER AND OCTAL).
C
C     DATA SMALL(1),SMALL(2) /    128,      0 /
C     DATA SMALL(3),SMALL(4) /      0,      0 /
C
C     DATA LARGE(1),LARGE(2) /  32767,     -1 /
C     DATA LARGE(3),LARGE(4) /     -1,     -1 /
C
C     DATA RIGHT(1),RIGHT(2) /   9344,      0 /
C     DATA RIGHT(3),RIGHT(4) /      0,      0 /
C
C     DATA DIVER(1),DIVER(2) /   9472,      0 /
C     DATA DIVER(3),DIVER(4) /      0,      0 /
C
C     DATA LOG10(1),LOG10(2) /  16282,   8346 /
C     DATA LOG10(3),LOG10(4) / -31493, -12296 /
C
C     DATA SMALL(1),SMALL(2) / O000200, O000000 /
C     DATA SMALL(3),SMALL(4) / O000000, O000000 /
C
C     DATA LARGE(1),LARGE(2) / O077777, O177777 /
C     DATA LARGE(3),LARGE(4) / O177777, O177777 /
C
C     DATA RIGHT(1),RIGHT(2) / O022200, O000000 /
C     DATA RIGHT(3),RIGHT(4) / O000000, O000000 /
C
C     DATA DIVER(1),DIVER(2) / O022400, O000000 /
C     DATA DIVER(3),DIVER(4) / O000000, O000000 /
C
C     DATA LOG10(1),LOG10(2) / O037632, O020232 /
C     DATA LOG10(3),LOG10(4) / O102373, O147770 /
C
C     MACHINE CONSTANTS FOR THE UNIVAC 1100 SERIES.
C
C     DATA SMALL(1),SMALL(2) / O000040000000, O000000000000 /
C     DATA LARGE(1),LARGE(2) / O377777777777, O777777777777 /
C     DATA RIGHT(1),RIGHT(2) / O170540000000, O000000000000 /
C     DATA DIVER(1),DIVER(2) / O170640000000, O000000000000 /
C     DATA LOG10(1),LOG10(2) / O177746420232, O411757177572 /
C
C     MACHINE CONSTANTS FOR THE VAX-11 WITH
C     FORTRAN IV-PLUS COMPILER
C
C     DATA SMALL(1),SMALL(2) / Z00000080, Z00000000 /
C     DATA LARGE(1),LARGE(2) / ZFFFF7FFF, ZFFFFFFFF /
C     DATA RIGHT(1),RIGHT(2) / Z00002480, Z00000000 /
C     DATA DIVER(1),DIVER(2) / Z00002500, Z00000000 /
C     DATA LOG10(1),LOG10(2) / Z209A3F9A, ZCFFA84FB /
C
      IF (I .LT. 1  .OR.  I .GT. 5) GOTO 100
C
      D1MACH = DMACH(I)
      RETURN
C
 100  IWUNIT = I1MACH(4)
      WRITE(IWUNIT, 99)
  99  FORMAT(24HD1MACH - I OUT OF BOUNDS)
      STOP
      END
```

```
C
C-----------------------------------------------------------------------
C  FUNCTION:  UNI
C  AUTHOR:    ALAN M. GROSS
C             BELL LABORATORIES, MURRAY HILL, NEW JERSEY 07974
C  PORTABLE RANDOM NUMBER GENERATOR
C-----------------------------------------------------------------------
C
      FUNCTION UNI(K)
      INTEGER IBYTE(4)
      DATA ICSEED/0/, ITSEED/0/, IFCN/1/
C
C  UNI IS RETURNED AS A SINGLE REAL RANDOM VARIATE
C  FROM THE UNIFORM DISTRIBUTION 0.0 .LE. UNI .LT. 1.0 .
C
C  IFCN = 1 IMPLIES THAT ICSEED, ITSEED, IBYTE, AND K ARE IGNORED.
C
      UNI=R1UNIF(ICSEED,ITSEED,IBYTE,IFCN)
      RETURN
      END
      SUBROUTINE RANSET(ICSEED,ITSEED)
      INTEGER IBYTE(4)
      DATA IFCN/0/
C
C  TO (RE)INITIALIZE THE UNIFORM RANDOM NUMBER GENERATOR, R1UNIF
C  (TO OTHER THAN THE DEFAULT INITIAL VALUES).
C
C    ICSEED IS THE NEW SEED FOR CONGRUENTIAL GENERATOR.
C    ITSEED IS THE NEW SEED FOR TAUSWORTHE GENERATOR.
C
C  ONE, BUT NOT BOTH, OF THE NEW SEEDS CAN BE ZERO
C
C
C  IFCN = 0 IMPLIES THAT UNI AND IBYTE ARE NOT COMPUTED.
C
      UNI=R1UNIF(ICSEED,ITSEED,IBYTE,IFCN)
      RETURN
      END
      SUBROUTINE RANBYT(UNI,IBYTE)
      DIMENSION IBYTE(4)
      DATA ICSEED/0/, ITSEED/0/, IFCN/2/
C
C  UNI IS RETURNED AS A SINGLE UNIFORM RANDOM VARIATE IN UNI.
C
C  IBYTE IS RETURNED WITH THE BITS OF UNI, 8 BITS PER WORD.
C  UNI=(IBYTE(1)*256**3+IBYTE(2)*256**2+IBYTE(3)*256+IBYTE(4))/2**32
C
C  IFCN = 2 IMPLIES THAT ICSEED AND ITSEED ARE IGNORED.
C
      UNI=R1UNIF(ICSEED,ITSEED,IBYTE,IFCN)
      RETURN
      END
      FUNCTION R1UNIF(ICSEED,ITSEED,IBYTE,IFCN)
C
C  R1UNIF - OUTPUT, THE UNIFORM RANDOM NUMBER IF IFCN .NE. 0
C  ICSEED - INPUT, THE NEW CONGRUENTIAL SEED IF IFCN = 0
C  ITSEED - INPUT, THE NEW TAUSWORTHE SEED IF IFCN = 0
C  IBYTE  - OUTPUT, THE BITS OF R1UNIF, 8 PER WORD, IF IFCN = 2
C  IFCN   - INPUT, = 0 FOR INITIALIZATION
C                  = 1 IF ONLY THE VALUE OF R1UNIF IS OF INTEREST
C                  = 2 IF BOTH R1UNIF AND IBYTE ARE OF INTEREST
C
C  THIS IS A PORTABLE FORTRAN IMPLEMENTATION OF UNI, A
C  UNIFORM RANDOM NUMBER GENERATOR ON (0.0, 1.0) DEVISED
C  BY MARSAGLIA, ET. AL., AND INCLUDED IN THEIR PACKAGE
C  CALLED "SUPER-DUPER".
C
C  TWO INDEPENDENT 32 BIT GENERATORS ARE MAINTAINED INTERNALLY AND
C  UPDATED FOR EACH CALL.
C
C  THE FIRST OF THESE IS A CONGRUENTIAL GENERATOR WITH
C  MULTIPLIER 69069 (=16*64**2 + 55*64 + 13).
C
C  THE SECOND IS A TAUSWORTHE OR SHIFT-REGISTER GENERATOR.
C  THIS GENERATOR TAKES THE SEED, SHIFTS IT RIGHT 15 BITS, EXCLUSIVE
C  ORS IT WITH ITSELF, SHIFTS THE RESULT 17 BITS TO THE LEFT, AND
C  EXCLUSIVE ORS THE SHIFTED RESULT WITH ITSELF (NOT WITH THE
C  ORIGINAL SEED).  THE OUTPUT OF THE PROCEDURE IS THE TAUSWORTHE
C  RANDOM NUMBER AND IS USED AS THE SEED FOR THE NEXT CALL.
C
C  FINALLY, THE OUTPUT FROM THE TWO GENERATORS IS
C  EXCLUSIVELY OR-ED TOGETHER.
C
C  THE FOLLOWING PROGRAM SHOULD WORK ON ANY 16+ BIT COMPUTER.
C
      LOGICAL  FIRST
      INTEGER CSEED(6), TSEED(32), XOR(29), IBYTE(4), ISCR(5)
      DATA XOR(1)/1/,XOR(2)/2/,XOR(3)/3/,XOR(4)/3/,XOR(5)/2/,
     1 XOR(6)/1/,XOR(7)/4/,XOR(8)/5/,XOR(9)/6/,XOR(10)/7/,XOR(11)/5/,
     2 XOR(12)/4/,XOR(13)/7/,XOR(14)/6/,XOR(15)/1/,XOR(16)/6/,
     3 XOR(17)/7/,XOR(18)/4/,XOR(19)/5/,XOR(20)/2/,XOR(21)/3/,
     4 XOR(22)/7/,XOR(23)/6/,XOR(24)/5/,XOR(25)/4/,XOR(26)/3/,
     5 XOR(27)/2/,XOR(28)/1/,XOR(29)/0/
      DATA FIRST/.TRUE./, JCSEED/12345/, JTSEED/1073/
C
C  INITIALIZE CSEED AND TSEED FOR PORTABILITY
C
      DATA CSEED(1)/0/,CSEED(2)/0/,CSEED(3)/0/,CSEED(4)/0/,
     1 CSEED(5)/0/,CSEED(6)/0/,TSEED(1)/0/,TSEED(2)/0/,TSEED(3)/0/,
     2 TSEED(4)/0/,TSEED(5)/0/,TSEED(6)/0/,TSEED(7)/0/,TSEED(8)/0/,
     3 TSEED(9)/0/,TSEED(10)/0/,TSEED(11)/0/,TSEED(12)/0/,
     4 TSEED(13)/0/,TSEED(14)/0/,TSEED(15)/0/,TSEED(16)/0/,
     5 TSEED(17)/0/,TSEED(18)/0/,TSEED(19)/0/,TSEED(20)/0/,
     6 TSEED(21)/0/,TSEED(22)/0/,TSEED(23)/0/,TSEED(24)/0/,
     7 TSEED(25)/0/,TSEED(26)/0/,TSEED(27)/0/,TSEED(28)/0/,
     8 TSEED(29)/0/,TSEED(30)/0/,TSEED(31)/0/,TSEED(32)/0/
C
      R1UNIF=0.0
      IF((.NOT.FIRST) .AND. (IFCN.GT.0)) GO TO 50
      IF(IFCN.GT.0) GO TO 10
C
C     TAKE USER VALUES AS SEEDS
C
      JCSEED=IABS(ICSEED)
      JTSEED=IABS(ITSEED)
   10 FIRST=.FALSE.
C
C.....DECODE SEEDS
C
      CSEED(1)=JCSEED
      DO 20 I=1,5
      CSEED(I+1)=CSEED(I)/64
   20 CSEED(I)=CSEED(I)-CSEED(I+1)*64
      CSEED(6)=MOD(CSEED(6),4)
C
C     ENSURE ODD UNLESS ZERO
C
      IF(JCSEED.NE.0 .AND. MOD(CSEED(1),2).EQ.0) CSEED(1)=CSEED(1)+1
      TSEED(1)=JTSEED
      DO 30 I=1,11
```

```
      TSEED(I+1)=TSEED(I)/2
   30 TSEED(I)=TSEED(I)-TSEED(I+1)*2
C
C     ONLY USE INITIAL VALUE MOD 2048
C
      DO 40 I=12,32
   40 TSEED(I)=0
C
C     ENSURE ODD UNLESS ZERO
C
      IF(JTSEED.NE.0) TSEED(1)=1
C
C     END OF INITIALIZATION
C
      IF(IFCN.EQ.0) RETURN
   50 CONTINUE
C
C.....TAUSWORTHE GENERATOR -- SHIFT RIGHT 15, THEN LEFT 17
C
      DO 60 I=1,17
   60 TSEED(I)=IABS(TSEED(I)-TSEED(I+15))
      DO 70 I=18,32
   70 TSEED(I)=IABS(TSEED(I)-TSEED(I-17))
C
C.....CONGRUENTIAL GENERATOR -- MULTIPLICATION IN BASE 64
C
C     MULTIPLY BASE 64
C
      CSEED(6)=13*CSEED(6)+55*CSEED(5)+16*CSEED(4)
      CSEED(5)=13*CSEED(5)+55*CSEED(4)+16*CSEED(3)
      CSEED(4)=13*CSEED(4)+55*CSEED(3)+16*CSEED(2)
      CSEED(3)=13*CSEED(3)+55*CSEED(2)+16*CSEED(1)
      CSEED(2)=13*CSEED(2)+55*CSEED(1)
      CSEED(1)=13*CSEED(1)
      K=-5
      ICARRY=0
      DO 80 I=1,5
      K=K+6
      CSEED(I)=CSEED(I)+ICARRY
      ICARRY=CSEED(I)/64
      CSEED(I)=CSEED(I)-64*ICARRY
      I2=CSEED(I)/8
      I1=CSEED(I)-8*I2
      J1=4*TSEED(K+2)+TSEED(K+1)+TSEED(K+1)+TSEED(K)
      J2=4*TSEED(K+5)+TSEED(K+4)+TSEED(K+4)+TSEED(K+3)
      IT1=28
      IF(I1.GT.J1) IT1=(I1*I1-I1)/2+J1
      IF(I1.LT.J1) IT1=(J1*J1-J1)/2+I1
      IT2=28
      IF(I2.GT.J2) IT2=(I2*I2-I2)/2+J2
      IF(I2.LT.J2) IT2=(J2*J2-J2)/2+I2
      ISCR(I)=8*XOR(IT2+1)+XOR(IT1+1)
   80 R1UNIF=(R1UNIF+FLOAT(ISCR(I)))/64.0
      CSEED(6)=MOD(CSEED(6)+ICARRY,4)
      J1=TSEED(31)+TSEED(32)+TSEED(32)
      IT1=IABS(CSEED(6)-J1)
      IF((IT1.EQ.1) .AND. (CSEED(6)+J1.EQ.3)) IT1=3
      R1UNIF=(R1UNIF+FLOAT(IT1))/4.0
      IF(IFCN.EQ.1) RETURN
      IBYTE(4)=ISCR(1)+MOD(ISCR(2),4)*64
      IBYTE(3)=ISCR(2)/4+MOD(ISCR(3),16)*16
      IBYTE(2)=ISCR(3)/16+ISCR(4)*4
      IBYTE(1)=ISCR(5)+IT1*64
      RETURN
      END
```

CHAPTER 1

Fast Fourier Transform Subroutines

J. W. Cooley and *M. T. Dolan*

Introduction

The following is but a small sample of the many excellent Discrete Fourier Transform (DFT) programs which were offered for inclusion in this book. Each of the programs selected for the book has some particular virtue in terms of ease of use, understanding, and efficiency in specific applications. Preceding each subroutine, there is included documentation which should be sufficient to enable the user to apply it. Comment cards in the subroutine listings give brief descriptions of the arguments and, to some extent, a description of the internal operations of the subroutine. The write-ups are complete and self-sufficient, so it only remains, in this introduction, to make a few remarks designed to enable the user to select, from among the subroutines, the one appropriate for his particular application, and to enable him to use it effectively. As a further aid to the user, Tables 1 to 5 described in more detail below, give storage requirements, execution time, and accuracy of test calculations for all subroutines.

FOUREA — A Short Demonstration Version of the FFT (1.1)

The subroutine FOUREA, by C. Rader, and others like it [1] have been written primarily to describe and demonstrate the FFT algorithm, and to be used as models for writing efficient special-purpose or microprocessor programs. In general, such programs are not intended to be used as operational programs. However, in some terminal-operated systems and interpreters, where the running time is less significant, these programs have been used effectively because they are short and easily understood.

Fast Fourier Transform Algorithms (1.2)

These efficient radix 2-4-8 subroutines, contributed by G. D. Bergland and M. T. Dolan, are written with a completely different objective. These are written to produce very fast programs at the expense of instruction memory and program complexity. The first DFT-IDFT pair, FAST-FSST, computes 1024-point transforms of real data in about one-fourth the time of the previous program, FOUREA, but requires 1940 words of instruction store on the Honeywell 6080N versus only 260 for FOUREA. The programs FFA-FFS do exactly the same thing but include the use of radix 8 DFT's internally and, as a result, run still faster, e.g., 134 ms. vs the 156 ms. for FAST-FSST for the 1024-point transform. The price for this is 3210 words vs 1940 words of instruction store. The third program in this package is FFT842, which computes the DFT of a complex input. For a 1024-point complex transform, this program takes 250 ms., which is almost twice as long as the time required for FAST. However, it is to be remembered that FAST and FFA do the calculation for real rather than complex input. Therefore, the amount of computing they do is a little more than half of that done by FOUREA and FFT842.

FFT Subroutines for Sequences with Special Properties (1.3)

The FFT subroutines for sequences with special symmetry properties by L. R. Rabiner are extremely useful for taking advantage of symmetries in the data to save computation. Such symmetries arise very often from the inherent nature of the data, such as in the case of the autocorrelation function, which is always real and even. Cosine series with only odd harmonics appear in certain types of Chebyshev approximations and sine series with odd harmonics are useful in solving linear partial differential equations with boundary conditions calling for zero values on one end and zero derivatives on the other. These are written in the style of FOUREA, i.e., to be short and easily read. Sines and cosines are computed as needed, rather than using tables, and calculations are done in "DO" loops without special cases for the simpler computations. The reason for this approach here is that a relatively insignificant amount of time is spent in the (1.3) subroutines; most of the time is spent in the DFT subroutine which is called internally. In the programs given here, the (1.3) subroutines call FAST and FSST but, it is to be emphasized that these calls may be replaced by calls to any other available DFT subroutines. For example, if N is not a power of two or if data is on some mass storage unit, one may want to use Singleton's (1.4) or Fraser's (1.5) FFT subroutines.

The first four programs, FFTSYM, IFTSYM, FFTASM, and IFTASM, call FAST and FSST to compute what are actually cosine and sine transforms. Such programs have been generally available in program libraries. Some recent contributions to the DFT literature referred to in the write-up, have shown how to take advantage of further symmetries which correspond to the existence of only odd harmonics in the DFT. These novel ideas are incorporated in the subroutines whose names include the letters "OH", meaning "odd harmonics". For example, the subroutine FFTOHM computes the DFT of real data having a two-fold redundancy: the top half is equal to minus the bottom half. Thus, only $N/2$ data need be stored and internally, a DFT of only $N/2$ real points is computed. Thus, a saving by a factor of 2 in storage and almost a factor of 2 in computation results. The last four subroutines, whose names contain "SOH" and "AOH" use additional two-fold redundancies due to symmetry and anti-symmetry about 0 in the data. These result in still further reductions in storage and computing time. The computing times given in Tables 2a and 2b show that as N increases, one approaches a saving of a factor of 2 for each two-fold redundancy in the data. If N is not a power of 2, the subroutine call statements can be changed so as to call the mixed radix subroutines in the next sub-section. However, if N is not a power of 2, it must be a multiple of 4 for the first six subroutines and a multiple of 8 for the last four subroutines.

Mixed Radix Fast Fourier Transforms (1.4)

All of the above subroutines require that N be a power of 2, 4, or 8 since these are the simplest and most efficient forms of the FFT algorithm. The mixed radix FFT by R. C. Singleton permits one to use values of N containing other factors than 2. In general, the larger the prime factors of N, the less efficient will be the calculation in terms of numbers of operations per output point.

Although the programs described above can be used for multidimensional transforms by simply computing the transforms for each vector of points in each dimension, the large number of subroutine calls would make this method quite inefficient. The second important feature of the mixed radix FFT is in the ease and efficiency with which it can be used for multidimensional complex and real transforms. One calls the subroutine once for each dimension and the subroutine computes the DFT for the number of points in that dimension, but, in the innermost loops, it repeats the calculation for all of the other dimensions. Table 2a shows that for powers of 2, FFT takes about the same amount of time as FFT842. The cost of the added flexibility is that FFT requires 2300 words of instruction storage while FFT842, with its subroutines, requires 1430.

Optimized Mass Storage FFT Program (1.5)

The sizes of the data arrays with which the above subroutines can be used depends, of course, upon the amount of core storage available. Depending upon the size of the machine, and the data arrays, it may sometimes be practical to keep the data array on a random access device such as a disk storage medium. The optimized mass storage FFT by D. Fraser makes this as easy and efficient as possible. The user defines the size of an available buffer store region of memory and provides subroutines which access the core storage where the data resides. The subroutine computes optimal size blocks and schedules the transfer of data between core storage and the mass storage medium in an optimal fashion according to algorithms which the author has designed and published. The array may have any number of dimensions and may be real or complex. The user will have to familiarize himself thoroughly with the documentation and will have to spend some effort on setting up read and write subroutines, but the result will be an extremely efficient program for very large Fourier transforms. If N is not a power of 2 or if one has two-dimensional arrays, then Singleton's 2-dimensional mass-storage mixed radix FFT (1.9) may be preferable.

Chirp z-Transform Algorithm Program (1.6)

The chirp z-transform program by L. Rabiner has been placed in this section because it is, in a sense, a generalization of a DFT. While the DFT may be regarded as an algorithm for evaluating the z-transform at N equidistant points on a unit circle, the chirp z-transform gives values of the z-transform at arbitrarily many equi-spaced points on an arbitrary spiral in the z-plane. In a sense, this is a generalized spectral analysis of discrete signals and plays the same role for discrete signals which the Laplace transform does in the case of continuous signals. The version given here uses FFT842 described above, but, as with many other programs in this book, FFT842 may be replaced by any other appropriate DFT subroutine.

Complex General-N Winograd Fourier Transform Algorithm (WFTA) (1.7)

The complex general-N Winograd algorithm by J. H. McClellan and H. Nawab uses an algorithm which is quite different from the FFT. As far as the use is concerned, if N is a power of 2, a subroutine using the FFT algorithm should be used, but, if there are important reasons to have other factors in N there may be as much as a 40% improvement in running time over the mixed radix DFT (1.4), for some values of N. There is no consistent rule for describing ranges of N-values for which one or the other algorithm or program is better. One must consult the tables included in the write-up. The WFTA subroutine with its required subroutines take quite a large amount of instruction store, as compared with any of the FFT programs. In judging time-wise efficiency, one must also consider that an initialization routine must be run for an N-value before computing the DFT. Therefore, the WFTA subroutine is in general preferable only when a number of computations with the same N are to be done.

Time-Efficient Radix-4 Fast Fourier Transform (1.8)

The time-efficient radix-4 fast Fourier transform by L. R. Morris is an extreme example of trading off memory for time-wise efficiency when many DFT's with the same N are to be performed. One is not likely to find this type of program in the usual program libraries, but since this situation is particularly typical in digital signal processing, it is regarded as a valuable contribution to this book. The program used here was produced by Morris's Autogen technique, which has been described in the literature. This technique essentially takes all data-independent parts of a given program, in this case, the radix-4 part of Singleton's program above, puts them in a preprocessing stage and effectively unwinds some inner loops, making special cases of certain simple operations. This produces customized optimal code for special situations. After reviewing this example, the reader may be interested in further optimization by contacting the author to produce other optimized forms of signal processing programs using the Autogen technique.

Two-Dimensional Mixed Radix Mass Storage Fourier Transform (1.9)

The mass-storage mixed radix 2-dimensional DFT by R. C. Singleton is essentially an adaptation of his mixed radix program described above, to do what Fraser's program does. The main differences are that the present program is slightly easier to use, it does not require the N to be a power of 2 and it requires less instruction store. However, it is limited in its utility to 2-dimensional arrays and may not be as optimal as Fraser's.

Storage and Timing

Table 1 contains a list of the user-called subroutines with sublists of the subroutines needed for each. Memory requirements are given for each subroutine along with totals for all subroutines needed by each user-called subroutine. These, of course, do not include storage required for data and other variables given as arguments in calling the subroutines. The run-totals include space for the sample calling programs provided by the authors, data arrays, FFT subroutines, and system subroutines on the Honeywell 6080N.

Tables 2a to 2d contain the average time in milliseconds for computing the transform and the inverse of the computed transform, starting with random sequences of length $N=2^m$, where $m=1,2,\ldots,15$, with some of the user-level subroutines. Tables 3a to 3d contain the standard deviation about the mean running time of each program. One may interpret this as saying that the computing time should differ from the given time by less than the standard deviation with a probability of 68%.

Accuracy

In Table 4, the average value, over several runs, of the difference between the initial random sequences and the result of computing the DFT and then the IDFT are given. Table 5 gives the standard deviation of the errors used for Table 4. According to the analysis of Kaneko and Liu [1], the variance of the error in the DFT will be

$$\sigma_{\delta a}^2 = H(m)\sigma_a^2 \ , \tag{1}$$

where σ_a^2 is the mean of the squares of the elements of $a(n)$, $n=0, 1,...,N-1$, and

$$H(m) = 0.5(5m-6)\sigma_\epsilon^2 + 0.25(25m^2-51m+18)\mu^2 \tag{2}$$

where $m = \log_2 N$ and μ and σ_ϵ^2 are the mean and variance, respectively, of the rounding error. From the definition of the DFT used in this book, we have $\sigma_a^2 = N\sigma_x^2$.

The error in the IDFT will be the sum of two errors: one due to the error in the computed DFT and the other resulting from the IDFT calculation. The former error will be the IDFT of the DFT error $\delta a(n)$ which, by Parseval's rule, make an additive contribution $\sigma_{\delta a}^2/N$ to the variance of the error. The error produced by the IDFT calculation itself is given by a formula like that given above, except with the a's replaced by x's. Assuming independence of all errors, we get, for the variance of the resulting $x(j)$'s,

$$\sigma_{\delta x}^2 = 2H(m)\sigma_x^2 \ , \tag{3}$$

With $\sigma_x^2 = 1/3$, this gives

$$\sigma_{\delta x}^2 = (2/3)H(m) \ . \tag{4}$$

Assuming m to be moderately large, machines with proper rounding, i.e., with $\mu = 0$, will produce an error with a standard deviation of approximately

$$\sigma_{\delta x} = \sqrt{5m/3}\ \sigma_\epsilon \ . \tag{5}$$

In machines which truncate instead of rounding, where $\mu \neq 0$, the term with μ in $H(m)$ dominates so that one obtains, approximately,

$$\sigma_{\delta x} = \frac{5}{\sqrt{6}}\ m_\mu \approx 2m\mu \tag{6}$$

It can be seen in Tables 4 and 5 that the errors agree roughly with these estimates.

References

1. J. W. Cooley, P. A. W. Lewis, and P. D. Welch, "The Fast Fourier Transform Algorithm and its Application", *IEEE Trans. on Education,* Vol. E-12, No. 1, pp. 27-34, March 1969.

2. T. Kaneko and B. Liu, "Accumulation of Round-Off Error in Fast Fourier Transforms", *J. Assoc. Comp. Mach.,* Vol. 17, No. 4, pp. 637-654, October 1970.

Table 1
Memory Requirements on Honeywell 6080N
with FORTRAN-Y Compiler

"Run-Total" includes all the FFT subroutines, the main test as prepared by authors, and the system subroutines required to run on the Honeywell 6080N computer.

"Total" is the memory requirement for all the FFT subroutines in the package.

"Subtotals" indicate memory necessary for independent subpackages.

1.1 FOUREA - A Short Demonstration Version of the FFT
C. M. Rader

Subroutine Name	Words
FOUREA	260
Run-Total=10K	Total= 260

1.2 Fast Fourier Transform Algorithms
G. D. Bergland and M. T. Dolan

Subroutine Name	Words
FAST	220
FSST	220
FR2TR	50
FR4TR	570
FR4SYN	570
FORD1	80
FORD2	230
FFA	260
FFS	260
R2TR	50
R4TR	90
R8TR	1060
R4SYN	90
R8SYN	1090
ORD1	80
ORD2	230
FFT842	500
R2TX	80
R4TX	150
R8TX	700
Run-Total=17K	Total= 6580

Subpackages	Subtotals
FAST,FSST,FR2TR,FR4TR,FR4SYN,FORD1,FORD2	1940
FFA,FFS,R2TR,R4TR,R8TR,R4SYN,R8SYN,ORD1,ORD2	3210
FFT842,R2TX,R4TX,R8TX	1430
FAST,FR2TR,FR4TR,FORD1,FORD2	1150
FSST,FR2TR,FR4SYN,FORD1,FORD2	1150
FFA,R2TR,R4TR,R8TR,ORD1,ORD2	1770
FFS,R2TR,R4SYN,R8SYN,ORD1,ORD2	1800

1.3 FFT Subroutines for Sequences with Special Properties

L. R. Rabiner

Subroutine Name	Words
FFTSYM	240
IFTSYM	300
FFTASM	210
IFTASM	220
FFTOHM	250
IFTOHM	200
FFTSOH	180
IFTSOH	270
FFTAOH	180
IFTAOH	240
FAST,FSST-Subpackage from 1.2	1940
Run-Total=24K Total=	4230

Subpackages	Subtotals
FFTSYM,FAST-Subpackage	1390
IFTSYM,FSST-Subpackage	1450
FFTSYM,IFTSYM,FAST-FSST-Subpackage	2480
FFTASM,FAST-Subpackage	1360
IFTASM,FSST-Subpackage	1370
FFTASM,IFTASM,FAST-FSST-Subpackage	2370
FFTOHM,FAST-Subpackage	1400
IFTOHM,FSST-Subpackage	1350
FFTOHM,IFTOHM,FAST-FSST-Subpackage	2390
FFTSOH,FFTOHM,FAST-Subpackage	1580
IFTSOH,IFTOHM,FSST-Subpackage	1620
FFTSOH,IFTSOH,FFTOHM,IFTOHM,FAST-FSST-Subpackage	2840
FFTAOH,FFTOHM,FAST-Subpackage	1580
IFTAOH,IFTOHM,FSST-Subpackage	1590
FFTAOH,IFTAOH,FFTOHM,IFTOHM,FAST-FSST-Subpackage	2810

1.4 Mixed Radix Fast Fourier Transforms

R. C. Singleton

Subroutine Name	Words
FFT	350
FFTMX	1690
REALS	260
REALT	370
SORTG	340
NORMAL	60
RMS	110
ISTKGT	110
ISTKRL	150
Run-Total=35K Total=	3440

Subpackages	Subtotals
FFT,FFTMX,ISTKGT,ISTKRL-Subpackage	2300
FFT,FFTMX,ISTKGT,ISTKRL,REALS-Subpackage	2560
FFT,FFTMX,ISTKGT,ISTKRL,REALT-Subpackage	2670
FFT,FFTMX,ISTKGT,ISTKRL,REALS,REALT-Subpackage	2930

1.5 Optimized Mass Storage FFT

D. Fraser

Subroutine Name	Words
RMFFT	120
CMFFT	270
MFCOMP	380
MFSORT	230
MFREV	320
MFLOAD	150
MFINDX	170
FMSUM	90
MFRCMP	830
MFRLOD	280
MFPAR	340
DMPERM	140
RANMF	60
NAIVE	350
MFREAD	50
MFWRIT	60
Run-Total=19K	Total= 3840

1.6 Chirp z-Transform Algorithm Program

L. R. Rabiner

Subroutine Name	Words
CZT	760
RECUR	140
DECUR	130
FFT842-Subpackage from 1.2	1430
Run-Total=17K	Total= 2460

1.7 Complex General-N Winograd Fourier Transform Algorithm (WFTA)

J. H. McClellan and H. Nawab

Subroutine Name	Words
INISHL	600
CONST	600
WFTA	42470
WE1AVE	1280
WEAVE2	1480
Run-Total=65K	Total= 46430

1.8 Time-Efficient Radix-4 Fast Fourier Transform

L. R. Morris

Subroutine Name	Words
RADIX4	6900
RAD4SB	90
Run-Total=21K	Total= 6990

1.9 Two-Dimensional Mixed Radix Mass Storage Fourier Transform

R. C. Singleton

Subroutine Name	Words
FFT2T	350
FFT2I	350
XFR	40
TRNSP	80
EXCH	50
MFREAD	50
MFWRIT	40
Run-Total=71K	Total= 960

Subpackages	Subtotals
FFT2T,XFR,TRNSP,MFWRIT,MFREAD,EXCH-Subpackage	610
FFT2I,XFR,TRNSP,MFWRIT,MFREAD,EXCH-Subpackage	610

Table 2a
Average Time in ms.

N	FOUREA	FAST-FSST	FFA-FFS	FFT842	FFT
2	0.767	1.020	0.998	0.820	0.888
4	1.510	1.510	1.100	1.060	1.100
8	3.190	1.810	1.740	2.040	2.370
16	6.790	2.960	2.360	3.370	3.110
32	14.700	4.820	4.110	6.000	6.230
64	31.200	8.770	7.370	12.000	11.800
128	66.900	17.400	14.500	25.600	28.800
256	143.000	35.200	29.200	52.600	54.700
512	304.000	75.000	60.900	110.000	122.000
1024	649.000	156.000	134.000	250.000	261.000
2048	1380.000	339.000	282.000	533.000	629.000
4096	2900.000	706.000	595.000	1120.000	1240.000
8192	6090.000	1520.000	1290.000	2470.000	2740.000
16384	12900.000	3180.000	2680.000	5140.000	5680.000
32768	27200.000	6790.000	5680.000	10800.000	12800.000

Table 2b
Average Time in ms.
Sequences with Special Properties

N	FFTSYM-IFTSYM	FFTASM-IFTASM	FFTOHM-IFTOHM	FFTSOH-IFTSOH	FFTAOH-IFTAOH
2	0.225	0.213	0.202	0.211	0.208
4	1.560	1.450	1.480	0.225	0.219
8	2.170	1.960	2.080	2.100	1.930
16	2.770	2.500	2.760	2.910	2.680
32	4.380	4.000	4.190	3.780	3.480
64	7.230	6.450	6.950	5.840	5.430
128	13.300	12.000	12.600	9.880	9.060
256	26.100	23.200	24.700	17.900	16.700
512	52.300	46.500	49.500	35.100	32.400
1024	109.000	97.400	103.000	69.800	64.800
2048	223.000	200.000	212.000	143.000	133.000
4096	467.000	421.000	445.000	286.000	266.000
8192	971.000	882.000	933.000	607.000	569.000
16384	2040.000	1860.000	1980.000	1250.000	1170.000
32768	4230.000	3870.000	4080.000	2600.000	2440.000

Table 2c
Average Time in ms.
WFTA

N	Initial	Subsequent
4	1.540	0.838
8	2.210	1.380
16	3.980	2.450
30	8.490	5.330
60	16.400	9.890
120	34.500	20.900
240	81.700	47.700
504	173.000	108.000
1008	416.000	252.000
2520	1040.000	728.000
5040	2360.000	1670.000

Table 2d
Average Time in ms.
Radix4

N	Initial	Subsequent
16	3.640	1.550
64	13.500	7.180
256	62.300	37.200
1024	295.000	188.000

Table 3a
Standard Deviation in ms.

N	FOUREA	FAST-FSST	FFA-FFS	FFT842	FFT
2	0.053	0.073	0.067	0.049	0.011
4	0.026	0.090	0.043	0.016	0.016
8	0.034	0.035	0.043	0.120	0.066
16	0.084	0.084	0.046	0.046	0.061
32	0.076	0.051	0.152	0.094	0.028
64	0.079	0.048	0.128	0.159	0.029
128	0.130	0.009	0.034	0.043	0.059
256	0.129	0.032	0.020	0.131	0.100
512	0.869	0.152	0.158	0.587	1.000
1024	1.740	0.401	0.170	1.300	0.694
2048	6.790	0.994	0.728	3.030	2.560
4096	48.400	8.750	7.750	16.400	16.400
8192	48.500	12.300	7.240	14.000	13.200
16384	105.000	11.700	27.400	41.200	19.400
32768	551.000	27.100	10.800	98.200	152.000

Table 3b
Standard Deviation in ms.
Sequences with Special Properties

N	FFTSYM-IFTSYM	FFTASM-IFTASM	FFTOHM-IFTOHM	FFTSOH-IFTSOH	FFTAOH-IFTAOH
2	0.006	0.012	0.006	0.005	0.006
4	0.070	0.079	0.030	0.018	0.007
8	0.045	0.031	0.034	0.096	0.041
16	0.038	0.043	0.126	0.139	0.096
32	0.055	0.131	0.045	0.158	0.076
64	0.061	0.056	0.092	0.062	0.067
128	0.045	0.186	0.026	0.091	0.060
256	0.089	0.042	0.035	0.035	0.120
512	0.162	0.045	0.053	0.144	0.065
1024	0.257	0.170	0.163	0.111	0.156
2048	0.138	1.730	0.258	0.185	0.413
4096	4.800	4.460	4.230	2.760	0.319
8192	9.510	10.600	11.300	6.030	6.070
16384	17.000	16.800	3.680	4.460	9.620
32768	45.500	29.600	29.000	26.400	28.400

Table 3c
Standard Deviation in ms.
WFTA

N	Initial	Subsequent
4	0.197	0.049
8	0.023	0.161
16	0.051	0.029
30	0.092	0.051
60	0.292	0.139
120	0.441	0.173
240	0.863	0.413
504	1.450	1.130
1008	3.630	1.670
2520	7.120	4.380
5040	11.000	6.350

Table 3d
Standard Deviation in ms.
Radix4

N	Initial	Subsequent
16	0.048	0.029
64	0.167	0.085
256	0.819	0.446
1024	3.000	1.500

Table 4
Sum of Squares Error

N	FOUREA	FAST-FSST	FFA-FFS	FFT842	FFT
2	0.	0.	0.	0.	0.
4	0.72164E-15	0.88818E-15	0.18128E-15	0.27756E-15	0.40246E-15
8	0.66613E-15	0.19429E-15	0.30618E-15	0.11241E-14	0.84655E-15
16	0.37331E-14	0.21272E-15	0.18501E-14	0.70499E-14	0.36533E-14
32	0.13850E-13	0.91221E-14	0.47967E-14	0.13177E-13	0.92981E-14
64	0.26326E-13	0.78466E-14	0.19558E-13	0.32644E-13	0.62922E-13
128	0.46768E-13	0.28238E-13	0.61767E-13	0.68565E-13	0.74431E-13
256	0.30520E-12	0.18360E-12	0.50912E-13	0.11421E-12	0.24460E-12
512	0.44593E-12	0.15395E-12	0.20711E-12	0.26937E-12	0.55185E-12
1024	0.89891E-12	0.34031E-12	0.35218E-12	0.62708E-12	0.14303E-11
2048	0.17115E-11	0.61706E-12	0.61415E-12	0.13545E-11	0.36801E-11
4096	0.36367E-11	0.19966E-11	0.20315E-11	0.50954E-11	0.82930E-11
8192	0.92585E-11	0.40613E-11	0.56809E-11	0.12832E-10	0.14287E-10
16384	0.25433E-10	0.14997E-10	0.95540E-11	0.17858E-10	0.33178E-10
32768	0.45252E-10	0.20410E-10	0.25955E-10	0.33980E-10	0.86588E-10

Table 5
Standard Deviation of Error

N	FOUREA	FAST-FSST	FFA-FFS	FFT842	FFT
2	0.	0.	0.	0.	0.
4	0.23264E-07	0.25810E-07	0.11660E-07	0.14428E-07	0.17374E-07
8	0.15805E-07	0.85357E-08	0.10715E-07	0.20531E-07	0.17817E-07
16	0.26457E-07	0.63155E-08	0.18625E-07	0.36357E-07	0.26172E-07
32	0.36034E-07	0.29244E-07	0.21206E-07	0.35147E-07	0.29525E-07
64	0.35128E-07	0.19178E-07	0.30278E-07	0.39118E-07	0.54309E-07
128	0.33108E-07	0.25726E-07	0.38048E-07	0.40087E-07	0.41767E-07
256	0.59805E-07	0.46385E-07	0.24426E-07	0.36584E-07	0.53539E-07
512	0.51116E-07	0.30035E-07	0.34836E-07	0.39728E-07	0.56863E-07
1024	0.51318E-07	0.31575E-07	0.32121E-07	0.42862E-07	0.64734E-07
2048	0.50071E-07	0.30065E-07	0.29994E-07	0.44544E-07	0.73422E-07
4096	0.51610E-07	0.38241E-07	0.38574E-07	0.61090E-07	0.77936E-07
8192	0.58229E-07	0.38565E-07	0.45611E-07	0.68552E-07	0.72333E-07
16384	0.68241E-07	0.52403E-07	0.41826E-07	0.57184E-07	0.77942E-07
32768	0.64366E-07	0.43227E-07	0.48747E-07	0.55776E-07	0.89036E-07

1.1

FOUREA - A Short Demonstration Version of the FFT

C. M. Rader

MIT Lincoln Laboratory
Lexington, MA 02173

1. Purpose

This program can be used to evaluate the discrete Fourier transform (DFT), $\{X(k);\ k=0,1,...,N-1\}$, of an N point sequence of complex numbers, $\{x(n);\ n=0,1,...,N-1\}$, where N must be a power of two, $N = 2^M$. It can also evaluate the inverse DFT. The DFT of $\{x(n)\}$ is defined as

$$X(k) = \sum_{n=0}^{N-1} x(n) e^{-j\frac{2\pi}{N}nk} ; \quad k = 0,1,...,N-1 \qquad (1)$$

The inverse DFT is defined as

$$x(n) = \frac{1}{N} \sum_{k=0}^{N-1} X(k) e^{j\frac{2\pi}{N}nk} ; \quad n = 0,1,...,N-1 \qquad (2)$$

Unlike other programs in this collection, FOUREA is not highly recommended for actual use. Other programs are provided which compute the DFT for more diverse values of N, for multidimensional sequences, and with shorter running time. FOUREA is intended mainly for educational and demonstration purposes.

2. Method

The program, which expects the input to be a complex array, is an implementation of the radix-2 Cooley-Tukey algorithm [1]. The input is first rearranged into "bit-reversed" order by the DO 80 loop, then $\log_2 N$ stages of "butterflies" are performed by statements 90 to 130. The computation is performed "in-place". The internal data structure is represented by Fig. 5 of Ref. 2.

3. Usage

The program consists of a subroutine, FOUREA, called by

CALL FOUREA (DATA,N,ISIGN).

- DATA — is the name of an array of N complex numbers (or $2N$ real numbers, in which case DATA(2 * I−1) + jDATA(2 * I) is treated as a complex number.)
- N — is the length of the sequence being transformed and must be a power of two.
- ISIGN — is an integer which should be −1 to compute the DFT defined by Eq. (1). If ISIGN is +1 , Eq. (2) is used to compute the inverse DFT. Note that the normalization by N is included.

The array DATA serves as both the input array and the output array, e.g., before the call to FOUREA, DATA holds the sequence to be transformed, whereas after the call it holds the transform. The indexing is displaced such that $x(n)$ is stored in array location $n+1$ and $X(k)$ is in array location $k+1$.

4. Test Problem

To test any DFT program, several approaches are available. We have chosen to compare the output of FOUREA with the predicted DFT of a sequence whose DFT has a simple algebraic form. The sequence we are using is

$$x(n) = Q^n \qquad n = 0, 1, \ldots, N-1 \tag{3}$$

and its DFT is

$$X(k) = (1-Q^N)/(1-QW^k) \qquad k = 0, 1, \ldots, N-1 \tag{4}$$

where

$$W = e^{-j\frac{2\pi}{N}}$$

and Q is the complex constant $0.9 + j0.3$. In the test we use $N = 2^5$. The test program computes and prints out $\{x(n)\}$, using Eq. (3), and $\{X(k)\}$, first using FOUREA and then using Eq. (4). Next, $\{x(n)\}$, computed as the inverse DFT of Eq. (4) using FOUREA, is printed. Finally the program prints the maximum absolute difference between $\{X(k)\}$ and the computed DFT of Eq. (3), and between $\{x(n)\}$ and the computed inverse DFT of Eq. (4). These differences should be small compared to unity -- otherwise the program has failed. The test was run on a Honeywell 6080N computer in the example reproduced in Table 1.

References

1. J. W. Cooley and J. W. Tukey, "An algorithm for the machine calculation of complex Fourier series", *Math. Comp.*, Vol. 19, No. 90, pp. 297-301, April 1965.

2. W. T. Cochran, et al, "What is the Fast Fourier Transform?", *IEEE Trans. Audio and Electroacoust.*, Vol. AU-15, No. 2, pp. 45-55, June 1967.

Table 1

COMPLEX INPUT SEQUENCE

(1)	0.100000E 01	0.	(2)	0.900000E 00	0.300000E 00
(3)	0.720000E 00	0.540000E 00	(4)	0.486000E 00	0.702000E 00
(5)	0.226800E 00	0.777600E 00	(6)	-0.291600E-01	0.767880E 00
(7)	-0.256608E 00	0.682344E 00	(8)	-0.435650E 00	0.537127E 00
(9)	-0.553224E 00	0.352719E 00	(10)	-0.603717E 00	0.151480E 00
(11)	-0.588789E 00	-0.447828E-01	(12)	-0.516476E 00	-0.216941E 00
(13)	-0.399746E 00	-0.350190E 00	(14)	-0.254714E 00	-0.435095E 00
(15)	-0.987144E-01	-0.467999E 00	(16)	0.515569E-01	-0.450814E 00
(17)	0.181645E 00	-0.390265E 00	(18)	0.280560E 00	-0.296745E 00
(19)	0.341528E 00	-0.182903E 00	(20)	0.362246E 00	-0.621539E-01
(21)	0.344667E 00	0.527352E-01	(22)	0.294380E 00	0.150862E 00
(23)	0.219684E 00	0.224090E 00	(24)	0.130488E 00	0.267586E 00
(25)	0.371637E-01	0.279974E 00	(26)	-0.505448E-01	0.263125E 00
(27)	-0.124428E 00	0.221649E 00	(28)	-0.178480E 00	0.162156E 00
(29)	-0.209279E 00	0.923965E-01	(30)	-0.216070E 00	0.203732E-01
(31)	-0.200575E 00	-0.464851E-01	(32)	-0.166572E 00	-0.102009E 00

FOUREA DFT

(1)	0.693973E 00	0.349972E 01	(2)	0.279227E 01	0.805046E 01
(3)	0.940296E 01	-0.913501E 01	(4)	0.186645E 01	-0.383383E 01
(5)	0.113182E 01	-0.223416E 01	(6)	0.904794E 00	-0.153463E 01
(7)	0.799557E 00	-0.113961E 01	(8)	0.739606E 00	-0.882314E 00
(9)	0.700862E 00	-0.698565E 00	(10)	0.673576E 00	-0.558478E 00
(11)	0.653109E 00	-0.446245E 00	(12)	0.636991E 00	-0.352689E 00
(13)	0.623788E 00	-0.272086E 00	(14)	0.612613E 00	-0.200642E 00
(15)	0.602883E 00	-0.135703E 00	(16)	0.594200E 00	-0.753136E-01
(17)	0.586277E 00	-0.179492E-01	(18)	0.578900E 00	0.376517E-01
(19)	0.571899E 00	0.926070E-01	(20)	0.565136E 00	0.147983E 00
(21)	0.558492E 00	0.204881E 00	(22)	0.551859E 00	0.264522E 00
(23)	0.545134E 00	0.328365E 00	(24)	0.538214E 00	0.398257E 00
(25)	0.531002E 00	0.476678E 00	(26)	0.523404E 00	0.567132E 00
(27)	0.515362E 00	0.674850E 00	(28)	0.506926E 00	0.808101E 00
(29)	0.498467E 00	0.980906E 00	(30)	0.491389E 00	0.121921E 01
(31)	0.490732E 00	0.157708E 01	(32)	0.517354E 00	0.218883E 01

Table 1
(Continued)

```
THEORETICAL DFT
 (  1)  0.693973E 00  0.349972E 01  (  2)  0.279227E 01  0.805046E 01
 (  3)  0.940296E 01 -0.913501E 01  (  4)  0.186645E 01 -0.383383E 01
 (  5)  0.113182E 01 -0.223416E 01  (  6)  0.904794E 00 -0.153463E 01
 (  7)  0.799557E 00 -0.113961E 01  (  8)  0.739606E 00 -0.882314E 00
 (  9)  0.700862E 00 -0.698565E 00  ( 10)  0.673576E 00 -0.558478E 00
 ( 11)  0.653109E 00 -0.446245E 00  ( 12)  0.636991E 00 -0.352689E 00
 ( 13)  0.623788E 00 -0.272086E 00  ( 14)  0.612613E 00 -0.200642E 00
 ( 15)  0.602883E 00 -0.135703E 00  ( 16)  0.594200E 00 -0.753137E-01
 ( 17)  0.586277E 00 -0.179492E-01  ( 18)  0.578900E 00  0.376517E-01
 ( 19)  0.571898E 00  0.926070E-01  ( 20)  0.565136E 00  0.147983E 00
 ( 21)  0.558492E 00  0.204881E 00  ( 22)  0.551859E 00  0.264522E 00
 ( 23)  0.545134E 00  0.328365E 00  ( 24)  0.538214E 00  0.398257E 00
 ( 25)  0.531002E 00  0.476678E 00  ( 26)  0.523404E 00  0.567132E 00
 ( 27)  0.515362E 00  0.674850E 00  ( 28)  0.506926E 00  0.808101E 00
 ( 29)  0.498467E 00  0.980906E 00  ( 30)  0.491389E 00  0.121921E 01
 ( 31)  0.490732E 00  0.157708E 01  ( 32)  0.517354E 00  0.218883E 01


FOUREA INVERSE DFT
 (  1)  0.100000E 01  0.372529E-08  (  2)  0.900000E 00  0.300000E 00
 (  3)  0.720000E 00  0.540000E 00  (  4)  0.486000E 00  0.702000E 00
 (  5)  0.226800E 00  0.777600E 00  (  6) -0.291600E-01  0.767880E 00
 (  7) -0.256608E 00  0.682344E 00  (  8) -0.435650E 00  0.537127E 00
 (  9) -0.553224E 00  0.352719E 00  ( 10) -0.603717E 00  0.151480E 00
 ( 11) -0.588789E 00 -0.447828E-01  ( 12) -0.516476E 00 -0.216941E 00
 ( 13) -0.399746E 00 -0.350190E 00  ( 14) -0.254714E 00 -0.435095E 00
 ( 15) -0.987144E-01 -0.467999E 00  ( 16)  0.515569E-01 -0.450814E 00
 ( 17)  0.181645E 00 -0.390265E 00  ( 18)  0.280560E 00 -0.296745E 00
 ( 19)  0.341528E 00 -0.182903E 00  ( 20)  0.362246E 00 -0.621539E-01
 ( 21)  0.344667E 00  0.527352E-01  ( 22)  0.294380E 00  0.150862E 00
 ( 23)  0.219684E 00  0.224090E 00  ( 24)  0.130488E 00  0.267586E 00
 ( 25)  0.371637E-01  0.279974E 00  ( 26) -0.505448E-01  0.263125E 00
 ( 27) -0.124428E 00  0.221649E 00  ( 28) -0.178480E 00  0.162156E 00
 ( 29) -0.209279E 00  0.923965E-01  ( 30) -0.216070E 00  0.203732E-01
 ( 31) -0.200575E 00 -0.464851E-01  ( 32) -0.166572E 00 -0.102009E 00

MAX DIFF BETWEEN THEOR AND FOUREA DFT IS     0.238E-06

MAX DIFF BETWEEN ORIGINAL DATA AND INVERSE DFT IS     0.263E-07
```

Appendix

```
C
C-----------------------------------------------------------------------
C MAIN PROGRAM: FOURSUBT
C AUTHOR:       C. M. RADER
C               MIT LINCOLN LABORATORY, LEXINGTON, MA 02173
C
C INPUT:        FUNCTION IS GENERATED BY THE PROGRAM TO TEST
C               SUBROUTINE FOUREA
C-----------------------------------------------------------------------
C
      COMPLEX W, C(32), D, E, B(32), QB(32), A
C
C SEQUENCE LENGTH N=2**MU; USES MU=5; THUS N=32
C METHOD IS TO COMPUTE DFT OF KNOWN FUNCTION; A**I, I=0,1,...,N-1
C OUTPUT OF PROGRAM IS LARGEST DIFFERENCE BETWEEN DFT COMPUTED TWO WAYS.
C WHEN RUN ON AN IBM 370, THE MAX DIFFERENCE WAS 0.389E-04
C WHEN RUN ON A HONEYWELL 6080N, THE MAX DIFFERENCE WAS 0.238E-06
C A MAX DIFFERENCE LARGE COMPARED TO THE ACCURACY OF THE COMPUTER
C WOULD PROBABLY INDICATE A PROGRAM ERROR.
C
      MU = 5
C
C SET UP MACHINE CONSTANT
C
      IOUTD = I1MACH(2)
      NN = 2**MU
      TPI = 8.*ATAN(1.)
      TPION = TPI/FLOAT(NN)
      W = CMPLX(COS(TPION),-SIN(TPION))
C
C GENERATE A**K AS TEST FUNCTION
C
      A = (.9,.3)
      B(1) = (1.,0.)
      QB(1) = B(1)
      DO 10 K=2,NN
        B(K) = A**(K-1)
        QB(K) = B(K)
  10  CONTINUE
C
C PRINT COMPLEX INPUT SEQUENCE
C
      WRITE (IOUTD,9999)
9999  FORMAT (1H1//////24H COMPLEX INPUT SEQUENCE )
      WRITE (IOUTD,9998) (I,QB(I),I=1,NN)
9998  FORMAT (2(2X, 1H(, I3, 1H), 2E14.6))
C
C B(1) CONTAINS A**0; B(K) CONTAINS A**K-1; ETC
C
C COMPUTE DFT OF B IN CLOSED FORM
C
      D = (1.,0.) - A**NN
      DO 20 K=1,NN
        E = (1.,0.) - A*W**(K-1)
        C(K) = D/E
  20  CONTINUE
C
C DFT OF B IS (1-A**NN)/(1-AW**K)
C
C NOW COMPUTE DFT OF B USING FOUREA
C
      CALL FOUREA(B, NN, -1)
C
C PRINT FOUREA DFT AND THEORETICAL DFT
C
      WRITE (IOUTD,9997)
9997  FORMAT (/12H FOUREA DFT )
      WRITE (IOUTD,9998) (I,B(I),I=1,NN)
      WRITE (IOUTD,9996)
9996  FORMAT (/17H THEORETICAL DFT )
      WRITE (IOUTD,9998) (I,C(I),I=1,NN)
C
C FIND MAX DIFFERENCE BETWEEN B AND C
C
      DD = 0.
      DO 30 I=1,NN
        DE = CABS(C(I)-B(I))
        IF (DD.GT.DE) GO TO 30
        DD = DE
  30  CONTINUE
C
C COMPUTE INVERSE DFT OF C
C
      CALL FOUREA(C, NN, 1)
C
C PRINT INVERSE DFT
C
      WRITE (IOUTD,9995)
9995  FORMAT (1H1//////20H FOUREA INVERSE DFT )
      WRITE (IOUTD,9998) (I,C(I),I=1,NN)
C
C COMPUTE MAX. DIFF. BETWEEN INPUT AND INVERSE OF THEOR. DFT
C
      GG = 0.
      DO 40 I=1,NN
        GG1 = CABS(QB(I)-C(I))
        IF (GG1.LE.GG) GO TO 40
        GG = GG1
  40  CONTINUE
C
C PRINT MAXIMUM DIFFERENCE
C
      WRITE (IOUTD,9994) DD
      WRITE (IOUTD,9993) GG
9994  FORMAT (/42H MAX DIFF BETWEEN THEOR AND FOUREA DFT IS , E12.3)
9993  FORMAT (/51H MAX DIFF BETWEEN ORIGINAL DATA AND INVERSE DFT IS ,
     *    E12.3)
      STOP
      END
C
C-----------------------------------------------------------------------
C SUBROUTINE: FOUREA
C PERFORMS COOLEY-TUKEY FAST FOURIER TRANSFORM
C-----------------------------------------------------------------------
C
      SUBROUTINE FOUREA(DATA, N, ISI)
C
C THE COOLEY-TUKEY FAST FOURIER TRANSFORM IN ANSI FORTRAN
C
C DATA IS A ONE-DIMENSIONAL COMPLEX ARRAY WHOSE LENGTH, N, IS A
C POWER OF TWO.  ISI IS +1 FOR AN INVERSE TRANSFORM AND -1 FOR A
C FORWARD TRANSFORM.  TRANSFORM VALUES ARE RETURNED IN THE INPUT
C ARRAY, REPLACING THE INPUT.
C TRANSFORM(J)=SUM(DATA(I)*W**((I-1)*(J-1))), WHERE I AND J RUN
C FROM 1 TO N AND W = EXP (ISI*2*PI*SQRT(-1)/N).  PROGRAM ALSO
C COMPUTES INVERSE TRANSFORM, FOR WHICH THE DEFINING EXPRESSION
C IS INVTR(J)=(1/N)*SUM(DATA(I)*W**((I-1)*(J-1))).
C RUNNING TIME IS PROPORTIONAL TO N*LOG2(N), RATHER THAN TO THE
C CLASSICAL N**2.
C AFTER PROGRAM BY BRENNER, JUNE 1967. THIS IS A VERY SHORT VERSION
```

```
C OF THE FFT AND IS INTENDED MAINLY FOR DEMONSTRATION. PROGRAMS
C ARE AVAILABLE IN THIS COLLECTION WHICH RUN FASTER AND ARE NOT
C RESTRICTED TO POWERS OF 2 OR TO ONE-DIMENSIONAL ARRAYS.
C SEE -- IEEE TRANS AUDIO (JUNE 1967), SPECIAL ISSUE ON FFT.
C
      COMPLEX DATA(1)
      COMPLEX TEMP, W
      IOUTD = I1MACH(2)
C
C CHECK FOR POWER OF TWO UP TO 15
C
      NN = 1
      DO 10 I=1,15
        M = I
        NN = NN*2
        IF (NN.EQ.N) GO TO 20
   10 CONTINUE
      WRITE (IOUTD,9999)
 9999 FORMAT (30H N NOT A POWER OF 2 FOR FOUREA)
      STOP
   20 CONTINUE
C
      PI = 4.*ATAN(1.)
      FN = N
C
C THIS SECTION PUTS DATA IN BIT-REVERSED ORDER
C
      J = 1
      DO 80 I=1,N
C
C AT THIS POINT, I AND J ARE A BIT REVERSED PAIR (EXCEPT FOR THE
C DISPLACEMENT OF +1)
C
        IF (I-J) 30, 40, 40
C
C EXCHANGE DATA(I) WITH DATA(J) IF I.LT.J.
C
   30   TEMP = DATA(J)
        DATA(J) = DATA(I)
        DATA(I) = TEMP
C
C IMPLEMENT J=J+1, BIT-REVERSED COUNTER
C
   40   M = N/2
   50   IF (J-M) 70, 70, 60
   60   J = J - M
        M = (M+1)/2
        GO TO 50
   70   J = J + M
   80 CONTINUE
C
C NOW COMPUTE THE BUTTERFLIES
C
      MMAX = 1
   90 IF (MMAX-N) 100, 130, 130
  100 ISTEP = 2*MMAX
      DO 120 M=1,MMAX
        THETA = PI*FLOAT(ISI*(M-1))/FLOAT(MMAX)
        W = CMPLX(COS(THETA),SIN(THETA))
        DO 110 I=M,N,ISTEP
          J = I + MMAX
          TEMP = W*DATA(J)
          DATA(J) = DATA(I) - TEMP
          DATA(I) = DATA(I) + TEMP
  110   CONTINUE
  120 CONTINUE
      MMAX = ISTEP
      GO TO 90
  130 IF (ISI) 160, 140, 140
C
C FOR INV TRANS -- ISI=1 -- MULTIPLY OUTPUT BY 1/N
C
  140 DO 150 I=1,N
        DATA(I) = DATA(I)/FN
  150 CONTINUE
  160 RETURN
      END
```

1.2

Fast Fourier Transform Algorithms

G. D. Bergland and *M. T. Dolan*
Bell Laboratories
Murray Hill, NJ 07974

1. Purpose

Fast Fourier analysis and fast Fourier synthesis algorithms are presented for computing both the discrete Fourier transform (DFT) of a real or complex sequence and the inverse DFT (IDFT) of a complex sequence.

2. Method

2.1 DFT (Real Input Sequence)

The subroutines FAST (radix 4-2) and FFA (radix 8-4-2) evaluate the equation

$$X(k) = \sum_{n=0}^{N-1} x(n) e^{-j\frac{2\pi}{N}nk} \tag{1}$$

where $x(n)$ is real, $X(k)$ is complex, and $k=0,1,...,N/2$. Only $(N/2+1)$ complex DFT values need be computed and stored because $x(n)$ is real and by symmetry

$$X(N-k)=X^*(k) \tag{2}$$

for $k=0,1,...,N/2$ and * denoting the complex conjugate.

The algorithms preserve the order and symmetry of the original Cooley-Tukey algorithm [1] but eliminate the operations which are redundant for real series, thus effecting a two-to-one reduction in the calculations and storage requirements [2].

The radix 8-4-2 routine performs as many base 8 iterations as possible and then performs one base 4 or a base 2 iteration, if necessary. Similarly, the radix 4-2 routine performs as many base 4 iterations as possible and then performs one base 2 iteration, if necessary. Thus, they can be used for any value of N which is an integral power of 2.

2.2 DFT (Complex Input Sequence)

The subroutine FFT842 (radix 8-4-2) evaluates (1) for a complex input sequence $x(n)$, $0 \leqslant n \leqslant N-1$, and for $0 \leqslant k \leqslant N-1$. With N factored into powers of 8, 4, and 2, the summed products are computed iteratively very efficiently by using equations which extend the recursive equations of the Cooley-Tukey algorithm [3,4].

2.3 Inverse DFT (Real Output Sequence)

The subroutines FSST (radix 4-2) and FFS (radix 8-4-2) evaluate the equation

$$x(n)=\frac{1}{N}\sum_{k=0}^{N-1} X(k) e^{j\frac{2\pi}{N}nk} \tag{3}$$

for $n=0,1,...,N-1$.

These subroutines use the fact that $X(k)$ has complex conjugate symmetry around $k=N/2$ in order to evaluate (3) in an efficient manner from the $(N/2+1)$ complex values of $X(k)$.

Effectively, these inverse algorithms are a specialization of the Sande-Tukey algorithm [5]. They are further characterized, as are all the algorithms, by performing in-place reordering and having no stored tables of sines or cosines.

2.4 Inverse DFT (Complex Output Sequence)

FFT842 computes the the inverse DFT by using the relationship

$$x(n)=\frac{1}{N}\left[\sum_{k=0}^{N-1} X(k)^* e^{-j\frac{2\pi}{N}nk}\right]^* \tag{4}$$

for $n=0,1,...,N-1$ and * denoting the complex conjugate operation. Thus to compute an inverse DFT, the sequence $X(k)$ is complex conjugated, a direct DFT is taken, and the resulting output is complex conjugated and normalized by the factor $1/N$.

3. Usage

3.1 For a DFT (Real Input Sequence X)

CALL FAST(X,N) (radix 4-2)
or
CALL FFA(X,N) (radix 8-4-2)

where

X Real array dimensioned to size N+2. On input, X contains the N-point sequence to be transformed (stored in locations 1 to N) and on output, X contains the (N/2+1) complex values of the transform. On output, the real components are stored in the odd locations of the array and the imaginary components are stored in the even locations.

N Size of the input sequence. N must be a power of 2 and $2 \leqslant N \leqslant 2^{15}$.

3.2 For a DFT (Complex Input Sequence X)

CALL FFT842(IN,N,X,Y) (radix 8-4-2)

where

IN Input parameter set to 0 for a DFT.

N Size of the input sequence. N must be a power of 2 and $2 \leqslant N \leqslant 2^{15}$.

X Real array dimensioned to size N. On input, X contains the real components of the sequence to be transformed. On output, X contains the real components of the transform. On input and output, the N values of X are stored in locations 1 to N.

Y Real array dimensioned to size N. On input, Y contains the imaginary components of the sequence to be transformed. On output, Y contains the imaginary components of the transform. On input and output, the N values of Y are stored in locations 1 to N.

3.3 For an Inverse DFT (Real Output Sequence X)

CALL FSST(X,N) (radix 4-2)
or
CALL FFS(X,N) (radix 8-4-2)

where

X Real array dimensioned to size N+2. On input, X contains the (N/2+1) complex values of the transform with the real components stored in the odd locations of the array and the imaginary components stored in the even locations. On output, X contains the sequence $x(n)$, $0 \leqslant n \leqslant N-1$, stored in locations 1 to N.

N Size of the sequence. N must be a power of 2 and $2 \leqslant N \leqslant 2^{15}$.

3.4 For an Inverse DFT (Complex Output Sequence X)

CALL FFT842(IN,N,X,Y) (radix 8-4-2)

where

IN Input parameter set to 1 for an inverse DFT.

N Size of the sequence. N must be a power of 2 and $2 \leqslant N \leqslant 2^{15}$.

X Real array dimensioned to size N. On input, X contains the real components of the direct transform. On output, X contains the real components of the inverse. On input and output, the N values of X are stored in locations 1 to N.

Y Real array dimensioned to size N. On input, Y contains the imaginary components of the direct transform. On output, Y contains the imaginary components of the inverse. On input and output, the N values of Y are stored in locations 1 to N.

4. Tests

The complete FAST-FSST package includes:

FAST and FSST (executive routines for DFT and IDFT)
FR2TR and FR4TR (radix 2 and radix 4 iteration routines)
FR4SYN (radix 4 synthesis routine)
FORD1 and FORD2 (in-place reordering routines)

The complete FFA-FFS package includes:

FFA and FFS (executive routines for DFT and IDFT)
R2TR, R4TR, and R8TR (radix 2, 4, and 8 iteration routines)
R4SYN and R8SYN (radix 4 and 8 synthesis routines)
ORD1 and ORD2 (in-place reordering routines)

The complete FFT842 package includes:

FFT842 (executive routine for both DFT and IDFT)
R2TX, R4TX, and R8TX (radix 2, 4, and 8 iteration routines)

The test program can be used to check out the three packages of code. The same 32-point real sequence of random numbers is used as data for each package. The random numbers are from a uniform distribution as generated by UNI (see Standards Section). The DFT and the inverse DFT are printed. In addition, a complex sequence of random numbers is used as input to FFT842. Complete output from the test program is in Table 1.

The three packages of code and the test program appear in the Appendix.

5. Acknowledgment

The contributions of Mrs. J. A. Malsbury in optimizing the FFT842 subroutine are gratefully acknowledged.

References

1. J. W. Cooley and J. W. Tukey, "An Algorithm for the Machine Calculation of Complex Fourier Series", *Math. Comp.*, Vol. 19, No. 90, pp. 297-301, April 1965.

2. G. D. Bergland, "A Radix-Eight Fast Fourier Transform Subroutine for Real-Valued Series", *IEEE Trans. on Audio and Electroacoust.*, Vol. AU-17, No. 2, pp. 138-144, June 1969.

3. G. D. Bergland, "The Fast Fourier Transform Recursive Equations for Arbitrary Length Records", *Math. Comp.*, Vol. 21, No. 98, pp. 236-238, April 1967.

4 G. D. Bergland, "A Fast Fourier Transform Algorithm Using Base 8 Iterations", *Math. Comp.*, Vol. 22, No. 102, pp. 275-279, April 1968.

5. C. Bingham, M. D. Godfrey, and J. W. Tukey, "Modern Techniques of Power Spectrum Estimation", *IEEE Trans. on Audio and Electroacoust.*, Vol. AU-15, No. 2, pp. 56-66, June 1967.

Table 1

```
TEST FAST AND FSST

REAL INPUT SEQUENCE
  0.22925607E 00   0.76687502E 00   0.68317685E 00   0.50919111E 00
  0.87455959E 00   0.64464101E 00   0.84746840E 00   0.35396343E 00
  0.39889160E 00   0.45709422E 00   0.23630936E 00   0.13318188E 00
  0.16605222E 00   0.22602680E 00   0.66245903E 00   0.25021175E 00
  0.61769668E 00   0.26246527E 00   0.51266762E 00   0.93920734E 00
  0.62402816E 00   0.42238195E 00   0.93970599E 00   0.28206823E 00
  0.46921754E 00   0.54879178E-01   0.51983086E 00   0.39682690E 00
  0.11315656E 00   0.60751725E 00   0.70150672E 00   0.88705479E 00

REAL COMPONENTS OF TRANSFORM
  0.15789569E 02   0.13026034E 01   0.11936771E 01   0.40612735E-01
 -0.46179910E 00   0.70122853E-01  -0.27579353E 00   0.46209661E 00
 -0.16102664E 01  -0.14930917E 01   0.21015245E 00  -0.93344460E 00
  0.33632980E 00  -0.12900518E 01  -0.12126616E 01  -0.12663724E 01
  0.14023971E 01

IMAG COMPONENTS OF TRANSFORM
  0.               0.19871481E 00  -0.28999118E 01   0.72163936E 00
  0.13081519E 01  -0.15931014E-01   0.10351044E 01   0.11709984E-01
  0.30982471E 00  -0.87044477E 00  -0.18331044E 00  -0.55190805E-01
 -0.10901590E 01   0.18219048E 01   0.75918928E 00  -0.10652235E 00
  0.

REAL INVERSE TRANSFORM
  0.22925607E 00   0.76687504E 00   0.68317685E 00   0.50919111E 00
  0.87455960E 00   0.64464101E 00   0.84746840E 00   0.35396343E 00
  0.39889160E 00   0.45709423E 00   0.23630936E 00   0.13318189E 00
  0.16605222E 00   0.22602681E 00   0.66245903E 00   0.25021176E 00
  0.61769669E 00   0.26246528E 00   0.51266762E 00   0.93920734E 00
  0.62402817E 00   0.42238196E 00   0.93970599E 00   0.28206824E 00
  0.46921754E 00   0.54879181E-01   0.51983087E 00   0.39682690E 00
  0.11315656E 00   0.60751726E 00   0.70150672E 00   0.88705480E 00

TEST FFA AND FFS

REAL INPUT SEQUENCE
  0.22925607E 00   0.76687502E 00   0.68317685E 00   0.50919111E 00
  0.87455959E 00   0.64464101E 00   0.84746840E 00   0.35396343E 00
  0.39889160E 00   0.45709422E 00   0.23630936E 00   0.13318188E 00
  0.16605222E 00   0.22602680E 00   0.66245903E 00   0.25021175E 00
  0.61769668E 00   0.26246527E 00   0.51266762E 00   0.93920734E 00
  0.62402816E 00   0.42238195E 00   0.93970599E 00   0.28206823E 00
  0.46921754E 00   0.54879178E-01   0.51983086E 00   0.39682690E 00
  0.11315656E 00   0.60751725E 00   0.70150672E 00   0.88705479E 00

REAL COMPONENTS OF TRANSFORM
  0.15789569E 02   0.13026034E 01   0.11936771E 01   0.40612705E-01
 -0.46179910E 00   0.70122856E-01  -0.27579353E 00   0.46209661E 00
 -0.16102664E 01  -0.14930917E 01   0.21015245E 00  -0.93344463E 00
  0.33632980E 00  -0.12900518E 01  -0.12126616E 01  -0.12663724E 01
  0.14023971E 01

IMAG COMPONENTS OF TRANSFORM
  0.               0.19871481E 00  -0.28999118E 01   0.72163937E 00
  0.13081519E 01  -0.15931029E-01   0.10351044E 01   0.11709988E-01
  0.30982471E 00  -0.87044477E 00  -0.18331045E 00  -0.55190814E-01
 -0.10901590E 01   0.18219048E 01   0.75918928E 00  -0.10652235E 00
  0.

REAL INVERSE TRANSFORM
  0.22925607E 00   0.76687504E 00   0.68317685E 00   0.50919111E 00
  0.87455960E 00   0.64464102E 00   0.84746840E 00   0.35396344E 00
  0.39889160E 00   0.45709423E 00   0.23630936E 00   0.13318189E 00
  0.16605222E 00   0.22602680E 00   0.66245904E 00   0.25021175E 00
  0.61769669E 00   0.26246528E 00   0.51266762E 00   0.93920734E 00
  0.62402817E 00   0.42238196E 00   0.93970599E 00   0.28206824E 00
  0.46921754E 00   0.54879183E-01   0.51983087E 00   0.39682691E 00
  0.11315656E 00   0.60751726E 00   0.70150673E 00   0.88705480E 00
```

Table 1
(Continued)

TEST FFT842 WITH REAL INPUT SEQUENCE

REAL COMPONENTS OF INPUT SEQUENCE			
0.22925607E 00	0.76687502E 00	0.68317685E 00	0.50919111E 00
0.87455959E 00	0.64464101E 00	0.84746840E 00	0.35396343E 00
0.39889160E 00	0.45709422E 00	0.23630936E 00	0.13318188E 00
0.16605222E 00	0.22602680E 00	0.66245903E 00	0.25021175E 00
0.61769668E 00	0.26246527E 00	0.51266762E 00	0.93920734E 00
0.62402816E 00	0.42238195E 00	0.93970599E 00	0.28206823E 00
0.46921754E 00	0.54879178E-01	0.51983086E 00	0.39682690E 00
0.11315656E 00	0.60751725E 00	0.70150672E 00	0.88705479E 00

IMAG COMPONENTS OF INPUT SEQUENCE			
0.	0.	0.	0.
0.	0.	0.	0.
0.	0.	0.	0.
0.	0.	0.	0.
0.	0.	0.	0.
0.	0.	0.	0.
0.	0.	0.	0.
0.	0.	0.	0.

REAL COMPONENTS OF TRANSFORM			
0.15789569E 02	0.13026034E 01	0.11936771E 01	0.40612735E-01
-0.46179911E 00	0.70122864E-01	-0.27579352E 00	0.46209662E 00
-0.16102664E 01	-0.14930917E 01	0.21015245E 00	-0.93344462E 00
0.33632981E 00	-0.12900518E 01	-0.12126616E 01	-0.12663724E 01
0.14023971E 01	-0.12663724E 01	-0.12126616E 01	-0.12900518E 01
0.33632981E 00	-0.93344463E 00	0.21015244E 00	-0.14930917E 01
-0.16102664E 01	0.46209661E 00	-0.27579353E 00	0.70122849E-01
-0.46179910E 00	0.40612698E-01	0.11936771E 01	0.13026034E 01

IMAG COMPONENTS OF TRANSFORM			
0.	0.19871481E 00	-0.28999118E 01	0.72163936E 00
0.13081520E 01	-0.15931033E-01	0.10351044E 01	0.11709969E-01
0.30982471E 00	-0.87044477E 00	-0.18331045E 00	-0.55190805E-01
-0.10901590E 01	0.18219048E 01	0.75918929E 00	-0.10652234E 00
0.	0.10652235E 00	-0.75918928E 00	-0.18219048E 01
0.10901590E 01	0.55190824E-01	0.18331043E 00	0.87044481E 00
-0.30982471E 00	-0.11709984E-01	-0.10351044E 01	0.15931014E-01
-0.13081520E 01	-0.72163937E 00	0.28999118E 01	-0.19871484E 00

REAL COMPONENTS OF INVERSE TRANSFORM			
0.22925607E 00	0.76687504E 00	0.68317685E 00	0.50919111E 00
0.87455961E 00	0.64464103E 00	0.84746840E 00	0.35396344E 00
0.39889160E 00	0.45709423E 00	0.23630936E 00	0.13318189E 00
0.16605222E 00	0.22602681E 00	0.66245903E 00	0.25021176E 00
0.61769669E 00	0.26246528E 00	0.51266762E 00	0.93920734E 00
0.62402817E 00	0.42238197E 00	0.93970599E 00	0.28206823E 00
0.46921754E 00	0.54879185E-01	0.51983087E 00	0.39682690E 00
0.11315656E 00	0.60751726E 00	0.70150672E 00	0.88705480E 00

IMAG COMPONENTS OF INVERSE TRANSFORM			
0.18626452E-08	0.74943591E-09	-0.45625189E-09	-0.12955597E-08
-0.65688967E-09	-0.21652834E-08	-0.13654090E-08	0.58739834E-09
-0.27939677E-08	-0.26120810E-08	-0.94073198E-09	0.37602551E-08
0.21702889E-08	0.16996221E-08	0.24131469E-08	-0.49095674E-09
0.18626452E-08	0.74943591E-09	0.14063933E-08	-0.68834951E-08
-0.18210429E-08	-0.30263824E-09	-0.66691706E-09	0.58739834E-09
-0.93132258E-09	0.11132092E-08	-0.18720546E-08	0.37602551E-08
0.30764370E-09	0.76829952E-09	0.14818243E-08	-0.25295450E-10

TEST FFT842 WITH COMPLEX INPUT SEQUENCE

REAL COMPONENTS OF INPUT SEQUENCE			
0.22925607E 00	0.76687502E 00	0.68317685E 00	0.50919111E 00
0.87455959E 00	0.64464101E 00	0.84746840E 00	0.35396343E 00
0.39889160E 00	0.45709422E 00	0.23630936E 00	0.13318188E 00
0.16605222E 00	0.22602680E 00	0.66245903E 00	0.25021175E 00
0.61769668E 00	0.26246527E 00	0.51266762E 00	0.93920734E 00
0.62402816E 00	0.42238195E 00	0.93970599E 00	0.28206823E 00
0.46921754E 00	0.54879178E-01	0.51983086E 00	0.39682690E 00
0.11315656E 00	0.60751725E 00	0.70150672E 00	0.88705479E 00

Table 1
(Continued)

```
IMAG COMPONENTS OF INPUT SEQUENCE
  0.33143324E 00    0.22096146E 00    0.16861611E 00    0.99224898E 00
  0.36706980E 00    0.49780032E 00    0.25410408E 00    0.79779523E 00
  0.13391190E 00    0.11614336E 00    0.21288166E 00    0.78336013E 00
  0.98293080E 00    0.78848629E 00    0.36542587E 00    0.47690938E 00
  0.82732023E 00    0.10252799E 00    0.94326350E 00    0.68425309E 00
  0.72906453E 00    0.72203852E 00    0.34003393E 00    0.41031617E 00
  0.38462142E 00    0.96184775E 00    0.48009393E 00    0.96515422E-01
  0.51983388E 00    0.76180377E 00    0.70444433E 00    0.34020287E 00

REAL COMPONENTS OF TRANSFORM
  0.15789569E 02    0.54187694E 00    0.10384466E 01    0.41030719E 00
 -0.91315009E 00   -0.15766885E 01   -0.42357683E 00    0.20724215E 00
 -0.20202583E 01   -0.60150058E 00   -0.67052441E 00   -0.14835396E 00
 -0.39671516E 00    0.22274646E 00   -0.47426426E 00   -0.16377531E 00
  0.14023971E 01   -0.23689694E 01   -0.19510589E 01   -0.28028500E 01
  0.10693748E 01   -0.17185353E 01    0.10908293E 01   -0.23846828E 01
 -0.12002746E 01    0.71695102E 00   -0.12801024E 00    0.17169342E 01
 -0.10448134E-01   -0.32908173E 00    0.13489076E 01    0.20633298E 01

IMAG COMPONENTS OF TRANSFORM
  0.16498260E 02   -0.12633017E 01   -0.22540791E 01    0.61337085E 00
 -0.95682339E 00    0.96735211E 00   -0.43692699E 00   -0.28613902E 01
  0.11171471E 01    0.51021537E 00    0.13039546E 01   -0.19500994E 00
 -0.66840816E 00    0.14368729E 01    0.26590036E 01   -0.14693253E 01
 -0.10081615E 01   -0.12562806E 01    0.11406250E 01   -0.22069367E 01
  0.15119098E 01   -0.84628310E-01    0.16705755E 01    0.22511049E 01
  0.49749768E 00   -0.28848102E 01   -0.25071358E 01    0.99921418E 00
 -0.35731273E 01   -0.82990787E 00    0.35457445E 01   -0.16607314E 01

REAL COMPONENTS OF INVERSE TRANSFORM
  0.22925607E 00    0.76687504E 00    0.68317686E 00    0.50919111E 00
  0.87455961E 00    0.64464102E 00    0.84746840E 00    0.35396344E 00
  0.39889160E 00    0.45709423E 00    0.23630936E 00    0.13318189E 00
  0.16605221E 00    0.22602681E 00    0.66245903E 00    0.25021175E 00
  0.61769669E 00    0.26246527E 00    0.51266762E 00    0.93920736E 00
  0.62402817E 00    0.42238196E 00    0.93970599E 00    0.28206823E 00
  0.46921755E 00    0.54879181E-01    0.51983086E 00    0.39682690E 00
  0.11315656E 00    0.60751726E 00    0.70150672E 00    0.88705480E 00

IMAG COMPONENTS OF INVERSE TRANSFORM
  0.33143324E 00    0.22096147E 00    0.16861611E 00    0.99224898E 00
  0.36706980E 00    0.49780032E 00    0.25410408E 00    0.79779525E 00
  0.13391190E 00    0.11614336E 00    0.21288165E 00    0.78336014E 00
  0.98293081E 00    0.78848629E 00    0.36542587E 00    0.47690939E 00
  0.82732023E 00    0.10252798E 00    0.94326352E 00    0.68425308E 00
  0.72906454E 00    0.72203852E 00    0.34003393E 00    0.41031619E 00
  0.38462142E 00    0.96184776E 00    0.48009393E 00    0.96515425E-01
  0.51983389E 00    0.76180378E 00    0.70444433E 00    0.34020288E 00
```

Appendix

```
C
C-----------------------------------------------------------------------
C MAIN PROGRAM: FASTMAIN  -  FAST FOURIER TRANSFORMS
C AUTHORS:      G. D. BERGLAND AND M. T. DOLAN
C               BELL LABORATORIES, MURRAY HILL, NEW JERSEY 07974
C
C INPUT:        THE PROGRAM CALLS ON A RANDOM NUMBER
C               GENERATOR FOR INPUT AND CHECKS DFT AND
C               IDFT WITH A 32-POINT SEQUENCE
C-----------------------------------------------------------------------
C
      DIMENSION X(32), Y(32), B(34)
C
C GENERATE RANDOM NUMBERS AND STORE ARRAY IN B SO
C THE SAME SEQUENCE CAN BE USED IN ALL TESTS.
C NOTE THAT B IS DIMENSIONED TO SIZE N+2.
C
C IW IS A MACHINE DEPENDENT WRITE DEVICE NUMBER
C
      IW = I1MACH(2)
C
      DO 10 I=1,32
        X(I) = UNI(0)
        B(I) = X(I)
  10  CONTINUE
      M = 5
      N = 2**M
      NP1 = N + 1
      NP2 = N + 2
      KNT = 1
C
C TEST FAST-FSST THEN FFA-FFS
C
      WRITE (IW,9999)
  20  WRITE (IW,9998) (B(I),I=1,N)
      IF (KNT.EQ.1) CALL FAST(B, N)
      IF (KNT.EQ.2) CALL FFA(B, N)
      WRITE (IW,9997) (B(I),I=1,NP1,2)
      WRITE (IW,9996) (B(I),I=2,NP2,2)
      IF (KNT.EQ.1) CALL FSST(B, N)
      IF (KNT.EQ.2) CALL FFS(B, N)
      WRITE (IW,9995) (B(I),I=1,N)
      KNT = KNT + 1
      IF (KNT.EQ.3) GO TO 40
C
      WRITE (IW,9994)
      DO 30 I=1,N
        B(I) = X(I)
  30  CONTINUE
      GO TO 20
C
C TEST FFT842 WITH REAL INPUT THEN COMPLEX
C
  40  WRITE (IW,9993)
      DO 50 I=1,N
        B(I) = X(I)
        Y(I) = 0.
  50  CONTINUE
  60  WRITE (IW,9992) (B(I),I=1,N)
      WRITE (IW,9991) (Y(I),I=1,N)
      CALL FFT842(0, N, B, Y)
      WRITE (IW,9997) (B(I),I=1,N)
      WRITE (IW,9996) (Y(I),I=1,N)
      CALL FFT842(1, N, B, Y)
      WRITE (IW,9990) (B(I),I=1,N)
      WRITE (IW,9989) (Y(I),I=1,N)
      KNT = KNT + 1
      IF (KNT.EQ.5) GO TO 80
C
      WRITE (IW,9988)
      DO 70 I=1,N
        B(I) = X(I)
        Y(I) = UNI(0)
  70  CONTINUE
      GO TO 60
C
9999  FORMAT (19H1TEST FAST AND FSST)
9998  FORMAT (20H0REAL INPUT SEQUENCE/(4E17.8))
9997  FORMAT (29H0REAL COMPONENTS OF TRANSFORM/(4E17.8))
9996  FORMAT (29H0IMAG COMPONENTS OF TRANSFORM/(4E17.8))
9995  FORMAT (23H0REAL INVERSE TRANSFORM/(4E17.8))
9994  FORMAT (17H1TEST FFA AND FFS)
9993  FORMAT (37H1TEST FFT842 WITH REAL INPUT SEQUENCE/(4E17.8))
9992  FORMAT (34H0REAL COMPONENTS OF INPUT SEQUENCE/(4E17.8))
9991  FORMAT (34H0IMAG COMPONENTS OF INPUT SEQUENCE/(4E17.8))
9990  FORMAT (37H0REAL COMPONENTS OF INVERSE TRANSFORM/(4E17.8))
9989  FORMAT (37H0IMAG COMPONENTS OF INVERSE TRANSFORM/(4E17.8))
9988  FORMAT (40H1TEST FFT842 WITH COMPLEX INPUT SEQUENCE)
  80  STOP
      END
C
C-----------------------------------------------------------------------
C SUBROUTINE:  FAST
C REPLACES THE REAL VECTOR B(K), FOR K=1,2,...,N,
C WITH ITS FINITE DISCRETE FOURIER TRANSFORM
C-----------------------------------------------------------------------
C
      SUBROUTINE FAST(B, N)
C
C THE DC TERM IS RETURNED IN LOCATION B(1) WITH B(2) SET TO 0.
C THEREAFTER THE JTH HARMONIC IS RETURNED AS A COMPLEX
C NUMBER STORED AS  B(2*J+1) + I B(2*J+2).
C THE N/2 HARMONIC IS RETURNED IN B(N+1) WITH B(N+2) SET TO 0.
C HENCE, B MUST BE DIMENSIONED TO SIZE N+2.
C THE SUBROUTINE IS CALLED AS  FAST(B,N) WHERE N=2**M AND
C B IS THE REAL ARRAY DESCRIBED ABOVE.
C
      DIMENSION B(2)
      COMMON /CONS/ PII, P7, P7TWO, C22, S22, PI2
C
C IW IS A MACHINE DEPENDENT WRITE DEVICE NUMBER
C
      IW = I1MACH(2)
C
      PII = 4.*ATAN(1.)
      PI8 = PII/8.
      P7 = 1./SQRT(2.)
      P7TWO = 2.*P7
      C22 = COS(PI8)
      S22 = SIN(PI8)
      PI2 = 2.*PII
      DO 10 I=1,15
        M = I
        NT = 2**I
        IF (N.EQ.NT) GO TO 20
  10  CONTINUE
      WRITE (IW,9999)
9999  FORMAT (33H N IS NOT A POWER OF TWO FOR FAST)
      STOP
  20  N4POW = M/2
```

```
C
C DO A RADIX 2 ITERATION FIRST IF ONE IS REQUIRED.
C
      IF (M-N4POW*2) 40, 40, 30
   30 NN = 2
      INT = N/NN
      CALL FR2TR(INT, B(1), B(INT+1))
      GO TO 50
   40 NN = 1
C
C PERFORM RADIX 4 ITERATIONS.
C
   50 IF (N4POW.EQ.0) GO TO 70
      DO 60 IT=1,N4POW
        NN = NN*4
        INT = N/NN
        CALL FR4TR(INT, NN, B(1), B(INT+1), B(2*INT+1), B(3*INT+1),
     *      B(1), B(INT+1), B(2*INT+1), B(3*INT+1))
   60 CONTINUE
C
C PERFORM IN-PLACE REORDERING.
C
   70 CALL FORD1(M, B)
      CALL FORD2(M, B)
      T = B(2)
      B(2) = 0.
      B(N+1) = T
      B(N+2) = 0.
      DO 80 IT=4,N,2
        B(IT) = -B(IT)
   80 CONTINUE
      RETURN
      END
C
C-----------------------------------------------------------------------
C SUBROUTINE:  FSST
C FOURIER SYNTHESIS SUBROUTINE
C-----------------------------------------------------------------------
C
      SUBROUTINE FSST(B, N)
C
C THIS SUBROUTINE SYNTHESIZES THE REAL VECTOR B(K), FOR
C K=1,2,...,N, FROM THE FOURIER COEFFICIENTS STORED IN THE
C B ARRAY OF SIZE N+2.  THE DC TERM IS IN B(1) WITH B(2) EQUAL
C TO  0.  THE JTH HARMONIC IS STORED AS B(2*J+1) + I B(2*J+2).
C THE N/2 HARMONIC IS IN B(N+1) WITH B(N+2) EQUAL TO 0.
C THE SUBROUTINE IS CALLED AS FSST(B,N) WHERE N=2**M AND
C B IS THE REAL ARRAY DISCUSSED ABOVE.
C
      DIMENSION B(2)
      COMMON /CONST/ PII, P7, P7TWO, C22, S22, PI2
C
C IW IS A MACHINE DEPENDENT WRITE DEVICE NUMBER
C
      IW = I1MACH(2)
C
      PII = 4.*ATAN(1.)
      PI8 = PII/8.
      P7 = 1./SQRT(2.)
      P7TWO = 2.*P7
      C22 = COS(PI8)
      S22 = SIN(PI8)
      PI2 = 2.*PII
      DO 10 I=1,15
        M = I
        NT = 2**I
        IF (N.EQ.NT) GO TO 20
   10 CONTINUE
      WRITE (IW,9999)
 9999 FORMAT (33H N IS NOT A POWER OF TWO FOR FSST)
      STOP
   20 B(2) = B(N+1)
      DO 30 I=4,N,2
        B(I) = -B(I)
   30 CONTINUE
C
C SCALE THE INPUT BY N
C
      DO 40 I=1,N
        B(I) = B(I)/FLOAT(N)
   40 CONTINUE
      N4POW = M/2
C
C SCRAMBLE THE INPUTS
C
      CALL FORD2(M, B)
      CALL FORD1(M, B)
C
      IF (N4POW.EQ.0) GO TO 60
      NN = 4*N
      DO 50 IT=1,N4POW
        NN = NN/4
        INT = N/NN
        CALL FR4SYN(INT, NN, B(1), B(INT+1), B(2*INT+1), B(3*INT+1),
     *      B(1), B(INT+1), B(2*INT+1), B(3*INT+1))
   50 CONTINUE
C
C DO A RADIX 2 ITERATION IF ONE IS REQUIRED
C
   60 IF (M-N4POW*2) 80, 80, 70
   70 INT = N/2
      CALL FR2TR(INT, B(1), B(INT+1))
   80 RETURN
      END
C
C-----------------------------------------------------------------------
C SUBROUTINE:  FR2TR
C RADIX 2 ITERATION SUBROUTINE
C-----------------------------------------------------------------------
C
      SUBROUTINE FR2TR(INT, B0, B1)
      DIMENSION B0(2), B1(2)
      DO 10 K=1,INT
        T = B0(K) + B1(K)
        B1(K) = B0(K) - B1(K)
        B0(K) = T
   10 CONTINUE
      RETURN
      END
C
C-----------------------------------------------------------------------
C SUBROUTINE:  FR4TR
C RADIX 4 ITERATION SUBROUTINE
C-----------------------------------------------------------------------
C
      SUBROUTINE FR4TR(INT, NN, B0, B1, B2, B3, B4, B5, B6, B7)
      DIMENSION L(15), B0(2), B1(2), B2(2), B3(2), B4(2), B5(2), B6(2),
     *    B7(2)
      COMMON /CONS/ PII, P7, P7TWO, C22, S22, PI2
      EQUIVALENCE (L15,L(1)), (L14,L(2)), (L13,L(3)), (L12,L(4)),
```

```
     *    (L11,L(5)), (L10,L(6)), (L9,L(7)), (L8,L(8)), (L7,L(9)),
     *    (L6,L(10)), (L5,L(11)), (L4,L(12)), (L3,L(13)), (L2,L(14)),
     *    (L1,L(15))
C
C JTHET IS A REVERSED BINARY COUNTER, JR STEPS TWO AT A TIME TO
C LOCATE THE REAL PARTS OF INTERMEDIATE RESULTS, AND JI LOCATES
C THE IMAGINARY PART CORRESPONDING TO JR.
C
      L(1) = NN/4
      DO 40 K=2,15
        IF (L(K-1)-2) 10, 20, 30
   10   L(K-1) = 2
   20   L(K) = 2
        GO TO 40
   30   L(K) = L(K-1)/2
   40 CONTINUE
C
      PIOVN = PII/FLOAT(NN)
      JI = 3
      JL = 2
      JR = 2
C
      DO 120 J1=2,L1,2
      DO 120 J2=J1,L2,L1
      DO 120 J3=J2,L3,L2
      DO 120 J4=J3,L4,L3
      DO 120 J5=J4,L5,L4
      DO 120 J6=J5,L6,L5
      DO 120 J7=J6,L7,L6
      DO 120 J8=J7,L8,L7
      DO 120 J9=J8,L9,L8
      DO 120 J10=J9,L10,L9
      DO 120 J11=J10,L11,L10
      DO 120 J12=J11,L12,L11
      DO 120 J13=J12,L13,L12
      DO 120 J14=J13,L14,L13
      DO 120 JTHET=J14,L15,L14
        TH2 = JTHET - 2
        IF (TH2) 50, 50, 90
   50   DO 60 K=1,INT
          T0 = B0(K) + B2(K)
          T1 = B1(K) + B3(K)
          B2(K) = B0(K) - B2(K)
          B3(K) = B1(K) - B3(K)
          B0(K) = T0 + T1
          B1(K) = T0 - T1
   60   CONTINUE
C
        IF (NN-4) 120, 120, 70
   70   K0 = INT*4 + 1
        KL = K0 + INT - 1
        DO 80 K=K0,KL
          PR = P7*(B1(K)-B3(K))
          PI = P7*(B1(K)+B3(K))
          B3(K) = B2(K) + PI
          B1(K) = PI - B2(K)
          B2(K) = B0(K) - PR
          B0(K) = B0(K) + PR
   80   CONTINUE
        GO TO 120
C
   90   ARG = TH2*PIOVN
        C1 = COS(ARG)
        S1 = SIN(ARG)
        C2 = C1**2 - S1**2
        S2 = C1*S1 + C1*S1
        C3 = C1*C2 - S1*S2
        S3 = C2*S1 + S2*C1
C
        INT4 = INT*4
        J0 = JR*INT4 + 1
        K0 = JI*INT4 + 1
        JLAST = J0 + INT - 1
        DO 100 J=J0,JLAST
          K = K0 + J - J0
          R1 = B1(J)*C1 - B5(K)*S1
          R5 = B1(J)*S1 + B5(K)*C1
          T2 = B2(J)*C2 - B6(K)*S2
          T6 = B2(J)*S2 + B6(K)*C2
          T3 = B3(J)*C3 - B7(K)*S3
          T7 = B3(J)*S3 + B7(K)*C3
          T0 = B0(J) + T2
          T4 = B4(K) + T6
          T2 = B0(J) - T2
          T6 = B4(K) - T6
          T1 = R1 + T3
          T5 = R5 + T7
          T3 = R1 - T3
          T7 = R5 - T7
          B0(J) = T0 + T1
          B7(K) = T4 + T5
          B6(K) = T0 - T1
          B1(J) = T5 - T4
          B2(J) = T2 - T7
          B5(K) = T6 + T3
          B4(K) = T2 + T7
          B3(J) = T3 - T6
  100   CONTINUE
C
        JR = JR + 2
        JI = JI - 2
        IF (JI-JL) 110, 110, 120
  110   JI = 2*JR - 1
        JL = JR
  120 CONTINUE
      RETURN
      END
C
C-----------------------------------------------------------------------
C SUBROUTINE:  FR4SYN
C RADIX 4 SYNTHESIS
C-----------------------------------------------------------------------
C
C
      SUBROUTINE FR4SYN(INT, NN, B0, B1, B2, B3, B4, B5, B6, B7)
      DIMENSION L(15), B0(2), B1(2), B2(2), B3(2), B4(2), B5(2), B6(2),
     *    B7(2)
      COMMON /CONST/ PII, P7, P7TWO, C22, S22, PI2
      EQUIVALENCE (L15,L(1)), (L14,L(2)), (L13,L(3)), (L12,L(4)),
     *    (L11,L(5)), (L10,L(6)), (L9,L(7)), (L8,L(8)), (L7,L(9)),
     *    (L6,L(10)), (L5,L(11)), (L4,L(12)), (L3,L(13)), (L2,L(14)),
     *    (L1,L(15))
C
      L(1) = NN/4
      DO 40 K=2,15
        IF (L(K-1)-2) 10, 20, 30
   10   L(K-1) = 2
   20   L(K) = 2
        GO TO 40
   30   L(K) = L(K-1)/2
```

```
   40   CONTINUE
C
        PIOVN = PII/FLOAT(NN)
        JI = 3
        JL = 2
        JR = 2
C
        DO 120 J1=2,L1,2
        DO 120 J2=J1,L2,L1
        DO 120 J3=J2,L3,L2
        DO 120 J4=J3,L4,L3
        DO 120 J5=J4,L5,L4
        DO 120 J6=J5,L6,L5
        DO 120 J7=J6,L7,L6
        DO 120 J8=J7,L8,L7
        DO 120 J9=J8,L9,L8
        DO 120 J10=J9,L10,L9
        DO 120 J11=J10,L11,L10
        DO 120 J12=J11,L12,L11
        DO 120 J13=J12,L13,L12
        DO 120 J14=J13,L14,L13
        DO 120 JTHET=J14,L15,L14
          TH2 = JTHET - 2
          IF (TH2) 50, 50, 90
   50     DO 60 K=1,INT
            T0 = B0(K) + B1(K)
            T1 = B0(K) - B1(K)
            T2 = B2(K)*2.0
            T3 = B3(K)*2.0
            B0(K) = T0 + T2
            B2(K) = T0 - T2
            B1(K) = T1 + T3
            B3(K) = T1 - T3
   60     CONTINUE
C
          IF (NN-4) 120, 120, 70
   70     K0 = INT*4 + 1
          KL = K0 + INT - 1
          DO 80 K=K0,KL
            T2 = B0(K) - B2(K)
            T3 = B1(K) + B3(K)
            B0(K) = (B0(K)+B2(K))*2.0
            B2(K) = (B3(K)-B1(K))*2.0
            B1(K) = (T2+T3)*P7TWO
            B3(K) = (T3-T2)*P7TWO
   80     CONTINUE
          GO TO 120
   90     ARG = TH2*PIOVN
          C1 = COS(ARG)
          S1 = -SIN(ARG)
          C2 = C1**2 - S1**2
          S2 = C1*S1 + C1*S1
          C3 = C1*C2 - S1*S2
          S3 = C2*S1 + S2*C1
C
          INT4 = INT*4
          J0 = JR*INT4 + 1
          K0 = JI*INT4 + 1
          JLAST = J0 + INT - 1
          DO 100 J=J0,JLAST
            K = K0 + J - J0
            T0 = B0(J) + B6(K)
            T1 = B7(K) - B1(J)
            T2 = B0(J) - B6(K)
            T3 = B7(K) + B1(J)
            T4 = B2(J) + B4(K)
            T5 = B5(K) - B3(J)
            T6 = B5(K) + B3(J)
            T7 = B4(K) - B2(J)
            B0(J) = T0 + T4
            B4(K) = T1 + T5
            B1(J) = (T2+T6)*C1 - (T3+T7)*S1
            B5(K) = (T2+T6)*S1 + (T3+T7)*C1
            B2(J) = (T0-T4)*C2 - (T1-T5)*S2
            B6(K) = (T0-T4)*S2 + (T1-T5)*C2
            B3(J) = (T2-T6)*C3 - (T3-T7)*S3
            B7(K) = (T2-T6)*S3 + (T3-T7)*C3
  100     CONTINUE
          JR = JR + 2
          JI = JI - 2
          IF (JI-JL) 110, 110, 120
  110     JI = 2*JR - 1
          JL = JR
  120   CONTINUE
        RETURN
        END
C
C-------------------------------------------------------------------
C SUBROUTINE:  FORD1
C IN-PLACE REORDERING SUBROUTINE
C-------------------------------------------------------------------
C
        SUBROUTINE FORD1(M, B)
        DIMENSION B(2)
C
        K = 4
        KL = 2
        N = 2**M
        DO 40 J=4,N,2
          IF (K-J) 20, 20, 10
   10     T = B(J)
          B(J) = B(K)
          B(K) = T
   20     K = K - 2
          IF (K-KL) 30, 30, 40
   30     K = 2*J
          KL = J
   40   CONTINUE
        RETURN
        END
C
C-------------------------------------------------------------------
C SUBROUTINE:  FORD2
C IN-PLACE REORDERING SUBROUTINE
C-------------------------------------------------------------------
C
        SUBROUTINE FORD2(M, B)
        DIMENSION L(15), B(2)
        EQUIVALENCE (L15,L(1)), (L14,L(2)), (L13,L(3)), (L12,L(4)),
     *     (L11,L(5)), (L10,L(6)), (L9,L(7)), (L8,L(8)), (L7,L(9)),
     *     (L6,L(10)), (L5,L(11)), (L4,L(12)), (L3,L(13)), (L2,L(14)),
     *     (L1,L(15))
        N = 2**M
        L(1) = N
        DO 10 K=2,M
          L(K) = L(K-1)/2
   10   CONTINUE
        DO 20 K=M,14
          L(K+1) = 2
   20   CONTINUE
```

```
      IJ = 2
      DO 40 J1=2,L1,2
      DO 40 J2=J1,L2,L1
      DO 40 J3=J2,L3,L2
      DO 40 J4=J3,L4,L3
      DO 40 J5=J4,L5,L4
      DO 40 J6=J5,L6,L5
      DO 40 J7=J6,L7,L6
      DO 40 J8=J7,L8,L7
      DO 40 J9=J8,L9,L8
      DO 40 J10=J9,L10,L9
      DO 40 J11=J10,L11,L10
      DO 40 J12=J11,L12,L11
      DO 40 J13=J12,L13,L12
      DO 40 J14=J13,L14,L13
      DO 40 JI=J14,L15,L14
        IF (IJ-JI) 30, 40, 40
  30    T = B(IJ-1)
        B(IJ-1) = B(JI-1)
        B(JI-1) = T
        T = B(IJ)
        B(IJ) = B(JI)
        B(JI) = T
  40    IJ = IJ + 2
      RETURN
      END
C
C-----------------------------------------------------------------------
C SUBROUTINE:  FFA
C FAST FOURIER ANALYSIS SUBROUTINE
C-----------------------------------------------------------------------
C
      SUBROUTINE FFA(B, NFFT)
C
C THIS SUBROUTINE REPLACES THE REAL VECTOR B(K),  (K=1,2,...,N),
C WITH ITS FINITE DISCRETE FOURIER TRANSFORM.  THE DC TERM IS
C RETURNED IN LOCATION B(1) WITH B(2) SET TO 0.  THEREAFTER, THE
C JTH HARMONIC IS RETURNED AS A COMPLEX NUMBER STORED AS
C B(2*J+1) + I B(2*J+2).  NOTE THAT THE N/2 HARMONIC IS RETURNED
C IN B(N+1) WITH B(N+2) SET TO 0.  HENCE, B MUST BE DIMENSIONED
C TO SIZE N+2.
C SUBROUTINE IS CALLED AS FFA (B,N) WHERE N=2**M AND B IS AN
C N TERM REAL ARRAY.  A REAL-VALUED, RADIX 8  ALGORITHM IS USED
C WITH IN-PLACE REORDERING AND THE TRIG FUNCTIONS ARE COMPUTED AS
C NEEDED.
C
      DIMENSION B(2)
      COMMON /CON/ PII, P7, P7TWO, C22, S22, PI2
C
C IW IS A MACHINE DEPENDENT WRITE DEVICE NUMBER
C
      IW = I1MACH(2)
C
      PII = 4.*ATAN(1.)
      PI8 = PII/8.
      P7 = 1./SQRT(2.)
      P7TWO = 2.*P7
      C22 = COS(PI8)
      S22 = SIN(PI8)
      PI2 = 2.*PII
      N = 1
      DO 10 I=1,15
        M = I
        N = N*2
        IF (N.EQ.NFFT) GO TO 20
  10  CONTINUE
      WRITE (IW,9999)
9999  FORMAT (30H NFFT NOT A POWER OF 2 FOR FFA)
      STOP
  20  CONTINUE
      N8POW = M/3
C
C DO A RADIX 2 OR RADIX 4 ITERATION FIRST IF ONE IS REQUIRED
C
      IF (M-N8POW*3-1) 50, 40, 30
  30  NN = 4
      INT = N/NN
      CALL R4TR(INT, B(1), B(INT+1), B(2*INT+1), B(3*INT+1))
      GO TO 60
  40  NN = 2
      INT = N/NN
      CALL R2TR(INT, B(1), B(INT+1))
      GO TO 60
  50  NN = 1
C
C PERFORM RADIX 8 ITERATIONS
C
  60  IF (N8POW) 90, 90, 70
  70  DO 80 IT=1,N8POW
        NN = NN*8
        INT = N/NN
        CALL R8TR(INT, NN, B(1), B(INT+1), B(2*INT+1), B(3*INT+1),
     *      B(4*INT+1), B(5*INT+1), B(6*INT+1), B(7*INT+1), B(1),
     *      B(INT+1), B(2*INT+1), B(3*INT+1), B(4*INT+1), B(5*INT+1),
     *      B(6*INT+1), B(7*INT+1))
  80  CONTINUE
C
C PERFORM IN-PLACE REORDERING
C
  90  CALL ORD1(M, B)
      CALL ORD2(M, B)
      T = B(2)
      B(2) = 0.
      B(NFFT+1) = T
      B(NFFT+2) = 0.
      DO 100 I=4,NFFT,2
        B(I) = -B(I)
 100  CONTINUE
      RETURN
      END
C
C-----------------------------------------------------------------------
C SUBROUTINE:  FFS
C FAST FOURIER SYNTHESIS SUBROUTINE
C RADIX 8-4-2
C-----------------------------------------------------------------------
C
      SUBROUTINE FFS(B, NFFT)
C
C THIS SUBROUTINE SYNTHESIZES THE REAL VECTOR B(K), WHERE
C K=1,2,...,N. THE INITIAL FOURIER COEFFICIENTS ARE PLACED IN
C THE B ARRAY OF SIZE N+2.  THE DC TERM IS IN B(1) WITH
C B(2) EQUAL TO 0.
C THE JTH HARMONIC IS STORED AS B(2*J+1) + I B(2*J+2).
C THE N/2 HARMONIC IS IN B(N+1) WITH B(N+2) EQUAL TO 0.
C THE SUBROUTINE IS CALLED AS FFS(B,N) WHERE N=2**M AND
C B IS THE N TERM REAL ARRAY DISCUSSED ABOVE.
C
      DIMENSION B(2)
      COMMON /CON1/ PII, P7, P7TWO, C22, S22, PI2
```

```
C
C IW IS A MACHINE DEPENDENT WRITE DEVICE NUMBER
C
      IW = I1MACH(2)
C
      PII = 4.*ATAN(1.)
      PI8 = PII/8.
      P7 = 1./SQRT(2.)
      P7TWO = 2.*P7
      C22 = COS(PI8)
      S22 = SIN(PI8)
      PI2 = 2.*PII
      N = 1
      DO 10 I=1,15
        M = I
        N = N*2
        IF (N.EQ.NFFT) GO TO 20
   10 CONTINUE
      WRITE (IW,9999)
 9999 FORMAT (30H NFFT NOT A POWER OF 2 FOR FFS)
      STOP
   20 CONTINUE
      B(2) = B(NFFT+1)
      DO 30 I=1,NFFT
        B(I) = B(I)/FLOAT(NFFT)
   30 CONTINUE
      DO 40 I=4,NFFT,2
        B(I) = -B(I)
   40 CONTINUE
      N8POW = M/3
C
C REORDER THE INPUT FOURIER COEFFICIENTS
C
      CALL ORD2(M, B)
      CALL ORD1(M, B)
C
      IF (N8POW.EQ.0) GO TO 60
C
C PERFORM THE RADIX 8 ITERATIONS
C
      NN = N
      DO 50 IT=1,N8POW
        INT = N/NN
        CALL R8SYN(INT, NN, B, B(INT+1), B(2*INT+1), B(3*INT+1),
     *      B(4*INT+1), B(5*INT+1), B(6*INT+1), B(7*INT+1), B(1),
     *      B(INT+1), B(2*INT+1), B(3*INT+1), B(4*INT+1), B(5*INT+1),
     *      B(6*INT+1), B(7*INT+1))
        NN = NN/8
   50 CONTINUE
C
C DO A RADIX 2 OR RADIX 4 ITERATION IF ONE IS REQUIRED
C
   60 IF (M-N8POW*3-1) 90, 80, 70
   70 INT = N/4
      CALL R4SYN(INT, B(1), B(INT+1), B(2*INT+1), B(3*INT+1))
      GO TO 90
   80 INT = N/2
      CALL R2TR(INT, B(1), B(INT+1))
   90 RETURN
      END
C
C-----------------------------------------------------------------------
C SUBROUTINE:  R2TR
C RADIX 2 ITERATION SUBROUTINE
C-----------------------------------------------------------------------
C
C
      SUBROUTINE R2TR(INT, B0, B1)
      DIMENSION B0(2), B1(2)
      DO 10 K=1,INT
        T = B0(K) + B1(K)
        B1(K) = B0(K) - B1(K)
        B0(K) = T
   10 CONTINUE
      RETURN
      END
C
C-----------------------------------------------------------------------
C SUBROUTINE:  R4TR
C RADIX 4 ITERATION SUBROUTINE
C-----------------------------------------------------------------------
C
      SUBROUTINE R4TR(INT, B0, B1, B2, B3)
      DIMENSION B0(2), B1(2), B2(2), B3(2)
      DO 10 K=1,INT
        R0 = B0(K) + B2(K)
        R1 = B1(K) + B3(K)
        B2(K) = B0(K) - B2(K)
        B3(K) = B1(K) - B3(K)
        B0(K) = R0 + R1
        B1(K) = R0 - R1
   10 CONTINUE
      RETURN
      END
C
C-----------------------------------------------------------------------
C SUBROUTINE: R8TR
C RADIX 8 ITERATION SUBROUTINE
C-----------------------------------------------------------------------
C
      SUBROUTINE R8TR(INT, NN, BR0, BR1, BR2, BR3, BR4, BR5, BR6, BR7,
     *    BI0, BI1, BI2, BI3, BI4, BI5, BI6, BI7)
      DIMENSION L(15), BR0(2), BR1(2), BR2(2), BR3(2), BR4(2), BR5(2),
     *    BR6(2), BR7(2), BI0(2), BI1(2), BI2(2), BI3(2), BI4(2),
     *    BI5(2), BI6(2), BI7(2)
      COMMON /CON/ PII, P7, P7TWO, C22, S22, PI2
      EQUIVALENCE (L15,L(1)), (L14,L(2)), (L13,L(3)), (L12,L(4)),
     *    (L11,L(5)), (L10,L(6)), (L9,L(7)), (L8,L(8)), (L7,L(9)),
     *    (L6,L(10)), (L5,L(11)), (L4,L(12)), (L3,L(13)), (L2,L(14)),
     *    (L1,L(15))
C
C SET UP COUNTERS SUCH THAT JTHET STEPS THROUGH THE ARGUMENTS
C OF W, JR STEPS THROUGH STARTING LOCATIONS FOR THE REAL PART OF THE
C INTERMEDIATE RESULTS AND JI STEPS THROUGH STARTING LOCATIONS
C OF THE IMAGINARY PART OF THE INTERMEDIATE RESULTS.
C
      L(1) = NN/8
      DO 40 K=2,15
        IF (L(K-1)-2) 10, 20, 30
   10   L(K-1) = 2
   20   L(K) = 2
        GO TO 40
   30   L(K) = L(K-1)/2
   40 CONTINUE
      PIOVN = PII/FLOAT(NN)
      JI = 3
      JL = 2
      JR = 2
      DO 120 J1=2,L1,2
      DO 120 J2=J1,L2,L1
```

```
      DO 120 J3=J2,L3,L2
      DO 120 J4=J3,L4,L3
      DO 120 J5=J4,L5,L4
      DO 120 J6=J5,L6,L5
      DO 120 J7=J6,L7,L6
      DO 120 J8=J7,L8,L7
      DO 120 J9=J8,L9,L8
      DO 120 J10=J9,L10,L9
      DO 120 J11=J10,L11,L10
      DO 120 J12=J11,L12,L11
      DO 120 J13=J12,L13,L12
      DO 120 J14=J13,L14,L13
      DO 120 JTHET=J14,L15,L14
        TH2 = JTHET - 2
        IF (TH2) 50, 50, 90
50      DO 60 K=1,INT
          T0 = BR0(K) + BR4(K)
          T1 = BR1(K) + BR5(K)
          T2 = BR2(K) + BR6(K)
          T3 = BR3(K) + BR7(K)
          T4 = BR0(K) - BR4(K)
          T5 = BR1(K) - BR5(K)
          T6 = BR2(K) - BR6(K)
          T7 = BR3(K) - BR7(K)
          BR2(K) = T0 - T2
          BR3(K) = T1 - T3
          T0 = T0 + T2
          T1 = T1 + T3
          BR0(K) = T0 + T1
          BR1(K) = T0 - T1
          PR = P7*(T5-T7)
          PI = P7*(T5+T7)
          BR4(K) = T4 + PR
          BR7(K) = T6 + PI
          BR6(K) = T4 - PR
          BR5(K) = PI - T6
60      CONTINUE
        IF (NN-8) 120, 120, 70
70      K0 = INT*8 + 1
        KL = K0 + INT - 1
        DO 80 K=K0,KL
          PR = P7*(BI2(K)-BI6(K))
          PI = P7*(BI2(K)+BI6(K))
          TR0 = BI0(K) + PR
          TI0 = BI4(K) + PI
          TR2 = BI0(K) - PR
          TI2 = BI4(K) - PI
          PR = P7*(BI3(K)-BI7(K))
          PI = P7*(BI3(K)+BI7(K))
          TR1 = BI1(K) + PR
          TI1 = BI5(K) + PI
          TR3 = BI1(K) - PR
          TI3 = BI5(K) - PI
          PR = TR1*C22 - TI1*S22
          PI = TI1*C22 + TR1*S22
          BI0(K) = TR0 + PR
          BI6(K) = TR0 - PR
          BI7(K) = TI0 + PI
          BI1(K) = PI - TI0
          PR = -TR3*S22 - TI3*C22
          PI = TR3*C22 - TI3*S22
          BI2(K) = TR2 + PR
          BI4(K) = TR2 - PR
          BI5(K) = TI2 + PI
          BI3(K) = PI - TI2
80      CONTINUE
        GO TO 120
90      ARG = TH2*PIOVN
        C1 = COS(ARG)
        S1 = SIN(ARG)
        C2 = C1**2 - S1**2
        S2 = C1*S1 + C1*S1
        C3 = C1*C2 - S1*S2
        S3 = C2*S1 + S2*C1
        C4 = C2**2 - S2**2
        S4 = C2*S2 + C2*S2
        C5 = C2*C3 - S2*S3
        S5 = C3*S2 + S3*C2
        C6 = C3**2 - S3**2
        S6 = C3*S3 + C3*S3
        C7 = C3*C4 - S3*S4
        S7 = C4*S3 + S4*C3
        INT8 = INT*8
        J0 = JR*INT8 + 1
        K0 = JI*INT8 + 1
        JLAST = J0 + INT - 1
        DO 100 J=J0,JLAST
          K = K0 + J - J0
          TR1 = BR1(J)*C1 - BI1(K)*S1
          TI1 = BR1(J)*S1 + BI1(K)*C1
          TR2 = BR2(J)*C2 - BI2(K)*S2
          TI2 = BR2(J)*S2 + BI2(K)*C2
          TR3 = BR3(J)*C3 - BI3(K)*S3
          TI3 = BR3(J)*S3 + BI3(K)*C3
          TR4 = BR4(J)*C4 - BI4(K)*S4
          TI4 = BR4(J)*S4 + BI4(K)*C4
          TR5 = BR5(J)*C5 - BI5(K)*S5
          TI5 = BR5(J)*S5 + BI5(K)*C5
          TR6 = BR6(J)*C6 - BI6(K)*S6
          TI6 = BR6(J)*S6 + BI6(K)*C6
          TR7 = BR7(J)*C7 - BI7(K)*S7
          TI7 = BR7(J)*S7 + BI7(K)*C7
C
          T0 = BR0(J) + TR4
          T1 = BI0(K) + TI4
          TR4 = BR0(J) - TR4
          TI4 = BI0(K) - TI4
          T2 = TR1 + TR5
          T3 = TI1 + TI5
          TR5 = TR1 - TR5
          TI5 = TI1 - TI5
          T4 = TR2 + TR6
          T5 = TI2 + TI6
          TR6 = TR2 - TR6
          TI6 = TI2 - TI6
          T6 = TR3 + TR7
          T7 = TI3 + TI7
          TR7 = TR3 - TR7
          TI7 = TI3 - TI7
C
          TR0 = T0 + T4
          TI0 = T1 + T5
          TR2 = T0 - T4
          TI2 = T1 - T5
          TR1 = T2 + T6
          TI1 = T3 + T7
          TR3 = T2 - T6
          TI3 = T3 - T7
          T0 = TR4 - TI6
          T1 = TI4 + TR6
```

```
            T4 = TR4 + TI6
            T5 = TI4 - TR6
            T2 = TR5 - TI7
            T3 = TI5 + TR7
            T6 = TR5 + TI7
            T7 = TI5 - TR7
            BR0(J) = TR0 + TR1
            BI7(K) = TI0 + TI1
            BI6(K) = TR0 - TR1
            BR1(J) = TI1 - TI0
            BR2(J) = TR2 - TI3
            BI5(K) = TI2 + TR3
            BI4(K) = TR2 + TI3
            BR3(J) = TR3 - TI2
            PR = P7*(T2-T3)
            PI = P7*(T2+T3)
            BR4(J) = T0 + PR
            BI3(K) = T1 + PI
            BI2(K) = T0 - PR
            BR5(J) = PI - T1
            PR = -P7*(T6+T7)
            PI = P7*(T6-T7)
            BR6(J) = T4 + PR
            BI1(K) = T5 + PI
            BI0(K) = T4 - PR
            BR7(J) = PI - T5
100       CONTINUE
          JR = JR + 2
          JI = JI - 2
          IF (JI-JL) 110, 110, 120
110       JI = 2*JR - 1
          JL = JR
120     CONTINUE
        RETURN
        END
C
C-----------------------------------------------------------------------
C SUBROUTINE:  R4SYN
C RADIX 4 SYNTHESIS
C-----------------------------------------------------------------------
C
        SUBROUTINE R4SYN(INT, B0, B1, B2, B3)
        DIMENSION B0(2), B1(2), B2(2), B3(2)
        DO 10 K=1,INT
          T0 = B0(K) + B1(K)
          T1 = B0(K) - B1(K)
          T2 = B2(K) + B2(K)
          T3 = B3(K) + B3(K)
          B0(K) = T0 + T2
          B2(K) = T0 - T2
          B1(K) = T1 + T3
          B3(K) = T1 - T3
10      CONTINUE
        RETURN
        END
C
C-----------------------------------------------------------------------
C SUBROUTINE:  R8SYN
C RADIX 8 SYNTHESIS SUBROUTINE
C-----------------------------------------------------------------------
C
        SUBROUTINE R8SYN(INT, NN, BR0, BR1, BR2, BR3, BR4, BR5, BR6, BR7,
     *      BI0, BI1, BI2, BI3, BI4, BI5, BI6, BI7)
        DIMENSION L(15), BR0(2), BR1(2), BR2(2), BR3(2), BR4(2), BR5(2),
     *      BR6(2), BR7(2), BI0(2), BI1(2), BI2(2), BI3(2), BI4(2),
     *      BI5(2), BI6(2), BI7(2)
        COMMON /CON1/ PII, P7, P7TWO, C22, S22, PI2
        EQUIVALENCE (L15,L(1)), (L14,L(2)), (L13,L(3)), (L12,L(4)),
     *      (L11,L(5)), (L10,L(6)), (L9,L(7)), (L8,L(8)), (L7,L(9)),
     *      (L6,L(10)), (L5,L(11)), (L4,L(12)), (L3,L(13)), (L2,L(14)),
     *      (L1,L(15))
        L(1) = NN/8
        DO 40 K=2,15
          IF (L(K-1)-2) 10, 20, 30
10        L(K-1) = 2
20        L(K) = 2
          GO TO 40
30        L(K) = L(K-1)/2
40      CONTINUE
        PIOVN = PII/FLOAT(NN)
        JI = 3
        JL = 2
        JR = 2
C
        DO 120 J1=2,L1,2
        DO 120 J2=J1,L2,L1
        DO 120 J3=J2,L3,L2
        DO 120 J4=J3,L4,L3
        DO 120 J5=J4,L5,L4
        DO 120 J6=J5,L6,L5
        DO 120 J7=J6,L7,L6
        DO 120 J8=J7,L8,L7
        DO 120 J9=J8,L9,L8
        DO 120 J10=J9,L10,L9
        DO 120 J11=J10,L11,L10
        DO 120 J12=J11,L12,L11
        DO 120 J13=J12,L13,L12
        DO 120 J14=J13,L14,L13
        DO 120 JTHET=J14,L15,L14
          TH2 = JTHET - 2
          IF (TH2) 50, 50, 90
50        DO 60 K=1,INT
            T0 = BR0(K) + BR1(K)
            T1 = BR0(K) - BR1(K)
            T2 = BR2(K) + BR2(K)
            T3 = BR3(K) + BR3(K)
            T4 = BR4(K) + BR6(K)
            T6 = BR7(K) - BR5(K)
            T5 = BR4(K) - BR6(K)
            T7 = BR7(K) + BR5(K)
            PR = P7*(T7+T5)
            PI = P7*(T7-T5)
            TT0 = T0 + T2
            TT1 = T1 + T3
            T2 = T0 - T2
            T3 = T1 - T3
            T4 = T4 + T4
            T5 = PR + PR
            T6 = T6 + T6
            T7 = PI + PI
            BR0(K) = TT0 + T4
            BR1(K) = TT1 + T5
            BR2(K) = T2 + T6
            BR3(K) = T3 + T7
            BR4(K) = TT0 - T4
            BR5(K) = TT1 - T5
            BR6(K) = T2 - T6
            BR7(K) = T3 - T7
60        CONTINUE
          IF (NN-8) 120, 120, 70
```

```
70      K0 = INT*8 + 1
        KL = K0 + INT - 1
        DO 80 K=K0,KL
          T1 = BI0(K) + BI6(K)
          T2 = BI7(K) - BI1(K)
          T3 = BI0(K) - BI6(K)
          T4 = BI7(K) + BI1(K)
          PR = T3*C22 + T4*S22
          PI = T4*C22 - T3*S22
          T5 = BI2(K) + BI4(K)
          T6 = BI5(K) - BI3(K)
          T7 = BI2(K) - BI4(K)
          T8 = BI5(K) + BI3(K)
          RR = T8*C22 - T7*S22
          RI = -T8*S22 - T7*C22
          BI0(K) = (T1+T5) + (T1+T5)
          BI4(K) = (T2+T6) + (T2+T6)
          BI1(K) = (PR+RR) + (PR+RR)
          BI5(K) = (PI+RI) + (PI+RI)
          T5 = T1 - T5
          T6 = T2 - T6
          BI2(K) = P7TWO*(T6+T5)
          BI6(K) = P7TWO*(T6-T5)
          RR = PR - RR
          RI = PI - RI
          BI3(K) = P7TWO*(RI+RR)
          BI7(K) = P7TWO*(RI-RR)
80      CONTINUE
        GO TO 120
90      ARG = TH2*PIOVN
        C1 = COS(ARG)
        S1 = -SIN(ARG)
        C2 = C1**2 - S1**2
        S2 = C1*S1 + C1*S1
        C3 = C1*C2 - S1*S2
        S3 = C2*S1 + S2*C1
        C4 = C2**2 - S2**2
        S4 = C2*S2 + C2*S2
        C5 = C2*C3 - S2*S3
        S5 = C3*S2 + S3*C2
        C6 = C3**2 - S3**2
        S6 = C3*S3 + C3*S3
        C7 = C3*C4 - S3*S4
        S7 = C4*S3 + S4*C3
        INT8 = INT*8
        J0 = JR*INT8 + 1
        K0 = JI*INT8 + 1
        JLAST = J0 + INT - 1
        DO 100 J=J0,JLAST
          K = K0 + J - J0
          TR0 = BR0(J) + BI6(K)
          TI0 = BI7(K) - BR1(J)
          TR1 = BR0(J) - BI6(K)
          TI1 = BI7(K) + BR1(J)
          TR2 = BR2(J) + BI4(K)
          TI2 = BI5(K) - BR3(J)
          TR3 = BI5(K) + BR3(J)
          TI3 = BI4(K) - BR2(J)
          TR4 = BR4(J) + BI2(K)
          TI4 = BI3(K) - BR5(J)
          T0 = BR4(J) - BI2(K)
          T1 = BI3(K) + BR5(J)
          TR5 = P7*(T0+T1)
          TI5 = P7*(T1-T0)
          TR6 = BR6(J) + BI0(K)
          TI6 = BI1(K) - BR7(J)
          T0 = BR6(J) - BI0(K)
          T1 = BI1(K) + BR7(J)
          TR7 = -P7*(T0-T1)
          TI7 = -P7*(T1+T0)
          T0 = TR0 + TR2
          T1 = TI0 + TI2
          T2 = TR1 + TR3
          T3 = TI1 + TI3
          TR2 = TR0 - TR2
          TI2 = TI0 - TI2
          TR3 = TR1 - TR3
          TI3 = TI1 - TI3
          T4 = TR4 + TR6
          T5 = TI4 + TI6
          T6 = TR5 + TR7
          T7 = TI5 + TI7
          TTR6 = TI4 - TI6
          TI6 = TR6 - TR4
          TTR7 = TI5 - TI7
          TI7 = TR7 - TR5
          BR0(J) = T0 + T4
          BI0(K) = T1 + T5
          BR1(J) = C1*(T2+T6) - S1*(T3+T7)
          BI1(K) = C1*(T3+T7) + S1*(T2+T6)
          BR2(J) = C2*(TR2+TTR6) - S2*(TI2+TI6)
          BI2(K) = C2*(TI2+TI6) + S2*(TR2+TTR6)
          BR3(J) = C3*(TR3+TTR7) - S3*(TI3+TI7)
          BI3(K) = C3*(TI3+TI7) + S3*(TR3+TTR7)
          BR4(J) = C4*(T0-T4) - S4*(T1-T5)
          BI4(K) = C4*(T1-T5) + S4*(T0-T4)
          BR5(J) = C5*(T2-T6) - S5*(T3-T7)
          BI5(K) = C5*(T3-T7) + S5*(T2-T6)
          BR6(J) = C6*(TR2-TTR6) - S6*(TI2-TI6)
          BI6(K) = C6*(TI2-TI6) + S6*(TR2-TTR6)
          BR7(J) = C7*(TR3-TTR7) - S7*(TI3-TI7)
          BI7(K) = C7*(TI3-TI7) + S7*(TR3-TTR7)
100       CONTINUE
          JR = JR + 2
          JI = JI - 2
          IF (JI-JL) 110, 110, 120
110       JI = 2*JR - 1
          JL = JR
120     CONTINUE
        RETURN
        END
C
C-----------------------------------------------------------------------
C SUBROUTINE:  ORD1
C IN-PLACE REORDERING SUBROUTINE
C-----------------------------------------------------------------------
C
        SUBROUTINE ORD1(M, B)
        DIMENSION B(2)
C
        K = 4
        KL = 2
        N = 2**M
        DO 40 J=4,N,2
          IF (K-J) 20, 20, 10
10        T = B(J)
          B(J) = B(K)
          B(K) = T
20        K = K - 2
          IF (K-KL) 30, 30, 40
```

```
   30    K = 2*J
         KL = J
   40  CONTINUE
       RETURN
       END
C
C-----------------------------------------------------------------------
C SUBROUTINE:  ORD2
C IN-PLACE REORDERING SUBROUTINE
C-----------------------------------------------------------------------
C
       SUBROUTINE ORD2(M, B)
       DIMENSION L(15), B(2)
       EQUIVALENCE (L15,L(1)), (L14,L(2)), (L13,L(3)), (L12,L(4)),
      *    (L11,L(5)), (L10,L(6)), (L9,L(7)), (L8,L(8)), (L7,L(9)),
      *    (L6,L(10)), (L5,L(11)), (L4,L(12)), (L3,L(13)), (L2,L(14)),
      *    (L1,L(15))
       N = 2**M
       L(1) = N
       DO 10 K=2,M
         L(K) = L(K-1)/2
   10  CONTINUE
       DO 20 K=M,14
         L(K+1) = 2
   20  CONTINUE
       IJ = 2
       DO 40 J1=2,L1,2
       DO 40 J2=J1,L2,L1
       DO 40 J3=J2,L3,L2
       DO 40 J4=J3,L4,L3
       DO 40 J5=J4,L5,L4
       DO 40 J6=J5,L6,L5
       DO 40 J7=J6,L7,L6
       DO 40 J8=J7,L8,L7
       DO 40 J9=J8,L9,L8
       DO 40 J10=J9,L10,L9
       DO 40 J11=J10,L11,L10
       DO 40 J12=J11,L12,L11
       DO 40 J13=J12,L13,L12
       DO 40 J14=J13,L14,L13
       DO 40 JI=J14,L15,L14
         IF (IJ-JI) 30, 40, 40
   30    T = B(IJ-1)
         B(IJ-1) = B(JI-1)
         B(JI-1) = T
         T = B(IJ)
         B(IJ) = B(JI)
         B(JI) = T
   40    IJ = IJ + 2
       RETURN
       END
C
C-----------------------------------------------------------------------
C SUBROUTINE:  FFT842
C FAST FOURIER TRANSFORM FOR N=2**M
C COMPLEX INPUT
C-----------------------------------------------------------------------
C
       SUBROUTINE FFT842(IN, N, X, Y)
C
C THIS PROGRAM REPLACES THE VECTOR Z=X+IY BY ITS  FINITE
C DISCRETE, COMPLEX FOURIER TRANSFORM IF IN=0.  THE INVERSE TRANSFORM
C IS CALCULATED FOR IN=1.  IT PERFORMS AS MANY BASE
C 8 ITERATIONS AS POSSIBLE AND THEN FINISHES WITH A BASE 4 ITERATION
C OR A BASE 2 ITERATION IF NEEDED.
C
C THE SUBROUTINE IS CALLED AS SUBROUTINE FFT842 (IN,N,X,Y).
C THE INTEGER N (A POWER OF 2), THE N REAL LOCATION ARRAY X, AND
C THE N REAL LOCATION ARRAY Y MUST BE SUPPLIED TO THE SUBROUTINE.
C
       DIMENSION X(2), Y(2), L(15)
       COMMON /CON2/ PI2, P7
       EQUIVALENCE (L15,L(1)), (L14,L(2)), (L13,L(3)), (L12,L(4)),
      *    (L11,L(5)), (L10,L(6)), (L9,L(7)), (L8,L(8)), (L7,L(9)),
      *    (L6,L(10)), (L5,L(11)), (L4,L(12)), (L3,L(13)), (L2,L(14)),
      *    (L1,L(15))
C
C
C IW IS A MACHINE DEPENDENT WRITE DEVICE NUMBER
C
       IW = I1MACH(2)
C
       PI2 = 8.*ATAN(1.)
       P7 = 1./SQRT(2.)
       DO 10 I=1,15
         M = I
         NT = 2**I
         IF (N.EQ.NT) GO TO 20
   10  CONTINUE
       WRITE (IW,9999)
 9999  FORMAT (35H N IS NOT A POWER OF TWO FOR FFT842)
       STOP
   20  N2POW = M
       NTHPO = N
       FN = NTHPO
       IF (IN.EQ.1) GO TO 40
       DO 30 I=1,NTHPO
         Y(I) = -Y(I)
   30  CONTINUE
   40  N8POW = N2POW/3
       IF (N8POW.EQ.0) GO TO 60
C
C RADIX 8 PASSES,IF ANY.
C
       DO 50 IPASS=1,N8POW
         NXTLT = 2**(N2POW-3*IPASS)
         LENGT = 8*NXTLT
         CALL R8TX(NXTLT, NTHPO, LENGT, X(1), X(NXTLT+1), X(2*NXTLT+1),
      *      X(3*NXTLT+1), X(4*NXTLT+1), X(5*NXTLT+1), X(6*NXTLT+1),
      *      X(7*NXTLT+1), Y(1), Y(NXTLT+1), Y(2*NXTLT+1), Y(3*NXTLT+1),
      *      Y(4*NXTLT+1), Y(5*NXTLT+1), Y(6*NXTLT+1), Y(7*NXTLT+1))
   50  CONTINUE
C
C IS THERE A FOUR FACTOR LEFT
C
   60  IF (N2POW-3*N8POW-1) 90, 70, 80
C
C GO THROUGH THE BASE 2 ITERATION
C
C
   70  CALL R2TX(NTHPO, X(1), X(2), Y(1), Y(2))
       GO TO 90
C
C GO THROUGH THE BASE 4 ITERATION
C
   80  CALL R4TX(NTHPO, X(1), X(2), X(3), X(4), Y(1), Y(2), Y(3), Y(4))
C
   90  DO 110 J=1,15
         L(J) = 1
         IF (J-N2POW) 100, 100, 110
```

```
100     L(J) = 2**(N2POW+1-J)
110   CONTINUE
      IJ = 1
      DO 130 J1=1,L1
      DO 130 J2=J1,L2,L1
      DO 130 J3=J2,L3,L2
      DO 130 J4=J3,L4,L3
      DO 130 J5=J4,L5,L4
      DO 130 J6=J5,L6,L5
      DO 130 J7=J6,L7,L6
      DO 130 J8=J7,L8,L7
      DO 130 J9=J8,L9,L8
      DO 130 J10=J9,L10,L9
      DO 130 J11=J10,L11,L10
      DO 130 J12=J11,L12,L11
      DO 130 J13=J12,L13,L12
      DO 130 J14=J13,L14,L13
      DO 130 JI=J14,L15,L14
        IF (IJ-JI) 120, 130, 130
120     R = X(IJ)
        X(IJ) = X(JI)
        X(JI) = R
        FI = Y(IJ)
        Y(IJ) = Y(JI)
        Y(JI) = FI
130     IJ = IJ + 1
      IF (IN.EQ.1) GO TO 150
      DO 140 I=1,NTHPO
        Y(I) = -Y(I)
140   CONTINUE
      GO TO 170
150   DO 160 I=1,NTHPO
        X(I) = X(I)/FN
        Y(I) = Y(I)/FN
160   CONTINUE
170   RETURN
      END
C
C-----------------------------------------------------------------------
C SUBROUTINE:  R2TX
C RADIX 2 ITERATION SUBROUTINE
C-----------------------------------------------------------------------
C
      SUBROUTINE R2TX(NTHPO, CR0, CR1, CI0, CI1)
      DIMENSION CR0(2), CR1(2), CI0(2), CI1(2)
      DO 10 K=1,NTHPO,2
        R1 = CR0(K) + CR1(K)
        CR1(K) = CR0(K) - CR1(K)
        CR0(K) = R1
        FI1 = CI0(K) + CI1(K)
        CI1(K) = CI0(K) - CI1(K)
        CI0(K) = FI1
  10  CONTINUE
      RETURN
      END
C
C-----------------------------------------------------------------------
C SUBROUTINE:  R4TX
C RADIX 4 ITERATION SUBROUTINE
C-----------------------------------------------------------------------
C
      SUBROUTINE R4TX(NTHPO, CR0, CR1, CR2, CR3, CI0, CI1, CI2, CI3)
      DIMENSION CR0(2), CR1(2), CR2(2), CR3(2), CI0(2), CI1(2), CI2(2),
     *    CI3(2)
      DO 10 K=1,NTHPO,4
        R1 = CR0(K) + CR2(K)
        R2 = CR0(K) - CR2(K)
        R3 = CR1(K) + CR3(K)
        R4 = CR1(K) - CR3(K)
        FI1 = CI0(K) + CI2(K)
        FI2 = CI0(K) - CI2(K)
        FI3 = CI1(K) + CI3(K)
        FI4 = CI1(K) - CI3(K)
        CR0(K) = R1 + R3
        CI0(K) = FI1 + FI3
        CR1(K) = R1 - R3
        CI1(K) = FI1 - FI3
        CR2(K) = R2 - FI4
        CI2(K) = FI2 + R4
        CR3(K) = R2 + FI4
        CI3(K) = FI2 - R4
  10  CONTINUE
      RETURN
      END
C
C-----------------------------------------------------------------------
C SUBROUTINE:  R8TX
C RADIX 8 ITERATION SUBROUTINE
C-----------------------------------------------------------------------
C
      SUBROUTINE R8TX(NXTLT, NTHPO, LENGT, CR0, CR1, CR2, CR3, CR4,
     *    CR5, CR6, CR7, CI0, CI1, CI2, CI3, CI4, CI5, CI6, CI7)
      DIMENSION CR0(2), CR1(2), CR2(2), CR3(2), CR4(2), CR5(2), CR6(2),
     *    CR7(2), CI1(2), CI2(2), CI3(2), CI4(2), CI5(2), CI6(2),
     *    CI7(2), CI0(2)
      COMMON /CON2/ PI2, P7
C
      SCALE = PI2/FLOAT(LENGT)
      DO 30 J=1,NXTLT
        ARG = FLOAT(J-1)*SCALE
        C1 = COS(ARG)
        S1 = SIN(ARG)
        C2 = C1**2 - S1**2
        S2 = C1*S1 + C1*S1
        C3 = C1*C2 - S1*S2
        S3 = C2*S1 + S2*C1
        C4 = C2**2 - S2**2
        S4 = C2*S2 + C2*S2
        C5 = C2*C3 - S2*S3
        S5 = C3*S2 + S3*C2
        C6 = C3**2 - S3**2
        S6 = C3*S3 + C3*S3
        C7 = C3*C4 - S3*S4
        S7 = C4*S3 + S4*C3
        DO 20 K=J,NTHPO,LENGT
          AR0 = CR0(K) + CR4(K)
          AR1 = CR1(K) + CR5(K)
          AR2 = CR2(K) + CR6(K)
          AR3 = CR3(K) + CR7(K)
          AR4 = CR0(K) - CR4(K)
          AR5 = CR1(K) - CR5(K)
          AR6 = CR2(K) - CR6(K)
          AR7 = CR3(K) - CR7(K)
          AI0 = CI0(K) + CI4(K)
          AI1 = CI1(K) + CI5(K)
          AI2 = CI2(K) + CI6(K)
          AI3 = CI3(K) + CI7(K)
          AI4 = CI0(K) - CI4(K)
          AI5 = CI1(K) - CI5(K)
          AI6 = CI2(K) - CI6(K)
```

```
            AI7 = CI3(K) - CI7(K)
            BR0 = AR0 + AR2
            BR1 = AR1 + AR3
            BR2 = AR0 - AR2
            BR3 = AR1 - AR3
            BR4 = AR4 - AI6
            BR5 = AR5 - AI7
            BR6 = AR4 + AI6
            BR7 = AR5 + AI7
            BI0 = AI0 + AI2
            BI1 = AI1 + AI3
            BI2 = AI0 - AI2
            BI3 = AI1 - AI3
            BI4 = AI4 + AR6
            BI5 = AI5 + AR7
            BI6 = AI4 - AR6
            BI7 = AI5 - AR7
            CR0(K) = BR0 + BR1
            CI0(K) = BI0 + BI1
            IF (J.LE.1) GO TO 10
            CR1(K) = C4*(BR0-BR1) - S4*(BI0-BI1)
            CI1(K) = C4*(BI0-BI1) + S4*(BR0-BR1)
            CR2(K) = C2*(BR2-BI3) - S2*(BI2+BR3)
            CI2(K) = C2*(BI2+BR3) + S2*(BR2-BI3)
            CR3(K) = C6*(BR2+BI3) - S6*(BI2-BR3)
            CI3(K) = C6*(BI2-BR3) + S6*(BR2+BI3)
            TR = P7*(BR5-BI5)
            TI = P7*(BR5+BI5)
            CR4(K) = C1*(BR4+TR) - S1*(BI4+TI)
            CI4(K) = C1*(BI4+TI) + S1*(BR4+TR)
            CR5(K) = C5*(BR4-TR) - S5*(BI4-TI)
            CI5(K) = C5*(BI4-TI) + S5*(BR4-TR)
            TR = -P7*(BR7+BI7)
            TI = P7*(BR7-BI7)
            CR6(K) = C3*(BR6+TR) - S3*(BI6+TI)
            CI6(K) = C3*(BI6+TI) + S3*(BR6+TR)
            CR7(K) = C7*(BR6-TR) - S7*(BI6-TI)
            CI7(K) = C7*(BI6-TI) + S7*(BR6-TR)
            GO TO 20
10          CR1(K) = BR0 - BR1
            CI1(K) = BI0 - BI1
            CR2(K) = BR2 - BI3
            CI2(K) = BI2 + BR3
            CR3(K) = BR2 + BI3
            CI3(K) = BI2 - BR3
            TR = P7*(BR5-BI5)
            TI = P7*(BR5+BI5)
            CR4(K) = BR4 + TR
            CI4(K) = BI4 + TI
            CR5(K) = BR4 - TR
            CI5(K) = BI4 - TI
            TR = -P7*(BR7+BI7)
            TI = P7*(BR7-BI7)
            CR6(K) = BR6 + TR
            CI6(K) = BI6 + TI
            CR7(K) = BR6 - TR
            CI7(K) = BI6 - TI
20        CONTINUE
30      CONTINUE
        RETURN
        END
```

1.3

FFT Subroutines for Sequences With Special Properties

L. R. Rabiner

Acoustics Research Dept.
Bell Laboratories
Murray Hill, NJ 07974

1. Purpose

The set of subroutines described in this paper computes the DFT and the IDFT of sequences with special properties. The special properties include:

1. Symmetrical, real sequences
2. Antisymmetrical, real sequences
3. Real sequences with only odd harmonics
4. Real, symmetrical sequences with only odd harmonics
5. Real, antisymmetrical sequences with only odd harmonics.

2. Method

2.1 DFT Routines

The subroutines for symmetrical and antisymmetrical sequences are based on the procedures described by Cooley, Lewis, and Welch [1]. If we call the input sequence $x(n)$, defined for $0 \leqslant n \leqslant N-1$, where N is a power of 2 (for this implementation), then if $x(n)$ is either symmetrical, i.e.,

$$x(n) = x(N-n) \qquad n = 1,2,\ldots,N/2-1 \tag{1}$$

or antisymmetrical, i.e.,

$$x(n) = -x(N-n) \qquad n = 1,2,\ldots,N/2-1 \tag{2}$$

then the N-Point DFT of $x(n)$ (call it $X(k)$, $k = 0,1,\ldots,N-1$), can be obtained from the $N/2$-point DFT of the sequence $y(n)$ (call it $Y(k)$, $k = 0,1,\ldots,N/2-1$) where $y(n)$ for symmetric sequences is obtained as

$$y(n) = x(2n) + (x(2n+1)-x(2n-1)) \qquad n = 1,2,\ldots,N/4-1 \tag{3a}$$
$$y(N/2-n) = x(2n) - (x(2n+1)-x(2n-1)) \qquad n = 1,2,\ldots,N/4-1 \tag{3b}$$
$$y(0) = x(0) \tag{3c}$$
$$y(N/4) = x(N/2) \tag{3d}$$

and for antisymmetric sequences, $y(n)$ is obtained as

$$y(n) = x(2n) + (x(2n+1)-x(2n-1)) \qquad n = 1,2,\ldots,N/4-1 \tag{4a}$$
$$y(N/2-n) = -x(2n) + (x(2n+1)-x(2n-1)) \qquad n = 1,2,\ldots,N/4-1 \tag{4b}$$
$$y(0) = 2x(1) \tag{4c}$$
$$y(N/4) = -2x(N/2-1) \tag{4d}$$

The relation for obtaining $X(k)$ from $Y(k)$ for symmetric sequences is [1,2]

$$X(k) = \mathrm{Re}[Y(k)] + \frac{\mathrm{Im}[Y(k)]}{2\sin(\frac{2\pi}{N}k)} \quad k = 1,2,...,N/4 \tag{5a}$$

$$X(N/2-k) = \mathrm{Re}[Y(k)] - \frac{\mathrm{Im}[Y(k)]}{2\sin(\frac{2\pi}{N}k)} \quad k = 1,2,...,N/4 \tag{5b}$$

$$X(N-k) = X^*(k) \quad k = 1,2,...,N/2-1 \tag{5c}$$

$$X(0) = B(0) + \mathrm{Re}[Y(0)] \tag{5d}$$

$$X(N/2) = -B(0) + \mathrm{Re}[Y(0)] \tag{5e}$$

where

$$B(0) = 2\sum_{n=0}^{\frac{N}{4}-1} x(2n+1) \tag{6}$$

and * denotes complex conjugate.

For antisymmetric sequences, the relation for obtaining $X(k)$ from $Y(k)$ is [1,2]

$$X(k) = \mathrm{Im}[Y(k)] - \frac{\mathrm{Re}[Y(k)]}{2\sin(\frac{2\pi}{N}k)} \quad k = 1,2,...,N/4 \tag{7a}$$

$$X(N/2-k) = -\mathrm{Im}[Y(k)] - \frac{\mathrm{Re}[Y(k)]}{2\sin(\frac{2\pi}{N}k)} \quad k = 1,2,...,N/4 \tag{7b}$$

$$X(0) = 0 \tag{7c}$$

$$X(N/2) = 0 \tag{7d}$$

It should be noted that for either symmetric or antisymmetric sequences, only $(N/2+1)$ (symmetric) or $(N/2)$ (antisymmetric) points of $x(n)$ need be specified.

For sequences containing only odd harmonics, i.e.,

$$X(2k) = 0 \quad \text{all } k \tag{8}$$

where $X(k)$ is the DFT of $x(n)$, the N-point DFT of $x(n)$ can be obtained from the $N/2$-point DFT of the sequence $y(n)$ (call it $Y(k)$, $k = 0,1,...,N/2-1$), where $y(n)$ is obtained as

$$y(n) = 4\sin(\frac{2\pi}{N}n)x(n) \quad n = 0,1,...,N/2-1 \tag{9}$$

A recursion relation is used to obtain $X(k)$ from $Y(k)$. The required initial values are

$$T_1 = \sum_{\substack{n=0 \\ n \text{ even}}}^{\frac{N}{2}-1} x(n)\cos(\frac{2\pi}{N}n) \tag{10a}$$

$$T_2 = \sum_{\substack{n=0 \\ n \text{ odd}}}^{\frac{N}{2}-1} x(n)\cos(\frac{2\pi}{N}n) \tag{10b}$$

giving

$$\mathrm{Re}[X(1)] = 2(T_1+T_2) \tag{10c}$$
$$\mathrm{Re}[X(N/2-1)] = 2(T_1-T_2) \tag{10d}$$

The initial condition

$$\mathrm{Im}[X(1)] = -\mathrm{Re}[Y(0)]/2 \tag{11}$$

along with Eq (10c) gives the recursion

$$\mathrm{Re}[X(2k+1)] = \mathrm{Im}[Y(k)] + \mathrm{Re}[X(2k-1)] \quad k = 1,2,...,N/4-2 \tag{12a}$$
$$\mathrm{Im}[X(2k+1)] = -\mathrm{Re}[Y(k)] + \mathrm{Im}[X(2k-1)] \quad k = 1,2,...,N/4-2 \tag{12b}$$

with final values

$$\mathrm{Im}[X(N/2-1)] = \mathrm{Re}[Y(N/4)]/2 \tag{13}$$

and Eq (10d) giving $\mathrm{Re}[X(N/2-1)]$. It should again be noted that only $N/2$ points of $x(n)$ are required to implement the procedure described above.

For real sequences that are both symmetrical, and possess only odd harmonics a combination of two of the above procedures can be used [2], to further reduce computation. For this case a total of $N/4$ values of $x(n)$ are required. The procedure is a cascade of the procedure for symmetric sequences with that for odd harmonic sequences. The procedure can be performed in either order (i.e., one can begin by accounting for symmetry and then odd harmonics, or vice versa). Details of the implementation are given in Ref. 5.

For real sequences that are both antisymmetrical, and possess only odd harmonics another combination of two of the above procedures can be used [5], to further reduce computation. For this case a total of $(N/4+1)$ values of $x(n)$ are required. The procedure is a cascade of the procedure for antisymmetric sequences with that for odd harmonic sequences. Again the procedure can be performed in either order (i.e. one can begin by accounting for antisymmetry, and then odd harmonics, or vice versa). Details of the implementation are given in Ref. 5.

2.2 IDFT Routines

The IDFT for sequences with the special properties discussed in Section 1 can be obtained from the DFT of these sequences by inverting the procedures described in Section 2.1. Throughout this section we denote the original sequence as $x(n)$ and its DFT as $X(k)$.

For symmetrical sequences we first precompute the sample $x(1)$, defined as

$$x(1) = X1 = X(0) - X(N/2) + 2\sum_{k=1}^{N/2-1} X(k)\cos(\frac{2\pi}{N}k) \tag{14}$$

The complex sequence $Y(k)$ is obtained from $X(k)$ via the relation

$$Y(0) = (X(0)+X(N/2))/2 \tag{15a}$$
$$\mathrm{Re}[Y(k)] = [X(k)+X(N/2-k)]/2 \quad k = 1,2,...,N/4 \tag{15b}$$
$$\mathrm{Im}[Y(k)] = [X(k)-X(N/2-k)]\,[\sin(\frac{2\pi}{N}k)] \quad k = 1,2,...,N/4 \tag{15c}$$

The $N/2$-point IDFT of $Y(k)$ is taken giving the $N/2$-point real sequence $y(n)$. The sequence $x(n)$ is obtained from $y(n)$ via the recursion relation

$$x(2n) = [y(n)+y(N/2-n)]/2 \quad n = 1,2,...,N/4-1 \tag{16a}$$
$$x(2n+1) = [y(n)-y(N/2-n)]/2 + x(2n-1) \quad n = 1,2,...,N/4-1 \tag{16b}$$

with initial conditions

$$x(0) = y(0) \tag{17a}$$
$$x(1) = X1 \tag{17b}$$

and final condition

$$x(N/2) = y(N/4) \tag{17c}$$

For antisymmetrical sequences the complex sequence $Y(k)$ is obtained from $X(k)$ via the relation

$$\mathrm{Im}[Y(k)] = [X(k)-X(N/2-k)]/2 \quad k = 1,2,...,N/4 \tag{18a}$$
$$\mathrm{Re}[Y(k)] = -\,[X(k)+X(N/2-k)]\,[\sin(\frac{2\pi}{N}k)] \quad k = 1,2,...,N/4 \tag{18b}$$

The $N/2$-point IDFT of $Y(k)$ is taken giving the $N/2$-point real sequence $y(n)$. The sequence $x(n)$ is obtained from $y(n)$ via the recursion relation

$$x(2n) = [y(n)-y(N/2-n)]/2 \qquad n = 1,2,...,N/4-1 \tag{19a}$$
$$x(2n+1) = [y(n)+y(N/2-n)]/2 + x(2n-1) \qquad n = 1,2,...,N/4-1 \tag{19b}$$

with initial conditions

$$x(0) = 0 \tag{20a}$$
$$x(1) = y(0)/2 \tag{20b}$$

and final condition

$$x(N/2-1) = -y(N/4)/2 \tag{20c}$$

For odd harmonic sequences we first precompute the value of $x(0)$ as

$$x(0) = \frac{2}{N} \sum_{k=0}^{N/4-1} \mathrm{Re}[X(2k+1)] \tag{21}$$

The complex sequence $Y(k)$ is obtained from $X(k)$ via the relation

$$\mathrm{Re}[Y(k)] = -\mathrm{Im}[X(2k+1)]+\mathrm{Im}[X(2k-1)] \quad k = 1,2,...,N/4-1 \tag{22a}$$
$$\mathrm{Im}[Y(k)] = \mathrm{Re}[X(2k+1)]-\mathrm{Re}[X(2k-1)] \quad k = 1,2,...,N/4-1 \tag{22b}$$

with initial and final values

$$\mathrm{Re}[Y(0) = -2\,\mathrm{Im}[X(1)] \tag{23a}$$
$$\mathrm{Im}[Y(0)] = 0 \tag{23b}$$
$$\mathrm{Re}[Y(N/4)] = 2\,\mathrm{Im}[X(N/2-1)] \tag{23c}$$
$$\mathrm{Im}[Y(N/4)] = 0 \tag{23d}$$

The $N/2$-point IDFT of $Y(k)$ is taken giving the $N/2$-point real sequence $y(n)$. The sequence $x(n)$ is obtained from $y(n)$ via the relation

$$x(n) = y(n)/[4\sin(\frac{2\pi}{N}n)] \qquad n = 1,2,...,N/2-1 \tag{24}$$

with $x(0)$ given from Eq. (21).

The IDFT for symmetric and antisymmetric, odd harmonic sequences is obtained by cascading the procedures given above. The specific implementations used here first account for the time symmetry (or antisymmetry) and then use the odd harmonics property of the sequence, as was done for the direct transforms. Details of the implementation are given in Ref. 5.

3. Program Description

3.1 Usage

The package consists of 10 subroutines; the DFT routines FFTSYM, FFTASM, FFTOHM, FFTSOH, FFTAOH, and the IDFT routines IFTSYM, IFTASM, IFTOHM, IFTSOH and IFTAOH. All the subroutines use the real input FFT subroutines FAST and FSST [3,4]. For the direct transforms, the user passes the required number of values of the data sequence ($(N/2+1)$ for FFTSYM, $N/2$ for FFTASM, $N/2$ for FFTOHM, $N/4$ for FFTSOH, $(N/4+1)$ for FFTAOH), the size of the full data sequence (N), and a scratch array of size $(N/2+2)$ for all subroutines except FFTOHM which doesn't require the scratch array. The subroutines return the DFT of $x(n)$, $X(k)$, in the same array in which the data is passed. The format in which $X(k)$ is returned is as follows:

FFTSYM - only the real part of $X(k)$, $k = 0,1,...,N/2$ is returned.
FFTASM - only the imaginary part of $X(k)$, $k = 0,1,...,N/2$ is returned.
FFTOHM - only the odd harmonics of $X(k)$, $k = 1,3,...,N/2-1$ are returned. The real part of $X(k)$ is followed by the imaginary part of $X(k)$.
FFTSOH - only the real parts of the odd harmonics of $X(k)$, $k = 1,3,...,N/2-1$ are returned.

FFTAOH - only the imaginary parts of the odd harmonics of $X(k)$ $k = 1,3,...,N/2 - 1$ are returned.

The FFT subroutine FAST is a real input, radix 4 and radix 2 FFT. Thus the subroutines only work for sequences whose length N is a power of 2. (If a more general FFT is used, these routines will work for sequences whose length N is a multiple of either 4 (for FFTSYM, FFTASM, and FFTOHM), or 8 (for FFTSOH and FFTAOH)).

For the inverse transforms, the user passes the required number of values of the transform ($N/2+1$ for IFTSYM and IFTASM, $N/2$ for IFTOHM, $N/4$ for IFTSOH, and $N/4$ for IFTAOH), in the format returned by the direct transform routines, the size of the full data sequence (N), and a scratch array of size ($N/2+2$) for all subroutines except IFTOHM which does the transform in place. The inverse subroutines return the IDFT of $X(k)$, i.e. the original sequence, $x(n)$, in the same array in which the data is passed, in the format required for the forward transforms.

The inverse FFT subroutine FSST is a real input, radix 4 and radix 2 inverse FFT. Thus the subroutines only work for sequences whose length N is a power of 2. (Again if a more general inverse FFT is used, these subroutines will work for values of N which are multiples of either 4 (for IFTSYM, IFTASM and IFTOHM), or 8 (for IFTSOH and IFTAOH).

3.2 Description of Parameters of FFT Subroutines

3.2.1 FFTSYM

This subroutine is used to give the DFT for real, symmetric, N-point sequences where N can be any power of 2, or a multiple of 4 consistent with the FFT subroutines. The call to FFTSYM is

CALL FFTSYM (X,N,Y)

where

X Array of size ($N/2+1$) points. On input X contains the first ($N/2+1$) values of $x(n)$. On output X contains the ($N/2+1$) real parts of the transform of the input.

N True size of input sequence.

Y Scratch array of size ($N/2+2$).

3.2.2 FFTASM

This subroutine is used to give the DFT for real, antisymmetric, N-point sequences where N can be any power of 2, or a multiple of 4 consistent with the FFT subroutine. The call to FFTASM is

CALL FFTASM (X,N,Y)

where

X Array of size ($N/2+1$) points. On input X contains the first ($N/2$) values of $x(n)$. On output X contains the ($N/2+1$) imaginary parts of the transform of the input.

N True size of input sequence.

Y Scratch array of size ($N/2+2$).

3.2.3 FFTOHM

This subroutine is used to give the DFT for real, N-point sequences having only odd harmonics where N can be any power of 2 or a multiple of 4 consistent with the FFT subroutine. The call to FFTOHM is

CALL FFTOHM (X,N)

where

X Array of size ($N/2+2$). On input X contains the first ($N/2$) values of $x(n)$. On output X contains the ($N/2$) values (complex) of the odd harmonics of the transform of the input. The complex values are stored as the real part of each harmonic followed by the imaginary part.

N True size of input sequence.

3.2.4 FFTSOH

This subroutine is used to give the DFT for real, symmetric N-point sequences having only odd harmonics where N is any power of 2 or a multiple of 8 consistent with the FFT subroutine. The call to FFTSOH is

CALL FFTSOH (X,N,Y)

where

X Array of size $(N/4)$. On input X contains the first $(N/4)$ values of $x(n)$. On output X contains the $(N/4)$ real parts of the odd harmonics of the transform of the input.

N True size of input sequence.

Y Scratch array of size $(N/4+2)$.

3.2.5 FFTAOH

This subroutine is used to give the DFT for real, antisymmetric N-point sequences having only odd harmonics, where N is any power of 2 or a multiple of 8 consistent with the FFT subroutine. The call to FFTAOH is

CALL FFTAOH (X,N,Y)

where

X Array of size $(N/4+1)$. On input X contains the first $(N/4+1)$ values of $x(n)$. On output X contains the $(N/4)$ imaginary parts of the odd harmonics of the transform of the input.

N True size of input sequence.

Y Scratch array of size $(N/4+2)$.

3.2.6 FAST

The subroutine FAST is a radix 4 and radix 2 FFT subroutine [3,4] which uses real inputs. The call to FAST is

CALL FAST (X,N)

where

X Array of size $(N+2)$. On input X contains the N values of the sequence to be transformed; on output X contains the $(N/2+1)$ complex values of the transform of the input.

N Size of the input sequence.

3.2.7 IFTSYM

This subroutine is used to give the inverse DFT (IDFT) for real, symmetric N-point sequences where N can be any power of 2, or a multiple of 4 consistent with the IFFT subroutine. The call to IFTSYM is

CALL IFTSYM (X,N,Y)

where

X Array of size $(N/2+1)$ points. On input X contains the $(N/2+1)$ real parts of the transform of $x(n)$; on output X contains the first $(N/2+1)$ values of $x(n)$.

N True size of input sequence.

Y Scratch array of size $(N/2+2)$.

3.2.8 IFTASM

This subroutine is used to give the IDFT for real, antisymmetric N-point sequences where N can be any power of 2, or a multiple of 4 consistent with the IFFT subroutine. The call to IFTASM is

CALL IFTASM (X,N,Y)

where

X Array of size $(N/2+1)$ points. On input X contains the $(N/2+1)$ imaginary parts of the transform of $x(n)$; on output X contains the first $(N/2)$ values of $x(n)$.

N True size of input sequence.

Y Scratch array of size $(N/2+2)$.

3.2.9 IFTOHM

This subroutine is used to give the IDFT for real, N-point sequences having only odd harmonics where N can be any power of 2, or a multiple of 4 consistent with the IFFT subroutine. The call to IFTOHM is

CALL IFTOHM (X,N)

where

X Array of size $(N/2+2)$. On input X contains the $(N/2)$ values (complex) of the odd harmonics of the transform of $x(n)$; on output X contains the first $(N/2)$ values of $x(n)$. On input, the complex values are stored as the real part of each harmonic followed by the imaginary part.

N True size of input sequence.

3.2.10 IFTSOH

This subroutine is used to give the IDFT for real, symmetric N-point sequences having only odd harmonics where N is any power of 2 or a multiple of 8 consistent with the IFFT subroutine. The call to IFTSOH is

CALL IFTSOH (X,N,Y)

where

X Array of size $(N/4)$. On input X contains the $(N/4)$ real parts of the odd harmonics of the transform of $x(n)$; on output X contains the first $(N/4)$ values of $x(n)$.

N True size of the input sequence.

Y Scratch array of size $(N/4+2)$.

3.2.11 IFTAOH

This subroutine is used to give the IDFT for real, antisymmetric N-point sequences having only odd harmonics, where N is any power of 2 or a multiple of 8 consistent with the IFFT subroutine. The call to IFTAOH is

CALL IFTAOH (X,N,Y)

where

X Array of size $(N/4+1)$. On input X contains the $(N/4)$ imaginary parts of the odd harmonics of the transform of $x(n)$; on output X contains the first $(N/4+1)$ values of $x(n)$.

N True size of input sequence.

Y Scratch array of size $(N/4+2)$.

3.2.12 FSST

The subroutine FSST is a radix 4 and radix 2 IFFT subroutine [3,4] for real input sequences. The call to FSST is

CALL FSST (X,N)

where

X Array of size $(N+2)$. On input X contains the $(N/2+1)$ complex values of the transform of $x(n)$; on output X contains the N real values of the sequence $x(n)$.

N Size of the input sequence.

3.3 Dimension Requirements

The DIMENSION statement in the calling program should be modified according to the requirements of each particular problem. The required dimensions of the arrays are

X $(N/2+2)$ values, or $(N/4+1)$ values, depending on the subroutine used, where N is the size of the data array.

Y $(N/2+2)$ values, or $(N/4+2)$ values depending on the subroutine used.

3.4 Summary of User Requirements

1. Specify the required number of values of $x(n)$, the input sequence, or $X(k)$ the transform of $x(n)$.
2. Call the appropriate FFT or IFFT subroutine.
3. Output DFT or IDFT is obtained in the same array in which the input was defined.

4. Test Program

The test program FFTSUBT tests each of the 10 subroutines in order. The program first asks the user to specify the value of N (the sequence length). It then creates an N-point random, real, symmetrical sequence, and then computes and prints out both the sequence and the entire $(N/2+1)$ point complex FFT with the real part of each FFT component preceding the imaginary part. The first $(N/2+1)$ time samples of $x(n)$ are then passed on to the subroutine FFTSYM and the $(N/2+1)$ points of the output (the real parts of the FFT) are printed. The output of FFTSYM is then passed to IFTSYM and the resulting $(N/2+1)$ values of $x(n)$ are printed.

The program next creates a new N-point random, real, antisymmetrical sequence and computes and prints out both the sequence and the entire $(N/2+1)$ point complex FFT. The subroutine FFTASM is used on the first $(N/2)$ time samples and the $(N/2+1)$ points of the output (the imaginary parts of the FFT) are printed. The output of FFTASM is then passed to IFTASM and the resulting $(N/2)$ values of $x(n)$ are printed.

The program next creates a new N-point random, real sequence with only odd harmonics. This sequence is created directly in the frequency domain by randomly specifying values for the complex odd harmonics, and setting the even harmonics to zero. Again the entire $(N/2+1)$ complex values of the FFT of the sequence are printed. The real sequence is obtained as the inverse FFT of the DFT (obtained using FSST [3,4]) and the N time values are printed, and the first $(N/2)$ time samples are passed on to the subroutine FFTOHM and the $(N/4)$ complex odd harmonics are printed. The output of FFTOHM is then passed to IFTOHM and the resulting $(N/2)$ values of $x(n)$ are printed.

Next, the program creates an N-point random, real, symmetric sequence with only odd harmonics. This sequence is again created directly in the frequency domain. The entire $(N/2+1)$ complex values of the FFT of the sequence are printed. The time sequence is obtained using the inverse FFT subroutine FSST and the N values of $x(n)$ are printed. The first $(N/4)$ time samples are passed on to the subroutine FFTSOH and the $(N/4)$ real values of the odd harmonics are printed. The output of FFTSOH is then passed to IFTSOH and the resulting $(N/4)$ values of $x(n)$ are printed.

Finally the program creates an N-point random, real, antisymmetric sequence with only odd harmonics. This sequence is again created directly in the frequency domain. The entire $(N/2+1)$ complex values of the FFT of the sequence are printed. The time sequence is again obtained using the inverse FFT subroutine FSST and the N values of $x(n)$ are printed. The first $(N/4+1)$ time samples are passed on to the subroutine FFTAOH and the $(N/4)$ imaginary values of the odd harmonics are printed. The output of FFTAOH is then passed to IFTAOH and the resulting $(N/4+1)$ values of $x(n)$ are printed.

The program then loops back and asks the user to specify a new value of N to repeat the tests. A value of $N = 0$ terminates the program.

The code for the test program and the subroutines FFTSYM, IFTSYM, FFTASM, IFTASM, FFTOHM, IFTOHM, FFTSOH, IFTSOH, FFTAOH, and IFTAOH is given in the Appendix. Table 1 gives output of the test program for a value of $N = 64$.

References

1. J. W. Cooley, P. A. Lewis, and P. D. Welch, "The Fast Fourier Transform Algorithm: Programming Considerations in the Calculation of Sine, Cosine and Laplace Transforms", *J. Sound Vib.*, Vol. 12, No. 3, pp. 315-337, July 1970.
2. L. R. Rabiner, "On the Use of Symmetry in FFT Computation", *IEEE Trans. on Acoustics, Speech, and Signal Processing,* 1979 (to appear).
3. G. D. Bergland, "Subroutines FAST and FSST, Radix 4 Routines for Computing the DFT of Real Sequences", unpublished work.
4. G. D. Bergland and M. T. Dolan, "Fast Fourier Transform Algorithms", Section 1.2, IEEE Press Book on *Programs for Digital Signal Processing,* 1979.
5. L. R. Rabiner, "Additional Special Routines for FFT and IFFT Computation", submitted for publication.

Table 1

```
TESTING N=    64 RANDOM SEQUENCES


ORIGINAL SYMMETRIC SEQUENCE
     .38705E 00   -.27074E 00    .26687E 00    .18318E 00    .91910E-02
     .37456E 00    .14464E 00    .34747E 00   -.14604E 00   -.10111E 00
    -.42906E-01   -.26369E 00   -.36682E 00   -.33395E 00   -.27397E 00
     .16246E 00   -.24979E 00    .11770E 00   -.23753E 00    .12667E-01
     .43921E 00    .12403E 00   -.77618E-01    .43971E 00   -.21793E 00
    -.30782E-01   -.44512E 00    .19831E-01   -.10317E 00   -.38684E 00
     .10752E 00    .20151E 00   -.16857E 00    .20151E 00    .10752E 00
    -.38684E 00   -.10317E 00    .19831E-01   -.44512E 00   -.30782E-01
    -.21793E 00    .43971E 00   -.77618E-01    .12403E 00    .43921E 00
     .12667E-01   -.23753E 00    .11770E 00   -.24979E 00    .16246E 00
    -.27397E 00   -.33395E 00   -.36682E 00   -.26369E 00   -.42906E-01
    -.10111E 00   -.14604E 00    .34747E 00    .14464E 00    .37456E 00
     .91910E-02    .18318E 00    .26687E 00   -.27074E 00


  64 POINT FFT OF SYMMETRIC SEQUENCE
    -.97649E 00    .00000E 00    .16830E 01    .12703E-05    .20771E 01
     .89407E-06    .38044E 01   -.33458E-06   -.56949E 00   -.83447E-06
    -.49675E 01   -.77672E-05    .31376E 00   -.53644E-06   -.20166E 01
    -.31777E-05    .64130E 00    .00000E 00    .10633E 01    .91270E-06
    -.50420E 00    .20862E-05   -.71593E 00    .15087E-05    .11459E 01
    -.47684E-06    .84500E 00    .14163E-05   -.35235E 00    .59232E-06
    -.25434E 01    .29653E-05    .64029E-01    .00000E 00    .40404E 00
    -.96858E-07   -.16805E 01   -.14342E-05    .31990E 01   -.82026E-06
    -.10552E 01   -.10729E-05    .52683E 00    .23060E-05    .38979E 00
    -.65565E-06    .19977E 01    .99465E-06   -.25730E 01    .00000E 00
     .32219E 01   -.13523E-05    .36140E 01   -.32783E-05    .67098E 00
    -.18887E-05    .22661E 01   -.14305E-05    .26874E 00    .18843E-05
     .18869E 01   -.11325E-05    .14484E 01   -.67428E-06   -.33604E 01
     .00000E 00
```

Table 1
(Continued)

```
OUTPUT OF FFTSYM
  -.97649E 00   .16830E 01   .20771E 01   .38044E 01  -.56949E 00
  -.49675E 01   .31375E 00  -.20166E 01   .64131E 00   .10633E 01
  -.50420E 00  -.71593E 00   .11459E 01   .84498E 00  -.35234E 00
  -.25434E 01   .64028E-01   .40406E 00  -.16805E 01   .31990E 01
  -.10552E 01   .52684E 00   .38979E 00   .19977E 01  -.25730E 01
   .32219E 01   .36141E 01   .67099E 00   .22661E 01   .26874E 00
   .18869E 01   .14484E 01  -.33604E 01


OUTPUT OF IFTSYM
   .38705E 00  -.27074E 00   .26687E 00   .18318E 00   .91908E-02
   .37456E 00   .14464E 00   .34747E 00  -.14604E 00  -.10110E 00
  -.42906E-01  -.26368E 00  -.36682E 00  -.33394E 00  -.27397E 00
   .16247E 00  -.24979E 00   .11770E 00  -.23753E 00   .12673E-01
   .43921E 00   .12403E 00  -.77618E-01   .43971E 00  -.21793E 00
  -.30778E-01  -.44512E 00   .19838E-01  -.10317E 00  -.38684E 00
   .10752E 00   .20151E 00  -.16857E 00

ORIGINAL ANTISYMMETRIC SEQUENCE
   .00000E 00  -.27904E 00  -.33138E 00   .49225E 00  -.13293E 00
  -.21999E-02  -.24590E 00   .29780E 00  -.36609E 00  -.38386E 00
  -.28712E 00   .28336E 00   .48293E 00   .28849E 00  -.13457E 00
  -.23091E-01   .32732E 00  -.39747E 00   .44326E 00   .18425E 00
   .22906E 00   .22204E 00  -.15997E 00  -.89684E-01  -.11538E 00
   .46185E 00  -.19906E-01  -.40348E 00   .19834E-01   .26180E 00
   .20444E 00  -.15980E 00   .00000E 00   .15980E 00  -.20444E 00
  -.26180E 00  -.19834E-01   .40348E 00   .19906E-01  -.46185E 00
   .11538E 00   .89684E-01   .15997E 00  -.22204E 00  -.22906E 00
  -.18425E 00  -.44326E 00   .39747E 00  -.32732E 00   .23091E-01
   .13457E 00  -.28849E 00  -.48293E 00  -.28336E 00   .28712E 00
   .38386E 00   .36609E 00  -.29780E 00   .24590E 00   .21999E-02
   .13293E 00  -.49225E 00   .33138E 00   .27904E 00


 64 POINT FFT OF ANTISYMMETRIC SEQUENCE
   .00000E 00   .00000E 00   .89407E-06  -.18801E 01  -.53644E-06
   .15215E 01   .77486E-06   .23367E 01   .35390E-06   .31046E 00
   .21458E-05  -.10940E 01   .53644E-06  -.73939E 00   .00000E 00
  -.24839E 01   .00000E 00   .90270E 00  -.23842E-06   .36229E 01
   .25034E-05   .32936E 01  -.50068E-05  -.24539E 01   .83819E-06
   .29557E 00   .17285E-05  -.44543E 00   .59605E-06   .50971E 00
   .48280E-05   .66126E 01   .00000E 00   .81998E 00  -.28014E-05
   .32414E 01   .83447E-06  -.17832E 01   .14901E-05   .10330E 01
  -.71153E-06   .17614E 01   .19073E-05  -.18068E 01   .15497E-05
  -.15702E 01  -.21458E-05   .26390E 00   .00000E 00   .14661E 01
  -.33379E-05  -.82171E 00  -.45896E-05  -.30256E 01   .23842E-06
   .18796E 01  -.48056E-06  -.14768E 01   .53644E-06  -.45427E 01
  -.89407E-06  -.22052E 01  -.10133E-05  -.44018E 00   .00000E 00
   .00000E 00


OUTPUT OF FFTASM
   .00000E 00  -.18801E 01   .15215E 01   .23367E 01   .31046E 00
  -.10940E 01  -.73940E 00  -.24839E 01   .90271E 00   .36229E 01
   .32936E 01  -.24539E 01   .29558E 00  -.44543E 00   .50970E 00
   .66127E 01   .81999E 00   .32414E 01  -.17832E 01   .10330E 01
   .17614E 01  -.18068E 01  -.15702E 01   .26392E 00   .14661E 01
  -.82172E 00  -.30256E 01   .18796E 01  -.14768E 01  -.45427E 01
  -.22052E 01  -.44018E 00   .00000E 00


OUTPUT OF IFTASM
   .00000E 00  -.27904E 00  -.33138E 00   .49225E 00  -.13293E 00
  -.22032E-02  -.24590E 00   .29780E 00  -.36609E 00  -.38386E 00
  -.28712E 00   .28336E 00   .48293E 00   .28849E 00  -.13457E 00
  -.23091E-01   .32732E 00  -.39747E 00   .44326E 00   .18425E 00
   .22906E 00   .22204E 00  -.15997E 00  -.89687E-01  -.11538E 00
   .46185E 00  -.19906E-01  -.40349E 00   .19834E-01   .26180E 00
   .20444E 00  -.15980E 00
```

Table 1
(Continued)

```
 64 POINT FFT OF ODD HARMONIC SEQUENCE
   .00000E 00    .00000E 00  -.39239E-01  -.24476E-01    .00000E 00
   .00000E 00    .24817E 00   .15425E 00   .00000E 00    .00000E 00
  -.78124E-01  -.58618E-01   .00000E 00   .00000E 00    .27441E 00
   .12580E 00    .00000E 00   .00000E 00   .24052E-01  -.40714E 00
   .00000E 00    .00000E 00   .48475E 00   .32958E 00    .00000E 00
   .00000E 00  -.30727E 00  -.20840E 00   .00000E 00    .00000E 00
  -.11457E 00  -.32150E-01   .00000E 00   .00000E 00   -.17971E 00
  -.25726E 00    .00000E 00   .00000E 00   .30139E 00   -.47430E 00
   .00000E 00    .00000E 00   .48588E 00   .40457E 00    .00000E 00
   .00000E 00  -.39448E 00  -.26213E 00   .00000E 00    .00000E 00
  -.24664E 00  -.36574E 00   .00000E 00   .00000E 00    .47930E 00
   .12915E 00    .00000E 00   .00000E 00  -.54238E-01    .41343E 00
   .00000E 00    .00000E 00  -.22918E 00  -.91972E-01    .00000E 00
   .00000E 00


ORIGINAL ODD HARMONIC SEQUENCE
   .20453E-01   .39408E-01  -.71683E-02  -.13413E-01  -.78217E-02
  -.29470E-01   .15314E-01   .57910E-02  -.27815E-01   .18272E-01
  -.20514E-01   .10145E-01   .10090E-01  -.54190E-01  -.32431E-01
   .11459E-01   .11932E-01   .73036E-01   .28193E-01  -.19987E-01
  -.53000E-01   .68412E-01  -.22913E-01   .21689E-02   .81134E-01
  -.87985E-01  -.46738E-02   .60175E-02   .33198E-01  -.23224E-01
  -.42196E-02   .91415E-02  -.20453E-01  -.39408E-01   .71683E-02
   .13413E-01   .78217E-02   .29470E-01  -.15314E-01  -.57910E-02
   .27815E-01  -.18272E-01   .20514E-01  -.10145E-01  -.10090E-01
   .54190E-01   .32431E-01  -.11459E-01  -.11932E-01  -.73036E-01
  -.28193E-01   .19987E-01   .53000E-01  -.68412E-01   .22913E-01
  -.21689E-02  -.81134E-01   .87985E-01   .46738E-02  -.60175E-02
  -.33198E-01   .23224E-01   .42196E-02  -.91415E-02


OUTPUT OF FFTOHM
  -.39239E-01  -.24476E-01   .24817E 00   .15425E 00  -.78123E-01
  -.58619E-01   .27441E 00   .12580E 00   .24053E-01  -.40714E 00
   .48475E 00   .32958E 00  -.30727E 00  -.20840E 00  -.11458E 00
  -.32151E-01  -.17971E 00  -.25726E 00   .30139E 00  -.47430E 00
   .48589E 00   .40457E 00  -.39448E 00  -.26213E 00  -.24664E 00
  -.36574E 00   .47931E 00   .12915E 00  -.54236E-01   .41343E 00
  -.22918E 00  -.91972E-01


OUTPUT OF IFTOHM
   .20453E-01   .39408E-01  -.71683E-02  -.13413E-01  -.78217E-02
  -.29470E-01   .15314E-01   .57910E-02  -.27815E-01   .18272E-01
  -.20514E-01   .10145E-01   .10090E-01  -.54190E-01  -.32432E-01
   .11459E-01   .11932E-01   .73036E-01   .28193E-01  -.19987E-01
  -.53001E-01   .68412E-01  -.22913E-01   .21689E-02   .81134E-01
  -.87986E-01  -.46738E-02   .60175E-02   .33198E-01  -.23224E-01
  -.42196E-02   .91415E-02


 64 POINT FFT OF ODD HARMONIC, SYMMETRIC SEQUENCE
   .00000E 00    .00000E 00  -.26539E 00   .00000E 00    .00000E 00
   .00000E 00    .14412E 00   .00000E 00   .00000E 00    .00000E 00
   .22329E 00    .00000E 00   .00000E 00   .00000E 00    .26810E 00
   .00000E 00    .00000E 00   .00000E 00  -.30587E 00    .00000E 00
   .00000E 00    .00000E 00   .13275E 00   .00000E 00    .00000E 00
   .00000E 00  -.34142E 00   .00000E 00   .00000E 00    .00000E 00
   .94914E-01    .00000E 00   .00000E 00   .00000E 00   -.34192E 00
   .00000E 00    .00000E 00   .00000E 00  -.14686E 00    .00000E 00
   .00000E 00    .00000E 00  -.44473E 00   .00000E 00    .00000E 00
   .00000E 00  -.69936E-01   .00000E 00   .00000E 00    .00000E 00
   .46172E-01    .00000E 00   .00000E 00   .00000E 00   -.78392E-03
   .00000E 00    .00000E 00   .00000E 00   .41555E 00    .00000E 00
   .00000E 00    .00000E 00   .12386E 00   .00000E 00    .00000E 00
   .00000E 00
```

Table 1
(Continued)

```
ORIGINAL ODD HARMONIC, SYMMETRIC SEQUENCE
  -.14629E-01  -.52831E-02   .43949E-01  -.30324E-01  -.75174E-02
  -.96356E-02  -.32010E-01  -.16822E-01  -.95453E-02  -.42268E-02
  -.90984E-02   .17533E-01  -.85548E-02   .19091E-01  -.24814E-01
  -.40847E-01   .00000E 00   .40847E-01   .24814E-01  -.19091E-01
   .85548E-02  -.17533E-01   .90984E-02   .42268E-02   .95453E-02
   .16822E-01   .32010E-01   .96356E-02   .75173E-02   .30324E-01
  -.43949E-01   .52831E-02   .14629E-01   .52831E-02  -.43949E-01
   .30324E-01   .75174E-02   .96356E-02   .32010E-01   .16822E-01
   .95453E-02   .42268E-02   .90984E-02  -.17533E-01   .85548E-02
  -.19091E-01   .24814E-01   .40847E-01   .00000E 00  -.40847E-01
  -.24814E-01   .19091E-01  -.85548E-02   .17533E-01  -.90984E-02
  -.42268E-02  -.95453E-02  -.16822E-01  -.32010E-01  -.96356E-02
  -.75173E-02  -.30324E-01   .43949E-01  -.52831E-02


OUTPUT OF FFTSOH
  -.26539E 00   .14412E 00   .22329E 00   .26811E 00  -.30587E 00
   .13275E 00  -.34142E 00   .94915E-01  -.34192E 00  -.14686E 00
  -.44473E 00  -.69939E-01   .46169E-01  -.78589E-03   .41555E 00
   .12386E 00


OUTPUT OF IFTSOH
  -.14630E-01  -.52827E-02   .43948E-01  -.30324E-01  -.75171E-02
  -.96353E-02  -.32011E-01  -.16822E-01  -.95451E-02  -.42265E-02
  -.90986E-02   .17533E-01  -.85547E-02   .19092E-01  -.24814E-01
  -.40847E-01


  64 POINT FFT OF ODD HARMONIC, ANTISYMMETRIC SEQUENCE
   .00000E 00   .00000E 00   .00000E 00   .13362E 00   .00000E 00
   .00000E 00   .00000E 00  -.45413E 00   .00000E 00   .00000E 00
   .00000E 00   .80915E-01   .00000E 00   .00000E 00   .00000E 00
  -.32781E 00   .00000E 00   .00000E 00   .00000E 00   .46997E 00
   .00000E 00   .00000E 00   .00000E 00   .41614E 00   .00000E 00
   .00000E 00   .00000E 00  -.99167E-01   .00000E 00   .00000E 00
   .00000E 00   .27775E 00   .00000E 00   .00000E 00   .00000E 00
   .68952E-01   .00000E 00   .00000E 00   .00000E 00  -.72483E-01
   .00000E 00   .00000E 00   .00000E 00  -.37515E 00   .00000E 00
   .00000E 00   .00000E 00   .27765E 00   .00000E 00   .00000E 00
   .00000E 00  -.45259E 00   .00000E 00   .00000E 00   .00000E 00
  -.23911E 00   .00000E 00   .00000E 00   .00000E 00  -.43002E 00
   .00000E 00   .00000E 00   .00000E 00   .28792E 00   .00000E 00
   .00000E 00


ORIGINAL ODD HARMONIC, ANTISYMMETRIC SEQUENCE
   .00000E 00  -.30825E-03  -.38045E-01   .40039E-01   .18682E-01
   .31662E-01   .20747E-02   .22736E-02  -.39392E-02  -.38762E-01
   .27449E-01   .23230E-01   .68283E-02  -.61149E-02  -.20351E-01
  -.65640E-01   .24044E-01  -.65640E-01  -.20351E-01  -.61149E-02
   .68284E-02   .23230E-01   .27449E-01  -.38762E-01  -.39392E-02
   .22736E-02   .20747E-02   .31662E-01   .18682E-01   .40039E-01
  -.38045E-01  -.30822E-03   .00000E 00   .30825E-03   .38045E-01
  -.40039E-01  -.18682E-01  -.31662E-01  -.20747E-02  -.22736E-02
   .39392E-02   .38762E-01  -.27449E-01  -.23230E-01  -.68283E-02
   .61149E-02   .20351E-01   .65640E-01  -.24044E-01   .65640E-01
   .20351E-01   .61149E-02  -.68284E-02  -.23230E-01  -.27449E-01
   .38762E-01   .39392E-02  -.22736E-02  -.20747E-02  -.31662E-01
  -.18682E-01  -.40039E-01   .38045E-01   .30822E-03


OUTPUT OF FFTAOH
   .13362E 00  -.45413E 00   .80916E-01  -.32781E 00   .46998E 00
   .41614E 00  -.99163E-01   .27776E 00   .68951E-01  -.72483E-01
  -.37515E 00   .27765E 00  -.45259E 00  -.23911E 00  -.43002E 00
   .28792E 00
```

Table 1
(Continued)

```
OUTPUT OF IFTAOH
   .00000E 00  -.30849E-03  -.38045E-01   .40039E-01   .18682E-01
   .31662E-01   .20747E-02   .22732E-02  -.39391E-02  -.38763E-01
   .27449E-01   .23230E-01   .68284E-02  -.61152E-02  -.20351E-01
  -.65640E-01   .24044E-01
```

Appendix

```
C
C-----------------------------------------------------------------------
C MAIN PROGRAM: TEST PROGRAM FOR FFT SUBROUTINES
C AUTHOR:       L R RABINER
C       BELL LABORATORIES, MURRAY HILL, NEW JERSEY, 07974
C INPUT:        RANDOMLY CHOSEN SEQUENCES TO TEST FFT SUBROUTINES
C       FOR SEQUENCES WITH SPECIAL PROPERTIES
C       N IS THE FFT LENGTH (N MUST BE A POWER OF 2)
C           2<= N <= 4096
C-----------------------------------------------------------------------
C
      DIMENSION X(4098), Y(4098)
C
C  DEFINE I/0 DEVICE CODES
C  INPUT: INPUT TO THIS PROGRAM IS USER-INTERACTIVE
C         THAT IS - A QUESTION IS WRITTEN ON THE USER
C         TERMINAL (IOUT1) AD THE USER TYPES IN THE ANSWER.
C
C OUTPUT: ALL OUTPUT IS WRITTEN ON THE STANDARD
C         OUTPUT UNIT (IOUT2).
C
      IND = I1MACH(1)
      IOUT1 = I1MACH(4)
      IOUT2 = I1MACH(2)
C
C READ IN ANALYSIS SIZE FOR FFT
C
  10  WRITE (IOUT1,9999)
9999  FORMAT (30H FFT SIZE(2.LE.N.LE.4096)(I4)=)
      READ (IND,9998) N
9998  FORMAT (I4)
      IF (N.EQ.0) STOP
      DO 20 I=1,12
        ITEST = 2**I
        IF (N.EQ.ITEST) GO TO 30
  20  CONTINUE
      WRITE (IOUT1,9997)
9997  FORMAT (45H N IS NOT A POWER OF 2 IN THE RANGE 2 TO 4096)
      GO TO 10
  30  WRITE (IOUT2,9996) N
9996  FORMAT (11H TESTING N=, I5, 17H RANDOM SEQUENCES)
      WRITE (IOUT2,9992)
      NP2 = N + 2
      NO2 = N/2
      NO2P1 = NO2 + 1
      NO4 = N/4
      NO4P1 = NO4 + 1
C
C CREATE SYMMETRICAL SEQUENCE OF SIZE N
C
      DO 40 I=2,NO2
        X(I) = UNI(0) - 0.5
        IND1 = NP2 - I
        X(IND1) = X(I)
  40  CONTINUE
      X(1) = UNI(0) - 0.5
      X(NO2P1) = UNI(0) - 0.5
      DO 50 I=1,NO2P1
        Y(I) = X(I)
  50  CONTINUE
      WRITE (IOUT2,9995)
9995  FORMAT (28H ORIGINAL SYMMETRIC SEQUENCE)
      WRITE (IOUT2,9993) (X(I),I=1,N)
      WRITE (IOUT2,9992)
C
C COMPUTE TRUE FFT OF N POINT SEQUENCE
C
      CALL FAST(X, N)
      WRITE (IOUT2,9994) N
9994  FORMAT (1H , I4, 32H POINT FFT OF SYMMETRIC SEQUENCE)
      WRITE (IOUT2,9993) (X(I),I=1,NP2)
9993  FORMAT (1H , 5E13.5)
      WRITE (IOUT2,9992)
9992  FORMAT (1H /1H )
C
C USE SUBROUTINE FFTSYM TO OBTAIN DFT FROM NO2 POINT FFT
C
      DO 60 I=1,NO2P1
        X(I) = Y(I)
  60  CONTINUE
      CALL FFTSYM(X, N, Y)
      WRITE (IOUT2,9991)
9991  FORMAT (17H OUTPUT OF FFTSYM)
      WRITE (IOUT2,9993) (X(I),I=1,NO2P1)
      WRITE (IOUT2,9992)
C
C USE SUBROUTINE IFTSYM TO OBTAIN ORIGINAL SEQUENCE FROM NO2 POINT DFT
C
      CALL IFTSYM(X, N, Y)
      WRITE (IOUT2,9990)
9990  FORMAT (17H OUTPUT OF IFTSYM)
      WRITE (IOUT2,9993) (X(I),I=1,NO2P1)
      WRITE (IOUT2,9992)
C
C CREATE ANTISYMMETRIC N POINT SEQUENCE
C
      DO 70 I=2,NO2
        X(I) = UNI(0) - 0.5
        IND1 = NP2 - I
        X(IND1) = -X(I)
  70  CONTINUE
      X(1) = 0.
      X(NO2P1) = 0.
      DO 80 I=1,NO2P1
        Y(I) = X(I)
  80  CONTINUE
      WRITE (IOUT2,9989)
9989  FORMAT (32H ORIGINAL ANTISYMMETRIC SEQUENCE)
      WRITE (IOUT2,9993) (X(I),I=1,N)
      WRITE (IOUT2,9992)
C
C OBTAIN N POINT DFT OF ANTISYMMETRIC SEQUENCE
C
      CALL FAST(X, N)
      WRITE (IOUT2,9988) N
9988  FORMAT (1H , I4, 36H POINT FFT OF ANTISYMMETRIC SEQUENCE)
      WRITE (IOUT2,9993) (X(I),I=1,NP2)
      WRITE (IOUT2,9992)
C
C USE SUBROUTINE FFTASM TO OBTAIN DFT FROM NO2 POINT FFT
C
      DO 90 I=1,NO2
        X(I) = Y(I)
  90  CONTINUE
      CALL FFTASM(X, N, Y)
      WRITE (IOUT2,9987)
9987  FORMAT (17H OUTPUT OF FFTASM)
      WRITE (IOUT2,9993) (X(I),I=1,NO2P1)
      WRITE (IOUT2,9992)
C
```

```
C USE SUBROUTINE IFTASM TO OBTAIN ORIGINAL SEQUENCE FROM NO2 POINT DFT
C
      CALL IFTASM(X, N, Y)
      WRITE (IOUT2,9986)
9986  FORMAT (17H OUTPUT OF IFTASM)
      WRITE (IOUT2,9993) (X(I),I=1,NO2)
      WRITE (IOUT2,9992)
C
C CREATE SEQUENCE WITH ONLY ODD HARMONICS--BEGIN IN FREQUENCY DOMAIN
C
      DO 100 I=1,NP2,2
        X(I) = 0.
        X(I+1) = 0.
        IF (MOD(I,4).EQ.1) GO TO 100
        X(I) = UNI(0) - 0.5
        X(I+1) = UNI(0) - 0.5
        IF (N.EQ.2) X(I+1) = 0.
 100  CONTINUE
      WRITE (IOUT2,9985) N
9985  FORMAT (1H , I4, 35H POINT FFT OF ODD HARMONIC SEQUENCE)
      WRITE (IOUT2,9993) (X(I),I=1,NP2)
      WRITE (IOUT2,9992)
C
C TRANSFORM BACK TO TIME SEQUENCE
C
      CALL FSST(X, N)
      WRITE (IOUT2,9984)
9984  FORMAT (31H ORIGINAL ODD HARMONIC SEQUENCE)
      WRITE (IOUT2,9993) (X(I),I=1,N)
      WRITE (IOUT2,9992)
C
C USE SUBROUTINE FFTOHM TO OBTAIN DFT FROM NO2 POINT FFT
C
      CALL FFTOHM(X, N)
      WRITE (IOUT2,9983)
9983  FORMAT (17H OUTPUT OF FFTOHM)
      WRITE (IOUT2,9993) (X(I),I=1,NO2)
      WRITE (IOUT2,9992)
C
C USE SUBROUTINE IFTOHM TO OBTAIN ORIGINAL SEQUENCE FROM NO2 POINT DFT
C
      CALL IFTOHM(X, N)
      WRITE (IOUT2,9982)
9982  FORMAT (17H OUTPUT OF IFTOHM)
      WRITE (IOUT2,9993) (X(I),I=1,NO2)
      WRITE (IOUT2,9992)
C
C CREATE SEQUENCE WITH ONLY REAL VALUED ODD HARMONICS
C
      DO 110 I=1,NP2,2
        X(I) = 0.
        X(I+1) = 0.
        IF (MOD(I,4).EQ.1) GO TO 110
        X(I) = UNI(0) - 0.5
 110  CONTINUE
      WRITE (IOUT2,9981) N
9981  FORMAT (1H , I4, 45H POINT FFT OF ODD HARMONIC, SYMMETRIC SEQUENC,
     *    1HE)
      WRITE (IOUT2,9993) (X(I),I=1,NP2)
      WRITE (IOUT2,9992)
C
C TRANSFORM BACK TO TIME SEQUENCE
C
      CALL FSST(X, N)
      WRITE (IOUT2,9980)
9980  FORMAT (42H ORIGINAL ODD HARMONIC, SYMMETRIC SEQUENCE)
      WRITE (IOUT2,9993) (X(I),I=1,N)
      WRITE (IOUT2,9992)
C
C USE SUBROUTINE FFTSOH TO OBTAIN DFT FROM NO4 POINT FFT
C
      CALL FFTSOH(X, N, Y)
      WRITE (IOUT2,9979)
9979  FORMAT (17H OUTPUT OF FFTSOH)
      WRITE (IOUT2,9993) (X(I),I=1,NO4)
      WRITE (IOUT2,9992)
C
C USE SUBROUTINE IFTSOH TO OBTAIN ORIGINAL SEQUENCE FROM NO4 POINT DFT
C
      CALL IFTSOH(X, N, Y)
      WRITE (IOUT2,9978)
9978  FORMAT (17H OUTPUT OF IFTSOH)
      WRITE (IOUT2,9993) (X(I),I=1,NO4)
      WRITE (IOUT2,9992)
C
C CREATE SEQUENCE WITH ONLY IMAGINARY VALUED ODD HARMONICS--BEGIN
C IN FREQUENCY DOMAIN
C
      DO 120 I=1,NP2,2
        X(I) = 0.
        X(I+1) = 0.
        IF (MOD(I,4).EQ.1) GO TO 120
        X(I+1) = UNI(0) - 0.5
 120  CONTINUE
      WRITE (IOUT2,9977) N
9977  FORMAT (1H , I4, 41H POINT FFT OF ODD HARMONIC, ANTISYMMETRIC,
     *    9H SEQUENCE)
      WRITE (IOUT2,9993) (X(I),I=1,NP2)
      WRITE (IOUT2,9992)
C
C TRANSFORM BACK TO TIME SEQUENCE
C
      CALL FSST(X, N)
      WRITE (IOUT2,9976)
9976  FORMAT (46H ORIGINAL ODD HARMONIC, ANTISYMMETRIC SEQUENCE)
      WRITE (IOUT2,9993) (X(I),I=1,N)
      WRITE (IOUT2,9992)
C
C USE SUBROUTINE FFTAOH TO OBTAIN DFT FROM NO4 POINT FFT
C
      CALL FFTAOH(X, N, Y)
      WRITE (IOUT2,9975)
9975  FORMAT (17H OUTPUT OF FFTAOH)
      WRITE (IOUT2,9993) (X(I),I=1,NO4)
      WRITE (IOUT2,9992)
C
C USE SUBROUTINE IFTAOH TO OBTAIN ORIGINAL SEQUENCE FROM N/4 POINT DFT
C
      CALL IFTAOH(X, N, Y)
      WRITE (IOUT2,9974)
9974  FORMAT (17H OUTPUT OF IFTAOH)
      WRITE (IOUT2,9993) (X(I),I=1,NO4P1)
      WRITE (IOUT2,9992)
C
C BEGIN A NEW PAGE
C
      WRITE (IOUT2,9973)
9973  FORMAT (1H1)
      GO TO 10
      END
```

```
C
C-----------------------------------------------------------------------
C SUBROUTINE: FFTSYM
C COMPUTE DFT FOR REAL, SYMMETRIC, N-POINT SEQUENCE X(M) USING
C N/2-POINT FFT
C SYMMETRIC SEQUENCE MEANS X(M)=X(N-M), M=1,...,N/2-1
C NOTE: INDEX M IS SEQUENCE INDEX--NOT FORTRAN INDEX
C-----------------------------------------------------------------------
C
      SUBROUTINE FFTSYM(X, N, Y)
      DIMENSION X(1), Y(1)
C
C X = REAL ARRAY WHICH ON INPUT CONTAINS THE N/2+1 POINTS OF THE
C     INPUT SEQUENCE (SYMMETRICAL)
C     ON OUTPUT X CONTAINS THE N/2+1 REAL POINTS OF THE TRANSFORM OF
C     THE INPUT--I.E. THE ZERO VALUED IMAGINARY PARTS ARE NOT RETURNED
C N = TRUE SIZE OF INPUT
C Y = SCRATCH ARRAY OF SIZE N/2+2
C
C
C FOR N = 2, COMPUTE DFT DIRECTLY
C
      IF (N.GT.2) GO TO 10
      T = X(1) + X(2)
      X(2) = X(1) - X(2)
      X(1) = T
      RETURN
  10  TWOPI = 8.*ATAN(1.0)
C
C FIRST COMPUTE B0 TERM, WHERE B0=SUM OF ODD VALUES OF X(M)
C
      NO2 = N/2
      NO4 = N/4
      NIND = NO2 + 1
      B0 = 0.
      DO 20 I=2,NIND,2
        B0 = B0 + X(I)
  20  CONTINUE
      B0 = B0*2.
C
C FOR N = 4 SKIP RECURSION LOOP
C
      IF (N.EQ.4) GO TO 40
C
C FORM NEW SEQUENCE, Y(M)=X(2*M)+(X(2*M+1)-X(2*M-1))
C
      DO 30 I=2,NO4
        IND = 2*I
        T1 = X(IND) - X(IND-2)
        Y(I) = X(IND-1) + T1
        IND1 = NO2 + 2 - I
        Y(IND1) = X(IND-1) - T1
  30  CONTINUE
  40  Y(1) = X(1)
      Y(NO4+1) = X(NO2+1)
C
C TAKE N/2 POINT (REAL) FFT OF Y
C
      CALL FAST(Y, NO2)
C
C FORM ORIGINAL DFT BY UNSCRAMBLING Y(K)
C USE RECURSION TO GIVE SIN(TPN*I) MULTIPLIER
C
      TPN = TWOPI/FLOAT(N)
      COSI = 2.*COS(TPN)
      SINI = 2.*SIN(TPN)
      COSD = COSI/2.
      SIND = SINI/2.
      NIND = NO4 + 1
      DO 50 I=2,NIND
        IND = 2*I
        BK = Y(IND)/SINI
        AK = Y(IND-1)
        X(I) = AK + BK
        NIND1 = N/2 + 2 - I
        X(NIND1) = AK - BK
        TEMP = COSI*COSD - SINI*SIND
        SINI = COSI*SIND + SINI*COSD
        COSI = TEMP
  50  CONTINUE
      X(1) = B0 + Y(1)
      X(NO2+1) = Y(1) - B0
      RETURN
      END
C
C-----------------------------------------------------------------------
C SUBROUTINE: IFTSYM
C COMPUTE IDFT FOR REAL, SYMMETRIC, N-POINT SEQUENCE X(M) USING
C N/2-POINT FFT
C SYMMETRIC SEQUENCE MEANS X(M)=X(N-M), M=1,...,N/2-1
C NOTE: INDEX M IS SEQUENCE INDEX--NOT FORTRAN INDEX
C-----------------------------------------------------------------------
C
      SUBROUTINE IFTSYM(X, N, Y)
      DIMENSION X(1), Y(1)
C
C X = REAL ARRAY WHICH ON INPUT CONTAINS THE N/2+1 REAL POINTS OF THE
C     TRANSFORM OF THE INPUT--I.E. THE ZERO VALUED IMAGINARY PARTS
C     ARE NOT GIVEN AS INPUT
C     ON OUTPUT X CONTAINS THE N/2+1 POINTS OF THE TIME SEQUENCE
C     (SYMMETRICAL)
C N = TRUE SIZE OF INPUT
C Y = SCRATCH ARRAY OF SIZE N/2+2
C
C
C FOR N = 2, COMPUTE IDFT DIRECTLY
C
      IF (N.GT.2) GO TO 10
      T = (X(1)+X(2))/2.
      X(2) = (X(1)-X(2))/2.
      X(1) = T
      RETURN
  10  TWOPI = 8.*ATAN(1.0)
C
C FIRST COMPUTE X1=X(1) TERM DIRECTLY
C USE RECURSION ON THE SINE COSINE TERMS
C
      NO2 = N/2
      NO4 = N/4
      TPN = TWOPI/FLOAT(N)
      COSD = COS(TPN)
      SIND = SIN(TPN)
      COSI = 2.
      SINI = 0.
      X1 = X(1) - X(NO2+1)
      DO 20 I=2,NO2
        TEMP = COSI*COSD - SINI*SIND
        SINI = COSI*SIND + SINI*COSD
        COSI = TEMP
        X1 = X1 + X(I)*COSI
```

```
   20 CONTINUE
      X1 = X1/FLOAT(N)
C
C SCRAMBLE ORIGINAL DFT (X(K)) TO GIVE Y(K)
C USE RECURSION RELATION TO GENERATE SIN(TPN*I) MULTIPLIER
C
      COSI = COS(TPN)
      SINI = SIN(TPN)
      COSD = COSI
      SIND = SINI
      Y(1) = (X(1)+X(NO2+1))/2.
      Y(2) = 0.
      NIND = NO4 + 1
      DO 30 I=2,NIND
        IND = 2*I
        NIND1 = NO2 + 2 - I
        AK = (X(I)+X(NIND1))/2.
        BK = (X(I)-X(NIND1))
        Y(IND-1) = AK
        Y(IND) = BK*SINI
        TEMP = COSI*COSD - SINI*SIND
        SINI = COSI*SIND + SINI*COSD
        COSI = TEMP
   30 CONTINUE
C
C TAKE N/2 POINT IDFT OF Y
C
      CALL FSST(Y, NO2)
C
C FORM X SEQUENCE FROM Y SEQUENCE
C
      X(1) = Y(1)
      X(2) = X1
      IF (N.EQ.4) GO TO 50
      DO 40 I=2,NO4
        IND = 2*I
        IND1 = NO2 + 2 - I
        X(IND-1) = (Y(I)+Y(IND1))/2.
        T1 = (Y(I)-Y(IND1))/2.
        X(IND) = T1 + X(IND-2)
   40 CONTINUE
   50 X(NO2+1) = Y(NO4+1)
      RETURN
      END
C
C-----------------------------------------------------------------------
C SUBROUTINE: FFTASM
C COMPUTE DFT FOR REAL, ANTISYMMETRIC, N-POINT SEQUENCE X(M) USING
C N/2-POINT FFT
C ANTISYMMETRIC SEQUENCE MEANS X(M)=-X(N-M), M=1,...,N/2-1
C NOTE: INDEX M IS SEQUENCE INDEX--NOT FORTRAN INDEX
C-----------------------------------------------------------------------
C
      SUBROUTINE FFTASM(X, N, Y)
      DIMENSION X(1), Y(1)
C
C X = REAL ARRAY WHICH ON INPUT CONTAINS THE N/2 POINTS OF THE
C     INPUT SEQUENCE (ASYMMETRICAL)
C     ON OUTPUT X CONTAINS THE N/2+1 IMAGINARY POINTS OF THE TRANSFORM
C     OF THE INPUT--I.E. THE ZERO VALUED REAL PARTS ARE NOT RETURNED
C N = TRUE SIZE OF INPUT
C Y = SCRATCH ARRAY OF SIZE N/2+2
C
C
C FOR N = 2, ASSUME X(1)=0, X(2)=0, COMPUTE DFT DIRECTLY
C
      IF (N.EQ.2) GO TO 30
      TWOPI = 8.*ATAN(1.0)
C
C FORM NEW SEQUENCE, Y(M)=X(2*M)+(X(2*M+1)-X(2*M-1))
C
      NO2 = N/2
      NO4 = N/4
      DO 10 I=2,NO4
        IND = 2*I
        T1 = X(IND) - X(IND-2)
        Y(I) = X(IND-1) + T1
        IND1 = NO2 + 2 - I
        Y(IND1) = -X(IND-1) + T1
   10 CONTINUE
      Y(1) = 2.*X(2)
      Y(NO4+1) = -2.*X(NO2)
C
C TAKE N/2 POINT (REAL) FFT OF Y
C
      CALL FAST(Y, NO2)
C
C FORM ORIGINAL DFT BY UNSCRAMBLING Y(K)
C USE RECURSION RELATION TO GENERATE SIN(TPN*I) MULTIPLIER
C
      TPN = TWOPI/FLOAT(N)
      COSI = 2.*COS(TPN)
      SINI = 2.*SIN(TPN)
      COSD = COSI/2.
      SIND = SINI/2.
      NIND = NO4 + 1
      DO 20 I=2,NIND
        IND = 2*I
        BK = Y(IND-1)/SINI
        AK = Y(IND)
        X(I) = AK - BK
        IND1 = NO2 + 2 - I
        X(IND1) = -AK - BK
        TEMP = COSI*COSD - SINI*SIND
        SINI = COSI*SIND + SINI*COSD
        COSI = TEMP
   20 CONTINUE
   30 X(1) = 0.
      X(NO2+1) = 0.
      RETURN
      END
C
C-----------------------------------------------------------------------
C SUBROUTINE: IFTASM
C COMPUTE IDFT FOR REAL, ANTISYMMETRIC, N-POINT SEQUENCE X(M) USING
C N/2-POINT FFT
C ANTISYMMETRIC SEQUENCE MEANS X(M)=-X(N-M), M=1,...,N/2-1
C NOTE: INDEX M IS SEQUENCE INDEX--NOT FORTRAN INDEX
C-----------------------------------------------------------------------
C
      SUBROUTINE IFTASM(X, N, Y)
      DIMENSION X(1), Y(1)
C
C X = IMAGINARY ARRAY WHICH ON INPUT CONTAINS THE N/2+1 REAL POINTS OF
C     THE TRANSFORM OF THE INPUT--I.E. THE ZERO VALUED REAL PARTS
C     ARE NOT GIVEN AS INPUT
C     ON OUTPUT X CONTAINS THE N/2 POINTS OF THE TIME SEQUENCE
C     (ANTISYMMETRICAL)
C N = TRUE SIZE OF INPUT
C Y = SCRATCH ARRAY OF SIZE N/2+2
```

```
C
C
C FOR N = 2, ASSUME X(1)=0, X(2)=0
C
      IF (N.GT.2) GO TO 10
      X(1) = 0
      X(2) = 0
      RETURN
  10  TWOPI = 8.*ATAN(1.0)
C
C FIRST COMPUTE X1=X(1) TERM DIRECTLY
C USE RECURSION ON THE SINE COSINE TERMS
C
      NO2 = N/2
      NO4 = N/4
      TPN = TWOPI/FLOAT(N)
C
C SCRAMBLE ORIGINAL DFT (X(K)) TO GIVE Y(K)
C USE RECURSION RELATION TO GIVE SIN(TPN*I) MULTIPLIER
C
      COSI = COS(TPN)
      SINI = SIN(TPN)
      COSD = COSI
      SIND = SINI
      NIND = NO4 + 1
      DO 20 I=2,NIND
        IND = 2*I
        IND1 = NO2 + 2 - I
        AK = (X(I)-X(IND1))/2.
        BK = -(X(I)+X(IND1))
        Y(IND) = AK
        Y(IND-1) = BK*SINI
        TEMP = COSI*COSD - SINI*SIND
        SINI = COSI*SIND + SINI*COSD
        COSI = TEMP
  20  CONTINUE
      Y(1) = 0.
      Y(2) = 0.
C
C TAKE N/2 POINT IDFT OF Y
C
      CALL FSST(Y, NO2)
C
C FORM X SEQUENCE FROM Y SEQUENCE
C
      X(2) = Y(1)/2.
      X(1) = 0.
      IF (N.EQ.4) GO TO 40
      DO 30 I=2,NO4
        IND = 2*I
        IND1 = NO2 + 2 - I
        X(IND-1) = (Y(I)-Y(IND1))/2.
        T1 = (Y(I)+Y(IND1))/2.
        X(IND) = T1 + X(IND-2)
  30  CONTINUE
  40  X(NO2) = -Y(NO4+1)/2.
      RETURN
      END
C
C-----------------------------------------------------------------------
C SUBROUTINE: FFTOHM
C COMPUTE DFT FOR REAL, N-POINT, ODD HARMONIC SEQUENCES USING AN
C N/2 POINT FFT
C ODD HARMONIC MEANS X(2*K)=0, ALL K WHERE X(K) IS THE DFT OF X(M)
C NOTE: INDEX M IS SEQUENCE INDEX--NOT FORTRAN INDEX
C-----------------------------------------------------------------------
C
      SUBROUTINE FFTOHM(X, N)
      DIMENSION X(1)
C
C X = REAL ARRAY WHICH ON INPUT CONTAINS THE FIRST N/2 POINTS OF THE
C     INPUT
C     ON OUTPUT X CONTAINS THE N/4 COMPLEX VALUES OF THE ODD
C     HARMONICS OF THE INPUT--STORED IN THE SEQUENCE RE(X(1)),IM(X(1)),
C     RE(X(2)),IM(X(2)),...
C ****NOTE: X MUST BE DIMENSIONED TO SIZE N/2+2 FOR FFT ROUTINE
C N = TRUE SIZE OF X SEQUENCE
C
C FIRST COMPUTE REAL(X(1)) AND REAL(X(N/2-1)) SEPARATELY
C ALSO SIMULTANEOUSLY MULTIPLY ORIGINAL SEQUENCE BY SIN(TWOPI*(M-1)/N)
C SIN AND COS ARE COMPUTED RECURSIVELY
C
C
C FOR N = 2, ASSUME X(1)=X0, X(2)=-X0, COMPUTE DFT DIRECTLY
C
      IF (N.GT.2) GO TO 10
      X(1) = 2.*X(1)
      X(2) = 0.
      RETURN
  10  TWOPI = 8.*ATAN(1.0)
      TPN = TWOPI/FLOAT(N)
C
C COMPUTE X1=REAL(X(1)) AND X2=IMAGINARY(X(N/2-1))
C X(N) = X(N)*4.*SIN(TWOPI*(I-1)/N)
C
      T1 = 0.
C
C COSD AND SIND ARE MULTIPLIERS FOR RECURSION FOR SIN AND COS
C COSI AND SINI ARE INITIAL CONDITIONS FOR RECURSION FOR SIN AND COS
C
      COSD = COS(TPN*2.)
      SIND = SIN(TPN*2.)
      COSI = 1.
      SINI = 0.
      NO2 = N/2
      DO 20 I=1,NO2,2
        T = X(I)*COSI
        X(I) = X(I)*4.*SINI
        TEMP = COSI*COSD - SINI*SIND
        SINI = COSI*SIND + SINI*COSD
        COSI = TEMP
        T1 = T1 + T
  20  CONTINUE
C
C RESET INITIAL CONDITIONS (COSI,SINI) FOR NEW RECURSION
C
      COSI = COS(TPN)
      SINI = SIN(TPN)
      T2 = 0.
      DO 30 I=2,NO2,2
        T = X(I)*COSI
        X(I) = X(I)*4.*SINI
        TEMP = COSI*COSD - SINI*SIND
        SINI = COSI*SIND + SINI*COSD
        COSI = TEMP
        T2 = T2 + T
  30  CONTINUE
      X1 = 2.*(T1+T2)
      X2 = 2.*(T1-T2)
C
```

```
C TAKE N/2 POINT (REAL) FFT OF PREPROCESSED SEQUENCE X
C
      CALL FAST(X, NO2)
C
C FOR N = 4--SKIP RECURSION AND INITIAL CONDITIONS
C
      IF (N.EQ.4) GO TO 50
C
C INITIAL CONDITIONS FOR RECURSION
C
      X(2) = -X(1)/2.
      X(1) = X1
C
C FOR N = 8, SKIP RECURSION
C
      IF (N.EQ.8) GO TO 50
C
C UNSCRAMBLE Y(K) USING RECURSION FORMULA
C
      NIND = NO2 - 2
      DO 40 I=3,NIND,2
        T = X(I)
        X(I) = X(I-2) + X(I+1)
        X(I+1) = X(I-1) - T
  40  CONTINUE
  50  X(NO2) = X(NO2+1)/2.
      X(NO2-1) = X2
      RETURN
      END
C
C-----------------------------------------------------------------------
C SUBROUTINE: IFTOHM
C COMPUTE IDFT FOR REAL, N-POINT, ODD HARMONIC SEQUENCES USING AN
C N/2 POINT FFT
C ODD HARMONIC MEANS X(2*K)=0, ALL K WHERE X(K) IS THE DFT OF X(M)
C NOTE: INDEX M IS SEQUENCE INDEX--NOT FORTRAN INDEX
C-----------------------------------------------------------------------
C
      SUBROUTINE IFTOHM(X, N)
      DIMENSION X(1)
C
C X = REAL ARRAY WHICH ON INPUT CONTAINS THE N/4 COMPLEX VALUES OF THE
C     ODD HARMONICS OF THE INPUT--STORED IN THE SEQUENCE RE(X(1)),
C     IM(X(1)),RE(X(2)),IM(X(2)),...
C     ON OUTPUT X CONTAINS THE FIRST N/2 POINTS OF THE INPUT
C ****NOTE: X MUST BE DIMENSIONED TO SIZE N/2+2 FOR FFT ROUTINE
C N = TRUE SIZE OF X SEQUENCE
C
C FIRST COMPUTE REAL(X(1)) AND REAL(X(N/2-1)) SEPARATELY
C ALSO SIMULTANEOUSLY MULTIPLY ORIGINAL SEQUENCE BY SIN(TWOPI*(M-1)/N)
C SIN AND COS ARE COMPUTED RECURSIVELY
C
C
C FOR N = 2, ASSUME X(1)=X0, X(2)=-X0, COMPUTE IDFT DIRECTLY
C
      IF (N.GT.2) GO TO 10
      X(1) = 0.5*X(1)
      X(2) = -X(1)
      RETURN
  10  TWOPI = 8.*ATAN(1.0)
      TPN = TWOPI/FLOAT(N)
      NO2 = N/2
      NO4 = N/4
      NIND = NO2
C
C SOLVE FOR X(0)=X0 DIRECTLY
C
      X0 = 0.
      DO 20 I=1,NO2,2
        X0 = X0 + 2.*X(I)
  20  CONTINUE
      X0 = X0/FLOAT(N)
C
C FORM Y(K)=J*(X(2K+1)-X(2K-1))
C OVERWRITE X ARRAY WITH Y SEQUENCE
C
      XPR = X(1)
      XPI = X(2)
      X(1) = -2.*X(2)
      X(2) = 0.
      IF (NO4.EQ.1) GO TO 40
      DO 30 I=3,NIND,2
        TI = X(I) - XPR
        TR = -X(I+1) + XPI
        XPR = X(I)
        XPI = X(I+1)
        X(I) = TR
        X(I+1) = TI
  30  CONTINUE
  40  X(NO2+1) = 2.*XPI
      X(NO2+2) = 0.
C
C TAKE N/2 POINT (REAL) IFFT OF PREPROCESSED SEQUENCE X
C
      CALL FSST(X, NO2)
C
C SOLVE FOR X(M) BY DIVIDING BY 4*SIN(TWOPI*M/N) FOR M=1,2,...,N/2-1
C FOR M=0 SUBSTITUTE PRECOMPUTED VALUE X0
C
      COSI = 4.
      SINI = 0.
      COSD = COS(TPN)
      SIND = SIN(TPN)
      DO 50 I=2,NO2
        TEMP = COSI*COSD - SINI*SIND
        SINI = COSI*SIND + SINI*COSD
        COSI = TEMP
        X(I) = X(I)/SINI
  50  CONTINUE
      X(1) = X0
      RETURN
      END
C
C-----------------------------------------------------------------------
C SUBROUTINE: FFTSOH
C COMPUTE DFT FOR REAL, SYMMETRIC, ODD HARMONIC, N-POINT SEQUENCE
C USING N/4-POINT FFT
C SYMMETRIC SEQUENCE MEANS X(M)=X(N-M), M=1,...,N/2-1
C ODD HARMONIC MEANS X(2*K)=0, ALL K, WHERE X(K) IS THE DFT OF X(M)
C X(M) HAS THE PROPERTY X(M)=-X(N/2-M), M=0,1,...,N/4-1,  X(N/4)=0
C NOTE: INDEX M IS SEQUENCE INDEX--NOT FORTRAN INDEX
C-----------------------------------------------------------------------
C
      SUBROUTINE FFTSOH(X, N, Y)
      DIMENSION X(1), Y(1)
C
C  X = REAL ARRAY WHICH ON INPUT CONTAINS THE N/4 POINTS OF THE
C      INPUT SEQUENCE (SYMMETRICAL)
C      ON OUTPUT X CONTAINS  THE N/4 REAL POINTS OF THE ODD HARMONICS
C      OF THE TRANSFORM OF THE INPUT--I.E. THE ZERO VALUED IMAGINARY
```

```
C       PARTS ARE NOT GIVEN NOR ARE THE ZERO-VALUED EVEN HARMONICS
C   N = TRUE SIZE OF INPUT
C   Y = SCRATCH ARRAY OF SIZE N/4+2
C
C
C  HANDLE N = 2 AND N = 4 CASES SEPARATELY
C
        IF (N.GT.4) GO TO 20
        IF (N.EQ.4) GO TO 10
C
C  FOR N=2, ASSUME X(1)=X0, X(2)=-X0, COMPUTE DFT DIRECTLY
C
        X(1) = 2.*X(1)
        RETURN
C
C  N = 4 CASE, COMPUTE DFT DIRECTLY
C
  10    X(1) = 2.*X(1)
        RETURN
  20    TWOPI = 8.*ATAN(1.0)
C
C  FORM NEW SEQUENCE, Y(M)=X(2*M)+(X(2*M+1)-X(2*M-1))
C
        NO2 = N/2
        NO4 = N/4
        NO8 = N/8
        IF (NO8.EQ.1) GO TO 40
        DO 30 I=2,NO8
          IND = 2*I
          T1 = X(IND) - X(IND-2)
          Y(I) = X(IND-1) + T1
          IND1 = N/4 + 2 - I
          Y(IND1) = -X(IND-1) + T1
  30    CONTINUE
  40    Y(1) = X(1)
        Y(NO8+1) = -2.*X(NO4)
C
C  THE SEQUENCE Y (N/4 POINTS) HAS ONLY ODD HARMONICS
C  CALL SUBROUTINE FFTOHM TO EXPLOIT ODD HARMONICS
C
        CALL FFTOHM(Y, NO2)
C
C  FORM ORIGINAL DFT FROM COMPLEX ODD HARMONICS OF Y(K)
C  BY UNSCRAMBLING Y(K)
C
        TPN = TWOPI/FLOAT(N)
        COSI = 2.*COS(TPN)
        SINI = 2.*SIN(TPN)
        COSD = COS(TPN*2.)
        SIND = SIN(TPN*2.)
        DO 50 I=1,NO8
          IND = 2*I
          BK = Y(IND)/SINI
          TEMP = COSI*COSD - SINI*SIND
          SINI = COSI*SIND + SINI*COSD
          COSI = TEMP
          AK = Y(IND-1)
          X(I) = AK + BK
          IND1 = N/4 + 1 - I
          X(IND1) = AK - BK
  50    CONTINUE
        RETURN
        END
C
C-----------------------------------------------------------------------
```

```
C  SUBROUTINE: IFTSOH
C  COMPUTE IDFT FOR REAL, SYMMETRIC, ODD HARMONIC, N-POINT SEQUENCE
C  USING N/4-POINT FFT
C  SYMMETRIC SEQUENCE MEANS X(M)=X(N-M), M=1,...,N/2-1
C  ODD HARMONIC MEANS X(2*K)=0, ALL K, WHERE X(K) IS THE DFT OF X(M)
C  X(M) HAS THE PROPERTY X(M)=-X(N/2-M), M=0,1,...,N/4-1,  X(N/4)=0
C  NOTE: INDEX M IS SEQUENCE INDEX--NOT FORTRAN INDEX
C-----------------------------------------------------------------------
C
        SUBROUTINE IFTSOH(X, N, Y)
        DIMENSION X(1), Y(1)
C
C   X = REAL ARRAY WHICH ON INPUT CONTAINS THE N/4 REAL POINTS OF
C       THE ODD HARMONICS OF THE TRANSFORM OF THE ORIGINAL TIME SEQUENCE
C       I.E. THE ZERO VALUED IMAGINARY PARTS ARE NOT GIVEN NOR ARE THE
C       ZERO VALUED EVEN HARMONICS
C       ON OUTPUT X CONTAINS THE FIRST N/4 POINTS OF THE ORIGINAL INPUT
C       SEQUENCE (SYMMETRICAL)
C   N = TRUE SIZE OF INPUT
C   Y = SCRATCH ARRAY OF SIZE N/4+2
C
C
C  HANDLE N = 2 AND N = 4 CASES SEPARATELY
C
        IF (N.GT.4) GO TO 10
C
C  FOR N=2, 4 ASSUME X(1)=X0, X(2)=-X0, COMPUTE IDFT DIRECTLY
C
        X(1) = X(1)/2.
        RETURN
C
C  CODE FOR VALUES OF N WHICH ARE MULTIPLES OF 8
C
  10    TWOPI = 8.*ATAN(1.0)
        NO2 = N/2
        NO4 = N/4
        NO8 = N/8
        TPN = TWOPI/FLOAT(N)
C
C  FIRST COMPUTE X1=X(1) TERM DIRECTLY
C  USE RECURSION ON THE SINE COSINE TERMS
C
        COSD = COS(TPN*2.)
        SIND = SIN(TPN*2.)
        COSI = 2.*COS(TPN)
        SINI = 2.*SIN(TPN)
        X1 = 0.
        DO 20 I=1,NO4
          X1 = X1 + X(I)*COSI
          TEMP = COSI*COSD - SINI*SIND
          SINI = COSI*SIND + SINI*COSD
          COSI = TEMP
  20    CONTINUE
        X1 = X1/FLOAT(N)
C
C  SCRAMBLE ORIGINAL DFT (X(K)) TO GIVE Y(K)
C  USE RECURSION RELATION TO GIVE SIN MULTIPLIERS
C
        COSI = COS(TPN)
        SINI = SIN(TPN)
        DO 30 I=1,NO8
          IND = 2*I
          IND1 = NO4 + 1 - I
          AK = (X(I)+X(IND1))/2.
          BK = (X(I)-X(IND1))
```

```
          Y(IND-1) = AK
          Y(IND) = BK*SINI
          TEMP = COSI*COSD - SINI*SIND
          SINI = COSI*SIND + SINI*COSD
          COSI = TEMP
   30   CONTINUE
C
C THE SEQUENCE Y(K) IS THE ODD HARMONICS DFT OUTPUT
C USE SUBROUTINE IFTOHM TO OBTAIN Y(M), THE INVERSE TRANSFORM
C
        CALL IFTOHM(Y, NO2)
C
C FORM X(M) SEQUENCE FROM Y(M) SEQUENCE
C USE X1 INITIAL CONDITION ON THE RECURSION
C
        X(1) = Y(1)
        X(2) = X1
        IF (NO8.EQ.1) RETURN
        DO 40 I=2,NO8
          IND = 2*I
          IND1 = NO4 + 2 - I
          T1 = (Y(I)+Y(IND1))/2.
          X(IND-1) = (Y(I)-Y(IND1))/2.
          X(IND) = T1 + X(IND-2)
   40   CONTINUE
        RETURN
        END
C
C-----------------------------------------------------------------------
C SUBROUTINE: FFTAOH
C COMPUTE DFT FOR REAL, ANTISYMMETRIC, ODD HARMONIC, N-POINT SEQUENCE
C USING N/4-POINT FFT
C ANTISYMMETRIC SEQUENCE MEANS X(M)=-X(N-M), M=1,...,N/2-1
C ODD HARMONIC MEANS X(2*K)=0, ALL K, WHERE X(K) IS THE DFT OF X(M)
C X(M) HAS THE PROPERTY X(M)=X(N/2-M), M=0,1,...,N/4-1,  X(0)=0
C NOTE: INDEX M IS SEQUENCE INDEX--NOT FORTRAN INDEX
C-----------------------------------------------------------------------
C
        SUBROUTINE FFTAOH(X, N, Y)
        DIMENSION X(1), Y(1)
C
C  X = REAL ARRAY WHICH ON INPUT CONTAINS THE (N/4+1) POINTS OF THE
C      INPUT SEQUENCE (ANTISYMMETRICAL)
C      ON OUTPUT X CONTAINS THE N/4 IMAGINARY POINTS OF THE ODD
C      HARMONICS OF THE TRANSFORM OF THE INPUT--I.E. THE ZERO
C      VALUED REAL PARTS ARE NOT GIVEN NOR ARE THE ZERO-VALUED
C      EVEN HARMONICS
C  N = TRUE SIZE OF INPUT
C  Y = SCRATCH ARRAY OF SIZE N/4+2
C
C
C HANDLE N = 2 AND N = 4 CASES SEPARATELY
C
        IF (N.GT.4) GO TO 20
        IF (N.EQ.4) GO TO 10
C
C FOR N=2, ASSUME X(1)=0, X(2)=0, COMPUTE DFT DIRECTLY
C
        X(1) = 0.
        RETURN
C
C N = 4 CASE, ASSUME X(1)=X(3)=0, X(2)=-X(4)=X0, COMPUTE DFT DIRECTLY
C
   10   X(1) = -2.*X(2)
        RETURN
   20   TWOPI = 8.*ATAN(1.0)
C
C FORM NEW SEQUENCE, Y(M)=X(2*M)+(X(2*M+1)-X(2*M-1))
C
        NO2 = N/2
        NO4 = N/4
        NO8 = N/8
        IF (NO8.EQ.1) GO TO 40
        DO 30 I=2,NO8
          IND = 2*I
          T1 = X(IND) - X(IND-2)
          Y(I) = X(IND-1) + T1
          IND1 = N/4 + 2 - I
          Y(IND1) = X(IND-1) - T1
   30   CONTINUE
   40   Y(1) = 2.*X(2)
        Y(NO8+1) = X(NO4+1)
C
C THE SEQUENCE Y (N/4 POINTS) HAS ONLY ODD HARMONICS
C CALL SUBROUTINE FFTOHM TO EXPLOIT ODD HARMONICS
C
        CALL FFTOHM(Y, NO2)
C
C FORM ORIGINAL DFT FROM COMPLEX ODD HARMONICS OF Y(K)
C BY UNSCRAMBLING Y(K)
C
        TPN = TWOPI/FLOAT(N)
        COSI = 2.*COS(TPN)
        SINI = 2.*SIN(TPN)
        COSD = COS(TPN*2.)
        SIND = SIN(TPN*2.)
        DO 50 I=1,NO8
          IND = 2*I
          BK = Y(IND-1)/SINI
          TEMP = COSI*COSD - SINI*SIND
          SINI = COSI*SIND + SINI*COSD
          COSI = TEMP
          AK = Y(IND)
          X(I) = AK - BK
          IND1 = N/4 + 1 - I
          X(IND1) = -AK - BK
   50   CONTINUE
        RETURN
        END
C
C-----------------------------------------------------------------------
C SUBROUTINE: IFTAOH
C COMPUTE IDFT FOR REAL, ANTISYMMETRIC, ODD HARMONIC, N-POINT SEQUENCE
C USING N/4-POINT FFT
C ANTISYMMETRIC SEQUENCE MEANS X(M)=-X(N-M), M=1,...,N/2-1
C ODD HARMONIC MEANS X(2*K)=0, ALL K, WHERE X(K) IS THE DFT OF X(M)
C X(M)HAS THE PROPERTY X(M)=X(N/2-M), M=0,1,...,N/4-1,  X(0)=0
C NOTE: INDEX M IS SEQUENCE INDEX--NOT FORTRAN INDEX
C-----------------------------------------------------------------------
C
        SUBROUTINE IFTAOH(X, N, Y)
        DIMENSION X(1), Y(1)
C
C  X = REAL ARRAY WHICH ON INPUT CONTAINS THE N/4 IMAGINARY POINTS
C      OF THE ODD HARMONICS OF THE TRANSFORM OF THE ORIGINAL TIME
C      SEQUENCE--I.E. THE ZERO VALUED REAL PARTS ARE NOT INPUT NOR
C      ARE THE ZERO-VALUED EVEN HARMONICS
C      ON OUTPUT X CONTAINS THE FIRST (N/4+1) POINTS OF THE ORIGINAL
C      TIME SEQUENCE (ANTISYMMETRICAL)
C  N = TRUE SIZE OF INPUT
```

```
C  Y = SCRATCH ARRAY OF SIZE N/4+2
C
C
C HANDLE N = 2 AND N = 4 CASES SEPARATELY
C
      IF (N.GT.4) GO TO 20
      IF (N.EQ.4) GO TO 10
C
C FOR N=2  ASSUME X(1)=0, X(2)=0, COMPUTE IDFT DIRECTLY
C
      X(1) = 0.
      RETURN
C
C FOR N=4, ASSUME X(1)=X(3)=0, X(2)=-X(4)=X0, COMPUTE IDFT DIRECTLY
C
  10  X(2) = -X(1)/2.
      X(1) = 0.
      RETURN
C
C CODE FOR VALUES OF N WHICH ARE MULTIPLES OF 8
C
  20  TWOPI = 8.*ATAN(1.0)
      NO2 = N/2
      NO4 = N/4
      NO8 = N/8
      TPN = TWOPI/FLOAT(N)
C
C SCRAMBLE ORIGINAL DFT (X(K)) TO GIVE Y(K)
C USE RECURSION TO GIVE SIN MULTIPLIERS
C
      COSI = COS(TPN)
      SINI = SIN(TPN)
      COSD = COS(TPN*2.)
      SIND = SIN(TPN*2.)
      DO 30 I=1,NO8
        IND = 2*I
        IND1 = NO4 + 1 - I
        AK = (X(I)-X(IND1))/2.
        BK = -(X(I)+X(IND1))
        Y(IND) = AK
        Y(IND-1) = BK*SINI
        TEMP = COSI*COSD - SINI*SIND
        SINI = COSI*SIND + SINI*COSD
        COSI = TEMP
  30  CONTINUE
C
C THE SEQUENCE Y(K) IS AN ODD HARMONIC SEQUENCE
C USE SUBROUTINE IFTOHM TO GIVE Y(M)
C
      CALL IFTOHM(Y, NO2)
C
C FORM X SEQUENCE FROM Y SEQUENCE
C
      X(2) = Y(1)/2.
      X(1) = 0.
      IF (N.EQ.8) RETURN
      DO 40 I=2,NO8
        IND = 2*I
        IND1 = NO4 + 2 - I
        X(IND-1) = (Y(I)+Y(IND1))/2.
        T1 = (Y(I)-Y(IND1))/2.
        X(IND) = T1 + X(IND-2)
  40  CONTINUE
      X(NO4+1) = Y(NO8+1)
      RETURN
      END
```

1.4

Mixed Radix Fast Fourier Transforms

Richard C. Singleton

SRI International
Menlo Park, CA 94025

1. Purpose

Mixed radix fast Fourier transform algorithms are presented, for both real and complex input sequences. These algorithms also perform multi-variate Fourier transforms. In addition, the program used to test the Fourier transform subroutine includes a fast in-place sort subroutine (with tag array) and a short bi-variate normal random number subroutine.

2. Method

The subroutine FFT evaluates the equation

$$X(k) = \sum_{n=0}^{N-1} x(n) e^{-j\frac{2\pi}{N}nk}$$

where $x(n)$ and $X(k)$ are complex, and $k = 0,1,...,N-1$. This subroutine also computes the $1/N$ scaled inverse Fourier transform. The subroutine REALT, when called following FFT, completes the final transform step for a real sequence $x(n)$ of length $2N$. Earlier versions of these algorithms were published in 1969 [1], and the methods used were described then in detail. This paper has also been reprinted [2], without program listings.

3. Usage

3.1 Single-Variate Transform of Complex Input Data

CALL FFT (A,B,1,N,1,−1) (transform, no scaling)

or

CALL FFT (A,B,1,N,1,1) (inverse, scaled by 1/N)

where

A Real array dimensioned to size N. On input, A contains the N real components of the data to be transformed. On output, the array A contains the Fourier cosine coefficients. The roles reverse for the inverse transform.

B Real array dimensioned to size N. On input, B contains the N real components of the data to be transformed. On output, the array B contains the Fourier sine coefficients. The roles reverse for the inverse transform.

N Size of input sequence, *N*.

3.2 Multi-variate Transform of Complex Input Data (Tri-variate Example)

```
CALL FFT(A,B,N2*N3,N1,1,-1)
CALL FFT(A,B,N3,N2,N1,-1)
CALL FFT(A,B,1,N3,N1*N2,-1)
```

where the input parameters are the same as in 3.1 above, except that arrays A and B are dimensioned (N1,N2,N3). The three calls to FFT can be arranged in any order. The inverse is computed by repeating the above calls, but with the sign of the sixth parameter changed to plus.

3.3 Single-variate Transform of 2*N Real Values, Stored Alternately A(1),B(1),...,A(N),B(N)

```
CALL FFT(A,B,1,N,1,-1)
CALL REALS(A,B,N,-1)
```

or

```
CALL FFT(A,B,1,N,1,-1)
CALL REALT(A,B,1,N,1,-1)
```

where the Fourier cosine coefficients are found in A(1),A(2),...,A(N),A(N+1) and the Fourier sine coefficients are found in B(1),B(2),...,B(N),B(N+1). The inverse is computed by changing the sign of the sixth parameter to plus, and reversing the order of the two calls; order is important. Scaling of the Fourier coefficients is the same as if FFT had been used to transform 2N real values in A, with B set to zero. As indicated above, either REALS or REALT can be used for a single-variate real transform; REALS is about ten percent faster than REALT.

3.4 Bi-variate Transform of Real Input Data, N=N1*N2

```
CALL FFT(A,B,N2,N1,1,-1)
CALL REALT(A,B,N2,N1,1,-1)
CALL FFT(A(N+1),B(N+1),1,N2,1,-1)
CALL FFT(A,B,1,N2,N1,-1)
```

where the real values are stored alternately in A and B. In this example, arrays A and B are dimensioned ((N1+1)*N2), and a 2*N1 by N2 transform is done. The call on FFT for the real dimension must be first, followed immediately by the corresponding call on REALT. (The folding frequency cosine coefficients are then in A(N+1),A(N+2),...,A(N+N2), and the corresponding locations in array B are set to zero.) The remaining calls may be in any order. If the sequence of calls is reversed and the sign of the sixth parameter is changed to plus, an inverse transform is done.

3.5 Multi-variate Transform of Real Input Data (Tri-variate Example N=N1*N2*N3)

```
CALL FFT(A,B,N3*N2,N1,1,-1)
CALL REALT(A,B,N3*N2,N1,1,-1)
CALL FFT(A(N+1),B(N+1),N3,N2,1,-1)
CALL FFT(A(N+1),B(N+1),1,N3,N2,-1)
CALL FFT(A,B,N3,N2,N1,-1)
CALL FFT(A,B,1,N3,N2*N1,-1)
```

where the real values are stored alternately in A and B. In this example, arrays A and B are dimensioned ((N1+1)*N2*N3). The order of the first two calls is fixed, but the remaining calls may be permuted in any order. If the sequence of calls is reversed and the sign of the sixth parameter is changed to plus, an inverse transform is done. Adding one more dimension adds two more calls on FFT, one for the main array and another for the folding frequency values. There is no limit on the number of dimensions.

The "real value" dimension need not be the column dimension, but the calls on FFT and REALT for the real value dimension must come first.

3.6 Interleaved Storage of Real and Imaginary Data Values

In any of the above examples, the array B can be replaced by A(2), and the magnitude of the sixth parameter set to 2. In this option, A(1),A(3),A(5)... hold the real values and A(2),A(4),A(6),... hold the imaginary values. The values of the third, fourth, and fifth parameters are unchanged. This method of calling is particularly useful when transforming real data, in which case the first dimension is the real dimension...as in the examples in 3.4 and 3.5 above.

3.7 Explanation of the Role of Parameters NSEG, N, NSPN and ISN

FFT(A,B,NSEG,N,NSPN,ISN)
REALT(A,B,NSEG,N,NSPN,ISN)

ISN The sign of ISN determines the transform direction, negative for forward and positive for inverse. The magnitude of ISN determines the indexing increment for arrays A and B.

N The dimension of the current variable.

NSPN The spacing of consecutive data values while indexing the current variable, in units determined by the magnitude of ISN.

NSEG NSEG*N*NSPN is the total number of complex data values.

Users familiar with the original 1969 version should note the changed meaning of the third and fifth parameters; to convert a program calling the earlier version of FFT, the user should replace NSPAN by NSPAN/N and should replace NTOT by NTOT/NSPAN. This change was made to eliminate two possible error conditions.

3.8 Error Conditions

The signs of N, NSPN, and NSEG are ignored, and any nonzero value can be valid. Any nonzero value of ISN can be valid, although it is not anticipated that magnitudes other than one or two will often be used. Thus the following error condition is recognized in FFT and REALT:

$$NSEG*N*NSPN*ISN = 0$$

This condition results in an error message and no transform. The corresponding error condition for REALS is N*ISN = 0.

If N has more than 15 "factors", FFT terminates with an error message. The smallest number with 16 "factors" is 12,754,584, so this error condition should not occur very often. (See [1] or [2].)

4. Tests

The subroutines included in this section have had extensive use in their earlier versions, and the modifications for inclusion in this publication have been held to a minimum. To enable the user to verify that the subroutines are working correctly on his system, a test program with sample output is included here.

The tests used here first do a transform-inverse pair on N = 32 real values, using FFT, REALS and REALT. The input value, the Fourier coefficient values, then the inverse result are printed; the final inverse values are identical with the input values. The sorted error values for arrays A and B are also listed.

The same sequence of N = 32 real values is then stored in array A, with array B set to zero, and a transform-inverse pair is done using FFT alone. The Fourier coefficients and final inverse values are printed; the Fourier coefficient values are seen to be the same as those obtained in the previous test. The first two tests are done using pseudo-random numbers on the unit interval. For the remaining tests, independent bi-variate random normal deviates with zero mean and unit standard deviation are used. The advantage of this approach is that the input data has an RMS value of one; the RMS error value for the final transform-inverse result then has a simple reference.

The next set of tests runs through a variety of test examples, with only the RMS error values for the transform-inverse pair printed. These examples are designed to test all branches of the subroutine.

The final test lists the Fourier coefficients for an 8 by 9 real transform, computed both by FFT alone (with zeros in the imaginary array) and by FFT and REALT. The results are identical, and illustrate the method used in REALT to store the folding frequency coefficients.

5. Acknowledgments

The helpful comments of many users over the years have been useful in developing the revisions included in the current versions of FFT, REALS and REALT. I particularly appreciate Jim Cooley's suggestions, including that of adding multi-variate indexing to REALT. Marie Dolan has been of great help in guiding me through the steps of preparing this material for publication.

References

1. R. C. Singleton, "An Algorithm for Computing the Mixed Radix Fast Fourier Transform", *IEEE Trans. Audio and Electroacoust.*, Vol. AU-17, No. 2, pp. 93-100, June 1969.

2. A. V. Oppenheim (ed.), *Papers on Digital Signal Processing,* The M.I.T. Press, Cambridge, Mass., pp. 163-170, 1969.

Table 1

TEST OF FFT, REALS AND REALT FOR REAL VALUES

REAL INPUT SEQUENCE ---

J	A(J)	B(J)	A(J+1)	B(J+1)
1	.22925607E+00	.76687502E+00	.68317685E+00	.50919111E+00
3	.87455959E+00	.64464100E+00	.84746840E+00	.35396343E+00
5	.39889159E+00	.45709421E+00	.23630936E+00	.13318189E+00
7	.16605222E+00	.22602680E+00	.66245903E+00	.25021174E+00
9	.61769668E+00	.26246527E+00	.51266762E+00	.93920734E+00
11	.62402816E+00	.42238195E+00	.93970599E+00	.28206823E+00
13	.46921754E+00	.54879178E-01	.51983086E+00	.39682690E+00
15	.11315656E+00	.60751725E+00	.70150672E+00	.88705479E+00

FOURIER COSINE AND SINE COEFFICIENTS ---

J	A(J)	B(J)	A(J+1)	B(J+1)
1	.15789569E+02	0.	.13026034E+01	.19871480E+00
3	.11936771E+01	-.28999118E+01	.40612714E-01	.72163937E+00
5	-.46179909E+00	.13081519E+01	.70122852E-01	-.15931004E-01
7	-.27579355E+00	.10351044E+01	.46209660E+00	.11709983E-01
9	-.16102664E+01	.30982473E+00	-.14930917E+01	-.87044475E+00
11	.21015246E+00	-.18331043E+00	-.93344462E+00	-.55190813E-01
13	.33632980E+00	-.10901590E+01	-.12900518E+01	.18219048E+01
15	-.12126616E+01	.75918926E+00	-.12663723E+01	-.10652233E+00
17	.14023971E+01	0.		

TRANSFORM-INVERSE RESULT ---

J	A(J)	B(J)	A(J+1)	B(J+1)
1	.22925607E+00	.76687502E+00	.68317685E+00	.50919111E+00
3	.87455959E+00	.64464100E+00	.84746840E+00	.35396343E+00
5	.39889159E+00	.45709421E+00	.23630936E+00	.13318189E+00
7	.16605222E+00	.22602680E+00	.66245903E+00	.25021174E+00
9	.61769668E+00	.26246527E+00	.51266762E+00	.93920734E+00
11	.62402816E+00	.42238195E+00	.93970599E+00	.28206823E+00
13	.46921754E+00	.54879178E-01	.51983086E+00	.39682690E+00
15	.11315656E+00	.60751725E+00	.70150672E+00	.88705479E+00

SORTED ERROR VALUES ---

J	A(J)	J	B(J)
16	-.36379788E-11	4	-.36379788E-11
6	-.36379788E-11	6	-.36379788E-11
11	-.18189894E-11	12	-.36379788E-11
12	0.	7	-.18189894E-11
10	0.	1	0.
14	0.	13	0.
9	0.	14	0.
8	0.	11	0.

Table 1
(Continued)

7	0.	9	0.
5	0.	5	0.
1	0.	3	0.
15	.22737368E-12	2	0.
3	.18189894E-11	8	.18189894E-11
13	.18189894E-11	15	.18189894E-11
4	.36379788E-11	10	.36379788E-11
2	.36379788E-11	16	.54569682E-11

TEST OF FFT FOR REAL VALUES

FOURIER COSINE AND SINE COEFFICIENTS ---

J	A(J)	B(J)	A(J+1)	B(J+1)
1	.15789569E+02	0.	.13026034E+01	.19871480E+00
3	.11936771E+01	-.28999118E+01	.40612714E-01	.72163937E+00
5	-.46179909E+00	.13081519E+01	.70122852E-01	-.15931004E-01
7	-.27579355E+00	.10351044E+01	.46209660E+00	.11709983E-01
9	-.16102664E+01	.30982473E+00	-.14930917E+01	-.87044475E+00
11	.21015246E+00	-.18331043E+00	-.93344462E+00	-.55190813E-01
13	.33632980E+00	-.10901590E+01	-.12900518E+01	.18219048E+01
15	-.12126616E+01	.75918926E+00	-.12663723E+01	-.10652233E+00
17	.14023971E+01	0.	-.12663723E+01	.10652233E+00
19	-.12126616E+01	-.75918926E+00	-.12900518E+01	-.18219048E+01
21	.33632980E+00	.10901590E+01	-.93344462E+00	.55190813E-01
23	.21015246E+00	.18331043E+00	-.14930917E+01	.87044475E+00
25	-.16102664E+01	-.30982473E+00	.46209660E+00	-.11709983E-01
27	-.27579355E+00	-.10351044E+01	.70122852E-01	.15931004E-01
29	-.46179909E+00	-.13081519E+01	.40612714E-01	-.72163937E+00
31	.11936771E+01	.28999118E+01	.13026034E+01	-.19871480E+00

TRANSFORM-INVERSE RESULT ---

J	A(J)	B(J)	A(J+1)	B(J+1)
1	.22925607E+00	0.	.76687502E+00	-.45474735E-12
3	.68317685E+00	.40946464E-12	.50919111E+00	-.15063506E-11
5	.87455959E+00	-.41774510E-13	.64464100E+00	.31852174E-12
7	.84746840E+00	-.42896152E-12	.35396343E+00	-.25880602E-11
9	.39889159E+00	.22737368E-12	.45709421E+00	.22737368E-12
11	.23630936E+00	.40946464E-12	.13318189E+00	-.82422957E-12
13	.16605222E+00	-.86772019E-12	.22602680E+00	.15004677E-11
15	.66245903E+00	-.98682366E-12	.25021174E+00	-.83580448E-13
17	.61769668E+00	0.	.26246527E+00	-.90949470E-12
19	.51266762E+00	-.45282706E-13	.93920734E+00	.25863756E-11
21	.62402816E+00	.22319622E-11	.42238195E+00	-.13622561E-12
23	.93970599E+00	-.59949178E-12	.28206823E+00	.10499186E-11
25	.46921754E+00	-.22737368E-12	.54879178E-01	.22737368E-12
27	.51983086E+00	.17737067E-11	.39682690E+00	.85265128E-13
29	.11315656E+00	-.13224675E-11	.60751725E+00	-.77326909E-12
31	.70150672E+00	-.53207631E-12	.88705479E+00	.12806616E-11

SINGLE-VARIATE TESTS OF SUBROUTINES

	N	RMS ERROR A ARRAY	RMS ERROR B ARRAY
FFT	2	.137E-10	0.
REALS	2	.386E-11	.103E-10
REALT	2	.386E-11	0.
FFT	3	.525E-11	.297E-11
REALS	3	.951E-11	.872E-11
REALT	3	.951E-11	.235E-11
FFT	5	.111E-10	.718E-11
REALS	5	.986E-11	.676E-11
REALT	5	.986E-11	.182E-11
FFT	210	.202E-10	.195E-10

Table 1
(Continued)

REALS	210	.913E-11	.810E-11
REALT	210	.913E-11	.804E-11
FFT	1000	.233E-10	.232E-10
REALS	1000	.836E-11	.815E-11
REALT	1000	.836E-11	.814E-11
FFT	2000	.233E-10	.224E-10
REALS	2000	.823E-11	.847E-11
REALT	2000	.823E-11	.846E-11
FFT	1024	.193E-10	.191E-10
REALS	1024	.850E-11	.781E-11
REALT	1024	.850E-11	.779E-11
FFT	2048	.195E-10	.191E-10
REALS	2048	.818E-11	.797E-11
REALT	2048	.818E-11	.796E-11
FFT	4096	.199E-10	.202E-10
REALS	4096	.842E-11	.833E-11
REALT	4096	.842E-11	.832E-11
FFT	2187	.272E-10	.259E-10
REALS	2187	.846E-11	.805E-11
REALT	2187	.846E-11	.804E-11
FFT	3125	.281E-10	.283E-10
REALS	3125	.853E-11	.821E-11
REALT	3125	.853E-11	.820E-11
FFT	2401	.285E-10	.284E-10
REALS	2401	.849E-11	.833E-11
REALT	2401	.849E-11	.832E-11
FFT	1331	.401E-10	.391E-10
REALS	1331	.879E-11	.825E-11
REALT	1331	.879E-11	.824E-11
FFT	2197	.293E-10	.288E-10
REALS	2197	.834E-11	.833E-11
REALT	2197	.834E-11	.833E-11
FFT	289	.233E-10	.226E-10
REALS	289	.875E-11	.769E-11
REALT	289	.875E-11	.764E-11
FFT	361	.220E-10	.221E-10
REALS	361	.886E-11	.847E-11
REALT	361	.886E-11	.843E-11
FFT	529	.425E-10	.402E-10
REALS	529	.917E-11	.862E-11
REALT	529	.917E-11	.860E-11

MULTI-VARIATE TESTS OF SUBROUTINES

		RMS ERROR	
	N	A ARRAY	B ARRAY
FFT	1080	.162E-10	.153E-10
FFT	1080	.119E-10	0.
REALT	1080	.921E-11	.912E-11
REALT	1080	.917E-11	0.

8 BY 9 REAL TRANSFORM --- FFT VS. FFT AND REALT

J	A(J)	B(J)	C(J)	D(J)
1	-.18509182E+02	0.	-.18509182E+02	0.
2	-.63947267E+01	.97863880E+01	-.63947267E+01	.97863880E+01

Table 1
(Continued)

3	-.78263712E+01	.25613030E+01	-.78263712E+01	.25613030E+01
4	-.28572332E+01	.10412241E+00	-.28572332E+01	.10412241E+00
5	-.10921541E+02	0.	.46916612E+01	.51173453E+01
6	-.28572332E+01	-.10412241E+00	-.70044347E+01	.40156402E+01
7	-.78263712E+01	-.25613030E+01	.25491851E+01	.37405183E+01
8	-.63947267E+01	-.97863880E+01	.57953108E+00	-.32320996E+01
9	.46916612E+01	.51173453E+01	-.89631614E+01	-.30857869E+01
10	-.70044347E+01	.40156402E+01	.30901046E+01	.92397343E+01
11	.25491851E+01	.37405183E+01	.21382018E+01	-.49035087E+01
12	.57953108E+00	-.32320996E+01	.67686785E+01	.76038599E+01
13	-.16600946E+01	-.23477400E+01	-.52371225E+01	.98372729E+01
14	-.56098061E+01	-.48183139E+01	-.56868554E+01	.61338098E+01
15	.73299048E+01	.85799253E+01	-.25913244E+01	.12416175E+02
16	.18873938E+01	.22840714E+01	-.25446027E+01	-.11353010E+01
17	-.89631614E+01	-.30857869E+01	.72919622E+00	-.24843916E+01
18	.30901046E+01	.92397343E+01	.42411487E+01	.11135565E+02
19	.21382018E+01	-.49035087E+01	.38718108E+01	.31550578E+01
20	.67686785E+01	.76038599E+01	.32136390E+01	-.21105050E+01
21	-.50517953E+01	.67468637E+01	.72919622E+00	.24843916E+01
22	-.16673708E+01	-.28199019E+01	.15985812E+01	.91010428E+01
23	.16322342E+02	-.10742899E+02	-.56624794E+01	-.14726845E+02
24	.38034461E+01	.13483020E+02	.61526515E+01	-.11118265E+02
25	-.52371225E+01	.98372729E+01	-.52371225E+01	-.98372729E+01
26	-.56868554E+01	.61338098E+01	-.35248262E+01	-.23698139E+01
27	-.25913244E+01	.12416175E+02	-.13925042E+01	.41322696E+01
28	-.25446027E+01	-.11353010E+01	.34296333E+01	.39694987E+01
29	-.89007166E+01	.30849262E+01	-.89631614E+01	.30857869E+01
30	.34296333E+01	-.39694987E+01	.38034461E+01	-.13483020E+02
31	-.13925042E+01	-.41322696E+01	.16322342E+02	.10742899E+02
32	-.35248262E+01	.23698139E+01	-.16673708E+01	.28199019E+01
33	.72919622E+00	-.24843916E+01	.46916612E+01	-.51173453E+01
34	.42411487E+01	.11135565E+02	.18873938E+01	-.22840714E+01
35	.38718108E+01	.31550578E+01	.73299048E+01	-.85799253E+01
36	.32136390E+01	-.21105050E+01	-.56098061E+01	.48183139E+01
37	.54606791E+00	-.22407029E+01	-.10921541E+02	0.
38	.61526515E+01	.11118265E+02	-.16600946E+01	-.23477400E+01
39	-.56624794E+01	.14726845E+02	-.50517953E+01	.67468637E+01
40	.15985812E+01	-.91010428E+01	-.89007166E+01	.30849262E+01
41	.72919622E+00	.24843916E+01	.54606791E+00	-.22407029E+01
42	.15985812E+01	.91010428E+01	.54606791E+00	.22407029E+01
43	-.56624794E+01	-.14726845E+02	-.89007166E+01	-.30849262E+01
44	.61526515E+01	-.11118265E+02	-.50517953E+01	-.67468637E+01
45	.54606791E+00	.22407029E+01	-.16600946E+01	.23477400E+01
46	.32136390E+01	.21105050E+01		
47	.38718108E+01	-.31550578E+01		
48	.42411487E+01	-.11135565E+02		
49	-.52371225E+01	-.98372729E+01		
50	-.35248262E+01	-.23698139E+01		
51	-.13925042E+01	.41322696E+01		
52	.34296333E+01	.39694987E+01		
53	-.89007166E+01	-.30849262E+01		
54	-.25446027E+01	.11353010E+01		
55	-.25913244E+01	-.12416175E+02		
56	-.56868554E+01	-.61338098E+01		
57	-.89631614E+01	.30857869E+01		
58	.38034461E+01	-.13483020E+02		
59	.16322342E+02	.10742899E+02		
60	-.16673708E+01	.28199019E+01		
61	-.50517953E+01	-.67468637E+01		
62	.67686785E+01	-.76038599E+01		
63	.21382018E+01	.49035087E+01		
64	.30901046E+01	-.92397343E+01		
65	.46916612E+01	-.51173453E+01		
66	.18873938E+01	-.22840714E+01		
67	.73299048E+01	-.85799253E+01		
68	-.56098061E+01	.48183139E+01		
69	-.16600946E+01	.23477400E+01		
70	.57953108E+00	.32320996E+01		
71	.25491851E+01	-.37405183E+01		
72	-.70044347E+01	-.40156402E+01		

MAXIMUM STACK SIZE = 251

Appendix

```
C
C-----------------------------------------------------------------------
C MAIN PROGRAM:  MXFFT
C AUTHOR:  RICHARD C. SINGLETON
C          SRI INTERNATIONAL, MENLO PARK, CALIFORNIA 94025
C INPUT:  NONE
C-----------------------------------------------------------------------
C
      DIMENSION A(4097), B(4097), C(4097), D(4097), NC(17)
      DIMENSION IA(64), IB(64), RA(64), RB(64)
      COMMON /CSTAK/ DSTAK(2500)
      DOUBLE PRECISION DSTAK
      INTEGER ISTAK(5000)
      EQUIVALENCE (DSTAK(1),ISTAK(1))
      EQUIVALENCE (IA(1),RA(1))
      EQUIVALENCE (IB(1),RB(1))
      EQUIVALENCE (ISTAK(1),LOUT)
      EQUIVALENCE (ISTAK(3),LUSED)
      DATA NC(1), NC(2), NC(3), NC(4), NC(5), NC(6), NC(7), NC(8),
     *    NC(9), NC(10), NC(11), NC(12), NC(13), NC(14), NC(15),
     *    NC(16), NC(17) /2,3,5,210,1000,2000,1024,2048,4096,2187,3125,
     *    2401,1331,2197,289,361,529/
C
C SET UP MACHINE CONSTANTS, USING PORT SUBPROGRAMS
C
      IOUTD = I1MACH(2)
C
      N = 16
      DO 10 J=1,N
        C(J) = UNI(K)
        D(J) = UNI(K)
   10 CONTINUE
 9999 FORMAT (/5X, 1HJ, 8X, 4HA(J), 12X, 4HB(J), 10X, 6HA(J+1), 10X,
     *    6HB(J+1))
 9998 FORMAT (I6, 4E16.8)
C
      DO 20 J=1,N
        A(J) = C(J)
        B(J) = D(J)
   20 CONTINUE
      WRITE (IOUTD,9997)
 9997 FORMAT (///12X, 44HTEST OF FFT, REALS AND REALT FOR REAL VALUES)
      WRITE (IOUTD,9996)
 9996 FORMAT (/24H REAL INPUT SEQUENCE ---)
      WRITE (IOUTD,9999)
      WRITE(IOUTD,9998) (J,A(J),B(J),A(J+1),B(J+1),J=1,N,2)
C
      CALL FFT(A, B, 1, N, 1, -1)
      CALL REALS(A, B, N, -1)
      WRITE (IOUTD,9995)
 9995 FORMAT (/41H FOURIER COSINE AND SINE COEFFICIENTS ---)
      WRITE (IOUTD,9999)
      WRITE(IOUTD,9998) (J,A(J),B(J),A(J+1),B(J+1),J=1,N,2)
      J = N + 1
      WRITE (IOUTD,9998) J, A(J), B(J)
C
C THE NEXT CALL ON REALT DOES THE SAME THING AS:  CALL REALS(A,B,N,1)
C
      CALL REALT(A, B, 1, N, 1, 1)
      CALL FFT(A, B, 1, N, 1, 1)
      WRITE (IOUTD,9994)
 9994 FORMAT (/29H TRANSFORM-INVERSE RESULT ---)
      WRITE (IOUTD,9999)
      WRITE(IOUTD,9998) (J,A(J),B(J),A(J+1),B(J+1),J=1,N,2)
C
      DO 30 J=1,N
        IA(J) = J
        IB(J) = J
        A(J) = A(J) - C(J)
        B(J) = B(J) - D(J)
   30 CONTINUE
      CALL SORTG(A, N, RA)
      CALL SORTG(B, N, RB)
      WRITE (IOUTD,9993)
 9993 FORMAT (/24H SORTED ERROR VALUES ---//9X, 1HJ, 11X, 4HA(J), 14X,
     *    1HJ, 11X, 4HB(J))
      WRITE (IOUTD,9992) (IA(J),A(J),IB(J),B(J),J=1,N)
 9992 FORMAT (I10, E20.8, I10, E20.8)
C
      DO 40 J=1,N
        A(2*J-1) = C(J)
        A(2*J) = D(J)
        B(2*J-1) = 0.0
        B(2*J) = 0.0
   40 CONTINUE
      WRITE (IOUTD,9991)
 9991 FORMAT (///16X, 27HTEST OF FFT FOR REAL VALUES)
C
      N = N + N
      CALL FFT(A, B, 1, N, 1, -1)
      WRITE (IOUTD,9995)
      WRITE (IOUTD,9999)
      WRITE(IOUTD,9998) (J,A(J),B(J),A(J+1),B(J+1),J=1,N,2)
C
      CALL FFT(A, B, 1, N, 1, 1)
      WRITE (IOUTD,9994)
      WRITE (IOUTD,9999)
      WRITE(IOUTD,9998) (J,A(J),B(J),A(J+1),B(J+1),J=1,N,2)
C
C GENERATE TEST DATA FOR TRANSFORM OF MAXIMUM SIZE 4096
C
      DO 50 J=1,4096
        CALL NORMAL(C(J), D(J))
   50 CONTINUE
C
      WRITE (IOUTD,9990)
 9990 FORMAT (///15X, 35HSINGLE-VARIATE TESTS OF SUBROUTINES)
      WRITE (IOUTD,9989)
 9989 FORMAT (/40X, 9HRMS ERROR/25X, 1HN, 9X, 7HA ARRAY, 5X, 7HB ARRAY)
C
      DO 90 I=1,17
        N = NC(I)
        WRITE (IOUTD,9998)
C
        DO 60 J=1,N
          A(J) = C(J)
          B(J) = D(J)
   60   CONTINUE
        CALL FFT(A, B, 1, N, 1, -1)
        CALL FFT(A, B, 1, N, 1, 1)
        CALL RMS(A, B, C, D, N, SS1, SS2)
        WRITE (IOUTD,9988) N, SS1, SS2
 9988   FORMAT (10X, 3HFFT, I13, 4X, 2E12.3)
C
        DO 70 J=1,N
          A(J) = C(J)
          B(J) = D(J)
   70   CONTINUE
        CALL REALS(A, B, N, -1)
        CALL REALS(A, B, N, 1)
```

```
      CALL RMS(A, B, C, D, N, SS1, SS2)
      WRITE (IOUTD,9987) N, SS1, SS2
9987  FORMAT (10X, 5HREALS, I11, 4X, 2E12.3)
C
      DO 80 J=1,N
        A(J) = C(J)
        B(J) = D(J)
 80   CONTINUE
      CALL REALT(A, B, 1, N, 1, -1)
      CALL REALT(A, B, 1, N, 1, 1)
      CALL RMS(A, B, C, D, N, SS1, SS2)
      WRITE (IOUTD,9986) N, SS1, SS2
9986  FORMAT (10X, 5HREALT, I11, 4X, 2E12.3)
C
 90   CONTINUE
C
      WRITE (IOUTD,9985)
9985  FORMAT (///15X, 34HMULTI-VARIATE TESTS OF SUBROUTINES)
      WRITE (IOUTD,9989)
C
      N1 = 4
      N2 = 30
      N3 = 9
      N = N1*N2*N3
C
      DO 100 J=1,N
        A(J) = C(J)
        B(J) = D(J)
100   CONTINUE
      CALL FFT(A, B, N2*N3, N1, 1, -1)
      CALL FFT(A, B, N3, N2, N1, -1)
      CALL FFT(A, B, 1, N3, N1*N2, -1)
      CALL FFT(A, B, 1, N3, N1*N2, 1)
      CALL FFT(A, B, N3, N2, N1, 1)
      CALL FFT(A, B, N2*N3, N1, 1, 1)
      CALL RMS(A, B, C, D, N, SS1, SS2)
      WRITE (IOUTD,9988) N, SS1, SS2
C
      M = N + N
      DO 110 J=1,M
        A(J) = C(J)
        B(J) = D(J)
110   CONTINUE
      CALL FFT(A, A(2), N3, N2, N1, -2)
      CALL FFT(A, A(2), N3, N2, N1, 2)
      CALL RMS(A, B, C, D, M, SS1, SS2)
      WRITE (IOUTD,9988) N, SS1, SS2
C
      DO 120 J=1,N
        A(J) = C(J)
        B(J) = D(J)
120   CONTINUE
      CALL REALT(A, B, N3, N2, N1, -1)
      CALL REALT(A, B, N3, N2, N1, 1)
      CALL RMS(A, B, C, D, N, SS1, SS2)
      WRITE (IOUTD,9986) N, SS1, SS2
C
      DO 130 J=1,M
        A(J) = C(J)
        B(J) = D(J)
130   CONTINUE
      CALL REALT(A, A(2), N3, N2, N1, -2)
      CALL REALT(A, A(2), N3, N2, N1, 2)
      CALL RMS(A, B, C, D, M, SS1, SS2)
      WRITE (IOUTD,9986) N, SS1, SS2
C
      WRITE (IOUTD,9984)
9984  FORMAT (///15X, 46H8 BY 9 REAL TRANSFORM --- FFT VS. FFT AND REAL,
     *    1HT)
      WRITE (IOUTD,9983)
9983  FORMAT (/5X, 1HJ, 8X, 4HA(J), 12X, 4HB(J), 12X, 4HC(J), 12X,
     *    4HD(J))
C
      N1 = 4
      N2 = 9
      N = N1*N2
C
      DO 140 J=1,N
        A(2*J-1) = C(J)
        B(2*J-1) = 0.0
        A(2*J) = D(J)
        B(2*J) = 0.0
140   CONTINUE
      CALL FFT(C, D, N2, N1, 1, -1)
      CALL REALT(C, D, N2, N1, 1, -1)
      CALL FFT(C, D, 1, N2, N1, -1)
      CALL FFT(C(N+1), D(N+1), 1, N2, 1, -1)
      N = N + N
      N1 = N1 + N1
      CALL FFT(A, B, N2, N1, 1, -1)
      CALL FFT(A, B, 1, N2, N1, -1)
C
      N1 = (N1/2+1)*N2
      N2 = N1 + 1
      WRITE (IOUTD,9998) (K,A(K),B(K),C(K),D(K),K=1,N1)
      WRITE (IOUTD,9982) (K,A(K),B(K),K=N2,N)
9982  FORMAT (I6, 2E16.8)
C
      WRITE (IOUTD,9981) LUSED
9981  FORMAT (/21H MAXIMUM STACK SIZE =, I6)
C
C DE-ALLOCATE WORKING STORAGE...AS A FINAL CHECK ON ARRAY BOUNDS
C
      J = LOUT
      IF (J.NE.0) CALL ISTKRL(J)
C
      STOP
      END
C
C-----------------------------------------------------------------------
C BLOCK DATA:  INITIALIZES LABELED COMMON
C-----------------------------------------------------------------------
C
      BLOCK DATA
C
      COMMON /CSTAK/ DSTAK(2500)
C
      DOUBLE PRECISION DSTAK
      INTEGER ISTAK(5000)
      INTEGER ISIZE(5)
C
      EQUIVALENCE (DSTAK(1),ISTAK(1))
      EQUIVALENCE (ISTAK(1),LOUT)
      EQUIVALENCE (ISTAK(2),LNOW)
      EQUIVALENCE (ISTAK(3),LUSED)
      EQUIVALENCE (ISTAK(4),LMAX)
      EQUIVALENCE (ISTAK(5),LBOOK)
      EQUIVALENCE (ISTAK(6),ISIZE(1))
C
      DATA ISIZE(1), ISIZE(2), ISIZE(3), ISIZE(4), ISIZE(5) /1,1,1,2,2/
```

```
      DATA LOUT, LNOW, LUSED, LMAX, LBOOK /0,10,10,5000,10/
C
      END
C
C-----------------------------------------------------------------------
C SUBROUTINE:  FFT
C MULTIVARIATE COMPLEX FOURIER TRANSFORM, COMPUTED IN PLACE
C USING MIXED-RADIX FAST FOURIER TRANSFORM ALGORITHM.
C-----------------------------------------------------------------------
C
      SUBROUTINE FFT(A, B, NSEG, N, NSPN, ISN)
C
C ARRAYS A AND B ORIGINALLY HOLD THE REAL AND IMAGINARY
C      COMPONENTS OF THE DATA, AND RETURN THE REAL AND
C      IMAGINARY COMPONENTS OF THE RESULTING FOURIER COEFFICIENTS.
C MULTIVARIATE DATA IS INDEXED ACCORDING TO THE FORTRAN
C      ARRAY ELEMENT SUCCESSOR FUNCTION, WITHOUT LIMIT
C      ON THE NUMBER OF IMPLIED MULTIPLE SUBSCRIPTS.
C      THE SUBROUTINE IS CALLED ONCE FOR EACH VARIATE.
C      THE CALLS FOR A MULTIVARIATE TRANSFORM MAY BE IN ANY ORDER.
C
C N IS THE DIMENSION OF THE CURRENT VARIABLE.
C NSPN IS THE SPACING OF CONSECUTIVE DATA VALUES
C      WHILE INDEXING THE CURRENT VARIABLE.
C NSEG*N*NSPN IS THE TOTAL NUMBER OF COMPLEX DATA VALUES.
C THE SIGN OF ISN DETERMINES THE SIGN OF THE COMPLEX
C      EXPONENTIAL, AND THE MAGNITUDE OF ISN IS NORMALLY ONE.
C      THE MAGNITUDE OF ISN DETERMINES THE INDEXING INCREMENT FOR A AND B.
C
C IF FFT IS CALLED TWICE, WITH OPPOSITE SIGNS ON ISN, AN
C      IDENTITY TRANSFORMATION IS DONE...CALLS CAN BE IN EITHER ORDER.
C      THE RESULTS ARE SCALED BY 1/N WHEN THE SIGN OF ISN IS POSITIVE.
C
C A TRI-VARIATE TRANSFORM WITH A(N1,N2,N3), B(N1,N2,N3)
C IS COMPUTED BY
C        CALL FFT(A,B,N2*N3,N1,1,-1)
C        CALL FFT(A,B,N3,N2,N1,-1)
C        CALL FFT(A,B,1,N3,N1*N2,-1)
C
C A SINGLE-VARIATE TRANSFORM OF N COMPLEX DATA VALUES IS COMPUTED BY
C        CALL FFT(A,B,1,N,1,-1)
C
C THE DATA MAY ALTERNATIVELY BE STORED IN A SINGLE COMPLEX
C      ARRAY A, THEN THE MAGNITUDE OF ISN CHANGED TO TWO TO
C      GIVE THE CORRECT INDEXING INCREMENT AND A(2) USED TO
C      PASS THE INITIAL ADDRESS FOR THE SEQUENCE OF IMAGINARY
C      VALUES, E.G.
C        CALL FFT(A,A(2),NSEG,N,NSPN,-2)
C
C ARRAY NFAC IS WORKING STORAGE FOR FACTORING N.  THE SMALLEST
C      NUMBER EXCEEDING THE 15 LOCATIONS PROVIDED IS 12,754,584.
C
      DIMENSION A(1), B(1), NFAC(15)
C
      COMMON /CSTAK/ DSTAK(2500)
      DOUBLE PRECISION DSTAK
      INTEGER ISTAK(5000)
      REAL RSTAK(5000)
C
      EQUIVALENCE (DSTAK(1),ISTAK(1))
      EQUIVALENCE (DSTAK(1),RSTAK(1))
C
C DETERMINE THE FACTORS OF N
C
      M = 0
      NF = IABS(N)
      K = NF
      IF (NF.EQ.1) RETURN
      NSPAN = IABS(NF*NSPN)
      NTOT = IABS(NSPAN*NSEG)
      IF (ISN*NTOT.NE.0) GO TO 20
      IERR = I1MACH(4)
      WRITE (IERR,9999) NSEG, N, NSPN, ISN
9999  FORMAT (31H ERROR - ZERO IN FFT PARAMETERS, 4I10)
      RETURN
C
  10  M = M + 1
      NFAC(M) = 4
      K = K/16
  20  IF (K-(K/16)*16.EQ.0) GO TO 10
      J = 3
      JJ = 9
      GO TO 40
  30  M = M + 1
      NFAC(M) = J
      K = K/JJ
  40  IF (MOD(K,JJ).EQ.0) GO TO 30
      J = J + 2
      JJ = J**2
      IF (JJ.LE.K) GO TO 40
      IF (K.GT.4) GO TO 50
      KT = M
      NFAC(M+1) = K
      IF (K.NE.1) M = M + 1
      GO TO 90
  50  IF (K-(K/4)*4.NE.0) GO TO 60
      M = M + 1
      NFAC(M) = 2
      K = K/4
C ALL SQUARE FACTORS OUT NOW, BUT K .GE. 5 STILL
  60  KT = M
      MAXP = MAX0(KT+KT+2,K-1)
      J = 2
  70  IF (MOD(K,J).NE.0) GO TO 80
      M = M + 1
      NFAC(M) = J
      K = K/J
  80  J = ((J+1)/2)*2 + 1
      IF (J.LE.K) GO TO 70
  90  IF (M.LE.KT+1) MAXP = M + KT + 1
      IF (M+KT.GT.15) GO TO 120
      IF (KT.EQ.0) GO TO 110
      J = KT
 100  M = M + 1
      NFAC(M) = NFAC(J)
      J = J - 1
      IF (J.NE.0) GO TO 100
C
 110  MAXF = M - KT
      MAXF = NFAC(MAXF)
      IF (KT.GT.0) MAXF = MAX0(NFAC(KT),MAXF)
      J = ISTKGT(MAXF*4,3)
      JJ = J + MAXF
      J2 = JJ + MAXF
      J3 = J2 + MAXF
      K = ISTKGT(MAXP,2)
      CALL FFTMX(A, B, NTOT, NF, NSPAN, ISN, M, KT, RSTAK(J),
     *    RSTAK(JJ), RSTAK(J2), RSTAK(J3), ISTAK(K), NFAC)
      CALL ISTKRL(2)
      RETURN
```

```
C
  120 IERR = I1MACH(4)
      WRITE (IERR,9998) N
 9998 FORMAT (50H ERROR - FFT PARAMETER N HAS MORE THAN 15 FACTORS-,
     *    I20)
      RETURN
      END
C
C-----------------------------------------------------------------------
C SUBROUTINE:  FFTMX
C CALLED BY SUBROUTINE 'FFT' TO COMPUTE MIXED-RADIX FOURIER TRANSFORM
C-----------------------------------------------------------------------
C
      SUBROUTINE FFTMX(A, B, NTOT, N, NSPAN, ISN, M, KT, AT, CK, BT,
     *    SK, NP, NFAC)
C
      DIMENSION A(1), B(1), AT(1), CK(1), BT(1), SK(1), NP(1), NFAC(1)
C
      INC = IABS(ISN)
      NT = INC*NTOT
      KS = INC*NSPAN
      RAD = ATAN(1.0)
      S72 = RAD/0.625
      C72 = COS(S72)
      S72 = SIN(S72)
      S120 = SQRT(0.75)
      IF (ISN.GT.0) GO TO 10
      S72 = -S72
      S120 = -S120
      RAD = -RAD
      GO TO 30
C
C SCALE BY 1/N FOR ISN .GT. 0
C
   10 AK = 1.0/FLOAT(N)
      DO 20 J=1,NT,INC
        A(J) = A(J)*AK
        B(J) = B(J)*AK
   20 CONTINUE
C
   30 KSPAN = KS
      NN = NT - INC
      JC = KS/N
C
C SIN, COS VALUES ARE RE-INITIALIZED EACH LIM STEPS
C
      LIM = 32
      KLIM = LIM*JC
      I = 0
      JF = 0
      MAXF = M - KT
      MAXF = NFAC(MAXF)
      IF (KT.GT.0) MAXF = MAX0(NFAC(KT),MAXF)
C
C COMPUTE FOURIER TRANSFORM
C
   40 DR = 8.0*FLOAT(JC)/FLOAT(KSPAN)
      CD = 2.0*SIN(0.5*DR*RAD)**2
      SD = SIN(DR*RAD)
      KK = 1
      I = I + 1
      IF (NFAC(I).NE.2) GO TO 110
C
C TRANSFORM FOR FACTOR OF 2 (INCLUDING ROTATION FACTOR)
C
      KSPAN = KSPAN/2
      K1 = KSPAN + 2
   50 K2 = KK + KSPAN
      AK = A(K2)
      BK = B(K2)
      A(K2) = A(KK) - AK
      B(K2) = B(KK) - BK
      A(KK) = A(KK) + AK
      B(KK) = B(KK) + BK
      KK = K2 + KSPAN
      IF (KK.LE.NN) GO TO 50
      KK = KK - NN
      IF (KK.LE.JC) GO TO 50
      IF (KK.GT.KSPAN) GO TO 350
   60 C1 = 1.0 - CD
      S1 = SD
      MM = MIN0(K1/2,KLIM)
      GO TO 80
   70 AK = C1 - (CD*C1+SD*S1)
      S1 = (SD*C1-CD*S1) + S1
C
C THE FOLLOWING THREE STATEMENTS COMPENSATE FOR TRUNCATION
C ERROR.  IF ROUNDED ARITHMETIC IS USED, SUBSTITUTE
C C1=AK
C
      C1 = 0.5/(AK**2+S1**2) + 0.5
      S1 = C1*S1
      C1 = C1*AK
   80 K2 = KK + KSPAN
      AK = A(KK) - A(K2)
      BK = B(KK) - B(K2)
      A(KK) = A(KK) + A(K2)
      B(KK) = B(KK) + B(K2)
      A(K2) = C1*AK - S1*BK
      B(K2) = S1*AK + C1*BK
      KK = K2 + KSPAN
      IF (KK.LT.NT) GO TO 80
      K2 = KK - NT
      C1 = -C1
      KK = K1 - K2
      IF (KK.GT.K2) GO TO 80
      KK = KK + JC
      IF (KK.LE.MM) GO TO 70
      IF (KK.LT.K2) GO TO 90
      K1 = K1 + INC + INC
      KK = (K1-KSPAN)/2 + JC
      IF (KK.LE.JC+JC) GO TO 60
      GO TO 40
   90 S1 = FLOAT((KK-1)/JC)*DR*RAD
      C1 = COS(S1)
      S1 = SIN(S1)
      MM = MIN0(K1/2,MM+KLIM)
      GO TO 80
C
C TRANSFORM FOR FACTOR OF 3 (OPTIONAL CODE)
C
  100 K1 = KK + KSPAN
      K2 = K1 + KSPAN
      AK = A(KK)
      BK = B(KK)
      AJ = A(K1) + A(K2)
      BJ = B(K1) + B(K2)
      A(KK) = AK + AJ
      B(KK) = BK + BJ
      AK = -0.5*AJ + AK
```

```
      BK = -0.5*BJ + BK
      AJ = (A(K1)-A(K2))*S120
      BJ = (B(K1)-B(K2))*S120
      A(K1) = AK - BJ
      B(K1) = BK + AJ
      A(K2) = AK + BJ
      B(K2) = BK - AJ
      KK = K2 + KSPAN
      IF (KK.LT.NN) GO TO 100
      KK = KK - NN
      IF (KK.LE.KSPAN) GO TO 100
      GO TO 290
C
C TRANSFORM FOR FACTOR OF 4
C
  110 IF (NFAC(I).NE.4) GO TO 230
      KSPNN = KSPAN
      KSPAN = KSPAN/4
  120 C1 = 1.0
      S1 = 0
      MM = MIN0(KSPAN,KLIM)
      GO TO 150
  130 C2 = C1 - (CD*C1+SD*S1)
      S1 = (SD*C1-CD*S1) + S1
C
C THE FOLLOWING THREE STATEMENTS COMPENSATE FOR TRUNCATION
C ERROR.  IF ROUNDED ARITHMETIC IS USED, SUBSTITUTE
C C1=C2
C
      C1 = 0.5/(C2**2+S1**2) + 0.5
      S1 = C1*S1
      C1 = C1*C2
  140 C2 = C1**2 - S1**2
      S2 = C1*S1*2.0
      C3 = C2*C1 - S2*S1
      S3 = C2*S1 + S2*C1
  150 K1 = KK + KSPAN
      K2 = K1 + KSPAN
      K3 = K2 + KSPAN
      AKP = A(KK) + A(K2)
      AKM = A(KK) - A(K2)
      AJP = A(K1) + A(K3)
      AJM = A(K1) - A(K3)
      A(KK) = AKP + AJP
      AJP = AKP - AJP
      BKP = B(KK) + B(K2)
      BKM = B(KK) - B(K2)
      BJP = B(K1) + B(K3)
      BJM = B(K1) - B(K3)
      B(KK) = BKP + BJP
      BJP = BKP - BJP
      IF (ISN.LT.0) GO TO 180
      AKP = AKM - BJM
      AKM = AKM + BJM
      BKP = BKM + AJM
      BKM = BKM - AJM
      IF (S1.EQ.0.0) GO TO 190
  160 A(K1) = AKP*C1 - BKP*S1
      B(K1) = AKP*S1 + BKP*C1
      A(K2) = AJP*C2 - BJP*S2
      B(K2) = AJP*S2 + BJP*C2
      A(K3) = AKM*C3 - BKM*S3
      B(K3) = AKM*S3 + BKM*C3
      KK = K3 + KSPAN
      IF (KK.LE.NT) GO TO 150
  170 KK = KK - NT + JC
      IF (KK.LE.MM) GO TO 130
      IF (KK.LT.KSPAN) GO TO 200
      KK = KK - KSPAN + INC
      IF (KK.LE.JC) GO TO 120
      IF (KSPAN.EQ.JC) GO TO 350
      GO TO 40
  180 AKP = AKM + BJM
      AKM = AKM - BJM
      BKP = BKM - AJM
      BKM = BKM + AJM
      IF (S1.NE.0.0) GO TO 160
  190 A(K1) = AKP
      B(K1) = BKP
      A(K2) = AJP
      B(K2) = BJP
      A(K3) = AKM
      B(K3) = BKM
      KK = K3 + KSPAN
      IF (KK.LE.NT) GO TO 150
      GO TO 170
  200 S1 = FLOAT((KK-1)/JC)*DR*RAD
      C1 = COS(S1)
      S1 = SIN(S1)
      MM = MIN0(KSPAN,MM+KLIM)
      GO TO 140
C
C TRANSFORM FOR FACTOR OF 5 (OPTIONAL CODE)
C
  210 C2 = C72**2 - S72**2
      S2 = 2.0*C72*S72
  220 K1 = KK + KSPAN
      K2 = K1 + KSPAN
      K3 = K2 + KSPAN
      K4 = K3 + KSPAN
      AKP = A(K1) + A(K4)
      AKM = A(K1) - A(K4)
      BKP = B(K1) + B(K4)
      BKM = B(K1) - B(K4)
      AJP = A(K2) + A(K3)
      AJM = A(K2) - A(K3)
      BJP = B(K2) + B(K3)
      BJM = B(K2) - B(K3)
      AA = A(KK)
      BB = B(KK)
      A(KK) = AA + AKP + AJP
      B(KK) = BB + BKP + BJP
      AK = AKP*C72 + AJP*C2 + AA
      BK = BKP*C72 + BJP*C2 + BB
      AJ = AKM*S72 + AJM*S2
      BJ = BKM*S72 + BJM*S2
      A(K1) = AK - BJ
      A(K4) = AK + BJ
      B(K1) = BK + AJ
      B(K4) = BK - AJ
      AK = AKP*C2 + AJP*C72 + AA
      BK = BKP*C2 + BJP*C72 + BB
      AJ = AKM*S2 - AJM*S72
      BJ = BKM*S2 - BJM*S72
      A(K2) = AK - BJ
      A(K3) = AK + BJ
      B(K2) = BK + AJ
      B(K3) = BK - AJ
      KK = K4 + KSPAN
      IF (KK.LT.NN) GO TO 220
```

```
      KK = KK - NN
      IF (KK.LE.KSPAN) GO TO 220
      GO TO 290
C
C TRANSFORM FOR ODD FACTORS
C
 230  K = NFAC(I)
      KSPNN = KSPAN
      KSPAN = KSPAN/K
      IF (K.EQ.3) GO TO 100
      IF (K.EQ.5) GO TO 210
      IF (K.EQ.JF) GO TO 250
      JF = K
      S1 = RAD/(FLOAT(K)/8.0)
      C1 = COS(S1)
      S1 = SIN(S1)
      CK(JF) = 1.0
      SK(JF) = 0.0
      J = 1
 240  CK(J) = CK(K)*C1 + SK(K)*S1
      SK(J) = CK(K)*S1 - SK(K)*C1
      K = K - 1
      CK(K) = CK(J)
      SK(K) = -SK(J)
      J = J + 1
      IF (J.LT.K) GO TO 240
 250  K1 = KK
      K2 = KK + KSPNN
      AA = A(KK)
      BB = B(KK)
      AK = AA
      BK = BB
      J = 1
      K1 = K1 + KSPAN
 260  K2 = K2 - KSPAN
      J = J + 1
      AT(J) = A(K1) + A(K2)
      AK = AT(J) + AK
      BT(J) = B(K1) + B(K2)
      BK = BT(J) + BK
      J = J + 1
      AT(J) = A(K1) - A(K2)
      BT(J) = B(K1) - B(K2)
      K1 = K1 + KSPAN
      IF (K1.LT.K2) GO TO 260
      A(KK) = AK
      B(KK) = BK
      K1 = KK
      K2 = KK + KSPNN
      J = 1
 270  K1 = K1 + KSPAN
      K2 = K2 - KSPAN
      JJ = J
      AK = AA
      BK = BB
      AJ = 0.0
      BJ = 0.0
      K = 1
 280  K = K + 1
      AK = AT(K)*CK(JJ) + AK
      BK = BT(K)*CK(JJ) + BK
      K = K + 1
      AJ = AT(K)*SK(JJ) + AJ
      BJ = BT(K)*SK(JJ) + BJ
      JJ = JJ + J
      IF (JJ.GT.JF) JJ = JJ - JF
      IF (K.LT.JF) GO TO 280
      K = JF - J
      A(K1) = AK - BJ
      B(K1) = BK + AJ
      A(K2) = AK + BJ
      B(K2) = BK - AJ
      J = J + 1
      IF (J.LT.K) GO TO 270
      KK = KK + KSPNN
      IF (KK.LE.NN) GO TO 250
      KK = KK - NN
      IF (KK.LE.KSPAN) GO TO 250
C
C MULTIPLY BY ROTATION FACTOR (EXCEPT FOR FACTORS OF 2 AND 4)
C
 290  IF (I.EQ.M) GO TO 350
      KK = JC + 1
 300  C2 = 1.0 - CD
      S1 = SD
      MM = MIN0(KSPAN,KLIM)
      GO TO 320
 310  C2 = C1 - (CD*C1+SD*S1)
      S1 = S1 + (SD*C1-CD*S1)
C
C THE FOLLOWING THREE STATEMENTS COMPENSATE FOR TRUNCATION
C ERROR.  IF ROUNDED ARITHMETIC IS USED, THEY MAY
C BE DELETED.
C
      C1 = 0.5/(C2**2+S1**2) + 0.5
      S1 = C1*S1
      C2 = C1*C2
 320  C1 = C2
      S2 = S1
      KK = KK + KSPAN
 330  AK = A(KK)
      A(KK) = C2*AK - S2*B(KK)
      B(KK) = S2*AK + C2*B(KK)
      KK = KK + KSPNN
      IF (KK.LE.NT) GO TO 330
      AK = S1*S2
      S2 = S1*C2 + C1*S2
      C2 = C1*C2 - AK
      KK = KK - NT + KSPAN
      IF (KK.LE.KSPNN) GO TO 330
      KK = KK - KSPNN + JC
      IF (KK.LE.MM) GO TO 310
      IF (KK.LT.KSPAN) GO TO 340
      KK = KK - KSPAN + JC + INC
      IF (KK.LE.JC+JC) GO TO 300
      GO TO 40
 340  S1 = FLOAT((KK-1)/JC)*DR*RAD
      C2 = COS(S1)
      S1 = SIN(S1)
      MM = MIN0(KSPAN,MM+KLIM)
      GO TO 320
C
C PERMUTE THE RESULTS TO NORMAL ORDER---DONE IN TWO STAGES
C PERMUTATION FOR SQUARE FACTORS OF N
C
 350  NP(1) = KS
      IF (KT.EQ.0) GO TO 440
      K = KT + KT + 1
      IF (M.LT.K) K = K - 1
      J = 1
```

```
      NP(K+1) = JC
  360 NP(J+1) = NP(J)/NFAC(J)
      NP(K) = NP(K+1)*NFAC(J)
      J = J + 1
      K = K - 1
      IF (J.LT.K) GO TO 360
      K3 = NP(K+1)
      KSPAN = NP(2)
      KK = JC + 1
      K2 = KSPAN + 1
      J = 1
      IF (N.NE.NTOT) GO TO 400
C
C PERMUTATION FOR SINGLE-VARIATE TRANSFORM (OPTIONAL CODE)
C
  370 AK = A(KK)
      A(KK) = A(K2)
      A(K2) = AK
      BK = B(KK)
      B(KK) = B(K2)
      B(K2) = BK
      KK = KK + INC
      K2 = KSPAN + K2
      IF (K2.LT.KS) GO TO 370
  380 K2 = K2 - NP(J)
      J = J + 1
      K2 = NP(J+1) + K2
      IF (K2.GT.NP(J)) GO TO 380
      J = 1
  390 IF (KK.LT.K2) GO TO 370
      KK = KK + INC
      K2 = KSPAN + K2
      IF (K2.LT.KS) GO TO 390
      IF (KK.LT.KS) GO TO 380
      JC = K3
      GO TO 440
C
C PERMUTATION FOR MULTIVARIATE TRANSFORM
C
  400 K = KK + JC
  410 AK = A(KK)
      A(KK) = A(K2)
      A(K2) = AK
      BK = B(KK)
      B(KK) = B(K2)
      B(K2) = BK
      KK = KK + INC
      K2 = K2 + INC
      IF (KK.LT.K) GO TO 410
      KK = KK + KS - JC
      K2 = K2 + KS - JC
      IF (KK.LT.NT) GO TO 400
      K2 = K2 - NT + KSPAN
      KK = KK - NT + JC
      IF (K2.LT.KS) GO TO 400
  420 K2 = K2 - NP(J)
      J = J + 1
      K2 = NP(J+1) + K2
      IF (K2.GT.NP(J)) GO TO 420
      J = 1
  430 IF (KK.LT.K2) GO TO 400
      KK = KK + JC
      K2 = KSPAN + K2
      IF (K2.LT.KS) GO TO 430
      IF (KK.LT.KS) GO TO 420
      JC = K3
  440 IF (2*KT+1.GE.M) RETURN
      KSPNN = NP(KT+1)
C
C PERMUTATION FOR SQUARE-FREE FACTORS OF N
C
      J = M - KT
      NFAC(J+1) = 1
  450 NFAC(J) = NFAC(J)*NFAC(J+1)
      J = J - 1
      IF (J.NE.KT) GO TO 450
      KT = KT + 1
      NN = NFAC(KT) - 1
      JJ = 0
      J = 0
      GO TO 480
  460 JJ = JJ - K2
      K2 = KK
      K = K + 1
      KK = NFAC(K)
  470 JJ = KK + JJ
      IF (JJ.GE.K2) GO TO 460
      NP(J) = JJ
  480 K2 = NFAC(KT)
      K = KT + 1
      KK = NFAC(K)
      J = J + 1
      IF (J.LE.NN) GO TO 470
C
C DETERMINE THE PERMUTATION CYCLES OF LENGTH GREATER THAN 1
C
      J = 0
      GO TO 500
  490 K = KK
      KK = NP(K)
      NP(K) = -KK
      IF (KK.NE.J) GO TO 490
      K3 = KK
  500 J = J + 1
      KK = NP(J)
      IF (KK.LT.0) GO TO 500
      IF (KK.NE.J) GO TO 490
      NP(J) = -J
      IF (J.NE.NN) GO TO 500
      MAXF = INC*MAXF
C
C REORDER A AND B, FOLLOWING THE PERMUTATION CYCLES
C
      GO TO 570
  510 J = J - 1
      IF (NP(J).LT.0) GO TO 510
      JJ = JC
  520 KSPAN = JJ
      IF (JJ.GT.MAXF) KSPAN = MAXF
      JJ = JJ - KSPAN
      K = NP(J)
      KK = JC*K + I + JJ
      K1 = KK + KSPAN
      K2 = 0
  530 K2 = K2 + 1
      AT(K2) = A(K1)
      BT(K2) = B(K1)
      K1 = K1 - INC
      IF (K1.NE.KK) GO TO 530
  540 K1 = KK + KSPAN
```

```
      K2 = K1 - JC*(K+NP(K))
      K = -NP(K)
 550  A(K1) = A(K2)
      B(K1) = B(K2)
      K1 = K1 - INC
      K2 = K2 - INC
      IF (K1.NE.KK) GO TO 550
      KK = K2
      IF (K.NE.J) GO TO 540
      K1 = KK + KSPAN
      K2 = 0
 560  K2 = K2 + 1
      A(K1) = AT(K2)
      B(K1) = BT(K2)
      K1 = K1 - INC
      IF (K1.NE.KK) GO TO 560
      IF (JJ.NE.0) GO TO 520
      IF (J.NE.1) GO TO 510
 570  J = K3 + 1
      NT = NT - KSPNN
      I = NT - INC + 1
      IF (NT.GE.0) GO TO 510
      RETURN
      END
C
C-----------------------------------------------------------------------
C SUBROUTINE:  REALS
C USED WITH 'FFT' TO COMPUTE FOURIER TRANSFORM OR INVERSE FOR REAL DATA
C-----------------------------------------------------------------------
C
      SUBROUTINE REALS(A, B, N, ISN)
C
C IF ISN=-1, THIS SUBROUTINE COMPLETES THE FOURIER TRANSFORM
C      OF 2*N REAL DATA VALUES, WHERE THE ORIGINAL DATA VALUES ARE
C      STORED ALTERNATELY IN ARRAYS A AND B, AND ARE FIRST
C      TRANSFORMED BY A COMPLEX FOURIER TRANSFORM OF DIMENSION N.
C      THE COSINE COEFFICIENTS ARE IN A(1),A(2),...A(N),A(N+1)
C      AND THE SINE COEFFICIENTS ARE IN B(1),B(2),...B(N),B(N+1).
C      NOTE THAT THE ARRAYS A AND B MUST HAVE DIMENSION N+1.
C      A TYPICAL CALLING SEQUENCE IS
C        CALL FFT(A,B,N,N,N,-1)
C        CALL REALS(A,B,N,-1)
C
C IF ISN=1, THE INVERSE TRANSFORMATION IS DONE, THE FIRST
C      STEP IN EVALUATING A REAL FOURIER SERIES.
C      A TYPICAL CALLING SEQUENCE IS
C        CALL REALS(A,B,N,1)
C        CALL FFT(A,B,N,N,N,1)
C      THE TIME DOMAIN RESULTS ALTERNATE IN ARRAYS A AND B,
C      I.E. A(1),B(1),A(2),B(2),...A(N),B(N).
C
C THE DATA MAY ALTERNATIVELY BE STORED IN A SINGLE COMPLEX
C      ARRAY A, THEN THE MAGNITUDE OF ISN CHANGED TO TWO TO
C      GIVE THE CORRECT INDEXING INCREMENT AND A(2) USED TO
C      PASS THE INITIAL ADDRESS FOR THE SEQUENCE OF IMAGINARY
C      VALUES, E.G.
C        CALL FFT(A,A(2),N,N,N,-2)
C        CALL REALS(A,A(2),N,-2)
C      IN THIS CASE, THE COSINE AND SINE COEFFICIENTS ALTERNATE IN A.
C
      DIMENSION A(1), B(1)
      INC = IABS(ISN)
      NF = IABS(N)
      IF (NF*ISN.NE.0) GO TO 10
      IERR = I1MACH(4)
      WRITE (IERR,9999) N, ISN
9999  FORMAT (33H ERROR - ZERO IN REALS PARAMETERS, 2I10)
      RETURN
C
  10  NK = NF*INC + 2
      NH = NK/2
      RAD = ATAN(1.0)
      DR = -4.0/FLOAT(NF)
      CD = 2.0*SIN(0.5*DR*RAD)**2
      SD = SIN(DR*RAD)
C
C SIN,COS VALUES ARE RE-INITIALIZED EACH LIM STEPS
C
      LIM = 32
      MM = LIM
      ML = 0
      SN = 0.0
      IF (ISN.GT.0) GO TO 40
      CN = 1.0
      A(NK-1) = A(1)
      B(NK-1) = B(1)
  20  DO 30 J=1,NH,INC
        K = NK - J
        AA = A(J) + A(K)
        AB = A(J) - A(K)
        BA = B(J) + B(K)
        BB = B(J) - B(K)
        RE = CN*BA + SN*AB
        EM = SN*BA - CN*AB
        B(K) = (EM-BB)*0.5
        B(J) = (EM+BB)*0.5
        A(K) = (AA-RE)*0.5
        A(J) = (AA+RE)*0.5
        ML = ML + 1
        IF (ML.EQ.MM) GO TO 50
        AA = CN - (CD*CN+SD*SN)
        SN = (SD*CN-CD*SN) + SN
C
C THE FOLLOWING THREE STATEMENTS COMPENSATE FOR TRUNCATION
C ERROR.  IF ROUNDED ARITHMETIC IS USED, SUBSTITUTE
C CN=AA
C
        CN = 0.5/(AA**2+SN**2) + 0.5
        SN = CN*SN
        CN = CN*AA
  30  CONTINUE
      RETURN
C
  40  CN = -1.0
      SD = -SD
      GO TO 20
C
  50  MM = MM + LIM
      SN = FLOAT(ML)*DR*RAD
      CN = COS(SN)
      IF (ISN.GT.0) CN = -CN
      SN = SIN(SN)
      GO TO 30
      END
C
C-----------------------------------------------------------------------
C SUBROUTINE:  REALT
C USED WITH 'FFT' OR ANY OTHER COMPLEX FOURIER TRANSFORM TO COMPUTE
C TRANSFORM OR INVERSE FOR REAL DATA
C THE DATA MAY BE EITHER SINGLE-VARIATE OR MULTI-VARIATE
```

```
C-----------------------------------------------------------------------
C
      SUBROUTINE REALT(A, B, NSEG, N, NSPN, ISN)
C
C IF ISN=-1, THIS SUBROUTINE COMPLETES THE FOURIER TRANSFORM
C      OF 2*N REAL DATA VALUES, WHERE THE ORIGINAL DATA VALUES ARE
C      STORED ALTERNATELY IN ARRAYS A AND B, AND ARE FIRST
C      TRANSFORMED BY A COMPLEX FOURIER TRANSFORM OF DIMENSION N.
C      THE COSINE COEFFICIENTS ARE IN A(1),A(2),...A(N),A(N+1)
C      AND THE SINE COEFFICIENTS ARE IN B(1),B(2),...B(N),B(N+1).
C      NOTE THAT THE ARRAYS A AND B MUST HAVE DIMENSION N+1.
C      A TYPICAL CALLING SEQUENCE IS
C        CALL FFT(A,B,1,N,1,-1)
C        CALL REALT(A,B,1,N,1,-1)
C
C IF ISN=1, THE INVERSE TRANSFORMATION IS DONE, THE FIRST
C      STEP IN EVALUATING A REAL FOURIER SERIES.
C      A TYPICAL CALLING SEQUENCE IS
C        CALL REALT(A,B,1,N,1,1)
C        CALL FFT(A,B,1,N,1,1)
C      THE TIME DOMAIN RESULTS ALTERNATE IN ARRAYS A AND B,
C      I.E. A(1),B(1),A(2),B(2),...A(N),B(N).
C
C THE DATA MAY ALTERNATIVELY BE STORED IN A SINGLE COMPLEX
C       ARRAY A, THEN THE MAGNITUDE OF ISN CHANGED TO TWO TO
C       GIVE THE CORRECT INDEXING INCREMENT AND A(2) USED TO
C       PASS THE INITIAL ADDRESS FOR THE SEQUENCE OF IMAGINARY
C       VALUES, E.G.
C         CALL FFT(A,A(2),1,N,1,-2)
C         CALL REALT(A,A(2),1,N,1,-2)
C      IN THIS CASE, THE COSINE AND SINE COEFFICIENTS ALTERNATE IN A.
C
C THIS SUBROUTINE IS SET UP TO DO THE ABOVE-DESCRIBED OPERATION ON
C      ALL SUB-VECTORS WITHIN ANY DIMENSION OF A MULTI-DIMENSIONAL
C      FOURIER TRANSFORM.  THE PARAMETERS NSEG, N, NSPN, AND INC
C      SHOULD AGREE WITH THOSE USED IN THE ASSOCIATED CALL OF 'FFT'.
C      THE FOLDING FREQUENCY COSINE COEFFICIENTS ARE STORED AT THE END
C      OF ARRAY A (WITH ZEROS IN THE CORRESPONDING LOCATIONS IN ARRAY B),
C      IN A SUB-MATRIX OF DIMENSION ONE LESS THAN THE MAIN ARRAY.  THE
C      DELETED DIMENSION IS THAT CORRESPONDING TO THE PARAMETER N IN
C      THE CALL OF REALT.  THUS ARRAYS A AND B MUST HAVE DIMENSION
C      NSEG*NSPN*(N+1).
C
      DIMENSION A(1), B(1)
      INC = IABS(ISN)
      KS = IABS(NSPN)*INC
      NF = IABS(N)
      NS = KS*NF
      NT = IABS(NS*NSEG)
      IF (ISN*NT.NE.0) GO TO 10
      IERR = I1MACH(4)
      WRITE (IERR,9999) NSEG, N, NSPN, ISN
9999  FORMAT (33H ERROR - ZERO IN REALT PARAMETERS, 3I10, I9)
      RETURN
C
   10 JC = KS
      K2 = IABS(KS*NSEG) - INC
      KD = NS
      NH = NS/2 + 1
      NN = NT - INC
      NT = NT + 1
      KK = 1
      RAD = ATAN(1.0)
      DR = -4.0/FLOAT(NF)
      CD = 2.0*SIN(0.5*DR*RAD)**2
      SD = SIN(DR*RAD)
C
C SIN,COS VALUES ARE RE-INITIALIZED EACH LIM STEPS
C
      LIM = 32
      KLIM = LIM*KS
      MM = MIN0(NH,KLIM)
      SN = 0.0
      IF (ISN.GT.0) GO TO 70
C
   20 AA = A(KK)
      BA = B(KK)
      B(KK) = 0
      A(KK) = AA + BA
      A(NT) = AA - BA
      B(NT) = 0
      NT = NT + JC
      KK = KK + NS
      IF (KK.LE.NN) GO TO 20
      NT = NT - K2
      KK = KK - NN
      IF (KK.LE.JC) GO TO 20
      CN = 1.0
   30 IF (NF.EQ.1) RETURN
C
   40 AA = CN - (CD*CN+SD*SN)
      SN = (SD*CN-CD*SN) + SN
C
C THE FOLLOWING THREE STATEMENTS COMPENSATE FOR TRUNCATION
C ERROR.  IF ROUNDED ARITHMETIC IS USED, SUBSTITUTE
C CN=AA
C
      CN = 0.5/(AA**2+SN**2) + 0.5
      SN = CN*SN
      CN = CN*AA
   50 JC = JC + KS
      KD = KD - KS - KS
   60 K2 = KK + KD
      AA = A(KK) + A(K2)
      AB = A(KK) - A(K2)
      BA = B(KK) + B(K2)
      BB = B(KK) - B(K2)
      RE = CN*BA + SN*AB
      EM = SN*BA - CN*AB
      B(K2) = (EM-BB)*0.5
      B(KK) = (EM+BB)*0.5
      A(K2) = (AA-RE)*0.5
      A(KK) = (AA+RE)*0.5
      KK = KK + NS
      IF (KK.LE.NN) GO TO 60
      KK = KK - NN
      IF (KK.LE.JC) GO TO 60
      IF (KK.LE.MM) GO TO 40
      IF (KK.GT.NH) RETURN
      SN = FLOAT(JC/KS)*DR*RAD
      CN = COS(SN)
      IF (ISN.GT.0) CN = -CN
      SN = SIN(SN)
      MM = MIN0(NH,MM+KLIM)
      GO TO 50
C
   70 AA = A(KK)
      BA = A(NT)
      A(KK) = (AA+BA)*0.5
      B(KK) = (AA-BA)*0.5
```

```
      NT = NT + JC
      KK = KK + NS
      IF (KK.LE.NN) GO TO 70
      NT = NT - K2
      KK = KK - NN
      IF (KK.LE.JC) GO TO 70
      CN = -1.0
      SD = -SD
      GO TO 30
      END
C
C-----------------------------------------------------------------------
C SUBROUTINE:  SORTG
C SORTS ARRAY A INTO INCREASING ORDER, FROM A(1) TO A(N)
C THE ARRAY TAG IS PERMUTED THE SAME AS ARRAY A
C-----------------------------------------------------------------------
C
      SUBROUTINE SORTG(A, N, TAG)
C
C TO SORT N ELEMENTS STARTING WITH A(K), CALL WITH A(K) AND TAG(K).
C AN EARLIER VERSION OF THIS ALGORITHM, WITHOUT THE TAG ARRAY, WAS
C PUBLISHED BY R.C. SINGLETON AS ACM ALGORITHM 347,
C COMM. ACM 12 (MARCH 1969), 1865-1866.  THE CURRENT VERSION
C SOLVES A MACHINE-DEPENDENT PROBLEM PRESENT IN THE EARLIER
C VERSION AND ALMOST ALL OTHER SORT SUBROUTINES.  ON MANY
C COMPUTERS, COMPARING A VERY LARGE NEGATIVE NUMBER WITH A
C VERY LARGE POSITIVE NUMBER GIVES A WRONG RESULT AND A BAD SORT.
C THIS PROBLEM WAS NOTED BY R. GRIFFIN AND K.A. REDISH, "REMARK
C ON ALGORITHM 347,", COMM. ACM 13 (JANUARY 1970), 54.
C THE PROBLEM IS AVOIDED HERE BY AN INITIAL SPLIT ON ZERO.
C TIME IS PROPORTIONAL TO N*LOG(N)
C AS FAR AS THE AUTHOR IS AWARE, NO FASTER IN-PLACE SORT METHOD HAS
C BEEN PUBLISHED SINCE THE ORIGINAL APPEARANCE OF THIS ALGORITHM.
C
C WORKING STORAGE ARRAYS IL AND IU SHOULD HAVE DIMENSION
C     INT(ALOG(FLOAT(N))/ALOG(2.0))
C     A DIMENSION OF 20 ALLOWS VALUES OF N UP TO 2**21-1
C
      DIMENSION A(1), IU(20), IL(20), TAG(1)
      M = 1
      I = 1
      J = N
      K = I
      L = J
      IF (I.GE.J) RETURN
      T = 0
      IF (A(I)) 30, 30, 10
  10  IF (A(L)) 90, 90, 20
  20  L = L - 1
      IF (L-I) 70, 70, 10
  30  IF (A(J)) 40, 110, 110
  40  IF (A(K)) 50, 90, 90
  50  K = K + 1
      IF (J-K) 70, 70, 40
  60  IF (I.GE.J) GO TO 140
  70  K = I
      IJ = (J+I)/2
      T = A(IJ)
      IF (A(I).LE.T) GO TO 80
      A(IJ) = A(I)
      A(I) = T
      T = A(IJ)
      TG = TAG(IJ)
      TAG(IJ) = TAG(I)
      TAG(I) = TG
  80  L = J
      IF (A(J).GE.T) GO TO 110
      A(IJ) = A(J)
      A(J) = T
      T = A(IJ)
      TG = TAG(IJ)
      TAG(IJ) = TAG(J)
      TAG(J) = TG
      IF (A(I).LE.T) GO TO 110
      A(IJ) = A(I)
      A(I) = T
      T = A(IJ)
      TG = TAG(IJ)
      TAG(IJ) = TAG(I)
      TAG(I) = TG
      GO TO 110
  90  TT = A(L)
 100  A(L) = A(K)
      A(K) = TT
      TG = TAG(L)
      TAG(L) = TAG(K)
      TAG(K) = TG
 110  L = L - 1
      IF (A(L).GT.T) GO TO 110
      TT = A(L)
 120  K = K + 1
      IF (A(K).LT.T) GO TO 120
      IF (K.LE.L) GO TO 100
      IF (L-I.LE.J-K) GO TO 130
      IL(M) = I
      IU(M) = L
      I = K
      M = M + 1
      GO TO 150
 130  IL(M) = K
      IU(M) = J
      J = L
      M = M + 1
      GO TO 150
 140  M = M - 1
      IF (M.EQ.0) RETURN
      I = IL(M)
      J = IU(M)
 150  IF (J-I.GT.10) GO TO 70
      IF (I.EQ.1) GO TO 60
      I = I - 1
 160  I = I + 1
      IF (I.EQ.J) GO TO 140
      T = A(I+1)
      IF (A(I).LE.T) GO TO 160
      TG = TAG(I+1)
      K = I
 170  A(K+1) = A(K)
      TAG(K+1) = TAG(K)
      K = K - 1
      IF (T.LT.A(K)) GO TO 170
      A(K+1) = T
      TAG(K+1) = TG
      GO TO 160
      END
C
C-----------------------------------------------------------------------
C SUBROUTINE:  NORMAL
C GENERATES AN INDEPENDENT PAIR OF RANDOM NORMAL DEVIATES
C METHOD DUE TO G. MARSAGLIA AND T.A. BRAY,
```

```
C SIAM REVIEW, VOL. 6, NO. 3, JULY 1964. 260-264
C-----------------------------------------------------------------------
C
      SUBROUTINE NORMAL(X, Y)
C
C OUTPUT:  X,Y = INDEPENDENT PAIR OF RANDOM NORMAL DEVIATES
C FUNCTION UNI GENERATES PSEUDO-RANDOM NUMBER BETWEEN 0.0 AND 1.0
C
   10 RX = UNI(NRM)*2.0 - 1.0
      RY = UNI(NRM)*2.0 - 1.0
      R = RX**2 + RY**2
      IF (R.GE.1.0) GO TO 10
      R = SQRT(-2.0*ALOG(R)/R)
      X = RX*R
      Y = RY*R
      RETURN
      END
C
C-----------------------------------------------------------------------
C SUBROUTINE:  RMS
C COMPUTES RMS ERROR FOR TRANSFORM-INVERSE PAIR
C-----------------------------------------------------------------------
C
      SUBROUTINE RMS(A, B, C, D, N, EA, EB)
C
C ARRAYS:  A,B = TRANSFORM, INVERSE RESULTS
C C,D = ORIGINAL DATA
C INPUT:  N = DIMENSION OF ARRAYS A, B, C AND D
C OUTPUT:  EA,EB = RMS ERRORS FOR A AND B ARRAYS
C
      DIMENSION A(1), B(1), C(1), D(1)
      SSA = 0.0
      SSB = 0.0
      DO 10 J=1,N
        SSA = (A(J)-C(J))**2 + SSA
        SSB = (B(J)-D(J))**2 + SSB
   10 CONTINUE
      EA = SQRT(SSA/FLOAT(N))
      EB = SQRT(SSB/FLOAT(N))
      RETURN
      END
C
C-----------------------------------------------------------------------
C FUNCTION:  ISTKGT(NITEMS,ITYPE)
C ALLOCATES WORKING STORAGE FOR NITEMS OF ITYPE, AS FOLLOWS
C
C 1 - LOGICAL
C 2 - INTEGER
C 3 - REAL
C 4 - DOUBLE PRECISION
C 5 - COMPLEX
C
C-----------------------------------------------------------------------
C
      INTEGER FUNCTION ISTKGT(NITEMS, ITYPE)
C
      COMMON /CSTAK/ DSTAK(2500)
C
      DOUBLE PRECISION DSTAK
      INTEGER ISTAK(5000)
      INTEGER ISIZE(5)
C
      EQUIVALENCE (DSTAK(1),ISTAK(1))
      EQUIVALENCE (ISTAK(1),LOUT)
      EQUIVALENCE (ISTAK(2),LNOW)
      EQUIVALENCE (ISTAK(3),LUSED)
      EQUIVALENCE (ISTAK(4),LMAX)
      EQUIVALENCE (ISTAK(5),LBOOK)
      EQUIVALENCE (ISTAK(6),ISIZE(1))
C
      ISTKGT = (LNOW*ISIZE(2)-1)/ISIZE(ITYPE) + 2
      I = ((ISTKGT-1+NITEMS)*ISIZE(ITYPE)-1)/ISIZE(2) + 3
      IF (I.GT.LMAX) GO TO 10
      ISTAK(I-1) = ITYPE
      ISTAK(I) = LNOW
      LOUT = LOUT + 1
      LNOW = I
      LUSED = MAX0(LUSED,LNOW)
      RETURN
C
   10 IERR = I1MACH(4)
      WRITE (IERR,9999) I
 9999 FORMAT (1H , 39HOVERFLOW OF COMMON ARRAY ISTAK --- NEED, I10)
      WRITE (IERR,9998) (ISTAK(J),J=1,10), ISTAK(LNOW-1), ISTAK(LNOW)
 9998 FORMAT (12I6)
      STOP
      END
C
C-----------------------------------------------------------------------
C SUBROUTINE:  ISTKRL(K)
C DE-ALLOCATES THE LAST K WORKING STORAGE AREAS
C-----------------------------------------------------------------------
C
      SUBROUTINE ISTKRL(K)
C
      COMMON /CSTAK/ DSTAK(2500)
C
      DOUBLE PRECISION DSTAK
      INTEGER ISTAK(5000)
C
      EQUIVALENCE (DSTAK(1),ISTAK(1))
      EQUIVALENCE (ISTAK(1),LOUT)
      EQUIVALENCE (ISTAK(2),LNOW)
      EQUIVALENCE (ISTAK(3),LUSED)
      EQUIVALENCE (ISTAK(4),LMAX)
      EQUIVALENCE (ISTAK(5),LBOOK)
C
      IN = K
C
      IF (LBOOK.LE.LNOW .AND. LNOW.LE.LUSED .AND. LUSED.LE.LMAX) GO TO
     *    10
      IERR = I1MACH(4)
      WRITE (IERR,9999)
 9999 FORMAT (53H WARNING...ISTAK(2),ISTAK(3),ISTAK(4) OR ISTAK(5) HIT)
      WRITE (IERR,9997) (ISTAK(J),J=1,10), ISTAK(LNOW-1), ISTAK(LNOW)
C
   10 IF (IN.LE.0) RETURN
      IF (LBOOK.GT.ISTAK(LNOW) .OR. ISTAK(LNOW).GE.LNOW-1) GO TO 20
      LOUT = LOUT - 1
      LNOW = ISTAK(LNOW)
      IN = IN - 1
      GO TO 10
C
   20 IERR = I1MACH(4)
      WRITE (IERR,9998)
 9998 FORMAT (45H WARNING...POINTER AT ISTAK(LNOW) OVERWRITTEN/11X,
     *    27HDE-ALLOCATION NOT COMPLETED)
      WRITE (IERR,9997) (ISTAK(J),J=1,10), ISTAK(LNOW-1), ISTAK(LNOW)
 9997 FORMAT (12I6)
      RETURN
C
      END
```

1.5

Optimized Mass Storage FFT Program

Donald Fraser

CSIRO, Division of Computing Research
Canberra City, ACT 2601, Australia

1. Purpose

The program computes the discrete Fourier transform (DFT) and inverse discrete Fourier transform (IDFT), in one or more dimensions, of a (large) array in mass store.

2. Method

The program is an implementation of the optimal sorting algorithm of the author [1] which allows a base 2 version of the Cooley-Tukey FFT algorithm [2,3,4] efficient access to a mass store array. Optimal sorting for the mass storage FFT has been determined independently by DeLotto and Dotti [5,6], but in the author's version the emphasis is on 'in-place' array modification. This results in slightly higher mass store I/O than the minimum, but requires no additional mass store working space. The method is also discussed in [7] and is a logical extension of the work of Singleton [8] and Brenner [9].

The program computes in-place the discrete Fourier transform of a one-dimensional or a multidimensional array. In the one-dimensional case the transform is defined by

$$X(k) = \text{SCAL} \sum_{n=0}^{N-1} x(n) e^{\pm j \frac{2\pi}{N} nk} \quad , \quad k = 0,1,...,N-1 \tag{1}$$

where SCAL is an arbitrary scaling factor, and $j = \sqrt{-1}$, the sign of the exponent being either minus (DFT) or plus (IDFT).

The definition Eq. (1) is easily generalized to cover more than one dimension; for example the two-dimensional case is given by

$$X(k_1,k_2) = \text{SCAL} \sum_{n_1=0}^{N_1-1} \sum_{n_2=0}^{N_2-1} x(n_1,n_2) e^{\pm j2\pi \left(\frac{n_1 k_1}{N_1} + \frac{n_2 k_2}{N_2} \right)} \tag{2}$$

$$\text{for} \quad k_1 = 0,1,...,N_1-1 \text{ and } k_2 = 0,1,...,N_2-1$$

The elements $x(n)$ or $x(n_1,n_2)$ in Eq. (1) or Eq. (2) are the initial complex data in a mass store array. These are replaced by the elements $X(k)$ or $X(k_1,k_2)$ as the final complex data.

Complex data is transformed by an in-place, base-2 algorithm using post-computation bit-reversal to sort the array [3]. The computation is handled by a modified, in-core FFT routine which does a sequence of partial transforms of the mass store array (the method is discussed in more detail in section 5). The sorting algorithm [1] calculates the most efficient way to access the mass store array for these computations. Finally the sorting algorithm is used to carry out an overall bit-reversed permutation of the array, again with as few accesses as possible (I/O efficiency is discussed in section 4).

Computation and sorting occur 'in-place', through a combination of 'virtual' permutations, where mass store blocks are accessed according to an indexing algorithm but are left physically unpermuted, and symmetric permutations, which interchange blocks according to a generalized index bit-reversal. In [1] it is shown that any unsymmetric permutation can be formed from two suitable symmetric permutations, each of which can be done in-place.

Multidimensional transforms are achieved automatically by making use of the indexing structure of the FFT algorithm itself. No change to the order of access of elements is necessary, so that the full advantage of the sorting algorithm and program simplicity is maintained.

For transforms in which the initial or final data is real, the usual time saving algorithm [4] is available to unscramble a half length complex transform of packed real data (or vice-versa). This requires an extra accessing and computing pass through the mass store array, but still results in a saving of nearly half the computation and I/O time. In this method, it is easy to allow a choice in the degree of redundancy in the final, complex result. The array may be expanded to full redundancy (twice the physical length of the original packed real array), or to partial redundancy, or maintained the same physical length by elimination of all redundancy.

Finally, we must define a number of terms used in the discussion. 'Core store' is used to describe a region where the elements of an array are equally accessible at random. 'Mass store' implies a region where elements are grouped into 'blocks' or 'records' which must be accessed as units, but which units are accessible efficiently at random. A 'pass' is an array operation which leaves the array elements in-place ready for another array operation. Thus, an 'I/O pass' reads and writes back once all the blocks of an array (this is a logical definition — sometimes in practice not all blocks need be physically accessed, or some may be accessed more than once).

3. Program Description

3.1 Usage

Figure 1 shows the structure of the internal subroutine calling network.

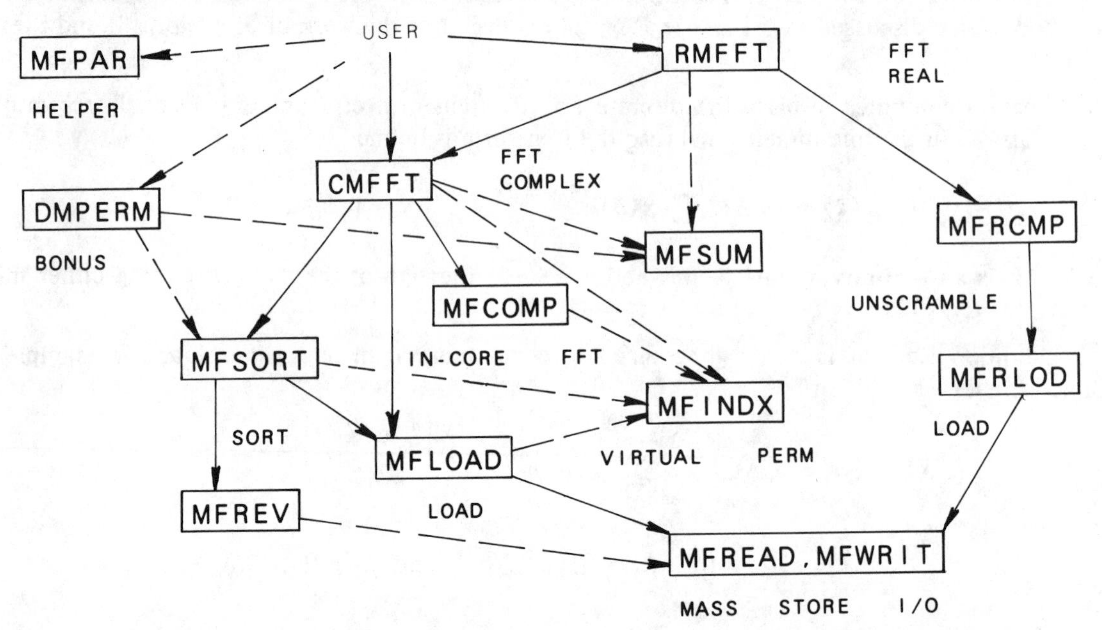

Fig. 1 Internal subroutine calling network. (Solid and dashed lines distinguish roughly between 'main path' and 'reference' calls).

The program consists of a set of subroutines written in ANSI FORTRAN, only two of which are called by a user program. If the data are always complex the user program calls subroutine CMFFT. If the input data or the results are real, the user program may call the faster subroutine RMFFT. Either subroutine replaces a one-dimensional or a multidimensional array in mass store by its discrete Fourier transform, as defined in section 2.

(If the elements of the mass store array are considered to be singly indexed by $i=0,1,2,...$, then the indices of the definitions map into $i=n$ in the one-dimensional case or $i=n_1+n_2 \cdot N_1$ in the two-dimensional case, resultant transposition giving $i=k_2+k_1 \cdot N_2$. In general, $i=n_1+n_2 \cdot N_1+n_3 \cdot N_2 \cdot N_1+...$ Remember also that for FORTRAN, indices are increased by 1.)

The mass store file is assumed to exist and to have been previously defined to the FORTRAN system and opened for random access by the user. The file is accessed through two system-dependent subroutines MFREAD and MFWRIT (see section 7). All other subroutines are system-independent.

The user has freedom to specify total array size, mass store block or record size, core store working space size and the dimensioning of the array, except that all sizes are to base 2 and are given by their binary exponents.

3.2 Calling Method

The complex FFT routine is called by

```
CALL CMFFT (MEXA,NDIM,ISGN,IDIR,SCAL,BUFA,IBEX,ICEX)
```

while the faster, real-to-complex (or vice-versa) FFT routine is called by

```
CALL RMFFT (MEXA,NDIM,ISGN,IDIR,SCAL,BUFA,IBEX,ICEX,IPAK)
```

3.3 Description of Parameters

MEXA Integer array of size NDIM (see below). MEXA consists of a list of dimension size binary exponents, defining the dimensioning of the mass store array. For example, a one-dimensional array has a size of 2**MEXA(1) elements. A two-dimensional array has 2**MEXA(2) sets of 2**MEXA(1) adjacent elements each (N of definition Eq. (1) or N_1 and N_2 of definition Eq. (2) are 2**MEXA(1) and 2**MEXA(2) in initial order).

Notes:

a) The FFT routines actively modify the MEXA list, if necessary, leaving it in the order corresponding to the final array dimensioning. In general, its order is reversed by the FFT, except in the special case with routine CMFFT discussed under parameter IDIR below. The MEXA list should therefore be linked to a unique mass store array so that it always indicates the current dimensioning of that array.

b) The MEXA list of exponents always refers to the dimensioning of a full array, as required by routine CMFFT or by routine RMFFT with IPAK=+1 (see below). The complex result of routine RMFFT operating on real data is truncated when IPAK=0 or −1, so that the MEXA list elements in these cases may refer to a virtual array length.

c) The first element of the list always gives the size 2**MEXA(1) of the set or sets of adjacent elements in the first dimension of the mass store array. If data are of type complex, this is the number of complex elements in each set while if data are of type real (e.g. before calling routine RMFFT in direction real-to-complex), this is the number of real elements.

d) Mass store array dimensioning is independent of I/O block transfer size and core working space size.

NDIM Number of dimensions in mass store array (size of MEXA list).

Range $1 \leqslant$ NDIM $\leqslant 4$ (RMFFT), to increase, see section 5.
or $1 \leqslant$ NDIM $\leqslant$ M (CMFFT).

(M) Not a call parameter, but defined here for convenience. M = Sum to NDIM of MEXA list, giving total mass store array size = 2**M elements (number of complex elements for routine CMFFT, real or complex elements for routine RMFFT, but see under IPAK packing parameter below).

ISGN The sign of ISGN is the sign of the complex exponent of the transform definition (e.g. equation 1 or 2). A positive or negative sign results in the inverse transform of the other, but see also below.

IDIR Transform direction (RMFFT), reversal (CMFFT) parameter. Routine RMFFT converts packed real data to complex data (or vice-versa) during transformation, so that IDIR is needed to determine the direction, independently of ISGN, thus:

IDIR=−1 Real-to-complex (RMFFT), Dimension reversal
IDIR=+1 Complex-to-real (RMFFT), Dimension reversal.

(Note that it is usually most convenient to use the same variable for both ISGN and IDIR so that a single negation results in transform inversion).

Routine CMFFT does not need a direction parameter, other than ISGN, and IDIR is used in this case to set dimension reversal (transposition in two dimensions) or not, as required, thus:

IDIR not zero, Dimension reversal (more efficient I/O)
IDIR=0 DO NOT USE (RMFFT), Suppress reversal (CMFFT).

SCAL Arbitrary type real scale factor of Eqs. (1) or (2). If SCAL=1.0 computation is fastest as no scaling occurs.

BUFA Array in core to be used as workspace by FFT routines. Note that internal FFT subroutines assume the following:

a) BUFA is either type real or complex, to suit local needs.

b) BUFA is given the trivial dimension BUFA(1) internally, since its actual size is known only as the exponent ICEX (see below).

c) Type complex data is assumed to be stored in the sequence real/imaginary/real/imaginary/.. in core store and mass store.

Some of these points may upset some FORTRAN compilers.

IBEX Mass store I/O block transfer size binary exponent. Block or record size = 2**IBEX real elements.

Limits are $2 \leqslant$ IBEX $\leqslant$ ICEX−2 (RMFFT),
or $1 \leqslant$ IBEX $\leqslant$ ICEX−1 (CMFFT).

ICEX Core store working space size binary exponent. Dimension of BUFA = 2**ICEX real elements.

Limits are IBEX+2 $\leqslant$ ICEX $\leqslant$ M (RMFFT),
or IBEX+1 $\leqslant$ ICEX $\leqslant$ M+1 (CMFFT).

IPAK Packing parameter (routine RMFFT) only.
Determines the degree of redundancy (discussed in more detail in section 5) desired in the complex result after a real-to-complex transform, thus:

IPAK=+1 gives a fully redundant complex result. The final mass store array is exactly twice the physical length of the initial real array, having the same number, 2**M, of complex elements as initial real elements. This is the same result that is obtained when calling CMFFT with the initial data occupying the real part of a complex array, zero imaginary.

IPAK=0 gives a partly redundant complex result, slightly longer than the initial real array. In this case there are 2**(M−1) + 2**(M−MEXA(NDIM)) complex elements in the final array, the MEXA list in the order after the transform (increased to an integral number of mass store blocks, if necessary). This is probably the most useful packing.

IPAK=−1 gives a result containing no redundancy, and having exactly the same physical length as the initial 2**M real elements, or 2**(M−1) complex elements. This is achieved by squeezing together those parts of the array having internal redundancy. The exact algorithm is discussed in section 5. No information is lost and, for one or two dimensions, only a little sorting is needed to access unsqueezed information. For example, in one dimension, the real value at the

Nyquist frequency becomes the imaginary part of the zero frequency element, which also must be real.

Notes:

a) For IPAK=+1 or 0 the mass store array file must be extendable. IPAK=−1 has the advantage that the mass store array file remains a fixed length, but has the disadvantage that some user effort is needed to access complex data correctly.

b) In the complex-to-real direction, IPAK=+1 or 0 are equivalent, since only part of a fully redundant complex array is accessed by RMFFT in this direction. But IPAK=−1 must be used in both directions to correctly handle the squeezed complex array.

3.4 Helper Routine

To help the user set up the arguments of section 3.3 and to determine the mass store file and block sizes, a helper routine MFPAR is included. Use of the routine is not essential, but is recommended. The routine is called by:

IERR = MFPAR (IRMF,ICOMP)

with IRMF=−1 if mass store data is currently packed real or +1 if data is currently complex when using routine RMFFT. When using routine CMFFT, IRMF=0. If ICOMP=0 the argument exponents MEXA( ), IBEX and ICEX are defined by the user while if ICOMP is not zero the exponents are to be computed by MFPAR.

Three COMMON blocks are used to transmit other data, thus:

```
COMMON/MFARG/ MEXA(4),NDIM,ISGN,IDIR,SCAL,IBEX,ICEX,IPAK
COMMON/MFVAL/ DIMA(4),TDM1,RDM1,FBLK,TBLK,RBLK,RCOR,SIZE
COMMON/MFINT/ NDMA(4),NTD1,NRD1,NFBK,NTBK,NRBK,NRCR,NSZE
```

MFARG holds a four element MEXA( ) list (for up to four dimensions) followed by the other mass store FFT call arguments (except BUFA). NDIM, IBEX, ICEX and IPAK must be present (unless ICOMP not zero), while ISGN, IDIR and SCAL are ignored here.

COMMON blocks MFVAL and MFINT return data computed by routine MFPAR. These include the four element arrays DIMA( ) and NDMA( ) corresponding element by element to MEXA( ) but containing the actual dimension sizes 2**MEXA( ). Similarly RBLK, NRBK and RCOR, NRCR hold the sizes 2**IBEX and 2**ICEX. Note that MFINT variables are one to one integer conversions of MFVAL variables, which are type real. A local variable FIXMAX, the biggest positive integer in the machine, determines whether a conversion is allowed. Any values not converted are set to −1 in MFINT and the function returns MFPAR=−1 to indicate this. IBEX and ICEX are forced to be within the limits defined in section 3.3, and MFPAR=+1 if IBEX is forced too small. Otherwise the function returns MFPAR=0 normally.

If the helper subroutine argument ICOMP is not zero, the computation is reversed and MEXA( ) exponents are computed from given DIMA( ) real sizes, IBEX and ICEX computed from RBLK and RCOR real sizes. Sizes are adjusted to be integral powers of 2, adjusted up or down to the closest power on a log scale. NDIM and IPAK must still be given in MFARG.

The most useful values computed by MFPAR are NTD1, NRD1, NFBK, NTBK and NRBK (in MFINT). NRBK is 2**IBEX and is the number of reals in the 'record' used for mass store access by the FFT. NTBK is the current total number of records of size NRBK, or the current file length. NFBK, on the other hand, is the maximum number of records of size NRBK to be expected, including any mass store array expansion by routine RMFFT (see IPAK=0 or 1 in section 3.3). NFBK is thus useful for defining the maximum file length to the operating system.

NRD1 is the number of reals in an equivalent 'record' of the current first dimension length ('first' defined by current MEXA(1)). NTD1 is the current number of such equivalent records and these two values are useful for logically accessing a multidimensional mass store array by the user (though not so useful if NDIM=1). Note that to do this the I/O routines MFREAD and MFWRIT must be able to handle correctly 'records' different in length from NRBK, otherwise NTBK and NRBK should be used. This will be system dependent (see section 7).

NSZE, SIZE is the effective total size of of the mass store array (2**M) and is useful, for example, in computing the scale factor SCAL.

Note that MFPAR should be called just prior to user file access, and with the correct value for IRMF, to ensure the computed variables reflect the current state of the mass store array.

3.5 Example

Suppose we wish to transform a two-dimensional real array having 512 rows (2**9) of 256 adjacent real elements each (2**8), or a total of $128K$ real elements (2**17), where $K = 1024$. Suppose we decide to allow a core store working space of $8K$ real elements (2**13) and to access the mass store array in blocks of 128 real elements (2**7) each, these being quite independent of array dimensioning. Then, using the helper routine MFPAR (section 3.4) to compute file parameters, the mass store file is defined and opened for random access, loadd with data and subroutine RMFFT is called with:

```
      COMMON/MFARG/ MEXA(4),NDIM,ISGN,IDIR,SCAL,IBEX,ICEX,IPAK
      COMMON/MFVAL/ DIMA(4),TDM1,RDM1,FBLK,TBLK,RBLK,RCOR,SIZE
      COMMON/MFINT/ NDMA(4),NTD1,NRD1,NFBK,NTBK,NRBK,NRCR,NSZE
C COMMON BLOCKS USED BY HELPER ROUTN MFPAR (NOT ESSENTIAL)
C
      REAL BUFA(8192)
      COMPLEX CBUFA(4096)
      EQUIVALENCE (BUFA(1),CBUFA(1))
C WORKING ARRAY IN CORE (EQUIVALENCED FOR USER ACCESS)
C
      MEXA(1)=8
      MEXA(2)=9
      NDIM=2
      ISGN=-1
      IDIR=-1
      SCAL=1.0
      IBEX=7
      ICEX=13
      IPAK=1
C MASS STORE FFT ARGUMENTS INITIALIZED
C
      IRMF=-1
      ICOMP=0
      IERR=MFPAR (IRMF,ICOMP)
C HELPER ROUTINE, DATA REAL, RMFFT, EXPONENTS DEFINED, IF(IERR.EQ.O) OK
C COMPUTES FILE PARAMETERS IN COMMON AREAS (NOT ESSENTIAL, BUT USEFUL)
C
      (open mass store file here)
C HERE CAN OPEN MASS STORE FILE, NFBK MAX 'RECDS' OF NRBK REALS
C
      DO 2 JB=1,NTD1                    (or NTBK)
      DO 1 I=1,NRD1                     (or NRBK)
1     BUFA(I)= (Enter a real variable here)
C
2     CALL MFWRIT (BUFA,NRD1,JB)        (or (BUFA,NRBK,JB))
C LOAD MASS STORE FILE WITH PACKED REAL DATA (NTD1=512, NRD1=256)
C (IF MFREAD/MFWRIT REQUIRE FIXED RECD LENGTH, USE NTBK,NRBK INSTEAD)
C
      CALL RMFFT (MEXA,NDIM,ISGN,IDIR,SCAL,BUFA,IBEX,ICEX,IPAK)
```

resulting in the DFT of Eq. (2), with SCAL=1.0, N_1=256 and N_2=512 and a negative complex exponent. Because the initial data are packed real and because the parameter IPAK=1 is chosen, the mass store array will be extended to $128K$ complex elements, or twice its initial physical size. This array is a fully redundant transform of the original, having 256 rows (2**8) of 512 adjacent complex elements (2**9), the dimensions being reversed or transposed. To indicate transposition the MEXA list will be reversed (MEXA(1)=9, MEXA(2)=8 after the transform).

To save unnecessary array extension the last parameter IPAK can be made 0 or −1. If IPAK=0 the result is a partially redundant transform, being the first 129 rows (2**MEXA(2)/2+1) of the transform, of 512 adjacent complex elements each. The first and last rows each have an internal conjugate symmetry, but other redundancy is deleted.

If IPAK=−1 the first and last rows above are 'squeezed' together, the second half of the last row becoming the second half of the first row. In addition, the first and middle real elements of the last row become the imaginary parts of the first and middle elements of the first row, resulting in an array of 128 rows of 512 complex elements, exactly the same physical length as the initial array, so that no file extension is necessary (see section 5).

The complex result is accessed, then an inverse transform is called by:

```
      IRMF=+1
      IERR=MFPAR (IRMF,ICOMP)
C HELPER ROUTINE AGAIN, DATA COMPLEX, ROUTINE RMFFT, IF(IERR.EQ.O) OK
C
      NCNT=NRD1/2                  (or NRBK/2)
C NCNT IS THE NUMBER OF COMPLEX ELEMENTS IN RECORD
      DO 4 JB=1,NTD1               (or NTBK)
      CALL MFREAD (BUFA,NRD1,JB)   (or (BUFA,NRBK,JB))
      DO 3 I=1,NCNT
3     (access each complex element CBUF(I) here)
C
4     CALL MFWRIT (BUFA,NRD1,JB)   (or BUFA,NRBK,JB))
C COMPLEX RESULT READ BY USER (WRITING NEW VALUES IF DESIRED)
C (IF MFREAD/MFWRIT REQUIRE FIXED RECD LENGTH, USE NTBK,NRBK INSTEAD)
C
      ISGN=-ISGN
      IDIR=-IDIR
      SCAL= 1.0/SIZE
      CALL RMFFT (MEXA,NDIM,ISGN,IDIR,SCAL,BUFA,IBEX,ICEX,IPAK)
```

the scale factor SCAL, using SIZE=2.**M computed by MFPAR, is chosen here to normalize the result to the same scale as the original data. The result is a real array having 512 rows of 256 adjacent real elements each, the MEXA list being restored to its initial order.

3.6 Addendum — Transposing, Dimension Shifting

The sorting algorithm used in the mass storage FFT may also be used for changing the order of dimensioning of a multidimensional mass store array. (This is similar to, but more general than, Eklundh's method [10]). As an example, and in addition to the FFT routines, a dimension shifting routine DMPERM is included for completeness, called thus:

```
      CALL DMPERM (MEXA,NDIM,NSHFT,IREX,BUFA,IBEX,ICEX)
```

where the arguments MEXA, NDIM, BUFA, IBEX, ICEX have the same meaning as in the FFT calls.

The other parameters are:

NSHFT Dimension shift count.
- NSHFT=0 — No shift or change occurs.
- NSHFT=1,2 etc. — First-to-next dimension, circular NSHFT place shift, Modulo (NDIM).
- NSHFT=−1 — Dimension order reversed.

IREX 'Element size' binary exponent (size =2** IREX reals), thus IREX=0 for real array, IREX=1 for complex array.

Most of the comments concerning the FFT routine parameters apply also to this routine. The mass store array is either real or complex (or 'elements' of multiple reals), with an MEXA list defining the current dimensioning (binary exponents). The MEXA list is actively modified by the routine if necessary. (Note that to sort arrays of elements smaller in size than type real, such as the type byte elements of some systems, it is necessary to alter type statements for BUFA and TEMP variables in all subroutines called).

The routine is quite trivial in design, consisting of only twenty statements. It operates by calling the generalized mass store bit-reversed sorting routine MFSORT, used internally by the FFT routines, and therefore has a similar I/O efficiency to the FFT routines (section 4).

4. Input/Output Efficiency

The number of I/O passes (see definitions, section 2) through the mass store array depends on the array size (2**M), the core store working space size (2**ICEX) and the I/O block or record size (2**IBEX). In general, the larger the working space and the smaller the block size the better, but block size should not be made too small because of other overheads. The FFT computation requires *I*((M−IBEX+1)/(ICEX−IBEX)) passes involving I/O (where *I*(x) is the smallest integer greater than or equal to x). Post computation sorting requires a similar number of passes (without sorting overlap), although DeLotto and Dotti [5] mention *I*((M−ICEX+1)/(ICEX−IBEX)) passes. The reason this is not achieved is that in-place sorting requires M−IBEX+1 virtual permutations to leave a physically unpermuted array. A change in the algorithm to a smaller number M−ICEX+1 virtual permutations requires an extra block-sorting pass of mass store, nullifying the advantage.

Figure 2 gives the maximum range of block size exponent IBEX necessary to keep the number of I/O passes, N, of the mass store array low, given ICEX and M. Solid lines bound the optimum $N=4$. Dashed lines are the boundaries for the next best $N=6$ passes, and it should not be difficult to operate with $N=4$ or 6 in most cases. When calling the half-length transform by subroutine RMFFT, increase diagram M and N by one to obtain the working M and N.

The hatched line cutting across the figure gives a bound below which all mass store passes access all blocks. In the area above the line, *I*((IBEX−1)/(ICEX−IBEX)) sorting passes are required with an additional pass of mass store involving a block reshuffle, in which only some of the blocks are accessed. That is, the optimum can approach $N=3$ passes. Given an I/O system which allows alteration of the index key of a block, the block shuffle can be replaced by an index shuffle, giving $N-1$ passes in the upper area. Note also that the diagram is only approximate, as integral rounding effects may increase or decrease the number of passes by one at some points.

Run time depends on two main factors, the computation time of the in-core FFT and the I/O time. A total elapsed time can be written approximately

$$T = TC*M*2**M + TM*N*2**M$$

where TC and TM are unit computation and unit mass store average access and transfer times corresponding to each complex element. As an example, consider two very different computer systems, a PDP 11/40 system and a CYBER 76 installation. Then we may expect TC=0.5 ms (PDP 11) or 2 μs (CYBER), and TM=1 ms (PDP 11, IBEX=7) or 0.5 ms (CYBER, IBEX=9) for routine CMFFT. Calling subroutine RMFFT with packed real data roughly halves the time, so that a 256×256 real array can be transformed in about 5 minutes (PDP 11) or 1 second (CYBER) CPU time and 8 minutes (PDP 11) or 1 minute (CYBER) elapsed time. But in machines such as the CYBER, where computing speed is very great compared to I/O latency, the routines are used to great advantage for transforming arrays in LCM (extended memory) in place of mass store, using high speed block-copy operations between this and main memory (see sections 6 and 7). Whether an array is multidimensional or not makes little difference to efficiency.

Both TC and TM are block size dependent, there being fixed overheads per block. Both therefore can be written

$$TC = TCO + TCA/2**(IBEX-1) \text{ and } TM = TMO + TMA/2**(IBEX-1)$$

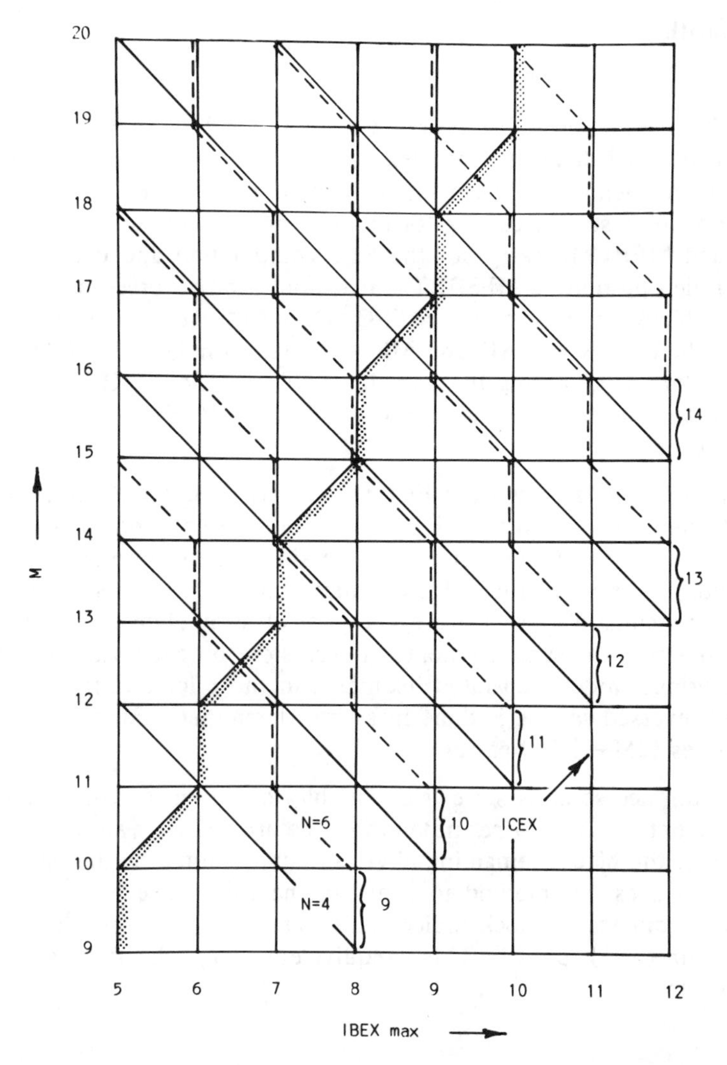

Fig. 2 Mass store I/O passes N related to binary exponents IBEX, ICEX and M, for routine CMFFT. Increase diagram values of M and N by one to obtain working M and N for routine RMFFT.

where TCA is the FFT computation overhead per block and TMA is the average mass store access time per block. TCO and TMO are the limits for large block size. TCA and TMA set a useful lower limit for block size. In the examples above TM is mainly the result of TMA, or system block access time. Increasing block size reduces TM but with a probable increase in N (depending on M and ICEX). TC is mainly due to TCO, TCA becoming important typically when IBEX < 5. To determine the most efficient parameter combination, begin by choosing ICEX as large as possible, use Fig. 2 as a guide in choosing IBEX but follow this up by timing comparisons over a range of IBEX.

The program was written to test the sorting algorithm [1] in practice. At the same time a useful and efficient Fourier program has been obtained, operating with a very general set of parameters. Some parameter combinations can lead to short cuts, not detected or implemented here (see section 5). Efficiency in a particular case can no doubt be improved by some recoding, at the expense of a loss in generality or increase in code complexity.

Hopefully, this program will stimulate designers not only of fast transforms but also of general purpose systems. Sorting by index bit manipulation should be considered an important concept for new machine architecture. Index bit-reversal is simple to achieve in hardware and, in its generalized form [1], is a symmetric permutation (allowing in-place operation) which forms the basis of a number of useful permutations, not exclusively associated with the FFT.

5. Discussion of Algorithm

5.1 General

The structure of the mass storage FFT program is shown in Fig. 1. When using real data, routine RMFFT calls CMFFT with a half length complex array before unscrambling by MFRCMP. If necessary, routines CMFFT and below may be overlaid with routines MFRCMP and MFRLOD. CMFFT controls the sequence of mass store accessing for in-core computation and sorting of the complex FFT. Routines MFCOMP and MFRCMP carry out the FFT computation and unscrambling of data in core store. MFREV, controlled by routine MFSORT, carries out a bit-reversed permutation on elements in core store or on whole blocks in mass store. MFLOAD and MFRLOD load and unload core store for the in-core routines, calling on MFREAD and MFWRIT (section 7). MFINDX defines a virtual permutation of the mass store array during the complex FFT operation. MFSUM sums elements in the dimensioning list MEXA( ), sorting them if necessary. Routine MFPAR is discussed in section 3.4 and routine DMPERM in section 3.6.

The main problem in writing a mass store FFT program is to devise an efficient means of accessing widely spaced data elements for use in the FFT computation kernel, and for sorting from bit-reversed order [3]. Singleton [8] describes an algorithm for accessing a mass store array two blocks at a time to carry out each computing pass of an FFT. He organizes mass store accesses in such a way that each I/O and computing pass results in a one place cyclic shift of array element index bits. This allows the FFT kernel program to access the same elements in core store on each pass, which not only simplifies the program but also brings 'widely separated' elements in the original array 'within reach' of the in-core computation. Bit-reversed sorting is done in a similar manner. For an array of 2**M elements, the basic method requires (2M−1) I/O passes.

With efficient, random access mass store it is possible to use an indexing algorithm to read widely scattered blocks, equivalent to a virtual permutation of the array. After an in-core operation the blocks are written back in-place, the blocks remaining physically unpermuted by the operation. In [1] an algorithm is described which uses this method to compute the FFT. The indexing subroutine MFINDX computes the required sequence of block indices using an algorithm which is a generalization of the Block Indexing Algorithm of [1], p. 303. This is equivalent to a cyclic shift (a number of places) of a set of the block index bits.

5.2 Complex Data

Unlike Singleton's algorithm, an attempt is made to load as many blocks together as possible in core store. In this way it is possible to include a number of FFT computing passes while the blocks are in core store, reducing the number of I/O passes by a factor ICEX−IBEX (see section 4). Routine CMFFT uses a slightly modified version of the algorithm of [1], p. 307 for this purpose, doing M−C instead of M−B initial computing passes with virtual permutations, followed by final C pass computations with direct access, instead of final B pass computations. I/O efficiency is unchanged, but the algorithm is neater (sizes being 2**B elements per block, 2**C in core store, 2**M total, all complex).

Routine MFCOMP computes the FFT in core store. But, unlike other FFT routines, the number of computing passes NPAS, and the effective initial pass number IPAS, are not fixed but are passed by the calling routine CMFFT. Weighting factors are computed using SIN, COS functions on mass store block boundaries but recursively within block boundaries, while array element accessing is ordered to minimize these computations. Multidimensional transforms are achieved by restarting the weighting factor sequence over passes corresponding to the exponent of each dimension, requiring no change to the order of access of elements.

Finally, routine MFSORT is called to sort the array from bit-reversed order. The algorithm is a slightly modified version of the sorting algorithm of [1], pp. 305-307, to allow the reversal of a more general set of index bits. Its structure is very similar to CMFFT, having a first stage using the virtual cyclic shift permutation for block access, with pairs of in-core bit-reversed permutations (MFREV) instead of FFT computations. (The pairs allow unsymmetric cyclic bit shift permutations to be done in-place, in-core). Only one in-core permutation is required on the first I/O pass. These are followed by a final bit-reversed permutation of whole blocks, if necessary.

In the special case when IDIR=0, MFSORT is called a number of times to individually reverse bits corresponding to each dimension exponent. This allows the complex FFT to be done without dimension reversal (but note that dimension reversal is required by RMFFT). The dimension shifting subroutine DMPERM, of section 3.6, also uses this feature. In both these cases, I/O is not fully optimized and specially written programs could reduce the number of I/O passes by combining some of the separate, in-core permutations.

Bit-reversed index pairs are generated by a very efficient recursion algorithm (routine MFREV). The algorithm maintains a hierarchy of 'reversed' integers of increasing number of bits, up to 2 less than the number being reversed. Incrementing a reversed integer then requires the alternate simple addition of a constant or replacement by the next lower incremented reversed integer in the hierarchy, recursively. For example, with a 3 bit reversal, the reversed set is (0,4,2,6,1,5,3,7) where the next value is obtained either by adding 4 to the last or by replacing it by one of the values (0,2,1,3) of a 2 bit reversed set.

This method of reversed series generation is in itself fast, as recursion depths are small on average. But, in addition, only quarter length series are generated (−2 bits) and the full length series is derived by scaling by 2 and adding offsets. This is equivalent to reversing an integer a(integer−2 bits)b to b(reversed integer−2 bits)a where a and b are the outer bits. There are four possible combinations for ab, but only those reversals greater numerically than before reversal are required (to prevent nullifying double swaps), leaving in general the three offsets 1...0, 0...0 and 1...1. In particular, only the first offset is required if the internal (reversed integer−2 bits) is smaller than or the same as before reversal. Thus, only valid swap index pairs are generated, saving the unnecessary reversed integer generation of some other methods.

5.3 Real Data

For transforms in which the initial or final data is real, a half length complex transform of packed real data must be unscrambled (or vice-versa) by routine MFRCMP. The method relies on complex-conjugate symmetry in the transform of real data [4]. Calculation of indices of multidimensional symmetry is more difficult than in the one-dimensional case. To do this a recursion algorithm is used, which is most easily described by the following FORTRAN program (for two dimensions, N1 by N2 elements):

```
          L2=N2/2 + 1
          DO 1 J2=1,L2
          K2=N2+2-J2
          IF(J2.EQ.1)K2=1
          L1=N1
          IF(J2.EQ.K2)L1=N1/2 + 1
          DO 1 J1=1,L1
          K1=N1+2-J1
          IF(J1.EQ.1)K1=1
C  (Now have A(K1,K2) and A(J1,J2) as a pair with conjugate symmetry)
1         CONTINUE
```

Note is taken of the special cases which exist when the J index is 1 (FORTRAN) and when the J and K indices are equal. The general case follows by repetition of the code between the two DO statements, replacing J2 by Jj etc. In routine MFRCMP arrays JAYA(4) etc. are used for this purpose. To increase the number of dimensions allowed in RMFFT (section 3.3), increase the array sizes. To help visualize the result, the index relationships (FORTRAN index −1) for an 8 by 4 array are as follows (r for real IF(index(K1,K2).EQ.index(J1,J2)), where index(J1,J2) = J1+(J2−1)*N1 for example, c for complex conjugate):

r00	01	02	03	r04	c03	c02	c01
10	11	12	13	14	15	16	17
r20	21	22	23	r24	c23	c22	c21
c10	c17	c16	c15	c14	c13	c12	c11

The array above represents the transformation and transposition of a 4 by 8 real array (imaginary part zero). However, we may also consider it to represent the scrambled transform of an array of initially packed real data, occupying alternately real and imaginary elements, that is initially 8 by 8 real values. In this case the result is only half length, element pairs such as 15 and c15 above containing information which can be unscrambled to give a new element 15 and an element 33 in place of c15. In addition, new elements c15 and c33 are derived. Thus the array is expanded to an 8 by 8 full transform of the real data:

r00	01	02	03	r04	c03	c02	c01
10	11	12	13	14	15	16	17
20	21	22	23	24	25	26	27
30	31	32	33	34	35	36	37
r40	41	42	43	r44	c43	c42	c41
c30	c37	c36	c35	c34	c33	c32	c31
c20	c27	c26	c25	c24	c23	c22	c21
c10	c17	c16	c15	c14	c13	c12	c11

Note that the same indexing algorithm applies in both examples (with different N2) but during unscrambling and expansion only the half-length symmetry is used ($N2=4$), corresponding expanded rows in the new array are obtained by a direct offset (4 in this case). Thus, by half-length symmetry, initial rows 1 and 3 are accessed together for unscrambling; these are replaced by unscrambled rows 1 and 3 of the full array while new rows 5 and 7 are also computed and added to the array. It is this ability to add data beyond the existing data which makes dimension reversal essential for in-place computation in the real mass store FFT routine RMFFT.

The complete array is 'fully redundant', nearly half the elements being complex conjugates of the other half. If the last three rows of the example are left out the result is 'partially redundant', since row 0 and row 4 still have some internal redundancy. To eliminate all redundancy, rows 0 and 4 can be merged (c43,c42,c41 replacing c03,c02,c01), and pairs of real elements combined as single complex elements (r40 as the imaginary part with r00 real, r44 with r04).

The exact algorithm by which an array, with IPAK=0 or −1, can be restored to full redundancy is given by the half-length symmetry program above, putting:

```
          A(J1,J2+N2) = CONJG( A(K1,K2))
and       A(K1,K2+N2) = CONJG( A(J1,J2)),     IF(J2.NE.1),
```

with special provision IF(J2.EQ.1) when IPAK=−1, below

```
          A(K1,K2+N2)  = A(K1,K2)
          A(J1,J2+N2)  = CONJG( A(K1,K2))
and       A(K1,K2)     = CONJG( A(J1,J2))
```

and IF(index(K1,K2).EQ.index(J1,J2)) when IPAK=−1, values are real, below

```
          A(J1,J2+N2)= CMPLX( AIMAG( A(J1,J2)), 0.)
and       A(J1,J2)   = CMPLX( REAL( A(J1,J2)), 0.)
```

the general case follows with Jj and Nj instead of J2 and N2 and as before with repetition of the code between DO loops.

6. Test Programs

A universal test program is provided which sets up, for a given M, exhaustive permutations and combinations of the different parameters IBEX, ICEX and NDIM from 1 to 3 dimensions. Mass store is simulated, through dummy routines MFREAD and MFWRIT, by a core store array so that the program is independent of system I/O.

The program stores a pseudo random number sequence in the simulated mass store, transforms this by the mass store FFT routines and compares the result with a discrete Fourier transform computed naively by routine NAIVE. An inverse mass store FFT is called and the result compared with the initial data. Maximum differences are noted and various levels of data and difference print-out are available through parameter IPRINT. The test is considered successful if maximum differences are explainable in terms of machine round-off errors (see Table 1).

Because the tests are exhaustive and because of the naive DFT computation the program is quite slow when the overall array size exponent M is other than very small. For example, in a CYBER 76 computer with M=5 the program takes 0.3 seconds while with M=10 the program takes about 800 seconds. Less exhaustive testing can be done by replacing program DO-loop variables by fixed variables (see program comments).

Example test programs are also included for accessing mass store in two specific computer systems. These are a CYBER 76 (true mass store and LCM 'static' store) and a PDP 11. Mass store I/O routines for other systems should be easily devised based on these examples and the comments of section 7.

7. System-Dependent I/O Routines

The system-dependent random access transfers between mass store and core store are handled by two subroutines called internally by the FFT routines, thus:

```
        CALL MFREAD (BUFA,NB,JB)    Read block into core store
and     CALL MFWRIT (BUFA,NB,JB)    Write block from core store
```

where BUFA is the core store address for the start of the transfer, NB is the number of real elements to be transferred and JB is the desired block or record number. The range of JB is:

for CMFFT or RMFFT if IPAK=+1

$1 \leqslant JB \leqslant$ 2**(M−IBEX+1),

for RMFFT if IPAK=0

$1 \leqslant JB \leqslant$ 2**(M−IBEX) + 2**(M−MEXA(NDIM)−IBEX+1), for MEXA(NDIM)+IBEX < M+1,

or $1 \leqslant JB \leqslant$ 2**(M−IBEX) + 1 for MEXA(NDIM)+IBEX $\geqslant$ M+1,
(the MEXA list in the order after the transform)

for RMFFT if IPAK=−1

$1 \leqslant JB \leqslant$ 2**(M−IBEX).

In addition, the I/O subroutines must know what file to access and any starting block offset, which may be given through COMMON variables, for example. The routines expect the mass store file to be defined and opened. The user program may, of course, call MFREAD and MFWRIT itself to load mass store with data or read the transform result.

For example, in PDP 11 FORTRAN a random access occurs with

```
        SUBROUTINE MFREAD (BUFA,NB,JB)
C READ BLOCK, INDEX JB, FROM MASS STORE TO BUFA, NB REAL VALUES
        REAL BUFA(NB)
        COMMON/FFTCOM/LUN
        READ (LUN'JB)(BUFA(I),I=1,NB)   or   READ (LUN'JB) BUFA
        RETURN
        END
```

and similarly for subroutine MFWRIT.

As discussed in sections 3.4 and 3.5 it may be useful to call MFREAD/MFWRIT with a 'record' length NB different from 2**IBEX. This can give a user more logical access to a multidimensional array, in 'records' of first dimension length, for example. Some systems allow a redefinition of the file structure, in which case the problem is solved. Others allow a 'word-addressable' file structure in which an I/O transfer may start at any word in the file. In this case, MFREAD/MFWRIT can include the following:

INDEX = (JB−1)*NB + 1

(Transfer between file real element indices (INDEX) and (INDEX+NB−1))

The dummy I/O routines of the universal test program of Section 6, and the CYBER Extended Core/LCM and PDP 11 Macro routines, include this feature.

Subroutines MFREAD and MFWRIT allow the user considerable scope for modifying the operation of the FFT routines. In the universal test driver program (section 6) the subroutines simply copy data from one core store array to another. In some systems external core store can be accessed efficiently in blocks, which allows this to be used as a fast, static mass store. Using different files on the first I/O pass for MFREAD and MFWRIT allows data to be copied automatically from a source file to a working and result file. In this case care must be taken to allow only a single read of each block of the source file, any subsequent read, even on the 'first' I/O pass, being from the working file (important during complex-to-real transformation by routine RMFFT).

References

1. Fraser, Donald, "Array permutation by index-digit permutation", *J. ACM,* Vol. 23, No. 2, pp. 298-309, April 1976.

2. Cooley, J. W. and Tukey, J. W., "An algorithm for the machine calculation of complex Fourier series", *Math of Comput.,* Vol. 19, pp. 297-301, April 1965.

3. Cochran, W. T. et al, "What is the fast Fourier transform?", *IEEE Trans. Audio and Electroacoust.,* Vol. AU-15, No. 2, pp. 45-55, June 1967.

4. Cooley, J. W. et al, "The fast Fourier transform algorithm; programming considerations in the calculation of sine, cosine and Laplace transforms", *J. Sound Vib.,* Vol. 12, No. 3, pp. 315-337, July 1970.

5. DeLotto, I. and Dotti, D., "A new procedure for optimum mass storage use in FT algorithms", *Alta Frequenza,* Vol. XLII, No. 8, pp. 379-384, August 1973.

6. DeLotto, I. and Dotti, D., "Two-dimensional transforms by minicomputer without matrix transposing", *Computer Graphics and Image Processing,* Vol. 4, pp. 271-278, October 1975.

7. Fraser, Donald, "An optimized mass storage FFT", *ACM Trans. on Mathematical Software,* to be published.

8. Singleton, Richard C., "A method for computing the fast Fourier transform with auxiliary memory and limited high-speed storage", *IEEE Trans. Audio and Electroacoust.,* Vol. AU-15, No. 2, pp. 91-98, June 1967.

9. Brenner, N. M., "FOR2D", Program No. 360D-13.4.007, SHARE Program Library Agency, 1969.

10. Eklundh, J. O., "A fast computer method for matrix transposing", *IEEE Trans. Computers,* Vol. C-21, No. 7, pp. 801-803, July 1972.

Table 1

```
MASS STORE FFT TEST
IPRINT =  1 (1=COPIOUS,0=MAX DIFFS,-1=OVERALL)
  IRMF = -1 (0=CMFFT TEST,-1=RMFFT TEST)
 NDIM =  1  ISGN = -1  IDIR = -1  IBEX =  2  ICEX =  4  IPAK =  1
 MEXA() =     5
  INDEX          FFT                        NAIVE                      DIFF
    0   0.160E 02  0.            0.160E 02  0.             0.         0.
    1  -0.132E 01 -0.638E 00   -0.132E 01 -0.638E 00   -0.10E-06  0.75E-07
    2  -0.971E 00 -0.417E 00   -0.971E 00 -0.417E 00    0.13E-06  0.86E-07
    3  -0.146E 00 -0.306E 00   -0.146E 00 -0.306E 00   -0.13E-07  0.34E-07
    4  -0.187E 00 -0.287E 00   -0.187E 00 -0.287E 00    0.37E-07  0.22E-07
    5   0.141E 01 -0.123E 01    0.141E 01 -0.123E 01    0.15E-07 -0.15E-07
    6  -0.997E 00  0.296E 00   -0.997E 00  0.296E 00   -0.22E-07 -0.67E-07
    7  -0.319E 00 -0.164E 01   -0.319E 00 -0.164E 01    0.18E-06 -0.36E-06
    8  -0.281E 00 -0.134E 01   -0.281E 00 -0.134E 01   -0.17E-06  0.30E-07
    9   0.181E 01  0.166E 01    0.181E 01  0.166E 01   -0.22E-06 -0.15E-07
   10  -0.111E 01 -0.104E 01   -0.111E 01 -0.104E 01   -0.75E-07 -0.36E-06
   11  -0.176E 01 -0.452E 00   -0.176E 01 -0.452E 00   -0.30E-07  0.22E-06
   12   0.874E 00 -0.210E 01    0.874E 00 -0.210E 01    0.13E-06  0.30E-07
   13   0.875E 00 -0.220E 01    0.875E 00 -0.220E 01   -0.89E-07 -0.18E-06
   14  -0.255E 01 -0.138E 01   -0.255E 01 -0.138E 01    0.15E-06  0.42E-06
   15   0.204E 00  0.738E 00    0.204E 00  0.738E 00    0.16E-06 -0.47E-06
   16  -0.844E 00  0.          -0.844E 00 -0.330E-07    0.         0.33E-07
   17   0.204E 00 -0.738E 00    0.204E 00 -0.738E 00    0.16E-06  0.22E-06
   18  -0.255E 01  0.138E 01   -0.255E 01  0.138E 01   -0.30E-07 -0.63E-06
   19   0.875E 00  0.220E 01    0.875E 00  0.220E 01   -0.27E-06 -0.60E-07
   20   0.874E 00  0.210E 01    0.874E 00  0.210E 01   -0.80E-06  0.60E-06
   21  -0.176E 01  0.452E 00   -0.176E 01  0.452E 00    0.55E-06  0.46E-06
   22  -0.111E 01  0.104E 01   -0.111E 01  0.104E 01    0.63E-06  0.39E-06
   23   0.181E 01 -0.166E 01    0.181E 01 -0.166E 01    0.60E-07 -0.31E-06
   24  -0.281E 00  0.134E 01   -0.281E 00  0.134E 01   -0.33E-06  0.30E-07
   25  -0.319E 00  0.164E 01   -0.319E 00  0.164E 01   -0.17E-05  0.83E-06
   26  -0.997E 00 -0.296E 00   -0.997E 00 -0.296E 00   -0.89E-07 -0.55E-06
   27   0.141E 01  0.123E 01    0.141E 01  0.123E 01    0.30E-06  0.28E-06
   28  -0.187E 00  0.287E 00   -0.187E 00  0.287E 00    0.11E-05  0.12E-05
   29  -0.146E 00  0.306E 00   -0.146E 00  0.306E 00   -0.75E-06 -0.21E-06
   30  -0.971E 00  0.417E 00   -0.971E 00  0.417E 00   -0.20E-05 -0.53E-06
   31  -0.132E 01  0.638E 00   -0.132E 01  0.638E 00   -0.72E-06  0.37E-07
MAX DIFF  0.1989E-05
 NDIM =  1  ISGN = -1  IDIR = -1  IBEX =  2  ICEX =  4  IPAK =  1
 MEXA() =     5

MASS STORE FFT TEST
IPRINT =  1 (1=COPIOUS,0=MAX DIFFS,-1=OVERALL)
  IRMF = -1 (0=CMFFT TEST,-1=RMFFT TEST)
 NDIM =  1  ISGN =  1  IDIR =  1  IBEX =  2  ICEX =  4  IPAK =  1
 MEXA() =     5
  INDEX  FFT/2**M     INPUT       DIFF
    0   0.194E 00  0.194E 00   0.373E-08
    1   0.958E 00  0.958E 00   0.745E-08
    2   0.135E 00  0.135E 00   0.745E-08
    3   0.696E 00  0.696E 00   0.149E-07
    4   0.233E 00  0.233E 00   0.186E-08
    5   0.465E 00  0.465E 00   0.745E-08
    6   0.737E 00  0.737E 00   0.
    7   0.516E 00  0.516E 00   0.149E-07
    8   0.897E 00  0.897E 00   0.745E-08
    9   0.981E-01  0.981E-01   0.373E-08
   10   0.963E 00  0.963E 00   0.149E-07
   11   0.461E 00  0.461E 00   0.
   12   0.186E 00  0.186E 00   0.745E-08
   13   0.856E 00  0.856E 00   0.745E-08
   14   0.815E 00  0.815E 00   0.745E-08
   15   0.532E 00  0.532E 00   0.149E-07
   16   0.100E 00  0.100E 00   0.373E-08
   17   0.739E 00  0.739E 00   0.745E-08
   18   0.292E 00  0.292E 00   0.
   19   0.727E 00  0.727E 00   0.149E-07
   20   0.639E 00  0.639E 00   0.
   21   0.747E 00  0.747E 00   0.745E-08
   22   0.393E 00  0.393E 00   0.
   23   0.474E-01  0.474E-01   0.373E-08
```

Table 1
(Continued)

```
    24   0.803E 00  0.803E 00   0.745E-08
    25   0.879E 00  0.879E 00   0.
    26   0.120E 00  0.120E 00   0.
    27   0.493E 00  0.493E 00   0.745E-08
    28   0.592E 00  0.592E 00   0.745E-08
    29   0.137E 00  0.137E 00   0.745E-08
    30   0.471E 00  0.471E 00   0.745E-08
    31   0.630E-01  0.630E-01   0.745E-08
MAX DIFF  0.1490E-07
 NDIM =  1  ISGN =  1  IDIR =  1  IBEX =  2  ICEX =  4  IPAK =  1
 MEXA() =    5
       .
       .
       .
```

Appendix

```
C
C-----------------------------------------------------------------------
C  PROGRAM:  AN OPTIMIZED MASS STORAGE FFT
C  AUTHOR:   DONALD FRASER
C            CSIRO, DIVISION OF COMPUTING RESEARCH
C            PO BOX 1800
C            CANBERRA CITY, ACT 2601, AUSTRALIA
C
C          REVISION DATE: JULY 1978.
C
C     THE SET INCLUDES A UNIVERSAL TEST PROGRAM, WHICH SIMULATES
C  MASS STORE THROUGH A FORTRAN ARRAY, SAMPLE PROGRAMS AND I/O
C  SUBROUTINES FOR CONTROL DATA 6000 AND CYBER COMPUTERS AND FOR
C  DEC PDP 11 MINICOMPUTERS.
C
C     I/O SUBROUTINES FOR OTHER SYSTEMS MAY BE EASILY CONSTRUCTED
C  FROM THE EXAMPLES AND WITH REFERENCE TO THE FORMAL PAPER.
C  BUT CARE SHOULD BE TAKEN WITH SOME FUSSY COMPILERS SINCE FFT
C  SUBROUTINES ASSIGN EITHER TYPE REAL OR TYPE COMPLEX TO THE SAME
C  ARRAY AS IT IS PASSED AS A FORMAL PARAMETER.  NOTE THAT COMPLEX
C  DATA IS ASSUMED TO BE STORED  REAL/IMAG/REAL/IMAG... IN BOTH
C  MASS STORE AND CORE STORE.
C
C     THE PROGRAM UNITS APPEAR IN THE FOLLOWING ORDER:
C
C         FIRST, THE FFT SUBROUTINE SET:
C  1  RMFFT        OPTIMIZED MASS STORAGE FFT (REAL DATA OR RESULT)
C  2  CMFFT        CALLED BY 1, OR MASS STORAGE FFT (ALL COMPLEX)
C  3  MFCOMP       IN-CORE FFT
C  4  MFSORT       MASS STORE SORTING
C  5  MFREV        IN-CORE SORTING OR WHOLE BLOCK SORTING
C  6  MFLOAD       LOADING/UNLOADING CORE STORE
C  7  MFINDX       BLOCK INDEXING ALGORITHM (VIRTUAL PERMUTATION)
C  8  MFSUM        MEXA() EXPONENT SUMMATIONS
C  9  MFRCMP       REAL-COMPLEX UNSCRAMBLING/SCRAMBLING
C 10  MFRLOD       LOADING/UNLOADING CORE STORE FOR MFRCMP
C 11  MFPAR        HELPER ROUTINE (NOT ESSENTIAL, BUT RECOMMENDED)
C 12  DMPERM       MASS STORE DIMENSION SHIFTING (BONUS SUBROUTINE)
C
C          THEN, TEST PROGRAMS AND SAMPLE I/O SUBROUTINES
C
C 13  UNIVERSAL TEST PROGRAM (SIMULATED MASS STORE, NEEDS 14 TO 17 ALSO)
C 14  RANMF        RANDOM NUMBER GENERATOR
C 15  NAIVE        DISCRETE FOURIER TRANSFORM COMPUTED NAIVELY
C 16  MFREAD       DUMMY I/O ROUTINE USING SIMULATED MASS STORE
C 17  MFWRIT          AS ABOVE
C
C 18  CYBER MASS STORE SAMPLE PROGRAM
C 19  MFREAD/MFWRIT  CYBER MASS STORE I/O ROUTINE
C
C 20  CYBER EXTENDED CORE/LCM SAMPLE PROGRAM
C 21  MFREAD/MFWRIT  CYBER EXTENDED CORE/LCM I/O ROUTINES
C
C 22  PDP 11 MASS STORE SAMPLE PROGRAM
C 23  MFREAD       PDP 11 STANDARD FORTRAN MASS STORE I/O ROUTINES
C 24  MFWRIT          AS ABOVE
C
C 25  PDP 11 FAST MACRO I/O SAMPLE PROGRAM (NEEDS MACRO OPEN SUBRTN)
C 26  MFREAD/MFWRIT  PDP 11 FAST MACRO I/O ROUTINE FOR RSX11M/RT11.
C-----------------------------------------------------------------------
C
      STOP
      END
C
C-----------------------------------------------------------------------
```

```
C SUBROUTINE:  RMFFT
C REAL-TO-COMPLEX FFT (OR VICE-VERSA) OF MULTI-DIMENSD MASS STORE ARRAY
C (FRASER, ACM TOMS - 1979, AND J.ACM, V.23,N.2, APRIL 76, PP. 298-309)
C MASS STORE ARRAY IS EITHER REAL OR COMPLEX DATA (SEE NOTE BELOW)
C-----------------------------------------------------------------------
C
      SUBROUTINE RMFFT(MEXA, NDIM, ISGN, IDIR, SCAL, BUFA, IBEX, ICEX,
     *    IPAK)
C
C NOTE WELL THAT TYPE COMPLEX DATA MUST EXIST AS ALTERNATING
C REAL/IMAG/REAL/IMAG... ELEMENTS, BOTH IN MASS STORE AND IN FORTRAN
C WORKING ARRAY BUFA;  IN THIS FFT, DIFFERENT SUBROUTINES WILL SET
C DIFFERENT TYPE (REAL OR COMPLEX) FOR ARRAY BUFA.
C
C MEXA(J) LIST OF DIMENSION SIZE EXPONS (BASE 2), ADJACENT FIRST
C NDIM IS NUMBER OF EXPONENTS IN LIST AND THUS THE NUMBER OF DIMENSIONS
C  SUM TO NDIM OF MEXA(J) = M, WHERE 2**M IS EFFECT SIZE OF TRANS.
C     THUS, 2**M PACKED REAL VALUES,
C     OR,   2**M COMPLEX VALUES, IF COMPLEX RESULT WITH IPAK=1
C RMFFT ALWAYS REVERSES DIMENSION ORDER AND MEXA LIST
C
C ISGN GIVES SIGN OF COMPLEX EXPONENT OF TRANSFORM (+ OR -), AND
C IDIR DETERMINES DIRECTION OF TRANSFORM, THUS:
C     IDIR=-1,  REAL-TO-COMPLEX
C     IDIR=+1,  COMPLEX-TO-REAL
C SCAL IS REAL MULTIPLIER OF RESULT (EG. SET SCAL=1. FWD, 1./2**M INV)
C
C BUFA IS CORE STORE WORKING ARRAY BASE ADDRESS (SEE NOTE ABOVE)
C IBEX, ICEX ARE BLOCK AND CORE SIZE EXPONENTS, THUS
C  2**IBEX IS NUMBER OF REAL ELEMENTS IN MASS STORE BLOCK
C  2**ICEX IS NUMBER OF REAL ELEMENTS IN CORE STORE BUFA
C
C IPAK IS ARRAY PACKING DETERMINATOR, THUS:
C  IPAK=+1 EXPANDS COMPLEX ARRAY TO FULL REDUNDANCY (SAME AS CMFFT)
C  IPAK=0  COMPUTES COMPLEX ARRAY OF JUST OVER HALF SIZE
C  IPAK=-1 HOLDS COMPLEX ARRAY AT EXACTLY HALF SIZE (2**(M-1) CMPLX
C
C MASS STORE ARRAY MUST BE OPEN FOR ACCESS BY SUBRTNS MFREAD/MFWRIT
C  EG. SUBRTN MFREAD(BUFA,NB,JB) AND MFWRIT(BUFA,NB,JB) TRANSFER ONE
C  BLOCK, INDEX JB, BETWEEN MASS STORE AND CORE STORE BUFA, NB REALS
C  (1.LE.JB.LE.2**(M-IBEX) IF REAL, OR 2**(M-IBEX+1) IF IPAK=1)
C
      COMPLEX BUFA(1)
      INTEGER B, MEXA(1)
C
      MH = MFSUM(MEXA,NDIM,99) - 1
      B = IBEX - 1
      IF (IDIR.GT.0) GO TO 10
C
C BELOW, REAL-TO-COMPLEX TRANSFORM
C
      MEXA(1) = MEXA(1) - 1
      CALL CMFFT(MEXA, NDIM, ISGN, IDIR, SCAL, BUFA, IBEX, ICEX)
      CALL MFRCMP(MEXA, NDIM, ISGN, IDIR, IPAK, BUFA, B, MH)
      MEXA(NDIM) = MEXA(NDIM) + 1
      RETURN
C
C BELOW, COMPLEX-TO-REAL TRANSFORM
C
   10 MEXA(NDIM) = MEXA(NDIM) - 1
      CALL MFRCMP(MEXA, NDIM, ISGN, IDIR, IPAK, BUFA, B, MH)
      CALL CMFFT(MEXA, NDIM, ISGN, IDIR, SCAL, BUFA, IBEX, ICEX)
      MEXA(1) = MEXA(1) + 1
      RETURN
      END
```

```
C
C-----------------------------------------------------------------------
C SUBROUTINE: CMFFT
C COMPLEX FFT OF MULTI-DIMENSD MASS STORE ARRAY, CALLED BY USER OR RMFFT
C FOR COMMENTS, SEE SUBRTN RMFFT; ARGUMENTS HAVE SAME MEANING EXCEPT
C FOR IDIR=+1 OR -1, ARRAY ALWAYS COMPLEX, DIMENSION ORDER REVERSED
C WHILE IDIR=0, DIMENSION ORDER (AND MEXA LIST) ARE NOT REVERSED
C-----------------------------------------------------------------------
C
      SUBROUTINE CMFFT(MEXA, NDIM, ISGN, IDIR, SCAL, BUFA, IBEX, ICEX)
C
      COMPLEX BUFA(1)
      INTEGER B, C, MEXA(1)
      DATA LSET /1/, LREAD /1/, LWRIT /2/
C
      M = MFSUM(MEXA,NDIM,99)
      MREAL = M + 1
      B = IBEX - 1
      C = ICEX - 1
      NC = 2**C
      IPAS = 0
      MPAS = M - C
      NPAS = C - B
C
C MOST EFFICIENT USE OF CORE STORE - TRIES TO DO C-B PASSES PER LOAD
C
      IDUM = MFINDX(LSET,B,M,M,NPAS)
C
C DUMMY CALL TO MFINDX TO SPECIFY VIRTUAL (B.'S'.M)**(C-B) PERMUTATION
C
C FIRST, PIECE-MEAL ATTACK ON FFT COMPUTATION FOLLOWS
C
  10  IF (MPAS.LE.0) GO TO 40
      IF (MPAS.LT.NPAS) NPAS = MPAS
  20  CALL MFLOAD(LREAD, BUFA, IBEX, ICEX, IFLG)
C
C LOAD CORE WORKING SPACE WITH 2**(C-B) BLOCKS ACCORDING TO MFINDX
C
      IF (IFLG.LT.0) GO TO 30
      CALL MFCOMP(MEXA, NDIM, ISGN, BUFA, B, C, M, NPAS, IPAS)
C
C DO MODIFIED IN-CORE FFT, REQUIRING NPAS PASSES STARTING WITH IPAS
C
      CALL MFLOAD(LWRIT, BUFA, IBEX, ICEX, IFLG)
C
C UNLOAD CORE AREA, WRITING BLOCKS BACK IN-PLACE TO MASS STORE
C
      GO TO 20
  30  IPAS = IPAS + NPAS
      MPAS = MPAS - NPAS
      GO TO 10
C
C END OF FIRST PART
C
C SPECIFY BLOCKS TO BE READ IN NEXT PART IN TRUE ORDER (NO PERM)
C
  40  IDUM = MFINDX(LSET,B,M,M,0)
C
C FINAL, CONCLUDING ATTACK ON FFT COMPUTATION FOLLOWS
C
  50  CALL MFLOAD(LREAD, BUFA, IBEX, ICEX, IFLG)
      IF (IFLG.LT.0) GO TO 80
      CALL MFCOMP(MEXA, NDIM, ISGN, BUFA, C, C, M, C, IPAS)
C
C DO FINAL, C-PASS IN-CORE FFT OF EACH CORE-LOAD
C
      IF (SCAL.EQ.1.) GO TO 70
      DO 60 J=1,NC
        BUFA(J) = BUFA(J)*SCAL
  60  CONTINUE
  70  CALL MFLOAD(LWRIT, BUFA, IBEX, ICEX, IFLG)
      GO TO 50
C
C BELOW, SORT ARRAY (FULL BIT-REVERSAL AND DIMEN REVERSAL IF IDIR.NE.0)
C
  80  IF (IDIR.EQ.0) GO TO 90
      CALL MFSORT(BUFA, IBEX, ICEX, 1, MREAL, MREAL)
      M = MFSUM(MEXA,NDIM,-1)
C
C DO FULL BIT-REVERSAL OF M BITS (AND REVERSE MEXA LIST)
C
      RETURN
C
C BELOW, REVERSE BITS OF EACH DIMEN SEPARATELY (NO DIMEN REVERSAL)
C
  90  IH = 1
      DO 100 J=1,NDIM
        IG = IH
        IH = IH + MEXA(J)
        CALL MFSORT(BUFA, IBEX, ICEX, IG, IH, MREAL)
 100  CONTINUE
      RETURN
      END
C
C-----------------------------------------------------------------------
C SUBROUTINE:  MFCOMP
C MODIFIED, IN-CORE FFT OF 2**C ELEMENTS, NPAS PASSES STARTING WITH IPAS
C MEXA, NDIM, ISGN ,BUFA AND M HAVE SAME MEANING AS IN RMFFT COMMENTS
C B,C EQUIVALENT TO IBEX,ICEX EXCEPT HERE REFER TO NUM COMPLEX ELMTS
C (2**C CMPLX ELMTS IN CORE STORE BUFA, IN BLOCKS OF 2**B CMPLX ELMTS)
C-----------------------------------------------------------------------
C
      SUBROUTINE MFCOMP(MEXA, NDIM, ISGN, BUFA, B, C, M, NPAS, IPAS)
C
C FFT W PHASE FACTOR COMPUTED RECURSIVELY EXCEPT ON BLOCK BOUNDS
C MULTIDIMEN FFT ACHIEVED BY REPEATING W SEQUENCES
C
      INTEGER B, C, SPAN, STEP, MEXA(1)
      COMPLEX TEMP, W, D, BUFA(1)
      DATA LINDX /0/, LREST /4/
C
      PI = 4.0D0 * DATAN(1.0D0)
      PIMOD = PI*2.0**(1-M)
      IF (ISGN.LT.0) PIMOD = -PIMOD
      NC = 2**C
C
C BELOW, NPAS COMPUTATION PASSES WHILE DATA IN-CORE
C KPAS IS GLOBAL COMPUTING PASS NUMBER (JPAS-1 IS LOCAL)
C KPEFF IS EFFECTIVE GLOBAL PASS FOR MULTIDIMEN. FFT W GENERATION
C D IS USED FOR RECURSIVE MODIFICATION OF W PHASE FACTOR
C SPAN SEPARATES VALUES IN FFT KERNEL, STEP TO NEXT PAIR, SAME W
C NRPT COUNTS REPETITION OF W FOR MULTIDIMEN. FFT
C IMOD AND NCLR ARE USED TO COMPUTE W WITHOUT EXCEEDING SMALL INTEGER
C
      DO 80 JPAS=1,NPAS
        KPAS = IPAS + JPAS - 1
        KDIFF = M - MFSUM(MEXA,NDIM,KPAS)
        KPEFF = KPAS + KDIFF
        D = CEXP(CMPLX(0.,PIMOD*2.0**KPEFF))
        ITEM = C - JPAS
```

```
      SPAN = 2**ITEM
      STEP = 2*SPAN
      IF (B.LT.ITEM) ITEM = B
      NB = 2**ITEM
      IF (ITEM.GT.KDIFF) ITEM = KDIFF
      NRPT = 2**ITEM
      IMOD = 2**(M-B-KPAS-1)
      IF (IMOD.LE.1) GO TO 20
      IDUM = MFINDX(LREST,0,0,0,0)
      MEXP = KPEFF
      ITEM = KDIFF - B
      IF (ITEM.GT.0) GO TO 10
      MEXP = B + KPAS
      ITEM = 0
10    NCLR = 2**ITEM
      PIMOD2 = PIMOD*2.0**MEXP
C
C BELOW, START OF ONE PASS THROUGH CORE, NOTING BLOCK BOUNDARIES
C
20    DO 70 I1=1,SPAN,NB
        W = (1.,0.)
        IF (IMOD.LE.1) GO TO 30
        INDWM = MOD(MFINDX(LINDX,0,0,0,0)-1,IMOD)/NCLR
        ANDWM = INDWM
        W = CEXP(CMPLX(0.,PIMOD2*ANDWM))
C
C NEW W COMPUTED DIRECTLY AT BEGINNING OF NEW BLOCK AREA
C
C BELOW, COMPUTATIONS WITHIN EACH BLOCK OF 2**B CMPLX ELMTS
C
30      DO 60 I2=1,NB,NRPT
C
C BELOW, REPETITION OF SAME W DUE TO MULTIDIMEN FFT
C
          DO 50 I3=1,NRPT
            I4 = I1 + I2 + I3 - 2
C
C BELOW, STEPPING THOUGH INDICES HAVING SAME W IN ONE DIMEN FFT
C
            DO 40 J=I4,NC,STEP
              K = J + SPAN
              TEMP = (BUFA(J)-BUFA(K))*W
              BUFA(J) = BUFA(J) + BUFA(K)
              BUFA(K) = TEMP
40          CONTINUE
50        CONTINUE
C
C FFT 2-POINT KERNEL ARITHMETIC (ALGORITHM BIT-REVERSAL FOLLOWS COMPUT)
C
          W = W*D
C
C RECURSIVE MODIFICATION OF W WITHIN BLOCK BOUNDARIES
C
60      CONTINUE
70    CONTINUE
80    CONTINUE
      RETURN
      END
C
C-----------------------------------------------------------------------
C SUBROUTINE:  MFSORT
C BIT-REVERSED PERMUTATION (IG.'R'.IH) OF MASS STORE REAL ARRAY
C REVERSES IH-IG BITS IN INDEX (M-1,...,IH-1,...,IG,...,0)
C NOTE THAT THIS IS MORE GENERAL THAN THE FULL M-BIT REVERSAL
C OF REFERENCE (FRASER, J.ACM, V.23, N.2, APR. 76, P. 306),
C BUT ALGORITHM IS LOGICALLY THE SAME, WITH ALTERED BIT LIMITS.
C BUFA,IBEX,ICEX AND M HAVE SAME MEANING AS IN COMMENTS IN RMFFT
C (BLOCKS 2**IBEX, CORE BUFA 2**ICEX, TOTAL 2**M, ALL REAL)
C-----------------------------------------------------------------------
C
      SUBROUTINE MFSORT(BUFA, IBEX, ICEX, IG, IH, M)
C
      REAL BUFA(1)
      DATA LSET /1/, LPERM /2/, LREAD /1/, LWRIT /2/
C
      IDUM = MFINDX(LSET,IBEX,M,M,0)
C
C DUMMY CALL TO INITIALISE MFINDX (INITIALLY UNPERMUTED ARRAY)
C
      IF (IH-IG.LE.1) RETURN
      IF (IG.GE.IBEX) GO TO 50
      IF (IH.LE.ICEX) GO TO 60
C
C CHECK FOR SPECIAL CASES, REQUIRING SIMPLER TREATMENT
C
C BELOW, MIXED PERMUTATION OF BOTH ELEMENTS AND BLOCKS
C MOST EFFICIENT USE OF CORE STORE - TRIES TO DO ICEX-IBEX PASSES
C PER LOAD
C
      IPAS = 0
      NPAS = ICEX - IBEX
      MPAS = IH - IBEX
      IF (IBEX-IG.LT.MPAS) MPAS = IBEX - IG
C
C BELOW, FIRST VIRTUAL 'S' PERMUTATIONS
C
10    IF (MPAS.LE.0) GO TO 40
      IF (MPAS.LT.NPAS) NPAS = MPAS
      IGCOR = IG + IPAS
      IF ((IGCOR.GT.IBEX-1) .AND. (IGCOR.GT.IBEX+NPAS-1)) GO TO 30
      IF ((IPAS.EQ.0) .AND. (IGCOR.GT.IBEX+NPAS-1)) GO TO 30
C
C BYPASS UNNECESSARY CORE LOAD IF TRIVIAL CASES
C
      IDUM = MFINDX(LPERM,IBEX,IH,M,NPAS)
C
C DUMMY CALL TO MFINDX TO SPECIFY VIRTUAL (IBEX.S.IH)**NPAS PERM
C
C BELOW, LOAD CORE ACCORDING TO VIRTUAL PERMUTATION AND PERM ELMTS
C
20    CALL MFLOAD(LREAD, BUFA, IBEX, ICEX, IFLG)
      IF (IFLG.LT.0) GO TO 30
      IF (IPAS.NE.0) CALL MFREV(BUFA, IGCOR, IBEX, ICEX, -1)
      CALL MFREV(BUFA, IGCOR, IBEX+NPAS, ICEX, -1)
C
C CARRY OUT IN-CORE, SYMMETRIC R PERMS. (ONE ONLY ON FIRST PASS)
C
      CALL MFLOAD(LWRIT, BUFA, IBEX, ICEX, IFLG)
C
C UNLOAD CORE AREA, WRITING BLOCKS BACK IN-PLACE TO MASS STORE
C
      GO TO 20
C
30    IPAS = IPAS + NPAS
      MPAS = MPAS - NPAS
      GO TO 10
C
C END OF FIRST PART
C
C BELOW, FINAL 'R' PERM. OF BLOCKS IN MASS STORE, IF (IH-2*IBEX+IG).GT.1
```

```
C
  40  CALL MFREV(BUFA, 0, IH-2*IBEX+IG, M-IBEX, IBEX)
      RETURN
C
C BELOW, PERMUTATION OF BLOCKS ONLY REQUIRED
C
  50  CALL MFREV(BUFA, IG-IBEX, IH-IBEX, M-IBEX, IBEX)
      RETURN
C
C BELOW, PERMUTATION OF ELEMENTS IN CORE ONLY REQUIRED
C
  60  CALL MFLOAD(LREAD, BUFA, IBEX, ICEX, IFLG)
      IF (IFLG.LT.0) RETURN
      CALL MFREV(BUFA, IG, IH, ICEX, -1)
      CALL MFLOAD(LWRIT, BUFA, IBEX, ICEX, IFLG)
      GO TO 60
      END
C
C-----------------------------------------------------------------------
C SUBROUTINE: MFREV
C BIT-REVERSED PERMUTATION OF RANDOMLY ADDRESSABLE ELEMENTS
C REVERSES IH-IG BITS IN INDEX (M-1,...,IH-1,...,IG,...,0)
C (GENERAL PERM IG.'R'.IH, FRASER, J.ACM, V.23, N.2, APR 1976, P. 300)
C IF IBEX.LT.0, SORTS 2**M REAL ELMTS IN CORE BUFA,
C IF IBEX.GE.0, SORTS BLOCKS IN MASS STORE
C (2**IBEX REAL ELMTS PER BLOCK AND 2**M BLOCKS IN SECOND CASE)
C-----------------------------------------------------------------------
C
      SUBROUTINE MFREV(BUFA, IG, IH, M, IBEX)
C
C THE ALGORITHM MAINTAINS A SET OF 'REVERSED' INTEGERS IN ARRAY IRA()
C OF INCREASING NUMBER OF BITS, UP TO 2 LESS THAN (IH-IG) BITS.
C INCREMENTING A REVERSED INTEGER THEN REQUIRES THE ALTERNATE
C ADDITION OF NRA() TO IRA(), OR REPLACEMENT BY THE NEXT LOWER
C INCREMENTED REVERSED INTEGER IN THE HIERARCHY, RECURSIVELY.
C THIS IN ITSELF IS FAST, AS RECURSION DEPTHS ARE ON AVERAGE SMALL.
C BUT, IN ADDITION, ONLY QUARTER LENGTH SERIES ARE GENERATED (-2 BITS)
C AND THE FULL LENGTH DERIVED BY SCALING BY 2 AND ADDING OFFSETS.
C IN THIS FINAL STAGE, ONLY VALID SWAP PAIRS ARE GENERATED (1 OR 3 EACH
C
C WITHIN THE INNER LOOPS, GROUPS OF 2**IG ELMTS ARE MOVED TOGETHER
C WHILE THIS IS REPEATED OVER 2**(M-IH) PARTS OF THE ARRAY,
C CORRESPONDING TO THE UNPERMUTED BITS M TO IH AND IG-1 TO 0.
C
      REAL BUFA(1), TEMP
      INTEGER IRA(16), NRA(16), IFOFA(3), IROFA(3)
C
      IHG = IH - IG - 3
      IF (IHG.LE.(-2)) RETURN
C
C NO PERMUTATION REQUIRED
C
      NB = 2**IBEX
      NB1 = NB + 1
      NG = 2**IG
      NGDB = NG*2
      NH = 2**IH
      NHHF = NH/2
C
C NG IS MOVEMENT GROUP SIZE, NH IS PERMUTATION REPLICATION SIZE
C
      NM = 2**M
      NPARS = NM - NH + 1
      NREV = NH/4
C
      DO 10 J=1,IHG
        IRA(J) = 0
        NREV = NREV/2
        NRA(J) = NREV
  10  CONTINUE
C
C REVERSED INTEGER RECURSION SETS INITIALISED
C
      NREV = NH/4
      IFOFA(1) = NG - 1
      IROFA(1) = NH/2 - 1
      IFOFA(2) = -1
      IROFA(2) = -1
      IFOFA(3) = NH/2 + NG - 1
      IROFA(3) = NH/2 + NG - 1
C
C THREE PAIRS OF OFFSETS TO CONVERT QUARTER TO FULL LENGTH SERIES
C
      IFOR = 0
      IREV = 0
C
C BELOW, GENERATE INDEX PAIRS AND SWAP (IREV IS 'TOP' OF IRA() SET)
C
  20  NOF = 3
      IF (IFOR.GE.IREV) NOF = 1
C
C SELECTS ONCE-ONLY SWAP PAIRS (EITHER 1 OR 3 PAIRS)
C
      DO 60 JOF=1,NOF
        IFOF = IFOFA(JOF)
        IROF = IROFA(JOF)
        DO 50 I1=1,NG
C
C REPETITION OVER GROUP OR SUPER ELEMENT OF NG ACTUAL ELEMENTS
C
          IN2F = IFOR + IFOF + I1
          IN2R = IREV + IROF + I1
          DO 40 I2=1,NPARS,NH
C
C REPETITION OF SAME PERMUTATION OVER ARRAY PARTS
C
            IN3F = IN2F + I2
            IN3R = IN2R + I2
            IF (IBEX.GE.0) GO TO 30
C
C BELOW, IN-CORE ELEMENT SORTING
C
            TEMP = BUFA(IN3R)
            BUFA(IN3R) = BUFA(IN3F)
            BUFA(IN3F) = TEMP
            GO TO 40
C
C BELOW, SORTING WHOLE BLOCKS IN MASS STORE
C
  30        CALL MFREAD(BUFA, NB, IN3F)
            CALL MFREAD(BUFA(NB1), NB, IN3R)
            CALL MFWRIT(BUFA, NB, IN3R)
            CALL MFWRIT(BUFA(NB1), NB, IN3F)
C
  40      CONTINUE
  50    CONTINUE
  60  CONTINUE
C
C END OF INNER, REPETITION LOOPS
C
```

```
      IFOR = IFOR + NGDB
C
C INCREMENT FORWARD QUARTER-LENGTH INTEGER (ALREADY SCALED BY NG*2)
C
      IF (IFOR.GE.NHHF) RETURN
      IF (IREV.GE.NREV) GO TO 70
C
C TEST FOR ALTERNATE METHODS OF REVERSE-INCREMENTING (SIMPLE BELOW)
C NOTE THAT REVERSE QUARTER-LENGTH INTEGER IS ALREADY SCALED BY NG*2
C
      IREV = IREV + NREV
      GO TO 20
C
C ALTERNATE RECURSIVE ALTERATION TO QUARTER-LENGTH REVERSED SERIES
C
  70  DO 80 J=1,IHG
        IF (IRA(J).LT.NRA(J)) GO TO 90
  80  CONTINUE
C
C BELOW, SIMPLE INCREMENT OF REVERSE INTEGER, LOWER IN HIERARCHY
C
  90  IRA(J) = IRA(J) + NRA(J)
      IREV = IRA(J)
 100  IF (J.EQ.1) GO TO 20
      J = J - 1
      IRA(J) = IREV
      GO TO 100
      END
C
C-----------------------------------------------------------------------
C SUBROUTINE:  MFLOAD
C LOADS, UNLOADS CORE STORE ARRAY BUFA, 2**ICEX REALS, 2**IBEX PER BLOCK
C RETURNS IFLG=+1 NORMALLY, IFLG=-1 WHEN FINISHED ONE PASS OF MASS STOR
C BLOCKS INDEXED ACCORDING TO VIRTUAL PERMUTATION FUNCTION MFINDX
C LOAD=1 (LREAD) READS BLOCKS FROM MASS STORE INTO CORE STORE BUFA
C LOAD=2 (LWRIT) WRITES BLOCKS BACK IN-PLACE TO MASS STORE
C-----------------------------------------------------------------------
C
      SUBROUTINE MFLOAD(LOAD, BUFA, IBEX, ICEX, IFLG)
C
      REAL BUFA(1)
      DATA LINDX /0/, LHOLD /3/, LREST /4/
C
      NB = 2**IBEX
      NCB = 2**(ICEX-IBEX)
      IF (LOAD.EQ.2) GO TO 30
      IFLG = +1
      IDUM = MFINDX(LHOLD,0,0,0,0)
C
C HOLDS CURRENT MFINDX VALUE FOR ENTRY 2 AND SUBRTN MFCOMP
C
      DO 10 J=1,NCB
        K = (J-1)*NB
        JB = MFINDX(LINDX,0,0,0,0)
        IF (JB.LT.0) GO TO 20
        CALL MFREAD(BUFA(K+1), NB, JB)
C
C READS BLOCK WITH NEXT VIRTUAL MFINDX INDEX
C
  10  CONTINUE
      RETURN
  20  IFLG = -1
      RETURN
C
C RESETS MFINDX TO START OF IN-PLACE BLOCK
C
  30  IDUM = MFINDX(LREST,0,0,0,0)
      DO 40 J=1,NCB
        K = (J-1)*NB
        JB = MFINDX(LINDX,0,0,0,0)
        CALL MFWRIT(BUFA(K+1), NB, JB)
C
C WRITES BLOCK WITH NEXT VIRTUAL MFINDX INDEX (REPEAT MFREAD SEQUENCE)
C
  40  CONTINUE
      RETURN
      END
C
C-----------------------------------------------------------------------
C FUNCTION:  MFINDX
C VIRTUAL 'S' PERMUTATION (FRASER, J.ACM, V.23, N.2, APR. 76, P.303)
C CYCLIC SHIFTS H-B BITS IN INDEX (M-1,...,H-1,...,B,...,0)
C COMPUTES NEXT INDEX FOR SEQUENTIAL CORE LOAD, PERM (B.'S'.H)**N
C BLOCK SIZE EXPON B, MASS STORE EXPON M (0.LE.B.LE.H.LE.M)
C N IS EFFECTIVE NUMBER OF LEFT SHIFTS PER I/O PASS (-N RIGHT SHIFTS)
C-----------------------------------------------------------------------
C
      FUNCTION MFINDX(LSPEC, B, H, M, N)
C
C NOTE VARIABLE NAMES AS FOLLOWS:
C IPERM IS 'P' OF ALGORITHM
C NPERM=N (ARGUMENT) IS 'N' OF ALGORITHM
C ISTEP IS 'Q**P' OF ALGORITHM
C JAY AND KAY ARE 'J' AND 'K' OF ALGORITHM
C NOTE UPPER BOUND H INSTEAD OF M, REQUIRING 2**(M-H) REPEATS
C
C LSPEC=0 (LINDX) RETURNS MFINDX FOR INDEX (B,H,M,N DUMMIES HERE)
C LSPEC=1 (LSET) SETS IPERM=0 (UNPERMED), ENTERS B,H,M,N PARAMS
C LSPEC=2 (LPERM) CHANGES THE B,H,M,N PARAMETERS
C LSPEC=3 (LHOLD) HOLDS CURRENT INDEXING STATE (B,H,M,N DUMMIES HERE)
C LSPEC=4 (LREST) RESTORES STATE TO LAST LHOLD (B,H,M,N DUMMIES HERE)
C
      COMMON /VARIAB/ JAY, JOFF, NRPT, IPERM, JAYH, JOFH, NRPTH, NHB,
     *    NMB, IHB, ISTEP, NPERM
      INTEGER B, H
C
      IF (LSPEC.EQ.1) GO TO 50
      IF (LSPEC.EQ.2) GO TO 60
      IF (LSPEC.EQ.3) GO TO 70
      IF (LSPEC.EQ.4) GO TO 90
C
      IF (ISTEP.NE.0) GO TO 10
C
C BELOW, PRECEDES FIRST MFINDX OF A PASS
C
      IF (NPERM.GT.0) IPERM = MOD(IPERM-NPERM,IHB)
      IF (IPERM.LT.0) IPERM = IHB + IPERM
      ISTEP = 2**IPERM
C
C 20 BELOW, NORMAL GENERATION OF NEXT MFINDX
C
  10  MFINDX = JAY + JOFF
      IF (MFINDX.GT.NMB) GO TO 30
      KAY = JAY + ISTEP
      JAY = MOD(KAY,NHB)
      IF (KAY.GE.NHB) JAY = JAY + 1
      NRPT = NRPT - 1
      IF (NRPT.GT.0) RETURN
C
C NRPT,JOFF REQUIRED TO REPEAT SEQUENCE ON 2**(M-H) PARTS OF ARRAY
```

```
C
      JOFF = JOFF + NHB
   20 NRPT = NHB
      JAY = 0
      RETURN
C
C 40 BELOW, END OF ONE PASS, PARS RESET, IPERM ALTERED IF INVERSE
C
   30 IF (NPERM.LT.0) IPERM = MOD(IPERM-NPERM,IHB)
      MFINDX = -1
   40 JOFF = 1
      ISTEP = 0
      GO TO 20
C
C LSPEC=1 (LSET) SETS IPERM=0 (UNPERMED), ENTERS B,H,M,N PARAMS
C
   50 IPERM = 0
C
C LSPEC=2 (LPERM) CHANGES THE B,H,M,N PARAMETERS (DUMMIES ELSEWHERE)
C
   60 IHB = H - B
      NPERM = N
      NHB = 2**IHB
      NMB = 2**(M-B)
      MFINDX = IPERM
      GO TO 40
C
C LSPEC=3 (LHOLD) HOLDS CURRENT MFINDX INDEXING PARAMETERS
C
   70 JAYH = JAY
      JOFH = JOFF
      NRPTH = NRPT
   80 MFINDX = IPERM
      RETURN
C
C LSPEC=4 (LREST) RESTORES PARAMETERS TO INDEX MFINDX AT LAST LHOLD
C
   90 JAY = JAYH
      JOFF = JOFH
      NRPT = NRPTH
      GO TO 80
      END
C
C-----------------------------------------------------------------------
C FUNCTION:  MFSUM
C SCANS MEXA LIST IN REVERSE ORDER, RETURNING (MFSUM.JUST GT.MLIM)
C (IF MLIM LARGE ENOUGH, RETURNS M TOTAL FOR NDIM VALUES)
C (IF MLIM NEGATIVE, RETURNS M TOTAL, REVERSES ORDER OF MEXA LIST)
C-----------------------------------------------------------------------
C
      FUNCTION MFSUM(MEXA, NDIM, MLIM)
C
      INTEGER MEXA(1)
C
      MFSUM = 0
      IF (NDIM.LE.0) RETURN
C
      DO 10 J=1,NDIM
        I = NDIM + 1 - J
        MFSUM = MFSUM + MEXA(I)
        IF ((MLIM.GE.0) .AND. (MLIM.LT.MFSUM)) RETURN
   10 CONTINUE
      IF (MLIM.GE.0) RETURN
C
C BELOW, REVERSE ORDER OF MEXA LIST
C
      NDIMH = NDIM/2
      DO 20 J=1,NDIMH
        K = NDIM + 1 - J
        MTEM = MEXA(J)
        MEXA(J) = MEXA(K)
        MEXA(K) = MTEM
   20 CONTINUE
      RETURN
      END
C
C-----------------------------------------------------------------------
C SUBROUTINE:  MFRCMP
C UNSCRAMBLES REAL-TO-COMPLEX FFT OR VICE-VERSA, CALLED BY SUBRTN RMFFT
C MOST ARGUMENTS HAVE SAME MEANING AS IN RMFFT COMMENTS
C BUT 2**B COMPLEX ELMTS IN MASS STORE BLOCK,
C USES (2**B)*4 CMPLX IN BUFA, 'LOWER', 'UPPER' PLUS EXPANSION AREAS
C TOTAL MASS STORE ARRAY SIZE OF 2**M COMPLEX ELMTS.
C-----------------------------------------------------------------------
C
      SUBROUTINE MFRCMP(MEXA, NDIM, ISGN, IDIR, IPAK, BUFA, B, M)
C
      COMPLEX ATEM, BTEM, TEMP, W, D, BUFA(1)
      INTEGER B, JAYA(4), KAYA(4), JWKA(4), KWKA(4), MEXA(1)
C
C JAYA,KAYA,JWKA,KWKA ALLOW UP TO 4 DIMENSIONS - INCREASE IF REQUIRED
C
      DATA LOWER /1/, LUPPR /2/, LCLR /-1/
      PI = 4.0D0 * DATAN(1.0D0)
C
      DO 10 IDIM=1,NDIM
        JAYA(IDIM) = 0
        KAYA(IDIM) = 0
   10 CONTINUE
C
C MULTIDIMEN. CONJUGATE-SYMMETRIC INDICES ZEROED
C
      IEXPND = 1
      NB = 2**B
      NBDB = NB*2
      JBOF = 2**(M-B)
      MAX = M - B - MEXA(NDIM)
      IF (IDIR*IPAK.LT.0) MAX = M - B
      JBMAX = 2**MAX
      IF (MAX.LT.0) JBMAX = 1
      IF (IPAK.LT.0) JBMAX = 0
C
C JBMAX IS MAXIMUM BLOCK INDEX REQUIRED (DEPENDS ON IPAK)
C
      IWFG = 0
      W = (1.,0.)
      D = CEXP(CMPLX(0.,PI*2.0**(-MEXA(NDIM))))
      IF (ISGN.LT.0) D = CONJG(D)
C
C W IS COMPLEX PHASE FACTOR, D IS RECURSIVE MODIFIER OF W
C
   20 JAY = 0
      KAY = 0
      IDIM = NDIM
   30 IF (IDIM.EQ.1) GO TO 40
      NUMD = 2**MEXA(IDIM)
      JWKA(IDIM) = JAY
      KWKA(IDIM) = KAY
      JAY = JAY*NUMD + JAYA(IDIM)
      KAY = KAY*NUMD + KAYA(IDIM)
```

```
      IDIM = IDIM - 1
      GO TO 30
C
C CONJUGATE-SYMMETRIC BASE INDICES COMPUTED FROM MULTIDIMEN. SET
C
   40 MEX1 = MEXA(1)
C
C 2**MEX1 IS NUMBER OF VALUES ADJACENT IN FIRST DIMENSION
C
      IFLG = -1
      IF (MEX1.LE.B) GO TO 160
C
C BELOW, FIRST DIMEN. GREATER THAN BLOCK SIZE, MULTIPLE BLOCKS
C
      NBPD1 = 2**(MEX1-B)
      NBLCNT = NBPD1
      KINC = NB
      KBINC = NBLCNT - 1
      IF (JAY.EQ.KAY) NBLCNT = NBLCNT/2
      JB = JAY*NBPD1 + 1
      KB = KAY*NBPD1 + 1
C
C JB AND KB ARE BLOCK INDEX PAIRS CONTAINING CONJUGATE ELEMENTS
C
      J1 = 0
      K1 = 0
C
   50 NCNT = NB + 1
   60 CALL MFRLOD(LOWER, IOF, BUFA, NB, JB, JBMAX, JBOF, IDIR, NCNT)
C
C LOWER BLOCK LOADED
C
      IF (JB.GT.JBMAX) IEXPND = -1
      J2 = J1 + IOF
      JB = JB + 1
      J3 = 0
      K3 = 0
      NEWBLK = NCNT - NB
      IF (IFLG.GE.0) GO TO 80
C
   70 CALL MFRLOD(LUPPR, IOF, BUFA, NB, KB, JBMAX, JBOF, IDIR, IFLG)
C
C UPPER BLOCK LOADED
C
      IFLG = IFLG + 1
      K2 = K1 + IOF
      KB = KB + KBINC
C
C FIRST TIME, UPPER BLOCK STEPS HIGH, FOLLOWING STEPS SMALL NEGATIVE
C
      KBINC = -1
C
C J AND K INDEX CONJUGATE-SYMMETRIC PAIRS IN CORE
C
   80 J = J2 + J3
      K = K2 + K3
      JJ = J + NBDB
      KK = K + NBDB
      IF (IDIR.GT.0) GO TO 200
C
C BELOW, UNSCRAMBLING FOR REAL-TO-COMPLEX FFT
C
      TEMP = (BUFA(J)+CONJG(BUFA(K)))*0.5
      BTEM = BUFA(K) - CONJG(BUFA(J))
      BTEM = (CMPLX(AIMAG(BTEM),REAL(BTEM)))*0.5*W
      ATEM = TEMP + BTEM
      BTEM = TEMP - BTEM
      BUFA(J) = ATEM
      IF (IEXPND.GT.0) BUFA(JJ) = BTEM
      IF (IWFG.EQ.0) GO TO 170
      BUFA(K) = CONJG(BTEM)
      IF (IEXPND.GT.0) BUFA(KK) = CONJG(ATEM)
C
   90 J3 = J3 + 1
      K3 = KINC - J3
C
C IN-CORE INDEX PAIRS STEPPED IN OPPOSING DIRECTIONS
C
      IF (IDIM.NE.NDIM) GO TO 100
      IWFG = 1
      W = W*D
C
C RECURSIVE MODIFICATION OF W IF UNIDIMEN. TRANSFORM
C
  100 NCNT = NCNT - 1
      IF (NCNT.LE.0) GO TO 110
C
C ENTER RECURSION ROUTINE IF OPERATION COMPLETE IN CURRENT DIMEN
C
      IF (J3.EQ.1) GO TO 70
      IF (NCNT.GT.NEWBLK) GO TO 80
C
C END OF INNER LOOP (NOTE SPECIAL CASE WHEN J3=1 ABOVE)
C
C BELOW, MAY REQUIRE TO READ NEW BLOCKS
C
      NBLCNT = NBLCNT - 1
      IF (NBLCNT.GT.0) GO TO 50
      IF (JAY.EQ.KAY) GO TO 60
C
C JAY.EQ.KAY NEEDS SYMMETRICAL MIDDLE, OTHERWISE CURRENT DIMEN COMPLT
C
C BELOW, RECURSION TO COMPUTE MULTIDIMEN. CONJUGATE-SYMMETRY
C
  110 JAYA(IDIM) = 0
      KAYA(IDIM) = 0
      IDIM = IDIM + 1
      IF (IDIM.GT.NDIM) GO TO 140
      NUMD = 2**MEXA(IDIM)
      IF (NUMD.LE.1) GO TO 140
      IF (IDIM.NE.NDIM) GO TO 120
      IWFG = 1
      W = W*D
C
C RECURSIVE MODIFICATION OF W IF MULTIDIMEN. FFT
C
  120 IF (JAYA(IDIM).EQ.0) GO TO 130
      IF ((JWKA(IDIM)*NUMD+JAYA(IDIM)).EQ.(KWKA(IDIM)*NUMD+KAYA(IDIM)))
     *    GO TO 110
      IF (KAYA(IDIM).EQ.1) GO TO 110
C
  130 JAYA(IDIM) = JAYA(IDIM) + 1
      KAYA(IDIM) = NUMD - JAYA(IDIM)
C
C RECURSIVE STEPPING OF MULTIDIMEN. CONJUGATE-SYMMETRIC INDEX PAIRS
C
      GO TO 20
C
C BELOW, OPERATION COMPLETE, TIDY UP AND RETURN FROM SUBROUTINE
C
```

```
 140  DO 150 IAREA=1,2
        CALL MFRLOD(IAREA, IOF, BUFA, NB, LCLR, JBMAX, JBOF, IDIR, IFLG)
C
C DUMMY CALL TO MFRLOD TO WRITE ANY UNWRITTEN BLOCKS TO MASS STORE
C
 150  CONTINUE
      RETURN
C
C RETURN FROM SUBROUTINE
C
C BELOW, FIRST DIMEN. LESS THAN BLOCK SIZE, INDEX PAIRS ALL IN-CORE
C
 160  NUMD1 = 2**MEX1
      NBPD1 = NB/NUMD1
      NCNT = NUMD1
      KINC = NCNT
      KBINC = 0
      IF (JAY.EQ.KAY) NCNT = NCNT/2 + 1
      JB = JAY/NBPD1 + 1
      KB = KAY/NBPD1 + 1
C
C JB AND KB ARE BLOCK INDEX PAIRS CONTAINING CONJUGATE ELEMENTS
C
      J1 = (JAY-(JB-1)*NBPD1)*NUMD1
      K1 = (KAY-(KB-1)*NBPD1)*NUMD1
      GO TO 60
C
C BELOW, UNSCRAMBLING WITH W0 (IWFG=0) MUST BE TREATED DIFFERENTLY
C
 170  IF (IEXPND.LT.0) GO TO 180
C
C BELOW, ARRAY EXPANSION (EITHER IPAK=+1 OR IPAK=0 AND STILL REDUNDANT)
C
      BUFA(K) = CONJG(ATEM)
      BUFA(KK) = CONJG(BTEM)
      GO TO 90
C
C BELOW, NO ARRAY EXPANSION (EITHER IPAK=-1 OR IPAK=0 NOT REDUNDANT)
C
 180  IF (J.EQ.K) GO TO 190
      BUFA(K) = CONJG(BTEM)
      GO TO 90
C
C BELOW, SPECIAL CASE IF IPAK=-1 AND ELEMENTS ARE SAME
C
 190  BUFA(J) = CMPLX(REAL(ATEM),REAL(BTEM))
      GO TO 90
C
C BELOW, SCRAMBLING FOR COMPLEX-TO-REAL FFT
C
 200  IF (IWFG.EQ.0) GO TO 220
      BTEM = CONJG(BUFA(K))
 210  ATEM = (BUFA(J)+BTEM)
      BTEM = (BUFA(J)-BTEM)*W
      BTEM = CMPLX(AIMAG(BTEM),REAL(BTEM))
      BUFA(J) = ATEM - CONJG(BTEM)
      BUFA(K) = CONJG(ATEM) + BTEM
      GO TO 90
C
C BELOW, SCRAMBLING WITH W0 (IWFG=0) MUST BE TREATED DIFFERENTLY
C
 220  IF (IEXPND.LT.0) GO TO 230
      BTEM = BUFA(JJ)
      GO TO 210
C
C BELOW, NO REDUNDANCY (EITHER IPAK=-1 OR IPAK=0 OR 1 NOT REDUND)
C
 230  IF (J.EQ.K) GO TO 240
      BTEM = CONJG(BUFA(K))
      GO TO 210
C
C BELOW, SPECIAL CASE IF IPAK=-1 AND ELEMENTS ARE SAME
C
 240  BTEM = CMPLX(AIMAG(BUFA(J)),0.)
      BUFA(J) = CMPLX(REAL(BUFA(J)),0.)
      GO TO 210
      END
C
C-----------------------------------------------------------------------
C SUBROUTINE:  MFRLOD
C LOADS, UNLOADS CORE STORE ARRAY BUFA, FOR REAL FFT UNSCRAMBLING ROUTN
C-----------------------------------------------------------------------
C
      SUBROUTINE MFRLOD(IAREA, IOF, BUFA, NB, JB, JBMAX, JBOF, IDIR,
     *    NCNT)
C
C BLOCK SIZE NB CMPLX, BLOCK NUMBER JB (JB=-1 DOES FINAL TIDY O/P)
C JBMAX IS MAX BLOCK INDX FOR EXPANSN, JBOF OFFSET TO EXPANDING BLOCKS
C IDIR=-1 DIRECTION REAL/CMPLX, +1 CMPLX/REAL
C IAREA=1 (LOWER) OR 2 (UPPER) OF TWO AREAS IN LOGICAL UNSCRAMBLING
C NOTE THAT BLOCK NORMALLY PHYSICALLY LOADED IN THESE AREAS,
C BUT, IF BLOCK ALREADY RESIDENT, MAY BE IN DIFFERENT AREA, SO
C IOF RETURNED AS ACTUAL OFFSET IN BUFA TO LOADED BLOCK.
C USES (2**B)*4 CMPLX IN BUFA, 'LOWER', 'UPPER' PLUS EXPANSION AREAS
C RETURNS IOF AS BUFFER OFFSET TO AREA (MAY NOT BE SAME, IF BLOCK RESID
C
C NCNT IS COUNT OF ELEMENTS TO BE ACCESSED IN THIS LOAD, TO ALLOW
C NOTE TO BE TAKEN OF ANY PARTLY FILLED BLOCKS DURING EXPANSION,
C PREVENTING THE READING OF 'NON-EXISTENT' BLOCKS.
C LISTS JBPFA(), NCPFA() OF SIZE NPARF HOLD THIS INFORMATION, DEFAULTS
C TO ALL-READ IF EXCEEDED, BUT INCREASE NPARF ETC, IF PROBLEM.
C
      COMPLEX BUFA(1)
      INTEGER JBAREA(2), NCAREA(2), JBPFA(5), NCPFA(5)
      DATA JBAREA(1) /-1/, JBAREA(2) /-1/, NPARF /5/
C
C JBAREA() HOLDS INDEX OF BLOCK LOADED IN AREA 1 OR 2 (FIRST TIME -1 BEL
C
      IF (JBAREA(1).GE.0) GO TO 20
      IEXIST = -1
      DO 10 I=1,NPARF
        JBPFA(I) = -1
  10  CONTINUE
C
C PRE-CLEARS PARTLY FILLED BLOCK LIST (ONCE BLOCK FILLED, ALSO CLEARED)
C
  20  NBDB = NB*2
      IF (JB.LT.0) GO TO 50
      IF (IAREA.EQ.2) GO TO 30
      NCLOW = NCNT
      IF (MOD(NCNT,2).NE.0) NCLOW = (NCLOW-1)*2
  30  NCHLD = NCLOW
      IF (IAREA.EQ.2) NCHLD = NCLOW - 1
      IF (NCNT.LT.0) NCHLD = 1
C
C NCHLD IS THE NUMBER OF ELEMENTS TO BE ACCESSED IN CURRENT READ
C
      DO 40 I=1,2
        IF (JB.EQ.JBAREA(I)) GO TO 140
  40  CONTINUE
```

```
C
C TEST DONE TO SEE IF REQUIRED BLOCK ALREADY IN CORE (TRIVIAL IF SO)
C
C OTHERWISE BELOW, FIRST WRITE OUT RESIDENT BLOCK, THEN READ IN NEW
C IOF IS BASE OFFSET OF CORE AREA WHERE BLOCK IS TO BE FOUND
C
   50  IOF = (IAREA-1)*NB + 1
       IOFDB = IOF + NBDB
       IF (JBAREA(IAREA).LT.0) GO TO 90
       CALL MFWRIT(BUFA(IOF), NBDB, JBAREA(IAREA))
C
C WRITE OUT BLOCK BEFORE READING NEW BLOCK
C
       IF (IDIR.GT.0) GO TO 90
       IF (JBAREA(IAREA).GT.JBMAX) GO TO 90
       IF (NCAREA(IAREA).GE.NB) GO TO 80
C
C BELOW, IF LAST BLOCK ONLY PART-FILLED, INDEX, ELMTS ACCESSED NOTED
C
       DO 60 I=1,NPARF
         IF (JBPFA(I).LT.0) GO TO 70
   60  CONTINUE
C
C NO ROOM IN LISTS, DEFAULTS TO ALL READ
C
       I = NPARF
       IEXIST = I
   70  JBPFA(I) = JBAREA(IAREA)
       NCPFA(I) = NCAREA(IAREA)
C
   80  CALL MFWRIT(BUFA(IOFDB), NBDB, JBOF+JBAREA(IAREA))
C
C SIMILARLY, WRITE OUT BLOCK PAIR IF EXPANDING
C
C BELOW, READ BLOCK NOTING BLOCK INDX (READ EXPANDED, IF PART FILLED
C
   90  JBAREA(IAREA) = JB
       IF (JB.LT.0) GO TO 130
       CALL MFREAD(BUFA(IOF), NBDB, JB)
C
C READ REQUIRED BLOCK AND NOTE ACCESS COUNT
C
       NCAREA(IAREA) = NCHLD
       IF (JB.GT.JBMAX) GO TO 130
       IF (IDIR.GT.0) GO TO 120
C
C BELOW, EXPANSION - DOES BLOCK EXIST TO READ
C
       DO 100 I=1,NPARF
         IF (JB.EQ.JBPFA(I)) GO TO 110
  100  CONTINUE
       IF (IEXIST.LT.0) GO TO 130
  110  JBPFA(I) = -1
       NCAREA(IAREA) = NCPFA(I) + NCHLD
C
C IF BLOCK TO BE READ WAS ONLY PART FILLED, THEN IT EXISTS TO READ
C
  120  CALL MFREAD(BUFA(IOFDB), NBDB, JBOF+JB)
C
C READ EXPANDED BLOCK IF REQUIRED
C
  130  RETURN
C
C RETURN FROM SUBROUTINE
C
C BELOW, TRIVIAL CASE - BLOCK ALREADY LOADED
C IOF IS BASE OFFSET OF CORE AREA WHERE BLOCK IS TO BE FOUND
C
  140  IOF = (I-1)*NB + 1
       IF (I.NE.IAREA) GO TO 150
       NCAREA(I) = NCAREA(I) + NCHLD
C
C INCREASE ACCESS COUNT IF CURRENT IAREA MATCHES ORIGINAL IAREA
C
  150  RETURN
       END
C
C-----------------------------------------------------------------------
C FUNCTION:  MFPAR
C HELPER ROUTINE TO CROSS-COMPUTE MASS STORE FFT FILE PARAMETERS
C PARAMETERS ARE HELD AND COMPUTED IN 3 COMMON AREAS (SEE BELOW)
C MFPAR RETURNS 0 NORMALLY, -1 IF NOT ALL MFINT CORRECT, +1 IBEX ERROR
C-----------------------------------------------------------------------
C
       FUNCTION MFPAR(IRMF, ICOMP)
C
C COMMON/MFARG/ HOLDS ARGUMENTS AS USED IN FFT CALLS, AS FOLLOWS:
C VARIABLE NAMES HAVE SAME MEANING AS COMMENTS, SUBROUTINE RMFFT
C MEXA() HOLDS EXPONENTS FOR UP TO 4 DIMENSIONS (R/T ZEROS EXCESS)
C NDIM NUM DIMENS, IBEX,ICEX BLOCK AND CORE EXPONS, IPAK RMFFT PACKI
C ISGN,IDIR,SCAL ARE IGNORED HERE, BUT INCLUDED FOR COMPLETENESS
C
C COMMON/MFVAL/,/MFINT/ RETURN COMPUTED VALUES, USEFUL FOR FILE ACCESS
C NDMA(4),DIMA(4) HOLD DIMENSION SIZES CORRESPONDING TO MEXA()
C (EG. NDMA(1)=DIMA(1)=2.**MEXA(1), ETC. AND =1. BEYOND NDIM)
C
C NTD1,TDM1  IS CURRENT TOTAL NUM OF 'RECDS' OF SIZE NRD1,RDM1
C NRD1,RDM1  IS NUM OF REALS IN CURRENT FIRST DIMENSION
C (USEFUL FOR ACCESSING DATA BY MFREAD/MFWRIT, NRD1,RDM1 REALS,
C ASSUMING THAT MFREAD/MFWRIT CAN HANDLE 'RECDS' OF DIFFERENT SIZES
C
C NFBK,FBLK  IS MAXIMUM FILE SIZE OF 'RECDS' OF SIZE NRBK,RBLK
C NTBK,TBLK  IS CURRENT TOTAL NUM OF 'RECDS' OF SIZE NRBK,RBLK
C NRBK,RBLK  IS NUM OF REALS IN FFT WORKING BLOCK (2**IBEX REALS)
C (GIVES MAX AND CURRENT FILE SIZE AND ACCESS BY FFT ROUTINES,
C NFBK.GT.NTBK ONLY WITH PACKED REAL DATA WHEN EXPANDING, IPAK=0 OR 1)
C
C NRCR,RCOR  IS NUM OF REALS IN FFT WORKING CORE  (2**ICEX REALS)
C NSZE=SIZE=2.**M, WHICH IS THE EFFECTIVE TOTAL SIZE OF TRANSFORM,
C WHERE M IS SUM TO NDIM OF MEXA() (SEE RMFFT COMMENTS)
C
C NOTE THAT ALL /MFARG/ ARE INTGS (EXCEPT SCAL), ALL /MFVAL/ REALS
C (/MFINT/ IS INTEGER CONVERSION OF /MFVAL/, ANY VALUE OF MFINT
C IS SET -1 IF TOO LARGE, BY FIXMAX, AND MFPAR RETURNED -1 AS FLAG,
C
C ROUTINE ARGUMENTS HAVE THE FOLLOWING EFFECT:
C IRMF=-1, DATA IS PACKED REAL, +1 DATA IS COMPLEX, ROUTINE RMFFT,
C IRMF=0,  DATA IS COMPLEX, ROUTINE CMFFT
C
C ICOMP=0, COMPUTES VALUES IN /MFVAL/ FROM VALUES GIVEN IN /MFARG/
C ICOMP=1 , REVERSE COMPUTES EXPONENTS IN /MFARG/ FROM /MFVAL/
C (DIMA(),RBLK,RCOR GIVEN INSTEAD OF MEXA(),IBEX,ICEX)
C
C NOTE, ROUTINE FORCES ICEX, IBEX TO CORRECT RANGE, MFPAR=+1 IF CANNOT
C
      COMMON /MFARG/ MEXA(4), NDIM, ISGN, IDIR, SCAL, IBEX, ICEX, IPAK
      COMMON /MFVAL/ DIMA(4), TDM1, RDM1, FBLK, TBLK, RBLK, RCOR, SIZE
      COMMON /MFINT/ NDMA(4), NTD1, NRD1, NFBK, NTBK, NRBK, NRCR, NSZE
      REAL VAL(11)
      INTEGER INT(11)
```

```
      EQUIVALENCE (VAL(1),DIMA(1)), (INT(1),NDMA(1))
C
      DATA NMAX /4/
      FIXMAX = FLOAT(I1MACH(9))
C
      MFPAR = 0
      IF (ICOMP.EQ.0) GO TO 20
      ALG2 = ALOG(2.)
      IBEX = IFIX(ALOG(RBLK)/ALG2+0.5)
      ICEX = IFIX(ALOG(RCOR)/ALG2+0.5)
C
      DO 10 I=1,NDIM
        MEXA(I) = IFIX(ALOG(DIMA(I))/ALG2+0.5)
   10 CONTINUE
C
   20 M = 0
      DO 30 I=1,NMAX
        IF (I.GT.NDIM) MEXA(I) = 0
        M = M + MEXA(I)
        DIMA(I) = 2.**MEXA(I)
   30 CONTINUE
      SIZE = 2.**M
C
      IF (IRMF.EQ.0) GO TO 90
      IF (ICEX.GT.M) ICEX = M
      IF (IBEX.GT.ICEX-2) IBEX = ICEX - 2
      IF (IBEX.LT.2) MFPAR = 1
C
C FORCES ICEX.NGT.M AND IBEX.NGT.ICEX-2, OR MFPAR=1 (IRMF=+/- 1)
C
   40 RBLK = 2.**IBEX
      RCOR = 2.**ICEX
C
      FADD = SIZE
      IF (IRMF.EQ.0 .OR. IPAK.GT.0) GO TO 50
C
C FADD IS ADDITIONAL FILE SIZE IN REALS, 'SIZE' IF CMFFT OR IPAK=1
C
      FADD = 0.
      IF (IPAK.LT.0) GO TO 50
C
C IPAK=-1 REQUIRES NO FILE EXPANSION
C
      IDIM = NDIM
      IF (IRMF.LT.0) IDIM = 1
      FADD = SIZE*2./DIMA(IDIM)
C
C FADD COMPUTED FOR PARTICULAR CASE OF IPAK=0, WHEN COMPLX
C
   50 FSIZ = SIZE + FADD
      ITEM = IFIX(FSIZ/RBLK+0.5)
      IF (FLOAT(ITEM)*RBLK+0.5.LT.FSIZ) ITEM = ITEM + 1
      FBLK = FLOAT(ITEM)
C
C FBLK IS MAXIMUM NUMBER OF 'RECDS', SIZE RBLK, POSSIBLE
C
      TBLK = FBLK
      IF (IRMF.GE.0) GO TO 60
C
C GENERALLY TBLK=FBLK, BUT FOR PACKED REAL NOT SO, BELOW
C
      FSIZ = SIZE
      TBLK = FSIZ/RBLK
C
   60 TDM1 = 1.
      RDM1 = FSIZ
      IF (NDIM.EQ.1) GO TO 70
C
C JOB COMPLETED IF NDIM=1
C
      RDM1 = DIMA(1)
      IF (IRMF.GE.0) RDM1 = RDM1*2.
      TDM1 = FSIZ/RDM1
C
C OTHERWISE COMPUTE TDM1 AS NUMBER OF 'RECDS', SIZE RDM1 REALS
C
   70 DO 80 I=1,11
        INT(I) = -1
        IF (VAL(I).LE.FIXMAX) INT(I) = IFIX(VAL(I)+0.5)
        IF (INT(I).LT.0 .AND. MFPAR.EQ.0) MFPAR = -1
   80 CONTINUE
C
C CONVERT VALUES IN /MFVAL/ TO INTEGERS IN /MFINT/ (-1 IF TOO LARGE)
C
      RETURN
C
   90 IF (ICEX.GT.M+1) ICEX = M + 1
      IF (IBEX.GT.ICEX-1) IBEX = ICEX - 1
      IF (IBEX.LT.1) MFPAR = 1
      GO TO 40
C
C FORCES ICEX.NGT.M+1 AND IBEX.NGT.ICEX-1, OR MFPAR=1 (IRMF=0)
C
      END
C
C-----------------------------------------------------------------------
C SUBROUTINE:  DMPERM
C SHIFTS ORDER OF DIMENSIONS OF REAL OR COMPLEX MASS STORE ARRAY
C NOTE, THIS IS NOT USED BY FFT SUBRTNS BUT IS INCLUDED FOR COMPLETENESS
C (FRASER, ACM TOMS - 1978/79, AND J.ACM, V.23,N.2, APRIL 76, PP. 298-3
C-----------------------------------------------------------------------
C
      SUBROUTINE DMPERM(MEXA, NDIM, NSHFT, IREX, BUFA, IBEX, ICEX)
C
C MEXA(J) LIST OF DIMENSION SIZE EXPONS (BASE 2), ADJACENT VARIABLES FIR
C NDIM IS NUMBER OF EXPONENTS IN LIST AND THUS THE NUMBER OF DIMENSIONS
C SUM TO NDIM: MEXA(J)=M, WHERE 2**M IS SIZE OF MASS STORE ARRAY (SEE B
C NSHFT IS DIMENSION SHIFT COUNT, THUS:
C NSHFT=0,  NO SHIFT OR CHANGE OCCURS
C NSHFT=1,2 ETC., FIRST TO NEXT DIMENSION, CIRC NSHFT PLACE SHIFT (MOD
C NSHFT=-1, REVERSES THE ORDER OF DIMENSIONS
C IREX=0 REAL, 1 COMPLEX (THAT IS, MOVEMENT GROUP IS 2**IREX REALS,
C AND TOTAL MASS STORE SIZE IS 2**(M+IREX) REAL ELEMENTS)
C
      REAL BUFA(1)
      INTEGER MEXA(1)
C
      NS = MOD(NSHFT,NDIM)
      IF (NS.EQ.0) RETURN
      M = MFSUM(MEXA,NDIM,-1) + IREX
C
C FINDS M TOTAL AND REVERSES MEXA LIST
C
      CALL MFSORT(BUFA, IBEX, ICEX, IREX, M, M)
C
C INITIAL OVERALL BIT-REVERSAL M BITS ABOVE IREX BITS
C
      IF (NSHFT.LT.0) GO TO 10
C
C BELOW, REVERSAL OF TWO PARTS, TO FORM REQUIRED SHIFT
```

```
C
      IH = MFSUM(MEXA,NS,-1) + IREX
      CALL MFSORT(BUFA, IBEX, ICEX, IREX, IH, M)
C
C REVERSE LOWER PART OF MEXA LIST AND LOWER PART OF ARRAY BITS
C
      CALL MFSORT(BUFA, IBEX, ICEX, IH, M, M)
C
C SEPARATELY REVERSE UPPER PART OF ARRAY BITS
C
      IH = MFSUM(MEXA(NSHFT+1),NDIM-NS,-1)
C
C REVERSE UPPER PART OF MEXA LIST
C
      RETURN
C
C RETURN FROM SUBROUTINE AFTER CYCLIC SHIFTS
C
C BELOW, SEPARATELY REVERSE OVER EACH DIMENSION (DIMEN REVERSAL)
C
   10 IH = IREX
      DO 20 J=1,NDIM
        IG = IH
        IH = IH + MEXA(J)
        CALL MFSORT(BUFA, IBEX, ICEX, IG, IH, M)
   20 CONTINUE
      RETURN
      END
C
C-----------------------------------------------------------------------
C TEST PROGRAM:  UNIVERSAL
C THIS PROGRAM TESTS THE MASS STORE FFT BY COMPARISON WITH NAIVE DFT
C FFT PARAMETERS MAY BE ALTERED AT WILL (SEE COMMENTS)
C MASS STORE IS SIMULATED BY FORTRAN ARRAYS (SEE DUMMY I/O SUBROUTINES)
C PRINTING MAY BE COPIOUS, OR ONLY MAX DIFFERENCES (SEE COMMENTS)
C TEST OK IF MAX DIFFERENCES ARE NEAR ORDER OF MACHINE ROUND-OFF
C-----------------------------------------------------------------------
C
C NOTE WELL THAT TYPE COMPLEX DATA MUST EXIST AS ALTERNATING
C REAL/IMAG/REAL/IMAG... ELEMENTS, BOTH IN MASS STORE AND IN FORTRAN
C WORKING ARRAY BUFA;  IN THIS FFT, DIFFERENT SUBROUTINES WILL SET
C DIFFERENT TYPE (REAL OR COMPLEX) FOR ARRAY BUFA.
C
      COMMON /MFARG/ MEXA(4), NDIM, ISGN, IDIR, SCAL, IBEX, ICEX, IPAK
      COMMON /MFVAL/ DIMA(4), TDM1, RDM1, FBLK, TBLK, RBLK, RCOR, SIZE
      COMMON /MFINT/ NDMA(4), NTD1, NRD1, NFBK, NTBK, NRBK, NRCR, NSZE
C
C COMMON AREAS /MFARG/,/MFVAL/,/MFINT/ USEFUL FOR RUNNING MASS STORE FFT
C /MFARG/ HOLDS ARGUMENTS USED IN FFT CALLES, MOSTLY EXPONENTS
C HELPER ROUTINE MFPAR COMPUTES VALUES FROM /MFARG/ INTO
C /MFVAL/ (REALS AS SOME LARGE), /MFINT/ (INTEGER EQUIVALENTS IF POSS)
C OR CAN REVERSE-COMPUTE SOME /MFARG/ FROM /MFVAL/
C SEE COMMENTS, ROUTINE MFPAR.
C
      COMMON /MASS/ RMAS(1024)
      DIMENSION XR(32)
      COMPLEX ANAIV(512), BNAIV(512), CBUFA(512), CDIF
      REAL RANDA(1024), BUFA(1024)
      EQUIVALENCE (CBUFA(1),BUFA(1))
C
C COMMON/MASS/RMAS() USED BY ROUTINES MFREAD/MFWRIT TO SIMULATE MASS STO
C ANAIV(), BNAIV()  HOLD RESULT OF NAIVE DFT FOR COMPARISON
C BUFA()=CBUFA()    IS WORKING AREA IN CORE STORE FOR FFT AND PROGRAM
C RANDA  HOLDS PSEUDO RANDOM DATA USED IN TEST
C
      LP = I1MACH(2)
      IPRINT = 1
C
C LP IS PRINTER LOGICAL UNIT, IPRINT=+1 FOR COPIOUS PRINT,
C IPRINT=0 FOR MAX DIFFERENCES ONLY, -1 OVERALL MAX DIFFERENCE ONLY
C
      M = 5
C
C M SETS THE OVERALL ARRAY SIZE FOR AUTO IBEX,ICEX,MEXA,NDIM STEPPING
C FIXED VALUES CAN BE USED (SEE COMMENTS BELOW DO LOOP)
C
      IRMF = -1
C
C IRMF=-1 REAL ROUTINE RMFFT TEST, 0 COMPLEX ROUTINE CMFFT TEST
C
      ISGN = -1
      IDIR = -1
      IPAK = 1
C
C FFT ARGS, ISGN=+/- 1, IDIR=-1 (RMFFT), IDIR=1 OR 0 (CMFFT)
C IPAK=1 OR 0 (RMFFT), -1 GIVES APPARENT FAILURES DUE TO SQUEEZED RESULT
C
C BELOW, PRINT HEADINGS FOR TEST OUTPUT
C
      IF (IPRINT.LE.0) WRITE (LP,9999) IPRINT, IRMF
9999  FORMAT (20H1MASS STORE FFT TEST/9H IPRINT =, I3, 13H (1=COPIOUS,0,
     *    22H=MAX DIFFS,-1=OVERALL)/9H   IRMF =, I3, 15H (0=CMFFT TEST,,
     *    14H-1=RMFFT TEST)/)
      DIFMG = 0.
C
C BELOW, COMPUTE ALL POSSIBLE MEXA FOR 1,2 AND 3 DIMENSIONS
C
      DO 200 NDIM1=1,3
        NDIM = NDIM1
        M2M = M - NDIM + 1
        IF (NDIM.EQ.1) M2M = 1
        DO 190 M2=1,M2M
          M3M = M2M - M2 + 1
          IF (NDIM.NE.3) M3M = 1
          DO 180 M3=1,M3M
            IF (NDIM.EQ.1) MEXA(1) = M
            IF (NDIM.EQ.2) MEXA(1) = M - M2
            IF (NDIM.EQ.3) MEXA(1) = M - M2 - M3
            MEXA(2) = M2
            MEXA(3) = M3
C
C REPLACE DO LOOP BY FIXED MEXA LIST, IF DESIRED
C
            IBEX = 2
            ICEX = 4
C
C DUMMY IBEX, ICEX SO NO ERROR IN MFPAR BELOW
C
            IERR = MFPAR(IRMF,0)
            IF (IERR.NE.0) GO TO 210
C
C CALL HELPER ROUTINE TO COMPUTE SIZES USED BY NAIVE DFT SUBRTN
C
            M = MFSUM(MEXA,NDIM,99)
            AR = RANMF(1)
C
C RESET RANDOM NUMBER GENERATOR AND COMPUTE M IN CASE NOT GIVEN
C
            DO 10 J=1,NSZE
              RANDA(J) = RANMF(-1)
```

```
            ANAIV(J) = CMPLX(RANDA(J),0.)
  10        CONTINUE
C
C LOAD RANDOM NUMBERS FOR FFT AND FOR NAIVE SUBROUTINE
C
            CALL NAIVE(ANAIV, BNAIV, NDMA(1), NDMA(2), NDMA(3), ISGN)
C
C NAIVE DFT SUBROUTINE CALLED TO COMPUTE 'SLOW' FOURIER TRANSFORM
C
            IF (IRMF.EQ.0) GO TO 20
C
C BELOW, SET LIMTS FOR STEPPING IBEX, ICEX, WHEN CALLING RMFFT
C
            IBEXL = 2
            ICEXL = 4
            ICEXM = M
            GO TO 30
C
C SETS DIFFERENT LIMITS FOR IBEX, ICEX FOR CMFFT BELOW (IRMF.EQ.0)
C
  20        IBEXL = 1
            ICEXL = 2
            ICEXM = M + 1
C
  30        IF (ICEXL.GT.ICEXM) GO TO 210
            DO 170 ICEX1=ICEXL,ICEXM
              ICEX = ICEX1
              IBEXM = ICEX - 2
              IF (IRMF.EQ.0) IBEXM = ICEX - 1
              DO 160 IBEX1=IBEXL,IBEXM
                IBEX = IBEX1
C
C IBEX AND ICEX COMPUTED; REPLACE DO LOOPS BY FIXED IBEX,ICEX IF REQ
C
                IERR = MFPAR(IRMF,0)
                IF (IERR.NE.0) GO TO 210
                NCNT = NRD1
C
C HELPER ROUTN, NTD1 TOTAL NUMB OF NRD1 REALS IN FIRST DIMENSION
C (NCNT IS NUMBER OF REAL ELMTS IN FIRST DIMENSION)
C
                SCAL = 1.
                IF (IRMF.EQ.0) GO TO 60
C
C SWITCH FOR COMPLEX ROUTINE CMFFT AT 50, REAL RMFFT BELOW
C
                DO 50 JB=1,NTD1
                  K = (JB-1)*NCNT
                  DO 40 I=1,NCNT
                    J = K + I
                    BUFA(I) = RANDA(J)
  40              CONTINUE
                  CALL MFWRIT(BUFA, NRD1, JB)
  50            CONTINUE
C
C LOAD RANDOM NUMBERS IN REAL ARRAY IN 'RECDS' OF FIRST DIMEN LENGTH
C (NOTE, THIS REQUIRES MFREAD/MFWRIT TO BE ABLE TO ACCEPT 'RECORDS'
C OF DIFFERENT LENGTH; OTHERWISE MUST USE NTBK AND NRBK HERE)
C
                CALL RMFFT(MEXA, NDIM, ISGN, IDIR, SCAL, BUFA, IBEX,
     *              ICEX, IPAK)
C
C REAL MASS STORE ROUTINE RMFFT TO TRANSFORM ARRAY TO COMPLEX RESULT
C
                GO TO 90
C
C BELOW, USE COMPLEX ROUTINE CMFFT,  NCNT NUM OF CMPLX ELMTS IN FIRST DI
C
  60            NCNT = NCNT/2
                DO 80 JB=1,NTD1
                  K = (JB-1)*NCNT
                  DO 70 I=1,NCNT
                    J = K + I
                    CBUFA(I) = CMPLX(RANDA(J),0.)
  70              CONTINUE
                  CALL MFWRIT(BUFA, NRD1, JB)
  80            CONTINUE
C
C LOAD REAL VALUES IN CMPLX ARRAY IN 'RECDS' OF FIRST DIMEN LENGTH
C (NOTE, THIS REQUIRES MFREAD/MFWRIT TO BE ABLE TO ACCEPT 'RECORDS'
C OF DIFFERENT LENGTH; OTHERWISE MUST USE NTBK AND NRBK HERE)
C
                CALL CMFFT(MEXA, NDIM, ISGN, IDIR, SCAL, BUFA, IBEX,
     *              ICEX)
C
C COMPLEX MASS STORE ROUTINE CMFFT TO TRANSFORM ARRAY TO COMPLEX RESULT
C
  90            IF (IPRINT.LE.0) GO TO 100
C
C BELOW, PRINT HEADINGS FOR TEST OUTPUT
C
                WRITE (LP,9999) IPRINT, IRMF
                WRITE (LP,9998) NDIM, ISGN, IDIR, IBEX, ICEX, IPAK,
     *              (MEXA(J),J=1,NDIM)
9998            FORMAT (8H  NDIM =, I3, 8H  ISGN =, I3, 8H  IDIR =, I3,
     *              8H  IBEX =, I3, 8H  ICEX =, I3, 8H  IPAK =,
     *              I3/10H  MEXA() =, 3I5)
                WRITE (LP,9997)
9997            FORMAT (8H   INDEX, 9X, 3HFFT, 19X, 5HNAIVE, 18X,
     *              4HDIFF)
C
 100            IERR = MFPAR(-IRMF,0)
                IF (IERR.NE.0) GO TO 210
                NCNT = NRD1/2
C
C HELPER ROUTN, NTD1 TOTAL NUMB OF NRD1 REALS IN FIRST DIMENSION
C (NOTE IRMF=+1 FOR DATA IN COMPLEX STATE, NCNT ELMTS)
C
C BELOW, COMPARES FFT COMPLEX RESULT WITH NAIVE RESULT
C
                DIFM = 0.
                DO 120 JB=1,NTD1
                  K = (JB-1)*NCNT
                  CALL MFREAD(BUFA, NRD1, JB)
C
C READ 'RECDS' OF FIRST DIMEN LENGTH (RECOMPUTED ABOVE BY MFPAR)
C (NOTE, THIS REQUIRES MFREAD/MFWRIT TO BE ABLE TO ACCEPT 'RECORDS'
C OF DIFFERENT LENGTH; OTHERWISE MUST USE NTBK AND NRBK HERE)
C
                  DO 110 I=1,NCNT
                    J = K + I
                    INDEX = J - 1
                    IF (IDIR.NE.0) CDIF = CBUFA(I) - BNAIV(J)
                    IF (IDIR.EQ.0) CDIF = CBUFA(I) - ANAIV(J)
                    DIF = ABS(REAL(CDIF))
                    IF (DIF.GT.DIFM) DIFM = DIF
                    DIF = ABS(AIMAG(CDIF))
                    IF (DIF.GT.DIFM) DIFM = DIF
                    IF (IPRINT.LE.0) GO TO 110
                    IF (IDIR.NE.0) WRITE (LP,9996) INDEX, CBUFA(I),
```

```
     *                  BNAIV(J), CDIF
                      IF (IDIR.EQ.0) WRITE (LP,9996) INDEX, CBUFA(I),
     *                  ANAIV(J), CDIF
 9996                 FORMAT (1X, I5, 2(1X, 2E11.3),1X,2E10.2)
  110               CONTINUE
  120             CONTINUE
C
C BELOW, PRINT INTERMEDIATE DIFFERENCES
C
                IF (IPRINT.GE.0) WRITE (LP,9995) DIFM, NDIM, ISGN,
     *              IDIR, IBEX, ICEX, IPAK, (MEXA(J),J=1,NDIM)
 9995           FORMAT (10H MAX DIFF , E11.4/8H  NDIM =, I3, 8H  ISGN =,
     *              I3, 8H  IDIR =, I3, 8H  IBEX =, I3, 8H  ICEX =, I3,
     *              8H  IPAK =, I3/10H  MEXA() =, 3I5)
                IF (DIFM.GT.DIFMG) DIFMG = DIFM
C
C BELOW, INVERT ISGN AND IDIR FOR INVERSE TRANSFORM (COMPLEX-TO-REAL)
C
                ISGN = -ISGN
                IDIR = -IDIR
                SCAL = 1./SIZE
                IF (IRMF.NE.0) CALL RMFFT(MEXA, NDIM, ISGN, IDIR, SCAL,
     *              BUFA, IBEX, ICEX, IPAK)
C
C EITHER ROUTINE RMFFT INVERSE TRANSFORMS ARRAY TO PACKED REAL
C
                IF (IRMF.EQ.0) CALL CMFFT(MEXA, NDIM, ISGN, IDIR, SCAL,
     *              BUFA, IBEX, ICEX)
C
C OR ROUTINE CMFFT INVERSE TRANSFORMS ARRAY
C
                IF (IPRINT.LE.0) GO TO 130
C
C BELOW, PRINT HEADINGS FOR TEST OUTPUT
C
                WRITE (LP,9999) IPRINT, IRMF
                WRITE (LP,9998) NDIM, ISGN, IDIR, IBEX, ICEX, IPAK,
     *              (MEXA(J),J=1,NDIM)
                WRITE (LP,9994)
 9994           FORMAT (8H   INDEX, 2X, 8HFFT/2**M, 4X, 5HINPUT, 7X,
     *              4HDIFF)
C
  130           IERR = MFPAR(IRMF,0)
                IF (IERR.NE.0) GO TO 210
                NCNT = NRD1
                IF (IRMF.EQ.0) NCNT = NCNT/2
C
C HELPER ROUTN, NTD1 TOTAL NUMB OF NRD1 REALS IN FIRST DIMENSION
C (NCNT IS NUMBER OF ELMTS IN FIRST DIMENSION, REAL OR CMPLX)
C
C BELOW, COMPARES FFT INVERSE RESULT WITH INITIAL RANDOM INPUT
C
                DIFM = 0.
                DO 150 JB=1,NTD1
                  K = (JB-1)*NCNT
                  CALL MFREAD(BUFA, NRD1, JB)
C
C READ 'RECDS' OF FIRST DIMEN LENGTH (RECOMPUTED ABOVE BY MFPAR)
C (NOTE, THIS REQUIRES MFREAD/MFWRIT TO BE ABLE TO ACCEPT 'RECORDS'
C OF DIFFERENT LENGTH; OTHERWISE MUST USE NTBK AND NRBK HERE)
C
                  DO 140 I=1,NCNT
                    J = K + I
                    INDEX = J - 1
                    IF (IRMF.NE.0) RM = BUFA(I)
                    IF (IRMF.EQ.0) RM = REAL(CBUFA(I))
                    DIF = ABS(RM-RANDA(J))
                    IF (DIF.GT.DIFM) DIFM = DIF
                    IF (IPRINT.GT.0) WRITE (LP,9996) INDEX, RM,
     *                RANDA(J), DIF
  140             CONTINUE
  150           CONTINUE
C
C PRINT INVERSE RESULTS (SHOULD BE SAME AS INITIAL RANDOM SET)
C
                IF (IPRINT.GE.0) WRITE (LP,9995) DIFM, NDIM, ISGN,
     *              IDIR, IBEX, ICEX, IPAK, (MEXA(J),J=1,NDIM)
                IF (DIFM.GT.DIFMG) DIFMG = DIFM
                ISGN = -ISGN
                IDIR = -IDIR
C
C RESTORE ISGN AND IDIR FOR FORWARD TRANSFORM
C
  160           CONTINUE
  170         CONTINUE
  180       CONTINUE
  190     CONTINUE
  200   CONTINUE
C
C BELOW, PRINT OVERALL MAXIMUM DIFFERENCE
C
        WRITE (LP,9993) DIFMG, M, ISGN, IDIR, IPAK
 9993   FORMAT (18H OVERALL MAX DIFF , E11.4/25H FOR ALL MEXA(1 TO 3 DIM),
     *    23H,IBEX,ICEX,MEXA FOR M =, I3/19H   ISGN,IDIR(+/-) =, 2I4,
     *    8H  IPAK =, I4)
        STOP
C
  210   WRITE (LP,9992) IERR
 9992   FORMAT (25H IBEX FORCED TOO SMALL OR, 24H NOT ALL /MFINT/ CORRECT,
     *    6H, IERR, I4)
        STOP
        END
C
C-----------------------------------------------------------------------
C FUNCTION:  RANMF
C RANDOM NUMBER GENERATOR FOR MASS STORE FFT TEST
C-----------------------------------------------------------------------
C
        FUNCTION RANMF(J)
        IF (J.GE.0) GO TO 20
C
C POSITIVE J CAUSES RESET OF INITIAL K
C NEGATIVE J MUST BE USED NORMALLY
C
        MODULO = 2048
        FLMOD = 2048.0
        DO 10 I=1,15
          K = MOD(5*K,MODULO)
   10   CONTINUE
        Z = FLOAT(K)/FLMOD
        RANMF = Z
        RETURN
C
   20   K = J
        RANMF = J
        RETURN
        END
C
C-----------------------------------------------------------------------
C SUBROUTINE:  NAIVE
```

```
C NAIVE DISCRETE FOURIER TRANSFORM - 1 TO 3 DIMENSIONS
C USED TO TEST MASS STORE FFT, INPUT ARRAY  ANAIV(NJ,NK,NL)
C RESULT RETURNED IN BOTH ARRAYS ANAIV AND BNAIV
C ANAIV DIMENSIONS IN INITIAL ORDER, BNAIV REVERSED ORDER
C NJ,NK,NL DIMENSIONING, ISGN SIGN OF COMPLEX EXPONENT OF FOURIER
C-----------------------------------------------------------------------
C
      SUBROUTINE NAIVE(ANAIV, BNAIV, NJ, NK, NL, ISGN)
      COMPLEX TEMP, ANAIV(NJ,NK,NL), BNAIV(NL,NK,NJ)
C
      PI = 4.0D0 * DATAN(1.0D0)
      PI2 = PI*2.0
      IF (ISGN.LT.0) PI2 = -PI2
C
      DO 60 JB=1,NJ
        AJB = FLOAT(JB-1)/FLOAT(NJ)
        DO 50 KB=1,NK
          AKB = FLOAT(KB-1)/FLOAT(NK)
          DO 40 LB=1,NL
            ALB = FLOAT(LB-1)/FLOAT(NL)
C
            TEMP = (0.,0.)
            DO 30 JA=1,NJ
              AJA = FLOAT(JA-1)*AJB
              DO 20 KA=1,NK
                AKA = FLOAT(KA-1)*AKB
                DO 10 LA=1,NL
                  ALA = FLOAT(LA-1)*ALB
                  TEMP = TEMP + ANAIV(JA,KA,LA)*CEXP(CMPLX(0.,PI2*
     *                (AJA+AKA+ALA)))
   10           CONTINUE
   20         CONTINUE
   30       CONTINUE
C
            BNAIV(LB,KB,JB) = TEMP
   40     CONTINUE
   50   CONTINUE
   60 CONTINUE
C
      DO 90 JA=1,NJ
        DO 80 KA=1,NK
          DO 70 LA=1,NL
            ANAIV(JA,KA,LA) = BNAIV(LA,KA,JA)
   70     CONTINUE
   80   CONTINUE
   90 CONTINUE
      RETURN
      END
C
C-----------------------------------------------------------------------
C SUBROUTINE:  MFREAD
C DUMMY SUBROUTINE TO SIMULATE RANDOM ACCESS MASS STORE READ
C READ BLOCK, INDEX JB, FROM MASS STORE TO BUFA, NB REAL VALUES
C COMMON ARRAY RMAS SIMULATES MASS STORE ARRAY
C-----------------------------------------------------------------------
C
      SUBROUTINE MFREAD(BUFA, NB, JB)
      COMMON /MASS/ RMAS(1024)
      REAL BUFA(NB)
C
      IOF = (JB-1)*NB
      DO 10 I=1,NB
        K = IOF + I
        BUFA(I) = RMAS(K)
   10 CONTINUE
      RETURN
      END
C
C-----------------------------------------------------------------------
C SUBROUTINE: MFWRIT
C DUMMY SUBROUTINE TO SIMULATE RANDOM ACCESS MASS STORE WRITE
C WRITE BLOCK, INDEX JB, FORM BUFA TO MASS STORE, NB REAL VALUES
C COMMON ARRAY RMAS SIMULATES MASS STORE ARRAY
C-----------------------------------------------------------------------
C
      SUBROUTINE MFWRIT(BUFA, NB, JB)
      COMMON /MASS/ RMAS(1024)
      REAL BUFA(NB)
C
      IOF = (JB-1)*NB
      DO 10 I=1,NB
        K = IOF + I
        RMAS(K) = BUFA(I)
   10 CONTINUE
      RETURN
      END
C
C-----------------------------------------------------------------------
C TEST PROGRAM:  MASTOM
C CONTROL DATA 6000 AND CYBER MASS STORE I/O FFT
C-----------------------------------------------------------------------
C
      PROGRAM MASTOM(TAPE1,INPUT,OUTPUT,TAPE60=INPUT,TAPE5=OUTPUT)
C
C NOTE WELL THAT TYPE COMPLEX DATA MUST EXIST AS ALTERNATING
C REAL/IMAG/REAL/IMAG... ELEMENTS, BOTH IN MASS STORE AND IN FORTRAN
C WORKING ARRAY BUFA;  IN THIS FFT, DIFFERENT SUBROUTINES WILL SET
C DIFFERENT TYPE (REAL OR COMPLEX) FOR ARRAY BUFA.
C
C

      COMMON /FFTCOM/ LUN, MINDX(512)
C
C COMMON /FFTCOM/ HOLDS LOGICAL UNIT NUMBER FOR MASS STORE I/O
C ARRAY MINDX HOLDS RECORD INDICES FOR CYBER MASS STORE I/O
C
      COMMON /MFARG/ MEXA(4), NDIM, ISGN, IDIR, SCAL, IBEX, ICEX, IPAK
      COMMON /MFVAL/ DIMA(4), TDM1, RDM1, FBLK, TBLK, RBLK, RCOR, SIZE
      COMMON /MFINT/ NDMA(4), NTD1, NRD1, NFBK, NTBK, NRBK, NRCR, NSZE
C
C COMMON AREAS /MFARG/,/MFVAL/,/MFINT/ USEFUL FOR RUNNING MASS STORE FFT
C /MFARG/ HOLDS ARGUMENTS USED IN FFT CALLES, MOSTLY EXPONENTS
C HELPER ROUTINE MFPAR COMPUTES VALUES FROM /MFARG/ INTO
C /MFVAL/ (REALS AS SOME LARGE), /MFINT/ (INTEGER EQUIVALENTS IF POSS)
C OR CAN REVERSE-COMPUTE SOME /MFARG/ FROM /MFVAL/
C SEE COMMENTS, ROUTINE MFPAR.
C
      COMPLEX CBUFA(4096)
      REAL BUFA(8192)
      EQUIVALENCE (CBUFA(1),BUFA(1))
C
C BUFA()=CBUFA()    IS WORKING AREA IN CORE STORE FOR FFT AND PROGRAM
C
      LUN = 1
C
C LUN IS LOGICAL UNIT FOR CYBER MASS STORE I/O
C
      LP = 5
      IPRINT = 0
C
C LP IS PRINTER LOGICAL UNIT, IPRINT=+1 COPIOUS, 0 MAX DIFFERENCES ONLY
```

```
C
      IRMF = -1
C
C IRMF=-1 REAL ROUTINE RMFFT TEST, 0 COMPLEX ROUTINE CMFFT TEST
C
      ISGN = -1
      IDIR = -1
      IPAK = 1
C
C FFT ARGS, ISGN=+/- 1, IDIR=-1 (RMFFT), IDIR=1 OR 0 (CMFFT)
C IPAK=1, 0 OR -1 (RMFFT), NOT USED (CMFFT)
C
C BELOW, PRINT HEADINGS FOR TEST OUTPUT
C
      IF (IPRINT.LE.0) WRITE (LP,9999) IPRINT, IRMF
9999  FORMAT (31H1MASS STORE FFT TEST - IPRINT =, I3, 14H (1=COPIOUS,0=,
     *    11HMAX DIFFS),, 9H   IRMF =, I3, 25H (0=CMFFT TEST,1=RMFFT TE,
     *    3HST)/)
      DIFMG = 0.
C
      MEXA(1) = 8
      MEXA(2) = 6
      NDIM = 2
      IBEX = 9
      ICEX = 13
C
C MORE FFT ARGS, 2**8 ROWS OF 2**6 ELMTS, 2 DIMEN,
C FFT MASS STORE BLOCKS 2**IBEX REALS, BUFA 2**ICEX REALS
C
      IERR = MFPAR(IRMF,0)
      IF (IERR.NE.0) GO TO 130
      CALL OPENMS(LUN, MINDX, NFBK+1, 0)
C
C CYBER 'READMS/WRITMS' MASS STORE OPENED WITH 'NUM REC'+1=NFBK+1
C (NOTE HELPER ROUTN MFPAR RETURNS NFBK AS MAXIMUM FILE SIZE)
C
      NCNT = NRBK
C
C HELPER ROUTINE, NCNT NUM OF ELMTS IN FFT BLOCK, NTBK TOTAL BLOCKS
C
      M = MFSUM(MEXA,NDIM,99)
      SCAL = 1.
      VALU = 0.
      VINC = 1./SIZE
      IF (IRMF.EQ.0) GO TO 30
C
C SWITCH FOR COMPLEX ROUTINE CMFFT AT 50, REAL RMFFT BELOW
C
      DO 20 JB=1,NTBK
        DO 10 I=1,NCNT
          BUFA(I) = VALU
          VALU = VALU + VINC
 10     CONTINUE
        CALL MFWRIT(BUFA, NRBK, JB)
 20   CONTINUE
C
C LOAD RAMP FUNCTION IN REAL ARRAY IN 'RECDS' OF NRBK LENGTH
C (NOTE, CYBER MFREAD/MFWRIT (USING READMS/WRITMS) CANNOT ACCEPT 'RECDS
C OF DIFFERENT LENGTH; OTHERWISE COULD USE NTD1 'RECDS' OF NRD1 HERE)
C
      CALL RMFFT(MEXA, NDIM, ISGN, IDIR, SCAL, BUFA, IBEX, ICEX, IPAK)
C
C REAL MASS STORE ROUTINE RMFFT TO TRANSFORM ARRAY TO COMPLEX RESULT
C
      GO TO 60
C
C BELOW, USE COMPLEX ROUTINE CMFFT,  NCNT NUM OF CMPLX ELMTS IN FFT BLOC
C
 30   NCNT = NCNT/2
      DO 50 JB=1,NTBK
        DO 40 I=1,NCNT
          CBUFA(I) = CMPLX(VALU,0.)
          VALU = VALU + VINC
 40     CONTINUE
        CALL MFWRIT(BUFA, NRBK, JB)
 50   CONTINUE
C
C LOAD RAMP FUNCTION IN COMPLEX ARRAY IN 'RECDS' OF NRBK LENGTH
C (NOTE, CYBER MFREAD/MFWRIT (USING READMS/WRITMS) CANNOT ACCEPT 'RECDS
C OF DIFFERENT LENGTH; OTHERWISE COULD USE NTD1 'RECDS' OF NRD1 HERE)
C
      CALL CMFFT(MEXA, NDIM, ISGN, IDIR, SCAL, BUFA, IBEX, ICEX)
C
C COMPLEX MASS STORE ROUTINE CMFFT TO TRANSFORM ARRAY TO COMPLEX RESULT
C
 60   IF (IPRINT.LE.0) GO TO 90
C
C BELOW, PRINT HEADINGS FOR TEST OUTPUT
C
      WRITE (LP,9999) IPRINT, IRMF
      WRITE (LP,9998) NDIM, ISGN, IDIR, IBEX, ICEX, IPAK,
     *    (MEXA(J),J=1,NDIM)
9998  FORMAT (8H  NDIM =, I3, 8H  ISGN =, I3, 8H  IDIR =, I3, 7H  IBEX ,
     *    1H=, I3, 8H  ICEX =, I3, 8H  IPAK =, I3, 10H  MEXA() =, 3I5)
      WRITE (LP,9997)
9997  FORMAT (8H  INDEX, 12X, 3HFFT)
C
      IERR = MFPAR(-IRMF,0)
      IF (IERR.NE.0) GO TO 130
      NCNT = NRBK/2
C
C HELPER ROUTN, NTBK TOTAL NUMB OF NRBK REALS IN FFT BLOCK
C (NOTE IRMF=+1 FOR DATA IN COMPLEX STATE, NCNT ELMTS)
C
C BELOW, PROGRAM CAN ACCESS FFT COMPLEX RESULT (AND FORM COMPARISON, IF
C
      DO 80 JB=1,NTBK
        K = (JB-1)*NCNT
        CALL MFREAD(BUFA, NRBK, JB)
C
C READ 'RECDS' OF NRBK LENGTH, TOTALING NTBK (RECOMPUTED BY MFPAR)
C (NOTE, CYBER MFREAD/MFWRIT (USING READMS/WRITMS) CANNOT ACCEPT 'RECDS
C OF DIFFERENT LENGTH; OTHERWISE COULD USE NTD1 'RECDS' OF NRD1 HERE)
C
        DO 70 I=1,NCNT
          INDEX = J - 1
          IF (IPRINT.LE.0) GO TO 70
          WRITE (LP,9996) INDEX, CBUFA(I)
9996      FORMAT (1X, I5, 2X, 2E13.4)
 70     CONTINUE
 80   CONTINUE
C
C BELOW, INVERT ISGN AND IDIR FOR INVERSE TRANSFORM (COMPLEX-TO-REAL)
C
 90   ISGN = -ISGN
      IDIR = -IDIR
      SCAL = 1./SIZE
      IF (IRMF.NE.0) CALL RMFFT(MEXA, NDIM, ISGN, IDIR, SCAL, BUFA,
     *    IBEX, ICEX, IPAK)
C
```

```
C EITHER ROUTINE RMFFT INVERSE TRANSFORMS ARRAY TO PACKED REAL
C
      IF (IRMF.EQ.0) CALL CMFFT(MEXA, NDIM, ISGN, IDIR, SCAL, BUFA,
     *    IBEX, ICEX)
C
C OR ROUTINE CMFFT INVERSE TRANSFORMS ARRAY
C
      IF (IPRINT.LE.0) GO TO 100
C
C BELOW, PRINT HEADINGS FOR TEST OUTPUT
C
      WRITE (LP,9999) IPRINT, IRMF
      WRITE (LP,9998) NDIM, ISGN, IDIR, IBEX, ICEX, IPAK,
     *     (MEXA(J),J=1,NDIM)
      WRITE (LP,9995)
9995  FORMAT (8H   INDEX, 4X, 8HFFT/2**M, 6X, 5HINPUT, 10X, 4HDIFF)
C
 100  IERR = MFPAR(IRMF,0)
      IF (IERR.NE.0) GO TO 130
      NCNT = NRBK
      IF (IRMF.EQ.0) NCNT = NCNT/2
C
C HELPER ROUTN, NTBK TOTAL NUMB OF NRBK REALS IN FFT BLOCK
C (NCNT IS NUMBER OF ELMTS IN FFT BLOCK, REAL OR CMPLX)
C
C BELOW, COMPARES FFT INVERSE RESULT WITH INITIAL RANDOM INPUT
C
      DIFM = 0.
      VALU = 0.
      DO 120 JB=1,NTBK
        CALL MFREAD(BUFA, NRBK, JB)
C
C READ 'RECDS' OF NRBK LENGTH, TOTALING NTBK (RECOMPUTED BY MFPAR)
C (NOTE, CYBER MFREAD/MFWRIT (USING READMS/WRITMS) CANNOT ACCEPT 'RECDS
C OF DIFFERENT LENGTH; OTHERWISE COULD USE NTD1 'RECDS' OF NRD1 HERE)
C
        DO 110 I=1,NCNT
          IF (IRMF.NE.0) RM = BUFA(I)
          IF (IRMF.EQ.0) RM = REAL(CBUFA(I))
          DIF = ABS(RM-VALU)
          IF (DIF.GT.DIFM) DIFM = DIF
          IF (IPRINT.GT.0) WRITE (LP,9994) I, RM, VALU, DIF
9994      FORMAT (1X, I5, 3(2X, E13.4))
          VALU = VALU + VINC
 110    CONTINUE
 120  CONTINUE
C
C PRINT INVERSE RESULTS (SHOULD BE SAME AS INITIAL RAMP)
C
      IF (IPRINT.GE.0) WRITE (LP,9993) DIFM, NDIM, ISGN, IDIR, IBEX,
     *    ICEX, IPAK, (MEXA(J),J=1,NDIM)
9993  FORMAT (10H MAX DIFF , E11.4, 1X, 8H  NDIM =, I3, 8H  ISGN =, I3,
     *    8H  IDIR =, I3, 8H  IBEX =, I3, 8H  ICEX =, I3, 8H  IPAK =,
     *    I3, 10H  MEXA() =, 3I5)
      IF (DIFM.GT.DIFMG) DIFMG = DIFM
C
      STOP
C
 130  WRITE (LP,9992) IERR
9992  FORMAT (25H IBEX FORCED TOO SMALL OR, 24H NOT ALL /MFINT/ CORRECT,
     *    6H, IERR, I4)
      STOP
      END
C
C-----------------------------------------------------------------------
C SUBROUTINES: MFREAD, MFWRIT
C CONTROL DATA 6000 AND CYBER MASS STORE I/O ROUTINES FOR FFT
C-----------------------------------------------------------------------
C
      SUBROUTINE MFREAD(BUFA, NB, JB)
C
C (LOGICAL UNIT LUN IN COMMON/FFTCOM/ MUST HAVE BEEN OPENED
C PREVIOUSLY; EG.     CALL OPENMS(LUN,INDXARRAY,NREC+1,0)
C WHERE NREC=2**(M-IBEX+1) IF IPAK=1 OR LESS IF IPAK=0 OR -1)
C
C SEE ALSO ALTERNATIVE SUBROUTINES USING EXTENDED CORE OR LCM
C (IN ADDITION, GETW, PUTW OR READM, WRITEM MACROS CAN BE USED)
C
C READ BLOCK, INDEX JB, FROM MASS STORE TO BUFA, NB REAL VALUES
C
      COMMON /FFTCOM/ LUN, MINDX(512)
      REAL BUFA(NB)
C
      CALL READMS(LUN, BUFA, NB, JB)
      RETURN
C
      ENTRY MFWRIT
C
C WRITE BLOCK, INDEX JB, FROM BUFA TO MASS STORE, NB REAL VALUES
C
      CALL WRITMS(LUN, BUFA, NB, JB, -1, 0)
      RETURN
      END
C
C-----------------------------------------------------------------------
C TEST PROGRAM: LCMTOM
C CONTROL DATA 6000 AND CYBER EXTENDED CORE/LCM FFT
C-----------------------------------------------------------------------
C
      PROGRAM LCMTOM(INPUT,OUTPUT,TAPE60=INPUT,TAPE5=OUTPUT)
C
C NOTE WELL THAT TYPE COMPLEX DATA MUST EXIST AS ALTERNATING
C REAL/IMAG/REAL/IMAG... ELEMENTS, BOTH IN MASS STORE AND IN FORTRAN
C WORKING ARRAY BUFA;  IN THIS FFT, DIFFERENT SUBROUTINES WILL SET
C DIFFERENT TYPE (REAL OR COMPLEX) FOR ARRAY BUFA.
C
C
      LEVEL3, LBUFA
      COMMON /FFTCOM/ LBUFA(32768)
C
C LBUFA IN EXTENDED CORE/LCM SIMULATES FAST MASS STORE
C DIMENSION LBUFA AS LARGE AS NECESSARY FOR RESULTANT ARRAY
C
      COMMON /MFARG/ MEXA(4), NDIM, ISGN, IDIR, SCAL, IBEX, ICEX, IPAK
      COMMON /MFVAL/ DIMA(4), TDM1, RDM1, FBLK, TBLK, RBLK, RCOR, SIZE
      COMMON /MFINT/ NDMA(4), NTD1, NRD1, NFBK, NTBK, NRBK, NRCR, NSZE
C
C COMMON AREAS /MFARG/,/MFVAL/,/MFINT/ USEFUL FOR RUNNING MASS STORE FFT
C /MFARG/ HOLDS ARGUMENTS USED IN FFT CALLES, MOSTLY EXPONENTS
C HELPER ROUTINE MFPAR COMPUTES VALUES FROM /MFARG/ INTO
C /MFVAL/ (REALS AS SOME LARGE), /MFINT/ (INTEGER EQUIVALENTS IF POSS)
C OR CAN REVERSE-COMPUTE SOME /MFARG/ FROM /MFVAL/
C SEE COMMENTS, ROUTINE MFPAR.
C
      COMPLEX CBUFA(4096)
      REAL BUFA(8192)
      EQUIVALENCE (CBUFA(1),BUFA(1))
C
C BUFA()=CBUFA()    IS WORKING AREA IN CORE STORE FOR FFT AND PROGRAM
C
```

```
      LP = 5
      IPRINT = 0
C
C LP IS PRINTER LOGICAL UNIT, IPRINT=+1 COPIOUS, 0 MAX DIFFERENCES ONLY
C
      IRMF = -1
C
C IRMF=-1 REAL ROUTINE RMFFT TEST, 0 COMPLEX ROUTINE CMFFT TEST
C
      ISGN = -1
      IDIR = -1
      IPAK = 1
C
C FFT ARGS, ISGN=+/- 1, IDIR=-1 (RMFFT), IDIR=1 OR 0 (CMFFT)
C IPAK=1, 0 OR -1 (RMFFT), NOT USED (CMFFT)
C
C BELOW, PRINT HEADINGS FOR TEST OUTPUT
C
      IF (IPRINT.LE.0) WRITE (L ,9999) IPRINT, IRMF
9999  FORMAT (31H1MASS STORE FFT TEST - IPRINT =, I3, 14H (1=COPIOUS,0=,
     *    11HMAX DIFFS),, 9H   IRMF =, I3, 25H (0=CMFFT TEST,1=RMFFT TE,
     *    3HST)/)
      DIFMG = 0.
C
      MEXA(1) = 8
      MEXA(2) = 6
      NDIM = 2
      IBEX = 7
      ICEX = 13
C
C MORE FFT ARGS, 2**8 ROWS OF 2**6 ELMTS, 2 DIMEN,
C FFT MASS STORE BLOCKS 2**IBEX REALS, BUFA 2**ICEX REALS
C
      IERR = MFPAR(IRMF   )
      IF (IERR.NE.0) GO TO 130
      NCNT = NRD1
C
C HELPER ROUTINE, NCNT NUM OF ELMTS IN FIRST DIMEN, NTD1 TOTAL
C
      M = MFSUM(MEXA,NDIM,99
      SCAL = 1.
      VALU = 0.
      VINC = 1./SIZE
      IF (IRMF.EQ.0) GO TO 3(
C
C SWITCH FOR COMPLEX ROUTINE      'T AT 50, REAL RMFFT BELOW
C
      DO 20 JB=1,NTD1
        DO 10 I=1,NCNT
          BUFA(I) = VALU
          VALU = VALU + VINC
   10   CONTINUE
        CALL MFWRIT(BUFA, NRD1, JB)
   20 CONTINUE
C
C LOAD RAMP FUNCTION IN REAL ARRAY IN 'RECDS' OF NRD1 LENGTH
C (NOTE, 'LCM' MFREAD/MFWRIT CAN ACCEPT 'RECORDS' OF DIFFERENT LENGTH;
C SO CAN USE NTD1,NRD1 HERE INSTEAD OF NTBK,NRBK AS IN CYBER MASS STOR
C
      CALL RMFFT(MEXA, NDIM, ISGN, IDIR, SCAL, BUFA, IBEX, ICEX, IPAK)
C
C REAL MASS STORE ROUTINE RMFFT TO TRANSFORM ARRAY TO COMPLEX RESULT
C
      GO TO 60
C
C BELOW, USE COMPLEX ROUTINE CMFFT,  NCNT NUM OF CMPLX ELMTS IN FFT BLOC
C
   30 NCNT = NCNT/2
      DO 50 JB=1,NTD1
        DO 40 I=1,NCNT
          CBUFA(I) = CMPLX(VALU,0.)
          VALU = VALU + VINC
   40   CONTINUE
        CALL MFWRIT(BUFA, NRD1, JB)
   50 CONTINUE
C
C LOAD RAMP FUNCTION IN COMPLEX ARRAY IN 'RECDS' OF NRD1 LENGTH
C (NOTE, 'LCM' MFREAD/MFWRIT CAN ACCEPT 'RECORDS' OF DIFFERENT LENGTH;
C SO CAN USE NTD1,NRD1 HERE INSTEAD OF NTBK,NRBK AS IN CYBER MASS STOR
C
      CALL CMFFT(MEXA, NDIM, ISGN, IDIR, SCAL, BUFA, IBEX, ICEX)
C
C COMPLEX MASS STORE ROUTINE CMFFT TO TRANSFORM ARRAY TO COMPLEX RESULT
C
   60 IF (IPRINT.LE.0) GO TO 90
C
C BELOW, PRINT HEADINGS FOR TEST OUTPUT
C
      WRITE (LP,9999) IPRINT, IRMF
      WRITE (LP,9998) NDIM, ISGN, IDIR, IBEX, ICEX, IPAK,
     *    (MEXA(J),J=1,NDIM)
9998  FORMAT (8H  NDIM =, I3, 8H  ISGN =, I3, 8H  IDIR =, I3, 7H  IBEX ,
     *    1H=, I3, 8H  ICEX =, I3, 8H  IPAK =, I3, 10H  MEXA() =, 3I5)
      WRITE (LP,9997)
9997  FORMAT (8H   INDEX, 12X, 3HFFT)
C
      IERR = MFPAR(-IRMF,0)
      IF (IERR.NE.0) GO TO 130
      NCNT = NRD1/2
C
C HELPER ROUTINE, NCNT NUM OF ELMTS IN FIRST DIMEN, NTD1 TOTAL
C (NOTE IRMF=+1 FOR DATA IN COMPLEX STATE, NCNT ELMTS)
C
C BELOW, PROGRAM CAN ACCESS FFT COMPLEX RESULT (AND FORM COMPARISON, IF
C
      DO 80 JB=1,NTD1
        K = (JB-1)*NCNT
        CALL MFREAD(BUFA, NRD1, JB)
C
C READ 'RECDS' OF NRD1 LENGTH, TOTALING NTD1 (RECOMPUTED BY MFPAR)
C (NOTE, 'LCM' MFREAD/MFWRIT CAN ACCEPT 'RECORDS' OF DIFFERENT LENGTH;
C SO CAN USE NTD1,NRD1 HERE INSTEAD OF NTBK,NRBK AS IN CYBER MASS STOR
C
        DO 70 I=1,NCNT
          INDEX = J - 1
          IF (IPRINT.LE.0) GO TO 70
          WRITE (LP,9996) INDEX, CBUFA(I)
9996      FORMAT (1X, I5, 2X, 2E13.4)
   70   CONTINUE
   80 CONTINUE
C
C BELOW, INVERT ISGN AND IDIR FOR INVERSE TRANSFORM (COMPLEX-TO-REAL)
C
   90 ISGN = -ISGN
      IDIR = -IDIR
      SCAL = 1./SIZE
      IF (IRMF.NE.0) CALL RMFFT(MEXA, NDIM, ISGN, IDIR, SCAL, BUFA,
     *    IBEX, ICEX, IPAK)
C
C EITHER ROUTINE RMFFT INVERSE TRANSFORMS ARRAY TO PACKED REAL
```

```
C
      IF (IRMF.EQ.0) CALL CMFFT(MEXA, NDIM, ISGN, IDIR, SCAL, BUFA,
     *     IBEX, ICEX)
C
C OR ROUTINE CMFFT INVERSE TRANSFORMS ARRAY
C
      IF (IPRINT.LE.0) GO TO 100
C
C BELOW, PRINT HEADINGS FOR TEST OUTPUT
C
      WRITE (LP,9999) IPRINT, IRMF
      WRITE (LP,9998) NDIM, ISGN, IDIR, IBEX, ICEX, IPAK,
     *    (MEXA(J),J=1,NDIM)
      WRITE (LP,9995)
9995  FORMAT (8H   INDEX, 4X, 8HFFT/2**M, 6X, 5HINPUT, 10X, 4HDIFF)
C
 100  IERR = MFPAR(IRMF,0)
      IF (IERR.NE.0) GO TO 130
      NCNT = NRD1
      IF (IRMF.EQ.0) NCNT = NCNT/2
C
C HELPER ROUTINE, NCNT NUM OF ELMTS IN FIRST DIMEN, NTD1 TOTAL
C (NCNT IS NUMBER OF ELMTS IN FFT BLOCK, REAL OR CMPLX)
C
C BELOW, COMPARES FFT INVERSE RESULT WITH INITIAL RANDOM INPUT
C
      DIFM = 0.
      VALU = 0.
      DO 120 JB=1,NTD1
        CALL MFREAD(BUFA, NRD1, JB)
C
C READ 'RECDS' OF NRD1 LENGTH, TOTALING NTD1 (RECOMPUTED BY MFPAR)
C (NOTE, 'LCM' MFREAD/MFWRIT CAN ACCEPT 'RECORDS' OF DIFFERENT LENGTH;
C SO CAN USE NTD1,NRD1 HERE INSTEAD OF NTBK,NRBK AS IN CYBER MASS STOR
C
        DO 110 I=1,NCNT
          IF (IRMF.NE.0) RM = BUFA(I)
          IF (IRMF.EQ.0) RM = REAL(CBUFA(I))
          DIF = ABS(RM-VALU)
          IF (DIF.GT.DIFM) DIFM = DIF
          IF (IPRINT.GT.0) WRITE (LP,9994) I, RM, VALU, DIF
9994      FORMAT (1X, I5, 3(2X, E13.4))
          VALU = VALU + VINC
 110    CONTINUE
 120  CONTINUE
C
C PRINT INVERSE RESULTS (SHOULD BE SAME AS INITIAL RAMP)
C
      IF (IPRINT.GE.0) WRITE (LP,9993) DIFM, NDIM, ISGN, IDIR, IBEX,
     *    ICEX, IPAK, (MEXA(J),J=1,NDIM)
9993  FORMAT (10H MAX DIFF , E11.4, 1X, 8H  NDIM =, I3, 8H  ISGN =, I3,
     *    8H  IDIR =, I3, 8H  IBEX =, I3, 8H  ICEX =, I3, 8H  IPAK =,
     *    I3, 10H  MEXA() =, 3I5)
      IF (DIFM.GT.DIFMG) DIFMG = DIFM
C
      STOP
C
 130  WRITE (LP,9992) IERR
9992  FORMAT (25H IBEX FORCED TOO SMALL OR, 24H NOT ALL /MFINT/ CORRECT,
     *    6H, IERR, I4)
      STOP
      END
C
C-----------------------------------------------------------------------
C SUBROUTINES: MFREAD, MFWRIT
C CONTROL DATA 6000 AND CYBER EXTENDED CORE STORE I/O ROUTINES FOR FFT
C-----------------------------------------------------------------------
C
      SUBROUTINE MFREAD(BUFA, NB, JB)
C
C (EXTENDED CORE OR LCM TAKES PLACE OF MASS STORE -
C DIMENSION LBUFA AS LARGE AS NECESSARY FOR LARGE ARRAYS)
C
C ALTERNATIVELY, EXTENDED CORE USE COULD BE COMBINED WITH
C READM/WRITEM MACROS TO ACCESS LARGER FILE WHEN NECESSARY
C (USE VIRTUAL MEMORY ALGORITHM - MANY SECTORS HELD IN
C EXTENDED CORE AND ONLY ACCESSED FROM DISC IF NOT PRESENT).
C
C READ BLOCK, INDEX JB, FROM EXTENDED CORE TO BUFA, NB REAL VALUES
C
      LEVEL3, LBUFA
      COMMON /FFTCOM/ LBUFA(32768)
      REAL BUFA(NB)
C
      CALL MOVLEV(LBUFA((JB-1)*NB+1), BUFA, NB)
      RETURN
C
      ENTRY MFWRIT
C
C WRITE BLOCK, INDEX JB, FROM BUFA TO EXTENDED CORE, NB REAL VALUES
C
      CALL MOVLEV(BUFA, LBUFA((JB-1)*NB+1), NB)
      RETURN
      END
C
C-----------------------------------------------------------------------
C TEST PROGRAM: PDP 11 MASS STORE I/O FFT SAMPLE
C-----------------------------------------------------------------------
C
C NOTE WELL THAT TYPE COMPLEX DATA MUST EXIST AS ALTERNATING
C REAL/IMAG/REAL/IMAG... ELEMENTS, BOTH IN MASS STORE AND IN FORTRAN
C WORKING ARRAY BUFA;  IN THIS FFT, DIFFERENT SUBROUTINES WILL SET
C DIFFERENT TYPE (REAL OR COMPLEX) FOR ARRAY BUFA.
C
C
      COMMON /FFTCOM/ LUN
C
C COMMON /FFTCOM/ HOLDS LOGICAL UNIT NUMBER FOR MASS STORE I/O
C
      COMMON /MFARG/ MEXA(4), NDIM, ISGN, IDIR, SCAL, IBEX, ICEX, IPAK
      COMMON /MFVAL/ DIMA(4), TDM1, RDM1, FBLK, TBLK, RBLK, RCOR, SIZE
      COMMON /MFINT/ NDMA(4), NTD1, NRD1, NFBK, NTBK, NRBK, NRCR, NSZE
C
C COMMON AREAS /MFARG/,/MFVAL/,/MFINT/ USEFUL FOR RUNNING MASS STORE FFT
C /MFARG/ HOLDS ARGUMENTS USED IN FFT CALLES, MOSTLY EXPONENTS
C HELPER ROUTINE MFPAR COMPUTES VALUES FROM /MFARG/ INTO
C /MFVAL/ (REALS AS SOME LARGE), /MFINT/ (INTEGER EQUIVALENTS IF POSS)
C OR CAN REVERSE-COMPUTE SOME /MFARG/ FROM /MFVAL/
C SEE COMMENTS, ROUTINE MFPAR.
C
      COMPLEX CBUFA(1024)
      REAL BUFA(2048)
      EQUIVALENCE (CBUFA(1),BUFA(1))
C
C BUFA()=CBUFA()    IS WORKING AREA IN CORE STORE FOR FFT AND PROGRAM
C
      LUN = 1
C
C LUN IS LOGICAL UNIT FOR PDP 11 MASS STORE I/O
C
```

```
      LP = 5
      IPRINT = 0
C
C LP IS PRINTER LOGICAL UNIT, IPRINT=+1 COPIOUS, 0 MAX DIFFERENCES ONLY
C
      IRMF = -1
C
C IRMF=-1 REAL ROUTINE RMFFT TEST, 0 COMPLEX ROUTINE CMFFT TEST
C
      ISGN = -1
      IDIR = -1
      IPAK = 1
C
C FFT ARGS, ISGN=+/- 1, IDIR=1 (RMFFT), IDIR=1 OR 0 (CMFFT)
C IPAK=1, 0 OR -1 (RMFFT), NOT USED (CMFFT)
C
C BELOW, PRINT HEADINGS FOR TEST OUTPUT
C
      IF (IPRINT.LE.0) WRITE (LP,9999) IPRINT, IRMF
9999  FORMAT (31H1MASS STORE FFT TEST - IPRINT =, I3, 14H (1=COPIOUS,0=,
     *    11HMAX DIFFS),, 9H   IRMF =, I3, 25H (0=CMFFT TEST,1=RMFFT TE,
     *    3HST)/)
      DIFMG = 0.
C
      MEXA(1) = 8
      MEXA(2) = 6
      NDIM = 2
      IBEX = 7
      ICEX = 11
C
C MORE FFT ARGS, 2**8 ROWS OF 2**6 ELMTS, 2 DIMEN,
C FFT MASS STORE BLOCKS 2**IBEX REALS, BUFA 2**ICEX REALS
C
      IERR = MFPAR(IRMF,0)
      IF (IERR.NE.0) GO TO 130
      CALL ASSIGN(LUN,'FFTEST.DAT',0)
      NWD = NRBK*2
      DEFINE FILE LUN(NFBK,NWD,U,INDX)
C
C PDP 11 MASS STORE OPENED WITH NUM REC=NFBK, NRBK*2 WORDS PER REC
C (NOTE HELPER ROUTN MFPAR RETURNS NFBK AS MAXIMUM FILE SIZE)
C
      NCNT = NRBK
C
C HELPER ROUTINE, NCNT NUM OF ELMTS IN FFT BLOCK, NTBK TOTAL BLOCKS
C
      M = MFSUM(MEXA,NDIM,99)
      SCAL = 1.
      VALU = 0.
      VINC = 1./SIZE
      IF (IRMF.EQ.0) GO TO 30
C
C SWITCH FOR COMPLEX ROUTINE CMFFT AT 50, REAL RMFFT BELOW
C
      DO 20 JB=1,NTBK
        DO 10 I=1,NCNT
          BUFA(I) = VALU
          VALU = VALU + VINC
 10     CONTINUE
        CALL MFWRIT(BUFA, NRBK, JB)
 20   CONTINUE
C
C LOAD RAMP FUNCTION IN REAL ARRAY IN 'RECDS' OF NRBK LENGTH
C (NOTE, PDP 11 MFREAD/MFWRIT (USING FORTRAN I/O) CANNOT ACCEPT 'RECDS'
C OF DIFFERENT LENGTH; OTHERWISE COULD USE NTD1 'RECDS' OF NRD1 HERE)
C
      CALL RMFFT(MEXA, NDIM, ISGN, IDIR, SCAL, BUFA, IBEX, ICEX, IPAK)
C
C REAL MASS STORE ROUTINE RMFFT TO TRANSFORM ARRAY TO COMPLEX RESULT
C
      GO TO 60
C
C BELOW, USE COMPLEX ROUTINE CMFFT,  NCNT NUM OF CMPLX ELMTS IN FFT BLOC
C
 30   NCNT = NCNT/2
      DO 50 JB=1,NTBK
        DO 40 I=1,NCNT
          CBUFA(I) = CMPLX(VALU,0.)
          VALU = VALU + VINC
 40     CONTINUE
        CALL MFWRIT(BUFA, NRBK, JB)
 50   CONTINUE
C
C LOAD RAMP FUNCTION IN COMPLEX ARRAY IN 'RECDS' OF NRBK LENGTH
C (NOTE, PDP 11 MFREAD/MFWRIT (USING FORTRAN I/O) CANNOT ACCEPT 'RECDS'
C OF DIFFERENT LENGTH; OTHERWISE COULD USE NTD1 'RECDS' OF NRD1 HERE)
C
      CALL CMFFT(MEXA, NDIM, ISGN, IDIR, SCAL, BUFA, IBEX, ICEX)
C
C COMPLEX MASS STORE ROUTINE CMFFT TO TRANSFORM ARRAY TO COMPLEX RESULT
C
 60   IF (IPRINT.LE.0) GO TO 90
C
C BELOW, PRINT HEADINGS FOR TEST OUTPUT
C
      WRITE (LP,9999) IPRINT, IRMF
      WRITE (LP,9998) NDIM, ISGN, IDIR, IBEX, ICEX, IPAK,
     *    (MEXA(J),J=1,NDIM)
9998  FORMAT (8H  NDIM =, I3, 8H  ISGN =, I3, 8H  IDIR =, I3, 7H  IBEX ,
     *    1H=, I3, 8H  ICEX =, I3, 8H  IPAK =, I3, 10H  MEXA() =, 3I5)
      WRITE (LP,9997)
9997  FORMAT (8H   INDEX, 12X, 3HFFT)
C
      IERR = MFPAR(-IRMF,0)
      IF (IERR.NE.0) GO TO 130
      IF (IERR.NE.0) GO TO 130
      NCNT = NRBK/2
C
C HELPER ROUTN, NTBK TOTAL NUMB OF NRBK REALS IN FFT BLOCK
C (NOTE IRMF=+1 FOR DATA IN COMPLEX STATE, NCNT ELMTS)
C
C BELOW, PROGRAM CAN ACCESS FFT COMPLEX RESULT (AND FORM COMPARISON, IF
C
      DO 80 JB=1,NTBK
        K = (JB-1)*NCNT
        CALL MFREAD(BUFA, NRBK, JB)
C
C READ 'RECDS' OF NRBK LENGTH, TOTALING NTBK (RECOMPUTED BY MFPAR)
C (NOTE, PDP 11 MFREAD/MFWRIT (USING FORTRAN I/O) CANNOT ACCEPT 'RECDS'
C OF DIFFERENT LENGTH; OTHERWISE COULD USE NTD1 'RECDS' OF NRD1 HERE)
C
        DO 70 I=1,NCNT
          INDEX = J - 1
          IF (IPRINT.LE.0) GO TO 70
          WRITE (LP,9996) INDEX, CBUFA(I)
9996      FORMAT (1X, I5, 2X, 2E13.4)
 70     CONTINUE
 80   CONTINUE
C
C BELOW, INVERT ISGN AND IDIR FOR INVERSE TRANSFORM (COMPLEX-TO-REAL)
```

```
C
   90  ISGN = -ISGN
       IDIR = -IDIR
       SCAL = 1./SIZE
       IF (IRMF.NE.0) CALL RMFFT(MEXA, NDIM, ISGN, IDIR, SCAL, BUFA,
     *     IBEX, ICEX, IPAK)
C
C EITHER ROUTINE RMFFT INVERSE TRANSFORMS ARRAY TO PACKED REAL
C
       IF (IRMF.EQ.0) CALL CMFFT(MEXA, NDIM, ISGN, IDIR, SCAL, BUFA,
     *     IBEX, ICEX)
C
C OR ROUTINE CMFFT INVERSE TRANSFORMS ARRAY
C
       IF (IPRINT.LE.0) GO TO 100
C
C BELOW, PRINT HEADINGS FOR TEST OUTPUT
C
       WRITE (LP,9999) IPRINT, IRMF
       WRITE (LP,9998) NDIM, ISGN, IDIR, IBEX, ICEX, IPAK,
     *     (MEXA(J),J=1,NDIM)
       WRITE (LP,9995)
9995   FORMAT (8H   INDEX, 4X, 8HFFT/2**M, 6X, 5HINPUT, 10X, 4HDIFF)
C
  100  IERR = MFPAR(IRMF,0)
       IF (IERR.NE.0) GO TO 130
       NCNT = NRBK
       IF (IRMF.EQ.0) NCNT = NCNT/2
C
C HELPER ROUTN, NTBK TOTAL NUMB OF NRBK REALS IN FFT BLOCK
C (NCNT IS NUMBER OF ELMTS IN FFT BLOCK, REAL OR CMPLX)
C
C BELOW, COMPARES FFT INVERSE RESULT WITH INITIAL RANDOM INPUT
C
       DIFM = 0.
       VALU = 0.
       DO 120 JB=1,NTBK
         CALL MFREAD(BUFA, NRBK, JB)
C
C READ 'RECDS' OF NRBK LENGTH, TOTALING NTBK (RECOMPUTED BY MFPAR)
C (NOTE, PDP 11 MFREAD/MFWRIT (USING FORTRAN I/O) CANNOT ACCEPT 'RECDS'
C OF DIFFERENT LENGTH; OTHERWISE COULD USE NTD1 'RECDS' OF NRD1 HERE)
C
         DO 110 I=1,NCNT
           IF (IRMF.NE.0) RM = BUFA(I)
           IF (IRMF.EQ.0) RM = REAL(CBUFA(I))
           DIF = ABS(RM-VALU)
           IF (DIF.GT.DIFM) DIFM = DIF
           IF (IPRINT.GT.0) WRITE (LP,9994) I, RM, VALU, DIF
9994       FORMAT (1X, I5, 3(2X, E13.4))
           VALU = VALU + VINC
  110    CONTINUE
  120  CONTINUE
C
C PRINT INVERSE RESULTS (SHOULD BE SAME AS INITIAL RAMP)
C
       IF (IPRINT.GE.0) WRITE (LP,9993) DIFM, NDIM, ISGN, IDIR, IBEX,
     *     ICEX, IPAK, (MEXA(J),J=1,NDIM)
9993   FORMAT (10H MAX DIFF , E11.4, 1X, 8H  NDIM =, I3, 8H  ISGN =, I3,
     *     8H  IDIR =, I3, 8H  IBEX =, I3, 8H  ICEX =, I3, 8H  IPAK =,
     *     I3, 10H  MEXA() =, 3I5)
       IF (DIFM.GT.DIFMG) DIFMG = DIFM
C
       STOP
C
  130  WRITE (LP,9992) IERR
9992   FORMAT (25H IBEX FORCED TOO SMALL OR, 24H NOT ALL /MFINT/ CORRECT,
     *     6H, IERR, I4)
       STOP
       END
C
C-----------------------------------------------------------------------
C SUBROUTINE:  MFREAD
C PDP 11 FORTRAN DIRECT ACCESS READ ROUTINE FOR FFT
C-----------------------------------------------------------------------
C
       SUBROUTINE MFREAD(BUFA, NB, JB)
C
C (LOGICAL UNIT LUN IN COMMON/FFTCOM/ MUST HAVE BEEN OPENED
C PREVIOUSLY; EG.      CALL ASSIGN(LUN,'FILENAME',0)
C AND                  DEFINE FILE LUN(NREC,NWD,U,INDX)
C WHERE NWD=2**(IBEX+1) AND NREC=2**(M-IBEX+1) OR LESS IF IPAK=0 OR -1)
C
C SEE ALSO ALTERNATIVE, FAST MACRO I/O SUBROUTINES
C (THESE ARE TO BE PREFERRED, SINCE SPEED 2 TO 10 TIMES BETTER)
C
C READ BLOCK, INDEX JB, FROM MASS STORE TO BUFA, NB REAL VALUES
C
       COMMON /FFTCOM/ LUN
       REAL BUFA(NB)
C
       READ(LUN'JB)BUFA
       RETURN
       END
C
C-----------------------------------------------------------------------
C SUBROUTINE: MFWRIT
C PDP 11 FORTRAN DIRECT ACCESS WRITE ROUTINE FOR FFT
C-----------------------------------------------------------------------
C
       SUBROUTINE MFWRIT(BUFA, NB, JB)
C
C (LOGICAL UNIT LUN IN COMMON/FFTCOM/ MUST HAVE BEEN OPENED
C PREVIOUSLY; EG.      CALL ASSIGN(LUN,'FILENAME',0)
C AND                  DEFINE FILE LUN(NREC,NWD,U,INDX)
C WHERE NWD=2**(IBEX+1) AND NREC=2**(M-IBEX+1) OR LESS IF IPAK=0 OR -1)
C
C SEE ALSO ALTERNATIVE, FAST MACRO I/O SUBROUTINES
C (THESE ARE TO BE PREFERRED, SINCE SPEED 2 TO 10 TIMES BETTER)
C
C WRITE BLOCK, INDEX JB, FROM BUFA TO MASS STORE, NB REAL VALUES
C
       COMMON /FFTCOM/ LUN
       REAL BUFA(NB)
C
       WRITE(LUN'JB)BUFA
       RETURN
       END
C
C-----------------------------------------------------------------------
C TEST PROGRAM: PDP 11 FAST MACRO I/O FFT SAMPLE
C (REQUIRES CALL TO SYSTEM MACRO TO OPEN FILE FOR I/O - SEE LINE 70)
C
C NOTE THAT FAST MACRO I/O IS MORE EFFICIENT THAN STANDARD FORTRAN I/O
C
C NOTE WELL THAT TYPE COMPLEX DATA MUST EXIST AS ALTERNATING
C REAL/IMAG/REAL/IMAG... ELEMENTS, BOTH IN MASS STORE AND IN FORTRAN
C WORKING ARRAY BUFA;  IN THIS FFT, DIFFERENT SUBROUTINES WILL SET
C DIFFERENT TYPE (REAL OR COMPLEX) FOR ARRAY BUFA.
C-----------------------------------------------------------------------
```

```
C
      COMMON /FFTCOM/ IFDB, IOERR
C
C COMMON /FFTCOM/ HOLDS FDB ADDRESS FOR MACRO I/O (RSX11M), CHANNEL (RT11)
C (SEE COMMENTS IN FAST MACRO ROUTINE MFREAD/MFWRIT)
C
      COMMON /MFARG/ MEXA(4), NDIM, ISGN, IDIR, SCAL, IBEX, ICEX, IPAK
      COMMON /MFVAL/ DIMA(4), TDM1, RDM1, FBLK, TBLK, RBLK, RCOR, SIZE
      COMMON /MFINT/ NDMA(4), NTD1, NRD1, NFBK, NTBK, NRBK, NRCR, NSZE
C
C COMMON AREAS /MFARG/,/MFVAL/,/MFINT/ USEFUL FOR RUNNING MASS STORE FFT
C /MFARG/ HOLDS ARGUMENTS USED IN FFT CALLES, MOSTLY EXPONENTS
C HELPER ROUTINE MFPAR COMPUTES VALUES FROM /MFARG/ INTO
C /MFVAL/ (REALS AS SOME LARGE), /MFINT/ (INTEGER EQUIVALENTS IF POSS)
C OR CAN REVERSE-COMPUTE SOME /MFARG/ FROM /MFVAL/
C SEE COMMENTS, ROUTINE MFPAR.
C
      COMPLEX CBUFA(1024)
      REAL BUFA(2048)
      EQUIVALENCE (CBUFA(1),BUFA(1))
C
C BUFA()=CBUFA()    IS WORKING AREA IN CORE STORE FOR FFT AND PROGRAM
C
      LUN = 1
C
C LUN IS LOGICAL UNIT FOR PDP 11 MASS STORE I/O
C
      LP = 5
      IPRINT = 0
C
C LP IS PRINTER LOGICAL UNIT, IPRINT=+1 COPIOUS, 0 MAX DIFFERENCES ONLY
C
      IRMF = -1
C
C IRMF=-1 REAL ROUTINE RMFFT TEST, 0 COMPLEX ROUTINE CMFFT TEST
C
      ISGN = -1
      IDIR = -1
      IPAK = 1
C
C FFT ARGS, ISGN=+/- 1, IDIR=-1 (RMFFT), IDIR=1 OR 0 (CMFFT)
C IPAK=1, 0 OR -1 (RMFFT), NOT USED (CMFFT)
C
C BELOW, PRINT HEADINGS FOR TEST OUTPUT
C
      IF (IPRINT.LE.0) WRITE (LP,9999) IPRINT, IRMF
9999  FORMAT (31H1MASS STORE FFT TEST - IPRINT =, I3, 14H (1=COPIOUS,0=,
     *     11HMAX DIFFS),, 9H   IRMF =, I3, 25H (0=CMFFT TEST,1=RMFFT TE,
     *     3HST)/)
      DIFMG = 0.
C
      MEXA(1) = 8
      MEXA(2) = 6
      NDIM = 2
      IBEX = 7
      ICEX = 11
C
C MORE FFT ARGS, 2**8 ROWS OF 2**6 ELMTS, 2 DIMEN,
C FFT MASS STORE BLOCKS 2**IBEX REALS, BUFA 2**ICEX REALS
C
      IERR = MFPAR(IRMF,0)
      IF (IERR.NE.0) GO TO 130
C
C (NOTE HELPER ROUTN MFPAR RETURNS NFBK AS MAXIMUM FILE SIZE)
C
      STOP 'ASSIGN AND OPEN FILE HERE'
C
C DELETE ABOVE LINE AND INVOKE SYSTEM MACRO 'OPEN$' (RSX) OR '.OPEN (RT1
C
C FOR EXAMPLE, A FORTRAN CALLABLE SUBROUTINE 'MFOPEN' COULD OPEN FILE
C 'FFTTEST.DAT', RETURNING IFDB=FDB ADDRESS (RSX11M) IN COMMON/FFTCOM/
C OF CHANNEL NUMBER (RT11), FOR USE BY MACRO MFREAD/MFWRIT, THUS:
C
C CALL MFOPEN(LUN,'FFTTEST.DAT',IFDB)
C
      IOERR = 0
C
C PRESET IOERR IN /FFTCOM/; ZERO IF NO ERRORS IN I/O
C
      NCNT = NRD1
C
C HELPER ROUTINE, NCNT NUM OF ELMTS IN 'FIRST' DIMEN, NTD1 TOTAL BLOCKS
C
      M = MFSUM(MEXA,NDIM,99)
      SCAL = 1.
      VALU = 0.
      VINC = 1./SIZE
      IF (IRMF.EQ.0) GO TO 30
C
C SWITCH FOR COMPLEX ROUTINE CMFFT AT 50, REAL RMFFT BELOW
C
      DO 20 JB=1,NTD1
        DO 10 I=1,NCNT
          BUFA(I) = VALU
          VALU = VALU + VINC
 10     CONTINUE
        CALL MFWRIT(BUFA, NRD1, JB)
 20   CONTINUE
C
C LOAD RAMP FUNCTION IN REAL ARRAY IN 'RECDS' OF NRD1 LENGTH
C (NOTE, 'MACRO' MFREAD/MFWRIT CAN ACCEPT 'RECORDS' OF DIFFERENT LENGTH
C SO CAN USE NTD1,NRD1 HERE INSTEAD OF NTBK,NRBK AS IN FORTRAN MASS ST
C
      CALL RMFFT(MEXA, NDIM, ISGN, IDIR, SCAL, BUFA, IBEX, ICEX, IPAK)
C
C REAL MASS STORE ROUTINE RMFFT TO TRANSFORM ARRAY TO COMPLEX RESULT
C
      GO TO 60
C
C BELOW, USE COMPLEX ROUTINE CMFFT,  NCNT NUM OF CMPLX ELMTS IN 'FIRST'
C
 30   NCNT = NCNT/2
      DO 50 JB=1,NTD1
        DO 40 I=1,NCNT
          CBUFA(I) = CMPLX(VALU,0.)
          VALU = VALU + VINC
 40     CONTINUE
        CALL MFWRIT(BUFA, NRD1, JB)
 50   CONTINUE
C
C LOAD RAMP FUNCTION IN COMPLEX ARRAY IN 'RECDS' OF NRD1 LENGTH
C (NOTE, 'MACRO' MFREAD/MFWRIT CAN ACCEPT 'RECORDS' OF DIFFERENT LENGTH
C SO CAN USE NTD1,NRD1 HERE INSTEAD OF NTBK,NRBK AS IN FORTRAN MASS ST
C
      CALL CMFFT(MEXA, NDIM, ISGN, IDIR, SCAL, BUFA, IBEX, ICEX)
C
C COMPLEX MASS STORE ROUTINE CMFFT TO TRANSFORM ARRAY TO COMPLEX RESULT
C
 60   IF (IPRINT.LE.0) GO TO 90
C
```

```
C BELOW, PRINT HEADINGS FOR TEST OUTPUT
C
      WRITE (LP,9999) IPRINT, IRMF
      WRITE (LP,9998) NDIM, ISGN, IDIR, IBEX, ICEX, IPAK,
     *    (MEXA(J),J=1,NDIM)
9998  FORMAT (8H  NDIM =, I3, 8H  ISGN =, I3, 8H  IDIR =, I3, 7H  IBEX ,
     *    1H=, I3, 8H  ICEX =, I3, 8H  IPAK =, I3, 10H  MEXA() =, 3I5)
      WRITE (LP,9997)
9997  FORMAT (8H   INDEX, 12X, 3HFFT)
C
      IERR = MFPAR(-IRMF,0)
      IF (IERR.NE.0) GO TO 130
      NCNT = NRD1/2
C
C HELPER ROUTN, NTD1 TOTAL NUMB OF NRD1 REALS IN 'FIRST' DIMEN
C (NOTE IRMF=+1 FOR DATA IN COMPLEX STATE, NCNT ELMTS)
C
C BELOW, PROGRAM CAN ACCESS FFT COMPLEX RESULT (AND FORM COMPARISON, IF
C
      DO 80 JB=1,NTD1
        K = (JB-1)*NCNT
        CALL MFREAD(BUFA, NRD1, JB)
C
C READ 'RECDS' OF NRD1 LENGTH, TOTALING NTD1 (RECOMPUTED BY MFPAR)
C (NOTE, 'MACRO' MFREAD/MFWRIT CAN ACCEPT 'RECORDS' OF DIFFERENT LENGTH
C SO CAN USE NTD1,NRD1 HERE INSTEAD OF NTBK,NRBK AS IN FORTRAN MASS ST
C
        DO 70 I=1,NCNT
          INDEX = J - 1
          IF (IPRINT.LE.0) GO TO 70
          WRITE (LP,9996) INDEX, CBUFA(I)
9996      FORMAT (1X, I5, 2X, 2E13.4)
  70    CONTINUE
  80  CONTINUE
C
C BELOW, INVERT ISGN AND IDIR FOR INVERSE TRANSFORM (COMPLEX-TO-REAL)
C
  90  ISGN = -ISGN
      IDIR = -IDIR
      SCAL = 1./SIZE
      IF (IRMF.NE.0) CALL RMFFT(MEXA, NDIM, ISGN, IDIR, SCAL, BUFA,
     *    IBEX, ICEX, IPAK)
C
C EITHER ROUTINE RMFFT INVERSE TRANSFORMS ARRAY TO PACKED REAL
C
      IF (IRMF.EQ.0) CALL CMFFT(MEXA, NDIM, ISGN, IDIR, SCAL, BUFA,
     *    IBEX, ICEX)
C
C OR ROUTINE CMFFT INVERSE TRANSFORMS ARRAY
C
      IF (IPRINT.LE.0) GO TO 100
C
C BELOW, PRINT HEADINGS FOR TEST OUTPUT
C
      WRITE (LP,9999) IPRINT, IRMF
      WRITE (LP,9998) NDIM, ISGN, IDIR, IBEX, ICEX, IPAK,
     *    (MEXA(J),J=1,NDIM)
      WRITE (LP,9995)
9995  FORMAT (8H   INDEX, 4X, 8HFFT/2**M, 6X, 5HINPUT, 10X, 4HDIFF)
C
 100  IERR = MFPAR(IRMF,0)
      IF (IERR.NE.0) GO TO 130
      NCNT = NRD1
      IF (IRMF.EQ.0) NCNT = NCNT/2
C
C HELPER ROUTN, NTD1 TOTAL NUMB OF NRD1 REALS IN 'FIRST' DIMEN
C (NCNT IS NUMBER OF ELMTS IN 'FIRST' DIMEN, REAL OR CMPLX)
C
C BELOW, COMPARES FFT INVERSE RESULT WITH INITIAL RANDOM INPUT
C
      DIFM = 0.
      VALU = 0.
      DO 120 JB=1,NTD1
        CALL MFREAD(BUFA, NRD1, JB)
C
C READ 'RECDS' OF NRD1 LENGTH, TOTALING NTD1 (RECOMPUTED BY MFPAR)
C (NOTE, 'MACRO' MFREAD/MFWRIT CAN ACCEPT 'RECORDS' OF DIFFERENT LENGTH
C SO CAN USE NTD1,NRD1 HERE INSTEAD OF NTBK,NRBK AS IN FORTRAN MASS ST
C
        DO 110 I=1,NCNT
          IF (IRMF.NE.0) RM = BUFA(I)
          IF (IRMF.EQ.0) RM = REAL(CBUFA(I))
          DIF = ABS(RM-VALU)
          IF (DIF.GT.DIFM) DIFM = DIF
          IF (IPRINT.GT.0) WRITE (LP,9994) I, RM, VALU, DIF
9994      FORMAT (1X, I5, 3(2X, E13.4))
          VALU = VALU + VINC
 110    CONTINUE
 120  CONTINUE
C
C PRINT INVERSE RESULTS (SHOULD BE SAME AS INITIAL RAMP)
C
      IF (IPRINT.GE.0) WRITE (LP,9993) DIFM, NDIM, ISGN, IDIR, IBEX,
     *    ICEX, IPAK, (MEXA(J),J=1,NDIM)
9993  FORMAT (10H MAX DIFF , E11.4, 1X, 8H  NDIM =, I3, 8H  ISGN =, I3,
     *    8H  IDIR =, I3, 8H  IBEX =, I3, 8H  ICEX =, I3, 8H  IPAK =,
     *    I3, 10H  MEXA() =, 3I5)
      IF (DIFM.GT.DIFMG) DIFMG = DIFM
C
      STOP
C
 130  WRITE (LP,9992) IERR
9992  FORMAT (25H IBEX FORCED TOO SMALL OR, 24H NOT ALL /MFINT/ CORRECT,
     *    6H, IERR, I4)
      STOP
      END
;
;-----------------------------------------------------------------
;  SUBROUTINE: MFREAD, MFWRIT
;  PDP11 RSX11M OR RT11 FAST MACRO I/O ROUTINES FOR FFT
;-----------------------------------------------------------------
;
;     SUBROUTINE MFREAD(BUFA,NB,JB)
;     SUBROUTINE MFWRIT(BUFA,NB,JB)
;
;C READ  BLOCK, INDEX JB, FROM MASS STORE TO BUFA, NB REAL VALUES
;C WRITE BLOCK, INDEX JB, FROM BUFA TO MASS STORE, NB REAL VALUES
;
;     COMMON/FFTCOM/IFDB,IOERR  (RSX11 FDB ADDRESS, ERROR)
;OR   COMMON/FFTCOM/ICHAN,IOERR (RT11  CHANNEL NUM, ERROR)
;
;NOTE1: FILE MUST BE PROPERLY OPEN FOR READ/WRITE ACCESS AND WITH
;       RSX11M FDB ADDRESS OR RT11 CHANNEL NUMBER IN FIRST WORD
;       OF COMMON/FFTCOM/ (IE. INVOKE RSX11M OR RT11 OPEN MACROS).
;
;NOTE2: I/O ERRORS, IF THEY OCCUR, ARE RETURNED IN SECOND WORD
;       OF COMMON/FFTCOM/.  THIS WORD (IOERR) REMAINS UNCHANGED
;       IF NO ERROR OCCURS - THUS, IF PRESET TO 0 BEFORE CALLING
;       MFREAD/MFWRIT (OR MASS STORE FFT), WILL FINALLY BE
;       0 IF NO ERRORS, OR NON ZERO=LAST ERROR (SEE FOLLOWING LABEL L4:)
```

```
;
;NOTE3: NB MUST BE AN INTEGRAL POWER OF 2; 128 OR MORE MOST EFFICIENT
;        (NB.LE.64 USES LOCAL 256 WORD JBUF TO HOLD SECTOR;
;         WRITE ALWAYS WRITES THIS SECTOR TO FILE; SLOW BUT SAFE).
;
;NOTE4: READY TO ASSEMBLE FOR RSX11M. FOR RT11, ERASE FOLLOWING LINE:
        RSX=0
;
        .IF     DF,RSX
; CONDITIONAL SECTION FOR RSX11M (V3)
        .TITLE  MFREAD(RSX11M)
        .MCALL  READ$,WRITE$,WAIT$
        .PSECT  FFTCOM,RW,D,GBL,REL,OVR
IFDB:   .WORD   0         ;FDB ADDRESS
IOERR:  .WORD   0         ;ERROR FLAG (UNCHANGED IF NO ERROR)
        .PSECT
        .ENDC
;
        .IF     NDF,RSX
; CONDITIONAL SECTION FOR RT11 (V3)
        .TITLE  MFREAD(RT11)
        .MCALL  .READW,.WRITW
        ERRBYT=52
        .PSECT  FFTCOM,RW,D,GBL,REL,OVR
ICHAN:  .WORD   0         ;CHANNEL NUMBER
IOERR:  .WORD   0         ;ERROR FLAG (UNCHANGED IF NO ERROR)
        .PSECT  MFREAD,RW,I,LCL,REL,OVR
        .ENDC
;
        .GLOBL  MFREAD,MFWRIT
;
MFREAD:  MOV     #1,R0              ;MFREAD SETS JSW=1
        BR      L1
;
MFWRIT:  MOV     #-1,R0             ;MFWRIT SETS JSW=-1
;
L1:     MOV     R0,JSW             ;HOLD JSW
        TST     (R5)+              ;IGNORE ARGUMENT COUNT
        MOV     (R5)+,R3           ;R3=ADDRESS OF 'BUFA'
        MOV     @(R5)+,R2          ;R2='NB'
        MOV     @(R5)+,R1          ;R1='JB'
; FINISH ACCESSING SUBROUTINE ARGUMENTS
;
        DEC     R1                 ;R1=JB-1
        ASL     R2                 ;R2=WORD COUNT, NB*2
        MOV     R2,R5              ;COPY TO R5
        BEQ     L6                 ;FINISH IF COUNT ZERO
        CLR     R0                 ;CLEAR R0 FOR WORD OFFSET
        CMP     R5,#256.           ;CHECK WHETHER WHOLE SECTORS
        BLT     L12                ;GO TO L12 IF NOT (MORE COMPLEX)
;
; BELOW, SCALE R1 TO SECTOR OFFSET (ASSUMES NWD POWER OF 2)
L10:    BIT     #256.,R5           ;LOOK FOR BIT AT 256.
        BNE     L11                ;R1 COMPLETE WHEN FOUND
        ASL     R1                 ;SCALE R1 ACCORDING TO R5
        ASR     R5
        BR      L10
;
; NOW HAVE      R1=SECTOR START (FIRST=0)
;               R2=WORD COUNT
;               R3=BUFA ADDRESS
;
L11:    MOV     JSW,JSWIO          ;SET UP FOR READ/WRITE
        BGT     L13
        MOV     #-1,SECHLD         ;DIRECT WRITE, ENSURE JBUF LOOKS EMPTY
L13:    JSR     PC,INOUT           ;CALL DIRECT INPUT/OUTPUT
L6:     RTS     PC                 ;RETURN FROM SUBROUTINE
; END OF SIMPLE DIRECT INPUT/OUTPUT
;
        .IF     DF,RSX
; CONDITIONAL SECTION FOR RSX11M
INOUT:  ASL     R2                 ;RSX11M NEEDS BYTE COUNT=NB*4
        INC     R1                 ;RSX11M SECTOR STARTS WITH 1
        MOV     R1,SECTL           ;HOLD AS LOWER SECTOR VALUE
        MOV     IFDB,R4            ;R4=FDB ADDRESS
        MOVB    F.LUN(R4),R1       ;USE LUN FOR EVENT FLAG
;
        TST     JSWIO
        BLT     L5                 ;BRANCH IF WRITE
;
        MOV     F.EFBK+2(R4),R0    ;GET LOW SECTOR OF EOF
        DEC     R0                 ;ALLOW FOR HEADER SECTOR
        CMP     R0,SECTL           ;TEST IF SECTOR EXISTS YET
        BLT     L7                 ;NO READ IF DOES NOT EXIST
;
        READ$   R4,R3,R2,#SECT,R1  ;FDB,BUF,BYTCNT,SECT,EVFLG
;
L4:     MOV     IFDB,R4            ;R4=FDB ADDRESS
        MOVB    F.LUN(R4),R1       ;USE LUN FOR EVENT FLAG
        WAIT$   R4,R1              ;WAIT FOR I/O FINISH
        BCC     L7                 ;NO ERROR IF CARRY CLEAR
;
        MOV     IFDB,R4            ;R4=FDB ADDRESS
        MOVB    F.ERR(R4),R4       ;HOLD NEGATIVE ERROR CODE
        MOV     R4,IOERR           ;IN IOERR IN COMMON/FFTCOM/
L7:     RTS     PC
;
L5:     WRITE$  R4,R3,R2,#SECT,R1  ;FDB,BUF,BYTCNT,SECT,EVFLG
        BR      L4
;
SECT:   .WORD   0         ;HOLDS SECTOR VALUE (HIGHER, ALWAYS 0)
SECTL:  .WORD   0         ;(LOWER, COMPUTED)
        .ENDC
;
        .IF     NDF,RSX
;CONDITIONAL SECTION FOR RT11
INOUT:  MOV     ICHAN,R4           ;R4=CHANNEL NUMBER
;
        TST     JSWIO
        BLT     L5                 ;BRANCH IF WRITE
;
        .READW  #AREA,R4,R3,R2,R1        ;AREA,CHAN,BUF,WDCNT,SECT
;
L4:     BCC     L7                 ;NO ERROR IF CARRY CLEAR
;
        MOVB    ERRBYT,R4          ;HOLD ERROR CODE + 1 (TO MAKE NON ZERO)
        INC     R4                  ;ERRORS 0,1,2 GO TO 1,2,3
        MOV     R4,IOERR           ;IN IOERR IN COMMON/FFTCOM/
L7:     RTS     PC
;
L5:     .WRITW  #AREA,R4,R3,R2,R1        ;AREA,CHAN,BUF,WDCNT,SECT
        BR      L4
AREA:   .BLKW   10        ;AREA FOR RT11 I/O MACROS USE
        .ENDC
;
; BELOW, MORE COMPLEX HANDLING OF NWD.LT.256 (PART SECTORS)
L12:    CLC
        ROR     R1                 ;SCALE R1/R0 RIGHT ACCORDING TO R5
        ROR     R0                 ;R0 HOLDS FLOW OUT OF R1
        ASL     R5
```

```
        BIT     #256.,R5        ;LOOK FOR BIT 256
        BNE     L20             ;R1/R0 COMPLETE IF FOUND
        BR      L12
;
L20:    SWAB    R0
        ASL     R0
;
; NOW HAVE      R0=BYTE OFFSET IN LOCAL BUFFER
;               R1=SECTOR REQUIRED
;               R2=WORD COUNT
;               R3=USER BUFA ADDRESS
;
; BELOW, TRANSFER BETWEEN LOCAL AND USER BUFFER
        ADD     #JBUF,R0        ;R0=JBUF ADDRESS + OFFSET
        CMP     SECHLD,R1               ;CHECK CURRENT SECTOR LOADED
        BNE     L24             ;MUST FIRST LOAD SECTO AT L24 IF NEC
L21:    TST     JSW             ;TEST WHETHER READ/WRITE
        BLT     L23
;
; BELOW, 'READ' PART SECTOR
L22:    MOV     (R0)+,(R3)+     ;COPY LOCAL TO USER BUFA
        DEC     R2
        BGT     L22
        JMP     L6              ;FINISH
;
; BELOW, 'WRITE' PART SECTOR
L23:    MOV     (R3)+,(R0)+     ;COPY USER TO LOCAL BUF
        DEC     R2
        BGT     L23
        MOV     #-1,JSWIO               ;SET UP I/O FOR WRITE
        JSR     PC,LINOUT       ;WRITE FROM LOCAL JBUF (SLOW BUT SAFE)
        JMP     L6              ;FINISH
;
; BELOW, LOAD LOCAL JBUF, IF POSSIBLE
L24:    MOV     R1,SECHLD       ;HOLD CURRENT SECTOR NUM
        MOV     #1,JSWIO        ;SET UP FOR READ
        JSR     PC,LINOUT       ;DO LOCAL READ TO JBUF
        BR      L21
;
; BELOW, LOCAL INPUT/OUTPUT ROUTINE
LINOUT: MOV     R0,-(SP)        ;SAVE REGISTERS R0 TO R3
        MOV     R1,-(SP)
        MOV     R2,-(SP)
        MOV     R3,-(SP)
        MOV     #256.,R2        ;R2=256. FOR WHOLE SECTOR
        MOV     #JBUF,R3        ;R3 IS LOCAL JBUF ADDRESS
        JSR     PC,INOUT        ;CALL INOUT ROUTINE
        MOV     (SP)+,R3        ;RESTORE REGISTERS R0 TO R3
        MOV     (SP)+,R2
        MOV     (SP)+,R1
        MOV     (SP)+,R0
        RTS     PC
;
; BELOW, LOCAL VARIABLES
JSW:    .WORD   0       ;JSW=+1 MFREAD, -1 MFWRIT
JSWIO:  .WORD   0       ;ACTUAL JSW USED FOR INOUT ROUTINE
SECHLD: .WORD   -1      ;CURRENT SECTOR LOADED IN JBUF
JBUF:   .BLKW   256.    ;JBUF 256 WORD LOCAL BUFFER FOR NWD.LT.256
;
        .END
```

1.6

Chirp z-Transform Algorithm Program

L. R. Rabiner

Acoustics Research Dept.
Bell Laboratories
Murray Hill, NJ 07974

1. Purpose

This program can be used to numerically evaluate the z-transform of a sequence along a circular or spiral contour beginning at any arbitrary point in the z-plane.

2. Method

The program is an implementation of the chirp z-transform algorithm of Rabiner, Schafer, and Rader [1]. If we call the input sequence $x(n)$, defined for $0 \leqslant n \leqslant N-1$, then the z-transform of $x(n)$, evaluated along the contour $z = AW^{\alpha}$ (with parameter α) at the set of points

$$z_k = AW^k \qquad k = 0,1,...,M-1 \tag{1}$$

is

$$X(z_k) = X_k = \sum_{n=0}^{N-1} x(n) A^{-n} W^{nk} \tag{2}$$

This program provides a computationally efficient algorithm for evaluating Eq. (2) at the set of M-points given by Eq. (1) by implementing it as a discrete convolution [1].

The general contour of Eq. (1) is that of a spiral in the z-plane. This is more clearly seen if we let the complex constants A and W be expressed in the form

$$A = A_0 e^{j2\pi\theta_0} \tag{3}$$
$$W = W_0 e^{j2\pi\phi_0} \tag{4}$$

As shown in Figure 1, the general z-plane contour begins with the point $z = A$ and, depending on the value of W, spirals in or out with respect to the origin. If $W_0 = 1$, the contour is an arc of a circle. The angular spacing of the samples is $2\pi\phi_0$. The equivalent s-plane contour with $s = (\ln z)/T$ begins with the point

$$s_0 = \sigma_0 + j\omega_0 = \frac{1}{T} \ln A \tag{5}$$

and the general point on the s-plane contour is

$$s_k = s_0 + k(\Delta\sigma + j\Delta\omega) = \frac{1}{T}(\ln A - k \ln W) \qquad 0 \leqslant k \leqslant M-1 \tag{6}$$

Since A and W are arbitrary complex numbers we see that the points s_k lie on an arbitrary straight line segment of arbitrary length and sampling density.

3. Program Description

3.1 Usage

The program consists of the CZT subroutine, two associated subroutines (RECUR and DECUR), and an FFT subroutine (FFT842). The user passes the data sequence and all the parameters required to specify the contour and the appropriate transform sizes for processing, and the subroutine returns the

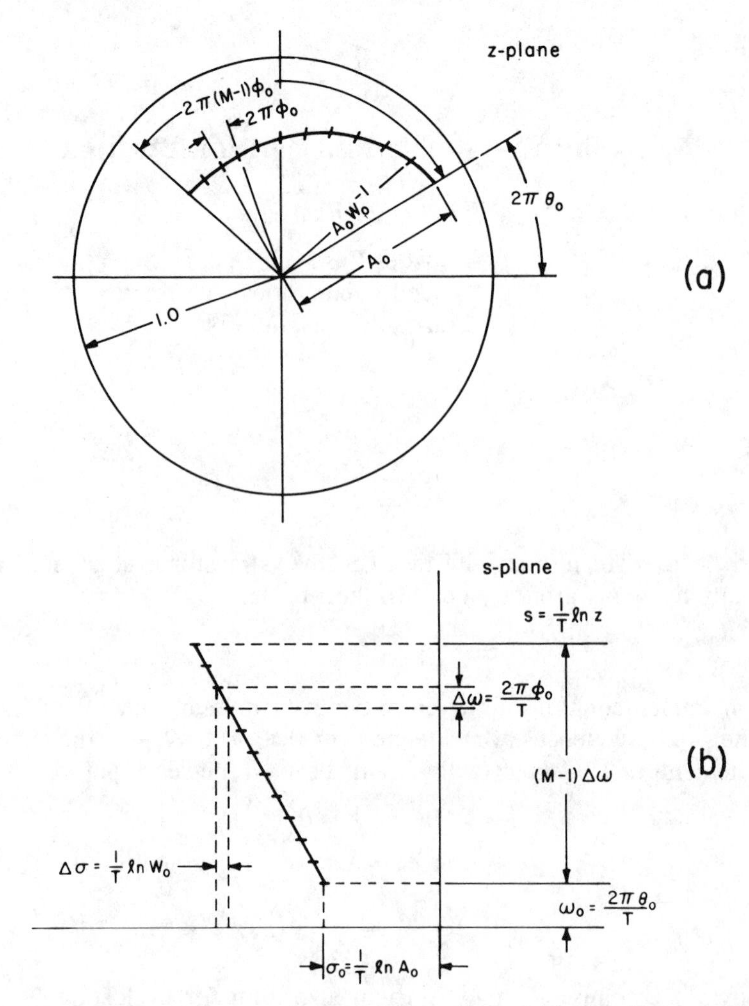

Fig. 1 An illustration of the independentent parameters of the CZT algorithm. (a) How the z-transform is evaluated on a spiral contour starting at the point $z = A$. (b) The corresponding straight line contour and independent parameters in the S-plane.

z-transform of the sequence in the same array in which the input sequence was passed. The FFT subroutine is a radix 2 complex input FFT [2]. As such the CZT routine requires that all relevant FFT's are for sequences whose length is a power of 2.

3.2 Description of Parameters of CZT Subroutine

The basic call to the CZT subroutine is of the form

CALL CZT (XR, XI, NDATA, NOPTS, DLTSIG, DLTOMG, WTR, WTI, SIG0, OME, NTR, NFFT, FS)

where

XR Array of size NFFT. On input XR contains the real part of the input data (Re$[x(n)]$); on output XR contains the real part of the transform (Re$[X_k]$).

XI Array of size NFFT. On input XI contains the imaginary part of the input data (Im$[x(n)]$); on output XI contains the imaginary part of the transform (Im$[X_k]$).

NDATA The number of points in the input sequence (N).

NOPTS The number of points in the output sequence (M).

DLTSIG Increment in sigma along the CZT contour in Hz. (DLTSIG $= \Delta\sigma/2\pi$ in Figure 1).

DLTOMG Increment in omega along the CZT contour in Hz. (DLTOMG $= \Delta\omega/2\pi$ in Figure 1).

WTR Array of size NFFT which holds the real part of the transform of $W^{(-n^2/2)}$.

WTI	Array of size NFFT which holds the imaginary part of the transform of $W^{(-n^2/2)}$.
SIG0	Initial value of σ of CZT contour in Hz. (SIG0 = $\sigma_0/2\pi$ in Figure 1).
OME	Initial value of ω of CZT contour in Hz. (OME = $\omega_0/2\pi$ in Figure 1).
NTR	Rotation factor on input data to make input appear to begin at n = −NTR instead of n = 0. (Set NTR = 0 to avoid using this option.)
NFFT	Size of FFT performed internally in CZT subroutine. NFFT is a power of 2 that is not less than (NDATA+NOPTS−1).
FS	Sampling frequency of data in Hz.
FFT842	FFT subroutine with calling sequence CALL FFT842(IT,NFFT,XR,XI) where

IT	= 0 for direct FFT
	= 1 for inverse FFT
NFFT	= Size of FFT
XR	= Array of length NFFT - real part of data
XI	= Array of length NFFT - imaginary part of data

3.3 Dimension Requirements

The DIMENSION statement in the calling program should be modified according to the requirements of each particular problem. The required dimensions of the arrays are

XR	must be a power of 2 no less than NDATA + NOPTS −1
XI	same as XR
WTR	same as XR
WTI	same as XR

3.4 Summary of User Requirements

(1) Specify input sequences XR and XI, and parameters NDATA, NOPTS, DLTSIG, DLTOMG, SIG0, OME, NTR, NFFT, FS and supply FFT routine FFT842.

(2) Call CZT.

(3) Output is obtained in arrays XR and XI.

4. Test Problem

Compute the z-transform of a delayed impulse along any CZT contour. The test program (See appendix) accepts as input from the teletype the CZT parameters NDATA, NOPTS, FS, SIG0, OME0, DLTSIG, and DLTOMG, as well as the delay parameter IDEL to form the sequence

$$x(n) = \begin{cases} 1 & n = \text{IDEL} \\ 0 & \text{otherwise} \end{cases}$$

The test program prints the input information, determines the FFT size, and prints the real and imaginary parts of the transform of the sequence.

The subroutines CZT, RECUR and DECUR are given in the appendix. Tables 1-4 give outputs of the test program for 4 test runs. Figures 2-5 show plots of the output sequences (both real and imaginary parts) for these test runs. The parameters used in these runs are printed in Tables 1-4.

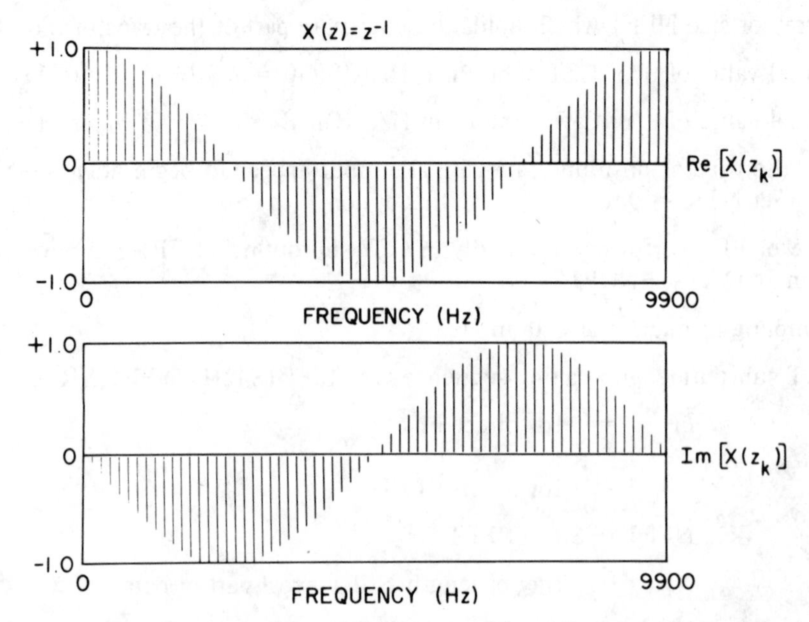

Fig. 2 Output sequences for the input signal $x(n) = 1$, $n = 1$ and $x(n) = 0$ for all other n, evaluated on the contour $z_k = e^{j\frac{2\pi}{100}k}$, $k = 0, 1, \ldots, 99$.

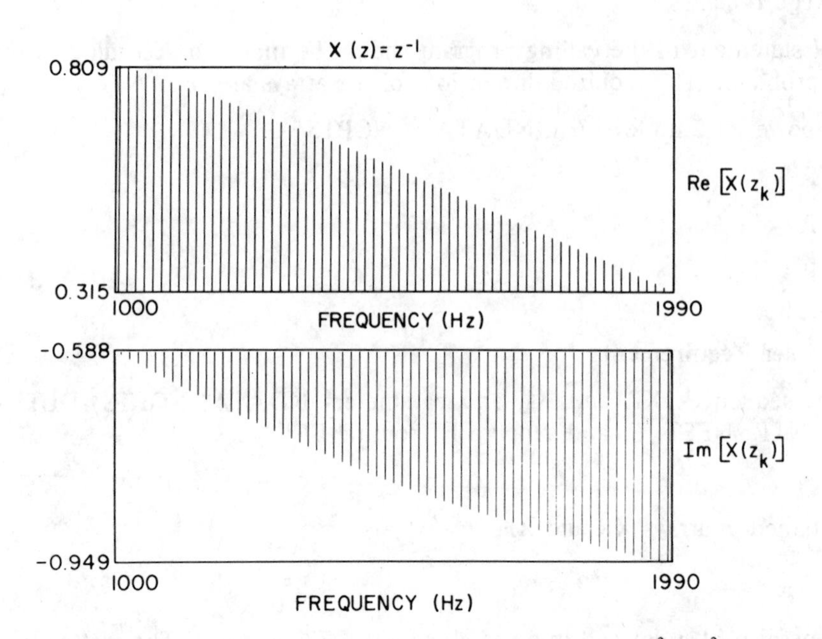

Fig. 3 Output sequences for the same input as Fig. 2, evaluated on the contour $z_k = e^{j\frac{2\pi}{100}} e^{j\frac{2\pi}{1000}k}$, $k = 0, 1, \ldots, 99$.

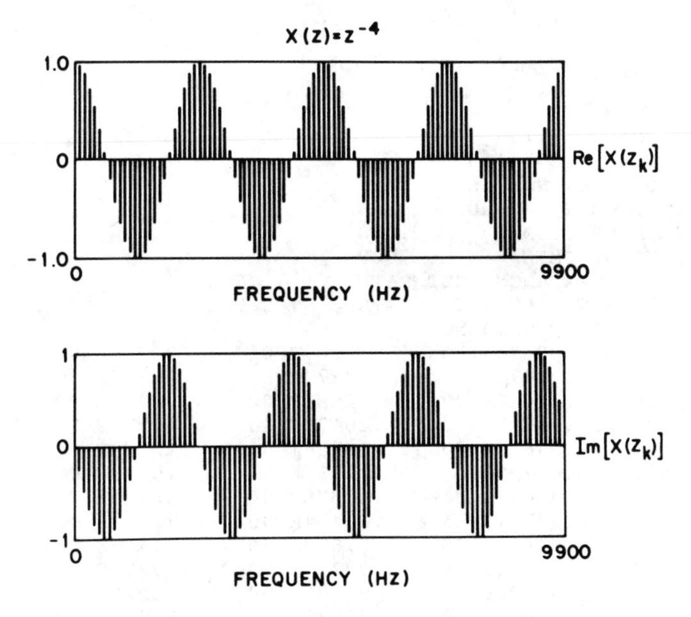

Fig. 4 Output sequences for the input signal $x(n) = 1$, $n = 4$, and $x(n) = 0$ for all other n, evaluated on the contour $z_k = e^{j\frac{2\pi}{100}k}$, $k = 0, 1, \ldots, 99$.

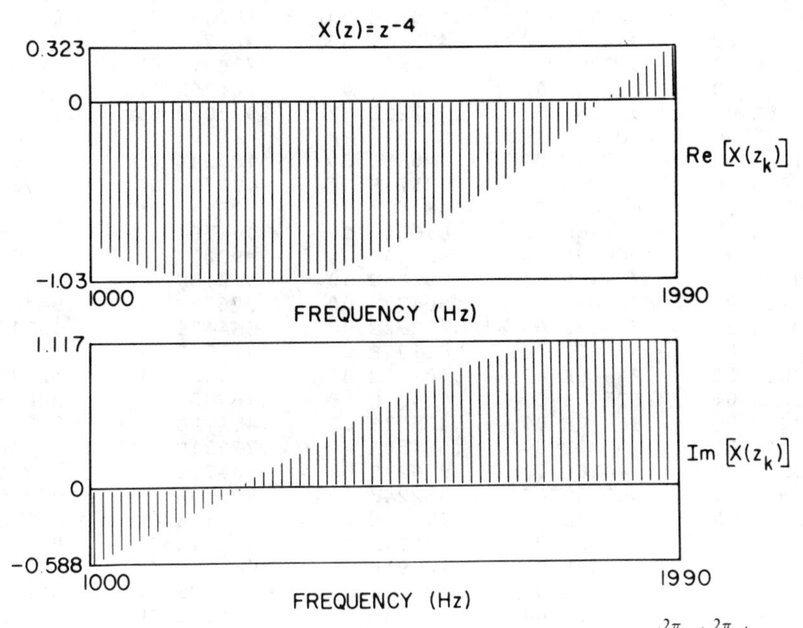

Fig. 5 Output sequences for the same input as Fig. 4, evaluated on the contour $z_k = e^{j\frac{2\pi}{10}} e^{j\frac{2\pi}{1000}k} e^{-k/10000}$, $k = 0, 1, \ldots, 99$.

References

1. L. R. Rabiner, R. W. Schafer, and C. M. Rader, "The Chirp z-Transform Algorithm", *IEEE Trans. Audio and Electroacoust.*, Vol. AU-17, No. 2, pp. 86-92, June 1969.

2. G. D. Bergland and M. T. Dolan, "Fast Fourier Transform Algorithms", in *Programs for Digital Signal Processing*, IEEE Press, 1979.

Table 1

```
NO OF DATA POINTS=  100
NO OF OUTPUT POINTS=  100
DELAYED IMPULSE AT SAMPLE NO=    1
SAMPLING FREQUENCY= 10000.000
INITIAL VALUE OF SIGMA=        .000
INITIAL VALUE OF OMEGA=        .000
INCREMENT IN SIGMA ALONG CZT PATH=         .000
INCREMENT IN OMEGA ALONG CZT PATH=    100.000
REAL PART OF OUTPUT SEQUENCE
    .99996E 00     .99804E 00     .99207E 00     .98224E 00     .96854E 00
    .95102E 00     .92971E 00     .90479E 00     .87627E 00     .84426E 00
    .80895E 00     .77048E 00     .72889E 00     .68450E 00     .63739E 00
    .58777E 00     .53579E 00     .48173E 00     .42575E 00     .36811E 00
    .30899E 00     .24869E 00     .18736E 00     .12529E 00     .62772E-01
  -.50576E-05   -.62791E-01   -.12531E 00   -.18733E 00   -.24863E 00
  -.30901E 00   -.36810E 00   -.42576E 00   -.48173E 00   -.53580E 00
  -.58779E 00   -.63741E 00   -.68454E 00   -.72891E 00   -.77045E 00
  -.80895E 00   -.84428E 00   -.87622E 00   -.90478E 00   -.92974E 00
  -.95097E 00   -.96852E 00   -.98225E 00   -.99207E 00   -.99798E 00
  -.99994E 00   -.99796E 00   -.99205E 00   -.98220E 00   -.96853E 00
  -.95101E 00   -.92968E 00   -.90479E 00   -.87625E 00   -.84426E 00
  -.80894E 00   -.77050E 00   -.72888E 00   -.68448E 00   -.63740E 00
  -.58777E 00   -.53580E 00   -.48173E 00   -.42573E 00   -.36810E 00
  -.30900E 00   -.24868E 00   -.18736E 00   -.12532E 00   -.62778E-01
  -.17870E-04    .62786E-01    .12533E 00    .18734E 00    .24866E 00
    .30901E 00     .36811E 00     .42574E 00     .48172E 00     .53580E 00
    .58775E 00     .63739E 00     .68450E 00     .72891E 00     .77047E 00
    .80895E 00     .84426E 00     .87620E 00     .90476E 00     .92973E 00
    .95097E 00     .96854E 00     .98226E 00     .99207E 00     .99799E 00
IMAGINARY PART OF OUTPUT SEQUENCE
  -.12159E-04   -.62797E-01   -.12532E 00   -.18734E 00   -.24865E 00
  -.30900E 00   -.36810E 00   -.42575E 00   -.48173E 00   -.53582E 00
  -.58777E 00   -.63742E 00   -.68451E 00   -.72894E 00   -.77043E 00
  -.80894E 00   -.84426E 00   -.87624E 00   -.90476E 00   -.92975E 00
  -.95097E 00   -.96853E 00   -.98224E 00   -.99207E 00   -.99795E 00
  -.99997E 00   -.99796E 00   -.99205E 00   -.98218E 00   -.96854E 00
  -.95099E 00   -.92967E 00   -.90477E 00   -.87628E 00   -.84427E 00
  -.80898E 00   -.77048E 00   -.72892E 00   -.68449E 00   -.63740E 00
  -.58776E 00   -.53582E 00   -.48172E 00   -.42576E 00   -.36808E 00
  -.30900E 00   -.24868E 00   -.18735E 00   -.12530E 00   -.62774E-01
  -.57220E-05    .62794E-01    .12531E 00    .18735E 00    .24864E 00
    .30902E 00     .36810E 00     .42573E 00     .48173E 00     .53580E 00
    .58774E 00     .63740E 00     .68450E 00     .72891E 00     .77046E 00
    .80899E 00     .84426E 00     .87627E 00     .90477E 00     .92974E 00
    .95098E 00     .96855E 00     .98224E 00     .99208E 00     .99796E 00
    .99994E 00     .99797E 00     .99208E 00     .98218E 00     .96853E 00
    .95100E 00     .92972E 00     .90477E 00     .87627E 00     .84424E 00
    .80895E 00     .77049E 00     .72891E 00     .68449E 00     .63741E 00
    .58775E 00     .53578E 00     .48171E 00     .42573E 00     .36807E 00
    .30898E 00     .24867E 00     .18735E 00     .12530E 00     .62779E-01
```

Table 2

```
NO OF DATA POINTS=  100
NO OF OUTPUT POINTS=  100
DELAYED IMPULSE AT SAMPLE NO=    1
SAMPLING FREQUENCY= 10000.000
INITIAL VALUE OF SIGMA=       .000
INITIAL VALUE OF OMEGA=  1000.000
INCREMENT IN SIGMA ALONG CZT PATH=       .000
INCREMENT IN OMEGA ALONG CZT PATH=    10.000
REAL PART OF OUTPUT SEQUENCE
   .80899E 00     .80530E 00     .80154E 00     .79779E 00     .79396E 00
   .79014E 00     .78624E 00     .78237E 00     .77843E 00     .77450E 00
   .77048E 00     .76648E 00     .76239E 00     .75833E 00     .75420E 00
   .75010E 00     .74593E 00     .74176E 00     .73750E 00     .73326E 00
   .72893E 00     .72463E 00     .72026E 00     .71593E 00     .71152E 00
   .70711E 00     .70261E 00     .69814E 00     .69359E 00     .68907E 00
   .68448E 00     .67993E 00     .67529E 00     .67067E 00     .66596E 00
   .66129E 00     .65654E 00     .65181E 00     .64701E 00     .64222E 00
   .63737E 00     .63254E 00     .62765E 00     .62275E 00     .61780E 00
   .61286E 00     .60789E 00     .60291E 00     .59789E 00     .59285E 00
   .58776E 00     .58266E 00     .57752E 00     .57240E 00     .56723E 00
   .56207E 00     .55684E 00     .55161E 00     .54633E 00     .54106E 00
   .53576E 00     .53046E 00     .52514E 00     .51981E 00     .51442E 00
   .50903E 00     .50360E 00     .49817E 00     .49270E 00     .48723E 00
   .48172E 00     .47622E 00     .47066E 00     .46512E 00     .45954E 00
   .45396E 00     .44835E 00     .44272E 00     .43709E 00     .43145E 00
   .42578E 00     .42007E 00     .41433E 00     .40861E 00     .40287E 00
   .39713E 00     .39136E 00     .38558E 00     .37976E 00     .37393E 00
   .36809E 00     .36224E 00     .35637E 00     .35051E 00     .34463E 00
   .33874E 00     .33278E 00     .32685E 00     .32091E 00     .31496E 00
IMAGINARY PART OF OUTPUT SEQUENCE
  -.58776E 00    -.59286E 00    -.59787E 00    -.60291E 00    -.60790E 00
  -.61290E 00    -.61782E 00    -.62275E 00    -.62764E 00    -.63255E 00
  -.63738E 00    -.64222E 00    -.64701E 00    -.65181E 00    -.65654E 00
  -.66128E 00    -.66597E 00    -.67068E 00    -.67531E 00    -.67995E 00
  -.68452E 00    -.68910E 00    -.69361E 00    -.69812E 00    -.70260E 00
  -.70710E 00    -.71151E 00    -.71592E 00    -.72027E 00    -.72462E 00
  -.72890E 00    -.73320E 00    -.73749E 00    -.74175E 00    -.74591E 00
  -.75009E 00    -.75421E 00    -.75834E 00    -.76238E 00    -.76645E 00
  -.77046E 00    -.77448E 00    -.77840E 00    -.78235E 00    -.78623E 00
  -.79011E 00    -.79392E 00    -.79774E 00    -.80152E 00    -.80529E 00
  -.80899E 00    -.81267E 00    -.81629E 00    -.81991E 00    -.82345E 00
  -.82702E 00    -.83054E 00    -.83405E 00    -.83747E 00    -.84089E 00
  -.84425E 00    -.84761E 00    -.85090E 00    -.85420E 00    -.85748E 00
  -.86072E 00    -.86388E 00    -.86703E 00    -.87012E 00    -.87321E 00
  -.87624E 00    -.87925E 00    -.88223E 00    -.88519E 00    -.88807E 00
  -.89094E 00    -.89376E 00    -.89657E 00    -.89932E 00    -.90206E 00
  -.90477E 00    -.90746E 00    -.91007E 00    -.91264E 00    -.91516E 00
  -.91769E 00    -.92015E 00    -.92261E 00    -.92500E 00    -.92739E 00
  -.92970E 00    -.93198E 00    -.93422E 00    -.93644E 00    -.93862E 00
  -.94078E 00    -.94292E 00    -.94501E 00    -.94702E 00    -.94901E 00
```

Table 3

```
NO OF DATA POINTS=  100
NO OF OUTPUT POINTS=  100
DELAYED IMPULSE AT SAMPLE NO=    4
SAMPLING FREQUENCY= 10000.000
INITIAL VALUE OF SIGMA=        .000
INITIAL VALUE OF OMEGA=        .000
INCREMENT IN SIGMA ALONG CZT PATH=        .000
INCREMENT IN OMEGA ALONG CZT PATH=    100.000
REAL PART OF OUTPUT SEQUENCE
    .99998E 00    .96855E 00    .87630E 00    .72893E 00    .53583E 00
    .30901E 00    .62805E-01   -.18737E 00   -.42578E 00   -.63738E 00
   -.80899E 00   -.92975E 00   -.99206E 00   -.99205E 00   -.92975E 00
   -.80894E 00   -.63739E 00   -.42578E 00   -.18739E 00    .62804E-01
    .30900E 00    .53579E 00    .72896E 00    .87624E 00    .96854E 00
    .99997E 00    .96853E 00    .87624E 00    .72895E 00    .53580E 00
    .30900E 00    .62806E-01   -.18736E 00   -.42577E 00   -.63741E 00
   -.80897E 00   -.92976E 00   -.99206E 00   -.99209E 00   -.92975E 00
   -.80899E 00   -.63739E 00   -.42578E 00   -.18738E 00    .62791E-01
    .30899E 00    .53581E 00    .72895E 00    .87627E 00    .96853E 00
    .99998E 00    .96853E 00    .87627E 00    .72894E 00    .53579E 00
    .30900E 00    .62789E-01   -.18735E 00   -.42580E 00   -.63737E 00
   -.80896E 00   -.92974E 00   -.99204E 00   -.99203E 00   -.92977E 00
   -.80898E 00   -.63740E 00   -.42577E 00   -.18737E 00    .62795E-01
    .30902E 00    .53581E 00    .72895E 00    .87625E 00    .96856E 00
    .99995E 00    .96855E 00    .87626E 00    .72895E 00    .53580E 00
    .30899E 00    .62789E-01   -.18738E 00   -.42577E 00   -.63740E 00
   -.80896E 00   -.92975E 00   -.99203E 00   -.99205E 00   -.92975E 00
   -.80897E 00   -.63737E 00   -.42579E 00   -.18737E 00    .62807E-01
    .30898E 00    .53582E 00    .72895E 00    .87628E 00    .96854E 00
IMAGINARY PART OF OUTPUT SEQUENCE
   -.17881E-05   -.24868E 00   -.48173E 00   -.68453E 00   -.84434E 00
   -.95103E 00   -.99799E 00   -.98225E 00   -.90481E 00   -.77047E 00
   -.58778E 00   -.36811E 00   -.12530E 00    .12533E 00    .36811E 00
    .58775E 00    .77050E 00    .90477E 00    .98225E 00    .99797E 00
    .95101E 00    .84428E 00    .68453E 00    .48172E 00    .24869E 00
    .54791E-06   -.24870E 00   -.48172E 00   -.68453E 00   -.84428E 00
   -.95100E 00   -.99792E 00   -.98227E 00   -.90479E 00   -.77048E 00
   -.58776E 00   -.36813E 00   -.12532E 00    .12534E 00    .36811E 00
    .58778E 00    .77047E 00    .90478E 00    .98223E 00    .99800E 00
    .95099E 00    .84430E 00    .68450E 00    .48173E 00    .24868E 00
   -.80466E-05   -.24869E 00   -.48173E 00   -.68450E 00   -.84428E 00
   -.95100E 00   -.99797E 00   -.98225E 00   -.90480E 00   -.77043E 00
   -.58778E 00   -.36810E 00   -.12531E 00    .12532E 00    .36811E 00
    .58776E 00    .77049E 00    .90478E 00    .98228E 00    .99798E 00
    .95103E 00    .84427E 00    .68452E 00    .48172E 00    .24868E 00
   -.63641E-05   -.24868E 00   -.48172E 00   -.68449E 00   -.84428E 00
   -.95104E 00   -.99796E 00   -.98227E 00   -.90478E 00   -.77047E 00
   -.58776E 00   -.36813E 00   -.12530E 00    .12532E 00    .36809E 00
    .58775E 00    .77045E 00    .90479E 00    .98222E 00    .99796E 00
    .95096E 00    .84429E 00    .68452E 00    .48174E 00    .24867E 00
```

Table 4

```
NO OF DATA POINTS=  100
NO OF OUTPUT POINTS=  100
DELAYED IMPULSE AT SAMPLE NO=   4
SAMPLING FREQUENCY= 10000.000
INITIAL VALUE OF SIGMA=        .000
INITIAL VALUE OF OMEGA=  1000.000
INCREMENT IN SIGMA ALONG CZT PATH=      1.000
INCREMENT IN OMEGA ALONG CZT PATH=     10.000
REAL PART OF OUTPUT SEQUENCE
  -.80902E 00   -.82454E 00   -.83963E 00   -.85418E 00   -.86828E 00
  -.88179E 00   -.89484E 00   -.90730E 00   -.91930E 00   -.93065E 00
  -.94151E 00   -.95174E 00   -.96147E 00   -.97053E 00   -.97907E 00
  -.98695E 00   -.99431E 00   -.10010E 01   -.10071E 01   -.10125E 01
  -.10174E 01   -.10215E 01   -.10251E 01   -.10280E 01   -.10303E 01
  -.10319E 01   -.10329E 01   -.10331E 01   -.10328E 01   -.10318E 01
  -.10302E 01   -.10278E 01   -.10250E 01   -.10213E 01   -.10170E 01
  -.10121E 01   -.10065E 01   -.10003E 01   -.99339E 00   -.98582E 00
  -.97768E 00   -.96883E 00   -.95939E 00   -.94927E 00   -.93859E 00
  -.92722E 00   -.91528E 00   -.90270E 00   -.88957E 00   -.87579E 00
  -.86146E 00   -.84651E 00   -.83102E 00   -.81492E 00   -.79828E 00
  -.78107E 00   -.76338E 00   -.74511E 00   -.72637E 00   -.70707E 00
  -.68732E 00   -.66703E 00   -.64630E 00   -.62509E 00   -.60350E 00
  -.58141E 00   -.55891E 00   -.53599E 00   -.51267E 00   -.48897E 00
  -.46491E 00   -.44048E 00   -.41573E 00   -.39062E 00   -.36525E 00
  -.33953E 00   -.31356E 00   -.28731E 00   -.26084E 00   -.23413E 00
  -.20720E 00   -.18009E 00   -.15279E 00   -.12531E 00   -.97641E-01
  -.69821E-01   -.41955E-01   -.13989E-01    .14026E-01    .42120E-01
   .70276E-01    .98476E-01    .12674E 00    .15491E 00    .18306E 00
   .21112E 00    .23918E 00    .26715E 00    .29500E 00    .32278E 00
IMAGINARY PART OF OUTPUT SEQUENCE
  -.58779E 00   -.56797E 00   -.54777E 00   -.52714E 00   -.50616E 00
  -.48477E 00   -.46306E 00   -.44096E 00   -.41856E 00   -.39581E 00
  -.37278E 00   -.34942E 00   -.32581E 00   -.30191E 00   -.27779E 00
  -.25341E 00   -.22882E 00   -.20401E 00   -.17903E 00   -.15386E 00
  -.12852E 00   -.10304E 00   -.77441E-01   -.51717E-01   -.25894E-01
   .64782E-05    .25956E-01    .51970E-01    .78026E-01    .10409E 00
   .13014E 00    .15618E 00    .18220E 00    .20816E 00    .23405E 00
   .25987E 00    .28559E 00    .31117E 00    .33663E 00    .36194E 00
   .38710E 00    .41204E 00    .43682E 00    .46136E 00    .48570E 00
   .50974E 00    .53357E 00    .55708E 00    .58035E 00    .60325E 00
   .62587E 00    .64813E 00    .67007E 00    .69160E 00    .71276E 00
   .73351E 00    .75385E 00    .77374E 00    .79323E 00    .81225E 00
   .83084E 00    .84888E 00    .86647E 00    .88351E 00    .90015E 00
   .91614E 00    .93165E 00    .94657E 00    .96098E 00    .97477E 00
   .98799E 00    .10006E 01    .10126E 01    .10240E 01    .10348E 01
   .10450E 01    .10545E 01    .10634E 01    .10716E 01    .10792E 01
   .10861E 01    .10923E 01    .10979E 01    .11028E 01    .11070E 01
   .11105E 01    .11133E 01    .11153E 01    .11168E 01    .11175E 01
   .11175E 01    .11168E 01    .11154E 01    .11132E 01    .11103E 01
   .11067E 01    .11025E 01    .10975E 01    .10919E 01    .10855E 01
```

Appendix

```
C
C-----------------------------------------------------------------------
C MAIN PROGRAM: TEST PROGRAM FOR CHIRP Z-TRANSFORM SUBROUTINE
C               USING A DELAY IMPULSE
C AUTHOR:       L R RABINER
C               BELL LABORATORIES, MURRAY HILL, NEW JERSEY 07974
C
C INPUT:        NDATA IS THE NUMBER OF DATA POINTS
C                   1 <= NDATA <= 512
C               NOPTS IS THE NUMBER OF OUTPUT POINTS
C                   1 <= NOPTS <= 512
C               IDEL IS THE DELAY OF THE IMPULSE IN SAMPLES
C               FS IS THE SAMPLING FREQUENCY IN HZ
C               SIG0 IS THE INITIAL VALUE OF SIGMA IN HZ
C               OME0 IS THE INITIAL VALUE OF OMEGA IN HZ
C               DLTSIG IS THE INCREMENT IN SIGMA IN HZ
C               DLTOMG IS THE INCREMENT IN OMEGA IN HZ
C               IMD REQUESTS ADDITIONAL RUNS
C                   IMD = 1  NEW RUN
C                   IMD = 0  TERMINATES PROGRAM
C-----------------------------------------------------------------------
C
      COMMON WTR(1024), WTI(1024), XR(1024), XI(1024)
      INTEGER TTI, TTO
C
C DEFINE I/O DEVICE CODES
C INPUT: INPUT TO THIS PROGRAM IS USER-INTERACTIVE
C        THAT IS - A QUESTION IS WRITTEN ON THE USER
C        TERMINAL (TTO) AND THE USER TYPES IN THE ANSWER.
C
C OUTPUT: ALL OUTPUT IS WRITTEN ON THE STANDARD
C         OUTPUT UNIT (LPT)
C
      TTI = I1MACH(1)
      TTO = I1MACH(4)
      LPT = I1MACH(2)
C
  10  WRITE (TTO,9999)
9999  FORMAT (23H NO OF DATA POINTS(I4)=)
      READ (TTI,9998) NDATA
9998  FORMAT (I4)
      WRITE (LPT,9997) NDATA
9997  FORMAT (19H NO OF DATA POINTS=, I5)
      WRITE (TTO,9996)
9996  FORMAT (25H NO OF OUTPUT POINTS(I4)=)
      READ (TTI,9998) NOPTS
      WRITE (LPT,9995) NOPTS
9995  FORMAT (21H NO OF OUTPUT POINTS=, I5)
      WRITE (TTO,9994)
9994  FORMAT (33H DATA INPUT DELAY IN SAMPLES(I4)=)
      READ (TTI,9998) IDEL
      WRITE (LPT,9993) IDEL
9993  FORMAT (30H DELAYED IMPULSE AT SAMPLE NO=, I4)
      WRITE (TTO,9992)
9992  FORMAT (33H SAMPLING FREQUENCY IN HZ(F10.0)=)
      READ (TTI,9991) FS
9991  FORMAT (F10.0)
      WRITE (LPT,9990) FS
9990  FORMAT (20H SAMPLING FREQUENCY=, F10.3)
      WRITE (TTO,9989)
9989  FORMAT (28H INITIAL SIGMA VALUE(F10.0)=)
      READ (TTI,9991) SIG0
      WRITE (LPT,9988) SIG0
9988  FORMAT (24H INITIAL VALUE OF SIGMA=, F10.3)
      WRITE (TTO,9987)
9987  FORMAT (28H INITIAL OMEGA VALUE(F10.0)=)
      READ (TTI,9991) OME0
      WRITE (LPT,9986) OME0
9986  FORMAT (24H INITIAL VALUE OF OMEGA=, F10.3)
      WRITE (TTO,9985)
9985  FORMAT (20H DELTA SIGMA(F10.0)=)
      READ (TTI,9991) DLTSIG
      WRITE (LPT,9984) DLTSIG
9984  FORMAT (35H INCREMENT IN SIGMA ALONG CZT PATH=, F10.3)
      WRITE (TTO,9983)
9983  FORMAT (20H DELTA OMEGA(F10.0)=)
      READ (TTI,9991) DLTOMG
      WRITE (LPT,9982) DLTOMG
9982  FORMAT (35H INCREMENT IN OMEGA ALONG CZT PATH=, F10.3)
C
C CREATE DELAYED INPUT SAMPLE WITH DELAY IDEL SAMPLES
C
      DO 20 I=1,NDATA
        XR(I) = 0.
        XI(I) = 0.
  20  CONTINUE
      XR(IDEL+1) = 1.
      NFFT = NDATA + NOPTS
      DO 30 I=1,10
        NTEST = 2**I
        IF (NTEST.GE.NFFT) GO TO 40
  30  CONTINUE
      WRITE (TTO,9981)
9981  FORMAT (18H N TOO BIG FOR FFT)
      STOP
  40  NFFT = NTEST
      CALL CZT(XR, XI, NDATA, NOPTS, DLTSIG, DLTOMG, WTR, WTI, SIG0,
     *    OME0, 0, NFFT, FS)
      WRITE (LPT,9980)
9980  FORMAT (//)
      WRITE (LPT,9979) (XR(I),I=1,NOPTS)
9979  FORMAT (29H REAL PART OF OUTPUT SEQUENCE/(5E14.5))
      WRITE (LPT,9978) (XI(I),I=1,NOPTS)
9978  FORMAT (34H IMAGINARY PART OF OUTPUT SEQUENCE/(5E14.5))
      WRITE (TTO,9977)
9977  FORMAT (27H MORE DATA(1=YES, 0=NO)(I1))
      READ (TTI,9976) IMD
9976  FORMAT (I1)
      WRITE (LPT,9975)
9975  FORMAT (1H1)
      IF (IMD.EQ.1) GO TO 10
      STOP
      END
C
C-----------------------------------------------------------------------
C SUBROUTINE: CZT
C CHIRP Z-TRANSFORM
C REFERENCE--L R RABINER, R W SCHAFER, C M RADER--BELL SYSTEM
C TECHNICAL JOURNAL--MAY 1969, PP 1249-1291
C-----------------------------------------------------------------------
C
      SUBROUTINE CZT(XR, XI, NDATA, NOPTS, DLTSIG, DLTOMG, WTR, WTI,
     *    SIGO, OME, NTR, NFFT, FS)
      DIMENSION WTR(1), WTI(1), XR(1), XI(1)
C
C     XR = ARRAY OF SIZE NFFT
C          ON INPUT XR CONTAINS REAL PART OF INPUT DATA IN LOCATIONS
C          1 TO NDATA
C          ON OUTPUT XR CONTAINS REAL PART OF OUTPUT DATA IN LOCATIONS
C          1 TO NOPTS
```

```
C      XI = ARRAY OF SIZE NFFT
C           ON INPUT XI CONTAINS IMAGINARY PART OF INPUT DATA
C           ON OUTPUT XI CONTAINS IMAGINARY PART OF OUTPUT DATA
C   NDATA = NUMBER OF INPUT POINTS BEING TRANSFORMED
C   NOPTS = NUMBER OF OUTPUT VALUES BEING COMPUTED
C  DLTSIG = INCREMENT IN SIGMA ALONG THE CZT CONTOUR--RELATIVE TO FS
C  DLTOMG = INCREMENT IN OMEGA ALONG THE CZT CONTOUR--RELATIVE TO FS
C           NOTE THAT IF DLTSIG IS LESS THAN ZERO, THE CONTOUR SPIRALS
C           OUTSIDE THE UNIT CIRCLE IN THE Z-PLANE
C           WTR IS AN ARRAY OF SIZE NFFT WHICH HOLDS THE REAL PART OF THE
C           TRANSFORM OF W**(-N**2/2)
C           WTI IS AN ARRAY OF SIZE NFFT WHICH HOLDS THE IMAGINARY PART
C           OF THE TRANSFORM OF W**(-N**2/2)
C    SIG0 = INITIAL VALUE OF SIGMA OF CZT CONTOUR, RELATIVE TO FS
C    OME0 = INITIAL VALUE OF OMEGA OF CZT CONTOUR, RELATIVE TO FS
C     NTR = ROTATION FACTOR ON INPUT DATA TO MAKE INPUT LOOK LIKE IT
C           BEGINS AT N = -NTR INSTEAD OF N=0. SET NTR=0 TO AVOID
C           THIS OPTION
C    NFFT = SIZE OF FFT PERFORMED INTERNALLY IN CZT PROGRAM--GENERALLY
C           NFFT IS A POWER OF 2--NFFT MUST BE GREATER THAN (NDATA+NOPTS)
C      FS = SAMPLING FREQUENCY IN HZ
C
      DOUBLE PRECISION A, B, PHINC, RO, XIO, PI, W1INC, WR, WI, W2INC,
     *    WOHR, WOHI, W3INC, W4INC, WPR, WPI, WPOHR, WPOHI, WNR, WNI,
     *    WNSQR, WNSQI, RAD, ANG, WAR, WAI
      DOUBLE PRECISION XRT, XADRT, WOPR, WIPR, WINR, WINI, WJNR, WJNI,
     *    XASQ, WKNR, WKNI, WLNR, WLNI
      DOUBLE PRECISION XTR
C
C SET INITIAL CONSTANTS
C
      N3 = NFFT
      NFT = N3/2 + 1
      N4 = N3/2
      X9 = N3
C
C N0 FACTOR FOR SYMMETRY OF W**(-N**2/2)
C
      NADRT = (NDATA-NOPTS)/2
      XADRT = NADRT + 1
      NRT = NADRT - NTR
      XRT = NRT
      PI = 4.0D0*DATAN(1.0D0)
      RO = SIGO
      XIO = OME
C
C COMPUTE W**(N**2/2) USING RECURSION RELATIONS
C OBTAIN INITIAL CONSTANTS
C
      A = DLTSIG
      B = DLTOMG
      PHINC = -PI*B*2.D0/FS
      W1INC = DEXP(A*PI/FS)
C
C W STORED IN WR AND WI
C
      WR = W1INC*DCOS(PHINC)
      WI = W1INC*DSIN(PHINC)
      W2INC = DSQRT(W1INC)
C
C W**(1/2) STORED IN WOHR AND WOHI
C
      WOHR = W2INC*DCOS(PHINC/2.0D0)
      WOHI = W2INC*DSIN(PHINC/2.0D0)
      W3INC = 1.0D0/W1INC
      W4INC = 1.0D0/W2INC
C
C W**(-1) STORED IN WPR AND WPI
C
      WPR = W3INC*DCOS(-PHINC)
      WPI = W3INC*DSIN(-PHINC)
C
C W**(-1/2) STORED IN WPOHR AND WPOHI
C
      WPOHR = W4INC*DCOS(-PHINC/2.0D0)
      WPOHI = W4INC*DSIN(-PHINC/2.0D0)
      DO 10 I=1,N4
        WTR(I) = 0.
        WTI(I) = 0.
 10   CONTINUE
C
C D0=W**(-1/2)   D0 STORED IN WNR AND WNI
C
      WNR = WPOHR
      WNI = WPOHI
C
C C0=1
C
      WNSQR = 1.0D0
      WNSQI = 0.0D0
C
C SOLVE RECURSION RELATION FOR W**(-N**2/2)
C
      CALL RECUR(WTR, WTI, WNR, WNI, WNSQR, WNSQI, WPR, WPI, N3)
C
C COMPUTE TRANSFORM OF W**(-N**2/2)
C
      CALL FFT842(0, N3, WTR, WTI)
C
C COMPUTE A**(-N)*W**(N**2/2)
C INITIAL CONSTANTS
C
      RAD = DEXP(RO*PI/FS)
      ANG = -XIO*PI*2.D0/FS
C
C A STORED IN WAR AND WAI
C
      WAR = RAD*DCOS(ANG)
      WAI = RAD*DSIN(ANG)
      XASQ = XADRT*XADRT/2.0D0
      WOPR = W1INC**(XASQ)
      WOPI = W1INC**(.5D0-XADRT)
      WKNR = WOPR*DCOS(PHINC*XASQ)
      WKNI = WOPR*DSIN(PHINC*XASQ)
      WLNR = WOPI*DCOS(PHINC*(.5D0-XADRT))
      WLNI = WOPI*DSIN(PHINC*(.5D0-XADRT))
      WNSQR = (WAR*WKNR+WAI*WKNI)/(RAD*RAD)
      WNSQI = (WAR*WKNI-WAI*WKNR)/(RAD*RAD)
      WNR = WAR*WLNR - WAI*WLNI
      WNI = WAI*WLNR + WAR*WLNI
C
C WEIGHT INPUT DATA BY A**(-N)*W**(N**2/2)
C
      CALL DECUR(XR, XI, WNR, WNI, WNSQR, WNSQI, WR, WI, NDATA)
C
C PAD DATA WITH ZEROS
C
      N5 = NDATA + 1
      DO 20 I=N5,N3
        XR(I) = 0.
```

```
        XI(I) = 0.
   20   CONTINUE
C
C TRANSFORM INPUT USING FFT
C
        CALL FFT842(0, N3, XR, XI)
C
C MULTIPLY FFTS
C
        DO 30 I=1,N3
          J = I
          IF (I.GT.NFT) J = N3 + 2 - I
          XT = XR(I)*WTR(J) - XI(I)*WTI(J)
          XI(I) = -XR(I)*WTI(J) - XI(I)*WTR(J)
          XR(I) = XT
   30   CONTINUE
C
C INVERSE TRANSFORM TO GET TIME SEQUENCE
C
        CALL FFT842(0, N3, XR, XI)
C
C SHUFFLE DATA
C
        IF (NADRT.EQ.0) GO TO 70
        IF (NADRT.LT.0) GO TO 50
        DO 40 I=1,NOPTS
          J = I + NADRT
          XR(I) = XR(J)
          XI(I) = XI(J)
   40   CONTINUE
        GO TO 70
   50   DO 60 I=1,NOPTS
          J = NOPTS + 1 - I
          K = J + NADRT
          IF (K.LE.0) K = K + N3
          XR(J) = XR(K)
          XI(J) = XI(K)
   60   CONTINUE
   70   CONTINUE
        DO 80 I=1,NOPTS
          XR(I) = XR(I)/X9
          XI(I) = -XI(I)/X9
   80   CONTINUE
C
C COMPUTE POST WEIGHTING W**(-K**2/2) INITIAL CONSTANTS FOR RECURSION
C
        WOPR = W1INC**(.5D0-XRT)
        WOPI = 1.0D0/WOPR
        WINR = WOPR*DCOS(PHINC*(.5D0-XRT))
        WINI = WOPR*DSIN(PHINC*(.5D0-XRT))
        WJNR = WOPI*DCOS(PHINC*(XRT-.5D0))
        WJNI = WOPI*DSIN(PHINC*(XRT-.5D0))
        XTR = NTR
        WOPR = RAD**(-XTR)
        WAR = WOPR*DCOS(-ANG*XTR)
        WAI = WOPR*DSIN(-ANG*XTR)
        WNR = WJNR
        WNI = WJNI
        WNSQR = WINR*WAR - WINI*WAI
        WNSQI = WINR*WAI + WINI*WAR
C
C POST WEIGHT DATA
C
        CALL DECUR(XR, XI, WNR, WNI, WNSQR, WNSQI, WR, WI, NOPTS)
        RETURN
        END
C
C-----------------------------------------------------------------------
C SUBROUTINE: RECUR
C RECURSION RELATION TO GIVE W**(-N**2/2)
C-----------------------------------------------------------------------
C
        SUBROUTINE RECUR(WR, WI, WNR, WNI, WNSQR, WNSQI, VR, VI, NFFT)
        DIMENSION WR(1), WI(1)
        DOUBLE PRECISION WNR, WNI, WNSQR, WNSQI, VR, VI
        DOUBLE PRECISION XT
C
C C(N+1) = C(N)*D(N)
C D(N+1) = D(N)*W**(-1)
C W**(-N**2/2) = D(N)
C WNR AND WNI ARE THE D TERMS
C WNSQR AND WNSQI ARE THE C TERMS
C WR AND WI STORE THE W**(-N**2/2) RESULTS
C
        WR(1) = 1.
        WI(1) = 0.
        N3 = NFFT/2 + 1
        DO 10 I=2,N3
C
C C(N+1)=C(N)*D(N)
C
          XT = WNSQR*WNR - WNSQI*WNI
          WNSQI = WNSQR*WNI + WNSQI*WNR
          WNSQR = XT
C
C D(N+1)=D(N)*W
C
          XT = WNR*VR - WNI*VI
          WNI = WNR*VI + WNI*VR
          WNR = XT
C
C STORE W**(-N**2/2)
C
          WR(I) = WNSQR
          WI(I) = WNSQI
          J = NFFT + 2 - I
          WR(J) = WR(I)
          WI(J) = WI(I)
   10   CONTINUE
        RETURN
        END
C
C-----------------------------------------------------------------------
C SUBROUTINE: DECUR
C RECURSION RELATION TO GIVE Y(N)=X(N)*A**(-N)*W**(N**2/2)
C-----------------------------------------------------------------------
C
        SUBROUTINE DECUR(XR, XI, WNR, WNI, WNSQR, WNSQI, VR, VI, N)
C
C Y(N) OVERWRITES X(N)
C X(N) STORED IN XR AND XI
C C(N+1) = C(N)*D(N)
C D(N+1) = D(N)*W
C WNR AND WNI STORE C(N)--INITIALLY SET TO C0
C WNSQR AND WNSQI STORE D(N)--INITIALLY SET TO D0
C
        DIMENSION XR(1), XI(1)
        DOUBLE PRECISION WNR, WNI, WNSQR, WNSQI, VR, VI
        DOUBLE PRECISION XT
        DO 10 J=1,N
```

```
C
C C(N+1)=C(N)*D(N)
C
         XT = WNSQR*WNR - WNSQI*WNI
         WNSQI = WNSQR*WNI + WNSQI*WNR
         WNSQR = XT
C
C D(N+1)=D(N)*W
C
         XT = WNR*VR - WNI*VI
         WNI = WNR*VI + WNI*VR
         WNR = XT
C
C WEIGHT INPUT DATA
C
         XS = XR(J)*WNSQR - XI(J)*WNSQI
         XI(J) = XR(J)*WNSQI + XI(J)*WNSQR
         XR(J) = XS
  10  CONTINUE
      RETURN
      END
```

1.7

Complex General-N Winograd Fourier Transform Algorithm (WFTA)*

J. H. McClellan
H. Nawab

Department of EECS
Mass. Inst. of Technology
Cambridge, MA 02139

1. Purpose

This program computes the discrete Fourier transform (DFT) of sequences whose length N is a product of relatively prime factors taken from the set {2,3,4,5,7,8,9,16}. The method used in this program for computing the DFT is a general-N implementation of Winograd's algorithm [1], referred to as the WFTA [2].

2. Method

In order to set up a common notational framework we first summarize the basic form of the WFTA. Consider the computation of the N-point DFT of the sequence $x[n]$, $n = 0, 1, \ldots, N-1$. Suppose the transform length factors into the product of μ mutually prime integers; $N = N_1 \times N_2 \times \cdots \times N_\mu$. The WFTA consists of five steps as illustrated in Fig. 1 for the case $N = 15$

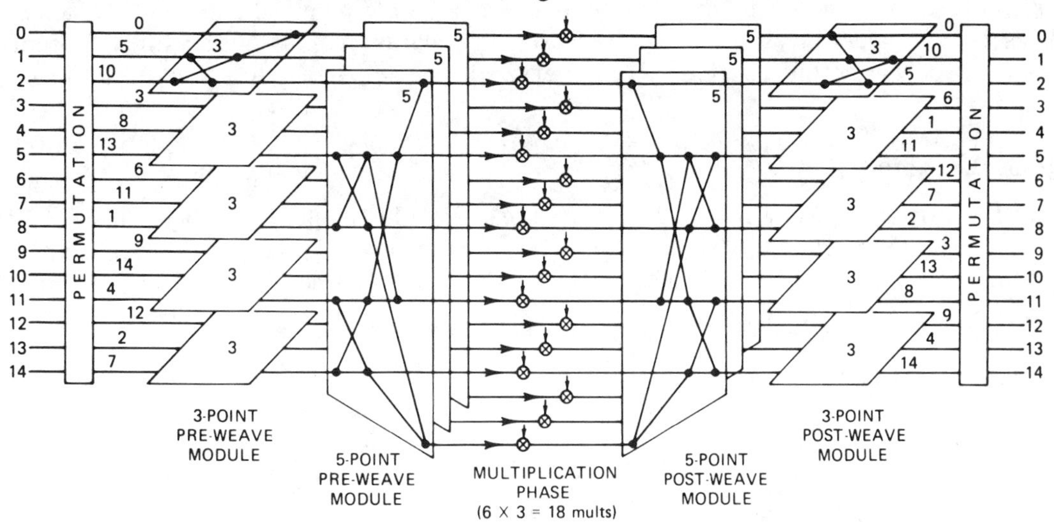

Fig. 1 Fifteen-point DFT algorithm decomposed using Good's mapping to obtain a two-dimensional (3×5)-point DFT and Winograd's algorithms for 3-point and 5-point DFTs. The algorithm requires 18=6×3 multiplications which are "nested" between the pre-weave and post-weave modules.

Step 1 The one-dimensional sequence $x[n]$ is mapped into a μ-dimensional array $s[n_1, n_2, \ldots, n_\mu]$ via the Sino correspondence [3,4]

$$n = < \sum_{i=1}^{\mu} \left(\frac{N}{N_i} \right) n_i >_N . \tag{1}$$

* An earlier version of this program can be found in Ref. [4].

where $<\cdot>_N$ denotes residue reduction modulo N.

Step 2 The *pre-weave* [6] modules are implemented, one for each factor N_i. Each pre-weave module contains only additions and subtractions. The size of the data array will expand, in general, when passing through the pre-weave elements.

Step 3 The data array is multiplied (point-by-point) by an array of real constants† derived from the multipliers of the small-N DFT algorithms for N_i, $i = 1, 2, ..., \mu$. These constants depend solely on the complex exponentials of the DFT, W_N, and are the only multiplications required in the algorithm.

Step 4 The *post-weave* [6] modules are implemented for each of the μ dimensions. Each of these modules contains only additions, subtractions and multiplications by j. A different module is required for each dimension. After passing through the post-weave section, the data array $S[k_1, k_2, ..., k_\mu]$ shrinks back to its original size, $N_1 \times N_2 \times \cdots \times N_\mu$. At this point in the computation we have obtained the μ-dimensional $N_1 \times N_2 \times \cdots \times N_\mu$ DFT of the array $s[n_1, n_2, ..., n_\mu]$.

Step 5 A mapping of the μ-dimensional array $S[k_1, k_2, ..., k_\mu]$ is performed in accordance with the Chinese Remainder theorem to obtain the correct one-dimensional DFT, $X[k]$. This correspondence is given by [3,4]

$$k = <s_1 k_1 + s_2 k_2 + \cdots + s_\mu k_\mu>_N \tag{2}$$

where the integers s_i satisfy

$$s_i \equiv \delta_{ij} \ mod \ N_j \quad ; \quad i,j = 1, 2, ..., \mu \tag{3}$$

2.1 Initialization

In order to have an efficient WFTA program several computations need be performed only once when DFTs of the same length are to be calculated over and over. First of all, the factors of N must be determined. Since the WFTA is composed of small-N DFT algorithms that have only been developed for certain lengths, a check against the allowable factors of N is made. Next, the constant multiplier array is formed by multiplying together the pre-stored constants for the individual small-N factors. Finally, permutation vectors for the input and output mappings are prepared based on the factors of N. This speeds up the actual mapping process because the congruences (3) of the Chinese Remainder theorem are solved during initialization. The expense, of course, is the storage to hold the mapping vectors.

3. Program Description

The WFTA program consists of four subroutines, each of which will be described in the following. The flow of control in the algorithm is given in Fig. 2. Documentation of the individual computational modules (pre-weave and post-weave) is presented using the annotated flowgraphs of Figs. 3 through 10.

The WFTA is invoked by calling the subroutine WFTA with the arguments

N - transform length

XR(N) - real part of array to be transformed

XI(N) - imaginary part of array to be transformed. The real and imaginary parts of the transform are returned in XR and XI.

INVRS - a flag to invoke the inverse transform. INVRS = 1 gives the inverse; INVRS ≠ 1 gives the forward transform. A factor of N^{-1} is included in the inverse. The inverse is actually obtained by modifying the permutation described in step 5 above to map to $<-k>_N$ instead of k.

INIT - a flag to specify whether the call to WFTA requires initialization. INIT=0 calls for initialization, INIT≠0 skips the initialization phase. When many DFTs of the same length are to be performed, initialization is needed only for the first DFT but not on succeeding calls. See subroutine INISHL for details.

† Note that the constants are not allowed to be imaginary as proposed by others; rather the post-weave modules contain "multiplications" by j to effectively obtain imaginary constants. Thus the need for a flag array is eliminated.

IERR - is an integer variable that contains, upon completion of a DFT, an error code. If the DFT was done successfully IERR=0; if an error occured IERR=−1 or −2. There are two causes of the error: (1) the transform length is illegal (i.e., it contains factors outside the set {2,3,4,5,7,8,9,16}), (IERR=−1), or (2) the program has not been initialized for the particular value of the transform length specified, (IERR=−2).

The subroutine WFTA calls in turn the subroutines INISHL (if needed), WEAVE1, and WEAVE2. The permutations and multiplications are performed within subroutine WFTA.

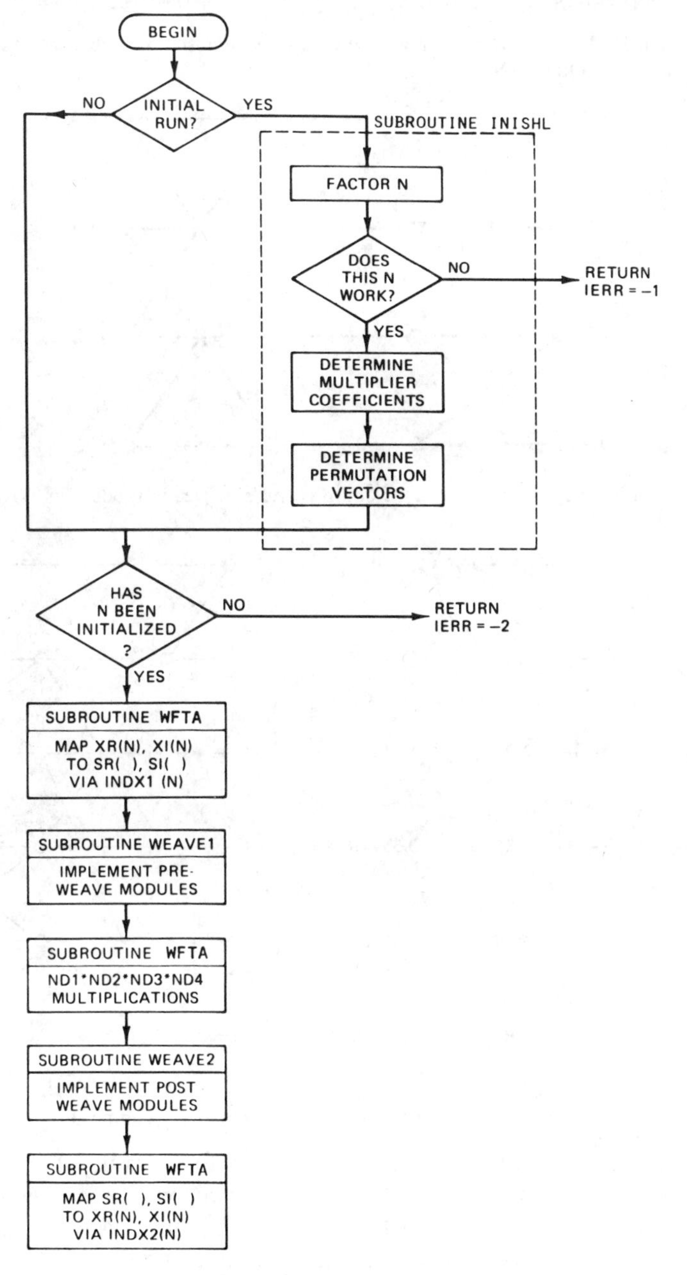

Fig. 2 Flow of control in program WFTA.

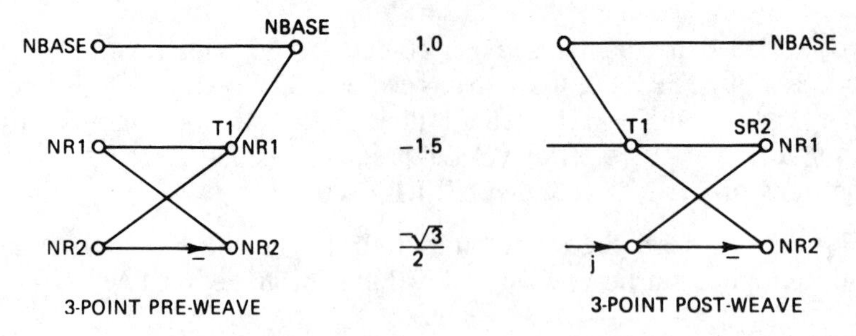

Fig. 3 Three-point DFT algorithm using Winograd's decomposition. Labels correspond to storage locations in FORTRAN program.

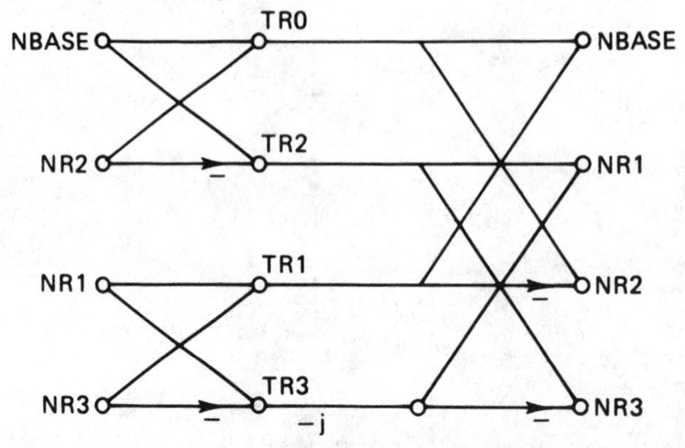

Fig. 4 Four-point DFT which is also used entirely as a post-weave module in WFTA program.

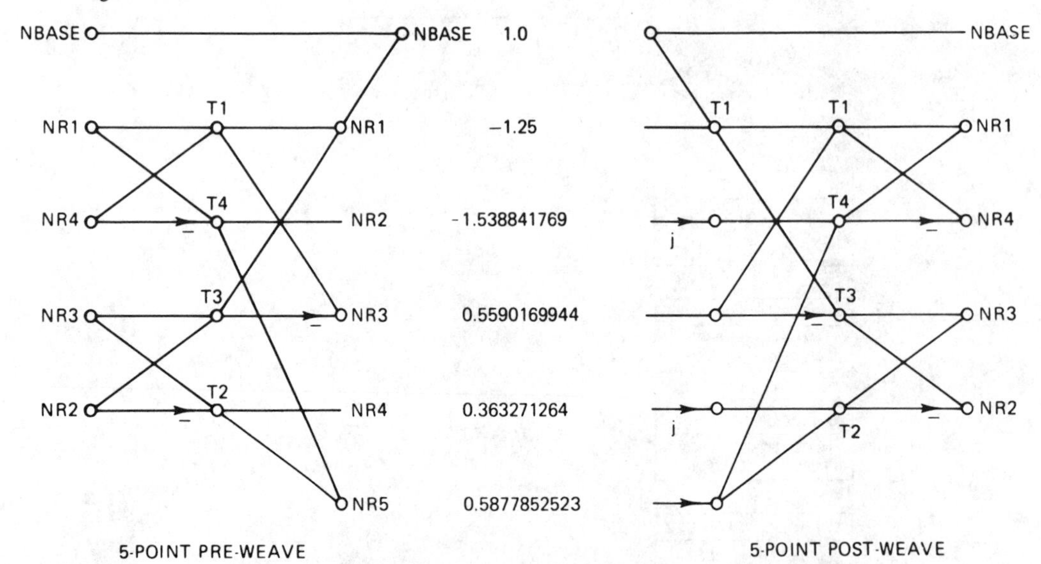

Fig. 5 Five-point DFT algorithm using Winograd's decomposition.

The subroutine INISHL performs those calculations, for a fixed value of N, that need not be done on each pass of the algorithm. These include factoring the transform length N into its mutually prime factors, calculating the constant coefficients to be used in the multiplication step, and calculating the mapping vectors for the pre- and post-permutations of the data to be used in the input and output permutations. The transform length N is factored into mutually prime factors taken from the set $\{2,3,4,5,7,8,9,16\}$. Since N can have at most four factors from this set, the factors are assigned to the integer variables NA, NB, NC and ND as follows:

NA is assigned the largest value from the set $\{1,2,4,8,16\}$ that divides N; NB the largest divisor of N from the set $\{1,3,9\}$; NC the largest from $\{1,7\}$; and ND the largest from $\{1,5\}$. The product NA*NB*NC*ND should equal N, if the transform length is permissible. In the following, NA will be referred to as the first (or outermost) factor, NB the second, NC the third, and ND the fourth (or innermost). This reference is with respect to the order of pre-weave modules in the calculations (i.e.,

NA is the size of the first pre-weave module and last post-weave module).†

From the factors of N we can now determine via Table 1 the amount of data expansion that will occur in the pre-weave phase and, equivalently, the number of multiplications that will be performed in the WFTA. For each of the four variables (NA, NB, NC, and ND) the number of multiplications is assigned to the variables ND1, ND2, ND3, and ND4, respectively. In order to accommodate the data expansion of the WFTA, the working arrays SR(·) and SI(·) must be dimensioned as large as ND1*ND2*ND3*ND4. If the present upper limit on the transform size of 5040 is to be reduced, the size of the working arrays can be reduced by using Table 1 to determine the data expansion for the largest allowable transform size.

Table 1
Number of multiplications, $M(N)$, needed for N-point WFTA

N	$M(N)$
2	2
3	3
4	4
5	6
7	9
8	8
9	11
16	18

Note: if $N = r_1 \times r_2 \times \cdots \times r_\mu$
then $M(N) = M(r_1)M(r_2)...M(r_\mu)$

The second step of the INISHL subroutine is to calculate the constant coefficients for the multiplication routine. These are generated from the coefficients of the individual factors as

$$\mathrm{COEF}(n_1,n_2,n_3,n_4) = \mathrm{COA}(n_1)*\mathrm{COB}(n_2)*\mathrm{COC}(n_3)*\mathrm{COD}(n_4)$$

where

$$n_1 = 1,2,...,\mathrm{ND1}$$
$$n_2 = 1,2,...,\mathrm{ND2}$$
$$n_3 = 1,2,...,\mathrm{ND3}$$
$$n_4 = 1,2,...,\mathrm{ND4}$$

The array COEF is actually treated as a one-dimensional array in the FORTRAN program, but can also be thought of as a multidimensional array in a mixed-radix index representation with the first index varying the most rapidly. Thus, the following addresses are equivalent:††

$$\mathrm{COEF}(n_1,n_2,n_3,n_4) = \mathrm{COEF}(n_1+\mathrm{ND1}*n_2+\mathrm{ND1}*\mathrm{ND2}*n_3+\mathrm{ND1}*\mathrm{ND2}*\mathrm{ND3}*n_4) .$$

This method of data addressing will also apply to the data arrays SR(·) and SI(·).

The third step of the INISHL subroutine is the calculation of the mapping vectors INDX1(·) and INDX2(·) for the pre- and post-permutations of the data. The pre-permutation of the data is done according to the Sino correspondence of Eq. (1). The mapping is implemented using the mapping vector INDX1(·) to obtain the highest speed possible. The post-permutation requires the inverse mapping of the Chinese Remainder theorem, as in Eqs. (2) and (3). Again, for maximum speed a mapping vector INDX2(·) is constructed.

† This order of pre-weave and post-weave modules will minimize the number of additions in the algorithm in all cases except N = 48, 240, 336 and 1680. The increases in additions for these cases is less than 4% compared with the optimal strategy [2]. The benefit of the order adopted here is a simpler program structure.

†† The multidimensional arrays were not used in the FORTRAN program because it was found that a typical compiler produced much more efficient code with the one-dimensional arrays.

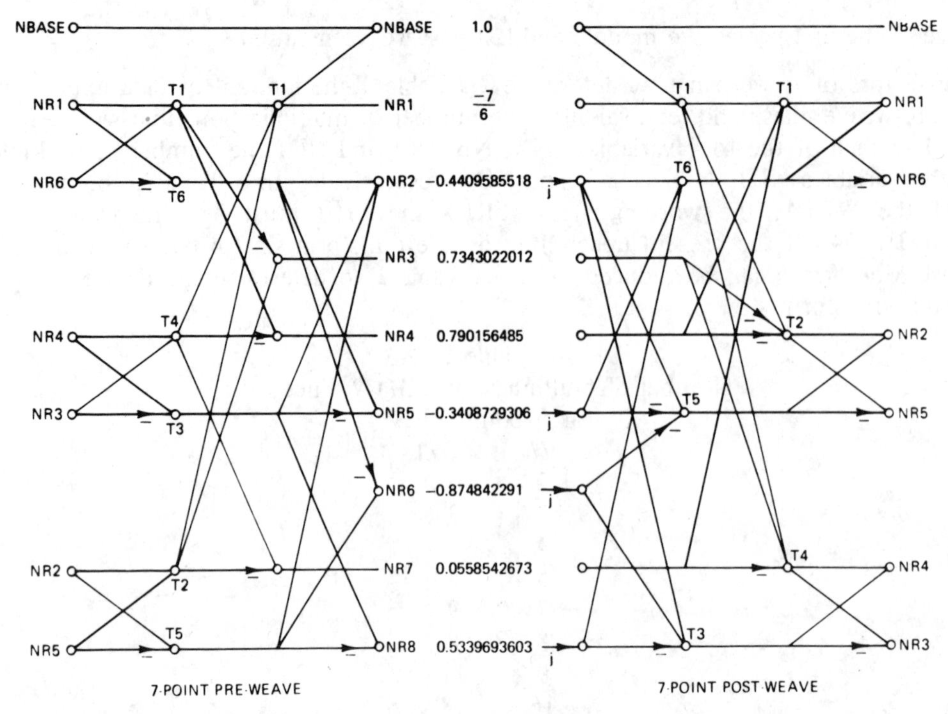

Fig. 6 Seven-point DFT using Winograd's decomposition. Nine multiplications are required.

The three functions of the INISHL subroutine need only be performed once for a given value of transform length *N*. The remaining three subroutines form the computational core of the algorithm. The subroutine WFTA maps the data input arrays XR and XI into the working arrays SR and SI using the pre-permutation vector INDX1. The arrays SR and SI are treated as one-dimensional arrays of size ND1*ND2*ND3*ND4 in subroutine WFTA. Note that the WFTA is NOT computed in-place because of the nature of the permutations.

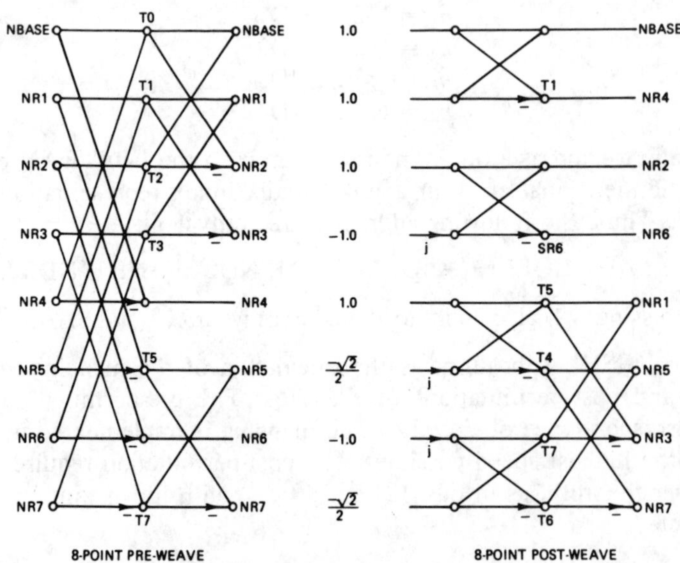

Fig. 7 Eight-point DFT algorithm using Winograd's decomposition.

Next the subroutine WEAVE1 is invoked to perform the pre-weave modules. The first pre-weave module performed is of length NA. If NA=1 the first module is skipped and the next factor, NB, is tested. Each factor NB, NC and ND is checked in turn. In WEAVE1 the arrays SR and SI are treated as one-dimensional arrays to gain speed in execution. There is no 4-point pre-weave module because the 4-point DFT is implemented entirely as a post-weave module. Likewise, the 2-point pre-weave module is a 2-point DFT and there is no 2-point post-weave module. Documentation of the code for each pre-weave module can be found in the flowgraphs of Figs. 3 - 10.

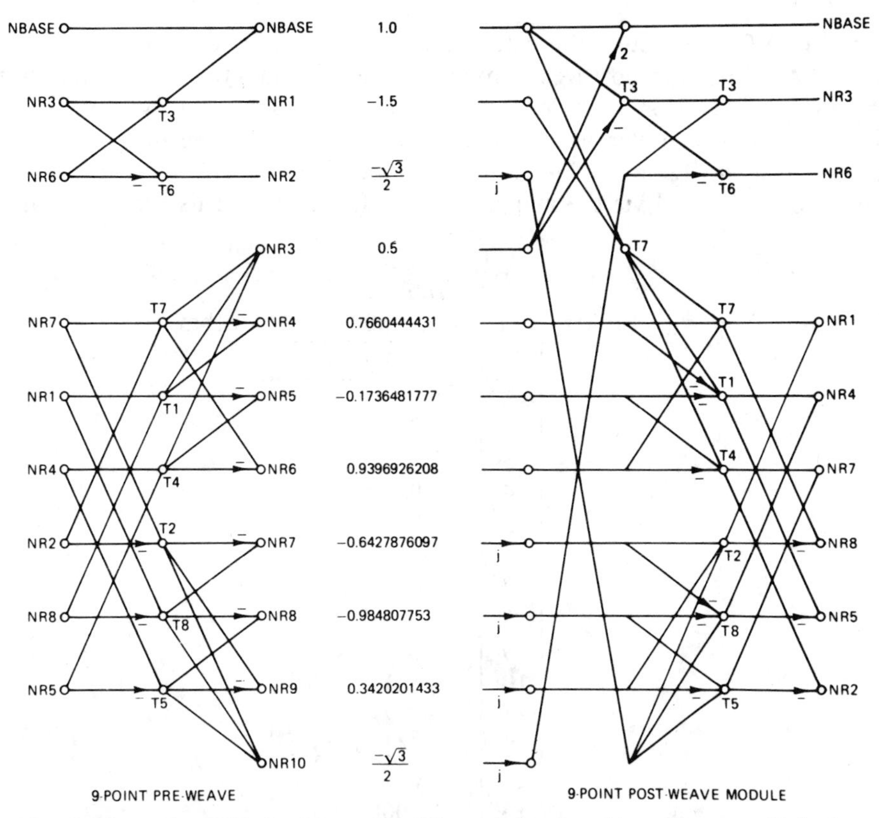

Fig. 8 Nine-point DFT algorithm using Winograd decomposition. Eleven multiplications are required.

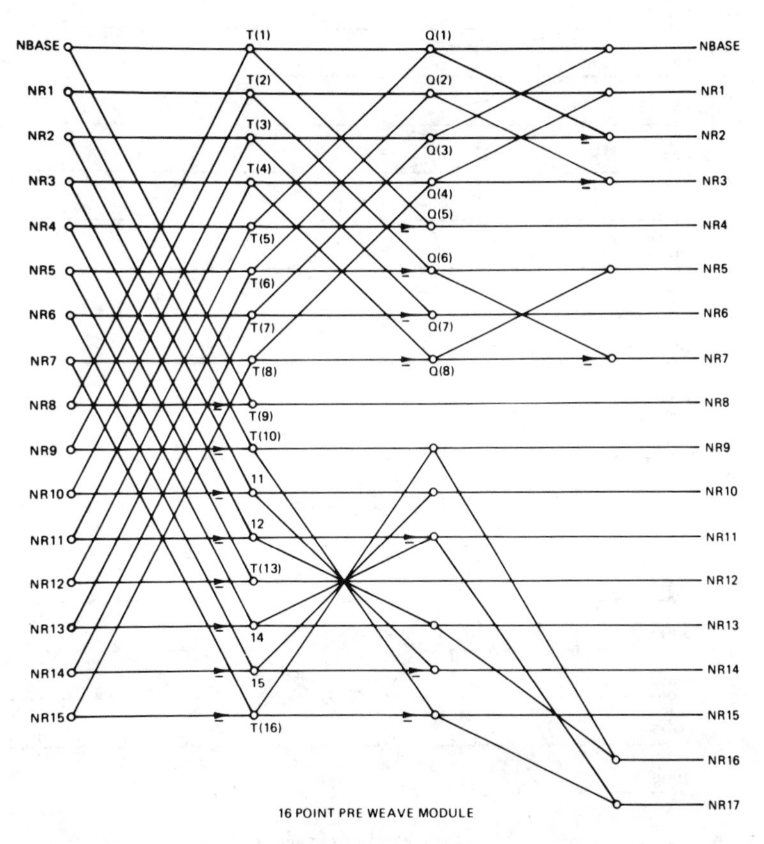

Fig. 9 Sixteen-point pre-weave module.

Following the WEAVE1 subroutine the data is multiplied point-by-point by the real array COEF in the subroutine WFTA. Then the post-weave modules are implemented in the subroutine WEAVE2. The module of length ND (=5 or 1) is done first, then NC, NB and finally NA. The flowgraphs of Figs. 3 - 10 give the post-weave algorithms for the various factors. The multipliers for the 16-point DFT algorithm are given in Table 2. Finally, subroutine WFTA performs the post-permutation by mapping SR and SI to XR and XI via the mapping vector INDX2(·). This completes the WFTA calculation.

Table 2
Multipliers for 16-point DFT Algorithm

I	CD16(I)
1	1.0
2	1.0
3	1.0
4	−1.0
5	1.0
6	$-\sqrt{2}/2$
7	−1.0
8	$\sqrt{2}/2$
9	1.0
10	0.5411961001
11	$-\sqrt{2}/2$
12	−0.5411961001
13	−1.0
14	−1.306562965
15	$\sqrt{2}/2$
16	1.306562965
17	−0.9238795325
18	0.3826834324

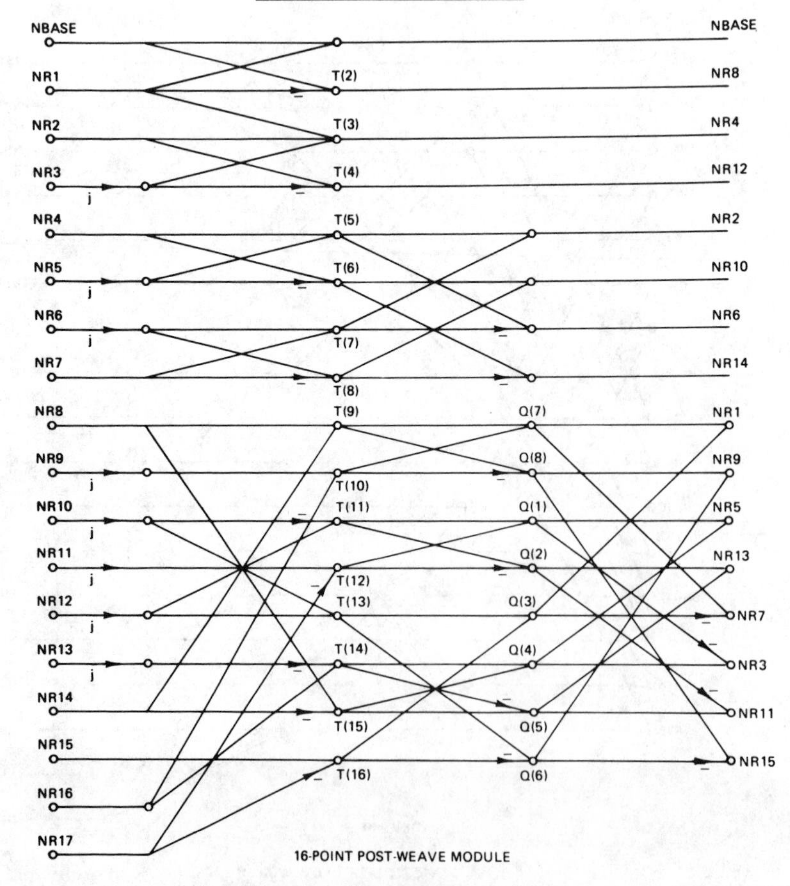

Fig. 10 Sixteen-point post-weave module.

4. Description of Parameters

CDA, CDB, CDC, CDD: real arrays that are assigned coefficients of the factors of N during initialization (CDA: for outermost factor, CDD: for innermost factor).

CO3, CO4, CO8, CO9, CO16: real arrays containing the short-DFT coefficients for factors 3,4,8,9,16 respectively.

COEF: real array of size ND1*ND2*ND3*ND4 to which the N-point DFT multiply coefficients are assigned during the initial run.

INDX1, INDX2: integer arrays of size N for storing initial and final permutation mappings respectively.

INIT: integer variable passed from user program to WFTA. It must be zero on the initial run and nonzero otherwise.

INVRS: integer variable to flag the inverse transform (INVRS = 1), otherwise (INVRS $\neq$ 1) the forward transform is calculated.

IERR: integer variable that returns error code indicating successful completion (IERR=0) or abnormal termination (IERR=−1 or −2).

N: transform length

NA, NB, NC, ND: integer variables for four factors of N (NA: outermost, ND: innermost).

ND1, ND2, ND3, ND4: integer variables representing the "expansion" of factors NA, NB, NC, ND, respectively. See Table 1.

NMULT: integer variable equal to product of ND1, ND2, ND3, and ND4.

SR, SI: real arrays of size ND1*ND2*ND3*ND4 to which XR, XI are transferred during initial permutation.

XR, XI: real and imaginary parts respectively of complex input data array. The transformed data are also placed in these arrays by the final permutation.

5. Summary of User Requirements

(1) Specify input sequences XR and XI and parameters N, INVRS, INIT and IERR.

(2) Call WFTA. For initial call use INIT=0; INVRS = 1 means the inverse DFT will be computed.

(3) Output is obtained in XR and XI. The error code IERR is zero for successful completion and −1 or −2 if the value of N is not permissible or was not initialized.

(4) After initial call, use INIT$\neq$0 for all other calls as long as N is not changed.

6. Test Example

A main program is provided to generate a test waveform whose DFT is known analytically. For this purpose a complex exponential signal was chosen

$$x[n] = \left(re^{j\frac{2\pi}{N}\phi}\right)^n \qquad n = 0,1,\ldots,N-1 \tag{4}$$

where N is the transform length to be tested. The DFT is easily calculated to be

$$X[k] = \frac{1 - \left(re^{j\frac{2\pi}{N}\phi}\right)^N}{1 - re^{j\frac{2\pi}{N}\phi}} \tag{5}$$

Thus the WFTA can be checked by comparing the results of a DFT computation with Eq. (5). The program will print out the maximum absolute and relative differences between the computed DFT and the ideal. The test example was run for the case $N = 28$, forward transform, $r = 0.98$, and $\phi = 0.5$ with the corresponding input and output given in Tables 3 and 4 respectively. By choosing different values for N the various pre-weave and post-weave modules can be exercised. Different inputs can be obtained by varying r and ϕ. The choice $r \approx 1$ and $\phi \equiv 0 \bmod N$ must be avoided since Eq. (5) does not hold in this case.

7. Summary

A FORTRAN program for the calculation of the DFT using Winograd's "nested" algorithm has been presented. While every effort has been made to provide a computationally efficient program, some compromises have been made to simplify the program structure. For example, the multiplications of the WFTA are all performed in a separate DO loop. A slightly more efficient implementation of the algorithm would provide code so that the innermost module of the pre-weave and post-weave computations is combined with the multiplications. For example, in Fig. 1 the innermost module is the 5-point factor. The advantage of such an approach is that it avoids some memory transfers (loads and stores) that are required in our implementation. Furthermore, it reduces slightly the amount of memory required for the arrays SR and SI. The disadvantage, of course, is that an additional block of code must be provided for each factor in the set {2,3,4,5,7,8,9,16} to allow for the possibility of each factor being the innermost. This could conceivably double the amount of program code.

Another possible improvement in the WFTA that would be quite helpful, would be the development of input and output permutations that could be done "in-place". If this were possible, the arrays SR and SI could share memory with XR and XI and only an additional (ND1*ND2*ND3*ND4 – N) locations would be needed to accommodate the data expansion of the WFTA.

If one only requires an algorithm to compute the DFT of real data, a rather simple modification of the existing program will exploit the real-valued nature of the input. The input permutation, subroutine WEAVE1 and the multiplications should be modified to remove all calculations involving the array SI which will be zero, until subroutine WEAVE2 is called. This will result in a slight improvement in the running time of the algorithm for real data.

As a final comment, it should be pointed out that this program is an initial attempt to produce an efficient WFTA program, and it is quite likely that future developments related to Winograd's algorithm will lead to more efficient implementations.

References

1. S. Winograd, "On Computing the Discrete Fourier Transform", *Proc. Nat. Acad. Sci. USA,* Vol. 73, No. 4, pp. 1005-1006, April 1976.
2. H. Silverman, "An Introduction to Programming the Winograd Fourier Transform Algorithm (WFTA)", *IEEE Trans. Acoust., Speech, and Signal Processing,* Vol. ASSP-25, No. 2, pp. 152-165, April 1977.
3. I. J. Good, "The Interaction Algorithm and Practical Fourier Analysis", *J. Royal Stat. Soc., Ser. B,* Vol. 20, pp. 361-372, 1958; Addendum, Vol. 22, 1960, pp. 372-375.
4. J. H. McClellan and C. M. Rader, *Number Theory in Digital Signal Processing,* Prentice-Hall, Inc., Englewood Cliffs, NJ, 1979.
5. L. R. Morris, "A Comparative Study of Time Efficient FFT and WFTA Programs for General Purpose Computers", *IEEE Trans. Acoust., Speech, and Signal Processing,* Vol. ASSP-26, No. 2, pp. 141-150, April 1978.

Table 3

```
   28
    0
0.98          .5
```

Table 4

```
LENGTH =     28
J =   1REAL =       0.10000000E 01IMAG =       0.
J =   2REAL =       0.97383796E 00IMAG =       0.10972519E 00
J =   3REAL =       0.93632076E 00IMAG =       0.21370910E 00
J =   4REAL =       0.88837543E 00IMAG =       0.31085601E 00
J =   5REAL =       0.83102498E 00IMAG =       0.40020054E 00
J =   6REAL =       0.76537160E 00IMAG =       0.48091486E 00
J =   7REAL =       0.69257944E 00IMAG =       0.55231368E 00
J =   8REAL =       0.61385744E 00IMAG =       0.61385744E 00
J =   9REAL =       0.53044206E 00IMAG =       0.66515329E 00
J =  10REAL =       0.44358054E 00IMAG =       0.70595438E 00
J =  11REAL =       0.35451459E 00IMAG =       0.73615713E 00
J =  12REAL =       0.26446479E 00IMAG =       0.75579695E 00
J =  13REAL =       0.17461589E 00IMAG =       0.76504221E 00
J =  14REAL =       0.86103185E-01IMAG =       0.76418691E 00
J =  15REAL =       0.22561488E-18IMAG =       0.75364190E 00
J =  16REAL =      -0.82693498E-01IMAG =       0.73392510E 00
J =  17REAL =      -0.16106014E 00IMAG =       0.70565056E 00
J =  18REAL =      -0.23427411E 00IMAG =       0.66951695E 00
J =  19REAL =      -0.30160790E 00IMAG =       0.62629525E 00
J =  20REAL =      -0.36243759E 00IMAG =       0.57681611E 00
J =  21REAL =      -0.41624673E 00IMAG =       0.52195689E 00
J =  22REAL =      -0.46262869E 00IMAG =       0.46262869E 00
J =  23REAL =      -0.50128740E 00IMAG =       0.39976336E 00
J =  24REAL =      -0.53203680E 00IMAG =       0.33430089E 00
J =  25REAL =      -0.55479886E 00IMAG =       0.26717705E 00
J =  26REAL =      -0.56960024E 00IMAG =       0.19931175E 00
J =  27REAL =      -0.57656787E 00IMAG =       0.13159785E 00
J =  28REAL =      -0.57592328E 00IMAG =       0.64890968E-01
J =   1REAL =       0.32239256E 01IMAG =       0.13521341E 02
J =   2REAL =       0.32239256E 01IMAG =      -0.13521341E 02
J =   3REAL =       0.10652301E 01IMAG =      -0.45975034E 01
J =   4REAL =       0.88719223E 00IMAG =      -0.27176643E 01
J =   5REAL =       0.83803070E 00IMAG =      -0.18913968E 01
J =   6REAL =       0.81782128E 00IMAG =      -0.14179013E 01
J =   7REAL =       0.80763211E 00IMAG =      -0.11045912E 01
J =   8REAL =       0.80182080E 00IMAG =      -0.87708147E 00
J =   9REAL =       0.79823032E 00IMAG =      -0.70048606E 00
J =  10REAL =       0.79589333E 00IMAG =      -0.55618385E 00
J =  11REAL =       0.79432575E 00IMAG =      -0.43323722E 00
J =  12REAL =       0.79326570E 00IMAG =      -0.32469963E 00
J =  13REAL =       0.79256435E 00IMAG =      -0.22583816E 00
J =  14REAL =       0.79213571E 00IMAG =      -0.13319096E 00
J =  15REAL =       0.79193212E 00IMAG =      -0.44023335E-01
J =  16REAL =       0.79193209E 00IMAG =       0.44023335E-01
J =  17REAL =       0.79213572E 00IMAG =       0.13319093E 00
J =  18REAL =       0.79256435E 00IMAG =       0.22583811E 00
J =  19REAL =       0.79326568E 00IMAG =       0.32469966E 00
J =  20REAL =       0.79432572E 00IMAG =       0.43323717E 00
J =  21REAL =       0.79589333E 00IMAG =       0.55618381E 00
J =  22REAL =       0.79823034E 00IMAG =       0.70048602E 00
J =  23REAL =       0.80182078E 00IMAG =       0.87708144E 00
J =  24REAL =       0.80763214E 00IMAG =       0.11045912E 01
J =  25REAL =       0.81782126E 00IMAG =       0.14179012E 01
J =  26REAL =       0.83803074E 00IMAG =       0.18913967E 01
J =  27REAL =       0.88719221E 00IMAG =       0.27176643E 01
J =  28REAL =       0.10652301E 01IMAG =       0.45975034E 01
ABSOLUTE DEVIATION =        0.11920929E-06 AT INDEX    0
RELATIVE DEVIATION =    0.0000025 PERCENT  AT INDEX   20
```

Appendix

```
C
C-----------------------------------------------------------------------
C MAIN PROGRAM: TEST PROGRAM TO EXERCISE THE WFTA SUBROUTINE
C   THE TEST WAVEFORM IS A COMPLEX EXPONENTIAL A**I WHOSE
C   TRANSFORM IS KNOWN ANALYTICALLY TO BE (1 - A**N)/(1 - A*W**K).
C
C AUTHORS:
C   JAMES H. MCCLELLAN     AND     HAMID NAWAB
C   DEPARTMENT OF ELECTRICAL ENGINEERING AND COMPUTER SCIENCE
C   MASSACHUSETTS OF TECHNOLOGY
C   CAMBRIDGE, MASS.  02139
C
C INPUTS:
C   N-- TRANSFORM LENGTH. IT MUST BE FORMED AS THE PRODUCT OF
C       RELATIVELY PRIME INTEGERS FROM THE SET:
C           2,3,4,5,7,8,9,16
C   INVRS IS THE FLAG FOR FORWARD OR INVERSE TRANSFORM.
C           INVRS = 1 YIELDS INVERSE TRANSFORM
C           INVRS .NE. 1 GIVES FORWARD TRANSFORM
C   RAD AND PHI ARE THE MAGNITUDE AND ANGLE (AS A FRACTION OF
C       2*PI/N) OF THE COMPLEX EXPONENTIAL TEST SIGNAL.
C          SUGGESTION: RAD = 0.98, PHI = 0.5.
C-----------------------------------------------------------------------
C
      DOUBLE PRECISION PI2,PIN,XN,XJ,XT
      DIMENSION XR(5040),XI(5040)
      COMPLEX CONE,CA,CAN,CNUM,CDEN
C
C   OUTPUT WILL BE PUNCHED
C
      IOUT=I1MACH(3)
      INPUT=I1MACH(1)
      CONE=CMPLX(1.0,0.0)
      PI2=8.0D0*DATAN(1.0D0)
50    CONTINUE
      READ(INPUT,130)N
130   FORMAT(I5)
      WRITE(IOUT,150) N
150   FORMAT(10H LENGTH = ,I5)
      IF(N.LE.0 .OR. N.GT.5040) STOP
C
C   ENTER A 1 TO PERFORM THE INVERSE
C
      READ(INPUT,130) INVRS
C
C   ENTER MAGNITUDE AND ANGLE (IN FRACTION OF 2*PI/N)
C   AVOID MULTIPLES OF N FOR THE ANGLE IF THE RADIUS IS
C   CLOSE TO ONE.  SUGGESTION: RAD = 0.98, PHI = 0.5.
C
      READ(INPUT,160) RAD,PHI
160   FORMAT(2F15.10)
      XN=FLOAT(N)
      PIN=PHI
      PIN=PIN*PI2/XN
C
C   GENERATE Z**J
C
      INIT=0
      DO 200 J=1,N
      AN=RAD**(J-1)
      XJ=J-1
      XJ=XJ*PIN
      XT=DCOS(XJ)
      XR(J)=XT
      XR(J)=XR(J)*AN
      XT=DSIN(XJ)
      XI(J)=XT
      XI(J)=XI(J)*AN
200   CONTINUE
      CAN=CMPLX(XR(N),XI(N))
      CA=CMPLX(XR(2),XI(2))
      CAN=CAN*CA
C
C   PRINT FIRST 50 VALUES OF INPUT SEQUENCE
C
      MAX=50
      IF(N.LT.50)MAX=N
      WRITE(IOUT,300)(J,XR(J),XI(J),J=1,MAX)
C
C   CALL THE WINOGRAD FOURIER TRANSFORM ALGORITHM
C
      CALL WFTA(XR,XI,N,INVRS,INIT,IERR)
C
C   CHECK FOR ERROR RETURN
C
      IF(IERR.LT.0) WRITE(IOUT,250) IERR
250   FORMAT(1X,5HERROR,I5)
      IF(IERR.LT.0) GO TO 50
C
C   PRINT FIRST 50 VALUES OF THE TRANSFORMED SEQUENCE
C
      WRITE(IOUT,300)(J,XR(J),XI(J),J=1,MAX)
300   FORMAT(1X,3HJ =,I3,6HREAL =,E20.12,6HIMAG =,E20.12)
C
C   CALCULATE ABSOLUTE AND RELATIVE DEVIATIONS
C
      DEVABS=0.0
      DEVREL=0.0         <-- { DEVMX1 = 0.0
      CNUM=CONE-CAN           { DEVMX2 = 0.0
      PIN=PI2/XN
      DO 350 J=1,N
      XJ=J-1
      XJ=-XJ*PIN
      IF(INVRS.EQ.1) XJ=-XJ
      TR=DCOS(XJ)
      TI=DSIN(XJ)
      CAN=CMPLX(TR,TI)
      CDEN=CONE-CA*CAN
      CDEN=CNUM/CDEN
C
C   TRUE VALUE OF THE TRANSFORM (1. - A**N)/(1. - A*W**K),
C   WHERE A = RAD*EXP(J*PHI*(2*PI/N)), W = EXP(-J*2*PI/N).
C   FOR THE INVERSE TRANSFORM THE COMPLEX EXPONENTIAL W
C   IS CONJUGATED.
C
      TR=REAL(CDEN)
      TI=AIMAG(CDEN)
      IF(INVRS.NE.1) GO TO 330
C
C   SCALE INVERSE TRANSFORM BY 1/N
C
      TR=TR/FLOAT(N)
      TI=TI/FLOAT(N)
330   TR=XR(J)-TR
      TI=XI(J)-TI
      DEVABS=SQRT(TR*TR+TI*TI)
      XMAG=SQRT(XR(J)*XR(J)+XI(J)*XI(J))
      DEVREL=100.0*DEVABS/XMAG
      IF(DEVABS.LE.DEVMX1)GO TO 340
      DEVMX1=DEVABS
```

```
      LABS=J-1
340   IF(DEVREL.LE.DEVMX2)GO TO 350
      DEVMX2=DEVREL
      LREL=J-1
350   CONTINUE
C
C   PRINT THE ABSOLUTE AND RELATIVE DEVIATIONS TOGETHER
C   WITH THEIR LOCATIONS.
C                     DEVMX1       DEVMX2
      WRITE(IOUT,380) DEVABS,LABS,DEVREL,LREL
380   FORMAT(1X,21HABSOLUTE DEVIATION = ,E20.12,9H AT INDEX,I5/
     1 1X,21HRELATIVE DEVIATION = ,F11.7,8H PERCENT,1X,9H AT INDEX,I5)
      GO TO 50
      END
C
C-----------------------------------------------------------------------
C SUBROUTINE: INISHL
C   THIS SUBROUTINE INITIALIZES THE WFTA ROUTINE FOR A GIVEN
C   VALUE OF THE TRANSFORM LENGTH N.  THE FACTORS OF N ARE
C   DETERMINED, THE MULTIPLICATION COEFFICIENTS ARE CALCULATED
C   AND STORED IN THE ARRAY COEF(.), THE INPUT AND OUTPUT
C   PERMUTATION VECTORS ARE COMPUTED AND STORED IN THE ARRAYS
C   INDX1(.) AND INDX2(.)
C
C-----------------------------------------------------------------------
C
      SUBROUTINE INISHL(N,COEF,XR,XI,INDX1,INDX2,IERR)
      DIMENSION COEF(1),XR(1),XI(1)
      INTEGER S1,S2,S3,S4,INDX1(1),INDX2(1),P1
      DIMENSION CO3(3),CO4(4),CO8(8),CO9(11),CO16(18),CDA(18),CDB(11),
     1CDC(9),CDD(6)
      COMMON NA,NB,NC,ND,ND1,ND2,ND3,ND4
C
C   DATA STATEMENTS ASSIGN SHORT DFT COEFFICIENTS.
C
      DATA CO4(1),CO4(2),CO4(3),CO4(4)/4*1.0/
C
      DATA CDA(1),CDA(2),CDA(3),CDA(4),CDA(5),CDA(6),CDA(7),
     1 CDA(8),CDA(9),CDA(10),CDA(11),CDA(12),CDA(13),CDA(14),
     2 CDA(15),CDA(16),CDA(17),CDA(18)/18*1.0/
C
      DATA CDB(1),CDB(2),CDB(3),CDB(4),CDB(5),CDB(6),CDB(7),CDB(8),
     1 CDB(9),CDB(10),CDB(11)/11*1.0/
C
      DATA IONCE/1/
C
C   GET MULTIPLIER CONSTANTS
C
      IF(IONCE.NE.1) GO TO 20
      CALL CONST(CO3,CO8,CO16,CO9,CDC,CDD)
20    IONCE=-1
C
C   FOLLOWING SEGMENT DETERMINES FACTORS OF N AND CHOOSES
C   THE APPROPRIATE SHORT DFT COEFFICIENTS.
C
      IOUT=I1MACH(2)
      IERR=0
      ND1=1
      NA=1
      NB=1
      ND2=1
      NC=1
      ND3=1
      ND=1
      ND4=1
      IF(N.LE.0 .OR. N.GT.5040) GO TO 190
      IF(16*(N/16).EQ.N) GO TO 30
      IF(8*(N/8).EQ.N) GO TO 40
      IF(4*(N/4).EQ.N) GO TO 50
      IF(2*(N/2).NE.N) GO TO 70
      ND1=2
      NA=2
      CDA(2)=1.0
      GO TO 70
30    ND1=18
      NA=16
      DO 31 J=1,18
31    CDA(J)=CO16(J)
      GO TO 70
40    ND1=8
      NA=8
      DO 41 J=1,8
41    CDA(J)=CO8(J)
      GO TO 70
50    ND1=4
      NA=4
      DO 51 J=1,4
51    CDA(J)=CO4(J)
70    IF(3*(N/3).NE.N) GO TO 120
      IF(9*(N/9).EQ.N) GO TO 100
      ND2=3
      NB=3
      DO 71 J=1,3
71    CDB(J)=CO3(J)
      GO TO 120
100   ND2=11
      NB=9
      DO 110 J=1,11
110   CDB(J)=CO9(J)
120   IF(7*(N/7).NE.N) GO TO 160
      ND3=9
      NC=7
160   IF(5*(N/5).NE.N) GO TO 190
      ND4=6
      ND=5
190   M=NA*NB*NC*ND
      IF(M.EQ.N) GO TO 250
      WRITE(IOUT,210)
210   FORMAT(21H THIS N DOES NOT WORK)
      IERR=-1
      RETURN
C
C   NEXT SEGMENT GENERATES THE DFT COEFFICIENTS BY
C   MULTIPLYING TOGETHER THE SHORT DFT COEFFICIENTS
C
250   J=1
      DO 300 N4=1,ND4
      DO 300 N3=1,ND3
      DO 300 N2=1,ND2
      DO 300 N1=1,ND1
      COEF(J)=CDA(N1)*CDB(N2)*CDC(N3)*CDD(N4)
      J=J+1
300   CONTINUE
C
C   FOLLOWING SEGMENT FORMS THE INPUT INDEXING VECTOR
C
      J=1
      NU=NB*NC*ND
      NV=NA*NC*ND
      NW=NA*NB*ND
```

```
      NY=NA*NB*NC
      K=1
      DO 440 N4=1,ND
      DO 430 N3=1,NC
      DO 420 N2=1,NB
      DO 410 N1=1,NA
405   IF(K.LE.N) GO TO 408
      K=K-N
      GO TO 405
408   INDX1(J)=K
      J=J+1
410   K=K+NU
420   K=K+NV
430   K=K+NW
440   K=K+NY
C
C   FOLLOWING SEGMENT FORMS THE OUTPUT INDEXING VECTOR
C
      M=1
      S1=0
      S2=0
      S3=0
      S4=0
      IF(NA.EQ.1) GO TO 530
520   P1=M*NU-1
      IF((P1/NA)*NA.EQ.P1) GO TO 510
      M=M+1
      GO TO 520
510   S1=P1+1
530   IF(NB.EQ.1) GO TO 540
      M=1
550   P1=M*NV-1
      IF((P1/NB)*NB.EQ.P1) GO TO 560
      M=M+1
      GO TO 550
560   S2=P1+1
540   IF(NC.EQ.1) GO TO 630
      M=1
620   P1=M*NW-1
      IF((P1/NC)*NC.EQ.P1) GO TO 610
      M=M+1
      GO TO 620
610   S3=P1+1
630   IF(ND.EQ.1) GO TO 660
      M=1
640   P1=M*NY-1
      IF((P1/ND)*ND.EQ.P1) GO TO 650
      M=M+1
      GO TO 640
650   S4=P1+1
660   J=1
      DO 810 N4=1,ND
      DO 810 N3=1,NC
      DO 810 N2=1,NB
      DO 810 N1=1,NA
      INDX2(J)=S1*(N1-1)+S2*(N2-1)+S3*(N3-1)+S4*(N4-1)+1
900   IF(INDX2(J).LE.N) GO TO 910
      INDX2(J)=INDX2(J)-N
      GO TO 900
910   J=J+1
810   CONTINUE
      RETURN
      END
C
C----------------------------------------------------------------------
C SUBROUTINE: CONST
C   COMPUTES THE MULTIPLIERS FOR THE VARIOUS MODULES
C----------------------------------------------------------------------
C
      SUBROUTINE CONST(CO3,CO8,CO16,CO9,CDC,CDD)
      DOUBLE PRECISION DTHETA,DTWOPI,DSQ32,DSQ2
      DOUBLE PRECISION DCOS1,DCOS2,DCOS3,DCOS4
      DOUBLE PRECISION DSIN1,DSIN2,DSIN3,DSIN4
      DIMENSION CO3(3),CO8(8),CO16(18),CO9(11),CDC(9),CDD(6)
      DTWOPI=8.0D0*DATAN(1.0D0)
      DSQ32=DSQRT(0.75D0)
      DSQ2=DSQRT(0.5D0)
C
C   MULTIPLIERS FOR THE THREE POINT MODULE
C
      CO3(1)=1.0
      CO3(2)=-1.5
      CO3(3)=-DSQ32
C
C   MULTIPLIERS FOR THE FIVE POINT MODULE
C
      DTHETA=DTWOPI/5.0D0
      DCOS1=DCOS(DTHETA)
      DCOS2=DCOS(2.0D0*DTHETA)
      DSIN1=DSIN(DTHETA)
      DSIN2=DSIN(2.0D0*DTHETA)
      CDD(1)=1.0
      CDD(2)=-1.25
      CDD(3)=-DSIN1-DSIN2
      CDD(4)=0.5*(DCOS1-DCOS2)
      CDD(5)=DSIN1-DSIN2
      CDD(6)=DSIN2
C
C
C   MULTIPLIERS FOR THE SEVEN POINT MODULE
C
      DTHETA=DTWOPI/7.0D0
      DCOS1=DCOS(DTHETA)
      DCOS2=DCOS(2.0D0*DTHETA)
      DCOS3=DCOS(3.0D0*DTHETA)
      DSIN1=DSIN(DTHETA)
      DSIN2=DSIN(2.0D0*DTHETA)
      DSIN3=DSIN(3.0D0*DTHETA)
      CDC(1)=1.0
      CDC(2)=-7.0D0/6.0D0
      CDC(3)=-(DSIN1+DSIN2-DSIN3)/3.0D0
      CDC(4)=(DCOS1+DCOS2-2.0D0*DCOS3)/3.0D0
      CDC(5)=(2.0D0*DCOS1-DCOS2-DCOS3)/3.0D0
      CDC(6)=-(2.0D0*DSIN1-DSIN2+DSIN3)/3.0D0
      CDC(7)=-(DSIN1+DSIN2+2.0D0*DSIN3)/3.0D0
      CDC(8)=(DCOS1-2.0D0*DCOS2+DCOS3)/3.0D0
      CDC(9)=-(DSIN1-2.0D0*DSIN2-DSIN3)/3.0D0
C
C   MULTIPLIERS FOR THE EIGHT POINT MODULE
C
      CO8(1)=1.0
      CO8(2)=1.0
      CO8(3)=1.0
      CO8(4)=-1.0
      CO8(5)=1.0
      CO8(6)=-DSQ2
      CO8(7)=-1.0
      CO8(8)=DSQ2
C
C   MULTIPLIERS FOR THE NINE POINT MODULE
```

```
C
      DTHETA=DTWOPI/9.0D0
      DCOS1=DCOS(DTHETA)
      DCOS2=DCOS(2.0D0*DTHETA)
      DCOS4=DCOS(4.0D0*DTHETA)
      DSIN1=DSIN(DTHETA)
      DSIN2=DSIN(2.0D0*DTHETA)
      DSIN4=DSIN(4.0D0*DTHETA)
      CO9(1)=1.0
      CO9(2)=-1.5
      CO9(3)=-DSQ32
      CO9(4)=0.5
      CO9(5)=(2.0D0*DCOS1-DCOS2-DCOS4)/3.0D0
      CO9(6)=(DCOS1-2.0D0*DCOS2+DCOS4)/3.0D0
      CO9(7)=(DCOS1+DCOS2-2.0D0*DCOS4)/3.0D0
      CO9(8)=-(2.0D0*DSIN1+DSIN2-DSIN4)/3.0D0
      CO9(9)=-(DSIN1+2.0D0*DSIN2+DSIN4)/3.0D0
      CO9(10)=-(DSIN1-DSIN2-2.0D0*DSIN4)/3.0D0
      CO9(11)=-DSQ32
C
C   MULTIPLIERS FOR THE SIXTEEN POINT MODULE
C
      DTHETA=DTWOPI/16.0D0
      DCOS1=DCOS(DTHETA)
      DCOS3=DCOS(3.0D0*DTHETA)
      DSIN1=DSIN(DTHETA)
      DSIN3=DSIN(3.0D0*DTHETA)
      CO16(1)=1.0
      CO16(2)=1.0
      CO16(3)=1.0
      CO16(4)=-1.0
      CO16(5)=1.0
      CO16(6)=-DSQ2
      CO16(7)=-1.0
      CO16(8)=DSQ2
      CO16(9)=1.0
      CO16(10)=-(DSIN1-DSIN3)
      CO16(11)=-DSQ2
      CO16(12)=-CO16(10)
      CO16(13)=-1.0
      CO16(14)=-(DSIN1+DSIN3)
      CO16(15)=DSQ2
      CO16(16)=-CO16(14)
      CO16(17)=-DSIN3
      CO16(18)=DCOS3
      RETURN
      END
C
C-----------------------------------------------------------------
C SUBROUTINE: WFTA
C WINOGRAD FOURIER TRANSFORM ALGORITHM
C-----------------------------------------------------------------
C
      SUBROUTINE WFTA(XR,XI,N,INVRS,INIT,IERR)
      DIMENSION XR(1),XI(1)
C
C INPUTS:
C   N-- TRANSFORM LENGTH.  MUST BE FORMED AS THE PRODUCT OF
C       RELATIVELY PRIME INTEGERS FROM THE SET:
C           2,3,4,5,7,8,9,16
C       THUS THE LARGEST POSSIBLE VALUE OF N IS 5040.
C   XR(.)-- ARRAY THAT HOLDS THE REAL PART OF THE DATA
C           TO BE TRANSFORMED.
C   XI(.)-- ARRAY THAT HOLDS THE IMAGINARY PART OF THE
C           DATA TO BE TRANSFORMED.
C   INVRS-- PARAMETER THAT FLAGS WHETHER OR NOT THE INVERSE
C           TRANSFORM IS TO BE CALCULATED.  A DIVISION BY N
C           IS INCLUDED IN THE INVERSE.
C           INVRS = 1 YIELDS INVERSE TRANSFORM
C           INVRS .NE. 1 GIVES FORWARD TRANSFORM
C   INIT-- PARAMETER THAT FLAGS WHETHER OR NOT THE PROGRAM
C          IS TO BE INITIALIZED FOR THIS VALUE OF N.  THE
C          INITIALIZATION IS PERFORMED ONLY ONCE IN ORDER TO
C          TO SPEED UP THE COMPUTATION ON SUCCEEDING CALLS
C          TO THE WFTA ROUTINE, WHEN N IS HELD FIXED.
C          INIT = 0 RESULTS IN INITIALIZATION.
C   IERR-- ERROR CODE THAT IS NEGATIVE WHEN THE WFTA
C          TERMINATES INCORRECTLY.
C           0 = SUCCESSFUL COMPLETION
C          -1 = THIS VALUE OF N DOES NOT FACTOR PROPERLY
C          -2 = AN INITIALIZATION HAS NOT BEEN DONE FOR
C               THIS VALUE OF N.
C
C
C   THE FOLLOWING TWO CARDS MAY BE CHANGED IF THE MAXIMUM
C   DESIRED TRANSFORM LENGTH IS LESS THAN 5040
C
C *******************************************************************
      DIMENSION SR(10692),SI(10692),COEF(10692)
      INTEGER INDX1(5040),INDX2(5040)
C *******************************************************************
C
      COMMON NA,NB,NC,ND,ND1,ND2,ND3,ND4
C
C   TEST FOR INITIAL RUN
C
      IF(INIT.EQ.0) CALL INISHL(N,COEF,XR,XI,INDX1,INDX2,IERR)
C
      IF(IERR.LT.0) RETURN
      M=NA*NB*NC*ND
      IF(M.EQ.N) GO TO 100
      IERR=-2
      RETURN
C
C  ERROR(-2)-- PROGRAM NOT INITIALIZED FOR THIS VALUE OF N
C
100   NMULT=ND1*ND2*ND3*ND4
C
C   THE FOLLOWING CODE MAPS THE DATA ARRAYS XR AND XI TO
C   THE WORKING ARRAYS SR AND SI VIA THE MAPPING VECTOR
C   INDX1(.).  THE PERMUTATION OF THE DATA FOLLOWS THE
C   SINO CORRESPONDENCE OF THE CHINESE REMAINDER THEOREM.
C
      J=1
      K=1
      INC1=ND1-NA
      INC2=ND1*(ND2-NB)
      INC3=ND1*ND2*(ND3-NC)
      DO 140 N4=1,ND
      DO 130 N3=1,NC
      DO 120 N2=1,NB
      DO 110 N1=1,NA
      IND=INDX1(K)
      SR(J)=XR(IND)
      SI(J)=XI(IND)
      K=K+1
110   J=J+1
120   J=J+INC1
130   J=J+INC2
140   J=J+INC3
```

```
C
C   DO THE PRE-WEAVE MODULES
C
       CALL WEAVE1(SR,SI)
C
C   THE FOLLOWING LOOP PERFORMS ALL THE MULTIPLICATIONS OF THE
C   WINOGRAD FOURIER TRANSFORM ALGORITHM.  THE MULTIPLICATION
C   COEFFICIENTS ARE STORED ON THE INITIALIZATION PASS IN THE
C   ARRAY COEF(.).
C
       DO 200 J=1,NMULT
       SR(J)=SR(J)*COEF(J)
       SI(J)=SI(J)*COEF(J)
  200  CONTINUE
C
C   DO THE POST-WEAVE MODULES
C
       CALL WEAVE2(SR,SI)
C
C
C   THE FOLLOWING CODE MAPS THE WORKING ARRAYS SR AND SI
C   TO THE DATA ARRAYS XR AND XI VIA THE MAPPING VECTOR
C   INDX2(.).  THE PERMUTATION OF THE DATA FOLLOWS THE
C   CHINESE REMAINDER THEOREM.
C
       J=1
       K=1
       INC1=ND1-NA
       INC2=ND1*(ND2-NB)
       INC3=ND1*ND2*(ND3-NC)
C
C   CHECK FOR INVERSE
C
       IF(INVRS.EQ.1) GO TO 400
       DO 340 N4=1,ND
       DO 330 N3=1,NC
       DO 320 N2=1,NB
       DO 310 N1=1,NA
       KNDX=INDX2(K)
       XR(KNDX)=SR(J)
       XI(KNDX)=SI(J)
       K=K+1
310    J=J+1
320    J=J+INC1
330    J=J+INC2
340    J=J+INC3
       RETURN
C
C   DIFFERENT PERMUTATION FOR THE INVERSE
C
400    FN=FLOAT(N)
       NP2=N+2
       INDX2(1)=N+1
       DO 440 N4=1,ND
       DO 430 N3=1,NC
       DO 420 N2=1,NB
       DO 410 N1=1,NA
       KNDX=NP2-INDX2(K)
       XR(KNDX)=SR(J)/FN
       XI(KNDX)=SI(J)/FN
       K=K+1
410    J=J+1
420    J=J+INC1
430    J=J+INC2
440    J=J+INC3
       RETURN
       END
C
C---------------------------------------------------------------------
C SUBROUTINE: WEAVE1
C   THIS SUBROUTINE IMPLEMENTS THE DIFFERENT PRE-WEAVE
C   MODULES OF THE WFTA.  THE WORKING ARRAYS ARE SR AND SI.
C   THE ROUTINE CHECKS TO SEE WHICH FACTORS ARE PRESENT
C   IN THE TRANSFORM LENGTH N = NA*NB*NC*ND AND EXECUTES
C   THE PRE-WEAVE CODE FOR THESE FACTORS.
C
C---------------------------------------------------------------------
C
       SUBROUTINE WEAVE1(SR,SI)
       COMMON NA,NB,NC,ND,ND1,ND2,ND3,ND4
       DIMENSION Q(8),T(16)
       DIMENSION SR(1),SI(1)
       IF(NA.EQ.1) GO TO 300
       IF(NA.NE.2) GO TO 800
C
C *******************************************************************
C
C      THE FOLLOWING CODE IMPLEMENTS THE 2 POINT PRE-WEAVE MODULE
C
C *******************************************************************
C
       NLUP2=2*(ND2-NB)
       NLUP23=2*ND2*(ND3-NC)
       NBASE=1
       DO 240 N4=1,ND
       DO 230 N3=1,NC
       DO 220 N2=1,NB
       NR1=NBASE+1
       T0=SR(NBASE)+SR(NR1)
       SR(NR1)=SR(NBASE)-SR(NR1)
       SR(NBASE)=T0
       T0=SI(NBASE)+SI(NR1)
       SI(NR1)=SI(NBASE)-SI(NR1)
       SI(NBASE)=T0
220    NBASE=NBASE+2
230    NBASE=NBASE+NLUP2
240    NBASE=NBASE+NLUP23
800    IF(NA.NE.8) GO TO 1600
C
C *******************************************************************
C
C      THE FOLLOWING CODE IMPLEMENTS THE 8 POINT PRE-WEAVE MODULE
C
C *******************************************************************
C
       NLUP2=8*(ND2-NB)
       NLUP23=8*ND2*(ND3-NC)
       NBASE=1
       DO 840 N4=1,ND
       DO 830 N3=1,NC
       DO 820 N2=1,NB
       NR1=NBASE+1
       NR2=NR1+1
       NR3=NR2+1
       NR4=NR3+1
       NR5=NR4+1
       NR6=NR5+1
       NR7=NR6+1
       T3=SR(NR3)+SR(NR7)
       T7=SR(NR3)-SR(NR7)
```

```
      T0=SR(NBASE)+SR(NR4)
      SR(NR4)=SR(NBASE)-SR(NR4)
      T1=SR(NR1)+SR(NR5)
      T5=SR(NR1)-SR(NR5)
      T2=SR(NR2)+SR(NR6)
      SR(NR6)=SR(NR2)-SR(NR6)
      SR(NBASE)=T0+T2
      SR(NR2)=T0-T2
      SR(NR1)=T1+T3
      SR(NR3)=T1-T3
      SR(NR5)=T5+T7
      SR(NR7)=T5-T7
      T3=SI(NR3)+SI(NR7)
      T7=SI(NR3)-SI(NR7)
      T0=SI(NBASE)+SI(NR4)
      SI(NR4)=SI(NBASE)-SI(NR4)
      T1=SI(NR1)+SI(NR5)
      T5=SI(NR1)-SI(NR5)
      T2=SI(NR2)+SI(NR6)
      SI(NR6)=SI(NR2)-SI(NR6)
      SI(NBASE)=T0+T2
      SI(NR2)=T0-T2
      SI(NR1)=T1+T3
      SI(NR3)=T1-T3
      SI(NR5)=T5+T7
      SI(NR7)=T5-T7
820   NBASE=NBASE+8
830   NBASE=NBASE+NLUP2
840   NBASE=NBASE+NLUP23
1600  IF(NA.NE.16) GO TO 300
C
C ********************************************************************
C
C     THE FOLLOWING CODE IMPLEMENTS THE 16 POINT PRE-WEAVE MODULE
C
C ********************************************************************
C
      NLUP2=18*(ND2-NB)
      NLUP23=18*ND2*(ND3-NC)
      NBASE=1
      DO 1640 N4=1,ND
      DO 1630 N3=1,NC
      DO 1620 N2=1,NB
      NR1=NBASE+1
      NR2=NR1+1
      NR3=NR2+1
      NR4=NR3+1
      NR5=NR4+1
      NR6=NR5+1
      NR7=NR6+1
      NR8=NR7+1
      NR9=NR8+1
      NR10=NR9+1
      NR11=NR10+1
      NR12=NR11+1
      NR13=NR12+1
      NR14=NR13+1
      NR15=NR14+1
      NR16=NR15+1
      NR17=NR16+1
      JBASE=NBASE
      DO 1645 J=1,8
      T(J)=SR(JBASE)+SR(JBASE+8)
      T(J+8)=SR(JBASE)-SR(JBASE+8)
      JBASE=JBASE+1
1645  CONTINUE
      DO 1650 J=1,4
      Q(J)=T(J)+T(J+4)
      Q(J+4)=T(J)-T(J+4)
1650  CONTINUE
      SR(NBASE)=Q(1)+Q(3)
      SR(NR2)=Q(1)-Q(3)
      SR(NR1)=Q(2)+Q(4)
      SR(NR3)=Q(2)-Q(4)
      SR(NR5)=Q(6)+Q(8)
      SR(NR7)=Q(6)-Q(8)
      SR(NR4)=Q(5)
      SR(NR6)=Q(7)
      SR(NR8)=T(9)
      SR(NR9)=T(10)+T(16)
      SR(NR15)=T(10)-T(16)
      SR(NR13)=T(14)+T(12)
      SR(NR11)=T(14)-T(12)
      SR(NR17)=SR(NR11)+SR(NR15)
      SR(NR16)=SR(NR9)+SR(NR13)
      SR(NR10)=T(11)+T(15)
      SR(NR14)=T(11)-T(15)
      SR(NR12)=T(13)
      JBASE=NBASE
      DO 1745 J=1,8
      T(J)=SI(JBASE)+SI(JBASE+8)
      T(J+8)=SI(JBASE)-SI(JBASE+8)
      JBASE=JBASE+1
1745  CONTINUE
      DO 1750 J=1,4
      Q(J)=T(J)+T(J+4)
      Q(J+4)=T(J)-T(J+4)
1750  CONTINUE
      SI(NBASE)=Q(1)+Q(3)
      SI(NR2)=Q(1)-Q(3)
      SI(NR1)=Q(2)+Q(4)
      SI(NR3)=Q(2)-Q(4)
      SI(NR5)=Q(6)+Q(8)
      SI(NR7)=Q(6)-Q(8)
      SI(NR4)=Q(5)
      SI(NR6)=Q(7)
      SI(NR8)=T(9)
      SI(NR9)=T(10)+T(16)
      SI(NR15)=T(10)-T(16)
      SI(NR13)=T(14)+T(12)
      SI(NR11)=T(14)-T(12)
      SI(NR17)=SI(NR11)+SI(NR15)
      SI(NR16)=SI(NR9)+SI(NR13)
      SI(NR10)=T(11)+T(15)
      SI(NR14)=T(11)-T(15)
      SI(NR12)=T(13)
1620  NBASE=NBASE+18
1630  NBASE=NBASE+NLUP2
1640  NBASE=NBASE+NLUP23
300   IF(NB.EQ.1) GO TO 700
      IF(NB.NE.3) GO TO 900
C
C ********************************************************************
C
C     THE FOLLOWING CODE IMPLEMENTS THE 3 POINT PRE-WEAVE MODULE
C
C ********************************************************************
C
      NLUP2=2*ND1
      NLUP23=3*ND1*(ND3-NC)
```

```
      NBASE=1
      NOFF=ND1
      DO 340 N4=1,ND
      DO 330 N3=1,NC
      DO 310 N2=1,ND1
      NR1=NBASE+NOFF
      NR2=NR1+NOFF
      T1=SR(NR1)+SR(NR2)
      SR(NBASE)=SR(NBASE)+T1
      SR(NR2)=SR(NR1)-SR(NR2)
      SR(NR1)=T1
      T1=SI(NR1)+SI(NR2)
      SI(NBASE)=SI(NBASE)+T1
      SI(NR2)=SI(NR1)-SI(NR2)
      SI(NR1)=T1
310   NBASE=NBASE+1
330   NBASE=NBASE+NLUP2
340   NBASE=NBASE+NLUP23
900   IF(NB.NE.9) GO TO 700
C
C **********************************************************************
C
C     THE FOLLOWING CODE IMPLEMENTS THE 9 POINT PRE-WEAVE MODULE
C
C **********************************************************************
C
      NLUP2=10*ND1
      NLUP23=11*ND1*(ND3-NC)
      NBASE=1
      NOFF=ND1
      DO 940 N4=1,ND
      DO 930 N3=1,NC
      DO 910 N2=1,ND1
      NR1=NBASE+NOFF
      NR2=NR1+NOFF
      NR3=NR2+NOFF
      NR4=NR3+NOFF
      NR5=NR4+NOFF
      NR6=NR5+NOFF
      NR7=NR6+NOFF
      NR8=NR7+NOFF
      NR9=NR8+NOFF
      NR10=NR9+NOFF
      T3=SR(NR3)+SR(NR6)
      T6=SR(NR3)-SR(NR6)
      SR(NBASE)=SR(NBASE)+T3
      T7=SR(NR7)+SR(NR2)
      T2=SR(NR7)-SR(NR2)
      SR(NR2)=T6
      T1=SR(NR1)+SR(NR8)
      T8=SR(NR1)-SR(NR8)
      SR(NR1)=T3
      T4=SR(NR4)+SR(NR5)
      T5=SR(NR4)-SR(NR5)
      SR(NR3)=T1+T4+T7
      SR(NR4)=T1-T7
      SR(NR5)=T4-T1
      SR(NR6)=T7-T4
      SR(NR10)=T2+T5+T8
      SR(NR7)=T8-T2
      SR(NR8)=T5-T8
      SR(NR9)=T2-T5
      T3=SI(NR3)+SI(NR6)
      T6=SI(NR3)-SI(NR6)
      SI(NBASE)=SI(NBASE)+T3
      T7=SI(NR7)+SI(NR2)
      T2=SI(NR7)-SI(NR2)
      SI(NR2)=T6
      T1=SI(NR1)+SI(NR8)
      T8=SI(NR1)-SI(NR8)
      SI(NR1)=T3
      T4=SI(NR4)+SI(NR5)
      T5=SI(NR4)-SI(NR5)
      SI(NR3)=T1+T4+T7
      SI(NR4)=T1-T7
      SI(NR5)=T4-T1
      SI(NR6)=T7-T4
      SI(NR10)=T2+T5+T8
      SI(NR7)=T8-T2
      SI(NR8)=T5-T8
      SI(NR9)=T2-T5
910   NBASE=NBASE+1
930   NBASE=NBASE+NLUP2
940   NBASE=NBASE+NLUP23
700   IF(NC.NE.7) GO TO 500
C
C **********************************************************************
C
C     THE FOLLOWING CODE IMPLEMENTS THE 7 POINT PRE-WEAVE MODULE
C
C **********************************************************************
C
      NOFF=ND1*ND2
      NBASE=1
      NLUP2=8*NOFF
      DO 740 N4=1,ND
      DO 710 N1=1,NOFF
      NR1=NBASE+NOFF
      NR2=NR1+NOFF
      NR3=NR2+NOFF
      NR4=NR3+NOFF
      NR5=NR4+NOFF
      NR6=NR5+NOFF
      NR7=NR6+NOFF
      NR8=NR7+NOFF
      T1=SR(NR1)+SR(NR6)
      T6=SR(NR1)-SR(NR6)
      T4=SR(NR4)+SR(NR3)
      T3=SR(NR4)-SR(NR3)
      T2=SR(NR2)+SR(NR5)
      T5=SR(NR2)-SR(NR5)
      SR(NR5)=T6-T3
      SR(NR2)=T5+T3+T6
      SR(NR6)=T5-T6
      SR(NR8)=T3-T5
      SR(NR3)=T2-T1
      SR(NR4)=T1-T4
      SR(NR7)=T4-T2
      T1=T1+T4+T2
      SR(NBASE)=SR(NBASE)+T1
      SR(NR1)=T1
      T1=SI(NR1)+SI(NR6)
      T6=SI(NR1)-SI(NR6)
      T4=SI(NR4)+SI(NR3)
      T3=SI(NR4)-SI(NR3)
      T2=SI(NR2)+SI(NR5)
      T5=SI(NR2)-SI(NR5)
      SI(NR5)=T6-T3
      SI(NR2)=T5+T3+T6
      SI(NR6)=T5-T6
```

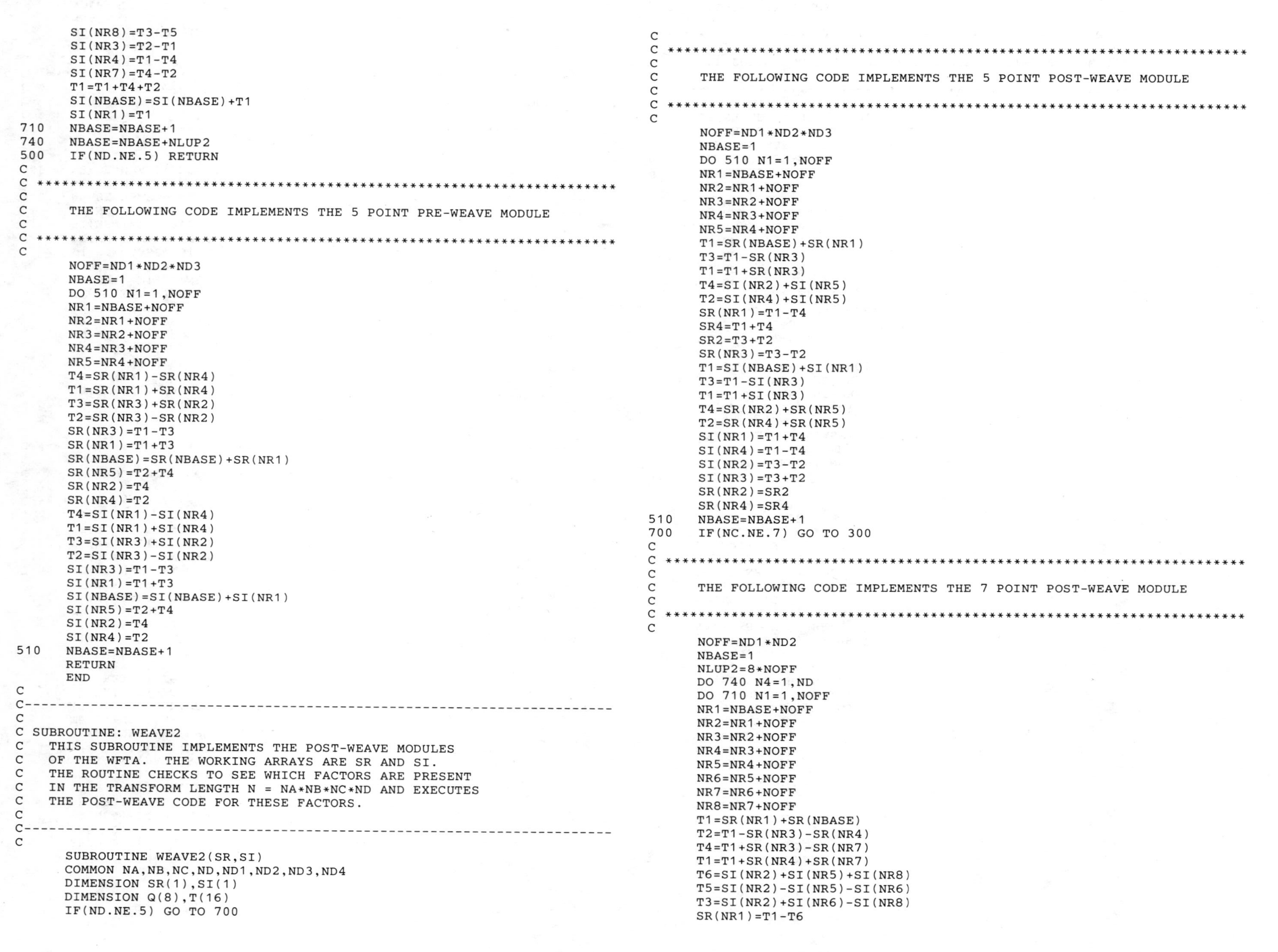

```
      SI(NR8)=T3-T5
      SI(NR3)=T2-T1
      SI(NR4)=T1-T4
      SI(NR7)=T4-T2
      T1=T1+T4+T2
      SI(NBASE)=SI(NBASE)+T1
      SI(NR1)=T1
710   NBASE=NBASE+1
740   NBASE=NBASE+NLUP2
500   IF(ND.NE.5) RETURN
C
C ***********************************************************************
C
C     THE FOLLOWING CODE IMPLEMENTS THE 5 POINT PRE-WEAVE MODULE
C
C ***********************************************************************
C
      NOFF=ND1*ND2*ND3
      NBASE=1
      DO 510 N1=1,NOFF
      NR1=NBASE+NOFF
      NR2=NR1+NOFF
      NR3=NR2+NOFF
      NR4=NR3+NOFF
      NR5=NR4+NOFF
      T4=SR(NR1)-SR(NR4)
      T1=SR(NR1)+SR(NR4)
      T3=SR(NR3)+SR(NR2)
      T2=SR(NR3)-SR(NR2)
      SR(NR3)=T1-T3
      SR(NR1)=T1+T3
      SR(NBASE)=SR(NBASE)+SR(NR1)
      SR(NR5)=T2+T4
      SR(NR2)=T4
      SR(NR4)=T2
      T4=SI(NR1)-SI(NR4)
      T1=SI(NR1)+SI(NR4)
      T3=SI(NR3)+SI(NR2)
      T2=SI(NR3)-SI(NR2)
      SI(NR3)=T1-T3
      SI(NR1)=T1+T3
      SI(NBASE)=SI(NBASE)+SI(NR1)
      SI(NR5)=T2+T4
      SI(NR2)=T4
      SI(NR4)=T2
510   NBASE=NBASE+1
      RETURN
      END
C
C-----------------------------------------------------------------------
C
C SUBROUTINE: WEAVE2
C   THIS SUBROUTINE IMPLEMENTS THE POST-WEAVE MODULES
C   OF THE WFTA.  THE WORKING ARRAYS ARE SR AND SI.
C   THE ROUTINE CHECKS TO SEE WHICH FACTORS ARE PRESENT
C   IN THE TRANSFORM LENGTH N = NA*NB*NC*ND AND EXECUTES
C   THE POST-WEAVE CODE FOR THESE FACTORS.
C
C-----------------------------------------------------------------------
C
      SUBROUTINE WEAVE2(SR,SI)
      COMMON NA,NB,NC,ND,ND1,ND2,ND3,ND4
      DIMENSION SR(1),SI(1)
      DIMENSION Q(8),T(16)
      IF(ND.NE.5) GO TO 700
C
C ***********************************************************************
C
C     THE FOLLOWING CODE IMPLEMENTS THE 5 POINT POST-WEAVE MODULE
C
C ***********************************************************************
C
      NOFF=ND1*ND2*ND3
      NBASE=1
      DO 510 N1=1,NOFF
      NR1=NBASE+NOFF
      NR2=NR1+NOFF
      NR3=NR2+NOFF
      NR4=NR3+NOFF
      NR5=NR4+NOFF
      T1=SR(NBASE)+SR(NR1)
      T3=T1-SR(NR3)
      T1=T1+SR(NR3)
      T4=SI(NR2)+SI(NR5)
      T2=SI(NR4)+SI(NR5)
      SR(NR1)=T1-T4
      SR4=T1+T4
      SR2=T3+T2
      SR(NR3)=T3-T2
      T1=SI(NBASE)+SI(NR1)
      T3=T1-SI(NR3)
      T1=T1+SI(NR3)
      T4=SR(NR2)+SR(NR5)
      T2=SR(NR4)+SR(NR5)
      SI(NR1)=T1+T4
      SI(NR4)=T1-T4
      SI(NR2)=T3-T2
      SI(NR3)=T3+T2
      SR(NR2)=SR2
      SR(NR4)=SR4
510   NBASE=NBASE+1
700   IF(NC.NE.7) GO TO 300
C
C ***********************************************************************
C
C     THE FOLLOWING CODE IMPLEMENTS THE 7 POINT POST-WEAVE MODULE
C
C ***********************************************************************
C
      NOFF=ND1*ND2
      NBASE=1
      NLUP2=8*NOFF
      DO 740 N4=1,ND
      DO 710 N1=1,NOFF
      NR1=NBASE+NOFF
      NR2=NR1+NOFF
      NR3=NR2+NOFF
      NR4=NR3+NOFF
      NR5=NR4+NOFF
      NR6=NR5+NOFF
      NR7=NR6+NOFF
      NR8=NR7+NOFF
      T1=SR(NR1)+SR(NBASE)
      T2=T1-SR(NR3)-SR(NR4)
      T4=T1+SR(NR3)-SR(NR7)
      T1=T1+SR(NR4)+SR(NR7)
      T6=SI(NR2)+SI(NR5)+SI(NR8)
      T5=SI(NR2)-SI(NR5)-SI(NR6)
      T3=SI(NR2)+SI(NR6)-SI(NR8)
      SR(NR1)=T1-T6
```

```
      SR6=T1+T6
      SR2=T2-T5
      SR5=T2+T5
      SR(NR4)=T4-T3
      SR(NR3)=T4+T3
      T1=SI(NR1)+SI(NBASE)
      T2=T1-SI(NR3)-SI(NR4)
      T4=T1+SI(NR3)-SI(NR7)
      T1=T1+SI(NR4)+SI(NR7)
      T6=SR(NR2)+SR(NR5)+SR(NR8)
      T5=SR(NR2)-SR(NR5)-SR(NR6)
      T3=SR(NR2)+SR(NR6)-SR(NR8)
      SI(NR1)=T1+T6
      SI(NR6)=T1-T6
      SI(NR2)=T2+T5
      SI(NR5)=T2-T5
      SI(NR4)=T4+T3
      SI(NR3)=T4-T3
      SR(NR2)=SR2
      SR(NR5)=SR5
      SR(NR6)=SR6
710   NBASE=NBASE+1
740   NBASE=NBASE+NLUP2
300   IF(NB.EQ.1) GO TO 400
      IF(NB.NE.3) GO TO 900
C
C *******************************************************************
C
C     THE FOLLOWING CODE IMPLEMENTS THE 3 POINT POST-WEAVE MODULE
C
C *******************************************************************
C
      NLUP2=2*ND1
      NLUP23=3*ND1*(ND3-NC)
      NBASE=1
      NOFF=ND1
      DO 340 N5=1,ND
      DO 330 N4=1,NC
      DO 310 N2=1,ND1
      NR1=NBASE+NOFF
      NR2=NR1+NOFF
      T1=SR(NBASE)+SR(NR1)
      SR(NR1)=T1-SI(NR2)
      SR2=T1+SI(NR2)
      T1=SI(NBASE)+SI(NR1)
      SI(NR1)=T1+SR(NR2)
      SI(NR2)=T1-SR(NR2)
      SR(NR2)=SR2
310   NBASE=NBASE+1
330   NBASE=NBASE+NLUP2
340   NBASE=NBASE+NLUP23
900   IF(NB.NE.9) GO TO 400
C
C *******************************************************************
C
C     THE FOLLOWING CODE IMPLEMENTS THE 9 POINT POST-WEAVE MODULE
C
C *******************************************************************
C
      NLUP2=10*ND1
      NLUP23=11*ND1*(ND3-NC)
      NBASE=1
      NOFF=ND1
      DO 940 N4=1,ND
      DO 930 N3=1,NC
      DO 910 N2=1,ND1
      NR1=NBASE+NOFF
      NR2=NR1+NOFF
      NR3=NR2+NOFF
      NR4=NR3+NOFF
      NR5=NR4+NOFF
      NR6=NR5+NOFF
      NR7=NR6+NOFF
      NR8=NR7+NOFF
      NR9=NR8+NOFF
      NR10=NR9+NOFF
      T3=SR(NBASE)-SR(NR3)
      T7=SR(NBASE)+SR(NR1)
      SR(NBASE)=SR(NBASE)+SR(NR3)+SR(NR3)
      T6=T3+SI(NR10)
      SR(NR3)=T3-SI(NR10)
      T4=T7+SR(NR5)-SR(NR6)
      T1=T7-SR(NR4)-SR(NR5)
      T7=T7+SR(NR4)+SR(NR6)
      SR(NR6)=T6
      T8=SI(NR2)-SI(NR7)-SI(NR8)
      T5=SI(NR2)+SI(NR8)-SI(NR9)
      T2=SI(NR2)+SI(NR7)+SI(NR9)
      SR(NR1)=T7-T2
      SR8=T7+T2
      SR(NR4)=T1-T8
      SR(NR5)=T1+T8
      SR7=T4-T5
      SR2=T4+T5
      T3=SI(NBASE)-SI(NR3)
      T7=SI(NBASE)+SI(NR1)
      SI(NBASE)=SI(NBASE)+SI(NR3)+SI(NR3)
      T6=T3-SR(NR10)
      SI(NR3)=T3+SR(NR10)
      T4=T7+SI(NR5)-SI(NR6)
      T1=T7-SI(NR4)-SI(NR5)
      T7=T7+SI(NR4)+SI(NR6)
      SI(NR6)=T6
      T8=SR(NR2)-SR(NR7)-SR(NR8)
      T5=SR(NR2)+SR(NR8)-SR(NR9)
      T2=SR(NR2)+SR(NR7)+SR(NR9)
      SI(NR1)=T7+T2
      SI(NR8)=T7-T2
      SI(NR4)=T1+T8
      SI(NR5)=T1-T8
      SI(NR7)=T4+T5
      SI(NR2)=T4-T5
      SR(NR2)=SR2
      SR(NR7)=SR7
      SR(NR8)=SR8
910   NBASE=NBASE+1
930   NBASE=NBASE+NLUP2
940   NBASE=NBASE+NLUP23
400   IF(NA.EQ.1) RETURN
      IF(NA.NE.4) GO TO 800
C
C *******************************************************************
C
C     THE FOLLOWING CODE IMPLEMENTS THE 4 POINT POST-WEAVE MODULE
C
C *******************************************************************
C
      NLUP2=4*(ND2-NB)
      NLUP23=4*ND2*(ND3-NC)
      NBASE=1
```

```
      DO 440 N4=1,ND
      DO 430 N3=1,NC
      DO 420 N2=1,NB
      NR1=NBASE+1
      NR2=NR1+1
      NR3=NR2+1
      TR0=SR(NBASE)+SR(NR2)
      TR2=SR(NBASE)-SR(NR2)
      TR1=SR(NR1)+SR(NR3)
      TR3=SR(NR1)-SR(NR3)
      TI1=SI(NR1)+SI(NR3)
      TI3=SI(NR1)-SI(NR3)
      SR(NBASE)=TR0+TR1
      SR(NR2)=TR0-TR1
      SR(NR1)=TR2+TI3
      SR(NR3)=TR2-TI3
      TI0=SI(NBASE)+SI(NR2)
      TI2=SI(NBASE)-SI(NR2)
      SI(NBASE)=TI0+TI1
      SI(NR2)=TI0-TI1
      SI(NR1)=TI2-TR3
      SI(NR3)=TI2+TR3
420   NBASE=NBASE+4
430   NBASE=NBASE+NLUP2
440   NBASE=NBASE+NLUP23
800   IF(NA.NE.8) GO TO 1600
C
C *********************************************************************
C
C     THE FOLLOWING CODE IMPLEMENTS THE 8 POINT POST-WEAVE MODULE
C
C *********************************************************************
C
      NLUP2=8*(ND2-NB)
      NLUP23=8*ND2*(ND3-NC)
      NBASE=1
      DO 840 N4=1,ND
      DO 830 N3=1,NC
      DO 820 N2=1,NB
      NR1=NBASE+1
      NR2=NR1+1
      NR3=NR2+1
      NR4=NR3+1
      NR5=NR4+1
      NR6=NR5+1
      NR7=NR6+1
      T1=SR(NBASE)-SR(NR1)
      SR(NBASE)=SR(NBASE)+SR(NR1)
      SR6=SR(NR2)+SI(NR3)
      SR(NR2)=SR(NR2)-SI(NR3)
      T4=SR(NR4)-SI(NR5)
      T5=SR(NR4)+SI(NR5)
      T6=SR(NR7)-SI(NR6)
      T7=SR(NR7)+SI(NR6)
      SR(NR4)=T1
      SR(NR1)=T4+T6
      SR3=T4-T6
      SR5=T5-T7
      SR(NR7)=T5+T7
      T1=SI(NBASE)-SI(NR1)
      SI(NBASE)=SI(NBASE)+SI(NR1)
      T3=SI(NR2)-SR(NR3)
      SI(NR2)=SI(NR2)+SR(NR3)
      T4=SI(NR4)+SR(NR5)
      T5=SI(NR4)-SR(NR5)
      SI(NR6)=T3
      T6=SR(NR6)+SI(NR7)
      T7=SR(NR6)-SI(NR7)
      SI(NR4)=T1
      SI(NR1)=T4+T6
      SI(NR3)=T4-T6
      SI(NR5)=T5+T7
      SI(NR7)=T5-T7
      SR(NR3)=SR3
      SR(NR5)=SR5
      SR(NR6)=SR6
820   NBASE=NBASE+8
830   NBASE=NBASE+NLUP2
840   NBASE=NBASE+NLUP23
1600  IF(NA.NE.16) RETURN
C
C *********************************************************************
C
C     THE FOLLOWING CODE IMPLEMENTS THE 16 POINT POST-WEAVE MODULE
C
C *********************************************************************
C
      NLUP2=18*(ND2-NB)
      NLUP23=18*ND2*(ND3-NC)
      NBASE=1
      DO 1640 N4=1,ND
      DO 1630 N3=1,NC
      DO 1620 N2=1,NB
      NR1=NBASE+1
      NR2=NR1+1
      NR3=NR2+1
      NR4=NR3+1
      NR5=NR4+1
      NR6=NR5+1
      NR7=NR6+1
      NR8=NR7+1
      NR9=NR8+1
      NR10=NR9+1
      NR11=NR10+1
      NR12=NR11+1
      NR13=NR12+1
      NR14=NR13+1
      NR15=NR14+1
      NR16=NR15+1
      NR17=NR16+1
      T(2)=SR(NBASE)-SR(NR1)
      SR(NBASE)=SR(NR1)+SR(NBASE)
      T(4)=SR(NR2)+SI(NR3)
      T(3)=SR(NR2)-SI(NR3)
      T(6)=SR(NR4)+SI(NR5)
      T(5)=SR(NR4)-SI(NR5)
      T(8)=-SI(NR6)-SR(NR7)
      T(7)=-SI(NR6)+SR(NR7)
      T(9)=SR(NR8)+SR(NR14)
      T(15)=SR(NR8)-SR(NR14)
      T(13)=-SI(NR10)-SI(NR12)
      T(11)=SI(NR10)-SI(NR12)
      T(16)=SR(NR15)-SR(NR17)
      T(12)=SR(NR11)-SR(NR17)
      T(10)=-SI(NR9)-SI(NR16)
      T(14)=-SI(NR16)+SI(NR13)
      SR(NR2)=T(5)+T(7)
      SR6=T(5)-T(7)
      SR10=T(6)+T(8)
      SR(NR14)=T(6)-T(8)
```

```
      Q(7)=T(9)+T(10)
      Q(8)=T(9)-T(10)
      Q(1)=T(11)+T(12)
      Q(2)=T(11)-T(12)
      Q(4)=T(14)+T(15)
      Q(5)=T(15)-T(14)
      Q(3)=T(13)+T(16)
      Q(6)=T(13)-T(16)
      SR(NR1)=Q(3)+Q(7)
      SR(NR7)=Q(7)-Q(3)
      SR9=Q(8)+Q(6)
      SR(NR15)=Q(8)-Q(6)
      SR5=Q(1)+Q(4)
      SR3=Q(4)-Q(1)
      SR13=Q(2)+Q(5)
      SR11=Q(5)-Q(2)
      SR(NR8)=T(2)
      SR(NR4)=T(3)
      SR12=T(4)
      T(2)=SI(NBASE)-SI(NR1)
      SI(NBASE)=SI(NR1)+SI(NBASE)
      T(4)=SI(NR2)-SR(NR3)
      T(3)=SI(NR2)+SR(NR3)
      T(6)=SI(NR4)-SR(NR5)
      T(5)=SI(NR4)+SR(NR5)
      T(8)=SR(NR6)-SI(NR7)
      T(7)=SR(NR6)+SI(NR7)
      T(9)=SI(NR8)+SI(NR14)
      T(15)=SI(NR8)-SI(NR14)
      T(13)=SR(NR10)+SR(NR12)
      T(11)=SR(NR12)-SR(NR10)
      T(16)=SI(NR15)-SI(NR17)
      T(12)=SI(NR11)-SI(NR17)
      T(10)=SR(NR9)+SR(NR16)        {T(14) = SR(NR16) - SR(NR13)
      SI(NR2)=T(5)+T(7)
      SI(NR6)=T(5)-T(7)
      SI(NR10)=T(6)+T(8)
      SI(NR14)=T(6)-T(8)
      Q(7)=T(9)+T(10)
      Q(8)=T(9)-T(10)
      Q(1)=T(11)+T(12)
      Q(2)=T(11)-T(12)
      Q(4)=T(14)+T(15)
      Q(5)=T(15)-T(14)
      Q(3)=T(13)+T(16)
      Q(6)=T(13)-T(16)
      SI(NR1)=Q(3)+Q(7)
      SI(NR7)=Q(7)-Q(3)
      SI(NR9)=Q(8)+Q(6)
      SI(NR15)=Q(8)-Q(6)
      SI(NR5)=Q(1)+Q(4)
      SI(NR3)=Q(4)-Q(1)
      SI(NR13)=Q(2)+Q(5)
      SI(NR11)=Q(5)-Q(2)
      SI(NR8)=T(2)
      SI(NR4)=T(3)
      SI(NR12)=T(4)
      SR(NR3)=SR3
      SR(NR5)=SR5
      SR(NR6)=SR6
      SR(NR9)=SR9
      SR(NR10)=SR10
      SR(NR11)=SR11
      SR(NR12)=SR12
      SR(NR13)=SR13
1620  NBASE=NBASE+18
1630  NBASE=NBASE+NLUP2
1640  NBASE=NBASE+NLUP23
      RETURN
      END
```

1.8

Time-Efficient Radix-4 Fast Fourier Transform

L. Robert Morris

Department of Systems Engineering and Computing Science
Carleton University
Ottawa, Canada

1. Purpose

This program uses the *autogen* technique [1-3] to implement a time efficient radix-4 fast Fourier transform. It is designed for those applications in which a fixed-size transform is performed repeatedly, such as speech analysis/synthesis (e.g., vocoders, spectrogram production) and two-dimensional picture processing. In such applications, speed optimization is especially profitable. Sample benchmark times for a 1K complex transform are:

Computer	*Compiler*	*Time*
DEC PDP-11/55	FORTRAN IV PLUS	250 msec
DEC VAX 11/780	FORTRAN IV PLUS	150 msec
IBM 370/168	FORTRAN H	32 msec

Finally, the program was derived from the radix-4 subset of Singleton's mixed-radix FFT [4] and thus illustrates the application of autogen techniques to speed optimization of other DSP software, such as that presented in this book.

2. Method

An autogen program is time efficient because most data-independent computation required for algorithm implementation is done prior to program execution. Such computation includes that necessary for data or coefficient access and algorithm flow control, including selection of the time optimum computational kernel (*ck*) applicable at any point in the algorithm. Conceptually, an autogen program for a selected size transform can be produced by unwinding the loops of any program whose flow is data-independent. The result is a customized program consisting essentially of a linear sequence of *ck*'s. Each *ck* may then be the optimum for that point in the algorithm and may contain precomputed constants which are pointers to data and/or coefficients necessary for execution of that *ck*. This code generation can in fact be automated by simply transforming an existing program into a generator program by replacing all sequences which operate on data points with "write" statements. When executed, the new program will then generate code, into a disc-based file for example. The file contents are then compiled/assembled to produce a customized, time efficient program [1]. However, excepting those cases when the *ck* size is "small" [2], in-line code is prohibitive insofar as memory storage requirements are concerned, and is a retrograde step for software speed enhancement on machines with cache memories. An alternative program structure for autogen software is "threaded code" [2,5]. In threaded code, the memory storage is divided into

(a) a set of computational kernels

(b) a data-point storage space

(c) a coefficient storage space (when necessary), and

(d) a driver array, or "thread".

The thread elements are accessed sequentially and consist essentially of precomputed pointers to *ck*'s, data, and coefficients. This means that no matter how complex the algorithm-dependent computation required at program generation time to determine the time optimum *ck* to invoke at any point during algorithm execution, the relevant *ck* can be invoked at run time by a single "computed go to" based on the value of a precomputed array element. Alternatively, using the premise that each *ck* must accept at least one nonzero data pointer per invokation, a time efficient "branch on nonzero (integer) pointer" can re-invoke the current *ck*. Because DSP *ck*'s are generally executed a number of times before transition to the next, this strategy -- "knotted code" [2] -- is to be preferred and is in fact used in this FFT autogen program.

In the radix-4 FFT, each butterfly includes three complex multiplications. Examination of the algorithm reveals that the multiplications required can be minimized by implementation of 5 different butterfly *ck* modules [6]. Further, by performing each nontrivial complex multiplication using a 3 multiply/3 add algorithm, possible when a table of $(\sin\theta_i - \cos\theta_i)$, $-(\sin\theta_i + \cos\theta_i)$ and $\cos\theta_i$ is stored, the 5 *ck*'s required have the following characteristics for an $N = 4^{**}M$ point complex transform:

ck	*# Multiplications*	*# Invokations*
1	0	$[4^{**}M-1]/3$
2	4	$[4^{**}(M-1)-1]/3$
3,4	8	$2[4^{**}(M-1)-1]/3$
5	9	$4^{**}(M-1)[M-7/3]+4/3$

The total number of butterfly invokations is then $M[4^{**}(M-1)] = MN/4$. These require 7,856 real multiplications and 28,341 real additions for $M = 5$, compared to 28,920 real additions and 15,112 real multiplications for Singleton's program. Using the DEC FORTRAN IV PLUS compiler, 59,725 floating-point register load/stores were executed (for $M = 5$) compared to 77,763 for Singleton's program.

In the documented program, thread and trig table generation are done during a first pass. This simplifies the program in terms of transportability (no disc file manipulation required) but means that the generator program is resident during FFT execution. Program storage requirements, in 16-bit words, are approximately $0.6NM + 5.2N + ck$ storage space. For the IBM 370 FORTRAN H and DEC FORTRAN IV PLUS compilers, *ck* storage space is about 1100 and 1500 16-bit words, respectively. An indication of the program time efficiency is that during execution of the program on a PDP-11/55, only about 10% of time is devoted to operations other than load/store and add/subtract/multiply of floating-point operands [6]. The relative speed increase as compared to conventional programs is very machine dependent. Time savings due to reduction of algorithm-dependent (and thus data-independent) operations will be greatest on modern computers which have hardware floating-point processors such that floating-point instructions require the same order of execution time as the nonfloating-point instructions eliminated. On machines with slow "software" floating-point arithmetic instructions, the nonfloating-point operations are, relatively, of less importance. However, on these machines, the reduction in "expensive" data-dependent operations via the ability of the autogen technique to *economically* invoke the time optimum *ck* (i.e., the one containing the fewest possible data-dependent operations), provides valuable time savings.

As noted earlier, the program was derived from the radix-4 subset of the well known Singleton mixed-radix program and implements

$$X(k) = \sum_{n=0}^{N-1} x(n) e^{-j\frac{2\pi}{N}nk}, \quad k=0,\ldots,N-1 \text{ (forward transform)}$$

or

$$x(n) = (1/N) \sum_{k=0}^{N-1} X(k) e^{j\frac{2\pi}{N}nk}, \quad n=0,\ldots,N-1 \text{ (inverse transform)}$$

The following table summarizes relative performance on the PDP-11/55. For completeness, execution times for an assembly language floating-point autogen FFT have been included [7]. As noted in [6], use of floating-point registers as temporary storage locations can be better optimized at the assembler level, thus reducing the load/store operations of floating-point data.

	Autogen	Autogen	Singleton		
N	*(a) Assembler*	*(b) FIV PLUS*	*(c) FIV PLUS*	*(c)/(a)*	*(c)/(b)*
16	0.90 msec	1.58 msec	3.23 msec	3.59	2.04
64	5.80 msec	9.07 msec	14.56 msec	2.51	1.60
256	33.00 msec	49.57 msec	71.51 msec	2.16	1.44
1024	168.70 msec	250.01 msec	343.88 msec	2.03	1.37

As per sec. 2, the proportion of time-favorable butterflies *decreases* with increasing N. Thus, use of this program for transforms of size 4 K or more is inappropriate from a time/space tradeoff point of view.

3. Program Description

3.1 Usage

The program consists of the basic subroutine RADIX4, which includes the 5 butterfly *ck*'s and the autogen code, and a small subroutine RAD4SB, used by the autogen. Much of the notation of the original Singleton program has been preserved. The user passes the data in an array "A", in labeled common block "AA". *This step is essential for ensuring "optimum" data addressing techniques used by optimizing compiler generated code.*

3.2 Description of Parameters

The basic call is of the form

```
CALL RADIX4(M,IFLAG,JFLAG)
```

where $N = 4^{**}M$ complex points are to be transformed and $2 \leqslant M \leqslant 5$

IFLAG = 1 on the first call
= 0 on subsequent calls with *the same value* of M.

JFLAG = −1 for a forward transform
= +1 for an inverse transform.

On input, the array A (passed in labeled common) contains the complex data points, real in odd array components, imaginary in even array components. On output, the array A contains the complex DFT coefficients $X(k)$ or $x(n)$ as per JFLAG.

3.3 Dimension Requirements

The DIMENSION statements in the main program and *both* subroutines can be modified to reflect the largest size transform to be implemented as follows:

	A	*IX*	*T*
	Array Dimensions		
$M = 2$	32	38	27
$M \leqslant 3$	128	144	135
$M \leqslant 4$	512	658	567
$M \leqslant 5$	2048	2996	2295

Arrays A and T are real (floating-point) and array IX is INTEGER*2, i.e., 16-bit.

3.4 Summary of User Requirements

On each call to RADIX4, load A with the input data and retrieve the output data from A. On the first call for a given size *M*, use IFLAG=1, on subsequent calls use IFLAG=0. For a DFT use JFLAG = −1, for an IDFT use JFLAG = +1.

4. Test Program

For $3 \leqslant M \leqslant 5$, $N = 4^{**}M$, compute the DFT of a complex sequence generated by UNI, then compute the IDFT of the result. The RMS error between DFT input and IDFT output is printed in each case. For $N = 64$, the DFT/IDFT input and output are listed. Note that the error will vary according to the format (e.g., bits for fraction) of floating-point numbers on any particular machine.

References

[1] L. R. Morris, "Automatic Generation of Time Efficient Digital Signal Processing Software", *IEEE Trans. Acoust., Speech, Signal Processing,* Vol. ASSP-25, No. 1, pp. 74-79, Feb. 1977.

[2] L. R. Morris, "Time/space efficiency of program structures for automatically generated digital signal processing software", *Conf. Rec.,* 1977 Int. Conf., ASSP, pp. 164-167, Hartford.

[3] L. R. Morris, J. C. Mudge, "Speed Enhancement of Digital Signal Processing Software via Microprogramming a General Purpose Minicomputer", *IEEE Trans. Acoust., Speech, Signal Processing,* Vol. ASSP-26, No. 2, pp. 135-140, April 1978.

[4] R. C. Singleton, "An Algorithm for Computing the Mixed Radix Fast Fourier Transform", *IEEE Trans. Audio Electroacoust.,* Vol. AU-17, No. 2, pp. 93-103, June 1969.

[5] J. Bell, "Threaded code", *Comm. ACM,* Vol. 16, No. 6, pp. 370-372, June 1973.

[6] L. R. Morris, "A Comparative Study of Time Efficient FFT and WFTA Programs for General Purpose Computers", *IEEE Trans. Acoust., Speech, Signal Processing,* Vol. ASSP-26, No. 2, pp. 141-150, April 1978.

[7] L. R. Morris, "Fast digital signal processing software package for the PDP-11", *DECUS Program Library,* No. 11-296, November 1976.

Table 1

DFT INPUT REAL	DFT INPUT IMAG	DFT OUTPUT REAL	DFT OUTPUT IMAG
0.229256	0.766875	31.377766	32.033501
0.683177	0.509191	-1.565459	-0.613354
0.874560	0.644641	-0.562676	-1.274236
0.847468	0.353963	-1.280178	0.779767
0.398892	0.457094	-3.942028	4.433803
0.236309	0.133182	-2.127302	0.555010
0.166052	0.226027	-0.364557	-6.019432
0.662459	0.250212	-0.179658	-0.879471
0.617697	0.262465	3.244458	-2.330645
0.512668	0.939207	1.399128	-1.712091
0.624028	0.422382	0.541036	-0.833928
0.939706	0.282068	-0.624628	-3.266858
0.469218	0.054879	-1.187343	1.579592
0.519831	0.396827	-1.564005	-2.370283
0.113157	0.607517	-0.138395	3.839718
0.701507	0.887055	-2.695570	4.165845
0.331433	0.220961	-1.135619	-5.030361
0.168616	0.992249	2.194667	1.753474
0.367070	0.497800	-0.264126	0.562021
0.254104	0.797795	-1.419245	-3.509271
0.133912	0.116143	2.231800	0.938169
0.212882	0.783360	1.686178	3.474952
0.982931	0.788486	-3.398292	0.256503
0.365426	0.476909	0.976109	2.695807
0.827320	0.102528	-0.999293	0.800101
0.943264	0.684253	-3.917623	-2.189232
0.729065	0.722039	1.067974	-0.777765
0.340034	0.410316	-4.502048	-0.440295
0.384621	0.961848	0.645658	-0.914912
0.480094	0.096515	-1.795084	0.368239
0.519834	0.761804	0.052220	-1.359995
0.704444	0.340203	2.705629	-0.261653
0.460761	0.475524	-1.924339	-2.254483
0.748171	0.654245	-1.059475	-1.224388
0.421876	0.441382	0.055644	4.737301
0.774413	0.625799	6.071876	0.011771
0.524052	0.092862	-4.694662	0.357120
0.984750	0.829579	2.227794	-0.556963
0.192729	0.291600	1.329456	1.476440
0.385426	0.467851	1.600311	-1.237184
0.320288	0.242745	-0.051197	-1.647760
0.801388	0.025701	-0.497059	0.466700
0.985884	0.904574	-0.116100	-1.350873
0.105523	0.237869	2.017721	3.162251
0.253360	0.134264	1.047104	1.395429
0.979304	0.629147	-3.381618	0.410234
0.445763	0.913431	0.277854	1.371036
0.270815	0.408028	2.875121	-1.404488
0.234614	0.644116	-2.133524	-3.313334
0.723286	0.768105	-0.886670	-1.253150
0.194135	0.632751	1.965688	2.067677
0.158584	0.594914	0.325368	3.580138
0.158079	0.353140	-0.403606	1.931254
0.055275	0.430064	-0.313278	1.078680
0.546173	0.499216	-4.102506	-1.050208
0.915547	0.623864	-1.932684	2.410821
0.633616	0.045868	0.861630	3.831643
0.580915	0.172191	0.113929	3.341182
0.969972	0.916135	1.158827	4.071988
0.400834	0.777754	0.789765	0.218662
0.568952	0.427517	-2.839772	1.910510
0.124854	0.777654	-0.988682	1.583505
0.047415	0.260892	4.481467	0.320888
0.069980	0.787919	-1.661487	0.184884

Table 1
(Continued)

DFT INPUT		IDFT OUTPUT	
REAL	IMAG	REAL	IMAG
0.229256	0.766875	0.229256	0.766875
0.683177	0.509191	0.683177	0.509191
0.874560	0.644641	0.874560	0.644641
0.847468	0.353963	0.847468	0.353963
0.398892	0.457094	0.398892	0.457094
0.236309	0.133182	0.236309	0.133182
0.166052	0.226027	0.166052	0.226027
0.662459	0.250212	0.662459	0.250212
0.617697	0.262465	0.617697	0.262465
0.512668	0.939207	0.512668	0.939207
0.624028	0.422382	0.624028	0.422382
0.939706	0.282068	0.939706	0.282068
0.469218	0.054879	0.469218	0.054879
0.519831	0.396827	0.519831	0.396827
0.113157	0.607517	0.113157	0.607517
0.701507	0.887055	0.701507	0.887055
0.331433	0.220961	0.331433	0.220961
0.168616	0.992249	0.168616	0.992249
0.367070	0.497800	0.367070	0.497800
0.254104	0.797795	0.254104	0.797795
0.133912	0.116143	0.133912	0.116143
0.212882	0.783360	0.212882	0.783360
0.982931	0.788486	0.982931	0.788486
0.365426	0.476909	0.365426	0.476909
0.827320	0.102528	0.827320	0.102528
0.943264	0.684253	0.943264	0.684253
0.729065	0.722039	0.729065	0.722039
0.340034	0.410316	0.340034	0.410316
0.384621	0.961848	0.384622	0.961848
0.480094	0.096515	0.480094	0.096515
0.519834	0.761804	0.519834	0.761804
0.704444	0.340203	0.704444	0.340203
0.460761	0.475524	0.460761	0.475524
0.748171	0.654245	0.748171	0.654246
0.421876	0.441382	0.421876	0.441382
0.774413	0.625799	0.774412	0.625799
0.524052	0.092862	0.524052	0.092862
0.984750	0.829579	0.984750	0.829579
0.192729	0.291600	0.192729	0.291600
0.385426	0.467851	0.385426	0.467851
0.320288	0.242745	0.320288	0.242745
0.801388	0.025701	0.801388	0.025701
0.985884	0.904574	0.985884	0.904574
0.105523	0.237869	0.105523	0.237869
0.253360	0.134264	0.253360	0.134263
0.979304	0.629147	0.979304	0.629147
0.445763	0.913431	0.445763	0.913431
0.270815	0.408028	0.270815	0.408028
0.234614	0.644116	0.234614	0.644116
0.723286	0.768105	0.723286	0.768105
0.194135	0.632751	0.194135	0.632751
0.158584	0.594914	0.158584	0.594914
0.158079	0.353140	0.158079	0.353140
0.055275	0.430064	0.055275	0.430064
0.546173	0.499216	0.546173	0.499216
0.915547	0.623864	0.915547	0.623864
0.633616	0.045868	0.633616	0.045868
0.580915	0.172191	0.580915	0.172191
0.969972	0.916135	0.969972	0.916135
0.400834	0.777754	0.400834	0.777754
0.568952	0.427517	0.568952	0.427517
0.124854	0.777654	0.124854	0.777654
0.047415	0.260892	0.047415	0.260892
0.069980	0.787919	0.069980	0.787919

```
RMS ERROR FOR M = 3 IS    0.114103E-06
RMS ERROR FOR M = 4 IS    0.221640E-06
RMS ERROR FOR M = 5 IS    0.566590E-06
```

Appendix

```
C
C-----------------------------------------------------------------------
C MAIN PROGRAM: TIME-EFFICIENT RADIX-4 FAST FOURIER TRANSFORM
C AUTHOR:       L. ROBERT MORRIS
C               DEPARTMENT OF SYSTEMS ENGINEERING AND COMPUTING SCIENCE
C               CARLETON UNIVERSITY, OTTAWA, CANADA K1S 5B6
C
C INPUT:        THE ARRAY "A" CONTAINS THE DATA TO BE TRANSFORMED
C-----------------------------------------------------------------------
C
C       TEST PROGRAM FOR AUTOGEN RADIX-4 FFT
C
        DIMENSION A(2048),B(2048)
        COMMON /AA/A
C
        IOUTD=I1MACH(2)
C
C       COMPUTE DFT AND IDFT FOR N = 64, 256, AND 1024 COMPLEX POINTS
C
        DO 1 MM=3,5
        N=4**MM
        DO 2 J=1,N
        A(2*J-1)=UNI(0)
        A(2*J  )=UNI(0)
        B(2*J-1)=A(2*J-1)
2       B(2*J  )=A(2*J)
C
C       FORWARD DFT
C
        CALL RADIX4(MM,1,-1)
C
        IF(MM.NE.3) GO TO 5
C
C       LIST DFT INPUT, OUTPUT FOR N = 64 ONLY
C
        WRITE(IOUTD,98)
        WRITE(IOUTD,100)
        DO 3 J=1,N
        WRITE(IOUTD,96) B(2*J-1),B(2*J),A(2*J-1),A(2*J)
3       CONTINUE
C
C       INVERSE DFT
C
5       CALL RADIX4(MM,0, 1)
C
C       LIST DFT INPUT, IDFT OUTPUT FOR N = 64 ONLY
C
        IF(MM.NE.3) GO TO 7
C
        WRITE(IOUTD,99)
        WRITE(IOUTD,100)
        DO 6 J=1,N
        WRITE(IOUTD,96) B(2*J-1),B(2*J),A(2*J-1),A(2*J)
6       CONTINUE
C
C       CALCULATE RMS ERROR
C
7       ERR=0.0
        DO 8 J=1,N
8       ERR=ERR+(A(2*J-1)-B(2*J-1))**2+(A(2*J)-B(2*J))**2
        ERR=SQRT(ERR/FLOAT(N))
        WRITE(IOUTD,97) MM,ERR
1       CONTINUE
C
96      FORMAT(1X,4(F10.6,2X))
97      FORMAT(1X,20H   RMS ERROR FOR M =,I2,4H IS ,E14.6/)
98      FORMAT(1X,43H       DFT INPUT                 DFT  OUTPUT/)
99      FORMAT(1X,43H       DFT INPUT                 IDFT OUTPUT/)
100     FORMAT(1X,44H   REAL        IMAG        REAL          IMAG/)
        STOP
        END
C
C-----------------------------------------------------------------------
C SUBROUTINE:  RADIX4
C COMPUTES FORWARD OR INVERSE COMPLEX DFT VIA RADIX-4 FFT.
C USES AUTOGEN TECHNIQUE TO YIELD TIME EFFICIENT PROGRAM.
C-----------------------------------------------------------------------
C
        SUBROUTINE RADIX4(MM,IFLAG,JFLAG)
C
C       INPUT:
C             MM = POWER OF 4 (I.E., N = 4**MM COMPLEX POINT TRANSFORM)
C             (MM.GE.2 AND MM.LE.5)
C
C          IFLAG = 1 ON FIRST PASS FOR GIVEN N
C                = 0 ON SUBSEQUENT PASSES FOR GIVEN N
C
C          JFLAG = -1 FOR FORWARD TRANSFORM
C                = +1 FOR INVERSE TRANSFORM
C
C       INPUT/OUTPUT:
C              A = ARRAY OF DIMENSIONS 2*N WITH REAL AND IMAGINARY PARTS
C                  OF DFT INPUT/OUTPUT IN ODD, EVEN ARRAY COMPONENTS.
C
C       FOR OPTIMAL TIME EFFICIENCY, COMMON IS USED TO PASS ARRAYS.
C       THIS MEANS THAT DIMENSIONS OF ARRAYS A, IX, AND T CAN BE
C       MODIFIED TO REFLECT MAXIMUM VALUE OF N = 4**MM TO BE USED. NOTE
C       THAT ARRAY "IX" IS ALSO DIMENSIONED IN SUBROUTINE "RAD4SB".
C
C    I.E.,     A(     ) IX(   ) T(    )
C
C       M =2       32      38      27
C       M<=3      128     144     135
C       M<=4      512     658     567
C       M<=5     2048    2996    2295
C
      DIMENSION A(2048),IX(2996),T(2295)
      DIMENSION NFAC(11),NP(209)
      COMMON NTYPL,KKP,INDEX,IXC
      COMMON /AA/A
      COMMON /XX/IX
C
C     CHECK FOR MM<2 OR MM>5
C
      IF(MM.LT.2.OR.MM.GT.5)STOP
C
C     INITIALIZE ON FIRST PASS """"""""""""""""""""""""""""""""""""""""
C
      IF(IFLAG.EQ.1) GO TO 9999
C
C     FAST FOURIER TRANSFORM START ##################################
C
8885  KSPAN=2*4**MM
      IF(JFLAG.EQ.1) GO TO 8887
C
C     CONJUGATE DATA FOR FORWARD TRANSFORM
C
      DO 8886 J=2,N2,2
8886  A(J)=-A(J)
      GO TO 8889
```

```
C
C       MULTIPLY DATA BY N**(-1) IF INVERSE TRANSFORM
C
8887    DO 8888 J=1,N2,2
        A(J)=A(J)*XP
8888    A(J+1)=A(J+1)*XP
8889    I=3
        IT=IX(I-1)
        GO TO (1,2,3,4,5,6,7,8),IT
C*********************************************************************
C
C                           8 MULTIPLY BUTTERFLY
C
1       KK=IX(I)
C
11      K1=KK+KSPAN
        K2=K1+KSPAN
        K3=K2+KSPAN
C
        AKP=A(KK)+A(K2)
        AKM=A(KK)-A(K2)
        AJP=A(K1)+A(K3)
        AJM=A(K1)-A(K3)
        A(KK)=AKP+AJP
C
        BKP=A(KK+1)+A(K2+1)
        BKM=A(KK+1)-A(K2+1)
        BJP=A(K1+1)+A(K3+1)
        BJM=A(K1+1)-A(K3+1)
        A(KK+1)=BKP+BJP
C
        BJP=BKP-BJP
C
        A(K2+1)=(AKP+BJP-AJP)*C707
        A(K2)=A(K2+1)+BJP*CM141
C
        BKP=BKM+AJM
        AKP=AKM-BJM
C
        AC0=(AKP+BKP)*C924
        A(K1+1)=AC0+AKP*CM541
        A(K1)   =AC0+BKP*CM131
C
        BKM=BKM-AJM
        AKM=AKM+BJM
C
        AC0=(AKM+BKM)*C383
        A(K3+1)=AC0+AKM*C541
        A(K3)   =AC0+BKM*CM131
C
        I=I+1
        KK=IX(I)
        IF (KK) 111,111,11
111     I=I+2
        IT=IX(I-1)
        GO TO (1,2,3,4,5,6,7,8), IT
C*********************************************************************
C
C                           4 MULTIPLY BUTTERFLY
C
2       KK=IX(I)
C
22      K1=KK+KSPAN
        K2=K1+KSPAN
        K3=K2+KSPAN
C
        AKP=A(KK)+A(K2)
        AKM=A(KK)-A(K2)
        AJP=A(K1)+A(K3)
        AJM=A(K1)-A(K3)
        A(KK)=AKP+AJP
C
        BKP=A(KK+1)+A(K2+1)
        BKM=A(KK+1)-A(K2+1)
        BJP=A(K1+1)+A(K3+1)
        BJM=A(K1+1)-A(K3+1)
        A(KK+1)=BKP+BJP
        A(K2)=-BKP+BJP
        A(K2+1)=AKP-AJP
C
        BKP=BKM+AJM
C
        A(K1+1)=(BKP+AKM-BJM)*C707
        A(K1)=A(K1+1)+BKP*CM141
C
        AKM=AKM+BJM
C
        A(K3+1)=(AKM+AJM-BKM)*C707
        A(K3)=A(K3+1)+AKM*CM141
C
        I=I+1
        KK=IX(I)
        IF (KK) 222,222,22
222     I=I+2
        IT=IX(I-1)
        GO TO (1,2,3,4,5,6,7,8), IT
C*********************************************************************
C
C                           8 MULTIPLY BUTTERFLY
C
3       KK=IX(I)
C
33      K1=KK+KSPAN
        K2=K1+KSPAN
        K3=K2+KSPAN
C
        AKP=A(KK)+A(K2)
        AKM=A(KK)-A(K2)
        AJP=A(K1)+A(K3)
        AJM=A(K1)-A(K3)
        A(KK)=AKP+AJP
C
        BKP=A(KK+1)+A(K2+1)
        BKM=A(KK+1)-A(K2+1)
        BJP=A(K1+1)+A(K3+1)
        BJM=A(K1+1)-A(K3+1)
        A(KK+1)=BKP+BJP
C
        AJP=AKP-AJP
C
        A(K2+1)=(AJP+BJP-BKP)*C707
        A(K2)=A(K2+1)+AJP*CM141
C
        BKP=BKM+AJM
        AKP=AKM-BJM
C
        AC0=(AKP+BKP)*C383
        A(K1+1)=AC0+AKP*C541
        A(K1)   =AC0+BKP*CM131
C
```

```
      BKM=BKM-AJM
      AKM=AKM+BJM
C
      AC0=(AKM+BKM)*CM924
      A(K3+1)=AC0+AKM*C541
      A(K3)  =AC0+BKM*C131
C
      I=I+1
      KK=IX(I)
      IF (KK) 333,333,33
333   I=I+2
      IT=IX(I-1)
      GO TO (1,2,3,4,5,6,7,8), IT
C**********************************************************************
C
C                                   GENERAL 9 MULTIPLY BUTTERFLY
C
4     KK=IX(I)
C
44    K1=KK+KSPAN
      K2=K1+KSPAN
      K3=K2+KSPAN
C
      AKP=A(KK)+A(K2)
      AKM=A(KK)-A(K2)
      AJP=A(K1)+A(K3)
      AJM=A(K1)-A(K3)
      A(KK)=AKP+AJP
C
      BKP=A(KK+1)+A(K2+1)
      BKM=A(KK+1)-A(K2+1)
      BJP=A(K1+1)+A(K3+1)
      BJM=A(K1+1)-A(K3+1)
      A(KK+1)=BKP+BJP
C
      AJP=AKP-AJP
      BJP=BKP-BJP
C
      J=IX(I+1)
C
      AC0=(AJP+BJP)*T(J+8)
      A(K2+1)=AC0+AJP*T(J+6)
      A(K2)  =AC0+BJP*T(J+7)
C
      BKP=BKM+AJM
      AKP=AKM-BJM
C
      AC0=(AKP+BKP)*T(J+5)
      A(K1+1)=AC0+AKP*T(J+3)
      A(K1)  =AC0+BKP*T(J+4)
C
      BKM=BKM-AJM
      AKM=AKM+BJM
C
      AC0=(AKM+BKM)*T(J+2)
      A(K3+1)=AC0+AKM*T(J)
      A(K3)  =AC0+BKM*T(J+1)
C
      I=I+2
      KK=IX(I)
      IF (KK) 444,444,44
444   I=I+2
      IT=IX(I-1)
      GO TO (1,2,3,4,5,6,7,8), IT
C**********************************************************************
C
C                                   0 MULTIPLY BUTTERFLY
C
5     KK=IX(I)
C
55    K1=KK+KSPAN
      K2=K1+KSPAN
      K3=K2+KSPAN
C
      AKP=A(KK)+A(K2)
      AKM=A(KK)-A(K2)
      AJP=A(K1)+A(K3)
      AJM=A(K1)-A(K3)
      A(KK)=AKP+AJP
      A(K2)=AKP-AJP
C
      BKP=A(KK+1)+A(K2+1)
      BKM=A(KK+1)-A(K2+1)
      BJP=A(K1+1)+A(K3+1)
      BJM=A(K1+1)-A(K3+1)
      A(KK+1)=BKP+BJP
      A(K2+1)=BKP-BJP
C
      A(K3+1)=BKM-AJM
      A(K1+1)=BKM+AJM
      A(K3)=AKM+BJM
      A(K1)=AKM-BJM
C
      I=I+1
      KK=IX(I)
      IF (KK) 555,555,55
555   I=I+2
      IT=IX(I-1)
      GO TO (1,2,3,4,5,6,7,8), IT
C**********************************************************************
C
C                                   OFFSET REDUCED
C
6     KSPAN=KSPAN/4
      I=I+2
      IT=IX(I-1)
      GO TO (1,2,3,4,5,6,7,8), IT
C**********************************************************************
C
C                                   BIT REVERSAL (SHUFFLING)
C
7     IP1=IX(I)
77    IP2=IX(I+1)
      T1=A(IP2)
      A(IP2)=A(IP1)
      A(IP1)=T1
      T1=A(IP2+1)
      A(IP2+1)=A(IP1+1)
      A(IP1+1)=T1
      I=I+2
      IP1=IX(I)
      IF (IP1) 777,777,77
777   I=I+2
      IT=IX(I-1)
      GO TO (1,2,3,4,5,6,7,8), IT
C**********************************************************************
8     IF(JFLAG.EQ.1) GO TO 888
C
C     CONJUGATE OUTPUT IF FORWARD TRANSFORM
C
```

```
      DO 88 J=2,N2,2
88    A(J)=-A(J)
888   RETURN
C
C     FAST FOURIER TRANSFORM ENDS #####################################
C
C     INITIALIZATION PHASE STARTS. DONE ONLY ONCE
C
9999  IXC=1
      N=4**MM
      XP=N
      XP=1./XP
      NTOT=N
      N2=N*2
      NSPAN=N
      N1TEST=N/16
      N2TEST=N/8
      N3TEST=(3*N)/16
      NSPAN4=NSPAN/4
      IBASE=0
      ISN=1
      INC=ISN
      RAD=8.0*ATAN(1.0)
      PI=4.*ATAN(1.0)
      C707=SIN(PI/4.)
      CM141=-2.*C707
      C383=SIN(PI/8.)
      C924=COS(PI/8.)
      CM924=-C924
      C541=C924-C383
      CM541=-C541
      C131=C924+C383
      CM131=-C131
10    NT=INC*NTOT
      KS=INC*NSPAN
      KSPAN=KS
      JC=KS/N
      RADF=RAD*FLOAT(JC)*.5
      I=0
C
C     DETERMINE THE FACTORS OF N
C     ALL FACTORS MUST BE 4 FOR THIS VERSION
C
      M=0
      K=N
15    M=M+1
      NFAC(M)=4
      K=K/4
20    IF(K-(K/4)*4.EQ.0) GO TO 15
      KT=1
      IF(N.GE.256) KT=2
      KSPAN0=KSPAN
      NTYPL=0
C
100   NDELTA=KSPAN0/KSPAN
      INDEX=0
      SD=RADF/FLOAT(KSPAN)
      CD=2.0*SIN(SD)**2
      SD=SIN(SD+SD)
      KK=1
      I=I+1
C
C     TRANSFORM FOR A FACTOR OF 4
C
      KSPAN=KSPAN/4
      IX(IXC)=0
      IX(IXC+1)=6
      IXC=IXC+2
C
410   C1=1.0
      S1=0.0
420   K1=KK+KSPAN
      K2=K1+KSPAN
      K3=K2+KSPAN
      IF(S1.EQ.0.0) GO TO 460
430   IF(KSPAN.NE.NSPAN4) GO TO 431
      T(IBASE+5)=-(S1+C1)
      T(IBASE+6)=C1
      T(IBASE+4)=S1-C1
      T(IBASE+8)=-(S2+C2)
      T(IBASE+9)=C2
      T(IBASE+7)=S2-C2
      T(IBASE+2)=-(S3+C3)
      T(IBASE+3)=C3
      T(IBASE+1)=S3-C3
      IBASE=IBASE+9
C
431   KKP=(KK-1)*2
      IF(INDEX.NE.N1TEST) GO TO 150
      CALL RAD4SB(1)
      GO TO 5035
150   IF(INDEX.NE.N2TEST) GO TO 160
      CALL RAD4SB(2)
      GO TO 5035
160   IF(INDEX.NE.N3TEST) GO TO 170
      CALL RAD4SB(3)
      GO TO 5035
170   CALL RAD4SB(4)
5035  KK=K3+KSPAN
      IF(KK.LE.NT) GO TO 420
440   INDEX=INDEX+NDELTA
      C2=C1-(CD*C1+SD*S1)
      S1=(SD*C1-CD*S1)+S1
      C1=C2
      C2=C1*C1-S1*S1
      S2=C1*S1+C1*S1
      C3=C2*C1-S2*S1
      S3=C2*S1+S2*C1
      KK=KK-NT+JC
      IF(KK.LE.KSPAN) GO TO 420
      KK=KK-KSPAN+INC
      IF(KK.LE.JC) GO TO 410
      IF(KSPAN.EQ.JC) GO TO 800
      GO TO 100
460   KKP=(KK-1)*2
      CALL RAD4SB(5)
5050  KK=K3+KSPAN
      IF(KK.LE.NT) GO TO 420
      GO TO 440
C
800   IX(IXC)=0
      IX(IXC+1)=7
      IXC=IXC+2
C
C     COMPUTE PARAMETERS TO PERMUTE THE RESULTS TO NORMAL ORDER
C     DONE IN TWO STEPS
C     PERMUTATION FOR SQUARE FACTORS OF N
C
      NP(1)=KS
      K=KT+KT+1
```

```
      IF(M.LT.K) K=K-1
      J=1
      NP(K+1)=JC
810   NP(J+1)=NP(J)/NFAC(J)
      NP(K)=NP(K+1)*NFAC(J)
      J=J+1
      K=K-1
      IF(J.LT.K) GO TO 810
      K3=NP(K+1)
      KSPAN=NP(2)
      KK=JC+1
      K2=KSPAN+1
      J=1
C
C     PERMUTATION FOR SINGLE VARIATE TRANSFORM
C
820   KKP=(KK-1)*2
      K2P=(K2-1)*2
      IX(IXC)=KKP+1
      IX(IXC+1)=K2P+1
      IXC=IXC+2
      KK=KK+INC
      K2=KSPAN+K2
      IF(K2.LT.KS) GO TO 820
830   K2=K2-NP(J)
      J=J+1
      K2=NP(J+1)+K2
      IF(K2.GT.NP(J)) GO TO 830
      J=1
840   IF(KK.LT.K2) GO TO 820
      KK=KK+INC
      K2=KSPAN+K2
      IF(K2.LT.KS) GO TO 840
      IF(KK.LT.KS) GO TO 830
      JC=K3
      IX(IXC)=0
      IX(IXC+1)=8
      GO TO 8885
      END
C
C-----------------------------------------------------------------------
C SUBROUTINE:  RAD4SB
C USED BY SUBROUTINE RADIX4. NEVER DIRECTLY ACCESSED BY USER.
C-----------------------------------------------------------------------
C
      SUBROUTINE RAD4SB(NTYPE)
C
C     INPUT: NTYPE = TYPE OF BUTTERFLY INVOKED
C     OUTPUT: PARAMETERS USED BY SUBROUTINE RADIX4
C
      DIMENSION IX(2996)
      COMMON /XX/IX
      COMMON NTYPL,KKP,INDEX,IXC
      IF(NTYPE.EQ.NTYPL) GO TO 7
      IX(IXC)=0
      IX(IXC+1)=NTYPE
      IXC=IXC+2
      IF(NTYPE.NE.4) GO TO 4
      INDEXP=(INDEX-1)*9
      IX(IXC)=KKP+1
      IX(IXC+1)=INDEXP+1
      IXC=IXC+2
      GO TO 6
4     IX(IXC)=KKP+1
      IXC=IXC+1
6     NTYPL=NTYPE
      RETURN
7     IF(NTYPE.NE.4) GO TO 8
      INDEXP=(INDEX-1)*9
      IX(IXC)=KKP+1
      IX(IXC+1)=INDEXP+1
      IXC=IXC+2
      RETURN
8     IX(IXC)=KKP+1
      IXC=IXC+1
      RETURN
      END
```

1.9

Two-Dimensional Mixed Radix Mass Storage Fourier Transform

Richard C. Singleton

SRI International
Menlo Park, CA 94025

1. Purpose

Two-dimensional mixed radix fast Fourier transform algorithms are presented, for use with a mass storage data file. These transforms operate on both real and complex input sequences. A transform (or inverse) is computed in two passes of the mass storage file. The method can be applied to any multi-dimensional transform that operates on the dimensions one by one, and is not limited to two dimensions.

Transforms of size 1024 by 1024 have been computed on the Burroughs 6700 and the CDC 6400; on both machines the program runs without problem in the normal mix, and the transform size is apparently limited only by the size of the mass store.

2. Method

The basic fast Fourier transform subroutines used here (FFT and REALT) are presented and tested in an earlier Section (1.4.). The new idea here is a permutation method to allow use of the basic FFT routines to do larger transforms on data on mass store.

If we assume a mass storage file to be an N1 by N2 array, stored with records written by columns, where N1 is an integer multiple of the record length, then the FFT by columns is simple. The problem is, how do we transform the rows?

This problem is solved here by following the FFT operation on the columns by a permutation of the input buffer so that the intermediate results can be written back to the input file in such a way as to partition the file into M1 by M2 sub-matrices, where M1*M2 is the record size of the file. After this is done, the file can be read back either by rows or by columns. M1 must be a multiple of 2, as the data are complex. N1 must be a multiple of M1*M2, the record size. However N2 need only be a multiple of M2.

On the first pass of the file, columns of length N1 are read into the input buffer in consecutive groups of M2. On the second pass of the file, N2/M2 sub-matrices of size M1 by M2 are read into the input buffer; the buffer then contains the values for M1 rows of length N2. Thus the input buffer must have dimension MAX0(N1*M2,N2*M1).

For a real transform, 2*M2 additional buffer locations are needed for the folding frequency Fourier coefficients, thus the buffer must have dimension MAX0((N1+2)*M2,N2*M1). Also, during a real transform or inverse, a working store of size 2*N2 is set up within the subroutine FFT2T or FFT2I.

This method can be generalized to higher-dimensional Fourier transforms, by sub-dividing either or both of the dimensions N1 and N2 used here. The method can also be applied to other transforms that operate independently on rows and columns.

3. Usage

3.1 Transform of Complex Input Data, N1/2 By N2

```
CALL FFT2T(A,J1,K2,I1,M2,0)        (transform, no scaling)
           or
CALL FFT2I(A,J1,K2,I1,M2,0)        (inverse, scaled)
```

where

A	Is real array dimensioned to MAX0(N1*M2,N2*M1)
M1=I1*2	Is the column dimension of the sub-matrix partitioning
M2	Is the row dimension of the sub-matrix partitioning
N1=J1*M2*M1	Is the column dimension of the data
N2=K2*M2	Is the row dimension of the data

On entry, the complex data are on the mass storage file in columns of length N1, or N1/2 complex values. Real and imaginary components alternate.

On completion of the transform, the data are left on the mass storage file, partitioned in sub-matricies of size M1 by M2, so that the results can be read either by columns or rows. The inverse transform expects input in this form. On completion of the inverse, the data are left on the mass storage file in sequential records by columns, scaled by 2/(N1*N2).

A call on FFT2T, followed by a call on FFT2I with the same input parameters, produces an identity transformation.

The mass storage file LUN should be declared by the user to have a logical record size of M1*M2; the unit number LUN is communicated through labeled common FFTCOM.

3.2 Transform of Real Input Data, N1 by N2

```
CALL FFT2T(A,J1,K2,I1,M2,1)        (transform, no scaling)
           or
CALL FFT2I(A,J1,K2,I1,M2,1)        (inverse, scaled by 1/N)
```

where the parameters are the same as in 3.1, except that any nonzero value for the sixth parameter causes the additional steps for a real transform or inverse to be done.

On completion of the transform, the Fourier coefficients are left on the mass storage file, partitioned in sub-matrices of dimension M1 by M2, with the cosine and sine coefficients alternating down columns. The N2 folding frequency cosine coefficients (alternating with zero sine coefficient values) are stored at the end of the main file in records of length M1*M2, with a possible final short record to bring the total length of the added records to 2*N2. The inverse transform expects data in this form. On completion of the inverse, the real data is in sequence down columns.

A call on FFT2T, followed by a call on FFT2I with the same input parameters, produces an identity transformation.

3.3 Other Subroutines Needed:

TRNSP, EXCH and XFT --- included here
FFT, FFTMX, REALT, ISTKGT and ISTKRL --- from Section 1.4

MFREAD and MFWRIT -- machine-dependent mass storage read and write subroutines. The versions used here were copied from "Optimized Mass Storage FFT Program," by Donald Fraser. His subroutines for the PDP 11 work on the Burroughs B6700 computer. They should also work on an IBM 360/370. For publication, the mass storage file LUN has been replaced by an array RMAS; this version was also tested.

3.4 Error Conditions

The signs of J1, K2, I1 and M2 are ignored. Thus the only error condition recognized by FFT2T or FFT2I is

$$J1*K2*I1*M2 = 0$$

This condition results in an error condition and no transform.

4. Testing

The subroutines included here were tested on a Burroughs B6700 computer. A real transform of size N1 = 240 by N2 = 225 was done with FFT2T, using consecutive integers from 0 to 53,999 as test data. The inverse was then done with FFT2I, and the result compared with the input. The square root of the sum of squares of the errors divided by the sum of squares about the mean for the input was computed as an accuracy check; the result was 0.147E-10 (see Table 1). The record size used was 240, allowing sub-matrices of size M1 = 16 by M2 = 15. The buffer size was thus 3630, and an additional 450 working storage locations were set up within FFT2T and FFT2I on a temporary basis, using ISTKGT and ISTKRL. Average CP time for the transform or inverse was 82 seconds. Most of this is time for FFT and REALT, as when the calls on these subroutines were removed and the permutation alone was tested, the average time each direction was 7 seconds.

Table 1

```
              TEST OF TWO-DIMENSIONAL MIXED RADIX FFT

RECORD SIZE = 240       BUFFER SIZE = 3630          240 BY  225 TRANSFORM

RMS ERROR NORM FOR TRANSFORM-INVERSE PAIR =  .147E-10

MAXIMUM STACK SIZE =  4147
```

```
C
C-----------------------------------------------------------------------
C MAIN PROGRAM:  FFT2D
C AUTHOR:  RICHARD C. SINGLETON
C          SRI INTERNATIONAL, MENLO PARK, CALIFORNIA 94025
C INPUT:  NONE
C-----------------------------------------------------------------------
C
      COMMON /FFTCOM/ LUN
C
C SET UP DYNAMIC STORAGE ALLOCATION, AS IN PORT
C
      COMMON /CSTAK/ DSTAK(2500)
C
      COMMON /MASS/ RMAS(54450)
C
C     NOTE:  MINICOMPUTERS MAY NOT BE ABLE TO ACCOMMODATE
C     THE LARGE ARRAY USED IN THIS TEST PROGRAM TO SIMULATE
C     MASS STORAGE.  IN SUCH CASES, TRUE MASS STORAGE SHOULD
C     BE USED.
C
      DOUBLE PRECISION DSTAK
      INTEGER ISTAK(5000)
      REAL RSTAK(5000)
C
      EQUIVALENCE (DSTAK(1),ISTAK(1))
      EQUIVALENCE (DSTAK(1),RSTAK(1))
      EQUIVALENCE (ISTAK(1),LOUT)
      EQUIVALENCE (ISTAK(3),LUSED)
C
C SET UP MACHINE CONSTANTS, USING PORT SUBPROGRAMS
C
      IOUTD = I1MACH(2)
C
C  THIS PROGRAM USES RANDOM DISK STORAGE, WITH A LOGICAL RECORD SIZE
C    OF M1*M2.  THE TOTAL FILE IS TREATED AS IF IT WERE AN ARRAY OF
C    SIZE N1=K1*M1 BY N2=K2*M2.
C  N1 MUST BE AN INTEGER NUMBER OF RECORDS, THUS K1 MUST BE A MULTIPLE
C    OF M2.  M1 MUST ALSO BE A MULTIPLE OF 2.
C
C  USES SUBROUTINES FFT, FFTMX, REALT, ISTKGT AND ISTKRL FROM SEC.1.4
C
C  USES DONALD FRASER'S SUBROUTINES MFREAD AND MFWRIT, WITH THE MASS
C    STORAGE UNIT NUMBER LUN COMMUNICATED THROUGH LABELED COMMON FFTCOM.
C
      LUN = 2
C
C THE FOLLOWING INPUT PARAMETERS SPECIFY A 240 BY 225 REAL FFT
C
      J1 = 1
      K2 = 15
      I1 = 8
      M2 = 15
C
      K1 = J1*M2
      KT = K1*K2
      M1 = 2*I1
      MT = M1*M2
      N1 = K1*M1
      N2 = K2*M2
      IB = MAX0((N1+2)*M2,N2*M1)
C
      WRITE (IOUTD,9999)
9999  FORMAT (///12X, 39HTEST OF TWO-DIMENSIONAL MIXED RADIX FFT)
      WRITE (IOUTD,9998) MT, IB, N1, N2
9998  FORMAT (/14H RECORD SIZE =, I4, 6X, 13HBUFFER SIZE =, I5, I12,
     *    3H BY, I5, 10H TRANSFORM)
C
C SET UP WORKING STORAGE FOR INPUT/OUTPUT
C
      IA = ISTKGT(IB,3)
C
C WRITE A TEST FILE
C
      IB = IA + MT - 1
      L = 0
      DO 20 J=1,KT
        DO 10 K=IA,IB
          RSTAK(K) = L
          L = L + 1
 10     CONTINUE
        CALL MFWRIT(RSTAK(IA), MT, J)
 20   CONTINUE
C
C TRANSFORM FROM TIME TO FREQUENCY
C
      CALL FFT2T(RSTAK(IA), J1, K2, I1, M2, 1)
C
C HERE WE ARE IN THE FREQUENCY DOMAIN --- ANY FILTERING TO DO?
C
C TRANSFORM BACK FROM FREQUENCY TO TIME DOMAIN
C
      CALL FFT2I(RSTAK(IA), J1, K2, I1, M2, 1)
C
C NOW BACK IN THE TIME DOMAIN...WILL CHECK TO SEE IF WE RETURNED SAFELY
C
      L = 0
      SSA = 0.0
      DO 40 J=1,KT
        CALL MFREAD(RSTAK(IA), MT, J)
        DO 30 K=IA,IB
          SSA = (RSTAK(K)-FLOAT(L))**2 + SSA
          L = L + 1
 30     CONTINUE
 40   CONTINUE
C
C COMPARE ERROR SUM OF SQUARES WITH SUM OF SQUARES ABOUT MEAN FOR DATA
C
      SSB = FLOAT(L)*FLOAT(L+1)*FLOAT(L-1)/12.0
      SSA = SQRT(SSA/SSB)
      WRITE (IOUTD,9997) SSA
9997  FORMAT (/44H RMS ERROR NORM FOR TRANSFORM-INVERSE PAIR =, E10.3)
C
      WRITE (IOUTD,9996) LUSED
9996  FORMAT (/21H MAXIMUM STACK SIZE =, I6)
C
C DE-ALLOCATE WORKING STORAGE...AS A FINAL CHECK ON ARRAY BOUNDS
C
      IF (LOUT.NE.0) CALL ISTKRL(LOUT)
C
      STOP
      END
C
C-----------------------------------------------------------------------
C BLOCK DATA --- INITIALIZES LABELED COMMON
C-----------------------------------------------------------------------
C
      BLOCK DATA
C
      COMMON /CSTAK/ DSTAK(2500)
```

```
C
      DOUBLE PRECISION DSTAK
      INTEGER ISTAK(5000)
      INTEGER ISIZE(5)
C
      EQUIVALENCE (DSTAK(1),ISTAK(1))
      EQUIVALENCE (ISTAK(1),LOUT)
      EQUIVALENCE (ISTAK(2),LNOW)
      EQUIVALENCE (ISTAK(3),LUSED)
      EQUIVALENCE (ISTAK(4),LMAX)
      EQUIVALENCE (ISTAK(5),LBOOK)
      EQUIVALENCE (ISTAK(6),ISIZE(1))
C
      DATA ISIZE(1), ISIZE(2), ISIZE(3), ISIZE(4), ISIZE(5) /1,1,1,2,2/
      DATA LOUT, LNOW, LUSED, LMAX, LBOOK /0,10,10,5000,10/
C
      END
C
C-----------------------------------------------------------------------
C SUBROUTINE:  FFT2T
C COMPUTES TWO-DIMENSIONAL FOURIER TRANSFORM FOR REAL OR COMPLEX DATA,
C IN TWO PASSES OF A MASS STORAGE UNIT LUN
C-----------------------------------------------------------------------
C
      SUBROUTINE FFT2T(A, JJ, KK, LL, MM, IR)
C
C  THE PARAMETER IR CONTROLS THE CALLING OF REALT.  IF IR .NE. 0,
C    COMPUTES AN N1 BY N2 FOURIER TRANSFORM OF REAL DATA STORED IN
C    RECORDS OF SIZE M1*M2 ON A MASS STORAGE FILE LUN.  THE FILE
C    NUMBER LUN MUST BE STORED BY THE USER IN COMMON FFTCOM.
C
C  ARRAY A IS THE INPUT/OUTPUT ARRAY FOR THE MASS STORAGE FILE LUN,
C    DIMENSIONED MAX0(N1*M2,N2*M1) FOR A COMPLEX TRANSFORM
C    OR MAX0((N1+2)*M2,N2*M1) FOR A REAL TRANSFORM.
C
C  THE PARAMETERS M1, M2, N1 AND N2 ARE COMMUNICATED AS FOLLOWS:
C
C        M1=IABS(LL)*2    I.E., M1 MUST BE EVEN
C        M2=IABS(MM)
C        N1=IABS(JJ)*M1*M2    I.E., N1 MUST BE A MULTIPLE OF M1*M2
C        N2=IABS(KK)*M2
C
C  ON ENTRY, THE REAL INPUT VALUES ARE ASSUMED TO BE STORED IN RECORDS
C    OF LENGTH M1*M2, ARRANGED IN SEQUENCE WITH N1 VALUES FOR THE
C    FIRST COLUMN, N1 VALUES FOR THE SECOND COLUMN, ETC.  THE COLUMNS
C    ARE TRANSFORMED, THEN THE DATA ARE PERMUTED TO M1 BY M2 SEGMENTS,
C    SO THAT THE FILE CAN BE READ BY ROWS TO TRANSFORM THE SECOND
C    DIMENSION.
C
C  ON EXIT, THE FOURIER COEFFICIENTS ARE ON MASS STORAGE FILE LUN,
C    WHICH IS LEFT PARTITIONED IN M1 BY M2 SUB-MATRICES...SO THAT
C    THE RESULTS CAN BE READ EITHER BY ROW OR COLUMN.  THE STRUCTURE OF
C    THE FIRST SEGMENT, FOR EXAMPLE, IS AS FOLLOWS:
C
C        0,0     0,1     0,2     ...     0,M2-1
C        1,0     1,1     1,2     ...     1,M2-1
C        2,0     2,1     2,2     ...     2,M2-1
C         .       .       .               .
C         .       .       .               .
C         .       .       .               .
C        M1-1,0 M1-1,1 M1-1,2 ...     M1-1,M2-1
C
C    THE COSINE AND SINE COEFFICIENTS ALTERNATE DOWN COLUMNS, EXCEPT
C    THAT THE FOLDING FREQUENCY COSINE COEFFICIENT VALUES (ALTERNATING
C    WITH ZERO SINE COEFFICIENT VALUES) ARE STORED AT THE END OF THE
C    MAIN FILE IN RECORDS OF LENGTH M1*M2, WITH A POSSIBLE SHORT
C    RECORD TO BRING THE TOTAL LENGTH OF THE ADDED RECORDS TO 2*N2.
C
C  TO DO AN N1/2 BY N2 COMPLEX FOURIER TRANSFORM, WHERE THE INPUT
C    VALUES ARE ARRANGED WITH REAL AND IMAGINARY COMPONENTS ALTERN-
C    ATING, CALL THIS SUBROUTINE WITH IR=0.
C
       COMMON /FFTCOM/ LUN
       DIMENSION A(1)
C
       COMMON /CSTAK/ DSTAK(2500)
       DOUBLE PRECISION DSTAK
       INTEGER ISTAK(5000)
       REAL RSTAK(5000)
C
       EQUIVALENCE (DSTAK(1),ISTAK(1))
       EQUIVALENCE (DSTAK(1),RSTAK(1))
C
       IF (JJ*KK*LL*MM.NE.0) GO TO 10
       IERR = I1MACH(4)
       WRITE (IERR,9999) JJ, KK, LL, MM
9999   FORMAT (33H ERROR - ZERO IN FFT2T PARAMETERS, 4I9)
       RETURN
C
  10   J1 = IABS(JJ)
       K2 = IABS(KK)
       M1H = IABS(LL)
       M1 = 2*M1H
       M2 = IABS(MM)
       M2C = 2*M2
       K1 = J1*M2
       KT = K1*K2
       MT = M1*M2
       N1H = K1*M1H
       N1 = K1*M1
       N2 = K2*M2
       L1 = K1*MT
       L2 = K2*MT
       JM = J1*MT
       IF (IR.EQ.0) GO TO 20
C
C SET UP WORKING STORAGE FOR FOLDING FREQUENCY COEFFICIENTS
C
       IC = ISTKGT(2*N2,3)
       JC = IC
C
  20   DO 60 J=1,KT,K1
         I = J
         DO 30 L=1,L1,MT
           CALL MFREAD(A(L), MT, I)
           I = I + 1
  30     CONTINUE
         CALL FFT(A, A(2), M2, N1H, 1, -2)
         IF (IR.EQ.0) GO TO 40
         CALL REALT(A, A(2), M2, N1H, 1, -2)
         CALL XFR(A(L1+1), RSTAK(JC), M2C)
         JC = JC + M2C
C
C THE FOLLOWING SECTION PARTITIONS THE N1 BY N2 MASS STORAGE INTO
C M1 BY M2 SUB-MATRICES.
C
  40     CALL TRNSP(A, N1, M1, M2)
         I = J
         L = 1
  50     CALL MFWRIT(A(L), MT, I)
```

```
         I = I + 1
         L = L + JM
         IF (L.LT.L1) GO TO 50
         L = L - L1 + MT
         IF (L.LT.JM) GO TO 50
C
   60  CONTINUE
C
       DO 90 J=1,K1
         I = J
         DO 70 L=1,L2,MT
           CALL MFREAD(A(L), MT, I)
           I = I + K1
   70    CONTINUE
         CALL FFT(A, A(2), 1, N2, M1H, -2)
         I = J
         DO 80 L=1,L2,MT
           CALL MFWRIT(A(L), MT, I)
           I = I + K1
   80    CONTINUE
   90  CONTINUE
C
       IF (IR.EQ.0) RETURN
       CALL FFT(RSTAK(IC), RSTAK(IC+1), 1, N2, 1, -2)
       J = KT + 1
       JC = IC
       K = 2*N2
  100  CALL MFWRIT(RSTAK(JC), MIN0(K,MT), J)
       J = J + 1
       JC = JC + MT
       K = K - MT
       IF (K.GT.0) GO TO 100
       CALL ISTKRL(1)
       RETURN
       END
C
C-----------------------------------------------------------------------
C SUBROUTINE:  FFT2I
C COMPUTES TWO-DIMENSIONAL FOURIER TRANSFORM INVERSE FOR REAL OR
C COMPLEX DATA, IN TWO PASSES OF A MASS STORAGE UNIT LUN
C-----------------------------------------------------------------------
C
       SUBROUTINE FFT2I(A, JJ, KK, LL, MM, IR)
C
C  THE PARAMETER IR CONTROLS THE CALLING OF REALT.  IF IR .NE. 0,
C    COMPUTES AN INVERSE FOURIER TRANSFORM, USING AS INPUT AN N1 BY N2
C    FILE OF FOURIER COEFFICIENTS ON MASS STORAGE FILE LUN.  THE FILE
C    NUMBER LUN MUST BE STORED BY THE USER IN COMMON FFTCOM.
C
C  ARRAY A IS THE INPUT/OUTPUT ARRAY FOR THE MASS STORAGE FILE LUN,
C    DIMENSIONED MAX0(N1*M2,N2*M1) FOR A COMPLEX TRANSFORM
C    OR MAX0((N1+2)*M2,N2*M1) FOR A REAL TRANSFORM.
C
C  THE PARAMETERS M1, M2, N1 AND N2 ARE COMMUNICATED AS FOLLOWS:
C
C        M1=IABS(LL)*2     I.E., M1 MUST BE EVEN
C        M2=IABS(MM)
C        N1=IABS(JJ)*M1*M2    I.E., N1 MUST BE A MULTIPLE OF M1*M2
C        N2=IABS(KK)*M2
C
C  ON ENTRY, THE FOURIER COEFFICIENTS ARE ON MASS STORAGE FILE LUN,
C    WHICH IS PARTITIONED IN M1 BY M2 SUB-MATRICES...SO THAT
C    THE RESULTS CAN BE READ EITHER BY ROW OR COLUMN.  THE STRUCTURE OF
C    THE FIRST SEGMENT, FOR EXAMPLE, IS AS FOLLOWS:
C
C        0,0    0,1    0,2    ...    0,M2-1
C        1,0    1,1    1,2    ...    1,M2-1
C        2,0    2,1    2,2    ...    2,M2-1
C         .      .      .             .
C         .      .      .             .
C         .      .      .             .
C        M1-1,0 M1-1,1 M1-1,2 ...    M1-1,M2-1
C
C    THE COSINE AND SINE COEFFICIENTS ALTERNATE DOWN COLUMNS, EXCEPT
C    THAT THE FOLDING FREQUENCY COSINE COEFFICIENT VALUES (ALTERNATING
C    WITH ZERO SINE COEFFICIENT VALUES) ARE STORED AT THE END OF THE
C    MAIN FILE IN RECORDS OF LENGTH M1*M2, WITH A POSSIBLE SHORT
C    RECORD TO BRING THE TOTAL LENGTH OF THE ADDED RECORDS TO 2*N2.
C
C  ON EXIT, THE REAL RESULT VALUES ARE STORED ON THE FILE LUN IN RECORDS
C    OF LENGTH M1*M2, ARRANGED IN SEQUENCE WITH N1 VALUES FOR THE
C    FIRST COLUMN, N1 VALUES FOR THE SECOND COLUMN, ETC.
C
C  THE FOLLOWING PAIR OF CALLS PRODUCES AN IDENTITY TRANSFORMATION:
C
C            CALL FFT2D(A,JJ,KK,LL,MM,IR)
C            CALL FFT2I(A,JJ,KK,LL,MM,IR)
C
C  TO DO AN N1/2 BY N2 COMPLEX INVERSE FOURIER TRANSFORM, WHERE THE
C    INPUT VALUES ARE ARRANGED WITH COSINE AND SINE COEFFICIENTS
C    ALTERNATING, CALL THIS SUBROUTINE WITH IR=0.
C
       COMMON /FFTCOM/ LUN
       DIMENSION A(1)
C
       COMMON /CSTAK/ DSTAK(2500)
       DOUBLE PRECISION DSTAK
       INTEGER ISTAK(5000)
       REAL RSTAK(5000)
C
       EQUIVALENCE (DSTAK(1),ISTAK(1))
       EQUIVALENCE (DSTAK(1),RSTAK(1))
C
       IF (JJ*KK*LL*MM.NE.0) GO TO 10
       IERR = I1MACH(4)
       WRITE (IERR,9999) JJ, KK, LL, MM
 9999  FORMAT (33H ERROR - ZERO IN FFT2I PARAMETERS, 4I9)
       RETURN
C
   10  J1 = IABS(JJ)
       K2 = IABS(KK)
       M1H = IABS(LL)
       M1 = 2*M1H
       M2 = IABS(MM)
       M2C = 2*M2
       K1 = J1*M2
       KT = K1*K2
       MT = M1*M2
       N1H = K1*M1H
       N1 = K1*M1
       N2 = K2*M2
       L1 = K1*MT
       L2 = K2*MT
       KM = M2*MT
       IF (IR.EQ.0) GO TO 30
C
C SET UP WORKING STORAGE FOR FOLDING FREQUENCY COEFFICIENTS, AND
C RETRIEVE THEM FROM THE END OF THE MAIN FILE.
C
       IC = ISTKGT(2*N2,3)
```

```
      J = KT + 1
      JC = IC
      K = 2*N2
   20 CALL MFREAD(RSTAK(JC), MIN0(K,MT), J)
      J = J + 1
      JC = JC + MT
      K = K - MT
      IF (K.GT.0) GO TO 20
      CALL FFT(RSTAK(IC), RSTAK(IC+1), 1, N2, 1, 2)
      JC = IC
C
   30 DO 60 J=1,K1
        I = J
        DO 40 L=1,L2,MT
          CALL MFREAD(A(L), MT, I)
          I = I + K1
   40   CONTINUE
        CALL FFT(A, A(2), 1, N2, M1H, 2)
        I = J
        DO 50 L=1,L2,MT
          CALL MFWRIT(A(L), MT, I)
          I = I + K1
   50   CONTINUE
   60 CONTINUE
C
      DO 100 J=1,KT,K1
C
C THE FOLLOWING SECTION RESTORES THE ORIGINAL ORDER OF THE N1 BY N2
C MASS STORAGE, ELIMINATING THE M1 BY M2 SUB-MATRICES.
C
        I = J
        L = 1
   70   CALL MFREAD(A(L), MT, I)
        I = I + 1
        L = L + KM
        IF (L.LT.L1) GO TO 70
        L = L - L1 + MT
        IF (L.LT.KM) GO TO 70
        CALL TRNSP(A, N1, M1, M2)
C
        IF (IR.EQ.0) GO TO 80
        CALL XFR(RSTAK(JC), A(L1+1), M2C)
        JC = JC + M2C
        CALL REALT(A, A(2), M2, N1H, 1, 2)
   80   CALL FFT(A, A(2), M2, N1H, 1, 2)
        I = J
        DO 90 L=1,L1,MT
          CALL MFWRIT(A(L), MT, I)
          I = I + 1
   90   CONTINUE
  100 CONTINUE
C
      IF (IR.NE.0) CALL ISTKRL(1)
      RETURN
      END
C
C-----------------------------------------------------------------------
C SUBROUTINE:  XFR
C MOVES A(J) TO B(J) FOR J=1,2,...,N
C-----------------------------------------------------------------------
C
      SUBROUTINE XFR(A, B, N)
C
      DIMENSION A(1), B(1)
      DO 10 J=1,N
        B(J) = A(J)
   10 CONTINUE
      RETURN
      END
C
C-----------------------------------------------------------------------
C SUBROUTINE:  TRNSP
C SELF-INVERSE PERMUTATION OF N1*M2 DATA VALUES IN ARRAY A
C-----------------------------------------------------------------------
C
      SUBROUTINE TRNSP(A, N1, M1, M2)
C
C SUPPOSE ARRAY A IS THE INPUT/OUTPUT BUFFER FOR A MASS STORE,
C WHERE THE RECORD SIZE IS M1*M2.  SUPPOSE THIS MASS STORE HOLDS
C A RECTANGULAR ARRAY, STORED BY COLUMNS OF LENGTH N1, WHERE N1
C IS A MULTIPLE OF M1*M2 AND THE NUMBER OF ROWS IS A MULTIPLE
C OF M2.  THIS SUBROUTINE REARRANGES THE BUFFER SO THAT ALL
C ELEMENTS BELONGING TO A ROW ARE IN THE SAME RECORD SEGMENT.
C WHILE THE SEGMENTS ARE NOT IN NORMAL ROW ORDER, THIS ORDER
C CAN BE OBTAINED BY REORDERING THE OUTPUT, AS IN 'FFT2T'...THEN
C BEFORE CALLING THIS SUBROUTINE A SECOND TIME TO RESTORE THE
C ORIGINAL ORDER, THE SEGMENTS MUST BE REORDERED ON INPUT, AS
C IN 'FFT2I'.
C
      DIMENSION A(1)
      NB = N1*M2
      MT = M1*M2
      MM = MT - M1
      J = 0
      K = 0
   10 J = J + M1
      K = K + N1
      IF (J.GE.K) GO TO 10
   20 CALL EXCH(A(J), A(K), M1)
      J = J + M1
      K = K + N1
      IF (K.LT.NB) GO TO 20
      K = K - NB + MT
      IF (K.LT.N1) GO TO 10
      K = K - N1 + M1
      IF (K.NE.MM) GO TO 10
      RETURN
      END
C
C-----------------------------------------------------------------------
C SUBROUTINE:  EXCH
C EXCHANGES A(J) AND B(J) FOR J=2,3,...,N+1
C-----------------------------------------------------------------------
C
      SUBROUTINE EXCH(A, B, N)
C
      DIMENSION A(1), B(1)
      J = 1
   10 J = J + 1
      T = A(J)
      A(J) = B(J)
      B(J) = T
      IF (J.LE.N) GO TO 10
      RETURN
      END
C
C-----------------------------------------------------------------------
C SUBROUTINE:  MFREAD
C READS RECORD JB FROM MASS STORE TO BUFA, NB REAL VALUES
C SOURCE:  DONALD FRASER, OPTIMIZED MASS STORAGE FFT PROGRAM, APPENDIX C
```

```
C-----------------------------------------------------------------------
C
      SUBROUTINE MFREAD(BUFA, NB, JB)
      REAL BUFA(NB)
      COMMON /FFTCOM/ LUN
C
C READ(LUN'JB) BUFA
C
      COMMON /MASS/ RMAS(54450)
      J = (JB-1)*NB + 1
      CALL XFR(RMAS(J), BUFA, NB)
      RETURN
      END
C
C-----------------------------------------------------------------------
C SUBROUTINE:  MFWRIT
C WRITES RECORD JB FROM BUFA TO MASS STORE, NB REAL VALUES
C SOURCE:  DONALD FRASER, OPTIMIZED MASS STORAGE FFT PROGRAM, APPENDIX C
C-----------------------------------------------------------------------
C
      SUBROUTINE MFWRIT(BUFA, NB, JB)
      REAL BUFA(NB)
      COMMON /FFTCOM/ LUN
C
C WRITE(LUN'JB) BUFA
C
      COMMON /MASS/ RMAS(54450)
      J = (JB-1)*NB + 1
      CALL XFR(BUFA, RMAS(J), NB)
      RETURN
      END
```

CHAPTER 2

Power Spectrum Analysis and Correlation

L. R. Rabiner

Introduction

The programs in this and the next section are various implementations of high speed convolution, correlation, and spectrum analysis algorithms. A wide variety of techniques have been proposed in this area and the four programs presented are only a small sampling of those available. However, they do serve to illustrate the types of techniques which have been applied in a wide variety of applications.

The first program by Rabiner, Schafer, and Dlugos is an implementation of the method of modified periodograms, as proposed by Welch [1], for directly estimating the power spectrum of a signal.

The second program, again by Rabiner, Schafer, and Dlugos, is an implementation of the correlation method, as proposed by Rader [2], for directly estimating the correlation function of a signal.

The third program in this section, by Carter and Ferrie, is one for estimating the coherence function of a signal. Such techniques are widely applied in the areas of sonar, radar, and other underwater applications of digital signal processing.

A recent IEEE Press volume [3] describes several other valuable algorithms for power spectrum analysis. Although programs are not included in that volume the interested reader should consult that reference for alternative methods of computing spectra and correlation functions of digital signals.

References

1. P. D. Welch, "The Use of the FFT for Estimation of Power Spectra: A Method Based on Averaging Over Short, Modified Periodograms", *IEEE Trans. on Audio and Electroacoustics,* Vol. AU-15, No. 2, pp. 70-73, June 1967.

2. C. M. Rader, "An Improved Algorithm for High Speed Autocorrelation with Applications to Spectral Estimation", *IEEE Trans. on Audio and Electroacoustics,* Vol. AU-18, No. 4, pp. 439-442, Dec. 1970.

3. D. G. Childers, Editor, *Modern Spectrum Analysis,* IEEE Press, 1978.

2.1

Periodogram Method for Power Spectrum Estimation

L. R. Rabiner

Acoustics Research Dept.
Bell Laboratories
Murray Hill, NJ 07974

R. W. Schafer and *D. Dlugos*

Dept. of Electrical Engineering
Georgia Institute of Technology
Atlanta, GA 30332

1. Purpose

This program can be used to obtain a direct estimate of the auto-power spectrum of a signal using the method of averaging periodograms.

2. Method

The use of the FFT in this manner was first described by Welch [1]. The essential features of the method are given below. Further details are available in Refs. [1-3].

A sequence $x(n)$ has a power spectrum denoted $P_{xx}(\omega)$. The modified periodogram spectrum estimate is obtained by dividing an available segment of $x(n)$, $0 \leqslant n \leqslant N-1$ into K overlapping segments of length L. In this implementation, the segments overlap by $L/2$ samples, giving the total number of segments as

$$K = [(N - L/2)/(L/2)]$$

where $[x]$ denotes the integer part of x. The ith segment of data is defined as

$$x_i(n) = x(iL/2+n)\, w_d(n) \quad 0 \leqslant n \leqslant L-1, \ \ 0 \leqslant i \leqslant K-1 \tag{1}$$

where, $w_d(n)$ is an L-point data window (e.g., rectangular, Hamming, etc.). The M-point $(M \geqslant L)$ DFTs of the windowed segments $x_i(n)$

$$X_i(k) = \sum_{n=0}^{M-1} x_i(n) e^{-j\frac{2\pi}{M}kn} \quad 0 \leqslant k \leqslant M-1, \quad 0 \leqslant i \leqslant K-1 \tag{2}$$

are computed using an FFT algorithm. (If $L < M$, the sequence $x_i(n)$ is augmented with $M-L$ zero valued samples.) The modified periodograms

$$S_i(k) = |X_i(k)|^2 \quad 0 \leqslant k \leqslant M-1, \quad 0 \leqslant i \leqslant K-1 \tag{3}$$

are averaged to produce the spectrum estimate at normalized radian frequency $2\pi k/M$

$$S_{xx}(2\pi k/M) = \frac{1}{KU} \sum_{i=0}^{K-1} S_i(k) \quad 0 \leqslant k \leqslant M-1. \tag{4}$$

where

$$U = \sum_{n=0}^{L-1} w_d^2(n)\ . \tag{5}$$

The quantity U is necessary for the spectrum estimate to be unbiased [1].

It can be shown [1,3] that the expected value of $S_{xx}(2\pi k/M)$ is

$$E[S_{xx}(2\pi k/M)] = \frac{1}{2\pi} \int_{-\pi}^{\pi} P_{xx}(\theta)\, W(2\pi k/M-\theta)\, d\theta \tag{6}$$

The effective spectrum smoothing "window" is

$$W(\omega) = \frac{1}{U} \, |W_d(\omega)|^2 , \tag{7}$$

where

$$W_d(\omega) = \sum_{n=0}^{L-1} w_d(n) e^{-j\omega n} \tag{8}$$

is the Fourier transform of the data window sequence, $w_d(n)$.

It can also be shown [1] that the variance of the spectrum estimate, $S_{xx}(2\pi k/M)$, is

$$var[S_{xx}(2\pi k/M)] \approx \frac{11}{9K} \, P_{xx}^2(2\pi k/M) \; . \tag{9}$$

Since the width of the spectrum window $W(e^{j\omega})$ is inversely proportional to L, and the variance of the spectrum estimate is inversely dependent on K, it is clear that for $N = (K+1)\,(L/2)$ fixed, good frequency resolution and low variance are conflicting requirements. Clearly resolution can be traded for low variance by using many short segments, and greater resolution can be achieved at the expense of greater statistical variance of the estimate.

3. Program Description

A flow chart of the program is shown in Fig. 1.

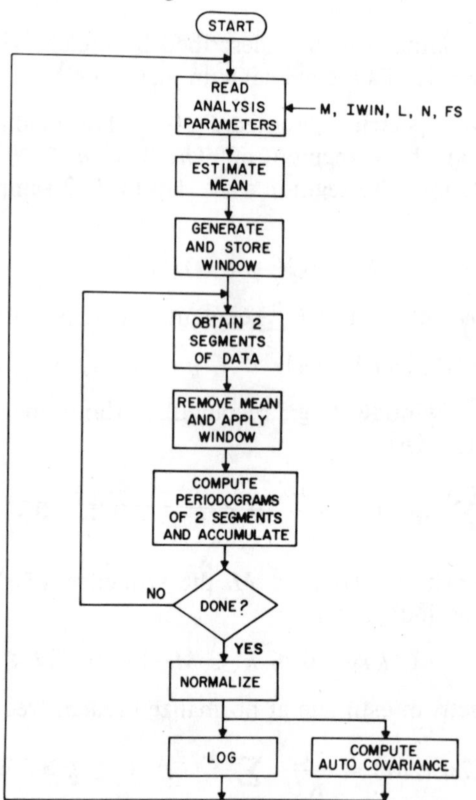

Fig. 1 Flow diagram of program, PMPSE.

First the program reads in parameters such as FFT size, window type, window length, maximum number of data samples available and sampling frequency of the data. Then the mean is estimated so that it can be subtracted from the data. (This is desirable so that the low frequency portion of the spectrum is not obscured.) Next the modified periodograms are computed for the available data, and accumulated according to Eq. (4).

A notable feature of the program is that the periodograms $|X_i(k)|^2$ are computed two at a time. It can be shown [1] that if two real segments $x_i(n)$ and $x_{i+1}(n)$ are used as real and imaginary parts of a sequence

$$x(n) = x_i(n) + jx_{i+1}(n) \tag{10}$$

then

$$X(k)X^*(k) + X(M-k)X^*(M-k) = 2[|X_i(k)|^2 + |X_{i+1}(k)|^2] \quad 0 \leqslant k \leqslant M/2 \tag{11}$$

where $X(k)$ is the M-point transform of the complex sequence, $x(n)$, and $X_i(k)$ and $X_{i+1}(k)$ are the M-point DFT's of $x_i(n)$ and $x_{i+1}(n)$ respectively. Thus the left hand side of Eq. (11) gives two periodograms from one FFT computation.

The final step in the program is the normalization by $K \cdot U$ and the computation of $20 \log_{10}[S_{xx}(2\pi k/M)]$. At this point the spectrum estimate is available for plotting, etc.

An estimate of the autocovariance of $x(n)$ is then obtained by taking the inverse DFT of the power spectral estimate. Since only a sampled version of the power spectrum is available, the resulting autocovariance is an aliased version of the true autocovariance of $x(n)$ [2].

3.1 Usage

The main program (PMPSE) requires two user supplied subroutines; GETX for accessing blocks of data, and FFT, a complex, radix 2, FFT subroutine. The main program reads in analysis parameters, and then computes the estimates of the mean and power spectrum and then does an inverse FFT to provide an autocovariance estimate of the input. The program prints out the log power spectrum (in dB), and the autocovariance function.

Although no graphical output capability is provided with the program, such can be supplied by the user at the places indicated in the main program. Typical graphical output is shown in the examples given below.

3.2 Description of Program Parameters

The main program PMPSE reads the following parameters:

M FFT length. M must be a power of 2 (for compatibility with the FFT subroutine) that is less than or equal to the maximum value MAXM specified in the main program.

IWIN Type of window used on data. If IWIN=1, a rectangular window is used; if IWIN=2, a Hamming window is used.

L Window length. Must be less than or equal to M.

N Maximum number of data samples available for use in the analysis. Because of the windowing applied to the data, it is preferrable that N be exactly divisible by L/2. Thus the program processes only NSECT = (N−L/2)/(L/2) overlapping segments of length L or a total of NP = (NSECT+1)(L/2) samples.

FS Sampling frequency in Hz. Used only for printing the power spectrum values.

The user must supply the routine GETX which provides values of $x(n)$ to the main program. The call to this routine is

CALL GETX (X,NRD,SS)

where the parameters X, NRD, and SS, are

X Array in which the samples of $x(n)$ are stored

NRD Number of samples of $x(n)$ to be read into array X.

SS Starting sample number of $x(n)$, i.e., samples $x(\mathrm{SS})$ to $x(\mathrm{SS} + \mathrm{NRD} - 1)$ are read into array X in the subroutine GETX. Note that SS is a floating point number to allow for maximum flexibility. The first input sample is assumed to be $x(1)$, i.e., SS = 1.

The main program PMPSE, and test subroutine GETX are given in the Appendix. The subroutine FFT is a complex, radix 2, FFT subroutine with calling sequence CALL FFT(X,M,INV), where

M = Size of FFT (a power of 2)
X = Complex array of length M
INV = 0 for direct FFT
= 1 for inverse FFT

3.3 Dimension Requirements

The real array XA is a data buffer and must be dimensioned to be the maximum window length; i.e., MAXM where MAXM is the largest FFT length. The complex array, X, (the working array for the FFT), and the real array, WD, (the window array) must be dimensioned MAXM. The real array SXX is used for accumulating the spectrum estimate. The real array XFR is used to generate a frequency scale corresponding to the DFT sample frequencies and the real array ILAG is used to generate a set of lag indices for printing. The arrays SXX, XFR and ILAG are all dimensioned (MAXM/2 + 1).

3.4 Summary of User Requirements

(1) Provide the subroutine GETX which specifies the input data $x(n)$.

(2) Provide the complex FFT subroutine FFT.

(3) Specify the analysis parameters, M, IWIN, L, N, and FS.

(4) Provide graphical output of the power spectrum and autocovariance estimates.

4. Test Problems

To illustrate the use of this program a simple test problem is provided in the form of the subroutine GETX which generates samples of the sequence

$$x(n) = \cos(2\pi n/10) \ ,$$

representing samples of a sinusoid of frequency FS/10, where FS is the sampling frequency.

Example #1

For this example, the parameters are M = 128, IWIN = 2, L = 64, N = 256, and FS = 10000. That is, a segment of length 256 is sectioned into 7 sections of length 64 each weighted with a Hamming window, and the 128-point periodograms are computed and averaged. Table 1 gives the numerical values of the power spectrum and autocovariance estimates for this example and Figure 2 is a plot of these values.

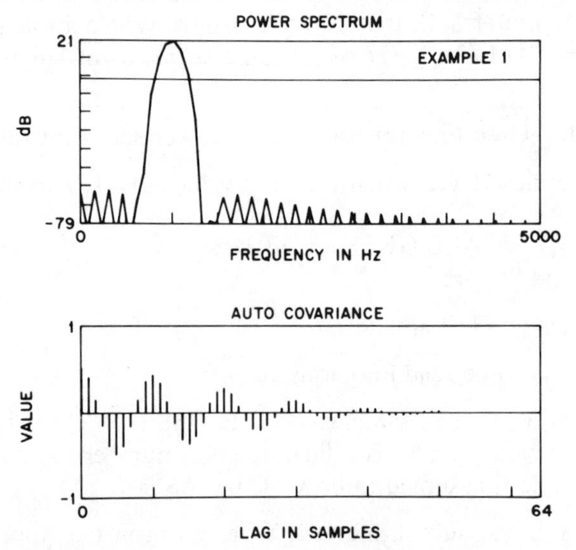

Fig. 2 Plots of the log power spectrum and the autocovariance for Example #1.

This example should be compared to Example #1 of the program CMPSE of section 2.2 in this book. In that example a 63-point Hamming window was applied to the *autocorrelation function.* In this case a 64-point Hamming window was applied to the *data.* According to Eq. (7) the spectral window for this example is essentially the square of the spectral window for Example #1 of CMPSE. Thus the width of the main lob (between the first zeros) is almost the same but the side lobes are twice as low (in dB). A comparison of Fig. 2 of PMPSE and Fig. 3 of CMPSE confirms this assertion.

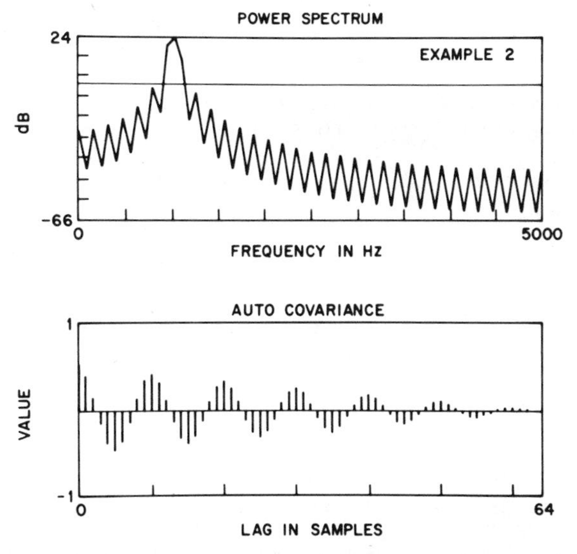

Fig. 3 Plots of the log power spectrum and the autocovariance for Example #2.

Example #2

For this example M = 128, IWIN = 1, L = 64, N = 256 and FS = 10000. That is, everything is the same as Example #1 except a rectangular window is used. The results are given in Table 2 and Fig. 3. Again, this example should be compared to Example #2 of CMPSE.

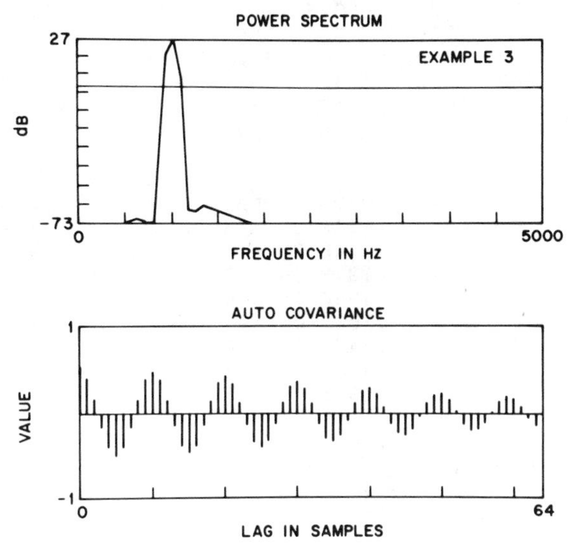

Fig. 4 Plots of the log power spectrum and the autocovariance for Example #3.

Example #3

It should be noted that the FFT length M can be greater than or equal to the window length L. Obviously the greatest efficiency results when M = L, however if a more detailed representation of the spectrum is desired, M can be larger than L as in Examples #1 and #2. In this example, M = L = 128, IWIN = 2, N = 256, and FS = 10000. In this case, about half as many segments are obtained from the 256 samples of data as in Example #1 (implying greater variance for a random signal) but the resolution is twice as good. The results are shown in Table 3 and Fig. 4.

References

1. P. D. Welch, "The Use of Fast Fourier Transform for the Estimation of Power Spectra: A Method Based on Time Averaging Over Short, Modified Periodograms", *IEEE Trans. Audio and Electroacoust.,* Vol. AU-15, No. 2, pp. 70-73, June 1967.

2. L. R. Rabiner, and B. Gold, *Theory and Application of Digital Signal Processing,* Chapter 6, pp. 399-419, Prentice-Hall, Inc., Englewood Cliffs, N.J., 1975.

3. A. V. Oppenheim and R. W. Schafer, *Digital Signal Processing,* Chapter 11, pp. 556-562, Prentice-Hall, Inc., Englewood Cliffs, N.J., 1975.

Table 1

WINDOW TYPE=HAMG
M= 128 NP= 256 L= 64 SAMPLING FREQUENCY=10000.0000
XMEAN= .69337E-06
LOG POWER SPECTRUM

FREQ	DB	FREQ	DB	FREQ	DB	FREQ	DB
.000	-61.4475	78.125	-82.8421	156.250	-61.3083	234.375	-83.4551
312.500	-61.2347	390.625	-86.3237	468.750	-63.3791	546.875	-92.0066
625.000	-67.5810	703.125	-51.5461	781.250	-9.4455	859.375	9.7459
937.500	19.1048	1015.625	21.1584	1093.750	16.3246	1171.875	3.5110
1250.000	-21.7077	1328.125	-89.0598	1406.250	-77.8186	1484.375	-78.8845
1562.500	-64.6289	1640.625	-78.5597	1718.750	-63.1345	1796.875	-80.5152
1875.000	-64.0265	1953.125	-82.6783	2031.250	-65.3884	2109.375	-84.6962
2187.500	-66.7943	2265.625	-86.5093	2343.750	-68.1207	2421.875	-88.1117
2500.000	-69.3376	2578.125	-89.5294	2656.250	-70.4361	2734.375	-90.7871
2812.500	-71.4257	2890.625	-91.9038	2968.750	-72.3136	3046.875	-92.8826
3125.000	-73.1069	3203.125	-93.7425	3281.250	-73.8075	3359.375	-94.5249
3437.500	-74.4394	3515.625	-95.1747	3593.750	-74.9815	3671.875	-95.7766
3750.000	-75.4666	3828.125	-96.2903	3906.250	-75.8803	3984.375	-96.7128
4062.500	-76.2405	4140.625	-97.0862	4218.750	-76.5297	4296.875	-97.3865
4375.000	-76.7718	4453.125	-97.6216	4531.250	-76.9536	4609.375	-97.7954
4687.500	-77.0851	4765.625	-97.9173	4843.750	-77.1607	4921.875	-97.9708
5000.000	-77.1911						

CORRELATION FUNCTION

LAG	CORR	LAG	CORR	LAG	CORR	LAG	CORR	LAG	CORR
0	.500E 00	1	.404E 00	2	.154E 00	3	-.153E 00	4	-.396E 00
5	-.483E 00	6	-.385E 00	7	-.144E 00	8	.141E 00	9	.362E 00
10	.435E 00	11	.342E 00	12	.127E 00	13	-.122E 00	14	-.308E 00
15	-.366E 00	16	-.283E 00	17	-.103E 00	18	.984E-01	19	.244E 00
20	.285E 00	21	.218E 00	22	.780E-01	23	-.733E-01	24	-.179E 00
25	-.206E 00	26	-.154E 00	27	-.544E-01	28	.504E-01	29	.121E 00
30	.136E 00	31	.100E 00	32	.346E-01	33	-.316E-01	34	-.742E-01
35	-.821E-01	36	-.592E-01	37	-.199E-01	38	.179E-01	39	.411E-01
40	.443E-01	41	.311E-01	42	.102E-01	43	-.904E-02	44	-.200E-01
45	-.210E-01	46	-.143E-01	47	-.449E-02	48	.393E-02	49	.834E-02
50	.840E-02	51	.547E-02	52	.163E-02	53	-.141E-02	54	-.281E-02
55	-.267E-02	56	-.164E-02	57	-.447E-03	58	.384E-03	59	.697E-03
60	.600E-03	61	.319E-03	62	.630E-04	63	-.490E-04	64	.000E 00

Table 2

WINDOW TYPE=RECT
M= 128 NP= 256 L= 64 SAMPLING FREQUENCY=10000.0000
XMEAN= .69337E-06
LOG POWER SPECTRUM

FREQ	DB	FREQ	DB	FREQ	DB	FREQ	DB
.000	-23.4182	78.125	-41.2622	156.250	-22.7620	234.375	-40.0745
312.500	-20.7564	390.625	-37.5837	468.750	-17.2393	546.875	-33.4572
625.000	-11.7112	703.125	-26.7759	781.250	-2.5602	859.375	-14.1903
937.500	19.2009	1015.625	23.7835	1093.750	12.2942	1171.875	-17.9184
1250.000	-4.5159	1328.125	-29.0914	1406.250	-12.6625	1484.375	-35.7182
1562.500	-17.9894	1640.625	-40.3775	1718.750	-21.8953	1796.875	-43.9277
1875.000	-24.9423	1953.125	-46.7622	2031.250	-27.4133	2109.375	-49.0958
2187.500	-29.4704	2265.625	-51.0590	2343.750	-31.2150	2421.875	-52.7352
2500.000	-32.7145	2578.125	-54.1828	2656.250	-34.0158	2734.375	-55.4436
2812.500	-35.1528	2890.625	-56.5475	2968.750	-36.1508	3046.875	-57.5154
3125.000	-37.0285	3203.125	-58.3657	3281.250	-37.8000	3359.375	-59.1143
3437.500	-38.4782	3515.625	-59.7669	3593.750	-39.0701	3671.875	-60.3393
3750.000	-39.5853	3828.125	-60.8339	3906.250	-40.0286	3984.375	-61.2544
4062.500	-40.4053	4140.625	-61.6092	4218.750	-40.7173	4296.875	-61.8999
4375.000	-40.9695	4453.125	-62.1289	4531.250	-41.1630	4609.375	-62.2989
4687.500	-41.3003	4765.625	-62.4118	4843.750	-41.3821	4921.875	-62.4676
5000.000	-41.4096						

CORRELATION FUNCTION

LAG	CORR	LAG	CORR	LAG	CORR	LAG	CORR	LAG	CORR
0	.499E 00	1	.397E 00	2	.149E 00	3	-.148E 00	4	-.379E 00
5	-.460E 00	6	-.365E 00	7	-.136E 00	8	.136E 00	9	.348E 00
10	.421E 00	11	.334E 00	12	.124E 00	13	-.124E 00	14	-.316E 00
15	-.382E 00	16	-.302E 00	17	-.112E 00	18	.112E 00	19	.284E 00
20	.343E 00	21	.271E 00	22	.100E 00	23	-.997E-01	24	-.253E 00
25	-.304E 00	26	-.239E 00	27	-.882E-01	28	.876E-01	29	.221E 00
30	.265E 00	31	.207E 00	32	.761E-01	33	-.755E-01	34	-.190E 00
35	-.226E 00	36	-.176E 00	37	-.641E-01	38	.635E-01	39	.158E 00
40	.187E 00	41	.144E 00	42	.520E-01	43	-.514E-01	44	-.126E 00
45	-.148E 00	46	-.113E 00	47	-.399E-01	48	.393E-01	49	.948E-01
50	.109E 00	51	.810E-01	52	.279E-01	53	-.272E-01	54	-.632E-01
55	-.696E-01	56	-.494E-01	57	-.158E-01	58	.152E-01	59	.316E-01
60	.306E-01	61	.178E-01	62	.371E-02	63	-.310E-02	64	-.358E-06

Table 3

WINDOW TYPE=HAMG
M= 128 NP= 256 L= 128 SAMPLING FREQUENCY=10000.0000
XMEAN= .69337E-06
LOG POWER SPECTRUM

FREQ	DB	FREQ	DB	FREQ	DB	FREQ	DB
.000	-80.3770	78.125	-80.1718	156.250	-79.5365	234.375	-78.5186
312.500	-77.1878	390.625	-75.5785	468.750	-73.8205	546.875	-72.0789
625.000	-70.8606	703.125	-72.1690	781.250	-91.4820	859.375	-34.5527
937.500	18.2261	1015.625	26.8289	1093.750	5.4495	1171.875	-65.6636
1250.000	-66.6186	1328.125	-63.7224	1406.250	-64.6737	1484.375	-66.2602
1562.500	-67.9136	1640.625	-69.4777	1718.750	-70.9325	1796.875	-72.2612
1875.000	-73.4819	1953.125	-74.6040	2031.250	-75.6304	2109.375	-76.5837
2187.500	-77.4638	2265.625	-78.2804	2343.750	-79.0422	2421.875	-79.7445
2500.000	-80.4112	2578.125	-81.0218	2656.250	-81.6119	2734.375	-82.1475
2812.500	-82.6574	2890.625	-83.1414	2968.750	-83.5880	3046.875	-84.0101
3125.000	-84.4137	3203.125	-84.7733	3281.250	-85.1392	3359.375	-85.4598
3437.500	-85.7784	3515.625	-86.0466	3593.750	-86.3405	3671.875	-86.5748
3750.000	-86.8187	3828.125	-87.0359	3906.250	-87.2378	3984.375	-87.4188
4062.500	-87.6011	4140.625	-87.7457	4218.750	-87.8939	4296.875	-88.0162
4375.000	-88.1343	4453.125	-88.2224	4531.250	-88.3248	4609.375	-88.3804
4687.500	-88.4489	4765.625	-88.4872	4843.750	-88.5287	4921.875	-88.5380
5000.000	-88.5555						

CORRELATION FUNCTION

LAG	CORR	LAG	CORR	LAG	CORR	LAG	CORR	LAG	CORR
0	.500E 00	1	.404E 00	2	.154E 00	3	-.154E 00	4	-.402E 00
5	-.496E 00	6	-.399E 00	7	-.151E 00	8	.152E 00	9	.394E 00
10	.484E 00	11	.388E 00	12	.146E 00	13	-.147E 00	14	-.380E 00
15	-.464E 00	16	-.370E 00	17	-.138E 00	18	.141E 00	19	.361E 00
20	.438E 00	21	.347E 00	22	.127E 00	23	-.134E 00	24	-.337E 00
25	-.406E 00	26	-.319E 00	27	-.114E 00	28	.127E 00	29	.312E 00
30	.371E 00	31	.287E 00	32	.981E-01	33	-.121E 00	34	-.285E 00
35	-.334E 00	36	-.253E 00	37	-.800E-01	38	.116E 00	39	.259E 00
40	.296E 00	41	.218E 00	42	.595E-01	43	-.114E 00	44	-.235E 00
45	-.259E 00	46	-.181E 00	47	-.368E-01	48	.115E 00	49	.215E 00
50	.225E 00	51	.145E 00	52	.117E-01	53	-.121E 00	54	-.199E 00
55	-.194E 00	56	-.110E 00	57	.157E-01	58	.131E 00	59	.190E 00
60	.168E 00	61	.769E-01	62	-.452E-01	63	-.147E 00	64	-.187E 00

Appendix

```
C
C-----------------------------------------------------------------------
C MAIN PROGRAM: MODIFIED PERIODOGRAM METHOD FOR POWER SPECTRUM
C               ESTIMATION-PMPSE
C AUTHORS:      L R RABINER
C               BELL LABORATORIES, MURRAY HILL, NEW JERSEY 07974
C               R W SCHAFER AND D DLUGOS
C               DEPT OF ELECTRICAL ENGINEERING
C               GEORGIA INSTITUTE OF TECHNOLOGY
C               ATLANTA, GEORGIA 30332
C
C METHOD BASED ON TECHNIQUE DESCRIBED BY P D WELCH, IEEE
C TRANS ON AUDIO AND ELECT, VOL 15, NO 2, PP 70-73, 1967.
C
C INPUT:        M IS THE FFT LENGTH (MUST BE A POWER OF 2)
C                   2<= M <= MAXM (1024)
C               IWIN IS THE WINDOW TYPE
C                   IWIN = 1  RECTANGULAR WINDOW
C                   IWIN = 2  HAMMING WINDOW
C               L IS THE WINDOW LENGTH
C                   L <= M
C               N IS THE MAXIMUM NUMBER OF SAMPLES AVAILABLE
C                   FOR ANALYSIS
C               FS IS THE SAMPLING FREQUENCY IN HZ
C               IMD REQUESTS ADDITIONAL RUNS
C                   IMD = 1   NEW RUN
C                   IMD = 0   TERMINATES PROGRAM\
C-----------------------------------------------------------------------
C
      DIMENSION XA(1024), XFR(513), SXX(513), WD(1024)
      DIMENSION JWIN(2,4)
      DIMENSION ILAG(513)
      COMPLEX X(1024), XMN
      DATA JWIN(1,1), JWIN(1,2), JWIN(1,3), JWIN(1,4) /1HR,1HE,1HC,1HT/
      DATA JWIN(2,1), JWIN(2,2), JWIN(2,3), JWIN(2,4) /1HH,1HA,1HM,1HG/
C
C DEFINE I/O DEVICE CODES
C INPUT: INPUT TO THIS IS USER-INTERACTIVE
C        THAT IS - A QUESTION IS WRITTEN ON THE USER
C        TERMINAL (IOUT1) AND THE USER TYPES IN THE ANSWER.
C
C OUTPUT: ALL OUTPUT IS WRITTEN ON THE STANDARD
C         OUTPUT UNIT (IOUT2).
C
      IND = I1MACH(1)
      IOUT1 = I1MACH(4)
      IOUT2 = I1MACH(2)
C
C SET MAXIMUM FFT SIZE MAXM DEPENDING ON DIMENSION WITHIN PROGRAM
C
      MAXM = 1024
      LHM = MAXM/2 + 1
C
C FILL LAG ARRAY FOR PRINTING
C
      DO 10 I=1,LHM
        ILAG(I) = I - 1
  10  CONTINUE
  20  CONTINUE
C
C READ IN ANALYSIS PARAMETERS M,IWIN,L,N,FS
C
      WRITE (IOUT1,9999)
9999  FORMAT (16H FFT LENGTH(I4)=)
      READ (IND,9997) M
      IF (M.GT.MAXM) WRITE (IOUT1,9998)
9998  FORMAT (27H M TOO LARGE--REENTER VALUE)
      IF (M.GT.MAXM) GO TO 20
9997  FORMAT (I4)
      WRITE (IOUT1,9996)
9996  FORMAT (43H WINDOW TYPE(I1)   1=RECTANGULAR, 2=HAMMING)
      READ (IND,9995) IWIN
9995  FORMAT (I1)
      WRITE (IOUT1,9994)
9994  FORMAT (19H WINDOW LENGTH(I4)=)
      READ (IND,9997) L
      WRITE (IOUT1,9993)
9993  FORMAT (40H MAXIMUM NUMBER OF ANALYSIS SAMPLES(I5)=)
      READ (IND,9992) N
9992  FORMAT (I5)
      WRITE (IOUT1,9991)
9991  FORMAT (33H SAMPLING FREQUENCY IN HZ(F10.4)=)
      READ (IND,9990) FS
9990  FORMAT (F10.4)
C
C NSECT = THE TOTAL NUMBER OF ANALYSIS SECTIONS
C    NP = THE TOTAL NUMBER OF SAMPLES ACTUALLY USED
C OVERLAP OF 2 TO 1 IS USED ON ADJACENT ANALYSIS SECTIONS
C    NP = N IF (N-L/2)/(L/2) = AN INTEGER
C
      MHLF1 = M/2 + 1
      NSECT = (N-L/2)/(L/2)
      NP = NSECT*(L/2) + L/2
      WRITE (IOUT2,9989) JWIN(IWIN,1), JWIN(IWIN,2), JWIN(IWIN,3),
     *    JWIN(IWIN,4)
9989  FORMAT (13H WINDOW TYPE=, 4A1)
      WRITE (IOUT2,9987)
      WRITE (IOUT2,9988) M, NP, L, FS
9988  FORMAT (3H M=, I4, 5H  NP=, I5, 4H  L=, I4, 18H  SAMPLING FREQUEN,
     *    3HCY=, F10.4)
C
C CALCULATE MEAN OF DATA.
C
      SS = 1.
      XSUM = 0.
      NS1 = NSECT + 1
      L1 = L/2
      DO 40 K=1,NS1
        CALL GETX(XA, L, SS)
        DO 30 I=1,L1
          XSUM = XSUM + XA(I)
  30    CONTINUE
        SS = SS + FLOAT(L1)
  40  CONTINUE
      XMEAN = XSUM/FLOAT(NP)
      XMN = CMPLX(XMEAN,XMEAN)
      WRITE (IOUT2,9987)
9987  FORMAT (//)
      WRITE (IOUT2,9986) XMEAN
9986  FORMAT (7H XMEAN=, E14.5)
C
C GENERATE WINDOW
C
      U = FLOAT(L)
      IF (IWIN.NE.2) GO TO 60
      U = 0.
      FL = FLOAT(L-1)
      TWOPI = 8.*ATAN(1.0)
      DO 50 I=1,L
        FI = FLOAT(I-1)
```

```
          WD(I) = .54 - .46*COS(TWOPI*FI/FL)
          U = U + WD(I)*WD(I)
   50   CONTINUE
   60   CONTINUE
C
C LOOP TO ACCUMULATE SPECTRA 2 AT A TIME
C
        SS = 1.
        DO 70 I=1,MHLF1
          SXX(I) = 0.
   70   CONTINUE
C
C READ L/2 SAMPLES TO INITIALIZE BUFFER
C
        NRD = L/2
        L2 = L/2 + 1
        CALL GETX(XA(L2), NRD, SS)
        SS = SS + FLOAT(NRD)
        IMN = L/2 + 1
        KMX = (NSECT+1)/2
        NSECTP = (NSECT+1)/2
        NRD = L
        DO 190 K=1,KMX
C
C MOVE DOWN UPPER HALF OF XA BUFFER
C
          DO 80 I=1,L1
            J = L1 + I
            X(I) = CMPLX(XA(J),0.)
   80     CONTINUE
          IF (K.NE.KMX .OR. NSECTP.EQ.NSECT) GO TO 100
          DO 90 I=IMN,NRD
            XA(I) = 0.0
   90     CONTINUE
          NRD = L/2
  100     CALL GETX(XA, NRD, SS)
          DO 110 I=1,L1
            J = I + L1
            X(J) = CMPLX(XA(I),XA(J)) - XMN
            X(I) = CMPLX(REAL(X(I)),XA(I)) - XMN
  110     CONTINUE
          IF (K.NE.KMX .OR. NSECTP.EQ.NSECT) GO TO 130
C
C AN ODD NUMBER OF SECTIONS--ZERO OUT THE SECOND PART
C
          DO 120 I=1,L
            X(I) = CMPLX(REAL(X(I)),0.)
  120     CONTINUE
  130     CONTINUE
          SS = SS + FLOAT(NRD)
          IF (IWIN.NE.2) GO TO 150
          DO 140 I=1,L
            X(I) = X(I)*WD(I)
  140     CONTINUE
  150     CONTINUE
          IF (L.EQ.M) GO TO 170
          LP1 = L + 1
          DO 160 I=LP1,M
            X(I) = (0.,0.)
  160     CONTINUE
  170     CONTINUE
          CALL FFT(X, M, 0)
          DO 180 I=2,MHLF1
            J = M + 2 - I
            SXX(I) = SXX(I) + REAL(X(I)*CONJG(X(I))+X(J)*CONJG(X(J)))
  180     CONTINUE
          SXX(1) = SXX(1) + REAL(X(1)*CONJG(X(1)))*2.
  190   CONTINUE
C
C NORMALIZE SPECTRAL ESTIMATE AND OBTAIN CORRELATION FUNCTION
C USING INVERSE FFT OF POWER SPECTRUM
C
        FNORM = 2.*U*FLOAT(NSECT)
        DO 200 I=1,MHLF1
          SXX(I) = SXX(I)/FNORM
          X(I) = CMPLX(SXX(I),0.)
          J = M + 2 - I
          X(J) = X(I)
  200   CONTINUE
        CALL FFT(X, M, 1)
        DO 210 I=1,MHLF1
          XA(I) = REAL(X(I))
  210   CONTINUE
C
C CORRELATION ESTIMATE IS IN XA FROM 1 TO MHLF1
C COMPUTE LOG OF POWER SPECTRUM ESTIMATE
C
        XFS = FS/FLOAT(M)
        DO 220 I=1,MHLF1
          XFR(I) = FLOAT(I-1)*XFS
          TMP = ALOG10(SXX(I))
          SXX(I) = 20.*TMP
  220   CONTINUE
C
C LOG POWER SPECTRUM (DB) IS IN ARRAY SXX
C IF DESIRED, THE USER MAY INSERT CODE AT THIS POINT TO PLOT THE LOG
C POWER SPECTRUM
C
        WRITE (IOUT2,9987)
        WRITE (IOUT2,9985)
 9985   FORMAT (19H LOG POWER SPECTRUM)
        WRITE (IOUT2,9987)
        WRITE (IOUT2,9984)
 9984   FORMAT (5X, 4HFREQ, 7X, 2HDB, 5X, 4HFREQ, 7X, 2HDB, 5X, 4HFREQ,
     *     7X, 2HDB, 5X, 4HFREQ, 7X, 2HDB)
        WRITE (IOUT2,9983) (XFR(I),SXX(I),I=1,MHLF1)
 9983   FORMAT (4(F9.3, F9.4))
C
C CORRELATION FUNCTION IS IN ARRAY XA
C IF DESIRED, THE USER MAY INSERT CODE AT THIS POINT TO PLOT THE
C CORRELATION FUNCTION
C
        WRITE (IOUT2,9987)
        WRITE (IOUT2,9982)
 9982   FORMAT (21H CORRELATION FUNCTION)
        WRITE (IOUT2,9987)
        WRITE (IOUT2,9981)
 9981   FORMAT (1X, 3HLAG, 2X, 4HCORR, 5X, 3HLAG, 2X, 4HCORR, 5X, 3HLAG,
     *     2X, 4HCORR, 5X, 3HLAG, 2X, 4HCORR, 5X, 3HLAG, 2X, 4HCORR)
        WRITE (IOUT2,9980) (ILAG(I),XA(I),I=1,MHLF1)
 9980   FORMAT (5(I4, E10.3))
        WRITE (IOUT2,9987)
        WRITE (IOUT2,9979)
 9979   FORMAT (////)
        WRITE (IOUT1,9978)
 9978   FORMAT (23H MORE DATA(1=YES,0=NO)=)
        READ (IND,9995) IMD
        IF (IMD.EQ.1) GO TO 20
        STOP
        END
```

SXX(1)/FNORM
X(1) = CMPLX(SXX(1), 0.)
DO 200 I=2, MHLF1

```
C
C-----------------------------------------------------------------------
C SUBROUTINE:  GETX
C READ DATA ROUTINE GENERATE X(N) FOR A SINE INPUT OF FREQUENCY FS/10.
C WHERE FS IS THE SAMPLING FREQUENCY IN HZ
C-----------------------------------------------------------------------
C
      SUBROUTINE GETX(X, NRD, SS)
      DIMENSION X(1)
C
C   X = ARRAY OF SIZE NRD TO HOLD GENERATOR OUTPUT DATA
C NRD = NUMBER OF SAMPLES TO BE CREATED
C  SS = STARTING SAMPLE OF GENERATOR OUTPUT
C SINE WAVE FREQUENCY IS 1000 HZ WITH AN ASSUMED SAMPLING FREQUENCY
C OF 10000 HZ
C
      TPI = 8.*ATAN(1.0)
      CF = 1000./10000.
      DO 10 I=1,NRD
        XSMP = (SS-1.) + FLOAT(I-1)
        X(I) = COS(TPI*CF*XSMP)
  10  CONTINUE
      RETURN
      END
C
C-----------------------------------------------------------------------
C SUBROUTINE:  FFT
C JIM COOLEY'S SIMPLE FFT PROGRAM--USES DECIMATION IN TIME ALGORITHM
C X IS AN N=2**M POINT COMPLEX ARRAY THAT INITIALLY CONTAINS THE INPUT
C AND ON OUTPUT CONTAINS THE TRANSFORM
C THE PARAMETER INV SPECIFIED DIRECT TRANSFORM IF 0 AND INVERSE IF 1
C-----------------------------------------------------------------------
C
      SUBROUTINE FFT(X, N, INV)
      COMPLEX X(1), U, W, T, CMPLX
C
C   X = COMPLEX ARRAY OF SIZE N--ON INPUT X CONTAINS
C       THE SEQUENCE TO BE TRANSFORMED
C       ON OUTPUT X CONTAINS THE DFT OF THE INPUT
C   N = SIZE OF FFT TO BE COMPUTED--N=2**M FOR 1.LE.M.LE.15
C INV = PARAMETER TO DETERMINE WHETHER TO DO A DIRECT TRANSFORM (INV=0)
C       OR AN INVERSE TRANSFORM (INV=1)
C
      M = ALOG(FLOAT(N))/ALOG(2.) + .1
      NV2 = N/2
      NM1 = N - 1
      J = 1
      DO 40 I=1,NM1
        IF (I.GE.J) GO TO 10
        T = X(J)
        X(J) = X(I)
        X(I) = T
  10    K = NV2
  20    IF (K.GE.J) GO TO 30
        J = J - K
        K = K/2
        GO TO 20
  30    J = J + K
  40  CONTINUE
      PI = 4.*ATAN(1.0)
      DO 70 L=1,M
        LE = 2**L
        LE1 = LE/2
        U = (1.0,0.0)
        W = CMPLX(COS(PI/FLOAT(LE1)),-SIN(PI/FLOAT(LE1)))
        IF (INV.NE.0) W = CONJG(W)
        DO 60 J=1,LE1
          DO 50 I=J,N,LE
            IP = I + LE1
            T = X(IP)*U
            X(IP) = X(I) - T
            X(I) = X(I) + T
  50      CONTINUE
          U = U*W
  60    CONTINUE
  70  CONTINUE
      IF (INV.EQ.0) RETURN
      DO 80 I=1,N
        X(I) = X(I)/CMPLX(FLOAT(N),0.)
  80  CONTINUE
      RETURN
      END
```

2.2

Correlation Method for Power Spectrum Estimation

L. R. Rabiner

Acoustics Research Dept.
Bell Laboratories
Murray Hill, NJ 07974

R. W. Schafer and *D. Dlugos*

Dept. of Electrical Engineering
Georgia Institute of Technology
Atlanta, GA 30332

1. Purpose

This program can be used to provide an estimate of either the autocorrelation function of a signal, or the cross-correlation between two signals using FFT techniques. An estimate of the power spectrum of the signal (obtained by computing the discrete Fourier transform of a windowed version of the correlation function) is also obtained in the program.

1.1 Method

The program is an implementation of the procedure originally described by Rader [1]. Figure 1 shows a block diagram of the signal

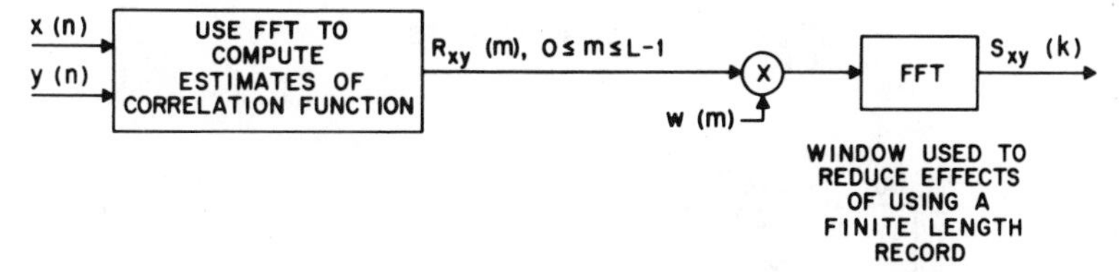

Fig. 1 Block diagram of the correlation method of power spectrum analysis.

processing within the program. In the general case there are two input signals, $x(n)$ and $y(n)$, which are defined for the interval $0 \leqslant n \leqslant N-1$. Outside this interval it is assumed that $x(n)$ and $y(n)$ are 0. We wish to compute the correlation function

$$R_{xy}(m) = \frac{1}{N} \sum_{n=0}^{N-1-m} (x(n)-\bar{x})(y(n+m)-\bar{y}) \qquad 0 \leqslant m \leqslant L-1, \tag{1}$$

where $\bar{x}$ and $\bar{y}$ are the estimated means of $x(n)$ and $y(n)$, i.e.,

$$\bar{x} = \frac{1}{N} \sum_{n=0}^{N-1} x(n) \tag{2a}$$

$$\bar{y} = \frac{1}{N} \sum_{n=0}^{N-1} y(n) \; . \tag{2b}$$

The program offers the option of subtracting the estimated means or not. If the means are subtracted an estimated *covariance function* is obtained. If the means are *not* subtracted, an estimated *correlation* function is obtained. For simplicity we shall use the term correlation function for both cases unless this would lead to confusion. Since, in general, $N >> L$, i.e., the number of samples of the signal is large compared to the number of correlation lags of interest, an efficient procedure for implementing Eq. (1) is to use FFT methods to accumulate partial correlation sums (being careful to avoid the aliasing effects of the circular correlation from the FFT), and then inverse transform the result to give the desired correlation function. Details of this procedure are given in Refs. [1-3]. It should be noted that for this method if the size of the FFT is M points, then $M/2+1$ values of the correlation function are obtained. Thus, the FFT size must be at least twice the desired number of correlation values, L.

To obtain the power spectrum estimate from the correlation function one of two procedures is used. If $y(n) = x(n)$, i.e., the correlation is an autocorrelation (autocovariance), the windowed autocorrelation function, $\tilde{R}_{xx}(m)$, is created as

$$\begin{aligned} \tilde{R}_{xx}(m) &= R_{xx}(m)w(m) && 0 \leqslant m \leqslant L-1 \\ &= R_{xx}(M-m)w(M-m) && M-L+1 \leqslant m \leqslant M-1 \\ &= 0 && \text{otherwise} \end{aligned} \tag{3}$$

where $w(m)$ is a symmetric finite duration window which tapers $R_{xx}(m)$ to 0 as m tends to $L-1$. Equation (3) shows that $\tilde{R}_{xx}(m)$ is a symmetric function so that the resulting DFT of the sequence of Eq. (3) is a real function which is a smoothed estimate of the power spectrum $S_{xx}(e^{j\omega})$.

If $y(n) \neq x(n)$, i.e., the correlation is a cross-correlation (cross-covariance), the signal $\tilde{R}_{xy}(m)$ is created as

$$\begin{aligned} \tilde{R}_{xy} &= R_{xy}(m)w(m) && 0 \leqslant m \leqslant L-1 \\ &= 0 && \text{otherwise} \end{aligned} \tag{4}$$

where $w(m)$ again represents a finite duration window. In general $w(m)$ is a rectangular window since the "origin" of $R_{xy}(m)$ need not be at $m = 0$; thus arbitrarily tapering $R_{xy}(m)$ to 0 can lead to erroneous estimates of the cross power spectrum. [For further discussion of this important issue see Ref. 2, p. 409-410]. The cross power spectrum estimate is again obtained as the DFT of the sequence of Eq. (4).

2. Program Description

The program flow is shown in Fig. 2.

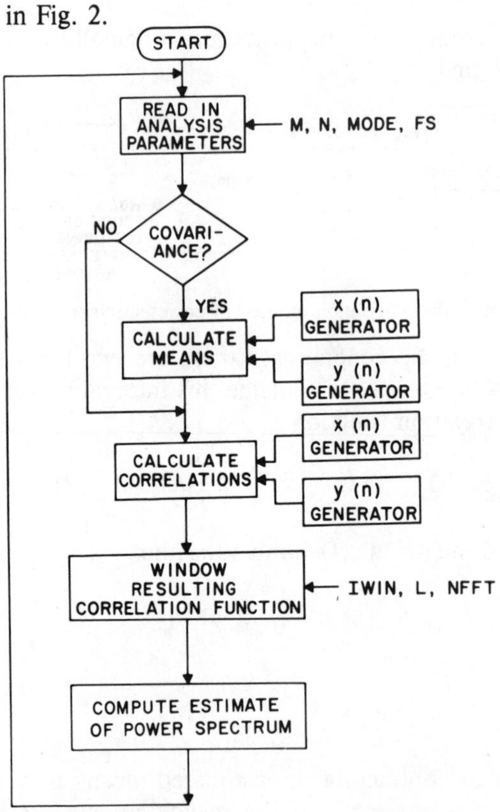

START
READ IN ANALYSIS PARAMETERS ← M, N, MODE, FS
COVARIANCE? (NO / YES)
CALCULATE MEANS ← x (n) GENERATOR, y (n) GENERATOR
CALCULATE CORRELATIONS ← x (n) GENERATOR, y (n) GENERATOR
WINDOW RESULTING CORRELATION FUNCTION ← IWIN, L, NFFT
COMPUTE ESTIMATE OF POWER SPECTRUM

Fig. 2 Flow diagram of program, CMPSE.

First the program reads in the parameters for performing the correlation analysis, namely the section size (FFT size), M, the total number of samples used in the estimate, N, and the type of correlation desired, i.e., autocorrelation, cross-correlation, autocovariance, or cross-covariance (as determined by the variable MODE). If a covariance function is desired, the next step is calculation of the means of $x(n)$ (and possibly $y(n)$) using the user provided subroutine(s) GETX (and GETY) which fill buffers with a block of samples of $x(n)$ (and $y(n)$) from a prescribed starting sample. (For simplicity we assume the N samples of $x(n)$ are numbered from 1 to N). Following computation of the means, the correlation function of Eq. (1) is determined by proper sectioning of $x(n)$ into blocks of M samples which overlap by M/2 samples. The implementation of Eq. (1) is done by computing M-point FFT's of the individual sections of $x(n)$ (and $y(n)$), accumulating sums in the frequency domain (for computational efficiency), and inverse transforming the final result back to the time domain.

Following the correlation computation the program prints the first (M/2+1) correlation values. To obtain a power spectrum estimate, window parameters are then read in and the appropriate windowed correlation function (either Eq. (3) or Eq. (4)) is computed. The window parameters are window type (IWIN - either rectangular or Hamming), window duration (L), and size of FFT used to obtain the spectral estimate (NFFT). The program then prints out the log spectral estimate in dB.

2.1 Usage

The program consists of a main program (CMPSE), two user supplied subroutines (GETX and GETY) for accessing blocks of data, and a complex FFT subroutine (FFT). The main program reads in the appropriate parameters for the analysis, and the program computes the correlation function and prints its values, and from this array it computes an estimate of the log power spectrum (dB), and prints its values. Since the FFT subroutine is a radix 2 complex FFT, the section size must be a power of 2. Although no graphical capability is provided with the program, it is suggested the user insert appropriate graphics displays at the points denoted in the program to facilitate use of the program. Typical graphical output is shown in the examples with this program description.

2.2 Description of Program Parameters

The main program reads the following parameters:

M Section size. M must be an integer that is a power of 2 (for compatibility with the FFT subroutine) and M must be less than or equal to a maximum section size MAXM as specified in the main program. (For this program MAXM is set to 512). Note that M/2+1 must be greater than or equal to the total number of lags desired.

N Total number of samples of $x(n)$ (and $y(n)$) used in the analysis. It is assumed that samples of $x(n)$ and $y(n)$ are numbered from 1 to N. (Note: For computers with 16-bit integer word lengths, N must be less than 32768.)

MODE Parameter which determines the type of correlation to be computed.

MODE = 0 Autocorrelation
MODE = 1 Cross-correlation
MODE = 2 Autocovariance
MODE = 3 Cross-covariance

FS Sampling frequency in Hz.

IWIN Type of window used on correlation function. If IWIN = 1 a rectangular window is used; if IWIN = 2 a Hamming window is used.

L Number of correlation points used in estimating the power spectrum. L must be less than or equal to M/2+1.

NFFT FFT size (power of 2) used to compute spectral estimate from windowed correlation. NFFT must be greater than or equal to 2*L − 1.

The user must supply the two routines GETX and GETY which provide values of $x(n)$ and $y(n)$ to the main program. The calls to these routines are

CALL GETX (X,NRD,SS)
CALL GETY (Y,NRD,SS)

where the parameters X, Y, NRD, and SS are

X Array in which the samples of $x(n)$ are stored

Y Array in which the samples of $y(n)$ are stored

NRD Number of samples of $x(n)$ (or $y(n)$) to be read into array X (or Y)

SS Starting sample number of $x(n)$ (or $y(n)$), i.e., samples $x(\mathrm{SS})$ to $x(\mathrm{SS} + \mathrm{NRD} - 1)$ are read into array X in the subroutine GETX. Note that SS is a floating point number to allow for maximum flexibility. The first input sample is assumed to be $x(1)$, i.e., SS = 1.

The main program, CMPSE, and test subroutines GETX and GETY are given in the Appendix. The subroutine FFT is a complex FFT subroutine with calling sequence CALL FFT(X,M,INV), where

M = Size of FFT (a power of 2)
X = Complex array of length M.
INV = 0 for direct FFT
= 1 for inverse FFT

2.3 Dimension Requirements

The real array XA and the complex array X are dimensioned to size MAXM, where MAXM is the largest size FFT to be used in any run. The array XA is a data buffer and the array X is the working array for the FFT computation. The complex array Z is used to accumulate the transform of the correlation estimate. The real array XFR is used to generate a frequency scale corresponding to the DFT sample frequencies; the real array ILAG is used to generate a set of lag indices for printing. The arrays Z, XFR, and ILAG are dimensioned to (MAXM/2+1).

2.4 Summary of User Requirements

(1) Provide the subroutines GETX and GETY which specify $x(n)$ and $y(n)$ for the analysis to be performed. (For accessing data from a disk file, etc., the user must perform any necessary initialization for reading the file.)

(2) Provide the complex FFT subroutine FFT.

(3) Specify the correlation parameters M, N, MODE, and FS, and the spectrum estimation parameters IWIN, L, and NFFT.

(4) Provide a graphical output to plot the correlation function and/or the power spectrum estimate.

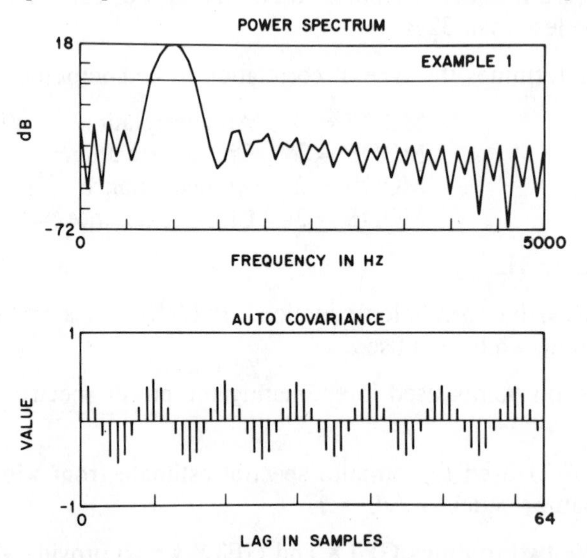

Fig. 3 Plots of the log power spectrum and the autocovariance for Example #1, using a Hamming window.

3. Test Problems

To illustrate the use of this program a simple test problem is provided in which

$$x(n) = \cos(2\pi n/10)$$

and

$$y(n) = \sin(2\pi n/10)$$

These sequences are sinusoids with frequency FS/10., where FS is the sampling frequency.

Example #1

For this example the program parameters are M = 128, N = 256, MODE = 2, FS = 10000, and the correlation parameters are IWIN = 2 (Hamming window), L = 32, and NFFT = 128. The numerical output of this example (i.e., the first M/2 + 1 points of the correlation function, and the M/2 + 1 points of the log power spectrum) are given in Table 1. Figure 3 shows the autocovariance function and the log power spectrum for this example. Note the slight linear taper in the autocovariance function estimate. This is due to the fact that only a finite number (256) of samples were used in the estimate.

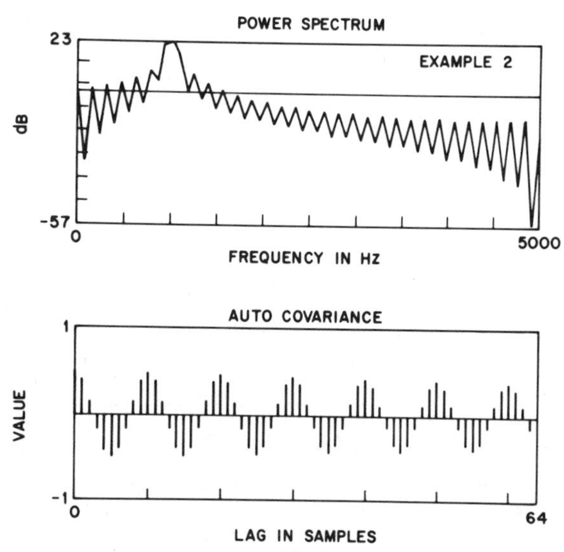

Fig. 4 Plots of the log power spectrum and the autocovariance for Example #2, using a rectangular window.

Example #2

For this example M = 128, N = 256, MODE = 2, FS = 10000, IWIN = 1 (rectangular window), L = 32 and NFFT = 128. The numerical results are in Table 2, and Figure 4 shows the autocovariance function and log power spectrum. A comparison of Figs. 3 and 4 shows the well known fact that the Hamming window has a "main lobe" about twice as wide as the same size rectangular window, but much lower side lobes.

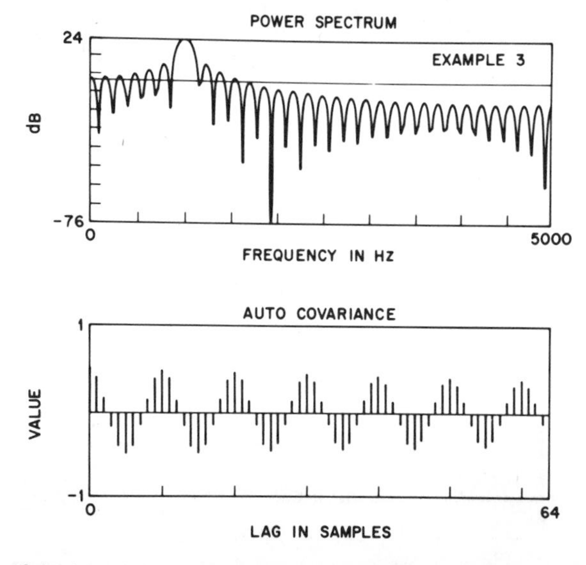

Fig. 5 Plots of the log power spectrum and the autocovariance for Example #3, using the same parameters as Example #2 except NFFT = 512.

Example #3

This case is identical to Example #2, except the FFT size, NFFT, was 512 instead of 128. Figure 5 shows the autocovariance and log power spectrum and Table 3 gives the numerical results. Note that a much smoother looking plot is obtained. However, it is well to point out that this is *not* a more accurate representation of the true spectrum, but rather a more detailed representation of the *estimate* of the spectrum.

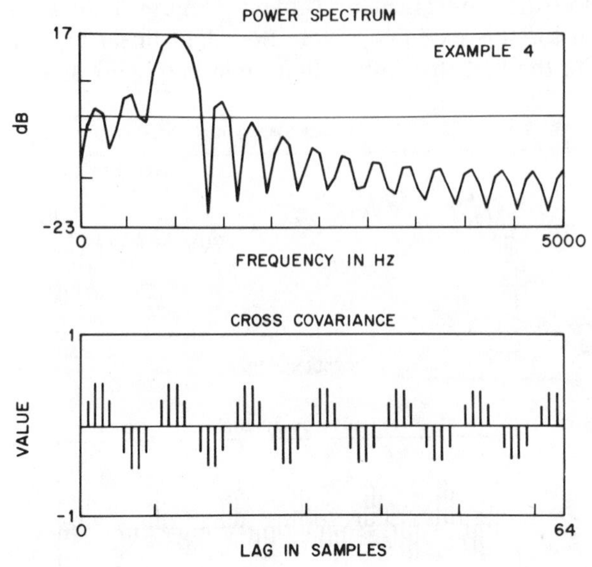

Fig. 6 Plots of the log power spectrum and the cross-covariance for Example #4 using a rectangular window.

Example #4

For this example M = 128, N = 256, MODE = 3, FS = 10000, IWIN = 1, L = 32, and NFFT = 128. The numerical results are in Table 4, and Figure 6 shows the cross-covariance function and log power spectrum. Note that $y(n) = x(n-2.5)$ so that $R_{xy}(m)$ should be roughly the same as the autocovariance function of Example #2 but shifted to the right by 2.5 samples. This assertion is supported by a comparison of Figs. 4 and 6 with an allowance for the slightly different linear tapering and the half sample shift (interpolation).

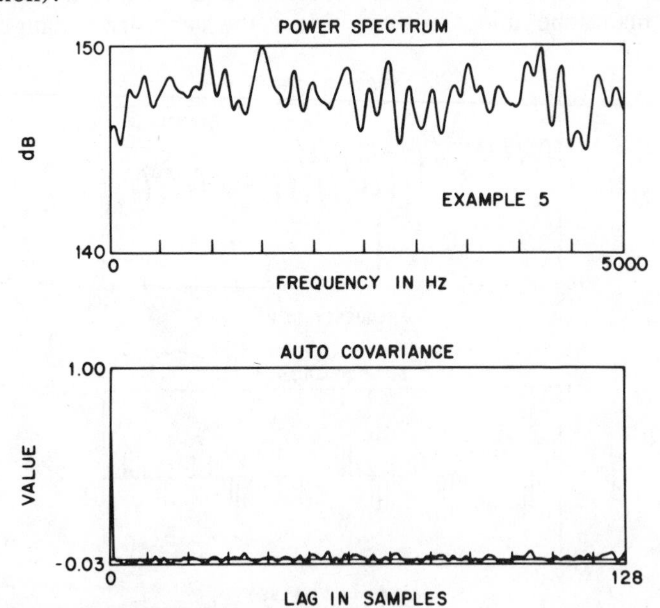

Fig. 7 Plots of the log power spectrum and the autocovariance for Example #5.

Example #5

For this example the signal generators were programmed to provide samples of a white Gaussian noise (program not provided). Figure 7 shows plots of the power spectrum and autocovariance for the noise with the following parameters:

$$M = 256, \quad N = 4096, \quad MODE = 2, \quad FS = 10000,$$
$$IWIN = 2, \quad L = 100, \quad NFFT = 1024.$$

A variation of about 5 dB is seen in the power spectrum for this theoretically white noise input.

References

1. C. M. Rader, "An Improved Algorithm for High Speed Autocorrelation with Applications to Spectral Estimation," *IEEE Trans. Audio Electroacoust.,* Vol. AU-18, No. 4, pp. 439-442, 1970.
2. L. R. Rabiner, and B. Gold, *Theory and Application of Digital Signal Processing,* Chapter 6, pp. 399-419, Prentice-Hall, Inc., Englewood Cliffs, N.J., 1975.
3. A. V. Oppenheim and R. W. Schafer, *Digital Signal Processing,* Chapter 11, pp. 556-562, Prentice-Hall, Inc., Englewood Cliffs, N.J., 1975.

Table 1

```
M= 128   N=  256  MODE=2  SAMPLING FREQUENCY=10000.0000
XMEAN=     .69337E-06  YMEAN=     .69337E-06
CORRELATION FUNCTION
LAG   CORR      LAG   CORR      LAG   CORR      LAG   CORR      LAG   CORR
  0   .502E 00    1   .403E 00    2   .151E 00    3 -.156E 00    4 -.401E 00
  5 -.492E 00    6 -.395E 00    7 -.148E 00    8   .153E 00    9   .393E 00
 10   .482E 00   11   .387E 00   12   .145E 00   13 -.150E 00   14 -.386E 00
 15 -.473E 00   16 -.379E 00   17 -.142E 00   18   .147E 00   19   .378E 00
 20   .463E 00   21   .371E 00   22   .139E 00   23 -.144E 00   24 -.370E 00
 25 -.453E 00   26 -.363E 00   27 -.136E 00   28   .141E 00   29   .362E 00
 30   .443E 00   31   .356E 00   32   .133E 00   33 -.138E 00   34 -.354E 00
 35 -.434E 00   36 -.348E 00   37 -.130E 00   38   .135E 00   39   .346E 00
 40   .424E 00   41   .340E 00   42   .127E 00   43 -.132E 00   44 -.338E 00
 45 -.414E 00   46 -.332E 00   47 -.124E 00   48   .129E 00   49   .330E 00
 50   .404E 00   51   .324E 00   52   .121E 00   53 -.126E 00   54 -.322E 00
 55 -.395E 00   56 -.316E 00   57 -.118E 00   58   .123E 00   59   .314E 00
 60   .385E 00   61   .308E 00   62   .115E 00   63 -.120E 00   64 -.307E 00
WINDOW TYPE=HAMG   NO OF WINDOW VALUES=   32   FFT SIZE= 128
LOG POWER SPECTRUM
     FREQ          DB       FREQ          DB       FREQ          DB       FREQ          DB
     .000   -21.4636     78.125   -52.2094    156.250   -21.9584    234.375   -51.8096
  312.500   -21.1400    390.625   -36.8481    468.750   -24.2275    546.875   -38.4634
  625.000   -29.4239    703.125   -11.0793    781.250     3.9527    859.375    12.6998
  937.500    17.1393   1015.625    18.1234   1093.750    15.8463   1171.875     9.9801
 1250.000     -.5387   1328.125   -17.6599   1406.250   -32.1351   1484.375   -42.3509
 1562.500   -38.4645   1640.625   -25.3551   1718.750   -24.1909   1796.875   -36.6632
 1875.000   -29.6829   1953.125   -29.4415   2031.250   -26.0121   2109.375   -38.0924
 2187.500   -30.1628   2265.625   -32.3922   2343.750   -27.7101   2421.875   -39.9763
 2500.000   -31.0733   2578.125   -34.7852   2656.250   -29.0482   2734.375   -41.9843
 2812.500   -31.8905   2890.625   -36.8850   2968.750   -30.0797   3046.875   -44.1559
 3125.000   -32.5535   3203.125   -38.8437   3281.250   -30.8688   3359.375   -46.6298
 3437.500   -33.0708   3515.625   -40.7724   3593.750   -31.4660   3671.875   -49.6887
 3750.000   -33.4647   3828.125   -42.7547   3906.250   -31.8997   3984.375   -54.1556
 4062.500   -33.7648   4140.625   -44.9129   4218.750   -32.2154   4296.875   -62.3889
 4375.000   -33.9483   4453.125   -47.4943   4531.250   -32.4179   4609.375   -67.1280
 4687.500   -34.0617   4765.625   -50.7562   4843.750   -32.5142   4921.875   -55.6432
 5000.000   -34.0996
```

Table 2

M= 128 N= 256 MODE=2 SAMPLING FREQUENCY=10000.0000
XMEAN= .69337E-06 YMEAN= .69337E-06
CORRELATION FUNCTION

LAG	CORR	LAG	CORR	LAG	CORR	LAG	CORR	LAG	CORR
0	.502E 00	1	.403E 00	2	.151E 00	3	-.156E 00	4	-.401E 00
5	-.492E 00	6	-.395E 00	7	-.148E 00	8	.153E 00	9	.393E 00
10	.482E 00	11	.387E 00	12	.145E 00	13	-.150E 00	14	-.386E 00
15	-.473E 00	16	-.379E 00	17	-.142E 00	18	.147E 00	19	.378E 00
20	.463E 00	21	.371E 00	22	.139E 00	23	-.144E 00	24	-.370E 00
25	-.453E 00	26	-.363E 00	27	-.136E 00	28	.141E 00	29	.362E 00
30	.443E 00	31	.356E 00	32	.133E 00	33	-.138E 00	34	-.354E 00
35	-.434E 00	36	-.348E 00	37	-.130E 00	38	.135E 00	39	.346E 00
40	.424E 00	41	.340E 00	42	.127E 00	43	-.132E 00	44	-.338E 00
45	-.414E 00	46	-.332E 00	47	-.124E 00	48	.129E 00	49	.330E 00
50	.404E 00	51	.324E 00	52	.121E 00	53	-.126E 00	54	-.322E 00
55	-.395E 00	56	-.316E 00	57	-.118E 00	58	.123E 00	59	.314E 00
60	.385E 00	61	.308E 00	62	.115E 00	63	-.120E 00	64	-.307E 00

WINDOW TYPE=RECT NO OF WINDOW VALUES= 32 FFT SIZE= 128
LOG POWER SPECTRUM

FREQ	DB	FREQ	DB	FREQ	DB	FREQ	DB
.000	1.2468	78.125	-29.4088	156.250	1.4597	234.375	-18.6579
312.500	2.2787	390.625	-13.7690	468.750	3.6083	546.875	-8.5367
625.000	6.1146	703.125	-4.2923	781.250	9.9020	859.375	5.4016
937.500	21.3157	1015.625	23.4804	1093.750	17.7880	1171.875	-.3154
1250.000	7.8799	1328.125	-2.6845	1406.250	4.0424	1484.375	-6.8817
1562.500	.7215	1640.625	-8.0628	1718.750	-1.1020	1796.875	-10.5720
1875.000	-3.0762	1953.125	-11.3458	2031.250	-4.1793	2109.375	-13.2565
2187.500	-5.5376	2265.625	-13.8561	2343.750	-6.2616	2421.875	-15.4983
2500.000	-7.2606	2578.125	-16.0150	2656.250	-7.7504	2734.375	-17.5438
2812.500	-8.5133	2890.625	-18.0259	2968.750	-8.8449	3046.875	-19.5409
3125.000	-9.4416	3203.125	-20.0211	3281.250	-9.6594	3359.375	-21.6114
3437.500	-10.1332	3515.625	-22.1209	3593.750	-10.2649	3671.875	-23.8939
3750.000	-10.6436	3828.125	-24.4750	3906.250	-10.7068	3984.375	-26.6121
4062.500	-11.0124	4140.625	-27.3203	4218.750	-11.0175	4296.875	-30.1657
4375.000	-11.2583	4453.125	-31.1973	4531.250	-11.2142	4609.375	-35.8501
4687.500	-11.4010	4765.625	-37.8602	4843.750	-11.3097	4921.875	-55.6854
5000.000	-11.4473						

Table 3

M= 128 N= 256 MODE=2 SAMPLING FREQUENCY=10000.0000
XMEAN= .69337E-06 YMEAN= .69337E-06
CORRELATION FUNCTION

LAG	CORR	LAG	CORR	LAG	CORR	LAG	CORR	LAG	CORR
0	.502E 00	1	.403E 00	2	.151E 00	3	-.156E 00	4	-.401E 00
5	-.492E 00	6	-.395E 00	7	-.148E 00	8	.153E 00	9	.393E 00
10	.482E 00	11	.387E 00	12	.145E 00	13	-.150E 00	14	-.386E 00
15	-.473E 00	16	-.379E 00	17	-.142E 00	18	.147E 00	19	.378E 00
20	.463E 00	21	.371E 00	22	.139E 00	23	-.144E 00	24	-.370E 00
25	-.453E 00	26	-.363E 00	27	-.136E 00	28	.141E 00	29	.362E 00
30	.443E 00	31	.356E 00	32	.133E 00	33	-.138E 00	34	-.354E 00
35	-.434E 00	36	-.348E 00	37	-.130E 00	38	.135E 00	39	.346E 00
40	.424E 00	41	.340E 00	42	.127E 00	43	-.132E 00	44	-.338E 00
45	-.414E 00	46	-.332E 00	47	-.124E 00	48	.129E 00	49	.330E 00
50	.404E 00	51	.324E 00	52	.121E 00	53	-.126E 00	54	-.322E 00
55	-.395E 00	56	-.316E 00	57	-.118E 00	58	.123E 00	59	.314E 00
60	.385E 00	61	.308E 00	62	.115E 00	63	-.120E 00	64	-.307E 00

Table 3
(Continued)

WINDOW TYPE=RECT NO OF WINDOW VALUES= 32 FFT SIZE= 512
LOG POWER SPECTRUM

FREQ	DB	FREQ	DB	FREQ	DB	FREQ	DB
.000	1.2468	19.531	.5363	39.063	-1.8771	58.594	-7.5362
78.125	-29.4091	97.656	-6.2786	117.188	-1.2692	136.719	.9062
156.250	1.4597	175.781	.5815	195.313	-2.1197	214.844	-8.7002
234.375	-18.6578	253.906	-4.5225	273.438	-.0885	292.969	1.8659
312.500	2.2787	332.031	1.2555	351.563	-1.7014	371.094	-9.2323
390.625	-13.7696	410.156	-2.5548	429.688	1.5086	449.219	3.3086
468.750	3.6082	488.281	2.4343	507.813	-.8658	527.344	-10.1205
546.875	-8.5368	566.406	.4372	585.938	4.1519	605.469	5.8471
625.000	6.1145	644.531	4.9176	664.063	1.5109	683.594	-8.7443
703.125	-4.2924	722.656	4.0842	742.188	7.7845	761.719	9.5558
781.250	9.9019	800.781	8.6839	820.313	4.7616	839.844	-14.9990
859.375	5.4015	878.906	12.5802	898.438	16.6793	917.969	19.4202
937.500	21.3156	957.031	22.5856	976.563	23.3390	996.094	23.6304
1015.625	23.4803	1035.156	22.8817	1054.688	21.7985	1074.219	20.1520
1093.750	17.7879	1113.281	14.3761	1132.813	9.0283	1152.344	-2.9158
1171.875	-.3156	1191.406	6.5389	1210.938	8.7332	1230.469	9.0190
1250.000	7.8799	1269.531	5.1494	1289.063	-.3682	1308.594	-23.3834
1328.125	-2.6846	1347.656	2.5766	1367.188	4.6145	1386.719	4.9980
1406.250	4.0424	1425.781	1.5587	1445.313	-3.5401	1464.844	-21.2599
1484.375	-6.8825	1503.906	-1.1278	1523.438	1.1186	1542.969	1.6235
1562.500	.7213	1582.031	-1.8174	1601.563	-7.3252	1621.094	-43.7521
1640.625	-8.0630	1660.156	-2.8873	1679.688	-.7555	1699.219	-.2531
1718.750	-1.1022	1738.281	-3.5264	1757.813	-8.7360	1777.344	-31.1360
1796.875	-10.5722	1816.406	-5.0576	1835.938	-2.8045	1855.469	-2.2432
1875.000	-3.0764	1894.531	-5.5508	1914.063	-11.0350	1933.594	-75.4598
1953.125	-11.3459	1972.656	-6.1529	1992.188	-3.9632	2011.719	-3.3955
2031.250	-4.1794	2050.781	-6.5441	2070.313	-11.7173	2089.844	-34.9787
2109.375	-13.2568	2128.906	-7.7161	2148.438	-5.4064	2167.969	-4.7792
2187.500	-5.5379	2207.031	-7.9203	2226.563	-13.2446	2246.094	-45.5196
2265.625	-13.8558	2285.156	-8.4974	2304.688	-6.2116	2324.219	-5.5614
2343.750	-6.2616	2363.281	-8.5254	2382.813	-13.5218	2402.344	-33.8661
2421.875	-15.4986	2441.406	-9.7425	2460.938	-7.3215	2480.469	-6.6007
2500.000	-7.2606	2519.531	-9.5116	2539.063	-14.5631	2558.594	-37.0923
2578.125	-16.0152	2597.656	-10.3452	2617.188	-7.9171	2636.719	-7.1583
2656.250	-7.7503	2675.781	-9.8752	2695.313	-14.5964	2714.844	-31.3533
2734.375	-17.5443	2753.906	-11.4013	2773.438	-8.8176	2792.969	-7.9756
2812.500	-8.5133	2832.031	-10.6013	2851.563	-15.3110	2871.094	-32.4305
2890.625	-18.0261	2910.156	-11.8908	2929.688	-9.2751	2949.219	-8.3826
2968.750	-8.8449	2988.281	-10.8027	3007.813	-15.1929	3027.344	-28.9483
3046.875	-19.5414	3066.406	-12.8282	3085.938	-10.0339	3105.469	-9.0455
3125.000	-9.4415	3144.531	-11.3450	3164.063	-15.6803	3183.594	-29.2885
3203.125	-20.0215	3222.656	-13.2412	3242.188	-10.3931	3261.719	-9.3441
3281.250	-9.6594	3300.781	-11.4318	3320.313	-15.4669	3339.844	-26.8918
3359.375	-21.6124	3378.906	-14.1040	3398.438	-11.0511	3417.969	-9.8941
3437.500	-10.1334	3457.031	-11.8391	3476.563	-15.7942	3496.094	-26.9421
3515.625	-22.1191	3535.156	-14.4615	3554.688	-11.3354	3574.219	-10.1094
3593.750	-10.2642	3613.281	-11.8403	3632.813	-15.5175	3652.344	-25.1417
3671.875	-23.8953	3691.406	-15.2804	3710.938	-11.9186	3730.469	-10.5723
3750.000	-10.6426	3769.531	-12.1432	3789.063	-15.7255	3808.594	-25.0527
3828.125	-24.4769	3847.656	-15.5995	3867.188	-12.1444	3886.719	-10.7210
3906.250	-10.7036	3925.781	-12.0761	3945.313	-15.4008	3964.844	-23.5957
3984.375	-26.6516	4003.906	-16.4137	4023.438	-12.6827	4042.969	-11.1259
4062.500	-11.0178	4082.031	-12.3111	4101.563	-15.5407	4121.094	-23.4887
4140.625	-27.3146	4160.156	-16.6869	4179.688	-12.8508	4199.219	-11.2141
4218.750	-11.0190	4238.281	-12.1907	4257.813	-15.1891	4277.344	-22.2846
4296.875	-30.1636	4316.406	-17.4862	4335.938	-13.3337	4355.469	-11.5516
4375.000	-11.2594	4394.531	-12.3462	4414.063	-15.2424	4433.594	-22.0957
4453.125	-31.1956	4472.656	-17.7521	4492.188	-13.4668	4511.719	-11.5964
4531.250	-11.2151	4550.781	-12.1842	4570.313	-14.8704	4589.844	-21.0559
4609.375	-35.8500	4628.906	-18.5720	4648.438	-13.9152	4667.969	-11.8858
4687.500	-11.4021	4707.031	-12.2841	4726.563	-14.8699	4746.094	-20.8460
4765.625	-37.8505	4785.156	-18.8174	4804.688	-14.0055	4824.219	-11.8830
4843.750	-11.3103	4863.281	-12.0794	4882.813	-14.4738	4902.344	-19.9108
4921.875	-55.6846	4941.406	-19.6813	4960.938	-14.4257	4980.469	-12.1289
5000.000	-11.4473						

Table 4

M= 128 N= 256 MODE=3 SAMPLING FREQUENCY=10000.0000
XMEAN= .69337E-06 YMEAN= .12023E-01
CORRELATION FUNCTION

LAG	CORR	LAG	CORR	LAG	CORR	LAG	CORR	LAG	CORR
0	-.286E-05	1	.293E 00	2	.472E 00	3	.470E 00	4	.289E 00
5	-.427E-04	6	-.287E 00	7	-.462E 00	8	-.461E 00	9	-.284E 00
10	-.588E-05	11	.281E 00	12	.453E 00	13	.451E 00	14	.278E 00
15	-.396E-04	16	-.276E 00	17	-.444E 00	18	-.442E 00	19	-.272E 00
20	-.903E-05	21	.270E 00	22	.435E 00	23	.433E 00	24	.266E 00
25	-.374E-04	26	-.264E 00	27	-.425E 00	28	-.423E 00	29	-.261E 00
30	-.120E-04	31	.258E 00	32	.416E 00	33	.414E 00	34	.255E 00
35	-.341E-04	36	-.253E 00	37	-.407E 00	38	-.405E 00	39	-.249E 00
40	-.150E-04	41	.247E 00	42	.397E 00	43	.396E 00	44	.243E 00
45	-.310E-04	46	-.241E 00	47	-.388E 00	48	-.386E 00	49	-.238E 00
50	-.168E-04	51	.235E 00	52	.379E 00	53	.377E 00	54	.232E 00
55	-.291E-04	56	-.230E 00	57	-.370E 00	58	-.368E 00	59	-.226E 00
60	-.199E-04	61	.224E 00	62	.360E 00	63	.358E 00	64	.220E 00

WINDOW TYPE=RECT NO OF WINDOW VALUES= 32 FFT SIZE= 128
LOG POWER SPECTRUM

FREQ	DB	FREQ	DB	FREQ	DB	FREQ	DB
.000	-9.1759	78.125	-1.5696	156.250	1.7312	234.375	.6426
312.500	-6.4814	390.625	-2.4477	468.750	3.7055	546.875	4.7017
625.000	-.0136	703.125	-1.1016	781.250	9.8122	859.375	14.5282
937.500	16.7107	1015.625	17.1139	1093.750	15.8456	1171.875	12.5509
1250.000	5.6713	1328.125	-19.9616	1406.250	2.0385	1484.375	3.3798
1562.500	-.6206	1640.625	-17.6371	1718.750	-3.7983	1796.875	-1.0490
1875.000	-3.8123	1953.125	-15.7870	2031.250	-7.6562	2109.375	-3.9764
2187.500	-5.8886	2265.625	-15.2234	2343.750	-10.5349	2421.875	-6.1414
2500.000	-7.3826	2578.125	-15.0337	2656.250	-12.7978	2734.375	-7.8359
2812.500	-8.5120	2890.625	-14.9642	2968.750	-14.6178	3046.875	-9.2064
3125.000	-9.3852	3203.125	-14.9200	3281.250	-16.0888	3359.375	-10.3365
3437.500	-10.0635	3515.625	-14.8582	3593.750	-17.2664	3671.875	-11.2788
3750.000	-10.5840	3828.125	-14.7564	3906.250	-18.1832	3984.375	-12.0686
4062.500	-10.9704	4140.625	-14.6016	4218.750	-18.8550	4296.875	-12.7286
4375.000	-11.2367	4453.125	-14.3844	4531.250	-19.2999	4609.375	-13.2785
4687.500	-11.3933	4765.625	-14.0969	4843.750	-19.5210	4921.875	-13.7311
5000.000	-11.4447						

Appendix

```
C
C-----------------------------------------------------------------------
C MAIN PROGRAM: CORRELATION METHOD FOR POWER SPECTRUM ESTIMATION - CMPSE
C AUTHORS:      L R RABINER
C               BELL LABORATORIES, MURRAY HILL, NEW JERSEY 07974
C               R W SCHAFER AND D DLUGOS
C               GEORGIA INSTITUTE OF TECHNOLOGY, ATLANTA, GEORGIA 30332
C
C THIS METHOD IS BASED ON THE TECHNIQUE DESCRIBED BY C M RADER, IN THE
C IEEE TRANS ON AUDIO AND ELECT, VOL 18, NO 4, PP 439-442, 1970.
C
C INPUT:        M IS THE SECTION SIZE(MUST BE A POWER OF 2)
C                   2 <= M <= 512
C               N IS THE NUMBER OF SAMPLES TO BE USED IS THE ANALYSIS
C               MODE IS THE DATA FORMAT TYPE
C                   MODE = 0   AUTO CORRELATION
C                   MODE = 1   CROSS CORRELATION
C                   MODE = 2   AUTO COVARIANCE
C                   MODE = 3   CROSS COVARIANCE
C               FS IS THE SAMPLING FREQUENCY IN HZ
C               IWIN IS THE WINDOW TYPE
C                   IWIN = 1   RECTANGULAR WINDOW
C                   IWIN = 2   HAMMING WINDOW
C               L IS THE NUMBER OF CORRELATION VALUES USED IN
C                   THE SPECTRAL ESTIMATE
C                   2 <= L <= M
C               NFFT IS THE SIZE FFT USED TO GIVE THE SPECTRAL ESTIMATE
C                   L <= NFFT <= MAXM
C               IMD REQUESTS ADDITIONAL RUNS
C                   IMD = 1   NEW RUN
C                   IMD = 0   TERMINATE PROGRAM
C-----------------------------------------------------------------------
C
      DIMENSION XA(512), XFR(257)
      DIMENSION JWIN(2,4)
      DIMENSION ILAG(257)
      COMPLEX X(512), Z(257), XMN, XI, YI
      INTEGER TTI, TTO
      DATA JWIN(1,1), JWIN(1,2), JWIN(1,3), JWIN(1,4) /1HR,1HE,1HC,1HT/
      DATA JWIN(2,1), JWIN(2,2), JWIN(2,3), JWIN(2,4) /1HH,1HA,1HM,1HG/
C
C DEFINE I/O DEVICE CODES
C INPUT: INPUT TO THIS PROGRAM IS USER-INTERACTIVE
C        THAT IS - A QUESTION IS WRITTEN ON THE USER
C        TERMINAL (TTO) AND THE USER TYPES IN THE ANSWER.
C
C OUTPUT: ALL OUTPUT IS WRITTEN ON THE STANDARD
C         OUTPUT UNIT (LPT)
C
      TTI = I1MACH(1)
      TTO = I1MACH(4)
      LPT = I1MACH(2)
      MAXM = 512
      MAXH = MAXM/2 + 1
C
C FILL LAG ARRAY FOR PRINTING
C
      DO 10 I=1,MAXH
        ILAG(I) = I - 1
  10  CONTINUE
C
C READ IN ANALYSIS PARAMETERS M,N
C
  20  WRITE (TTO,9999)
9999  FORMAT (18H SECTION SIZE(I4)=)
      READ (TTI,9998) M
9998  FORMAT (I4)
      IF (M.GT.0 .AND. M.LE.MAXM) GO TO 30
      WRITE (TTO,9997) M
9997  FORMAT (30HILLEGAL INPUT -- REENTER VALUE)
      GO TO 20
  30  WRITE (TTO,9996)
9996  FORMAT (38H TOTAL NUMBER OF ANALYSIS SAMPLES(I5)=)
      READ (TTI,9995) N
9995  FORMAT (I5)
C
C NSECT IS THE TOTAL NUMBER OF ANALYSIS SECTIONS
C LSHFT IS THE SHIFT BETWEEN ADJACENT ANALYSIS SAMPLES
C
      LSHFT = M/2
      MHLF1 = LSHFT + 1
      NSECT = (FLOAT(N)+FLOAT(LSHFT)-1.)/FLOAT(LSHFT)
C
C READ IN MODE DATA TYPE FORMAT
C
      WRITE (TTO,9994)
9994  FORMAT (10H MODE(I1)=)
      READ (TTI,9993) MODE
9993  FORMAT (I1)
      WRITE (TTO,9992)
9992  FORMAT (33H SAMPLING FREQUENCY IN HZ(F10.4)=)
      READ (TTI,9991) FS
9991  FORMAT (F10.4)
      WRITE (LPT,9990) M, N, MODE, FS
9990  FORMAT (3H M=, I4, 4H  N=, I5, 7H  MODE=, I1, 16H  SAMPLING FREQU,
     *    5HENCY=, F10.4)
      IF (MODE.LT.2) GO TO 80
C
C SS IS GENERATOR SAMPLE NUMBER
C NRD IS NUMBER OF SAMPLES OF GENERATOR OUTPUT TO BE COMPUTED
C
      SS = 1.
      NRD = LSHFT
      XSUM = 0.
      YSUM = 0.
C
C LOOP TO CALCULATE MEANS OF X AND Y DATA
C USE GETX TO READ NRD SAMPLES FROM X GENERATOR STARTING AT SAMPLE SS
C USE GETY TO READ NRD SAMPLES FROM Y GENERATOR IF CROSS VARIANCE
C
      DO 70 K=1,NSECT
        IF (K.EQ.NSECT) NRD = N - (K-1)*NRD
        CALL GETX(XA, NRD, SS)
        DO 40 I=1,NRD
          XSUM = XSUM + XA(I)
  40    CONTINUE
        IF (MODE.EQ.2) GO TO 60
        CALL GETY(XA, NRD, SS)
        DO 50 I=1,NRD
          YSUM = YSUM + XA(I)
  50    CONTINUE
  60    SS = SS + FLOAT(NRD)
  70  CONTINUE
      XMEAN = XSUM/FLOAT(N)
      YMEAN = YSUM/FLOAT(N)
      IF (MODE.EQ.2) YMEAN = XMEAN
      XMN = CMPLX(XMEAN,YMEAN)
      WRITE (LPT,9989)
9989  FORMAT (//)
      WRITE (LPT,9988) XMEAN, YMEAN
```

```
9988  FORMAT (7H XMEAN=, E14.5, 8H  YMEAN=, E14.5)
C
C LOOP TO ACCUMULATE CORRELATIONS
C
  80  SS = 1.
      NRDY = M
      NRDX = LSHFT
      DO 90 I=1,MHLF1
        Z(I) = (0.,0.)
  90  CONTINUE
      DO 190 K=1,NSECT
        NSECT1 = NSECT - 1
        IF (K.LT.NSECT1) GO TO 110
        NRDY = N - (K-1)*LSHFT
        IF (K.EQ.NSECT) NRDX = NRDY
        IF (NRDY.EQ.M) GO TO 110
        NRDY1 = NRDY + 1
        DO 100 I=NRDY1,M
          X(I) = (0.,0.)
 100    CONTINUE
C
C READ NRDY SAMPLES FROM X GENERATOR STARTING AT SAMPLE SS
C
 110    CALL GETX(XA, NRDY, SS)
        DO 120 I=1,NRDY
          X(I) = CMPLX(XA(I),XA(I))
 120    CONTINUE
        IF (MODE.EQ.0 .OR. MODE.EQ.2) GO TO 140
C
C READ NRDY SAMPLES FROM Y GENERATOR IF CROSS COR. OR CROSS COV.
C
        CALL GETY(XA, NRDY, SS)
        DO 130 I=1,NRDY
          X(I) = CMPLX(REAL(X(I)),XA(I))
 130    CONTINUE
 140    IF (MODE.LT.2) GO TO 160
        DO 150 I=1,NRDY
          X(I) = X(I) - XMN
 150    CONTINUE
 160    NRDX1 = NRDX + 1
        DO 170 I=NRDX1,M
          X(I) = CMPLX(0.,AIMAG(X(I)))
 170    CONTINUE
C
C CORRELATE X AND Y SECTIONS
C DO EVEN-ODD SEPARATION AND ACCUMULATE  CONJG(X)*Y
C
        CALL FFT(X, M, 0)
        DO 180 I=2,LSHFT
          J = M + 2 - I
          XI = (X(I)+CONJG(X(J)))*.5
          YI = (X(J)-CONJG(X(I)))*.5
          YI = CMPLX(AIMAG(YI),REAL(YI))
          Z(I) = Z(I) + CONJG(XI)*YI
 180    CONTINUE
        XI = X(1)
        Z(1) = Z(1) + CMPLX(REAL(XI)*AIMAG(XI),0.)
        XI = X(MHLF1)
        Z(MHLF1) = Z(MHLF1) + CMPLX(REAL(XI)*AIMAG(XI),0.)
        SS = SS + FLOAT(LSHFT)
 190  CONTINUE
C
C INVERSE DFT TO GIVE CORRELATION
C
      DO 200 I=2,LSHFT
        J = M + 2 - I
        X(I) = Z(I)
        X(J) = CONJG(Z(I))
 200  CONTINUE
      X(1) = Z(1)
      X(MHLF1) = Z(MHLF1)
      CALL FFT(X, M, 1)
      FN = FLOAT(N)
      DO 210 I=1,MHLF1
        XA(I) = REAL(X(I))/FN
 210  CONTINUE
C
C IF DESIRED, THE USER MAY INSERT CODE AT THIS POINT TO PLOT
C THE CORRELATION FUNCTION WHICH IS IN THE ARRAY XA.
C
C PRINT THE CORRELATION FUNCTION
C
      WRITE (LPT,9989)
      WRITE (LPT,9987)
9987  FORMAT (21H CORRELATION FUNCTION)
      WRITE (LPT,9989)
      WRITE (LPT,9986)
9986  FORMAT (1X, 3HLAG, 2X, 4HCORR, 5X, 3HLAG, 2X, 4HCORR, 5X, 3HLAG,
     *    2X, 4HCORR, 5X, 3HLAG, 2X, 4HCORR, 5X, 3HLAG, 2X, 4HCORR)
      WRITE (LPT,9985) (ILAG(I),XA(I),I=1,MHLF1)
9985  FORMAT (5(I4, E10.3))
      WRITE (LPT,9989)
C
C WINDOW CORRELATION USING L VALUES TO GIVE SPECTRAL ESTIMATE
C CREATE SYMMETRICAL ARRAY IF MODE=0
C READ IN WINDOW TYPE AND WINDOW LENGTH
C NOTE SPECTRAL ESTIMATE MAY NOT BE MEANINGFUL IF X NOT EQUAL TO Y
C
      WRITE (TTO,9984)
9984  FORMAT (43H WINDOW TYPE(I1)-  1=RECTANGULAR, 2=HAMMING)
      READ (TTI,9983) IWIN
9983  FORMAT (I1)
      WRITE (TTO,9982)
9982  FORMAT (35H NO OF CORRELATION VALUES USED(I4)=)
      READ (TTI,9981) L
9981  FORMAT (I4)
      WRITE (TTO,9980)
9980  FORMAT (14H FFT SIZE(I4)=)
      READ (TTI,9981) NFFT
      NHLF1 = NFFT/2 + 1
      WRITE (LPT,9979) JWIN(IWIN,1), JWIN(IWIN,2), JWIN(IWIN,3),
     *    JWIN(IWIN,4), L, NFFT
9979  FORMAT (13H WINDOW TYPE=, 4A1, 22H  NO OF WINDOW VALUES=, I4,
     *    11H  FFT SIZE=, I4)
C
C WINDOW CORRELATION FUNCTION--BEWARE IF X NOT EQUAL TO Y
C
      PI = 4.0*ATAN(1.0)
      DO 230 I=2,L
        IF (IWIN.EQ.1) GO TO 220
        XA(I) = XA(I)*(0.54+0.46*COS(PI*FLOAT(I-1)/FLOAT(L-1)))
 220    IF (MODE.EQ.1 .OR. MODE.EQ.3) GO TO 230
        J = NFFT + 2 - I
        XA(J) = XA(I)
 230  CONTINUE
      NLAST = NFFT + 1 - L
      IF (MODE.EQ.1 .OR. MODE.EQ.3) NLAST = NFFT
      L1 = L + 1
      DO 240 I=L1,NLAST
        XA(I) = 0.
```

```
 240  CONTINUE
      DO 250 I=1,NFFT
        X(I) = CMPLX(XA(I),0.)
 250  CONTINUE
      CALL FFT(X, NFFT, 0)
C
C OBTAIN LOG POWER SPECTRUM  IN DB
C
      XFS = FS/FLOAT(NFFT)
      NHF = NFFT/2
      NHF1 = NHF + 1
      DO 260 I=1,NHF1
        XFR(I) = FLOAT(I-1)*XFS
        T = ALOG10(CABS(X(I)))
        XA(I) = 20.*T
 260  CONTINUE
C
C LOG POWER SPECTRUM (DB) IS IN XA
C IF DESIRED, THE USER MAY INSERT CODE AT THIS POINT TO
C PLOT LOG POWER SPECTRUM AT THIS POINT
C
      WRITE (LPT,9989)
      WRITE (LPT,9978)
9978  FORMAT (19H LOG POWER SPECTRUM)
      WRITE (LPT,9989)
      WRITE (LPT,9977)
9977  FORMAT (5X, 4HFREQ, 7X, 2HDB, 5X, 4HFREQ, 7X, 2HDB, 5X, 4HFREQ,
     *    7X, 2HDB, 5X, 4HFREQ, 7X, 2HDB)
      WRITE (LPT,9976) (XFR(I),XA(I),I=1,NHLF1)
9976  FORMAT (4(F9.3, F9.4))
      WRITE (LPT,9975)
9975  FORMAT (////)
      WRITE (TTO,9974)
9974  FORMAT (22HMORE DATA(1=YES,0=NO)=)
      READ (TTI,9993) IMD
      IF (IMD.EQ.1) GO TO 20
      STOP
      END
C
C-----------------------------------------------------------------------
C SUBROUTINE: GETX
C GENERATE X(N) FOR A SINE INPUT OF FREQUENCY 1000 HZ WITH AN ASSUME
C SAMPLING FREQUENCY OF 10000 HZ
C-----------------------------------------------------------------------
C
      SUBROUTINE GETX(X, NRD, SS)
      DIMENSION X(1)
C
C   X = ARRAY OF SIZE NRD TO HOLD GENERATOR OUTPUT DATA
C NRD = NUMBER OF SAMPLES TO BE CREATED
C  SS = STARTING SAMPLE OF GENERATOR OUTPUT
C
      TPI = 8.*ATAN(1.0)
      CF = 1000./10000.
      DO 10 I=1,NRD
        XSMP = (SS-1.) + FLOAT(I-1)
        X(I) = COS(TPI*CF*XSMP)
  10  CONTINUE
      RETURN
      END
C
C-----------------------------------------------------------------------
C SUBROUTINE: GETY
C GENERATE Y(N) FOR A SINE INPUT OF FREQUENCY 1000 HZ WITH AN
C ASSUMED SAMPLING FREQUENCY OF 10000 HZ
C-----------------------------------------------------------------------
C
      SUBROUTINE GETY(Y, NRD, SS)
      DIMENSION Y(1)
C
C   Y = ARRAY OF SIZE NRD TO HOLD GENERATOR OUTPUT DATA
C NRD = NUMBER OF SAMPLES TO BE CREATED
C  SS = STARTING SAMPLE OF GENERATOR OUTPUT
C
      TPI = 8.*ATAN(1.0)
      CF = 1000./10000.
      DO 10 I=1,NRD
        XSMP = (SS-1.) + FLOAT(I-1)
        Y(I) = SIN(TPI*CF*XSMP)
  10  CONTINUE
      RETURN
      END
C
C-----------------------------------------------------------------------
C SUBROUTINE: FFT
C JIM COOLEY'S SIMPLE FFT PROGRAM USING DECIMATION IN TIME ALGORITHM
C-----------------------------------------------------------------------
C
      SUBROUTINE FFT(X, N, INV)
C
C   X = 2**M COMPLEX ARRAY THAT INITIALLY CONTAINS INPUT
C       AND ON RETURN CONTAINS TRANSFORM
C   N = 2**M POINTS
C INV = 0, DIRECT TRANSFORM
C INV = 1, INVERSE TRANSFORM
C
      COMPLEX X(1), U, W, T, CMPLX
      M = ALOG(FLOAT(N))/ALOG(2.) + .1
      NV2 = N/2
      NM1 = N - 1
      J = 1
      DO 40 I=1,NM1
        IF (I.GE.J) GO TO 10
        T = X(J)
        X(J) = X(I)
        X(I) = T
  10    K = NV2
  20    IF (K.GE.J) GO TO 30
        J = J - K
        K = K/2
        GO TO 20
  30    J = J + K
  40  CONTINUE
      PI = 4.0*ATAN(1.0)
      DO 70 L=1,M
        LE = 2**L
        LE1 = LE/2
        U = (1.0,0.0)
        W = CMPLX(COS(PI/FLOAT(LE1)),-SIN(PI/FLOAT(LE1)))
        IF (INV.NE.0) W = CONJG(W)
        DO 60 J=1,LE1
          DO 50 I=J,N,LE
            IP = I + LE1
            T = X(IP)*U
            X(IP) = X(I) - T
            X(I) = X(I) + T
  50      CONTINUE
          U = U*W
  60    CONTINUE
  70  CONTINUE
```

```
      IF (INV.EQ.0) RETURN
      DO 80 I=1,N
        X(I) = X(I)/CMPLX(FLOAT(N),0.)
80    CONTINUE
      RETURN
      END
```

2.3

A Coherence and Cross Spectral Estimation Program*

G. Clifford Carter and *James F. Ferrie*

Naval Underwater Systems Center
New London, CT 06320

1. Purpose

This program can be used to estimate various second order statistics between two real sequences. In particular the auto and cross power spectral density functions together with coherence and generalized cross correlation functions can be estimated.

2. Method

The method uses the overlapped fast Fourier transform (FFT) technique discussed by Carter, Knapp and Nuttall [1]. The method discussed is sometimes referred to as the direct method (as opposed to the indirect correlation methods) and has been discussed in part by Welch [2], Bingham, Godfrey and Tukey [3], Benignus [4], Nuttall [5], Williams [6], and more fully by Carter, Knapp and Nuttall [1].

The method begins with two (one from each process) digital waveforms (or with analog waveforms that have been lowpass filtered and digitized). Briefly, there are four steps in the estimation procedure: First, each time series is segmented into N segments, each having P data points. Second, each segment is multiplied by a smooth weighting function. Third, the z transform of the weighted P-point sequence is evaluated on the unit circle in the z-plane. Finally, the discrete Fourier coefficients obtained in the third step are used to estimate the elements of the power spectral density matrix by averaging "raw" power spectral estimates over all the N segments. A summary flow chart is given in Figure 1.

More explicitly, two random processes that are jointly stationary over N data segments are processed as follows (Carter and Knapp [7]):

1. Each of the two time series is segmented into N segments of P points. The segments may either be disjoint or overlapped. Then one segment of P data points with the same (or aligned) time origin is selected from each of two time records. Even if each of the N data segments is large, P should be selected to ensure that the sampling frequency divided by P will afford sufficiently fine spectral resolution.

2. Each of the two P point segments is multiplied by a smooth weighting function. Here smooth means that as many higher order derivatives be continuous as reasonable over the full interval of data points. The smoother the weighting function, the more rapidly the side lobes of its Fourier transform, or window function, will decay, and thus the less leakage there will be extraneous power, which corrupts spectral measurements. Hence, smooth weighting functions result in better spectral estimates. Smooth weighting functions however, result in poorer frequency resolution when P is held fixed. If better resolution is desired, more data points per segment will be required. From a stability point of view, increasing P decreases the available number of independent data segments when the data duration is finite.

The application of some smooth weighting function, has been shown to be *necessary* to reduce errors due to side lobe leakage. (Carter and Knapp [7]). However, weighting functions have the apparent disadvantage of wasting the available data. This apparent wastage can be overcome through the added

---

* This program, originally written and documented by C. R. Arnold, G. C. Carter, and J. F. Ferrie, has been rewritten and tested by J. C. Sikorski, G. C. Carter, R. G. Williams, and J. F. Ferrie.

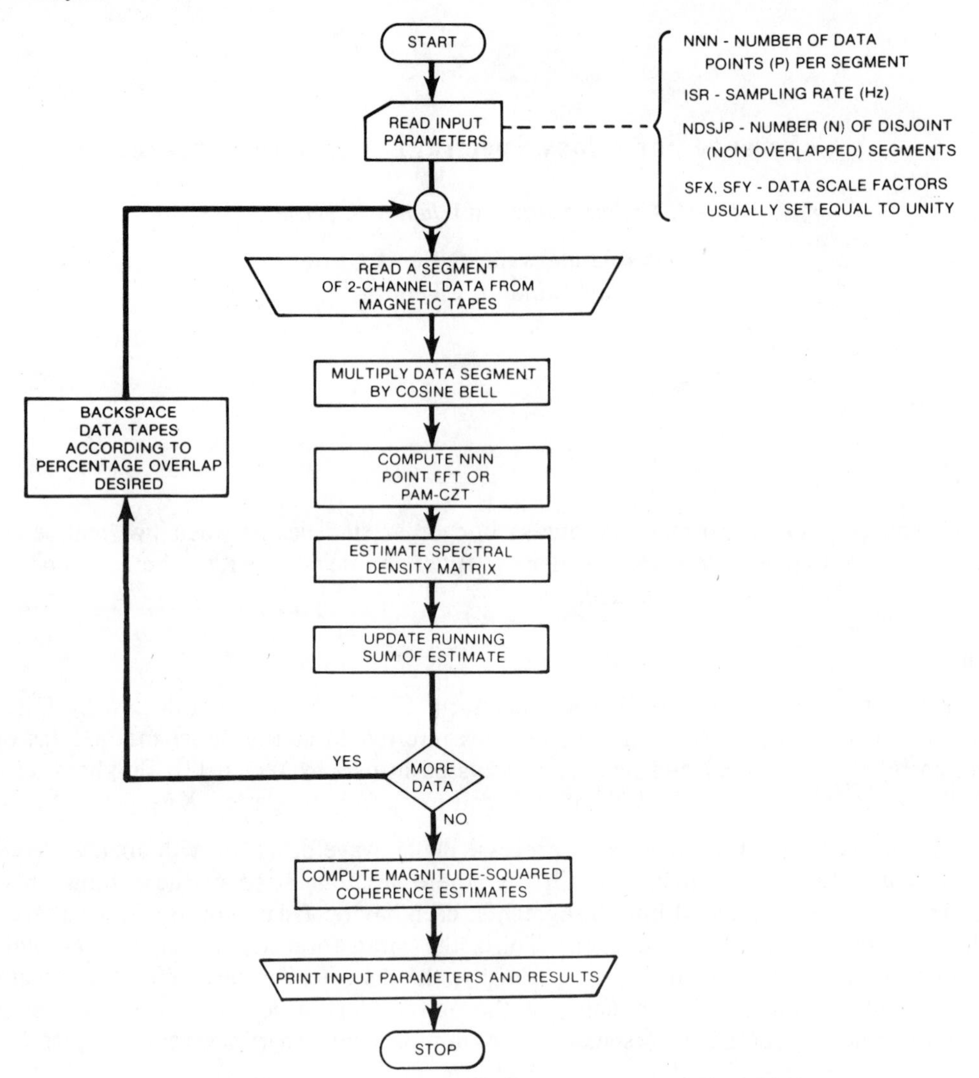

Fig. 1 Summary flow chart for coherence and spectral estimation program.

cost of overlapped processing. Nuttall [5] has shown that the same equivalent degrees of freedom can be obtained from a fixed amount of data via overlapped processing as with indirect correlation processing for both auto and cross spectral density estimation.

3. The transform of the weighted P-point sequence is evaluated on the unit circle in the z plane. The two sided z-transform of an infinite sequence is defined by Oppenheim and Schafer [8] as

$$X_n(z) = \sum_{p=-\infty}^{\infty} x_n(p) z^{-p} , \quad n = 1,2,\ldots,N , \tag{1}$$

where z equals any complex variable.

Similarly, $Y_n(z)$ is defined as the z-transform of $y_n(p)$. When $x_n(p)$ and $y_n(p)$ are finite in duration, the infinite series Eq. (1) becomes finite. Evaluation of the z-transform at P equally spaced points around the circle yields the discrete Fourier transform or DFT:

$$X_n(k) = \sum_{p=0}^{P-1} x_n(p) e^{-j2\pi pk/P} . \tag{2}$$

Similarly, $Y_n(k)$ is the DFT of the nth weighted data segment $y_n(p)$, $p = 0, 1,\ldots,P-1$. The DFT can *rapidly* be evaluated by the FFT. Having computed the DFT, Eq. (2), by an FFT, we are ready to proceed with the fourth step in the spectral estimation algorithm.

4. The spectral estimates are

$$\hat{G}_{xx}(k) = c_g \sum_{n=1}^{N} |X_n(k)|^2 , \tag{3a}$$

$$\hat{G}_{yy}(k) = c_g \sum_{n=1}^{N} |Y_n(k)|^2 , \tag{3b}$$

$$\hat{G}_{xy}(k) = c_g \sum_{n=1}^{N} X_n^*(k)\, Y_n(k) , \tag{3c}$$

with the constant gain

$$c_g = \frac{1}{N \cdot f_s \cdot P} , \tag{3d}$$

f_s = sampling frequency and * indicates complex conjugate. The estimate of magnitude squared coherence (MSC)

$$\hat{C}_{xy}(k) = \frac{|\hat{G}_{xy}(k)|^2}{\hat{G}_{xx}(k)\, \hat{G}_{yy}(k)} . \tag{4}$$

Without some understanding of the statistics of the estimator in Eq. (4) it would be of little practical value. Under certain assumptions the statistical characteristics of $\hat{C}$ can be evaluated. These characteristics are based on the derivation by Goodman [9] of an analytical expression for the probability density function of the magnitude coherence estimate and the subsequent extensions to magnitude squared coherence by Carter, Knapp and Nuttall [1]. These results are based on two zero-mean stochastic processes that were jointly *stationary, Gaussian,* and had been segmented into *N independent* segments.[1] Each segment was assumed large enough to ensure *adequate spectral resolution.* Further, each segment was assumed *perfectly weighted* (windowed), in the sense that the Fourier coefficient at some k^{th} frequency was to have "leaked" no power from other bins. The statistics do not hold at the zero-th or folding frequencies (Hannan [10]). Extensions to Goodman's work are given by Haubrich [11], Nettheim [12], Wahba [13], Carter, Knapp and Nuttall [1], and Halvorsen and Bendat [14].

3. Program Description

The main program estimates the auto and cross spectral density functions. The programs listed are intended to be general FORTRAN IV programs; they have been compiled and executed on the PDP-11/70, System Engineering Laboratory (SEL) 32/55, Univac 1108, Control Data Corporation (CDC) 6600 and International Business Machine (IBM) 360. The spectral estimation programs have been used for research projects by: Williams [6], Brady [15], Carter, Knapp and Nuttall [1], Carter, Nuttall and Cable [16], Santopietro [17], Carter and Knapp [7] and Carter [18]. These research projects were conducted entirely on the Univac 1108 and a significant program rewrite was undertaken to make the programs more transferable from one computer system to another. The programs as a complete data processing system consists of input, computations and display. We have concentrated our rewrite efforts on the computations; both the input and display programs are expected to contain peculiarities of the particular computer being used. The input and display subroutines are modular so that only a minimum rewrite is required to transfer the program to another installation. The function of the input subroutine LOAD is to load the XX and YY arrays each with a real data segment of P (equivalent to the FORTRAN variable NNN) data points. If the data were stored on logical magnetic tape number 6 in binary format the call to LOAD could be replaced by the FORTRAN statement

```
READ 6, XX(I), YY(I), I=1, NNN.
```

The subroutine LOAD listed in the Appendix is used to generate synthetic data for a suitable test case.

---

1. Despite the fact that it is mathematically tractable to obtain analytic expressions only when the segments are independent, we would in practice use some overlapped processing to regain the apparent data wastage incurred by the necessity of data weighting. Carter, Knapp and Nuttall [1] report the results of an empirical study that demonstrates how bias and variance decrease as a function of increased data segment overlap. Fifty percent overlap is recommended with cosine weighting.

The main program also calls (in addition to LOAD): COPLT, HICMP, FFT842, PRCES, PRPRT, LREMV and ZERO. The subroutine ZERO is used to store zeroes in an array. The subroutines PRPRT and COPLT are display subroutines. The subroutine LREMV estimates (and optionally removes) the linear trend and mean value for the input time waveforms. These computations are performed for every time segment and are printed out by the main program as in aid to detecting nonstationarities or digitizing errors.

The subroutine FFT842 computes the FFT (see, for example, Cooley and Tukey [19]) coded and listed by Bergland and Dolan [20]. FFT842 can be replaced with a different FFT program such as the mixed radix FFT of Singleton [21]. Note that the subroutine PRCES and the main program presume that the FFT output array is subscripted from 1 to NPFFT and *not* from 0 to (NPFFT−1), where NPFFT is the number of point FFT to be performed.

The subroutine PRCES implements the generalized cross correlation functions discussed in Knapp and Carter [22]. The subroutine PRCES calls on the subroutines FFT842 and COPLT. Given the FORTRAN variable NNN, the subroutine HICMP finds the highly composite number closest to (but greater than or equal to) NNN. The output of HICMP is NEWNN. For FFT842, NEWNN is 2 to an integer power. (For Singleton's FFT see Carter [18].) For some applications, the program user will want NEWNN to be twice as large as NNN; this is because the main program fills the data arrays with zero from NNN+1 to NEWNN. Such zero filling is (theoretically) required to inhibit the effect of circular convolution; in practice, though, (with stochastic data) zero filling often does *not* warrant the added computational cost. If it is desired, zero filling can simply be achieved by adding one line to HICMP: "NEWNN = 2*NEWNN". In addition to calling several critical subroutines, the main program performs computations necessary to estimate the spectral characteristics of the two waveforms under investigation.

When the two input waveforms are complex, one FFT of each waveform segment is required as specified in earlier. In most (though not all) practical data collection facilities, the input waveforms are real (not complex). When $x(t)$ and $y(t)$ are real, one FFT of the complex waveform $x(t) + j\,y(t)$ can be computed and quickly be manipulated to form the FFT of $x(t)$ and the FFT of $y(t)$. (See pp. 333-334 of Oppenheim and Schafer [8]; see also pp. 271-293 of Rabiner and Rader [23]). These observations, combined with Eq. (3) give rise to the FORTRAN statements used to estimate the spectral characteristics of $x(t)$ and $y(t)$. The application of this theory reduces the computation time for two real waveforms by a factor of two.

The final comment necessary before presenting the computer listings is to describe the input FORTRAN variables. NNN is the number of data points per segment. ISR is the integer sampling rate (Hz). NDSJP is the number of disjoint segments in the total time waveform. SFX and SFY are scale factors used to adjust the level of the input waveform to correct for frequency independent attenuations in the data collection and digitizing process. (When no correction is desired, the user sets SFX=SFY=1.0. When the user desires the spectral estimates to appear twice as large in power, he sets SFX=SFY=2.0.) With these five sample inputs, the input time data are processed.

4. Test Problem

The test problem was run with NNN = 32, ISR = 256, NDSJP = 8, SFX = 1.0 and SFY = 1.0. The LOAD subroutine used to generate the synthetic data is the one given in the Appendix. Selected output is given in the following tables.

References

1. G. C. Carter, C. H. Knapp and A. H. Nuttall, "Estimation of the Magnitude-Squared Coherence Function via Overlapped Fast Fourier Transform Processing", *IEEE Trans. Audio Electroacoust.*, Vol. AU-21, No. 4, pp. 337-344, Aug. 1973.

2. P. D. Welch, "The Use of Fast Fourier Transform for the Estimation of Power Spectra: A Method Based on Time Averaging Over Short, Modified Periodograms", *IEEE Trans. Audio Electroacoust.*, Vol. AU-15, No. 2, pp. 70-73, June 1967.

3. C. Bingham, M. D. Godfrey, and J. W. Tukey, "Modern Techniques of Power Spectrum Estimation", *IEEE Trans. Audio Electroacoust.*, Vol. AU-15, No. 2, pp. 56-66, June 1967.

4. V. A. Benignus, "Estimation of the Coherence Spectrum and its Confidence Interval Using the Fast Fourier Transform", *IEEE Trans. Audio Electroacoust.*, Vol. AU-17, No. 2, pp. 145-150, June 1969.

5. A. H. Nuttall, "Spectral Estimation by Means of Overlapped FFT Processing of Windowed Data", Naval Underwater Systems Center Report No. 4169, New London, CT. (and supplement NUSC TR-4169S), 1971.

6. R. G. Williams, "Estimating Ocean Wind Wave Spectra by Means of Underwater Sound", Ph.D. Dissertation, New York University, N.Y., (also published in *J. Acoust. Soc. Am.*, Vol. 53, No. 3, pp. 910-920, 1973), 1971.

7. G. C. Carter and C. H. Knapp, "Coherence and Its Estimation via the Partitioned Modified Chip-z Transform", *IEEE Trans. Acoust., Speech, Signal Processing,* Vol. ASSP-23, No. 3, pp. 257-264, June 1975.

8. A. V. Oppenheim and R. W. Schafer, *Digital Signal Processing,* Prentice-Hall, Inc., Englewood Cliffs, N.J., 1975.

9. N. R. Goodman, "On the Joint Estimation of the Spectra, Cospectrum, and Quadrature Spectrum of a Two-Dimensional Stationary Guassian Process," Scientific Paper 10, New York University, New York, 1957.

10. E. J. Hannan, *Multiple Time Series,* John Wiley and Sons, Inc., New York, 1970.

11. R. W. Haubrich, "Earth Noise, 5 to 500 Millicycles per Second, 1. Spectral Stationarity, Normality, and Non-Linearity", *J. Geophysical Res.*, Vol. 70, No. 6, pp. 1415-1427, Mar. 1965.

12. N. Nettheim, "The Estimation of Coherence", Stanford University Department of Stat., TR No. 5, Stanford, CA., 1966.

13. G. Wahba, "Cross Spectral Distribution Theory for Mixed Spectra and Estimation of Prediction Filter Coefficients", Stanford University Department of Stat., TR No. 15, Stanford, CA., 1966.

14. W. G. Halvorsen and J. S. Bendat, "Noise Source Identification Using Coherent Output Power Spectra", *J. Sound and Vibration,* Vol. 9, No. 8, pp. 15-24, Aug. 1975.

15. J. F. Brady, "An Experimental Study of Vibration Noise, and Drag of a Cylinder Rotating in Water and Certain Polymer Solutions", University of Rhode Island Ph.D. Thesis, Kingston, R.I., 1973.

16. G. C. Carter, A. H. Nuttall, and P. G. Cable, "The Smoothed Coherence Transform", *Proc. IEEE,* Vol. 61, No. 10, pp. 1497-1498, Oct. 1973.

17. R. F. Santopietro, "Measurement, Analysis and Reduction of Noise in the High Frequency", University of Pennsylvania, Ph.D. Thesis, Philadelphia, PA., 1973. (Also published in *Proc. IEEE,* Vol. 65, No. 5, pp. 707-713, May 1977.)

18. G. C. Carter, "Time Delay Estimation", University of Connecticut, Ph.D. Thesis, Storrs, CT, 1976. (Also NUSC TR-5335 AD-A025408.)

19. J. W. Cooley and J. W. Tukey, "An Algorithm for the Machine Calculation of Complex Fourier Series", *Mathematics of Computation,* Vol. 19, No. 90, pp. 297-301, April 1965.

20. G. D. Bergland, and M. T. Dolan, "Fast Fourier Transform Algorithms", Section 1.2 this volume, 1979.

21. R. C. Singleton, "An Algorithm for Computing the Mixed Radix Fast Fourier Transform", *IEEE Trans. Audio Electroacoust.*, Vol. AU-17, No. 2, pp. 93-102, June 1969.

22. C. H. Knapp and G. C. Carter, "The Generalized Correlation Method for Estimation of Time Delay", *IEEE Trans. Acoustic, Speech, Signal Processing,* Vol. ASSP-24, No. 4, pp. 320-327, Aug. 1976.

23. L. R. Rabiner and C. M. Rader, (ed.), *Digital Signal Processing,* IEEE Press, New York, 1972.

Table 1

```
32  256    8   1.0         1.0
```

Table 2

```
NNN =      32      ISR =      256      NDSJP =         8

SFX = 0.10000000E 01           SFY = 0.10000000E 01
          THE    8 DISJOINT PIECES COMPRISE     1.00 SECONDS OF DATA
          NUMBER OF POINT FFT =    32
1         PRINTOUT OF FIRST 50 VALUES OF INPUT DATA


      1      0.92138798      3.00000000
      2     -0.06441755     -0.12499915
      3      0.80970714      1.00000000
      4     -0.54630972      3.00000000
      5      1.00000000      3.00000000
      6     -0.25065174      3.00000000
      7     -1.00000000      1.00000000
      8      1.00000000      1.00000000
      9      1.00000000      1.00000000
     10      1.00000000      1.00000000
     11     -1.00000000      1.54018403
     12     -1.00000000      0.83475567
     13     -1.00000000      1.34758016
     14     -1.00000000      1.00000000
     15      0.66246481      0.71841439
     16      0.44273215      2.02324235
     17      0.60807347      1.00000000
     18     -1.00000000      2.28788233
     19      0.39938986      0.08707205
     20      0.78639817     -0.00627428
     21     -1.00000000      0.76997684
     22      0.84838115      3.00000000
     23     -0.57563173      3.00000000
     24     -0.00635505      1.66902228
     25     -0.91894575      3.00000000
     26      1.00000000      0.16576968
     27      1.00000000     -0.10451246
     28      0.69710672      1.00000000
     29      1.00000000      0.96631333
     30     -0.63129364      3.00000000
     31     -0.14878848      1.00000000
     32     -1.00000000     -0.12384148

1
  1 DX= 0.63539E-01 DY= 0.14078E 01 SX=-0.69924E-02 SY=-0.20702E-01
VX= 0.69209E 00 VY= 0.32296E 01     0
  2 DX=-0.41329E-01 DY= 0.12917E 01 SX=-0.22229E-02 SY=-0.16657E-01
VX= 0.74174E 00 VY= 0.29968E 01     0
  3 DX=-0.21002E-01 DY= 0.14245E 01 SX=-0.54174E-02 SY=-0.23483E-01
VX= 0.69332E 00 VY= 0.33341E 01     0
  4 DX= 0.41618E-01 DY= 0.14021E 01 SX=-0.96131E-02 SY=-0.67735E-02
VX= 0.63633E 00 VY= 0.37315E 01     0
  5 DX= 0.10607E 00 DY= 0.14050E 01 SX= 0.12152E-01 SY= 0.27676E-01
VX= 0.68438E 00 VY= 0.35066E 01     0
  6 DX=-0.55038E-01 DY= 0.12663E 01 SX= 0.47732E-02 SY=-0.69094E-02
VX= 0.76118E 00 VY= 0.26802E 01     0
  7 DX=-0.12450E 00 DY= 0.13886E 01 SX=-0.82681E-02 SY=-0.18342E-01
VX= 0.67879E 00 VY= 0.35007E 01     0
  8 DX= 0.41899E-01 DY= 0.13454E 01 SX=-0.15682E-01 SY=-0.15694E-01
VX= 0.76327E 00 VY= 0.32313E 01     0

          AVERAGE VARIANCES ARE, VX=0.706388E 00 VY=0.327634E 01


          INTEGRATED VARIANCES ARE, VX=0.751587E 00 VY=0.455639E 01


1
                PLOT OF AUTO SPECTRUM GXX
```

Table 2
(Continued)

```
     FMIN = -32.49 DB     PEAK = -22.76 DB

 INDEX     FREQ       DB
     1    0.       -28.69 *
     2    8.000 -22.76 ****
     3   16.000 -24.14 ****
     4   24.000 -23.85 ****
     5   32.000 -25.94 ***
     6   40.000 -23.66 ****
     7   48.000 -23.82 ****
     8   56.000 -30.54 *
     9   64.000 -28.08 **
    10   72.000 -22.88 ****
    11   80.000 -28.47 **
    12   88.000 -30.59 *
    13   96.000 -32.49 *
    14  104.000 -24.22 ****
    15  112.000 -23.31 ****
    16  120.000 -30.09 *
    17  128.000 -28.35 **
1
                         PLOT OF AUTOCORRELATION RXX


     FMIN =-0.17620937E 00          PEAK = 0.10000000E 01

 INDEX     TIME       VALUE
     1    0.          1.0000 ********************************************
     2    0.00391     0.1584 ************
     3    0.00781     0.0703 *********
     4    0.01172    -0.0385 *****
     5    0.01563    -0.0151 ******
     6    0.01953     0.0123 *******
     7    0.02344    -0.1025 **
     8    0.02734     0.1921 **************
     9    0.03125    -0.1762 *
    10    0.03516    -0.1458 *
    11    0.03906     0.0042 ******
    12    0.04297     0.0480 ********
    13    0.04688     0.0086 *******
    14    0.05078    -0.1496 *
    15    0.05469     0.0149 *******
    16    0.05859    -0.0865 ***
    17    0.06250    -0.1143 **
1
                    PLOT OF AUTO SPECTRUM GYY


     FMIN = -27.55 DB     PEAK =   -7.77 DB

 INDEX     FREQ       DB
     1    0.        -7.77 *********
     2    8.000 -13.28 *******
     3   16.000 -20.87 ***
     4   24.000 -22.16 **
     5   32.000 -22.86 **
     6   40.000 -21.48 ***
     7   48.000 -22.34 **
     8   56.000 -23.57 *
     9   64.000 -23.81 *
    10   72.000 -23.12 **
    11   80.000 -26.30 *
    12   88.000 -25.31 *
    13   96.000 -26.21 *
    14  104.000 -21.52 ***
    15  112.000 -22.17 **
    16  120.000 -27.55 *
    17  128.000 -26.22 *
1
                         PLOT OF AUTOCORRELATION RYY
```

Table 2
(Continued)

```
      FMIN = 0.18801017E 00          PEAK = 0.10000000E 01


 INDEX     TIME     VALUE
     1  0.          1.0000 ********************************************
     2  0.00391     0.6667 **************************
     3  0.00781     0.6123 ***********************
     4  0.01172     0.5594 ********************
     5  0.01563     0.5295 ******************
     6  0.01953     0.5461 *******************
     7  0.02344     0.4426 **************
     8  0.02734     0.4713 ***************
     9  0.03125     0.3698 **********
    10  0.03516     0.3382 ********
    11  0.03906     0.3196 *******
    12  0.04297     0.2811 *****
    13  0.04688     0.2830 *****
    14  0.05078     0.2195 *
    15  0.05469     0.2356 **
    16  0.05859     0.2086 *
    17  0.06250     0.1880 *
1             DUMP OF CONTINUOUS PHASE VALUES IN DEGREES
   FREQUENCY
       0.             0.
       8.000         46.59
      16.000         48.43
      24.000        139.80
      32.000        140.38
      40.000        244.10
      48.000        251.48
      56.000        297.47
      64.000        402.70
      72.000        413.14
      80.000        426.34
      88.000        486.45
      96.000        495.55
     104.000        610.33
     112.000        616.86
     120.000        637.78
     128.000        720.00
1             DUMP OF THE MAGNITUDE SQUARED COHERENCE
   FREQUENCY
       0.             0.02
       8.000          0.17
      16.000          0.46
      24.000          0.66
      32.000          0.42
      40.000          0.51
      48.000          0.66
      56.000          0.33
      64.000          0.05
      72.000          0.60
      80.000          0.49
      88.000          0.54
      96.000          0.44
     104.000          0.66
     112.000          0.73
     120.000          0.08
     128.000          0.13
1
                  PLOT OF CROSS SPECTRUM GXY


      FMIN = -34.40 DB     PEAK = -21.84 DB

 INDEX     FREQ      DB
     1    0.      -27.20 ***
     2    8.000   -21.84 ******
     3   16.000   -24.21 *****
     4   24.000   -23.92 *****
     5   32.000   -26.28 ****
     6   40.000   -24.05 *****
```

Table 2
(Continued)

```
       7  48.000 -23.99 *****
       8  56.000 -29.45 **
       9  64.000 -32.38 *
      10  72.000 -24.10 *****
      11  80.000 -28.93 **
      12  88.000 -29.30 **
      13  96.000 -31.14 *
      14 104.000 -23.76 *****
      15 112.000 -23.44 *****
      16 120.000 -34.40 *
      17 128.000 -31.74 *
1
                         PLOT OF SCOT FUNCTION


      FMIN =-0.27144101E 00        PEAK = 0.10000000E 01


  INDEX    TIME      VALUE
    -15 -0.05859  -0.0367 ********
    -14 -0.05469   0.0685 ************
    -13 -0.05078  -0.0173 ********
    -12 -0.04688  -0.0309 ********
    -11 -0.04297  -0.0413 ********
    -10 -0.03906  -0.0666 *******
     -9 -0.03516   0.0150 **********
     -8 -0.03125  -0.1832 ***
     -7 -0.02734  -0.0906 ******
     -6 -0.02344  -0.0721 *******
     -5 -0.01953  -0.1263 *****
     -4 -0.01563   1.0000 ********************************************
     -3 -0.01172   0.0884 ************
     -2 -0.00781  -0.0508 *******
     -1 -0.00391   0.0785 ************
      0  0.       -0.0964 ******
      1  0.00391  -0.0254 ********
      2  0.00781   0.0076 *********
      3  0.01172   0.1127 *************
      4  0.01563  -0.2714 *
      5  0.01953   0.0138 **********
      6  0.02344   0.0477 ***********
      7  0.02734   0.0947 ************
      8  0.03125  -0.0031 *********
      9  0.03516  -0.2503 *
     10  0.03906   0.0520 ***********
     11  0.04297  -0.2190 *
     12  0.04688  -0.0850 ******
     13  0.05078   0.1026 *************
     14  0.05469   0.0808 ************
     15  0.05859   0.0955 ************
     16  0.06250   0.0334 **********
1
                         PLOT OF PHAT FUNCTION


      FMIN =-0.20107339E 00        PEAK = 0.10000000E 01


  INDEX    TIME      VALUE
    -15 -0.05859  -0.1132 ***
    -14 -0.05469   0.0576 *********
    -13 -0.05078   0.0087 *******
    -12 -0.04688   0.1063 ***********
    -11 -0.04297  -0.0956 ***
    -10 -0.03906   0.0106 *******
     -9 -0.03516   0.0342 ********
     -8 -0.03125  -0.0108 *******
     -7 -0.02734  -0.1228 **
     -6 -0.02344   0.0016 *******
     -5 -0.01953  -0.0710 ****
     -4 -0.01563   1.0000 *********************************************
```

Table 2
(Continued)

```
   -3 -0.01172   0.0834 **********
   -2 -0.00781   0.0005 *******
   -1 -0.00391   0.1309 ************
    0  0.         0.0364 ********
    1  0.00391  -0.0259 ******
    2  0.00781   0.0367 ********
    3  0.01172   0.1267 ************
    4  0.01563  -0.1574 *
    5  0.01953   0.0259 ********
    6  0.02344  -0.0032 *******
    7  0.02734   0.1261 ************
    8  0.03125  -0.0266 ******
    9  0.03516  -0.2011 *
   10  0.03906  -0.0047 *******
   11  0.04297  -0.1278 **
   12  0.04688  -0.0534 *****
   13  0.05078   0.0720 **********
   14  0.05469   0.0483 *********
   15  0.05859   0.1496 *************
   16  0.06250   0.0708 **********
1
                        PLOT OF CROSS CORRELATION


      FMIN =-0.34397384E 00          PEAK = 0.99999999E 00


 INDEX     TIME     VALUE
   -15 -0.05859   0.0363 ************
   -14 -0.05469   0.0346 ************
   -13 -0.05078  -0.0797 ********
   -12 -0.04688  -0.0962 ********
   -11 -0.04297   0.1154 ***************
   -10 -0.03906  -0.0943 ********
    -9 -0.03516   0.0692 *************
    -8 -0.03125  -0.0789 ********
    -7 -0.02734  -0.0351 **********
    -6 -0.02344   0.0645 *************
    -5 -0.01953   0.1599 ****************
    -4 -0.01563   1.0000 *********************************************
    -3 -0.01172   0.2454 *******************
    -2 -0.00781   0.1985 ******************
    -1 -0.00391   0.0785 **************
     0  0.         0.0137 ***********
     1  0.00391   0.1357 ****************
     2  0.00781  -0.0413 **********
     3  0.01172   0.2720 ********************
     4  0.01563  -0.3031 *
     5  0.01953  -0.1149 *******
     6 -0.02344   0.0263 ************
     7  0.02734   0.0278 ************
     8  0.03125   0.0109 ***********
     9  0.03516  -0.3440 *
    10  0.03906  -0.0676 *********
    11  0.04297  -0.3070 *
    12  0.04688  -0.2010 ****
    13  0.05078   0.0053 ***********
    14  0.05469   0.0484 *************
    15  0.05859  -0.0226 **********
    16  0.06250  -0.0096 ***********
1
                        PLOT OF IMPULSE RESPONSE


      FMIN =-0.21953384E 00          PEAK = 0.10000000E 01


 INDEX     TIME     VALUE
   -15 -0.05859  -0.0803 *****
   -14 -0.05469   0.0746 **********
   -13 -0.05078   0.0486 *********
```

Table 2
(Continued)

```
  -12 -0.04688   0.0800 ***********
  -11 -0.04297  -0.0617 *****
  -10 -0.03906  -0.0118 *******
   -9 -0.03516   0.1547 *************
   -8 -0.03125  -0.1106 ****
   -7 -0.02734   0.0197 ********
   -6 -0.02344   0.0170 ********
   -5 -0.01953  -0.0603 *****
   -4 -0.01563   1.0000 *********************************************
   -3 -0.01172   0.1841 **************
   -2 -0.00781   0.0015 ********
   -1 -0.00391   0.1916 ***************
    0  0.       -0.0267 *******
    1  0.00391   0.0846 ***********
    2  0.00781   0.1157 ************
    3  0.01172   0.0311 *********
    4  0.01563  -0.1551 **
    5  0.01953   0.1101 ************
    6  0.02344   0.0212 ********
    7  0.02734   0.0761 **********
    8  0.03125  -0.0276 *******
    9  0.03516  -0.2195 *
   10  0.03906   0.0860 ***********
   11  0.04297  -0.1769 *
   12  0.04688  -0.0984 ****
   13  0.05078   0.0904 ***********
   14  0.05469   0.0731 **********
   15  0.05859   0.1506 *************
   16  0.06250   0.0278 *********
1
                     PLOT OF H-T(I) FUNCTION


    FMIN =-0.41350540E 00        PEAK = 0.99999999E 00

INDEX     TIME     VALUE
  -15 -0.05859   0.1616 ******************
  -14 -0.05469   0.1164 ****************
  -13 -0.05078   0.0293 **************
  -12 -0.04688  -0.2013 ******
  -11 -0.04297   0.0462 **************
  -10 -0.03906  -0.1862 *******
   -9 -0.03516   0.0473 **************
   -8 -0.03125  -0.2488 *****
   -7 -0.02734  -0.0879 **********
   -6 -0.02344  -0.0597 ***********
   -5 -0.01953  -0.1886 *******
   -4 -0.01563   1.0000 *********************************************
   -3 -0.01172  -0.0182 ************
   -2 -0.00781  -0.0176 ************
   -1 -0.00391  -0.0216 ************
    0  0.       -0.2124 ******
    1  0.00391   0.0345 **************
    2  0.00781  -0.1292 *********
    3  0.01172   0.2419 ********************
    4  0.01563  -0.4135 *
    5  0.01953   0.0438 **************
    6  0.02344   0.1171 ****************
    7  0.02734   0.1649 ******************
    8  0.03125   0.1535 ******************
    9  0.03516  -0.3226 **
   10  0.03906   0.1061 ****************
   11  0.04297  -0.2780 ****
   12  0.04688  -0.0597 ***********
   13  0.05078   0.0888 ***************
   14  0.05469   0.0750 ***************
   15  0.05859  -0.0103 ************
   16  0.06250   0.0463 **************
1
                     PLOT OF ECKART FUNCTION
```

Table 2
(Continued)

```
       FMIN =-0.63544157E 00         PEAK = 0.10000000E 01


   INDEX     TIME     VALUE
     -15 -0.05859   0.5846 *********************************
     -14 -0.05469   0.1941 **********************
     -13 -0.05078   0.2290 ***********************
     -12 -0.04688  -0.5636 *
     -11 -0.04297  -0.3422 ********
     -10 -0.03906  -0.6354 *
      -9 -0.03516  -0.4095 ******
      -8 -0.03125  -0.3856 ******
      -7 -0.02734  -0.3801 *******
      -6 -0.02344   0.2161 ***********************
      -5 -0.01953   0.2143 ***********************
      -4 -0.01563   1.0000 *********************************************
      -3 -0.01172   0.4729 ******************************
      -2 -0.00781   0.4576 ******************************
      -1 -0.00391  -0.0076 *****************
       0  0.       -0.1843 ************
       1  0.00391  -0.0194 ****************
       2  0.00781  -0.3895 ******
       3  0.01172  -0.0398 ****************
       4  0.01563  -0.5450 **
       5  0.01953   0.0318 ******************
       6  0.02344  -0.0044 *****************
       7  0.02734   0.1967 **********************
       8  0.03125   0.1180 ********************
       9  0.03516  -0.3017 *********
      10  0.03906  -0.0853 ***************
      11  0.04297  -0.3988 ******
      12  0.04688   0.2159 ***********************
      13  0.05078  -0.0369 ****************
      14  0.05469   0.2568 ************************
      15  0.05859   0.2253 ***********************
      16  0.06250   0.3553 ***************************
1            PLOT OF THE TRANSFER FUNCTION


       FMIN =   -8.64 DB     PEAK =     2.97 DB

   INDEX      FREQ        DB
       1    0.          2.97 *****
       2    8.000       1.84 *****
       3   16.000      -0.14 ****
       4   24.000      -0.14 ****
       5   32.000      -0.69 ***
       6   40.000      -0.77 ***
       7   48.000      -0.33 ****
       8   56.000       2.18 *****
       9   64.000      -8.60 *
      10   72.000      -2.44 ***
      11   80.000      -0.92 ***
      12   88.000       2.60 *****
      13   96.000       2.70 *****
      14  104.000       0.91 ****
      15  112.000      -0.26 ****
      16  120.000      -8.64 *
      17  128.000      -6.78 *
```

Appendix

```
C
C-----------------------------------------------------------------------
C MAIN PROGRAM:  A COHERENCE AND CROSS SPECTRAL ESTIMATION PROGRAM
C AUTHORS:       G. C. CARTER, J. F. FERRIE
C                NAVAL UNDERWATER SYSTEMS CENTER
C                NEW LONDON, CONNECTICUT 06320
C
C INPUT:         NNN IS THE NUMBER OF DATA POINTS PER SEGMENT
C                  4 < NNN < 1025
C                ISR IS THE SAMPLING RATE
C                NDSJP IS THE NUMBER OF DISJOINT SEGMENTS
C                SFX IS THE SCALE FACTOR FOR THE INPUT DATA STORED IN
C                  THE XX ARRAY
C                SFY IS THE SCALE FACTOR FOR THE INPUT DATA STORED IN
C                  THE YY ARRAY
C-----------------------------------------------------------------------
C
C SPECIFICATION AND TYPE STATEMENTS
C
      DIMENSION XX(1024), YY(1024)
      DIMENSION GXX(513), GYY(513), GXYRE(513), GXYIM(513)
      DIMENSION WEGHT(513), PHI(513)
      DIMENSION LINE(50)
      EQUIVALENCE (WEGHT(1),PHI(1))
C
C SET UP MACHINE CONSTANTS
C
      IOIN1 = I1MACH(1)
      IPRTR = I1MACH(2)
      SMALL = R1MACH(1)
C
C READ INPUT CONTROL PARAMETERS FROM COMPUTER DATA CARD
C
      READ (IOIN1,9999) NNN, ISR, NDSJP, SFX, SFY
C NNN IS THE NUMBER OF DATA POINTS PER SEGMENT
C ISR IS THE SAMPLING RATE
C NDSJP IS THE NUMBER OF DISJOINT SEGMENTS
C SFX AND SFY ARE SCALE FACTORS FOR THE INPUT DATA
9999  FORMAT (3I5, 2F10.5)
      NFFTS = NDSJP
C
C PRINT INPUT CONTROL PARAMETERS
C
      WRITE (IPRTR,9998) NNN, ISR, NDSJP, SFX, SFY
9998  FORMAT (/1X, 5HNNN =, I6, 5X, 5HISR =, I7, 5X, 7HNDSJP =, I7,
     *    5X//1X, 5HSFX =, E15.8, 8X, 5HSFY =, E15.8/)
C
C CALCULATE CONSTANTS
C
      TPI = 8.0*ATAN(1.0)
      DEG = 360.0/TPI
      IF (NNN.GT.0 .AND. NNN.LE.1024) GO TO 10
      WRITE (IPRTR,9997)
9997  FORMAT (10X, 9HNNN ERROR)
      STOP
  10  CONTINUE
      VARX = 0.0
      VARY = 0.0
      DT = 1.0/FLOAT(ISR)
      SF = SQRT(ABS(SFX*SFY))
C
C PRINT OUT USER INFORMATION
C
      TIME = FLOAT(NDSJP*NNN)*DT
      WRITE (IPRTR,9996) NDSJP, TIME
9996  FORMAT (10X, 3HTHE, I4, 25H DISJOINT PIECES COMPRISE, F8.2,
     *    16H SECONDS OF DATA)
C
C COMPUTE NEW COMPOSITE NUMBER NNN
C
      CALL HICMP(NNN, NPFFT)
      IF (NPFFT.GT.1024) STOP
      WRITE (IPRTR,9995) NPFFT
9995  FORMAT (10X, 21HNUMBER OF POINT FFT =, I5/)
C
C CALCULATE CONSTANTS
C
      NNNP1 = NNN + 1
      NNND2 = NNN/2
      NND21 = NNND2 + 1
      NP2 = NPFFT + 2
      ND2 = NPFFT/2
      ND2P1 = ND2 + 1
      DF = 1.0/(DT*FLOAT(NPFFT))
      FNYQ = FLOAT(ISR)/2.0
      CONST = 0.25*DT/FLOAT(NNN)
      FLOW = 0.0
      FHIGH = FNYQ
      ISTRT = IFIX(FLOW/DF) + 1
      ISTOP = IFIX(FHIGH/DF) + 1
C
C COMPUTE AND SAVE WEIGHTING FUNCTION
C
      TEMP = TPI/FLOAT(NNN+1)
      SCL = SQRT(2.0/3.0)
      DO 20 I=1,NNND2
        WEGHT(I) = SCL*(1.0-COS(TEMP*FLOAT(I)))
  20  CONTINUE
C
C STORE ZEROS IN THE SUMMING ARRAYS
C
      CALL ZERO(GXX, ND2P1)
      CALL ZERO(GYY, ND2P1)
      CALL ZERO(GXYRE, ND2P1)
      CALL ZERO(GXYIM, ND2P1)
C
C COMPUTE AND SUM NPFFT ESTIMATES
C
      DO 80 KOUNT=1,NFFTS
C
        CALL ZERO(XX, NPFFT)
        CALL ZERO(YY, NPFFT)
C
C LOAD XX AND YY ARRAYS WITH NNN DATA POINTS
C
        CALL LOAD(XX, YY, NNN, KOUNT, ISR)
C
C PRINT OF FIRST 50 INPUT VALUES
C
        IF (KOUNT.NE.1) GO TO 40
        WRITE (IPRTR,9994)
9994    FORMAT (1H1, 9X, 41HPRINTOUT OF FIRST 50 VALUES OF INPUT DATA///
     *    )
        LPMAX = MIN0(NPFFT,50)
        DO 30 I=1,LPMAX
          WRITE (IPRTR,9993) I, XX(I), YY(I)
9993      FORMAT (I6, 1X, 2F15.8, 6X)
  30    CONTINUE
        WRITE (IPRTR,9992)
9992    FORMAT (/1H1)
```

```
  40    CONTINUE
C
C REMOVE THE LINEAR TREND AND COMPUTE THE VARIANCE
C IF IS3 = 0 DO NOT REMOVE DC COMPONENT OR SLOPE
C = 1 REMOVE THE DC COMPONENT
C > 1 REMOVE THE DC COMPONENT AND SLOPE
C
        IS3 = 0
        CALL LREMV(XX, NNN, IS3, DX, SX)
        CALL LREMV(YY, NNN, IS3, DY, SY)
        VARXI = 0.0
        VARYI = 0.0
        DO 50 I=1,NNN
          VARXI = VARXI + XX(I)*XX(I)
          VARYI = VARYI + YY(I)*YY(I)
  50    CONTINUE
        VARXI = VARXI/FLOAT(NNN-1)
        VARYI = VARYI/FLOAT(NNN-1)
        WRITE (IPRTR,9991) KOUNT, DX, DY, SX, SY, VARXI, VARYI, IS3
9991    FORMAT (1X, I3, 4H DX=, E12.5, 4H DY=, E12.5, 4H SX=, E12.5,
     *     4H SY=, E12.5/4H VX=, E12.5, 4H VY=, E12.5, I5)
        VARX = VARX + VARXI
        VARY = VARY + VARYI
C
C WEIGHT THE INPUT DATA WITH COSINE WINDOW
C
        DO 60 I=1,NNND2
          ITMP = NNNP1 - I
          XX(I) = XX(I)*WEGHT(I)
          YY(I) = YY(I)*WEGHT(I)
          XX(ITMP) = XX(ITMP)*WEGHT(I)
          YY(ITMP) = YY(ITMP)*WEGHT(I)
  60    CONTINUE
C
C COMPUTE FORWARD FFT
C
        CALL FFT842(0, NPFFT, XX, YY)
C
C COMPUTE SPECTRA
C
        GXX(1) = GXX(1) + 4.0*XX(1)**2
        DO 70 K=2,ND2P1
          J = NP2 - K
          GXX(K) = GXX(K) + (XX(K)+XX(J))**2 + (YY(K)-YY(J))**2
          GYY(K) = GYY(K) + (YY(K)+YY(J))**2 + (XX(J)-XX(K))**2
          GXYRE(K) = GXYRE(K) + XX(K)*YY(J) + XX(J)*YY(K)
          GXYIM(K) = GXYIM(K) + XX(J)**2 + YY(J)**2 - XX(K)**2 -
     *       YY(K)**2
  70    CONTINUE
        GYY(1) = GYY(1) + 4.0*YY(1)**2
        GXYRE(1) = GXYRE(1) + 2.0*(XX(1)*YY(1))
        GXYIM(1) = 0.0
C
C GO BACK FOR NEXT SEGMENT
C
  80  CONTINUE
C
C NORMALIZE ESTIMATES
C
      FNSG = FLOAT(NFFTS)
      OFNSG = 1.0/FNSG
      VARX = VARX*OFNSG
      VARY = VARY*OFNSG
      TEMP1 = CONST*OFNSG*SFX
      TEMP2 = CONST*OFNSG*SFY
      TEMP4 = CONST*OFNSG*SF
      TEMP3 = 2.0*TEMP4
      DO 90 K=1,ND2P1
        GXX(K) = GXX(K)*TEMP1
        GYY(K) = GYY(K)*TEMP2
        GXYRE(K) = GXYRE(K)*TEMP3
        GXYIM(K) = GXYIM(K)*TEMP4
  90  CONTINUE
C
C PRINT OUT VARIANCES
C
      WRITE (IPRTR,9990) VARX, VARY
9990  FORMAT (/10X, 26HAVERAGE VARIANCES ARE, VX=, E12.6, 4H VY=,
     *   E12.6//)
      VARX = 0.0
      VARY = 0.0
      DO 100 K=1,ND2P1
        VARX = VARX + GXX(K)
        VARY = VARY + GYY(K)
 100  CONTINUE
      VARX = VARX*DF*2.0/SFX
      VARY = VARY*DF*2.0/SFY
      WRITE (IPRTR,9989) VARX, VARY
9989  FORMAT (/10X, 29HINTEGRATED VARIANCES ARE, VX=, E12.6, 4H VY=,
     *   E12.6//)
C
C CONVERT GXX TO DB AND PLOT
C
      DO 110 I=1,ND2P1
        XX(I) = GXX(I)
        PHI(I) = 10.0*ALOG10(AMAX1(GXX(I),SMALL))
 110  CONTINUE
      WRITE (IPRTR,9988)
9988  FORMAT (1H1/15X, 25HPLOT OF AUTO SPECTRUM GXX)
      CALL PRPRT(PHI, LINE, DF, FLOW, FHIGH, IPRTR)
C
C COMPUTE AND DISPLAY AUTOCORRELATION FUNCTION OF INPUT SIGNAL XX
C
      CALL ZERO(YY, NPFFT)
      DO 120 K=2,ND2P1
        J = NP2 - K
        XX(J) = XX(K)
 120  CONTINUE
      CALL FFT842(1, NPFFT, XX, YY)
      ONDRO = 1.0/XX(1)
      DO 130 I=1,ND2P1
        XX(I) = XX(I)*ONDRO
 130  CONTINUE
      WRITE (IPRTR,9987)
9987  FORMAT (1H1/20X, 27HPLOT OF AUTOCORRELATION RXX)
      CALL COPLT(XX, ND2P1, DT, 1, IPRTR, LINE)
C
C CONVERT GYY TO DB AND PLOT
C
      DO 140 I=1,ND2P1
        XX(I) = GYY(I)
        PHI(I) = 10.0*ALOG10(AMAX1(GYY(I),SMALL))
 140  CONTINUE
      WRITE (IPRTR,9986)
9986  FORMAT (1H1/15X, 25HPLOT OF AUTO SPECTRUM GYY)
      CALL PRPRT(PHI, LINE, DF, FLOW, FHIGH, IPRTR)
C
C COMPUTE AND DISPLAY AUTOCORRELATION FUNCTION OF INPUT SIGNAL YY
C
      CALL ZERO(YY, NPFFT)
```

```
      DO 150 K=2,ND2P1
        J = NP2 - K
        XX(J) = XX(K)
  150 CONTINUE
      CALL FFT842(1, NPFFT, XX, YY)
      ONDRO = 1.0/XX(1)
      DO 160 I=1,ND2P1
        XX(I) = XX(I)*ONDRO
  160 CONTINUE
      WRITE (IPRTR,9985)
 9985 FORMAT (1H1/20X, 27HPLOT OF AUTOCORRELATION RYY)
      CALL COPLT(XX, ND2P1, DT, 1, IPRTR, LINE)
C
C COMPUTE AND DISPLAY PHASE FROM AVERAGED GXYRE AND GXYIM SPECTRUM
C
      GXYIM(1) = 0.0
      PHI(1) = 0.0
      DO 200 K=2,ND2P1
        XXK = GXYRE(K)
        IF (XXK) 190, 170, 190
  170   IF (GXYIM(K)) 190, 180, 190
  180   XXK = 1.0
  190   PHI(K) = DEG*ATAN2(GXYIM(K),XXK)
  200 CONTINUE
C
C PLOT PHASE FROM -PHLIM TO PHLIM
C
      PHLIM = 1800.0
      DO 210 K=2,ND2P1
        X = PHI(K) - PHI(K-1)
        PHI(K) = PHI(K) - SIGN(360.,X)*AINT(0.5+ABS(X)/360.0)
        IF (PHI(K).GT.PHLIM) PHI(K) = PHI(K) - PHLIM
        IF (PHI(K).LT.(-PHLIM)) PHI(K) = PHI(K) + PHLIM
  210 CONTINUE
      WRITE (IPRTR,9984)
 9984 FORMAT (1H1, 10X, 42HDUMP OF CONTINUOUS PHASE VALUES IN DEGREES/)
      WRITE (IPRTR,9983)
 9983 FORMAT (3X, 9HFREQUENCY)
      DO 220 I=ISTRT,ISTOP
        TEMP = DF*FLOAT(I-1)
        WRITE (IPRTR,9982) TEMP, PHI(I)
 9982   FORMAT (4X, F8.3, 2X, F10.2)
  220 CONTINUE
C
C COMPUTE CROSS SPECTRUM AND MAGNITUDE SQUARED COHERENCE
C
      DO 230 K=1,ND2P1
        PHI(K) = GXYRE(K)**2 + GXYIM(K)**2
        XX(K) = PHI(K)/(GXX(K)*GYY(K))
  230 CONTINUE
      WRITE (IPRTR,9981)
 9981 FORMAT (1H1, 10X, 39HDUMP OF THE MAGNITUDE SQUARED COHERENCE/)
      WRITE (IPRTR,9983)
      DO 240 I=ISTRT,ISTOP
        TEMP = DF*FLOAT(I-1)
        WRITE (IPRTR,9982) TEMP, XX(I)
  240 CONTINUE
C
C CONVERT GXY TO DB AND ?LOT
C
      DO 250 I=1,ND2P1
        PHI(I) = 5.0*ALOG10(AMAX1(PHI(I),SMALL))
  250 CONTINUE
      WRITE (IPRTR,9980)
 9980 FORMAT (1H1/15X, 26HPLOT OF CROSS SPECTRUM GXY)
      CALL PRPRT(PHI, LINE, DF, FLOW, FHIGH, IPRTR)
C
C COMPUTE SIX GENERALIZED CROSS CORRELATION FUNCTIONS
C
      CALL PRCES(GXX, GYY, GXYRE, GXYIM, NPFFT, XX, YY, IPRTR, DT, LINE)
C
C COMPUTE MODULUS OF TRANSFER FUNCTION IN DB AND PLOT
C
      DO 260 K=1,ND2P1
        TEMP = (GXYRE(K)**2+GXYIM(K)**2)/GXX(K)**2
        PHI(K) = 10.0*ALOG10(AMAX1(TEMP,SMALL))
  260 CONTINUE
      WRITE (IPRTR,9979)
 9979 FORMAT (1H1, 10X, 29HPLOT OF THE TRANSFER FUNCTION)
      CALL PRPRT(PHI, LINE, DF, FLOW, FHIGH, IPRTR)
C
C TERMINATE PROGRAM
C
      STOP
      END
C
C-----------------------------------------------------------------------
C SUBROUTINE:  COPLT
C PLOT A CORRELOGRAM ON A 72 COLUMN PRINTER
C-----------------------------------------------------------------------
C
      SUBROUTINE COPLT(DATA, N, DT, ISWCH, IPRTR, LINE)
C
C INPUT:  DATA = ARRAY OF N VALUES WHICH CONTAIN THE CORRELOGRAM
C         N = NUMBER OF CORRELOGRAM POINTS TO PLOT
C         DT = TIME BETWEEN CORRELOGRAM POINTS
C         ISWCH = 1 FOR AUTO CORRELATION PLOT
C               = 2 FOR CROSS CORRELATION PLOT
C         IPRTR = LOGICAL UNIT NUMBER OF 72 COLUMN PRINTER
C         LINE = INTEGER SCRATCH ARRAY IN CALLING ROUTINE OF AT LEAST
C                45 WORDS
C
      DIMENSION DATA(1), LINE(1)
      DATA ISTAR /1H*/
C
C FIND PEAK AND MINIMUM VALUES OF ARRAY DATA
C
      FMIN = 10000.0
      PEAK = -10000.0
      DO 10 K=1,N
        PEAK = AMAX1(PEAK,DATA(K))
        FMIN = AMIN1(FMIN,DATA(K))
   10 CONTINUE
      WRITE (IPRTR,9999) FMIN, PEAK
 9999 FORMAT (///5X, 6HFMIN =, E15.8, 8X, 6HPEAK =, E15.8///1X, 5HINDEX,
     *    4X, 4HTIME, 4X, 5HVALUE/)
C
C PLOT CORRELOGRAM ON A 72 COLUMN PRINTER
C ALL VALUES OF ARRAY DATA ARE SCALED TO FIT ON PRINTER
C
      DO 20 K=1,45
        LINE(K) = ISTAR
   20 CONTINUE
      ND2 = N/2
      ND2P1 = ND2 + 1
      DELTA = 45.0/(PEAK-FMIN)
      DO 50 K=1,N
        IF (ISWCH.EQ.2) GO TO 30
        TIME = DT*FLOAT(K-1)
        J = K
```

```
      GO TO 40
 30   TIME = DT*FLOAT(K-ND2)
      J = K - ND2
 40   INDEX = IFIX((DATA(K)-FMIN)*DELTA)
      IF (INDEX.LT.1) INDEX = 1
      IF (INDEX.GT.45) INDEX = 45
      WRITE (IPRTR,9998) J, TIME, DATA(K), (LINE(I),I=1,INDEX)
9998  FORMAT (I6, F9.5, F9.4, 1X, 45A1)
 50 CONTINUE
C
    RETURN
    END
C
C-----------------------------------------------------------------------
C SUBROUTINE:  HICMP
C THIS SUBROUTINE COMPUTES A NEW COMPOSITE NUMBER
C-----------------------------------------------------------------------
C
    SUBROUTINE HICMP(NNN, NEWNN)
C
C INPUT:  NNN = NUMBER OF DATA POINTS
C OUTPUT: NEWNN = A NEW COMPOSITE NUMBER ( A POWER OF 2 ) > OR = TO NNN
C
    DO 10 I=1,15
      M = I
      NT = 2**I
      IF (NNN.LE.NT) GO TO 20
 10 CONTINUE
C
 20 NEWNN = 2**M
C
    RETURN
    END
C
C-----------------------------------------------------------------------
C SUBROUTINE:  LOAD
C THIS SUBROUTINE GENERATES TWO CHANNELS OF SYNTHETIC DATA FOR THE
C MAIN PROGRAM. THIS SUBROUTINE CAN BE REPLACED BY A DISC OR A
C MAGNETIC TAPE READ.
C-----------------------------------------------------------------------
C
    SUBROUTINE LOAD(XX, YY, NNN, KOUNT, ISR)
C
C INPUT:  NNN = NUMBER OF DATA POINTS TO BE GENERATED PER CHANNEL
C         KOUNT = NUMBER OF CURRENT SPECTRAL ESTIMATE
C         ISR = INTEGER SAMPLING RATE
C OUTPUT: XX = FIRST CHANNEL OF TIME DATA TO BE PROCESSED. THIS
C              BROADBAND SIGNAL IS GENERATED BY NON LINEARLY DISTORTING
C              A SIGNAL CONSISTING OF THE SUM OF FIVE SINUSOIDS. (THUS
C              GXX CONSISTS NOT ONLY OF THE FIVE SINE WAVES BUT MANY
C              INTERMODULATION PRODUCTS.
C         YY = SECOND CHANNEL OF TIME DATA TO BE PROCESSED. THIS
C              BROADBAND SIGNAL IS DETERMINISTICALLY RELATED TO THE XX
C              ARRAY WITH BOTH A LINEAR AND INCOHERENT COMPONENT
C              ADVANCED (DELAYED) BY ND8 UNITS.
C
    DIMENSION XX(1), YY(1)
    DIMENSION PHASE(5), FREQ(5)
C
    DT = 1.0/FLOAT(ISR)
    TPI = 8.0*ATAN(1.0)
    FREQ(1) = 10.0
    FREQ(2) = 27.0
    FREQ(3) = 43.9
    FREQ(4) = 71.8
    FREQ(5) = 108.31
    TPID = TPI/10.0
    DO 10 K=1,5
      PHASE(K) = FLOAT(K)*TPI*0.2
      FREQ(K) = TPI*FREQ(K)
 10 CONTINUE
C
    ND8 = NNN/8
    NLOOP = NNN + ND8
    DO 30 I=1,NLOOP
      TIME = FLOAT((KOUNT-1)*NNN+I)*DT
      SUM = 0.0
      DO 20 K=1,5
        SUM = SUM + SIN(FREQ(K)*TIME+PHASE(K))
 20   CONTINUE
      IF (SUM.GT.1.0) SUM = 1.0
      IF (SUM.LT.(-1.0)) SUM = -1.0
      IF (I.LE.NNN) XX(I) = SUM
      TEMP = SUM + 2.0*(SUM**2)
      J = I - ND8
      IF (I.GT.ND8) YY(J) = TEMP
 30 CONTINUE
C
    RETURN
    END
C
C-----------------------------------------------------------------------
C SUBROUTINE:  LREMV
C THIS SUBROUTINE CAN REMOVE THE DC COMPONENT AND SLOPE OF AN ARRAY OF
C DATA IF DESIRED
C-----------------------------------------------------------------------
C
    SUBROUTINE LREMV(XX, NNN, ISWCH, DC, SLOPE)
C
C INPUT:  XX = INPUT DATA ARRAY
C         NNN = NUMBER OF POINTS IN DATA ARRAY
C         ISWCH = 0 DO NOT REMOVE DC COMPONENT OR SLOPE
C               = 1 REMOVE THE DC COMPONENT
C               > 1 REMOVE THE DC COMPONENT AND SLOPE
C OUTPUT: DC = DC COMPONENT OF DATA
C         SLOPE = SLOPE OF DATA
C
    DIMENSION XX(1)
C
C ESTABLISH CONSTANTS
C
    FLN = FLOAT(NNN)
    DC = 0.0
    SLOPE = 0.0
C
    DO 10 I=1,NNN
      DC = DC + XX(I)
      SLOPE = SLOPE + XX(I)*FLOAT(I)
 10 CONTINUE
C
C COMPUTE STATISTICS
C
    DC = DC/FLN
    SLOPE = 12.0*SLOPE/(FLN*(FLN*FLN-1.0)) - 6.0*DC/(FLN-1.0)
C
C DETERMINE KIND OF TREND REMOVAL
C
    IF (ISWCH-1) 60, 40, 20
C
C REMOVE TREND (MEAN AND SLOPE)
```

```
C
  20  CONTINUE
      FLN = DC - 0.5*(FLN+1.0)*SLOPE
      DO 30 I=1,NNN
        XX(I) = XX(I) - FLOAT(I)*SLOPE - FLN
  30  CONTINUE
      GO TO 60
C
C REMOVE THE DC COMPONENT
C
  40  CONTINUE
      DO 50 I=1,NNN
        XX(I) = XX(I) - DC
  50  CONTINUE
C
  60  RETURN
      END
C
C-------------------------------------------------------------------
C SUBROUTINE:  PRCES
C THIS SUBROUTINE COMPUTES AND PLOTS SIX GENERALIZED CROSS CORRELATION
C FUNCTIONS
C-------------------------------------------------------------------
C
      SUBROUTINE PRCES(GXX, GYY, GXYRE, GXYIM, NPFFT, XX, YY, IPRTR,
     *    DT, LINE)
C
C INPUT:  GXX = ARRAY OF AUTO SPECTRAL VALUES OF XX DATA CHANNEL
C         GYY = ARRAY OF AUTO SPECTRAL VALUES OF YY DATA CHANNEL
C         GXYRE = ARRAY REPRESENTING REAL PART OF CROSS SPECTRAL
C                 DENSITY FUNCTION
C         GXYIM = ARRAY REPRESENTING IMAGINARY PART OF CROSS SPECTRAL
C                 DENSITY FUNCTION
C         NPFFT = NUMBER REPRESENTING FOURIER TRANSFORM SIZE
C         XX = SCRATCH ARRAY OF LENGTH NPFFT
C         YY = SCRATCH ARRAY OF LENGTH NPFFT
C         IPRTR = LOGICAL UNIT NUMBER OF 80 COLUMN PRINTER
C         LINE = INTEGER SCRATCH ARRAY IN CALLING ROUTINE OF AT LEAST
C                50 WORDS
C
      DIMENSION GXX(1), GYY(1), GXYRE(1), GXYIM(1), XX(1), YY(1)
      DIMENSION LINE(1)
C
C CALCULATE CONSTANTS
C
      SMALL = R1MACH(1)
      SMALL = AMAX1(0.0001,SMALL)
      NP2 = NPFFT + 2
      ND2P1 = (NPFFT/2) + 1
      ND2 = NPFFT/2
      ND2M1 = ND2 - 1
C
C PROCESS SIX GENERALIZED CROSS CORRELATION FUNCTIONS
C
      DO 90 NTIME=1,6
C
        DO 10 K=1,ND2P1
C
          IF (NTIME.EQ.1) TEMP = 1.0/SQRT(GXX(K)*GYY(K))
          IF (NTIME.EQ.2) TEMP = 1.0/SQRT(GXYRE(K)**2+GXYIM(K)**2)
          IF (NTIME.EQ.3) TEMP = 1.0
          IF (NTIME.EQ.4) TEMP = 1.0/GXX(K)
          GXYMG = SQRT(GXYRE(K)**2+GXYIM(K)**2)
          COHR2 = GXYMG**2/(GXX(K)*GYY(K))
          COHR2 = AMIN1(COHR2,1.0-SMALL)
          IF (NTIME.EQ.5) TEMP = COHR2/((1.-COHR2)*GXYMG)
          TEMP1 = GXX(K) - GXYMG
          H = 1.0
          IF (ABS(TEMP1).LT.SMALL) TEMP1 = SMALL*SIGN(H,TEMP1)
          IF (NTIME.EQ.6) TEMP = GXYMG/(TEMP1**2)
C
          XX(K) = GXYRE(K)*TEMP
          YY(K) = GXYIM(K)*TEMP
C
  10    CONTINUE
C
        DO 20 K=2,ND2P1
          J = NP2 - K
          XX(J) = XX(K)
          YY(J) = -YY(K)
  20    CONTINUE
        YY(ND2P1) = 0.0
C
C COMPUTE INVERSE FFT
C
        CALL FFT842(1, NPFFT, XX, YY)
C
        TEMP = 0.0
        DO 30 K=1,NPFFT
          IF (TEMP.GE.ABS(XX(K))) GO TO 30
          KOFMX = K
          TEMP = ABS(XX(K))
  30    CONTINUE
C
        TEMP = 1.0/TEMP
        DO 40 K=1,NPFFT
          XX(K) = XX(K)*TEMP
  40    CONTINUE
C
        DO 50 I=1,ND2P1
          ITMP1 = I + ND2M1
          YY(ITMP1) = XX(I)
  50    CONTINUE
        DO 60 I=1,ND2M1
          ITMP1 = ND2P1 + I
          YY(I) = XX(ITMP1)
  60    CONTINUE
        NLAG = 100
        ITMP1 = ND2 - NLAG
        ITMP2 = ND2 + NLAG
        IF (ITMP1.GE.1) GO TO 70
        ITMP1 = 1
  70    IF (ITMP2.LE.NPFFT) GO TO 80
        ITMP2 = NPFFT
  80    CONTINUE
        XMIN = -DT*FLOAT(1-NLAG)
        XMAX = DT*FLOAT(1+NLAG)
C
C PLOT GENERALIZED CROSS CORRELATION FUNCTIONS
C
        IF (NTIME.EQ.1) WRITE (IPRTR,9999)
        IF (NTIME.EQ.2) WRITE (IPRTR,9998)
        IF (NTIME.EQ.3) WRITE (IPRTR,9997)
        IF (NTIME.EQ.4) WRITE (IPRTR,9996)
        IF (NTIME.EQ.5) WRITE (IPRTR,9995)
        IF (NTIME.EQ.6) WRITE (IPRTR,9994)
9999    FORMAT (1H1/20X, 21HPLOT OF SCOT FUNCTION)
9998    FORMAT (1H1/20X, 21HPLOT OF PHAT FUNCTION)
9997    FORMAT (1H1/20X, 25HPLOT OF CROSS CORRELATION)
9996    FORMAT (1H1/20X, 24HPLOT OF IMPULSE RESPONSE)
```

```
9995   FORMAT (1H1/20X, 23HPLOT OF H-T(I) FUNCTION)
9994   FORMAT (1H1/20X, 23HPLOT OF ECKART FUNCTION)
       CALL COPLT(YY, NPFFT, DT, 2, IPRTR, LINE)
C
  90  CONTINUE
C
      RETURN
      END
C
C----------------------------------------------------------------------
C SUBROUTINE:  PRPRT
C THIS SUBROUTINE PLOTS A POWER SPECTRUM ON A 72 COLUMN PRINTER FROM
C AN ARRAY OF DB VALUES
C----------------------------------------------------------------------
C
      SUBROUTINE PRPRT(POWER, LINE, DF, FLOW, FHIGH, IPRTR)
C
C INPUT:  POWER = AN ARRAY OF POWER SPECTRAL VALUES IN DB
C          LINE = INTEGER SCRATCH ARRAY IN CALLING ROUTINE OF AT LEAST
C                 50 WORDS WHICH IS USED TO STORE THE CHARACTER *
C            DF = FREQUENCY RESOLUTION IN HERTZ
C          FLOW = STARTING FREQUENCY OF SIGNAL TO BE PLOTTED
C                 MINIMUM VALUE = 0.0 HERTZ
C         FHIGH = ENDING FREQUENCY OF SIGNAL TO BE PLOTTED
C                  MAXIMUM VALUE = FLOAT(ISR/2) HERTZ
C         IPRTR = LOGICAL UNIT NUMBER OF 72 COLUMN PRINTER
C
      DIMENSION POWER(1), LINE(1)
      DATA ISTAR /1H*/
C
C FIND PEAK AND MINIMUM DB VALUES OF ARRAY POWER BETWEEN FLOW AND FHIGH
C
      ISTRT = IFIX(FLOW/DF) + 1
      ISTOP = IFIX(FHIGH/DF) + 1
      FMIN = 10000.0
      PEAK = -10000.0
      DO 10 K=ISTRT,ISTOP
        PEAK = AMAX1(PEAK,POWER(K))
        FMIN = AMIN1(FMIN,POWER(K))
  10  CONTINUE
      WRITE (IPRTR,9999) FMIN, PEAK
9999  FORMAT (///5X, 6HFMIN =, F7.2, 3H DB, 4X, 6HPEAK =, F7.2, 3H DB//
     *    1X, 5HINDEX, 4X, 4HFREQ, 5X, 2HDB/)
C
C PLOT SPECTRUM ON PRINTER
C
      DO 20 K=1,50
        LINE(K) = ISTAR
  20  CONTINUE
C
      FBEG = FLOAT(IFIX(FLOW/DF))*DF
      DO 30 K=ISTRT,ISTOP
        FREQ = FBEG + DF*FLOAT(K-ISTRT)
        INDEX = IFIX(POWER(K)-FMIN)/2
        IF (INDEX.LT.1) INDEX = 1
        IF (INDEX.GT.50) INDEX = 50
        WRITE (IPRTR,9998) K, FREQ, POWER(K), (LINE(I),I=1,INDEX)
9998    FORMAT (I6, F8.3, F7.2, 1X, 50A1)
  30  CONTINUE
C
      RETURN
      END
C
C----------------------------------------------------------------------
C SUBROUTINE:  ZERO
C THIS SUBROUTINE STORES ZEROES IN A FLOATING POINT ARRAY
C----------------------------------------------------------------------
C
      SUBROUTINE ZERO(ARRAY, NUMBR)
C
C INPUT:   ARRAY = AN ARRAY OF FLOATING POINT VALUES TO BE
C                  ZERO FILLED
C          NUMBR = NUMBER OF ARRAY VALUES
C
      DIMENSION ARRAY(1)
C
      DO 10 K=1,NUMBR
        ARRAY(K) = 0.0
  10  CONTINUE
C
      RETURN
      END
```

CHAPTER 3

Fast Convolution

L. R. Rabiner

Introduction

The only program in this chapter is an implementation of a high speed (FFT) convolution as proposed by Stockham [1], and Helms [2]. The program was provided by J. Allen and is a direct implementation of the method of overlap-add. The program is designed to handle efficiently large amounts of data assumed available on a file or disk.

References

1. T. G. Stockham, "High Speed Convolution and Correlation", *Proc. AFIPS Spring Joint Computer Conf.,* Vol. 28, pp. 229-233, 1966.

2. H. D. Helms, "Fast Fourier Transform Method of Computing Difference Equations and Simulating Filters", *IEEE Trans. on Audio and Electroacoustics,* Vol. AU-15, No. 2, pp. 85-90, June 1967.

3.1

FASTFILT -- An FFT Based Filtering Program

Jont B. Allen

Acoustics Research Dept.
Bell Laboratories
Murray Hill, NJ 07974

1. Purpose

FASTFILT is intended to be a general purpose utility program for filtering large amounts of data from disk. It consists of a MAIN program called FASTFILT and a subroutine RFILT which may be used independently if desired. The input data sequence is assumed to exist in a file on disk (as binary data). The function of FASTFILT is to sequentially read in blocked segments of data, filter them (by a call to RFILT), and store the results on disk in a second file. The filter is always represented by its impulse response.[1] (A test program for RFILT called TESTFILT has also been provided.)

2. Method

The filter technique is the well known method of overlap-add [1,2].

3. Program Description

When specifying the filter impulse response to FASTFILT the user may elect to either read a precomputed impulse response from disk or input an impulse response from the teletype terminal input. In the following, disk files will be given as names in quotations, i.e., "DISKFILEIN", "DKOUT". The teletype terminal name on input will always be represented as "TTYI" and the teletype name on output as "TTYO".

When entering the filter impulse response data from "TTYI" the program recursively asks for delay and gain values. The delay requested IDEL is the index of the impulse response array sample F(IDEL). The gain value requested is the value of the impulse response at delay IDEL. For example if IDEL = 1024 and G = .8, then F(1024) = .8. Any delay values not specified will have zero gain. IDEL = 1 represents no delay, while a gain of one represents no gain change. When IDEL is set to zero, the loop is terminated and the impulse response F(I) is stored on disk in a file called "FILT-DATA", thereby making it available for subsequent use. The filter length must be a power of 2, but this represents no restriction of course, since one may append any number of zeros to the impulse response.

The length of the filter, NP, is limited by the maximum FFT size NMAX. The length of the FFT used NPT2 must be twice as long as the impulse response, NP. Thus, NP $\leqslant$ NMAX/2. The total array space required is two real arrays NMAX long and a real array NMAX/2 long. Thus the total space required is 5*NMAX/2 reals. If one real word is stored in two integer locations, the total integer array space required is therefore 5*NMAX or 10*NP. These numbers assume that a real to complex FFT is used, such as Bergland's FAST, FSST, or FFA, FFS (see section 1.2). On our Data General Eclipse, each user has 32K words of 16 bit core. In this case NMAX may assume a maximum of 4096 and the maximum filter length NP is 2048 points. Under these conditions the total integer array requirement is 20,480 words (leaving 12,288 words for the program code.)

---

1. If the user has a filter design package he wishes to use with this program he must compute the filter impulse response using his program and store it on disk.

It is an important fact that the amount of time to perform the filtering per sample of input data increases only as $\log_2$(NPT2), the logarithm base 2 of the FFT length. Since the disk overhead reads and writes usually decrease as NPT2 increases, *the time required for filtering is nearly independent of the filter length.*

The first question asked by our program is the filter length. This question is asked in order to keep the size of files stored on disk as small as possible: if disk storage were not a consideration, a standard 2048 long impulse response could be assumed with no significant loss in speed.

3.1 Data I/O

This program has not been intended for a batch environment. (The batch user must use his own I/O and RFILT.) We assume that the user may respond to questions through keyboard input.

Besides TTY I/O, it is necessary to read and write data from the disk. DISK I/O is done in one Routine DKIO. We assume that the user has named disk files and that commands are available to open named disk file on a FORTRAN channel for reading and writing.

The subroutine DKIO must be supplied by the user. The arguments which must be supplied are (in order as they appear in the call)

NAME = An array containing the name of the disk file to be accessed (defined by call GNAME)

ICN = Fortran channel number of named file

IB0 = Relative block number to be read or written

IARRAY = Integer array for core data storage

NBLK = Size of IARRAY, in disk blocks

IER = Error Flag (not used here)

IOPER = Operation Code

The final argument determines the operation to be performed according to the following table

IOPER	Operation
0	open disk file
1	close disk file
2	read from disk file
3	write onto disk file

Alternatively the user may rewrite the four calls FOPEN, CLOSE, RBLK, and WBLK.

A remaining complication is that of reading the file names from the TTY keyboard. The problem is that a file name may be any number of characters long and thus some form of free format alphanumeric READ on "TTYI" must be available. In order to handle this we assume that the user has a subroutine called GNAME. When GNAME is called, the user types in a file name of any (reasonable) number of characters, and this name is returned in an array NAME, formatted so that it may be used directly when the file is opened. For example

```
DIMENSION NAME (10)
CALL GNAME(NAME)
CALL DKIO(NAME,0,...,0)
```

opens FORTRAN channel 0 on the disk with a disk file name defined by the user in response to the GNAME call.

3.2 Speed

The following times are for the DATA GENERAL ECLIPSE computer. Our 1024 point real FFT time is 288 (milliseconds, FFA time). The time required to filter 25,600 data points with a 2048 point filter is 40 sec. Thus

$$\frac{\text{time}}{\text{point}} = \frac{40\text{ sec}}{25600} = 1.17\ \frac{\text{ms}}{\text{point}}$$

These figures may be used to compensate for the speed of the FFT on the users machine by proportional adjustment of this time.

3.3 Accuracy

The accuracy will be close to the accuracy of the FFT (see Tables 4 and 5 of Section 1.0).

3.4 RFILT

Subroutine RFILT is called as follows

CALL RFILT(R,F,S,NP)

where S is a scratch array NP points long, F is the filter frequency response 2*NP+2 real (NP + 1 complex) points long. R(1) to R(NP) are the input (prior to the call) and output data points (after return from the call), while R(NP+1) to R(2*NP) are data points left from the previous call to RFILT. Prior to the first call to RFILT, R(NP+1) through R(2*NP) should be zeroed. After the initial call to RFILT, this portion of R *should not be changed in any way.* The array S may be used in the main program for any purpose since it is only used as a temporary storage array. For example it may be EQUIVALENCE'd to an integer array for the disk read and writes. The array F is assumed to be the output of FAST (or FFA) corresponding to the filter input impulse response. The filter impulse response must be zero in the range of time samples n between NP+1 through 2*NP, i.e., the second half of F, prior to calling FAST (FFA). The user should study the listings for further details.

3.5 FAST, FSST

These FFT subroutines are those given in this book in Section 1.2. They may be replaced by FFA and FFS for greater speed.

3.6 Further Constant Definitions Are:

ITTO	TTYO output channel number
ITTI	TTYI input channel number
IDSK0, IDSK1	allowed disk channel numbers
NBLKL	length in words of minimum disk block size (power of 2 only).
NMAX	maximum FFT size allowed
SCALE	A positive number equal to the maximum magnitude of data to be stored or read from disk (This number may be D/A and A/D dependent).
NP	Filter length

3.7 Required External Calls and Functions

CALL GNAME	CALL FAST	CALL DKIO
	CALL FSST	I1MACH

4. Test Examples

In the following examples, computer responses are in **BOLD FACE** while user responses are in *ITALICS*. Comments are in normal type.

Example 1

For the following dialogue we assume that the filter impulse response is on disk in a file named "FILTER". We wish to filter some speech data also on disk in file "SPEECHIN". The output, the filtered speech data, will go to disk file "DKOUT".

DO YOU WISH TO DO FILTERING (=0)
OR CORRELATION (MATCHED FILTERING) (=1) ?
0
PROGRAM WILL DO FILTERING
FILTER LENGTH IN 256 WORD DISK BLOCKS (1,2,4,OR8) =
8
FILTER WILL BE 2048 POINTS LONG
INPUT FILTER FROM DISK(1) OR TTY(2) =
1
FILTER FILENAME =
FILTER
SYSTEM GAIN = (F10.1)
1.
INPUT DATA FILENAME =
SPEECHIN
OUTPUT DATA FILENAME =
DKOUT
 # OF DATA BLOCKS TO BE FILTERED = (I5)
00100
STOP

Example 2

As a second example we input the filter impulse response from the keyboard. All other conditions remain the same.

DO YOU WISH TO DO FILTERING (=0)
OR CORRELATION (MATCHED FILTERING) (=1) ?
0
PROGRAM WILL DO FILTERING
FILTER LENGTH IN 256 WORD DISK BLOCKS (1,2,4,OR8) =
8
FILTER WILL BE 2048 POINTS LONG
INPUT FILTER FROM DISK(1) OR TTY(2) =
2
DELAY = 0 TO TERMINATE INPUT MODE
DELAY = (I4)
0002 delay by one sample
GAIN = (F10.1)
-1. change sign of data
DELAY = 0 TO TERMINATE INPUT MODE
DELAY = (I4)
0000
IMPULSE RESPONSE IS ON DISK IN FILE "FILTDATA"
INPUT DATA FILENAME =
SPEECHIN
OUTPUT DATA FILENAME =
DKOUT
 # OF DATA BLOCKS TO BE FILTERED = (I5)
00100
STOP

The data will be delayed by one sample and inverted as a result of these commands.

5. TESTFILT

TESTFILT is a main program written to test subroutine RFILT. It also provides an excellent test of the FFT routines. The program requires no input and upon completion prints a pass/fail message on the output device. It filters an internally generated function with the impulse response (0.,–1.,0.). It then compares the result to the original input function. Required external routines are FAST, FSST, RFILT, I1MACH, and R1MACH. TESTFILT should be run before attempting to run FASTFILT.

References

1. T. G. Stockham, "High Speed Convolution and Correlation", *Proc. AFIPS Spring Joint Computer Conf.,* Vol. 28, pp. 229-233, 1966.

2. H. D. Helms, "Fast Fourier Transform Method of Computing Difference Equations and Simulating Filters", *IEEE Trans. Audio Electroacoust.,* Vol. AU-15, No. 2, pp. 85-90, June 1967.

Appendix

```
C
C-----------------------------------------------------------------------
C MAIN PROGRAM: FASTFILT - 8/1/75
C AUTHOR:       JONT B. ALLEN
C               BELL LABS,MURRAY HILL N.J., 07974
C
C INPUT:
C               A)0=FILTERING;1=CORRELATION
C               B)LENGTH OF FILTER SPECIFIED IN DISK BLOCKS
C               C)IMPULSE RESPONSE INPUT FILENAME
C               THIS MAY BE 'TTI' OR ANY DISK FILENAME
C               D)IF INPUT IS FROM 'TTI',THE IMPULSE RESP
C               IS NOW TYPED IN AS DELAY AND GAIN.
C               GAIN MUST BE LESS THAN OR EQUAL TO 1. IN
C               MAGNITUDE.
C               D)IF INPUT IS FROM DISK FILE, AN OVERALL GAIN
C               VALUE IS REQUESTED; THIS WILL BE USED TO SCALE
C               THE IMPULSE RESPONSE AFTER IT IS READ FROM DISK.
C               E)THE INPUT DISK FILENAME WHERE THE SIGNAL TO BE
C               FILTERED IS STORED IN INTEGER (BINARY) FORMAT.
C               F)OUTPUT DISK FILENAME WHERE FILTERED OUTPUT IS TO
C               G)NUMBER OF DISK BLOCKS TO BE FILTERED
C
C  IF THE USER OPTS THE CORRELATION OPTION, THE IMPULSE RESP. IS
C      TIME REVERSED.
C
C
C      USES BERGLAND REAL FFT TO INCREASE SIZE OF MAXIMUM FILTER LENGTH
C      PGM TO FILTER DATA OFF OF DISK BY FFT OVERLAP-SAVE FILTERING
C      THE FILTER IMPULSE RESPONSE DATA IS EITHER SPECIFIED IN A DISK
C      FILE, OR FROM THE TTY
C
C SUBROUTINES:  RFILT,RBLK,WBLK,FOPEN,CLOSE,DKIO,GNAME,I1MACH,FAST,FSST
C
C RFILT:        DOES OVERLAP-ADD FILTERING;MAY BE USED ALONE
C RBLK,WBLK,FOPEN,CLOSE,DKIO:   DISK FILE HANDELING ROUTINES.
C
C*** THERE ARE 2 MACHINE DEPENDENT SUBROUTINES WHICH MUST BE
C*** SUPPLIED BY THE USER - DKIO AND GNAME - COMPLETE INSTRUCTIONS
C*** FOR WRITING THEM ARE GIVEN IN THE ROUTINES. FORTRAN 5
C*** CODE IS SUPPLIED FOR A DATA GENERAL COMPUTER.
C***     DKIO - OPENS, CLOSES, READS AND WRITES A DISK FILE
C***     GNAME - READS AN ASCII FILENAME FROM THE INPUT DEVICE(TTY)
C       I1MACH: MACHINE DEPENDENT; SETS MACHINE CONSTANTS
C       FAST,FSST: BERGLAND RADIX-4 REAL TO COMPLEX FFT'S
C-----------------------------------------------------------------------
C
      DIMENSION IS(4096), F(4098), R(4098), S(2049)
      DIMENSION IFIN(10), IFOUT(10)
      LOGICAL COR
      EQUIVALENCE (S(1),IS(1))
      DATA IDSK0 /0/, IDSK1 /1/, IB0 /0/
      DATA NBLKL /256/, NMAX /4096/, COR /.FALSE./
      ITTO = I1MACH(2)
      ITTI = I1MACH(1)
      SCALE = I1MACH(9)
      NMAXD2 = NMAX/2
C
C INPUT INFO FROM TTY
C
      WRITE (ITTO,9999)
9999  FORMAT (33H DO YOU WISH TO DO FILTERING (=0)/17H OR CORRELATION (,
     *    25HMATCHED FILTERING) (=1) =)
      READ (ITTI,9998) I
9998  FORMAT (I1)
      IF (I.EQ.1) COR = .TRUE.
      IF (COR) WRITE (ITTO,9997)
9997  FORMAT (28H PROGRAM WILL DO CORRELATION)
      IF (.NOT.COR) WRITE (ITTO,9996)
9996  FORMAT (26H PROGRAM WILL DO FILTERING)
  10  CONTINUE
C
C INPUT FILTER LENGTH
C
      WRITE (ITTO,9995) NBLKL
9995  FORMAT (18H FILTER LENGTH IN , I3/28H WORD DISK BLOCKS (1,2,4 OR ,
     *    3H8)=)
      READ (ITTI,9994) NB
9994  FORMAT (I1)
      IF (NB.LE.0) GO TO 10
      NP = NB*NBLKL
      IF (NP.GT.NMAXD2) WRITE (ITTO,9993)
9993  FORMAT (23H FILTER LENGTH TOO LONG)
      IF (NP.GT.NMAXD2) GO TO 10
      WRITE (ITTO,9992) NP
9992  FORMAT (16H FILTER WILL BE , I4, 12H POINTS LONG)
      NPT2 = NP*2
      NPP1 = NP + 1
C
C ZERO FILTER
C
      DO 20 I=1,NPT2
        F(I) = 0
  20  CONTINUE
      WRITE (ITTO,9991)
9991  FORMAT (37H INPUT FILTER FROM DISK(1) OR TTY(2)=)
      READ (ITTI,9990) IRESP
9990  FORMAT (I1)
      IF (IRESP.EQ.1) GO TO 60
C
C TTY INPUT OF IMPULSE RESPONSE
C
  30  CONTINUE
      WRITE (ITTO,9989)
9989  FORMAT (32H DELAY=0 TO TERMINATE INPUT MODE/12H DELAY= (I4))
      READ (ITTI,9988) IDEL
9988  FORMAT (I4)
C
C CHECK FOR INPUT TERMINATION
C
      IF (IDEL.LE.0) GO TO 40
C
C CHECK FOR TO LARGE A DELAY
C
      IF (IDEL.GT.NP) GO TO 30
C
C READ IN GAIN VALUE OF IMPULSE RESP. AT DELAY IDEL
C
      WRITE (ITTO,9987)
9987  FORMAT (36H GAIN=(F10.1), ABS(GAIN) LESS THAN 1)
      READ (ITTI,9986) F(IDEL)
9986  FORMAT (F10.1)
      IF (ABS(F(IDEL)).GT.1.) WRITE (ITTO,9985)
9985  FORMAT (13HGAIN TO LARGE)
      IF (ABS(F(IDEL)).GT.1.) F(IDEL) = 0
C
C GET A NEW RESP VALUE
C
      GO TO 30
  40  CONTINUE
```

```
C
C WRITE IMPULSE RESPONSE DATA ON DISK IN FILE "FILTDATA"
C OPEN A DISK FILE CALLED "FILTDATA" ON CHANNEL IDSK0
C AND SCALE DATA
C
      CALL FOPEN(IDSK0, 8HFILTDATA)
      DO 50 I=1,NP
        IS(I) = SCALE*F(I)
  50  CONTINUE
C
C PUT DATA ON DISK IN BINARY FORMAT
C WRITE NB BLOCKS STARTING AT BLOCK IB0 ON CHANNEL IDSK0
C FROM ARRAY IS; IER IS UNUSED ERROR FLAG
C
      CALL WBLK(IDSK0, IB0, IS, NB, IER)
      CALL CLOSE(IDSK0)
      WRITE (ITTO,9984)
9984  FORMAT (47H IMPULSE RESPONSE IS ON DISK IN FILE - FILTDATA)
      GO TO 80
C
C READ FILTER H(T) FROM DISK AND PUT IN ARRAY F
C
  60  CONTINUE
      WRITE (ITTO,9983)
9983  FORMAT (17H FILTER FILENAME=)
      CALL GNAME(IFIN)
      WRITE (ITTO,9982)
9982  FORMAT (20H SYSTEM GAIN=(F10.1))
      READ (ITTI,9981) GAIN
9981  FORMAT (F10.1)
C
C OPEN DISK ON CHANNEL IDSK0 ; NAME OF FILE IS IN ARRAY IFIN
C AND READ IN IMPULSE RESPONSE
C
      CALL FOPEN(IDSK0, IFIN)
      CALL RBLK(IDSK0, IB0, IS, NB, IER)
      CALL CLOSE(IDSK0)
C
C SCALE H(T) TO 'GAIN' MAXIMUM
C
      DO 70 I=1,NP
        F(I) = GAIN*FLOAT(IS(I))/SCALE
  70  CONTINUE
C
C FILTER RESP HAS BEEN DEFINED
C
  80  CONTINUE
      WRITE (ITTO,9980)
9980  FORMAT (21H INPUT DATA FILENAME=)
      CALL GNAME(IFIN)
      WRITE (ITTO,9979)
9979  FORMAT (22H OUTPUT DATA FILENAME=)
      CALL GNAME(IFOUT)
      CALL FOPEN(IDSK0, IFIN)
      CALL FOPEN(IDSK1, IFOUT)
      WRITE (ITTO,9978)
9978  FORMAT (43H NUMBER OF DATA BLOCKS TO BE FILTERED= (I5))
      READ (ITTI,9977) NBLKS
9977  FORMAT (I5)
C
C FILTER DATA FROM FILE IFIN AND PUT IN FILE IFOUT
C IF COR IS TRUE, REVERSE IMPULSE RESP
C
      IF (.NOT.COR) GO TO 100
      NPD2 = NP/2
      DO 90 I=1,NPD2
        J = NP + 1 - I
        F0 = F(J)
        F(J) = F(I)
        F(I) = F0
  90  CONTINUE
 100  CONTINUE
C
C FFT FILTER H(T) TO GET FILTER SPECTRUM
C FORCE SECOND HALF OF FILTER TO ZERO; ZERO R SECOND HALF
C
      DO 110 I=NPP1,NPT2
        F(I) = 0
        R(I) = 0
 110  CONTINUE
      CALL FAST(F, NPT2)
C
C START LOOP OVER FRAMES OF DATA
C
      IFRAME = 0
      DO 140 IB=1,NBLKS,NB
C
C READ IN FRAME OF DATA FROM DISK CHANNEL IDSK0
C AND STORE FLOATED DATA IN R ARRAY
C
        CALL RBLK(IDSK0, IFRAME, IS, NB, IER)
        DO 120 I=1,NP
          R(I) = IS(I)
 120    CONTINUE
C
C FILTER FRAME OF DATA
C
        CALL RFILT(R, F, S, NP)
C
C OUTPUT PRESENT FRAME OF DATA TO OUTPUT DISK FILE IDSK1
C
        DO 130 I=1,NP
          IF (ABS(R(I)).GT.SCALE) WRITE (ITTO,9976) IB, I
9976      FORMAT (32H OUTPUT OVERFLOW BLOCK AND WORD , 2I5)
          IS(I) = R(I)
 130    CONTINUE
C
C WRITE DATA ON DISK IN BINARY FORMAT
C
        CALL WBLK(IDSK1, IFRAME, IS, NB, IER)
        WRITE (ITTO,9975) IB
9975    FORMAT (7H BLOCK , I6, 12H IS FINISHED)
        IFRAME = IFRAME + NB
 140  CONTINUE
C
C WRITE OUT LAST FRAME
C
      DO 150 I=NPP1,NPT2
        IF (ABS(R(I)).GT.SCALE) WRITE (ITTO,9976) IB, I
        J = I - NP
        IS(J) = R(I)
 150  CONTINUE
      CALL WBLK(IDSK1, IFRAME, IS, NB, IER)
      CALL CLOSE(IDSK0)
      CALL CLOSE(IDSK1)
      STOP
      END
C
C-----------------------------------------------------------------------
C SUBROUTINE: RFILT
```

```
C FILTER ONE FRAME (I.E., NP POINTS) OF DATA
C
C PROGRAM ASSUMES:
C      1. R(NP+1) TO R(2*NP) HAS NOT BEEN USED SINCE LAST CALL TO
C      'RFILT' AND WAS ZERO BEFORE FIRST CALL.
C      2. INPUT AND OUTPUT DATA (TIME SERIES) ARE IN R(1) TO R(NP)
C      3. F(1) TO F(2*NP+2) IS THE FILTER FREQUENCY RESP.
C      4. DATA ARE STORED IN F AS: DC,0.,F1REAL,F1IMAG,
C      F2REAL,F2IMAG,...,FSAMPLING/2.,0.
C      5. S(1) TO S(NP) IS A SCRATCH ARRAY; IT MAY BE USED IN 'MAIN
C      6. IMPULSE RESP. OF FILTER F IS ZERO FOR NP+1 THRU 2*NP
C      (I.E. SECOND HALF OF H(T)=0, WHERE F(W)=FFT(H) )
C-----------------------------------------------------------------------
C
      SUBROUTINE RFILT(R, F, S, NP)
      DIMENSION S(1), R(1), F(1)
      NPT2 = NP*2
C
C STORE PREVIOUS TAIL IN SCRATCH ARRAY S(.);
C ZERO SECOND HALF OF R--FIRST HALF OF R IS NEW DATA TO BE FILTERED
C
      DO 10 K=1,NP
        NPPK = NP + K
        S(K) = R(NPPK)
        R(NPPK) = 0.
   10 CONTINUE
C
C TAKE FFT OF DATA
C
      CALL FAST(R, NPT2)
C
C HANDLE VALUES AS COMPLEX VALUES
C
      KMAX = NPT2 + 1
      DO 20 K=1,KMAX,2
        X = F(K)*R(K) - F(K+1)*R(K+1)
        Y = F(K)*R(K+1) + F(K+1)*R(K)
        R(K) = X
        R(K+1) = Y
   20 CONTINUE
C
C INVERSE TRANSFORM PRODUCT
C
      CALL FSST(R, NPT2)
C
C ADD IN TAIL FROM PREVIOUS FRAME WHICH WAS STORED IN S(.)
C
      DO 30 K=1,NP
        R(K) = R(K) + S(K)
   30 CONTINUE
      RETURN
      END
C
C-----------------------------------------------------------------------
C PROGRAM: RBLK
C PGM TO READ DATA FROM DISK
C-----------------------------------------------------------------------
C
      SUBROUTINE RBLK(ICN, IB0, IARRAY, NBLK, IER)
      DIMENSION NAME(1), IARRAY(1)
      CALL DKIO(NAME, ICN, IB0, IARRAY, NBLK, IER, 2)
      RETURN
      END
C
C-----------------------------------------------------------------------
C SUBROUTINE: WBLK
C PGM TO WRITE DATA ON DISK
C-----------------------------------------------------------------------
C
      SUBROUTINE WBLK(ICN, IB0, IARRAY, NBLK, IER)
      DIMENSION NAME(1), IARRAY(1)
      CALL DKIO(NAME, ICN, IB0, IARRAY, NBLK, IER, 3)
      RETURN
      END
C
C-----------------------------------------------------------------------
C SUBROUTINE:  FOPEN
C PGM TO OPEN A DISC FILE ON CHANNEL ICN WITH NAME IN ARRAY "NAME"
C-----------------------------------------------------------------------
C
      SUBROUTINE FOPEN(ICN, NAME)
      DIMENSION NAME(1), IARRAY(1)
      CALL DKIO(NAME, ICN, 0, IARRAY, 0, IER, 0)
      RETURN
      END
C
C-----------------------------------------------------------------------
C SUBROUTINE: CLOSE
C PGM TO CLOSE A DISK FILE ON CHANNEL ICN
C-----------------------------------------------------------------------
C
      SUBROUTINE CLOSE(ICN)
      DIMENSION NAME(1), IARRAY(1)
      CALL DKIO(NAME, ICN, 0, IARRAY, 0, IER, 1)
      RETURN
      END
C
C-----------------------------------------------------------------------
C SUBROUTINE: DKIO
C *** THIS PROGRAM IS MACHINE DEPENDENT
C *** FORTRAN CODE HAS BEEN SUPPLIED FOR A DATA GENERAL COMPUTER
C *** WITH A FORTRAN 5 COMPILER
C ***
      SUBROUTINE DKIO(NAME, ICN, IB0, IARRAY, NBLK, IER, IOPER)
      DIMENSION NAME(10), IARRAY(1)
C PGM TO OPEN, CLOSE, READ, AND WRITE DATA FROM BULK STORAGE
C DEVICE.
C
C..    OPER    OPERATION
C      0       OPEN
C      1       CLOSE DISC
C      2       READ FROM DISK
C      3       WRITE ON DISC
C
C      ICN=FORTRAN CHANNEL NUMBER FOR BULK STORAGE
C      NAME=FORTRAN ARRAY CONTAINING ASCII NAME OF DISC FILE
C      IB0=BLOCK TO BE READ OR WRITTEN FROM DISK FILE
C      IARRAY=ARRAY WHERE DATA IS TO BE TRANSFERED TO OR FROM
C      NBLK=NUMBER OF BLOCKS TO BE TRANSFERED EACH CALL
C      IER=ERROR FLAG (PARAMETER NOT USED IN THIS PROGRAM)
C-----------------------------------------------------------------------
C
      LENBLK = 256
      IOPER1 = IOPER + 1
      GO TO (10, 20, 30, 40), IOPER1
C
C *** OPEN - INTRODUCE A FILE ON A SPECIFIC CHANNEL TO THE
C ***        OPERATING SYSTEM PRIOR TO READING OR WRITING DATA IN IT
C
   10 CONTINUE
```

```
      OPEN ICN,NAME
      RETURN
C
C *** CLOSE - RELEASE THE FILE AND CHANNEL NUMBER FROM THE
C ***         OPERATING SYSTEM
C
20     CLOSE ICN
      RETURN
C
C *** RDBLK - READS NBLK BLOCKS (1 BLOCK=256 WORDS; 1WORD=16 BITS) OF
C ***         DATA STARTING AT BLOCK IB0 INTO ARRAY IARRAY (FIRST
C ***         BLOCK IN FILE IS NUMBER 0)
C
  30  CALL RDBLK(ICN, IB0, IARRAY, NBLK, IER)
      RETURN
C
C *** WRBLK - WRITES NBLK BLOCKS OF DATA STARTING AT BLOCK IB0 FROM
C ***         ARRAY IARRAY INTO DISK FILE OPENED ON CHANNEL ICN
C
  40  CALL WRBLK(ICN, IB0, IARRAY, NBLK, IER)
      RETURN
      END
C
C-----------------------------------------------------------------------
C SUBROUTINE: GNAME
C ***
C ***  THIS PROGRAM IS MACHINE DEPENDENT
C ***  FORTRAN CODE HAS BEEN SUPPLIED FOR A DATA GENERAL COMPUTER
C *** WITH A FORTRAN 5 COMPILER
C ***
C THIS PROGRAM READS ASCII DATA INTO AN ARRAY 'NAME(I)'
C IN A FORMAT THAT MAY BE USED BY : CALL FOPEN(ICN,NAME)
C WHICH OPENS DISK FILE 'NAME' ON FORTRAN CHANNEL ICN
C-----------------------------------------------------------------------
C
      SUBROUTINE GNAME(NAME)
      DIMENSION NAME(10)
      ITTI = I1MACH(1)
C
C READ UP TO 10 CHARACTERS FROM DEVICE ITTI IN S (STRING) FORMAT
C THE CHARACTERS ARE PACKED 2 PER 16 BIT WORD AND ARE LEFT
C JUSTIFIED IN THE ARRAY NAME.
C
      READ (ITTI,9999) NAME(1)
9999  FORMAT (S10)
      RETURN
      END
C
C-----------------------------------------------------------------------
C MAIN PROGRAM: TESTFILT - TEST PROGRAM FOR RFILT SUBROUTINE - 1/8/79
C SUBROUTINES:  RFILT,FAST,FSST,R1MACH,I1MACH
C-----------------------------------------------------------------------
C
      DIMENSION Y(4096), F(258), R(258), S(128)
C
C DEFINE FUNCTION STATEMENT FOR INPUT "FILE" X(*)
C
      X(K) = (AMOD(FLOAT(K-1),10.) - 5.) / 5.
C
C DEFINE CONSTANTS
C
      ITTO = I1MACH(2)
      NPTS = 4096
      NFILT = 128
      NFFT = 2*NFILT
      NBLKL = 64
      NBLKS = NPTS/NBLKL
      NB = NFILT/NBLKL
C
C INITIALIZE FILTER: DELAY OF 1 SAMPLE,CHANGE OF SIGN
C
      DO 10 I=1,NFFT
        F(I) = 0.
        R(I) = 0.
  10  CONTINUE
      F(2) = -1.
C
C FFT FILTER
C
      CALL FAST(F, NFFT)
C
C START LOOP ON DATA
C
      DO 40 IB=1,NBLKS,NB
C
C
C GET BLK OF DATA FROM INPUT "FILE" FUNCTION X(I)
C
        DO 20 I=1,NFILT
          J = I + (IB-1)*NBLKL
          R(I) = X(J)
  20    CONTINUE
C
C FILTER DATA
C
        CALL RFILT(R, F, S, NFILT)
C
C WRITE OUTPUT INTO OUTPUT "FILE" ARRAY Y(I)
C
        DO 30 I=1,NFILT
          J = I + (IB-1)*NBLKL
          Y(J) = R(I)
  30    CONTINUE
  40  CONTINUE
C
C CHECK RESULTS
C
      RMACH4 = R1MACH(4)
      ERMAX = 0.
      FNFFT = NFFT
      RMAX = SQRT(FNFFT)*RMACH4
      NPTS1 = NPTS - 1
      DO 50 I=1,NPTS1
        ERR = X(I) + Y(I+1)
        ERMAX = AMAX1(ERMAX,ABS(ERR))
  50  CONTINUE
      WRITE (ITTO,9999) RMACH4, ERMAX
9999  FORMAT (18H MACHINE ACCURACY=, E12.5, 11H MAX ERROR=, E12.5)
      IF (RMAX.GT.ERMAX) WRITE (ITTO,9998)
9998  FORMAT (12H TEST PASSES)
      IF (RMAX.LE.ERMAX) WRITE (ITTO,9997)
9997  FORMAT (11H TEST FAILS)
      STOP
      END
```

CHAPTER 4

Linear Prediction Analysis of Speech Signals

B. S. Atal

Introduction

Linear prediction is one of the most widely used techniques in speech analysis. It provides a simple and effective method of representing speech signals in terms of a small number of slowly-varying parameters. The linear predictability of speech signals is based directly on a linear filter model of speech production. In this model, the filtering properties of the glottal flow, the vocal tract, and the radiation at the lips are represented by a single time-varying all-pole filter. The linear prediction techniques provide a simple method for estimating the parameters of the all-pole filter.

Linear prediction parameters have been found useful in a variety of applications, such as speech coding, speech recognition, speech synthesis, and speaker verification. Although the technique of linear prediction is applicable to a much broader class of signals, the programs presented in this section have developed from applications to speech signals. For applications to other classes of signals, great care must be exercised in choosing the right program and in the appropriate selection of the variables of the programs. Even for speech signals, there is still a large number of unresolved issues: How many predictor parameters should be used? What is the ideal analysis interval? Where should the analysis interval be located with reference to the pitch period? What is an ideal anti-aliasing filter for analog-to-digital conversion of the speech signal prior to LPC analysis? The answers to such questions often depend on the particular application for which the LPC analysis is used. However, it is fair to say that we still lack a clear understanding of the constraints which a physical signal must satisfy for the LPC analysis to work satisfactorily on that signal.

The programs presented in this section are of two types: 1) a set of programs (AUTO, COVAR, CLHARM, and COVLAT) related to different methods of LPC analysis, and 2) another program, LPTRAN, that transforms one LPC representation to another.

Several alternate ways of doing LPC analysis have evolved over the past ten years or so, mainly to take care of the complications introduced by the nonstationary nature of speech signals. For a nonstationary signal, it is important to perform a "short-time analysis" [1] using short segments of the signal during which it could be considered to be nearly stationary. The differences in the various LPC analysis methods lie in the manner in which they treat the signal outside the analysis interval in the process of determining predictor coefficients.

The correlations between different samples of a signal play a key role in the LPC analysis. All of the LPC analysis methods require that such correlations be computed in the process of determining LPC parameters. In the program AUTO (based on the *autocorrelation* method of LPC analysis), these correlations are expressed in terms of a short-time autocorrelation function [1,2]. This method requires that the signal be set to zero outside the analysis interval and a suitable window, such as a Hamming window [2], be used to reduce the abrupt change in signal values occurring at the beginning and at the end of the analysis interval. The use of an autocorrelation function leads to a particularly simple and efficient algorithm for determining LPC parameters. The program COVAR (patterned after the *covariance* method of LPC analysis) in the same section computes the LPC parameters by minimizing the mean-squared prediction error over the analysis interval. This method avoids truncation of the signal, but does require that an entire matrix of covariances be computed from the speech signal. In its original form [3], this method did not always produce predictor coefficients which could be used for a later synthesis of the signal without producing unstable outputs. However, the program COVAR can be modified to produce a set of reflection coefficients, $k_1, k_2, \ldots, k_M$, which can be transformed to a set of predictor coefficients corresponding to a stable all-pole filter [4,5]. This transformation can be performed, if desired, by the program LPTRAN (to be discussed later). The programs COVLAT and CLHARM described in Sec. 4.2 follow a somewhat different approach to linear prediction based on the lattice structure of the all-pole filter. The lattice formulation does not need a knowledge of the short-time autocorrelation function for determining predictor coefficients and therefore avoids windowing of the signal. Also, the resulting all-pole filter is always stable. Thus, the lattice method may be preferable for certain applications.

As expected, the LPC analysis produces accurate results if a signal has been generated as the output of an all-pole filter excited either by a delta function or white noise. However, it is rare that a physical signal can be modelled exactly as stated above. The performance of LPC techniques suffers whenever any one of the assumptions underlying the linear prediction model is violated. For example, in the prsence of spectral zeros (which could be introduced by the vocal tract, by the glottal excitation, or even by the anti-aliasing filter used in the A/D conversion), the linear prediction techniques not only skip over the spectral zeros but often introduce errors in formant frequencies and their bandwidths. The periodic nature of voiced excitation also introduces similar errors [3]. These errors must be kept in mind while interpreting the results of LPC analysis.

There are several equivalent ways of describing the characteristics of an all-pole filter. The program LPTRAN described in Sec. 4.3 allows one to go from one representation to another in an efficient fashion. Although all these representations are equivalent and are uniquely related to each other, some of them may be more suitable than others for certain applications [2,3]. For example, the area function representation has been found to be well-suited for speech synthesis-by-rule applications where a continuous speech message is created by concatenating and then smoothing a number of prestored elements. Similarly, the cepstral parameters have been found to be most effective for speaker recognition tasks. Another important transformation is the set of partial correlation coefficients (PARCOR) or reflection coefficients. Table 1 of Sec. 4.3 lists the different LPC representations which can be transformed by LPTRAN. It should be noted that the transformations between alternate LPC representations are valid only for all-pole filters.

References

1. J. L. Flanagan, *Speech Analysis, Synthesis and Perception,* Springer-Verlag, 1972.
2. L. R. Rabiner and R. W. Schafer, *Digital Processing of Speech Signals,* Prentice-Hall, 1978.
3. J. D. Markel and A. H. Gray, Jr., *Linear Prediction of Speech,* Springer-Verlag, New York, 1976.
4. B. S. Atal, "On Determining Partial Correlation Coefficients by the Covariance Method of Linear Prediction", *J. Acoust. Soc. Amer.,* Vol. 62, Supplement No. 1, Paper BB12, page S64, Fall 1977.
5. B. W. Dickinson and J. M. Turner, "Reflection Coefficient Estimation Using Cholesky Decomposition", *IEEE Trans. Acoust., Speech, and Signal Processing,* Vol. ASSP-27, No. 2, pp. 146-149, April 1979.

4.1

Linear Prediction Analysis Programs (AUTO-COVAR)

A. H. Gray, Jr.

Department of Electrical Eng. & Computer Science
University of California at Santa Barbara
Santa Barbara, CA 93106
and
Signal Technology, Inc.*

J. D. Markel

*Signal Technology, Inc.
15 De La Guerra Street
Santa Barbara, CA 93101

1. Purpose

The purpose of these programs is to implement the autocorrelation (AUTO) and covariance (COVAR) methods of linear prediction analysis.

2. COVARiance and AUTOcorrelation Methods

Two cases will be treated here, commonly termed the covariance and autocorrelation methods. The Fortran subroutines which carry out the solution are called COVAR and AUTO. For derivations, detailed discussion, motivation, and references the reader is referred to [1]--only a summary of the algorithms is presented here.

The general problem can be stated as follows. Given a data sequence $\{x(0),x(1),...,x(N-1)\}$ and an integer M, find the coefficients $\{a_1,a_2,...,a_M\}$ which minimize the summation

$$\alpha = \sum_{n=n_0}^{n_1} [x(n) + \sum_{k=1}^{M} a_k x(n-k)]^2 . \tag{1}$$

This minimization is carried out by solving the equations

$$\sum_{i=1}^{M} a_i c_{ik} = - c_{0k} \quad \text{for} \quad k = 1,2,...,M , \tag{2}$$

where

$$c_{ik} = \sum_{n=n_0}^{n_1} x(n-i)x(n-k) . \tag{3}$$

In the covariance method the limits n_0 and n_1 are defined so that no data points $\{x(n)\}$ outside of the range $0 \leqslant n < N$ are needed for the evaluation of Eq. (3):

$$n_0 = M, \quad n_1 = N - 1 . \tag{4}$$

In this case, the coefficient set $\{c_{ik}\}$ forms a symmetric, positive semi-definite matrix. The matrix will be singular if the input data sequence $\{x(n)\}$ satisfies a linear homogeneous difference equation of order M or less.

In the autocorrelation method, the data sequence is treated as though it were zero outside of the interval from $n = 0$ through $n = N - 1$:

$$x(n) = 0 \quad \text{for} \quad n < 0 \quad \text{and for} \quad n > N - 1 \quad (AUTO) . \tag{5}$$

In this case n_0 and n_1 are defined as $-\infty$ and $+\infty$ respectively. The coefficient set $\{c_{ik}\}$ forms the elements of a symmetric positive definite Toeplitz matrix, and the coefficients can be expressed in terms of an autocorrelation sequence as

$$c_{ik} = r(|i-k|) \qquad (AUTO) \tag{6}$$

where

$$r(k) = \sum_{n=0}^{N-1-k} x(n)x(n+k) \quad \text{for} \quad k = 0,1,...,N \quad (AUTO) \ . \tag{7}$$

Only values of $r(k)$ for $k = 0,1,...,M$ are needed for the solution.

The two methods give identical results when the data sequence is truncated so that $x(n) = 0$ for $n < M$ and for $n > N-1-M$. The two methods give similar results when $N >> M$.

3. The Algorithms

The subroutine COVAR implements a form of Cholesky decomposition, and the subroutine AUTO implements a form of Robinson's recursion. As many of the steps are common to both, a common notation is used in this discussion.

3.1 Initialization (step zero)

For COVAR Eqs. (3) and (4) are applied directly to obtain the covariance coefficients c_{00}, c_{10}, and c_{11}. For AUTO all of the correlation coefficients are obtained by directly applying Eq. (7), though only $r(0)$ and $r(1)$ are needed for initialization:

$$c_{00} = c_{11} = r(0) \quad \text{and} \quad c_{10} = r(1) \quad (AUTO) \ .$$

The following parameters are then defined:

$$\begin{aligned} a_{00} &= 1, & b_{01} &= 1, \\ \alpha_0 &= c_{00}, & \beta_0 &= c_{11}, \\ k_1 &= -c_{10}/c_{11}, \\ a_{10} &= 1, & a_{11} &= k_1, \\ \alpha_1 &= \alpha_0 - k_1^2 \beta_0 \ . \end{aligned}$$

3.2 Recursion (step m for $m=1,2,...,M-1$)

For COVAR, $c_{m+1,0}$ is directly evaluated from Eq. (3) and Eq. (4). The coefficients $c_{m+1,k}$ for $k = 1,2,...,m+1$ are found from

$$c_{m+1,k} = c_{m,k-1} + x(M-m-1)x(M-k) - x(N-m-1)x(N-k)$$

For AUTO, the correlation sequence $\{r(k)\}$ is was evaluated in the initialization step.

For COVAR, one evaluates then

$$\gamma_{mn} = \frac{1}{\beta_n} \sum_{j=1}^{n+1} c_{m+1,j}\, b_{nj} \quad \text{for} \quad n = 0,1,...,m-1 \ , \tag{8a}$$

$$b_{mj} = -\sum_{i=j-1}^{m-1} \gamma_{mi}\, b_{ij} \quad \text{for} \quad j = 1,2,...,m \ , \tag{8b}$$

$$b_{m,m+1} = 1 \ , \tag{8c}$$

$$\beta_m = \sum_{j=1}^{m+1} c_{m+1,j}\, b_{mj} \ . \tag{8d}$$

While these equations are correct for the AUTO case, they are not necessary since

$$b_{mj} = a_{m,m+1-j} \quad \text{for} \quad j = 1,2,...,m+1 \quad (AUTO). \tag{9a}$$

and

$$\beta_m = \alpha_m \ . \tag{9b}$$

Finally, defining

$$k_{m+1} = -\frac{1}{\beta_m} \sum_{i=0}^{m} c_{m+1,i}\, a_{mi} \,, \tag{10}$$
$$a_{m+1,0} = 1 \,,$$
$$a_{m+1,i} = a_{mi} + k_{m+1} \quad b_{mi} \quad \text{for} \quad i = 1,2,\ldots,m \,,$$
$$a_{m+1,m+1} = k_{m+1} \,,$$

and

$$\alpha_{m+1} = \alpha_m - k_{m+1}^2 \beta_m \,,$$

step m is completed.

3.3 Termination (end of step $M-1$)

At the end of step $M - 1$, the results

$$a_k = a_{Mk} \quad \text{for} \quad k = 1,2,\ldots,M$$

and

$$\alpha = \alpha_M \tag{11}$$

are obtained. In addition, the intermediate parameter set $\{k_1, k_2, \ldots k_M\}$ is supplied as an output. These parameters are often called reflection coefficients for the AUTO case. In the AUTO case, they will all have a magnitude less than one, and represent reflection coefficients in an acoustic tube model of speech production. What meaning they may have in the COVAR case is not clear.

3.4 Numerical Accuracy

In essence, the solution is an indirect inversion of a matrix whose elements are the $\{c_{ik}\}$ coefficients, whose determinant is the product $\beta_0 \beta_1 \beta_2 \cdots \beta_{M-1}$. Theoretically, the coefficients $\{\beta_k\}$ are non-negative, and in fact are positive for the AUTO case. A negative value for one of the coefficients $\{\beta_k\}$ can only result from a numerical error. A zero value could result in the COVAR case.

Any very small value of one of the coefficients $\{\beta_k\}$ will indicate problems, for they are used as divisors in Eq. (8b) for COVAR and as divisors in Eq. (10) for COVAR and AUTO, thus amplifying the effects of any numerical errors in evaluating the numerators of Eq. (8a) and Eq. (10). This problem is *not* relieved by resorting to other solution methods such as Gauss-Siedell, and can only be handled by increasing the number of significant figures, lowering the value of M, or by properly pre-emphasizing the data [1, Chapter 9].

Both subroutines have test statements to test for a zero or negative α_k or β_k. A print statement is placed at the appropriate point in the program to warn the user that a singular matrix has been defined.

4. Program Description

4.1 Usage

Inputs and outputs to the subroutines COVAR and AUTO are in the argument lists, COVAR(N,X,M,A,ALPHA,GRC) and AUTO(N,X,M,A,ALPHA,RC). The inputs are

- N — the number, N, of data points,
- X(·) — the input data with $X(k) = x(k-1)$ for $k = 1,2,\ldots,N$,
- M — the order, M, of the filter.

The main outputs are the coefficients $\{a_k\}$ contained in the dimensioned array with

$$A(1) = 1$$
$$A(k) = a_{k-1} \quad \text{for} \quad k = 2,3,\ldots,M+1 \,.$$

An auxiliary set of outputs are the coefficients $\{k_m\}$ with

$$GRC(m) = k_m \quad \text{for} \quad m = 1,2,\ldots,M \qquad (COVAR)$$
$$RC(m) = k_m \quad \text{for} \quad m = 1,2,\ldots,M \qquad (AUTO)$$

The dimensioned variables X(·), A(·), and GRC(·) or RC(·) are dimensioned by the calling program. The dimension of X(·) must be at least N, the dimension of A(·) must be at least $M+1$, and the dimension of GRC(·) or RC(·) must be at least M. ALPHA defines the residual energy of Eq. (11).

Table 1: Program Variables

TEXT	*COVAR*	*AUTO*
N	N	N
$x(n)$	X(n+1)	X(n+1)
M	M	M
a_{mi}	A(i+1)	A(i+1)
α_{mi}	ALPHA	ALPHA
k_m	GRC(m)	RC(m)
$c_{m+1,j}$	CC(j+1)	R(m+2−j)
β_m	BETA(m+1)	ALPHA
γ_{mi}	GAM	--
b_{mi}	B(·)	A(·)

Table 2: Minimum Dimensions for Variables

DIMENSIONED VARIABLE	*MIN. DIM. FOR COVAR*	*MIN. DIM. FOR AUTO*
X(·)	N	N
A(·)	$M+1$	$M+1$
CC(·)	$M+1$	
R(·)		$M+1$
GRC(·)	M	
RC(·)		M
BETA(·)	M	
B(·)	$M(M-1)/2$	

4.2 Program Variables

Table 1 shows a listing of the variables in this discussion. While some of the variables in the discussion have double subscripts, only singly dimensioned variables are used in the program.

The coefficients $\{b_{mk}\}$ do not appear explicitly in AUTO, for they are related to the coefficients $\{a_{mk}\}$ by Eq. (9a). In COVAR these coefficients are stored in the singly dimensioned array B(·) in the order

$$(b_{11}, b_{21}, b_{22}, b_{31}, b_{32}, b_{33}, b_{41}, \ldots b_{M-1,M-1}) .$$

Table 2 shows the minimum dimension needed for each of the dimensioned variables. The subroutines are presently set to handle values of M up to 20. For values of M larger than 20, the dimension statements in the subroutines must be changed according to Table 2.

5. Test Program

A sample test program and its output is given. This particular program first generates a 40 point data sequence which is the unit sample response of the filter

$$1/[1 - 1.44z^{-1} + 1.26z^{-2} - 0.81z^{-3}] .$$

As the filter is third order and the numerator is of lower order than the denominator, the covariance approach with M=3 theoretically reproduces the denominator and gives a total squared error of zero. The test program gives the results of both covariance and autocorrelation methods for M=2 and M=3, and for M=3 indicates this, within the limits of computer accuracy, giving a residual energy of 0.6×10^{-5} for M=3 as opposed to 1.7 for M=2.

The data sequence is then shifted by three steps, with three zeros appended at the start and end for the new 46 point sequence. Running both the methods on this sequence theoretically gives identical results to the autocorrelation method for the 40 point sequence with M=3. This is seen in the final results for the test program printout.

References

1. Markel, J. D. and Gray, A. H., Jr., *Linear Prediction of Speech,* Springer Verlag, New York, 1976.

Table 3

```
        OUTPUT DATA FROM AUTO/COVAR TEST PROGRAM

  1    0.100000E 01   -0.891115E 00
  2   -0.108019E 01    0.197712E 00
  3    0.197712E 00    0.000000E 00
0.172095E 01

  1    0.100000E 01   -0.905548E 00
  2   -0.116199E 01    0.283199E 00
  3    0.283199E 00    0.000000E 00
0.308386E 01

  1    0.100000E 01   -0.935486E 00
  2   -0.143999E 01    0.313706E 00
  3    0.125999E 01   -0.809997E 00
  4   -0.809997E 00    0.000000E 00
0.619888E-05

  1    0.100000E 01   -0.905548E 00
  2   -0.13 308E 01    0.283199E 00
  3    0.110826E 01   -0.710041E 00
  4   -0.710041E 00    0.000000E 00
0.152910E 01

  1    0.100000E 01   -0.905548E 00
  2   -0.136308E 01    0.283199E 00
  3    0.110826E 01   -0.710042E 00
  4   -0.710042E 00    0.000000E 00
0.152910E 01

  1    0.100000E 01   -0.905548E 00
  2   -0.136308E 01    0.283199E 00
  3    0.110826E 01   -0.710041E 00
  4   -0.710041E 00    0.000000E 00
0.152910E 01
```

Appendix

```
C
C-------------------------------------------------------------------
C  MAIN PROGRAM: TEST PROGRAM FOR COVAR AND AUTO
C  AUTHORS:      A. H. GRAY, JR.   AND J. D. MARKEL
C      GRAY -    UNIV. OF CALIF., SANTA BARBARA, CA 93109
C      BOTH -    SIGNAL TECHNOLOGY, INC., 15 W. DE LA GUERRA STREET
C                SANTA BARBARA, CA 93101
C-------------------------------------------------------------------
C
      DIMENSION X(100), A(21), RC(21)
      IOUTD = I1MACH(2)
      X(1) = 1.
      X(2) = 1.44
      X(3) = 1.44*X(2) - 1.26
      DO 10 J=4,40
        X(J) = 1.44*X(J-1) - 1.26*X(J-2) + .81*X(J-3)
   10 CONTINUE
 9999 FORMAT (I3, 2E15.6)
 9998 FORMAT (E12.6/)
      N = 40
      DO 20 M=2,3
        CALL COVAR(N, X, M, A, ALPHA, RC)
        MP = M + 1
        WRITE (IOUTD,9999) (I,A(I),RC(I),I=1,MP)
        WRITE (IOUTD,9998) ALPHA
        CALL AUTO(N, X, M, A, ALPHA, RC)
        WRITE (IOUTD,9999) (I,A(I),RC(I),I=1,MP)
        WRITE (IOUTD,9998) ALPHA
   20 CONTINUE
      DO 30 JB=1,40
        J = 44 - JB
        X(J) = X(J-3)
   30 CONTINUE
      DO 40 J=1,3
        JB = 43 + J
        X(J) = 0.
        X(JB) = 0.
   40 CONTINUE
      M = 3
      N = 46
      CALL COVAR(N, X, M, A, ALPHA, RC)
      MP = M + 1
      WRITE (IOUTD,9999) (I,A(I),RC(I),I=1,MP)
      WRITE (IOUTD,9998) ALPHA
      CALL AUTO(N, X, M, A, ALPHA, RC)
      WRITE (IOUTD,9999) (I,A(I),RC(I),I=1,MP)
      WRITE (IOUTD,9998) ALPHA
      STOP
      END
C
C-------------------------------------------------------------------
C SUBROUTINE:  COVAR
C A SUBROUTINE FOR IMPLEMENTING THE COVARIANCE
C METHOD OF LINEAR PREDICTION ANALYSIS
C-------------------------------------------------------------------
C
      SUBROUTINE COVAR(N, X, M, A, ALPHA, GRC)
C
C   INPUTS:     N - NO. OF DATA POINTS
C               X(N) - INPUT DATA SEQUENCE
C               M - ORDER OF FILTER (M<21, SEE NOTE*)
C  OUTPUTS:     A - FILTER COEFFICIENTS
C               ALPHA - RESIDUAL "ENERGY"
C               A - FILTER COEFFICIENTS
C               GRC - "GENERALIZED REFLECTION COEFFICIENTS",
C
C *PROGRAM LIMITED TO M=20, BECAUSE OF THE DIMENSIONS
C B(M*(M+1)/2), BETA(M), AND CC(M+1)
C
      DIMENSION X(1), A(1), GRC(1)
      DIMENSION B(190), BETA(20), CC(21)
      MP = M + 1
      MT = MP*M/2
      MT = (MP*M)/2
      DO 10 J=1,MT
        B(J) = 0.
   10 CONTINUE
      ALPHA = 0.
      CC(1) = 0.
      CC(2) = 0.
      DO 20 NP=MP,N
        NP1 = NP - 1
        ALPHA = ALPHA + X(NP)*X(NP)
        CC(1) = CC(1) + X(NP)*X(NP1)
        CC(2) = CC(2) + X(NP1)*X(NP1)
   20 CONTINUE
      B(1) = 1.
      BETA(1) = CC(2)
      GRC(1) = -CC(1)/CC(2)
      A(1) = 1.
      A(2) = GRC(1)
      ALPHA = ALPHA + GRC(1)*CC(1)
      MF = M
      DO 130 MINC=2,MF
        DO 30 J=1,MINC
          JP = MINC + 2 - J
          N1 = MP + 1 - JP
          N2 = N + 1 - MINC
          N3 = N + 2 - JP
          N4 = MP - MINC
          CC(JP) = CC(JP-1) + X(N4)*X(N1) - X(N2)*X(N3)
   30   CONTINUE
        CC(1) = 0.
        DO 40 NP=MP,N
          N1 = NP - MINC
          CC(1) = CC(1) + X(N1)*X(NP)
   40   CONTINUE
        MSUB = (MINC*MINC-MINC)/2
        MM1 = MINC - 1
        N1 = MSUB + MINC
        B(N1) = 1.
        DO 80 IP=1,MM1
          ISUB = (IP*IP-IP)/2
          IF (BETA(IP)) 150, 150, 50
   50     GAM = 0.
          DO 60 J=1,IP
            N1 = ISUB + J
            GAM = GAM + CC(J+1)*B(N1)
   60     CONTINUE
          GAM = GAM/BETA(IP)
          DO 70 JP=1,IP
            N1 = MSUB + JP
            N2 = ISUB + JP
            B(N1) = B(N1) - GAM*B(N2)
   70     CONTINUE
   80   CONTINUE
        BETA(MINC) = 0.
        DO 90 J=1,MINC
          N1 = MSUB + J
          BETA(MINC) = BETA(MINC) + CC(J+1)*B(N1)
```

```
 90     CONTINUE
        IF (BETA(MINC)) 150, 150, 100
100     S = 0.
        DO 110 IP=1,MINC
          S = S + CC(IP)*A(IP)
110     CONTINUE
        GRC(MINC) = -S/BETA(MINC)
        DO 120 IP=2,MINC
          M2 = MSUB + IP - 1
          A(IP) = A(IP) + GRC(MINC)*B(M2)
120     CONTINUE
        A(MINC+1) = GRC(MINC)
        S = GRC(MINC)*GRC(MINC)*BETA(MINC)
        ALPHA = ALPHA - S
        IF (ALPHA) 150, 150, 130
130   CONTINUE
140   RETURN
150   CONTINUE
C
C WARNING - SINGULAR MATRIX
C
      IOUTD = I1MACH(2)
      WRITE (IOUTD,9999)
9999  FORMAT (34H WARNING - SINGULAR MATRIX - COVAR)
      GO TO 140
      END
C
C-----------------------------------------------------------------------
C SUBROUTINE:  AUTO
C A SUBROUTINE FOR IMPLEMENTING THE AUTOCORRELATION
C METHOD OF LINEAR PREDICTION ANALYSIS
C-----------------------------------------------------------------------
C
      SUBROUTINE AUTO(N, X, M, A, ALPHA, RC)
C
C   INPUTS:   N - NO. OF DATA POINTS
C             X(N) - INPUT DATA SEQUENCE
C             M - ORDER OF FILTER (M<21, SEE NOTE*)
C  OUTPUTS:   A - FILTER COEFFICIENTS
C             ALPHA - RESIDUAL "ENERGY"
C             RC - REFLECTION COEFFICIENTS
C
C *PROGRAM LIMITED TO M<21 BECAUSE OF DIMENSIONS OF R(.)
C
      DIMENSION X(1), A(1), RC(1)
      DIMENSION R(21)
      MP = M + 1
      DO 20 K=1,MP
        R(K) = 0.
        NK = N - K + 1
        DO 10 NP=1,NK
          N1 = NP + K - 1
          R(K) = R(K) + X(NP)*X(N1)
 10     CONTINUE
 20   CONTINUE
      RC(1) = -R(2)/R(1)
      A(1) = 1.
      A(2) = RC(1)
      ALPHA = R(1) + R(2)*RC(1)
      DO 50 MINC=2,M
        S = 0.
        DO 30 IP=1,MINC
          N1 = MINC - IP + 2
          S = S + R(N1)*A(IP)
 30     CONTINUE
        RC(MINC) = -S/ALPHA
        MH = MINC/2 + 1
        DO 40 IP=2,MH
          IB = MINC - IP + 2
          AT = A(IP) + RC(MINC)*A(IB)
          A(IB) = A(IB) + RC(MINC)*A(IP)
          A(IP) = AT
 40     CONTINUE
        A(MINC+1) = RC(MINC)
        ALPHA = ALPHA + RC(MINC)*S
        IF (ALPHA) 70, 70, 50
 50   CONTINUE
 60   RETURN
 70   CONTINUE
C
C WARNING - SINGULAR MATRIX
C
      IOUTD = I1MACH(2)
      WRITE (IOUTD,9999)
9999  FORMAT (33H WARNING - SINGULAR MATRIX - AUTO)
      GO TO 60
      END
C
```

4.2

Efficient Lattice Methods For Linear Prediction*

R. Viswanathan and *J. Makhoul*

Bolt Beranek and Newman, Inc.
50 Moulton Street
Cambridge, MA 02138

1. Introduction and Purpose

The lattice formulation for linear prediction was introduced in speech processing by Itakura [1]. Recently, Makhoul [2] showed the existence of a class of such lattice methods all of which have the following properties in common: (1) windowing of the signal is not required; (2) the resulting all-pole linear prediction filter is guaranteed to be stable; (3) stability is less sensitive to finite wordlength computations; and (4) quantization of the lattice model parameters (for the purpose of data compression) can be accomplished within the recursion for retention of accuracy in representation. However, the computation for these lattice methods is about four times more expensive than the traditional autocorrelation and covariance methods. To overcome this drawback, Makhoul introduced the covariance lattice (CL) methods: these compute the lattice model parameters directly from the covariance of the signal, and thus require about the same order of computational complexity as the traditional autocorrelation and covariance methods [2]. (For additional references on lattice methods, see [3]-[5].)

Given the covariance matrix of the signal, the subroutines COVLAT and CLHARM described below compute the parameters of the lattice model, known as the reflection (or partial correlation) coefficients,† and the predictor coefficients of the all-pole filter. COVLAT is a general routine in that it allows, through proper specification of input parameters, the choice of any one of a large class of CL methods. On the other hand, CLHARM uses a specific method, namely the harmonic-mean method [2]; consequently, the routine CLHARM requires less computation than COVLAT as it capitalizes on the specific structure of the harmonic-mean method.

We also give two other subroutines, COVAR1 and COVAR2, for computing the covariance of the signal. When using COVAR1, one computes at the outset the covariance of the signal up to the maximum desired lag, and employs this *fixed* covariance in the estimation of all the reflection coefficients. COVAR2, on the other hand, recomputes for each new lattice stage the covariance up to that stage in a manner that makes maximum use of the signal data [2]. To allow the use of CL methods with either type of covariance computation, the subroutine COVLAT and CLHARM have been organized for computing the reflection coefficient of only one stage at a time.

Covariance computation with COVAR1 is more common among speech researchers. It can be shown that when the routine COVAR2 for computing the covariance is used in conjunction with the harmonic-mean method, the results obtained are identical to those using Burg's technique [3] as described in [6], the latter technique being more popular in the geophysics area. However, the use of the covariance lattice routine CLHARM given below results in about a four-fold savings in computation.

2. Brief Review of the Methods

For a detailed exposition of covariance lattice methods, the reader is referred to [2]. The purpose of this review is to aid the reader in understanding the description of program parameters given in Section 3 below, and to point out some changes in the algorithm that are necessitated by the ANSI FORTRAN convention of nonzero subscripts.

---

* This work was supported by the Information Processing Techniques Branch of the Advanced Research Projects Agency.

† Partial correlation coefficients are actually the negative of the reflection coefficients as defined here.

2.1 Lattice Formulation

Linear prediction methods model the signal spectrum by an all-pole spectrum with a transfer function given by

$$H(z) = G/A(z) \tag{1a}$$

where

$$A(z) = 1 + \sum_{k=1}^{p} a_k z^{-k} \tag{1b}$$

is the inverse filter, G is a gain factor, a_k are the predictor coefficients, and p is the number of poles or predictor coefficients in the model. If $H(z)$ is stable (minimum phase), $A(z)$ can be implemented as a lattice filter [1], as shown in Fig. 1.

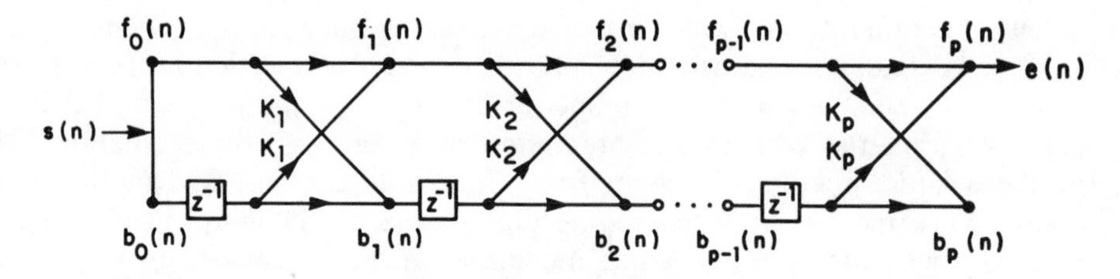

Fig. 1 Lattice inverse filter $A(z)$

The quantities $f_m(n)$ and $b_m(n)$ denote the forward and backward errors or residuals at the output of the mth lattice stage at time n. From Fig. 1, we have:

$$f_0(n) = b_0(n) = s(n) \tag{2a}$$
$$f_{m+1}(n) = f_m(n) + K_{m+1} b_m(n-1) \tag{2b}$$
$$b_{m+1}(n) = K_{m+1} f_m(n) + b_m(n-1) \tag{2c}$$

where $s(n)$ is the input signal, K_{m+1} is the reflection coefficient at stage $m+1$, and $e(n) = f_p(n)$ is the output residual.

The reflection coefficients K_m in the lattice are uniquely related to the predictor coefficients. We give below the recursive relations for obtaining a_k from K_m. These relations are employed by both subroutines COVLAT and CLHARM:

$$a_m^{(m)} = K_m$$
$$a_j^{(m)} = a_j^{(m-1)} + K_m a_{m-j}^{(m-1)}, \quad 1 \leqslant j \leqslant m-1, \tag{3}$$

where $a_j^{(m)}$, $1 \leqslant j \leqslant m$, are the predictor coefficients at stage m. The equations in (3) are computed recursively for $m = 1,2,...,p$; the final solution is given by $a_j = a_j^{(p)}$, $1 \leqslant j \leqslant p$. The necessary and sufficient condition for the stability of $H(z)$ is given by

$$|K_m| < 1, \quad 1 \leqslant m \leqslant p. \tag{4}$$

2.2 Covariance Lattice Recursions

For easy reference, major relations used by the two subroutines COVLAT and CLHARM are given below, employing only nonzero subscripts. (For example, the covariance matrix element $\phi(0,0)$ in [2] is denoted here by $\phi(1,1)$.)

Assume that the input signal $s(n)$ of the lattice in Fig. 1 is given in the range $1 \leqslant n \leqslant N$. We then define the following four quantities:

$$F_m = \sum_{n=L}^{N} f_m^2(n) \tag{5a}$$

$$B_m = \sum_{n=L}^{N} b_m^2(n) \tag{5b}$$

$$B_m' = \sum_{n=L}^{N} b_m^2(n-1) \tag{5c}$$

$$C_m = \sum_{n=L}^{N} f_m(n)\, b_m(n-1). \tag{5d}$$

The lower limit $L = p + 1$ if COVAR1 is used for the covariance computation, and $L = m + 1$ if COVAR2 is employed instead. The time averages in Eq. (5) approximate expected values considered in reference [2], where the above quantities F_m, B_m, B_m', and C_m are denoted, respectively, by $F_m(n)$, $B_m(n)$, $B_m(n-1)$ and $C_m(n)$. Below, we shall call F_m and B_m as the mean-square value (or the variance) of the forward and backward residuals, respectively.

From Eqs. (2) and (5), one can show that the residual variances at stage m+1 are given in terms of F_m, B_m', C_m and K_{m+1}, as follows:

$$F_{m+1} = F_m + 2K_{m+1}C_m + K_{m+1}^2 B_m' \tag{6a}$$

$$B_{m+1} = K_{m+1}^2 F_m + 2K_{m+1}C_m + B_m' \,. \tag{6b}$$

We employ the symbols K_{m+1}^f and K_{m+1}^b to denote the reflection coefficient at stage $m + 1$ produced by minimizing, respectively, the forward residual variance F_{m+1} (forward method), or the backward residual variance B_{m+1} (backward method). K_{m+1} or K_{m+1}^r shall denote the rth mean between K_{m+1}^f and K_{m+1}^b (see below). K^f and K^b have the same sign [2], which shall be denoted by S:

$$S = \text{sign } K^f = \text{sign } K^b \,.$$

2.2.1 General Covariance Lattice Method (COVLAT)

Three quantities, F_m, B_m' and C_m, are needed for computing the reflection coefficient at stage $m + 1$. The three quantities are computed in the subroutine from the following equations:

$$F_m = \phi(1,1) + 2\sum_{i=1}^{m} a_i^{(m)}\phi(1,i+1) + \sum_{i=1}^{m} [a_i^{(m)}]^2\phi(i+1,i+1) + 2\sum_{i=1}^{m-1}\sum_{j=i+1}^{m} a_i^{(m)}a_j^{(m)}\phi(i+1,j+1) \tag{7}$$

$$B_m' = \phi(m+2,m+2) + 2\sum_{i=1}^{m} a_i^{(m)}\phi(m+2,m+2-i) + \sum_{i=1}^{m} [a_i^{(m)}]^2\phi(m+2-i,m+2-i) + 2\sum_{i=1}^{m-1}\sum_{j=i+1}^{m} a_i^{(m)}a_j^{(m)}\phi(m+2-i,m+2-j) \tag{8}$$

$$C_m = \phi(1,m+2) + \sum_{i=1}^{m} a_i^{(m)}[\phi(1,m+2-i) + \phi(i+1,m+2)] + \sum_{i=1}^{m} [a_i^{(m)}]^2\phi(i+1,m+2-i) + \sum_{i=1}^{m-1}\sum_{j=i+1}^{m} a_i^{(m)}a_j^{(m)}[\phi(i+1,m+2-j) + \phi(j+1,m+2-i)] \tag{9}$$

We then have

$$K_{m+1}^f = -\, C_m/B_m' \tag{10}$$

$$K_{m+1}^b = -\, C_m/F_m \tag{11}$$

$$K_{m+1}^r = S[\tfrac{1}{2}(|K_{m+1}^f|^r + |K_{m+1}^b|^r)]^{1/r}, \tag{12}$$

where r is the order of the generalized mean. The three specific choices $r = 0, -1$, and $-\infty$ correspond to the three methods: geometric-mean method corresponding to that of Itakura [1], harmonic-mean method corresponding to that of Burg [3], and the minimum method [2]. The reflection coefficient estimates for the three methods are given below:

$$K_{m+1}^{0} = -\, C_m/\sqrt{F_m B_m'} = S\sqrt{K_{m+1}^{f} K_{m+1}^{b}} \tag{13}$$

$$K_{m+1}^{-1} = -\, 2C_m/(F_m + B_m') \tag{14}$$

$$K_{m+1}^{-\infty} = S \min\left(|K_{m+1}^{f}|,\ |K_{m+1}^{b}|\right) \tag{15}$$

Stability of the all-pole filter $H(z)$ can be guaranteed if $r \leqslant 0$ [2]. Given F_m, B_m', C_m and the computed value for K_{m+1}, COVLAT computes the residual variance F_{m+1} and B_{m+1} using the recursive relations (6).

2.2.2 Harmonic-Mean Method (CLHARM)

For this method, substantial savings can be achieved by computing the sum of the forward and backward residual variances directly as follows:

$$\begin{aligned} F_m + B_m' = {} & \phi(1,1) + \phi(m+2,m+2) \\ & + 2\sum_{i=1}^{m} a_i^{(m)}[\phi(1,i+1) + \phi(m+2,m+2-i)] \\ & + \sum_{i=1}^{m} [a_i^{(m)}]^2[\phi(i+1,i+1) + \phi(m+2-i,m+2-i)] \\ & + 2\sum_{i=1}^{m-1}\sum_{j=i+1}^{m} a_i^{(m)} a_j^{(m)}[\phi(i+1,j+1) + \phi(m+2-i,m+2-j)] \end{aligned} \tag{16}$$

The expression for C_m stays the same as given above in Eq. (9). Then, K_{m+1}^{-1} is computed using Eq. (14).

For the special case of the harmonic-mean method, the value of K_{m+1}^{-1} given in Eq. (14) minimizes the sum of the forward and backward variances $F_{m+1} + B_{m+1}$. One can show from Eqs. (6) and (14) that

$$F_{m+1} + B_{m+1} = [1 - (K_{m+1}^{-1})^2](F_m + B_m'). \tag{17}$$

Subroutine CLHARM uses Eq. (17) instead of Eq. (6).

2.3 Covariance Computation (COVAR1 and COVAR2)

Given the signal $s(n)$, $1 \leqslant n \leqslant N$, its covariance is computed using one of the two methods given below.

2.3.1 Method 1 (COVAR1)

$$\phi(i,j) = \sum_{k=p+1}^{N} s(k+1-i)s(k+1-j), \quad 1 \leqslant i,\ j \leqslant p+1, \tag{18}$$

where p is the order of the predictor or the number of lattice stages. Only the first row of $p+1$ elements (i=1, $1 \leqslant j \leqslant p+1$) of the covariance matrix is computed using the above expression. All other p rows are computed using the symmetry property of the covariance matrix $\phi(i,j) = \phi(j,i)$ and the following recursive relations:

$$\begin{aligned} \phi(i,j) = {} & \phi(i-1,j-1) + s(p+2-i)s(p+2-j) \\ & - s(N+2-i)s(N+2-j), \quad 2 \leqslant i,\ j \leqslant p+1. \end{aligned} \tag{19}$$

The elements $\phi(i,j)$ are computed only once and remain fixed throughout the solution.

2.3.2 Method 2 (COVAR2)

Here, all covariance terms up to stage m are recomputed for each new stage. The covariance for stage m is defined as follows:

$$\phi_m(i,j) = \sum_{k=m+1}^{N} s(k+1-i)s(k+1-j), \quad 1 \leqslant i,j \leqslant m+1 \tag{20}$$

The first m rows of the covariance matrix are computed from the covariance for stage $m-1$, using the symmetry property and the following recursive relations:

$$\phi_m(i,j) = \phi_{m-1}(i,j) - s(m+1-i)s(m+1-j), \quad 1 \leqslant i, \; j \leqslant m. \tag{21}$$

The elements of the $(m+1)^{\text{th}}$ row are computed as follows:

$$\phi_m(m+1,1) = \sum_{k=m+1}^{N} s(k)s(k-m) \tag{22a}$$

$$\phi_m(m+1,j) = \phi_{m-1}(m,j-1) - s(N+1-m)s(N+2-j), \quad 2 \leqslant j \leqslant m+1. \tag{22b}$$

2.4 Number of Computations

For computing the $(p+1)\times(p+1)$ covariance matrix for an N-sample signal, the number of multiplies and adds required by the programs COVAR1 and COVAR2 are given in Table 1.

Method	Number of Multiplies and Adds
COVAR1	$N(p+1)$
COVAR2	$N(p+1) + \frac{p^3}{3} + \frac{p^2}{2} + \frac{p}{6} - 1$

Table 1. Number of multiplies and adds required for the covariance computation.

The total number of computations required for computing p reflection coefficients using COVLAT and CLHARM are given in Table 2.

Method	Multiplies	Adds	Divides	Shifts
COVLAT	$\frac{2}{3}p^3 + 2p^2 + \frac{4}{3}p$	$\frac{2}{3}p^3 + 2p^2 + \frac{31}{3}p - 9$	$2p$	$3p-2$
CLHARM	$\frac{p^3}{2} + \frac{3}{2}p^2$	$\frac{2}{3}p^3 + \frac{5}{2}p^2 + \frac{29}{6}p - 6$	p	$2p-1$

Table 2. Number of computations for getting p reflection coefficients.

Number of shift operations given in Table 2 refers to multiplication by 2. Computations given there include the evaluation of forward and backward errors (see Eq. (6)) for COVLAT ($4p$ multiplies, p shifts and $4p$ adds) and the evaluation of forward-plus-backward error (see Eq. (17)) for CLHARM ($2p$ multiplies and p adds). In counting the adds, we did not consider the adds that figure in the subscripts. In addition to the computations given in Table 2, use of COVLAT requires p evaluations of the function GMEAN (Section 3.2), which is the generalized mean between the reflection coefficient values given by the forward method and the backward method (see Eq. (12)).

The total number of computations required for computing the reflection coefficients from the signal data is then found by adding the number of computations for obtaining the signal covariance (Table 1) to that for computing the reflection coefficients (Table 2).

3. Program Description

3.1 Main Program

The main program listed in the Appendix is provided primarily to show how the subroutines are called and to produce results for two test cases. For any practical problem, the user should write his own main program.

For each test example, the main program computes and prints three sets of reflection coefficients obtained from: (1) the harmonic-mean method using CLHARM, (2) the geometric-mean method using COVLAT with $r = 0$ and (3) the minimum method using COVLAT with $r = -\infty$. For case (1), the main program normalizes the forward-plus-backward error at the output of the last (p^{th}) stage, $F_p + B_p$, by dividing by $2\phi(1,1)$ and prints this normalized error. The three CL methods are demonstrated using first Method 1 (COVAR1) for the covariance computation and then Method 2 (COVAR2).

3.2 Subroutines

(a) COVLAT(PHI,NMAX,M,R,IFLAG,A,K(M),FERROR,BERROR,SCR)
Called from the main program; the parameters R and IFLAG define a class of CL methods.

(b) GMEAN(KF,KB,R,IFLAG)
Called by COVLAT; computes the generalized mean between the estimate KF of the reflection coefficient produced by the forward method and the estimate KB by the backward method.

(c) CLHARM(PHI,NMAX,M,A,K(M),ERROR,SCR)
Called from the main program; uses the harmonic-mean method.

(d) COVAR1(SIG,NSIG,NSTAGE,PHI,NMAX)
Called from the main program. Since the covariance computed by this method does not depend on the stage number, COVAR1 is called only once before computing the reflection coefficients in a stage-by-stage fashion.

(e) COVAR2(SIG,NSIG,M,PHI,NMAX)
Called from the main program. Since the covariance computed by this method depends on the lattice stage number, COVAR2 is called for each new stage before the reflection coefficient is computed.

3.3 Description of Parameters

PHI — Covariance matrix

NMAX — Maximum dimension of PHI, or one plus the maximum number of lattice stages. (Actually, PHI needs to have the dimension NSTAGE+1 only. However, the user may be interested in experimenting with different values for NSTAGE: for example, to drive the prediction error(s) sufficiently low, or for any other reason. NMAX in this case is then one plus the maximum value of NSTAGE that the user may want to consider. Notice that the maximum dimension of each of the vectors A, K and SCR defined below may be set to NMAX−1; however, we have set those dimensions to NMAX for convenience.)

M — Lattice stage number

NSTAGE — Number of lattice stages or reflection coefficients, or the order of the all-pole model (NSTAGE is the same as p used in Section 2.)

R — Order of the generalized mean; filter stability is guaranteed for $R \leqslant 0$.

IFLAG — $= -1$ means $R = -\infty$ (Minimum method)
$= 1$ means $R = +\infty$ (Maximum method)
$= 0$ means R is finite

A — A(1) through A(NSTAGE) are the predictor coefficients of the all-pole filter

K — Reflection coefficient vector (declared to be *real*)

FERROR — Forward error F_m at m=NSTAGE

BERROR	Backward error B_m at m=NSTAGE
SCR	Scratch vector
KF	Reflection coefficient given by the forward method (declared to be *real*)
KB	Reflection coefficient given by the backward method (declared to be *real*)
ERROR	Forward-plus-backward error, (F_m+B_m) at m=NSTAGE
SIG	SIG(1) through SIG(NSIG) are the signal samples
NSIG	Number of signal samples. (NSIG is the same as N used in Section 2.)
NO	Printer unit number (This is defined in the main program through the integer FUNCTION I1MACH.)

3.4 DIMENSION Requirements

The DIMENSION statement in the main program should be modified according to the requirements of each particular problem. The parameters included in the following DIMENSION statement conform to the above parameter definitions:

```
DIMENSION PHI(NMAX,NMAX),A(NMAX),SCR(NMAX),SIG(NSIG)
REAL K(NMAX)
```

3.5 Output

The main program prints out NSTAGE reflection coefficients and the errors FERROR, BERROR and ERROR, all defined above.

3.6 Summary of User Requirements

(a) Determine the value of NMAX and adjust DIMENSION statement.

(b) (Optional) Specify a subroutine for quantizing each reflection coefficient as it is computed. (Comments are provided in COVLAT and CLHARM as to where to insert a call to the user's own quantization routine. For details on quantization, refer to [7] and [8]. No modification of the subroutines COVLAT and CLHARM is necessary if the user's application does not require quantization or if the user does not want to perform quantization within the recursion.)

4. Test Problems

The main program has two 4th order (i.e., NSTAGE=4) test problems: (1) Non-Toeplitz covariance matrix case; and (2) Stationary case where the covariance matrix reduces to a Toeplitz autocorrelation matrix. For the first problem, the main program defines a 32-sample signal vector SIG in data statements. This signal segment was taken from 10 kHz sampled real speech data for which the traditional covariance method produced an unstable filter. (The "reflection coefficient" values obtained using the covariance method were: −1.023, −0.168, −1.684, 1.03.)

As seen from the listing of the main program output (Table 3), the CL methods produce stable filters. For the second problem, which requires a stationary (or windowed) signal, the main program appends NSTAGE zero samples on both sides of the above-mentioned 32-sample signal vector. This guarantees that the resulting covariance matrix is Toeplitz. For the stationary case, all CL methods should give the same values for the reflection coefficients. Also, the forward error FERROR and the backward error BERROR should be equal. These results are readily verified from the output listing given in Table 3.

References

1. F. Itakura and S. Saito, "Digital Filtering Techniques for Speech Analysis and Synthesis", *Proc. 7th Int. Cong. Acoust.* (Budapest), 25-C-1, pp. 261-264, 1971.
2. J. Makhoul, "Stable and Efficient Lattice Methods for Linear Prediction", *IEEE Trans. Acoustics, Speech, and Signal Processing,* Vol. ASSP-25, No. 5, pp. 423-428, October 1977. Also, "New Lattice Methods for Linear Prediction", *Proc. 1976 IEEE Int'l. Conf. Acoustics, Speech, and Signal Processing,* pp. 462-465, April 1976.
3. J. Burg, "Maximum Entropy Spectral Analysis", Ph.D. dissertation, Stanford Univ., Stanford, CA, May 1975.
4. A. H. Gray and J. D. Markel, "Digital Lattice and Ladder Filter Synthesis", *IEEE Trans. Audio and Electroacoust.,* Vol. AU-21, No. 6, pp. 491-500, December 1973.
5. R. Viswanathan and J. Makhoul, "Sequential Lattice Methods for Stable Linear Prediction", *Proc. EASCON '76,* pp. 155A-155H, September 1976.
6. D. E. Smylie, G. K. C. Clarke, and T. J. Ulrych, "Analysis of Irregularities in the Earth's Rotation", in *Methods in Computational Physics,* Vol. 13, Academic Press, New York, pp. 391-430, 1973.
7. R. Viswanathan and J. Makhoul, "Quantization Properties of Transmission Parameters in Linear Predictive Systems", *IEEE Trans. Acoustics, Speech, and Signal Processing,* Vol. ASSP-23, No. 3, pp. 309-321, June 1975.
8. J. D. Markel and A. H. Gray, *Linear Prediction of Speech,* Springer-Verlag, Berlin, 1976.

Table 3

```
***FIRST TEST EXAMPLE. NON-TOEPLITZ COVARIANCE MATRIX CASE***

     SOLUTION USING COVARIANCE MATRIX COMPUTED VIA METHOD 1

REFLECTION COEFFICIENTS FROM PROGRAM CLHARM
-0.98576966E+00   0.75438073E+00  -0.16509753E+00   0.77495836E+00
FORWARD-PLUS-BACKWARD ERROR = 0.39345233E-02

REFLECTION COEFFICIENTS FROM PROGRAM COVLAT (R=0, GEOMETRIC MEAN)
-0.98578447E+00   0.75686465E+00  -0.16748354E+00   0.79276231E+00
FORWARD ERROR = 0.46530087E-02, BACKWARD ERROR = 0.31714242E-02

REFLECTION COEFFS. FROM PROGRAM COVLAT (R=-INFINITY, MINIMUM METHOD)
-0.98039611E+00   0.69331092E+00  -0.12956089E+00   0.60512754E+00
FORWARD ERROR = 0.63612880E-02, BACKWARD ERROR = 0.37597384E-02

     SOLUTION USING COVARIANCE MATRIX COMPUTED VIA METHOD 2

REFLECTION COEFFICIENTS FROM PROGRAM CLHARM
-0.98586581E+00   0.48746297E+00   0.27865216E+00   0.26831956E+00
FORWARD-PLUS-BACKWARD ERROR = 0.39385329E-01

REFLECTION COEFFICIENTS FROM PROGRAM COVLAT (R=0, GEOMETRIC MEAN)
-0.98589755E+00   0.50014503E+00   0.31662578E+00   0.35155942E+00
FORWARD ERROR = 0.61631529E-01, BACKWARD ERROR = 0.17051948E-01

REFLECTION COEFFS. FROM PROGRAM COVLAT (R=-INFINITY, MINIMUM METHOD)
-0.97801873E+00   0.39491358E+00   0.18140856E+00   0.14482672E+00
FORWARD ERROR = 0.65123382E-01, BACKWARD ERROR = 0.14874170E-01
```

Table 3
(Continued)

```
***SECOND TEST EXAMPLE. STATIONARY OR TOEPLITZ MATRIX CASE***

       SOLUTION USING COVARIANCE MATRIX COMPUTED VIA METHOD 1

REFLECTION COEFFICIENTS FROM PROGRAM CLHARM

-0.97392271E+00   0.22170082E+00   0.20598015E+00   0.14759705E+00

FORWARD-PLUS-BACKWARD ERROR = 0.45846863E-01

REFLECTION COEFFICIENTS FROM PROGRAM COVLAT (R=0, GEOMETRIC MEAN)

-0.97392271E+00   0.22170082E+00   0.20598015E+00   0.14759705E+00

FORWARD ERROR = 0.45846863E-01, BACKWARD ERROR = 0.45846863E-01

REFLECTION COEFFS. FROM PROGRAM COVLAT (R=-INFINITY, MINIMUM METHOD)

-0.97392271E+00   0.22170082E+00   0.20598015E+00   0.14759705E+00

FORWARD ERROR = 0.45846863E-01, BACKWARD ERROR = 0.45846863E-01

       SOLUTION USING COVARIANCE MATRIX COMPUTED VIA METHOD 2

REFLECTION COEFFICIENTS FROM PROGRAM CLHARM

-0.97392271E+00   0.22170082E+00   0.20598015E+00   0.14759705E+00

FORWARD-PLUS-BACKWARD ERROR = 0.45846863E-01

REFLECTION COEFFICIENTS FROM PROGRAM COVLAT (R=0, GEOMETRIC MEAN)

-0.97392271E+00   0.22170082E+00   0.20598015E+00   0.14759705E+00

FORWARD ERROR = 0.45846863E-01, BACKWARD ERROR = 0.45846863E-01

REFLECTION COEFFS. FROM PROGRAM COVLAT (R=-INFINITY, MINIMUM METHOD)

-0.97392271E+00   0.22170082E+00   0.20598015E+00   0.14759705E+00

FORWARD ERROR = 0.45846863E-01, BACKWARD ERROR = 0.45846863E-01
```

```
C
C-----------------------------------------------------------------
C MAIN PROGRAM: EFFICIENT LATTICE PROGRAMS FOR LINEAR PREDICTION
C AUTHORS:      R.VISWANATHAN AND J.MAKHOUL
C               BOLT BERANEK AND NEWMAN INC
C               50 MOULTON STREET
C               CAMBRIDGE, MA 02138
C
C INPUT:        ALL INPUT DATA ARE PROVIDED IN DATA STATEMENTS
C-----------------------------------------------------------------
C
      DIMENSION PHI(20,20), A(20), SCR(20), SIG(100)
      REAL K(20)
      DATA NSTAGE, NMAX, NSIG, NFLAG /4,20,32,0/
C
C SIGNAL DATA FOR THE TEST EXAMPLES
C
      DATA SIG(1), SIG(2), SIG(3), SIG(4) /-16.0,-18.0,-21.0,-24.0/
      DATA SIG(5), SIG(6), SIG(7), SIG(8) /-27.0,-30.0,-31.0,-33.0/
      DATA SIG(9), SIG(10), SIG(11), SIG(12) /-36.0,-39.0,-42.0,-44.0/
      DATA SIG(13), SIG(14), SIG(15), SIG(16) /-46.0,-49.0,-51.0,-55.0/
      DATA SIG(17), SIG(18), SIG(19), SIG(20) /-58.0,-61.0,-64.0,-67.0/
      DATA SIG(21), SIG(22), SIG(23), SIG(24) /-71.0,-73.0,-68.0,-53.0/
      DATA SIG(25), SIG(26), SIG(27), SIG(28) /-38.0,-36.0,-31.0,-13.0/
      DATA SIG(29), SIG(30), SIG(31), SIG(32) /6.0,15.0,19.0,35.0/
C
C SET UP MACHINE CONSTANT (OUTPUT UNIT NUMBER)
C
      NO = I1MACH(2)
C
C FIRST TEST EXAMPLE: NON-TOEPLITZ COVARIANCE MATRIX CASE
C
      WRITE (NO,9999)
 9999 FORMAT (1H1, 4X, 45H***FIRST TEST EXAMPLE. NON-TOEPLITZ COVARIANC,
     *    16HE MATRIX CASE***)
C
C COVARIANCE MATRIX COMPUTATION USING METHOD 1
C
   10 WRITE (NO,9998)
 9998 FORMAT (//8X, 48HSOLUTION USING COVARIANCE MATRIX COMPUTED VIA ME,
     *    6HTHOD 1)
      WRITE (NO,9997)
 9997 FORMAT (//44H REFLECTION COEFFICIENTS FROM PROGRAM CLHARM)
      CALL COVAR1(SIG, NSIG, NSTAGE, PHI, NMAX)
      DO 20 M=1,NSTAGE
        CALL CLHARM(PHI, NMAX, M, A, K(M), ERROR, SCR)
   20 CONTINUE
      ERROR = ERROR/(2.*PHI(1,1))
      WRITE (NO,9996) (K(I),I=1,NSTAGE)
 9996 FORMAT (/1X, 4(E15.8, 2X))
      WRITE (NO,9995) ERROR
 9995 FORMAT (/30H FORWARD-PLUS-BACKWARD ERROR =, E15.8)
      WRITE (NO,9994)
 9994 FORMAT (//44H REFLECTION COEFFICIENTS FROM PROGRAM COVLAT,
     *    22H (R=0, GEOMETRIC MEAN))
      R = 0
      IFLAG = 0
      DO 30 M=1,NSTAGE
        CALL COVLAT(PHI, NMAX, M, R, IFLAG, A, K(M), FERROR, BERROR,
     *      SCR)
   30 CONTINUE
      FERROR = FERROR/PHI(1,1)
      BERROR = BERROR/PHI(1,1)
      WRITE (NO,9996) (K(I),I=1,NSTAGE)
      WRITE (NO,9993) FERROR, BERROR
 9993 FORMAT (/16H FORWARD ERROR =, E15.8, 18H, BACKWARD ERROR =, E15.8)
      WRITE (NO,9992)
 9992 FORMAT (//39H REFLECTION COEFFS. FROM PROGRAM COVLAT, 9H (R=-INFI,
     *    21HNITY, MINIMUM METHOD))
      IFLAG = -1
      DO 40 M=1,NSTAGE
        CALL COVLAT(PHI, NMAX, M, R, IFLAG, A, K(M), FERROR, BERROR,
     *      SCR)
   40 CONTINUE
      FERROR = FERROR/PHI(1,1)
      BERROR = BERROR/PHI(1,1)
      WRITE (NO,9996) (K(I),I=1,NSTAGE)
      WRITE (NO,9993) FERROR, BERROR
C
C COVARIANCE MATRIX COMPUTATION USING METHOD 2
C
      WRITE (NO,9991)
 9991 FORMAT (//8X, 48HSOLUTION USING COVARIANCE MATRIX COMPUTED VIA ME,
     *    6HTHOD 2)
      WRITE (NO,9997)
      DO 50 M=1,NSTAGE
        CALL COVAR2(SIG, NSIG, M, PHI, NMAX)
        CALL CLHARM(PHI, NMAX, M, A, K(M), ERROR, SCR)
   50 CONTINUE
      ERROR = ERROR/(2.*PHI(1,1))
      WRITE (NO,9996) (K(I),I=1,NSTAGE)
      WRITE (NO,9995) ERROR
      WRITE (NO,9994)
      R = 0
      IFLAG = 0
      DO 60 M=1,NSTAGE
        CALL COVAR2(SIG, NSIG, M, PHI, NMAX)
        CALL COVLAT(PHI, NMAX, M, R, IFLAG, A, K(M), FERROR, BERROR,
     *      SCR)
   60 CONTINUE
      FERROR = FERROR/PHI(1,1)
      BERROR = BERROR/PHI(1,1)
      WRITE (NO,9996) (K(I),I=1,NSTAGE)
      WRITE (NO,9993) FERROR, BERROR
      WRITE (NO,9992)
      IFLAG = -1
      DO 70 M=1,NSTAGE
        CALL COVAR2(SIG, NSIG, M, PHI, NMAX)
        CALL COVLAT(PHI, NMAX, M, R, IFLAG, A, K(M), FERROR, BERROR,
     *      SCR)
   70 CONTINUE
      FERROR = FERROR/PHI(1,1)
      BERROR = BERROR/PHI(1,1)
      WRITE (NO,9996) (K(I),I=1,NSTAGE)
      WRITE (NO,9993) FERROR, BERROR
      IF (NFLAG.EQ.1) GO TO 110
C
C SECOND TEST EXAMPLE: STATIONARY  OR TOEPLITZ MATRIX CASE
C
      WRITE (NO,9990)
 9990 FORMAT (1H1, 4X, 45H***SECOND TEST EXAMPLE. STATIONARY OR TOEPLIT,
     *    16HZ MATRIX CASE***)
C
C AUGMENT THE SIGNAL VECTOR OF THE FIRST EXAMPLE WITH NSTAGE ZEROS
C SO THAT THE RESULTING COVARIANCE MATRIX IS TOEPLITZ
C
      N1 = NSIG + NSTAGE
      NSTP1 = NSTAGE + 1
      N1P1 = N1 + 1
      N = N1
```

```
   80 NMNST = N - NSTAGE
      SIG(N) = SIG(NMNST)
      N = N - 1
      IF (N.GE.NSTP1) GO TO 80
      DO 90 N=1,NSTAGE
        SIG(N) = 0
   90 CONTINUE
      N2 = N1 + NSTAGE
      DO 100 N=N1P1,N2
        SIG(N) = 0
  100 CONTINUE
      NSIG = N2
      NFLAG = 1
      GO TO 10
C
  110 STOP
      END
C
C------------------------------------------------------------------
C SUBROUTINE: COVLAT
C GENERAL COVARIANCE LATTICE ROUTINE; IT COMPUTES THE REFLECTION
C COEFFICIENT OF STAGE M,GIVEN THE COVARIANCE MATRIX OF THE SIGNAL AND
C PREDICTOR COEFFICIENTS UP TO STAGE M-1.
C------------------------------------------------------------------
C
      SUBROUTINE COVLAT(PHI, NMAX, M, R, IFLAG, A, K, FERROR, BERROR,
     *    SCR)
      DIMENSION PHI(NMAX,NMAX), A(NMAX), SCR(NMAX)
      REAL K, KF, KB
C
      IF (M.GT.1) GO TO 20
C
C EXPLICIT COMPUTATION OF THE FIRST STAGE REFLECTION COEFFICIENT
C
      F = PHI(1,1)
      B = PHI(2,2)
      C = PHI(1,2)
      K = 0
      IF (C.EQ.0.0) GO TO 10
      KF = -C/B
      KB = -C/F
      K = GMEAN(KF,KB,R,IFLAG)
   10 A(1) = K
      GO TO 90
C
C RECURSIVE COMPUTATION OF THE M-TH STAGE (M.GE.2) REFLECTION
C COEFFICIENT
C
   20 MP1 = M + 1
      MM1 = M - 1
      SUM1 = 0
      SUM2 = 0
      SUM3 = 0
      SUM4 = 0
      SUM5 = 0
      SUM6 = 0
      DO 30 I=1,MM1
        IP1 = I + 1
        SCR(I) = A(I)
        SUM1 = SUM1 + A(I)*PHI(1,IP1)
        MP1MI = MP1 - I
        SUM2 = SUM2 + A(I)*PHI(MP1,MP1MI)
        SUM3 = SUM3 + A(I)*(PHI(1,MP1MI)+PHI(IP1,MP1))
        Y = A(I)**2
        SUM4 = SUM4 + Y*PHI(IP1,IP1)
        SUM5 = SUM5 + Y*PHI(MP1MI,MP1MI)
        SUM6 = SUM6 + Y*PHI(IP1,MP1MI)
   30 CONTINUE
      SUM7 = 0
      SUM8 = 0
      SUM9 = 0
      IF (M.EQ.2) GO TO 60
      MM2 = M - 2
      DO 50 I=1,MM2
        IP1 = I + 1
        MP1MI = MP1 - I
        DO 40 J=IP1,MM1
          Y = A(I)*A(J)
          SUM7 = SUM7 + Y*PHI(IP1,J+1)
          MP1MJ = MP1 - J
          SUM8 = SUM8 + Y*PHI(MP1MI,MP1MJ)
          SUM9 = SUM9 + Y*(PHI(IP1,MP1MJ)+PHI(J+1,MP1MI))
   40   CONTINUE
   50 CONTINUE
   60 F = PHI(1,1) + 2.*(SUM1+SUM7) + SUM4
      B = PHI(MP1,MP1) + 2.*(SUM2+SUM8) + SUM5
      C = PHI(1,MP1) + SUM3 + SUM6 + SUM9
      K = 0
      IF (C.EQ.0.0) GO TO 70
      KF = -C/B
      KB = -C/F
      K = GMEAN(KF,KB,R,IFLAG)
   70 CONTINUE
C
C INSERT CALL TO USER'S OWN SUBROUTINE TO QUANTIZE THE REFLECTION
C COEFFICIENT K
C
C RECURSION TO CONVERT REFLECTION COEFFICIENTS TO PREDICTOR
C COEFFICIENTS
C
      DO 80 I=1,MM1
        MMI = M - I
        A(I) = SCR(I) + K*SCR(MMI)
   80 CONTINUE
      A(M) = K
C
C COMPUTE FORWARD AND BACKWARD ERRORS AT THE OUTPUT OF THE M-TH
C LATTICE STAGE
C
   90 X = K**2
      Y = 2.*K*C
      FERROR = F + X*B + Y
      BERROR = B + X*F + Y
C
      RETURN
      END
C
C------------------------------------------------------------------
C FUNCTION: GMEAN
C FUNCTION TO COMPUTE THE R-TH GENERALIZED MEAN BETWEEN KF AND KB
C------------------------------------------------------------------
C
      FUNCTION GMEAN(KF, KB, R, IFLAG)
      REAL KF, KB
C
      AKF = ABS(KF)
      AKB = ABS(KB)
      S = KF/AKF
      IF (IFLAG) 10, 20, 50
C
```

```
C MINIMUM METHOD
C
  10  GMEAN = AKF
      IF (AKB.LT.AKF) GMEAN = AKB
      GMEAN = GMEAN*S
      RETURN
C
C GENERAL FINITE R-TH MEAN
C
  20  IF (R.EQ.(-1.0)) GO TO 30
      IF (R.EQ.0.0) GO TO 40
      X = (AKF)**R + (AKB)**R
      GMEAN = S*((X/2.)**(1./R))
      RETURN
C
C HARMONIC MEAN METHOD
C
  30  GMEAN = 2.*KF*KB/(KF+KB)
      RETURN
C
C GEOMETRIC MEAN METHOD
C
  40  GMEAN = S*SQRT(AKF*AKB)
      RETURN
C
C MAXIMUM METHOD
C
  50  GMEAN = AKF
      IF (AKB.GT.AKF) GMEAN = AKB
      GMEAN = GMEAN*S
C
      RETURN
      END
C
C-----------------------------------------------------------------------
C SUBROUTINE: CLHARM
C COVARIANCE LATTICE ROUTINE FOR HARMONIC MEAN METHOD; IT COMPUTES THE
C REFLECTION COEFFICIENT OF STAGE M, GIVEN THE COVARIANCE MATRIX OF THE
C SIGNAL AND PREDICTOR COEFFICIENTS UP TO STAGE M-1.
C-----------------------------------------------------------------------
C
      SUBROUTINE CLHARM(PHI, NMAX, M, A, K, ERROR, SCR)
      DIMENSION PHI(NMAX,NMAX), A(NMAX), SCR(NMAX)
      REAL K
C
      IF (M.GT.1) GO TO 20
C
C EXPLICIT COMPUTATION OF THE FIRST STAGE REFLECTION COEFFICIENT
C
      FPLUSB = PHI(1,1) + PHI(2,2)
      C = PHI(1,2)
      K = 0
      IF (C.EQ.0.0) GO TO 10
      K = -2.*C/FPLUSB
  10  A(1) = K
      GO TO 90
C
C RECURSIVE COMPUTATION OF THE M-TH STAGE (M.GE.2) REFLECTION
C COEFFICIENT
C
  20  MP1 = M + 1
      MM1 = M - 1
      SUM1 = 0
      SUM3 = 0
      SUM4 = 0
      SUM6 = 0
      DO 30 I=1,MM1
        IP1 = I + 1
        SCR(I) = A(I)
        MP1MI = MP1 - I
        SUM1 = SUM1 + A(I)*(PHI(1,IP1)+PHI(MP1,MP1MI))
        SUM3 = SUM3 + A(I)*(PHI(1,MP1MI)+PHI(IP1,MP1))
        Y = A(I)**2
        SUM4 = SUM4 + Y*(PHI(IP1,IP1)+PHI(MP1MI,MP1MI))
        SUM6 = SUM6 + Y*PHI(IP1,MP1MI)
  30  CONTINUE
      SUM7 = 0
      SUM9 = 0
      IF (M.EQ.2) GO TO 60
      MM2 = M - 2
      DO 50 I=1,MM2
        IP1 = I + 1
        MP1MI = MP1 - I
        DO 40 J=IP1,MM1
          Y = A(I)*A(J)
          MP1MJ = MP1 - J
          SUM7 = SUM7 + Y*(PHI(IP1,J+1)+PHI(MP1MI,MP1MJ))
          SUM9 = SUM9 + Y*(PHI(IP1,MP1MJ)+PHI(J+1,MP1MI))
  40    CONTINUE
  50  CONTINUE
  60  FPLUSB = PHI(1,1) + PHI(MP1,MP1) + 2.*(SUM1+SUM7) + SUM4
      C = PHI(1,MP1) + SUM3 + SUM6 + SUM9
      K = 0
      IF (C.EQ.0.0) GO TO 70
      K = -2.*C/FPLUSB
  70  CONTINUE
C
C INSERT CALL TO USER'S OWN SUBROUTINE TO QUANTIZE THE REFLECTION
C COEFFICIENT K
C
C RECURSION TO CONVERT REFLECTION COEFFICIENTS TO PREDICTOR
C COEFFICIENTS
C
      DO 80 I=1,MM1
        MMI = M - I
        A(I) = SCR(I) + K*SCR(MMI)
  80  CONTINUE
      A(M) = K
C
C COMPUTE MINIMUM FORWARD-PLUS-BACKWARD ERROR AT THE OUTPUT OF
C THE M-TH LATTICE STAGE
C
  90  ERROR = FPLUSB*(1.-K**2)
C
      RETURN
      END
C
C-----------------------------------------------------------------------
C SUBROUTINE: COVAR1
C GIVEN THE SIGNAL, THIS ROUTINE GENERATES ITS COVARIANCE MATRIX
C USING THE RELATION: PHI(I,J)=SUM[SIG(K+1-I)*SIG(K+1-J)] OVER THE
C RANGE FROM K=NSTAGE+1 TO K=NSIG.
C-----------------------------------------------------------------------
C
      SUBROUTINE COVAR1(SIG, NSIG, NSTAGE, PHI, NMAX)
      DIMENSION SIG(NSIG), PHI(NMAX,NMAX)
C
      NP1 = NSTAGE + 1
      NP2 = NSTAGE + 2
      NSP2 = NSIG + 2
```

```
      DO 20 J=1,NP1
        TEMP = 0
        DO 10 K=NP1,NSIG
          KP1MJ = K + 1 - J
          TEMP = TEMP + SIG(K)*SIG(KP1MJ)
 10     CONTINUE
        PHI(1,J) = TEMP
 20   CONTINUE
      DO 50 I=2,NP1
        IM1 = I - 1
        DO 30 J=1,IM1
          PHI(I,J) = PHI(J,I)
 30     CONTINUE
        NP2MI = NP2 - I
        NSP2MI = NSP2 - I
        DO 40 J=I,NP1
          NP2MJ = NP2 - J
          NSP2MJ = NSP2 - J
          PHI(I,J) = PHI(I-1,J-1) + SIG(NP2MI)*SIG(NP2MJ) -
     *        SIG(NSP2MI)*SIG(NSP2MJ)
 40     CONTINUE
 50   CONTINUE
C
      RETURN
      END
C
C-----------------------------------------------------------------------
C SUBROUTINE: COVAR2
C GIVEN THE SIGNAL, THIS ROUTINE COMPUTES THE (M+1) X (M+1) COVARIANCE
C MATRIX CORRESPONDING TO THE LATTICE STAGE M.  FOR M LARGER THAN
C 1, THIS COMPUTATION IS DONE EFFICIENTLY, USING THE COVARIANCE
C MATRIX FOR STAGE (M-1).
C-----------------------------------------------------------------------
C
      SUBROUTINE COVAR2(SIG, NSIG, M, PHI, NMAX)
      DIMENSION SIG(NSIG), PHI(NMAX,NMAX)
C
      IF (M.GT.1) GO TO 20
C
C FIRST STAGE: M=1
C
      TEMP1 = 0
      TEMP2 = 0
      DO 10 K=2,NSIG
        TEMP1 = TEMP1 + SIG(K)**2
        TEMP2 = TEMP2 + SIG(K)*SIG(K-1)
 10   CONTINUE
      PHI(1,1) = TEMP1
      PHI(1,2) = TEMP2
      PHI(2,1) = TEMP2
      PHI(2,2) = TEMP1 + SIG(1)**2 - SIG(NSIG)**2
      RETURN
C
C M-TH STAGE, M.GE.2
C
 20   MP1 = M + 1
      NSP1 = NSIG + 1
      NSM = NSP1 - M
      DO 30 J=2,MP1
        NSJ = NSP1 + 1 - J
        PHI(MP1,J) = PHI(M,J-1) - SIG(NSM)*SIG(NSJ)
        PHI(J,MP1) = PHI(MP1,J)
 30   CONTINUE
      TEMP1 = 0
      DO 40 K=MP1,NSIG
        KMM = K - M
        TEMP1 = TEMP1 + SIG(K)*SIG(KMM)
 40   CONTINUE
      PHI(MP1,1) = TEMP1
      PHI(1,MP1) = TEMP1
      DO 60 I=1,M
        MP1MI = MP1 - I
        DO 50 J=1,M
          MP1MJ = MP1 - J
          PHI(I,J) = PHI(I,J) - SIG(MP1MI)*SIG(MP1MJ)
          PHI(J,I) = PHI(I,J)
 50     CONTINUE
 60   CONTINUE
C
      RETURN
      END
```

4.3

Linear Predictor Coefficient Transformations
Subroutine LPTRN

A. H. Gray

Department of Electrical Eng. & Computer Science
University of California at Santa Barbara
Santa Barbara, CA 93106

and

J. D. Markel

Signal Technology, Inc.
15 De La Guerra Street
Santa Barbara, CA 93101

1. Purpose

The purpose of the subroutine LPTRN is to allow the user to transform between the different parameter sets often used in linear predictive speech analysis to describe the synthesis filter

$$H(z) = \sqrt{\alpha}/A(z), \tag{1}$$

where

$$A(z) = \sum_{k=0}^{M} a_k z^{-k} \quad \text{with} \quad a_0 = 1 \, . \tag{2}$$

The program is a slight modification of the program LPTRAN in [1], where a more detailed discussion may be found.

2. Parameter Set Choices

Table 1 shows the choices of parameter sets to be used as inputs, along with their FORTRAN labels. The column labeled I, where I = 1, 2,..., 6, defines which parameter set is to be used as an input to the subroutine. All other parameter sets are then outputs.

I	Name	Text	FORTRAN	Subscript Range
1	filter coefficients	$a_0 = 1$	A(1) = 1	
		a_j	A(j+1) = SA(j)	$j = 1, 2, \ldots, M$
	gain	α	ALPHA	
2	cepstral coefficients	$c(j)$	C(j+1)	$j = 0, 1, \ldots, M$
3	autocorrelation coefficients	$r(j)$	R(j+1)	$j = 0, 1, \ldots, M$
4	reflection coefficients	k_j	RC(j)	$j = 1, 2, \ldots, M$
	input energy	$r(0)$	R(1)	
5	log area ratios	$\ln(S_j/S_{j-1})$	ALAR(j)	$j = 1, 2, \ldots M$
	input energy	$r(0)$	R(1)	
6	area functions	S_j	AREA(j+1)	$j = 0, 1, \ldots, M$
	input energy	$r(0)$	R(1)	

Table 1. Definitions of Text and Fortran Variables for subroutine LPTRN

In each case, there are a total of $M + 1$ parameters which uniquely describe the filter $H(z)$. When the array A($\cdot$) for the filter coefficients is used as an input (I=1), then the value of A(1) is ignored, since $a_0 = 1$ by definition. When A($\cdot$) is an output (I$\neq$1), then A(1) is set equal to one by the program. When the array representing the are functions, AREA($\cdot$), is used as an input (I=6), then $M + 2$

parameters are actually entered into the subroutine. When AREA(·) is an output (I≠6), then AREA(M+1) is always set equal to one for normalization.

2.1 Filter Coefficients

This is the parameter set $\{a_0, a_1, ..., a_M\}$ defining the filter denominator as in Eq. (2). The coefficient a_0 equals one by definition. On input A(1) is ignored (I=1). When A(·) is an output (I≠1), then A(1) = 1. During the calculations another dimensioned array, SA( ) is used, where

$$\mathrm{SA}(j) = a_j \quad \text{for} \quad j = 1, 2, ..., M.$$

This is to avoid changing any values of A(·) when it is used as an input (I=1).

2.2 Cepstral Coefficients

When $H(z)$ is stable and causal, or $A(z)$ is minimum phase, a set of cepstral coefficients $\{c(k)\}$ is defined by

$$\ln[\sqrt{\alpha}/A(z)] = \sum_{k=0}^{\infty} c(k) z^{-k}, \tag{3}$$

or equivalently,

$$c(0) = \ln[\sqrt{\alpha}] \tag{4a}$$

$$c(1) + a_1 = 0 \tag{4b}$$

$$n\,c(n) + n\,a_n = -\sum_{k=1}^{n-1} (n-k) c(n-k) a_k, \tag{4c}$$

$$\text{for} \quad n = 2, ..., M.$$

Equation (4c) can also be used for $n > M$, provided one defines

$$a_k = 0 \quad \text{for} \quad k > M. \tag{5}$$

Equations (4) can be used directly to find the cepstral coefficients from the filter coefficients or vice versa. If the filter $H(z)$ is not stable, the results are meaningless.

2.3 Autocorrelation Coefficients

When the filter $H(z)$ is stable, the autocorrelation coefficients $\{r(k)\}$ are defined by

$$H(z)H(1/z) = \sum_{k=-\infty}^{\infty} r(k) z^{-k}, \tag{6}$$

and are even functions of k,

$$r(k) = r(-k). \tag{7}$$

The autocorrelation coefficients $\{r(0), r(1), ..., r(M)\}$ can be used as inputs or outputs in the subroutine. When used as inputs, the standard Levinson recursion is used to define a "step-up" procedure. This is abbreviated as follows. First there is an initialization of the form

$$\alpha_0 = r(0), \qquad k_1 = -r(1)/r(0), \tag{9a}$$

$$a_{10} = 1, \qquad a_{11} = k_1, \tag{9b}$$

$$\alpha_1 = \alpha_0(1 - k_1^2). \tag{9c}$$

This is followed by the recursive steps

$$k_m = -\left[\sum_{j=0}^{m-1} r(m-j)a_{m-1,j}\right]/\alpha_{m-1}, \tag{10a}$$

$$a_{m0} = 1 \tag{10b}$$

$$a_{mj} = a_{m-1,j} + k_m a_{m-1,m-j} \quad \text{for} \quad j = 1,2,...,m-1 \tag{10c}$$

$$a_{mm} = k_m, \tag{10d}$$

$$\alpha_m = \alpha_{m-1}(1-k_m^2), \tag{10e}$$

for $m = 2,3,...,M$. At the completion of the recursion,

$$a_k = a_{Mk} \quad \text{for} \quad k = 0,1,...,M, \tag{11a}$$

and

$$\alpha = \alpha_M. \tag{11b}$$

A necessary and sufficient condition for stability of $H(z)$ is that

$$|k_j| < 1 \quad \text{for} \quad j = 1,2,...,M. \tag{12}$$

Therefore a division by zero cannot theoretically occur in Eq. (10a) for a stable filter.

If the autocorrelation sequence is not an input (I≠3), then it is found from the same step-up approach, but with the input being the reflection coefficient set $\{k_1,k_2,...,k_M\}$ and the energy term $r(0)$. The only changes in the formulas lie in replacing Eq. (9a) by

$$\alpha_0 = r(0), \qquad r(1) = -k_1 r(0) \tag{13}$$

removing Eq. (10a), and inserting at the end of step m the expression

$$r(m) = -\sum_{j=1}^{m} r(m-j)a_{mj}. \tag{14}$$

2.4 Reflection Coefficients

The reflection coefficients $\{k_1,k_2,...,k_M\}$ represent the reflection coefficients for an equivalent acoustic tube model for $H(z)$. When used as inputs along with the filter energy $r(0)$, the filter coefficients and autocorrelation coefficients are generated as described above.

When the filter coefficients are used as an input (I=1), the reflection coefficients are generated through a "step-down" procedure as follows. First, the initialization of Eq. (11a) is performed. Then taking $m = M,M-1,...,1$,

$$k_m = a_{mm}, \tag{15a}$$

$$a_{m-1,0} = 1, \tag{15b}$$

and

$$a_{m-1,j} = \frac{a_{m,j} - k_m a_{m,m-j}}{1 - k_m^2} \tag{15c}$$

for $j = 1,2,...,m-1$.

2.5 Log Area Ratios

The log area ratios are a nonlinear transformation of the reflection coefficients, given by

$$\ln(S_j/S_{j-1}) = \ln[(1-k_j)/(1+k_j)] \quad \text{for} \quad j = 1,2,...,M. \tag{16}$$

Scaling of all area functions by the same scale factor does not change the filter, for only the ratio of area functions enters into the calculations. When area functions are calculated as an output of the subroutine (I≠6) the value of S_M is always normalized to one.

2.6 Area Functions:

The area functions $\{S_0,S_1, \ldots, S_M\}$ are areas of an acoustic tube model of $H(z)$. They are related to the reflection coefficients by the equation

$$S_j/S_{j-1} = (1-k_j)/(1+k_j) \quad \text{for} \quad j = 1,2,...,M. \tag{17}$$

Scaling of all area functions by the same scale factor does not change the filter, for only the ratio of area functions enters into the calculations. When area functions are calculated as an output of the subroutine (I≠6) the value of S_M is always normalized to one.

3. Filter Stability

The various transformations are meaningless if the filter $H(z)$ is not a stable causal filter. Causality of the filter $H(z)$ will be assumed, so that stability of $H(z)$ is equivalent to $H(z)$ being minimum phase. Stability can be guaranteed in advance for the cases I = 4, 5, and 6. When the reflection coefficients are input (I=4), their magnitudes must be less than one. When the log area functions are input (I=5) stability is guaranteed. When the area functions are input (I=6), they must all be positive to guarantee stability.

When the parameters represent a stable $H(z)$ there will be no error messages from the subroutine. Should parameters for an unstable $H(z)$ be used then there will be attempts to take the logarithm of a nonpositive number in the evaluation of the log area functions, and there may be a division by zero.

A short stability test, DO loop 330, is inserted in the subroutine to test the magnitudes of the reflection coefficients. If they do not satisfy Eq. (12) an error message is printed out. This subroutine provides a stability test for any rational digital filter whose denominator is expressed in the form of Eq. (2), and is equivalent to the Schur-Cohn method [2,3].

4. A Test Example

The program can be used to test itself by changing the index I after each use. An example test program is given in the Appendix. Here the program generates the first parameter set to be used as an input (I=1) using $M = 3$, $a_0 = 1$, $a_1 = -.45$, $a_2 = .81$, $a_3 = 0$, and $\alpha = 10$. By incrementing I from 1 through 6, the roles of inputs are changed according to Table 1. Each printout will be the same, except for possible computer roundoff errors and the index I. A sample printout is shown in Table 2 for I = 1.

The example used here illustrates how one can get additional information about autocorrelation values and cepstral coefficients. Note that the filter is actually only of second order. If $M = 2$ had been used, then the autocorrelation values and cepstral coefficients would only have been evaluated up to a subscript of 2. By setting $M = 3$ and $a_3 = 0$ they are evaluated up to a subscript of 3.

References

1. J. D. Markel and A. H. Gray, Jr., *Linear Prediction of Speech,* Springer Verlag, New York, 1976.

2. A. Cohn, "Uber die Anzahl deter Wurzeln einer Algebraischen Gleichung in einem Kreise", *Math. Z,* Vol. 14, pp. 110-148, 1922.

3. E. I. Jury, *Theory and Application of the z-Transform Method,* Wiley, New York, 1964.

Table 1

```
OUTPUT DATA FROM LPTRN TEST PROGRAM

   1  10.000001
  1.000000   2.302585  30.993995  -0.248618   0.507879   5.732649
 -0.450000   0.450000   7.705689   0.810000  -2.254057   9.526313
  0.810000  -0.708750 -21.637573   0.000000   0.000000   1.000000
  0.000000  -0.334124 -15.978511   0.000000   0.000000   1.000000

   2   9.999994

  1.000000   2.302585  30.993957  -0.248618   0.507879   5.732645
 -0.450000   0.450000   7.705680   0.809999  -2.254056   9.526308
  0.809999  -0.708750 -21.637542  -0.000000   0.000000   1.000000
 -0.000000  -0.334124 -15.978488   0.000000   0.000000   1.000000

   3  09.999999

  1.000000   2.302584  30.993957  -0.248618   0.507879   5.732640
 -0.449999   0.449999   7.705680   0.809999  -2.254055   9.526298
  0.809999  -0.708749 -21.637542  -0.000000   0.000000   0.999999
 -0.000000  -0.334124 -15.978488   0.000000   0.000000   1.000000

   4  09.999999

  1.000000   2.302584  30.993957  -0.248618   0.507879   5.732640
 -0.449999   0.449999   7.705679   0.809999  -2.254055   9.526298
  0.809999  -0.708749 -21.637542  -0.000000   0.000000   0.999999
 -0.000000  -0.334124 -15.978483   0.000000   0.000000   1.000000

   5  10.000011

  1.000000   2.302586  30.993957  -0.248618   0.507879   5.732637
 -0.449999   0.449999   7.705673   0.809999  -2.254055   9.526290
  0.809999  -0.708749 -21.637538  -0.000000   0.000000   0.999999
 -0.000000  -0.334124 -15.978466   0.000000   0.000000   1.000000

   6  10.000011

  1.000000   2.302586  30.993957  -0.248618   0.507879   5.732637
 -0.449999   0.449999   7.705673   0.809999  -2.254055   9.526290
  0.809999  -0.708749 -21.637538  -0.000000   0.000000   0.999999
 -0.000000  -0.334124 -15.978466   0.000000   0.000000   1.000000
```

```
C
C-----------------------------------------------------------------------
C  MAIN PROGRAM:  TEST PROGRAM FOR LPTRN
C  AUTHORS:       A. H. GRAY, JR.  AND J. D. MARKEL
C       GRAY -    UNIV. OF CALIF., SANTA BARBARA, CA 93106
C       BOTH -    SIGNAL TECHNOLOGY INC., 15 W. DE LA GUERRA AVE.
C                 SANTA BARBARA, CA 93101
C-----------------------------------------------------------------------
C
      DIMENSION A(21), C(21), R(21), RC(21), ALAR(21)
      DIMENSION AREA(21)
      COMMON IOUTD
      IOUTD = I1MACH(2)
      DO 10 J=4,21
        A(J) = 0.
   10 CONTINUE
      A(1) = 1.
      A(2) = -.45
      A(3) = .81
      M = 3
      ALPHA = 10.
      DO 30 I=1,6
        CALL LPTRN(I, M, A, C, R, RC, ALAR, AREA, ALPHA)
        MP = M + 1
        WRITE (IOUTD,9999) I, ALPHA
9999    FORMAT (/I10, F11.6/)
        DO 20 J=1,MP
          WRITE (IOUTD,9998) A(J), C(J), R(J), RC(J), ALAR(J), AREA(J)
   20   CONTINUE
   30 CONTINUE
9998  FORMAT (5X, 6F11.6)
      STOP
      END
C
C-----------------------------------------------------------------------
C SUBROUTINE:  LPTRN
C THIS SUBROUTINE CARRIES OUT THE TRANSFORMATIONS BETWEEN
C VARIOUS PARAMETER SETS USED IN LINEAR PREDICTION
C-----------------------------------------------------------------------
C
      SUBROUTINE LPTRN(I, M, A, C, R, RC, ALAR, AREA, ALPHA)
C
C  INPUTS:  I  -  VARIABLE IDENTIFYING WHICH ARE INPUTS
C           M  -  FILTER ORDER (M<51, SEE NOTE BELOW*)
C
C     I=1     INPUT=A(.) & ALPHA
C     I=2     INPUT=C(.)
C     I=3     INPUT=R(.)
C     I=4     INPUT=RC(.) & R(1)
C     I=5     INPUT=ALAR(.) & R(1)
C     I=6     INPUT=AREA(.) & R(1)
C
C        A(.)    =  FILTER COEF.
C        ALPHA   =  GAIN
C        C(.)    =  CEPSTRAL COEF.
C        R(.)    =  AUTOCORRELATION COEF.
C        R(1)    =  FIRST AUTO. COEF. (ENERGY)
C        RC(.)   =  REFLECTION COEF.
C        ALAR(.) =  LOG AREA RATIOS
C        AREA(.) =  AREA FUNCTIONS
C
C    * PROGRAM LIMITED TO M=50 BY DIMENSION SA(50),
C      A TEMPORARY STORAGE FOR FILTER COEFFICIENTS
C
      DIMENSION A(1), C(1), R(1), RC(1), ALAR(1), AREA(1)
      DIMENSION SA(50)
      COMMON IOUTD
C
C TEST FOR M OUT OF RANGE
C
      MTEST = (M-1)*(50-M)
      IF (MTEST) 340, 340, 10
   10 MP = M + 1
      IF (I-2) 50, 20, 50
C
C ..GENERATES A(.) ,ALPHA, FROM C(.)
C
   20 ALPHA = EXP(C(1))
      A(1) = 1.
      DO 40 K=1,M
        KP = K + 1
        SUM = 0.
        DO 30 J=1,K
          JB = K - J + 2
          SUM = SUM + A(J)*C(JB)*FLOAT(JB-1)
   30   CONTINUE
        A(KP) = -SUM/FLOAT(K)
   40 CONTINUE
   50 GO TO (110, 110, 160, 60, 200, 220), I
C
C ..GENERATES SA(.),R(.), & ALPHA FROM RC(.) & R(1)
C
   60 DO 70 J=1,M
        SA(J) = RC(J)
   70 CONTINUE
      R(2) = -RC(1)*R(1)
      ALPHA = R(1)*(1.-RC(1)*RC(1))
      DO 100 J=2,M
        MH = J/2
        Q = RC(J)
        ALPHA = ALPHA*(1.-Q*Q)
        DO 80 K=1,MH
          KB = J - K
          AT = SA(K) + Q*SA(KB)
          SA(KB) = SA(KB) + Q*SA(K)
          SA(K) = AT
   80   CONTINUE
        SUM = 0.
        DO 90 L=1,J
          LB = J + 1 - L
          SUM = SUM + SA(L)*R(LB)
   90   CONTINUE
        R(J+1) = -SUM
  100 CONTINUE
      IF (I-4) 240, 240, 260
  110 DO 120 J=1,M
        SA(J) = A(J+1)
  120 CONTINUE
C
C ..GENERATES RC(.),R(1), FROM A(.) & ALPHA
C
      DO 130 J=1,M
        RC(J) = SA(J)
  130 CONTINUE
      ALT = ALPHA
      DO 150 J=2,M
        JB = M + 1 - J
        MH = (JB+1)/2
        RCT = RC(JB+1)
        D = 1. - RCT*RCT
```

JB = M+2-J
MH = JB/2
RCT = RC(JB)
DR = 1./(1. - RCT * RCT)

```
      DO 140 K=1,MH
        KB = JB - K + 1
        Q = (RC(K)-RCT*RC(K))/D
        RC(K) = Q
140   CONTINUE
      ALT = ALT/D
150   CONTINUE
      R(1) = ALT/(1.-RC(1)*RC(1))
      GO TO 60
C
C ..GENERATES RC(.),SA(.),& ALPHA FROM R(.)
C
160   RC(1) = -R(2)/R(1)
      SA(1) = RC(1)
      ALPHA = R(1)*(1.-RC(1)*RC(1))
      DO 190 J=2,M
        MH = J/2
        JM = J - 1
        Q = R(J+1)
        DO 170 L=1,JM
          LB = J + 1 - L
          Q = Q + SA(L)*R(LB)
170     CONTINUE
        Q = -Q/ALPHA
        RC(J) = Q
        DO 180 K=1,MH
          KB = J - K
          AT = SA(K) + Q*SA(K)
          SA(K) = AT
180     CONTINUE
        SA(J) = Q
        ALPHA = ALPHA*(1.-Q*Q)
190   CONTINUE
      GO TO 240
C
C ..GENERATES RC(.) & AREA(.) FROM ALAR(.)
C
200   AREA(MP) = 1.
      DO 210 J=1,M
        JB = M + 1 - J
        AR = EXP(ALAR(JB))
        RC(JB) = (1.-AR)/(1.+AR)
        AREA(JB) = AREA(JB+1)/AR
210   CONTINUE
      GO TO 60
C
C ..GENERATES ALRAR(.) & RC(.) FROM AREA(.)
C
220   DO 230 J=1,M
        AR = AREA(J+1)/AREA(J)
        ALAR(J) = ALOG(AR)
        RC(J) = (1.-AR)/(1.+AR)
230   CONTINUE
      GO TO 60
C
C ..GENERATES AREA(.) & ALAR(.) FROM RC(.)
C
240   AREA(MP) = 1.
      DO 250 J=1,M
        JB = M + 1 - J
        AR = (1.-RC(JB))/(1.+RC(JB))
        ALAR(JB) = ALOG(AR)
        AREA(JB) = AREA(JB+1)/AR
250   CONTINUE
      IF (I-2) 280, 310, 260
260   DO 270 J=2,MP
        A(J) = SA(J-1)
270   CONTINUE
      A(1) = 1.
C
C ..GENERATE C(.) FROM A(.) & ALPHA
C
280   C(1) = ALOG(ALPHA)
      C(2) = -A(2)
      DO 300 L=2,M
        LP = L + 1
        SUM = FLOAT(L)*A(LP)
        DO 290 J=2,L
          JB = L - J + 2
          SUM = SUM + A(J)*C(JB)*FLOAT(JB-1)
290     CONTINUE
        C(LP) = -SUM/FLOAT(L)
300   CONTINUE
310   DO 330 J=1,M
        IF (ABS(RC(J))-1.) 330, 320, 320
320     WRITE (IOUTD,9999)
330   CONTINUE
9999  FORMAT (19H FILTER IS UNSTABLE)
      RETURN
340   WRITE (IOUTD,9998) M
9998  FORMAT (3H M=, I6, 13H OUT OF RANGE)
      RETURN
      END
```

CHAPTER 5

FIR Filter Design and Synthesis

J. H. McClellan

Introduction

The problem of designing finite impulse response (FIR) digital filters experienced great activity in the early 1970's. Most of this work was directed at the problem of optimal (in the weighted Chebyshev sense) filter design. The first program in this section, by McClellan, Parks, and Rabiner, will design the optimal Chebyshev approximation for linear phase filters. The Chebyshev optimization is done very efficiently by means of the Remez exchange algorithm, and this particular FIR filter design program has found wide use. Interestingly enough, this usage seems to have come as such from the designers of CCD (charge-coupled device) and SAW (surface acoustic wave) filters as from the digital filtering community. Of course, any FIR design program can be used for these other problems since the underlying mathematics is the same.

The second program in this section implements the classical method of windowed linear-phase filter design. Seven window types can be specified including Chebyshev and Kaiser windows. The third program designs maximally flat linear-phase lowpass filters. In both of these programs (5.2 and 5.3) the design can be expressed in closed form; no optimization is required.

The final program, by Heute, is a subroutine for the synthesis of a finite wordlength (fixed-point rounding of coefficients) realization of a linear phase FIR filter. The program determines the minimum possible wordlength to meet the prescribed filter specifications. This technique requires, as a starting point, an exact (infinite-precision) design as would be obtained from a program such as the Remez exchange algorithm. Program 5.4 then seeks to fulfill the given specs. As a final comment, we note that the problem of optimizing the finite wordlength design of FIR digital filters remains an unsolved problem.

The four programs contained in this section are but some of the programs (and techniques) that have been proposed for the design and implementation of linear-phase FIR digital filters. Other methods for Chebyshev approximation of linear-phase filters include Linear Programming [1,7]. Herrmann and Hofstetter's method [2,5], and McCallig's CONRIP program [6]. Among other error norms, the weighted least-squares norm is a straightforward method that is not restricted to linear-phase filters.

One important difference among various methods is the manner in which the desired filter is specified. The filter design process is always a compromise among filter length, transition width, and passband and stopband deviation (in the bandpass case). Not all of these specs can be chosen arbitrarily. In connection with program 5.1 the work of Herrmann, Rabiner and Chan [4] and Rabiner [8], describes tradeoff relationships among the various parameters. Similar results are available for windowing and the maximally flat designs.

The design of nonlinear phase FIR digital filters has proven to be a more difficult problem. A particular case of interest is the design of minimum phase filters. A polynomial rooting technique proposed by Herrmann and Schuessler [3] can be used to convert a linear-phase filter to a minimum phase design. The more general problem of optimal magnitude and phase approximation still remains as an unsolved problem.

References

1. H. D. Helms, "Digital Filters with Equiripple or Minimax Responses", *IEEE Trans. Audio Electroacoust.*, Vol. AU-19, No. 1, pp. 87-93, March 1971.

2. O. Herrmann, "Design of Nonrecursive Digital Filters with Linear Phase", *Electronics Letters*, Vol. 6, No. 11, pp. 328-329, May 28, 1970.

3. O. Herrmann and H. W. Schuessler, "Design of Nonrecursive Digital Filters with Minimum Phase", *Electronics Letters*, Vol. 6, No. 11, pp. 329-330, May 28, 1970.

4. O. Herrmann, L. R. Rabiner and D. S. K. Chan, "Practical Design Rules or Optimum Finite Impulse Response Lowpass Digital Filters", *Bell System Technical Journal*, Vol. 52, No. 6, pp. 769-799, July-August 1973.

5. E. Hofstetter, A. Oppenheim and J. Siegel, "A New Technique for the Design of Nonrecursive Digital Filters", in *Proceedings of 5th Annual Princeton Conference on Information Sciences and Systems*, pp. 64-72, 1971.

6. M. T. McCallig and B. J. Leon, "Constrained Ripple Design of FIR Digital Filters", *IEEE Trans. Circuits and Systems*, Vol. CAS-25, No. 11, pp. 893-902, Nov. 1978.

7. L. R. Rabiner, "Linear Program Design of Finite Impulse Response (FIR) Digital Filters", *IEEE Trans. Audio, Electroacoust.*, Vol. AU-20, No. 5, pp. 280-288, Oct. 1972.

8. L. R. Rabiner, "Approximate Design Relationships for Lowpass FIR Digital Filters", *IEEE Trans. Audio, Electroacoust.*, Vol. AU-21, No. 5, pp. 456-460, Oct. 1973.

5.1

FIR Linear Phase Filter Design Program

J. H. McClellan

Dept. EECS
M.I.T.
Cambridge, MA 02139

T. W. Parks

EE Dept.
Rice University
Houston, TX 77001

L. R. Rabiner

Acoustics Research Dept.
Bell Laboratories
Murray Hill, NJ 07974

1. Purpose

This program uses the Remez exchange algorithm [1], [2] to design linear phase FIR digital filters with minimum weighted Chebyshev error in approximating a desired ideal frequency response. A unified treatment of the theory of this method of filter approximation can be found in Refs. [3]-[7]. The program has a special built-in section for specifying the more common ideal filter types such as multi-band, bandpass filters, Hilbert transform filters, and differentiators. It is also possible to approximate an arbitrary ideal frequency response specified by the user, but two user-written subroutines must be provided to replace two of the existing subroutines [6].

2. Method

The frequency response of an FIR digital filter with an N-point impulse response $\{h(n), n=0,1,...,N-1\}$ is the z-transform of the sequence evaluated on the unit circle, i.e.,

$$H(f)^1 = H(z)\big|_{z=e^{j2\pi f}} = \sum_{n=0}^{N-1} h(n)e^{-j2\pi nf} . \tag{1}$$

The frequency responses of a linear phase filter can be written as

$$H(f) = G(f)\exp\left[j\left(\frac{L\pi}{2} - \left(\frac{N-1}{2}\right)2\pi f\right)\right] \tag{2}$$

where $G(f)$ is a real valued function and $L = 0$ or 1. It is possible to show that there are exactly four cases of linear phase FIR filters to consider [3,5]. These four cases differ in the length of the impulse response (even or odd) and the symmetry of the impulse response [positive (L=0) or negative (L=1)]. By positive symmetry we mean $h(n) = h(N-1-n)$, and by negative symmetry $h(n) = -h(N-1-n)$.

In all cases, the real-valued function $G(f)$ will be used to approximate the desired ideal magnitude specifications since the linear phase term in Eq. (2) has no effect on the magnitude response of the filter. The form of $G(f)$ depends on which of the four cases is being used. Using the appropriate

1 For convenience, throughout this paper the notation $H(f)$ rather than $H(e^{j2\pi f})$ is used to denote the frequency response of the digital filter.

symmetry relations, $G(f)$ can be expressed as follows:

Case 1: Positive symmetry, odd length:

$$G(f) = \sum_{n=0}^{N1} a(n) \cos(2\pi n f) \tag{3}$$

where $N1 = (N-1)/2$, $a(0) = h(N1)$, and $a(n) = 2h(N1-n)$ for $n = 1,2,\ldots,N1$.

Case 2: Positive symmetry, even length:

$$G(f) = \sum_{n=1}^{N1} b(n) \cos[2\pi(n - \frac{1}{2})f] \tag{4}$$

where $N1 = N/2$ and $b(n) = 2h(N1-n)$ for $n = 1,\ldots,N1$.

Case 3: Negative symmetry, odd length:

$$G(f) = \sum_{n=1}^{N1} c(n) \sin(2\pi n f) \tag{5}$$

where $N1 = (N-1)/2$ and $c(n) = 2h(N1-n)$ for $n = 1,2,\ldots,N1$ and $h(N1) = 0$.

Case 4: Negative symmetry, even length:

$$G(f) = \sum_{n=1}^{N1} d(n) \sin[2\pi(n - \frac{1}{2})f] \tag{6}$$

where $N1 = N/2$ and $d(n) = 2h(N1-n)$ for $n = 1,\ldots,N1$.

All four cases can be combined into one algorithm. This is accomplished by noting that $G(f)$ can be rewritten as $G(f) = Q(f)P(f)$ where $P(f)$ is a linear combination of cosine functions.

It is convenient to express the summations in (4)-(6) as a sum of cosines directly. Simple manipulations of (4)-(6) yield the expressions

Case 2:

$$\sum_{n=1}^{N1} b(n) \cos[2\pi(n - \frac{1}{2})f] = \cos(\pi f) \sum_{n=0}^{N1-1} \tilde{b}(n)\cos(2\pi n f) \tag{7}$$

Case 3:

$$\sum_{n=1}^{N1} c(n)\sin(2\pi n f) = \sin(2\pi f) \sum_{n=0}^{N1-1} \tilde{c}(n)\cos(2\pi n f) \tag{8}$$

Case 4:

$$\sum_{n=1}^{N1} d(n)\sin[2\pi(n - \frac{1}{2})f] = \sin(\pi f) \sum_{n=0}^{N1-1} \tilde{d}(n)\cos(2\pi n f) \tag{9}$$

where

$$\text{Case 2:} \quad \begin{cases} b(1) = \tilde{b}(0) + \frac{1}{2}\,\tilde{b}(1) \\ b(n) = \frac{1}{2}\,[\tilde{b}(n-1) + \tilde{b}(n)], \\ \qquad n = 2,3,\ldots,N1-1 \\ b(N1) = \frac{1}{2}\,\tilde{b}(N1-1) \end{cases} \tag{10}$$

$$\textit{Case 3:}\begin{cases} c(1) = \tilde{c}(0) - \frac{1}{2}\tilde{c}(2) \\ c(n) = \frac{1}{2}[\tilde{c}(n-1) - \tilde{c}(n+1)], \\ \qquad n = 2,3,\ldots,N1-2 \\ c(N1-1) = \frac{1}{2}\tilde{c}(N1-2) \\ c(N1) = \frac{1}{2}\tilde{c}(N1-1) \end{cases} \tag{11}$$

$$\textit{Case 4:}\begin{cases} d(1) = \tilde{d}(0) - \frac{1}{2}\tilde{d}(1) \\ d(n) = \frac{1}{2}[\tilde{d}(n-1) - \tilde{d}(n)], \\ \qquad n = 2,3,\ldots,N1-1 \\ d(N1) = \frac{1}{2}\tilde{d}(N1-1). \end{cases} \tag{12}$$

The motivation for rewriting the four cases in a common form is that a single central computation routine (based on the Remez exchange method) can be used to calculate the best approximation in each of the four cases. This is accomplished by modifying both the desired magnitude function and the weighting function to formulate a new equivalent approximation problem.

The original approximation problem can be stated as follows: given a desired magnitude response $D(f)$ and a positive weight function $W(f)$, both of which are continuous on a compact subset of $[0,\frac{1}{2}]$ (note that the sampling frequency is 1.0) and given one of the four cases of linear phase filters [i.e., the forms of $G(f)$], then one wishes to minimize the maximum absolute weighted error, defined as

$$||E(f)|| = \max_{f \in F} W(f)|D(f)-G(f)| \tag{13}$$

over the set of coefficients of $G(f)$.

The error function $E(f)$ can be rewritten in the form

$$E(f) = W(f)[D(f)-G(f)] = W(f)Q(f)\left[\frac{D(f)}{Q(f)} - P(f)\right] \tag{14}$$

if one is careful to omit those endpoint(s) where $Q(f) = 0$. Letting $\hat{D}(f) = D(f)/Q(f)$ and $\hat{W}(f) = W(f)Q(f)$, an equivalent approximation problem is to minimize the quantity

$$||E(f)|| = \max_{f \in F'} \hat{W}(f)|\hat{D}(f)-P(f)| \tag{15}$$

by choice of the coefficients of $P(f)$. The set F has been replaced by $F' = F$ - {endpoints where $Q(f)=0$}.

The net effect of this reformulation of the problem is a unification of the four cases of linear phase FIR filters from the point of view of the approximation problem. Furthermore, Eq. (15) provides a simplified viewpoint from which it is easy to see the necessary and sufficient conditions that must be satisfied by the best approximation. Finally, Eq. (15) shows how to calculate this best approximation using an algorithm which can do only cosine approximations (Case 1). The necessary and sufficient conditions for this best approximation are given in the following alternation theorem [1].

Alternation Theorem: If $P(f)$ is a linear combination of r cosine functions i.e.,

$$P(f) = \sum_{n=0}^{r-1} \alpha(n)\cos(2\pi nf) ,$$

then a necessary and sufficient condition that $P(f)$ be the unique best weighted Chebyshev approximation to a continuous function $\hat{D}(f)$ on F' is that the weighted error function $E(F) = \hat{W}(f)$ $[\hat{D}(f)-P(f)]$ exhibit *at least* $r+1$ extremal frequencies in F'.

These extremal frequencies are a set of $r+1$ points $\{F_i\}$, $i = 1,2,...,r+1$ such that $F_1 < F_2 \cdots < F_r < F_{r+1}$, with $E(F_i) = -E(F_{i+1})$, $i = 1,2,...,r$ and $|E(F_i)| = \max_{f \in F'} |E(f)|$.

An algorithm can now be designed to make the error function of the filter satisfy the set of necessary and sufficient conditions for optimality as stated in the alternation theorem. The next section describes such an algorithm along with details of its implementation.

3. Program Description

The program consists of a main program that handles the input, sets up the appropriate approximation problem, calls the Remez exchange subroutine and handles the output of the optimal filter coefficients and the value of the minimum error. The program contains seven subroutines the most important of which are REMEZ, EFF and WATE. Subroutine REMEZ implements the Remez exchange algorithm for Chebyshev approximation. Subroutines EFF and WATE define the desired ideal frequency response, $D(f)$, and the weight function, $W(f)$, respectively. It is possible for the user to modify these subroutines to suit his own choice of ideal frequency response and tolerance scheme [6]. Figure 1 provides a summary of the overall program structure. Note that the program yields the impulse response of the optimum filter, but does *not* produce a frequency response plot. Such a plot can be generated easily by the user with an FFT subroutine and some straightforward graphical programs.

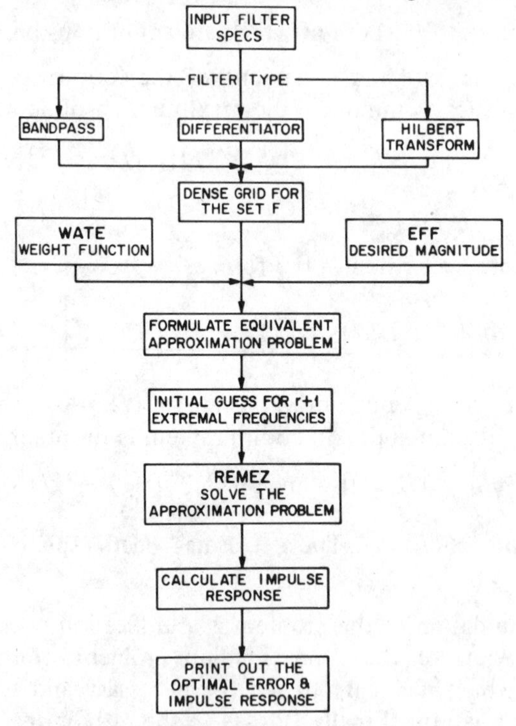

Fig. 1 Overall flowchart of filter design algorithm.

The specification of the filter design problem requires the following input parameters:

NFILT — The filter length in samples. NFILT must satisfy $3 \leqslant$ NFILT $\leqslant$ NFMAX (the upper limit can be changed by the programmer). In the present implementation NFMAX has been arbitrarily chosen to be 128.

JTYPE — The type of filter
(a) Multiple passband/stopband (JTYPE=1)

(b) Differentiator (JTYPE=2)
(c) Hilbert transformer (JTYPE=3).

EDGE An array of size 20. The frequency bands, specified by upper and lower cutoff frequencies, up to a maximum of 10 bands.

FX An array of size 10. The desired frequency response in each band.

WTX An array of size 10. A positive weight function in each band.

LGRID The grid density, set to be 16 unless altered by the user by specifying a positive value for LGRID.

The EDGE array specifies the set F to be of the form $F = UB_i$ where each frequency band B_i is a closed subinterval of the frequency axis $[0, 1/2]$. The arrays FX and WTX then specify the ideal response and weight function in each band. These input arrays are interpreted in different manners by the program depending on whether the filter type is a differentiator (JTYPE=2) or not. In the case of a differentiator FX(I) specifies the slope of $D(f)$ in the frequency band B_i while WTX(I) gives W(f) as 1/WTX(I), resulting in a *relative* error tolerance scheme [6].

The set F must be replaced by a finite set of points for implementation on a computer. A dense grid of points is used with the spacing between points being $0.5/(\text{LGRID}\cdot r)$ where r is the number of cosine basis functions. Both $D(f)$ and $W(f)$ are evaluated on the grid and stored in the arrays DES and WT by the subroutines EFF and WATE, respectively. Then the equivalent approximation problem is set up by forming $\hat{D}(f)$ and $\hat{W}(f)$ as in Eq. (14), and an initial guess of the extremal frequencies is made by taking $r + 1$ equally spaced frequency values. The subroutine REMEZ is called to perform the calculation of the best approximation for the equivalent problem. The mechanics of the Remez algorithm will not be discussed here since they are treated elsewhere for the particular case of low-pass filters [4].

The appropriate equations ((3)-(12)) are used to recover the impulse response from the coefficients of the best cosine approximation obtained in the REMEZ subroutine. The outputs of the program are the coefficients of the best impulse response, the optimal error $(min||E(f)||)$, and the $r + 1$ extremal frequencies where $E(f) = \pm ||E(f)||$.

It is possible that one might want to design a filter to approximate a magnitude specification which is not included among the possibilities given above, or change the weight function to get a new tolerance scheme. In such cases, the user must code new functions EFF and WATE to calculate $D(f)$ and $W(f)$ as a function of f. Note that the value of f is passed to each of these functions via the parameter FREQ. The input is the same as before, except that there are only two types of filters, depending on whether the impulse symmetry is positive or negative (JTYPE=1 or 2). One example of the design of a nonstandard filter using this program is given in Ref. [6].

4. Design Examples

This design example illustrates the multiband capability of the program, for an $N = 55$ five-band filter with three stopbands and two passbands. The weighting in each of the stopbands is different, making the peak approximation error differ in each of these bands. The required input for this example is given in Table 1. Table 2 shows the computer output listing, and Figure 2 shows a plot of the filter frequency response on a log magnitude scale.

An extensive set of design examples which illustrate the versatility of this program is given in Ref. [6].

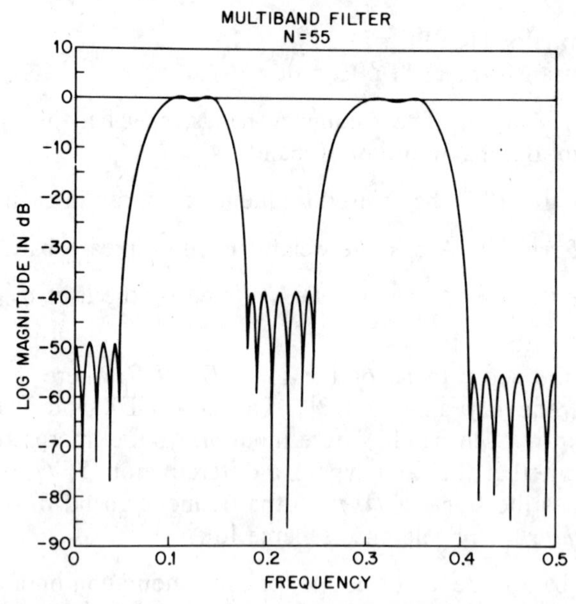

Fig. 2 Log magnitude response for an $N = 55$ multiband filter.

References

1. E. W. Cheney, *Introduction to Approximation Theory,* New York: McGraw-Hill, pp. 77-100, 1966.

2. E. Ya. Remez, "General Computational Methods of Tchebycheff Approximation", Kiev, (Atomic Energy Translation 4491), pp. 1-85, 1957.

3. J. H. McClellan and T. W. Parks, "A Unified Approach to the Design of Optimum FIR Linear Phase Digital Filters", *IEEE Trans. on Circuit Theory,* Vol. CT-20, pp. 697-701, November 1973.

4. T. W. Parks and J. H. McClellan, "Chebyshev Approximation for Nonrecursive Digital Filters with Linear Phase", *IEEE Trans. on Circuit Theory,* Vol. CT-19, pp. 189-194, March 1972.

5. L. R. Rabiner and B. Gold, *Theory and Application of Digital Signal Processing,* Prentice-Hall, Inc., Englewood Cliffs, New Jersey, Chapter 3, pp. 123-204, 1975.

6. J. H. McClellan, T. W. Parks, and L. R. Rabiner, "A Computer Program for Designing Optimum FIR Linear Phase Digital Filters", *IEEE Trans. on Audio and Electroacoustics,* Vol. AU-21, No. 6, pp. 506-526, December 1973. (Also reprinted in *Digital Signal Processing, II,* IEEE Press, pp. 97-117, 1975.)

7. L. R. Rabiner, J. H. McClellan, and T. W. Parks, "FIR Digital Filter Design Techniques Using Weighted Chebyshev Approximation", *Proc. IEEE,* Vol. 63, No. 4, pp. 595-610, April 1975.

Table 1

```
  55    1     5     0
0.            .05               .1             .15
0.18          .25               .3             .36
0.41          .5
0.            1.                0.             1.
0.
10.           1.                3.             1.
20.
   0
```

Table 2 Output listing for an $N = 55$ multiband filter.

```
DEVIATION =  0.000734754
DEVIATION =  0.006315947
DEVIATION =  0.021567374
DEVIATION =  0.026203127
DEVIATION = -0.032680369
DEVIATION = -0.034435446
DEVIATION = -0.034448378
DEVIATION = -0.034448593

**********************************************************************

             FINITE IMPULSE RESPONSE (FIR)
           LINEAR PHASE DIGITAL FILTER DESIGN
               REMEZ EXCHANGE ALGORITHM
                    BANDPASS FILTER
                  FILTER LENGTH =  55
             ***** IMPULSE RESPONSE *****
           H( 1) =  0.10662652E-02 = H( 55)
           H( 2) =  0.63777615E-02 = H( 54)
           H( 3) =  0.35755609E-02 = H( 53)
           H( 4) = -0.90677854E-02 = H( 52)
           H( 5) = -0.90906978E-02 = H( 51)
           H( 6) =  0.29155630E-02 = H( 50)
           H( 7) =  0.39637965E-02 = H( 49)
           H( 8) =  0.11172051E-01 = H( 48)
           H( 9) =  0.11646759E-01 = H( 47)
           H(10) = -0.99630785E-02 = H( 46)
           H(11) = -0.92384245E-02 = H( 45)
           H(12) = -0.20406392E-01 = H( 44)
           H(13) = -0.19460483E-01 = H( 43)
           H(14) =  0.31243014E-01 = H( 42)
           H(15) =  0.63045568E-02 = H( 41)
           H(16) = -0.20482803E-01 = H( 40)
           H(17) =  0.65740513E-02 = H( 39)
           H(18) = -0.11202127E-02 = H( 38)
           H(19) =  0.41956986E-01 = H( 37)
           H(20) =  0.35784266E-01 = H( 36)
           H(21) =  0.34744803E-01 = H( 35)
           H(22) =  0.71496359E-01 = H( 34)
           H(23) = -0.17138831E 00 = H( 33)
           H(24) = -0.18255044E 00 = H( 32)
           H(25) =  0.74059024E-01 = H( 31)
           H(26) = -0.10317421E 00 = H( 30)
           H(27) =  0.25716721E-01 = H( 29)
           H(28) =  0.37813546E 00 = H( 28)

                       BAND  1        BAND  2        BAND  3        BAND  4
 LOWER BAND EDGE     0.             0.1000000      0.1800000      0.3000000
 UPPER BAND EDGE     0.0500000      0.1500000      0.2500000      0.3600000
 DESIRED VALUE       0.             1.0000000      0.             1.0000000
 WEIGHTING          10.0000000      1.0000000      3.0000000      1.0000000
 DEVIATION           0.0034449      0.0344486      0.0114829      0.0344486
 DEVIATION IN DB   -49.2565703      0.2941783    -38.7989955      0.2941783

                       BAND  5
 LOWER BAND EDGE     0.4100000
 UPPER BAND EDGE     0.5000000
 DESIRED VALUE       0.
 WEIGHTING          20.0000000
 DEVIATION           0.0017224
 DEVIATION IN DB   -55.2771702

 EXTREMAL FREQUENCIES--MAXIMA OF THE ERROR CURVE
    0.          0.0167411   0.0323661   0.0446429   0.0500000
    0.1000000   0.1089286   0.1267857   0.1424107   0.1500000
    0.1800000   0.1855804   0.1978571   0.2134821   0.2302232
    0.2436160   0.2500000   0.3000000   0.3122768   0.3323661
    0.3502232   0.3600000   0.4100000   0.4155804   0.4289732
    0.4457143   0.4635714   0.4814285   0.5000000

**********************************************************************
```

Appendix

```
C
C-----------------------------------------------------------------------
C MAIN PROGRAM: FIR LINEAR PHASE FILTER DESIGN PROGRAM
C
C AUTHORS: JAMES H. MCCLELLAN
C          DEPARTMENT OF ELECTRICAL ENGINEERING AND COMPUTER SCIENCE
C          MASSACHUSETTS INSTITUTE OF TECHNOLOGY
C          CAMBRIDGE, MASS. 02139
C
C          THOMAS W. PARKS
C          DEPARTMENT OF ELECTRICAL ENGINEERING
C          RICE UNIVERSITY
C          HOUSTON, TEXAS 77001
C
C          LAWRENCE R. RABINER
C          BELL LABORATORIES
C          MURRAY HILL, NEW JERSEY 07974
C
C INPUT:
C  NFILT-- FILTER LENGTH
C  JTYPE-- TYPE OF FILTER
C          1 = MULTIPLE PASSBAND/STOPBAND FILTER
C          2 = DIFFERENTIATOR
C          3 = HILBERT TRANSFORM FILTER
C  NBANDS-- NUMBER OF BANDS
C  LGRID-- GRID DENSITY, WILL BE SET TO 16 UNLESS
C          SPECIFIED OTHERWISE BY A POSITIVE CONSTANT.
C
C  EDGE(2*NBANDS)-- BANDEDGE ARRAY, LOWER AND UPPER EDGES FOR EACH BAND
C                   WITH A MAXIMUM OF 10 BANDS.
C
C  FX(NBANDS)-- DESIRED FUNCTION ARRAY (OR DESIRED SLOPE IF A
C               DIFFERENTIATOR) FOR EACH BAND.
C
C  WTX(NBANDS)-- WEIGHT FUNCTION ARRAY IN EACH BAND.  FOR A
C                DIFFERENTIATOR, THE WEIGHT FUNCTION IS INVERSELY
C                PROPORTIONAL TO F.
C
C  SAMPLE INPUT DATA SETUP:
C       32,1,3,0
C       0.0,0.1,0.2,0.35,0.425,0.5
C       0.0,1.0,0.0
C       10.0,1.0,10.0
C     THIS DATA SPECIFIES A LENGTH 32 BANDPASS FILTER WITH
C     STOPBANDS 0 TO 0.1 AND 0.425 TO 0.5, AND PASSBAND FROM
C     0.2 TO 0.35 WITH WEIGHTING OF 10 IN THE STOPBANDS AND 1
C     IN THE PASSBAND.  THE GRID DENSITY DEFAULTS TO 16.
C     THIS IS THE FILTER IN FIGURE 10.
C
C     THE FOLLOWING INPUT DATA SPECIFIES A LENGTH 32 FULLBAND
C     DIFFERENTIATOR WITH SLOPE 1 AND WEIGHTING OF 1/F.
C     THE GRID DENSITY WILL BE SET TO 20.
C       32,2,1,20
C       0,0.5
C       1.0
C       1.0
C
C-----------------------------------------------------------------------
C
      COMMON PI2,AD,DEV,X,Y,GRID,DES,WT,ALPHA,IEXT,NFCNS,NGRID
      COMMON /OOPS/NITER,IOUT
      DIMENSION IEXT(66),AD(66),ALPHA(66),X(66),Y(66)
      DIMENSION H(66)
      DIMENSION DES(1045),GRID(1045),WT(1045)
      DIMENSION EDGE(20),FX(10),WTX(10),DEVIAT(10)
      DOUBLE PRECISION PI2,PI
      DOUBLE PRECISION AD,DEV,X,Y
      DOUBLE PRECISION GEE,D
      INTEGER BD1,BD2,BD3,BD4
      DATA BD1,BD2,BD3,BD4/1HB,1HA,1HN,1HD/
      INPUT=I1MACH(1)
      IOUT=I1MACH(2)
      PI=4.0*DATAN(1.0D0)
      PI2=2.0D00*PI
C
C  THE PROGRAM IS SET UP FOR A MAXIMUM LENGTH OF 128, BUT
C  THIS UPPER LIMIT CAN BE CHANGED BY REDIMENSIONING THE
C  ARRAYS IEXT, AD, ALPHA, X, Y, H TO BE NFMAX/2 + 2.
C  THE ARRAYS DES, GRID, AND WT MUST DIMENSIONED
C  16(NFMAX/2 + 2).
C
      NFMAX=128
  100 CONTINUE
      JTYPE=0
C
C  PROGRAM INPUT SECTION
C
      READ(INPUT,110) NFILT,JTYPE,NBANDS,LGRID
      IF(NFILT.EQ.0)STOP
  110 FORMAT(4I5)
      IF(NFILT.LE.NFMAX.OR.NFILT.GE.3) GO TO 115
      CALL ERROR
      STOP
  115 IF(NBANDS.LE.0) NBANDS=1
C
C  GRID DENSITY IS ASSUMED TO BE 16 UNLESS SPECIFIED
C  OTHERWISE
C
      IF(LGRID.LE.0) LGRID=16
      JB=2*NBANDS
      READ(INPUT,120) (EDGE(J),J=1,JB)
  120 FORMAT(4F15.9)
      READ(INPUT,120) (FX(J),J=1,NBANDS)
      READ(INPUT,120) (WTX(J),J=1,NBANDS)
      IF(JTYPE.GT.0.AND.JTYPE.LE.3) GO TO 125
      CALL ERROR
      STOP
  125 NEG=1
      IF(JTYPE.EQ.1) NEG=0
      NODD=NFILT/2
      NODD=NFILT-2*NODD
      NFCNS=NFILT/2
      IF(NODD.EQ.1.AND.NEG.EQ.0) NFCNS=NFCNS+1
C
C  SET UP THE DENSE GRID.  THE NUMBER OF POINTS IN THE GRID
C  IS (FILTER LENGTH + 1)*GRID DENSITY/2
C
      GRID(1)=EDGE(1)
      DELF=LGRID*NFCNS
      DELF=0.5/DELF
      IF(NEG.EQ.0) GO TO 135
      IF(EDGE(1).LT.DELF) GRID(1)=DELF
  135 CONTINUE
      J=1
      L=1
      LBAND=1
  140 FUP=EDGE(L+1)
  145 TEMP=GRID(J)
C
C  CALCULATE THE DESIRED MAGNITUDE RESPONSE AND THE WEIGHT
```

```
C  FUNCTION ON THE GRID
C
      DES(J)=EFF(TEMP,FX,WTX,LBAND,JTYPE)
      WT(J)=WATE(TEMP,FX,WTX,LBAND,JTYPE)
      J=J+1
      GRID(J)=TEMP+DELF
      IF(GRID(J).GT.FUP) GO TO 150
      GO TO 145
  150 GRID(J-1)=FUP
      DES(J-1)=EFF(FUP,FX,WTX,LBAND,JTYPE)
      WT(J-1)=WATE(FUP,FX,WTX,LBAND,JTYPE)
      LBAND=LBAND+1
      L=L+2
      IF(LBAND.GT.NBANDS) GO TO 160
      GRID(J)=EDGE(L)
      GO TO 140
  160 NGRID=J-1
      IF(NEG.NE.NODD) GO TO 165
      IF(GRID(NGRID).GT.(0.5-DELF)) NGRID=NGRID-1
  165 CONTINUE
C
C  SET UP A NEW APPROXIMATION PROBLEM WHICH IS EQUIVALENT
C  TO THE ORIGINAL PROBLEM
C
      IF(NEG) 170,170,180
  170 IF(NODD.EQ.1) GO TO 200
      DO 175 J=1,NGRID
      CHANGE=DCOS(PI*GRID(J))
      DES(J)=DES(J)/CHANGE
  175 WT(J)=WT(J)*CHANGE
      GO TO 200
  180 IF(NODD.EQ.1) GO TO 190
      DO 185 J=1,NGRID
      CHANGE=DSIN(PI*GRID(J))
      DES(J)=DES(J)/CHANGE
  185 WT(J)=WT(J)*CHANGE
      GO TO 200
  190 DO 195 J=1,NGRID
      CHANGE=DSIN(PI2*GRID(J))
      DES(J)=DES(J)/CHANGE
  195 WT(J)=WT(J)*CHANGE
C
C  INITIAL GUESS FOR THE EXTREMAL FREQUENCIES--EQUALLY
C  SPACED ALONG THE GRID
C
  200 TEMP=FLOAT(NGRID-1)/FLOAT(NFCNS)
      DO 210 J=1,NFCNS
      XT=J-1
  210 IEXT(J)=XT*TEMP+1.0
      IEXT(NFCNS+1)=NGRID
      NM1=NFCNS-1
      NZ=NFCNS+1
C
C  CALL THE REMEZ EXCHANGE ALGORITHM TO DO THE APPROXIMATION
C  PROBLEM
C
      CALL REMEZ
C
C  CALCULATE THE IMPULSE RESPONSE.
C
      IF(NEG) 300,300,320
  300 IF(NODD.EQ.0) GO TO 310
      DO 305 J=1,NM1
      NZMJ=NZ-J
  305 H(J)=0.5*ALPHA(NZMJ)
      H(NFCNS)=ALPHA(1)
      GO TO 350
  310 H(1)=0.25*ALPHA(NFCNS)
      DO 315 J=2,NM1
      NZMJ=NZ-J
      NF2J=NFCNS+2-J
  315 H(J)=0.25*(ALPHA(NZMJ)+ALPHA(NF2J))
      H(NFCNS)=0.5*ALPHA(1)+0.25*ALPHA(2)
      GO TO 350
  320 IF(NODD.EQ.0) GO TO 330
      H(1)=0.25*ALPHA(NFCNS)
      H(2)=0.25*ALPHA(NM1)
      DO 325 J=3,NM1
      NZMJ=NZ-J
      NF3J=NFCNS+3-J
  325 H(J)=0.25*(ALPHA(NZMJ)-ALPHA(NF3J))
      H(NFCNS)=0.5*ALPHA(1)-0.25*ALPHA(3)
      H(NZ)=0.0
      GO TO 350
  330 H(1)=0.25*ALPHA(NFCNS)
      DO 335 J=2,NM1
      NZMJ=NZ-J
      NF2J=NFCNS+2-J
  335 H(J)=0.25*(ALPHA(NZMJ)-ALPHA(NF2J))
      H(NFCNS)=0.5*ALPHA(1)-0.25*ALPHA(2)
C
C  PROGRAM OUTPUT SECTION.
C
  350 WRITE(IOUT,360)
  360 FORMAT(1H1, 70(1H*)//15X,29HFINITE IMPULSE RESPONSE (FIR)/
     113X,34HLINEAR PHASE DIGITAL FILTER DESIGN/
     217X,24HREMEZ EXCHANGE ALGORITHM/)
      IF(JTYPE.EQ.1) WRITE(IOUT,365)
  365 FORMAT(22X,15HBANDPASS FILTER/)
      IF(JTYPE.EQ.2) WRITE(IOUT,370)
  370 FORMAT(22X,14HDIFFERENTIATOR/)
      IF(JTYPE.EQ.3) WRITE(IOUT,375)
  375 FORMAT(20X,19HHILBERT TRANSFORMER/)
      WRITE(IOUT,378) NFILT
  378 FORMAT(20X,16HFILTER LENGTH = ,I3/)
      WRITE(IOUT,380)
  380 FORMAT(15X,28H***** IMPULSE RESPONSE *****)
      DO 381 J=1,NFCNS
      K=NFILT+1-J
      IF(NEG.EQ.0) WRITE(IOUT,382) J,H(J),K
      IF(NEG.EQ.1) WRITE(IOUT,383) J,H(J),K
  381 CONTINUE
  382 FORMAT(13X,2HH(,I2,4H) = ,E15.8,5H = H(,I3,1H))
  383 FORMAT(13X,2HH(,I2,4H) = ,E15.8,6H = -H(,I3,1H))
      IF(NEG.EQ.1.AND.NODD.EQ.1) WRITE(IOUT,384) NZ
  384 FORMAT(13X,2HH(,I2,8H) =  0.0)
      DO 450 K=1,NBANDS,4
      KUP=K+3
      IF(KUP.GT.NBANDS) KUP=NBANDS
      WRITE(IOUT,385) (BD1,BD2,BD3,BD4,J,J=K,KUP)
  385 FORMAT(/24X,4(4A1,I3,7X))
      WRITE(IOUT,390) (EDGE(2*J-1),J=K,KUP)
  390 FORMAT(2X,15HLOWER BAND EDGE,5F14.7)
      WRITE(IOUT,395) (EDGE(2*J),J=K,KUP)
  395 FORMAT(2X,15HUPPER BAND EDGE,5F14.7)
      IF(JTYPE.NE.2) WRITE(IOUT,400) (FX(J),J=K,KUP)
  400 FORMAT(2X,13HDESIRED VALUE,2X,5F14.7)
      IF(JTYPE.EQ.2) WRITE(IOUT,405) (FX(J),J=K,KUP)
  405 FORMAT(2X,13HDESIRED SLOPE,2X,5F14.7)
      WRITE(IOUT,410) (WTX(J),J=K,KUP)
```

```
410 FORMAT(2X,9HWEIGHTING,6X,5F14.7)
    DO 420 J=K,KUP
420 DEVIAT(J)=DEV/WTX(J)
    WRITE(IOUT,425) (DEVIAT(J),J=K,KUP)
425 FORMAT(2X,9HDEVIATION,6X,5F14.7)
    IF(JTYPE.NE.1) GO TO 450
    DO 430 J=K,KUP
430 DEVIAT(J)=20.0*ALOG10(DEVIAT(J)+FX(J))
    WRITE(IOUT,435) (DEVIAT(J),J=K,KUP)
435 FORMAT(2X,15HDEVIATION IN DB,5F14.7)
450 CONTINUE
    DO 452 J=1,NZ
    IX=IEXT(J)
452 GRID(J)=GRID(IX)
    WRITE(IOUT,455) (GRID(J),J=1,NZ)
455 FORMAT(/2X,47HEXTREMAL FREQUENCIES--MAXIMA OF THE ERROR CURVE/
   1 (2X,5F12.7))
    WRITE(IOUT,460)
460 FORMAT(/1X,70(1H*)/1H1)
    GO TO 100
    END
C
C-----------------------------------------------------------------------
C FUNCTION: EFF
C   FUNCTION TO CALCULATE THE DESIRED MAGNITUDE RESPONSE
C   AS A FUNCTION OF FREQUENCY.
C   AN ARBITRARY FUNCTION OF FREQUENCY CAN BE
C   APPROXIMATED IF THE USER REPLACES THIS FUNCTION
C   WITH THE APPROPRIATE CODE TO EVALUATE THE IDEAL
C   MAGNITUDE.  NOTE THAT THE PARAMETER FREQ IS THE
C   VALUE OF NORMALIZED FREQUENCY NEEDED FOR EVALUATION.
C-----------------------------------------------------------------------
C
      FUNCTION EFF(FREQ,FX,WTX,LBAND,JTYPE)
      DIMENSION FX(5),WTX(5)
      IF(JTYPE.EQ.2) GO TO 1
      EFF=FX(LBAND)
      RETURN
    1 EFF=FX(LBAND)*FREQ
      RETURN
      END
C
C-----------------------------------------------------------------------
C FUNCTION: WATE
C   FUNCTION TO CALCULATE THE WEIGHT FUNCTION AS A FUNCTION
C   OF FREQUENCY.  SIMILAR TO THE FUNCTION EFF, THIS FUNCTION CAN
C   BE REPLACED BY A USER-WRITTEN ROUTINE TO CALCULATE ANY
C   DESIRED WEIGHTING FUNCTION.
C-----------------------------------------------------------------------
C
      FUNCTION WATE(FREQ,FX,WTX,LBAND,JTYPE)
      DIMENSION FX(5),WTX(5)
      IF(JTYPE.EQ.2) GO TO 1
      WATE=WTX(LBAND)
      RETURN
    1 IF(FX(LBAND).LT.0.0001) GO TO 2
      WATE=WTX(LBAND)/FREQ
      RETURN
    2 WATE=WTX(LBAND)
      RETURN
      END
C
C-----------------------------------------------------------------------
C SUBROUTINE: ERROR
C   THIS ROUTINE WRITES AN ERROR MESSAGE IF AN
C   ERROR HAS BEEN DETECTED IN THE INPUT DATA.
C-----------------------------------------------------------------------
C
      SUBROUTINE ERROR
      COMMON /OOPS/NITER,IOUT
      WRITE(IOUT,1)
    1 FORMAT(44H ************ ERROR IN INPUT DATA **********)
      RETURN
      END
C
C-----------------------------------------------------------------------
C SUBROUTINE: REMEZ
C   THIS SUBROUTINE IMPLEMENTS THE REMEZ EXCHANGE ALGORITHM
C   FOR THE WEIGHTED CHEBYSHEV APPROXIMATION OF A CONTINUOUS
C   FUNCTION WITH A SUM OF COSINES.  INPUTS TO THE SUBROUTINE
C   ARE A DENSE GRID WHICH REPLACES THE FREQUENCY AXIS, THE
C   DESIRED FUNCTION ON THIS GRID, THE WEIGHT FUNCTION ON THE
C   GRID, THE NUMBER OF COSINES, AND AN INITIAL GUESS OF THE
C   EXTREMAL FREQUENCIES.  THE PROGRAM MINIMIZES THE CHEBYSHEV
C   ERROR BY DETERMINING THE BEST LOCATION OF THE EXTREMAL
C   FREQUENCIES (POINTS OF MAXIMUM ERROR) AND THEN CALCULATES
C   THE COEFFICIENTS OF THE BEST APPROXIMATION.
C-----------------------------------------------------------------------
C
      SUBROUTINE REMEZ
      COMMON PI2,AD,DEV,X,Y,GRID,DES,WT,ALPHA,IEXT,NFCNS,NGRID
      COMMON /OOPS/NITER,IOUT
      DIMENSION IEXT(66),AD(66),ALPHA(66),X(66),Y(66)
      DIMENSION DES(1045),GRID(1045),WT(1045)
      DIMENSION A(66),P(65),Q(65)
      DOUBLE PRECISION PI2,DNUM,DDEN,DTEMP,A,P,Q
      DOUBLE PRECISION DK,DAK
      DOUBLE PRECISION AD,DEV,X,Y
      DOUBLE PRECISION GEE,D
C
C  THE PROGRAM ALLOWS A MAXIMUM NUMBER OF ITERATIONS OF 25
C
      ITRMAX=25
      DEVL=-1.0
      NZ=NFCNS+1
      NZZ=NFCNS+2
      NITER=0
  100 CONTINUE
      IEXT(NZZ)=NGRID+1
      NITER=NITER+1
      IF(NITER.GT.ITRMAX) GO TO 400
      DO 110 J=1,NZ
      JXT=IEXT(J)
      DTEMP=GRID(JXT)
      DTEMP=DCOS(DTEMP*PI2)
  110 X(J)=DTEMP
      JET=(NFCNS-1)/15+1
      DO 120 J=1,NZ
  120 AD(J)=D(J,NZ,JET)
      DNUM=0.0
      DDEN=0.0
      K=1
      DO 130 J=1,NZ
      L=IEXT(J)
      DTEMP=AD(J)*DES(L)
      DNUM=DNUM+DTEMP
      DTEMP=FLOAT(K)*AD(J)/WT(L)
      DDEN=DDEN+DTEMP
  130 K=-K
      DEV=DNUM/DDEN
```

```
      WRITE(IOUT,131) DEV
  131 FORMAT(1X,12HDEVIATION = ,F12.9)
      NU=1
      IF(DEV.GT.0.0) NU=-1
      DEV=-FLOAT(NU)*DEV
      K=NU
      DO 140 J=1,NZ
      L=IEXT(J)
      DTEMP=FLOAT(K)*DEV/WT(L)
      Y(J)=DES(L)+DTEMP
  140 K=-K
      IF(DEV.GT.DEVL) GO TO 150
      CALL OUCH
      GO TO 400
  150 DEVL=DEV
      JCHNGE=0
      K1=IEXT(1)
      KNZ=IEXT(NZ)
      KLOW=0
      NUT=-NU
      J=1
C
C  SEARCH FOR THE EXTREMAL FREQUENCIES OF THE BEST
C  APPROXIMATION
C
  200 IF(J.EQ.NZZ) YNZ=COMP
      IF(J.GE.NZZ) GO TO 300
      KUP=IEXT(J+1)
      L=IEXT(J)+1
      NUT=-NUT
      IF(J.EQ.2) Y1=COMP
      COMP=DEV
      IF(L.GE.KUP) GO TO 220
      ERR=GEE(L,NZ)
      ERR=(ERR-DES(L))*WT(L)
      DTEMP=FLOAT(NUT)*ERR-COMP
      IF(DTEMP.LE.0.0) GO TO 220
      COMP=FLOAT(NUT)*ERR
  210 L=L+1
      IF(L.GE.KUP) GO TO 215
      ERR=GEE(L,NZ)
      ERR=(ERR-DES(L))*WT(L)
      DTEMP=FLOAT(NUT)*ERR-COMP
      IF(DTEMP.LE.0.0) GO TO 215
      COMP=FLOAT(NUT)*ERR
      GO TO 210
  215 IEXT(J)=L-1
      J=J+1
      KLOW=L-1
      JCHNGE=JCHNGE+1
      GO TO 200
  220 L=L-1
  225 L=L-1
      IF(L.LE.KLOW) GO TO 250
      ERR=GEE(L,NZ)
      ERR=(ERR-DES(L))*WT(L)
      DTEMP=FLOAT(NUT)*ERR-COMP
      IF(DTEMP.GT.0.0) GO TO 230
      IF(JCHNGE.LE.0) GO TO 225
      GO TO 260
  230 COMP=FLOAT(NUT)*ERR
  235 L=L-1
      IF(L.LE.KLOW) GO TO 240
      ERR=GEE(L,NZ)
      ERR=(ERR-DES(L))*WT(L)
      DTEMP=FLOAT(NUT)*ERR-COMP
      IF(DTEMP.LE.0.0) GO TO 240
      COMP=FLOAT(NUT)*ERR
      GO TO 235
  240 KLOW=IEXT(J)
      IEXT(J)=L+1
      J=J+1
      JCHNGE=JCHNGE+1
      GO TO 200
  250 L=IEXT(J)+1
      IF(JCHNGE.GT.0) GO TO 215
  255 L=L+1
      IF(L.GE.KUP) GO TO 260
      ERR=GEE(L,NZ)
      ERR=(ERR-DES(L))*WT(L)
      DTEMP=FLOAT(NUT)*ERR-COMP
      IF(DTEMP.LE.0.0) GO TO 255
      COMP=FLOAT(NUT)*ERR
      GO TO 210
  260 KLOW=IEXT(J)
      J=J+1
      GO TO 200
  300 IF(J.GT.NZZ) GO TO 320
      IF(K1.GT.IEXT(1)) K1=IEXT(1)
      IF(KNZ.LT.IEXT(NZ)) KNZ=IEXT(NZ)
      NUT1=NUT
      NUT=-NU
      L=0
      KUP=K1
      COMP=YNZ*(1.00001)
      LUCK=1
  310 L=L+1
      IF(L.GE.KUP) GO TO 315
      ERR=GEE(L,NZ)
      ERR=(ERR-DES(L))*WT(L)
      DTEMP=FLOAT(NUT)*ERR-COMP
      IF(DTEMP.LE.0.0) GO TO 310
      COMP=FLOAT(NUT)*ERR
      J=NZZ
      GO TO 210
  315 LUCK=6
      GO TO 325
  320 IF(LUCK.GT.9) GO TO 350
      IF(COMP.GT.Y1) Y1=COMP
      K1=IEXT(NZZ)
  325 L=NGRID+1
      KLOW=KNZ
      NUT=-NUT1
      COMP=Y1*(1.00001)
  330 L=L-1
      IF(L.LE.KLOW) GO TO 340
      ERR=GEE(L,NZ)
      ERR=(ERR-DES(L))*WT(L)
      DTEMP=FLOAT(NUT)*ERR-COMP
      IF(DTEMP.LE.0.0) GO TO 330
      J=NZZ
      COMP=FLOAT(NUT)*ERR
      LUCK=LUCK+10
      GO TO 235
  340 IF(LUCK.EQ.6) GO TO 370
      DO 345 J=1,NFCNS
      NZZMJ=NZZ-J
      NZMJ=NZ-J
  345 IEXT(NZZMJ)=IEXT(NZMJ)
      IEXT(1)=K1
```

```
      GO TO 100
  350 KN=IEXT(NZZ)
      DO 360 J=1,NFCNS
  360 IEXT(J)=IEXT(J+1)
      IEXT(NZ)=KN
      GO TO 100
  370 IF(JCHNGE.GT.0) GO TO 100
C
C  CALCULATION OF THE COEFFICIENTS OF THE BEST APPROXIMATION
C  USING THE INVERSE DISCRETE FOURIER TRANSFORM
C
  400 CONTINUE
      NM1=NFCNS-1
      FSH=1.0E-06
      GTEMP=GRID(1)
      X(NZZ)=-2.0
      CN=2*NFCNS-1
      DELF=1.0/CN
      L=1
      KKK=0
      IF(GRID(1).LT.0.01.AND.GRID(NGRID).GT.0.49) KKK=1
      IF(NFCNS.LE.3) KKK=1
      IF(KKK.EQ.1) GO TO 405
      DTEMP=DCOS(PI2*GRID(1))
      DNUM=DCOS(PI2*GRID(NGRID))
      AA=2.0/(DTEMP-DNUM)
      BB=-(DTEMP+DNUM)/(DTEMP-DNUM)
  405 CONTINUE
      DO 430 J=1,NFCNS
      FT=J-1
      FT=FT*DELF
      XT=DCOS(PI2*FT)
      IF(KKK.EQ.1) GO TO 410
      XT=(XT-BB)/AA
      XT1=SQRT(1.0-XT*XT)
      FT=ATAN2(XT1,XT)/PI2
  410 XE=X(L)
      IF(XT.GT.XE) GO TO 420
      IF((XE-XT).LT.FSH) GO TO 415
      L=L+1
      GO TO 410
  415 A(J)=Y(L)
      GO TO 425
  420 IF((XT-XE).LT.FSH) GO TO 415
      GRID(1)=FT
      A(J)=GEE(1,NZ)
  425 CONTINUE
      IF(L.GT.1) L=L-1
  430 CONTINUE
      GRID(1)=GTEMP
      DDEN=PI2/CN
      DO 510 J=1,NFCNS
      DTEMP=0.0
      DNUM=J-1
      DNUM=DNUM*DDEN
      IF(NM1.LT.1) GO TO 505
      DO 500 K=1,NM1
      DAK=A(K+1)
      DK=K
  500 DTEMP=DTEMP+DAK*DCOS(DNUM*DK)
  505 DTEMP=2.0*DTEMP+A(1)
  510 ALPHA(J)=DTEMP
      DO 550 J=2,NFCNS
  550 ALPHA(J)=2.0*ALPHA(J)/CN
      ALPHA(1)=ALPHA(1)/CN
      IF(KKK.EQ.1) GO TO 545
      P(1)=2.0*ALPHA(NFCNS)*BB+ALPHA(NM1)
      P(2)=2.0*AA*ALPHA(NFCNS)
      Q(1)=ALPHA(NFCNS-2)-ALPHA(NFCNS)
      DO 540 J=2,NM1
      IF(J.LT.NM1) GO TO 515
      AA=0.5*AA
      BB=0.5*BB
  515 CONTINUE
      P(J+1)=0.0
      DO 520 K=1,J
      A(K)=P(K)
  520 P(K)=2.0*BB*A(K)
      P(2)=P(2)+A(1)*2.0*AA
      JM1=J-1
      DO 525 K=1,JM1
  525 P(K)=P(K)+Q(K)+AA*A(K+1)
      JP1=J+1
      DO 530 K=3,JP1
  530 P(K)=P(K)+AA*A(K-1)
      IF(J.EQ.NM1) GO TO 540
      DO 535 K=1,J
  535 Q(K)=-A(K)
      NF1J=NFCNS-1-J
      Q(1)=Q(1)+ALPHA(NF1J)
  540 CONTINUE
      DO 543 J=1,NFCNS
  543 ALPHA(J)=P(J)
  545 CONTINUE
      IF(NFCNS.GT.3) RETURN
      ALPHA(NFCNS+1)=0.0
      ALPHA(NFCNS+2)=0.0
      RETURN
      END
C
C-----------------------------------------------------------------------
C FUNCTION: D
C   FUNCTION TO CALCULATE THE LAGRANGE INTERPOLATION
C   COEFFICIENTS FOR USE IN THE FUNCTION GEE.
C-----------------------------------------------------------------------
C
      DOUBLE PRECISION FUNCTION D(K,N,M)
      COMMON PI2,AD,DEV,X,Y,GRID,DES,WT,ALPHA,IEXT,NFCNS,NGRID
      DIMENSION IEXT(66),AD(66),ALPHA(66),X(66),Y(66)
      DIMENSION DES(1045),GRID(1045),WT(1045)
      DOUBLE PRECISION AD,DEV,X,Y
      DOUBLE PRECISION Q
      DOUBLE PRECISION PI2
      D=1.0
      Q=X(K)
      DO 3 L=1,M
      DO 2 J=L,N,M
      IF(J-K)1,2,1
    1 D=2.0*D*(Q-X(J))
    2 CONTINUE
    3 CONTINUE
      D=1.0/D
      RETURN
      END
C
C-----------------------------------------------------------------------
C FUNCTION: GEE
C   FUNCTION TO EVALUATE THE FREQUENCY RESPONSE USING THE
C   LAGRANGE INTERPOLATION FORMULA IN THE BARYCENTRIC FORM
C-----------------------------------------------------------------------
```

```
C
      DOUBLE PRECISION FUNCTION GEE(K,N)
      COMMON PI2,AD,DEV,X,Y,GRID,DES,WT,ALPHA,IEXT,NFCNS,NGRID
      DIMENSION IEXT(66),AD(66),ALPHA(66),X(66),Y(66)
      DIMENSION DES(1045),GRID(1045),WT(1045)
      DOUBLE PRECISION P,C,D,XF
      DOUBLE PRECISION PI2
      DOUBLE PRECISION AD,DEV,X,Y
      P=0.0
      XF=GRID(K)
      XF=DCOS(PI2*XF)
      D=0.0
      DO 1 J=1,N
      C=XF-X(J)
      C=AD(J)/C
      D=D+C
    1 P=P+C*Y(J)
      GEE=P/D
      RETURN
      END
C
C-----------------------------------------------------------------------
C SUBROUTINE: OUCH
C   WRITES AN ERROR MESSAGE WHEN THE ALGORITHM FAILS TO
C   CONVERGE.  THERE SEEM TO BE TWO CONDITIONS UNDER WHICH
C   THE ALGORITHM FAILS TO CONVERGE: (1) THE INITIAL
C   GUESS FOR THE EXTREMAL FREQUENCIES IS SO POOR THAT
C   THE EXCHANGE ITERATION CANNOT GET STARTED, OR
C   (2) NEAR THE TERMINATION OF A CORRECT DESIGN,
C   THE DEVIATION DECREASES DUE TO ROUNDING ERRORS
C   AND THE PROGRAM STOPS.  IN THIS LATTER CASE THE
C   FILTER DESIGN IS PROBABLY ACCEPTABLE, BUT SHOULD
C   BE CHECKED BY COMPUTING A FREQUENCY RESPONSE.
C-----------------------------------------------------------------------
C
      SUBROUTINE OUCH
      COMMON /OOPS/NITER,IOUT
      WRITE(IOUT,1)NITER
    1 FORMAT(44H ************ FAILURE TO CONVERGE **********/
     141H0PROBABLE CAUSE IS MACHINE ROUNDING ERROR/
     223H0NUMBER OF ITERATIONS =,I4/
     339H0IF THE NUMBER OF ITERATIONS EXCEEDS 3,/
     462H0THE DESIGN MAY BE CORRECT, BUT SHOULD BE VERIFIED WITH AN FFT)
      RETURN
      END
```

5.2

FIR Windowed Filter Design Program - WINDOW

L. R. Rabiner and *C. A. McGonegal*

Acoustics Research Dept.
Bell Laboratories
Murray Hill, NJ 07974

D. Paul

MIT Lincoln Laboratory
Lexington, MA 02173

1. Purpose

This program can be used to design FIR digital filters using the window method [1-3]. The program is capable of designing lowpass, bandpass, bandstop, and highpass filters for both even and odd values of N, (the impulse response duration in samples), using either a rectangular, a triangular, a Hamming, a Hanning, a Chebyshev, or a Kaiser window.

2. Method

This program uses the well known method of window design for FIR digital filters. If we denote the N-point windows as $w(n)$, for $0 \leqslant n \leqslant N-1$, and we denote the impulse response of the ideal digital filter (obtained as the inverse Fourier transform of the ideal frequency response of the filter) as $h(n)$, $-\infty < n < \infty$, then the windowed digital filter is given as

$$\begin{aligned} \hat{h}(n) &= w(n)h(n) \quad & 0 \leqslant n \leqslant N-1 \\ &= 0 & \text{otherwise} \end{aligned} \qquad (1)$$

In the discussion above it is assumed that $h(n)$ incorporates an ideal delay of $(N-1)/2$ samples, and that $w(n)$ is symmetric around the point $(N-1)/2$.

In order to design windowed digital filters requires specification of

1. Type of ideal filter - i.e., lowpass, highpass, bandpass or bandstop filter.
2. Filter cutoff frequencies.
3. Type of window - i.e., rectangular, triangular, Hamming, Hanning, Chebyshev, Kaiser.
4. Window duration in samples.

From the filter specifications the sequences $h(n)$ and $w(n)$ of Eq. (1) are computed, and the windowed filter is obtained as the final output.

2.1 Form of the Windows

The seven windows used in this program are

1. Rectangular Window
2. Triangular Window
3. Hamming Window
4. Hanning Window
5. Generalized Hamming Window
6. Kaiser (I_0-Sinh) Window
7. Chebyshev Window

The form of each of these windows (assuming they are symmetric around $n=0$) is:

Rectangular Window

For N odd $\quad w(n) = 1 \qquad -(N-1)/2 \leqslant n \leqslant (N-1)/2$ (2a)

For N even, $\quad w(n) = 1 \qquad -(N/2) \leqslant n \leqslant (N/2-1)$ (2b)

Triangular Window

For N odd, $\quad w(n) = 1 - |2n|/(N+1) \qquad -(N-1)/2 \leqslant n \leqslant (N-1)/2$ (3a)

For N even, $\quad w(n) = 1 - |2n+1|/N \qquad -(N/2) \leqslant n \leqslant (N/2-1)$ (3b)

Hamming Window

For N odd, $\quad w(n) = 0.54 + 0.46 \cos[2\pi n/(N-1)] \qquad -(N-1)/2 \leqslant n \leqslant (N-1)/2$ (4a)

For N even, $\quad w(n) = 0.54 + 0.46 \cos\left[\frac{2\pi(2n+1)}{2(N-1)}\right] \qquad -(N/2) \leqslant n \leqslant (N/2-1)$ (4b)

Hanning Window

For N odd, $\quad w(n) = 0.5 + 0.5 \cos[2\pi n/(N+1)] \qquad -(N-1)/2 \leqslant n \leqslant (N-1)/2$ (5a)

For N even, $\quad w(n) = 0.5 + 0.5 \cos\left[\frac{2\pi(2n+1)}{2(N+1)}\right] \qquad -(N/2) \leqslant n \leqslant (N/2-1)$ (5b)

Generalized Hamming Window

For N odd, $\quad w(n) = \alpha + (1-\alpha)\cos\,[2\pi n/(N-1)] \qquad -(N-1)/2 \leqslant n \leqslant (N-1)/2$ (6a)

For N even, $\quad w(n) = \alpha + (1-\alpha)\cos\left[\frac{2\pi(2n+1)}{2(N-1)}\right] \qquad -(N/2) \leqslant n \leqslant (N/2-1)$ (6b)

where α is variable parameter.

Kaiser (I_0-Sinh) Window

For N odd, $\quad w(n) = \dfrac{I_0\left[\beta\sqrt{1 - \dfrac{4n^2}{(N-1)^2}}\right]}{I_0(\beta)} \qquad -(N-1)/2 \leqslant n \leqslant (N-1)/2$ (7a)

For N even, $\quad w(n) = \dfrac{I_0\left[\beta\sqrt{1 - \dfrac{4(n+1/2)^2}{(N-1)^2}}\right]}{I_0(\beta)} \qquad -(N/2) \leqslant n \leqslant (N/2-1)$ (7b)

where β is window parameter related to desired minimum stopband attenuation [4].

Chebyshev Window

$w(n)$ is obtained as the inverse DFT of the Chebyshev polynomial, evaluated at N equally spaced frequencies around the unit circle [2,3]. The parameters of the Chebyshev window are the ripple, δ_p, the filter length, N, and the normalized transition width, ΔF. Figure 1 shows a plot of the frequency response of a Chebyshev window illustrating how δ_p and ΔF are measured.

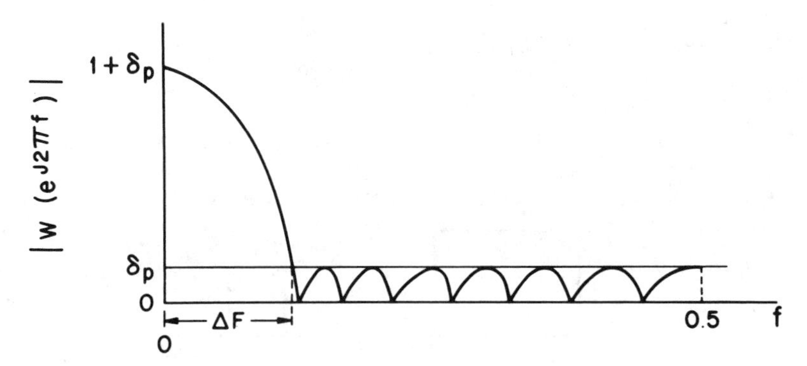

Fig. 1 Definition of Chebyshev window parameters in the frequency domain.

Only 2 of the 3 parameters N, δ_p and ΔF can be independently specified. If the parameters δ_p and ΔF are specified, then N is determined by the equation

$$N \geqslant 1 + \left[\frac{\cosh^{-1}\left(\frac{1+\delta_p}{\delta_p}\right)}{\cosh^{-1}\left(\frac{1}{\cos(\pi \Delta F)}\right)} \right] \tag{8}$$

If the parameters N and δ_p are specified, then ΔF is determined as

$$\Delta F = \frac{1}{\pi} \cos^{-1} \left[\frac{1}{\cosh\left[\frac{\cosh^{-1}\left(\frac{1+\delta_p}{\delta_p}\right)}{N-1} \right]} \right] \tag{9}$$

Finally if the parameters N and ΔF are specified, then δ_p is determined as

$$\delta_p = \frac{1}{\cosh\left[(N-1)\cosh^{-1}\left(\frac{1}{\cos(\pi \Delta F)}\right)\right] - 1} \tag{10}$$

3. Program Description

3.1 Usage

The program consists of a main program and eleven subroutines. The main program reads the input data (i.e., the filter parameters discussed above), and the program prints the filter parameters, the window coefficients, and the resulting impulse response of the digital filter (stored in array G). User prompts are given to aid in the input stage of the program.

3.2 Description of Input Parameters

CARD 1: NF, ITYPE, JTYPE

FORMAT: (I4,I2,I2)

NF — The filter length in samples. NF must satisfy the relation $3 \leqslant$ NF $\leqslant 1024$. For highpass or bandpass filters NF *must* be odd.

ITYPE — The window type. Values of ITYPE from 1 to 7 correspond to rectangular, triangular, Hamming, generalized Hamming, Hanning, Kaiser, and Chebyshev windows, respectively.

JTYPE — The filter type. Values of JTYPE from 1 to 4 correspond to LP (lowpass), HP (highpass), BP (bandpass), and BS (bandstop) designs, respectively.

If JTYPE = 1 or JTYPE = 2, CARD 2: FC

If JTYPE = 3 or JTYPE = 4, CARD 2: FL, FH

FORMAT: (2F14.7)

FC — Normalized filter cutoff frequency for lowpass and highpass filters. FC is normalized to the range $0 \leqslant FC \leqslant 0.5$. (See Figure 2).

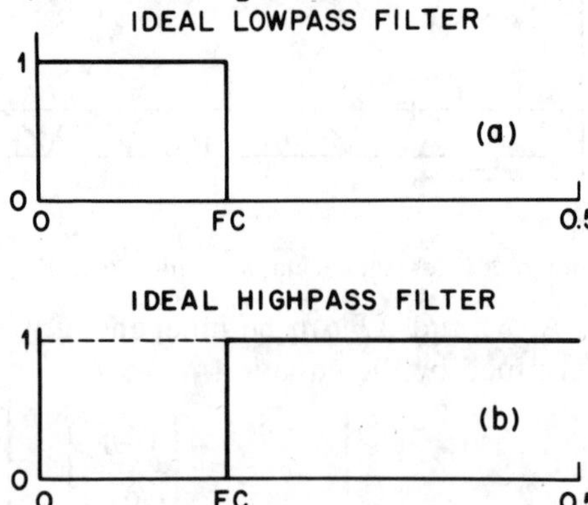

Fig. 2 Definition of normalized cutoff frequency, FC, for (a) an ideal lowpass and, (b) an ideal highpass filter.

FL,FH — Normalized filter cutoff frequencies for bandpass filters and bandstop filters. FL and FH are normalized so that $0 \leqslant FL < FH \leqslant 0.5$. (See Figure 3).

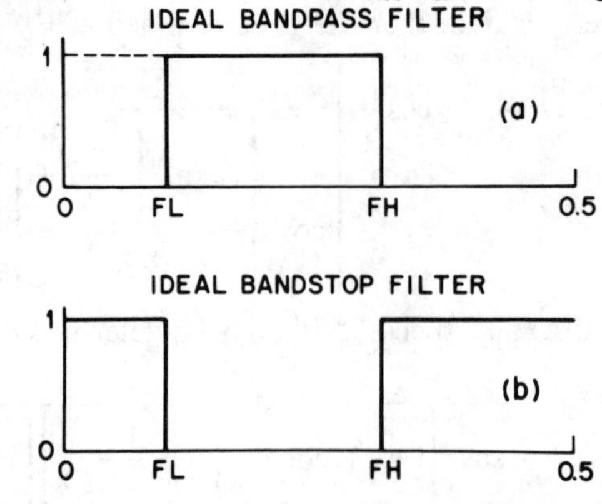

Fig. 3 Definition of normalized cutoff frequencies, FL and FH, for (a) an ideal bandpass, and, (b) an ideal bandstop filter.

If ITYPE = 4, CARD 3: ALPHA
If ITYPE = 6, CARD 3: ATT
If ITYPE = 7, CARD 3: DPLOG,DF

FORMAT: (2F14.7)

ALPHA — Parameter of the generalized Hamming window.

ATT — The desired minimum stopband attenuation in dB (from which the parameter β of the Kaiser window is determined).

DPLOG,DF — Ripple and transition width parameters for Chebyshev window designs. DPLOG is the desired filter ripple (on a decibel scale) and DF is the normalized transition width of the filter ($0 \leqslant DF \leqslant 0.5$). For Chebyshev windows exactly 2 of the 3 parameters NF,DP, and DF must be specified. The unspecified parameter (input set to 0) is determined analytically.

CARD 4: IWIN

FORMAT: (I1)

IWIN — IWIN is set to 1 to print $w(n)$.

CARD 5: IMD

FORMAT: (I1)

IMD — IMD is set to 1 to design a new filter, IMD is set to 0 to terminate the program.

3.3 Description of Main Program (WINDOW) and Subroutines

Main Program (WINDOW) — This program reads in filter parameters, determines the window and the ideal lowpass filter response, computes the windowed digital filter coefficients, transforms the lowpass filter to the specified filter type, writes out the filter coefficients and determines filter bandedges and ripples within each band. As an option it will also write out the window coefficients.

TRIANG — This subroutines creates a triangular window.

HAMMIN — This subroutine creates a generalized Hamming window.

KAISER — This subroutine creates a Kaiser window.

INO — This function evaluates $I_o(x)$ [4].

CHEBC — This subroutine computes the unspecified parameter of a Chebyshev window.

CHEBY — This subroutine creates a Chebyshev window.

FLCHAR — This subroutine computes a high resolution frequency response of the resulting filter using a DFT algorithm. For all window designs (except the triangular windows for which no ripple or bandedges are readily obtained) each ideal filter band is searched for the peak ripple and the bandedge. The bandedge is defined as the point where the frequency response crosses the ripple bound. A more precise estimate of the bandedge is determined by linear interpolation of the frequency response of the filter in the vicinity of the bandedge. Finally the ripple (i.e., peak deviation from the ideal response) is determined as

$$\text{Ripple} = 20 \log (1+\delta_p) \text{ (dB) for passbands}$$
$$\text{Ripple} = 20 \log (\delta_p) \text{ (dB) for stopbands.}$$

RIPPLE — This subroutine determines the filter ripples and bandedges from the filter parameters and the high resolution frequency response.

COSH — This function evaluates $\cosh(x)$.

COSHIN — This function evaluates $\cosh^{-1}(x)$.

ARCCOS — This function evaluates $\cos^{-1}(x)$.

3.4 Dimension Requirements

The dimensioned arrays are

W — This array holds half of the window. (The second half is obtained by symmetry.) The size of W must be (NF+1)/2.

G — This array holds half the filter coefficients. (Symmetry is again used for the second half.) Its size must also be (NF+1)/2.

PR — This array is only dimensioned in the subroutine CHEBY. It holds the real part of the DFT of the window. The size must be NF.

PI — This array is only dimensioned in the subroutine CHEBY. It holds the imaginary part of the DFT of the window. The size must be NF.

RESP — This array is only dimensioned in the subroutine FLCHAR. It holds the high resolution frequency response of the filter. The size of RESP is 2048 points.

3.5 Summary of User Requirements

(1) Specify filter parameters NF, ITYPE, JTYPE, FC (or FL,FH), and special window parameters DPLOG, DF (for Chebyshev windows), ALPHA (for generalized Hamming windows) and ATT (for Kaiser windows).

(2) Design $w(n)$, $h(n)$ and $\hat{h}(n)$ using the main program which prints $\hat{h}(n)$, the filter coefficients, which are in array G, and optionally prints $w(n)$ in array W.

4. Test Problems

The input data for a set of 7 windowed filter designs which exercises some of the types of windows and filters is given in Table 1. The output for each of these 7 cases is given in Tables 2-8, respectively. Test example 1 is a 25 point lowpass filter with ideal cutoff frequency of 0.3. A rectangular window is used for this case. Example 2 is the same as example 1 except a Hamming window is used, and the value of NF is increased to 30. Example 3 is a 25 point lowpass filter with ideal cutoff frequency of 0.3. This case uses a triangular window. Example 4 is a 56 point bandpass filter designed using a Kaiser window. The ideal bandedges of the filter are 0.15 and 0.35, and the window is designed for 60 dB minimum attenuation. Example 5 is a 55 point bandpass filter designed using a Chebyshev window. The ideal bandedges are 0.15 and 0.35, and the peak stopband ripple is specified to be 60 dB (0.001 on a linear scale). The transition width of the window is unspecified (set to 0.) as this is the free parameter of the Chebyshev window for this run. Example 6 is a bandstop filter designed using a Chebyshev window. The filter length is unspecified (set to 0) and is determined from the specified ripple (60 dB) and transition width (0.05) for the window. The bandedges of the filter are 0.15 and 0.35. The last example is a 55 point highpass filter designed using a Kaiser window. The minimum stopband attenuation is 60 dB, and the ideal cutoff frequency of the filter is 0.35. Figures 4 to 7 show plots of the window (part a), the window frequency response (part b), the filter impulse response (part c) and the filter frequency response (part d) for examples 2, 5, 6, and 7.

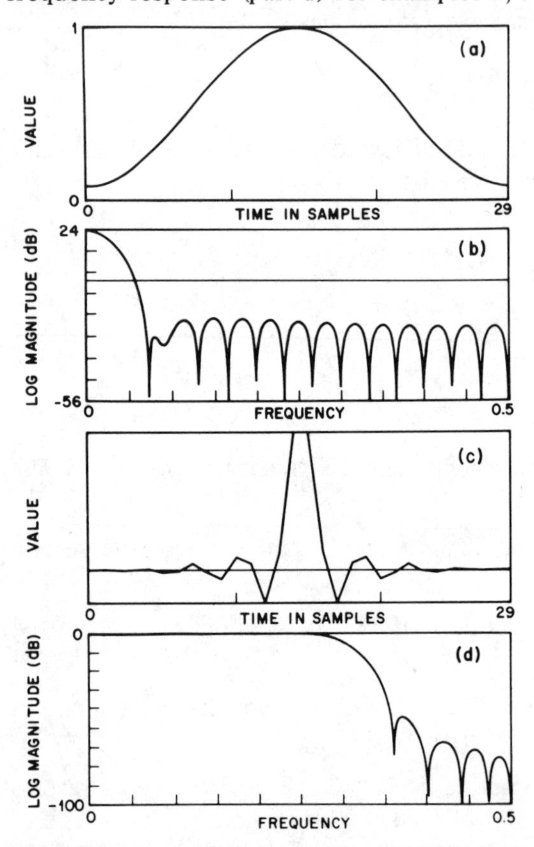

Fig. 4 Responses of the window (parts a and b) and the resulting filter (parts c and d) for example 2.

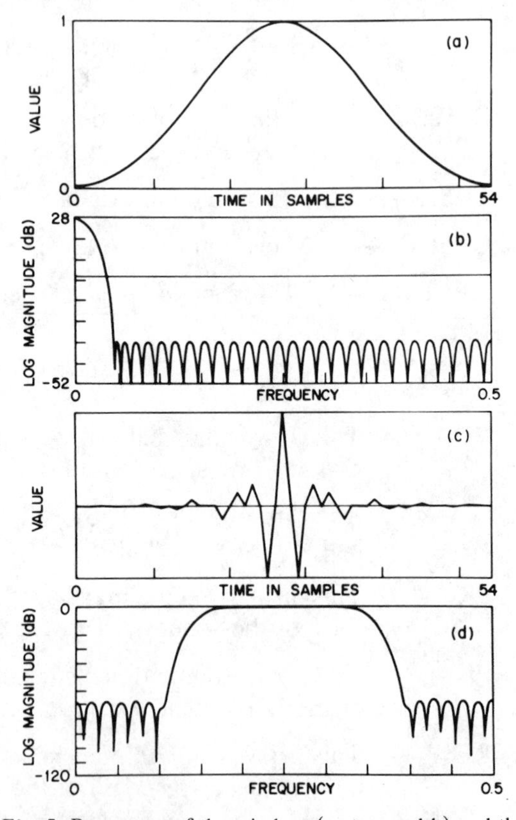

Fig. 5 Responses of the window (parts a and b) and the resulting filter (parts c and d) for example 5.

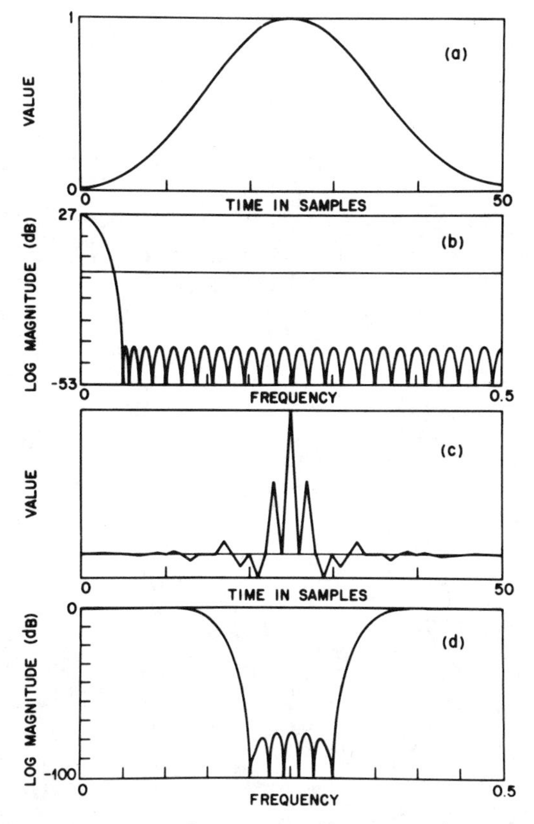

Fig. 6 Responses of the window (parts a and b) and the resulting filter (parts c and d) for example 6.

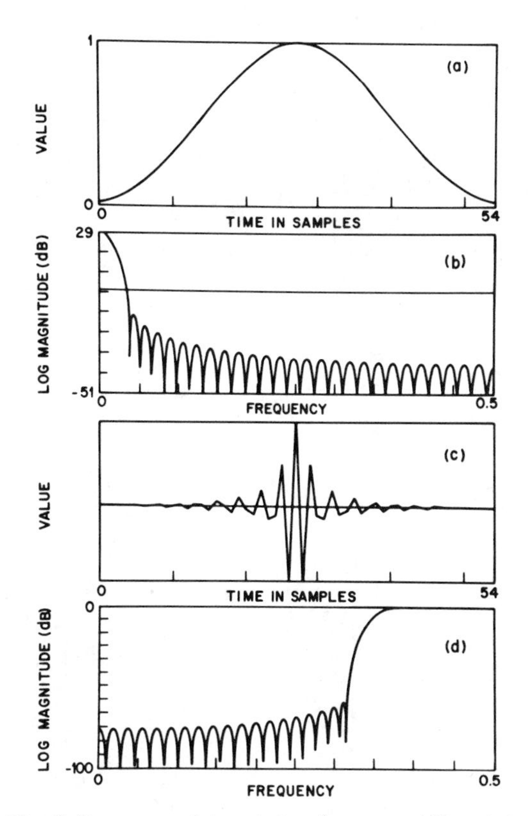

Fig. 7 Responses of the window (parts a and b) and the resulting filter (parts c and d) for example 7.

The listing of the main program and subroutines is given in the Appendix.

References

1. J. F. Kaiser, "Digital Filters", Chapter 7 in *System Analysis by Digital Computer* (F.F. Kuo and J. F. Kaiser, ed.), John Wiley and Sons, Inc. New York, NY, 1966.

2. H. D. Helms, "Nonrecursive Digital Filters: Design Methods for Achieving Specifications on Frequency Response", *IEEE Trans. Audio Electroacoust.,* Vol. AU-16, No. 3, pp. 336-342, September 1968.

3. L. R. Rabiner and B. Gold, *Theory and Application of Digital Signal Processing,* Chapter 3, pp. 88-105, Prentice-Hall, Inc., Englewood Cliffs, NJ, 1975.

4. J. F. Kaiser, "Nonrecursive Digital Filter Design Using the I_0-Sinh Window Function", *Proc. 1974 IEEE Symp. Circuits and Syst.,* pp. 20-23, April 1974; also in *Digital Signal Processing II,* IEEE Press, New York, NY, 1975.

Table 1

```
  25  1  1
0.3
1
1
  30  3  1
0.3
1
1
  25  2  1
0.3
1
1
  56  6  3
0.15     ,       .35
60.
1
1
  55  7  3
0.15             .35
60.              0.
1
1
   0  7  4
0.15             .35
60.              .05
1
1
  55  6  2
0.35
60.
1
0
```

Table 2

```
RECTANGULAR WINDOW-NF=  25
WINDOW VALUES
          W(  1)=  .10000000E 01 =W(  25)
          W(  2)=  .10000000E 01 =W(  24)
          W(  3)=  .10000000E 01 =W(  23)
          W(  4)=  .10000000E 01 =W(  22)
          W(  5)=  .10000000E 01 =W(  21)
          W(  6)=  .10000000E 01 =W(  20)
          W(  7)=  .10000000E 01 =W(  19)
          W(  8)=  .10000000E 01 =W(  18)
          W(  9)=  .10000000E 01 =W(  17)
          W( 10)=  .10000000E 01 =W(  16)
          W( 11)=  .10000000E 01 =W(  15)
          W( 12)=  .10000000E 01 =W(  14)
          W( 13)=  .10000000E 01 =W(  13)
**LOWPASS FILTER DESIGN**
IDEAL LOWPASS CUTOFF=          .3000000
          H(  1)= -.15591270E-01 =H(  25)
          H(  2)=  .27521070E-01 =H(  24)
          H(  3)= -.27820390E-06 =H(  23)
          H(  4)= -.33636670E-01 =H(  22)
          H(  5)=  .23387470E-01 =H(  21)
          H(  6)=  .26728050E-01 =H(  20)
          H(  7)= -.50455220E-01 =H(  19)
          H(  8)=  .27820390E-06 =H(  18)
          H(  9)=  .75682580E-01 =H(  17)
          H( 10)= -.62366210E-01 =H(  16)
          H( 11)= -.93548720E-01 =H(  15)
          H( 12)=  .30273090E 00 =H(  14)
          H( 13)=  .60000000E 00 =H(  13)
PASSBAND CUTOFF        .2807 HZ  RIPPLE       .643 DB
STOPBAND CUTOFF        .3176 HZ  RIPPLE    -19.677 DB
```

Table 3

```
HAMMING WINDOW-NF=  30
ALPHA=       .5400000
WINDOW VALUES
          W(  1)=  .80000100E-01 =W(  30)
          W(  2)=  .90754870E-01 =W(  29)
          W(  3)=  .12251590E 00 =W(  28)
          W(  4)=  .17379830E 00 =W(  27)
          W(  5)=  .24220310E 00 =W(  26)
          W(  6)=  .32453290E 00 =W(  25)
          W(  7)=  .41693830E 00 =W(  24)
          W(  8)=  .51509700E 00 =W(  23)
          W(  9)=  .61442080E 00 =W(  22)
          W( 10)=  .71026440E 00 =W(  21)
          W( 11)=  .79814640E 00 =W(  20)
          W( 12)=  .87395820E 00 =W(  19)
          W( 13)=  .93415440E 00 =W(  18)
          W( 14)=  .97592040E 00 =W(  17)
          W( 15)=  .99730340E 00 =W(  16)
**LOWPASS FILTER DESIGN**
IDEAL LOWPASS CUTOFF=        .3000000
          H(  1)=  .14208050E-02 =H(  30)
          H(  2)=  .66123040E-03 =H(  29)
          H(  3)= -.31198420E-02 =H(  28)
          H(  4)=  .14865990E-02 =H(  27)
          H(  5)=  .59401210E-02 =H(  26)
          H(  6)= -.87972210E-02 =H(  25)
          H(  7)= -.48247580E-02 =H(  24)
          H(  8)=  .21861400E-01 =H(  23)
          H(  9)= -.92980680E-02 =H(  22)
          H( 10)= -.33255520E-01 =H(  21)
          H( 11)=  .45675080E-01 =H(  20)
          H( 12)=  .24561290E-01 =H(  19)
          H( 13)= -.11894020E 00 =H(  18)
          H( 14)=  .63996670E-01 =H(  17)
          H( 15)=  .51364740E 00 =H(  16)
PASSBAND CUTOFF      .2413 HZ  RIPPLE     .020 DB
STOPBAND CUTOFF      .3563 HZ  RIPPLE  -50.398 DB
```

Table 4

```
TRIANGULAR WINDOW-NF=  25
WINDOW VALUES
          W(  1)=  .76923130E-01 =W(  25)
          W(  2)=  .15384620E 00 =W(  24)
          W(  3)=  .23076930E 00 =W(  23)
          W(  4)=  .30769230E 00 =W(  22)
          W(  5)=  .38461540E 00 =W(  21)
          W(  6)=  .46153850E 00 =W(  20)
          W(  7)=  .53846160E 00 =W(  19)
          W(  8)=  .61538460E 00 =W(  18)
          W(  9)=  .69230770E 00 =W(  17)
          W( 10)=  .76923080E 00 =W(  16)
          W( 11)=  .84615390E 00 =W(  15)
          W( 12)=  .92307690E 00 =W(  14)
          W( 13)=  .10000000E 01 =W(  13)
**LOWPASS FILTER DESIGN**
IDEAL LOWPASS CUTOFF=        .3000000
          H(  1)= -.11993290E-02 =H(  25)
          H(  2)=  .42340080E-02 =H(  24)
          H(  3)= -.64200890E-07 =H(  23)
          H(  4)= -.10349750E-01 =H(  22)
          H(  5)=  .89951790E-02 =H(  21)
          H(  6)=  .12336020E-01 =H(  20)
          H(  7)= -.27168200E-01 =H(  19)
          H(  8)=  .17120240E-06 =H(  18)
          H(  9)=  .52395630E-01 =H(  17)
          H( 10)= -.47974010E-01 =H(  16)
          H( 11)= -.79156580E-01 =H(  15)
          H( 12)=  .27944390E 00 =H(  14)
          H( 13)=  .60000000E 00 =H(  13)
```

Table 5

```
 KAISER WINDOW-NF=  56
  ATT=    60.0000000  BETA=      5.6532570
 WINDOW VALUES
           W(  1)=  .20388090E-01 =W(  56)
           W(  2)=  .33787730E-01 =W(  55)
           W(  3)=  .50466800E-01 =W(  54)
           W(  4)=  .70650820E-01 =W(  53)
           W(  5)=  .94510080E-01 =W(  52)
           W(  6)=  .12215050E 00 =W(  51)
           W(  7)=  .15360570E 00 =W(  50)
           W(  8)=  .18883060E 00 =W(  49)
           W(  9)=  .22769680E 00 =W(  48)
           W( 10)=  .26999030E 00 =W(  47)
           W( 11)=  .31541080E 00 =W(  46)
           W( 12)=  .36357260E 00 =W(  45)
           W( 13)=  .41401090E 00 =W(  44)
           W( 14)=  .46618500E 00 =W(  43)
           W( 15)=  .51949130E 00 =W(  42)
           W( 16)=  .57326480E 00 =W(  41)
           W( 17)=  .62680230E 00 =W(  40)
           W( 18)=  .67936970E 00 =W(  39)
           W( 19)=  .73021390E 00 =W(  38)
           W( 20)=  .77858810E 00 =W(  37)
           W( 21)=  .82375820E 00 =W(  36)
           W( 22)=  .86502400E 00 =W(  35)
           W( 23)=  .90173230E 00 =W(  34)
           W( 24)=  .93329350E 00 =W(  33)
           W( 25)=  .95919230E 00 =W(  32)
           W( 26)=  .97900370E 00 =W(  31)
           W( 27)=  .99239710E 00 =W(  30)
           W( 28)=  .99915160E 00 =W(  29)
 **BANDPASS FILTER DESIGN**
 IDEAL CUTOFF FREQUENCIES=       .1500000      .3500000
           H(  1)= -.33373340E-03 =H(  56)
           H(  2)=  .46434790E-03 =H(  55)
           H(  3)=  .27529500E-03 =H(  54)
           H(  4)=  .40115490E-03 =H(  53)
           H(  5)=  .14646240E-02 =H(  52)
           H(  6)= -.24439120E-02 =H(  51)
           H(  7)= -.26018560E-02 =H(  50)
           H(  8)=  .12813520E-02 =H(  49)
           H(  9)= -.16243030E-02 =H(  48)
           H( 10)=  .53150390E-02 =H(  47)
           H( 11)=  .81133020E-02 =H(  46)
           H( 12)= -.80248120E-02 =H(  45)
           H( 13)= -.37155130E-02 =H(  44)
           H( 14)= -.44724530E-02 =H(  43)
           H( 15)= -.14014040E-01 =H(  42)
           H( 16)=  .20645000E-01 =H(  41)
           H( 17)=  .19849570E-01 =H(  40)
           H( 18)= -.90004760E-02 =H(  39)
           H( 19)=  .10692340E-01 =H(  38)
           H( 20)= -.33359160E-01 =H(  37)
           H( 21)= -.49442590E-01 =H(  36)
           H( 22)=  .48466250E-01 =H(  35)
           H( 23)=  .22806540E-01 =H(  34)
           H( 24)=  .28850660E-01 =H(  33)
           H( 25)=  .99806550E-01 =H(  32)
           H( 26)= -.17628290E 00 =H(  31)
           H( 27)= -.24094420E 00 =H(  30)
           H( 28)=  .27797700E 00 =H(  29)
 STOPBAND CUTOFF       .1171 HZ  RIPPLE  -59.714 DB
 PASSBAND CUTOFFS       .1825 HZ       .3174 HZ  RIPPLE      .012 DB
 STOPBAND CUTOFF       .3832 HZ  RIPPLE  -61.655 DB
```

Table 6

```
CHEBYSHEV WINDOW-NF=  55
DP=      .0010000  DF=       .0446622
WINDOW VALUES
          W(  1)=  .22114040E-01 =W(  55)
          W(  2)=  .23350940E-01 =W(  54)
          W(  3)=  .34994060E-01 =W(  53)
          W(  4)=  .49832560E-01 =W(  52)
          W(  5)=  .68227770E-01 =W(  51)
          W(  6)=  .90503870E-01 =W(  50)
          W(  7)=  .11691700E 00 =W(  49)
          W(  8)=  .14763970E 00 =W(  48)
          W(  9)=  .18275320E 00 =W(  47)
          W( 10)=  .22222850E 00 =W(  46)
          W( 11)=  .26590410E 00 =W(  45)
          W( 12)=  .31350470E 00 =W(  44)
          W( 13)=  .36460550E 00 =W(  43)
          W( 14)=  .41867210E 00 =W(  42)
          W( 15)=  .47501970E 00 =W(  41)
          W( 16)=  .53289100E 00 =W(  40)
          W( 17)=  .59138560E 00 =W(  39)
          W( 18)=  .64956160E 00 =W(  38)
          W( 19)=  .70639290E 00 =W(  37)
          W( 20)=  .76085360E 00 =W(  36)
          W( 21)=  .81189510E 00 =W(  35)
          W( 22)=  .85851200E 00 =W(  34)
          W( 23)=  .89975800E 00 =W(  33)
          W( 24)=  .93478490E 00 =W(  32)
          W( 25)=  .96284500E 00 =W(  31)
          W( 26)=  .98333280E 00 =W(  30)
          W( 27)=  .99580760E 00 =W(  29)
          W( 28)=  .10000000E 01 =W(  28)
**BANDPASS FILTER DESIGN**
IDEAL CUTOFF FREQUENCIES=      .1500000      .3500000
          H(  1)=  .10587540E-07 =H(  55)
          H(  2)=  .33606800E-03 =H(  54)
          H(  3)=  .73327410E-13 =H(  53)
          H(  4)=  .77696800E-03 =H(  52)
          H(  5)= -.32665510E-07 =H(  51)
          H(  6)= -.24907520E-02 =H(  50)
          H(  7)=  .34595170E-07 =H(  49)
          H(  8)= -.15649260E-07 =H(  48)
          H(  9)=  .54076300E-07 =H(  47)
          H( 10)=  .74750520E-02 =H(  46)
          H( 11)= -.12730700E-06 =H(  45)
          H( 12)= -.73319790E-02 =H(  44)
          H( 13)= -.45840200E-12 =H(  43)
          H( 14)= -.11190420E-01 =H(  42)
          H( 15)=  .22742580E-06 =H(  41)
          H( 16)=  .26887070E-01 =H(  40)
          H( 17)= -.17498860E-06 =H(  39)
          H( 18)=  .68851140E-07 =H(  38)
          H( 19)= -.20901980E-06 =H(  37)
          H( 20)= -.57583410E-01 =H(  36)
          H( 21)=  .38871140E-06 =H(  35)
          H( 22)=  .53541870E-01 =H(  34)
          H( 23)=  .37707510E-12 =H(  33)
          H( 24)=  .87448120E-01 =H(  32)
          H( 25)= -.46098210E-06 =H(  31)
          H( 26)= -.29768500E 00 =H(  30)
          H( 27)=  .29465610E-06 =H(  29)
          H( 28)=  .40000010E 00 =H(  28)
STOPBAND CUTOFF        .1073 HZ   RIPPLE   -67.961 DB
PASSBAND CUTOFFS        .1915 HZ        .3082 HZ  RIPPLE       .004 DB
STOPBAND CUTOFF        .3924 HZ   RIPPLE   -67.961 DB
```

Table 7

```
NF MUST BE ODD INTEGER FOR HP OR BS FILTERS--NF IS BEING INCREASED BY 1
CHEBYSHEV WINDOW-NF=  51
DP=      .0010000  DF=       .0500000
WINDOW VALUES
          W(  1)=  .16471760E-01 =W(  51)
          W(  2)=  .20148630E-01 =W(  50)
          W(  3)=  .31728420E-01 =W(  49)
          W(  4)=  .46980540E-01 =W(  48)
          W(  5)=  .66402670E-01 =W(  47)
          W(  6)=  .90445820E-01 =W(  46)
          W(  7)=  .11946230E 00 =W(  45)
          W(  8)=  .15370090E 00 =W(  44)
          W(  9)=  .19324890E 00 =W(  43)
          W( 10)=  .23803970E 00 =W(  42)
          W( 11)=  .28783570E 00 =W(  41)
          W( 12)=  .34218940E 00 =W(  40)
          W( 13)=  .40046500E 00 =W(  39)
          W( 14)=  .46183130E 00 =W(  38)
          W( 15)=  .52529320E 00 =W(  37)
          W( 16)=  .58969080E 00 =W(  36)
          W( 17)=  .65375500E 00 =W(  35)
          W( 18)=  .71612070E 00 =W(  34)
          W( 19)=  .77539310E 00 =W(  33)
          W( 20)=  .83018530E 00 =W(  32)
          W( 21)=  .87916110E 00 =W(  31)
          W( 22)=  .92110100E 00 =W(  30)
          W( 23)=  .95491870E 00 =W(  29)
          W( 24)=  .97973990E 00 =W(  28)
          W( 25)=  .99489920E 00 =W(  27)
          W( 26)=  .10000000E 01 =W(  26)
**BANDSTOP FILTER DESIGN**
IDEAL CUTOFF FREQUENCIES=      .1500000      .3500000
          H(  1)= -.34515350E-13 =H(  51)
          H(  2)= -.31414860E-03 =H(  50)
          H(  3)=  .15190660E-07 =H(  49)
          H(  4)=  .12929490E-02 =H(  48)
          H(  5)= -.19648240E-07 =H(  47)
          H(  6)=  .95869230E-08 =H(  46)
          H(  7)= -.35348630E-07 =H(  45)
          H(  8)= -.51700030E-02 =H(  44)
          H(  9)=  .92521820E-07 =H(  43)
          H( 10)=  .55670700E-02 =H(  42)
          H( 11)=  .36188270E-12 =H(  41)
          H( 12)=  .91461610E-02 =H(  40)
          H( 13)= -.19173110E-06 =H(  39)
          H( 14)= -.23301740E-01 =H(  38)
          H( 15)=  .15543210E-06 =H(  37)
          H( 16)= -.62505020E-07 =H(  36)
          H( 17)=  .19344440E-06 =H(  35)
          H( 18)=  .54197900E-01 =H(  34)
          H( 19)= -.37123540E-06 =H(  33)
          H( 20)= -.51775250E-01 =H(  32)
          H( 21)= -.36844320E-12 =H(  31)
          H( 22)= -.86167990E-01 =H(  30)
          H( 23)=  .45718720E-06 =H(  29)
          H( 24)=  .29659740E 00 =H(  28)
          H( 25)= -.29438730E-06 =H(  27)
          H( 26)=  .59999990E 00 =H(  26)
PASSBAND CUTOFF       .1030 HZ  RIPPLE       .002 DB
STOPBAND CUTOFFS       .1974 HZ       .3024 HZ  RIPPLE  -73.147 DB
PASSBAND CUTOFF       .3968 HZ  RIPPLE       .002 DB
```

Table 8

```
KAISER WINDOW-NF=  55
 ATT=    60.0000000  BETA=      5.6532570
WINDOW VALUES
          W(  1)=  .20388090E-01 =W(  55)
          W(  2)=  .34066010E-01 =W(  54)
          W(  3)=  .51150610E-01 =W(  53)
          W(  4)=  .71878610E-01 =W(  52)
          W(  5)=  .96427080E-01 =W(  51)
          W(  6)=  .12490300E 00 =W(  50)
          W(  7)=  .15733500E 00 =W(  49)
          W(  8)=  .19366690E 00 =W(  48)
          W(  9)=  .23375280E 00 =W(  47)
          W( 10)=  .27735250E 00 =W(  46)
          W( 11)=  .32413550E 00 =W(  45)
          W( 12)=  .37367830E 00 =W(  44)
          W( 13)=  .42547510E 00 =W(  43)
          W( 14)=  .47893880E 00 =W(  42)
          W( 15)=  .53341600E 00 =W(  41)
          W( 16)=  .58819190E 00 =W(  40)
          W( 17)=  .64251240E 00 =W(  39)
          W( 18)=  .69559720E 00 =W(  38)
          W( 19)=  .74664920E 00 =W(  37)
          W( 20)=  .79488300E 00 =W(  36)
          W( 21)=  .83953300E 00 =W(  35)
          W( 22)=  .87987460E 00 =W(  34)
          W( 23)=  .91524270E 00 =W(  33)
          W( 24)=  .94504490E 00 =W(  32)
          W( 25)=  .96877100E 00 =W(  31)
          W( 26)=  .98601800E 00 =W(  30)
          W( 27)=  .99648990E 00 =W(  29)
          W( 28)=  .99999900E 00 =W(  28)
**HIGHPASS FILTER DESIGN**
IDEAL HIGHPASS CUTOFF=       .3500000
          H(  1)= -.74281390E-04 =H(  55)
          H(  2)= -.24513290E-03 =H(  54)
          H(  3)=  .65126990E-03 =H(  53)
          H(  4)= -.56036520E-03 =H(  52)
          H(  5)= -.41235770E-03 =H(  51)
          H(  6)=  .17187140E-02 =H(  50)
          H(  7)= -.19293910E-02 =H(  49)
          H(  8)=  .59010880E-07 =H(  48)
          H(  9)=  .31681480E-02 =H(  47)
          H( 10)= -.46646410E-02 =H(  46)
          H( 11)=  .18755650E-02 =H(  45)
          H( 12)=  .43695570E-02 =H(  44)
          H( 13)= -.90288630E-02 =H(  43)
          H( 14)=  .64007230E-02 =H(  42)
          H( 15)=  .40358830E-02 =H(  41)
          H( 16)= -.14838590E-01 =H(  40)
          H( 17)=  .15041810E-01 =H(  39)
          H( 18)= -.21195050E-06 =H(  38)
          H( 19)= -.21363840E-01 =H(  37)
          H( 20)=  .30079520E-01 =H(  36)
          H( 21)= -.11797260E-01 =H(  35)
          H( 22)= -.27436900E-01 =H(  34)
          H( 23)=  .58266170E-01 =H(  33)
          H( 24)= -.44204220E-01 =H(  32)
          H( 25)= -.31763530E-01 =H(  31)
          H( 26)=  .14924880E 00 =H(  30)
          H( 27)= -.25661440E 00 =H(  29)
          H( 28)=  .30000070E 00 =H(  28)
STOPBAND CUTOFF       .3164 HZ  RIPPLE   -59.842 DB
PASSBAND CUTOFF       .3834 HZ  RIPPLE      .008 DB
```

Appendix

```
C
C-----------------------------------------------------------------------
C MAIN PROGRAM: WINDOW DESIGN OF LINEAR PHASE, LOWPASS, HIGHPASS
C               BANDPASS, AND BANDSTOP FIR DIGITAL FILTERS
C AUTHOR:       LAWRENCE R. RABINER AND CAROL A. MCGONEGAL
C               BELL LABORATORIES, MURRAY HILL, NEW JERSEY, 07974
C MODIFIED JAN. 1978 BY DOUG PAUL, MIT LINCOLN LABORATORIES
C TO INCLUDE SUBROUTINES FOR OBTAINING FILTER BAND EDGES AND RIPPLES
C
C INPUT:        NF IS THE FILTER LENGTH IN SAMPLES
C                   3 <= NF <= 1024
C
C               ITYPE IS THE WINDOW TYPE
C                   ITYPE = 1      RECTANGULAR WINDOW
C                   ITYPE = 2      TRIANGULAR WINDOW
C                   ITYPE = 3      HAMMING WINDOW
C                   ITYPE = 4      GENERALIZED HAMMING WINDOW
C                   ITYPE = 5      HANNING WINDOW
C                   ITYPE = 6      KAISER (I0-SINH) WINDOW
C                   ITYPE = 7      CHEBYSHEV WINDOW
C
C               JTYPE IS THE FILTER TYPE
C                   JTYPE = 1      LOWPASS FILTER
C                   JTYPE = 2      HIGHPASS FILTER
C                   JTYPE = 3      BANDPASS FILTER
C                   JTYPE = 4      BANDSTOP FILTER
C
C               FC IS THE NORMALIZED CUTOFF FREQUENCY
C                   0 <= FC <= 0.5
C               FL AND FH ARE THE NORMALIZED FILTER CUTOFF FREQUENCIES
C                   0 <= FL <= FH <= 0.5
C               IWP OPTIONALLY PRINTS OUT THE WINDOW VALUES
C                   IWP = 0  DO NOT PRINT
C                   IWP = 1  PRINT
C               IMD REQUESTS ADDITIONAL RUNS
C                   IMD = 1   NEW RUN
C                   IMD = 0   TERMINATES PROGRAM
C-----------------------------------------------------------------------
C
      DIMENSION W(512), G(512)
      INTEGER OTCD1, OTCD2
C
      PI = 4.0*ATAN(1.0)
      TWOPI = 2.0*PI
C
C DEFINE I/O DEVICE CODES
C INPUT: INPUT TO THIS PROGRAM IS USER-INTERACTIVE
C        THAT IS - A QUESTION IS WRITTEN ON THE USER
C        TERMINAL (OTCD1) AND THE USER TYPES IN THE ANSWER.
C
C OUTPUT: ALL OUTPUT IS WRITTEN ON THE STANDARD
C         OUTPUT UNIT (OTCD2)
C
      INCOD = I1MACH(1)
      OTCD1 = I1MACH(4)
      OTCD2 = I1MACH(2)
C
C INPUT THE FILTER LENGTH(NF), WINDOW TYPE(ITYPE) AND FILTER TYPE(JTYPE)
C
   10 WRITE (OTCD1,9999)
 9999 FORMAT (44H SPECIFY FILTER LENGTH(I4), WINDOW TYPE(I2),, 6H FILTE,
     *    10HR TYPE(I2))
      READ (INCOD,9998) NF, ITYPE, JTYPE
 9998 FORMAT (I4, I2, I2)
      IF (NF.LE.1024) GO TO 30
   20 WRITE (OTCD2,9997) NF
 9997 FORMAT (4H NF=, I4, 17H IS OUT OF BOUNDS)
      GO TO 10
   30 IF (ITYPE.NE.7 .AND. NF.LT.3) GO TO 20
      IF (ITYPE.EQ.7 .AND. (NF.EQ.1 .OR. NF.EQ.2)) GO TO 20
C
C N IS HALF THE LENGTH OF THE SYMMETRIC FILTER
C
      N = (NF+1)/2
      IF (JTYPE.NE.1 .AND. JTYPE.NE.2) GO TO 50
C
C FOR THE IDEAL LOWPASS OR HIGHPASS DESIGN - INPUT FC
C
   40 WRITE (OTCD1,9996)
 9996 FORMAT (38H SPECIFY IDEAL CUTOFF FREQUENCY(F14.7))
      READ (INCOD,9993) FC
      IF (FC.GT.0.0 .AND. FC.LT.0.5) GO TO 60
      WRITE (OTCD1,9995) FC
 9995 FORMAT (13H VALUE OF FC=, F14.7, 29H IS OUT OF BOUNDS, REENTER DA,
     *    2HTA)
      GO TO 40
C
C FOR THE IDEAL BANDPASS OR BANDSTOP DESIGN - INPUT FL AND FH
C
   50 WRITE (OTCD1,9994)
 9994 FORMAT (43H SPECIFY LOWER AND UPPER CUTOFF FREQUENCIES, 7H(2F14.7,
     *    1H))
      READ (INCOD,9993) FL, FH
 9993 FORMAT (2F14.7)
      IF (FL.GT.0.0 .AND. FL.LT.0.5 .AND. FH.GT.0.0 .AND. FH.LT.0.5
     *    .AND. FH.GT.FL) GO TO 60
      IF (FL.LT.0. .OR. FL.GT.0.5) WRITE (OTCD1,9995) FL
      IF (FH.LT.0. .OR. FH.GT.0.5) WRITE (OTCD1,9995) FH
      IF (FH.LT.FL) WRITE (OTCD1,9992) FH, FL
 9992 FORMAT (4H FH=, F14.7, 20H IS SMALLER THAN FL=, F14.7, 8H REENTER,
     *    5H DATA)
      GO TO 50
   60 IF (ITYPE.NE.7) GO TO 70
C
C INPUT FOR CHEBYSHEV WINDOW--2 OF THE 3 PARAMETERS NF, DPLOG, AND DF
C MUST BE SPECIFIED, WHERE DPLOG IS THE DESIRED FILTER RIPPLE(DB SCALE),
C DF IS THE TRANSITION WIDTH (NORMALIZED) OF THE FILTER,
C AND NF IS THE FILTER LENGTH.  THE UNSPECIFIED PARAMETER
C IS READ IN WITH THE ZERO VALUE.
C
      WRITE (OTCD1,9991)
 9991 FORMAT (46H SPECIFY CHEBYSHEV RIPPLE IN DB (F14.7) AND/OR,
     *24H TRANSITION WIDTH(F14.7))
      READ (INCOD,9993) DPLOG, DF
      DP = 10.0**(-DPLOG/20.0)
      CALL CHEBC(NF, DP, DF, N, X0, XN)
C
C IEO IS AN EVEN, ODD INDICATOR, IEO = 0 FOR EVEN, IEO = 1 FOR ODD
C
   70 IEO = MOD(NF,2)
      IF (IEO.EQ.1 .OR. JTYPE.EQ.1 .OR. JTYPE.EQ.3) GO TO 80
      WRITE (OTCD1,9990)
 9990 FORMAT (48H NF MUST BE ODD INTEGER FOR HP OR BS FILTERS--NF,
     *    24H IS BEING INCREASED BY 1)
      NF = NF + 1
      N = (1+NF)/2
      IEO = 1
   80 CONTINUE
C
C COMPUTE IDEAL (UNWINDOWED) IMPULSE RESPONSE FOR FILTER
```

```
C
      C1 = FC
      IF (JTYPE.EQ.3 .OR. JTYPE.EQ.4) C1 = FH - FL
      IF (IEO.EQ.1) G(1) = 2.*C1
      I1 = IEO + 1
      DO 90 I=I1,N
        XN = I - 1
        IF (IEO.EQ.0) XN = XN + 0.5
        C = PI*XN
        C3 = C*C1
        IF (JTYPE.EQ.1 .OR. JTYPE.EQ.2) C3 = 2.*C3
        G(I) = SIN(C3)/C
        IF (JTYPE.EQ.3 .OR. JTYPE.EQ.4) G(I) = G(I)*2.*COS(C*(FL+FH))
   90 CONTINUE
C
C COMPUTE A RECTANGULAR WINDOW
C
      IF (ITYPE.EQ.1) WRITE (OTCD2,9989) NF
 9989 FORMAT (23H RECTANGULAR WINDOW-NF=, I4)
      DO 100 I=1,N
        W(I) = 1.
  100 CONTINUE
C
C DISPATCH ON WINDOW TYPE
C
      GO TO (200, 110, 120, 140, 150, 160, 170), ITYPE
C
C TRIANGULAR WINDOW
C
  110 CALL TRIANG(NF, W, N, IEO)
      WRITE (OTCD2,9988) NF
 9988 FORMAT (22H TRIANGULAR WINDOW-NF=, I4)
      GO TO 180
C
C HAMMING WINDOW
C
  120 ALPHA = 0.54
      WRITE (OTCD2,9987) NF
 9987 FORMAT (19H HAMMING WINDOW-NF=, I4)
  130 BETA = 1. - ALPHA
      CALL HAMMIN(NF, W, N, IEO, ALPHA, BETA)
      WRITE (OTCD2,9986) ALPHA
 9986 FORMAT (7H ALPHA=, F14.7)
      GO TO 180
C
C GENERALIZED HAMMING WINDOW
C FORM OF WINDOW IS W(M)=ALPHA+BETA*COS((TWOPI*M)/(NF-1))
C BETA IS AUTOMATICALLY SET TO 1.-ALPHA
C READ IN ALPHA
C
  140 WRITE (OTCD1,9985)
 9985 FORMAT (45H SPECIFY ALPHA FOR GENERALIZED HAMMING WINDOW)
      READ (INCOD,9993) ALPHA
      WRITE (OTCD2,9984) NF
 9984 FORMAT (31H GENERALIZED HAMMING WINDOW-NF=, I4)
      GO TO 130
C
C HANNING WINDOW
C
  150 ALPHA = 0.5
      WRITE (OTCD2,9983) NF
 9983 FORMAT (19H HANNING WINDOW-NF=, I4)
C
C INCREASE NF BY 2 AND N BY 1 FOR HANNING WINDOW SO ZERO
C ENDPOINTS ARE NOT PART OF WINDOW
C
      NF = NF + 2
      N = N + 1
      GO TO 130
C
C KAISER (I0-SINH) WINDOW
C NEED TO SPECIFY PARAMETER ATT=STOPBAND ATTENUATION IN DB
C
  160 WRITE (OTCD1,9982)
 9982 FORMAT (33H SPECIFY ATTENUATION IN DB(F14.7), 16H FOR KAISER WIND,
     *    2HOW)
      READ (INCOD,9993) ATT
      IF (ATT.GT.50.) BETA = 0.1102*(ATT-8.7)
      IF (ATT.GE.20.96 .AND. ATT.LE.50.) BETA = 0.58417*(ATT-20.96)**
     *    0.4 + 0.07886*(ATT-20.96)
      IF (ATT.LT.20.96) BETA = 0.
      CALL KAISER(NF, W, N, IEO, BETA)
      WRITE (OTCD2,9981) NF
 9981 FORMAT (18H KAISER WINDOW-NF=, I4)
      WRITE (OTCD2,9980) ATT, BETA
 9980 FORMAT (6H  ATT=, F14.7, 7H  BETA=, F14.7)
      GO TO 180
C
C CHEBYSHEV WINDOW
C
  170 CALL CHEBY(NF, W, N, IEO, DP, DF, X0, XN)
      WRITE (OTCD2,9979) NF
 9979 FORMAT (21H CHEBYSHEV WINDOW-NF=, I4)
      WRITE (OTCD2,9978) DP, DF
 9978 FORMAT (4H DP=, F14.7, 5H  DF=, F14.7)
C
C WINDOW IDEAL FILTER RESPONSE
C CHANGE BACK NF AND N FOR HANNING WINDOW
C
  180 IF (ITYPE.EQ.5) NF = NF - 2
      IF (ITYPE.EQ.5) N = N - 1
      DO 190 I=1,N
        G(I) = G(I)*W(I)
  190 CONTINUE
C
C PRINT OUT RESULTS
C
  200 WRITE (OTCD1,9977)
 9977 FORMAT (36H PRINT OUT WINDOW VALUES(1=YES,0=NO))
      READ (INCOD,9976) IWP
 9976 FORMAT (I1)
      IF (IWP.EQ.0) GO TO 220
      WRITE (OTCD2,9975)
 9975 FORMAT (14H WINDOW VALUES)
      DO 210 I=1,N
        J = N + 1 - I
        K = NF + 1 - I
        WRITE (OTCD2,9974) I, W(J), K
 9974   FORMAT (10X, 3H W(, I3, 2H)=, E15.8, 4H =W(, I4, 1H))
  210 CONTINUE
  220 IF (JTYPE.EQ.1) WRITE (OTCD2,9973)
 9973 FORMAT (26H **LOWPASS FILTER DESIGN**)
      IF (JTYPE.EQ.2) WRITE (OTCD2,9972)
 9972 FORMAT (27H **HIGHPASS FILTER DESIGN**)
      IF (JTYPE.EQ.3) WRITE (OTCD2,9971)
 9971 FORMAT (27H **BANDPASS FILTER DESIGN**)
      IF (JTYPE.EQ.4) WRITE (OTCD2,9970)
 9970 FORMAT (27H **BANDSTOP FILTER DESIGN**)
      IF (JTYPE.EQ.1) WRITE (OTCD2,9969) FC
 9969 FORMAT (22H IDEAL LOWPASS CUTOFF=, F14.7)
```

```
      IF (JTYPE.EQ.2) WRITE (OTCD2,9968) FC
9968  FORMAT (23H IDEAL HIGHPASS CUTOFF=, F14.7)
      IF (JTYPE.EQ.3 .OR. JTYPE.EQ.4) WRITE (OTCD2,9967) FL, FH
9967  FORMAT (26H IDEAL CUTOFF FREQUENCIES=, 2F14.7)
      IF (JTYPE.EQ.1 .OR. JTYPE.EQ.3) GO TO 240
      DO 230 I=2,N
        G(I) = -G(I)
 230  CONTINUE
      G(1) = 1.0 - G(1)
C
C WRITE OUT IMPULSE RESPONSE
C
 240  DO 250 I=1,N
        J = N + 1 - I
        K = NF + 1 - I
        WRITE (OTCD2,9966) I, G(J), K
9966    FORMAT (10X, 3H H(, I3, 2H)=, E15.8, 4H =H(, I4, 1H))
 250  CONTINUE
      CALL FLCHAR(NF, ITYPE, JTYPE, FC, FL, FH, N, IEO, G, OTCD2)
      WRITE (OTCD2,9965)
9965  FORMAT (1H /1H /1H /1H )
      WRITE (OTCD2,9964)
9964  FORMAT (1H1)
      WRITE (OTCD1,9963)
9963  FORMAT (26H MORE DATA(1=YES,0=NO)(I1))
      READ (INCOD,9962) IMD
9962  FORMAT (I1)
      IF (IMD.EQ.1) GO TO 10
      STOP
      END
C
C-----------------------------------------------------------------------
C SUBROUTINE:  TRIANG
C TRIANGULAR WINDOW
C-----------------------------------------------------------------------
C
      SUBROUTINE TRIANG(NF, W, N, IEO)
C
C  NF = FILTER LENGTH IN SAMPLES
C   W = WINDOW COEFFICIENTS FOR HALF THE WINDOW
C   N = HALF WINDOW LENGTH=(NF+1)/2
C IEO = EVEN - ODD INDICATION--IEO=0 FOR NF EVEN
C
      DIMENSION W(1)
      FN = N
      DO 10 I=1,N
        XI = I - 1
        IF (IEO.EQ.0) XI = XI + 0.5
        W(I) = 1. - XI/FN
  10  CONTINUE
      RETURN
      END
C
C-----------------------------------------------------------------------
C SUBROUTINE:  HAMMIN
C GENERALIZED HAMMING WINDOW ROUTINE
C WINDOW IS W(N) = ALPHA + BETA * COS( TWOPI*(N-1) / (NF-1) )
C-----------------------------------------------------------------------
C
      SUBROUTINE HAMMIN(NF, W, N, IEO, ALPHA, BETA)
C
C    NF = FILTER LENGTH IN SAMPLES
C     W = WINDOW ARRAY OF SIZE N
C     N = HALF LENGTH OF FILTER=(NF+1)/2
C   IEO = EVEN ODD INDICATOR--IEO=0 IF NF EVEN
C ALPHA = CONSTANT OF WINDOW
C  BETA = CONSTANT OF WINDOW--GENERALLY BETA=1-ALPHA
C
      DIMENSION W(1)
      PI2 = 8.0*ATAN(1.0)
      FN = NF - 1
      DO 10 I=1,N
        FI = I - 1
        IF (IEO.EQ.0) FI = FI + 0.5
        W(I) = ALPHA + BETA*COS((PI2*FI)/FN)
  10  CONTINUE
      RETURN
      END
C
C-----------------------------------------------------------------------
C SUBROUTINE:  KAISER
C KAISER WINDOW
C-----------------------------------------------------------------------
C
      SUBROUTINE KAISER(NF, W, N, IEO, BETA)
C
C   NF = FILTER LENGTH IN SAMPLES
C    W = WINDOW ARRAY OF SIZE N
C    N = FILTER HALF LENGTH=(NF+1)/2
C  IEO = EVEN ODD INDICATOR--IEO=0 IF NF EVEN
C BETA = PARAMETER OF KAISER WINDOW
C
      DIMENSION W(1)
      REAL INO
      BES = INO(BETA)
      XIND = FLOAT(NF-1)*FLOAT(NF-1)
      DO 10 I=1,N
        XI = I - 1
        IF (IEO.EQ.0) XI = XI + 0.5
        XI = 4.*XI*XI
        W(I) = INO(BETA*SQRT(1.-XI/XIND))
        W(I) = W(I)/BES
  10  CONTINUE
      RETURN
      END
C
C-----------------------------------------------------------------------
C FUNCTION:  INO
C BESSEL FUNCTION FOR KAISER WINDOW
C-----------------------------------------------------------------------
C
      REAL FUNCTION INO(X)
      Y = X/2.
      T = 1.E-08
      E = 1.
      DE = 1.
      DO 10 I=1,25
        XI = I
        DE = DE*Y/XI
        SDE = DE*DE
        E = E + SDE
        IF (E*T-SDE) 10, 10, 20
  10  CONTINUE
  20  INO = E
      RETURN
      END
C
C-----------------------------------------------------------------------
C SUBROUTINE:  CHEBC
C SUBROUTINE TO GENERATE CHEBYSHEV WINDOW PARAMETERS WHEN
```

```
C ONE OF THE THREE PARAMETERS NF,DP AND DF IS UNSPECIFIED
C-----------------------------------------------------------------------
C
      SUBROUTINE CHEBC(NF, DP, DF, N, X0, XN)
C
C NF = FILTER LENGTH (IN SAMPLES)
C DP = FILTER RIPPLE (ABSOLUTE SCALE)
C DF = NORMALIZED TRANSITION WIDTH OF FILTER
C  N = (NF+1)/2 = FILTER HALF LENGTH
C X0 = (3-C0)/(1+C0) WITH C0=COS(PI*DF) = CHEBYSHEV WINDOW CONSTANT
C XN = NF-1
C
      PI = 4.*ATAN(1.0)
      IF (NF.NE.0) GO TO 10
C
C DP,DF SPECIFIED, DETERMINE NF
C
      C1 = COSHIN((1.+DP)/DP)
      C0 = COS(PI*DF)
      X = 1. + C1/COSHIN(1./C0)
C
C INCREMENT BY 1 TO GIVE NF WHICH MEETS OR EXCEEDS SPECS ON DP AND DF
C
      NF = X + 1.0
      N = (NF+1)/2
      XN = NF - 1
      GO TO 30
  10  IF (DF.NE.0.0) GO TO 20
C
C NF,DP SPECIFIED, DETERMINE DF
C
      XN = NF - 1
      C1 = COSHIN((1.+DP)/DP)
      C2 = COSH(C1/XN)
      DF = ARCCOS(1./C2)/PI
      GO TO 30
C
C NF,DF SPECIFIED, DETERMINE DP
C
  20  XN = NF - 1
      C0 = COS(PI*DF)
      C1 = XN*COSHIN(1./C0)
      DP = 1./(COSH(C1)-1.)
  30  X0 = (3.-COS(2.*PI*DF))/(1.+COS(2.*PI*DF))
      RETURN
      END
C
C-----------------------------------------------------------------------
C SUBROUTINE:  CHEBY
C DOLPH CHEBYSHEV WINDOW DESIGN
C-----------------------------------------------------------------------
C
      SUBROUTINE CHEBY(NF, W, N, IEO, DP, DF, X0, XN)
C
C  NF = FILTER LENGTH IN SAMPLES
C   W = WINDOW ARRAY OF SIZE N
C   N = HALF LENGTH OF FILTER = (NF+1)/2
C IEO = EVEN-ODD INDICATOR--IEO=0 FOR NF EVEN
C  DP = WINDOW RIPPLE ON AN ABSOLUTE SCALE
C  DF = NORMALIZED TRANSITION WIDTH OF WINDOW
C  X0 = WINDOW PARAMETER RELATED TO TRANSITION WIDTH
C  XN = NF-1
C
      DIMENSION W(1)
      DIMENSION PR(1024), PI(1024)
      PIE = 4.*ATAN(1.0)
      XN = NF - 1
      FNF = NF
      ALPHA = (X0+1.)/2.
      BETA = (X0-1.)/2.
      TWOPI = 2.*PIE
      C2 = XN/2.
      DO 40 I=1,NF
        XI = I - 1
        F = XI/FNF
        X = ALPHA*COS(TWOPI*F) + BETA
        IF (ABS(X)-1.) 10, 10, 20
  10    P = DP*COS(C2*ARCCOS(X))
        GO TO 30
  20    P = DP*COSH(C2*COSHIN(X))
  30    PI(I) = 0.
        PR(I) = P
C
C FOR EVEN LENGTH FILTERS USE A ONE-HALF SAMPLE DELAY
C ALSO THE FREQUENCY RESPONSE IS ANTISYMMETRIC IN FREQUENCY
C
        IF (IEO.EQ.1) GO TO 40
        PR(I) = P*COS(PIE*F)
        PI(I) = -P*SIN(PIE*F)
        IF (I.GT.(NF/2+1)) PR(I) = -PR(I)
        IF (I.GT.(NF/2+1)) PI(I) = -PI(I)
  40  CONTINUE
C
C USE DFT TO GIVE WINDOW
C
      TWN = TWOPI/FNF
      DO 60 I=1,N
        XI = I - 1
        SUM = 0.
        DO 50 J=1,NF
          XJ = J - 1
          SUM = SUM + PR(J)*COS(TWN*XJ*XI) + PI(J)*SIN(TWN*XJ*XI)
  50    CONTINUE
        W(I) = SUM
  60  CONTINUE
      C1 = W(1)
      DO 70 I=1,N
        W(I) = W(I)/C1
  70  CONTINUE
      RETURN
      END
C
C-----------------------------------------------------------------------
C FUNCTION:  COSHIN
C FUNCTION FOR HYPERBOLIC INVERSE COSINE OF X
C-----------------------------------------------------------------------
C
      REAL FUNCTION COSHIN(X)
      COSHIN = ALOG(X+SQRT(X*X-1.))
      RETURN
      END
C
C-----------------------------------------------------------------------
C FUNCTION:  ARCCOS
C FUNCTION FOR INVERSE COSINE OF X
C-----------------------------------------------------------------------
C
      FUNCTION ARCCOS(X)
      IF (X) 30, 20, 10
  10  A = SQRT(1.-X*X)/X
```

```
      ARCCOS = ATAN(A)
      RETURN
   20 ARCCOS = 2.*ATAN(1.0)
      RETURN
   30 A = SQRT(1.-X*X)/X
      ARCCOS = ATAN(A) + 4.*ATAN(1.0)
      RETURN
      END
C
C-----------------------------------------------------------------------
C FUNCTION:  COSH
C FUNCTION FOR HYPERBOLIC COSINE OF X
C-----------------------------------------------------------------------
C
      REAL FUNCTION COSH(X)
      COSH = (EXP(X)+EXP(-X))/2.
      RETURN
      END
C
C-----------------------------------------------------------------------
C SUBROUTINE:  FLCHAR
C SUBROUTINE TO DETERMINE FILTER CHARACTERISTICS
C-----------------------------------------------------------------------
C
      SUBROUTINE FLCHAR(NF, ITYPE, JTYPE, FC, FL, FH, N, IEO, G, OTCD2)
C
C     NF = FILTER LENGTH IN SAMPLES
C  ITYPE = WINDOW TYPE
C  JTYPE = FILTER TYPE
C     FC = IDEAL CUTOFF OF LP OR HP FILTER
C     FL = LOWER CUTOFF OF BP OR BS FILTER
C     FH = UPPER CUTOFF OF BP OR BS FILTER
C      N = FILTER HALF LENGTH = (NF+1) / 2
C    IEO = EVEN ODD INDICATOR
C      G = FILTER ARRAY OF SIZE N
C  OTCD2 = OUTPUT CODE FOR LINE PRINTER USED IN WRITE STATEMENTS
C
      DIMENSION G(1)
      DIMENSION RESP(2048)
      INTEGER OTCD2
C
C NOT FOR FOR TRIANGULAR WINDOW
C
      IF (ITYPE.EQ.2) RETURN
C
C DFT TO GET FREQ RESP
C
      PI = 4.*ATAN(1.0)
C
C UP TO 4096 PT DFT
C
      NR = 8*NF
      IF (NR.GT.2048) NR = 2048
      XNR = NR
      TWN = PI/XNR
      SUMI = -G(1)/2.
      IF (IEO.EQ.0) SUMI = 0.
      DO 20 I=1,NR
        XI = I - 1
        TWNI = TWN*XI
        SUM = SUMI
        DO 10 J=1,N
          XJ = J - 1
          IF (IEO.EQ.0) XJ = XJ + .5
          SUM = SUM + G(J)*COS(XJ*TWNI)
   10   CONTINUE
        RESP(I) = 2.*SUM
   20 CONTINUE
C
C DISPATCH ON FILTER TYPE
C
      GO TO (30, 40, 50, 60), JTYPE
C
C LOWPASS
C
   30 CALL RIPPLE(NR, 1., 0., FC, RESP, F1, F2, DB)
      WRITE (OTCD2,9999) F2, DB
 9999 FORMAT (17H PASSBAND CUTOFF , F6.4, 9H  RIPPLE , F8.3, 3H DB)
      CALL RIPPLE(NR, 0., FC, .5, RESP, F1, F2, DB)
      WRITE (OTCD2,9998) F1, DB
 9998 FORMAT (17H STOPBAND CUTOFF , F6.4, 9H  RIPPLE , F8.3, 3H DB)
      RETURN
C
C HIGHPASS
C
   40 CALL RIPPLE(NR, 0., 0., FC, RESP, F1, F2, DB)
      WRITE (OTCD2,9998) F2, DB
      CALL RIPPLE(NR, 1., FC, .5, RESP, F1, F2, DB)
      WRITE (OTCD2,9999) F1, DB
      RETURN
C
C BANDPASS
C
   50 CALL RIPPLE(NR, 0., 0., FL, RESP, F1, F2, DB)
      WRITE (OTCD2,9998) F2, DB
      CALL RIPPLE(NR, 1., FL, FH, RESP, F1, F2, DB)
      WRITE (OTCD2,9997) F1, F2, DB
 9997 FORMAT (18H PASSBAND CUTOFFS , F6.4, 2X, F6.4, 8H  RIPPLE, F9.3,
     *    3H DB)
      CALL RIPPLE(NR, 0., FH, .5, RESP, F1, F2, DB)
      WRITE (OTCD2,9998) F1, DB
      RETURN
C
C STOPBAND
C
   60 CALL RIPPLE(NR, 1., 0., FL, RESP, F1, F2, DB)
      WRITE (OTCD2,9999) F2, DB
      CALL RIPPLE(NR, 0., FL, FH, RESP, F1, F2, DB)
      WRITE (OTCD2,9996) F1, F2, DB
 9996 FORMAT (18H STOPBAND CUTOFFS , F6.4, 2X, F6.4, 8H  RIPPLE, F9.3,
     *    3H DB)
      CALL RIPPLE(NR, 1., FH, .5, RESP, F1, F2, DB)
      WRITE (OTCD2,9999) F1, DB
      RETURN
      END
C
C-----------------------------------------------------------------------
C SUBROUTINE:  RIPPLE
C FINDS LARGEST RIPPLE IN BAND AND LOCATES BAND EDGES BASED ON THE
C POINT WHERE THE TRANSITION REGION CROSSES THE MEASURED RIPPLE BOUND
C-----------------------------------------------------------------------
C
      SUBROUTINE RIPPLE(NR, RIDEAL, FLOW, FHI, RESP, F1, F2, DB)
C
C     NR = SIZE OF RESP
C RIDEAL = IDEAL FREQUENCY RESPONSE
C   FLOW = LOW EDGE OF IDEAL BAND
C    FHI = HIGH EDGE OF IDEAL BAND
C   RESP = FREQUENCY RESPONSE OF SIZE NR
C     F1 = COMPUTED LOWER BAND EDGE
```

```
C     F2 = COMPUTED UPPER BAND EDGE
C     DB = DEVIATION FROM IDEAL RESPONSE IN DB
C
      DIMENSION RESP(1)
      XNR = NR
C
C BAND LIMITS
C
      IFLOW = 2.*XNR*FLOW + 1.5
      IFHI = 2.*XNR*FHI + 1.5
      IF (IFLOW.EQ.0) IFLOW = 1
      IF (IFHI.GE.NR) IFHI = NR - 1
C
C FIND MAX AND MIN PEAKS IN BAND
C
      RMIN = RIDEAL
      RMAX = RIDEAL
      DO 20 I=IFLOW,IFHI
        IF (RESP(I).LE.RMAX .OR. RESP(I).LT.RESP(I-1) .OR.
     *      RESP(I).LT.RESP(I+1)) GO TO 10
        RMAX = RESP(I)
 10     IF (RESP(I).GE.RMIN .OR. RESP(I).GT.RESP(I-1) .OR.
     *      RESP(I).GT.RESP(I+1)) GO TO 20
        RMIN = RESP(I)
 20   CONTINUE
C
C PEAK DEVIATION FROM IDEAL
C
      RIPL = AMAX1(RMAX-RIDEAL,RIDEAL-RMIN)
C
C SEARCH FOR LOWER BAND EDGE
C
      F1 = FLOW
      IF (FLOW.EQ.0.0) GO TO 50
      DO 30 I=IFLOW,IFHI
        IF (ABS(RESP(I)-RIDEAL).LE.RIPL) GO TO 40
 30   CONTINUE
 40   XI = I - 1
C
C LINEAR INTERPOLATION OF BAND EDGE FREQUENCY TO IMPROVE ACCURACY
C
      X1 = .5*XI/XNR
      X0 = .5*(XI-1.)/XNR
      Y1 = ABS(RESP(I)-RIDEAL)
      Y0 = ABS(RESP(I-1)-RIDEAL)
      F1 = (X1-X0)/(Y1-Y0)*(RIPL-Y0) + X0
C
C SEARCH FOR UPPER BAND EDGE
C
 50   F2 = FHI
      IF (FHI.EQ.0.5) GO TO 80
      DO 60 I=IFLOW,IFHI
        J = IFHI + IFLOW - I
        IF (ABS(RESP(J)-RIDEAL).LE.RIPL) GO TO 70
 60   CONTINUE
 70   XI = J - 1
C
C LINEAR INTERPOLATION OF BAND EDGE FREQUENCY TO IMPROVE ACCURACY
C
      X1 = .5*XI/XNR
      X0 = .5*(XI+1.)/XNR
      Y1 = ABS(RESP(J)-RIDEAL)
      Y0 = ABS(RESP(J+1)-RIDEAL)
      F2 = (X1-X0)/(Y1-Y0)*(RIPL-Y0) + X0
C
C DEVIATION FROM IDEAL IN DB
C
 80   DB = 20.*ALOG10(RIPL+RIDEAL)
      RETURN
      END
```

Design Subroutine (MXFLAT) for Symmetric FIR Low Pass Digital Filters with Maximally-Flat Pass and Stop Bands

J. F. Kaiser

Digital Systems Research Dept.
Bell Laboratories
Murray Hill, NJ 07974

1. Purpose

The program is designed to give the coefficients of a maximally-flat pass and stop band symmetric FIR low pass digital filter with an odd total number of terms. The design proceeds from the initial specification of the width and the center of the transition band.

2. Method

In those situations where a low pass (smoothing) filter with a monotonic frequency response characteristic [1] is necessary the MXFLAT routine gives the coefficients of the desired filter. The MXFLAT designs are characterized by a tangency of the filter magnitude-frequency characteristic of L'th order at zero frequency and K'th order at half sampling frequency. The combined order is thus $2(K+L-1)$.

The frequency response characteristic of a symmetric FIR low pass digital filter having a maximally-flat magnitude characteristic is given [2-5] by

$$H(y) = (1-y)^K \sum_{n=0}^{L-1} \frac{(K-1+n)!}{(K-1)!n!} y^n \tag{1}$$

where

$$y = [1 - \cos(2\pi f/f_s)]/2 \tag{2}$$

and

K = order of tangency at $f = f_s/2$
L = order of tangency at $f = 0$
f_s = sampling frequency in Hz

Equation (1) is rewritten in a form more convenient for calculation as

$$H(y) = (1-y)^K \left[1 + \sum_{n=1}^{N_T-K} \left\{ \prod_{i=1}^{K-1} (1 + \frac{n}{i}) \right\} y^n \right] \tag{3}$$

where

$N_T = K+L-1$ = the half-order of the filter.

To find the actual coefficients of the filter having the magnitude characteristic given by Eq. (1) with the substitution Eq. (2) it is necessary to rewrite $H(y)$ in the form

$$H(f) = \mathrm{B}(1) + \sum_{n=1}^{N_T} 2\,\mathrm{B}(n+1) \cos(2\pi nf/f_s) \tag{4}$$

The resulting filter coefficients can then be used directly to filter data; the filtered output data y_n is calculated from the original input data x_n as

$$y_n = B(1)x_n + B(2)(x_{n+1} + x_{n-1}) + B(3)(x_{n+2} + x_{n-2}) + \cdots + B(NP)(x_{n+NP-1} + x_{n-NP+1}) \quad (5)$$

where B(k) is the array of filter coefficients and NP $= N_T + 1$.

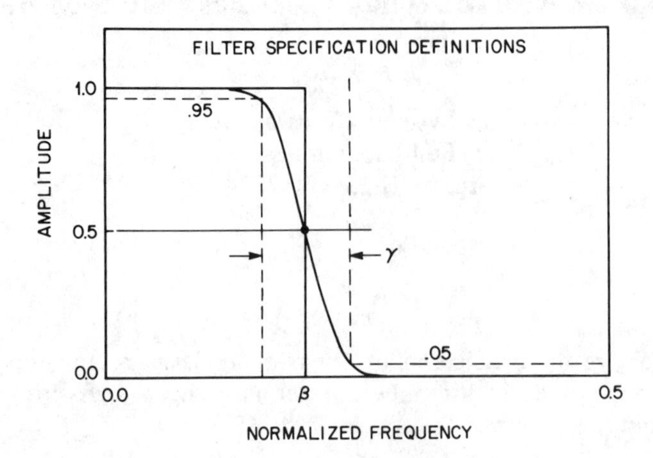

Fig. 1 Definitions of β and γ used in the design of maximally-flat FIR filters.

Referring to Fig. 1 the specification of N_T determines the sharpness of cutoff of the filter as measured by γ, the normalized width of the transition region defined as the region where the filter magnitude response varies from 95% (passband edge) down to 5% (stopband edge). The location of the center, β, of the transition region is determined by the ratio K/N_T. For these filters all of the K zeros of transmission are located at $f_s/2$ rather than being distributed throughout the stopband region as in the equiripple FIR designs.

The design equation relating the width of the transition region to filter order is not known in closed form for this filter class. However it can be approximated [1] by

$$N - 1 = 2 * N_T \approx \frac{1}{2\gamma^2} \quad (6)$$

where $N-1$ is the order of the filter and γ is the normalized transition band width as measured by the region where the filter response magnitude varies from 95% (passband edge) down to 5% (stopband edge). Normalization is with respect to the sampling frequency. These filter parameters are shown in Fig. 1.

Note that in this class of filters the number of terms required is inversely proportional to the square of the transition width whereas in the equiripple pass and stop band designs it is inversely proportional to only the first power of the transition width. These filter designs thus tend to require more terms for a given transition band width.

3. Program Description

The package consists of a main program and three subroutines - MXFLAT, PRCOEF, and RATPRX. The code for these routines is given in the Appendix.

After reading in the values of β and γ, the main program calls up MXFLAT which computes the filter coefficients after determining the proper filter order through RATPRX. The filter coefficients are then printed out by PRCOEF. The program is dimensioned to handle up to a 398'th order filter. The dimension can be increased by changing the value of the variable LIMIT.

The detailed design proceeds as follows:

1. Specify β, the normalized center of the transition band, and γ, the normalized width of the transition band; the normalization is with respect to the sampling frequency, i.e. $0<\beta<0.5$, $0<\gamma<$ *minimum*$(2\beta,\ 1-2\beta)$. If β and γ lie outside these limits the design routine prints out a

diagnostic to this effect and execution stops. Using the approximate design formula, Eq. (6), a lower estimate of the required filter order, $N-1$, is determined.

2. The degree of tangency at zero frequency is then determined such as to give the desired location for the center, β, of the transition region. The filter order is permitted to change by up to a factor of two over that determined in Step 1 in order to adjust the transition band center, β, to as close to the prescribed value as possible.

3. The filter coefficients, B(I), are then determined by first evaluating the frequency response characteristic at an equally spaced set of frequency points and then performing an inverse discrete Fourier transform on this point set. Advantage is taken of the even symmetry in both the frequency response and the weighting coefficient sequence. This method of obtaining the coefficients simplifies the coding considerably.

4. Test Problem

An example of the use of this procedure is shown in Fig. 2. Here beginning with the desired specifications of $\beta = 0.2$ (center of transition band) and $\gamma = 0.1$ (transition band width) a filter design was produced by MXFLAT with NP=26, N_T=25, i.e. a filter with a span of 50 samples. The program printed output is displayed in Table 1. The filter coefficients are used according to Eq. (5). The actual resulting β was 0.198 and the γ was 0.102 as could be determined from an evaluation of the frequency response characteristic.

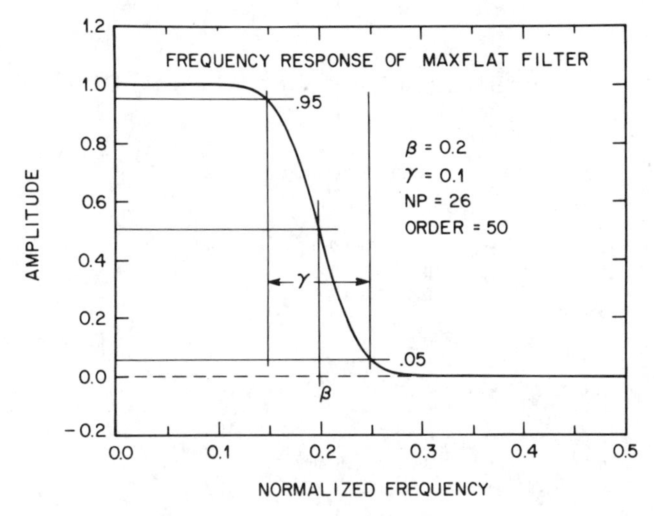

Fig. 2 Frequency response characteristic of maximally flat (MXFLAT) filter.

Often in the resulting designs the last few filter coefficients will be virtually zero. This is especially true for s mall values of γ. These coefficients can be set to zero and the order of the filter reduced accordingly. The net effect is to distribute the transmission zeros in the upper part of the stopband from their clustered position at half the sampling frequency.

As a general rule the use of this program should be restricted to values of γ not much smaller than 0.05. For example with β=0.2 and γ=0.05 the resulting filter had 110 coefficients (218 order), the last 52 of which were smaller than 0.0000002 in magnitude.

References

1. J. F. Kaiser and W. A. Reed, "Data smoothing using low-pass digital filters", *Rev. Sci. Instrum.*, Vol. 48, No. 11, pp. 1447-1457, November 1977.
2. J. F. Kaiser and R. W. Hamming, "Sharpening the Response of a Symmetric Nonrecursive Filter by Multiple Use of the Same Filter", *IEEE Trans. Acoustics, Speech, and Signal Processing,* Vol. ASSP-25, No. 5, pp. 415-422, October 1977.
3. O. Herrmann, "On the Approximation Problem in Nonrecursive Digital Filter Design", *IEEE Trans. Circuit Theory,* Vol. CT-18, No. 3, pp. 411-413, May 1971; reprinted in [5], pp. 202-203.
4. R. W. Hamming, *Digital Filters,* Prentice-Hall, Inc., Englewood Cliffs, NJ, 1977.
5. L. R. Rabiner and C. M. Rader, *Digital Signal Processing,* IEEE Press, New York, pp. 202-203, 1972.

Table 1

Program output for $\beta = 0.2$, $\gamma = 0.1$

```
FOR BETA = .200 AND GAMMA = .100
NO. OF COEFS.=   26
 B(  1) =  .39847448
 B(  2) =  .29650429
 B(  3) =  .08785310
 B(  4) = -.05124769
 B(  5) = -.05604429
 B(  6) = -.00136329
 B(  7) =  .02472394
 B(  8) =  .01120456
 B(  9) = -.00592278
 B( 10) = -.00709112
 B( 11) = -.00061605
 B( 12) =  .00232969
 B( 13) =  .00113164
 B( 14) = -.00028640
 B( 15) = -.00043241
 B( 16) = -.00008700
 B( 17) =  .00006961
 B( 18) =  .00004320
 B( 19) =  .00000222
 B( 20) = -.00000624
 B( 21) = -.00000241
 B( 22) = -.00000007
 B( 23) =  .00000016
 B( 24) =  .00000002
 B( 25) = -.00000002
 B( 26) = -.00000001
```

Appendix

```
C
C-----------------------------------------------------------------------
C MAIN PROGRAM: MAXIMALLY FLAT FIR FILTER DESIGN
C AUTHOR:       J. F. KAISER
C               BELL LABORATORIES, MURRAY HILL, NEW JERSEY 07974
C
C INPUT:        BETA IS THE CENTER OF THE TRANSITION REGION,
C                    0 < BETA < 0.5
C               GAMMA IS THE WIDTH OF THE TRANSITION REGION,
C                    0 < GAMMA < MIN( 2*BETA, 1-2*BETA )
C-----------------------------------------------------------------------
C
      DIMENSION A(200), B(200), C(200)
      COMMON IOUTD
C
C SET UP MACHINE CONSTANTS
C
      IND = I1MACH(1)
      IOUTD = I1MACH(2)
C
      LIMIT = 200
      READ (IND,9999) BETA
      READ (IND,9999) GAMMA
9999  FORMAT (E15.7)
      CALL MXFLAT(BETA, GAMMA, NP, A, B, C, LIMIT, IERR)
C
C PRINT RESULTS
C
      IF (IERR.GT.1) WRITE (IOUTD,9998) BETA, GAMMA
9998  FORMAT (9HFOR BETA=, F5.3, 11H AND GAMMA=, F5.3)
      GO TO (10, 20, 30, 40), IERR
  10  CALL PRCOEF(BETA, GAMMA, NP, B, LIMIT)
      STOP
  20  WRITE (IOUTD,9997)
9997  FORMAT (24H BETA NOT IN RANGE 0-0.5)
      STOP
  30  WRITE (IOUTD,9996)
9996  FORMAT (19H GAMMA NOT IN RANGE)
      STOP
  40  WRITE (IOUTD,9995)
9995  FORMAT (29H GAMMA TOO SMALL, MIN IS 0.4+)
      STOP
      END
C
C-----------------------------------------------------------------------
C SUBROUTINE:  PRCOEF
C PRINT THE FILTER COEFFICIENTS
C-----------------------------------------------------------------------
C
      SUBROUTINE PRCOEF(BETA, GAMMA, NP, B, LIMIT)
      DIMENSION B(LIMIT)
      COMMON IOUTD
      WRITE (IOUTD,9999) BETA, GAMMA, NP
9999  FORMAT (9HFOR BETA=, F5.3, 11H AND GAMMA=, F5.3, 13H NO. OF COEFS,
     *    2H.=, I4)
      DO 10 I=1,NP
        WRITE (IOUTD,9998) I, B(I)
9998    FORMAT (3H B(, I3, 3H) =, F11.8)
  10  CONTINUE
      RETURN
      END
C
C-----------------------------------------------------------------------
C SUBROUTINE:  MXFLAT
C COMPUTE THE COEFFICIENTS OF A MAXIMALLY FLAT FIR LINEAR PHASE FILTER
C WITH ODD NUMBER OF TERMS AND EVEN SYMMETRY IN FILTER COEFFICIENTS
C-----------------------------------------------------------------------
C
      SUBROUTINE MXFLAT(BE,GA,NP,A,B,C,LIMIT,IERR)
C
C INPUT: BE = CENTER OF TRANSITION REGION
C           (BETA), 0<BE<0.5
C        GA = WIDTH OF TRANSITION REGION (GAMMA)
C          WIDTH IS REGION WHERE 5% < MAG < 95%
C           0 < GAMMA < MIN(2*BE, 1-2*BE)
C        LIMIT = DIMENSION OF COEF ARRAY
C OUTPUT: NP = NO. OF FILTER COEFS, = <LIMIT
C           (TOTAL FILTER LENGTH = 2*NP-1)
C         B = ARRAY OF NP FILTER COEFFICIENTS
C          B(1),B(2),...B(NP)
C         IERR = 1, NORMAL RETURN
C         IERR = 2, BETA NOT IN RANGE, 0 TO 0.5
C         IERR = 3, GAMMA NOT IN RANGE
C         IERR = 4, GAMMA TOO SMALL, MIN IS 0.04+
C A = WORKING ARRAY OF SAME SIZE AS B
C C = WORKING ARRAY OF SAME SIZE AS B
C K = NO. OF ZEROS AT NYQUIST FREQ.
C L = NO. OF ZERO DERIVATIVES AT ZERO FREQ.
C NT = FILTER HALF ORDER = NP-1
C
      DIMENSION A(LIMIT), B(LIMIT), C(LIMIT)
      IERR = 1
      NP = 0
      TWOPI = 8.0*ATAN(1.0)
      IF (BE.LE.0. .OR. BE.GE.0.5) GO TO 80
      BM = AMIN1(2.0*BE,1.0-2.0*BE)
      IF (GA.LE.0. .OR. GA.GE.BM) GO TO 90
      NT = INT(1.0/(4.0*GA*GA))
      IF (NT.GT.160) GO TO 100
      AC = (1.0+COS(TWOPI*BE))/2.0
      QLIM = LIMIT
      CALL RATPRX(AC, NT, K, NP, QLIM)
      N = 2*NP - 1
      IF (K.EQ.0) K = 1
C
C COMPUTE MAGNITUDE AT NP PTS.
C
      C(1) = 1.0
      A(1) = 1.0
      LL = NT - K
      L = LL + 1
      DO 40 I=2,NP
        FF = FLOAT(I-1)/FLOAT(N)
        C(I) = COS(TWOPI*FF)
        X = (1.0-C(I))/2.0
        SUM = 1.0
        IF (K.EQ.NT) GO TO 40
        Y = X
        DO 30 J=1,LL
          FJ = J
          JL = K - 1
          Z = Y
          IF (K.EQ.1) GO TO 20
          DO 10 JJ=1,JL
            AJ = JJ
            Z = Z*(1.0+FJ/AJ)
  10      CONTINUE
  20      Y = Y*X
          SUM = SUM + Z
  30    CONTINUE
```

```
        A(I) = SUM*(1.0-X)**K
   40 CONTINUE
C
C CALCULATE WEIGHTING COEFS BY
C AN N-POINT IDFT
C
      DO 70 I=1,NP
        B(I) = A(1)/2.0
        DO 60 J=2,NP
          M = MOD((I-1)*(J-1),N)
          IF (M.LE.NT) GO TO 50
          M = N - M
   50     B(I) = B(I) + C(M+1)*A(J)
   60   CONTINUE
        B(I) = 2.0*B(I)/FLOAT(N)
   70 CONTINUE
      RETURN
   80 IERR = 2
      RETURN
   90 IERR = 3
      RETURN
  100 IERR = 4
      RETURN
      END
C
C-----------------------------------------------------------------------
C SUBROUTINE:  RATPRX
C COMPUTE RATIONAL FRACTION APPROXIMATION, K/NP, TO NUMBER A WITHIN
C THE LIMIT OF N <= NP <= 2*N FOR THE DENOMINATOR
C-----------------------------------------------------------------------
C
      SUBROUTINE RATPRX(A, N, K, NP, QLIM)
C
C INPUT: A = DESIRED NUMBER
C N = INTEGER MAX LOWER LIMIT ON NP
C OUTPUT: K = INTEGER NUMERATOR
C NP = INTEGER DENOMINATOR
C N RETURNS AS NP-1
C K/NP IS NEAREST TO A IN THE ALGEBRAIC
C SENSE, N < LIMIT
C
      IF(N .LE. 0)GOTO 3
      AA = ABS(A)
      AI = IFIX(AA)
      AF = AMOD(AA,1.0)
      QMAX = 2*N
      IF(QMAX .GT. QLIM)QMAX = QLIM
      Q = N-1
      EM = 1.0
    1 Q = Q+1.0
      IF(Q .GT. QMAX)GOTO 2
      PS = Q*AF
      IP = PS+0.5
      E = ABS((PS-FLOAT(IP))/Q)
      IF(E .GE. EM)GOTO 1
      EM = E
      PP = IP
      QQ = Q
      GOTO 1
    2 K = SIGN(AI*QQ+PP,A)
      NP = QQ
      N = NP-1
      IF(K .EQ. NP)GOTO 4
      RETURN
    3 K = 0
      N = -1
      NP = 0
      RETURN
    4 NP = QMAX
      K = NP-1
      N = K
      RETURN
      END
```

5.4

A Subroutine for Finite Wordlength FIR Filter Design

Ulrich Heute

Institut fuer Nachrichtentechnik
Universitaet Erlangen-Nuernberg
Cauerstr. 7
D-8520 Erlangen, W. Germany

1. Purpose

FIR filters in direct form have a transfer function

$$H(z) = \sum_{\nu=0}^{n_F-1} h_o(\nu) z^{-\nu}$$

with the coefficients $h_o(\nu)$ being identical to the impulse response of the finite length n_F.

Linear-phase filters have a symmetrical impulse response, i.e., $h_o(\nu) = \pm h_o(n_F-1-\nu)$; thus, only $n_1 = [\frac{n_F+1}{2}]$ different elements $h_o(\nu)$, $\nu \in \{0, 1, ..., (n_1-1)\}$ are needed for their description. The frequency response, in this case, becomes purely real or purely imaginary, if a constant delay term is neglected.

The subroutine IDEFIR *I*terative *D*esign of Linear-Phase *FIR* Filters") finds these n_1 coefficients in a way that the frequency response with a rounded fixed-point version, $R[h_o(\nu)]$, of the coefficients fulfills a prescribed tolerance scheme. Rounding is carried out to the minimum possible wordlength.

The tolerance schemes can be of the following types:

- lowpass filters,
- highpass filters,
- symmetrical bandpass filters,
- symmetrical bandstop filters,
- HILBERT transformers.

The tolerance schemes are described by sets of at most six parameters $\{\delta_1, \delta_2; \Omega_{p1}, \Omega_{s1}; \Omega_{p2}, \Omega_{s2}\}$, the meaning of which is to be seen from the figures in [3]. All tolerances have to be constant; the symmetry in the bandfilter cases includes identical tolerances in all passbands or all stopbands, respectively, and a center frequency at $\Omega = 0.25 \mathrel{\hat{=}} \frac{\pi}{2} \mathrel{\hat{=}} f = f_s/4$.

2. Method

2.1 Principal Idea

The subroutine IDEFIR realizes a "modified iterative design" procedure as proposed in [1],[2].

The main idea is a CHEBYSHEV design of a linear-phase lowpass filter with *exact* (i.e., computer accuracy) coefficients $h_o(\nu)$ for a *transformed* and *reduced* tolerance scheme such that the *original* requirements are fulfilled with *rounded* coefficients $R[h_o(\nu)]$.

The appropriate reduction is found with the aid of a well-known filter length estimation [3],[4], an approximate statistical bound for the effective coefficient rounding error [1],[2],[5], and a minimum wordlength estimation derived from this error bound [1]; the "exact design" is carried out by calling any convenient lowpass design procedure. Iterations, with modified tolerance reductions, are provided because of the statistical and approximate character of the estimations applied.

2.2 Design Process

The design process can easily be understood from the block diagram and the following explanations, concerning the numbered stages of the diagram:

(1) Only the data of the tolerance scheme $\{\delta_1, \delta_2; \Omega_{p1}, \Omega_{s1}; (\Omega_{p2}, \Omega_{s2})\}$ are needed as input parameters.

(2) Filters of the types named in Section 1 can all be derived from equivalent lowpass designs by simple transformations (e.g., [4]).

(3) The minimum wordlength $w_{c_{min}}$ for the (transformed) specifications is estimated.

(4) The filter length n_F is estimated from the specifications.

(5) The rounding error effects S_1, S_2 in the passband and the stopband are estimated, taking into account the filter length, the wordlength, and the size of "ripples" δ_1', δ_2' of the "exact coefficients" frequency response. This calculation, itself, is carried out iteratively [1].

(6) The *estimated* total errors, $F_1 = \delta_1' + S_1$, $F_2 = \delta_2' + S_2$, are compared to the prescribed tolerances. If the assumption results that the tolerance scheme is *fulfilled,* a CHEBYSHEV design procedure is called. If the tolerances are assumed to be *violated,* the tolerances δ_1', δ_2' are reduced; the program jumps back to label* 20, and the estimations of n_F, S_1, and S_2 are repeated.

(7) Any convenient design routine can be applied, here. The filter length n_{Fo} of the first design iteration is stored for later comparisons (see point (9)!).

(8) After coefficient rounding, the *real* maximum total errors E_{max1}, E_{max2} are compared to the specifications. If the tolerances are *fulfilled,* further checks follow (see comment (9a)!). If they are *violated,* a further reduction of δ_1', δ_2' is necessary, the program jumps back to label 20, (see point (6)!), and the (inner) design iteration loop is re-entered.

(9) (a) The "exploitation" of the tolerances is checked: if less than a certain percentage $p \cdot 100$ (e.g., $p = 0.9 \hat{=} 90\%$) is used and the filter length exceeds the value n_{Fo} (see comments (9b) and (7)!), the tolerances δ_1', δ_2' may be enlarged; in this case, the program jumps back to label 20, as well.
(b) If no solution was found for the filter length estimated in the first iteration (see point (7)!), the length for the first solution found at all takes the place of n_{Fo}.

(10) Tolerance reductions and increases are carried out with different "damping factors" C_R, C_I.

(11) After a change of the design tolerances δ_1', δ_2', it has to be checked, whether the scheme is realizable, at all. If negative tolerances result, the wordlength is too small. With an increased wordlength, the whole (outer) design iteration loop is re-entered, and the program jumps back to label 10.

(12) The (exact and rounded) coefficients of the lowpass fulfilling the requirements are transformed to the desired filter type (see comment (2)!). Finally, their "actual wordlength" is determined, i.e., the number of bits needed for the sign and for the magnitude of the largest absolute coefficient value without leading zeros.

3. Program Description

3.1 Main Program

A main program is needed calling the iterative design routine:

```
CALL IDEFIR (DEL1, DEL2, OMGP1, OMGS1, OMGP2, OMGS2, ITYP, NFOO, NF, IWCMIN,
             IWC, IWACT, DEL11, DEL21, EMAX1, EMAX2, H0, RH0, ITERA, IERR).
```

The meanings of the parameters in the list are to be found in the "Comment"-part of the subroutine

* The labels cited here and marked by encircled numbers in the block diagram, correspond to labels in the program listing.

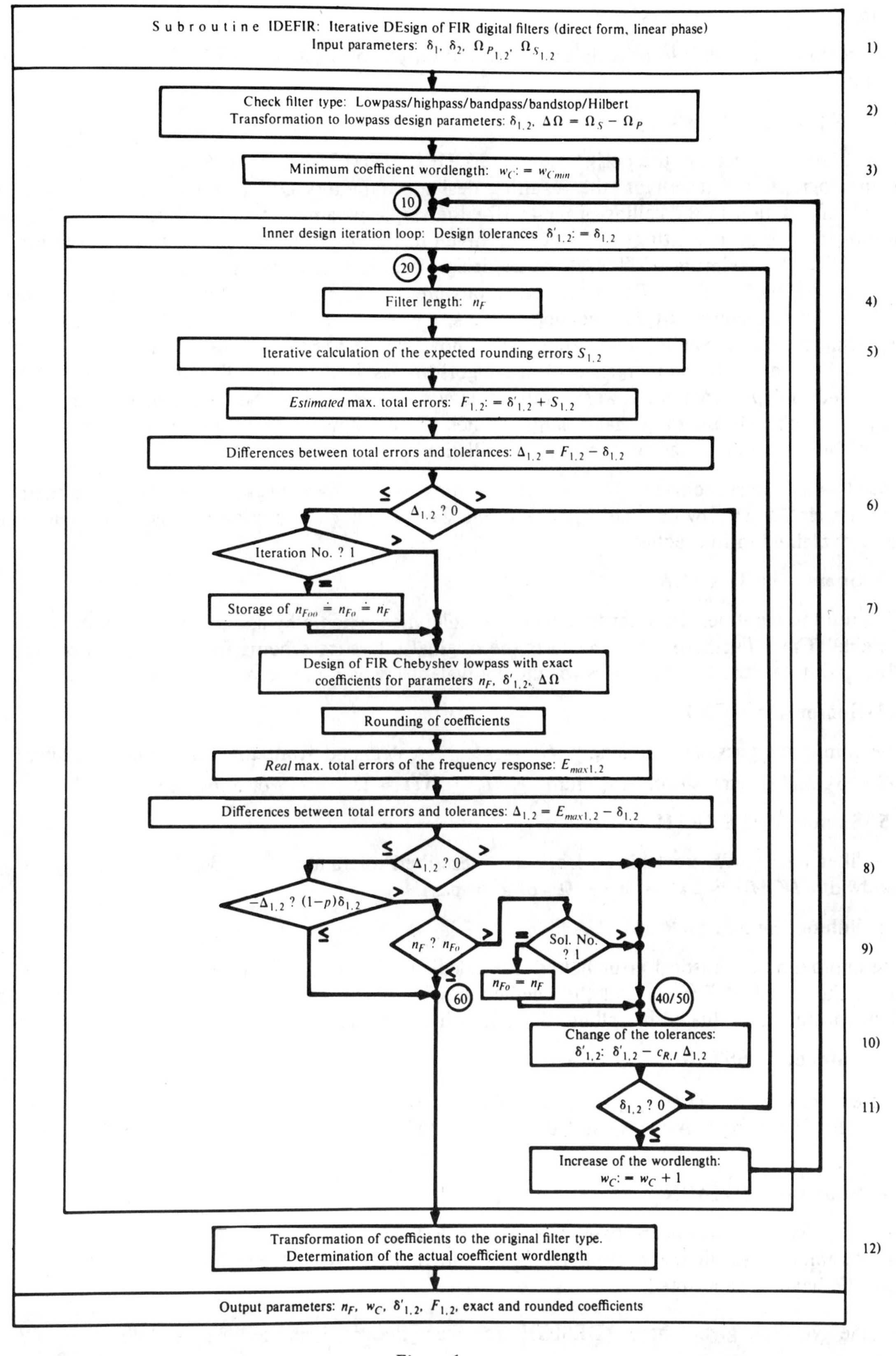
Subroutine IDEFIR: Iterative DEsign of FIR digital filters (direct form, linear phase)
Input parameters: δ_1, δ_2, $\Omega_{P_{1,2}}$, $\Omega_{S_{1,2}}$
1)
Check filter type: Lowpass/highpass/bandpass/bandstop/Hilbert
Transformation to lowpass design parameters: $\delta_{1,2}$, $\Delta\Omega = \Omega_S - \Omega_P$
2)
Minimum coefficient wordlength: $w_C := w_{C_{min}}$
3)
10
Inner design iteration loop: Design tolerances $\delta'_{1,2} := \delta_{1,2}$
20
Filter length: n_F
4)
Iterative calculation of the expected rounding errors $S_{1,2}$
5)
Estimated max. total errors: $F_{1,2} := \delta'_{1,2} + S_{1,2}$
Differences between total errors and tolerances: $\Delta_{1,2} = F_{1,2} - \delta_{1,2}$
$\Delta_{1,2}$? 0
6)
Iteration No. ? 1
Storage of $n_{F_{oo}} \doteq n_{F_o} \doteq n_F$
7)
Design of FIR Chebyshev lowpass with exact coefficients for parameters n_F, $\delta'_{1,2}$, $\Delta\Omega$
Rounding of coefficients
Real max. total errors of the frequency response: $E_{max1,2}$
Differences between total errors and tolerances: $\Delta_{1,2} = E_{max1,2} - \delta_{1,2}$
$\Delta_{1,2}$? 0
8)
$-\Delta_{1,2}$? $(1-p)\delta_{1,2}$
n_F ? n_{F_o}
Sol. No. ? 1
9)
$n_{F_o} = n_F$
60
40/50
Change of the tolerances:
$\delta'_{1,2}$: $\delta'_{1,2} - c_{R,I}\,\Delta_{1,2}$
10)
$\delta_{1,2}$? 0
11)
Increase of the wordlength:
$w_C := w_C + 1$
Transformation of coefficients to the original filter type.
Determination of the actual coefficient wordlength
12)
Output parameters: n_F, w_C, $\delta'_{1,2}$, $F_{1,2}$, exact and rounded coefficients

Figure 1

listing.

This main program has to provide the input and output of all data; no READ/WRITE - commands are included in IDEFIR.

3.2 Subroutine IDEFIR

The subroutine receives the design parameters DEL1...OMGS2 defining the tolerance scheme from the main program, and it delivers the resulting design parameters DEL1...OMGS1, coefficients, and additional information to the calling program (for the single parameters and variables, see COMMENT section of the subroutine listing). The version given in the printout is prepared to handle filters of lengths $n_F \leqslant 256$. A maximum of ITMAX = 20 design iterations is allowed. A $p = 90\%$ -tolerance usage is aimed at (PROZ = 0.9). Tolerance reductions are carried out by 75% of the tolerance violation value (CR = 0.75), increases by 25% of the unused space (CI = 0.25). Changes of p, C_R, C_I are possible by replacing the corresponding three statements; however, it has to be considered that the choice of these values influences the convergence of the algorithm as well as the efficiency of the results: with C_R increased and p decreased, fewer iterations may be needed, but the resulting solution may have too large a length! On the other hand, larger values C_I and p may provide a more efficient use of the tolerances, but they might lead to a program "oscillation"!

The steps of the subroutine IDEFIR corresponding to the block diagram (Fig. 1) are marked in the listing in a similar way by comments to be found easily. They are carried out by appropriate subroutines to be explained in the sequel.

3.3 Subroutine TYPTRA

This routine identifies the filter type from the tolerance scheme by assigning an appropriate value to the variable ITYP. Furthermore, it receives the original tolerance scheme from the calling subroutine IDEFIR, and it delivers the tolerances for the equivalent lowpass problem.

3.4 Subroutine WCMIN

The minimum possible wordlength $IW \hat{=} w_{c_{\min}}$ is calculated from the prescribed tolerances $D1 \hat{=} \delta_1$, $D2 \hat{=} \delta_2$, and the transition bandwidth $DOMG \hat{=} \Delta\Omega = \Omega_S - \Omega_P$ of a lowpass.

3.5 Subroutine LENGTH

The filter length NF is calculated from the prescribed tolerances $D1 \hat{=} \delta_1$, $D2 \hat{=} \delta_2$, and the transition bandwidth $DOMG \hat{=} \Delta\Omega = \Omega_S - \Omega_P$ of a lowpass.

3.6 Subroutine EFFERR

The approximate statistical error bound S is calculated from the tolerance $D \hat{=} \delta$ of an exact-coefficient filter, the length $NF \hat{=} n_F$, and the coefficient wordlength $IW \hat{=} w_c$ in an iterating loop. The actual error formula is evaluated by calling the subroutine SSTAT.

3.7 Subroutine SSTAT

The expected error $SQ = S/Q$ (i.e., normalized to the quantization step-size $Q = 2^{-(w_c-1)}$) is calculated from the filter length $NF \hat{=} n_F$ and the ratio $ALPHA \hat{=} \alpha = \delta/S$ of original "ripple" size and rounding error.

3.8 Subroutine DESIGN

Here, any convenient linear-phase, direct-form, FIR lowpass design routine can be applied. Of course, the parameter lists in the routine DESIGN and in the CALL DESIGN statement in the subroutine IDEFIR have to be adapted.

In the version given here, DESIGN receives the equivalent lowpass tolerances $DEL11 \hat{=} \delta_1'$, $DEL21 \hat{=} \delta_2'$, $OMGP1 \hat{=} \Omega_{p1}$, $OMGS1 \hat{=} \Omega_{s1}$, and the filter length $NF \hat{=} n_F$; it delivers to the calling routine IDEFIR the coefficient vector $HO = h_o(\nu)$, $\nu \in \{0, 1, ..., (n_1-1)\}$.

3.9 Subroutine MAXERR

The maximum deviation $EMAX \triangleq E_{max}$ of a linear-phase FIR lowpass from the corresponding ideal lowpass is calculated. The parameters used are the coefficient vector $RHO = R[h_o(\nu)]$, $\nu \in \{0,1,..., (n_1-1)\}$, the length $N1 \triangleq n_1$ of this vector, the filter length $NF \triangleq n_F$, and the frequency range $[OMGA, OMGE] \triangleq [\Omega_A, \Omega_E]$, for which the maximum error is wanted.

The actual error calculation is carried out by calling the subroutine ERROR

3.10 Subroutine ERROR

The difference E between an ideal lowpass and a linear-phase FIR lowpass at the frequency point $OMG \triangleq \Omega$ is calculated from the $N1 \triangleq n_1$ coefficients $RHO = R[h_o(\nu)]$, $\nu \in \{0,1,..., (n_1-1)\}$ of a length $NF \triangleq n_F$ filter. The routine assumes that a fairly close approximation is given: $E < 0.5$ is necessary.

3.11 Subroutine BACTRA

According to the information ITYP on the desired filter type, delivered by SUBROUTINE TYPTRA (see part 3!), the exact and rounded coefficients, $HO \triangleq h_o(\nu)$, $RHO \triangleq R[h_o(\nu)]$, of the lowpass filter are transformed. The *lowpass* filter length $NF \triangleq n_F$ is received and the length $NF \triangleq n_F$ of the *desired* filter is delivered to the calling program.

3.12 Subroutine WCACT

The $N1 = n_1$ rounded coefficients $RHO \triangleq R[h_o(\nu)]$, $\nu \in \{0,1,..., (n_1-1)\}$ are quantized to a stepsize $Q = 2^{-(w_c-1)}$; thus, with a sign bit, $IWC \triangleq w_c$ bits are used to completely define any rounded coefficients, in general.

In special realizations, however, omitting leading zeros behind the binary point might decrease the filter wordlength. Therefore, the "actual wordlength" $IWACT \triangleq w_c$ - (minimum number of leading zeros in all coefficients) is determined.

3.13 DIMENSION and COMMON Requirements

The routines IDEFIR, DESIGN, MAXERR, ERROR, WCACT contain array specifications for the coefficient vectors $HO \triangleq h_o(\nu)$ or $RHO \triangleq R[h_o(\nu)]$ or both. In the state presented here, a maximum of $n_1 = 128$ different coefficients is allowed, meaning a maximum filter length of $n_F = 256$. COMMON specifications are not contained. If, however, the subroutine DESIGN to be applied needs such definitions, they must be inserted into the main program and the subroutine IDEFIR as well. In this case, it has to be checked, whether any variables contained in the COMMON definition, are already used in subroutine parameter lists.

4. Summary of User Requirements

1. Write MAIN program with input/output commands, the subroutine call
CALL IDEFIR (DEL1, DEL2, OMGP1, OMGS1, OMGP2, OMGS2, ITYP, NFOO, NF, IWCMIN,
IWC, IWACT, DEL11, DEL21, EMAX1, EMAX2, HO, RHO, ITERA, IERR),
and the proper DIMENSION and COMMON specifications according to those contained in IDEFIR, DESIGN, MAXERR, ERROR, WCACT.

2. Adapt a convenient DESIGN routine to the subroutine IDEFIR requirements; especially, check the parameter lists, DIMENSION, and COMMON specifications.

5. Test Program and Test Problems

The routines described above were linked to a main program "IDEFIX" (see the listing below), meeting the requirements given in Section 3.1. As a possible design routine, a subroutine-package version of the PARKS-McCLELLAN program [6], capable of designing odd-length filters only, was applied.

Five examples were computed by supplying to the computer the following five data cards:

	DEL1	DEL2	OMGP1	OMGS1	OMGP2	OMGS2
1)	0.01	0.001	0.3	0.35	0.0	0.0
2)	0.01	0.001	0.2	0.15	0.0	0.0
3)	0.01	0.001	0.175	0.15	0.325	0.35
4)	0.01	0.001	0.12	0.15	0.38	0.35
5)	0.005	0.005	0.025	0.0	0.0	0.0

The resulting line printer outputs are given below in the tables for the five examples:

- Lowpass
- Highpass
- Bandpass
- Bandstop
- HILBERT-Transform

References

1. Heute, U., "Ueber Realisierungsprobleme bei nichtrekursiven Digitalfiltern", Ausgewaehlte Arbeiten ueber Nachrichtensysteme, No. 20, ed. H. W. Schuessler, Universitaet Erlangen, 1975.

2. Heute, U., "Koeffizienten-Empfindlichkeit nicht-rekursiver Digital-Filter in direkter Struktur", DFG-Colloq., *Digitale Systeme zur Signalverarbeitung,* Universitaet Erlangen, 1974.

3. Herrmann, O., Rabiner, L. R., Chan, D. S. K., "Practical Design Rules for Optimum FIR Low-Pass Digital Filters", *Bell Syst. Tech. J.,* Vol. 52, No. 6, pp. 769-799, July-August 1973.

4. Schuessler, H. W., *Digitale Systeme zur Signalverarbeitung,* Springer-Verlag, Berlin-Heidelberg-New York, 1973.

5. Heute, U., "Necessary and Efficient Expenditure for Nonrecursive Digital Filters in Direct Structure", Europ. Conf. Circuit Theory and Design, London, IEE Publ. No. 116, pp. 13-19, 1974.

6. Parks. T. W., McClellan, J. H., "A Program for the Design of Linear Phase Finite Impulse Response Digital Filters", *IEEE Trans. Audio Electroacoust.,* Vol. AU-20, No. 3, pp. 195-199, Aug. 1972.

Table 1

```
ITERATIVE DESIGN OF LINEAR PHASE FIR FILTERS IN DIRECT FORM
WITH MINIMUM COEFFICIENT WORDLENGTH

LOWPASS WITH DELTA 1 = 0.010000,DELTA 2 = 0.001000
             OMEGA P = 0.300000,OMEGA S = 0.350000
EST. MIN. WORDLENGTH WC MIN = 15 BITS (SIGN+BITS BEHIND THE DUAL POINT)
ESTIMATED LOWPASS LENGTH NF =  55
COEFFICIENT WORDLENGTH APPLIED        WC = 15
ACTUAL WORDLENGTH WITHOUT LEADING ZEROS      WC ACT = 15
ACTUAL FILTER LENGTH        NF =  57
REDUCED LOWPASS TOLERANCE SCHEME       DELTA 1 PRIME = 0.008796
                                       DELTA 2 PRIME = 0.000593
EXACT FILTER COEFFICIENTS H0 (I), I = 0...((NF-1)/2)
-0.001865     -0.000806      0.003469      0.001039
-0.002183      0.003388      0.001655     -0.004896
 0.003558      0.003980     -0.008106      0.002993
 0.008183     -0.011719      0.000822      0.014755
-0.015467     -0.004170      0.024699     -0.018991
-0.014353      0.040839     -0.021880     -0.037062
 0.075071     -0.023781     -0.122956      0.286443
 0.642227

ROUNDED FILTER COEFFICIENTS H0 (I)
-0.001831     -0.000793      0.003418      0.001038
-0.002136      0.003357      0.001648     -0.004883
 0.003540      0.003967     -0.008057      0.002991
```

Table 1
(Continued)

```
 0.008179     -0.011719      0.000793      0.014709
-0.015442     -0.004150      0.024658     -0.018982
-0.014343      0.040833     -0.021851     -0.037048
 0.075012     -0.023743     -0.122925      0.286438
 0.642212

ACTUAL TOLERANCES OF THE EQUIVALENT LOWPASS WITH ROUNDED COEFFICIENTS
DELTA 1 = 0.008044,DELTA 2 = 0.000671
NUMBER OF DESIGN ITERATIONS =   4
```

Table 2

```
ITERATIVE DESIGN OF LINEAR PHASE FIR FILTERS IN DIRECT FORM
WITH MINIMUM COEFFICIENT WORDLENGTH

HIGHPASS WITH DELTA 1 = 0.010000,DELTA 2 = 0.001000
              OMEGA P = 0.200000,OMEGA S = 0.150000
EQUIVALENT LOWPASS WITH DELTA 1 = 0.010000,DELTA 2 = 0.001000
                        OMEGA P = 0.300000,OMEGA S = 0.350000
EST. MIN. WORDLENGTH WC MIN = 15 BITS (SIGN+BITS BEHIND THE DUAL POINT)
ESTIMATED LOWPASS LENGTH NF =  55
COEFFICIENT WORDLENGTH APPLIED      WC = 15
ACTUAL WORDLENGTH WITHOUT LEADING ZEROS      WC ACT = 15
ACTUAL FILTER LENGTH       NF =  57
REDUCED LOWPASS TOLERANCE SCHEME      DELTA 1 PRIME = 0.008796
                                      DELTA 2 PRIME = 0.000593
EXACT FILTER COEFFICIENTS H0 (I), I = 0...((NF-1)/2)
-0.001865      0.000806      0.003469     -0.001039
-0.002183     -0.003388      0.001655      0.004896
 0.003558     -0.003980     -0.008106     -0.002993
 0.008183      0.011719      0.000822     -0.014755
-0.015467      0.004170      0.024699      0.018991
-0.014353     -0.040839     -0.021880      0.037062
 0.075071      0.023781     -0.122956     -0.286443
 0.642227

ROUNDED FILTER COEFFICIENTS H0 (I)
-0.001831      0.000793      0.003418     -0.001038
-0.002136     -0.003357      0.001648      0.004883
 0.003540     -0.003967     -0.008057     -0.002991
 0.008179      0.011719      0.000793     -0.014709
-0.015442      0.004150      0.024658      0.018982
-0.014343     -0.040833     -0.021851      0.037048
 0.075012      0.023743     -0.122925     -0.286438
 0.642212

ACTUAL TOLERANCES OF THE EQUIVALENT LOWPASS WITH ROUNDED COEFFICIENTS
DELTA 1 = 0.008044,DELTA 2 = 0.000671
NUMBER OF DESIGN ITERATIONS =   4
```

Table 3

```
ITERATIVE DESIGN OF LINEAR PHASE FIR FILTERS IN DIRECT FORM
WITH MINIMUM COEFFICIENT WORDLENGTH

BANDPASS WITH DELTA 1 = 0.010000,DELTA 2 = 0.001000
              OMEGAP1 = 0.175000,OMEGAS1 = 0.150000
              OMEGAP2 = 0.325000,OMEGAS2 = 0.350000
EQUIVALENT LOWPASS WITH DELTA 1 = 0.010000,DELTA 2 = 0.001000
                        OMEGA P = 0.150000,OMEGA S = 0.200000
EST. MIN. WORDLENGTH WC MIN = 15 BITS (SIGN+BITS BEHIND THE DUAL POINT)
ESTIMATED LOWPASS LENGTH NF =  55
COEFFICIENT WORDLENGTH APPLIED      WC = 15
ACTUAL WORDLENGTH WITHOUT LEADING ZEROS      WC ACT = 14
ACTUAL FILTER LENGTH       NF = 117
REDUCED LOWPASS TOLERANCE SCHEME      DELTA 1 PRIME = 0.008362
                                      DELTA 2 PRIME = 0.000430
EXACT FILTER COEFFICIENTS H0 (I), I = 0...((NF-1)/2)
 0.000607      0.           -0.000572      0.
-0.000520      0.            0.002465      0.
-0.003505      0.            0.001855      0.
```

Table 3
(Continued)

```
 0.001906       0.              -0.004329       0.
 0.002120       0.               0.003647       0.
-0.007141       0.               0.003230       0.
 0.005906       0.              -0.010919       0.
 0.004360       0.               0.009573       0.
-0.016369       0.               0.005519       0.
 0.015509       0.              -0.024539       0.
 0.006580       0.               0.025888       0.
-0.038408       0.               0.007432       0.
 0.047958       0.              -0.070128       0.
 0.007984       0.               0.132416       0.
-0.279054       0.               0.341510

 ROUNDED FILTER COEFFICIENTS H0 (I)
 0.000549       0.              -0.000549       0.
-0.000488       0.               0.002441       0.
-0.003479       0.               0.001831       0.
 0.001892       0.              -0.004272       0.
 0.002075       0.               0.003601       0.
-0.007141       0.               0.003174       0.
 0.005859       0.              -0.010864       0.
 0.004333       0.               0.009521       0.
-0.016357       0.               0.005493       0.
 0.015503       0.              -0.024536       0.
 0.006531       0.               0.025879       0.
-0.038391       0.               0.007385       0.
 0.047913       0.              -0.070068       0.
 0.007935       0.               0.132385       0.
-0.279053       0.               0.341492

 ACTUAL TOLERANCES OF THE EQUIVALENT LOWPASS WITH ROUNDED COEFFICIENTS
 DELTA 1 = 0.009576,DELTA 2 = 0.000987
 NUMBER OF DESIGN ITERATIONS =  12
```

Table 4

```
 ITERATIVE DESIGN OF LINEAR PHASE FIR FILTERS IN DIRECT FORM
 WITH MINIMUM COEFFICIENT WORDLENGTH

 BANDSTOP WITH DELTA 1 = 0.010000,DELTA 2 = 0.001000
               OMEGAP1 = 0.120000,OMEGAS1 = 0.150000
               OMEGAP2 = 0.380000,OMEGAS2 = 0.350000
 EQUIVALENT LOWPASS WITH DELTA 1 = 0.010000,DELTA 2 = 0.001000
                         OMEGA P = 0.240000,OMEGA S = 0.300000
 EST. MIN. WORDLENGTH WC MIN = 15 BITS (SIGN+BITS BEHIND THE DUAL POINT)
 ESTIMATED LOWPASS LENGTH NF =  47
 COEFFICIENT WORDLENGTH APPLIED      WC = 15
 ACTUAL WORDLENGTH WITHOUT LEADING ZEROS      WC ACT = 15
 ACTUAL FILTER LENGTH      NF =  93
 REDUCED LOWPASS TOLERANCE SCHEME       DELTA 1 PRIME = 0.009881
                                        DELTA 2 PRIME = 0.000676
 EXACT FILTER COEFFICIENTS H0 (I), I = 0...((NF-1)/2)
-0.001520       0.              -0.003752       0.
-0.000894       0.               0.003676       0.
 0.000481       0.              -0.005578       0.
 0.000340       0.               0.008082       0.
-0.001983       0.              -0.011180       0.
 0.005010       0.               0.014701       0.
-0.010025       0.              -0.018449       0.
 0.017968       0.               0.022139       0.
-0.030583       0.              -0.025456       0.
 0.052231       0.               0.028100       0.
-0.098901       0.              -0.029799       0.
 0.315849       0.               0.530385

 ROUNDED FILTER COEFFICIENTS H0 (I)
-0.001465       0.              -0.003723       0.
-0.000854       0.               0.003662       0.
 0.000427       0.              -0.005554       0.
 0.000305       0.               0.008057       0.
-0.001953       0.              -0.011169       0.
 0.005005       0.               0.014648       0.
```

Table 4
(Continued)

```
-0.010010      0.             -0.018433      0.
 0.017944      0.              0.022095      0.
-0.030579      0.             -0.025452      0.
 0.052185      0.              0.028076      0.
-0.098877      0.             -0.029785      0.
 0.315796      0.              0.530334

 ACTUAL TOLERANCES OF THE EQUIVALENT LOWPASS WITH ROUNDED COEFFICIENTS
 DELTA 1 = 0.009237,DELTA 2 = 0.000885
 NUMBER OF DESIGN ITERATIONS =    1
```

Table 5

```
ITERATIVE DESIGN OF LINEAR PHASE FIR FILTERS IN DIRECT FORM
WITH MINIMUM COEFFICIENT WORDLENGTH

HILBERT-TRANSFORMER WITH DELTA = 0.005000,OMEGAP = 0.025000
EQUIVALENT LOWPASS WITH DELTA 1 = 0.002500,DELTA 2 = 0.002500
                        OMEGA P = 0.225000,OMEGA S = 0.275000
EST. MIN. WORDLENGTH WC MIN = 14 BITS (SIGN+BITS BEHIND THE DUAL POINT)
ESTIMATED LOWPASS LENGTH NF =   59
COEFFICIENT WORDLENGTH APPLIED      WC = 14
ACTUAL WORDLENGTH WITHOUT LEADING ZEROS      WC ACT = 14
ACTUAL FILTER LENGTH      NF =   59
REDUCED LOWPASS TOLERANCE SCHEME       DELTA 1 PRIME = 0.002137
                                       DELTA 2 PRIME = 0.002137
EXACT FILTER COEFFICIENTS H0 (I), I = 0...((NF-1)/2)
0.002960       0.000055       0.003471       0.000055
0.005015       0.000055       0.007564       0.000055
0.010351       0.000055       0.014415       0.000055
0.019086       0.000055       0.025550       0.000055
0.033402       0.000055       0.044355       0.000055
0.059084       0.000055       0.081828       0.000055
0.120482       0.000055       0.208221       0.000055
0.635134       0.

ROUNDED FILTER COEFFICIENTS H0 (I)
0.002930       0.             0.003418       0.
0.004883       0.             0.007324       0.
0.010254       0.             0.014404       0.
0.019043       0.             0.025391       0.
0.033203       0.             0.044189       0.
0.059082       0.             0.081787       0.
0.120361       0.             0.208008       0.
0.635010       0.

ACTUAL TOLERANCES OF THE EQUIVALENT LOWPASS WITH ROUNDED COEFFICIENTS
DELTA 1 = 0.002485,DELTA 2 = 0.002422
NUMBER OF DESIGN ITERATIONS =    1
```

Appendix

```
C
C-----------------------------------------------------------------------
C MAIN PROGRAM:  IDEFIX - TEST PROGRAM FOR THE SUBROUTINE IDEFIR
C
C AUTHOR:  ULRICH HEUTE
C INSTITUT FUER NACHRICHTENTECHNIK,UNIV.ERLANGEN-NUERNBERG
C D-8520 ERLANGEN, W-GERMANY
C INPUT:
C     FORMAT...6F8.6 (1 CARD).
C     DEL1....PASSBAND TOLERANCE;
C     DEL2....STOPBAND TOLERANCE;
C     OMGP1...FIRST PASSBAND EDGE FREQUENCY;
C     OMGS1...FIRST STOPBAND EDGE FREQUENCY;
C     OMGP2...SECOND PASSBAND EDGE FREQUENCY;
C     OMGS2...SECOND STOPBAND EDGE FREQUENCY.
C
C     NORMALIZED FREQUENCY OMG = F / SAMPLING RATE.
C     IN LOWPASS AND HIGHPASS CASES, THE LAST TWO PARAMETERS ARE ZERO
C     IN BANDFILTER CASES, THE LAST TWO PARAMETERS HAVE THE VALUES
C               OMGP(S)2 = 0.5 - OMGP(S)1
C     IN HILBERT CASES, DEL2 = DEL1 IS VALID;OMGP1 IS THE HILBERT BAND
C     EDGE FREQUENCY; THE LAST T H R E E PARAMETERS ARE ZERO.
C
C OUTPUT:
C     ALL DATA CONCERNING THE DESIGN ARE DELIVERED TO THE STANDARD
C     OUTPUT UNIT.
C-----------------------------------------------------------------------
C
      COMMON /CALC/ X(131), AF(2096), AD(131), Y(131), SCALE
      COMMON /CHEB/ N, NGRID, KGRID, KFP, KFS, IFR(131)
      COMMON /CHEK/ NL, NG, KA, KB
C
C THE COMMON-DEFINITIONS HAVE TO BE ADAPTED TO THE REQUIREMENTS OF
C THE SUBROUTINE 'DESIGN' CALLED BY THE SUBROUTINE 'IDEFIR'.
C
      DIMENSION H0(256), RH0(128)
C
C SET UP MACHINE CONSTANTS.
C
C IRCHN...CHANNEL ASSIGNED TO INPUT DEVICE;
C IWCHN...CHANNEL ASSIGNED TO OUTPUT DEVICE;
C
      IRCHN = I1MACH(1)
      IWCHN = I1MACH(2)
C
C INPUT OF THE TOLERANCE SCHEME.
C
      READ (IRCHN,9999) DEL1, DEL2, OMGP1, OMGS1, OMGP2, OMGS2
C
C A MAXIMUM OF 20 DESIGN ITERATIONS IS ALLOWED.
C
      ITMAX = 20
C
      DP = DEL1
      OP = OMGP1
      OS = OMGS1
C
      WRITE (IWCHN,9998)
C
      WRITE (IWCHN,9997)
C
      CALL IDEFIR(DEL1, DEL2, OMGP1, OMGS1, OMGP2, OMGS2, ITYP, NF00,
     *    NF, IWCMIN, IWC, IWACT, DEL11, DEL21, EMAX1, EMAX2, H0, RH0,
     *    ITERA, IERR)
C
C OUTPUT OF FILTER TYPE AND DESIGN PARAMETERS.
C
      IF (ITYP.EQ.0) WRITE (IWCHN,9995) DEL1, DEL2, OP, OS
      IF (ITYP.EQ.1) WRITE (IWCHN,9994) DEL1, DEL2, OP, OS
      IF (ITYP.EQ.2) WRITE (IWCHN,9993) DEL1, DEL2, OP, OS, OMGP2, OMGS2
      IF (ITYP.EQ.3) WRITE (IWCHN,9992) DEL1, DEL2, OP, OS, OMGP2, OMGS2
      IF (ITYP.EQ.4) WRITE (IWCHN,9991) DP, OP
C
C TEST FOR POSSIBLE ERROR RETURNS.
C
      GO TO (40, 10, 20, 30), IERR
C
C OUTPUT FOR THE CASE OF INCONSISTENT SPECIFICATIONS.
C
   10 WRITE (IWCHN,9981)
      STOP
C
C OUTPUT, IF THE MAXIMUM ALLOWED FILTER ORDER IS EXCEEDED.
C
   20 WRITE (IWCHN,9980)
      STOP
C
C OUTPUT,IF NO SOLUTION WAS FOUND.
C
   30 WRITE (IWCHN,9982) ITERA
   40 CONTINUE
C
C OUTPUT OF REALIZATION PARAMETERS.
C
      IF (ITYP.NE.0) WRITE (IWCHN,9990) DEL1, DEL2, OMGP1, OMGS1
      WRITE (IWCHN,9989) IWCMIN, NF00
      WRITE (IWCHN,9996) IWC, IWACT, NF
C
C OUTPUT OF THE REDUCED TOLERANCE SCHEME.
C
      WRITE (IWCHN,9988) DEL11, DEL21
C
C OUTPUT OF EXACT AND ROUNDED COEFFICIENTS.
C
      N1 = (NF+1)/2
      WRITE (IWCHN,9985)
      WRITE (IWCHN,9987) (H0(I),I=1,N1)
      WRITE (IWCHN,9984)
      WRITE (IWCHN,9987) (RH0(I),I=1,N1)
C
C OUTPUT OF ACTUAL LOWPASS TOLERANCES.
C
      WRITE (IWCHN,9986) EMAX1, EMAX2
      WRITE (IWCHN,9983) ITERA
C
      STOP
C
 9999 FORMAT (6F8.6)
C
 9998 FORMAT (1H1)
 9997 FORMAT (54H ITERATIVE DESIGN OF LINEAR PHASE FIR FILTERS IN DIREC,
     *    6HT FORM/36H WITH MINIMUM COEFFICIENT WORDLENGTH//)
 9996 FORMAT (41H COEFFICIENT WORDLENGTH APPLIED      WC = , I2/6H ACTUA,
     *    48HL WORDLENGTH WITHOUT LEADING ZEROS     WC ACT = ,
     *    I2/31H ACTUAL FILTER LENGTH      NF = , I3/)
 9995 FORMAT (24H LOWPASS WITH DELTA 1 = , F8.6, 11H,DELTA 2 = ,
     *    F8.6/14X, 10HOMEGA P = , F8.6, 11H,OMEGA S = , F8.6/)
 9994 FORMAT (25H HIGHPASS WITH DELTA 1 = , F8.6, 11H,DELTA 2 = ,
     *    F8.6/15X, 10HOMEGA P = , F8.6, 11H,OMEGA S = , F8.6/)
 9993 FORMAT (25H BANDPASS WITH DELTA 1 = , F8.6, 11H,DELTA 2 = ,
```

```
     *    F8.6/15X, 10HOMEGAP1 = , F8.6, 11H,OMEGAS1 = , F8.6/15X,
     *    10HOMEGAP2 = , F8.6, 11H,OMEGAS2 = , F8.6/)
 9992 FORMAT (25H BANDSTOP WITH DELTA 1 = , F8.6, 11H,DELTA 2 = ,
     *    F8.6/15X, 10HOMEGAP1 = , F8.6, 11H,OMEGAS1 = , F8.6/15X,
     *    10HOMEGAP2 = , F8.6, 11H,OMEGAS2 = , F8.6/)
 9991 FORMAT (34H HILBERT-TRANSFORMER WITH DELTA = , F8.6, 9H,OMEGAP =,
     *    1H , F8.6/)
 9990 FORMAT (35H EQUIVALENT LOWPASS WITH DELTA 1 = , F8.6, 9H,DELTA 2 ,
     *    2H= , F8.6/25X, 10HOMEGA P = , F8.6, 11H,OMEGA S = , F8.6/)
 9989 FORMAT (31H EST. MIN. WORDLENGTH WC MIN = , I2,      9H BITS (SI,
     *    30HGN+BITS BEHIND THE DUAL POINT)/  22H ESTIMATED LOWPASS LEN,
     *    9HGTH NF = , I3/)
 9988 FORMAT (54H REDUCED LOWPASS TOLERANCE SCHEME      DELTA 1 PRIME = ,
     *    F8.6/ 54H                                       DELTA 2 PRIME =
     *    , F8.6/)
 9987 FORMAT (4(F9.6, 5X))
 9986 FORMAT (1H /50H ACTUAL TOLERANCES OF THE EQUIVALENT LOWPASS WITH ,
     *    21HROUNDED COEFFICIENTS /11H DELTA 1 = , F8.6, 11H,DELTA 2 = ,
     *    F8.6/)
 9985 FORMAT (54H EXACT FILTER COEFFICIENTS H0 (I), I = 0...((NF-1)/2)
     *    /)
 9984 FORMAT (1H /37H ROUNDED FILTER COEFFICIENTS H0 (I)  /)
 9983 FORMAT (31H NUMBER OF DESIGN ITERATIONS = , I3)
 9982 FORMAT (34H NO SOLUTION FOUND WITHIN ALLOWED , I2, 11H DESIGN ITE,
     *    8HRATIONS-/47H IN THE LAST TRIAL, THE FOLLOWING PARAMETERS WE,
     *    9HRE USED /)
 9981 FORMAT (1H ///48H INCONSISTENT SPECIFICATIONS,CHECK PARAMETER ORD,
     *    6HERING )
 9980 FORMAT (1H ///48H FILTER REQUIREMENTS EXCEED MAX. ALLOWED FILTER ,
     *    6HORDER )
      END
C
C-----------------------------------------------------------------------
C SUBROUTINE: I D E F I R
C
C     DESIGN OF LINEAR-PHASE FIR FILTERS IN DIRECT FORM WITH MINIMUM
C     COEFFICIENT WORDLENGTH. POSSIBLE FILTER TYPES INCLUDE LOWPASS
C     HIGHPASS,SYMMETRICAL BANDPASS AND BANDSTOP FILTERS (WITH CENTER
C     FREQUENCY AT OMEGA = 0.25, WHERE OMEGA = 0.5 IS HALF THE SAMPLING
C     FREQUENCY), AND HILBERT TRANSFORMERS.
C     THE DESIGN IS CARRIED OUT IN THE FOLLOWING WAY - -
C     PROBLEM TRANSFORMATION TO AN EQUIVALENT LOWPASS PROBLEM;
C     FILTER LENGTH,WORDLENGTH, AND ROUNDING ERROR ESTIMATION;
C     LOWPASS DESIGN WITH APPROPRIATELY REDUCED TOLERANCES;
C     COEFFICIENT ROUNDING, TOLERANCE CONTROL; ITERATION IF NECESSARY;
C     COEFFICIENT TRANSFORMATION BACK TO THE DESIRED FILTER TYPE.
C
C  I N P U T
C     I N P U T   P A R A M E T E R S = TOLERANCE SCHEME WITH
C     DEL1...PASSBAND TOLERANCE DELTA 1;
C     DEL2...STOPBAND TOLERANCE DELTA 2;
C     OMGP1..FIRST PASSBAND EDGE FREQUENCY;
C     OMGS1..FIRST STOPBAND EDGE FREQUENCY;
C     OMGP2..SECOND PASSBAND EDGE FREQUENCY;
C     OMGS2..SECOND STOPBAND EDGE FREQUENCY.
C     NORMALIZED FREQUENCY OMG = F / SAMPLING RATE.
C     IN LOWPASS AND HIGHPASS CASES,THE LAST TWO PARAMETERS ARE ZERO.
C     IN BANDFILTER CASES,THE LAST TWO PARAMETERS HAVE THE VALUES
C                 OMGP(S)2 = 0.5 - OMGP(S)1
C     IN HILBERT CASES, DEL2 = DEL1 IS VALID;OMGP1 IS THE HILBERT BAND
C     EDGE FREQUENCY; THE LAST T H R E E  PARAMETERS ARE ZERO.
C
C  O U T P U T
C     O U T P U T   P A R A M E T E R S
C     DEL1,DEL2,OMGP1,OMGS1...TOLERANCE SCHEME OF THE EQUIVALENT LOWPASS
C     PROBLEM; THE ORIGINAL TOLERANCE SCHEME IS REPLACED
C     ITYP...FILTER TYPE INFORMATION- ITYP =0..LOWPASS;
C                                           1..HIGHPASS;
C                                           2..BANDPASS;
C                                           3..BANDSTOP;
C                                           4..HILBERT-TRANSFORMER;
C     NF00...FILTER LENGTH ACCORDING TO TOLERANCES AND ERROR ESTIMATION;
C     NF.....FILTER LENGTH ACTUALLY APPLIED;
C     IWCMIN...ESTIMATED MIN.COEFF.WORDLENGTH (ACCORDING TO MAX.QUANTI-
C              ZATION STEPSIZE + SIGN BIT);
C     IWC......WORDLENGTH USED FOR THE DESIGN (IN THE SAME SENSE);
C     IWACT....ACTUAL WORDLENGTH (SIGN + NECESSARY BITS BEHIND THE DUAL
C              POINT (WITHOUT LEADING ZEROS) FOR THE LARGEST ABS.VALUE);
C     DEL11,DEL21...REDUCED TOLERANCES FOR THE EQUIVALENT LOWPASS DESIGN
C     EMAX1,2..TOLERANCES (I.E. MAX. ERRORS) OF THE EQUIVALENT LOWPASS
C              WITH ROUNDED COEFFICIENTS;
C     H0.......COEFFICIENTS BEFORE ROUNDING;
C     RH0......ROUNDED COEFFICIENTS;
C     ITERA....NUMBER OF DESIGN ITERATIONS;
C     IERR.....ERROR PARAMETER -
C              IERR= 1...NO ERROR;
C                    2...INCONSISTENT SPECIFICATIONS;
C                    3...MAX. ALLOWED FILTER ORDER EXCEEDED;
C                    4...MAX. ALLOWED NUMBER OF ITERATIONS EXCEEDED
C-----------------------------------------------------------------------
C
      SUBROUTINE IDEFIR(DEL1, DEL2, OMGP1, OMGS1, OMGP2, OMGS2, ITYP,
     *    NF00, NF, IWCMIN, IWC, IWACT, DEL11, DEL21, EMAX1, EMAX2, H0,
     *    RH0, ITERA, IERR)
C
C COMMON-DEFINITIONS HAVE TO BE INSERTED HERE ACCORDING TO THE SUB-
C ROUTINE 'DESIGN' CALLED BY THIS SUBROUTINE.
C
      COMMON /CALC/ X(131), AF(2096), AD(131), Y(131), SCALE
      COMMON /CHEB/ N, NGRID, KGRID, KFP, KFS, IFR(131)
      COMMON /CHEK/ NL, NG, KA, KB
C
      DIMENSION H0(256), RH0(128)
      IERR = 1
      SMALL = 2.*R1MACH(3)
C
C THE FOLLOWING PARAMETERS CONCERN THE MAX. ALLOWED NUMBER OF DESIGN
C ITERATIONS (ITMAX), THE MINIMUM USAGE OF THE TOLERANCES (PROZ),THE
C REDUCTION (CR) AND INCREASE (CI) FACTORS FOR THE TOLERANCE CHANGE.
C
      ITMAX = 20
      PROZ = 0.9
      CR = 0.75
      CI = 0.25
C
C TRANSFORMATION TO LOWPASS PROBLEM.
C
      CALL TYPTRA(DEL1, DEL2, OMGP1, OMGS1, OMGP2, OMGS2, ITYP, IERR)
C
C IERR = 2 IN THE CASE OF INCONSISTENT SPECIFICATIONS.
C
      IF (IERR.EQ.2) RETURN
C
      DELOMG = OMGS1 - OMGP1
C
C CHECKING THE FILTER SYMMETRY.
C
      OM = ABS(OMGP1+OMGS1-.5)
      ISYM = 0
      IF ((DEL1.EQ.DEL2) .AND. (OM.LT.SMALL)) ISYM = 1
```

```
      ITERA = 0
      ISOLUT = 0
C
C DETERMINATION OF MINIMUM COEFFICIENT WORDLENGTH.
C
      CALL WCMIN(DEL1, DEL2, DELOMG, IWCMIN)
      IWC = IWCMIN
C
C O U T E R   D E S I G N   I T E R A T I O N   L O O P
C
   10 CONTINUE
      DEL11 = DEL1
      DEL21 = DEL2
C
C I N N E R   D E S I G N   I T E R A T I O N   L O O P
C
   20 CONTINUE
C
C STOP AFTER MAXIMUM NUMBER OF ITERATIONS.
C
      IF (ITERA.LT.ITMAX) GO TO 30
      IERR = 4
      GO TO 80
   30 CONTINUE
C
C DETERMINATION OF FILTER LENGTH.
C
      CALL LENGTH(DEL11, DEL21, DELOMG, NF)
      NN = (NF/2)*2
      IF (NN.EQ.NF) NF = NF + 1
C
C IERR = 3 , IF THE MAX. ALLOWED FILTER ORDER IS EXCEEDED.
C
      IF (NF.LE.128) GO TO 40
      IF ((ITYP.NE.2) .AND. (ITYP.NE.3) .AND. (NF.LE.256)) GO TO 40
      IERR = 3
      RETURN
   40 CONTINUE
C
C EXPECTED EFFICIENT ROUNDING ERROR.
C
      IF (ISYM.EQ.1) NF = ((NF+2)/4)*2
      CALL EFFERR(DEL11, NF, IWC, S1)
      CALL EFFERR(DEL21, NF, IWC, S2)
      IF (ISYM.EQ.1) NF = 2*NF - 1
C
C COMPARISON OF ESTIMATED TOTAL ERRORS WITH TOLERANCES.
C
      F1 = DEL11 + S1
      DD1 = F1 - DEL1
      F2 = DEL21 + S2
      DD2 = F2 - DEL2
      IF ((DD1.GT.SMALL).OR.(DD2.GT.SMALL)) GO TO 70
      ITERA = ITERA + 1
      IF (ITERA.EQ.1) NF0 = NF
      IF (ITERA.EQ.1) NF00 = NF0
C
C EQUIVALENT LOWPASS DESIGN FOR REDUCED TOLERANCES.
C
      N1 = (NF+1)/2
      N = N1
      CALL DESIGN(DEL11, DEL21, OMGP1, OMGS1, H0, NF)
C
C COEFFICIENT ROUNDING TO (IWC-1) BITS BEHIND THE DUAL POINT.
C
      Q = 2.**(1-IWC)
      DO 50 I=1,N1
        HR = H0(I)/Q
        RH0(I) = Q*AINT(HR)
   50 CONTINUE
C
C DETERMINATION OF REAL MAX. ERRORS.
C
      CALL MAXERR(RH0, N1, NF, 0., OMGP1, EMAX1)
      CALL MAXERR(RH0, N1, NF, OMGS1, .5, EMAX2)
C
C COMPARISON WITH TOLERANCES.
C
      DD1 = EMAX1 - DEL1
      DD2 = EMAX2 - DEL2
      IF ((DD1.GT.SMALL).OR.(DD2.GT.SMALL)) GO TO 60
      ISOLUT = ISOLUT + 1
C
C CHECKING FOR SUFFICIENT USAGE OF TOLERANCES.
C
      IF ((DD1/DEL1).GE.(PROZ-1.)) DD1 = 0.
      IF ((DD2/DEL2).GE.(PROZ-1.)) DD2 = 0.
      IF ((DD1.EQ.0.) .AND. (DD2.EQ.0.)) GO TO 80
      IF (NF.LE.NF0) GO TO 80
      IF (ISOLUT.EQ.1) NF0 = NF
C
C TOLERANCE CHANGE.
C
   60 CONTINUE
      IF (DD1.LE.0.) DEL11 = DEL11 - CI*DD1
      IF (DD2.LE.0.) DEL21 = DEL21 - CI*DD2
   70 CONTINUE
      IF (DD1.GT.0.) DEL11 = DEL11 - CR*DD1
      IF (DD2.GT.0.) DEL21 = DEL21 - CR*DD2
      IF (ISYM.EQ.1) DEL11 = AMIN1(DEL11,DEL21)
      IF (ISYM.EQ.1) DEL21 = DEL11
C
C CHECK FOR REALIZABILITY.
C
      IF ((DEL11.GT.0.) .AND. (DEL21.GT.0.)) GO TO 20
C
C E N D   O F   I N N E R   D E S I G N   I T E R A T I O N   L O O P .
C
C
C INCREASE OF WORDLENGTH IN CASE OF UNREALIZABILITY.
C
      IWC = IWC + 1
      GO TO 10
C
C E N D   O F   O U T E R   D E S I G N   I T E R A T I O N   L O O P .
C
C
C TRANSFORMATION TO DESIRED FILTER TYPE.
C
   80 CONTINUE
      IF (ITYP.EQ.0) GO TO 90
      CALL BACTRA(ITYP, H0, RH0, NF)
      N1 = (NF+1)/2
   90 CONTINUE
C
C DETERMINATION OF ACTUAL COEFFICIENT WORDLENGTH.
C
      CALL WCACT(RH0, N1, IWC, IWACT)
      RETURN
      END
```

```
C
C-----------------------------------------------------------------------
C SUBROUTINE: TYPTRA
C TRANSFORMATION TO EQUIVALENT LOWPASS;SPECIFICATIONS TEST.
C-----------------------------------------------------------------------
C
      SUBROUTINE TYPTRA(DEL1, DEL2, OMGP1, OMGS1, OMGP2, OMGS2, ITYP,
     *    IERR)
C
C     TRANSFORMATION TO EQUIVALENT LOWPASS PROBLEM.
C     IDENTIFICATION OF FILTER TYPES- ITYP = 0..LOWPASS)
C                                     ITYP = 1..HIGHPASS)
C                                     ITYP = 2..BANDPASS)
C                                     ITYP = 3..BANDSTOP)
C                                     ITYP = 4..HILBERT-TRANSFORMER
C
C L O W P A S S
C
      IF ((OMGP2.NE.0.) .OR. (OMGS2.NE.0.)) GO TO 20
      IF (OMGP1.GT.OMGS1) GO TO 10
      ITYP = 0
      GO TO 60
C
C H I G H P A S S
C
   10 CONTINUE
      IF (OMGS1.EQ.0.) GO TO 40
      ITYP = 1
      OMGP1 = .5 - OMGP1
      OMGS1 = .5 - OMGS1
      GO TO 60
   20 CONTINUE
C
C CHECKING FOR INCONSISTENT SPECIFICATIONS.
C SET UP A MACHINE CONSTANT.
C
      EPSMAC = 2.*R1MACH(3)
      DOP = ABS(OMGP1+OMGP2-.5)
      DOS = ABS(OMGS1+OMGS2-.5)
      DO = AMAX1(DOP,DOS)
      IF (DO.GE.EPSMAC) GO TO 50
C
C B A N D P A S S
C
      IF (OMGP1.LT.OMGS1) GO TO 30
      ITYP = 2
      OMGP1 = .5 - 2.*OMGP1
      OMGS1 = .5 - 2.*OMGS1
      GO TO 60
C
C B A N D S T O P
C
   30 CONTINUE
      ITYP = 3
      OMGP1 = 2.*OMGP1
      OMGS1 = 2.*OMGS1
      GO TO 60
C
C H I L B E R T - T R A N S F O R M E R
C
   40 CONTINUE
C
C CHECKING FOR INCONSISTENT SPECIFICATIONS.
C
      IF ((DEL1.NE.DEL2) .AND. (DEL2.NE.0.)) GO TO 50
      ITYP = 4
      OMGS1 = .25 + OMGP1
      QMGP1 = .25 - OMGP1
      DEL1 = .5*DEL1
      DEL2 = DEL1
      GO TO 60
C
C ERROR - INCONSISTENT SPECIFICATIONS.
C
   50 IERR = 2
C
   60 RETURN
      END
C
C-----------------------------------------------------------------------
C SUBROUTINE: BACTRA
C TRANSFORMATION OF THE DESIGNED LOWPASS TO THE DESIRED FILTER TYPE.
C-----------------------------------------------------------------------
C
      SUBROUTINE BACTRA(ITYP, H0, RH0, NF)
C
      DIMENSION H0(256), RH0(128)
C
      N1 = (NF+1)/2
      IF (ITYP-3) 10, 30, 60
   10 CONTINUE
C
C H I G H P A S S  AND B A N D P A S S
C
      N = N1 - 1
      DO 20 I=1,N,2
        J = N1 - I
        H0(J) = -H0(J)
        RH0(J) = -RH0(J)
   20 CONTINUE
      IF (ITYP.EQ.1) GO TO 80
   30 CONTINUE
C
C B A N D P A S S   AND   B A N D S T O P
C
      N11 = N1
      N1 = NF
      NF = 2*NF - 1
      NN = 2*N11
      I = N1
      J = N11
      IF (NN-N1) 40, 40, 50
   40 CONTINUE
C
C LOWPASS FILTER LENGTH EVEN.
C
      H0(I) = 0.
      RH0(I) = 0.
      I = I - 1
   50 CONTINUE
C
C LOWPASS FILTER LENGTH EVEN OR ODD.
C
      H0(I) = H0(J)
      H0(I-1) = 0.
      RH0(I) = RH0(J)
      RH0(I-1) = 0.
      I = I - 2
      J = J - 1
      IF (J-1) 80, 80, 50
```

```
   60 CONTINUE
C
C H I L B E R T - T R A N S F O R M E R
C
      I = N1
      H0(I) = 0.
      RH0(I) = 0.
      V = -2.
      I = I - 1
   70 CONTINUE
      V = -V
      H0(I) = V*H0(I)
      RH0(I) = V*RH0(I)
      I = I - 2
      IF (I) 80, 80, 70
   80 RETURN
      END
C
C-----------------------------------------------------------------------
C SUBROUTINE: WCMIN
C MINIMUM COEFFICIENT WORDLENGTH ESTIMATION BY AN EMPIRICAL FORMULA.
C-----------------------------------------------------------------------
C
      SUBROUTINE WCMIN(D1, D2, DOMG, IW)
C
      ALD = ALOG10(D1)
      ALS = ALOG10(D2)
      ALO = ALOG10(DOMG*360.)
      W = 7.22116 - 0.589268*ALD
      W = W + (0.112354*ALD-2.328539)*ALO
      W = W - 3.559051*ALS
      IW = IFIX(W)
      RETURN
      END
C
C-----------------------------------------------------------------------
C SUBROUTINE: LENGTH
C FILTER LENGTH ESTIMATION BY AN EMPIRICAL FORMULA.
C-----------------------------------------------------------------------
C
      SUBROUTINE LENGTH(D1, D2, DOMG, NF)
C
      DL1 = ALOG10(D1)
      DL2 = ALOG10(D2)
      DQ1 = DL1*DL1
      D = .005309*DQ1 + .07114*DL1 - .4761
      D = D*DL2
      D = D - .00266*DQ1 - .5941*DL1 - .4278
      D = 2. + D/DOMG
      NF = IFIX(D)
      RETURN
      END
C
C-----------------------------------------------------------------------
C SUBROUTINE: EFFERR
C ITERATIVE CALCULATION OF THE APPROXIMATE STATISTICAL ERROR BOUND S
C FOR A GIVEN TOLERANCE D,FILTER LENGTH NF,COEFF.WORDLENGTH IW.
C-----------------------------------------------------------------------
C
      SUBROUTINE EFFERR(D, NF, IW, S)
C
      QH = 2.**(-IW)
      CALL SSTAT(NF, 0., SQ)
      ALPHA = D/(SQ*QH)
   10 CONTINUE
      CALL SSTAT(NF, ALPHA, SQ)
      ALPHA1 = D/(SQ*QH)
      DALPHA = ALPHA1 - ALPHA
      IF (DALPHA-.05) 30, 30, 20
   20 CONTINUE
      ALPHA = ALPHA1
      GO TO 10
   30 CONTINUE
      S = QH*SQ
      RETURN
      END
C
C-----------------------------------------------------------------------
C SUBROUTINE: SSTAT
C EVALUATION OF THE APPROXIMATE STATISTICAL ERROR FORMULA.
C-----------------------------------------------------------------------
C
      SUBROUTINE SSTAT(NF, ALPHA, SQ)
C
      PI = 4.*ATAN(1.0)
      PIQ = PI*PI
C
C WEIGHTING FUNCTION M(ALPHA).
C
      AM = 1./(1.+SQRT(ALPHA))
C
C ERROR BOUND.
C
      ANZ = FLOAT(NF-1)
      SQ = (ANZ/PI+.5)*AM
      X = (4.-24.*AM/PIQ)*ANZ
      X = X + 4. - 3.*AM
      X = SQRT(X*AM/3.)
      SQ = SQ + X
      RETURN
      END
C
C-----------------------------------------------------------------------
C SUBROUTINE: MAXERR
C SEARCH FOR THE MAX.ABS.ERROR OF A LINEAR-PHASE FIR FILTER.
C-----------------------------------------------------------------------
C
      SUBROUTINE MAXERR(RH0, N1, NF, OMGA, OMGE, EMAX)
C
      DIMENSION RH0(128)
      EMAX = 0.
      OMG = OMGA
      IA = 1
      ISTG = 1
      DOM1 = 1./(4.*FLOAT(NF))
      DOM2 = DOM1/2.
      DOM3 = DOM2/2.
      DFR = .05
      DFRH = .025
      CALL ERROR(RH0, N1, NF, OMG, E)
      BF1 = ABS(E)
      IF (BF1.GT.EMAX) EMAX = BF1
      DELOMG = DOM3
      OMG = OMG + DELOMG
   10 CONTINUE
      BF2 = BF1
   20 CONTINUE
      CALL ERROR(RH0, N1, NF, OMG, E)
      BF1 = ABS(E)
      IF (BF1.GT.EMAX) EMAX = BF1
```

```
      IF (IA) 30, 110, 30
   30 CONTINUE
      DELFR = 2.*(BF1-BF2)/(BF1+BF2)
      IF (ISTG) 90, 90, 40
   40 CONTINUE
      IF (DELFR) 80, 80, 50
   50 CONTINUE
      ISTG = 1
      DELOMG = DOM2
      IF (DELFR.LT.DFRH) DELOMG = DOM3
      IF (DELFR.GT.DFR) DELOMG = DOM1
      OMG = OMG + DELOMG
   60 CONTINUE
      IF (OMG-OMGE) 10, 70, 70
   70 CONTINUE
      OMG = OMGE
      IA = 0
      GO TO 10
   80 CONTINUE
      ISTG = -1
      IF (DELOMG.EQ.DOM1) DELOMG = -3.*DOM3
      IF (DELOMG.EQ.DOM2) DELOMG = -DOM3
      IF (DELOMG.EQ.DOM3) DELOMG = DOM1
      OMG = OMG + DELOMG
      IF (DELOMG) 20, 60, 60
   90 CONTINUE
      IF (DELFR) 100, 100, 50
  100 CONTINUE
      DELOMG = DOM1
      OMG = OMG + DELOMG
      GO TO 60
  110 RETURN
      END
C
C-----------------------------------------------------------------------
C SUBROUTINE: ERROR
C ERROR E OF A LINEAR-PHASE FIR LOWPASS TRANSFER FUNCTION WITH
C RESPECT TO THE IDEAL LOWPASS.
C-----------------------------------------------------------------------
C
      SUBROUTINE ERROR(RH0, N1, NF, OMG, E)
C
      DIMENSION RH0(128)
C
      PI = 4.*ATAN(1.0)
      PI2 = PI*2.
      NN = 2*N1
      IF (NF-NN) 10, 30, 30
   10 CONTINUE
C
C LENGTH NF ODD.
C
      H = RH0(N1)
      N = N1 - 1
      DO 20 I=1,N
        ANNU = FLOAT(N1-I)
        H = H + 2.*RH0(I)*COS(ANNU*OMG*PI2)
   20 CONTINUE
      GO TO 50
   30 CONTINUE
C
C LENGTH NF EVEN.
C
      H = 0.
      DO 40 I=1,N1
        ANNU = FLOAT(N1-I) + .5
        H = H + RH0(I)*COS(ANNU*OMG*PI2)
   40 CONTINUE
      H = H*2.
   50 CONTINUE
C
C ERROR VALUE E.
C
      E = H
      E1 = H - 1.
      EA = ABS(E)
      EA1 = ABS(E1)
      IF (EA1.LT.EA) E = E1
      RETURN
      END
C
C-----------------------------------------------------------------------
C SUBROUTINE: WCACT
C DETERMINATION OF ACTUAL WORDLENGTH.
C-----------------------------------------------------------------------
C
      SUBROUTINE WCACT(RH0, N1, IWC, IWACT)
C
      DIMENSION RH0(128)
      HMAX = ABS(RH0(1))
      DO 10 I=2,N1
        H = ABS(RH0(I))
        HMAX = AMAX1(HMAX,H)
   10 CONTINUE
      L = -2
   20 CONTINUE
      L = L + 1
      IF (HMAX-1.) 30, 40, 40
   30 CONTINUE
      HMAX = HMAX*2.
      GO TO 20
   40 CONTINUE
      IWACT = IWC - L
      RETURN
      END
C
C-----------------------------------------------------------------------
C SUBROUTINE: DESIGN
C AUXILIARY ROUTINE CALLING THE ODD-LENGTH LOWPASS DESIGN ROUTINE
C 'TOMSB2'.
C-----------------------------------------------------------------------
C
      SUBROUTINE DESIGN(DEL11, DEL21, OMGP1, OMGS1, H0, NF)
C
      COMMON /CALC/ X(131), AF(2096), AD(131), Y(131), SCALE
      COMMON /CHEB/ N, NGRID, KGRID, KFP, KFS, IFR(131)
      COMMON /CHEK/ NL, NG, KA, KB
      DIMENSION H0(256)
C
C AUXILIARY VARIABLES FOR 'TOMSB2'.
C
      FP = OMGP1
      FS = OMGS1
      CALL TOMSB2(DEL11, DEL21, FP, FS, H0, NF)
      RETURN
      END
C
C-----------------------------------------------------------------------
C SUBROUTINE: TOMSB2
C     SUBROUTINE ACCORDING TO "PROGRAM TOM" WITH  THE
```

```
C     EXCHANGE ALGORITHM FOR DESIGNING LOW PASS FINITE LENGTH IMPULSE
C     RESPONSE DIGITAL FILTERS WITH LINEAR PHASE.
C
C     J.H. MCCLELLAN AND T.W. PARKS---RICE UNIVERSITY---OCTOBER 26, 1971
C
C     THE FILTER LENGTH (NF), THE PASSBAND CUTOFF FREQUENCY (FP),
C     THE STOPBAND FREQUENCY (FS), AND THE WEIGHTING FACTOR (AA) ARE
C     INPUTS TO THE ALGORITHM.  THEN THE STOPBAND DEVIATION (DS) IS
C     MINIMIZED AND THUS THE PASSBAND DEVIATION (DP=AA*DS) IS ALSO
C     MINIMIZED. THE PROGRAM SEKS TO DETERMINE THE BEST LOCATION OF
C     THE PEAKS OF THE ERROR CURVE IN ORDER TO MINIMIZE THE DEVIATIONS.
C
C     REFERENCE--PARKS,T.W. AND J.H. MCCLELLAN,*CHEBYCHEV APPROXIMATION
C     OF NON-RECURSIVE DIGITAL FILTERS WITH LINEAR PHASE*, IEEE TRANS.
C     ON CIRCUIT THEORY, VOL. CT-19, MARCH, 1972.
C
C     SPECIAL PARAMETERS FOR CHANGE OF FS, IF STOPBAND RIPPLE IS
C     SPECIFIED
C     D1D   =  STOPBAND RIPPLE D2
C     STEP  =  INITIAL STEPSIZE FOR CHANGING FS& STEP DECREASES WITH
C              EXP.& IF SOLUTION  FAR FROM OPTIMUM TRY 0.02
C     XAC   =  ACCURACY OF D2
C              1000. MEANS ACCURACY TO 0.1 PERCENT
C              100.  MEANS ACCURACY TO 1 PERCENT ETC.
C     IF XAC = D1D  NO ACTION IS TAKEN AND FS REMAINS UNCHANGED
C
C        DIMENSION OF X, AD, Y, IFR, FOPT = N*/2+3 WHERE N* IS
C        MAXIMUM LENGTH OF N
C        DIMENSION OF AF = 16(N*/2+3)+1
C-----------------------------------------------------------------------
C
      SUBROUTINE TOMSB2(DEL1, DEL2, FP, FS, H0, NF)
C
      COMMON /CALC/ X(131), AF(2096), AD(131), Y(131), SCALE
      COMMON /CHEB/ N, NGRID, KGRID, KFP, KFS, IFR(131)
      COMMON /CHEK/ NL, NG, KA, KB
      DIMENSION H0(256)
C
      IWCHN = I1MACH(2)
      DO 5 I=1,131
        X(I) = 0.
    5 CONTINUE
C
C INITIALIZATION OF THE GRID OF 16N POINTS.  THE CONSTANT LGRID
C CAN BE CHANGED IF A DIFFERENT GRID DENSITY IS DESIRED.
C
      LGRID = 16
      XAC = DEL2
      AA = DEL1/DEL2
      D1D = DEL2
      STEP = 0.02
      ICH = 1
      KITER = 0
      D1C = D1D/XAC
   10 CONTINUE
      PI2 = 8.0*ATAN(1.0)
      KCOUNT = 0
      TW = FS - FP
      NGRID = LGRID*(N-1)
      NGRID = NGRID + LGRID/2
      KGRID = 2
      DELF = 0.5/FLOAT(NGRID)
      AF(1) = 0.0
      DO 20 J=1,NGRID
        AF(J+1) = FLOAT(J)*DELF
   20 CONTINUE
      KFP = FP/DELF - 1.
      KFS = FS/DELF
   30 CONTINUE
      IF (FP.LT.AF(KFP+1)) GO TO 40
      KFP = KFP + 1
      GO TO 30
   40 CONTINUE
      IF (FS.LE.AF(KFS)) GO TO 50
      KFS = KFS + 1
      GO TO 40
   50 CONTINUE
      AF(KFP) = FP
      AF(KFS) = FS
      NGRID = NGRID + 1
      DO 60 J=1,NGRID
        AF(J) = COS(PI2*AF(J))
   60 CONTINUE
C
C INITIAL GUESS FOR THE N+1 OPTIMAL FREQUENCIES
C
      KP = (KFP-1)/2
      KP = 2*KP + 1
      KS = KFS/2
      KS = 2*KS + 1
      XPP = KP - 1
      XSS = NGRID - KS
      XAA = XPP + XSS
      PT = XAA/FLOAT(N-1)
      IF (PT.GT.3.) GO TO 70
      WRITE (IWCHN,9999)
 9999 FORMAT (20H ERROR IN INPUT DATA)
      GO TO 560
   70 KR = XPP/PT
      IF (KR.EQ.0) KR = KR + 1
      PT = XPP/FLOAT(KR)
      IFR(1) = 1
      DO 80 J=1,KR
        XTW = FLOAT(J)*PT
        NTW = IFIX(XTW) + 1
        NTW = NTW/2
        IFR(J+1) = 2*NTW + 1
   80 CONTINUE
      KR = KR + 1
      IFR(KR) = KP
      IFR(KR+1) = KS
      NTW = N - KR
      IF (NTW.GE.0) GO TO 90
      WRITE (IWCHN,9999)
      GO TO 560
   90 IF (NTW.EQ.0) GO TO 110
      PT = XSS/FLOAT(NTW)
      DO 100 J=1,NTW
        XTW = FLOAT(J)*PT + FLOAT(KS) - 1.0
        NB = XTW
        NB = NB/2
        K111 = KR + 1 + J
        IFR(K111) = 2*NB - 1
  100 CONTINUE
  110 CONTINUE
C
C MAIN CALCULATION ROUTINE
C CALCULATE RHO ON THE OPTIMAL SET
C THE FIRST 4 PASSES ARE MADE ON A GRID OF 8N POINTS THEN THE
C FINER GRID OF 16N POINTS IS USED.
```

```
C
 120  CONTINUE
      N111 = N + 1
      DO 130 J=1,N111
        J111 = IFR(J)
        X(J) = AF(J111)
 130  CONTINUE
      M = (N-1)/15 + 1
      M = (N-1)/20 + 1
      N111 = N + 1
      DO 140 J=1,N111
        AD(J) = D(J,N+1,M)
 140  CONTINUE
      R1 = 0.0
      DO 150 J=1,KR
        R1 = R1 + AD(J)
 150  CONTINUE
      R2 = 0.0
      K = 1
      DO 160 J=1,KR
        R2 = R2 + AD(J)*AA*FLOAT(K)
        K = -K
 160  CONTINUE
      K111 = KR + 1
      N111 = N + 1
      DO 170 J=K111,N111
        R2 = R2 + FLOAT(K)*AD(J)
        K = -K
 170  CONTINUE
      RHO = R1/R2
      NU = 1
      IF (RHO.GT.0.0) NU = -1
      RHO = ABS(RHO)
      K = NU
      XRAY = X(KR+1)
      DO 180 J=1,KR
        AD(J) = AD(J)*(XRAY-X(J))
        Y(J) = 1.0 + AA*FLOAT(K)*RHO
        K = -K
 180  CONTINUE
      K = -K
      K111 = KR + 1
      DO 190 J=K111,N
        X(J) = X(J+1)
        AD(J) = AD(J+1)*(XRAY-X(J))
        Y(J) = FLOAT(K)*RHO
        K = -K
 190  CONTINUE
      Y(N+1) = FLOAT(K)*RHO
      NL = 0
      NG = 0
      TEST = GEE(IFR(KR+1))
      RO = ABS(TEST-RHO)/RHO
      IF (RO.LT.0.1 .OR. KCOUNT.LT.4) GO TO 200
      WRITE (IWCHN,9998)
9998  FORMAT (20H INTERPOLATION ERROR)
      GO TO 560
 200  CONTINUE
C
C SEARCH FOR THE NEW EXTREMAL FREQUENCIES
C
      NTOT = 0
      NPT = IFR(1)
      NG = IFR(2)
      KA = 1
      KB = 1
      KSIGN = NU
      XT = Y(1)
      NFIR = IFR(1)
      NOLD = IFR(N+1)
      IF (KR.EQ.1) GO TO 330
      IF (NPT-1) 210, 260, 210
 210  ZT = GEE(NPT-KGRID)
      IF (FLOAT(KSIGN)*(ZT-XT)) 260, 260, 220
 220  KPT = NPT - KGRID
 230  KPT = KPT - KGRID
      XT = ZT
      IF (KPT-1) 250, 240, 240
 240  ZT = GEE(KPT)
      IF (FLOAT(KSIGN)*(ZT-XT)) 250, 250, 230
 250  IFR(1) = KPT + KGRID
      NTOT = NTOT + 1
      KSIGN = -KSIGN
      GO TO 310
 260  ZT = GEE(NPT+KGRID)
      IF (FLOAT(KSIGN)*(ZT-XT)) 300, 300, 270
 270  KPT = NPT
 280  KPT = KPT + KGRID
      XT = ZT
      ZT = GEE(KPT+KGRID)
      IF (FLOAT(KSIGN)*(ZT-XT)) 290, 290, 280
 290  IFR(1) = KPT
      NTOT = NTOT + 1
 300  KSIGN = -KSIGN
 310  XFIR = ABS(1.0-XT)/AA
      KA = 2
 320  IF (KA.GT.(N+1)) GO TO 420
      NL = NPT
      NG = NGRID + 1
      IF (KA.LE.N) NG = IFR(KA+1)
      NPT = IFR(KA)
      XT = Y(KA)*FLOAT(KB)
      IF (KA.EQ.KR) GO TO 330
      IF (KA.EQ.(KR+1)) GO TO 340
      GO TO 350
 330  IF ((KFP-NPT).LT.KGRID) GO TO 410
      NPT = KFP
      NTOT = 1
      GO TO 410
 340  KB = -1
      IF ((NPT-KFS).LT.KGRID) GO TO 410
      NPT = KFS
      NTOT = 1
      GO TO 410
 350  KPT = NPT
      ZT = GEE(KPT-KGRID)
      IF (FLOAT(KSIGN)*(ZT-XT)) 380, 380, 360
 360  XT = ZT
      KPT = KPT - KGRID
      ZT = GEE(KPT-KGRID)
      IF (FLOAT(KSIGN)*(ZT-XT)) 370, 370, 360
 370  IFR(KA) = KPT
      KSIGN = -KSIGN
      NTOT = NTOT + 1
      KA = KA + 1
      GO TO 320
 380  IF ((KPT+KGRID).GT.NGRID) GO TO 410
      ZT = GEE(KPT+KGRID)
      IF (FLOAT(KSIGN)*(ZT-XT)) 410, 410, 390
 390  XT = ZT
```

```
      KPT = KPT + KGRID
      IF ((KGRID+KPT).GT.NGRID) GO TO 400
      ZT = GEE(KPT+KGRID)
      IF (FLOAT(KSIGN)*(ZT-XT)) 400, 400, 390
 400  IFR(KA) = KPT
      KSIGN = -KSIGN
      NTOT = NTOT + 1
      KA = KA + 1
      GO TO 320
 410  IFR(KA) = NPT
      KSIGN = -KSIGN
      KA = KA + 1
      GO TO 320
 420  CONTINUE
      XOLD = ABS(XT)*AA
      NL = 0
      NG = 0
      IF (IFR(1).EQ.1) GO TO 450
      IF (IFR(N+1).EQ.NGRID) GO TO 430
      WRITE (IWCHN,9997)
9997  FORMAT (38H ERROR-NEITHER ENDPOINT IS AN EXTREMUM)
      GO TO 560
 430  IF (NFIR.EQ.1) GO TO 470
      TEST = GEE(1)
      TEST = -FLOAT(NU)*(TEST-1.0) - XOLD
      IF (TEST.LE.0.0) GO TO 470
      NTOT = 1
      DO 440 J=1,N
        N111 = N + 2 - J
        M111 = N + 1 - J
        IFR(N111) = IFR(M111)
 440  CONTINUE
      IFR(1) = 1
      KR = KR + 1
      GO TO 470
 450  IF (IFR(N+1).EQ.NGRID) GO TO 470
      IF (NOLD.EQ.NGRID) GO TO 470
      TEST = GEE(NGRID)
      TEST = FLOAT(KSIGN)*TEST - XFIR
      IF (TEST.LE.0.0) GO TO 470
      NTOT = 1
      DO 460 J=1,N
        IFR(J) = IFR(J+1)
 460  CONTINUE
      IFR(N+1) = NGRID
      KR = KR - 1
 470  CONTINUE
      IF (KR.GT.0) GO TO 480
      WRITE (IWCHN,9996)
9996  FORMAT (6H ERROR)
      GO TO 560
 480  CONTINUE
      KCOUNT = KCOUNT + 1
      IF (KCOUNT.GT.50) GO TO 500
      IF (KCOUNT.EQ.4 .AND. KGRID.EQ.2) GO TO 490
      IF (NTOT.NE.0) GO TO 120
      IF (KGRID.EQ.2) GO TO 490
      GO TO 500
 490  KGRID = 1
      GO TO 120
 500  CONTINUE
      IF (ABS(RHO-D1D).LT.D1C) GO TO 520
      KITER = KITER + 1
      IF (KITER.GT.1000) GO TO 520
      IF (RHO.GT.D1D) GO TO 510
      IF (ICH.EQ.1) STEP = STEP/2.
      FS = FS - STEP
      ICH = -1
      GO TO 10
 510  IF (ICH.EQ.(-1)) STEP = STEP/2.
      FS = FS + STEP
      ICH = 1
      GO TO 10
 520  CONTINUE
      KRP = N - KR
      R1 = AA*RHO
      J = N
 530  CONTINUE
      IF (X(J)) 540, 550, 540
 540  J = J + 1
      GO TO 530
 550  X(J) = -1.
      CALL PREF2(H0, NF)
      GO TO 570
 560  CONTINUE
      STOP
 570  CONTINUE
      RETURN
      END
C
C-----------------------------------------------------------------------
C FUNCTION: D
C THE SUBROUTINE D(,,,) CALCULATES THE LAGRANGIAN INTERPOLATION
C COEFFICIENTS FOR USE IN THE BARYCENTRIC INTERPOLATION FORMULA
C-----------------------------------------------------------------------
C
      FUNCTION D(K, N, M)
C
      COMMON /CALC/ X(131), AF(2096), AD(131), Y(131), SCALE
      D = 1.0
      Q = X(K)
      DO 30 L=1,M
        DO 20 J=L,N,M
          IF (J-K) 10, 20, 10
 10       D = 2.0*D*(X(J)-Q)
 20     CONTINUE
 30   CONTINUE
      D = 1.0/D
      RETURN
      END
C
C-----------------------------------------------------------------------
C FUNCTION: GEE
C THE SUBROUTINE GEE EVALUATES THE FREQUENCY RESPONSE USING THE
C INTERPOLATION COEFFICIENTS WHICH ARE CALCULATED IN THE ROUTINE D.
C-----------------------------------------------------------------------
C
      FUNCTION GEE(K)
C
      COMMON /CALC/ X(131), AF(2096), AD(131), Y(131), SCALE
      COMMON /CHEB/ N, NGRID, KGRID, KFP, KFS, IFR(131)
      COMMON /CHEK/ NL, NG, KA, KB
      IF (K-NL) 10, 40, 10
 10   IF (K-NG) 20, 40, 20
 20   P = 0.0
      XF = AF(K)
      D = 0.0
      DO 30 J=1,N
        C = AD(J)/(XF-X(J))
        D = D + C
```

```
      P = P + C*Y(J)
 30   CONTINUE
      GEE = P/D
      RETURN
 40   K111 = KA + KB
      GEE = Y(K111)
      RETURN
      END
C
C-----------------------------------------------------------------------
C SUBROUTINE: PREF2
C THE SUBROUTINE PREF SAMPLES THE FREQUENCY RESPONSE AT 2**M POINTS
C SO THAT THE FFT (R.C. SINGLETON) CAN BE CALLED TO INVERSE TRANSFORM
C THIS DATA TO OBTAIN THE IMPULSE RESPONSE.
C-----------------------------------------------------------------------
C
      SUBROUTINE PREF2(A, NF)
C
      COMMON /CALC/ X(131), AF(2096), AD(131), Y(131), SCALE
      COMMON /CHEB/ N, NGRID, KGRID, KFP, KFS, IFR(131)
      COMPLEX Z
      DIMENSION A(256), B(256)
      DIMENSION Z(256)
      EQUIVALENCE (B(1),AF(257)), (Z(1),AF(520))
C
C SET UP A MACHINE CONSTANT.
C
      FSH = R1MACH(1)
C
      PI2 = 8.0*ATAN(1.0)
      NX = 2
      MX = 1
 10   IF (NX.GE.NF) GO TO 20
      NX = 2*NX
      MX = MX + 1
      GO TO 10
 20   NN = NX
      MM = MX
      NX = NX/2
      XNN = NN
      DF = 1.0/XNN
      IF (IFR(1).EQ.1) GO TO 30
      P = GOO(1.0)
      GO TO 40
 30   P = Y(1)
 40   A(1) = P
      IF (IFR(N+1).EQ.NGRID) GO TO 50
      P = GOO(-1.0)
      GO TO 60
 50   P = Y(N)
 60   A(NX+1) = P
      L = 1
      N111 = NX - 1
      DO 110 J=1,N111
        AT = DF*FLOAT(J)
        AT = COS(PI2*AT)
 70     AS = X(L)
        IF (AT.GT.AS) GO TO 90
        IF ((AS-AT).LT.FSH) GO TO 80
        L = L + 1
        GO TO 70
 80     A(J+1) = Y(L)
        GO TO 100
 90     IF ((AT-AS).LT.FSH) GO TO 80
        A(J+1) = GOO(AT)
 100    L111 = NN + 1 - J
        A(L111) = A(J+1)
        B(J+1) = 0.0
        IF (L.GT.1) L = L - 1
        M111 = NN + 1 - J
        B(M111) = 0.
 110  CONTINUE
      B(1) = 0.0
      NX1 = NX + 1
      B(NX1) = 0.0
      DO 120 J=1,NN
        Z(J) = CMPLX(A(J),B(J))
 120  CONTINUE
      CALL BITREV(Z, MM)
      CALL FFTFT(Z, MM)
      DO 130 J=1,NX
        B(J) = REAL(Z(J))
 130  CONTINUE
      DO 140 J=1,N
        JJ = N + 1 - J
        A(JJ) = B(J)
 140  CONTINUE
      RETURN
      END
C
C-----------------------------------------------------------------------
C FUNCTION: GOO
C THE SUBROUTINE GOO IS THE SAME AS GEE EXCEPT THAT IT IS CALLED BY
C A FREQUENCY VALUE RATHER THAN BY BY A GRID INDEX VALUE.
C-----------------------------------------------------------------------
C
      FUNCTION GOO(F)
C
      COMMON /CALC/ X(131), AF(2096), AD(131), Y(131), SCALE
      COMMON /CHEB/ N, NGRID, KGRID, KFP, KFS, IFR(131)
      XF = F
      P = 0.0
      D = 0.0
      DO 20 J=1,N
        IF (XF.NE.X(J)) GO TO 10
        GOO = Y(J)
        GO TO 30
 10     C = AD(J)/(XF-X(J))
        D = D + C
        P = P + C*Y(J)
 20   CONTINUE
      GOO = P/D
 30   RETURN
      END
C
C-----------------------------------------------------------------------
C SUBROUTINE: FFTFT
C FREQUENCY-TO-TIME FFT ROUTINE; INPLACE ALGORITHM; COMPLEX DATA IN
C ARRAY A; INPUT DATA ASSUMED TO BE IN BITREVERSED ORDER.
C-----------------------------------------------------------------------
C
      SUBROUTINE FFTFT(A, M)
C
      COMPLEX A, U, W, T
      DIMENSION A(1)
      PI = 4.0*ATAN(1.)
      N = 2**M
      DO 30 L=1,M
        LE = 2**L
        LE1 = LE/2
```

```
      U = (1.0,0.0)
      W = CMPLX(COS(PI/FLOAT(LE1)),SIN(PI/FLOAT(LE1)))
      DO 20 J=1,LE1
        DO 10 I=J,N,LE
          IP = I + LE1
          T = A(IP)*U
          A(IP) = A(I) - T
          A(I) = A(I) + T
10      CONTINUE
        U = U*W
20    CONTINUE
30  CONTINUE
    DO 40 K=1,N
      A(K) = A(K)/FLOAT(N)
40  CONTINUE
    RETURN
    END
C
C-----------------------------------------------------------------------
C SUBROUTINE: BITREV
C REORDERING OF 2**M DATA IN COMPLEX ARRAY A IN BITREVERSED ORDER.
C-----------------------------------------------------------------------
C
    SUBROUTINE BITREV(A, M)
C
    DIMENSION A(1)
    COMPLEX A, T
C
    N = 2**M
    NV2 = N/2
    NM1 = N - 1
    J = 1
    DO 40 I=1,NM1
      IF (I.GE.J) GO TO 10
      T = A(J)
      A(J) = A(I)
      A(I) = T
10    K = NV2
20    IF (K.GE.J) GO TO 30
      J = J - K
      K = K/2
      GO TO 20
30    J = J + K
40  CONTINUE
    RETURN
    END
```

CHAPTER 6

IIR Filter Design and Synthesis

J. F. Kaiser and *K. Steiglitz*

The problems of designing infinite impulse response (IIR) and finite impulse response (FIR) digital filters differ in a number of significant ways. First, in the important special cases of lowpass, highpass, and bandpass magnitude specifications, we have available in the IIR case the classical closed-form designs associated with the names of Butterworth (maximally-flat pass and stop bands), Chebyshev (equiripple pass or stop band), and Cauer (elliptic, equiripple pass and stop bands) originally developed for the continuous case.

The first program in this chapter, DOREDI by G. F. Dehner, is devoted to the design, implementation, and analysis of finite wordlength IIR digital filters based on such closed form designs. This very comprehensive program includes the capability of designing the classic lowpass, highpass, bandpass, and bandstop filters from specifications on cutoff frequencies and passband and stopband tolerances. The program has the facility to determine from the extended degree infinite-precision coefficient design the minimum coefficient wordlength assuming coefficient rounding necessary to meet the filter specifications. There is also the facility to optimize the noise performance of the filter by changing the pairing and ordering of the poles and zeros of the filter being realized as a cascade of second-order sections. These last two powerful features can also be used directly on filter designs input separately. Thus the Dehner program provides a very comprehensive IIR filter design capability.

When the design of an IIR digital filter with arbitrary magnitude or phase characteristics is needed, we encounter another important difference between the IIR and FIR cases. The general IIR design problem is highly nonlinear in the filter coefficients, whereas the FIR problem is linear. This means that general purpose design programs are necessarily more complicated and time consuming than their FIR counterparts. Two such programs are included here: LPIIR by A. Deczky, based on the minimum p-error criterion [1], and OPTIIR by M. T. Dolan and J. F. Kaiser, which uses the penalty function method. LPIIR minimizes the pth power of the error, and so includes the least-square-error case (when $p=2$) and the minimax criterion (when $p \to \infty$). Typically $p=40$ is sufficient for the minimax criterion. LPIIR also provides an option for phase equalization.

Two different general purpose IIR design programs are included here because these optimization problems can be quite difficult numerically. Any one given program may fail to converge on some problem, especially one of high order. Also, we have no mathematical guarantee that the solutions obtained are globally optimal, since the convergence criterion is usually that the gradient of the error be very small in norm - a necessary but not sufficient condition for global optimality in general nonlinear design problems. The reader may therefore wish to try a given problem on both programs, and compare the solutions if they both converge within the allotted execution time. Even in the cases where convergence is obtained and the specifications are satisfied, different computers may give different results. These differences can be attributed to different computer word lengths, different rounding rules, different conversion routines, and different FORTRAN compilers.

The last program, FWIIR by K. Steiglitz and B. D. Ladendorf, deals with the problem of choosing finite word-length coefficients in a cascade implementation of an IIR filter. The program requires the user to provide an "infinite-precision" design in this form. This problem is doubly difficult, since it is not only nonlinear in its parameters, but its parameter space is discrete. No algorithm is known which is practical for problems with as many as 12-16 parameters and which also guarantees optimality, so it is necessary to use a heuristic. FWIIR uses a randomized version of the Hooke-and-Jeeves search algorithm [2,3], which attempts to build a pattern of successful moves on the grid in parameter space. The fact that it is randomized means that different runs may produce different local optima; the user can then select the best.

FWIIR yields results which are very close to those reported by Avenhaus and Schuessler [4], who also use a search on a discrete grid. Recently Brglez [5] presented a method which improves on some published examples by 1 or 2 bits. He does a sequence of discrete local optimizations interspersed with continuous optimizations, thus allowing excursions in parameter space larger than would be obtained by search on a discrete grid only. Brglez's paper appeared too late for inclusion of his program here. The reader is also referred to related work on the IIR problem [6,7,8] and the FIR problem [9,10].

References

1. A. G. Deczky, "Synthesis of Recursive Digital Filters Using the Minimum p-Error Criterion", *IEEE Trans. Audio Electroacoust.*, Vol. AU-20, No. 4, pp. 257-263, Oct. 1972.

2. K. Steiglitz, "Designing Short-Word Recursive Digital Filters", *Proc. Ninth Annual Allerton Conf. Circuit and Syst. Theory*, pp. 778-785, Oct. 6-8, 1971; also in *Digital Signal Processing II*, IEEE Press, New York, 1975.

3. R. Hooke and T. A. Jeeves, "'Direct Search' Solution of Numerical and Statistical Problems", *Jour. Assoc. Comput. Mach.*, Vol. 8, No. 2, pp. 212-229, April 1961.

4. E. Avenhaus and H. W. Schuessler, "On the Approximation Problem in the Design of Digital Filters with Limited Word-Length", *Archiv der Elecktrischen Uebertragung*, Vol. 24, No. 12, pp. 571-572, Dec. 1970.

5. F. Brglez, "Digital Filter Design with Short Word-Length Coefficients", *IEEE Trans. Circuits and Syst.*, Vol. CAS-25, No. 12, pp. 1044-1050, Dec. 1978.

6. M. Suk and S. K. Mitra, "Computer-Aided Design of Digital Filters with Finite Word Lengths", *IEEE Trans. Audio Electroacoust.*, Vol. AU-20, No. 5, pp. 356-363, Dec. 1972.

7. C. Charalambous and M. J, Best, "Optimization of Recursive Digital Filters with Finite Word Lengths", *IEEE Trans. Audio Electroacoust.*, Vol. ASSP-22, No. 6, pp. 424-431, Dec. 1974.

8. J. W. Bandler, B. L. Bardakjian, and J. H. K. Chen, "Design of Recursive Digital Filters with Optimum Word Length Coefficients", *Proc. Eighth Annual Princeton Conf. Inf. Sci. Syst.*, pp. 126-131, March 1974.

9. Y. Chen, S. M. Kang, and T. G. Marshall, "The Optimal Design of CCD Transversal Filters Using Mixed-Integer Programming Techniques", *Proc. 1978 Int. Symp. Circuits and Syst.*, pp. 748-751.

10. D. Kodek, "Design of Optimal Finite Word Length FIR Digital Filters Using Integer Programming Techniques", submitted for publication.

6.1

Program for the Design of Recursive Digital Filters

G. F. Dehner

Institut für Nachrichtentechnik
Univ. Erlangen-Nürnberg, Cauerstrasse 7
D-8520 Erlangen, Germany

1. Purpose and Method

The general purpose program, DOREDI, an acronym for **D**esign and **O**ptimization of **RE**cursive **DI**gital filters, is designed to solve the classic approximation problem for recursive digital filters, to minimize the coefficient wordlength and then to minimize the noise performance by scaling and pairing and ordering the poles and zeros for a number of different structures, and finally to analyze the performance of the resulting filter.

A general view of the design possibilities of DOREDI is given in the block diagram in Fig. 1.

The analysis and noise optimization procedures not only can be applied to filters generated within the design program, but also can be used for the investigation of other filter designs by inputting separately their filter coefficients.

1.1 Design of selective IIR-filters

The filter types that can be handled are the lowpass, highpass, symmetrical bandpass and bandstop filters. The types of approximation available are the maximally-flat magnitude (Butterworth), the Chebyshev types I and II, and the elliptic (Cauer) forms. The method used is the transformation of the tolerance scheme into the continuous domain, the approximation, and then the transformation back to the discrete domain [1], [2].

1.2 Optimization of the coefficient wordlength

The optimization is obtained by choosing an appropriate filter degree and, thus, a certain margin inside the given tolerance scheme (necessary extension of the filter degree, see [3], [4]). This margin, unused for filters with a high coefficient wordlength, is distributed between passband and stopband such that the required wordlength of the rounded coefficients is small (trial and error procedure) [3], [4].

1.3 Optimization of the state variable wordlength

For cascade structures with given second-order blocks, the output noise and consequently the wordlength of the state variables [1] is optimized by pairing poles and zeros and ordering the subsystems; a heuristic method is used called "optimization with limited storage" and is described in [4]. For a system of lower order, this procedure is equivalent to the "dynamic programming" approach [5] and leads to an exact solution for systems including independent white noise sources, only.

In order to reduce storage and computing time only the S_{max} best intermediate results are stored and handled in the consecutive step. For $S_{max} = 1$, the procedure equals the "direct procedure" explained in [4] and [6]. For general noise sources (see Appendix B) the optimization procedure considers the correlations within the second-order blocks, only.

1.4 Analysis of the optimized filter

Input: Coefficients designed by the program or specified independently with input data cards (class 01, 03, 04, and 05, see sect. 2.2).

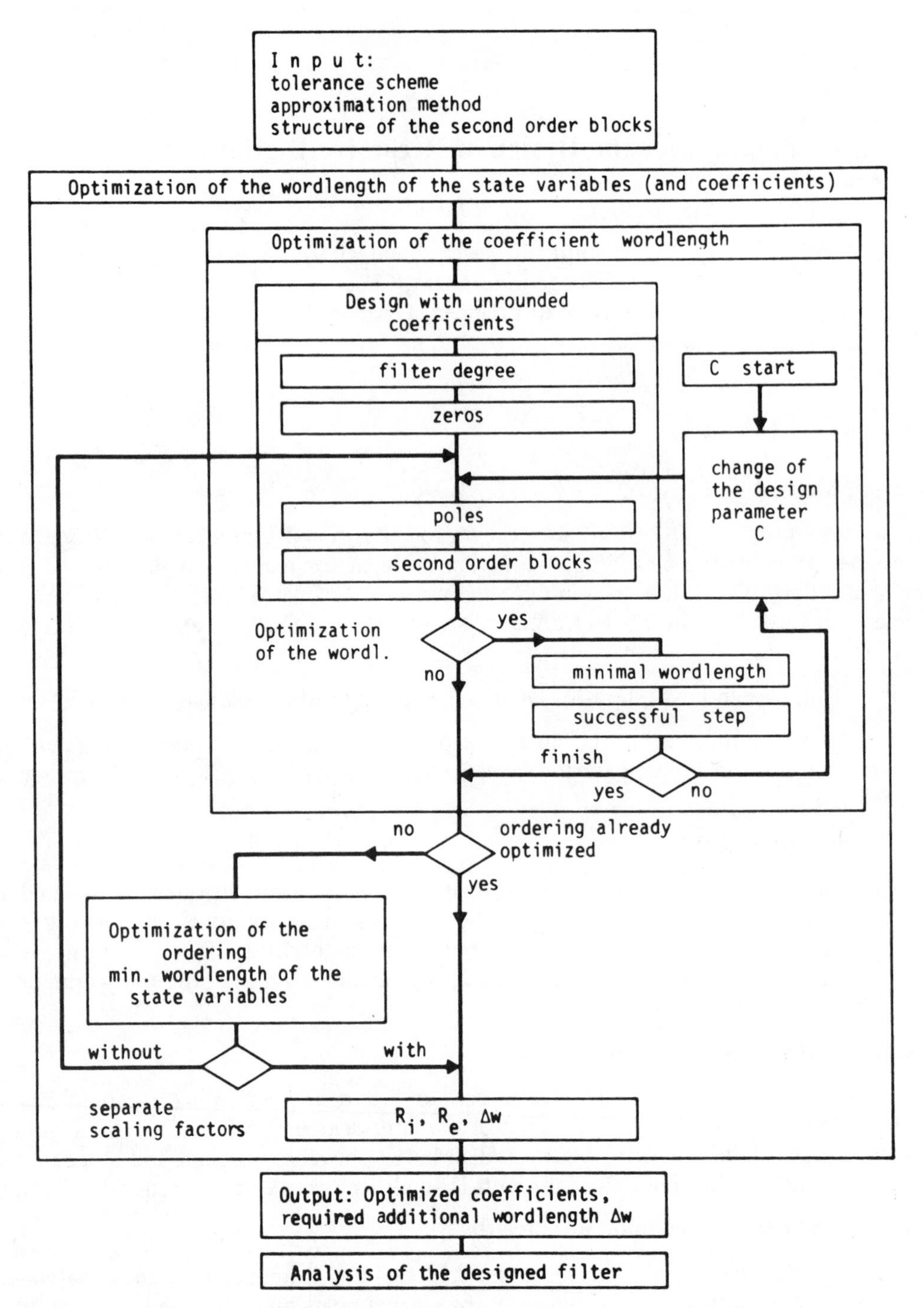

Fig. 1 Block diagram of DOREDI.

(a) Test of the necessary coefficient wordlength.

(b) Calculations of the output noise and the additional wordlength of the state variables; all correlations of the noise sources within the total filter are taken into account.

(c) Printer plots of the magnitude and the phase of the frequency response as well as the impulse response.

1.5 Realization of the transfer function

In the cascade realization, the transfer function is given by

$$H(z) = \prod_{\lambda=1}^{l} \frac{b_{2\lambda} z^2 + b_{1\lambda} z + b_{0\lambda}}{z^2 + c_{1\lambda} z + c_{0\lambda}} \tag{1}$$

The coefficient and state variable wordlengths can be optimized for different hardware realizations.

1.5.1 Coefficients

The options available are:

(a) fixed-point sign-magnitude realization
(b) "pseudo"-floating-point realization
(c) realization by differences

For a detailed description of the different realizations see Appendix A and [4].

1.5.2 State Variables

Fixed-point two's complement representation: The necessary reduction of the wordlength inside the filter can be done by different arithmetics and at different points within the filter structures. For a detailed description see Appendix B and [4], [6], [7], [8].

1.5.3 Structures of Second-Order Blocks

The structures available are variations of the first and second canonic forms (Fig. 2) with or without separate scaling factors.

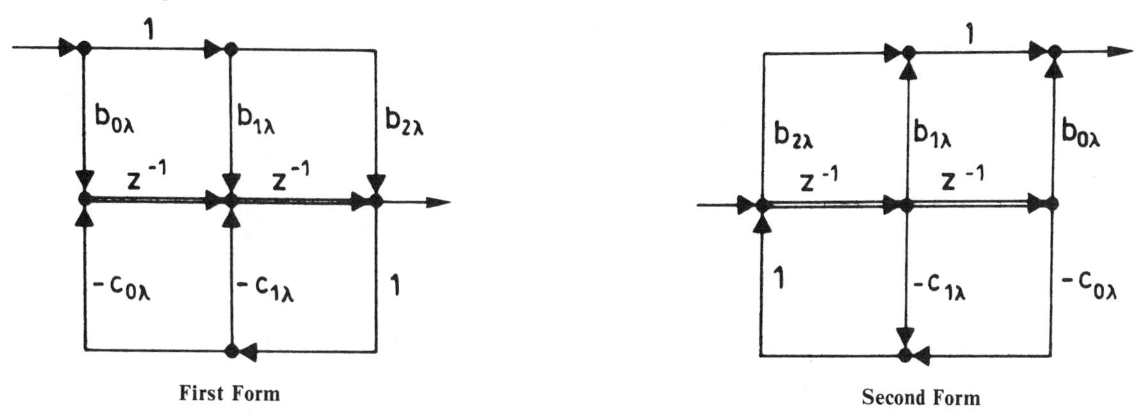

Fig. 2 Signal-flow graph of the canonic forms.

Scaling can be realized by a multiplication with an arbitrary value or by a power of two [4], [6], [10].

2. How to Use DOREDI - Description of the Input Data

The whole design and optimization program is controlled by the main program calling the on-line input data compiler and the proper program sections sequentially for the different tasks. All input data are read in via the on-line input compiler, allowing an almost free ordering of the data cards within one program section.

2.1 Call of the program sections and the input compiler

Normal program section card format

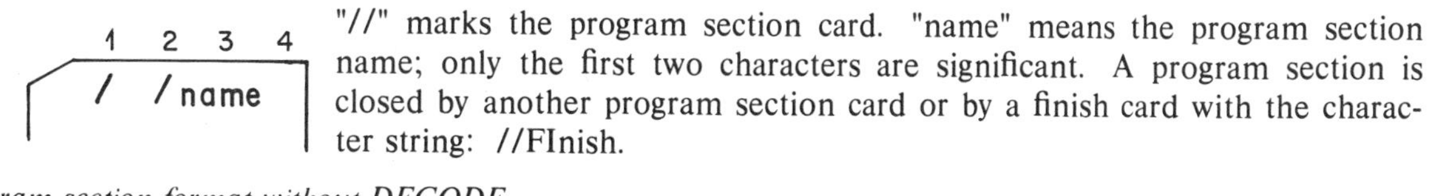

"//" marks the program section card. "name" means the program section name; only the first two characters are significant. A program section is closed by another program section card or by a finish card with the character string: //FInish.

Program section format without DECODE

Normally the input data card is processed by the "DECODE" statement. If "DECODE" does not conform to the compiler in use, an alternative program section input can be installed by compilation (see sect. 4.1). The original program section card is split up into a special control card "//" and a second card with the program section name beginning in column 3.

1 2 3 4
name

Description of the program sections

//SYnthesis: Design of recursive digital filters with or without optimized coefficients

//SEquence: Optimization of the pairing and ordering of second-order blocks achieving minimum output noise

//SS (Synthesis-Sequence)
This task includes the sections //SY and //SE. For a realization of second-order blocks without separate scaling factor, another optimization of the coefficient wordlength is done in a third step taking into account the sequence of the second-order blocks fixed in the step before (Fig. 1).

//ANalysis: Analysis of the digital filter designed before or defined by input data cards.

2.2 Control parameters and filter data input

Normal input formats

Two formats are used for the input of the design parameters (tolerance scheme, program control parameters, filter realization, etc.):

(a) The first one is the *control card* with a unified format for one LOGICAL, one INTEGER and three REAL parameters:

FORMAT(1A1,2I2,2X,L1,1X,I5,1X,3E15.0,12A1)

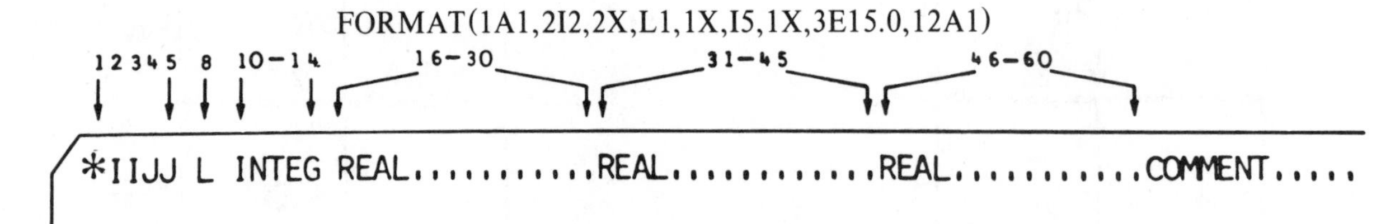

In the following, the different input variables are described by a short notation:

*IIJJ L,INTEG,REAL,REAL,REAL;COMMENT

If any data arrays within one set of data are superfluous, they are marked by "bb"; following superfluous arrays are not marked.

Description of the data arrays

*IIJJ: Input code
II, Input class
JJ, Number within a class
Input cards to be used alternatively belong to the same decade (e.g., JJ=41 *or* JJ=42).

L LOGICAL input
FORMAT: L1 T ≙ TRUE
F, any other character, or blank ≙ FALSE

INTEG INTEGER input
FORMAT: I5, right-justified

REAL REAL input
FORMAT: unformatted floating-point read statement within the specified array, if necessary with a right-justified exponent x.xxE±ee

(b) For the input of INTEGER arrays (e.g., a predetermined pairing and ordering, or the exponents of the "pseudo"-floating point coefficients) a second type of format, the INTEGER data card, is necessary. This data card can only be used as a continuation card of a control card.

FORMAT (1A1,2I2,3X,16I4)

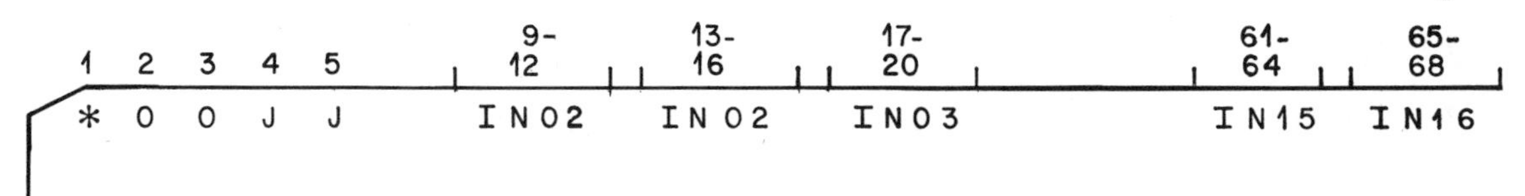

Description of the data arrays

*00JJ — Input code for the INTEGER data card. II has to be 00 because this format is only used for data input following a control card. JJ Number of the leading control card

IN01...IN16 — INTEGER-array
FORMAT: 16I4, right-justified

Input formats without "DECODE"

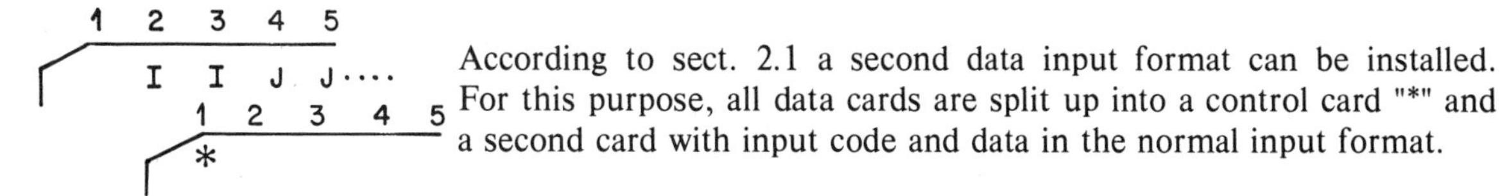

According to sect. 2.1 a second data input format can be installed. For this purpose, all data cards are split up into a control card "*" and a second card with input code and data in the normal input format.

Description of the input parameters

In the following, the different input data are explained. They are classified according to their functions:

Class 00 - data input (only used in connection with a control card)

Class 01 - general input/output control

Class 02 - parameters for the optimizations

Class 03 - tolerance scheme

Class 04 - realization of the coefficients

Class 05 - realization of the blocks of second order

Class 06 - special analysis of the designed filter.

The use of some of the input data classes makes no sense, if called in an individual program section. Therefore, Table 1 shows (x) the principally allowed input data classes for the different program sections.

As far as it is useful, default values are assigned to the input parameters; in addition to the definition of the tolerance scheme, the input is reduced to those data (or data cards, respectively) disagreeing with the default values.

At the beginning of the program all parameters agree with their default values. When a new program section within a program run is reopened, only the parameters within the input data classes, marked by "d" in Table 1, are redefined by their default values.

CLASS	SY	SE	SS	AN
01	x	x	x	x
02	x d	x d	x d	
03	x d		x d	x
04	x d	x	x d	x
05		x d	x d	x
06				x d

Table 1 Allowed input data classes (x) and redefinition of the default values (d).

2.2.1 General Input/Output Control — Data Cards of Class 01

2.2.1.1 Input of filter coefficients (for the program sections SE and AN)

*0110 bb,NINP

NINP Input parameter

NINP=1: Unrounded coefficients and, if not explicitly specified by input data, the realization parameters from the previous program section

NINP=2: See NINP=1, but from a disk file (channel number KA4)

NINP=3: Input of second-order blocks from the card reader; the following data cards are expected to immediately follow the control card†):

```
*0010 bb,NB,FACT
*0010 bb,bb,B2(01),B1(01),B0(01)
  :
*0010 bb,bb,B2(NB),B1(NB),B0(NB)
*0010 bb,bb,bb,C1(01),C0(01)
  :
*0010 bb,bb,bb,C1(NB),C0(NB)
```

NINP=4: Input of the zeros and poles; the following data cards are expected immediately behind the control card †):

```
*0010 bb,NB,FACT
*0010 bb,bb,ZZR(01),ZZI(01)
  :
*0010 bb,bb,ZZR(NB),ZZI(NB)
*0010 bb,bb,ZPR(01),ZPI(01)
  :
*0010 bb,bb,ZPR(NB),ZPI(NB)
```

only the zeros and poles in the closed upper complex plane have to be declared.

Default value: NINP=1

2.2.1.2 Output of the results on the line printer

*0120 bb,NOUT

NOUT INTEGER parameter for the output on the line printer (channel number KA2)

NOUT=0: No line printer output

to

NOUT=5: Complete line printer output with all intermediate results

For the intermediate results, see output example in section 3.

Default value: NOUT=3

2.2.1.3 Output of the results on the disk

*0130 LDOUT

LDOUT LOGICAL parameter for the output on the disk (channel number KA4)

True: An output is provided according to the ASCII format given below

False: No disk output

† For the meaning of the names see section 2.2.1.3.

Default value: LDOUT=.FALSE.

List of formatted records for the output on the disk

```
' === DOREDI === VERSION V005 ==='
'FILTER DESCRIPTION'
ITYP,IAPRO,NDEG
SF,OM(1),...,OM(4),ADELP,ADELS
AC,ROM(1),...,ROM(4),RDELP,RDELS
NZM(1),...,NZM(4),M
ZM(1,1),...,ZM(1,4)
 :
ZM(M,1),...,ZM(M,4)
'DATA'
NB
FACT
B2(1),B1(1),B0(1),C1(1),C0(1)
 :
B2(NB),B1(NB),B0(NB),C1(NB),C0(NB)
IRCO(1),...,IRCO(5)
IECO(1,1),...,IECO(1,5)
 :
IECO(NB,1),...,IECO(NB,5)
IDCO(1,1),...,IDCO(1,5)
 :
IDCO(NB,1),...,IDCO(NB,5)
IWL,IECOM,JMAXV,ITRB2
'END OF DATA'
```

All INTEGER records in the format 10I5
All REAL records in the format 4E15.7

Description of the parameters

ITYP	= filter type
IAPRO	= kind of approximation
NDEG	= filter degree
SF	= sampling frequency
OM(I)	= cutoff frequencies Ω in radians
ADELP	= ripple in the passband(s)
ADELS	= ripple in the stopband(s)
AC	= design parameter C
ROM(I)	= changed cutoff frequencies
RDELP	= changed ripple in the passband(s)
RDELS	= changed ripple in the stopband(s)
NZM(I)	= number of the different extrema
M	= max {NZM(J)}; J = 1,2,3,4
ZM(I,1)	= location of the maxima in the passband
ZM(I,2)	= location of the minima in the passband

ZM(I,3)	= location of the maxima in the stopband
ZM(I,4)	= location of the zeros in the stopband
NB	= number of blocks of second order
FACT	= gain factor
B2(I),B1(I),B0(I)	= numerator coefficients
C1(I),C0(I)	= denominator coefficients
ZZR(I)	= real part of the zeros in the z-plane
ZZI(I)	= imaginary part of the zeros in the z-plane (ZZI(I)=0, real zero)
ZPR(I)	= real part of the poles in the z-plane
ZPI(I)	= imaginary part of the poles in the z-plane (ZPI(I)=0, real pole)

IRCO(N),IFCO(L,N),IDCO(L,N),IWL,IECOM,JMAXV and JTRB2 are described in sect. 2.2.4.

2.2.1.4 Special output

*0140 LSPOUT,ISPOUT

LSPOUT LOGICAL parameter for the special output

True: Output is provided; a special output procedure "OUTSPE" has to be defined and linked to the program by the user (see sect. 4).

False: No special output

ISPOUT: Free parameter for the special output

Default values: LSPOUT=.FALSE.
ISPOUT=0

2.2.2 Parameters for the Optimization — Data Cards of Class 02

2.2.2.1 Optimization of the coefficient wordlength

*0210 LWLF,ITERM,ACX,ACXMI,ACXMA

LWLF LOGICAL parameter for the optimization of the tolerance scheme utilization with a fixed wordlength

True: Optimization of the parameters ACX for a fixed coefficient wordlength IWL; IWL can be defined by the control card *0410 (see sect. 2.2.4.1)

False: No fixed wordlength; coefficient wordlength will be minimized

ITERM Maximum number of optimization cycles for the final coefficient optimization

ACX See section 2.2.3.2 (for the alternative input format see section 4.1)

ACXMI Lower bound for the design parameter ACX: $0. \leqslant ACXMI < 1.$

ACXMA Upper bound of the design parameter ACX: $0. < ACXMA \leqslant 1.$

Default values: LWLF=.FALSE.
ITERM=10
ACX, see *0320 in section 2.2.3.2 and Appendix D
ACXMI=0.
ACXMA=1.

*0220 bb,ITERM1

ITERM1 Maximum number of optimization cycles for the pre-optimization (see sect. 2.1, //SS)

Default value: ITERM1=4

2.2.2.2 Optimization of the output noise

*0230 LPAIRF,ISTOR

The program section call //SE or //SS starts the output noise optimization automatically. This control card is only necessary to alter the default values.

LPAIRF LOGICAL parameter for a fixed pairing

True: Optimization of the ordering of blocks of second order with a fixed pairing of the poles and zeros, according to a rule of thumb [1], [4], [6], or to the input by the control card *0240. For structures 21, 23, 25, and 27 the input *0240 provides fixed pairing of the auxiliary blocks (Appendix C). **ATTENTION:** fixed pairing is not possible for the structures 22, 24, 26, and 28.

False: Free optimization of the pairing and ordering

ISTOR Maximum number of intermediate storage S_{max} for the "optimization with limited storage" [4], ISTOR $\leqslant$ 100

Default values: LPAIRF=.FALSE.
ISTOR=100

*0240 LSEQ,ISCAL,SCALM

LSEQ LOGICAL parameter for the input of a fixed pairing and ordering (e.g., as a start ordering for the heuristic optimization)

True: Input of a fixed pairing of the zeros and poles. Two input data cards have to follow immediately:
*0040 ISEQN(1),...,ISEQN(16) } INTEGER data format
*0040 ISEQD(1),...,ISEQD(16) }
The numerator and denominator are numbered in an ascending order. ISEQN(I)=J means the J-th numerator out of the input sequence is taken as the I-th numerator in the final realization. The same is true for the denominator declaration ISEQD(I).

False: No input

ISCAL INTEGER parameter for the kind of scaling [1], [4]

ISCAL=1: absolute criterion

$$\sum_{k=0}^{\infty} |f_{0_i}(k)| \leqslant \text{SCALM} \dagger$$

ISCAL=2: criterion for sinusoid input

$$L_\infty = \max|F_i(\Omega)| \leqslant \text{SCALM} \dagger \ddagger$$

ISCAL=3: criterion for the power of a white noise source

$$L_2^2 = \frac{1}{\pi}\int_0^{\pi} |F_i(\Omega)|^2 \, d\Omega \leqslant \text{SCALM} \dagger \ddagger$$

SCALM chosen maximum of the overflow points

Default values: LSEQ=.FALSE.
ISCAL=2
SCALM=1.

† The criterion has to be satisfied for all overflow points within the realized structure.

‡ $L_2 \triangleq L_2$ norm; $L_\infty \triangleq L_\infty$ norm; F_i see Appendix B.

2.2.3 Description of the Desired Filter — Data Cards of Class 03

2.2.3.1 Type of the filter; sampling frequency

*0310 bb,ITYP,SF

ITYP Filter type

ITYP=1: lowpass filter

ITYP=2: highpass filter

ITYP=3: symmetrical bandpass filter

ITYP=4: symmetrical bandstop filter

SF Sampling frequency (input may as well be executed with the data cards *0341 or *0342)

Default values: ITYP=1
SF no default value provided

2.2.3.2 Type of approximation; design parameter

*0320 bb,IAPRO,ACX

IAPRO Type of approximation

IAPRO=1: Butterworth (maximally flat)

IAPRO=2: Chebyshev I (passband)

IAPRO=3: Chebyshev II (stopband)

IAPRO=4: Elliptic filter (Cauer)

ACX Normalized design parameter C for the utilization of the tolerances in the passband and the stopband [3],[4] (start value for the optimization) 0.⩽ACX⩽1. (See Appendix D)

ACX=0.: the passband tolerance is used completely

to

ACX=1.: the stopband tolerance is used completely

Default values: IAPRO=4 (elliptic filter)
ACX, such that passband and stopband are used as equally as possible. (For the alternative input format see section 4.1; in this case for the default input, ACX has to be set to ACX=1000.)

2.2.3.3 Filter degree (order of the recursive system)

*0330 LDEGF,NDEG,EDEG

LDEGF LOGICAL parameter related to the specification of a filter degree. For the check of the filter order, the degree extension [4] is introduced into the calculation.

True: If the tolerance scheme cannot be fulfilled with the prescribed degree, the program run is terminated.

False: If the given degree is not sufficient, the order is increased automatically.

NDEG Prescribed filter order

EDEG Relative extension of the degree [4].

Default values: LDEGF=.FALSE.
NDEG see remarks
EDEG=0.2 (for the alternative input format see section 4.1)

Remarks:

a) Normally, the degree is determined automatically. The minimum degree n_{min} found in this way is not necessarily an integer value. Taking this into account as well as including the degree extension for the purpose of a minimized coefficient expenditure [3], [4], yields the order to be used as

$$n = MAX \begin{cases} [n_{min}(1 + \frac{\Delta n}{n_{min}}|_{opt})]_{rounded} \\ \\ < n_{min} >_{half-adjusted} \end{cases}$$

$$\frac{\Delta n}{n_{min}}|_{opt} \hat{=} EDEG$$

b) maximum filter degree is set to $n_{max} = 32$

c) To obtain the minimum filter degree for a prescribed tolerance scheme, the degree extension EDEG has to be set explicitly to EDEG=0.0.

2.2.3.4 Cutoff frequencies for the passband(s) and stopband(s)

The cutoff frequencies are read in by one (and only one) of the following sets of data:

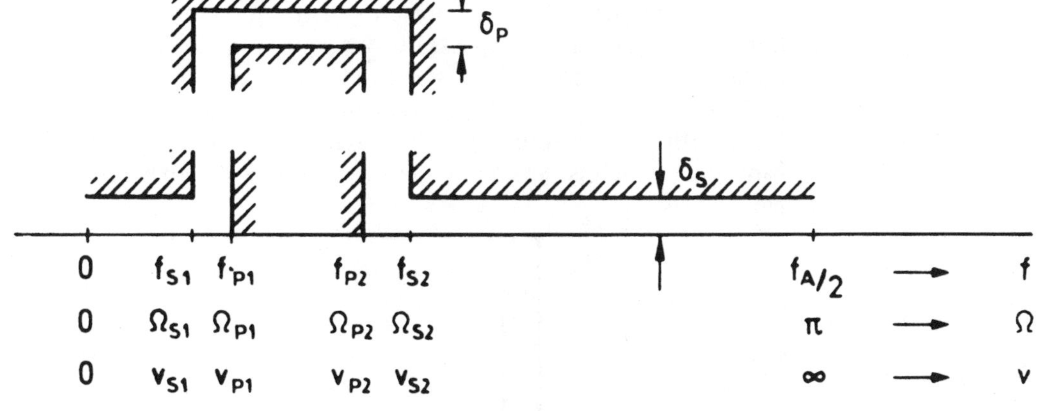

Fig. 3. Cutoff frequencies (e.g., for a bandpass).

(a) *0341 bb,bb,FR(1),FR(2),<SF>

and for bandpass and bandstop filters in addition

*0342 bb,bb,FR(3),FR(4),<SF>

FR(I) Unnormalized cutoff frequencies

	lowpass	highpass	bandpass	bandstop
FR(1)	f_P	f_S	f_{S1}	f_{P1}
FR(2)	f_S	f_P	f_{P1}	f_{S1}
FR(3)			f_{P2}	f_{S2}
FR(4)			f_{S2}	f_{P2}

SF Sampling frequency; superfluous, if already defined in data card *0310. The units of FR(I) and SF have to be identical - i.e., both Hz or kHz. (<> means that this parameter may be omitted.)

(b) *0343 bb,bb,OM(1),OM(2)

and additionally for bandpass or bandstop filters

	*0344	bb,bb,OM(3),OM(4)
	OM(I)	Cutoff frequencies in radians, normalized by the sampling frequency
(c)	*0345	bb,bb,S(1),S(2)
		and additionally for bandpass and bandstop
	*0346	bb,bb,S(3),S(4)
	S(I)	Cutoff frequencies of the corresponding analog system (input if analog filters are to be designed)
(d)	*0347	bb,bb,VSN,VD,A
	VSN	Normalized stopband cutoff frequency ν_s' of the corresponding analog lowpass filter (actual design parameter for the transition from passband to stopband [1])
	VD	Passband edge frequency of the corresponding analog lowpass filter; VD $\hat{=}$ ν_P
	A	Parameter for the transformations from the normalized lowpass to the bandpass and bandstop filters.

Remarks:

a) Prescription of only *three* cutoff frequencies for symmetrical band filters. If *four* frequency values are prescribed, the requirements for the calculation of the reference lowpass are overdetermined. Without further data input, the tolerance scheme is utilized such that a minimal filter degree is achieved (see section 2.2.3.6). If only *three* frequency values are given, the fourth one is automatically chosen in a manner yielding an analog band filter which is symmetric with respect to its center frequency. Example:

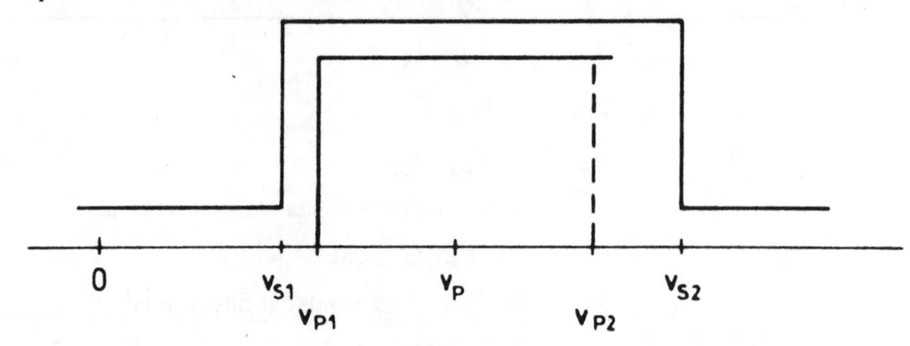

Fig. 4. Free choice of the cutoff frequency ν_{P2} for a symmetric bandpass.

b) Determination of the transition band. The filter order and the tolerances δ_P and δ_S being given, the transition bandwidth of the filter can be determined automatically. A required degree extension for the optimization of the coefficient wordlength is taken into account. Lowpass example:

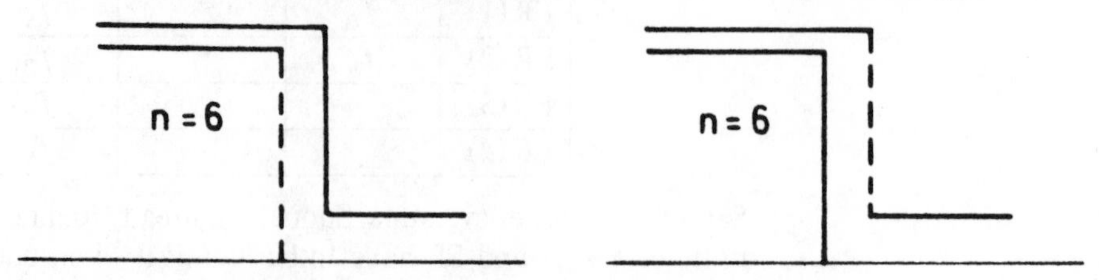

Fig. 5. Free choice of one cutoff frequency for a lowpass.

For a band filter, remark a) can be regarded in addition; then, only two cutoff frequencies have to be prescribed. Bandpass example:

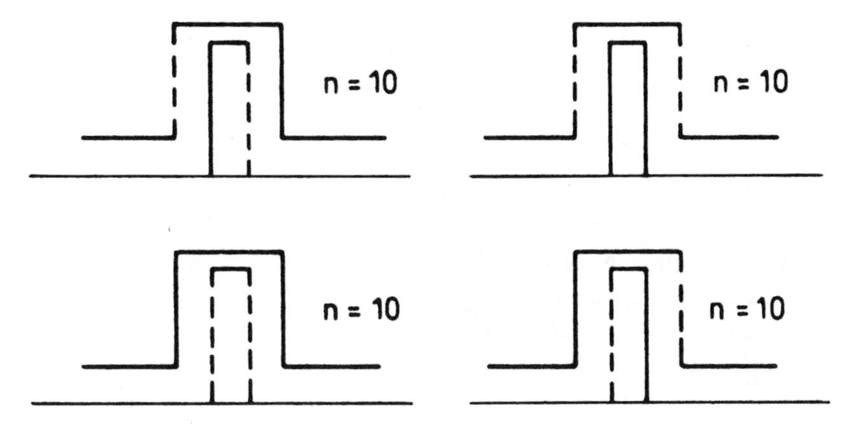

Fig. 6. Free choice of two cutoff frequencies for a bandpass.

2.2.3.5 Tolerances in passband and stopband

The passband and stopband tolerances are read in by one of the following data cards, optionally:

(a) *0351 bb,bb,ADELP,ADELS

ADELP — Tolerance δ_P in the passband

ADELS — Tolerance δ_S in the stopband

(b) *0352 bb,bb,AP,AS

AP — Tolerated passband attenuation a_P (in dB)

AS — Stopband attenuation a_S (in dB)

(c) *0353 bb,bb,P,AS

P — Reflection coefficient ρ

$P \hat{=} \rho = \sqrt{1 - (1-\delta_P)^2}$

Remarks: The filter order, the transition band, and one of the δ tolerances being given, the other one can be determined. A required degree extension for the optimization of the coefficient wordlength can also be regarded. Examples are given in Fig. 7.

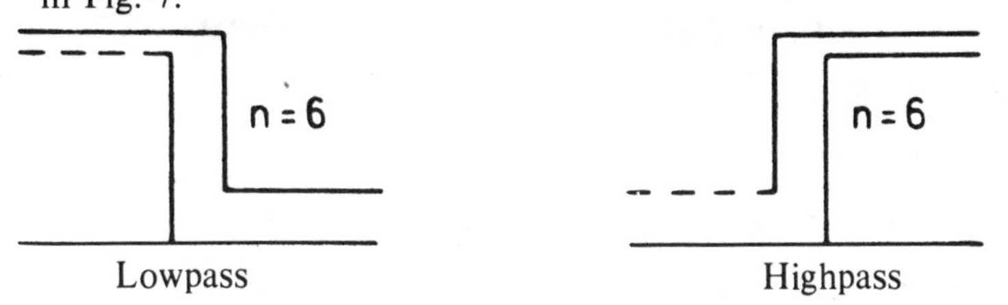

Bandpass

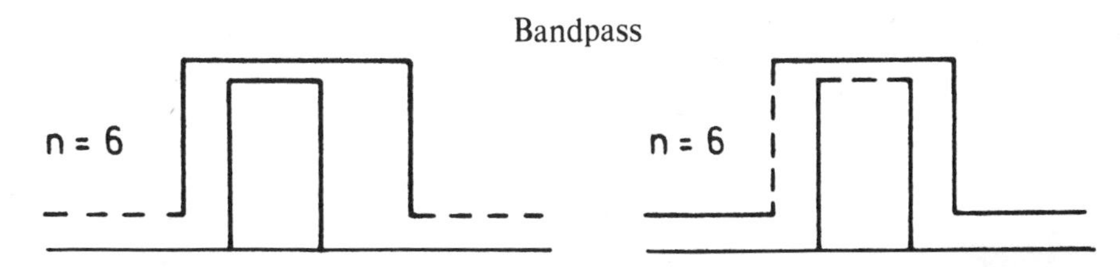

Fig. 7. Free choice of one of the tolerances.

2.2.3.6 'Choice' of the symmetrical band filter

*0360 bb,NORMA

NORMA — Parameter for the determination of the reference lowpass of symmetrical bandfilters

NORMA=0: calculation of ν_P such that $n_{\min}$ is minimized;

NORMA=1: calculation of ν_P from the inner cutoff frequencies;

NORMA=2: calculation of ν_P from the outer cutoff frequencies;

NORMA=3: calculation of ν_P as the geometric mean of all cutoff frequencies. Examples for a bandpass filter are shown in Fig. 8.

Default value: NORMA=0

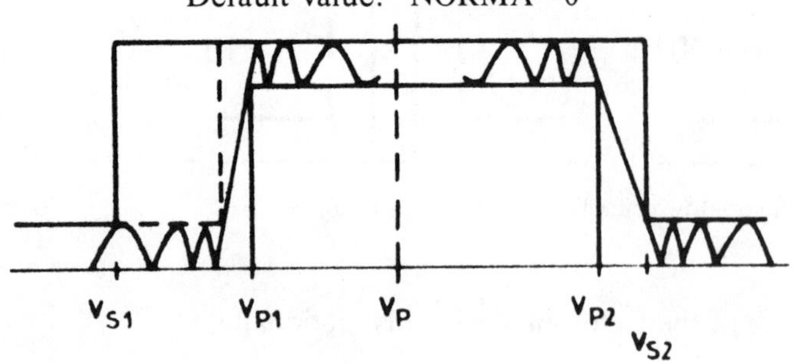

NORMA = 1; $\nu_P = \sqrt{\nu_{P1}\nu_{P2}}$

VSN = 1.18

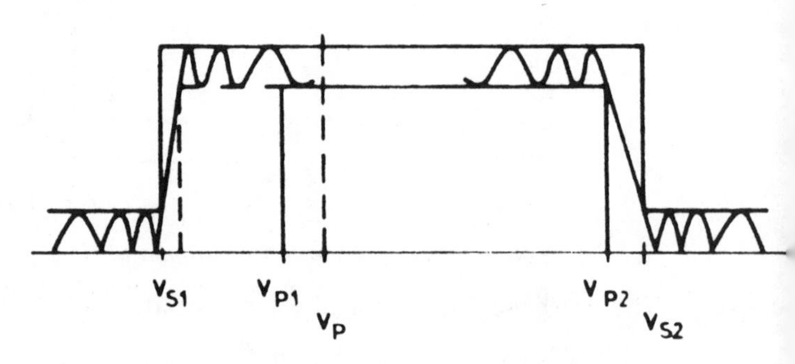

NORMA = 2; $\nu_P = \sqrt{\nu_{S1}\nu_{S2}}$

VSN = 1.12

here, this is equivalent to NORMA = 0

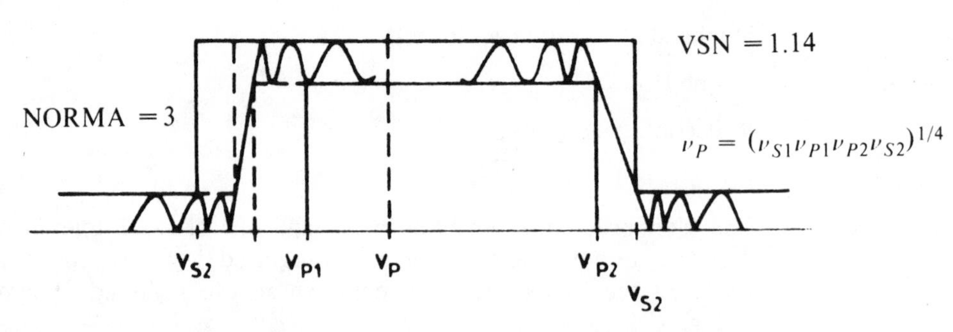

Fig. 8. Different choice of the parameter NORMA

2.2.3.7 Output of the analog filter

*0370	LSOUT
LSOUT	LOGICAL parameter for the special output
True:	Output of poles, zeros, and 2nd order blocks of the corresponding analog filter
False:	No special output
Default value:	LSOUT=.FALSE.

2.2.4 Realization of the Coefficients — Data Cards of Class 04

General coefficient representation (see Appendix A)
ACO(L,N)=ACOS(L,N)*2.**(IRCO(N) − IECO(L,N))
−SIGN(2.**IDCO(L,N),ACOS(L,N))

Description of the parameters

ACO(L,1) ≙ B2(L) = $b_{2\lambda}$
ACO(L,2) ≙ B1(L) = $b_{1\lambda}$
ACO(L,3) ≙ B0(L) = $b_{0\lambda}$
ACO(L,4) ≙ C1(L) = $c_{1\lambda}$
ACO(L,5) ≙ C0(L) = $c_{0\lambda}$

ACOS(L,N) = sign + mantissa of the "pseudo"-floating-point coefficients
IRCO(N) = exponent of the number range
IECO(L,N) = "pseudo"-floating-point exponent
IDCO(L,N) = exponent of the difference branch

2.2.4.1 Coefficient wordlength

*0410	LSTAB,IWL
LSTAB	LOGICAL parameter for the stability test
True:	Check only whether the system with rounded coefficients is stable; *no* correction.
False:	Check whether the system with rounded coefficients is stable and change, if necessary.
IWL	Wordlength of sign and mantissa; quantization $Q_c = 2^{-(IWL-1)}$ IWLMI ⩽ IWL ⩽ IWLMA; for IWLMI and IWLMA see sect. 4.
Default value:	LSTAB=.FALSE. IWL = 16
Attention:	IWL is only used for the optimization of the coefficient errors. For other inputs apply IWLR in *0530.

2.2.4.2 Number range of the coefficients

*0420	RCOINP,JRCO
RCOINP	LOGICAL parameter for the input of a predetermined array of coefficient ranges
True:	One additional INTEGER-data card is necessary: *0020 IRCO(1),...,IRCO(5)
False:	No extra input.
JRCO	Parameter for the calculation of the necessary coefficient ranges
JRCO<0	ranges IRCO(I)<0 allowed
JRCO>0	ranges IRCO(I)⩾0 only
|JRCO|=1	separate ranges for B2(L),B1(L),B0(L),C1(L),C0(L)
|JRCO|=2	one common range for (B2(L),B1(L),B0(L)); separate range for C1(L) and for C0(L)

|JRCO|=3 one common range each for (B2(L),B1(L),B0(L)) and for (C1(L),C0(L))

|JRCO|=4 common ranges for (B1(L),C1(L)) and (B0(L),C0(L))

|JRCO|=5 one common range for all coefficients

Default values: RCOINP=.FALSE.
JRCO=2

2.2.4.3 'Pseudo'-floating-point coefficients

*0430 ECOINP,JECO

ECOINP LOGICAL parameter for the input of predetermined exponents

True: Five additional INTEGER-data cards are necessary:

*0030 IECO(1,1),...,IECO(NB,1)
⋮
*0030 IECO(1,5),...,IECO(NB,5)
NB = number of second-order blocks used

False: No extra input

JECO Parameter for the calculation of the exponents

JECO=0 no "pseudo"-floating-point realization

JECO≠0 value of the "pseudo"-floating-point exponent must be less or equal to JECO

Default values: ECOINP=.FALSE.
JECO=0

2.2.4.4 Realization by differences

*0440 DCOINP,JDCO

DCOINP LOGICAL parameter for the input of predetermined difference branches

True: Five additional INTEGER-data cards are necessary:

*0040 IDCO(1,1),...,IDCO(NB,1)
⋮
*0040 IDCO(1,5),...,IDCO(NB,5)
NB = number of second-order blocks used
IDCO(L,N) = −100 means no difference branch for this coefficient

False: No extra input

JDCO Parameter for the calculation of the difference branches

JDCO=0: no realization with differences

JDCO≠0: realization with differences is possible. JDCO is used as a code word; the digits are assigned to the control parameters JJDCO(N) of the different coefficient classes:

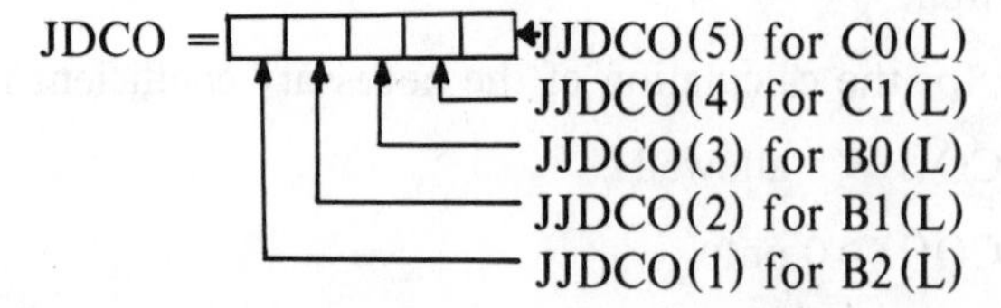

JJDCO(N)=1: common difference branch realization for all blocks (block multiplexing)

JJDCO(N)=2: separate realization; only for those blocks with coefficients less than half the original number range

Default values: DCOINP=.FALSE
JDCO=00000

2.2.4.5 Realization of the largest numbers within the range

*0450 bb,JMAXV

JMAXV Parameter for the control of the largest numbers

JMAXV=1:	$-2^{i_R} \leqslant x \leqslant 2^{i_R}$	
JMAXV=2:	$-2^{i_R} \leqslant x < 2^{i_R}$	two's complement representation
JMAXV=−2:	$-2^{i_R} \leqslant x < 2^{i_R}$ and $2^{i_R} \rightarrow 2^{i_R}-LSB$	
JMAXV=3:	$-2^{i_R} < x < 2^{i_R}$	sign-magnitude representation
JMAXV=−3	$-2^{i_R} < x < 2^{i_R}$ and $-2^{i_R} \rightarrow -2^{i_R}+LSB$, $2^{i_R} \rightarrow 2^{i_R}-LSB$	

$LSB \hat{=}$ Least significant bit

Remark: JMAXV=−2,−3 is useful to reduce the necessary number range

Default values: JMAXV=−3

*0460 bb,JTRB2

JTRB2 Parameter to control the rounding of the scaling factors; correction of the scaling error

JTRB2<0: normal rounding of the coefficients with an influence on scaling; violation of the scaling is possible

JTRB2>0: truncation of these coefficients

|JTRB2|=1 only the total gain factor FACT is corrected

|JTRB2|=2 the scaling of the next block is corrected (useful for structures with separate scaling factors (see sect. 2.2.5.1))

|JTRB2|=3: in addition to JTRB2=2, the coefficients B1(L) and B0(L) of the same block are corrected (useful for structures without separate scaling factors)

Default values: JTRB2=3

2.2.5 Realization of the Cascade Structures with Blocks of Second Order — Data Cards of Class 05

The implemented structures for second-order blocks with five or four (zeros on the unit circle) coefficients are shown in Figs. 9 and 10. The nonlinearities are marked by a square box with an N over it. The characteristics of these nonlinearities are declared by a control card. For scaling, because of the two's complement realization of the state variables, only the overflow points marked by → have to be checked.

2.2.5.1 Structures of second-order blocks

*0510 LPOT2,ISTRU

LPOT2 LOGICAL parameter for the choice of the scaling factor

True: Scaling by powers of two (bit shifts)

False: Scaling with any factor

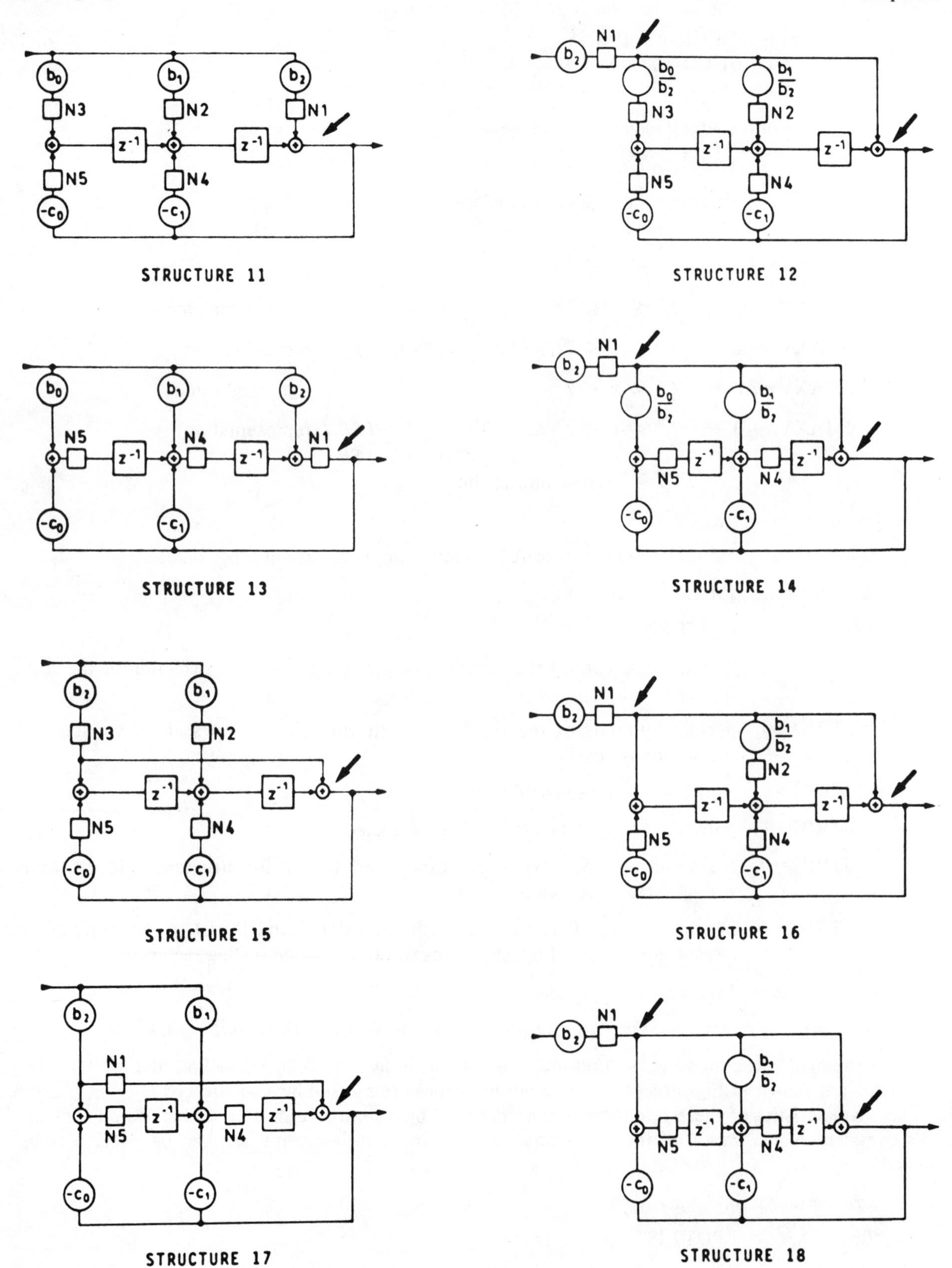

Fig. 9. Blocks of second order in the first canonic form.

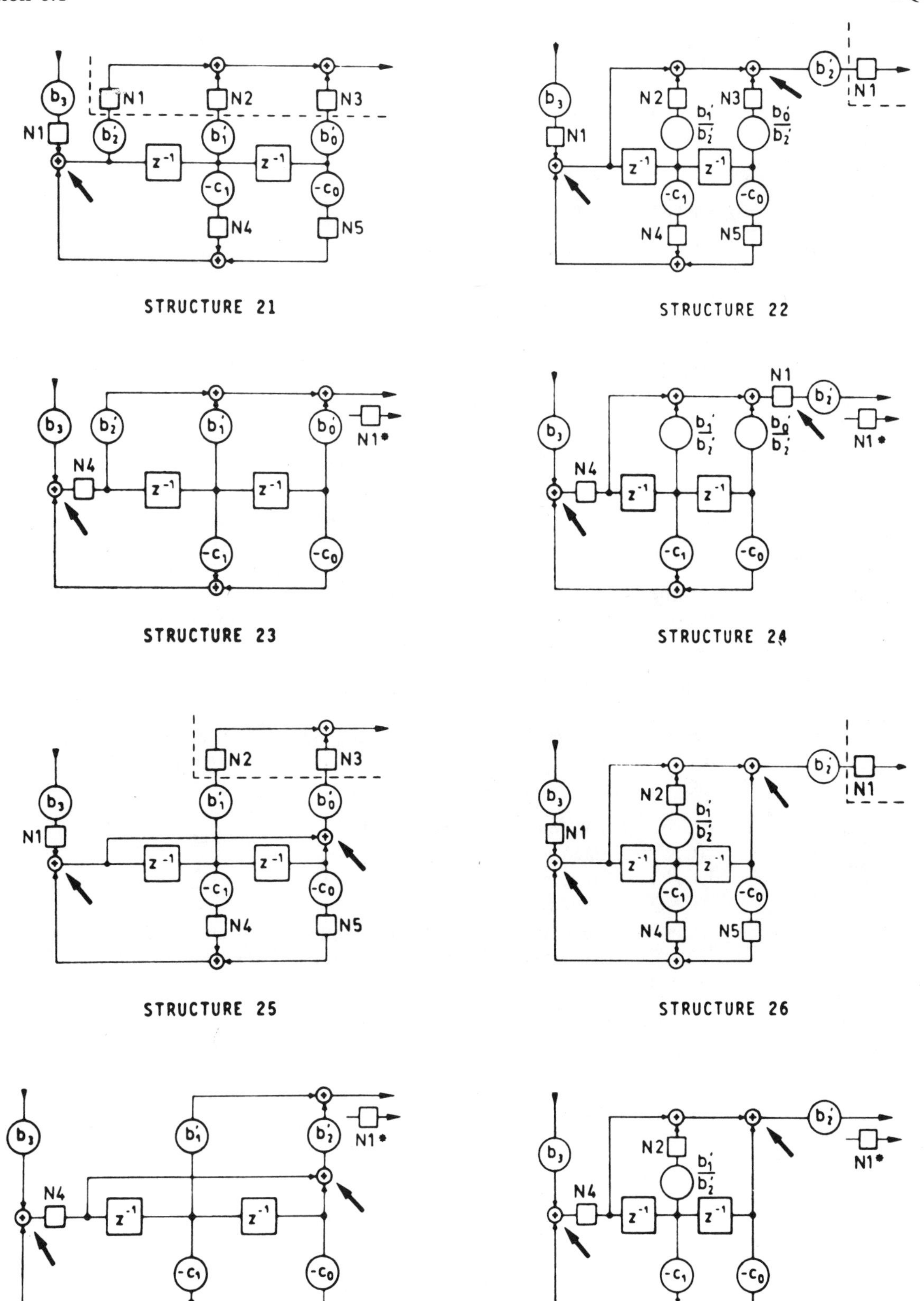

Fig. 10 Blocks of second order in the second canonic form. The coefficient b_3 is only realized in the first block of a cascade; for the calculation of the noise, the part, separated by ⌊ - - - - is attached to the following block; $b_i' = b_i/b_3$.

* only realized for the last block in the cascade

ISTRU — Number of the structure of the second-order blocks to be realized (see catalog in Figs. 9 and 10).

Default value: LPOT2=.FALSE.
ISTRU=13

2.2.5.2 Characteristics of the nonlinearities

*0520 — CNOINP,JCNO

CNOINP — LOGICAL parameter for the input of inhomogeneous nonlinearity characteristics for the different second-order blocks

True: Five additional input data cards are necessary to characterize the nonlinearities

*0020 ICNO(1,1),...,ICNO(NB,1)
⋮
*0020 ICNO(1,5),...,ICNO(NB,5)
NB = number of second-order blocks

ICNO(L,N) = control parameter according to JJCNO (see next section); in this case JCNO is unaffected

False: No extra input

JCNO — parameter for the input of the nonlinearity characteristics, unified for all blocks JCNO is used as a code word; the digits are assigned to the five nonlinearities, according to JJCNO

JCNO = | N1 | N2 | N3 | N4 | N5 |

Zeros are replaced by the values of the nearest succeeding nonzero digit; e.g., 00205 $\hat{=}$ 22255. The different characteristics are described by the parameter JJCNO (see Appendix B).

JJCNO=1: two's complement rounding
JJCNO=2: sign magnitude rounding
JJCNO=3: mathematical rounding
JJCNO=4: two's complement truncation
JJCNO=5: unsymmetrical sign magnitude truncation
JJCNO=6: symmetrical sign magnitude truncation
JJCNO=7: random rounding

Especially for small coefficients, these different characteristics are distinct in their generation of random noise, DC offset, and errors correlated with the signal [4]. For the error values see Appendix B.

Default value: CNOINP=.FALSE.
JCNO=33333

*0530 — LCNO,IWLR

LCNO — LOGICAL parameter for the consideration of the coefficient quantization by the calculation of the output noise (see Appendix B).

True: No coefficient quantization is considered

False: Noise parameters will be adjusted (see Appendix B, Table B-1).

IWLR — parameter for the specification of the realized coefficient wordlength. In the noise analysis, the wordlength may be different from that found by the coefficient optimization.

Default value: LCNO=.FALSE.
IWLR=100, (no coefficient rounding) or result of the coefficients optimization

2.2.6 Analysis of the Designed Filter — Data Cards of Class 06

2.2.6.1 Coefficient analysis

*0610	LWLM,IWL
LWLM	LOGICAL parameter for the search of minimum wordlength
True:	For a given set of coefficients and realization the necessary coefficient wordlength will be found.
False:	Error analysis is done for a fixed wordlength IWL.
IWL	Coefficient wordlength
Default value:	LWLM=.FALSE. IWL=16

2.2.6.2 Noise analysis

For a given realization, the output noise and the necessary additional wordlength Δw are calculated.

*0620	LSEQ,ISCAL,SCALM
	For description of the parameters see control card *0240, sect. 2.2.2.2. In addition if ISCAL=−1 no scaling is done.

2.2.6.3 Printer plot of the magnitude of the frequency response

*0630	LNOR,IPAG,OMLO,OMUP,RMAX
	Response in linear scale
LNOR	LOGICAL parameter for the normalization
True:	Normalization to the extreme values in the given range
False:	No normalization
IPAG	Number of line printer pages to be used for one plot
OMLO	Lower limit for the abscissa (normalized frequency in radians)
OMUP	Upper limit for the abscissa (normalized frequency in radians)
RMAX	Maximum value for the ordinate
Default values:	LNOR=.FALSE. (range from 0. to RMAX) IPAG=2 OMLO=0. (for the alternative input format see sect. 4.1) OMUP=π RMAX=1.

2.2.6.4 Printer plot of the attenuation of the frequency response

*0640	LNOR,IPAG,OMLO,OMUP,RMAX
	Response in dB, for parameter description see sect. 2.2.6.3.
Default values:	LNOR=.FALSE. (range from 0. to RMAX) IPAG=2 OMLO=0.; (for the alternative input format see sect. 4.1) OMUP=π RMAX=100 dB

2.2.6.5 Printer plot of the phase response

*0650	LNOR,IPAG,OMLO,OMUP

For parameter description see sect. 2.2.6.3.

Default values: LNOR=.FALSE. (range from $-\pi$ to $+\pi$)
IPAG=2
OMLO=0.; (for the alternative input format see sect. 4.1)
OMUP=π

2.2.6.6 Printer plot of the impulse response

*0660 LNOR,IPAG,RMIN,RMAX

LNOR Parameter for the normalization, see sect. 2.2.6.3.

IPAG Number of line printer pages

RMIN Minimum value for the ordinate of the unnormalized plot range

RMAX Maximum value for the ordinate of the unnormalized plot range

Default values: LNOR=.FALSE. (range from RMIN to RMAX)
IPAG=2
RMIN=−1.
RMAX=1.

3. Examples for the Filter Design and Optimization

3.1 Design of a filter without optimization

3.1.1 Specifications

Lowpass:

sampling frequency $f_A = 20.0$ kHz

cutoff frequency in the passband $f_P = 2.0$ kHz

cutoff frequency in the stopband $f_S = 4.0$ kHz

passband tolerance $\delta_P = 0.02$

stopband tolerance $\delta_S = 0.001$

3.1.2 Execution of the Design

extension of the filter degree $\frac{\Delta n}{n_{\min}} = 0.2$

equal use of passband and stopband

3.1.3 Input Data Cards

```
//SY
*0310          20.
*0341          2.          4.
*0351          0.02        0.001
//FI
```

3.1.4 Output on the Line Printer

3.1.4.1 Normal output, according to the default output code NOUT = 3

```
    ===  DOREDI  ===  VERSION  V005 ===


DATA INPUT
//SY
*    3  10          0     0.200000E 02   0.              0.
*    3  41          0     0.200000E 01   0.400000E 01    0.
*    3  51          0     0.200000E-01   0.100000E-02    0.
//FI


TOLERANCE SCHEME


FILTER-TYPE
                        LOWPASS

APPROXIMATION
                        ELLIPTIC

SAMPLING FREQ.         20.000000
CUTOFF FREQUENCIES      2.000000      4.000000

NORM. CUTOFF FREQ.      0.628319      1.256637

CUTOFF FREQ. S-DOM.     0.324920      0.726543


PASSBAND RIPPLE(S)        0.020000       0.1755 DB    P =       0.1990
STOPBAND RIPPLE(S)        0.001000      60.0000 DB


MIN. FILTER DEGREE       4.6285

DEGREE EXTENSION         0.9257

CHOSEN FILTER DEG.       6


BOUND PAIR OF THE DESIGN PARAMETER C     3.2927  .LE.  C  .LE.   61.6686

REALIZED
NORM. CUTOFF FREQ.      0.628319      1.256637

CHOSEN DESIGN PAR.    CX = 0.6632134   (=)  C =      22.9879777

UTILIZATION OF THE PASSBAND  DELTA P  = 0.0028525  =  14.26252 PERCENT
                   STOPBAND  DELTA S  = 0.0001432  =  14.32376 PERCENT



POLES AND ZEROS IN THE Z-DOMAIN

CONSTANT GAIN FACTOR =    0.1559554E-02


  NUM.           POLES                    NUM.          ZEROS
    2     0.717153 +-J*   0.590872          2     0.280642 +-J*   0.959812
    2     0.675963 +-J*   0.412069          2     0.000000 +-J*   1.000000
    2     0.666428 +-J*   0.150211          2    -0.753474 +-J*   0.657478


BLOCKS OF SECOND ORDER

CONSTANT GAIN FACTOR =    0.1559554E-02

 L      B2(L)         B1(L)         B0(L)           C1(L)         C0(L)
 1    1.00000000   -0.56128475   1.00000000     -1.43430551   0.86343788
 2    1.00000000   -0.00000001   1.00000000     -1.35192616   0.62672679
 3    1.00000000    1.50694796   1.00000000     -1.33285540   0.46668934
```

3.1.4.2 Complete line printer output, according to the greatest output, code NOUT = 5

NOUT

```
    ===  DOREDI  ===  VERSION  V005 ===


DATA INPUT
//SY
*    1   20        5    0.               0.               0.
*    3   10        0    0.200000E 02     0.               0.
*    3   41        0    0.200000E 01     0.400000E 01     0.
*    3   51        0    0.200000E-01     0.100000E-02     0.
//FI


TOLERANCE SCHEME


FILTER-TYPE
                             LOWPASS

APPROXIMATION
                             ELLIPTIC

SAMPLING FREQ.              20.000000
CUTOFF FREQUENCIES           2.000000        4.000000

NORM. CUTOFF FREQ.           0.628319        1.256637

CUTOFF FREQ. S-DOM.          0.324920        0.726543


PASSBAND RIPPLE(S)             0.020000        0.1755 DB    P =       0.1990
STOPBAND RIPPLE(S)             0.001000       60.0000 DB


NORMALIZED PARAMETER IN THE S-DOMAIN

VD        =      0.324920
VSN       =      2.236068


MIN. FILTER DEGREE          4.6285

DEGREE EXTENSION            0.9257

CHOSEN FILTER DEG.          6

CAP. DELTA                     0.003293


ZEROS OF THE CHARACTERISTIC FUNCTION   /K(J*V)/**2
                                         2     0.969399 +-J*    0.
                                         2     0.726543 +-J*    0.
                                         2     0.272422 +-J*    0.


BOUND PAIR OF THE DESIGN PARAMETER C     3.2927  .LE.  C  .LE.   61.6686




POLES AND ZEROS OF THE NORMALIZED REFERENCE LOWPASS IN THE S-DOMAIN


  NUM.           POLES                  NUM.            ZEROS
                                         2    0.          +-J*    2.306654
                                         2    0.          +-J*    3.077684
                                         2    0.          +-J*    8.208093
```

≥3

≥4

≥3

=5

≥3

=5

```
POLES AND ZEROS OF THE REFERENCE FILTER IN THE S-DOMAIN

 NUM.           POLES                  NUM.            ZEROS                       ≥4
                                       2     0.          +-J*    0.749477
                                       2     0.          +-J*    1.000000
                                       2     0.          +-J*    2.666971

REALIZED
NORM. CUTOFF FREQ.        0.628319        1.256637                                 ≥3

CHOSEN DESIGN PAR.    CX = 0.6632134  (=)  C =       22.9879777

UTILIZATION OF THE PASSBAND  DELTA P   = 0.0028525   =   14.26252 PERCENT
                   STOPBAND  DELTA S   = 0.0001432   =   14.32376 PERCENT


POLES AND ZEROS OF THE NORMALIZED REFERENCE LOWPASS IN THE S-DOMAIN

CONSTANT GAIN FACTOR =    0.1432376E-03                                           =5

 NUM.           POLES                  NUM.            ZEROS
   2   -0.127449 +-J*    1.102886
   2   -0.385683 +-J*    0.851538
   2   -0.586296 +-J*    0.330270


POLES AND ZEROS OF THE REFERENCE FILTER IN THE S-DOMAIN

CONSTANT GAIN FACTOR =    0.1432376E-03                                           ≥4

 NUM.           POLES                  NUM.            ZEROS
   2   -0.041411 +-J*    0.358349       2     0.          +-J*    0.749477
   2   -0.125316 +-J*    0.276681       2     0.          +-J*    1.000000
   2   -0.190499 +-J*    0.107311       2     0.          +-J*    2.666971


POLES AND ZEROS IN THE Z-DOMAIN

CONSTANT GAIN FACTOR =    0.1559554E-02                                           ≥3

 NUM.           POLES                  NUM.            ZEROS
   2    0.717153 +-J*    0.590872       2    0.280642 +-J*    0.959812
   2    0.675963 +-J*    0.412069       2    0.000000 +-J*    1.000000
   2    0.666428 +-J*    0.150211       2   -0.753474 +-J*    0.657478

BLOCKS OF SECOND ORDER

CONSTANT GAIN FACTOR =    0.1559554E-02                                           ≥2

L     B2(L)        B1(L)        B0(L)          C1(L)        C0(L)
1   1.00000000  -0.56128475  1.00000000    -1.43430551   0.86343788
2   1.00000000  -0.00000001  1.00000000    -1.35192616   0.62672679
3   1.00000000   1.50694796  1.00000000    -1.33285540   0.46668934
```

```
EXTREMES OF THE MAGNITUDE OF THE TRANSFER FUNCTION (COEFS NOT ROUNDED)

MAXIMA IN THE PASSBAND (P)
MAXIMA IN THE STOPBAND (S)
BAND   S-DOMAIN          Z-DOMAIN              MAGNITUDE
                   IN RAD     IN DEGREE
P      0.0885      0.1766      10.1168          1.000000
S      0.7265      1.2566      72.0000          0.000143
P      0.2361      0.4636      26.5651          1.000000
S      0.8277      1.3828      79.2305          0.000143
P      0.3150      0.6103      34.9664          1.000000
S      1.3948      1.8975     108.7215          0.000143
S0.4254E 38        3.1416     180.0000          0.000143


MINIMA IN THE PASSBAND (P)
MINIMA IN THE STOPBAND (S)
BAND   S-DOMAIN          Z-DOMAIN              MAGNITUDE
                   IN RAD     IN DEGREE
P      0.          0.           0.              0.997147
S      0.7495      1.2863      73.7015          0.
P      0.1693      0.3353      19.2128          0.997147
S      1.0000      1.5708      90.0000          0.000000
P      0.2852      0.5557      31.8366          0.997147
S      2.6670      2.4241     138.8922          0.
P      0.3249      0.6283      36.0000          0.997147
```

≥4

3.2 Optimized design of a lowpass filter

3.2.1 Specifications (see sect. 3.1.1)

3.2.2 Chosen Realization

Structure 16; first canonic form with separate scaling factors; mathematical rounding after each multiplier; scaling by powers of two.

3.2.3 Necessary Input Cards

```
//SS
*0210        20 0.2
*0230  T
*0310           20.
*0341           2.              4.
*0351           0.02            0.001
*0420         1
*0450         1
*0510  T     16
//FI
```

3.2.4 Line Printer Output of the Designed Filter

```
     ===  DOREDI  ===  VERSION  V005 ===


DATA INPUT
//SS
*   2  10       20   0.200000E 00   0.              0.
*   2  30  T     0   0.             0.              0.
*   3  10        0   0.200000E 02   0.              0.
*   3  41        0   0.200000E 01   0.400000E 01    0.
*   3  51        0   0.200000E-01   0.100000E-02    0.
*   4  20        1   0.             0.              0.
*   4  50        1   0.             0.              0.
*   5  10  T    16   0.             0.              0.
//FI
```

```
TOLERANCE SCHEME


FILTER-TYPE
                          LOWPASS

APPROXIMATION
                          ELLIPTIC

SAMPLING FREQ.           20.000000
CUTOFF FREQUENCIES        2.000000      4.000000

NORM. CUTOFF FREQ.        0.628319      1.256637

CUTOFF FREQ. S-DOM.       0.324920      0.726543


PASSBAND RIPPLE(S)          0.020000      0.1755 DB   P =       0.1990
STOPBAND RIPPLE(S)          0.001000     60.0000 DB


MIN. FILTER DEGREE         4.6285

DEGREE EXTENSION           0.9257

CHOSEN FILTER DEG.         6


BOUND PAIR OF THE DESIGN PARAMETER C     3.2927  .LE.  C  .LE.   61.6686

REALIZED
NORM. CUTOFF FREQ.        0.628319      1.256637



OPTIMIZATION IS TERMINATED AFTER THE    10. STEP
THREE UNSUCCESSFUL STEPS

CHOSEN DESIGN PAR.   CX = 0.0031250  (=)  C =        3.3230284

UTILIZATION OF THE PASSBAND  DELTA P  = 0.0000599  =   0.29930 PERCENT
                   STOPBAND  DELTA S  = 0.0009909  =  99.08854 PERCENT



POLES AND ZEROS IN THE Z-DOMAIN

CONSTANT GAIN FACTOR =   0.6750014E-02


  NUM.          POLES                    NUM.           ZEROS
    2    0.621103 +-J*   0.646824         2     0.280642 +-J*   0.959812
    2    0.532684 +-J*   0.440233         2     0.000000 +-J*   1.000000
    2    0.481727 +-J*   0.159623         2    -0.753474 +-J*   0.657478

BLOCKS OF SECOND ORDER

CONSTANT GAIN FACTOR =   0.6750014E-02

 L      B2(L)          B1(L)          B0(L)            C1(L)          C0(L)
 1   1.00000000   -0.56128475    1.00000000     -1.24220650    0.80415063
 2   1.00000000   -0.00000001    1.00000000     -1.06536840    0.47755731
 3   1.00000000    1.50694796    1.00000000     -0.96345360    0.25754030
```

```
 EXTREMES OF THE MAGNITUDE OF THE TRANSFER FUNCTION (COEFS NOT ROUNDED)

 MAXIMA IN THE PASSBAND (P)
 MAXIMA IN THE STOPBAND (S)
BAND   S-DOMAIN         Z-DOMAIN            MAGNITUDE
                   IN RAD    IN DEGREE
P      0.0885      0.1766      10.1168       1.000000
S      0.7265      1.2566      72.0000       0.000991
P      0.2361      0.4636      26.5651       1.000000
S      0.8277      1.3828      79.2305       0.000991
P      0.3150      0.6103      34.9664       1.000000
S      1.3948      1.8975     108.7215       0.000991
S0.4254E 38        3.1416     180.0000       0.000991

 MINIMA IN THE PASSBAND (P)
 MINIMA IN THE STOPBAND (S)
BAND   S-DOMAIN         Z-DOMAIN            MAGNITUDE
                   IN RAD    IN DEGREE
P      0.          0.           0.           0.999940
S      0.7495      1.2863      73.7015       0.
P      0.1693      0.3353      19.2128       0.999940
S      1.0000      1.5708      90.0000       0.000000
P      0.2852      0.5557      31.8366       0.999940
S      2.6670      2.4241     138.8922       0.
P      0.3249      0.6283      36.0000       0.999940


 LAYOUT OF THE ROUNDED COEFFICIENTS

 WORDLENGTH      8


               COEF            COEFS             IR    IE            ID      OCTAL

B2( 1)     1.00000000=    1.00000000*2**(   0 -   0)      2**-100 000000000000
B2( 2)     1.00000000=    1.00000000*2**(   0 -   0)      2**-100 000000000000
B2( 3)     1.00000000=    1.00000000*2**(   0 -   0)      2**-100 000000000000

B1( 1)    -0.56250000=   -0.28125000*2**(   1 -   0)     -2**-100 156000000000
B1( 2)     0.         =   0.        *2**(   1 -   0)      2**-100 100000000000
B1( 3)     1.50000000=    0.75000000*2**(   1 -   0)      2**-100 060000000000

B0( 1)     1.00000000=    1.00000000*2**(   0 -   0)      2**-100 000000000000
B0( 2)     1.00000000=    1.00000000*2**(   0 -   0)      2**-100 000000000000
B0( 3)     1.00000000=    1.00000000*2**(   0 -   0)      2**-100 000000000000

C1( 1)    -1.25000000=   -0.62500000*2**(   1 -   0)     -2**-100 130000000000
C1( 2)    -1.06250000=   -0.53125000*2**(   1 -   0)     -2**-100 136000000000
C1( 3)    -0.96875000=   -0.48437500*2**(   1 -   0)     -2**-100 141000000000

C0( 1)     0.80468750=    0.80468750*2**(   0 -   0)      2**-100 063400000000
C0( 2)     0.47656250=    0.47656250*2**(   0 -   0)      2**-100 036400000000
C0( 3)     0.25781250=    0.25781250*2**(   0 -   0)      2**-100 020400000000

 SEARCH OF MINIMUM WORDLENGTH

 IWL       EPS-P       EPS-S       PMAX
   8     0.289635    0.979014    1.023068

 BLOCKS OF SECOND ORDER

 CONSTANT GAIN FACTOR =    0.6750014E-02

  L     B2(L)        B1(L)        B0(L)           C1(L)         C0(L)
  1   1.00000000  -0.56250000   1.00000000     -1.25000000   0.80468750
  2   1.00000000   0.           1.00000000     -1.06250000   0.47656250
  3   1.00000000   1.50000000   1.00000000     -0.96875000   0.25781250

 OPTIMIZATION OF THE PAIRING AND ORDERING

 STORAGE FOR THE INTERMEDIATE RESULTS  ISTOR = 100

 FIXED PAIRING                                   YES
```

```
REALIZED STRUCTURE                          ISTRU =  16

SCALING  OPTION                             ISCAL =   2
         BY A FACTOR OF POWER TWO                   YES
         CHOSEN MAXIMUM OF THE
         OVERFLOW POINTS                    SCALM = 1.000

REALIZATION OF THE COEFFICIENTS
    CONSIDERATION OF THE QUANTIZATION               YES
    WORDLENGTH                              IWLR  =   8
                   (FOR IWLR=100 NO ROUNDING)

                                                             [1]   [2]
NUM  DEN   SCALING   UNCOR. NOISE   COR. NOISE      DC-OFFSET CHA CO-QUAN

  2    2   0.125000   0.28031E 02  0.            0.             3    3 BIT
                                                                0    0 BIT
                                                                0    0 BIT
                                                                3    4 BIT
                                                                3    7 BIT

  3    3   0.125000   0.65466E 02  0.            0.             3    3 BIT
                                                                3    1 BIT
                                                                0    0 BIT
                                                                3    5 BIT
                                                                3    7 BIT

  1    1   0.250000   0.64145E 02  0.            0.             3    2 BIT
                                                                3    4 BIT
                                                                0    0 BIT
                                                                3    2 BIT
                                                                3    7 BIT
                                                  --------------
                                                 0.
TOTAL NOISE           0.15764E 03  0.            0.           = 0.1576E 03

ABSOLUTE OUTPUT NOISE        ANP   =  0.46048E 01  (Q*Q)     =  6.63 DB

SCALING AT THE OUTPUT        SCA   =  0.59205E 00

RELATIVE OUTPUT NOISE        RNP   =  0.13137E 02  (Q*Q)     , 11.18 DB

RELATED TO MAX /H/ = 1

INNER NOISE FIGURE           RIN   =  0.15764E 03 (Q*Q)/12 =  21.98 DB

RELATED TO MAX /H/ = 1

ENTRANCE NOISE FIGURE        REN   =  0.25892E 00

RELATED TO MAX /H/ = 1

ADDITIONAL WORDLENGTH       DELTA W = /   3.484/ =     4

STRUCTURE WITH SEPARATE SCALING FACTORS B2(L)

FOR STRUCTURES 21 TO 28  B2(0) = GAIN FACTOR

BLOCKS OF SECOND ORDER

CONSTANT GAIN FACTOR =   0.1000000E 01
 L      B2(L)          B1(L)          B0(L)           C1(L)          C0(L)
 1   0.12500000   0.            1.00000000     -1.06250000    0.47656250
 2   0.12500000   1.50000000    1.00000000     -0.96875000    0.25781250
 3   0.25000000  -0.56250000    1.00000000     -1.25000000    0.80468750
```

---

[1] CHA chosen characteristic for the nonlinearities N1...N5.

[2] CO-QUAN actual wordlength reduction in the nonlinearities N1...N5; this number of bits to be cut off is determined by the minimum actual coefficient quantization step size.

3.3 Optimized design of a bandpass filter

The following examples illustrate some further possibilities of DOREDI. First, the input for the optimization of a bandpass filter is given.

3.3.1 Specification of the Tolerance Scheme

*0310	Bandpass	
*0330	Order	$n = 10$
*0343	Normalized passband cutoff frequencies;	$\Omega_{P_1} = \pi/10$; $\Omega_{P_2} = 2\pi/10$
*0344	The normalized stopband cutoff frequencies are chosen automatically, taking into account the default degree extension of 0.2;	
*0352	Tolerated passband attenuation	$a_P = 0.2$ dB
	Minimum stop band attenuation	$a_S = 40$ dB

3.3.2 Chosen Realization

*0420	For all the coefficients one common range for the fixed point representation is chosen, according to the greatest magnitude of the actual coefficients.
*0240	Pairing and ordering of the numerators and denominators of the second-order blocks will be done by dynamic programming. To avoid small overflows, evoked by the rounding error of the state variables, the chosen maximum of the scaling is set to SCALM=0.99.
*0510	The output noise will be optimized for the second canonic form with only one nonlinearity after the summation (structure 23).
*0520	To avoid limit cycles, random rounding is used within the recursive loop (JJCNO=7). Only at the end of the cascade two's complement rounding is assigned (JJCNO=1)

3.3.3 Analysis of the Designed Filter

*0630 *0640	After the optimization of the described filter a printer plot of the magnitude and the attenuation will show the behavior of the frequency response for the coefficient set with final wordlength.

3.3.4 Necessary Input Data Cards

```
//SS
*0240          0.99
*0310        3
*0330       10
*0343                          0.31416
*0344          0.62832
*0352          0.2             40.
*0420        5
*0510       23
*0520      107
//AN
*0630  T
*0640  T
//FI
```

3.3.5 Line Printer Output of the Designed Filter

```
    ===  DOREDI  ===  VERSION  V005 ===


DATA INPUT
//SS
*    2  40         0    0.990000E 00   0.             0.
*    3  10         3    0.             0.             0.
*    3  30        10    0.             0.             0.
*    3  43         0    0.             0.314160E 00   0.
*    3  44         0    0.628320E 00   0.             0.
*    3  52         0    0.200000E 00   0.400000E 02   0.
*    4  20         5    0.             0.             0.
*    5  10        23    0.             0.             0.
*    5  20       107    0.             0.             0.
//AN


TOLERANCE SCHEME


FILTER-TYPE
                        BANDPASS

APPROXIMATION
                        ELLIPTIC

NORM. CUTOFF FREQ.      0.             0.314160      0.628320      0.

CUTOFF FREQ. S-DOM.     0.             0.158385      0.324921      0.


PASSBAND RIPPLE(S)        0.022763      0.2000 DB    P =       0.2121
STOPBAND RIPPLE(S)        0.010000     40.0000 DB


FILTER DEGREE ASSIGNED TO ORDER 10


MIN. FILTER DEGREE       4.1667

DEGREE EXTENSION         0.8333

CHOSEN FILTER DEG.       5
FOR THE REFERENCE LOWPASS

ACTUAL FILTER DEGREE    10


BOUND PAIR OF THE DESIGN PARAMETER C     2.1967  .LE.  C  .LE.    9.8821

REALIZED
NORM. CUTOFF FREQ.      0.251463      0.314160      0.628320      0.773302



OPTIMIZATION IS TERMINATED AFTER THE     4. STEP
MAXIMUM NUMBER OF STEPS


BLOCKS OF SECOND ORDER

CONSTANT GAIN FACTOR =    0.8861927E-02

 L      B2(L)         B1(L)          B0(L)            C1(L)         C0(L)
 1   1.00000000  -1.41015625   1.00000000     -1.52148437   0.93359375
 2   1.00000000  -1.09375000   1.00000000     -1.46679687   0.78515625
 3   1.00000000   0.          -1.00000000     -1.57226563   0.74218750
 4   1.00000000  -1.96484375   1.00000000     -1.77539063   0.87500000
 5   1.00000000  -1.93945313   1.00000000     -1.88085937   0.96679687


OPTIMIZATION OF THE PAIRING AND ORDERING

STORAGE FOR THE INTERMEDIATE RESULTS  ISTOR = 100
```

```
FIXED PAIRING                                     NO

REALIZED STRUCTURE                      ISTRU =  23

SCALING  OPTION                         ISCAL =   2
         BY A FACTOR OF POWER TWO                NO
         CHOSEN MAXIMUM OF THE
         OVERFLOW POINTS                SCALM = 0.990

REALIZATION OF THE COEFFICIENTS
    CONSIDERATION OF THE QUANTIZATION            YES
    WORDLENGTH                          IWLR  =  11
                        (FOR IWLR=100 NO ROUNDING)

NUM  DEN   SCALING   UNCOR. NOISE   COR. NOISE     DC-OFFSET CHA CO-QUAN

  1    3   0.103516   0.43090E 02  0.           0.             0    0 BIT
                                                               0    0 BIT
                                                               0    0 BIT
                                                               7    9 BIT
                                                               0    0 BIT

  2    1   0.558594   0.60455E 02  0.           0.             0    0 BIT
                                                               0    0 BIT
                                                               0    0 BIT
                                                               7   10 BIT
                                                               0    0 BIT

  5    2   0.250000   0.32107E 02  0.           0.             0    0 BIT
                                                               0    0 BIT
                                                               0    0 BIT
                                                               7    9 BIT
                                                               0    0 BIT

  4    5   0.550781   0.51710E 03  0.           0.             0    0 BIT
                                                               0    0 BIT
                                                               0    0 BIT
                                                               7   10 BIT
                                                               0    0 BIT

  3    4   0.679688   0.16365E 03  0.           0.             0    0 BIT
                                                               0    0 BIT
                                                               0    0 BIT
                                                               7   10 BIT
                                                               0    0 BIT

  0    0   1.583008   0.10204E 01  0.           0.49324E-03    1   10 BIT
                                                               0    0 BIT
                                                               0    0 BIT
                                                               0    0 BIT
                                                               0    0 BIT
                                             --------------
                                                0.49324E-03
TOTAL NOISE              0.81743E 03  0.        0.24329E-06= 0.8174E 03

ABSOLUTE OUTPUT NOISE        ANP   =  0.66756E 02  (Q*Q)     = 18.24 DB

SCALING AT THE OUTPUT        SCA   =  0.98994E 00

RELATIVE OUTPUT NOISE        RNP   =  0.68119E 02  (Q*Q)     , 18.33 DB

RELATED TO MAX /H/ = 1

INNER NOISE FIGURE           RIN   =  0.81743E 03 (Q*Q)/12 =  29.12 DB

RELATED TO MAX /H/ = 1

ENTRANCE NOISE FIGURE        REN   =  0.11543E 00

RELATED TO MAX /H/ = 1

ADDITIONAL WORDLENGTH       DELTA W = /  4.759/ =     5
```

BLOCKS OF SECOND ORDER

CONSTANT GAIN FACTOR = 0.1035156E 00

L	B2(L)	B1(L)	B0(L)	C1(L)	C0(L)
1	0.55859375	-0.78770447	0.55859375	-1.57226563	0.74218750
2	0.25000000	-0.27343750	0.25000000	-1.52148437	0.93359375
3	0.55078125	-1.06821442	0.55078125	-1.46679687	0.78515625
4	0.67968750	-1.33547974	0.67968750	-1.88085937	0.96679687
5	1.58300781	0.	-1.58300781	-1.77539063	0.87500000

OPTIMIZATION IS TERMINATED AFTER THE 6. STEP
THREE UNSUCCESSFUL STEPS

CHOSEN DESIGN PAR. CX = 0.0617819 (=) C = 2.4105675

UTILIZATION OF THE PASSBAND DELTA P = 0.0013992 = 6.14692 PERCENT
STOPBAND DELTA S = 0.0091129 = 91.12871 PERCENT

POLES AND ZEROS IN THE Z-DOMAIN

CONSTANT GAIN FACTOR = 0.9106737E-02

NUM.	POLES		NUM.	ZEROS	
2	0.760564 +-J*	0.594928	2	0.704866 +-J*	0.709340
2	0.731746 +-J*	0.498722	2	0.546626 +-J*	0.837377
2	0.784508 +-J*	0.351874	1	-1.000000 +-J*	0.
2	0.887864 +-J*	0.292069	1	1.000000 +-J*	0.
2	0.940917 +-J*	0.286555	2	0.982093 +-J*	0.188399
			2	0.969864 +-J*	0.243648

BLOCKS OF SECOND ORDER

CONSTANT GAIN FACTOR = 0.9106737E-02

L	B2(L)	B1(L)	B0(L)	C1(L)	C0(L)
1	1.00000000	-1.40973199	1.00000000	-1.52112896	0.93239814
2	1.00000000	-1.09325187	1.00000000	-1.46349257	0.78417628
3	1.00000000	0.	-1.00000000	-1.56901532	0.73926776
4	1.00000000	-1.96418508	1.00000000	-1.77572700	0.87360604
5	1.00000000	-1.93972760	1.00000000	-1.88183418	0.96743870

EXTREMES OF THE MAGNITUDE OF THE TRANSFER FUNCTION (COEFS NOT ROUNDED)

MAXIMA IN THE PASSBAND (P)
MAXIMA IN THE STOPBAND (S)

BAND	S-DOMAIN	Z-DOMAIN IN RAD	Z-DOMAIN IN DEGREE	MAGNITUDE
P	0.1606	0.3184	18.2449	1.000001
S	0.4071	0.7733	44.3070	0.009113
P	0.1804	0.3571	20.4577	1.000000
S	0.4495	0.8449	48.4102	0.009113
P	0.2269	0.4462	25.5628	1.000000
S	0.8845	1.4483	82.9829	0.009113
P	0.2852	0.5556	31.8355	1.000000
S	0.0582	0.1162	6.6600	0.009113
P	0.3205	0.6203	35.5398	1.000000
S	0.1145	0.2280	13.0618	0.009113
S	0.1264	0.2515	14.4078	0.009113

```
 MINIMA IN THE PASSBAND (P)
 MINIMA IN THE STOPBAND (S)
BAND  S-DOMAIN        Z-DOMAIN           MAGNITUDE
                 IN RAD   IN DEGREE
P     0.1584     0.3142    18.0000        0.998602
S     0.4161     0.7886    45.1813        0.
P     0.1676     0.3320    19.0251        0.998601
S     0.5414     0.9925    56.8642        0.
P     0.2003     0.3954    22.6550        0.998601
S0.8507E 38      3.1416   180.0000        0.000000
P     0.2569     0.5029    28.8158        0.998601
S     0.         0.         0.            0.
P     0.3071     0.5959    34.1450        0.998601
S     0.0951     0.1895    10.8594        0.
P     0.3249     0.6283    36.0001        0.998601
S     0.1237     0.2461    14.1019        0.000000
```

```
 LAYOUT OF THE ROUNDED COEFFICIENTS

 WORDLENGTH    11


            COEF            COEFS            IR   IE          ID      OCTAL

B2( 1)    0.54687500=   0.27343750*2**(   1 -  0)     2**-100 021400000000
B2( 2)    0.25000000=   0.12500000*2**(   1 -  0)     2**-100 010000000000
B2( 3)    0.55468750=   0.27734375*2**(   1 -  0)     2**-100 021600000000
B2( 4)    0.69140625=   0.34570312*2**(   1 -  0)     2**-100 026100000000
B2( 5)    1.60546875=   0.80273438*2**(   1 -  0)     2**-100 063300000000

B1( 1)   -0.77148437=  -0.38574219*2**(   1 -  0)    -2**-100 147240000000
B1( 2)   -0.27343750=  -0.13671875*2**(   1 -  0)    -2**-100 167200000000
B1( 3)   -1.07617188=  -0.53808594*2**(   1 -  0)    -2**-100 135440000000
B1( 4)   -1.35742187=  -0.67871094*2**(   1 -  0)    -2**-100 124440000000
B1( 5)    0.         =   0.          *2**(   1 -  0)     2**-100 100000000000

B0( 1)    0.54687500=   0.27343750*2**(   1 -  0)     2**-100 021400000000
B0( 2)    0.25000000=   0.12500000*2**(   1 -  0)     2**-100 010000000000
B0( 3)    0.55468750=   0.27734375*2**(   1 -  0)     2**-100 021600000000
B0( 4)    0.69140625=   0.34570312*2**(   1 -  0)     2**-100 026100000000
B0( 5)   -1.60546875=  -0.80273438*2**(   1 -  0)    -2**-100 114500000000

C1( 1)   -1.56835937=  -0.78417969*2**(   1 -  0)    -2**-100 115640000000
C1( 2)   -1.52148437=  -0.76074219*2**(   1 -  0)    -2**-100 117240000000
C1( 3)   -1.46289063=  -0.73144531*2**(   1 -  0)    -2**-100 121140000000
C1( 4)   -1.88085937=  -0.94042969*2**(   1 -  0)    -2**-100 103640000000
C1( 5)   -1.77539063=  -0.88769532*2**(   1 -  0)    -2**-100 107140000000

C0( 1)    0.74023438=   0.37011719*2**(   1 -  0)     2**-100 027540000000
C0( 2)    0.93164062=   0.46582032*2**(   1 -  0)     2**-100 035640000000
C0( 3)    0.78320312=   0.39160156*2**(   1 -  0)     2**-100 031040000000
C0( 4)    0.96679687=   0.48339844*2**(   1 -  0)     2**-100 036740000000
C0( 5)    0.87304687=   0.43652344*2**(   1 -  0)     2**-100 033740000000
```

```
 SEARCH OF MINIMUM WORDLENGTH

 IWL      EPS-P       EPS-S       PMAX
  11    0.557585    0.991198    0.984173

 BLOCKS OF SECOND ORDER

 CONSTANT GAIN FACTOR =    0.1058192E 00

 L      B2(L)        B1(L)        B0(L)          C1(L)        C0(L)
 1   0.54687500  -0.77148437   0.54687500    -1.56835937   0.74023438
 2   0.25000000  -0.27343750   0.25000000    -1.52148437   0.93164062
 3   0.55468750  -1.07617188   0.55468750    -1.46289063   0.78320312
 4   0.69140625  -1.35742187   0.69140625    -1.88085937   0.96679687
 5   1.60546875   0.          -1.60546875    -1.77539063   0.87304687
```

```
REALIZED STRUCTURE                          ISTRU =  23

SCALING  OPTION                             ISCAL =   0
         BY A FACTOR OF POWER TWO                    NO
         CHOSEN MAXIMUM OF THE
         OVERFLOW POINTS                    SCALM = 0.990

REALIZATION OF THE COEFFICIENTS
    CONSIDERATION OF THE QUANTIZATION                YES

    WORDLENGTH                              IWLR  =  11
                         (FOR IWLR=100 NO ROUNDING)

NUM  DEN   SCALING   UNCOR. NOISE   COR. NOISE    DC-OFFSET CHA CO-QUAN

  1    3   1.000000   0.42382E 02  0.            0.            0    0 BIT
                                                               0    0 BIT
                                                               0    0 BIT
                                                               7    9 BIT
                                                               0    0 BIT

  2    1   1.000000   0.63044E 02  0.            0.            0    0 BIT
                                                               0    0 BIT
                                                               0    0 BIT
                                                               7    9 BIT
                                                               0    0 BIT

  5    2   1.000000   0.33679E 02  0.            0.            0    0 BIT
                                                               0    0 BIT
                                                               0    0 BIT
                                                               7    9 BIT
                                                               0    0 BIT

  4    5   1.000000   0.52348E 03  0.            0.            0    0 BIT
                                                               0    0 BIT
                                                               0    0 BIT
                                                               7    9 BIT
                                                               0    0 BIT

  3    4   1.000000   0.16769E 03  0.            0.            0    0 BIT
                                                               0    0 BIT
                                                               0    0 BIT
                                                               7    9 BIT
                                                               0    0 BIT

  0    0   1.000000   0.10324E 01  0.            0.19845E-02   1    8 BIT
                                                               0    0 BIT
                                                               0    0 BIT
                                                               0    0 BIT
                                                               0    0 BIT
                                            --------------
                                             0.19845E-02
TOTAL NOISE               0.83131E 03  0.    0.39384E-05= 0.8313E 03

ABSOLUTE OUTPUT NOISE        ANP   =  0.67100E 02  (Q*Q)    = 18.27 DB

SCALING AT THE OUTPUT        SCA   =  0.98417E 00

RELATIVE OUTPUT NOISE        RNP   =  0.69276E 02  (Q*Q)    , 18.41 DB

RELATED TO MAX /H/ = 1

INNER NOISE FIGURE           RIN   =  0.83131E 03 (Q*Q)/12 =  29.20 DB

RELATED TO MAX /H/ = 1

ENTRANCE NOISE FIGURE        REN   =  0.11864E 00

RELATED TO MAX /H/ = 1

ADDITIONAL WORDLENGTH      DELTA W = /  4.769/ =    5
```

BLOCKS OF SECOND ORDER

CONSTANT GAIN FACTOR =   0.1058192E 00

L	B2(L)	B1(L)	B0(L)	C1(L)	C0(L)
1	0.54687500	-0.77148437	0.54687500	-1.56835937	0.74023438
2	0.25000000	-0.27343750	0.25000000	-1.52148437	0.93164062
3	0.55468750	-1.07617188	0.55468750	-1.46289063	0.78320312
4	0.69140625	-1.35742187	0.69140625	-1.88085937	0.96679687
5	1.60546875	0.	-1.60546875	-1.77539063	0.87304687

```
 DATA INPUT
 *   6  30  T     0    0.            0.            0.

                             MAGNITUDE OF THE FREQUENCY RESPONSE
MAGNITUDE  X    0.                0.400               0.800
0.        0.    +------------------------------------------------------
0.279E-02 0.026 +    .    .    .    .    .    .    .    .    .    .
0.536E-02 0.051 +    .    .    .    .    .    .    .    .    .    .
0.749E-02 0.077 +    .    .    .    .    .    .    .    .    .    .
0.884E-02 0.102 +    .    .    .    .    .    .    .    .    .    .
0.897E-02 0.128 +    .    .    .    .    .    .    .    .    .    .
0.729E-02 0.153 +    .    .    .    .    .    .    .    .    .    .
0.315E-02 0.179 +    .    .    .    .    .    .    .    .    .    .
0.339E-02 0.204 +    .    .    .    .    .    .    .    .    .    .
0.779E-02 0.230 +    .    .    .    .    .    .    .    .    .    .
0.196E-01 0.255 .+   .    .    .    .    .    .    .    .    .    .
0.273E 00 0.281 .    .    .  + .    .    .    .    .    .    .    .
0.942E 00 0.306 .    .    .    .    .    .    .    .    .  +  .
0.972E 00 0.332 .    .    .    .    .    .    .    .    .   + .
0.973E 00 0.358 .    .    .    .    .    .    .    .    .   + .
0.973E 00 0.383 .    .    .    .    .    .    .    .    .   + .
0.976E 00 0.409 .    .    .    .    .    .    .    .    .    +.
0.980E 00 0.434 .    .    .    .    .    .    .    .    .    +.
0.982E 00 0.460 .    .    .    .    .    .    .    .    .    +.
0.983E 00 0.485 .    .    .    .    .    .    .    .    .    +.
0.983E 00 0.511 .    .    .    .    .    .    .    .    .    +.
0.984E 00 0.536 .    .    .    .    .    .    .    .    .    +.
0.984E 00 0.562 .    .    .    .    .    .    .    .    .    +.
0.982E 00 0.587 .    .    .    .    .    .    .    .    .    +.
0.982E 00 0.613 .    .    .    .    .    .    .    .    .    +.
0.965E 00 0.639 .    .    .    .    .    .    .    .    .   + .
0.761E 00 0.664 .    .    .    .    .    .    .  + .    .    .
0.371E 00 0.690 .    .    .    .  + .    .    .    .    .    .
0.147E 00 0.715 .    . +  .    .    .    .    .    .    .    .
0.532E-01 0.741 . +  .    .    .    .    .    .    .    .    .
0.144E-01 0.766 +    .    .    .    .    .    .    .    .    .    .
0.160E-02 0.792 +    .    .    .    .    .    .    .    .    .    .
0.757E-02 0.817 +    .    .    .    .    .    .    .    .    .    .
0.898E-02 0.843 +    .    .    .    .    .    .    .    .    .    .
0.833E-02 0.868 +    .    .    .    .    .    .    .    .    .    .
0.680E-02 0.894 +    .    .    .    .    .    .    .    .    .    .
0.496E-02 0.919 +    .    .    .    .    .    .    .    .    .    .
0.311E-02 0.945 +    .    .    .    .    .    .    .    .    .    .
0.135E-02 0.971 +    .    .    .    .    .    .    .    .    .    .
0.234E-03 0.996 +    .    .    .    .    .    .    .    .    .    .
0.164E-02 1.022 +    .    .    .    .    .    .    .    .    .    .
0.287E-02 1.047 +    .    .    .    .    .    .    .    .    .    .
0.394E-02 1.073 +    .    .    .    .    .    .    .    .    .    .
0.485E-02 1.098 +    .    .    .    .    .    .    .    .    .    .
0.563E-02 1.124 +    .    .    .    .    .    .    .    .    .    .
0.629E-02 1.149 +    .    .    .    .    .    .    .    .    .    .
0.685E-02 1.175 +    .    .    .    .    .    .    .    .    .    .
0.732E-02 1.200 +    .    .    .    .    .    .    .    .    .    .


0.288E-03 3.091 +    .    .    .    .    .    .    .    .    .    .
0.144E-03 3.116 +    .    .    .    .    .    .    .    .    .    .
0.112E-10 3.142 +    .    .    .    .    .    .    .    .    .    .
```

```
 DATA INPUT
 *   6  40  T     0    0.              0.              0.


                                    ATTENUATION OF THE FREQUENCY RESPONSE
MAGNITUDE  X    0.                        40.000                  80.000
0.163E 03 0.    ------------------------------------------------------I
0.511E 02 0.026 .    .    .    .    .    +    .    .    .    .    .
0.454E 02 0.051 .    .    .    .    . +  .    .    .    .    .    .
0.425E 02 0.077 .    .    .    .    .+   .    .    .    .    .    .
0.411E 02 0.102 .    .    .    .    +    .    .    .    .    .    .
0.409E 02 0.128 .    .    .    .    +    .    .    .    .    .    .
0.427E 02 0.153 .    .    .    .    .+   .    .    .    .    .    .
0.500E 02 0.179 .    .    .    .    .    +    .    .    .    .    .
0.494E 02 0.204 .    .    .    .    .   +.    .    .    .    .    .
0.422E 02 0.230 .    .    .    .    .+   .    .    .    .    .    .
0.342E 02 0.255 .    .    .    . +  .    .    .    .    .    .    .
0.113E 02 0.281 .    +    .    .    .    .    .    .    .    .    .
0.516E 00 0.306 +    .    .    .    .    .    .    .    .    .    .
0.244E 00 0.332 +    .    .    .    .    .    .    .    .    .    .
0.240E 00 0.358 +    .    .    .    .    .    .    .    .    .    .
0.236E 00 0.383 +    .    .    .    .    .    .    .    .    .    .
0.213E 00 0.409 +    .    .    .    .    .    .    .    .    .    .
0.179E 00 0.434 +    .    .    .    .    .    .    .    .    .    .
0.158E 00 0.460 +    .    .    .    .    .    .    .    .    .    .
0.152E 00 0.485 +    .    .    .    .    .    .    .    .    .    .
0.148E 00 0.511 +    .    .    .    .    .    .    .    .    .    .
0.140E 00 0.536 +    .    .    .    .    .    .    .    .    .    .
0.141E 00 0.562 +    .    .    .    .    .    .    .    .    .    .
0.156E 00 0.587 +    .    .    .    .    .    .    .    .    .    .
0.158E 00 0.613 +    .    .    .    .    .    .    .    .    .    .
0.310E 00 0.639 +    .    .    .    .    .    .    .    .    .    .
0.237E 01 0.664 .+   .    .    .    .    .    .    .    .    .    .
0.860E 01 0.690 .   +.    .    .    .    .    .    .    .    .    .
0.167E 02 0.715 .    .   +.    .    .    .    .    .    .    .    .
0.255E 02 0.741 .    .    . +  .    .    .    .    .    .    .    .
0.368E 02 0.766 .    .    .    .  + .    .    .    .    .    .    .
0.559E 02 0.792 .    .    .    .    .    .   +.    .    .    .    .
0.424E 02 0.817 .    .    .    .    .+   .    .    .    .    .    .
0.409E 02 0.843 .    .    .    .    +    .    .    .    .    .    .
0.416E 02 0.868 .    .    .    .    .+   .    .    .    .    .    .
0.434E 02 0.894 .    .    .    .    .+   .    .    .    .    .    .
0.461E 02 0.919 .    .    .    .    .   + .    .    .    .    .    .
0.502E 02 0.945 .    .    .    .    .    +    .    .    .    .    .
0.574E 02 0.971 .    .    .    .    .    .   +.    .    .    .    .
0.726E 02 0.996 .    .    .    .    .    .    .    .+   .    .    .
0.557E 02 1.022 .    .    .    .    .    .   +.    .    .    .    .
0.508E 02 1.047 .    .    .    .    .    +    .    .    .    .    .
0.481E 02 1.073 .    .    .    .    .   +.    .    .    .    .    .
0.463E 02 1.098 .    .    .    .    .   + .    .    .    .    .    .
0.450E 02 1.124 .    .    .    .    . +  .    .    .    .    .    .
0.440E 02 1.149 .    .    .    .    . +  .    .    .    .    .    .
0.433E 02 1.175 .    .    .    .    .+   .    .    .    .    .    .
0.427E 02 1.200 .    .    .    .    .+   .    .    .    .    .    .
0.423E 02 1.226 .    .    .    .    .+   .    .    .    .    .    .
0.419E 02 1.252 .    .    .    .    .+   .    .    .    .    .    .
0.416E 02 1.277 .    .    .    .    .+   .    .    .    .    .    .
0.414E 02 1.303 .    .    .    .    +    .    .    .    .    .    .
0.413E 02 1.328 .    .    .    .    +    .    .    .    .    .    .
0.411E 02 1.354 .    .    .    .    +    .    .    .    .    .    .
0.411E 02 1.379 .    .    .    .    +    .    .    .    .    .    .
0.410E 02 1.405 .    .    .    .    +    .    .    .    .    .    .
0.410E 02 1.430 .    .    .    .    +    .    .    .    .    .    .
0.410E 02 1.456 .    .    .    .    +    .    .    .    .    .    .


0.708E 02 3.091 .    .    .    .    .    .    .    +    .    .    .
0.768E 02 3.116 .    .    .    .    .    .    .    .  + .    .    .
0.163E 03 3.142 .    .    .    .    .    .    .    .    .    .    .    I


 DATA INPUT
 //FI
```

3.4 Analysis of a given digital filter

*0110 — The coefficients of the given cascade of second-order blocks are read in directly by the data input mode 3 (results of example 3.2).

*0620 — For the noise analysis, pairing and ordering will be done in the sequence of the coefficient input. For the realization structure 16 (Fig. 9) is assumed again.

*0510
*0520 — Yet, instead of mathematical rounding, unsymmetrical sign magnitude truncation is used for the nonlinearities.

*0630 — After the noise analysis a printer plot of the frequency response in the region of $\Omega=0.0$ to 0.8 is given in the same program section.

The input data cards for this analysis are given here.

```
//AN
*0110        3
*0010        3 0.00675
*0010          1.0            -0.5625        1.0
*0010          1.0             0.0           1.0
*0010          1.0             1.5           1.0
*0010                         -1.25          0.8046875
*0010                         -1.0625        0.4765625
*0010                         -0.96875       0.2578125
*0510       16
*0520        5
*0620
*0630  T       0.0             0.8
*0650
*0660
//FI
```

3.5 Design and analysis with different parameters

3.5.1 Filter Design

*0130 — In the first program activity //SY a lowpass filter will be designed, according to the specification in sect. 3.1.1. The results are output on the line printer and additionally stored on the disk (channel number KA4) to save the original (unrounded) coefficients for further program sections. The input data are shown in sect. 3.5.4.

3.5.2 Noise Analysis of the Designed Filter

*0530 — In the second program activity //AN the filter parameters and coefficients are read in from the disk. Besides the default parameter for the filter realization the actual coefficient wordlength is assigned to IWLR=12 bit by the control card *0530. (Attention: Don't mix up the parameter IWLR with IWL for the optimization.)

Scaling is done by the absolute criterion, and the chosen maximum is set to SCALM=0.99 (see sect. 3.3.2). For the noise analysis the ordering of the second-order blocks will be arranged according to the special input.

3.5.3 Frequency Response for Rounded Coefficients

*0530
*0630
*0640 — The actual coefficient wordlength is defined by the parameter IWLR. To avoid a double rounding of the coefficients, the printer plots for frequency responses with different coefficient wordlength have to be done in separate program sections.

*0110 — At the beginning of each activity the original coefficients are read in from the disk, initialized by the parameter NIMP=2.

*0630 Then, the normalized printer plots (length one page) are output on the line
*0640 printer for the magnitude and the attenuation of the frequency response.

3.5.4 Necessary Input Data Cards

Four separate activities are necessary only in computers where it is illegal to follow a write-disc with a read-disc.

```
//SY
*0130  T
*0310         20.
*0341         2.              4.
*0351         0.02            0.001
//FI
//AN
*0110       2
*0530      12
*0620  T    1 0.99
*0040     3   1   2
*0040     3   1   2
*0630  T    1
*0640  T    1
//FI
//AN
*0110       2
*0530      10
*0630  T    1
*0640  T    1
//FI
//AN
*0110       2
*0530       8
*0630  T    1
*0640  T    1
//FI
```

4. Considerations for the Installation of the Program on a Special Computer

4.1 Language

DOREDI is written in FORTRAN and follows ANSI FORTRAN conventions very closely. The parameters and variable names are limited to six characters. Only the DECODE statement, permissible for most of the compilers, is used in addition. If DECODE or a corresponding statement is not available on the compiler in use, the alternative input format should be chosen for the program section and the input data cards (see sect. 2.2). Therefore, the input subroutines INP001, INP002, and INP003 have to be changed according to the remarks in the beginning of these subroutines before compilation. Consequently, for the alternative input format the parameters ACX, EDEG, OMLO and RMIN cannot have automatic default values. The defined default values are active as long as the parameter is not explicitly redefined by the use of an input card including this parameter; e.g., for the definition of a fixed filter degree NDEG=10, with input in the normal format

```
*0330        10
```

the default degree extension EDEG=0.2 has to be set, for the alternative input format, too.

```
*0330        10    0.2
```

4.2 Declarations, depending on the computer used

All declarations depending on the computer precision and special I/O-handling have to be defined by the PORT Mathematical Subroutine Library (see Standards chapter of this book). The functions I1MACH, R1MACH, and D1MACH are all required.

4.2.1 Format for Input of Hollerith Characters

The format must be modified to accommodate the machine in use. The following example applies *only* to the PDP 11/45.

<table>
<tr><td>IOFO</td><td>FORMAT statement for the input of 80 Hollerith characters

(n A m) n = number of INTEGER
m = number of characters encoded by one INTEGER

n * m = 80</td></tr>
<tr><td>ION</td><td>n</td></tr>
<tr><td colspan="2">For the PDP 11/45

DATA IOFO/2H(4,2H0A,2H2)/,ION/40/</td></tr>
</table>

4.2.2 Special Parameters for the Design and Optimization

		PDP 11/45
MAXDEG	maximum filter degree †	32
MBL	maximum number of blocks of second order †	16
IWLMI	minimum coefficient wordlength for the optimization	2
IWLMA	maximum coefficient wordlength for the optimization ($\leqslant$ wordlength of REAL mantissa)	24

4.2.3 Subroutine for SPECIAL OUTPUT

In order to implement a special output, like a paper tape running on a "hardware filter", the user has to replace the dummy subroutine "OUTSPE" by his own. The parameter ISPOUT, see sect. 2.2.1.4, can be used to generate several output options. The dummy subroutine is listed as an example below:

```
C
C-------------------------------------------------------------------
C SUBROUTINE:    OUTSPE
C       DUMMY SUBROUTINE FOR 'SPECIAL OUTPUT'
C-------------------------------------------------------------------
C
      SUBROUTINE OUTSPE
C
      COMMON /CONTR / IPRUN,IPCON,NINP,NOUT,NDOUT,LSPOUT,ISPOUT
      COMMON /CANPAR/ KA1,KA2,KA3,KA4,KA5,LINE
C
C     DATA TRANSFER BY NAMED COMMON BLOCKS:
C
      COMMON /FILT   / NB,FACT,B2(16),B1(16),B0(16),C1(16),C0(16)
C
C     ISPMAX: MAXIMUM NUMBER OF IMPLEMENTED OPTIONS
      ISPMAX=0
C
      IF(ISPOUT.LE.0) RETURN
      IF(ISPOUT.GT.ISPMAX) GO TO 900
C
C     EXAMPLE FOR ISPMAX=3
C
```

† A change of the parameters MAXDEG and MBL entails a change of all COMMON- and DIMENSION-statements, depending on these parameters.

```
C     GO TO (100,200,300),ISPOUT
C 100 CONTINUE
C     .
C     OPTION NO. 1
C     .
C     RETURN
C
C 200 CONTINUE
C     .
C     OPTION NO. 2
C     .
C     RETURN
C
C 300 CONTINUE
C     .
C     OPTION NO. 3
C     .
C     RETURN
C
  900 CALL ERROR (8)
      RETURN
      END
```

5. Acknowledgment

The fundamental optimization algorithms and a first program version were developed within the author's Ph.D-thesis. He wishes to express his sincere thanks to Prof. Dr. -Ing. H. W. Schuessler for the initialization of this interesting work, for many discussions, and for valuable hints. As well, the author wants to express his thanks to the reviewers, especially to Dr. J. F. Kaiser and Marie T. Dolan for their help in testing the program and achieving a portable code, and to Dr. -Ing. U. Heute for his aid in writing the manual.

References

1. H. W. Schuessler, *Digitale Systeme zur Signalverarbeitung,* Springer-Verlag, Berlin, 1973.

2. B. Gold, C. M. Rader, *Digital Processing of Signals,* McGraw-Hill Book Co., Inc., New York, 1969.

3. G. Dehner, "On the design of digital Cauer filters with coefficients of limited wordlength", *AEU,* Vol. 29, No. 4, pp. 165-168, April 1975.

4. G. Dehner, "Ein Beitrag zum rechnergestützten Entwurf rekursiver digitaler Filter minimalen Aufwandes", Ausgewählte Arbeiten über Nachrichtensysteme, No. 23, edited by H. W. Schuessler, Univ. Erlangen-Nürnberg, 1976.

5. S. Y. Hwang, "On optimization of cascade fixed-point digital filters", *IEEE Trans. Circuits and Systems,* Vol. CAS-21, No. 1, pp. 163-166, Jan. 1974.

6. G. Dehner, "A contribution to the optimization of roundoff-noise in recursive digital filters", *AEU,* Vol. 29, No. 12, pp. 505-510, Dec. 1975.

7. B. Eckhardt, "Untersuchungen des Multipliziererfehlers in digitalen Filtern", Ausgewählte Arbeiten über Nachrichtensysteme, No. 21, edited by H. W. Schuessler, Univ. Erlangen-Nürnberg, 1975.

8. T. A. C. M. Claasen, W. F. G. Mecklenbrauker, and J. B. H. Peek, "Quantization noise analysis for fixed-point digital filters using magnitude truncation for quantization", *IEEE Trans. Circuits and Systems,* Vol. CAS-22, No. 11, pp. 887-895, Nov. 1975.

9. M. Buttner, "A novel approach to eliminate limit cycles in digital filters with a minimum increase in the quantization noise", *Proc. of the 1976 IEEE Int. Symp. on Circuits and Systems,* Munchen, pp. 291-294, 1976.

10. G. Dehner, "On the noise behavior of a digital filter in cascade structure", *Proc. of the 1976 IEEE Int. Symp. on Circuits and Systems,* Munchen, pp. 348-351, 1976.

Appendix A

A.1 General description of the coefficient realization

For the realization of coefficients with finite wordlength a general description, including the fixed-point and "pseudo"-floating-point realization as well as the realization by differences, is useful. The difference realization seems to be appropriate for coefficients in the region of the largest possible numbers, especially in connection with the "pseudo"-floating-point implementation [4]. The coefficients of the transfer function,

$$H(z) = b_{20} \prod_{\lambda=1}^{l} \frac{b_{2\lambda} z^2 + b_{1\lambda} z + b_{0\lambda}}{z^2 + c_{1\lambda} z + c_{0\lambda}} \tag{A1}$$

are renamed by

$$\{b_{2\lambda}, b_{1\lambda}, b_{0\lambda}, c_{1\lambda}, c_{0\lambda}\} = a_{\nu\lambda}; \quad \nu = 1(1)5 . \tag{A2}$$

The $a_{\nu\lambda}$ are divided by Eq. (A3) according to Fig. A1.

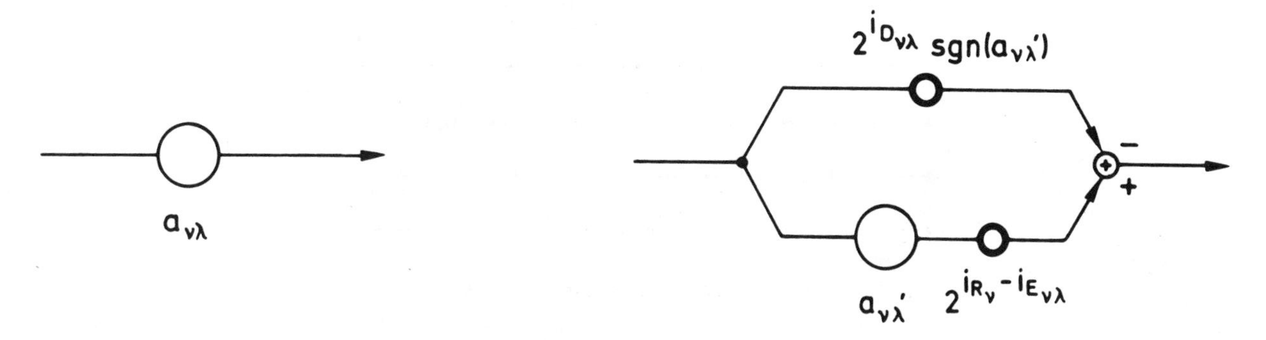

Fig. A1 "Pseudo"-floating point and difference realization of coefficients.

$$a_{\nu\lambda} = a'_{\nu\lambda} 2^{i_{R_\nu} - i_{E_{\nu\lambda}}} - \mathrm{sgn}(a'_{\nu\lambda}) 2^{i_{D_{\nu\lambda}}} . \tag{A3a}$$

"pseudo"-floating-point

$$a'_{\nu\lambda} = (a_{\nu\lambda} - \mathrm{sgn}(a_{\nu\lambda}) 2^{i_{D_{\nu\lambda}}}) 2^{i_{E_{\nu\lambda}} - i_{R_\nu}} \tag{A3b}$$

For the difference realization the exponent i_{D_ν} is calculated as

$$i_{D_{\nu\lambda}} = <\log_2 |a_{\nu\lambda}|> \tag{A4}$$

otherwise the exponent is fixed to $i_{D_{\nu\lambda}} = -100$; with the definition $2^{-100} \mathrel{\hat=} 0$. $<x>$ means the smallest integer greater than or equal to x.

Since the realized coefficient $a'_{\nu\lambda}$ has to be in the range $-1 \leqslant a'_{\nu\lambda} < 1$, i_{R_ν} describes the number range for the coefficients with the same index ν,

$$i_{R_\nu} = <\log_2 \max_{\lambda=1}^{l} \{|a_{\nu\lambda} - 2^{i_{D_\nu}} \mathrm{sgn}(a_{\nu\lambda})|\}> \tag{A5}$$

The "pseudo"-floating-point realization with no forced normalization is represented by only the positive exponent $i_{E_{\nu\lambda}}$,

$$i_{E_{\nu\lambda}} = \min \begin{cases} -[\log_2 (a_{\nu\lambda} - \mathrm{sgn}(a_{\nu\lambda}) 2^{i_{D_{\nu\lambda}}}) 2^{-i_{R_{\nu\lambda}}}] \\ \\ i_{E_{\max}} \end{cases} \tag{A6}$$

with the maximum exponent $i_{E_{\max}}$ and the definition $[x]$ meaning the largest integer less than or equal to x. In this description with $i_{D_{\nu\lambda}} = -100$ and $i_{E_{\nu\lambda}} = 0$ the simple fixed-point implementation is included.

Appendix B

B.1 Characteristics of the nonlinearities

Both the character and the power of the error at the output of a general transfer system (see Fig. B1) depend upon the characteristics of the nonlinearities applied [1], [4]. A nonlinearity may produce errors of variance $\sigma_i^{2(R)}$ which exhibit no correlation with the signal, DC-offsets of size Δ_i, and errors, described by the value $e_i^{(S)}$, which are correlated with the signal (see Fig. B2). Furthermore, DC-offsets and correlated errors are correlated with each other.

The partial errors can be determined following [4], [7], [8]. The quantization characteristics and the error values of the different nonlinearities are given in Table B1.

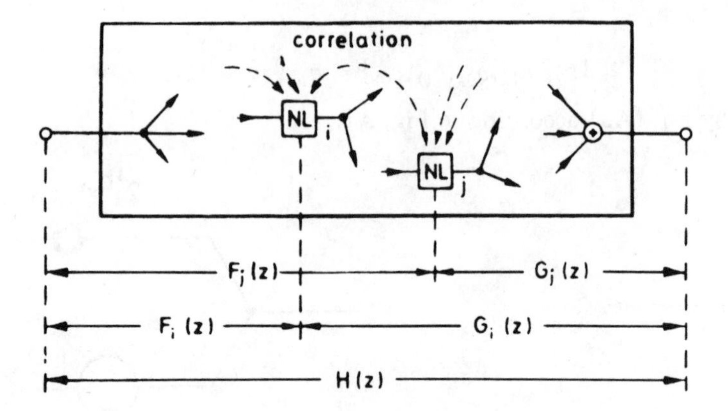

Fig. B1 General transfer function system.

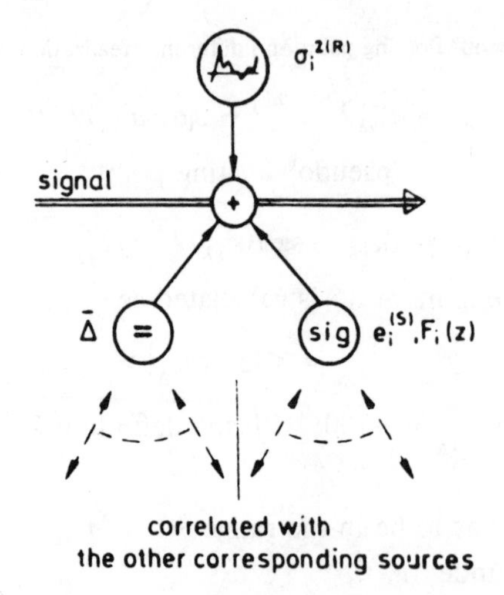

Fig. B2 Model of the nonlinearity.

B.2 Limitation of the error analysis

For reasons of computation time, the correlated errors are calculated, here, according to the simpler linearized model [8]. Furthermore, during the optimization only error correlations *within* the second-order blocks can be taken into account. In the final analysis, however, all correlations are regarded.

Table B1 Characteristics of the nonlinearities.

Nonlinearity	Characteristic	DC-offset $\overline{\Delta}_i$	correlated error value $e^{(S)}$	uncorrelated noise $\sigma_i^{2(iR)}$	mean power $\overline{\Delta_i^2}$
two's complement rounding	$[x]_{TR} = Q_S[Q_S^{-1}x+0.5]$	$\frac{Q_S Q_C}{2}$	0	$\frac{Q_S^2}{12}(1-Q_C^2)$	$\frac{Q_S^2}{12}(1+2Q_C^2)$
sign-magnitude rounding	$[x]_R = \text{sgn}(x)\,Q_S[Q_S^{-1}\|x\|+0.5]$	0	$\frac{Q_S Q_C}{2}$	$\frac{Q_S^2}{12}(1-Q_C^2)$	$\frac{Q_S^2}{12}(1+2Q_C^2)$
mathematical rounding	$[x]_{MR} = \begin{cases} Q_S[Q_S^{-1}x+0.5] & x \neq x_\nu \\ 2Q_S[Q_S^{-1}/2+0.5] & x = x_\nu \end{cases}$ [1]	0	0	$\frac{Q_S^2}{12}(1+2Q_C^2)$	$\frac{Q_S^2}{12}(1+2Q_C^2)$
two's complement truncation	$[x]_{TT} = Q_S[Q_S^{-1}x]$	$-\frac{Q_S}{2}(1-Q_C)$	0	$\frac{Q_S^2}{12}(1-Q_C^2)$	$\frac{Q_S^2}{3}(1-\frac{3}{2}Q_C+\frac{Q_C^2}{2})$
unsymmetrical sign-magnitude truncation	$[x]_{UT} = Q_S([Q_S^{-1}x]+\xi_1)$ [2]	$\frac{Q_S Q_C}{2}$	$-\frac{Q_S}{2}$	$\frac{Q_S^2}{12}(1-Q_C^2)$	$\frac{Q_S^2}{3}(1+\frac{Q_C^2}{2})$
symmetrical sign-magnitude truncation	$[x]_{ST} = Q_S([Q_S^{-1}x]+\xi_2)$ [3]	0	$-\frac{Q_S}{2}(1-Q_C)$	$\frac{Q_S^2}{12}(1-Q_C^2)$	$\frac{Q_S^2}{3}(1-\frac{3}{2}Q_C+\frac{Q_C^2}{2})$
random rounding	$[x]_{RR} = Q_S([Q_S^{-1}x]+\xi_3)$ [4]	$\frac{Q_S Q_C}{2}$	0	$\frac{Q_S^2}{3}(1-\frac{Q_C^2}{4})$	$\frac{Q_S^2}{3}(1+\frac{Q_C^2}{2})$

[1] $x_\nu = \pm(2\nu-1)\frac{Q_S}{2}\ ;\ \nu = 1(1)Q_S^{-1}$

[2] $\xi_1 = \begin{cases} 0 & \text{if } x \geq 0 \\ 1 & \text{if } x < 0 \end{cases}$

[3] $\xi_2 = \begin{cases} 0 & \text{if } x \geq 0 \text{ or } x = -\nu Q_S \\ 1 & \text{if } x < 0 \text{ and } x \neq -\nu Q_S \end{cases}$

[4] $\xi_3 = 0, 1$ randomly

x x x x x x	x x	0 0 0
q_S	q_C	

$Q_C = 2^{-q_C}$

$Q_S = 2^{-q_S}$

state variable with $q = q_S + q_C$ bits truncated to q_S bits.

Appendix C

C.1 Fixed pairing

To reduce the necessary computing time for the optimization of the state variable wordlength, a simple heuristic method can be used to get a predetermined combination of poles and zeros [1],[4],[6]. To get the fixed pairing, in a first step those poles and zeros are combined which are situated most closely to each other. Then, without taking into account poles and zeros regarded before, the following steps repeat this procedure. The output of the filter synthesis obeys this rule.

For the first canonical form (Fig. 2) the paired poles and zeros are ordered directly in the blocks of second order (Fig. C1). By means of a simple consideration this rule can be extended for the second canonical structure. The numerator function of one block may be thought of as being combined with the denominator function of the following block in an auxiliary block [6]; if the rule, mentioned above, is applied to these auxiliary blocks, a predetermined combination for the second canonical form is possible, too. The scheme in Fig. C1 explains the different pairings for the two canonical forms.

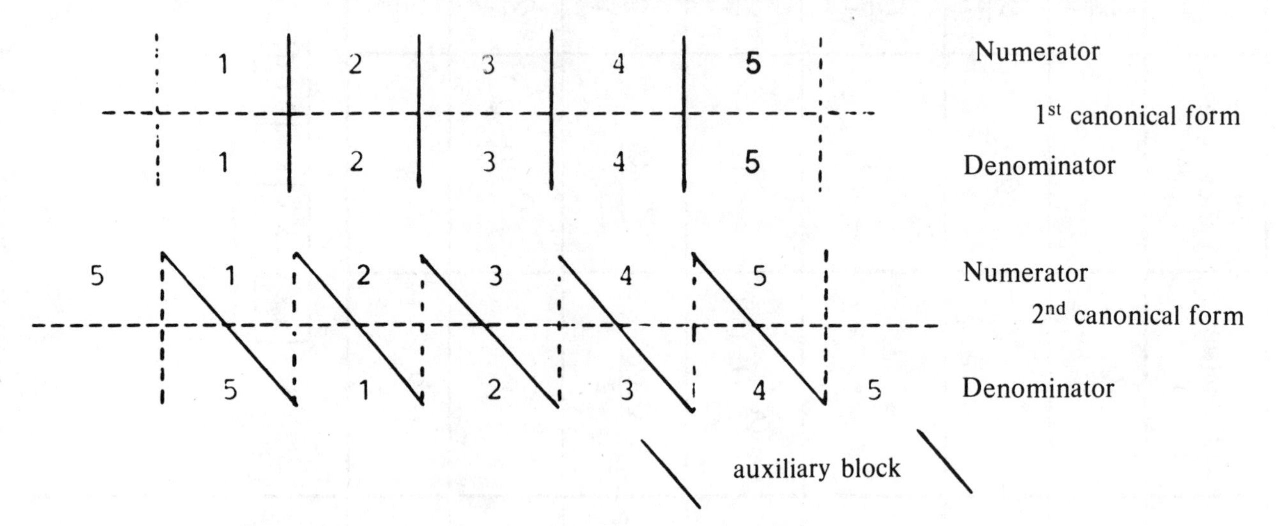

Fig. C1 Pairing of poles and zeros.

Appendix D

D.1 Parameters for the coefficient optimization

For a tolerance scheme, given by Ω_P, Ω_S, δ_P and δ_S, a minimum filter degree $n_{\min}$ is found by

$$n_{\min} = f(\Omega_P,\ \Omega_S,\ \delta_P,\ \delta_S)\ ,$$

see [1], [2] and [4]. This $n_{\min}$ is not necessarily an integer value. Taking the smallest integer n on condition that $n \geqslant n_{\min}$ and utilizing the total transition band, an unused margin in the passband or stopband tolerances remains. According to the choice of the design parameter C, this margin can be partitioned to the passband and to the stopband. For $C = C_{\min}$, the total stopband tolerances are utilized for the approximation, and the passband ripple receives a minimum. On the other hand, for $C = C_{\max}$ the total passband tolerances are utilized [1],[4].

$$C_{\min} = f_1(\Omega_P,\ \Omega_S,\ \delta_P,\ \delta_S,\ n)$$

$$C_{\max} = f_2(\Omega_P,\ \Omega_S,\ \delta_P,\ \delta_S,\ n)$$

Rounding the coefficients of the transfer function leads to errors not proportional to the passband or stopband ripples, in general. For this reason, the parameter C can be used as an additional variable for the design of digital filters with rounded coefficients.

In DOREDI, the optimization of C ($C_{\min} \leqslant C \leqslant C_{\max}$) is done by a trial and error procedure such that the required wordlength of the rounded coefficients is small [3],[4].

Finally, the parameter C is normalized to the border values $C_{\min}$ and $C_{\max}$ in order to get a uniform variation range in different filter designs:

$$\mathrm{ACX} = c = \frac{\log_{10} C - \log_{10} C_{\min}}{\log_{10} C_{\max} - \log_{10} C_{\min}}\ ;\quad 0 \leqslant c < 1\ .$$

Appendix E

Error diagnostics

Error code:	Description
1	Unexpected input code
2	Input data card out of sequence
3	Undefined program section name
4	Illegal input data
5	Parameter out of defined range
6	Sampling frequency not defined
7	Sampling frequency multi defined
8	Not implemented
9	Meaningless input data
10	Lower bound greater than upper bound
11	Missing //FI card
12	No input data available
15	Filter degree given by input data too small → filter degree chosen automatically or program aborted
16	Coefficient wordlength out of range
20	Not defined filter type
21	Not defined approximation type
22	Cutoff frequencies not in ascending order
23	Missing or illegal definition of the cutoff frequencies
24	Missing definition of the tolerances
25	Necessary filter degree too high
26	Missing filter coefficients
30	System not stable
31	No fixed pairing possible
32	Zeros not on the unit circle
33	More than one pole at $z = 0$
38	Option not implemented → Stop

Appendix F

Control data input cards

CLASS 01	*CLASS 04*
*0110 bb,NINP	*0410 LSTAB,IWL
*0120 bb,NOUT	*0420 RCOINP,JRCO
*0130 LDOUT	*0430 ECOINP,JECO
*0140 LSPOUT,ISPOUT	*0440 DCOINP,JDCO
	*0450 bb,JMAXV
	*0460 bb,JTRB2
CLASS 02	*CLASS 05*
*0210 LWLF,ITERM,ACX,ACXMI,ACXMA	*0510 LPOT2,ISTRU
*0220 bb,ITERM1	*0520 CNOINP,JCNO
*0230 LPAIRF,ISTOR	*0530 LNCO,IWLR
*0240 LSEQ,ISCAL,SCALM	
CLASS 03	*CLASS 06*
*0310 bb,ITYP,SF	*0610 LWL,IWL
*0320 bb,IAPRO,ACX	*0620 LSEQ,ISCAL,SCALM
*0330 LDEGF,NDEG,EDEG	*0630 LNOR,IPAG,OMLO,OMUP,RMAX
*0341 bb,bb,FR(1),FR(2),SF	*0640 LNOR,IPAG,OMLO,OMUP,RMAX
*0342 bb,bb,FR(3),FR(4),SF	*0650 LNOR,IPAG,OMLO,OMUP
*0343 bb,bb,OM(1),OM(2)	*0660 LNOR,IPAG,RMIN,RMAX
*0344 bb,bb,OM(3),OM(4)	
*0345 bb,bb,S(1),S(2)	
*0346 bb,bb,S(3),S(4)	
*0347 bb,bb,VSN,VD,A	
*0351 bb,bb,ADELP,ADELS	
*0352 bb,bb,AP,AS	
*0353 bb,bb,P,AS	
*0360 bb,NORMA	
*0370 LSOUT	

Appendix G

```
C
C-----------------------------------------------------------------------
C MAIN PROGRAM:          D O R E D I
C
C               DESIGN AND OPTIMIZATION OF RECURSIVE DIGITAL FILTERS
C
C      BUTTERWOTH-, CHEBYSHEV- AND ELLIPTIC-APROXIMATION
C      OPTIMIZATION OF THE COEFFICIENT WORDLENGTH AND THE OUTPUT NOISE
C      ANALYSIS OF THE DESIGNED FILTER
C
C VERSION:        V005
C
C AUTHOR:         DR.-ING. GUENTER F. DEHNER
C
C                 INSTITUT FUER NACHRICHTENTECHNIK
C                 UNIVERSITAET ERLANGEN-NUERNBERG
C
C                 CAUERSTRASSE 7
C                 D-8520 ERLANGEN, GERMANY
C
C INPUT:          MANUAL, SECTION 2
C-----------------------------------------------------------------------
C
      DOUBLE PRECISION DPI, DOMI
C
      COMPLEX ZP, ZPS
C
      COMMON /CONTR/ IPRUN, IPCON, NINP, NOUT, NDOUT, LSPOUT, NSPOUT
      COMMON /CANPAR/ KA1, KA2, KA3, KA4, KA5, LINE
      COMMON /CONST/ PI, FLMA, FLMI, FLER
      COMMON /CONST1/ MAXDEG, IWLMI, IWLMA, MBL
      COMMON /CONST2/ DPI, DOMI
      COMMON /CPLOT/ IPLOT, LNOR, IPAG, OMLO, OMUP, RMAX, RMIN
C
      COMMON /TOL/ ITYP, IAPRO, NDEG, OM(4), SF, ADELP, ADELS
      COMMON /TOLSN/ VSN, VD, A
      COMMON /DESIGN/ NDEGF, EDEG, ACX, NORMA, LSOUT, LVSN, LSYM
      COMMON /RES/ AC, ROM(4), RDELP, RDELS, NZM(4)
      COMMON /RESZ/ ZFA, ZM(18,4), ZZR(16), ZZI(16), ZPR(16), ZPI(16)
      COMMON /FILT/ NB, FACT, B2(16), B1(16), B0(16), C1(16), C0(16)
      COMMON /SFILT/ NBS, SFACT, SB2(16), SB1(16), SB0(16), SC1(16),
     *    SC0(16)
      COMMON /FILTRE/ IRCO(5), IECO(16,5), IDCO(16,5), IECOM
      COMMON /CGRID/ GR(64), NGR(12)
      COMMON /CRECO/ JRCO, JECO, JDCO, JJDCO(5), JMAXV, JTRB2, LREF
      COMMON /CCOEFW/ IWL, IWLG, IWLD, IWLL, ADEPSG, ADEPSD, ADEPSL,
     *    ISTAB, IDEPSL, IDEPSD
      COMMON /COPTCO/ LOPTW, LSTAB, ACXMI, ACXMA, ITER, ITERM, ITERM1
      COMMON /COPTST/ LOPTS, ISTOR
      COMMON /CRENO/ LCNO, ICNO(16,5)
      COMMON /CREST/ ISTRU, ISCAL, SCALM, ISEQ(16,2), LSEQ, IWLR,
     *    LPOT2, JSTRU
      COMMON /CREST1/ JSTRUS, JSTRUD
      COMMON /CREST2/ KSEQ(16,2)
      COMMON /CNOISE/ RI, RIN, RE, REN, FAC
C
      COMMON /SCRAT/ ADUM(32)
      COMMON /CPOL/ ZP(16,2), ZPS(16,2)
      COMMON /COPST2/ PN(100,2), TF(100,2), TFA(100,2)
      COMMON /CFFUNC/ PHI(5), BF2(5), BF1(5), BF0(5), IBB(5), ICOR
      COMMON /CNFUNC/ AQC(5), BN2(5), BN1(5), BN0(5)
      COMMON /CPOW/ PNU, PNC, AND, ITCORP
      COMMON /OUTDAT/ IP, PRE, PIM, IZ, ZRE, ZIM
      COMMON /COPTSP/ IB, JSEQN(16), JSEQD(16), AMAX, SCA, ALSBI
C
      CALL DEFIN0
C
      NDDEL = -1
      CALL HEAD(KA2)
C
      IPRUN = 0
      IPCON = 0
      IPLOT = 0
   10 CALL DORINP
C
      IF (NDOUT.NE.(-1) .OR. NDDEL.NE.(-1)) GO TO 20
      CALL HEAD(KA4)
      NDDEL = 0
   20 IWLRR = IWLR
      GO TO (30, 40, 30, 80), IPRUN
C
   30 NGR(8) = 0
      CALL SYNTHE
      NINP = 1
      IF (IWLRR.EQ.100) IWLR = IWL
      IF (IPRUN.LT.0) GO TO 50
      IF (IPRUN.NE.3) GO TO 90
C
   40 IF (NINP.EQ.0) GO TO 100
      IF (NINP.GE.3) NGR(8) = 0
C
      CALL SEQSTA
C
      GO TO 60
   50 LOPTS = 0
      ISCAL = 0
      LSEQ = 0
C
   60 CALL SEQOPT
C
      IF (IPRUN.EQ.(-3)) GO TO 90
      IF (IPRUN.EQ.3) GO TO 70
      IF (IWLR.LT.100) CALL SEQSTA
      GO TO 90
C
C JSTRU .EQ. 1        SECOND COEFFICIENT OPTIMIZATION
C
   70 IF (JSTRU.EQ.2) GO TO 90
      IPRUN = -3
      GO TO 30
   80 IF (NINP.EQ.0) GO TO 100
      IF (NINP.GE.3) NGR(8) = 0
C
      CALL ANALYS
C
C >>>>>>>>>>>>>>>>>>>>>>>>>>>>>>>>>>>>>>>>>>>>>>>>>>>>>>>>>>>>>>>>>>>>>>
C HERE INSERT THE CALLS TO YOUR OWN DESIGN AND ANALYSIS PROCEDURES
C DEFINE NEW LABELS AND EXTEND THE STATEMENT '100+1'
C IPRUN HAS TO BE EQUIVALENT TO THE DEFINITIONS IN 'DORINP'
C >>>>>>>>>>>>>>>>>>>>>>>>>>>>>>>>>>>>>>>>>>>>>>>>>>>>>>>>>>>>>>>>>>>>>>
C
   90 IF (LSPOUT.EQ.(-1)) CALL OUTSPE
      IF (NINP.NE.2) GO TO 10
      ENDFILE KA4
      GO TO 10
C
  100 CALL ERROR(26)
C
C THE PROGRAM WILL COME TO THE END IN THE SUBROUTINE DORINP
C OR IN THE SUBROUTINE ERROR
```

```
C
      STOP
      END
C
C-----------------------------------------------------------------------
C SUBROUTINE:   DEFIN0
C DEFINITION OF THE MACHINE PARAMETERS
C-----------------------------------------------------------------------
C
      SUBROUTINE DEFIN0
C
      DOUBLE PRECISION DPI, DOMI, D1MACH
C
      COMMON /CANPAR/ KA1, KA2, KA3, KA4, KA5, LINE
      COMMON /CONST/ PI, FLMA, FLMI, FLER
      COMMON /CONST1/ MAXDEG, IWLMI, IWLMA, MBL
      COMMON /CONST2/ DPI, DOMI
      COMMON /CIOFOR/ IOFO(3), ION
C
      DIMENSION JOFO(3)
C
C >>>>>>>>>>>>>>>>>>>>>>>>>>>>>>>>>>>>>>>>>>>>>>>>>>>>>>>>>>>>>>>>>>>>
      DATA JOFO /2H(4,2H0A,2H2)/, JON /40/
C
C THIS STATEMENT IS WRITTEN FOR THE PDP11/45 AND IS
C MACHINE DEPENDENT (SEE MANUAL 4.2.2.)
C E.G.:   CDC CYBER 172
C DATA IOFO /10H(8A10)     ,2*10H           /,ION/8/
C >>>>>>>>>>>>>>>>>>>>>>>>>>>>>>>>>>>>>>>>>>>>>>>>>>>>>>>>>>>>>>>>>>>>
C
      KA1 = I1MACH(1)
      KA2 = I1MACH(2)
      KA3 = 6
      KA4 = 1
      KA5 = 9
      LINE = 65
C
      PI = 4.*ATAN(1.0)
      FLMA = 2.**(I1MACH(13)-2)
      FLMI = R1MACH(3)   <------  FLMI = 2.0 * R1MACH (4)
      FLER = 1.0E02*FLMI
C
      DPI = 4.00*DATAN(1.0D00)
      DOMI = D1MACH(3)   <------  DOMI = 2.0D0 * D1MACH (4)
C
      MAXDEG = 32
      IWLMI = 2
      IWLMA = I1MACH(11)
      MBL = 16
C
      IOFO(1) = JOFO(1)
      IOFO(2) = JOFO(2)
      IOFO(3) = JOFO(3)
      ION = JON
C
      RETURN
C
      END
C
C =======================================================================
C
C DOREDI - SUBROUTINES:  ROOT ELEMENTS FOR OVERLAYS
C
C =======================================================================
C
C
C-----------------------------------------------------------------------
C SUBROUTINE:   SYNTHE
C SUBROUTINE TO HANDLE THE NECESSARY COMMONS FOR SYNNOR AND SYNOPT
C FOR OVERLAYS
C-----------------------------------------------------------------------
C
      SUBROUTINE SYNTHE
C
      DOUBLE PRECISION DK, DKS, DCAP02, DCAP04
      COMMON /TOLCHA/ GD1, GD2, ACAP12, ADELTA, ADEG
      COMMON /TOLNOR/ VSNN, NDEGN, NBN
      COMMON /RESS/ SFA, SM(18,4), NZERO(16), SPR(16), SPI(16)
      COMMON /RESIN1/ PREN(16), PIMN(16), UGC, OGC, ACK, NJ, NH
      COMMON /RESIN2/ DK, DKS, DCAP02, DCAP04
      COMMON /COPTCO/ LOPTW, LSTAB, ACXMI, ACXMA, ITER, ITERM, ITERM1
      COMMON /OUTDAT/ IP, PRE, PIM, IZ, ZRE, ZIM
      COMMON /COPTSP/ IB, JSEQN(16), JSEQD(16), AMAX, SCA, ALSBI
C
      CALL SYNNOR
      IF (LOPTW.EQ.0) RETURN
      CALL SYNOPT
      RETURN
      END
C
C-----------------------------------------------------------------------
C SUBROUTINE:   SYNNOR
C SYNTHESIS OF RECURSIVE DIGITAL FILTERS
C CALCULATION OF THE ZEROS AND THE POLES
C-----------------------------------------------------------------------
C
      SUBROUTINE SYNNOR
C
      COMMON /CONTR/ IPRUN, IPCON, NINP, NOUT, NDOUT, LSPOUT, NSPOUT
      COMMON /COPTCO/ LOPTW, LSTAB, ACXMI, ACXMA, ITER, ITERM, ITERM1
      COMMON /OUTDAT/ IP, PRE, PIM, IZ, ZRE, ZIM
C
      MOUT = NOUT
      IF (IPRUN.EQ.(-3)) NOUT = 0
      CALL DESIA
      NOUT = MOUT
      IF (LOPTW.NE.0) RETURN
      CALL DESIB
      IF (NDOUT.EQ.(-1)) CALL OUT012
      RETURN
      END
C
C-----------------------------------------------------------------------
C SUBROUTINE:   SYNOPT
C SYNTHESIS OF RECURSIVE DIGITAL FILTERS WITH OPTIMIZED COEFFICIENTS
C CALCULATION OF THE POLES AND THE OPTIMIZATION PARAMETER
C-----------------------------------------------------------------------
C
      SUBROUTINE SYNOPT
C
      COMMON /CONTR/ IPRUN, IPCON, NINP, NOUT, NDOUT, LSPOUT, NSPOUT
      COMMON /CONST/ PI, FLMA, FLMI, FLER
      COMMON /CONST1/ MAXDEG, IWLMI, IWLMA, MBL
      COMMON /COPTCO/ LOPTW, LSTAB, ACXMI, ACXMA, ITER, ITERM, ITERM1
      COMMON /CCOEFW/ IWL, IWLG, IWLD, IWLL, ADEPSG, ADEPSD, ADEPSL,
     *    ISTAB, IDEPSL, IDEPSD
      COMMON /DESIGN/ NDEGF, EDEG, ACX, NORMA, LSOUT, LVSN, LSYM
      COMMON /COPTCP/ IS1, IS2, IABO, IWLGS, IWLGN, IWLGP, ADEPSS,
     *    ADEPSN, ADEPSP, ACXS, ACXN, ACXP
      COMMON /CREST/ ISTRU, ISCAL, SCALM, ISEQ(16,2), LSEQ, IWLR,
```

```
     *    LPOT2, JSTRU
      COMMON /CREST1/ JSTRUS, JSTRUD
      COMMON /SFILT/ NBS, SFACT, SB2(16), SB1(16), SB0(16), SC1(16),
     *    SC0(16)
      COMMON /OUTDAT/ IP, PRE, PIM, IZ, ZRE, ZIM
      COMMON /COPTSP/ IB, JSEQN(16), JSEQD(16), AMAX, SCA, ALSBI
C
      FLAB = 10.*FLMI
      ACXN = ACXMI
      ACXP = ACXMA
      ADEPSS = FLMA
      ADEPSP = FLMA
      ADEPSN = -FLMA
      ITER = 0
      IWLGS = IWLMA
      IWLGP = IWLMA
      IWLGN = IWLMA
      IWLL = IWLMA
      IDEPSL = 0
      IS1 = 0
      IS2 = 0
      IABO = 0
      ITERM2 = ITERM
      IF (IPRUN.EQ.3 .AND. JSTRU.EQ.1) ITERM = ITERM1
      MOUT = NOUT
      IF (NOUT.LE.3) NOUT = NOUT - 2
C
   10 ITER = ITER + 1
      IF (IPRUN.EQ.(-3)) CALL BLNUMZ
      IF (NOUT.GE.2) CALL OUT023
      CALL DESIB
      IF (ITER.GT.1) GO TO 20
      CALL GRID01
      CALL GRID04
      IF (NOUT.EQ.5) CALL OUT042
      ACXS = ACX
   20 IF (LOPTW.EQ.0) GO TO 60
      CALL COEFW
      CALL COPY02
      CALL OPTPAR
      IF (NOUT.GE.2) CALL OUT045
      IF (ITER.GE.ITERM) GO TO 30
      IF (ACX-ACXMI.LT.FLAB) GO TO 40
      IF (ACXMA-ACX.LT.FLAB) GO TO 40
      IF (IABO.LT.3) GO TO 10
      IABO = 1
      GO TO 50
   30 IABO = 2
      GO TO 50
   40 IABO = 3
   50 NOUT = MOUT
      IF (NOUT.GE.1) CALL OUT022
      ACX = ACXS
      IF (IPRUN.EQ.3 .AND. JSTRU.EQ.1) GO TO 80
      IF (IPRUN.EQ.(-3)) CALL BLNUMZ
      CALL DESIB
      IF (NOUT.EQ.1) CALL OUT011
      IF (NOUT.EQ.2) CALL OUT038
   60 IF (NOUT/2.EQ.1) CALL OUT039
      IWL = IWLGS
      IF (IPRUN.EQ.3 .AND. JSTRUS.EQ.1) GO TO 70
      IF (NDOUT.EQ.(-1)) CALL OUT012
   70 LOPTW = 1
      MOUT = NOUT
      IF (NOUT.NE.0) NOUT = NOUT + 2
      CALL COEFW
      IF (IPRUN.NE.(-3)) CALL COPY02
      NOUT = MOUT
   80 ITERM = ITERM2
      RETURN
      END
C
C-----------------------------------------------------------------------
C SUBROUTINE:   SEQSTA
C START NOISE ANALISIS AND OPTIMIZATION
C-----------------------------------------------------------------------
C
      SUBROUTINE SEQSTA
C
      COMMON /CONTR/ IPRUN, IPCON, NINP, NOUT, NDOUT, LSPOUT, NSPOUT
      COMMON /CONST1/ MAXDEG, IWLMI, IWLMA, MBL
      COMMON /CREST/ ISTRU, ISCAL, SCALM, ISEQ(16,2), LSEQ, IWLR,
     *    LPOT2, JSTRU
      COMMON /CREST1/ JSTRUS, JSTRUD
      COMMON /CCOEFW/ IWL, IWLG, IWLD, IWLL, ADEPSG, ADEPSD, ADEPSL,
     *    ISTAB, IDEPSL, IDEPSD
      COMMON /COPTCO/ LOPTW, LSTAB, ACXMI, ACXMA, ITER, ITERM, ITERM1
      COMMON /COPTST/ LOPTS, ISTOR
C
      IF (LSEQ.NE.(-1)) GO TO 10
      IF (NOUT.GE.3) CALL OUT011
      CALL COPY01
      CALL ALLO02
      GO TO 20
C
   10 IF (NINP.GE.3 .AND. LOPTS.EQ.1) CALL FIXPAR
   20 L = LSTAB
      LSTAB = 0
      CALL STABLE
      IF (ISTAB.NE.(-1)) GO TO 50
      LSTAB = L
      IF (ISTRU.GE.30) GO TO 30
      CALL TSTR01
      GO TO 40
C
C >>>>>>>>>>>>>>>>>>>>>>>>>>>>>>>>>>>>>>>>>>>>>>>>>>>>>>>>>>>>>>>>>>>>>
C HERE INSERT TEST'S FOR YOUR OWN STRUCTURE IMPLEMENTATIONS
C
   30 CALL ERROR(8)
C
C >>>>>>>>>>>>>>>>>>>>>>>>>>>>>>>>>>>>>>>>>>>>>>>>>>>>>>>>>>>>>>>>>>>>>
C
   40 IF (IWLR.EQ.100) RETURN
      IF (IWLR.GT.IWLMA) RETURN
      L = IWL
      IWL = IWLR
      CALL RECO
      CALL ROUND(1, 5)
      CALL STABLE
      IWL = L
      RETURN
   50 CALL ERROR(30)
      RETURN
      END
C
C-----------------------------------------------------------------------
C SUBROUTINE:   SEQOPT
C OPTIMIZATION OF THE PAIRING AND ORDERING OF THE BLOCKS OF SECOND
C ORDER
C OPTIMIZATION PROCEDURE WITH LIMITED STORAGE
```

```
C-----------------------------------------------------------------------
C
      SUBROUTINE SEQOPT
C
      COMPLEX ZP, ZPS
      COMMON /CONTR/ IPRUN, IPCON, NINP, NOUT, NDOUT, LSPOUT, NSPOUT
      COMMON /CREST/ ISTRU, ISCAL, SCALM, ISEQ(16,2), LSEQ, IWLR,
     *    LPOT2, JSTRU
      COMMON /CREST1/ JSTRUS, JSTRUD
      COMMON /COPTST/ LOPTS, ISTOR
      COMMON /CGRID/ GR(64), NGR(12)
      COMMON /SFILT/ NBS, SFACT, SB2(16), SB1(16), SB0(16), SC1(16),
     *    SC0(16)
      COMMON /CPOL/ ZP(16,2), ZPS(16,2)
      COMMON /CNOISE/ RI, RIN, RE, REN, FAC
      COMMON /COPST2/ PN(100,2), TF(100,2), TFA(100,2)
      COMMON /CFFUNC/ PHI(5), BF2(5), BF1(5), BF0(5), IBB(5), ICOR
      COMMON /CNFUNC/ AQC(5), BN2(5), BN1(5), BN0(5)
      COMMON /CPOW/ PNU, PNC, AND, ITCORP
C
      IF (ISCAL.EQ.0) GO TO 10
      CALL DESCAL
   10 CALL COPY01
      IF (NOUT.GE.3) CALL OUT011
      CALL POLLOC
      IF (NGR(8).EQ.0) CALL GRID02
      IF (NOUT.GE.5) CALL OUT042
      IF (NOUT.GE.3) CALL OUT025
      IF (LOPTS.NE.0) CALL OPTLST
      CALL ANANOI
      IF (NOUT.EQ.0) RETURN
      CALL OUT027
      IF (NOUT.EQ.1) RETURN
      IF (JSTRUS.EQ.2) CALL OUT010
      CALL OUT011
      IF (JSTRUS.EQ.2) CALL DENORM
      RETURN
      END
C
C-----------------------------------------------------------------------
C SUBROUTINE:   ANALYS
C ANALYSIS OF THE DESIGNED RECURSIVE DIGITAL FILTER
C-----------------------------------------------------------------------
C
      SUBROUTINE ANALYS
C
      COMPLEX ZP, ZPS
      COMMON /CPLOT/ IPLOT, LNOR, IPAG, OMLO, OMUP, RMAX, RMIN
      COMMON /COPTCO/ LOPTW, LSTAB, ACXMI, ACXMA, ITER, ITERM, ITERM1
      COMMON /CCOEFW/ IWL, IWLG, IWLD, IWLL, ADEPSG, ADEPSD, ADEPSL,
     *    ISTAB, IDEPSL, IDEPSD
      COMMON /COPTST/ LOPTS, ISTOR
      COMMON /COPTSP/ IB, JSEQN(16), JSEQD(16), AMAX, SCA, ALSBI
      COMMON /CREST/ ISTRU, ISCAL, SCALM, ISEQ(16,2), LSEQ, IWLR,
     *    LPOT2, JSTRU
      COMMON /CREST1/ JSTRUS, JSTRUD
      COMMON /CPOL/ ZP(16,2), ZPS(16,2)
      COMMON /CNOISE/ RI, RIN, RE, REN, FAC
C
      CALL ANAL01
      IF (LOPTS.EQ.0) GO TO 10
      CALL ANAL02
   10 IF (IPLOT.EQ.0) GO TO 20
      CALL PLOT00
C
   20 RETURN
      END
C
C-----------------------------------------------------------------------
C SUBROUTINE:   ANAL01
C FIRST PART OF ANALYSIS
C-----------------------------------------------------------------------
C
      SUBROUTINE ANAL01
C
      COMMON /CONTR/ IPRUN, IPCON, NINP, NOUT, NDOUT, LSPOUT, NSPOUT
      COMMON /CONST1/ MAXDEG, IWLMI, IWLMA, MBL
      COMMON /COPTCO/ LOPTW, LSTAB, ACXMI, ACXMA, ITER, ITERM, ITERM1
      COMMON /CCOEFW/ IWL, IWLG, IWLD, IWLL, ADEPSG, ADEPSD, ADEPSL,
     *    ISTAB, IDEPSL, IDEPSD
      COMMON /COPTST/ LOPTS, ISTOR
      COMMON /CREST/ ISTRU, ISCAL, SCALM, ISEQ(16,2), LSEQ, IWLR,
     *    LPOT2, JSTRU
      COMMON /CREST1/ JSTRUS, JSTRUD
      COMMON /CGRID/ GR(64), NGR(12)
      COMMON /CPLOT/ IPLOT, LNOR, IPAG, OMLO, OMUP, RMAX, RMIN
      COMMON /COPTSP/ IB, JSEQN(16), JSEQD(16), AMAX, SCA, ALSBI
C
      IF (LSEQ.NE.(-1)) GO TO 10
      IF (NOUT.GE.3) CALL OUT011
      CALL COPY01
      CALL ALLO02
C
   10 L = LSTAB
      LSTAB = 0
      CALL STABLE
      IF (ISTAB.NE.(-1)) GO TO 50
      LSTAB = L
      IF (LOPTW.EQ.0) GO TO 30
      IF (NGR(1).NE.0) GO TO 20
      IF (NINP.GE.3) CALL GRID03
      CALL GRID01
      IF (NINP.LE.2) CALL GRID04
   20 IF (NOUT.EQ.5) CALL OUT042
      CALL COEFW
      CALL COPY02
      LOPTW = 0
   30 IF (LOPTS.NE.(-1) .AND. IPLOT.EQ.0) GO TO 60
      IF (IWLR.EQ.100) GO TO 60
      IWL = IWLR
      IF (IWL.GT.IWLMA) GO TO 40
      CALL RECO
      CALL ROUND(1, 5)
      CALL STABLE
      GO TO 60
   40 CALL ERROR(16)
      GO TO 60
   50 CALL ERROR(30)
   60 RETURN
      END
C
C-----------------------------------------------------------------------
C SUBROUTINE:   ANAL02
C SECOND PART OF THE ANALYSIS
C-----------------------------------------------------------------------
C
      SUBROUTINE ANAL02
C
      COMMON /CONTR/ IPRUN, IPCON, NINP, NOUT, NDOUT, LSPOUT, NSPOUT
      COMMON /COPTST/ LOPTS, ISTOR
```

```
      COMMON /CREST/ ISTRU, ISCAL, SCALM, ISEQ(16,2), LSEQ, IWLR,
     *    LPOT2, JSTRU
      COMMON /CREST1/ JSTRUS, JSTRUD
      COMMON /CNOISE/ RI, RIN, RE, REN, FAC
      COMMON /COPST2/ PN(100,2), TF(100,2), TFA(100,2)
      COMMON /CFFUNC/ PHI(5), BF2(5), BF1(5), BF0(5), IBB(5), ICOR
      COMMON /CNFUNC/ AQC(5), BN2(5), BN1(5), BN0(5)
      COMMON /CPOW/ PNU, PNC, AND, ITCORP
C
      IF (ISTRU.GE.30) GO TO 10
      CALL TSTR01
      GO TO 20
C
C >>>>>>>>>>>>>>>>>>>>>>>>>>>>>>>>>>>>>>>>>>>>>>>>>>>>>>>>>>>>>>>>>>>>
C HERE INSERT TEST'S FOR YOUR OWN STRUCTURE IMPLEMENTATIONS
C
  10  CALL ERROR(8)
C
C >>>>>>>>>>>>>>>>>>>>>>>>>>>>>>>>>>>>>>>>>>>>>>>>>>>>>>>>>>>>>>>>>>>>
C
  20  IF (ISCAL.NE.0) CALL DESCAL
      CALL COPY01
      IF (NOUT.GE.3) CALL OUT011
      LOPTS = 0
      CALL POLLOC
      CALL GRID02
      IF (NOUT.GE.3) CALL OUT025
      CALL ANANOI
      CALL OUT027
      IF (JSTRUS.EQ.2) CALL OUT010
      CALL OUT011
      IF (JSTRUS.EQ.2) CALL DENORM
      RETURN
      END
C
C-----------------------------------------------------------------------
C SUBROUTINE:   PLOT00
C PRINTER PLOTS
C-----------------------------------------------------------------------
C
      SUBROUTINE PLOT00
C
      COMMON /CONST/ PI, FLMA, FLMI, FLER
      COMMON /CPLOT/ IPLOT, LNOR, IPAG, OMLO, OMUP, RMAX, RMIN
      COMMON /CANPAR/ KA1, KA2, KA3, KA4, KA5, LINE
      COMMON /CSTATV/ X1(16), X2(16)
      COMMON /CLINE/ Y, X, IZ(111)
C
      DIMENSION A(6)
      EQUIVALENCE (A(1),IZ(1))
C
      DATA ICHI, ICHP, ICHC, ICHM, ICHB /1HI,1H+,1H.,1H-,1H /
C
C NUMBER OF PRINTED LINES
C
      IF (IPLOT.GT.4) GO TO 230
      ML = IPAG*LINE - 6
      ADOM = (OMUP-OMLO)/FLOAT(ML-1)
C
C SCALING
C
      IF (LNOR.NE.(-1)) GO TO 60
C
      RRMAX = -FLMA
      RRMIN = FLMA
C
      IF (IPLOT.EQ.4) Y = RESP(1.,1)
      DO 30 L=1,ML
        IF (IPLOT.EQ.4) GO TO 20
        X = ADOM*FLOAT(L-1) + OMLO
        IF (IPLOT.EQ.3) GO TO 10
        Y = AMAGO(X)
        IF (IPLOT.EQ.1) GO TO 20
        IF (Y.LT.FLMI) Y = FLMI
        Y = -20.*ALOG10(Y)
        IF (Y.GT.RMAX) Y = RMAX
        GO TO 20
  10    Y = PHASE(X)
  20    RRMAX = AMAX1(RRMAX,Y)
        RRMIN = AMIN1(RRMIN,Y)
        IF (IPLOT.NE.4) GO TO 30
        Y = RESP(0.,0)
  30  CONTINUE
C
C SCALING OF THE PLOTS
C
      RR = ABS(RRMAX-RRMIN)
      RN = 0.005*RR
      RRMAX = RRMAX - RN
      RRMIN = RRMIN + RN
      RR = ABS(RRMAX-RRMIN)
      IF (RR.LE.0.) RR = ABS(RRMAX)
      IQ = INT(ALOG10(RR))
      RN = 10.**FLOAT(IQ)
      IF (RN.LT.RR) RN = RN*10.
      AN = RR/RN
      IF (AN.GT.0.0) AAN = 0.1
      IF (AN.GT.0.1) AAN = 0.2
      IF (AN.GT.0.2) AAN = 0.5
      IF (AN.GT.0.5) AAN = 1.0
      RR = AAN*RN
C
  40  RN = RR*0.2
      RMA = AINT(RRMAX/RN)*RN
      IF (RMA.LT.RRMAX) RMA = RMA + RN
      RMI = RMA - RR
      IF (RMI.LE.RRMIN) GO TO 50
      RMI = AINT(RRMIN/RN)*RN
      IF (RMI.GT.RRMIN) RMI = RMI - RN
      RMA = RMI + RR
      IF (RMA.GE.RRMAX) GO TO 50
      RR = RR*2.
      IF (AAN.EQ.0.2) RR = RR*1.25
      GO TO 40
C
  50  RRMAX = RMA
      RRMIN = RMI
      GO TO 110
C
  60  GO TO (70, 70, 80, 90), IPLOT
  70  RRMAX = RMAX
      RRMIN = 0.
      GO TO 100
  80  RRMAX = PI
      RRMIN = -RRMAX
      GO TO 100
  90  RRMAX = RMAX
      RRMIN = RMIN
 100  RR = RRMAX - RRMIN
C
```

```
  110 CALL OUT060
      NULL = INT(-RRMIN*100./RR+1.5)
      IF (NULL.LT.1 .OR. NULL.GT.111) NULL = 0
      Q = RR/5.
      DO 120 I=1,6
        A(I) = Q*FLOAT(I-1) + RRMIN
  120 CONTINUE
      CALL OUT061
      DO 130 I=1,111
        IZ(I) = ICHM
  130 CONTINUE
      IF (IPLOT.EQ.4) Y = RESP(1.,1)
      X = 1.
      DO 220 L=1,ML
        IF (IPLOT.EQ.4) GO TO 150
        X = ADOM*FLOAT(L-1) + OMLO
        IF (IPLOT.EQ.3) GO TO 140
        Y = AMAGO(X)
        IF (IPLOT.NE.2) GO TO 150
        IF (Y.LT.FLMI) Y = FLMI
        Y = -20.*ALOG10(Y)
        GO TO 150
  140   Y = PHASE(X)
  150   DO 160 I=1,101,10
          IZ(I) = ICHC
  160   CONTINUE
        IF (NULL.NE.0) IZ(NULL) = ICHC
        Y1 = (Y-RRMIN)/RR
        IY = INT(100.*Y1+1.5)
        IF (IY.GT.0) GO TO 170
        IZ(1) = ICHI
        GO TO 190
  170   IF (IY.LE.111) GO TO 180
        IZ(111) = ICHI
        GO TO 190
  180   IZ(IY) = ICHP
  190   CALL OUT062
        IF (IPLOT.NE.4) GO TO 200
        Y = RESP(0.,0)
        X = FLOAT(L)
  200   DO 210 I=1,111
          IZ(I) = ICHB
  210   CONTINUE
  220 CONTINUE
      GO TO 240
  230 CALL ERROR(8)
  240 RETURN
      END
C
C-----------------------------------------------------------------------
C SUBROUTINE:   DORINP
C INPUT DATA COMPILER FOR THE DESIGN PROGRAM 'DOREDI'
C-----------------------------------------------------------------------
C
      SUBROUTINE DORINP
C
      COMMON /CONTR/ IPRUN, IPCON, NINP, NOUT, NDOUT, LSPOUT, NSPOUT
      COMMON /CARD/ LCODE, IZ(40)
      COMMON /INPDAT/ ICODE, JCODE, LINP, IINP, AINP(3), NPAR
      COMMON /INTDAT/ JINP(16)
      COMMON /CPLOT/ IPLOT, LNOR, IPAG, OMLO, OMUP, RMAX, RMIN
      COMMON /COPTCO/ LOPTW, LSTAB, ACXMI, ACXMA, ITER, ITERM, ITERM1
      COMMON /COPTST/ LOPTS, ISTOR
C
C
      DIMENSION IPT1(5), IPT2(5)
      DATA IPT1(1), IPT1(2), IPT1(3), IPT1(4), IPT1(5) /1HS,1HS,1HS,1HA,
     *     1HF/
      DATA IPT2(1), IPT2(2), IPT2(3), IPT2(4), IPT2(5) /1HY,1HE,1HS,1HN,
     *     1HI/
      DATA IPMAX /5/
C
C >>>>>>>>>>>>>>>>>>>>>>>>>>>>>>>>>>>>>>>>>>>>>>>>>>>>>>>>>>>>>>>>>>>>
C HERE INSERT YOUR OWN PROGRAM DEFINITIONS
C CHANGE DIMENSION AND DATA STATEMENT INFRONT
C 'FI' HAS TO BE LAST DEFINITION
C >>>>>>>>>>>>>>>>>>>>>>>>>>>>>>>>>>>>>>>>>>>>>>>>>>>>>>>>>>>>>>>>>>>>
C
      LCODE = -1
      IF (IPLOT.NE.0) GO TO 50
      IF (IPCON.NE.0) GO TO 10
      CALL DEFIN1
      CALL DEFIN2
      CALL DEFIN3
      CALL DEFIN4
      CALL DEFIN5
      CALL DEFIN6
      GO TO 60
C
   10 IF (IPCON.EQ.99) STOP
      IPRUN = IPCON
      IF (IPCON.EQ.4) GO TO 40
      CALL DEFIN2
      IF (IPCON.EQ.2) GO TO 20
      CALL DEFIN3
      CALL DEFIN4
      IF (IPCON.EQ.1) GO TO 60
   20 CALL DEFIN5
   30 IF (IPRUN.EQ.2 .OR. IPRUN.EQ.3) LOPTS = -1
      IF (IPRUN.EQ.3) LOPTW = -1
      GO TO 60
   40 CALL DEFIN6
      GO TO 60
C
   50 IPLOT = 0
   60 CALL INP001
      IF (LCODE.EQ.0) GO TO 110
   70 DO 80 IP=1,IPMAX
        IF (ICODE.EQ.IPT1(IP) .AND. JCODE.EQ.IPT2(IP)) GO TO 90
   80 CONTINUE
      GO TO 190
   90 IF (IPRUN.EQ.0) GO TO 100
      IF (IP.EQ.IPMAX) IP = 99
      IPCON = IP
      IF (IPRUN.EQ.1 .OR. IPRUN.EQ.3) CALL INP031
      RETURN
C
  100 IPRUN = IP
      GO TO 30
C
  110 IF (IPRUN.EQ.0) GO TO 190
      CALL INP002
      IF (ICODE.EQ.0) GO TO 210
      GO TO (120, 130, 140, 150, 160, 170), ICODE
  120 CALL INP010
      GO TO 180
  130 IF (IPRUN.GE.4) GO TO 200
      CALL INP020
      GO TO 60
  140 IF (IPRUN.EQ.2) GO TO 200
```

```
      CALL INP030
      GO TO 60
  150 CALL INP040
      GO TO 180
  160 IF (IPRUN.LE.1) GO TO 200
      CALL INP050
      GO TO 180
  170 IF (IPRUN.NE.4) GO TO 200
      CALL INP060
      IF (IPLOT.EQ.0) GO TO 60
      RETURN
C
  180 IF (LCODE.NE.0) GO TO 70
      GO TO 60
C
  190 I = 3
      GO TO 220
  200 I = 4
      GO TO 220
  210 I = 2
  220 CALL ERROR(I)
      GO TO 60
      END
C
C ======================================================================
C
C DOREDI - SUBROUTINES: PART I
C
C ======================================================================
C
C
C-----------------------------------------------------------------------
C SUBROUTINE:   DESIA
C FILTER DESIGN -- FIRST SECTION
C-----------------------------------------------------------------------
C
      SUBROUTINE DESIA
C
      COMMON /CONTR/ IPRUN, IPCON, NINP, NOUT, NDOUT, LSPOUT, NSPOUT
      COMMON /TOL/ ITYP, IAPRO, NDEG, OM(4), SF, ADELP, ADELS
      COMMON /TOLSN/ VSN, VD, A
      COMMON /TOLCHA/ GD1, GD2, ACAP12, ADELTA, ADEG
      COMMON /TOLNOR/ VSNN, NDEGN, NBN
      COMMON /DESIGN/ NDEGF, EDEG, ACX, NORMA, LSOUT, LVSN, LSYM
      COMMON /RESS/ SFA, SM(18,4), NZERO(16), SPR(16), SPI(16)
      COMMON /RESIN1/ PREN(16), PIMN(16), UGC, OGC, ACK, NJ, NH
      COMMON /OUTDAT/ IP, PRE, PIM, IZ, ZRE, ZIM
C
      SFA = 0.
      IF (NOUT.GE.3) CALL OUT030
      CALL DESI00
      IF (NOUT.GE.4) CALL OUT031
      CALL DESI01
      IF (NOUT.GE.3) CALL OUT033
      GO TO (10, 20, 20, 30), IAPRO
   10 CALL DESI11
      GO TO 40
   20 CALL DESI12
      GO TO 40
   30 CALL DESI14
C
   40 VSNN = VSN
      NDEGN = NDEG
      NBN = NJ
      IF (NOUT.GE.5) CALL OUT034
      IF (NOUT.GE.3) CALL OUT035
      IF (NOUT.LT.4) GO TO 50
      IZ = 0
      IP = 0
      CALL OUT036
   50 CALL TRANZE
      IF (NOUT.LT.4) GO TO 60
      IZ = 1
      IP = 0
      CALL OUT036
   60 CALL TRBIZE
      CALL BLNUMZ
      CALL ROMEG
      IF (NOUT.GE.3) CALL OUT032
      RETURN
      END
C
C-----------------------------------------------------------------------
C SUBROUTINE:   DESIB
C FILTER DESIGN  --  SECOND SECTION
C-----------------------------------------------------------------------
C
      SUBROUTINE DESIB
C
C
      COMMON /CONTR/ IPRUN, IPCON, NINP, NOUT, NDOUT, LSPOUT, NSPOUT
      COMMON /TOL/ ITYP, IAPRO, NDEG, OM(4), SF, ADELP, ADELS
      COMMON /TOLSN/ VSN, VD, A
      COMMON /TOLNOR/ VSNN, NDEGN, NBN
      COMMON /DESIGN/ NDEGF, EDEG, ACX, NORMA, LSOUT, LVSN, LSYM
      COMMON /OUTDAT/ IP, PRE, PIM, IZ, ZRE, ZIM
      COMMON /RESIN1/ PREN(16), PIMN(16), UGC, OGC, ACK, NJ, NH
C
      VSN = VSNN
      NDEG = NDEGN
      NJ = NBN
      GO TO (10, 20, 20, 30), IAPRO
   10 CALL DESI21
      GO TO 40
   20 CALL DESI22
      GO TO 40
   30 CALL DESI24
C
   40 IF (NOUT.GE.3) CALL OUT037
      IF (NOUT.LT.4) GO TO 50
      IZ = 0
      IP = 1
      CALL OUT036
   50 CALL TRANPO
      IF (LSOUT.EQ.(-1)) GO TO 60
      IF (NOUT.LT.4) GO TO 70
   60 IZ = 1
      IP = 1
      CALL OUT036
      IF (LSOUT.EQ.(-1)) CALL OUT016
   70 CALL TRBIPO
      IF (NOUT.GE.3) CALL OUT038
      CALL BLDENZ
      IF (NOUT.GE.2) CALL OUT011
      IF (NOUT.GE.4) CALL OUT039
      RETURN
      END
C
C-----------------------------------------------------------------------
C SUBROUTINE:   DESI00
```

```
C TRANSFORM TOLERANCE SCHEME
C-----------------------------------------------------------------------
C
      SUBROUTINE DESI00
C
      COMMON /TOL/ ITYP, IAPRO, NDEG, OM(4), SF, ADELP, ADELS
      COMMON /TOLCHA/ GD1, GD2, ACAP12, ADELTA, ADEG
      COMMON /DESIGN/ NDEGF, EDEG, ACX, NORMA, LSOUT, LVSN, LSYM
C
      IF (ITYP.GE.3) NDEG = (NDEG+1)/2
      IF (NDEG.NE.0) ADEG = FLOAT(NDEG)/(1.+EDEG)
      IF (LVSN.LT.0) GO TO 20
      IF (LVSN.EQ.0) GO TO 10
      CALL PARCHA
      CALL VSNMAX
   10 CALL TRANSN
   20 RETURN
      END
C
C-----------------------------------------------------------------------
C SUBROUTINE:   DESI01
C DESIGN OF BUTTERWORTH-, CHEBYSHEV (PASSBAND OR STOPBAND -, AND
C ELLIPTIC-FILTERS
C-----------------------------------------------------------------------
C
      SUBROUTINE DESI01
C
      COMMON /CONST1/ MAXDEG, IWLMI, IWLMA, MBL
      COMMON /TOL/ ITYP, IAPRO, NDEG, OM(4), SF, ADELP, ADELS
      COMMON /TOLSN/ VSN, VD, A
      COMMON /TOLCHA/ GD1, GD2, ACAP12, ADELTA, ADEG
      COMMON /DESIGN/ NDEGF, EDEG, ACX, NORMA, LSOUT, LVSN, LSYM
C
      IF (LVSN.NE.0) GO TO 20
      CALL PARCHA
      CALL DEGREE
      Q = ADEG*(1.+EDEG) + 0.5
      N = INT(Q)
      M = INT(ADEG)
      IF (FLOAT(M).LT.ADEG) M = M + 1
      N = MAX0(M,N)
      IF (NDEG.EQ.0) GO TO 10
      IF (NDEG.GE.N) GO TO 20
      CALL ERROR(15)
      IF (NDEGF.EQ.(-1)) STOP
C
   10 NDEG = N
C
   20 IF (NDEG.LE.MAXDEG) RETURN
      CALL ERROR(25)
      RETURN
      END
C
C-----------------------------------------------------------------------
C SUBROUTINE:   PARCHA
C COMPUTATION OF THE PARAMETERS OF THE CHARACTERISTIC FUNCTION
C-----------------------------------------------------------------------
C
      SUBROUTINE PARCHA
C
      COMMON /TOL/ ITYP, IAPRO, NDEG, OM(4), SF, ADELP, ADELS
      COMMON /TOLSN/ VSN, VD, A
      COMMON /TOLCHA/ GD1, GD2, ACAP12, ADELTA, ADEG
      COMMON /DESIGN/ NDEGF, EDEG, ACX, NORMA, LSOUT, LVSN, LSYM
C
      GD1 = 0.
      GD2 = -1.
      IF (ADELP.GT.0.) GD1 = SQRT((2.-ADELP)*ADELP)/(1.-ADELP)
      IF (ADELS.GT.0.) GD2 = SQRT(1.-ADELS*ADELS)/ADELS
      ACAP12 = GD1/GD2
      IF (ACAP12.GT.0.) GO TO 60
      GO TO (10, 20, 20, 30), IAPRO
   10 ACAP12 = VSN**(-ADEG)
      GO TO 40
   20 Q = ARCOSH(VSN)*ADEG
      ACAP12 = 1./COSH(Q)
      GO TO 40
   30 CALL BOUND
   40 IF (GD2.EQ.(-1.)) GO TO 50
      GD1 = ACAP12*GD2
      ADELP = 1. - 1./SQRT(1.+GD1*GD1)
      GO TO 60
   50 GD2 = GD1/ACAP12
      ADELS = 1./SQRT(1.+GD2*GD2)
   60 RETURN
      END
C
C-----------------------------------------------------------------------
C SUBROUTINE:   VSNMAX
C COMPUTATION OF THE NORMALIZED CUTOFF FREQUENCY FOR A FILTER
C DEGREE AND THE RIPPLES GIVEN
C-----------------------------------------------------------------------
C
      SUBROUTINE VSNMAX
C
      COMMON /TOL/ ITYP, IAPRO, NDEG, OM(4), SF, ADELP, ADELS
      COMMON /TOLSN/ VSN, VD, A
      COMMON /TOLCHA/ GD1, GD2, ACAP12, ADELTA, ADEG
C
      AC = 1./ACAP12
      E = 1./ADEG
      GO TO (10, 20, 20, 30), IAPRO
   10 VSN = AC**E
      GO TO 40
   20 VSN = COSH(E*ARCOSH(AC))
      GO TO 40
   30 CALL BOUND
   40 RETURN
      END
C
C-----------------------------------------------------------------------
C SUBROUTINE:   BOUND
C CALCULATION OF A BOUND FOR VSN OR ACAP12 FOR ELLIPTIC FILTERS
C-----------------------------------------------------------------------
C
      SUBROUTINE BOUND
C
      DOUBLE PRECISION DPI, DOMI
      DOUBLE PRECISION DE, DKK, DDEG, DCAP12, DEG, DCAP14, DQ, DK1,
     *    DAB, DMAX, DDE
      DOUBLE PRECISION DK, DF, DD
      DOUBLE PRECISION DFF, DELLK
C
      COMMON /CONST2/ DPI, DOMI
      COMMON /TOLSN/ VSN, VD, A
      COMMON /TOLCHA/ GD1, GD2, ACAP12, ADELTA, ADEG
C
      DIMENSION DK(3), DF(3)
C
      DATA DE /1.D00/
```

```
      DFF(DD) = (DELLK(DD)*DKK/DELLK(DSQRT(DE-DD*DD)))**II - DDEG
      IF (ACAP12.LE.0.) GO TO 10
      DCAP12 = DBLE(ACAP12)
      DEG = DBLE(1./ADEG)
      II = 1
      GO TO 20
   10 DCAP12 = DE/DBLE(VSN)
      DEG = DBLE(ADEG)
      II = -1
   20 DCAP14 = DSQRT(DE-DCAP12*DCAP12)
      DKK = DELLK(DCAP14)/DELLK(DCAP12)
      DQ = DEXP(-DPI*DKK*DEG)
      DK1 = 4.D00*DSQRT(DQ)
      IF (DK1.LT.DE) GO TO 30
      DQ = 2.D00*DQ
      DQ = (DE-DQ)/(DE+DQ)
      DQ = DQ*DQ
      DK1 = DSQRT(DE-DQ*DQ)
   30 DK(1) = DK1
      DK(2) = (DE+DK(1))/2.D00
      DDEG = DBLE(ADEG)
      DF(1) = DFF(DK(1))
      DF(2) = DFF(DK(2))
   40 DK(3) = DK(1) - DF(1)*(DK(1)-DK(2))/(DF(1)-DF(2))
      DF(3) = DFF(DK(3))
      IF (DABS(DF(3)).LT.1.D-6) GO TO 60
      DMAX = 0.D00
      DO 50 J=1,3
        DAB = DABS(DF(J))
        IF (DMAX.GT.DAB) GO TO 50
        JJ = J
        DMAX = DAB
   50 CONTINUE
      IF (JJ.EQ.3) GO TO 40
      DK(JJ) = DK(3)
      DF(JJ) = DF(3)
      GO TO 40
   60 IF (ACAP12.LE.0.) GO TO 70
      DDE = DE/DK(3)
      VSN = SNGL(DDE)
      RETURN
   70 ACAP12 = SNGL(DK(3))
      RETURN
      END
C
C-----------------------------------------------------------------------
C SUBROUTINE:   DEGREE
C COMPUTATION OF THE MINIMUM FILTER DEGREE (ADEG)
C-----------------------------------------------------------------------
C
      SUBROUTINE DEGREE
C
      DOUBLE PRECISION DE, DCAP02, DCAP04, DCAP12, DCAP14, DADEG
      DOUBLE PRECISION DELLK
      COMMON /TOL/ ITYP, IAPRO, NDEG, OM(4), SF, ADELP, ADELS
      COMMON /TOLSN/ VSN, VD, A
      COMMON /TOLCHA/ GD1, GD2, ACAP12, ADELTA, ADEG
C
      GO TO (10, 20, 20, 30), IAPRO
C
   10 ADEG = ALOG(1./ACAP12)/ALOG(VSN)
      RETURN
C
   20 ADEG = ARCOSH(1./ACAP12)/ARCOSH(VSN)
      RETURN
C
   30 DE = 1.D00
      DCAP02 = DE/DBLE(VSN)
      DCAP04 = DSQRT(DE-DCAP02*DCAP02)
      DCAP12 = DBLE(ACAP12)
      DCAP14 = DSQRT(DE-DCAP12*DCAP12)
      DADEG = (DELLK(DCAP02)*DELLK(DCAP14))/(DELLK(DCAP04)*DELLK(DCAP12)
     *    )
      ADEG = SNGL(DADEG)
      RETURN
      END
C
C-----------------------------------------------------------------------
C SUBROUTINE:   DESI11
C COMPUTATION OF THE ZEROS AND LOCATIONS OF THE EXTREMAS
C FOR BUTTERWORTH FILTER
C-----------------------------------------------------------------------
C
      SUBROUTINE DESI11
C
      COMMON /CONST/ PI, FLMA, FLMI, FLER
      COMMON /TOL/ ITYP, IAPRO, NDEG, OM(4), SF, ADELP, ADELS
      COMMON /TOLSN/ VSN, VD, A
      COMMON /TOLCHA/ GD1, GD2, ACAP12, ADELTA, ADEG
      COMMON /DESIGN/ NDEGF, EDEG, ACX, NORMA, LSOUT, LVSN, LSYM
      COMMON /RES/ AC, ROM(4), RDELP, RDELS, NZM(4)
      COMMON /RESS/ SFA, SM(18,4), NZERO(16), SPR(16), SPI(16)
      COMMON /RESIN1/ PREN(16), PIMN(16), UGC, OGC, ACK, NJ, NH
C
      ADELTA = VSN**NDEG
C
      NH = NDEG/2
      NJ = (NDEG+1)/2
      FDEG = FLOAT(NDEG)
      FN = PI/2./FDEG
C
      DO 10 I=1,NJ
        NZERO(I) = 0
        III = I + I - 1
        Q = FN*FLOAT(III)
        PREN(I) = SIN(Q)
        PIMN(I) = COS(Q)
   10 CONTINUE
C
      FN = 2.*FN
      NZERO(1) = NDEG
      NZM(1) = 1
      SM(1,1) = 0.
      NZM(2) = 1
      SM(1,2) = 1.
      NZM(3) = 1
      SM(1,3) = VSN
      NZM(4) = 1
      SM(1,4) = FLMA
C
      UGC = GD2/ADELTA
      OGC = GD1
      SM(17,4) = 1.
      RETURN
      END
C
C-----------------------------------------------------------------------
C SUBROUTINE:   DESI12
C CHEBYSHEV-FILTER (PASSBAND OR STOPBAND)
C COMPUTATION OF THE ZEROS AND THE LOCATIONS OF THE EXTREMAS
```

```
C-----------------------------------------------------------------------
C
      SUBROUTINE DESI12
C
      COMMON /CONST/ PI, FLMA, FLMI, FLER
      COMMON /TOL/ ITYP, IAPRO, NDEG, OM(4), SF, ADELP, ADELS
      COMMON /TOLSN/ VSN, VD, A
      COMMON /TOLCHA/ GD1, GD2, ACAP12, ADELTA, ADEG
      COMMON /DESIGN/ NDEGF, EDEG, ACX, NORMA, LSOUT, LVSN, LSYM
      COMMON /RES/ AC, ROM(4), RDELP, RDELS, NZM(4)
      COMMON /RESS/ SFA, SM(18,4), NZERO(16), SPR(16), SPI(16)
      COMMON /RESIN1/ PREN(16), PIMN(16), UGC, OGC, ACK, NJ, NH
C
      ADELTA = COSH(FLOAT(NDEG)*ARCOSH(VSN))
C
      FA = 1.
      NH = NDEG/2
      NJ = (NDEG+1)/2
      FN = PI/(2.*FLOAT(NDEG))
C
      DO 10 I=1,NJ
        NZERO(I) = 0
        INJ = I + I - 1
        Q = FN*FLOAT(INJ)
        PREN(I) = SIN(Q)
        PIMN(I) = COS(Q)
   10 CONTINUE
C
      FN = 2.*FN
C
      IF (IAPRO.EQ.3) GO TO 40
C
      M = NJ + 1
      DO 20 I=1,NJ
        J = M - I
        SM(I,1) = PIMN(J)
   20 CONTINUE
      NZM(1) = NJ
      M = NH + 1
      DO 30 I=1,M
        MI = M - I
        Q = FLOAT(MI)*FN
        SM(I,2) = COS(Q)
   30 CONTINUE
      NZM(2) = M
      SM(1,3) = VSN
      NZM(3) = 1
      SM(1,4) = FLMA
      NZM(4) = 1
      NZERO(1) = NDEG
C
      UGC = GD2/ADELTA
      OGC = GD1
      GO TO 80
C
   40 SM(1,1) = 0.
      NZM(1) = 1
      SM(1,2) = 1.
      NZM(2) = 1
      DO 50 I=1,NJ
        INJ = NJ - I
        Q = FLOAT(INJ)*FN
        SM(INJ+1,3) = VSN/COS(Q)
   50 CONTINUE
      NZM(3) = NJ
      DO 60 I=1,NH
        NZERO(I) = 2
        Q = PIMN(I)
        FA = FA*Q*Q
        SM(I,4) = VSN/Q
   60 CONTINUE
      IF (NH.EQ.NJ) GO TO 70
      NZERO(NJ) = 1
      SM(NJ,4) = FLMA
   70 NZM(4) = NJ
C
      UGC = GD2
      OGC = GD1*ADELTA
   80 ACK = FA
      SM(17,4) = 1.
      RETURN
      END
C
C-----------------------------------------------------------------------
C SUBROUTINE:   DESI14
C ELLIPTIC-FILTER
C COMPUTATION OF THE ZEROS AND THE LOCATIONS OF THE EXTREMAS
C-----------------------------------------------------------------------
C
      SUBROUTINE DESI14
C
      DOUBLE PRECISION DPI, DOMI
      DOUBLE PRECISION DK, DKS, DCAP02, DCAP04, DE
      DOUBLE PRECISION DSK(16)
      DOUBLE PRECISION DCAP01, DQ, DN, DU, DM
      DOUBLE PRECISION DELLK, DSN2, DEL1, DEL2, DDE, DDELTA, DDELT
C
      COMMON /CONST/ PI, FLMA, FLMI, FLER
      COMMON /CONST2/ DPI, DOMI
      COMMON /TOL/ ITYP, IAPRO, NDEG, OM(4), SF, ADELP, ADELS
      COMMON /TOLSN/ VSN, VD, A
      COMMON /TOLCHA/ GD1, GD2, ACAP12, ADELTA, ADEG
      COMMON /DESIGN/ NDEGF, EDEG, ACX, NORMA, LSOUT, LVSN, LSYM
      COMMON /RES/ AC, ROM(4), RDELP, RDELS, NZM(4)
      COMMON /RESS/ SFA, SM(18,4), NZERO(16), SPR(16), SPI(16)
      COMMON /RESIN1/ PREN(16), PIMN(16), UGC, OGC, ACK, NJ, NH
      COMMON /RESIN2/ DK, DKS, DCAP02, DCAP04
      EQUIVALENCE (PREN(1),DSK(1))
C
      DATA DE /1.D00/
C
      DCAP02 = DE/DBLE(VSN)
      DCAP01 = DSQRT(DCAP02)
      DCAP04 = DSQRT(DE-DCAP02*DCAP02)
      DK = DELLK(DCAP02)
      DKS = DELLK(DCAP04)
C
      DQ = DEXP(-DPI*DKS/DK)
C
      NH = NDEG/2
      M = NDEG + 1
      NJ = M/2
      MH = NH + 1
C
      DN = DK/DBLE(FLOAT(NDEG))
C
      DEL1 = DE
      IF (NH.EQ.0) GO TO 20
      DO 10 I=1,NH
        INH = M - I - I
```

```
      DU = DN*DBLE(FLOAT(INH))
      DM = DSN2(DU,DK,DQ)
      DEL1 = DEL1*DM*DCAP01
      DSK(I) = DM
      J = MH - I
      SM(J,1) = SNGL(DM)
      NZERO(I) = 2
      DDE = DE/(DCAP02*DM)
      SM(I,4) = SNGL(DDE)
   10 CONTINUE
      GO TO 30
   20 SM(1,1) = 0.
   30 KJ = NJ - 1
      MJ = NJ + 1
      DEL2 = DE
      IF (KJ.EQ.0) GO TO 50
      DO 40 I=1,KJ
        NDEGI = NDEG - I - I
        DU = DN*DBLE(FLOAT(NDEGI))
        DM = DSN2(DU,DK,DQ)
        J = MJ - I
        SM(J,2) = SNGL(DM)
        DDE = DE/(DCAP02*DM)
        SM(I+1,3) = SNGL(DDE)
        DEL2 = DEL2*DM*DCAP01
   40 CONTINUE
      GO TO 60
   50 SM(NDEG,2) = 1.
      SM(1,3) = VSN
C
   60 DDELT = DEL1*DEL1
      ADELTA = SNGL(DDELT)
      ACK = 1./ADELTA
      IF (NH.EQ.NJ) GO TO 80
      ACK = ACK*SNGL(DCAP01)
      DDELTA = DEL2*DEL2*DCAP01
      ADELTA = SNGL(DDELTA)
      DSK(NJ) = 0.D00
      NZERO(NJ) = 1
      SM(NJ,4) = FLMA
C
      IF (NH.EQ.0) GO TO 90
      DO 70 I=1,NH
        J = MJ - I
        SM(J,1) = SM(J-1,1)
        SM(I,2) = SM(I+1,2)
   70 CONTINUE
      SM(1,1) = 0.
      GO TO 90
C
   80 SM(MH,3) = FLMA
      SM(1,2) = 0.
C
   90 NZM(1) = NJ
      NZM(4) = NJ
      NZM(2) = MH
      NZM(3) = MH
      SM(MH,2) = 1.
      SM(1,3) = VSN
      UGC = GD2*ADELTA
      OGC = GD1/ADELTA
      SM(17,4) = 1.
      RETURN
      END
C
C-----------------------------------------------------------------------
C SUBROUTINE:   DESI21
C BUTTERWORTH-FILTER
C COMPUTATION OF THE POLES
C-----------------------------------------------------------------------
C
      SUBROUTINE DESI21
C
      COMMON /CONST/ PI, FLMA, FLMI, FLER
      COMMON /TOL/ ITYP, IAPRO, NDEG, OM(4), SF, ADELP, ADELS
      COMMON /TOLCHA/ GD1, GD2, ACAP12, ADELTA, ADEG
      COMMON /DESIGN/ NDEGF, EDEG, ACX, NORMA, LSOUT, LVSN, LSYM
      COMMON /RES/ AC, ROM(4), RDELP, RDELS, NZM(4)
      COMMON /RESS/ SFA, SM(18,4), NZERO(16), SPR(16), SPI(16)
      COMMON /RESIN1/ PREN(16), PIMN(16), UGC, OGC, ACK, NJ, NH
C
C COMPUTATION OF CONSTANT C AND REDUCED TOLERANCE SCHEME
C
      IF (ACX.LT.999.) GO TO 20
      IF ((OGC-UGC).LT.FLMI) GO TO 10
      AC = (2.*ADELP/(ADELTA*ADELS))**0.33333
      ACX = ALOG10(AC/UGC)/ALOG10(OGC/UGC)
      IF (ACX.GE.0. .AND. ACX.LE.1.) GO TO 30
   10 ACX = 0.5
   20 AC = UGC*(OGC/UGC)**ACX
   30 RDELP = 1. - SQRT(1./(1.+AC*AC))
      Q = AC*ADELTA
      RDELS = SQRT(1./(1.+Q*Q))
C
C COMPUTATION OF FACTOR SFA AND POLES
C
      SFA = 1./AC
      Q = AC**(-1./FLOAT(NDEG))
C
      DO 40 I=1,NJ
        SPR(I) = -Q*PREN(I)
        SPI(I) = Q*PIMN(I)
   40 CONTINUE
      RETURN
      END
C
C-----------------------------------------------------------------------
C SUBROUTINE:   DESI22
C CHEBYSHEV-FILTER (PASSBAND OR STOPBAND)
C COMPUTATION OF THE POLES
C-----------------------------------------------------------------------
C
      SUBROUTINE DESI22
C
      COMMON /CONST/ PI, FLMA, FLMI, FLER
      COMMON /TOL/ ITYP, IAPRO, NDEG, OM(4), SF, ADELP, ADELS
      COMMON /TOLSN/ VSN, VD, A
      COMMON /TOLCHA/ GD1, GD2, ACAP12, ADELTA, ADEG
      COMMON /DESIGN/ NDEGF, EDEG, ACX, NORMA, LSOUT, LVSN, LSYM
      COMMON /RES/ AC, ROM(4), RDELP, RDELS, NZM(4)
      COMMON /RESS/ SFA, SM(18,4), NZERO(16), SPR(16), SPI(16)
      COMMON /RESIN1/ PREN(16), PIMN(16), UGC, OGC, ACK, NJ, NH
C
      IF (ACX.LT.999.) GO TO 20
      IF ((OGC-UGC).LT.FLMI) GO TO 10
      IF (IAPRO.EQ.2) Q = 1./ADELTA
      IF (IAPRO.EQ.3) Q = ADELTA*ADELTA
      AC = (2.*ADELP*Q/ADELS)**0.33333
      ACX = ALOG10(AC/UGC)/ALOG10(OGC/UGC)
      IF (ACX.GE.0. .AND. ACX.LE.1.) GO TO 30
```

```
   10 ACX = 0.5
   20 AC = UGC*(OGC/UGC)**ACX
C
C COMPUTATION OF THE REDUCED TOLERANCE SCHEME
C
   30 Q = AC
      IF (IAPRO.EQ.3) Q = Q/ADELTA
      Q = 1. + Q*Q
      RDELP = 1. - SQRT(1./Q)
      Q = AC
      IF (IAPRO.EQ.2) Q = Q*ADELTA
      Q = 1. + Q*Q
      RDELS = SQRT(1./Q)
C
C COMPUTATION OF THE FACTOR SFA AND THE POLES
C
      IF (IAPRO.EQ.3) GO TO 40
      SFA = 2./(AC*2.**NDEG)
      Q = -1./AC
      GO TO 50
C
   40 SFA = ACK
      Q = AC
   50 Q = ARSINH(Q)/FLOAT(NDEG)
      QR = SINH(Q)
      QI = COSH(Q)
      IF (IAPRO.EQ.3) GO TO 70
      DO 60 I=1,NJ
        SPR(I) = QR*PREN(I)
        SPI(I) = QI*PIMN(I)
   60 CONTINUE
      RETURN
C
   70 DO 80 I=1,NH
        Q = PIMN(I)*QI
        QA = PREN(I)*QR
        QQ = Q*Q
        QQA = QA*QA
        SFA = SFA/(QQ+QQA)
        SPR(I) = -VSN/(QQ/QA+QA)
        SPI(I) = VSN/(Q+QQA/Q)
   80 CONTINUE
      IF (NH.EQ.NJ) RETURN
      SPI(NJ) = 0.
      Q = VSN/QR
      SFA = SFA*Q
      SPR(NJ) = -Q
      RETURN
      END
C
C-----------------------------------------------------------------------
C SUBROUTINE:  DESI24
C ELLIPTIC FILTER
C COMPUTATION OF THE REDUCED TOLERANCE SCHEME, THE FACTOR SFA AND
C THE POLES
C-----------------------------------------------------------------------
C
      SUBROUTINE DESI24
C
      DOUBLE PRECISION DPI, DOMI
      DOUBLE PRECISION DK, DKS, DCAP02, DCAP04, DE
      DOUBLE PRECISION DSK(16)
      DOUBLE PRECISION DU, DR, DQ, DUD, DUC, DRC, DRD, DM
      DOUBLE PRECISION DELLK, DSN2
C
      COMMON /CONST/ PI, FLMA, FLMI, FLER
      COMMON /CONST2/ DPI, DOMI
      COMMON /TOL/ ITYP, IAPRO, NDEG, OM(4), SF, ADELP, ADELS
      COMMON /TOLSN/ VSN, VD, A
      COMMON /TOLCHA/ GD1, GD2, ACAP12, ADELTA, ADEG
      COMMON /DESIGN/ NDEGF, EDEG, ACX, NORMA, LSOUT, LVSN, LSYM
      COMMON /RES/ AC, ROM(4), RDELP, RDELS, NZM(4)
      COMMON /RESS/ SFA, SM(18,4), NZERO(16), SPR(16), SPI(16)
      COMMON /RESIN1/ PREN(16), PIMN(16), UGC, OGC, ACK, NJ, NH
      COMMON /RESIN2/ DK, DKS, DCAP02, DCAP04
      EQUIVALENCE (PREN(1),DSK(1))
C
      DATA DE /1.D00/
C
C IF ACX NOT DEFINED COMPUTE A SYMMETRICAL USAGE OF THE TOLERANCE
C SCHEME
C
      IF (ACX.LT.999.) GO TO 20
      IF ((OGC-UGC).LT.FLMI) GO TO 10
      AC = (2.*ADELP/(ADELTA*ADELS))**0.333333333
      ACX = ALOG10(AC/UGC)/ALOG10(OGC/UGC)
      IF (ACX.GE.0. .AND. ACX.LE.1.) GO TO 30
   10 ACX = 0.5
C
   20 AC = UGC*(OGC/UGC)**ACX
C
C COMPUTATION OF THE REDUCED TOLERANCE SCHEME
C
   30 Q = AC*ADELTA
      DU = DE/DBLE(Q)
      RDELP = 1. - SQRT(1./(1.+Q*Q))
      Q = 1. + AC*AC/(ADELTA*ADELTA)
      RDELS = SQRT(1./Q)
C COMPUTATION OF THE FACTOR SFA AND THE POLES
C
      Q = AC*ACK
      IF (NH.EQ.NJ) Q = SQRT(1.+Q*Q)
      SFA = 1./Q
C
      DR = DBLE(ADELTA)
      DR = DR*DR
      DQ = DBLE(Q)
      CALL DELI1(DQ, DU, DR)
      DU = DQ
      DQ = DSQRT(DE-DR*DR)
      DQ = DELLK(DR)
      DU = DK*DU/(DQ*DBLE(FLOAT(NDEG)))
      DQ = DEXP(-DPI*DK/DKS)
      DU = -DSN2(DU,DKS,DQ)
      DQ = DU*DU
      DUD = DE - DCAP04*DCAP04*DQ
      DUD = DSQRT(DUD)
      DUC = DSQRT(DE-DQ)
      DO 40 I=1,NJ
        DR = DSK(I)
        DRC = DR*DR
        DRD = DE - DCAP02*DCAP02*DRC
        DRC = DSQRT(DE-DRC)
        DM = DE - DQ*DRD
        DRD = DSQRT(DRD)
        DRD = DRD*DU*DUC*DRC/DM
        SPR(I) = SNGL(DRD)
        DR = DR*DUD/DM
        SPI(I) = SNGL(DR)
   40 CONTINUE
```

```
      RETURN
      END
C
C-----------------------------------------------------------------------
C SUBROUTINE:   TRANZE
C REACTANCE TRANSFORMATION OF THE ZEROS AND THE LOCATIONS OF THE
C EXTREMES
C-----------------------------------------------------------------------
C
      SUBROUTINE TRANZE
C
      DOUBLE PRECISION DR, DQI
C
      COMMON /CONST/ PI, FLMA, FLMI, FLER
      COMMON /TOL/ ITYP, IAPRO, NDEG, OM(4), SF, ADELP, ADELS
      COMMON /TOLSN/ VSN, VD, A
      COMMON /RES/ AC, ROM(4), RDELP, RDELS, NZM(4)
      COMMON /RESS/ SFA, SM(18,4), NZERO(16), SPR(16), SPI(16)
C
      DIMENSION MSM(4)
      FA = 1.
      IF (ITYP.EQ.1) GO TO 190
      IF (ITYP.EQ.3) GO TO 60
C
      ME = NZM(4)
      DO 10 I=1,ME
        Q = SM(I,4)
        IF (Q.LT.FLMA) FA = FA*Q
   10 CONTINUE
C
      FA = FA*FA
C LOWPASS - HIGHPASS
C
      DO 50 J=1,4
        ME = NZM(J)
        DO 40 I=1,ME
          QI = SM(I,J)
          IF (ABS(QI).LT.FLMI) GO TO 20
          QI = 1./QI
          GO TO 30
   20     QI = FLMA
   30     SM(I,J) = QI
   40   CONTINUE
   50 CONTINUE
      GO TO 90
   60 DO 80 J=1,2
        ME = NZM(J)
        MA = ME + 1
        ME = ME/2
        DO 70 I=1,ME
          QI = SM(I,J)
          II = MA - I
          SM(I,J) = SM(II,J)
          SM(II,J) = QI
   70   CONTINUE
   80 CONTINUE
C
   90 IF (ITYP.EQ.2) GO TO 190
C
C LOWPASS - BANDPASS TRANSFORMATION
C
      QA = 2.*A
      NN = NDEG + 1
      IF (ITYP.EQ.4) GO TO 110
C
      MSM(1) = 1
      IF (NZM(1).NE.1) MSM(1) = NDEG
      MSM(2) = 2
      IF (NZM(2).NE.1) MSM(2) = NN
      DO 100 J=3,4
        MSM(J) = 2*NZM(J)
  100 CONTINUE
      GO TO 130
C
  110 DO 120 J=1,2
        MSM(J) = 2*NZM(J)
  120 CONTINUE
      MSM(3) = 2
      IF (NZM(3).NE.1) MSM(3) = NN
      MSM(4) = 1
      IF (NZM(4).NE.1) MSM(4) = NDEG
C
  130 S = 1.
      DO 180 J=1,4
        ME = NZM(J)
        MA = MSM(J)
        NZM(J) = MA
        IF (J.EQ.3) S = -1.
        DO 170 I=1,ME
          QR = SM(I,J)
          NU = NZERO(I)
          IF (ABS(QR).LT.FLMA) GO TO 150
          IF (J.NE.4) GO TO 140
          FA = FA*(VD/A)**NU
  140     QI = QR
          GO TO 160
C
  150     QR = QR/QA
          DR = DBLE(QR)
          DQI = DSQRT(DR*DR+1.D00)
          QI = SNGL(DQI)
  160     SM(I,J) = QI - S*QR
          II = MA - I + 1
          IF (ABS(QR).LT.FLMI) NU = 2*NU
          IF (J.EQ.4) NZERO(II) = NU
          SM(II,J) = QI + S*QR
  170   CONTINUE
  180 CONTINUE
C
  190 DO 220 J=1,4
        ME = NZM(J)
        DO 210 I=1,ME
          Q = SM(I,J)
          IF (Q.LT.FLMA) GO TO 200
          IF (J.NE.4 .OR. ITYP.GE.3) GO TO 210
          NU = NZERO(I)
          FA = FA*VD**NU
          GO TO 210
  200     SM(I,J) = Q*VD
  210   CONTINUE
  220 CONTINUE
      SM(17,4) = SM(17,4)*FA
      RETURN
      END
C
C-----------------------------------------------------------------------
C SUBROUTINE:   TRANPO
C REACTANCE TRANSFORMATION OF THE POLES
C-----------------------------------------------------------------------
C
```

```
      SUBROUTINE TRANPO
C
      DOUBLE PRECISION DR, DI, DQ
C
      COMMON /CONST/ PI, FLMA, FLMI, FLER
      COMMON /TOL/ ITYP, IAPRO, NDEG, OM(4), SF, ADELP, ADELS
      COMMON /TOLSN/ VSN, VD, A
      COMMON /RESS/ SFA, SM(18,4), NZERO(16), SPR(16), SPI(16)
      COMMON /RESIN1/ PREN(16), PIMN(16), UGC, OGC, ACK, NJ, NH
C
      IF (ITYP.EQ.1) GO TO 90
      IF (ITYP.EQ.3) GO TO 40
C
      DO 30 I=1,NJ
        QR = SPR(I)
        QI = SPI(I)
        QH = QR*QR + QI*QI
        IF (ABS(QI).GT.FLMI) GO TO 10
        SFA = -SFA/QR
        GO TO 20
 10     SFA = SFA/QH
 20     QI = QI/QH
        IF (ABS(QI).LT.FLMI) QI = 0.
        SPI(I) = QI
        SPR(I) = QR/QH
 30   CONTINUE
      IF (ITYP.EQ.2) GO TO 90
C
 40   QA = 2.*A
      NN = NJ
      NJ = NDEG
      NDEG = 2*NDEG
C
      ME = NJ
      DO 80 I=1,NN
        QR = SPR(I)/QA
        QI = SPI(I)/QA
        IF (ABS(QI).GE.FLMA) GO TO 70
        DR = DBLE(QR)
        DI = DBLE(QI)
        DQ = DI*DI
        DI = DI*DR*2.D00
        DR = DR*DR - DQ - 1.D00
        CALL DSQRTC(DR, DI, DR, DI)
        QZ = SNGL(DR)
        QN = SNGL(DI)
        IF (ABS(QN).GT.FLMI) GO TO 60
        JJ = NJ + ME
        DO 50 II=ME,NJ
          J = JJ - II
          SPR(J+1) = SPR(J)
          SPI(J+1) = SPI(J)
 50     CONTINUE
        NJ = NJ + 1
        ME = ME + 1
 60     SPR(I) = QR + QZ
        SPI(I) = QI + QN
        SPR(ME) = QR - QZ
        SPI(ME) = QN - QI
        ME = ME - 1
        GO TO 80
C
 70     SPR(I) = QR
        SPI(I) = FLMA
        NJ = NJ + 1
        SPR(NJ) = QR
        SPI(NJ) = 0.
 80   CONTINUE
C
 90   DO 100 I=1,NJ
        SPR(I) = SPR(I)*VD
        SPI(I) = SPI(I)*VD
 100  CONTINUE
C
      SFA = SFA*SM(17,4)
      RETURN
      END
C
C-----------------------------------------------------------------------
C SUBROUTINE:   TRBIZE
C BILINEAR TRANSFORMATION OF THE ZEROS AND THE LOCATIONS OF THE
C EXTREMAS
C-----------------------------------------------------------------------
C
      SUBROUTINE TRBIZE
C
      COMMON /CONST/ PI, FLMA, FLMI, FLER
      COMMON /RES/ AC, ROM(4), RDELP, RDELS, NZM(4)
      COMMON /RESS/ SFA, SM(18,4), NZERO(16), SPR(16), SPI(16)
      COMMON /RESZ/ ZFA, ZM(18,4), ZZR(16), ZZI(16), ZPR(16), ZPI(16)
C
      FA = 1.
      DO 50 J=1,4
        ME = NZM(J)
        DO 40 I=1,ME
          QI = SM(I,J)
          ZM(I,J) = 2.*ATAN(QI)
          IF (J.NE.4) GO TO 40
          IF (QI.GE.FLMA) GO TO 10
          IF (QI.LT.FLMI) GO TO 20
          QQI = QI*QI
          Q = 1. + QQI
          ZZR(I) = (1.-QQI)/Q
          ZZI(I) = 2.*QI/Q
          NU = NZERO(I)/2
          FA = FA*Q**NU
          GO TO 40
C
 10       ZZR(I) = -1.
          GO TO 30
C
 20       ZZR(I) = 1.
 30       ZZI(I) = 0.
 40     CONTINUE
 50   CONTINUE
C
      SM(17,1) = FA
      RETURN
      END
C
C-----------------------------------------------------------------------
C SUBROUTINE:   TRBIPO
C BILINEAR TRANSFORMATION OF THE POLES
C-----------------------------------------------------------------------
C
      SUBROUTINE TRBIPO
C
      COMMON /CONST/ PI, FLMA, FLMI, FLER
      COMMON /RESS/ SFA, SM(18,4), NZERO(16), SPR(16), SPI(16)
      COMMON /RESZ/ ZFA, ZM(18,4), ZZR(16), ZZI(16), ZPR(16), ZPI(16)
```

```
      COMMON /RESIN1/ PREN(16), PIMN(16), UGC, OGC, ACK, NJ, NH
C
      ZFA = SFA*SM(17,1)
C
      DO 20 I=1,NJ
        QR = SPR(I)
        Q = 1. - QR
        QI = SPI(I)
        IF (ABS(QI).LT.FLMI) GO TO 10
        QQR = QR*QR
        QQI = QI*QI
        ZFA = ZFA/(Q-QR+QQR+QQI)
        Q = 1./(Q*Q+QQI)
        ZPR(I) = (1.-QQR-QQI)*Q
        ZPI(I) = 2.*QI*Q
        GO TO 20
C
   10   ZPR(I) = (1.+QR)/Q
        ZPI(I) = 0.
        ZFA = ZFA/Q
   20 CONTINUE
      RETURN
      END
C
C-----------------------------------------------------------------------
C SUBROUTINE:   BLNUMZ
C BUILD NUMERATOR BLOCKS OF SECOND ORDER
C-----------------------------------------------------------------------
C
      SUBROUTINE BLNUMZ
C
      COMMON /RES/ AC, ROM(4), RDELP, RDELS, NZM(4)
      COMMON /RESS/ SFA, SM(18,4), NZERO(16), SPR(16), SPI(16)
      COMMON /RESZ/ ZFA, ZM(18,4), ZZR(16), ZZI(16), ZPR(16), ZPI(16)
      COMMON /FILT/ NB, FACT, B2(16), B1(16), B0(16), C1(16), C0(16)
C
      DIMENSION NZE(16)
      COMMON /SCRAT/ ADUM(32)
      EQUIVALENCE (ADUM(1),NZE(1))
C
      N = 0
      ME = NZM(4)
      DO 10 I=1,ME
        NZE(I) = NZERO(I)
   10 CONTINUE
C
      DO 70 I=1,ME
        QR = ZZR(I)
        NZ = NZE(I)
C
   20   IF (NZ.EQ.0) GO TO 70
        N = N + 1
        B2(N) = 1.
        IF (NZ.EQ.1) GO TO 30
        B1(N) = -2.*QR
        B0(N) = 1.
        NZ = NZ - 2
        IF (NZ.GT.0) GO TO 20
        GO TO 70
C
   30   IF (I.EQ.ME) GO TO 60
        MA = I + 1
        DO 40 II=MA,ME
          IF (ZZI(II).EQ.0.) GO TO 50
   40   CONTINUE
        GO TO 60
C
   50   QRR = ZZR(II)
        B1(N) = -QR - QRR
        B0(N) = QR*QRR
        NZE(II) = NZE(II) - 1
        GO TO 70
C
   60   B1(N) = -QR
        B0(N) = 0.
C
   70 CONTINUE
      RETURN
      END
C
C-----------------------------------------------------------------------
C SUBROUTINE:   BLDENZ
C BUILD DENOMINATOR BLOCKS OF SECOND ORDER
C Z-DOMAIN
C-----------------------------------------------------------------------
C
      SUBROUTINE BLDENZ
C
      COMMON /CONST/ PI, FLMA, FLMI, FLER
      COMMON /TOL/ ITYP, IAPRO, NDEG, OM(4), SF, ADELP, ADELS
      COMMON /RESS/ SFA, SM(18,4), NZERO(16), SPR(16), SPI(16)
      COMMON /RESZ/ ZFA, ZM(18,4), ZZR(16), ZZI(16), ZPR(16), ZPI(16)
      COMMON /RESIN1/ PREN(16), PIMN(16), UGC, OGC, ACK, NJ, NH
      COMMON /FILT/ NB, FACT, B2(16), B1(16), B0(16), C1(16), C0(16)
C
      NB = (NDEG+1)/2
      N = 0
      FACT = ZFA
C
      DO 40 I=1,NB
        N = N + 1
        QR = ZPR(N)
        QI = ZPI(N)
        IF (ABS(QI).LT.FLMI) GO TO 10
        C1(I) = -2.*QR
        C0(I) = QR*QR + QI*QI
        GO TO 40
C
   10   IF (N.GE.NJ) GO TO 20
        IF (ABS(ZPI(N+1)).LT.FLMI) GO TO 30
   20   C1(I) = -QR
        C0(I) = 0.
        GO TO 40
C
   30   N = N + 1
        QI = ZPR(N)
        C1(I) = -QR - QI
        C0(I) = QR*QI
   40 CONTINUE
      RETURN
      END
C
C-----------------------------------------------------------------------
C SUBROUTINE:   ROMEG
C REALIZED FREQUENCIES OMEGA
C-----------------------------------------------------------------------
C
      SUBROUTINE ROMEG
C
      COMMON /TOL/ ITYP, IAPRO, NDEG, OM(4), SF, ADELP, ADELS
```

```
      COMMON /RES/ AC, ROM(4), RDELP, RDELS, NZM(4)
      COMMON /RESZ/ ZFA, ZM(18,4), ZZR(16), ZZI(16), ZPR(16), ZPI(16)
C
      N2 = NZM(2)
      N3 = NZM(3)
      GO TO (10, 20, 30, 40), ITYP
   10 ROM(1) = ZM(N2,2)
      ROM(2) = ZM(1,3)
      GO TO 50
   20 ROM(1) = ZM(1,3)
      ROM(2) = ZM(N2,2)
      GO TO 50
   30 ROM(1) = ZM(N3,3)
      ROM(2) = ZM(1,2)
      ROM(3) = ZM(N2,2)
      ROM(4) = ZM(1,3)
      GO TO 50
   40 N2 = N2/2
      ROM(1) = ZM(N2,2)
      ROM(4) = ZM(N2+1,2)
      ROM(3) = ZM(1,3)
      ROM(2) = ZM(N3,3)
   50 RETURN
      END
C
C-----------------------------------------------------------------------
C SUBROUTINE:   TRANSN
C COMPUTATION OF THE PARAMETERS OF THE NORMALIZED LOWPASS
C-----------------------------------------------------------------------
C
      SUBROUTINE TRANSN
C
      COMMON /TOL/ ITYP, IAPRO, NDEG, OM(4), SF, ADELP, ADELS
      COMMON /TOLSN/ VSN, VD, A
      COMMON /DESIGN/ NDEGF, EDEG, ACX, NORMA, LSOUT, LVSN, LSYM
C
      TAN2(AA) = SIN(AA/2.)/COS(AA/2.)
C
      V1 = TAN2(OM(1))
      V2 = TAN2(OM(2))
      IF (ITYP.LE.2) GO TO 210
      V3 = TAN2(OM(3))
      V4 = TAN2(OM(4))
      IF (ITYP.EQ.3) GO TO 10
      Q = V1
      V1 = -V4
      V4 = -Q
      Q = V2
      V2 = -V3
      V3 = -Q
      IF (LSYM.NE.0) LSYM = 5 - LSYM
      IF (LVSN.EQ.0) GO TO 10
      J = 5 - LVSN
      LVSN = LSYM
      LSYM = J
C
   10 J = LSYM + 1
      JJ = LVSN/2 + 1
      GO TO (20, 80, 120, 160, 180), J
   20 J = NORMA + 1
      GO TO (30, 30, 40, 70), J
   30 VDQ1 = V2*V3
      VSN1 = VDQ1/V1 - V1
      Q = V4 - VDQ1/V4
      IF (Q.LT.VSN1) VSN1 = Q
      A1 = 1./(V3-V2)
      VSN1 = VSN1*A1
      GO TO (40, 50, 40), J
   40 VDQ = V1*V4
      A = V2/(VDQ-V2*V2)
      Q = V3/(V3*V3-VDQ)
      IF (Q.LT.A) A = Q
      VSN = A*(V4-V1)
      IF (NORMA.EQ.2) GO TO 200
      IF (VSN.GE.VSN1) GO TO 200
   50 VDQ = VDQ1
   60 VSN = VSN1
      A = A1
      GO TO 200
C
   70 VDQ = SQRT(V1*V2*V3*V4)
      A1 = V3/(V3*V3-VDQ)
      VSN1 = (V4-VDQ/V4)*A1
      A = V2/(VDQ-V2*V2)
      VSN = (VDQ/V1-V1)*A
      IF (VSN.GE.VSN1) GO TO 200
      GO TO 60
C
   80 GO TO (90, 100, 110), JJ
   90 VDQ = V2*V3
      VSN = V4 - VDQ/V4
      GO TO 190
  100 V3 = V4*(V4+VSN*V2)/(V2+VSN*V4)
  110 VDQ = V2*V3
      A = 1./(V3-V2)
      GO TO 200
  120 GO TO (130, 150, 140), JJ
  130 VDQ = V1*V4
      A = V3/(V3*V3-VDQ)
      GO TO 170
  140 V4 = V3*(V1+VSN*V3)/(V3+VSN*V1)
  150 VDQ = V1*V4
      A = VSN/(V4-V1)
      GO TO 200
  160 VDQ = V1*V4
      A = V2/(VDQ-V2*V2)
  170 VSN = (V4-V1)*A
      GO TO 200
  180 VDQ = V2*V3
      VSN = VDQ/V1 - V1
  190 A = 1./(V3-V2)
      VSN = VSN*A
C
  200 VD = SQRT(VDQ)
      A = A*VD
      IF (ITYP.LE.3) GO TO 270
      A = A/VSN
      GO TO 270
C
  210 J = LVSN*2 + ITYP
      GO TO (220, 220, 230, 240, 250, 260), J
  220 VSN = V2/V1
      GO TO (250, 240), J
  230 VD = V2/VSN
      GO TO 270
  240 VD = V2
      GO TO 270
  250 VD = V1
      GO TO 270
  260 VD = V1*VSN
```

```
C
 270  RETURN
      END
C
C ======================================================================
C
C DOREDI - SUBROUTINES: PART II
C
C ======================================================================
C
C
C-----------------------------------------------------------------------
C SUBROUTINE:   GRID01
C BUILD A START GRID FOR THE SEARCH OF THE GLOBAL EXTREMES OF THE
C TRANSFER FUNCTION WITH ROUNDED COEFFICIENTS
C-----------------------------------------------------------------------
C
      SUBROUTINE GRID01
C
      COMMON /CONST/ PI, FLMA, FLMI, FLER
      COMMON /TOL/ ITYP, IAPRO, NDEG, OM(4), SF, ADELP, ADELS
      COMMON /RES/ AC, ROM(4), RDELP, RDELS, NZM(4)
      COMMON /RESZ/ ZFA, ZM(18,4), ZZR(16), ZZI(16), ZPR(16), ZPI(16)
      COMMON /CGRID/ GR(64), NGR(12)
C
      M = NDEG
      IF (NDEG.GT.22) M = 22
      IF (ITYP.GE.3) M = M/2
      M1 = M/2
      FM = PI/FLOAT(2*M)
      M11 = 2*M1 + 1
      FM1 = PI/FLOAT(M11)
      M = M + 1
      PH2 = 1.5*PI
C
      NGR(8) = 1
C
C PASS BAND
C
      N = 0
      GO TO (70, 10, 20, 40), ITYP
C
C HIGH PASS
C
  10  IA = 1
      IB = NZM(2)
      GO TO 30
C
C BAND PASS
C
  20  IA = NZM(3)
      IB = 1
  30  N = N + 1
      GR(N) = 0.
      GO TO 50
C
C BAND STOP
C
  40  IA = 1
      IB = NZM(2)/2 + 1
  50  J = 1
      PH1 = 0.
      JA = 3
      JB = 2
      GO TO 350
  60  NGR(1) = N + 1
  70  K = 2
      L = 1
      MAK = 1
      MAL = 1
      MEK = NZM(K)
      MEL = NZM(L)
      J = 1
      GO TO (80, 90, 100, 110), ITYP
C LOW PASS
  80  ID = 1
      NGR(1) = 1
      PH = 0.
      IF (MEK.GT.MEL) GO TO 370
      IQ = MEK
      MEK = MEL
      MEL = IQ
      K = 1
      L = 2
      GO TO 370
C
C HIGH PASS
C
  90  ID = -1
      GO TO 130
C BAND PASS
 100  MEK = MEK/2
      MEL = MEK
      GO TO 120
C BAND STOP
 110  MAK = MEK/2 + 1
      MAL = MEL/2 + 1
 120  ID = 1
 130  PH = PH2
      GO TO 370
C
 140  NGR(2) = N
      GO TO (150, 160, 160, 160), ITYP
C LOW PASS
 150  IA = NZM(2)
      IB = 1
      JA = 2
      JB = 3
      PH1 = PH2
      J = 2
      GO TO 350
 160  NGR(3) = N
      IF (ITYP.LE.2) GO TO 230
      K = 2
      L = 1
      MEK = NZM(K)
      MEL = NZM(L)
      IF (ITYP.EQ.3) GO TO 180
      IF (MEK.GT.MEL) GO TO 170
      K = 1
      L = 2
 170  MEK = MEK/2
      MEL = MEL/2
      MAK = 1
      MAL = 1
      GO TO 190
 180  MAK = MEK/2 + 1
      MAL = MAK
 190  ID = 1
      PH = 0.
```

```
      J = 2
      GO TO 370
C
 200  NGR(4) = N
      IF (ITYP.EQ.3) GO TO 210
C
C BAND STOP
C
      IA = NZM(2)/2
      IB = NZM(3)
      GO TO 220
C
C BAND PASS
C
 210  IA = NZM(2)
      IB = 1
 220  JA = 2
      JB = 3
      PH1 = PH2
      J = 3
C
      GO TO 350
 230  IF (ABS(GR(N)-PI).LT.FLMI) GO TO 240
      N = N + 1
      GR(N) = PI
 240  NGR(9) = N
      NGR(5) = N
C
C
C STOP BAND
C
      K = 3
      L = 4
      MEK = NZM(K)
      MEL = NZM(L)
      MAK = 1
      MAL = 1
      J = 3
      ID = 1
      GO TO (280, 250, 270, 250), ITYP
C
C HIGH PASS
C
 250  IF (MEK.GT.MEL) GO TO 260
      IQ = MEK
      MEK = MEL
      MEL = IQ
      K = 4
      L = 3
 260  PH = 0.
      ID = -1
      IF (ITYP.NE.4) GO TO 330
C
C BAND STOP
C
      MAK = MEK/2 + 1
      MAL = MEL/2 + 1
      GO TO 280
C BAND PASS
 270  MEK = MEK/2
      MEL = MEL/2
 280  PH = PH2
      GO TO 380
C
 290  NGR(6) = N
      J = 4
      GO TO (500, 500, 300, 320), ITYP
C
C BAND PASS
C
 300  IF (MEK.GT.MEL) GO TO 310
      IQ = MEK
      MEK = MEL
      MEL = IQ
      K = 4
      L = 3
 310  MAK = MEK + 1
      MAL = MEL + 1
      MEK = NZM(K)
      MEL = NZM(L)
      GO TO 330
C
C BAND STOP
C
 320  MEK = MEK/2
      MEL = (MEL+1)/2
      K = 4
      L = 3
      MAK = 1
      MAL = 1
 330  PH = 0.
      GO TO 380
C
 340  NGR(7) = N
      GO TO 500
C
C
 350  QA = ZM(IA,JA)
      QD = ZM(IB,JB) - QA
      IF (PH1.NE.0.) QA = QA + QD
      DO 360 I=1,M1
        N = N + 1
        Q = FLOAT(I)*FM1 + PH1
        GR(N) = QA + SIN(Q)*QD
 360  CONTINUE
      GO TO (60, 160, 230), J
C
 370  GO TO (460, 390, 460, 390), IAPRO
 380  GO TO (460, 460, 390, 390), IAPRO
C MULTIPLE EXTREMES
 390  IF (ID.LT.0) GO TO 400
      IK = MAK
      IL = MAL
      GO TO 410
C
 400  IK = MEK
      IL = MEL
 410  DO 450 I=MAK,MEK
        N = N + 1
        GR(N) = ZM(IK,K)
        IF (IL.GT.MEL .OR. IL.LT.MAL) GO TO 440
        IF (NDEG.LE.22) GO TO 420
        IF (IL.NE.MEL .AND. IL.NE.MAL) GO TO 430
 420    N = N + 1
        GR(N) = ZM(IL,L)
 430    IL = IL + ID
 440    IK = IK + ID
 450  CONTINUE
      GO TO 490
C
```

```
C SINGLE EXTREMES
C
 460  QA = ZM(MEK,K)
      QD = ZM(MEL,L)
      IF (QD.GT.QA) GO TO 470
      Q = QA
      QA = QD
      QD = Q
 470  QD = QD - QA
      IF (PH.NE.0.) QA = QA + QD
      DO 480 I=1,M
        N = N + 1
        II = I - 1
        Q = FLOAT(II)*FM + PH
        GR(N) = QA + SIN(Q)*QD
 480  CONTINUE
 490  GO TO (140, 200, 290, 340), J
C
 500  NGR(10) = NGR(3) + 1
      NGR(11) = NGR(5) + 1
      NGR(12) = NGR(6) + 1
      RETURN
      END
C
C-----------------------------------------------------------------------
C SUBROUTINE:   GRID03
C START VALUES FOR THE CALCULATION OF A GRID
C-----------------------------------------------------------------------
C
      SUBROUTINE GRID03
C
      COMMON /CONST/ PI, FLMA, FLMI, FLER
      COMMON /TOL/ ITYP, IAPRO, NDEG, OM(4), SF, ADELP, ADELS
      COMMON /RESZ/ ZFA, ZM(18,4), ZZR(16), ZZI(16), ZPR(16), ZPI(16)
      COMMON /RES/ AC, ROM(4), RDELP, RDELS, NZM(4)
C
      IF (ITYP.LE.0 .OR. ITYP.GT.4) GO TO 70
      IF (NDEG.LE.0) GO TO 70
      IAPRO = 1
      DO 10 I=1,4
        NZM(I) = (ITYP+1)/2
  10  CONTINUE
      ME = 2
      IF (ITYP.GE.3) ME = 4
      DO 20 I=1,ME
        IF (OM(I).LE.0.) GO TO 80
  20  CONTINUE
C
      GO TO (30, 40, 50, 60), ITYP
  30  ZM(1,1) = 0.
      ZM(1,2) = OM(1)
      ZM(1,3) = OM(2)
      ZM(1,4) = PI
      GO TO 100
  40  ZM(1,1) = PI
      ZM(1,2) = OM(2)
      ZM(1,3) = OM(1)
      ZM(1,4) = 0.
      GO TO 100
  50  NZM(1) = 1
      ZM(1,1) = (OM(2)+OM(3))/2.
      ZM(1,2) = OM(2)
      ZM(2,2) = OM(3)
      ZM(1,3) = OM(4)
      ZM(2,3) = OM(1)
      ZM(1,4) = PI
      ZM(2,4) = 0.
      GO TO 100
  60  ZM(1,1) = 0.
      ZM(2,1) = PI
      ZM(1,2) = OM(1)
      ZM(2,2) = OM(4)
      ZM(1,3) = OM(3)
      ZM(2,3) = OM(2)
      NZM(4) = 1
      ZM(1,4) = (OM(2)+OM(3))/2.
      GO TO 100
C
  70  IE = 20
      GO TO 90
  80  IE = 23
  90  CALL ERROR(IE)
 100  RETURN
      END
C
C-----------------------------------------------------------------------
C SUBROUTINE:   GRID04
C CHANGE GRID FOR UNSYM. TOLERANCE SCHEME
C-----------------------------------------------------------------------
C
      SUBROUTINE GRID04
C
      COMMON /CONST/ PI, FLMA, FLMI, FLER
      COMMON /TOL/ ITYP, IAPRO, NDEG, OM(4), SF, ADELP, ADELS
      COMMON /RES/ AC, ROM(4), RDELP, RDELS, NZM(4)
      COMMON /CGRID/ GR(64), NGR(12)
C
      IF (ITYP.LE.2) GO TO 130
      DO 10 I=1,4
        IF (OM(I).EQ.0.) GO TO 130
  10  CONTINUE
C
      I = 1
      MA = NGR(1)
      ME = NGR(2)
      IF (ITYP.EQ.3) OMM = OM(2)
      IF (ITYP.EQ.4) OMM = OM(4)
      N = 1
      GO TO 50
C
  20  MA = NGR(10)
      ME = NGR(4)
      IF (ITYP.EQ.3) OMM = OM(3)
      IF (ITYP.EQ.4) OMM = OM(1)
      N = 4
      GO TO 90
C
  30  I = 2
      MA = NGR(11)
      ME = NGR(6)
      IF (ITYP.EQ.3) OMM = OM(4)
      IF (ITYP.EQ.4) OMM = OM(2)
      N = 11
      GO TO 50
C
  40  MA = NGR(12)
      ME = NGR(7)
      IF (ITYP.EQ.3) OMM = OM(1)
      IF (ITYP.EQ.4) OMM = OM(3)
      N = 7
```

```
      GO TO 90
C
   50 DO 60 II=MA,ME
        J = ME + MA - II
        IF ((GR(J)-OMM).LE.FLER) GO TO 70
   60 CONTINUE
      GO TO 80
   70 GR(J) = OMM
      NGR(N) = J
   80 IF (I.EQ.1) GO TO 20
      GO TO 40
C
   90 DO 100 II=MA,ME
        IF ((GR(II)-OMM).GE.(-FLER)) GO TO 110
  100 CONTINUE
      GO TO 120
  110 GR(II) = OMM
      NGR(N) = II
  120 IF (I.EQ.1) GO TO 30
C
  130 RETURN
      END
C
C-----------------------------------------------------------------------
C SUBROUTINE:   COEFW
C SEARCH OF THE MINIMUM COEFFICIENT WORDLENGTH
C COMPUTATION OF THE OPTIMIZATION CRITERION
C-----------------------------------------------------------------------
C
      SUBROUTINE COEFW
C
      COMMON /CONTR/ IPRUN, IPCON, NINP, NOUT, NDOUT, LSPOUT, NSPOUT
      COMMON /CONST/ PI, FLMA, FLMI, FLER
      COMMON /CONST1/ MAXDEG, IWLMI, IWLMA, MBL
      COMMON /COPTCO/ LOPTW, LSTAB, ACXMI, ACXMA, ITER, ITERM, ITERM1
      COMMON /CCOEFW/ IWL, IWLG, IWLD, IWLL, ADEPSG, ADEPSD, ADEPSL,
     *    ISTAB, IDEPSL, IDEPSD
      COMMON /CEPSIL/ EPS(2), PMAX
      COMMON /CREST/ ISTRU, ISCAL, SCALM, ISEQ(16,2), LSEQ, IWLR,
     *    LPOT2, JSTRU
      COMMON /CREST1/ JSTRUS, JSTRUD
      COMMON /COPTSP/ IB, JSEQN(16), JSEQD(16), AMAX, SCA, ALSBI
      COMMON /FILT/ NB, FACT, B2(16), B1(16), B0(16), C1(16), C0(16)
      COMPLEX ZP, ZPS
      COMMON /CPOL/ ZP(16,2), ZPS(16,2)
C
      IWLRR = IWLR
      IWLR = 100
      IWLG = IWLMA
      IWLD = 0
      IWLM = IWL
      II1 = 1
      II2 = 0
      II3 = 0
      CALL COPY01
      IF (IPRUN.GT.0) GO TO 20
      IF (LSEQ.NE.(-1)) GO TO 10
      CALL ALLO02
      CALL COPY01
      IF (NOUT.GE.2) CALL OUT011
   10 CALL ALLO01
      CALL COPY01
   20 IF (NOUT.EQ.4) CALL OUT043
      CALL RECO
   30 CALL COPY02
      IF (IPRUN.GT.0) GO TO 70
      CALL ROUND(4, 5)
      IF (ISCAL.NE.3) GO TO 40
      CALL POLLOC
      CALL COPY03
   40 AMAX = 1.
      FACT = 1.
      DO 60 J=1,NB
        IB = J
   50   CALL SCAL01
        CALL SCAL02
        IF (JSTRUD.EQ.1) GO TO 60
        IF (IB.NE.NB) GO TO 60
        IB = IB + 1
        GO TO 50
   60 CONTINUE
      CALL RECO
      CALL ROUND(1, 3)
      GO TO 80
   70 CALL ROUND(1, 5)
   80 CALL STABLE
      IF (NOUT.GE.5) CALL OUT046
      IF (ISTAB.NE.(-1)) GO TO 250
      CALL EPSILO
      IDEPS = 0
      DO 90 K=1,2
        IF (EPS(K).GT.1.) IDEPS = IDEPS + 1
   90 CONTINUE
      IF (NOUT.GE.5) CALL OUT043
      IF (NOUT.GE.4) CALL OUT044
      ADEPS = EPS(1) - EPS(2)
      IF (LOPTW.NE.(-1)) GO TO 260
      IF (IWL.NE.IWLL) GO TO 100
      ADEPSL = ADEPS
      IDEPSL = IDEPS
  100 IF (IDEPS.EQ.0) GO TO 110
      IF (IWL.LE.IWLD .OR. IWL.GT.IWLG) GO TO 120
      IWLD = IWL
      ADEPSD = ADEPS
      IDEPSD = IDEPS
      GO TO 120
  110 IF (IWL.GE.IWLG) GO TO 120
      IWLG = IWL
      ADEPSG = ADEPS
      IF (IWLD.LE.IWLG) GO TO 120
      IWLD = 0
      II3 = 0
  120 GO TO (130, 140, 180), II1
  130 II1 = 2
      IF (IDEPS.EQ.2) II1 = 3
      GO TO 120
  140 IF (IDEPS.NE.0) GO TO 170
  150 II2 = II2 + 1
  160 IWL = IWL - 1
      GO TO 220
  170 II3 = II3 + 1
      IF (II3.LT.2) GO TO 160
      IF (II2.GE.2) GO TO 270
      IWL = IWLM + 1
      II1 = 3
      GO TO 220
  180 IF (IDEPS.EQ.0) GO TO 210
  190 II3 = II3 + 1
  200 IWL = IWL + 1
      GO TO 220
```

```
 210  II2 = II2 + 1
      IF (II2.LT.2) GO TO 200
      IF (II3.GE.2) GO TO 270
      IWL = IWLM - 1
      II1 = 2
 220  IF (IWL.GE.IWLMI .AND. IWL.LE.IWLMA) GO TO 30
 230  IF (II1.EQ.3) GO TO 240
      IF (IDEPS.EQ.0) GO TO 270
      II1 = 3
      II2 = 2
      GO TO 190
 240  II1 = 2
      IF (IDEPS.NE.0) GO TO 270
      II3 = 2
      GO TO 150
 250  IF (LOPTW.NE.(-1)) GO TO 270
      IF (II1.NE.1) GO TO 230
      IWL = IWL + 1
      IF (IWL.LE.IWLMA) GO TO 30
 260  ADEPSD = ADEPS
      IWLG = IWL
      IWLD = IWL
 270  RETURN
      END
C
C-----------------------------------------------------------------------
C SUBROUTINE:   RECO
C TRANSFORM THE COEFFICIENTS FOR THE CHOSEN REALIZATION
C-----------------------------------------------------------------------
C
      SUBROUTINE RECO
C
      COMMON /CONST/ PI, FLMA, FLMI, FLER
      COMMON /CRECO/ JRCO, JECO, JDCO, JJDCO(5), JMAXV, JTRB2, LREF
      COMMON /FILT/ NB, FACT, B2(16), B1(16), B0(16), C1(16), C0(16)
      COMMON /FILTRE/ IRCO(5), IECO(16,5), IDCO(16,5), IECOM
C
      DIMENSION ACO(16,5)
      EQUIVALENCE (ACO(1,1),B2(1))
      EQUIVALENCE (IRCO(1),IRB2), (IRCO(2),IRB1), (IRCO(3),IRB0)
      EQUIVALENCE (IRCO(4),IRC1), (IRCO(5),IRC0)
C
      ALOG2 = ALOG(2.)
C
C REPRESENTATION OF THE COEFFICIENTS WITH DIFFERENCES
C
      IF (JDCO.EQ.0) GO TO 70
      DO 60 J=1,5
        ACOM = 0.
        DO 10 I=1,NB
          ACOM = AMAX1(ABS(ACO(I,J)),ACOM)
  10    CONTINUE
        IF (ACOM.GE.FLMI) GO TO 20
        ID = -100
        GO TO 30
  20    ID = ALOG(ACOM)/ALOG2
        Q = 2.**ID
        IF (Q.GE.ACOM) GO TO 30
        ID = ID + 1
        Q = Q*2.
  30    JJ = JJDCO(J)
        IF (JJ.EQ.0) GO TO 60
        BCOM = 0.5*Q
        DO 50 I=1,NB
          AC = ACO(I,J)
          ACA = ABS(AC)
          IF (ACA.GE.BCOM) GO TO 40
          IF (JJ.NE.1) GO TO 50
  40      IDCO(I,J) = ID
  50    CONTINUE
  60  CONTINUE
C
C RANGE OF COEFFICIENT NUMBERS
C
  70  IF (JRCO.EQ.0) GO TO 180
      DO 120 J=1,5
        ACOM = 0.
        BCOM = 0.
        DO 80 I=1,NB
          ACS = ACOEFS(ACO(I,J),0,IDCO(I,J),0)
          ACOM = AMIN1(ACOM,ACS)
          BCOM = AMAX1(BCOM,ACS)
  80    CONTINUE
        ACOM = AMAX1(ABS(ACOM),BCOM)
        IF (ACOM.GE.FLMI) GO TO 90
        IQ = -100
        GO TO 110
  90    IQ = ALOG(ACOM)/ALOG2
        Q = 2.**IQ
        IF (Q.LT.ACOM) GO TO 100
        IF (JMAXV.LE.1) GO TO 110
        IF (JMAXV.EQ.2 .AND. Q.GT.BCOM) GO TO 110
        IF (JMAXV.EQ.3 .AND. Q.GT.ACOM) GO TO 110
 100    IQ = IQ + 1
 110    IF (JRCO.GT.0) IQ = MAX0(IQ,0)
        IRCO(J) = IQ
 120  CONTINUE
      JJ = IABS(JRCO)
      GO TO (180, 130, 130, 150, 160), JJ
 130  IQ = MAX0(IRB2,IRB1,IRB0)
      DO 140 J=1,3
        IRCO(J) = IQ
 140  CONTINUE
      IF (JJ.EQ.2) GO TO 180
      IQ = MAX0(IRC1,IRC0)
      IRC1 = IQ
      IRC0 = IQ
      GO TO 180
 150  IQ = MAX0(IRB1,IRC1)
      IRB1 = IQ
      IRC1 = IQ
      GO TO 180
 160  IQ = MAX0(IRB2,IRB1,IRB0,IRC1,IRC0)
      DO 170 J=1,5
        IRCO(J) = IQ
 170  CONTINUE
C
C COMPUTE "PSEUDO" FLOAT POINT EXPONENTS
C
 180  IF (JECO.EQ.0) GO TO 230
      IECOM = 0
      DO 220 J=1,5
        DO 210 I=1,NB
          QA = ACOEFS(ACO(I,J),0,IDCO(I,J),IRCO(J))
          QA = ABS(QA)
          IF (QA.GE.FLMI) GO TO 190
          IQ = JECO
          GO TO 200
 190      IQ = -ALOG(QA)/ALOG2
          IF (IQ.GT.JECO) IQ = JECO
```

```
200         IECO(I,J) = IQ
            IECOM = MAX0(IECOM,IQ)
210       CONTINUE
220     CONTINUE
230     RETURN
        END
C
C-----------------------------------------------------------------------
C SUBROUTINE:   ROUND
C ROUNDING OF THE CHANGED COEFFICIENTS
C-----------------------------------------------------------------------
C
      SUBROUTINE ROUND(IRA, IRB)
C
      COMMON /FILT/ NB, FACT, B2(16), B1(16), B0(16), C1(16), C0(16)
      COMMON /CRECO/ JRCO, JECO, JDCO, JJDCO(5), JMAXV, JTRB2, LREF
      COMMON /FILTRE/ IRCO(5), IECO(16,5), IDCO(16,5), IECOM
      COMMON /CCOEFW/ IWL, IWLG, IWLD, IWLL, ADEPSG, ADEPSD, ADEPSL,
     *    ISTAB, IDEPSL, IDEPSD
C
      DIMENSION ACO(16,5)
      EQUIVALENCE (ACO(1,1),B2(1))
C
      ALSB = 2.**(1-IWL)
      ELSB = 1. - ALSB
      DO 120 I=1,NB
        DO 110 J=IRA,IRB
          AC = ACO(I,J)
          IF (J.NE.2 .AND. J.NE.3) GO TO 10
          IF (ACO(I,1).EQ.AC) GO TO 110
10        ACS = ACOEFS(AC,IECO(I,J),IDCO(I,J),IRCO(J))
          ACSA = ABS(ACS)
          IF (ACSA.LT.ALSB .AND. J.EQ.1) GO TO 110
          IF (ACSA.GT.ELSB) GO TO 30
          R = 0.5
          IF (J.NE.1 .OR. JTRB2.LE.0) GO TO 20
          R = 0.
          IF (IDCO(I,J).GT.(-100)) R = 1.
20        ACSA = AINT(ACSA/ALSB+R)
          ACSA = ACSA*ALSB
          GO TO 50
30        IF (JMAXV.GE.1) GO TO 40
          IF (JMAXV.EQ.(-2) .AND. AC.LT.0.) GO TO 40
          ACSA = ELSB
          GO TO 50
40        ACSA = 1.
50        BC = SIGN(ACSA,ACS)
          BC = ACOEF(BC,IECO(I,J),IDCO(I,J),IRCO(J))
          ACO(I,J) = BC
          IF (J.NE.1) GO TO 110
          F = BC/AC
          JJ = IABS(JTRB2)
          GO TO (100, 80, 60), JJ
60        DO 70 K=2,3
            ACO(I,K) = ACO(I,K)*F
70        CONTINUE
80        II = I + 1
          IF (II.GT.NB) GO TO 100
          DO 90 K=1,3
            ACO(II,K) = ACO(II,K)/F
90        CONTINUE
          GO TO 110
100       FACT = FACT/F
110     CONTINUE
120   CONTINUE
        RETURN
        END
C
C-----------------------------------------------------------------------
C SUBROUTINE:   STABLE
C CHECK STABILITY OF THE ROUNDED SYSTEM
C-----------------------------------------------------------------------
C
      SUBROUTINE STABLE
C
      COMMON /FILT/ NB, FACT, B2(16), B1(16), B0(16), C1(16), C0(16)
      COMMON /SFILT/ NBS, SFACT, SB2(16), SB1(16), SB0(16), SC1(16),
     *    SC0(16)
      COMMON /FILTRE/ IRCO(5), IECO(16,5), IDCO(16,5), IECOM
      COMMON /CCOEFW/ IWL, IWLG, IWLD, IWLL, ADEPSG, ADEPSD, ADEPSL,
     *    ISTAB, IDEPSL, IDEPSD
      COMMON /COPTCO/ LOPTW, LSTAB, ACXMI, ACXMA, ITER, ITERM, ITERM1
C
      DO 120 I=1,NB
        IT = 0
        CC0 = C0(I)
        CC1 = C1(I)
        IF (CC0.GE.1.) GO TO 20
        AC1 = ABS(CC1)
10      IF (AC1.GE.1.+CC0) GO TO 30
        IF (IT.EQ.0) GO TO 120
        GO TO 110
20      IT = IT + 1
        GO TO 40
30      IT = IT + 2
40      IF (LSTAB.NE.(-1)) GO TO 130
        ALSB = 2.**(1-IWL)
        Q0 = ACOEF(ALSB,IECO(I,5),-100,IRCO(5))
        IF (IT.GE.2) GO TO 50
        CC0 = CC0 - Q0
        GO TO 10
50      IF (IT.GE.3) GO TO 80
        AC0 = CC0
60      AC0 = AC0 + Q0
        IF (AC0.GE.1.) GO TO 70
        IF (AC1.GE.1.+AC0) GO TO 60
        Q3 = AC1 - ABS(SC1(I))
        Q2 = AC0 - SC0(I)
        A = Q3*Q3 + Q2*Q2
        A = SQRT(A)
        GO TO 80
70      IT = 3
80      Q1 = ACOEF(ALSB,IECO(I,4),-100,IRCO(4))
        BC1 = AC1
90      BC1 = BC1 - Q1
        IF (BC1.GE.1.+AC0) GO TO 90
        IF (IT.EQ.3) GO TO 100
        Q3 = BC1 - ABS(SC1(I))
        Q2 = CC0 - SC0(I)
        B = Q3*Q3 + Q2*Q2
        B = SQRT(B)
        IF (A.GE.B) GO TO 100
        BC1 = AC1
        CC0 = AC0
100     C1(I) = SIGN(BC1,CC1)
110     C0(I) = CC0
120   CONTINUE
      ISTAB = -1
      RETURN
130   ISTAB = 0
```

```
      RETURN
      END
C
C-----------------------------------------------------------------------
C SUBROUTINE:   EPSILO
C COMPUTATION OF THE GLOBL EXTREMES OF THE TRANSFER FUNCTION
C WITH ROUNDED COEFFICIENTS
C COMPUTATION OF THE ERROR VALUES EPSILON
C-----------------------------------------------------------------------
C
      SUBROUTINE EPSILO
C
      COMMON /TOL/ ITYP, IAPRO, NDEG, OM(4), SF, ADELP, ADELS
      COMMON /CGRID/ GR(64), NGR(12)
      COMMON /CEPSIL/ EPS(2), PMAX
C
C
C MAXIMUM IN THE PASS BAND(S)
C
      ME = 3
      IF (ITYP.EQ.3) ME = 5
      CALL SMAX(PMAX, 8, ME, 1)
      IF (ITYP.NE.4) GO TO 10
      CALL SMAX(QMAX, 10, 5, 1)
      PMAX = AMAX1(PMAX,QMAX)
C
C MINIMUM IN THE PASS BAND(S)
C
  10  ME = 2
      IF (ITYP.EQ.3) ME = 4
      CALL SMAX(PMIN, 1, ME, -1)
      IF (ITYP.NE.4) GO TO 20
      CALL SMAX(QMIN, 10, 4, -1)
      PMIN = AMIN1(PMIN,QMIN)
C
C MAXIMUM IN THE STOP BAND(S)
C
  20  ME = 6
      IF (ITYP.EQ.4) ME = 7
      CALL SMAX(STMAX, 11, ME, 1)
      IF (ITYP.NE.3) GO TO 30
      CALL SMAX(QMAX, 12, 7, 1)
      STMAX = AMAX1(STMAX,QMAX)
C
C COMPUTATION OF EPSILON
C
  30  EPS(1) = (1.-PMIN/PMAX)/ADELP
      EPS(2) = STMAX/PMAX/ADELS
      RETURN
      END
C
C-----------------------------------------------------------------------
C SUBROUTINE:   OPTPAR
C OPTIMIZATION OF THE COEFFICIENTS WORDLENGTH BY VARIATION OF THE
C DESIGN PARAMETER ACX
C-----------------------------------------------------------------------
C
      SUBROUTINE OPTPAR
C
      COMMON /CONST/ PI, FLMA, FLMI, FLER
      COMMON /DESIGN/ NDEGF, EDEG, ACX, NORMA, LSOUT, LVSN, LSYM
      COMMON /CCOEFW/ IWL, IWLG, IWLD, IWLL, ADEPSG, ADEPSD, ADEPSL,
     *    ISTAB, IDEPSL, IDEPSD
      COMMON /COPTCO/ LOPTW, LSTAB, ACXMI, ACXMA, ITER, ITERM, ITERM1
      COMMON /COPTCP/ IS1, IS2, IABO, IWLGS, IWLGN, IWLGP, ADEPSS,
     *    ADEPSN, ADEPSP, ACXS, ACXN, ACXP
C
      IF (LOPTW.NE.(-1)) GO TO 40
      IF (IWLD.GT.IWLL .AND. IDEPSL.LT.2) GO TO 20
      IF (IDEPSD.NE.2) GO TO 10
      ADEPSD = ADEPSG
      IWLD = IWLG
  10  IWLL = IWLD
      GO TO 30
  20  ADEPSD = ADEPSL
      IWLD = IWLL
  30  IWL = IWLD
  40  IABO = IABO + 1
      IF (ADEPSD.LT.0.) GO TO 80
      IF (IS1.NE.(-1)) GO TO 50
      IF (IS2.NE.(-1)) GO TO 70
      GO TO 60
  50  IF (IS2.NE.(-1)) GO TO 60
      IS1 = -1
  60  IS2 = -((IS2+2)/2)
      ACXA = (ACX+ACXN)/2.
      GO TO 120
  70  ACXA = (ACX+ACXMI)/2.
      GO TO 120
C
  80  IF (IS1.EQ.(-1)) GO TO 90
      IF (IS2.NE.(-1)) GO TO 110
      GO TO 100
  90  IF (IS2.NE.(-1)) GO TO 100
      IS1 = 0
 100  IS2 = -((IS2+2)/2)
      ACXA = (ACX+ACXP)/2.
      GO TO 120
 110  ACXA = (ACX+ACXMA)/2.
C
 120  IF (IWLG.LT.IWLGS) GO TO 130
      IF (ADEPSS-ABS(ADEPSD).LE.FLMI .OR. IWLG.GT.IWLGS) GO TO 140
 130  IWLGS = IWLG
      ADEPSS = ABS(ADEPSD)
      ACXS = ACX
 140  IF (ADEPSD.LT.0.) GO TO 160
      IF (IWLG.LT.IWLGP) GO TO 150
      IF (ADEPSP-ADEPSD.LE.FLMI .OR. IWLG.GT.IWLGP) GO TO 190
 150  IWLGP = IWLG
      ADEPSP = ADEPSD
      ACXP = ACX
      GO TO 180
 160  IF (IWLG.LT.IWLGN) GO TO 170
      IF (ADEPSN-ADEPSD.GE.FLMI .OR. IWLG.GT.IWLGN) GO TO 190
 170  IWLGN = IWLG
      ADEPSN = ADEPSD
      ACXN = ACX
 180  IABO = 0
 190  ACX = ACXA
      RETURN
      END
C
C =====================================================================
C
C DOREDI - SUBROUTINES: PART III
C
C =====================================================================
C
C
C-----------------------------------------------------------------------
```

```
C SUBROUTINE:   GRID02
C BUILD A START GRID OUT OF THE POLE LOCATIONS FOR THE SEARCH OF THE
C GLOBL EXTREMES OF THE TRANSFER FUNCTION
C-----------------------------------------------------------------------
C
      SUBROUTINE GRID02
C
      COMMON /CONST/ PI, FLMA, FLMI, FLER
      COMMON /FILT/ NB, FACT, B2(16), B1(16), B0(16), C1(16), C0(16)
      COMMON /CGRID/ GR(64), NGR(12)
C
      NGR(1) = 0
      J = 1
      GR(1) = 0.
      DO 10 I=1,NB
        Q = -C1(I)/2.
        QN = C0(I)
        IF (ABS(QN).LT.FLMI) GO TO 10
        QN = QN - Q*Q
        IF (QN.LE.0.) GO TO 10
        J = J + 1
        GR(J) = ATAN2(SQRT(QN),Q)
  10  CONTINUE
      J = J + 1
      GR(J) = PI
      NGR(8) = 1
      NGR(9) = J
C
C SORTING OF THE LOCATIONS
C
      J = J - 1
  20  JJ = 0
      DO 30 I=1,J
        Q1 = GR(I)
        Q2 = GR(I+1)
        IF (Q1.LT.Q2) GO TO 30
        JJ = 1
        GR(I) = Q2
        GR(I+1) = Q1
  30  CONTINUE
      IF (JJ.NE.0) GO TO 20
      RETURN
      END
C
C-----------------------------------------------------------------------
C SUBROUTINE:   ANANOI
C NOISE ANALYSIS OF A GIVEN CASCADE
C-----------------------------------------------------------------------
C
      SUBROUTINE ANANOI
C
      COMMON /CONTR/ IPRUN, IPCON, NINP, NOUT, NDOUT, LSPOUT, NSPOUT
      COMMON /CONST1/ MAXDEG, IWLMI, IWLMA, MBL
      COMMON /FILT/ NB, FACT, B2(16), B1(16), B0(16), C1(16), C0(16)
      COMMON /COPTST/ LOPTS, ISTOR
      COMMON /COPTSP/ IB, JSEQN(16), JSEQD(16), AMAX, SCA, ALSBI
      COMMON /CREST/ ISTRU, ISCAL, SCALM, ISEQ(16,2), LSEQ, IWLR,
     *    LPOT2, JSTRU
      COMMON /CREST1/ JSTRUS, JSTRUD
      COMMON /CPOW/ PNU, PNC, AND, ITCORP
      COMMON /CNFUNC/ AQC(5), BN2(5), BN1(5), BN0(5)
      COMMON /CFFUNC/ PHI(5), BF2(5), BF1(5), BF0(5), IBB(5), ICOR
      COMMON /CNOISE/ RI, RIN, RE, REN, FAC
      COMMON /COPST2/ PN(100,2), TF(100,2), TFA(100,2)
C
      IQ = IWLR
      IF (IQ-1.GT.IWLMA) IQ = IWLMA + 1
      IQ = IQ - 2
      ALSBI = 2.**IQ
      IB = NB
      IF (LOPTS.NE.0) GO TO 20
      DO 10 I=1,NB
        JSEQN(I) = I
        JSEQD(I) = I
  10  CONTINUE
      GO TO 40
  20  DO 30 I=1,NB
        JSEQN(I) = ISEQ(I,1)
        JSEQD(I) = ISEQ(I,2)
  30  CONTINUE
  40  SAND = 0.
      SPNU = 0.
      SPNC = 0.
      CALL ALLOND
      CALL SMAX(Q, 8, 9, 1)
      FAC1 = FACT/Q
      IF (ISCAL.NE.0 .OR. JSTRUD.EQ.1) FACT = 1.
      AMAX = 1.
      ITCORP = -1
C
      DO 80 I=1,NB
        IB = I
        IF (ISTRU.GE.30) GO TO 60
  50    IF (ISCAL.EQ.0) SCA = 1.
        IF (ISCAL.NE.0) CALL SCAL01
        CALL ALNS01
        CALL ALNS02
        GO TO 70
  60    CONTINUE
C
C >>>>>>>>>>>>>>>>>>>>>>>>>>>>>>>>>>>>>>>>>>>>>>>>>>>>>>>>>>>>>>>>>>>>>
C HERE INSERT THE CALLS TO YOUR OWN STRUCTURE IMPLEMENTATIONS
C SEE SUBROUTINE OPTLST
C
        CALL ERROR(8)
C
C >>>>>>>>>>>>>>>>>>>>>>>>>>>>>>>>>>>>>>>>>>>>>>>>>>>>>>>>>>>>>>>>>>>>>
C
  70    Q = FAC1/FACT
        CALL POWER
        AND = AND*Q
        PNC = Q
        Q = Q*Q
        PNU = PNU*Q
        CALL PCORP
        PNC = PNC*Q
        SAND = SAND + AND
        SPNU = SPNU + PNU
        SPNC = SPNC + PNC
        IF (NOUT.GE.3) CALL OUT026
        IF (ISCAL.NE.0) CALL SCAL02
        IF (JSTRUD.EQ.1 .OR. IB.NE.1) FAC1 = FAC1/SCA
        IF (JSTRUD.EQ.1) GO TO 80
        IF (IB.NE.NB) GO TO 80
        IB = IB + 1
        GO TO 50
  80  CONTINUE
      CALL TCORP
      RIN = SPNU + SAND*SAND + PNC
      FAC = FACT/FAC1
```

```
      Q = FAC*FAC
      RI = RIN*Q
      IF (NOUT.LT.3) GO TO 90
      IB = 0
      AND = SAND
      PNU = SPNU
      CALL OUT026
   90 CALL SPOW(RE)
      REN = RE/Q
      RETURN
      END
C
C-----------------------------------------------------------------------
C SUBROUTINE:   OPTLST
C OPTIMIZATION OF THE PAIRING AND ORDERING BY THE PROCEDURE WITH
C LIMITED STORAGE
C DYNAMIC PROGRAMMING FOR SYSTEMS OF AN ORDER LESS OR EQUAL TO TEN
C DIRECT PROCEDURE FOR THE PARAMETER ISTOR EQUAL TO ONE
C-----------------------------------------------------------------------
C
      SUBROUTINE OPTLST
C
      COMMON /CONST/ PI, FLMA, FLMI, FLER
      COMMON /CONST1/ MAXDEG, IWLMI, IWLMA, MBL
      COMMON /FILT/ NB, FACT, B2(16), B1(16), B0(16), C1(16), C0(16)
      COMMON /CREST/ ISTRU, ISCAL, SCALM, ISEQ(16,2), LSEQ, IWLR,
     *    LPOT2, JSTRU
      COMMON /CREST1/ JSTRUS, JSTRUD
      COMMON /COPTST/ LOPTS, ISTOR
      COMMON /COPTSP/ IB, JSEQN(16), JSEQD(16), AMAX, SCA, ALSBI
      COMMON /COPST1/ LSEQN(16,100,2), LSEQD(16,100,2)
      COMMON /COPST2/ PN(100,2), TF(100,2), TFA(100,2)
      COMMON /CPOW/ PNU, PNC, AND, ITCORP
      COMMON /CNFUNC/ AQC(5), BN2(5), BN1(5), BN0(5)
      COMMON /CFFUNC/ PHI(5), BF2(5), BF1(5), BF0(5), IBB(5), ICOR
      COMMON /CNOISE/ RI, RIN, RE, REN, FAC
C
      DIMENSION ISEQN(16), ISEQD(16)
      EQUIVALENCE (ISEQN(1),ISEQ(1,1)), (ISEQD(1),ISEQ(1,2))
C
      ITCORP = 0
      ALSBI = 2.**(IWLMA-1)
      II = 1
      PN(1,2) = 0.
      TF(1,2) = 1.
      TFA(1,2) = 1.
      CALL SMAX(Q, 8, 9, 1)
      FAC1 = FACT/Q
      JS = 1
      IS = 2
      IN = 1
      PNTM = FLMA
      MA = NB
      IF (LOPTS.NE.(-1)) MA = 1
C
      DO 220 IK=1,NB
        IIK = IK
        MK = IK - 1
        IQ = IS
        IS = JS
        JS = IQ
        JN = IN
        IN = 0
C
C EXTENSION OF THE ASSIGNED ORDERING
C
        DO 210 JK=1,JN
          IF (IK.EQ.1) GO TO 10
          CALL CODE4(JSEQD, LSEQD(1,JK,JS), MK)
          IF (LOPTS.EQ.(-1)) CALL CODE4(JSEQN, LSEQN(1,JK,JS), MK)
   10     DO 200 I=1,NB
            IF (IK.EQ.1) GO TO 30
            DO 20 II=1,MK
              IF (JSEQD(II).EQ.I) GO TO 200
   20       CONTINUE
   30       JSEQD(IK) = I
            DO 190 J=1,MA
              IF (LOPTS.NE.(-1)) GO TO 60
              IF (IK.EQ.1) GO TO 50
              DO 40 II=1,MK
                IF (JSEQN(II).EQ.J) GO TO 190
   40         CONTINUE
   50         JSEQN(IK) = J
C
   60         FACT = TF(JK,JS)
              AMAX = TFA(JK,JS)
              PNT = PN(JK,JS)
              IB = IK
              CALL ALLOND
   70         IF (ISTRU.GE.30) GO TO 80
              CALL SCAL01
              CALL ALNS01
              CALL ALNS02
              GO TO 90
   80         CONTINUE
C
C >>>>>>>>>>>>>>>>>>>>>>>>>>>>>>>>>>>>>>>>>>>>>>>>>>>>>>>>>>>>>>>>>>>>>>>
C HERE INSERT THE CALLS TO YOUR OWN STRUCTURE IMPLEMENTATIONS
C SCALING OF THE IB-TH BLOCK
C ALLOCATIONS OF THE NOISE SOURCES IN THE IB-TH BLOCK
C ALLOCATION OF THE INPUT  TRANSFERFUNCTIONS
C FACT = PRODUCT OF THE IB-1 SCALING FACTORS INFRONT
C AMAX = SCALING OF THE (IB-1)-TH BLOCK
C JSTRU  = 1 : SECOND COEFFICIENT OPTIMIZATION
C JSTRU  = 2 : NO SECOND OPTIMIZATION
C JSTRUS = 1 : NO SEPARATE SCALING FACTOR
C JSTRUS = 2 : SEPARATE SCALING FACTOR
C JSTRUD = 1 : NUMERATOR FIRST
C JSTRUD = 2 : DENOMINATOR FIRST
C BN2(I),BN1(I),BN0(I) : NOISE TRANSFER FUNCTION IN THE IB-TH BLOCK
C BF2(I),BF1(I),BF0(I) : TRANSFER FUNCTIONS IN THE IB-TH BLOCK
C FROM THE INPUT TO THE NONLINEARITIES
C IBB(I) .NE. IB : BLOCK IB ONLY NUMERATOR
C PHI(I) : VARIANCE OF THE INPUT TRANSFER FUNCTIONS
C
C REMOVE THE FOLLOWING STATEMENT
C
              CALL ERROR(8)
C
C >>>>>>>>>>>>>>>>>>>>>>>>>>>>>>>>>>>>>>>>>>>>>>>>>>>>>>>>>>>>>>>>>>>>>>>
C
   90         Q = FAC1/FACT
              CALL POWER
              CALL PCORP
              PNT = (PNU+PNC+AND*AND)*Q*Q + PNT
              IF (JSTRUD.EQ.1) GO TO 100
              IF (IB.NE.NB) GO TO 100
              IB = IB + 1
              GO TO 70
  100         IF (IK.EQ.NB) GO TO 170
```

```
C
C STORE INTERMEDIAT RESULT
C CHECK IF COMBINATION IS IN STORAGE
C
            IF (IK.EQ.1) GO TO 130
            IF (IN.EQ.0) GO TO 150
            DO 120 IL=1,IN
              CALL CODE5(ISEQD, LSEQD(1,IL,IS), IIK, JSEQD, II)
              IF (II.EQ.2) GO TO 120
              IF (LOPTS.NE.(-1)) GO TO 110
              CALL CODE5(ISEQN, LSEQN(1,IL,IS), IIK, JSEQN, II)
              IF (II.EQ.2) GO TO 120
 110          CONTINUE
              IF (PN(IL,IS).LE.PNT) GO TO 190
              CALL CODE3(JSEQD, LSEQD(1,IL,IS), IIK)
              IF (LOPTS.EQ.(-1)) CALL CODE3(JSEQN, LSEQN(1,IL,IS),
     *            IIK)
              PN(IL,IS) = PNT
              TF(IL,IS) = FACT
              TFA(IL,IS) = AMAX
              GO TO 190
 120        CONTINUE
 130        IF (IN.LT.ISTOR) GO TO 150
C
C SEARCH IN THE STORAGE FOR THE ALLOCATION WITH  GREATEST NOISE
C POWER
C
            PNM = PNT
            IQ = 0
            DO 140 II=1,ISTOR
              IF (PN(II,IS).LT.PNM) GO TO 140
              IQ = II
              PNM = PN(II,IS)
 140        CONTINUE
            IF (IQ.EQ.0) GO TO 190
            GO TO 160
 150        IN = IN + 1
            IQ = IN
 160        PN(IQ,IS) = PNT
            TF(IQ,IS) = FACT
            TFA(IQ,IS) = AMAX
            CALL CODE3(JSEQD, LSEQD(1,IQ,IS), IIK)
            IF (LOPTS.EQ.(-1)) CALL CODE3(JSEQN, LSEQN(1,IQ,IS), IIK)
            GO TO 190
C
C CHECK RESULT
C
 170        IF (PNT.GE.PNTM) GO TO 190
            PNTM = PNT
            FAC = FACT
            DO 180 II=1,NB
              ISEQD(II) = JSEQD(II)
              ISEQN(II) = JSEQN(II)
 180        CONTINUE
 190      CONTINUE
 200    CONTINUE
        IF (IK.EQ.1) GO TO 220
 210  CONTINUE
 220  CONTINUE
      FAC = FAC/FAC1
      RIN = PNTM
      RI = RIN*FAC*FAC
      RETURN
      END
C
C-----------------------------------------------------------------------
C SUBROUTINE:   ALLOND
C ALLOCATION OF THE NUMERATOR AND DENOMINATOR BLOCKS
C-----------------------------------------------------------------------
C
      SUBROUTINE ALLOND
C
      COMPLEX ZP, ZPS
C
      COMMON /FILT/ NB, FACT, B2(16), B1(16), B0(16), C1(16), C0(16)
      COMMON /SFILT/ NBS, SFACT, SB2(16), SB1(16), SB0(16), SC1(16),
     *    SC0(16)
      COMMON /CPOL/ ZP(16,2), ZPS(16,2)
      COMMON /COPTST/ LOPTS, ISTOR
      COMMON /CREST/ ISTRU, ISCAL, SCALM, ISEQ(16,2), LSEQ, IWLR,
     *    LPOT2, JSTRU
      COMMON /CREST1/ JSTRUS, JSTRUD
      COMMON /COPTSP/ IB, JSEQN(16), JSEQD(16), AMAX, SCA, ALSBI
C
      DO 40 I=1,IB
        J = JSEQD(I)
        C1(I) = SC1(J)
        C0(I) = SC0(J)
        ZP(I,1) = ZPS(J,1)
        ZP(I,2) = ZPS(J,2)
        IF (LOPTS.NE.1) GO TO 20
        IF (JSTRUD.EQ.1) GO TO 30
        IF (I.NE.IB) GO TO 10
        J = JSEQD(1)
        GO TO 30
 10     J = JSEQD(I+1)
        GO TO 30
 20     J = JSEQN(I)
 30     B2(I) = SB2(J)
        B1(I) = SB1(J)
        B0(I) = SB0(J)
        JSEQN(I) = J
 40   CONTINUE
      IF (IB.GE.NB) RETURN
      II = IB
      JJ = IB
      DO 80 I=1,NB
        DO 50 J=1,IB
          IF (JSEQD(J).EQ.I) GO TO 60
 50     CONTINUE
        II = II + 1
        C1(II) = SC1(I)
        C0(II) = SC0(I)
        ZP(II,1) = ZPS(I,1)
        ZP(II,2) = ZPS(I,2)
 60     DO 70 J=1,IB
          IF (JSEQN(J).EQ.I) GO TO 80
 70     CONTINUE
        JJ = JJ + 1
        B2(JJ) = SB2(I)
        B1(JJ) = SB1(I)
        B0(JJ) = SB0(I)
 80   CONTINUE
      RETURN
      END
C
C-----------------------------------------------------------------------
C SUBROUTINE:   SCAL01
C SCALING OF THE IB-TH BLOCK
C-----------------------------------------------------------------------
```

```
C
      SUBROUTINE SCAL01
C
      COMMON /FILT/ NB, FACT, B2(16), B1(16), B0(16), C1(16), C0(16)
      COMMON /CREST/ ISTRU, ISCAL, SCALM, ISEQ(16,2), LSEQ, IWLR,
     *    LPOT2, JSTRU
      COMMON /CREST1/ JSTRUS, JSTRUD
      COMMON /COPTSP/ IB, JSEQN(16), JSEQD(16), AMAX, SCA, ALSBI
C
      NNB = NB
      BMAX = 0.
      J = 1
      IF (IB.GT.NB) GO TO 30
      NB = IB
      IF (JSTRUD.EQ.1) GO TO 30
      J = 2
      BB1 = B1(IB)
      BB0 = B0(IB)
      B1(IB) = 0.
      B0(IB) = 0.
      GO TO 30
   10 IF (JSTRUS.EQ.1) GO TO 20
      J = 3
      B1(IB) = BB1
      B0(IB) = BB0
      GO TO 30
   20 IF (ISTRU.LT.25) GO TO 80
      J = 3
      B1(IB) = 0.
      B0(IB) = 1.
   30 GO TO (40, 50, 60), ISCAL
   40 CALL SMIMP(QMAX)
      GO TO 70
   50 CALL SMAX(QMAX, 8, 9, 1)
      GO TO 70
   60 CALL SPOW(QMAX)
      QMAX = SQRT(QMAX)
   70 BMAX = AMAX1(BMAX,QMAX)
      GO TO (90, 10, 80), J
   80 B1(IB) = BB1
      B0(IB) = BB0
   90 SCA = SCALM/BMAX
      IF (JSTRUS.EQ.1) GO TO 100
      IF (JSTRUD.EQ.2) GO TO 100
      IF (AMAX*SCA.GT.SCALM) SCA = SCALM/AMAX
C
C SCALING WITH A POWER OF TWO
C
  100 IF (LPOT2.NE.(-1)) GO TO 110
      I = ALOG(SCA)/ALOG(2.)
      Q = 2.**I
      IF (Q.GT.SCA) Q = Q/2.
      SCA = Q
      GO TO 120
  110 IF (IWLR.GE.100) GO TO 120
      BB0 = 2.*ALSBI
      Q = SCA*BB0
      SCA = AINT(Q)/BB0
  120 AMAX = BMAX*SCA
      FACT = FACT*SCA
      NB = NNB
      RETURN
      END
C
C-----------------------------------------------------------------------
C SUBROUTINE:   SCAL02
C CALCULATE SCALING FACTOR INTO NUMERATOR
C-----------------------------------------------------------------------
C
      SUBROUTINE SCAL02
C
      COMMON /FILT/ NB, FACT, B2(16), B1(16), B0(16), C1(16), C0(16)
      COMMON /COPTSP/ IB, JSEQN(16), JSEQD(16), AMAX, SCA, ALSBI
      COMMON /CREST/ ISTRU, ISCAL, SCALM, ISEQ(16,2), LSEQ, IWLR,
     *    LPOT2, JSTRU
      COMMON /CREST1/ JSTRUS, JSTRUD
C
      I = IB
      IF (JSTRUD.EQ.1) GO TO 10
      IF (I.EQ.1) GO TO 20
      I = I - 1
   10 B2(I) = SCA
      B1(I) = B1(I)*SCA
      B0(I) = B0(I)*SCA
      FACT = FACT/SCA
   20 RETURN
      END
C
C-----------------------------------------------------------------------
C SUBROUTINE:   ALNS01
C ALLOCATION OF THE NOISE SOURCES FOR THE MODIFIED STRUCTURES
C OF THE FIRST AND SECOND CANONIC FORM
C-----------------------------------------------------------------------
C
      SUBROUTINE ALNS01
C
      COMMON /CREST/ ISTRU, ISCAL, SCALM, ISEQ(16,2), LSEQ, IWLR,
     *    LPOT2, JSTRU
      COMMON /CREST1/ JSTRUS, JSTRUD
      COMMON /FILT/ NB, FACT, B2(16), B1(16), B0(16), C1(16), C0(16)
      COMMON /COPTSP/ IB, JSEQN(16), JSEQD(16), AMAX, SCA, ALSBI
      COMMON /CNFUNC/ AQC(5), BN2(5), BN1(5), BN0(5)
C
C
      DO 10 I=1,5
        AQC(I) = 1.
   10 CONTINUE
C
      BB2 = SCA
      IF (IB.GT.NB) GO TO 120
      CALL QUAN(C1(IB), AQC1)
      CALL QUAN(C0(IB), AQC0)
      BB1 = B1(IB)
      BB0 = B0(IB)
C
      J = ISTRU/10
      IF (J.EQ.2) GO TO 130
      BB2 = BB2*B2(IB)
      CALL QUAN(BB2, AQB2)
      IF (JSTRU.EQ.2) GO TO 20
      BB1 = BB1*SCA
      BB0 = BB0*SCA
   20 CALL QUAN(BB1, AQB1)
      CALL QUAN(BB0, AQB0)
      J = ISTRU - 10
      IIS = 1
      GO TO (30, 30, 50, 50, 60, 70, 90, 100), J
   30 AQC(5) = AQC0
      AQC(4) = AQC1
      AQC(3) = AQB0
```

```
      AQC(2) = AQB1
40    AQC(1) = AQB2
      GO TO 270
50    AQC(4) = AMIN1(AQB1,AQC1)
      AQC(5) = AMIN1(AQB0,AQC0)
      GO TO 40
60    AQC(3) = AQB0
      GO TO 80
70    AQC(1) = AQB2
80    AQC(2) = AQB1
      AQC(4) = AQC1
      AQC(5) = AQC0
      GO TO 270
90    AQC(5) = AMIN1(AQB0,AQC0)
      GO TO 110
100   AQC(5) = AQC0
110   AQC(4) = AMIN1(AQB1,AQC1)
      GO TO 40
C
120   AQC1 = 1.
      AQC0 = 1.
      AQB1 = 1.
      AQB0 = 1.
      GO TO 140
130   J = ISTRU - 20
      CALL QUAN(BB1, AQB1)
      CALL QUAN(BB0, AQB0)
140   IIS = 2
      IF (IB.NE.1) BB2 = BB2*B2(IB-1)
      CALL QUAN(BB2, AQB2)
      GO TO (170, 150, 210, 200, 190, 160, 210, 220), J
150   AQC(3) = AQB0
160   AQC(2) = AQB1
170   AQC(1) = AQB2
180   AQC(4) = AQC1
      AQC(5) = AQC0
      GO TO 230
190   IF (IB.EQ.1) GO TO 170
      GO TO 180
200   AQC(2) = AMIN1(AQB1,AQB0)
210   AQC(4) = AMIN1(AQB2,AQC1,AQC0)
      IF (IB.LE.NB) GO TO 230
      AQC(1) = AQC(4)
      AQC(4) = 1.
      GO TO 270
220   AQC(2) = AQB1
      GO TO 210
C
230   IF (IB.EQ.1) GO TO 270
      BB1 = B1(IB-1)
      BB0 = B0(IB-1)
      IF (JSTRUS.EQ.2) GO TO 240
      BB1 = BB1*SCA
      BB0 = BB0*SCA
240   CALL QUAN(BB1, AQB1)
      CALL QUAN(BB0, AQB0)
      GO TO (250, 270, 260, 270, 250, 270, 260, 270), J
250   AQC(3) = AQB0
      AQC(2) = AQB1
      GO TO 270
260   AQC(4) = AMIN1(AQC(4),AQB1,AQB0)
C
270   DO 460 I=1,5
        BN2(I) = 0.
        BN1(I) = 0.
        BN0(I) = 0.
        IF (AQC(I).EQ.1.) GO TO 460
        IF (IB.GT.NB) GO TO 420
        IF (IIS.EQ.2) GO TO 340
        GO TO (280, 290, 300, 310, 320, 330, 300, 310), J
280     GO TO (420, 430, 440, 430, 440), I
290     GO TO (400, 430, 440, 430, 440), I
300     GO TO (420, 460, 460, 430, 440), I
310     GO TO (400, 460, 460, 430, 440), I
320     GO TO (460, 430, 410, 430, 440), I
330     GO TO (400, 430, 460, 430, 440), I
C
340     GO TO (390, 350, 360, 370, 390, 380, 360, 370), J
350     GO TO (390, 450, 450, 390, 390), I
360     GO TO (460, 460, 460, 390, 460), I
370     GO TO (460, 450, 460, 390, 460), I
380     GO TO (390, 450, 460, 390, 390), I
C
390     BN2(I) = B2(IB)
        BN1(I) = B1(IB)
        BN0(I) = B0(IB)
        GO TO 460
400     BN2(I) = 1.
        BN1(I) = B1(IB)/B2(IB)
        BN0(I) = B0(IB)/B2(IB)
        GO TO 460
410     BN0(I) = 1.
420     BN2(I) = 1.
        GO TO 460
430     BN1(I) = 1.
        GO TO 460
440     BN0(I) = 1.
        GO TO 460
450     BN2(I) = B2(IB)
        BN1(I) = B2(IB)*C1(IB)
        BN0(I) = B2(IB)*C0(IB)
460   CONTINUE
C
      RETURN
      END
C
C-----------------------------------------------------------------------
C SUBROUTINE: ALNS02
C ALLOCATION OF THE TRANSFER FUNCTIONS FROM THE INPUT
C TO THE NONLINEARITIES
C-----------------------------------------------------------------------
C
      SUBROUTINE ALNS02
C
      COMMON /CONST/ PI, FLMA, FLMI, FLER
      COMMON /FILT/ NB, FACT, B2(16), B1(16), B0(16), C1(16), C0(16)
      COMMON /COPTSP/ IB, JSEQN(16), JSEQD(16), AMAX, SCA, ALSBI
      COMMON /CREST/ ISTRU, ISCAL, SCALM, ISEQ(16,2), LSEQ, IWLR,
     *    LPOT2, JSTRU
      COMMON /CREST1/ JSTRUS, JSTRUD
      COMMON /CRENO/ LCNO, ICNO(16,5)
      COMMON /CNFUNC/ AQC(5), BN2(5), BN1(5), BN0(5)
      COMMON /CFFUNC/ PHI(5), BF2(5), BF1(5), BF0(5), IBB(5), ICOR
C
      NNB = NB
      IF (IB.GT.NNB) GO TO 10
      BB2 = B2(IB)
      BB1 = B1(IB)
      BB0 = B0(IB)
10    ICOR = 0
```

```
      IS = ISTRU - 10
      IIS = 1
      IF (IS.LE.8) GO TO 20
      IIS = 2
      IS = IS - 10
C
   20 DO 450 N=1,5
        BF2(N) = 0.
        BF1(N) = 0.
        BF0(N) = 0.
        IBB(N) = 0
        AQ = AQC(N)
        IF (AQ.EQ.1.) GO TO 400
        IF (LCNO.EQ.(-1)) AQ = 0.
        K = ICNO(IB,N)
        GO TO (400, 30, 400, 400, 40, 50, 400), K
   30   AQ = SQRT(3.)*AQ
        GO TO 60
   40   AQ = -SQRT(3.)
        GO TO 60
   50   AQ = SQRT(3.)*(AQ-1.)
   60   IF (AQ.EQ.0.) GO TO 440
        IBB(N) = IB
        IF (IB.GT.NNB) GO TO 380
        IF (IIS.EQ.2) GO TO 120
        GO TO (70, 70, 80, 90, 100, 110, 90, 90), IS
   70   GO TO (200, 210, 220, 240, 250), N
   80   GO TO (260, 400, 400, 280, 290), N
   90   GO TO (200, 400, 400, 280, 290), N
  100   GO TO (400, 210, 220, 240, 250), N
  110   GO TO (200, 210, 400, 240, 250), N
C
  120   GO TO (130, 140, 150, 160, 170, 180, 150, 190), IS
  130   GO TO (300, 310, 320, 340, 350), N
  140   GO TO (330, 360, 370, 340, 350), N
  150   GO TO (330, 400, 400, 380, 400), N
  160   GO TO (330, 390, 400, 380, 400), N
  170   GO TO (330, 310, 320, 340, 350), N
  180   GO TO (330, 360, 400, 340, 350), N
  190   GO TO (330, 360, 400, 380, 400), N
C
  200   Q = BB2
        GO TO 230
  210   Q = BB1
        GO TO 230
  220   Q = BB0
  230   BF2(N) = Q
        BF1(N) = Q*C1(IB)
        BF0(N) = Q*C0(IB)
        GO TO 410
  240   Q = -C1(IB)
        GO TO 270
  250   Q = -C0(IB)
        GO TO 270
  260   Q = 1.
  270   BF2(N) = Q*BB2
        BF1(N) = Q*BB1
        BF0(N) = Q*BB0
        GO TO 410
  280   BF2(N) = BB1 - BB2*C1(IB)
        BF1(N) = BB0 - BB2*C0(IB)
        GO TO 410
  290   BF2(N) = BB0 - BB2*C0(IB)
        BF1(N) = BB0*C1(IB) - BB1*C0(IB)
        GO TO 410
  300   IF (IB.EQ.1) GO TO 330
        IQ = IB - 1
        IBB(N) = IQ
        BF2(N) = B2(IQ)
        GO TO 410
  310   IQ = IB - 1
        IBB(N) = IQ
        BF1(N) = B1(IQ)
        GO TO 410
  320   IQ = IB - 1
        IBB(N) = IQ
        BF0(N) = B0(IQ)
        GO TO 410
  330   Q = 1.
        GO TO 230
  340   BF1(N) = -C1(IB)
        GO TO 410
  350   BF0(N) = -C0(IB)
        GO TO 410
  360   BF1(N) = BB1/BB2
        GO TO 410
  370   BF0(N) = BB0/BB2
        GO TO 410
  380   BF2(N) = 1.
        GO TO 410
  390   BF1(N) = BB1/BB2
        GO TO 370
C
  400   AQ = 0.
        GO TO 440
C
  410   NB = IBB(N)
        IF (IB.GT.NNB) GO TO 430
        IF (NB.EQ.IB) GO TO 420
        BB22 = B2(NB)
        BB12 = B1(NB)
        BB02 = B0(NB)
  420   B2(NB) = BF2(N)
        B1(NB) = BF1(N)
        B0(NB) = BF0(N)
  430   CALL SPOW(Q)
        PHI(N) = SQRT(Q)
        ICOR = 1
        IF (NB.EQ.IB) GO TO 440
        B2(NB) = BB22
        B1(NB) = BB12
        B0(NB) = BB02
C
  440   BF2(N) = BF2(N)*AQ
        BF1(N) = BF1(N)*AQ
        BF0(N) = BF0(N)*AQ
  450 CONTINUE
C
      IF (IB.GT.NNB) GO TO 460
      B2(IB) = BB2
      B1(IB) = BB1
      B0(IB) = BB0
  460 NB = NNB
C
      RETURN
      END
C
C-----------------------------------------------------------------------
C SUBROUTINE:   TSTR01
C TEST ROUTINE FOR THE STRUCTURES NO 11-28
```

```
C-----------------------------------------------------------------------
C
      SUBROUTINE TSTR01
C
      COMMON /CONST/ PI, FLMA, FLMI, FLER
      COMMON /COPTST/ LOPTS, ISTOR
      COMMON /CREST/ ISTRU, ISCAL, SCALM, ISEQ(16,2), LSEQ, IWLR,
     *    LPOT2, JSTRU
      COMMON /CREST1/ JSTRUS, JSTRUD
      COMMON /FILT/ NB, FACT, B2(16), B1(16), B0(16), C1(16), C0(16)
C
      IF (LOPTS.NE.1) GO TO 10
      IF (ISTRU.GT.20 .AND. JSTRUS.EQ.2) GO TO 30
   10 L = ISTRU
      IF (L.GT.20) L = L - 10
      IF (L.LT.15) GO TO 60
      DO 20 L=1,NB
        Q = ABS(B0(L))
        IF (Q.LT.FLMI) Q = ABS(B1(L))
        IF ((B2(L)-Q).GT.FLMI) GO TO 40
   20 CONTINUE
      GO TO 60
   30 L = 31
      GO TO 50
   40 L = 32
   50 CALL ERROR(L)
   60 RETURN
      END
C
C-----------------------------------------------------------------------
C SUBROUTINE:   PCORP
C COMPUTATION OF THE CORRELATED NOISE OF THE BLOCK IB BY A
C LINEARIZED MODELL
C-----------------------------------------------------------------------
C
      SUBROUTINE PCORP
C
      COMPLEX ZBL1, ZQQ, ZOM, Z, ZA, ZAA
      COMPLEX ZPHI(300)
C
      COMMON /CONST/ PI, FLMA, FLMI, FLER
      COMMON /FILT/ NB, FACT, B2(16), B1(16), B0(16), C1(16), C0(16)
      COMMON /CPOW/ PNU, PNC, AND, ITCORP
      COMMON /COPTSP/ IB, JSEQN(16), JSEQD(16), AMAX, SCA, ALSBI
      COMMON /CNFUNC/ AQC(5), BN2(5), BN1(5), BN0(5)
      COMMON /CFFUNC/ PHI(5), BF2(5), BF1(5), BF0(5), IBB(5), ICOR
      COMMON /COPST2/ PN(100,2), TF(100,2), TFA(100,2)
C
      EQUIVALENCE (ZPHI(1),PN(1,1))
C
      ZBL1(ZQQ,A2,A1,A0) = (ZQQ*A2+A1)*ZQQ + A0
C
      FAC = PNC
      IF (ITCORP.GE.0) GO TO 20
      DO 10 I=1,300
        ZPHI(I) = CMPLX(0.,0.)
   10 CONTINUE
      ITCORP = 1
C
   20 PNC = 0.
      IF (ICOR.EQ.0) GO TO 80
      ADOM = PI/299.
      DO 70 LL=1,300
        OM = ADOM*FLOAT(LL-1)
        ZOM = CMPLX(COS(OM),SIN(OM))
        Z = CMPLX(FACT,0.)
C
        DO 30 I=1,NB
          IF (I.NE.IB) Z = Z*ZBL1(ZOM,B2(I),B1(I),B0(I))
          Z = Z/ZBL1(ZOM,1.,C1(I),C0(I))
   30   CONTINUE
C
        IF (IB.LE.NB) Z = Z/ZBL1(ZOM,1.,C1(IB),C0(IB))
        ZA = CMPLX(0.,0.)
        DO 60 N=1,5
          IF (IBB(N).EQ.0) GO TO 60
          IF (IB.GT.NB) GO TO 50
          ZAA = ZBL1(ZOM,BF2(N),BF1(N),BF0(N))*ZBL1(ZOM,BN2(N),BN1(N),
     *        BN0(N))
          ZAA = ZAA/PHI(N)
          IF (IBB(N).EQ.IB) GO TO 40
          ZAA = ZAA*ZBL1(ZOM,1.,C1(IB),C0(IB))/ZBL1(ZOM,B2(IB),B1(IB),
     *        B0(IB))
   40     ZA = ZA + ZAA
          GO TO 60
   50     ZA = ZA + BF2(N)
   60   CONTINUE
C
        Z = Z*ZA
        IF (ITCORP.EQ.1) ZPHI(LL) = ZPHI(LL) + Z*FAC
        Q = CABS(Z)
        PNC = PNC + Q*Q
   70 CONTINUE
C
      PNC = PNC/PI/150.
   80 RETURN
      END
C
C-----------------------------------------------------------------------
C SUBROUTINE:   TCORP
C COMPUTATION OF THE TOTAL CORRELATED NOISE BY A LINEARIZED
C MODELL
C-----------------------------------------------------------------------
C
      SUBROUTINE TCORP
C
      COMPLEX ZPHI(300)
C
      COMMON /CONST/ PI, FLMA, FLMI, FLER
      COMMON /CPOW/ PNU, PNC, AND, ITCORP
      COMMON /COPST2/ PN(100,2), TF(100,2), TFA(100,2)
      COMMON /CRENO/ LCNO, ICNO(16,5)
      COMMON /FILT/ NB, FACT, B2(16), B1(16), B0(16), C1(16), C0(16)
C
      EQUIVALENCE (ZPHI(1),PN(1,1))
C
      PNC = 0.
C
      DO 20 I=1,NB
        DO 10 J=1,5
          K = ICNO(I,J)
          IF (K.EQ.2) GO TO 30
          IF (K.EQ.5) GO TO 30
          IF (K.EQ.6) GO TO 30
   10   CONTINUE
   20 CONTINUE
      GO TO 50
C
   30 DO 40 I=1,300
        Q = CABS(ZPHI(I))
```

```
      PNC = PNC + Q*Q
   40 CONTINUE
C
      PNC = PNC/PI/150.
   50 RETURN
      END
C
C-----------------------------------------------------------------------
C SUBROUTINE:   DESCAL
C REMOVE SCALING OF SECOND ORDER BLOCKS
C-----------------------------------------------------------------------
C
      SUBROUTINE DESCAL
C
      COMMON /FILT/ NB, FACT, B2(16), B1(16), B0(16), C1(16), C0(16)
C
      DO 10 I=1,NB
        Q = B2(I)
        B2(I)= 1.
        B1(I) = B1(I)/Q
        B0(I) = B0(I)/Q
        FACT = FACT*Q
   10 CONTINUE
      RETURN
      END
C
C-----------------------------------------------------------------------
C SUBROUTINE:  DENORM
C TRANSFORM STRUCTURE WITH  TO STRUCTURE WITHOUT SEPARATE
C SCALING FACTOR
C-----------------------------------------------------------------------
C
      SUBROUTINE DENORM
C
      COMMON /FILT/ NB, FACT, B2(16), B1(16), B0(16), C1(16), C0(16)
C
      DO 10 I=1,NB
        Q = B2(I)
        B1(I) = B1(I)*Q
        B0(I) = B0(I)*Q
   10 CONTINUE
      RETURN
      END
C
C-----------------------------------------------------------------------
C SUBROUTINE:   ALLO01
C ALLOCATION OF NUMERATORS AND DENOMINATORS ACCORDING TO THE
C ALLOCATION FIELD  ISEQ(I,J)
C-----------------------------------------------------------------------
C
      SUBROUTINE ALLO01
C
      COMMON /FILT/ NB, FACT, B2(16), B1(16), B0(16), C1(16), C0(16)
      COMMON /SFILT/ NBS, SFACT, SB2(16), SB1(16), SB0(16), SC1(16),
     *   SC0(16)
      COMMON /CREST/ ISTRU, ISCAL, SCALM, ISEQ(16,2), LSEQ, IWLR,
     *    LPOT2, JSTRU
C
      NB = NBS
      FACT = SFACT
      DO 10 I=1,NB
        J = ISEQ(I,1)
        B2(I) = SB2(J)
        B1(I) = SB1(J)
        B0(I) = SB0(J)
        J = ISEQ(I,2)
        C1(I) = SC1(J)
        C0(I) = SC0(J)
   10 CONTINUE
      RETURN
      END
C
C-----------------------------------------------------------------------
C SUBROUTINE:   ALLO02
C ALLOCATION OF NUMERATORS AND DENOMINATORS ACCORDING TO THE
C ALLOCATION FIELD  KSEQ(I,J), FOR A GIVEN START ORDERING
C-----------------------------------------------------------------------
C
      SUBROUTINE ALLO02
C
      COMMON /FILT/ NB, FACT, B2(16), B1(16), B0(16), C1(16), C0(16)
      COMMON /SFILT/ NBS, SFACT, SB2(16), SB1(16), SB0(16), SC1(16),
     *   SC0(16)
      COMMON /CREST2/ KSEQ(16,2)
C
      NB = NBS
      FACT = SFACT
      DO 10 I=1,NB
        J = KSEQ(I,1)
        B2(I) = SB2(J)
        B1(I) = SB1(J)
        B0(I) = SB0(J)
        J = KSEQ(I,2)
        C1(I) = SC1(J)
        C0(I) = SC0(J)
   10 CONTINUE
      RETURN
      END
C
C-----------------------------------------------------------------------
C SUBROUTINE:   FIXPAR
C FIXED PAIRIND OF POLES AND ZEROS
C USED TOGETHER WITH A FREE INPUT OF THE COEFFICIENTS
C-----------------------------------------------------------------------
C
      SUBROUTINE FIXPAR
C
      COMMON /CONST/ PI, FLMA, FLMI, FLER
      COMMON /FILT/ NB, FACT, B2(16), B1(16), B0(16), C1(16), C0(16)
      COMMON /SFILT/ NBS, SFACT, SB2(16), SB1(16), SB0(16), SC1(16),
     *   SC0(16)
      COMMON /SCRAT/ ADUM(32)
C
      DIMENSION IPOL(16,2)
      EQUIVALENCE (ADUM(1),IPOL(1,1))
C
      DO 10 I=1,NB
        IPOL(I,1) = 0
        IPOL(I,2) = 0
   10 CONTINUE
C
      DO 30 I=1,NB
        PM = 0.
        DO 20 J=1,NB
          IF (IPOL(J,1).NE.0) GO TO 20
          P2O = POW2O(1.0,0.,0.,C1(J),C0(J))
          IF (P2O.LE.PM) GO TO 20
          PM = P2O
          IPM = J
   20   CONTINUE
```

```
         IPOL(IPM,1) = I
   30  CONTINUE
C
C
       DO 70 I=1,NB
         DO 40 K=1,NB
           IF (IPOL(K,1).NE.I) GO TO 40
           KK = K
           GO TO 50
   40    CONTINUE
C
   50    CC0 = C0(KK)
         CC1 = C1(KK)
         DO 60 J=1,NB
           IF (IPOL(J,2).NE.0) GO TO 60
           P2O = POW2O(1.0,BB1,BB0,CC1,CC0)
           PM = P2O
           IZM = J
   60    CONTINUE
         IPOL(IZM,2) = KK
   70  CONTINUE
C
       CALL COPY01
       DO 80 I=1,NB
         IPM = IPOL(I,2)
         B2(IPM) = SB2(I)
         B1(IPM) = SB1(I)
         B0(IPM) = SB0(I)
   80  CONTINUE
       RETURN
       END
C
C-----------------------------------------------------------------------
C SUBROUTINE:   POLLOC
C CALCULATE THE POLES IN THE CLOSED UPPER Z-DOMAIN
C-----------------------------------------------------------------------
C
       SUBROUTINE POLLOC
C
       COMPLEX ZP, ZPS
C
       COMMON /FILT/ NB, FACT, B2(16), B1(16), B0(16), C1(16), C0(16)
       COMMON /CPOL/ ZP(16,2), ZPS(16,2)
C
       DO 20 I=1,NB
         Q = -C1(I)/2.
         QN = C0(I)
         QN = Q*Q - QN
         IF (QN.LT.0.) GO TO 10
         QN = SQRT(QN)
         ZPS(I,1) = CMPLX(Q+QN,0.)
         ZPS(I,2) = CMPLX(Q-QN,0.)
         GO TO 20
   10    QN = SQRT(-QN)
         ZPS(I,1) = CMPLX(Q,QN)
         ZPS(I,2) = CMPLX(Q,-QN)
   20  CONTINUE
       RETURN
       END
C
C-----------------------------------------------------------------------
C FUNCTION:     POW2O
C NOISE POWER OF A SECOND ORDER BLOCK
C-----------------------------------------------------------------------
C
       FUNCTION POW2O(B2, B1, B0, C1, C0)
C
       BB0 = B2*B2 + B1*B1 + B0*B0
       BB1 = 2.*B1*(B2+B0)
       BB2 = 2.*B2*B0
       E1 = 1. + C0
       Q = BB0*E1 - BB1*C1 + BB2*(C1*C1-C0*E1)
       Q = Q/((1.-C0*C0)*E1-(C1-C1*C0)*C1)
       POW2O = Q
       RETURN
       END
C
C =======================================================================
C
C DOREDI - SUBROUTINES: PART IV
C
C =======================================================================
C
C
C-----------------------------------------------------------------------
C SUBROUTINE:   DEFIN1
C DEFAULT VALUES FOR CLASS 1 INPUT DATA
C-----------------------------------------------------------------------
C
       SUBROUTINE DEFIN1
C
       COMMON /CONTR/ IPRUN, IPCON, NINP, NOUT, NDOUT, LSPOUT, NSPOUT
       COMMON /RES/ AC, ROM(4), RDELP, RDELS, NZM(4)
       COMMON /CGRID/ GR(64), NGR(12)
C
       NINP = 0
       NOUT = 3
       NDOUT = 0
       LSPOUT = 0
       NSPOUT = 0
       DO 10 I=1,12
         NGR(I) = 0
   10  CONTINUE
       DO 20 I=1,4
         ROM(I) = 0.
   20  CONTINUE
       RETURN
       END
C
C-----------------------------------------------------------------------
C SUBROUTINE:   INP010
C COMMAND CARDS FOR INPUT AND OUTPUT DIRECTIVES
C PROCESSING OF THE COMMAND CARDS CLASS 1
C-----------------------------------------------------------------------
C
       SUBROUTINE INP010
C
       COMMON /CONTR/ IPRUN, IPCON, NINP, NOUT, NDOUT, LSPOUT, NSPOUT
       COMMON /INPDAT/ ICODE, JCODE, LINP, IINP, AINP(3), NPAR
       COMMON /CGRID/ GR(64), NGR(12)
C
       J = JCODE/10
       IF (J.EQ.0) GO TO 110
       IF (J.GT.4) GO TO 110
       GO TO (10, 80, 90, 100), J
C
C *0110  B,NINP
C
   10  IF (IINP.GT.4) GO TO 120
       GO TO (20, 30, 40, 50), IINP
```

```
   20  IF (NINP.EQ.0) GO TO 130
       GO TO 70
   30  CALL INP012
       GO TO 70
   40  NINP = IINP
       CALL INP013
       GO TO 60
   50  NINP = IINP
       CALL INP014
   60  NGR(1) = 0
       RETURN
   70  NINP = IINP
       RETURN
C
C *0120  B,NOUT
C
   80  IF (IINP.LT.0 .OR. IINP.GT.5) GO TO 120
       NOUT = IINP
       RETURN
C
C *0130  NDOUT
C
   90  NDOUT = LINP
       RETURN
C
C *0140  LSPOUT,ISPOUT
C
  100  LSPOUT = LINP
       NSPOUT = IINP
       RETURN
C
  110  I = 1
       GO TO 140
  120  I = 4
       GO TO 140
  130  I = 12
  140  CALL ERROR(I)
       RETURN
       END
C
C-----------------------------------------------------------------------
C SUBROUTINE:   INP012
C INPUT OF THE FILTER DESCRIPTION AND DATA FROM DISK:(CHANNEL 4)
C-----------------------------------------------------------------------
C
      SUBROUTINE INP012
C
      COMMON /CANPAR/ KA1, KA2, KA3, KA4, KA5, LINE
      COMMON /TOL/ ITYP, IAPRO, NDEG, OM(4), SF, ADELP, ADELS
      COMMON /RES/ AC, ROM(4), RDELP, RDELS, NZM(4)
      COMMON /RESZ/ ZFA, ZM(18,4), ZZR(16), ZZI(16), ZPR(16), ZPI(16)
      COMMON /FILT/ NB, FACT, B2(16), B1(16), B0(16), C1(16), C0(16)
      COMMON /FILTRE/ IRCO(5), IECO(16,5), IDCO(16,5), IECOM
      COMMON /CRECO/ JRCO, JECO, JDCO, JJDCO(5), JMAXV, JTRB2, LREF
      COMMON /CCOEFW/ IWL, IWLG, IWLD, IWLL, ADEPSG, ADEPSD, ADEPSL,
     *    ISTAB, IDEPSL, IDEPSD
C
9999  FORMAT (10I5)
9998  FORMAT (4E15.7)
C
      DO 10 I=1,3
        READ (KA4,9999)
   10 CONTINUE
      READ (KA4,9999) ITYP, IAPRO, NDEG
      READ (KA4,9998) SF, OM, ADELP, ADELS
      READ (KA4,9998) AC, ROM, RDELP, RDELS
      READ (KA4,9999) NZM, M
      DO 20 I=1,M
        READ (KA4,9998) (ZM(I,J),J=1,4)
   20 CONTINUE
      READ (KA4,9999)
      READ (KA4,9999) NB
      READ (KA4,9998) FACT
      DO 30 I=1,NB
        READ (KA4,9998) B2(I), B1(I), B0(I), C1(I), C0(I)
   30 CONTINUE
      READ (KA4,9999) IRCO
      DO 40 I=1,NB
        READ (KA4,9999) (IECO(I,J),J=1,5)
   40 CONTINUE
      DO 50 I=1,NB
        READ (KA4,9999) (IDCO(I,J),J=1,5)
   50 CONTINUE
      READ (KA4,9999) IWL, IECOM, JMAXV, JTRB2
      RETURN
      END
C
C-----------------------------------------------------------------------
C SUBROUTINE:   INP013
C INPUT OF SECOND ORDER BLOCKS
C-----------------------------------------------------------------------
C
      SUBROUTINE INP013
C
      COMMON /CARD/ LCODE, IZ(40)
      COMMON /INPDAT/ ICODE, JCODE, LINP, IINP, AINP(3), NPAR
      COMMON /FILT/ NB, FACT, B2(16), B1(16), B0(16), C1(16), C0(16)
C
      CALL INP001
      IF (LCODE.NE.0) RETURN
      CALL INP002
      IF (ICODE.NE.0 .OR. JCODE.NE.10) CALL ERROR(2)
      NB = IINP
      FACT = AINP(1)
C
      MA = 2*NB
      DO 20 I=1,MA
        CALL INP001
        IF (LCODE.NE.0) RETURN
        CALL INP002
        IF (ICODE.NE.0 .OR. JCODE.NE.10) CALL ERROR(2)
        IF (I.GT.NB) GO TO 10
        B2(I) = AINP(1)
        B1(I) = AINP(2)
        B0(I) = AINP(3)
        GO TO 20
   10   J = I - NB
        C1(J) = AINP(2)
        C0(J) = AINP(3)
   20 CONTINUE
      RETURN
      END
C
C-----------------------------------------------------------------------
C SUBROUTINE:   INP014
C INPUT OF POLES AND ZEROS
C-----------------------------------------------------------------------
C
      SUBROUTINE INP014
C
```

```
      COMMON /CARD/ LCODE, IZ(40)
      COMMON /INPDAT/ ICODE, JCODE, LINP, IINP, AINP(3), NPAR
      COMMON /FILT/ NB, FACT, B2(16), B1(16), B0(16), C1(16), C0(16)
C
      CALL INP001
      IF (LCODE.NE.0) RETURN
      CALL INP002
      IF (ICODE.NE.0 .OR. JCODE.NE.10) CALL ERROR(2)
      NB = IINP
      FACT = AINP(1)
      NN = 0
      JJ = NB
C
   10 CALL INP001
      IF (LCODE.NE.0) RETURN
      CALL INP002
      IF (ICODE.NE.0 .OR. JCODE.NE.10) CALL ERROR(2)
      QR = AINP(1)
      QI = AINP(2)
      IF (QI.EQ.0.) GO TO 20
      A0 = QR*QR + QI*QI
      A1 = -2.*QR
      GO TO 40
   20 IF (JJ.NE.NB) GO TO 30
      JJ = NB - 1
      R = QR
      GO TO 10
   30 JJ = NB
      A0 = R*QR
      A1 = -R - QR
   40 NN = NN + 1
      IF (K.EQ.2) GO TO 50
      B2(NN) = 1.
      B1(NN) = A1
      B0(NN) = A0
      GO TO 60
   50 C1(NN) = A1
      C0(NN) = A0
   60 IF (NN.LT.JJ) GO TO 10
      IF (JJ.EQ.NB) GO TO 70
      A1 = -R
      A0 = 0.
      JJ = JJ + 1
      GO TO 40
   70 IF (K.EQ.2) RETURN
      K = 2
      NN = 0
      GO TO 10
C
      END
C
C-----------------------------------------------------------------------
C SUBROUTINE:   DEFIN2
C DEFAULT VALUES FOR THE COMMAND CARDS OF CLASS 02
C-----------------------------------------------------------------------
C
      SUBROUTINE DEFIN2
C
      COMMON /CONST1/ MAXDEG, IWLMI, IWLMA, MBL
      COMMON /COPTCO/ LOPTW, LSTAB, ACXMI, ACXMA, ITER, ITERM, ITERM1
      COMMON /CREST/ ISTRU, ISCAL, SCALM, ISEQ(16,2), LSEQ, IWLR,
     *    LPOT2, JSTRU
      COMMON /CREST2/ KSEQ(16,2)
      COMMON /COPTST/ LOPTS, ISTOR
C
      LOPTW = 0
      LSTAB = -1
      IWLR = 100
      ACXMI = 0.
      ACXMA = 1.
      LOPTS = 0
      ISTOR = 100
      LSEQ = 0
      ISCAL = 2
      SCALM = 1.
      ITERM = 10
      ITERM1 = 4
      DO 20 J=1,2
        DO 10 I=1,MBL
          KSEQ(I,J) = I
          ISEQ(I,J) = I
   10   CONTINUE
   20 CONTINUE
      RETURN
      END
C
C-----------------------------------------------------------------------
C SUBROUTINE:   INP020
C COMMAND CARDS FOR CONTROLLING THE OPTIMIZATION
C PROCESSING OF THE COMMAND CARDS  CLASS 02
C-----------------------------------------------------------------------
C
      SUBROUTINE INP020
C
      COMMON /CONST1/ MAXDEG, IWLMI, IWLMA, MBL
      COMMON /INPDAT/ ICODE, JCODE, LINP, IINP, AINP(3), NPAR
      COMMON /INTDAT/ JINP(16)
      COMMON /DESIGN/ NDEGF, EDEG, ACX, NORMA, LSOUT, LVSN, LSYM
      COMMON /CCOEFW/ IWL, IWLG, IWLD, IWLL, ADEPSG, ADEPSD, ADEPSL,
     *    ISTAB, IDEPSL, IDEPSD
      COMMON /COPTCO/ LOPTW, LSTAB, ACXMI, ACXMA, ITER, ITERM, ITERM1
      COMMON /CREST/ ISTRU, ISCAL, SCALM, ISEQ(16,2), LSEQ, IWLR,
     *    LPOT2, JSTRU
      COMMON /CREST2/ KSEQ(16,2)
      COMMON /COPTST/ LOPTS, ISTOR
C
      IF (JCODE.LT.10) GO TO 110
      J = JCODE/10
      IF (J.GT.4) GO TO 110
      GO TO (10, 50, 60, 70), J
C
C *0210  LWLF,ITERM,ACX,ACXMI,ACXMA
C
   10 LOPTW = -1
      IF (LINP.EQ.(-1)) LOPTW = 1
      IF (IINP.EQ.0) GO TO 20
      IF (IINP.LT.0) GO TO 120
      ITERM = IINP
   20 IF (NPAR.EQ.0) GO TO 30
      Q = AINP(1)
      IF (Q.LT.0. .OR. Q.GT.1.) GO TO 120
      ACX = Q
   30 Q = AINP(2)
      IF (Q.EQ.0.) GO TO 40
      IF (Q.LT.0. .OR. Q.GE.1.) GO TO 120
      ACXMI = Q
   40 Q = AINP(3)
      IF (Q.EQ.0.) RETURN
      IF (Q.LE.0. .OR. Q.GT.1.) GO TO 120
      ACXMA = Q
```

```
      RETURN
C
C *0220  B,ITERM1
C
   50 IF (IINP.LT.0) GO TO 120
      ITERM1 = IINP
      RETURN
C
C *0230  PAIRF,ISTOR
C
   60 LOPTS = -1
      IF (LINP.EQ.(-1)) LOPTS = 1
      IF (IINP.GT.100) GO TO 120
      IF (IINP.LT.0) GO TO 120
      IF (IINP.NE.0) ISTOR = IINP
      RETURN
C
C *0240  LSEQ,ISCAL,SCALM
C
   70 IF (LINP.NE.(-1)) GO TO 100
      LSEQ = -1
      DO 90 J=1,2
        CALL INP001
        IF (ICODE.NE.0) GO TO 130
        IF (JCODE.NE.40) GO TO 130
        CALL INP003
        DO 80 I=1,MBL
          KSEQ(I,J) = JINP(I)
   80   CONTINUE
   90 CONTINUE
C
  100 IF (IINP.NE.0) ISCAL = IINP
      IF (ISCAL.LT.0) ISCAL = 0
      IF (AINP(1).NE.0.) SCALM = AINP(1)
      RETURN
C
  110 I = 4
      GO TO 140
  120 I = 5
      GO TO 140
  130 I = 2
  140 CALL ERROR(I)
      RETURN
      END
C
C-----------------------------------------------------------------------
C SUBROUTINE:   DEFIN3
C SET DEFAULT VALUES
C-----------------------------------------------------------------------
C
      SUBROUTINE DEFIN3
C
      COMMON /CONST/ PI, FLMA, FLMI, FLER
      COMMON /TOL/ ITYP, IAPRO, NDEG, OM(4), SF, ADELP, ADELS
      COMMON /TOLSN/ VSN, VD, A
      COMMON /DESIGN/ NDEGF, EDEG, ACX, NORMA, LSOUT, LVSN, LSYM
C
      ITYP = 1
      IAPRO = 4
      NDEG = 0
      NDEGF = 0
      MDEG = 0
      NORMA = 0
      LSOUT = 0
C
      DO 10 I=1,4
        OM(I) = 0.
   10 CONTINUE
      SF = 0.
      A = 0.
      ADELP = 0.
      ADELS = 1./FLMA
      VSN = 0.
      VD = 0.
      ACX = 1000.
      EDEG = ACX
C
      RETURN
      END
C
C-----------------------------------------------------------------------
C SUBROUTINE:   INP030
C INPUT ROUTINE FOR THE FILTER TYPE AND FOR THE TOLERANCE SCHEME
C
C PROCESSING OF THE COMMAND CARDS CLASS 3
C-----------------------------------------------------------------------
C
      SUBROUTINE INP030
C
      COMMON /CONST/ PI, FLMA, FLMI, FLER
      COMMON /INPDAT/ ICODE, JCODE, LINP, IINP, AINP(3), NPAR
      COMMON /TOL/ ITYP, IAPRO, NDEG, OM(4), SF, ADELP, ADELS
      COMMON /TOLSN/ VSN, VD, A
      COMMON /DESIGN/ NDEGF, EDEG, ACX, NORMA, LSOUT, LVSN, LSYM
C
      SOM(AA) = 2.*ATAN(AA)
C
      J = JCODE/10
      IF (J.GT.7) GO TO 240
      GO TO (10, 20, 30, 40, 170, 220, 230), J
C
C *0310  B,FILTER TYPE8SAMPLING FREQUENCY
C
   10 IF (IINP.NE.0) ITYP = IINP
      IF (AINP(1).NE.0.) SF = AINP(1)
      RETURN
C
C *0320  B8IAPRO88ACX
C
   20 IF (IINP.NE.0) IAPRO = IINP
      IF (NPAR.NE.0) ACX = AINP(1)
      RETURN
C
C *0330  NDEGF,EDEG
C
   30 IF (LINP.EQ.(-1)) NDEGF = -1
      IF (IINP.NE.0) NDEG = IINP
      IF (NPAR.NE.0) EDEG = AINP(1)
      RETURN
C
C CUTOFF FREQUENCIES OF THE TOLERANCE SCHEME
C
   40 K = JCODE - J*10
      IF (K.GT.7) GO TO 250
      GO TO (50, 50, 100, 110, 120, 140, 160), K
C
C *0341  B,B,FR(1),FR(2),<SF>     ::  FREQUENCY DOMAIN
C *0342  B,B,FR(3),FR(4),<SF>
C
   50 IF (AINP(3).EQ.0.) GO TO 70
```

```
      IF (SF.NE.0.) GO TO 260
   60 SF = AINP(3)
   70 IF (SF.EQ.0.) GO TO 270
      Q = 2.*PI/SF
      GO TO (80, 90), K
   80 OM(1) = AINP(1)*Q
      OM(2) = AINP(2)*Q
      RETURN
C
   90 OM(3) = AINP(1)*Q
      OM(4) = AINP(2)*Q
      RETURN
C
C *0343  B,B,OM(1),OM(2)           ::  Z-DOMAIN
C
  100 OM(1) = AINP(1)
      OM(2) = AINP(2)
      RETURN
C
C *0344  B,B,OM(3),OM(4)           ::  Z-DOMAIN SEC. CARD
C
  110 OM(3) = AINP(1)
      OM(4) = AINP(2)
      RETURN
C
C *0345  B,B,S(1),S(2)             ::  S-DOMAIN
C
  120 DO 130 I=1,2
        OM(I) = SOM(AINP(I))
  130 CONTINUE
      RETURN
C
C *0346  B,B,S(3),S(4)             :: S-DOMAIN  SEC. CARD
C
  140 DO 150 I=1,2
        OM(I+2) = SOM(AINP(I))
  150 CONTINUE
      RETURN
C
C *0347  B,B,VSN,VD,A              ::  NORMALIZED PARAMETER
C
  160 VSN = AINP(1)
      VD = AINP(2)
      A = AINP(3)
      RETURN
C
C DELTA P , DELTA S
C
  170 K = JCODE - J*10
      IF (K.GT.3) GO TO 250
      GO TO (180, 190, 210), K
C *0351  B,B,DELP,DELS
C
  180 ADELP = AINP(1)
      ADELS = AINP(2)
      RETURN
C
C *0352  B,B,AP,AS (IN DB)
C
  190 ADELP = 1. - 10.**(-0.05*AINP(1))
  200 ADELS = 10.**(-0.05*AINP(2))
      RETURN
C
C *0353  B,B,P,AS     (AS IN DB)
C
  210 Q = AINP(1)
      ADELP = 1. - SQRT(1.-Q*Q)
      GO TO 200
C
C *0360  B,NORMA
C
  220 IF (IINP.LT.0 .OR. IINP.GT.3) GO TO 250
      NORMA = IINP
      RETURN
C
C *0370  LSOUT
C
  230 LSOUT = LINP
      RETURN
C
C OUTPUT OF ERROR MESSAGES
C
  240 I = 4
      GO TO 280
  250 I = 5
      GO TO 280
  260 I = 7
      GO TO 280
  270 I = 6
  280 CALL ERROR(I)
      IF (I.EQ.7) GO TO 60
      RETURN
      END
C
C-----------------------------------------------------------------------
C SUBROUTINE:   INP031
C CHECK ROUTINE FOR THE FILTER INPUT DATA
C CHECK DATA CARDS OF CLASS 3
C-----------------------------------------------------------------------
C
      SUBROUTINE INP031
C
      COMMON /CONST/ PI, FLMA, FLMI, FLER
      COMMON /CONST1/ MAXDEG, IWLMI, IWLMA, MBL
      COMMON /TOL/ ITYP, IAPRO, NDEG, OM(4), SF, ADELP, ADELS
      COMMON /TOLSN/ VSN, VD, A
      COMMON /DESIGN/ NDEGF, EDEG, ACX, NORMA, LSOUT, LVSN, LSYM
C
      IF (ITYP.GT.4) GO TO 90
      IF (IAPRO.GT.4) GO TO 100
C
      IF (EDEG.GT.999.) EDEG = 0.2
C
C CHECK OM(1) ... OM(4)
C
      LVSN = 0
      LSYM = 0
      J = 0
      Q = 0.
      ME = 2
      IF (ITYP.GE.3) ME = 4
      DO 20 I=1,ME
        QQ = OM(I)
        IF (QQ.EQ.0.) GO TO 10
        IF (QQ.GT.PI) GO TO 120
        IF (QQ.LT.Q) GO TO 110
        Q = QQ
        GO TO 20
   10   LSYM = LVSN
        LVSN = I
```

```
         J = J + 1
   20  CONTINUE
       IF (J.EQ.0) GO TO 80
       IF (ITYP.GT.2) GO TO 30
       IF (J.EQ.1) GO TO 70
       GO TO 50
   30  GO TO (40, 60, 50, 50), J
   40  LSYM = LVSN
       LVSN = 0
       J = 0
       GO TO 80
C
   50  IF (VSN.EQ.0.) GO TO 120
       IF (VD.EQ.0.) GO TO 120
       IF (ITYP.GT.2 .AND. A.EQ.0.) GO TO 110
       LVSN = -1
       GO TO 80
C
   60  J = 1
       IQ = LSYM + LVSN
       IF (IQ.LE.3 .OR. IQ.GE.7) GO TO 120
   70  IF (NDEG.EQ.0) GO TO 120
C
C CECK FOR PASS BAND AND STOP BAND RIPPLE
C
   80  IF (ADELP.EQ.0.) J = J + 1
       IF (ADELS.LE.(1./FLMA)) J = J + 1
       IF (J.EQ.0) RETURN
       IF (J.NE.1 .OR. NDEG.EQ.0) GO TO 130
       RETURN
C
C
C OUTPUT OF ERROR MESSAGE
C
   90  I = 20
       GO TO 140
  100  I = 21
       GO TO 140
  110  I = 22
       GO TO 140
  120  I = 23
       GO TO 140
  130  I = 24
  140  CALL ERROR(I)
       RETURN
       END
C
C-----------------------------------------------------------------------
C SUBROUTINE:   DEFIN4
C SET DEFAULT VALUES OF CLASS 4
C-----------------------------------------------------------------------
C
       SUBROUTINE DEFIN4
C
       COMMON /CONST1/ MAXDEG, IWLMI, IWLMA, MBL
       COMMON /COPTCO/ LOPTW, LSTAB, ACXMI, ACXMA, ITER, ITERM, ITERM1
       COMMON /CRECO/ JRCO, JECO, JDCO, JJDCO(5), JMAXV, JTRB2, LREF
       COMMON /FILTRE/ IRCO(5), IECO(16,5), IDCO(16,5), IECOM
       COMMON /CCOEFW/ IWL, IWLG, IWLD, IWLL, ADEPSG, ADEPSD, ADEPSL,
     *     ISTAB, IDEPSL, IDEPSD
C
       LSTAB = 0
       LREF = 0
       IWL = 16
       JRCO = 2
       JECO = 0
       JDCO = 0
       JMAXV = -3
       JTRB2 = 3
       DO 20 J=1,5
         IRCO(J) = 0
         DO 10 I=1,MBL
           IECO(I,J) = 0
           IDCO(I,J) = -100
   10    CONTINUE
   20  CONTINUE
       RETURN
       END
C
C-----------------------------------------------------------------------
C SUBROUTINE:   INP040
C INPUT ROUTINE FOR THE SPECIFICATION OF THE COEFFICIENT REALI-
C SATION
C-----------------------------------------------------------------------
C
       SUBROUTINE INP040
C
       COMMON /CONST1/ MAXDEG, IWLMI, IWLMA, MBL
       COMMON /CARD/ LCODE, IZ(40)
       COMMON /INPDAT/ ICODE, JCODE, LINP, IINP, AINP(3), NPAR
       COMMON /INTDAT/ JINP(16)
       COMMON /COPTCO/ LOPTW, LSTAB, ACXMI, ACXMA, ITER, ITERM, ITERM1
       COMMON /CRECO/ JRCO, JECO, JDCO, JJDCO(5), JMAXV, JTRB2, LREF
       COMMON /FILTRE/ IRCO(5), IECO(16,5), IDCO(16,5), IECOM
       COMMON /CCOEFW/ IWL, IWLG, IWLD, IWLL, ADEPSG, ADEPSD, ADEPSL,
     *     ISTAB, IDEPSL, IDEPSD
C
       J = JCODE/10
       IF (J.EQ.0) GO TO 170
       IF (J.GT.6) GO TO 170
       GO TO (10, 20, 50, 90, 150, 160), J
C
C *0410  LSTAB,IWL
C
   10  LSTAB = LINP
       IF (IINP.NE.0) IWL = IINP
       RETURN
C
C *0420  RCOINP,JRCO
C
   20  IF (LINP.NE.(-1)) GO TO 40
       CALL INP001
       IF (LCODE.EQ.1) RETURN
       CALL INP003
       IF (ICODE.NE.0 .OR. JCODE.NE.20) CALL ERROR(2)
       DO 30 I=1,5
         IRCO(I) = JINP(I)
   30  CONTINUE
       JRCO = 0
       IF (IINP.NE.0) GO TO 180
   40  JRCO = IINP
       RETURN
C
C *0430  ECOINP,JECO
C
   50  IF (LINP.NE.(-1)) GO TO 80
       JECO = 0
       IECOM = 0
       DO 70 J=1,5
         CALL INP001
```

```
      IF (LCODE.EQ.1) RETURN
      CALL INP003
      IF (ICODE.NE.0 .OR. JCODE.NE.30) CALL ERROR(2)
      DO 60 I=1,MBL
        IECOM = MAX0(IECOM,JINP(I))
        IECO(I,J) = JINP(I)
 60     CONTINUE
 70   CONTINUE
      RETURN
C
 80   JECO = IINP
      RETURN
C
C *0440  DCOINP,JDCO
C
 90   IF (LINP.EQ.(-1)) GO TO 120
      DO 110 J=1,5
        II = IINP/10
        M = 6 - J
        JJ = IINP - 10*II
        IINP = II
        JJDCO(M) = JJ
        IF (II.NE.0) GO TO 110
        DO 100 I=1,16
          IDCO(I,J) = -100
 100    CONTINUE
 110  CONTINUE
      JDCO = -1
      RETURN
C
 120  JDCO = 0
      DO 140 J=1,5
        CALL INP001
        IF (LCODE.EQ.1) RETURN
        CALL INP003
        IF (ICODE.NE.0 .OR. JCODE.NE.40) CALL ERROR(2)
        DO 130 I=1,16
          IDCO(I,J) = JINP(I)
 130    CONTINUE
 140  CONTINUE
      RETURN
C
C *0450  B,JMAXV
C
 150  JMAXV = IINP
      RETURN
C
C *0460  B,JTRB2
C
 160  JTRB2 = IINP
      RETURN
C
C OUTPUT OF ERROR MESSAGES
C
 170  I = 1
      GO TO 190
 180  I = 9
 190  CALL ERROR(I)
      RETURN
      END
C
C-----------------------------------------------------------------------
C SUBROUTINE:   DEFIN5
C DEFAULT VALUES FOR COMMAND CARDS OF CLASS 05
C-----------------------------------------------------------------------
C
      SUBROUTINE DEFIN5
C
      COMMON /CONST1/ MAXDEG, IWLMI, IWLMA, MBL
      COMMON /CREST/ ISTRU, ISCAL, SCALM, ISEQ(16,2), LSEQ, IWLR,
     *    LPOT2, JSTRU
      COMMON /CREST1/ JSTRUS, JSTRUD
      COMMON /CRENO/ LCNO, ICNO(16,5)
C
      LCNO = 0
      LPOT2 = 0
      ISTRU = 13
      JSTRU = 1
      JSTRUS = 1
      JSTRUD = 1
      IWLR = 100
      DO 20 J=1,5
        DO 10 I=1,MBL
          ICNO(I,J) = 3
 10     CONTINUE
 20   CONTINUE
      RETURN
      END
C
C-----------------------------------------------------------------------
C SUBROUTINE:   INP050
C INPUT DATA FOR THE STRUCTURE REALISATION
C COMMAND CARDS CLASS 5
C-----------------------------------------------------------------------
C
      SUBROUTINE INP050
C
      COMMON /CONST1/ MAXDEG, IWLMI, IWLMA, MBL
      COMMON /CARD/ LCODE, IZ(40)
      COMMON /INPDAT/ ICODE, JCODE, LINP, IINP, AINP(3), NPAR
      COMMON /INTDAT/ JINP(16)
      COMMON /CREST/ ISTRU, ISCAL, SCALM, ISEQ(16,2), LSEQ, IWLR,
     *    LPOT2, JSTRU
      COMMON /CREST1/ JSTRUS, JSTRUD
      COMMON /CRENO/ LCNO, ICNO(16,5)
C
      J = JCODE/10
      IF (J.EQ.0) GO TO 100
      IF (J.GT.3) GO TO 100
      GO TO (10, 20, 90), J
C
C *0510  LPOT2,ISTRU
C
 10   LPOT2 = LINP
      IF (IINP.EQ.0) RETURN
      IF (IINP.GE.30) GO TO 110
C
C >>>>>>>>>>>>>>>>>>>>>>>>>>>>>>>>>>>>>>>>>>>>>>>>>>>>>>>>>>>>>>>>>>>>>
C HERE INSERT  THE PARAMETERS JSTRU,JSTRUS,JSTRUD FOR YOUR
C STRUCTRUES
C >>>>>>>>>>>>>>>>>>>>>>>>>>>>>>>>>>>>>>>>>>>>>>>>>>>>>>>>>>>>>>>>>>>>>
C
      ISTRU = IINP
      JSTRU = 1
      IF (ISTRU/2.EQ.(ISTRU+1)/2) JSTRU = 2
      JSTRUS = JSTRU
      JSTRUD = 1
      IF (ISTRU.GE.20) JSTRUD = 2
      RETURN
C
```

```
C *0520  CNOINP,JCNO
C
   20  IF (LINP.NE.(-1)) GO TO 50
       DO 40 J=1,5
         CALL INP001
         IF (LCODE.NE.0) RETURN
         CALL INP003
         IF (ICODE.NE.0 .OR. JCODE.NE.20) CALL ERROR(2)
         DO 30 I=1,MBL
           ICNO(I,J) = JINP(I)
   30    CONTINUE
   40  CONTINUE
       IF (IINP.NE.0) GO TO 130
       RETURN
C
   50  IF (IINP.EQ.0) GO TO 120
       DO 80 J=1,5
         II = IINP/10
         M = 6 - J
         JJ = IINP - 10*II
         IINP = II
         IF (JJ.NE.0) GO TO 60
         IF (J.EQ.1) GO TO 130
         JJ = ICNO(1,M+1)
   60    DO 70 I=1,16
           ICNO(I,M) = JJ
   70    CONTINUE
   80  CONTINUE
       RETURN
C
C *0530  LCNO,IWLR
C
   90  LCNO = LINP
       IF (IINP.LT.0) GO TO 110
       IF (IINP.NE.0) IWLR = IINP
       RETURN
C
  100  I = 1
       GO TO 140
  110  I = 5
       GO TO 140
  120  I = 9
       GO TO 140
  130  I = 4
  140  CALL ERROR(I)
       RETURN
       END
C
C-----------------------------------------------------------------------
C SUBROUTINE:   DEFIN6
C DEFAULT VALUES FOR THR COMMAND CARDS OF CLASS 06
C-----------------------------------------------------------------------
C
       SUBROUTINE DEFIN6
C
       COMMON /CONST/ PI, FLMA, FLMI, FLER
       COMMON /COPTCO/ LOPTW, LSTAB, ACXMI, ACXMA, ITER, ITERM, ITERM1
       COMMON /COPTST/ LOPTS, ISTOR
       COMMON /CPLOT/ IPLOT, LNOR, IPAG, OMLO, OMUP, RMAX, RMIN
C
       LOPTS = 0
       LOPTW = 0
       IPLOT = 0
       LNOR = 0
       IPAG = 2
       OMLO = 0.
       OMUP = PI
       RMAX = 1.
       RMIN = -1.
       RETURN
       END
C
C-----------------------------------------------------------------------
C SUBROUTINE:   INP060
C COMMANND CARDS FOR THE ANALYSIS
C PROCESSING OF COMMAND CARDS CLASS 06
C-----------------------------------------------------------------------
C
       SUBROUTINE INP060
C
       COMMON /INPDAT/ ICODE, JCODE, LINP, IINP, AINP(3), NPAR
       COMMON /COPTST/ LOPTS, ISTOR
       COMMON /COPTCO/ LOPTW, LSTAB, ACXMI, ACXMA, ITER, ITERM, ITERM1
       COMMON /CCOEFW/ IWL, IWLG, IWLD, IWLL, ADEPSG, ADEPSD, ADEPSL,
      *    ISTAB, IDEPSL, IDEPSD
       COMMON /CPLOT/ IPLOT, LNOR, IPAG, OMLO, OMUP, RMAX, RMIN
C
       J = JCODE/10
       IF (J.EQ.0) GO TO 90
       IF (J.GT.6) GO TO 90
       GO TO (10, 20, 30, 60, 70, 80), J
C
C *0610  LWLM,IWL
C
   10  LOPTW = 1
       IF (LINP.EQ.(-1)) LOPTW = -1
       IF (IINP.LT.0) GO TO 100
       IF (IINP.NE.0) IWL = IINP
       RETURN
C
C *0620  LSEQ,LSCAL,SCALM
C
   20  LOPTS = -1
       JCODE = 40
       CALL INP020
       RETURN
C
C *0630  LNOR,IPAG,OMLO,OMUP,RMAX
C
   30  IPLOT = 1
   40  LNOR = LINP
       IF (IINP.GT.0) IPAG = IINP
       IF (IPLOT.EQ.4) RETURN
       IF (NPAR.NE.0) OMLO = AINP(1)
       Q = AINP(2)
       IF (Q.EQ.0.) GO TO 50
       IF (Q.LE.OMLO) GO TO 110
       OMUP = Q
   50  IF (AINP(3).NE.0.) RMAX = AINP(3)
       RETURN
C
C *0640  LNOR,IPAG,OMLO,OMUP,RMAX
C
   60  IPLOT = 2
       RMAX = 100.
       GO TO 40
C
C *0650  LNOR,IPAG,OMLO,OMUP
C
   70  IPLOT = 3
```

```
      GO TO 40
C
C *0660  LNOR,IPAG,RMIN,RMAX
C
   80 IPLOT = 4
      RMAX = 1.
      IF (NPAR.NE.0) RMIN = AINP(1)
      Q = AINP(2)
      IF (Q.EQ.0.) GO TO 40
      IF (Q.LE.RMIN) GO TO 110
      RMAX = Q
      GO TO 40
C
   90 I = 1
      GO TO 120
  100 I = 5
      GO TO 120
  110 I = 10
      IPLOT = 0
  120 CALL ERROR(I)
C
      RETURN
      END
C
C-----------------------------------------------------------------------
C SUBROUTINE:   INP001
C READ COMMAND CARD
C
C IF 'DECODE' DOES NOT CONFORM TO YOUR COMPILER,
C SET THE ALTERNATIVE INPUT VERSION
C
C ELIMINATE THE STATEMENTS ENCLOSED BY THE COMMENT LINES C111
C
C AND REMOVE THE COMMENT MARKS C222
C
C-----------------------------------------------------------------------
C
      SUBROUTINE INP001
C
      COMMON /CANPAR/ KA1, KA2, KA3, KA4, KA5, LINE
      COMMON /CARD/ LCODE, IZ(40)
      COMMON /INPDAT/ ICODE, JCODE, LINP, IINP, AINP(3), NPAR
C111
      COMMON /CIOFOR/ IOFO(3), ION
C111
C
      DIMENSION IZ1(2)
      DATA IPR, ISL /1H*,1H//
C
9999  FORMAT (1X, 4A1)
9998  FORMAT (1X//11H DATA INPUT)
C
      IF (LCODE.EQ.(-1)) WRITE (KA2,9998)
C
   10 CONTINUE
C111
      READ (KA1,IOFO) (IZ(I),I=1,ION)
C
C >>>>>>>>>>>>>>>>>>>>>>>>>>>>>>>>>>>>>>>>>>>>>>>>>>>>>>>>>>>>>>>>>>>>>
C DECODE IS A MACHINE-DEPENDENT FUNCTION
C >>>>>>>>>>>>>>>>>>>>>>>>>>>>>>>>>>>>>>>>>>>>>>>>>>>>>>>>>>>>>>>>>>>>>
C
      DECODE(2,21,IZ) IZ1(1),IZ1(2)
C111
C
C222  READ (KA1,21) (IZ1(I),I=1,2)
   21 FORMAT (2A1)
      IF (IZ1(1).EQ.IPR) GO TO 30
      IF (IZ1(1).EQ.ISL .AND. IZ1(2).EQ.ISL) GO TO 20
      CALL ERROR(1)
      GO TO 10
   20 CONTINUE
C111
C
C >>>>>>>>>>>>>>>>>>>>>>>>>>>>>>>>>>>>>>>>>>>>>>>>>>>>>>>>>>>>>>>>>>>>>
C DECODE IS A MACHINE-DEPANDENT FUNCTION
C >>>>>>>>>>>>>>>>>>>>>>>>>>>>>>>>>>>>>>>>>>>>>>>>>>>>>>>>>>>>>>>>>>>>>
C
      DECODE (4,22,IZ) ICODE,JCODE
C111
C
C222  READ (KA1,22) ICODE,JCODE
   22 FORMAT (2X, 2A1)
      WRITE (KA2,9999) IZ1(1), IZ1(2), ICODE, JCODE
      LCODE = 1
      RETURN
   30 LCODE = 0
      ICODE = 0
      RETURN
C
      END
C
C-----------------------------------------------------------------------
C SUBROUTINE:   INP002
C DECODE COMMAND CARD
C
C IF 'DECODE' DOES NOT CONFORM TO YOUR COMPILER, SEE INP001
C
C-----------------------------------------------------------------------
C
      SUBROUTINE INP002
C
      COMMON /CANPAR/ KA1, KA2, KA3, KA4, KA5, LINE
      COMMON /CARD/ LCODE, IZ(40)
      COMMON /INPDAT/ ICODE, JCODE, LINP, IINP, AINP(3), NPAR
      COMMON /SCRAT/ ADUM(32)
      DIMENSION IZZ(15)
      EQUIVALENCE (ADUM(1),IZZ(1))
C
      DATA IBL, IT /1H ,1HT/
C
      NPAR = 0
C111
C
C >>>>>>>>>>>>>>>>>>>>>>>>>>>>>>>>>>>>>>>>>>>>>>>>>>>>>>>>>>>>>>>>>>>>>
C DECODE IS A MACHINE-DEPENDENT FUNCTION
C >>>>>>>>>>>>>>>>>>>>>>>>>>>>>>>>>>>>>>>>>>>>>>>>>>>>>>>>>>>>>>>>>>>>>
C
      DECODE (30,11,IZ) IZZ
   11 FORMAT (15X, 15A1)
      DO 10 I=1,15
        IF (IZZ(I).NE.IBL) NPAR = 1
   10 CONTINUE
C
C >>>>>>>>>>>>>>>>>>>>>>>>>>>>>>>>>>>>>>>>>>>>>>>>>>>>>>>>>>>>>>>>>>>>>
C DECODE IS A MACHINE-DEPENDENT FUNCTION
C >>>>>>>>>>>>>>>>>>>>>>>>>>>>>>>>>>>>>>>>>>>>>>>>>>>>>>>>>>>>>>>>>>>>>
C
      DECODE (60,12,IZ) ICODE,JCODE,JINP,IINP,AINP
C111
```

```
C
C222  READ (KA1,12) ICODE,JCODE,JINP,IINP,(AINP(I),I=1,3)
C222  NPAR=1
C
      WRITE (KA2,9999) ICODE, JCODE, JINP, IINP, AINP
C
  12  FORMAT (1X, 2I2, 2X, 1A1, I6, 1X, 3E15.0)
9999  FORMAT (2H *, 2I4, 2X, 1A1, I6, 1X, 3E15.6)
C
      LINP = 0
      IF (JINP.EQ.IT) LINP = -1
      RETURN
      END
C
C-----------------------------------------------------------------------
C SUBROUTINE:   INP003
C INTEGER DATA INPUT
C
C
C IF 'DECODE' DOES NOT CONFORM TO YOUR COMPILER, SEE INP001
C-----------------------------------------------------------------------
C
      SUBROUTINE INP003
C
      COMMON /CANPAR/ KA1, KA2, KA3, KA4, KA5, LINE
      COMMON /CARD/ LCODE, IZ(40)
      COMMON /INTDAT/ JINP(16)
C
  12  FORMAT (1X, 2I2, 3X, 16I4)
9999  FORMAT (1H*, 2I3, 1X, 16I4)
C111
C
C >>>>>>>>>>>>>>>>>>>>>>>>>>>>>>>>>>>>>>>>>>>>>>>>>>>>>>>>>>>>>>>>>>>>
C DECODE IS A MACHINE-DEPENDENT FUNCTION
C >>>>>>>>>>>>>>>>>>>>>>>>>>>>>>>>>>>>>>>>>>>>>>>>>>>>>>>>>>>>>>>>>>>>
C
      DECODE (72,12,IZ) ICODE,JCODE,(JINP(I),I=1,16)
C111
C
C222  READ (KA1,12) ICODE,JCODE,(JINP(I),I=1,16)
C
      WRITE (KA2,9999) ICODE, JCODE, JINP
      RETURN
      END
C
C =====================================================================
C
C DOREDI - SUBROUTINES: PART V
C
C =====================================================================
C
C
C-----------------------------------------------------------------------
C SUBROUTINE:   OUT010
C LEADER TO OUT011 FOR STRUCTURES WITH SEPARATE SCALING
C-----------------------------------------------------------------------
C
      SUBROUTINE OUT010
C
      COMMON /CANPAR/ KA1, KA2, KA3, KA4, KA5, LINE
      COMMON /FILT/ NB, FACT, B2(16), B1(16), B0(16), C1(16), C0(16)
C
      WRITE (KA2,9999)
9999  FORMAT (1X//46H STRUCTURE WITH SEPARATE SCALING FACTORS B2(L)//
     *    45H FOR STRUCTURES 21 TO 28  B2(0) = GAIN FACTOR)
C
      DO 10 I=1,NB
        Q = B2(I)
        B1(I) = B1(I)/Q
        B0(I) = B0(I)/Q
  10  CONTINUE
      RETURN
      END
C
C-----------------------------------------------------------------------
C SUBROUTINE:   OUT011
C OUTPUT OF BLOCKS OF SECOND ORDER
C-----------------------------------------------------------------------
C
      SUBROUTINE OUT011
C
      COMMON /CANPAR/ KA1, KA2, KA3, KA4, KA5, LINE
      COMMON /FILT/ NB, FACT, B2(16), B1(16), B0(16), C1(16), C0(16)
C
      WRITE (KA2,9999) FACT
9999  FORMAT (//23H BLOCKS OF SECOND ORDER//24H CONSTANT GAIN FACTOR = ,
     *    E15.7//3H  L, 5X, 5HB2(L), 8X, 5HB1(L), 8X, 5HB0(L), 10X,
     *    5HC1(L), 8X, 5HC0(L)/)
      WRITE (KA2,9998) (I,B2(I),B1(I),B0(I),C1(I),C0(I),I=1,NB)
9998  FORMAT (I3, 3F13.8, 2X, 2F13.8)
      RETURN
      END
C
C-----------------------------------------------------------------------
C SUBROUTINE:   OUT012
C OUTPUT OF THE RESULTS TO THE DISK: (CHANNEL 4)
C-----------------------------------------------------------------------
C
      SUBROUTINE OUT012
C
      COMMON /CANPAR/ KA1, KA2, KA3, KA4, KA5, LINE
      COMMON /TOL/ ITYP, IAPRO, NDEG, OM(4), SF, ADELP, ADELS
      COMMON /RES/ AC, ROM(4), RDELP, RDELS, NZM(4)
      COMMON /RESZ/ ZFA, ZM(18,4), ZZR(16), ZZI(16), ZPR(16), ZPI(16)
      COMMON /FILT/ NB, FACT, B2(16), B1(16), B0(16), C1(16), C0(16)
      COMMON /FILTRE/ IRCO(5), IECO(16,5), IDCO(16,5), IECOM
      COMMON /CRECO/ JRCO, JECO, JDCO, JJDCO(5), JMAXV, JTRB2, LREF
      COMMON /CCOEFW/ IWL, IWLG, IWLD, IWLL, ADEPSG, ADEPSD, ADEPSL,
     *    ISTAB, IDEPSL, IDEPSD
C
9999  FORMAT (10I5)
9998  FORMAT (4E15.7)
C
      WRITE (KA4,9997)
9997  FORMAT (19H FILTER DESCRIPTION)
      WRITE (KA4,9999) ITYP, IAPRO, NDEG
      WRITE (KA4,9998) SF, OM, ADELP, ADELS
      WRITE (KA4,9998) AC, ROM, RDELP, RDELS
      M = 0
      DO 10 I=1,4
        M = MAX0(M,NZM(I))
  10  CONTINUE
      WRITE (KA4,9999) NZM, M
      DO 20 I=1,M
        WRITE (KA4,9998) (ZM(I,J),J=1,4)
  20  CONTINUE
      WRITE (KA4,9996)
9996  FORMAT (5H DATA)
      WRITE (KA4,9999) NB
      WRITE (KA4,9998) FACT
```

```
      DO 30 I=1,NB
        WRITE (KA4,9998) B2(I), B1(I), B0(I), C1(I), C0(I)
   30 CONTINUE
      WRITE (KA4,9999) IRCO
      DO 40 I=1,NB
        WRITE (KA4,9999) (IECO(I,J),J=1,5)
   40 CONTINUE
      DO 50 I=1,NB
        WRITE (KA4,9999) (IDCO(I,J),J=1,5)
   50 CONTINUE
      WRITE (KA4,9999) IWL, IECOM, JMAXV, JTRB2
      WRITE (KA4,9995)
 9995 FORMAT (12H END OF DATA)
      RETURN
      END
C
C-----------------------------------------------------------------------
C SUBROUTINE:   OUT016
C OUTPUT OF BLOCKS OF SECOND ORDER IN THE S-DOMAIN
C-----------------------------------------------------------------------
C
      SUBROUTINE OUT016
C
      COMMON /CONST/ PI, FLMA, FLMI, FLER
      COMMON /CANPAR/ KA1, KA2, KA3, KA4, KA5, LINE
      COMMON /TOL/ ITYP, IAPRO, NDEG, OM(4), SF, ADELP, ADELS
      COMMON /RES/ AC, ROM(4), RDELP, RDELS, NZM(4)
      COMMON /RESS/ SFA, SM(18,4), NZERO(16), SPR(16), SPI(16)
      COMMON /RESIN1/ PREN(16), PIMN(16), UGC, OGC, ACK, NJ, NH
C
      COMMON /SCRAT/ ADUM(32)
      DIMENSION NZE(16)
      EQUIVALENCE (ADUM(1),NZE(1))
C
      WRITE (KA2,9999) SFA
 9999 FORMAT (//39H BLOCKS OF SECOND ORDER IN THE S-DOMAIN//9H CONSTANT,
     *    15H GAIN FACTOR = , E15.7//3H  L, 5X, 5HA2(L), 8X, 5HA1(L),
     *    8X, 5HA0(L), 10X, 5HD1(L), 8X, 5HD0(L)/)
C
      N = 0
      ME = NZM(4)
      DO 10 I=1,ME
        NZE(I) = NZERO(I)
   10 CONTINUE
      NB = (NDEG+1)/2
      II = 1
      NZ = NZE(II)
C
      DO 180 I=1,NB
   20   IF (NZ.GT.0) GO TO 30
        II = II + 1
        NZ = NZE(II)
        GO TO 20
   30   QI = SM(II,4)
        IF (NZ.EQ.1) GO TO 70
        IF (QI.GE.FLMA) GO TO 50
   40   B2 = 1.
        B1 = 0.
        B0 = QI*QI
        GO TO 60
   50   B2 = 0.
        B1 = 0.
        B0 = 1.
   60   NZ = NZ - 2
        GO TO 130
   70   IF (II.EQ.ME) GO TO 120
        MA = II + 1
        DO 80 J=MA,ME
          IF (SM(J,4).GE.FLMA) GO TO 90
          IF (SM(J,4).LE.FLMI) GO TO 100
   80   CONTINUE
        GO TO 120
   90   NZE(J) = NZE(J) - 1
        IF (QI.GE.FLMA) GO TO 50
        GO TO 110
  100   NZE(J) = NZE(J) - 1
        IF (QI.LE.FLMI) GO TO 40
  110   B2 = 0.
        B1 = 1.
        B0 = 0.
        NZ = NZ - 1
        GO TO 130
  120   IF (QI.GE.FLMA) GO TO 50
        GO TO 110
C
  130   N = N + 1
        QR = SPR(N)
        QI = SPI(N)
        IF (ABS(QI).LT.FLMI) GO TO 140
        C1 = -2.*QR
        C0 = QR*QR + QI*QI
        GO TO 170
  140   IF (N.GE.NJ) GO TO 150
        IF (ABS(SPI(N+1)).LT.FLMI) GO TO 160
  150   C1 = -QR
        C0 = 0.
        GO TO 170
  160   N = N + 1
        QI = SPR(N)
        C1 = -QR - QI
        C0 = QR*QI
C
  170   WRITE (KA2,9998) I, B2, B1, B0, C1, C0
 9998   FORMAT (I3, 3F13.8, 2X, 2F13.8)
  180 CONTINUE
      RETURN
      END
C
C-----------------------------------------------------------------------
C SUBROUTINE:   OUT022
C OUTPUT OF THE TERMINATION KIND
C-----------------------------------------------------------------------
C
      SUBROUTINE OUT022
C
      COMMON /CANPAR/ KA1, KA2, KA3, KA4, KA5, LINE
      COMMON /COPTCO/ LOPTW, LSTAB, ACXMI, ACXMA, ITER, ITERM, ITERM2
      COMMON /COPTCP/ IS1, IS2, IABO, IWLGS, IWLGN, IWLGP, ADEPSS,
     *    ADEPSN, ADEPSP, ACXS, ACXN, ACXP
C
      WRITE (KA2,9999) ITER
 9999 FORMAT (///38H OPTIMIZATION IS TERMINATED AFTER THE , I4,
     *    6H. STEP)
      GO TO (10, 20, 30), IABO
   10 WRITE (KA2,9998)
 9998 FORMAT (25H THREE UNSUCCESSFUL STEPS)
      GO TO 40
   20 WRITE (KA2,9997)
 9997 FORMAT (24H MAXIMUM NUMBER OF STEPS)
      GO TO 40
```

```
   30  WRITE (KA2,9996)
 9996  FORMAT (54H OPTIMIZATION PARAMETER IS COMING TO THE END OF THE RE,
     *     4HGION)
   40  RETURN
       END
C
C-----------------------------------------------------------------------
C SUBROUTINE:   OUT023
C OUTPUT OF THE ITERATION NUMBER
C-----------------------------------------------------------------------
C
       SUBROUTINE OUT023
C
       COMMON /CANPAR/ KA1, KA2, KA3, KA4, KA5, LINE
       COMMON /COPTCO/ LOPTW, LSTAB, ACXMI, ACXMA, ITER, ITERM, ITERM2
C
       WRITE (KA2,9999) ITER
 9999  FORMAT (1X////1X, I3, 13H-TH ITERATION)
       RETURN
       END
C
C-----------------------------------------------------------------------
C SUBROUTINE:   OUT025
C OUTPUT OF THE OPTIMIZATION OPTIONS
C-----------------------------------------------------------------------
C
       SUBROUTINE OUT025
C
       COMMON /CANPAR/ KA1, KA2, KA3, KA4, KA5, LINE
       COMMON /COPTST/ LOPTS, ISTOR
       COMMON /CREST/ ISTRU, ISCAL, SCALM, ISEQ(16,2), LSEQ, IWLR,
     *     LPOT2, JSTRU
       COMMON /CRENO/ LCNO, ICNO(16,5)
C
       DIMENSION IES(4), NO(4)
       DATA IES(1), IES(2), IES(3), IES(4) /1H ,1HY,1HE,1HS/
       DATA NO(1), NO(2), NO(3), NO(4) /1H ,1H ,1HN,1HO/
C
       IF (LOPTS.EQ.0) GO TO 10
       WRITE (KA2,9999)
 9999  FORMAT (1X//41H OPTIMIZATION OF THE PAIRING AND ORDERING//)
       WRITE (KA2,9998) ISTOR
 9998  FORMAT (46H STORAGE FOR THE INTERMEDIATE RESULTS  ISTOR =, I4/)
       IF (LOPTS.EQ.(-1)) WRITE (KA2,9997) NO
       IF (LOPTS.NE.(-1)) WRITE (KA2,9997) IES
 9997  FORMAT (/14H FIXED PAIRING, 32X, 4A1)
C
   10  WRITE (KA2,9996) ISTRU
 9996  FORMAT (1X/20H REALIZED STRUCTURE , 19X, 7HISTRU =, I4/)
       WRITE (KA2,9995) ISCAL
 9995  FORMAT (/16H SCALING  OPTION, 23X, 7HISCAL =, I4)
       IF (LPOT2.EQ.(-1)) WRITE (KA2,9994) IES
       IF (LPOT2.NE.(-1)) WRITE (KA2,9994) NO
 9994  FORMAT (10X, 24HBY A FACTOR OF POWER TWO, 12X, 4A1)
       WRITE (KA2,9993) SCALM
 9993  FORMAT (10X, 21HCHOSEN MAXIMUM OF THE/10X, 15HOVERFLOW POINTS,
     *     14X, 7HSCALM =, F6.3/)
       WRITE (KA2,9992)
 9992  FORMAT (/33H REALIZATION OF THE COEFFICIENTS )
       IF (LCNO.EQ.(-1)) WRITE (KA2,9991) NO, IWLR
       IF (LCNO.NE.(-1)) WRITE (KA2,9991) IES, IWLR
 9991  FORMAT (5X, 33HCONSIDERATION OF THE QUANTIZATION, 8X, 4A1/5X,
     *     10HWORDLENGTH, 24X, 7HIWLR  =, I4/24X, 19H(FOR IWLR=100 NO RO,
     *     7HUNDING)/)
       RETURN
       END
C
C-----------------------------------------------------------------------
C SUBROUTINE:   OUT026
C OUTPUT OF THE NOISE VALUES OF THE SECOND ORDER BLOCKS
C-----------------------------------------------------------------------
C
       SUBROUTINE OUT026
C
       COMMON /CANPAR/ KA1, KA2, KA3, KA4, KA5, LINE
       COMMON /CREST/ ISTRU, ISCAL, SCALM, ISEQ(16,2), LSEQ, IWLR,
     *     LPOT2, JSTRU
       COMMON /COPTSP/ IB, JSEQN(16), JSEQD(16), AMAX, SCA, ALSBI
       COMMON /CNFUNC/ AQC(5), BN2(5), BN1(5), BN0(5)
       COMMON /CRENO/ LCNO, ICNO(16,5)
       COMMON /CPOW/ PNU, PNC, AND, ITCORP
       COMMON /FILT/ NB, FACT, B2(16), B1(16), B0(16), C1(16), C0(16)
C
       DIMENSION NCHA(5), NQUAN(5)
C
       IF (IB.LE.0) GO TO 50
       IF (IB.EQ.1) WRITE (KA2,9999)
 9999  FORMAT (1X//47H NUM  DEN   SCALING   UNCOR. NOISE   COR. NOISE,
     *     14H     DC-OFFSET, 12H CHA CO-QUAN)
       ALOG2 = ALOG(2.)
       IF (IB.GT.NB) GO TO 10
       II = IB
       IN = ISEQ(IB,1)
       ID = ISEQ(IB,2)
       GO TO 20
   10  IN = 0
       ID = 0
       II = NB
   20  DO 40 J=1,5
         Q = AQC(J)
         IF (Q.EQ.1.) GO TO 30
         Q = ALOG(Q)/ALOG2
         NQUAN(J) = -IFIX(Q-0.5)
         NCHA(J) = ICNO(II,J)
         GO TO 40
   30    NQUAN(J) = 0
         NCHA(J) = 0
   40  CONTINUE
       WRITE (KA2,9998) IN, ID, SCA, PNU, PNC, AND, (NCHA(J),NQUAN(J),J=
     *     1,5)
 9998  FORMAT (/1X, 2(I3, 2X), F9.6, 1X, 3E13.5, 2I4, 4H BIT/(60X, 2I4,
     *     4H BIT))
       RETURN
C
   50  WRITE (KA2,9997) AND
 9997  FORMAT (46X, 14(1H-)/46X, E14.5/)
       Q = AND*AND
       PNT = PNU + PNC + Q
       WRITE (KA2,9996) PNU, PNC, Q, PNT
 9996  FORMAT (12H TOTAL NOISE, 9X, 3E13.5, 1H=, E11.4)
       RETURN
       END
C
C-----------------------------------------------------------------------
C SUBROUTINE:   OUT027
C OUTPUT OF THE DIFFERENT NOISE VALUES
C-----------------------------------------------------------------------
C
       SUBROUTINE OUT027
C
```

```
      COMMON /CANPAR/ KA1, KA2, KA3, KA4, KA5, LINE
      COMMON /CNOISE/ RI, RIN, RE, REN, FAC
C
      ANP = RI/12.
      ALANP = 10.*ALOG10(ANP)
      WRITE (KA2,9999) ANP, ALANP
 9999 FORMAT (1H //22H ABSOLUTE OUTPUT NOISE, 7X, 3HANP, 3X, 1H=,
     *    E13.5, 12H  (Q*Q)    =, F6.2, 3H DB/)
      WRITE (KA2,9998) FAC
 9998 FORMAT (/22H SCALING AT THE OUTPUT, 7X, 3HSCA, 3X, 1H=, E13.5/)
      RNP = RIN/12.
      ALRNP = 10.*ALOG10(RNP)
      WRITE (KA2,9997) RNP, ALRNP
 9997 FORMAT (/22H RELATIVE OUTPUT NOISE, 7X, 3HRNP, 3X, 1H=, E13.5,
     *    12H  (Q*Q)    ,F6.2, 3H DB)
      WRITE (KA2,9995)
      ALRIN = 10.*ALOG10(RIN)
      WRITE (KA2,9996) RIN, ALRIN
 9996 FORMAT (/19H INNER NOISE FIGURE, 10X, 3HRIN, 3X, 1H=, E13.5,
     *    12H (Q*Q)/12 = , F6.2, 3H DB)
      WRITE (KA2,9995)
 9995 FORMAT (/23H RELATED TO MAX /H/ = 1/)
      WRITE (KA2,9994) REN
 9994 FORMAT (/22H ENTRANCE NOISE FIGURE, 7X, 3HREN, 3X, 1H=, E13.5/)
      WRITE (KA2,9995)
      ADW = RIN/(1.+REN)
      ADW = ALOG(ADW)/(2.*ALOG(2.))
      IDW = IFIX(ADW)
      Q = FLOAT(IDW)
      IF (Q.LT.ADW) IDW = IDW + 1
      WRITE (KA2,9993) ADW, IDW
 9993 FORMAT (/22H ADDITIONAL WORDLENGTH, 6X, 11HDELTA W = /, F7.3,
     *    4H/ = , I4/)
      RETURN
      END
C
C-----------------------------------------------------------------------
C SUBROUTINE:   OUT030
C OUTPUT OF THE TOLERANCE SCHEME
C-----------------------------------------------------------------------
C
      SUBROUTINE OUT030
C
      COMMON /CANPAR/ KA1, KA2, KA3, KA4, KA5, LINE
      COMMON /CONST/ PI, FLMA, FLMI, FLER
      COMMON /TOL/ ITYP, IAPRO, NDEG, OM(4), SF, ADELP, ADELS
C
      DIMENSION FE(4)
C
      AL(ADEL) = -20.*ALOG10(ADEL)
C
      ME = 2
      IF (ITYP.GT.2) ME = 4
      WRITE (KA2,9999)
 9999 FORMAT (//17H TOLERANCE SCHEME)
      WRITE (KA2,9998)
 9998 FORMAT (//13H FILTER-TYPE )
      GO TO (10, 20, 30, 40), ITYP
   10 WRITE (KA2,9997)
 9997 FORMAT (25X, 7HLOWPASS)
      GO TO 50
   20 WRITE (KA2,9996)
 9996 FORMAT (25X, 8HHIGHPASS)
      GO TO 50
   30 WRITE (KA2,9995)
 9995 FORMAT (25X, 8HBANDPASS)
      GO TO 50
   40 WRITE (KA2,9994)
 9994 FORMAT (25X, 8HBANDSTOP)
   50 WRITE (KA2,9993)
 9993 FORMAT (/15H APPROXIMATION )
      GO TO (60, 70, 80, 90), IAPRO
   60 WRITE (KA2,9992)
 9992 FORMAT (25X, 11HBUTTERWORTH)
      GO TO 100
   70 WRITE (KA2,9991)
 9991 FORMAT (25X, 11HCHEBYSHEV I)
      GO TO 100
   80 WRITE (KA2,9990)
 9990 FORMAT (25X, 12HCHEBYSHEV II)
      GO TO 100
   90 WRITE (KA2,9989)
 9989 FORMAT (25X, 8HELLIPTIC)
C
  100 IF (SF.EQ.0.) GO TO 120
      WRITE (KA2,9988) SF
 9988 FORMAT (/18H SAMPLING FREQ.    , F15.6)
      Q = SF*0.5/PI
      DO 110 I=1,ME
        FE(I) = OM(I)*Q
  110 CONTINUE
      WRITE (KA2,9987) (FE(I),I=1,ME)
 9987 FORMAT (20H CUTOFF FREQUENCIES  , 4F13.6)
C
  120 WRITE (KA2,9986) (OM(I),I=1,ME)
 9986 FORMAT (/20H NORM. CUTOFF FREQ.  , 4F13.6)
      DO 130 I=1,ME
        Q = OM(I)*0.5
        FE(I) = SIN(Q)/COS(Q)
  130 CONTINUE
      WRITE (KA2,9985) (FE(I),I=1,ME)
 9985 FORMAT (/20H CUTOFF FREQ. S-DOM., 4F13.6)
      P = 1. - ADELP
      Q = AL(P)
      P = SQRT(1.-P*P)
      WRITE (KA2,9984) ADELP, Q, P
 9984 FORMAT (//20H PASSBAND RIPPLE(S)  , F15.6, F12.4, 9H DB   P =,
     *    F12.4)
      Q = FLMA
      IF (ADELS.GT.0.) Q = AL(ADELS)
      WRITE (KA2,9983) ADELS, Q
 9983 FORMAT (20H STOPBAND RIPPLE(S)  , F15.6, F12.4, 3H DB)
      IF (NDEG.EQ.0) RETURN
      WRITE (KA2,9982) NDEG
 9982 FORMAT (//32H FILTER DEGREE ASSIGNED TO ORDER, I3)
      RETURN
      END
C
C-----------------------------------------------------------------------
C SUBROUTINE:   OUT031
C OUTPUT OF THE NORMALIZED PARAMETER IN THE S-DOMAIN
C-----------------------------------------------------------------------
C
      SUBROUTINE OUT031
C
      COMMON /CANPAR/ KA1, KA2, KA3, KA4, KA5, LINE
      COMMON /TOL/ ITYP, IAPRO, NDEG, OM(4), SF, ADELP, ADELS
      COMMON /TOLSN/ VSN, VD, A
C
      WRITE (KA2,9999)
```

```
9999  FORMAT (//37H NORMALIZED PARAMETER IN THE S-DOMAIN)
      WRITE (KA2,9998) VD, VSN
9998  FORMAT (/3H VD, 6X, 1H=, F13.6/4H VSN, 5X, 1H=, F13.6)
      IF (ITYP.LE.2) RETURN
C
      WRITE (KA2,9997) A
9997  FORMAT (2H A, 7X, 1H=, F13.6)
      RETURN
      END
C
C-----------------------------------------------------------------------
C SUBROUTINE:   OUT032
C OUTPUT OF THE REALIZED CUTOFF FREQUENCIES
C-----------------------------------------------------------------------
C
      SUBROUTINE OUT032
C
      COMMON /CANPAR/ KA1, KA2, KA3, KA4, KA5, LINE
      COMMON /TOL/ ITYP, IAPRO, NDEG, OM(4), SF, ADELP, ADELS
      COMMON /RES/ AC, ROM(4), RDELP, RDELS, NZM(4)
C
      ME = 2
      IF (ITYP.GE.3) ME = 4
      WRITE (KA2,9999) (ROM(I),I=1,ME)
9999  FORMAT (/9H REALIZED/20H NORM. CUTOFF FREQ. , 4F13.6)
      RETURN
      END
C
C-----------------------------------------------------------------------
C SUBROUTINE:   OUT033
C OUTPUT OF THE FILTER DEGREE
C-----------------------------------------------------------------------
C
      SUBROUTINE OUT033
C
      COMMON /CANPAR/ KA1, KA2, KA3, KA4, KA5, LINE
      COMMON /TOL/ ITYP, IAPRO, NDEG, OM(4), SF, ADELP, ADELS
      COMMON /TOLSN/ VSN, VD, A
      COMMON /TOLCHA/ GD1, GD2, ACAP12, ADELTA, ADEG
      COMMON /DESIGN/ NDEGF, EDEG, ACX, NORMA, LSOUT, LVSN, LSYM
C
      Q = ADEG*EDEG
      WRITE (KA2,9999) ADEG, Q, NDEG
9999  FORMAT (//20H MIN. FILTER DEGREE , F12.4//18H DEGREE EXTENSION ,
     *    F14.4//20H CHOSEN FILTER DEG. , I7)
      IF (ITYP.LE.2) RETURN
      IQ = 2*NDEG
      WRITE (KA2,9998) IQ
9998  FORMAT (26H FOR THE REFERENCE LOWPASS//21H ACTUAL FILTER DEGREE,
     *    I6)
      RETURN
      END
C
C-----------------------------------------------------------------------
C SUBROUTINE:   OUT034
C OUTPUT OF THE CHARACTERISTIC FUNCTION  /K(JV)/**2
C-----------------------------------------------------------------------
C
      SUBROUTINE OUT034
C
      DOUBLE PRECISION DSK(16), DK, DKS, DCAP02, DCAP04
C
      COMMON /CANPAR/ KA1, KA2, KA3, KA4, KA5, LINE
      COMMON /OUTDAT/ IP, PRE, PIM, IZ, ZRE, ZIM
      COMMON /TOL/ ITYP, IAPRO, NDEG, OM(4), SF, ADELP, ADELS
      COMMON /TOLCHA/ GD1, GD2, ACAP12, ADELTA, ADEG
      COMMON /RESIN1/ PREN(16), PIMN(16), UGC, OGC, ACK, NJ, NH
      COMMON /RESIN2/ DK, DKS, DCAP02, DCAP04
      EQUIVALENCE (PREN(1),DSK(1))
C
      WRITE (KA2,9999) ADELTA
9999  FORMAT (/11H CAP. DELTA, 10X, F15.6)
      IF (IAPRO.LT.4) RETURN
      WRITE (KA2,9998)
9998  FORMAT (//50H ZEROS OF THE CHARACTERISTIC FUNCTION  /K(J*V)/**2/)
      IP = 0
      IZ = 2
      ZIM = 0.
      DO 10 I=1,NJ
        ZRE = SNGL(DSK(I))
        CALL OUT002
  10  CONTINUE
      RETURN
      END
C
C-----------------------------------------------------------------------
C SUBROUTINE:   OUT035
C OUTPUT OF THE TOLERANCE
C OUTPUT OF THE BOUND PAIR OF THE DESIGN PARAMETER AC
C-----------------------------------------------------------------------
C
      SUBROUTINE OUT035
C
      COMMON /CANPAR/ KA1, KA2, KA3, KA4, KA5, LINE
      COMMON /RESIN1/ PREN(16), PIMN(16), UGC, OGC, ACK, NJ, NH
C
      WRITE (KA2,9999) UGC, OGC
9999  FORMAT (//38H BOUND PAIR OF THE DESIGN PARAMETER C , F9.4,
     *    16H  .LE.  C  .LE. , F9.4)
      RETURN
      END
C
C-----------------------------------------------------------------------
C SUBROUTINE:   OUT036
C OUTPUT OF THE ZEROS AND POLES OF THE NORMALIZED REFERENCE LOWPASS
C-----------------------------------------------------------------------
C
      SUBROUTINE OUT036
C
      COMMON /CANPAR/ KA1, KA2, KA3, KA4, KA5, LINE
      COMMON /OUTDAT/ IP, PRE, PIM, IZ, ZRE, ZIM
      COMMON /CONST/ PI, FLMA, FLMI, FLER
      COMMON /RES/ AC, ROM(4), RDELP, RDELS, NZM(4)
      COMMON /RESS/ SFA, SM(18,4), NZERO(16), SPR(16), SPI(16)
      COMMON /RESIN1/ PREN(16), PIMN(16), UGC, OGC, ACK, NJ, NH
C
C
      NZ = NZM(4)
      IF (IZ.NE.0) GO TO 10
      WRITE (KA2,9999)
9999  FORMAT (////50H POLES AND ZEROS OF THE NORMALIZED REFERENCE LOWPA,
     *    18HSS IN THE S-DOMAIN)
      IF (IP.NE.0) NZ = 0
      GO TO 20
  10  WRITE (KA2,9998)
9998  FORMAT (////50H POLES AND ZEROS OF THE REFERENCE FILTER IN THE S-,
     *    6HDOMAIN)
  20  CONTINUE
      PRE = SFA
      CALL OUT001
```

```
      ME = MAX0(NZ,NJ)
      DO 60 I=1,ME
        IF (IP.EQ.0) GO TO 30
        IP = 2
        Q = SPI(I)
        IF (ABS(Q).LT.FLMI) IP = 1
        PRE = SPR(I)
        PIM = Q
        IF (NZ.LT.I) GO TO 40
        IF (NJ.LT.I) IP = 0
  30    IZ = NZERO(I)
        ZRE = 0.
        ZIM = SM(I,4)
        GO TO 50
  40    IZ = 0
  50    CALL OUT002
  60  CONTINUE
C
      RETURN
      END
C
C-----------------------------------------------------------------------
C SUBROUTINE:   OUT037
C OUTPUT OF THE CHOSEN DESIGN PARAMETER
C-----------------------------------------------------------------------
C
      SUBROUTINE OUT037
C
      COMMON /CANPAR/ KA1, KA2, KA3, KA4, KA5, LINE
      COMMON /TOL/ ITYP, IAPRO, NDEG, OM(4), SF, ADELP, ADELS
      COMMON /DESIGN/ NDEGF, EDEG, ACX, NORMA, LSOUT, LVSN, LSYM
      COMMON /RES/ AC, ROM(4), RDELP, RDELS, NZM(4)
C
      WRITE (KA2,9999) ACX, AC
9999  FORMAT (/26H CHOSEN DESIGN PAR.   CX =, F10.7, 10H  (=)  C =,
     *    F15.7)
      QP = 100.*RDELP/ADELP
      QS = 100.*RDELS/ADELS
      WRITE (KA2,9998) RDELP, QP, RDELS, QS
9998  FORMAT (/40H UTILIZATION OF THE PASSBAND  DELTA P  =, F10.7,
     *    3H  =, F10.5, 8H PERCENT/20X, 20HSTOPBAND  DELTA S  =, F10.7,
     *    3H  =, F10.5, 8H PERCENT)
      RETURN
      END
C
C-----------------------------------------------------------------------
C SUBROUTINE:   OUT038
C OUTPUT OF THE POLES AND ZEROS IN THE Z-DOMAIN
C-----------------------------------------------------------------------
C
      SUBROUTINE OUT038
C
      COMMON /CANPAR/ KA1, KA2, KA3, KA4, KA5, LINE
      COMMON /OUTDAT/ IP, PRE, PIM, IZ, ZRE, ZIM
      COMMON /CONST/ PI, FLMA, FLMI, FLER
      COMMON /RES/ AC, ROM(4), RDELP, RDELS, NZM(4)
      COMMON /RESS/ SFA, SM(18,4), NZERO(16), SPR(16), SPI(16)
      COMMON /RESZ/ ZFA, ZM(18,4), ZZR(16), ZZI(16), ZPR(16), ZPI(16)
      COMMON /RESIN1/ PREN(16), PIMN(16), UGC, OGC, ACK, NJ, NH
C
C
      WRITE (KA2,9999)
9999  FORMAT (////32H POLES AND ZEROS IN THE Z-DOMAIN)
      NZ = NZM(4)
      PRE = ZFA
      CALL OUT001
      ME = MAX0(NJ,NZ)
C
      DO 30 I=1,ME
        IP = 2
        Q = ZPI(I)
        IF (ABS(Q).LT.FLMI) IP = 1
        PRE = ZPR(I)
        PIM = Q
        IF (NJ.LT.I) IP = 0
        IF (NZ.LT.I) GO TO 10
        IZ = NZERO(I)
        ZRE = ZZR(I)
        ZIM = ZZI(I)
        GO TO 20
  10    IZ = 0
  20    CALL OUT002
  30  CONTINUE
      RETURN
      END
C
C-----------------------------------------------------------------------
C SUBROUTINE:   OUT039
C OUTPUT OF THE EXTREMES OF THE MAGNITUDE OF THE TRANSFER FUNCTION
C-----------------------------------------------------------------------
C
      SUBROUTINE OUT039
C
      COMMON /CANPAR/ KA1, KA2, KA3, KA4, KA5, LINE
      COMMON /CONST/ PI, FLMA, FLMI, FLER
      COMMON /RES/ AC, ROM(4), RDELP, RDELS, NZM(4)
      COMMON /RESS/ SFA, SM(18,4), NZERO(16), SPR(16), SPI(16)
      COMMON /RESZ/ ZFA, ZM(18,4), ZZR(16), ZZI(16), ZPR(16), ZPI(16)
C
C
      WRITE (KA2,9999)
9999  FORMAT (///51H EXTREMES OF THE MAGNITUDE OF THE TRANSFER FUNCTION,
     *    1H , 20H(COEFS NOT ROUNDED) )
      FN = 180./PI
      WRITE (KA2,9998)
9998  FORMAT (/1X, 26HMAXIMA IN THE PASSBAND (P)/1X, 15HMAXIMA IN THE S,
     *    11HTOPBAND (S)/14HBAND  S-DOMAIN, 7X, 8HZ-DOMAIN, 10X,
     *    9HMAGNITUDE/17X, 6HIN RAD, 3X, 9HIN DEGREE/)
C
      II = 1
      JJ = 3
  10  I = NZM(II)
      J = NZM(JJ)
      K = MAX0(I,J)
      DO 40 L=1,K
        IF (I.LT.L) GO TO 20
        OMEG1 = FN*ZM(L,II)
        AMAG1 = AMAGO(ZM(L,II))
        IF (J.LT.L) GO TO 30
        OMEG2 = FN*ZM(L,JJ)
        AMAG2 = AMAGO(ZM(L,JJ))
        WRITE (KA2,9997) SM(L,II), ZM(L,II), OMEG1, AMAG1, SM(L,JJ),
     *      ZM(L,JJ), OMEG2, AMAG2
9997    FORMAT (1HP, 3(F10.4, 2X), F12.6/1HS, 3(F10.4, 2X), F12.6)
9996    FORMAT (1HP, 3(F10.4, 2X), F12.6)
        GO TO 40
  20    OMEG2 = FN*ZM(L,JJ)
        AMAG2 = AMAGO(ZM(L,JJ))
        WRITE (KA2,9995) SM(L,JJ), ZM(L,JJ), OMEG2, AMAG2
9995    FORMAT (1HS, 3(F10.4, 2X), F12.6)
```

```
      GO TO 40
30    WRITE (KA2,9996) SM(L,II), ZM(L,II), OMEG1, AMAG1
40  CONTINUE
    IF (II.EQ.2) RETURN
    WRITE (KA2,9994)
9994 FORMAT (/1X, 26HMINIMA IN THE PASSBAND (P)/1X, 15HMINIMA IN THE S,
   *    11HTOPBAND (S)/14HBAND  S-DOMAIN, 7X, 8HZ-DOMAIN, 10X,
   *    9HMAGNITUDE/17X, 6HIN RAD, 3X, 9HIN DEGREE/)
    II = 2
    JJ = 4
    GO TO 10
C
    END
C
C-----------------------------------------------------------------------
C SUBROUTINE:   OUT042
C OUTPUT OF THE START GRID FOR THE SEARCH OF THE GLOBL EXTREMES
C OF THE TRANSFER FUNCTION
C-----------------------------------------------------------------------
C
      SUBROUTINE OUT042
C
      COMMON /CANPAR/ KA1, KA2, KA3, KA4, KA5, LINE
      COMMON /CGRID/ GR(64), NGR(12)
C
      WRITE (KA2,9999)
9999  FORMAT (///36H GRID FOR THE SEARCH OF THE EXTREMES)
      J = 1
      IF (NGR(1).NE.0) GO TO 10
      MA = NGR(8)
      ME = NGR(9)
      J = 4
  10  GO TO (20, 30, 40, 50, 80), J
  20  WRITE (KA2,9998)
9998  FORMAT (/10H PASS BAND/)
      MA = 1
      ME = NGR(3)
      GO TO 60
  30  MA = ME + 1
      ME = NGR(5)
      IF (ME.LT.MA) GO TO 70
      GO TO 60
  40  WRITE (KA2,9997)
9997  FORMAT (/10H STOP BAND/)
      MA = MAX0(NGR(3),NGR(5)) + 1
      ME = NGR(6)
      GO TO 60
  50  MA = ME + 1
      ME = NGR(7)
      IF (ME.LT.MA) GO TO 80
  60  WRITE (KA2,9996) (GR(I),I=MA,ME)
9996  FORMAT (7F10.4)
  70  J = J + 1
      GO TO 10
  80  RETURN
      END
C
C-----------------------------------------------------------------------
C SUBROUTINE:   OUT043
C HEAD LINE FOR THE SEARCH OF THE MIN. WORDLENGTH
C-----------------------------------------------------------------------
C
      SUBROUTINE OUT043
C
      COMMON /CANPAR/ KA1, KA2, KA3, KA4, KA5, LINE
C
      WRITE (KA2,9999)
9999  FORMAT (//29H SEARCH OF MINIMUM WORDLENGTH//4H IWL, 6X, 5HEPS-P,
     *    7X, 5HEPS-S, 7X, 4HPMAX)
      RETURN
      END
C
C-----------------------------------------------------------------------
C SUBROUTINE:   OUT044
C OUTPUT OF ERROR EPSILON AND MAX. MAGNITUDE
C-----------------------------------------------------------------------
C
      SUBROUTINE OUT044
C
      COMMON /CANPAR/ KA1, KA2, KA3, KA4, KA5, LINE
      COMMON /CCOEFW/ IWL, IWLG, IWLD, IWLL, ADEPSG, ADEPSD, ADEPSL,
     *    ISTAB, IDEPSL, IDEPSD
      COMMON /CEPSIL/ EPS(2), PMAX
C
9999  FORMAT (I4, 3F12.6)
C
      WRITE (KA2,9999) IWL, EPS, PMAX
      RETURN
      END
C
C-----------------------------------------------------------------------
C SUBROUTINE:   OUT045
C OUTPUT OF MINIMUM WORDLENGTH
C-----------------------------------------------------------------------
C
      SUBROUTINE OUT045
C
      COMMON /CANPAR/ KA1, KA2, KA3, KA4, KA5, LINE
      COMMON /CCOEFW/ IWL, IWLG, IWLD, IWLL, ADEPSG, ADEPSD, ADEPSL,
     *    ISTAB, IDEPSL, IDEPSD
C
      WRITE (KA2,9999) IWLG, ADEPSG, IWLD, ADEPSD
9999  FORMAT (/19H MINIMUM WORDLENGTH, I4, 13H  DELTA EPS =,
     *    F12.6/19H TEST    WORDLENGTH, I4, 13H  DELTA EPS =, F12.6)
      RETURN
      END
C
C-----------------------------------------------------------------------
C SUBROUTINE:   OUT046
C OUTPUT OF THE LAYOUT OF THE ROUNDED COEFFICIENTS
C-----------------------------------------------------------------------
C
      SUBROUTINE OUT046
C
      COMMON /CANPAR/ KA1, KA2, KA3, KA4, KA5, LINE
      COMMON /CONST/ PI, FLMA, FLMI, FLER
      COMMON /CCOEFW/ IWL, IWLG, IWLD, IWLL, ADEPSG, ADEPSD, ADEPSL,
     *    ISTAB, IDEPSL, IDEPSD
      COMMON /FILT/ NB, FACT, B2(16), B1(16), B0(16), C1(16), C0(16)
      COMMON /FILTRE/ IRCO(5), IECO(16,5), IDCO(16,5), IECOM
      COMMON /SCRAT/ ADUM(32)
C
      DIMENSION ACO(16,5)
      DIMENSION ITEXT(5), JTEXT(5)
      DIMENSION IOCT(12)
      EQUIVALENCE (B2(1),ACO(1,1))
      EQUIVALENCE (ADUM(1),IOCT(1))
C
      DATA ITEXT(1), ITEXT(2), ITEXT(3), ITEXT(4), ITEXT(5) /1HB,1HB,
     *    1HB,1HC,1HC/
```

```
      DATA JTEXT(1), JTEXT(2), JTEXT(3), JTEXT(4), JTEXT(5) /1H2,1H1,
     *     1H0,1H1,1H0/
C
      WRITE (KA2,9999) IWL
 9999 FORMAT (///35H LAYOUT OF THE ROUNDED COEFFICIENTS//11H WORDLENGTH,
     *    2H  , I4///11X, 4HCOEF, 10X, 5HCOEFS, 10X, 2HIR, 3X, 2HIE,
     *    10X, 2HID, 5X, 5HOCTAL/)
      DO 20 J=1,5
        WRITE (KA2,9998)
 9998   FORMAT (/)
        DO 10 I=1,NB
          AC = ACO(I,J)
          ACS = ACOEFS(AC,IECO(I,J),IDCO(I,J),IRCO(J))
          ID = SIGN(2.,AC)
          Q = ACS
          CALL OCTAL(Q, IOCT, 12)
          WRITE (KA2,9997) ITEXT(J), JTEXT(J), I, AC, ACS, IRCO(J),
     *        IECO(I,J), ID, IDCO(I,J), (IOCT(II),II=1,12)
 9997     FORMAT (2A1, 1H(, I2, 2H) , F13.8, 1H=, F13.8, 5H*2**(, I3,
     *        2H -, I3, 1H), 2X, I3, 2H**, I4, 1X, 12I1)
   10   CONTINUE
   20 CONTINUE
      RETURN
      END
C
C-----------------------------------------------------------------------
C SUBROUTINE:   OUT060
C NEW PAGE
C-----------------------------------------------------------------------
C
      SUBROUTINE OUT060
C
      COMMON /CANPAR/ KA1, KA2, KA3, KA4, KA5, LINE
      WRITE (KA2,9999)
 9999 FORMAT (1H1)
      RETURN
      END
C
C-----------------------------------------------------------------------
C SUBROUTINE:   OUT061
C HEAD OF PRINTER PLOT
C-----------------------------------------------------------------------
C
      SUBROUTINE OUT061
C
      COMMON /CANPAR/ KA1, KA2, KA3, KA4, KA5, LINE
      COMMON /CPLOT/ IPLOT, LNOR, IPAG, OMLO, OMUP, RMAX, RMIN
      COMMON /CLINE/ Y, X, IZ(111)
      DIMENSION A(6)
      EQUIVALENCE (A(1),IZ(1))
      GO TO (10, 20, 30, 40), IPLOT
   10 WRITE (KA2,9999)
 9999 FORMAT (1H /30X, 35HMAGNITUDE OF THE FREQUENCY RESPONSE/)
      GO TO 50
   20 WRITE (KA2,9998)
 9998 FORMAT (1H /28X, 37HATTENUATION OF THE FREQUENCY RESPONSE/)
      GO TO 50
   30 WRITE (KA2,9997)
 9997 FORMAT (1H /31X, 14HPHASE RESPONSE/)
      GO TO 50
   40 WRITE (KA2,9996)
 9996 FORMAT (1H /31X, 16HIMPULSE RESPONSE/)
   50 WRITE (KA2,9995) A(1), A(3), A(5)
 9995 FORMAT (9HMAGNITUDE, 2X, 1HX, F8.3, 2(11X, F9.3)/)
      RETURN
      END
C
C-----------------------------------------------------------------------
C SUBROUTINE:   OUT062
C OUTPUT OF THE PLOTTER LINE
C-----------------------------------------------------------------------
C
      SUBROUTINE OUT062
C
      COMMON /CANPAR/ KA1, KA2, KA3, KA4, KA5, LINE
      COMMON /CPLOT/ IPLOT, LNOR, IPAG, OMLO, OMUP, RMAX, RMIN
      COMMON /CLINE/ Y, X, IZ(111)
      DIMENSION IZDUM(56)
      DATA ICHP, ICHB /1H+,1H /
      DO 10 I=1,110,2
        J = (I+1)/2
        IZDUM(J) = ICHB
        IF (IZ(I).NE.ICHB) IZDUM(J) = IZ(I)
        IF (IZ(I+1).NE.ICHB) IZDUM(J) = IZ(I+1)
        IF ((IZ(I).EQ.ICHP) .OR. (IZ(I+1).EQ.ICHP)) IZDUM(J) = ICHP
   10 CONTINUE
      IZDUM(56) = IZ(111)
      IF (IPLOT.EQ.4) GO TO 20
      WRITE (KA2,9999) Y, X, IZDUM
 9999 FORMAT (E9.3, F6.3, 1X, 56A1)
      GO TO 30
   20 IX = INT(X)
      WRITE (KA2,9998) Y, IX, IZDUM
 9998 FORMAT (E9.3, I6, 1X, 56A1)
   30 RETURN
      END
C
C-----------------------------------------------------------------------
C SUBROUTINE:   HEAD
C OUTPUT OF A HEAD LINE
C-----------------------------------------------------------------------
C
      SUBROUTINE HEAD(KAN)
C
      WRITE (KAN,9999)
 9999 FORMAT (41H     ===  DOREDI  ===   VERSION  V005 === )
C
C HERE ON LINE FOR THE OUTPUT OF DATE AND TIME;
C REMOVE THE FOLLOWING STATEMENT
      WRITE (KAN,9998)
 9998 FORMAT (1H )
      RETURN
      END
C
C-----------------------------------------------------------------------
C SUBROUTINE:   OUT001
C HEADLINE OF THE POLES AND ZERO OUTPUT
C-----------------------------------------------------------------------
C
      SUBROUTINE OUT001
C
      COMMON /CANPAR/ KA1, KA2, KA3, KA4, KA5, LINE
      COMMON /OUTDAT/ IP, PRE, PIM, IZ, ZRE, ZIM
C
      IF (PRE.NE.0.) WRITE (KA2,9999) PRE
 9999 FORMAT (/24H CONSTANT GAIN FACTOR = , E15.7)
      WRITE (KA2,9998)
 9998 FORMAT (//3X, 4HNUM., 11X, 5HPOLES, 15X, 4HNUM., 11X, 5HZEROS/)
      RETURN
      END
```

```
C
C-----------------------------------------------------------------------
C SUBROUTINE:   OUT002
C OUTPUT OF POLE AND ZERO
C-----------------------------------------------------------------------
C
      SUBROUTINE OUT002
C
      COMMON /CANPAR/ KA1, KA2, KA3, KA4, KA5, LINE
      COMMON /OUTDAT/ IP, PRE, PIM, IZ, ZRE, ZIM
C
      IF (IP.EQ.0) GO TO 20
      IF (IZ.EQ.0) GO TO 10
      WRITE (KA2,9999) IP, PRE, PIM, IZ, ZRE, ZIM
9999  FORMAT (2(1X, I5, 1X, F11.6, 5H +-J*, F11.6, 1X))
      RETURN
  10  WRITE (KA2,9999) IP, PRE, PIM
      RETURN
  20  IF (IZ.EQ.0) RETURN
      WRITE (KA2,9998) IZ, ZRE, ZIM
9998  FORMAT (36X, I5, 1X, F11.6, 5H +-J*, F11.6, 1X)
      RETURN
      END
C
C-----------------------------------------------------------------------
C SUBROUTINE:   OCTAL
C OCTAL CONVERSION
C-----------------------------------------------------------------------
C
      SUBROUTINE OCTAL(AC, IOCT, N)
C
      COMMON /CONST/ PI, FLMA, FLMI, FLER
C
      DIMENSION IOCT(1)
C
      BC = AC
      IF (BC.LT.(-1.0)) BC = -1.0
      IF (BC.LT.0.) BC = -BC
      IF (BC.GE.1.) BC = 0.
C
      IOCT(1) = 0
      DO 10 J=2,N
        Q = 8.*BC
        IQ = INT(Q)
        IOCT(J) = IQ
        BC = Q - FLOAT(IQ)
  10  CONTINUE
C
      IF (AC.GE.FLMI) RETURN
      IA = 1
      IOCT(1) = 1
      IQ = N - 1
      DO 20 J=1,IQ
        JJ = N - J + 1
        IF (IA.EQ.1 .AND. IOCT(JJ).EQ.0) GO TO 20
        IOCT(JJ) = 7 - IOCT(JJ) + IA
        IA = 0
  20  CONTINUE
C
      RETURN
      END
C
C-----------------------------------------------------------------------
C SUBROUTINE:   OUTSPE
C     DUMMY SUBROUTINE FOR 'SPECIAL OUTPUT'
C-----------------------------------------------------------------------
C
      SUBROUTINE OUTSPE
C
      COMMON /CONTR / IPRUN,IPCON,NINP,NOUT,NDOUT,LSPOUT,ISPOUT
      COMMON /CANPAR/ KA1,KA2,KA3,KA4,KA5,LINE
C
C     DATA TRANSFER BY NAMED COMMON BLOCKS:
C
      COMMON /FILT  / NB,FACT,B2(16),B1(16),B0(16),C1(16),C0(16)
C
C     ISPMAX: MAXIMUM NUMBER OF IMPLEMENTED OPTIONS
      ISPMAX=0
C
      IF(ISPOUT.LE.0) RETURN
      IF(ISPOUT.GT.ISPMAX) GO TO 900
C
C     EXAMPLE FOR ISPMAX=3
C
C     GO TO (100,200,300),ISPOUT
C 100 CONTINUE
C     .
C     OPTION NO.  1
C     .
C     RETURN
C
C 200 CONTINUE
C     .
C     OPTION NO.  2
C     .
C     RETURN
C
C 300 CONTINUE
C     .
C     OPTION NO.  3
C     .
C     RETURN
C
 900  CALL ERROR (8)
      RETURN
      END
C
C =======================================================================
C
C DOREDI - SUBROUTINES: PART VI
C
C =======================================================================
C
C
C-----------------------------------------------------------------------
C SUBROUTINE:   SPOW
C CALCULATION OF THE L-2 NORM
C-----------------------------------------------------------------------
C
      SUBROUTINE SPOW(BMAX)
C
      COMMON /CNFUNC/ AQC(5), BN2(5), BN1(5), BN0(5)
      COMMON /CPOW/ PNU, PNC, AND, ITCORP
      COMMON /FILT/ NB, FACT, B2(16), B1(16), B0(16), C1(16), C0(16)
      COMMON /CRENO/ LCNO, ICNO(16,5)
      COMMON /COPTSP/ IB, JSEQN(16), JSEQD(16), AMAX, SCA, ALSBI
C
      BB2 = BN2(1)
      BB1 = BN1(1)
      BB0 = BN0(1)
```

```
      BN2(1) = B2(1)
      BN1(1) = B1(1)
      BN0(1) = B0(1)
      LQ = LCNO
      LCNO = 2
      IQ = IB
      IB = 1
      Q = PNU
      CALL POWER
      BMAX = PNU*FACT*FACT
      PNU = Q
      IB = IQ
      LCNO = LQ
      BN2(1) = BB2
      BN1(1) = BB1
      BN0(1) = BB0
      RETURN
      END
C
C-----------------------------------------------------------------------
C SUBROUTINE:   POWER
C CALCULATION OF THE NOISE POWER
C GENERATED IN THE IB-TH BLOCK
C-----------------------------------------------------------------------
C
      SUBROUTINE POWER
C
      COMPLEX ZA, ZB, ZP, ZPS, ZQ, ZQI, Z0, Z1, ZBL, ZBL1, ZZ, ZZ1
C
      COMMON /CONST/ PI, FLMA, FLMI, FLER
      COMMON /CPOL/ ZP(16,2), ZPS(16,2)
      COMMON /CRENO/ LCNO, ICNO(16,5)
      COMMON /CNFUNC/ AQC(5), BN2(5), BN1(5), BN0(5)
      COMMON /CPOW/ PNU, PNC, AND, ITCORP
      COMMON /FILT/ NB, FACT, B2(16), B1(16), B0(16), C1(16), C0(16)
      COMMON /COPTSP/ IB, JSEQN(16), JSEQD(16), AMAX, SCA, ALSBI
C
      DIMENSION EU(5)
C
      DATA Z0,Z1 /(0.,0.),(1.,0.)/
C
      ZBL(ZZ,A1,A0) = (ZZ+A1)*ZZ + A0
      ZBL1(ZZ1,A21,A11,A01) = (A21*ZZ1+A11)*ZZ1 + A01
C
      ZA = Z0
      PNU = 0.
      IF (LCNO.NE.2) GO TO 20
      EU(1) = 1.
      DO 10 I=2,5
        EU(I) = 0.
  10  CONTINUE
      GO TO 160
  20  DO 130 N=1,5
        AQ = AQC(N)
        IF (AQ.EQ.1.) GO TO 120
        IF (LCNO.EQ.(-1)) AQ = 0.
        KK = IB
        IF (KK.GT.NB) KK = NB
        K = ICNO(KK,N)
        QS = 1.
        QC1 = 0.
        GO TO (30, 40, 50, 60, 70, 80, 90), K
  30    QC2 = -1.
        Q = AQ/2.
        GO TO 100
  40    QC2 = 2.*(1.-3./PI)
        GO TO 110
  50    QC2 = 2.
        GO TO 110
  60    QC2 = -1.
        Q = (AQ-1.)/2.
        GO TO 100
  70    QS = 4. - 6./PI
        GO TO 30
  80    QC1 = -6.*(1.-2./PI)
        GO TO 40
  90    QS = 4.
        GO TO 30
 100    ZA = ZA + Q*ZBL1(Z1,BN2(N),BN1(N),BN0(N))
 110    EU(N) = QS + (QC2*AQ+QC1)*AQ
        GO TO 130
 120    EU(N) = 0.
 130  CONTINUE
      IF (IB.GT.NB) GO TO 340
      IF (CABS(ZA).EQ.0.) GO TO 150
      ZA = ZA/ZBL(Z1,C1(IB),C0(IB))
      IF (IB.EQ.NB) GO TO 150
      MA = IB + 1
      DO 140 I=MA,NB
        ZA = ZA*ZBL1(Z1,B2(I),B1(I),B0(I))/ZBL(Z1,C1(I),C0(I))
 140  CONTINUE
 150  AND = REAL(ZA)
C
 160  MARK = 0
      R1 = 0.
      DO 170 N=1,5
        IF (EU(N).EQ.0.) GO TO 170
        R1 = R1 + BN2(N)*BN0(N)*EU(N)
 170  CONTINUE
      R2 = 1.
      DO 190 J=IB,NB
        IF (J.NE.IB) R2 = R2*B2(J)*B0(J)
        Q = C0(J)
        IF (ABS(Q).GT.FLMI) GO TO 180
        IF (MARK.NE.0) GO TO 360
        MARK = J
        Q = C1(J)
 180    R2 = R2/Q
 190  CONTINUE
      IF (MARK.NE.0) GO TO 200
      PNU = PNU + R1*R2
      GO TO 280
C
 200  DO 270 I=IB,NB
        IF (I.EQ.MARK) GO TO 210
        Q1 = C0(I)
        Q2 = C1(I)
        GO TO 220
 210    Q1 = C1(I)
        Q2 = 1.
 220    IF (I.NE.IB) GO TO 240
        R = 0.
        DO 230 N=1,5
          IF (EU(N).EQ.0.) GO TO 230
          QQ1 = BN2(N)*(BN1(N)*Q1-BN0(N)*Q2)/Q1
          QQ2 = BN0(N)*(BN1(N)-BN2(N)*C1(IB))
          R = R + (QQ1+QQ2)*EU(N)
 230    CONTINUE
        R = R*R2
        GO TO 260
```

```
240     R = R1*(B2(I)*(B1(I)*Q1-B0(I)*Q2)/Q1+B0(I)*(B1(I)-C1(I)*B2(I)))
        DO 250 J=IB,NB
          IF (J.NE.I .AND. J.NE.IB) R = R*B2(J)*B0(J)
          Q = C0(J)
          IF (J.EQ.MARK) Q = C1(J)
          R = R/Q
250     CONTINUE
260     PNU = PNU + R
270   CONTINUE
C
280   DO 330 J=IB,NB
        DO 320 K=1,2
          ZQ = ZP(J,K)
          IF (CABS(ZQ).LT.FLMI) GO TO 320
          KK = 3 - K
          ZQI = ZP(J,KK)
          ZB = Z1/(ZQ-ZQI)
          ZQI = Z1/ZQ
          ZB = ZB*ZQI
          ZA = Z0
          DO 290 N=1,5
            IF (EU(N).EQ.0.) GO TO 290
            ZA = ZA + EU(N)*ZBL1(ZQ,BN2(N),BN1(N),BN0(N))*ZBL1(ZQI,
     *          BN2(N),BN1(N),BN0(N))
290       CONTINUE
          ZB = ZB*ZA
          DO 310 JJ=IB,NB
            IF (JJ.EQ.IB) GO TO 300
            ZB = ZB*ZBL1(ZQ,B2(JJ),B1(JJ),B0(JJ))*ZBL1(ZQI,B2(JJ),B1(JJ)
     *          ,B0(JJ))
300         IF  (JJ.NE.J) ZB = ZB/ZBL(Q,C1(JJ),C0(JJ))
            ZB = ZB/ZBL(ZQI,C1(JJ),C0(JJ))
310       CONTINUE
          PNU = PNU + REAL(ZB)
320     CONTINUE
330   CONTINUE
      RETURN
C
340   AND = REAL(ZA)
      DO 350 N=1,5
        PNU = PNU + EU(N)
350   CONTINUE
      RETURN
C
360   CALL ERROR(33)
      RETURN
      END
C
C-----------------------------------------------------------------------
C SUBROUTINE:   SMAX
C COMPUTE THE MAXIMUM (IC=1) OR MINIMUM (IC=-1) OF THE TRANSFER
C FUNCTION
C-----------------------------------------------------------------------
C
      SUBROUTINE SMAX(BMAX, MMA, MME, IC)
C
      COMMON /CONST/ PI, FLMA, FLMI, FLER
      COMMON /CGRID/ GR(64), NGR(12)
C
      DIMENSION O(5), B(5)
C
      FLM = 1./FLMA
      FLER1 = FLER*0.05
      MA = NGR(MMA)
      ME = NGR(MME)
      BMAX = 0.
      DO 30 I=MA,ME
        QR = AMAGO(GR(I))
        IF (IC.GT.0) GO TO 10
        IF (QR.LT.FLM) GO TO 220
        QR = 1./QR
10      IF (QR.LT.BMAX) GO TO 20
        BMAX1 = BMAX
        MAX1 = MAX
        BMAX = QR
        MAX = I
        GO TO 30
20      IF (QR.LT.BMAX1) GO TO 30
        BMAX1 = QR
        MAX1 = I
30    CONTINUE
      JJ = 1
      IF ((ME-MA).GT.1) GO TO 40
      O(1) = GR(MA)
      O(3) = (GR(MA)+GR(ME))/2.
      O(5) = GR(ME)
      JJ = 3
      GO TO 60
40    IF (IABS(MAX-MAX1).LT.2) JJ = 3
50    IF (MAX.EQ.MA) MAX = MA + 1
      IF (MAX.EQ.ME) MAX = MAX - 1
      O(1) = GR(MAX-1)
      O(3) = GR(MAX)
      O(5) = GR(MAX+1)
60    DO 80 I=1,5,2
        Q = AMAGO(O(I))
        IF (IC.GT.0) GO TO 70
        IF (Q.LT.FLM) GO TO 220
        Q = 1./Q
70      B(I) = Q
80    CONTINUE
      GO TO 150
C
90    EMAX = 0.
      DO 100 I=1,5
        Q = B(I)
        IF (Q.LT.EMAX) GO TO 100
        MAX = I
        EMAX = Q
100   CONTINUE
      IF (EMAX.EQ.BMAX) GO TO 110
      IF (EMAX/BMAX-1..LT.FLER) GO TO 180
110   IF (MAX.EQ.3) GO TO 130
      IF (MAX.LT.3) GO TO 120
      O(1) = O(3)
      B(1) = B(3)
      O(3) = O(4)
      B(3) = B(4)
      GO TO 140
120   O(5) = O(3)
      B(5) = B(3)
      O(3) = O(2)
      B(3) = B(2)
      GO TO 140
130   O(1) = O(2)
      B(1) = B(2)
      O(5) = O(4)
      B(5) = B(4)
140   BMAX = EMAX
150   O(2) = (O(3)+O(1))/2.
```

```
      Q = AMAGO(O(2))
      IF (IC.GT.0) GO TO 160
      IF (Q.LT.FLM) GO TO 220
      Q = 1./Q
  160 B(2) = Q
      O(4) = (O(3)+O(5))/2.
      Q = AMAGO(O(4))
      IF (IC.GT.0) GO TO 170
      IF (Q.LT.FLM) GO TO 220
      Q = 1./Q
  170 B(4) = Q
      IF (O(5)-O(1).GE.FLER1) GO TO 90
  180 GO TO (190, 200, 210), JJ
  190 EMAX = BMAX
      BMAX = BMAX1
      BMAX1 = EMAX
      MAX = MAX1
      JJ = 2
      GO TO 50
  200 BMAX = AMAX1(BMAX,BMAX1)
  210 IF (IC.LT.0) BMAX = 1./BMAX
      RETURN
  220 BMAX = 0.
      RETURN
      END
C
C-----------------------------------------------------------------------
C SUBROUTINE:   SMIMP
C CALCULATION OF THE SUM OF THE IMPULSE RESPONSE MAGNITUDES
C ABSOLUT SCALING CRETERION
C-----------------------------------------------------------------------
C
      SUBROUTINE SMIMP(SUM)
C
      COMMON /CONST/ PI, FLMA, FLMI, FLER
      COMMON /FILT/ NB, FACT, B2(16), B1(16), B0(16), C1(16), C0(16)
C
      DIMENSION X1(16), X2(16)
      COMMON /SCRAT/ ADUM(32)
      EQUIVALENCE (ADUM(1),X1(1)), (ADUM(17),X2(1))
C
      DO 10 I=1,NB
        X1(I) = 0.
        X2(I) = 0.
   10 CONTINUE
C
      BMAX = 0.
      SUM = 0.
      IN = 0
      JN = 0
      Y = FACT
      IM = 0
      GO TO 30
   20 Y = 0.
   30 DO 40 I=1,NB
        U = Y
        Y = U*B2(I) + X1(I)
        X1(I) = U*B1(I) - Y*C1(I) + X2(I)
        X2(I) = U*B0(I) - Y*C0(I)
   40 CONTINUE
      JM = SIGN(1.,Y)
      Y = ABS(Y)
      SUM = SUM + Y
      BMAX = AMAX1(BMAX,Y)
      IF (IM.NE.0) GO TO 50
      EMAX = BMAX
      IM = JM
      GO TO 20
   50 IF (IM.EQ.JM) GO TO 70
      IF (JN.GT.3) RETURN
      IN = 0
      Q = EMAX/BMAX
      EMAX = Y
      IM = JM
      IF (Q.LT.FLER) GO TO 60
      JN = 0
      GO TO 20
   60 JN = JN + 1
      GO TO 20
   70 IF (IN.GT.3) RETURN
      EMAX = AMAX1(EMAX,Y)
      IF (Y/EMAX.LT.FLER) IN = IN + 1
      GO TO 20
      END
C
C-----------------------------------------------------------------------
C FUNCTION:     AMAGO
C COMPUTATION OF THE MAGNITUDE OF THE TRANSFER FUNCTION
C PARAMETER OMEGA IN RADIANS
C-----------------------------------------------------------------------
C
      FUNCTION AMAGO(OMEG)
C
      COMMON /FILT/ NB, FACT, B2(16), B1(16), B0(16), C1(16), C0(16)
C
      COMPLEX C, CH
C
      C = CMPLX(COS(OMEG),SIN(OMEG))
      CH = CMPLX(FACT,0.)
C
      DO 10 I=1,NB
        CH = CH*((B2(I)*C+B1(I))*C+B0(I))/((C+C1(I))*C+C0(I))
   10 CONTINUE
      AMAGO = CABS(CH)
      RETURN
      END
C
C-----------------------------------------------------------------------
C FUNCTION:     PHASE
C PHASE OF OMEGA
C-----------------------------------------------------------------------
C
      FUNCTION PHASE(OM)
C
      COMPLEX CQ, CF
      COMMON /FILT/ NB, FACT, B2(16), B1(16), B0(16), C1(16), C0(16)
      CF = CMPLX(FACT,0.)
      CQ = CMPLX(COS(OM),SIN(OM))
      DO 10 I=1,NB
        CF = CF*((B2(I)*CQ+B1(I))*CQ+B0(I))/((CQ+C1(I))*CQ+C0(I))
   10 CONTINUE
      Y = AIMAG(CF)
      X = REAL(CF)
      PHASE = ATAN2(Y,X)
      RETURN
      END
C
C-----------------------------------------------------------------------
C FUNCTION:     RESP
C RESPONSE OF AN INPUT SIGNAL
```

```
C-----------------------------------------------------------------------
C
      FUNCTION RESP(X, IS)
C
      COMMON /FILT/ NB, FACT, B2(16), B1(16), B0(16), C1(16), C0(16)
      COMMON /CSTATV/ X1(16), X2(16)
C
      IF (IS.EQ.0) GO TO 20
C CLEAR STATA VARIABLES
      DO 10 I=1,NB
        X1(I) = 0.
        X2(I) = 0.
  10  CONTINUE
C
  20  Y = X*FACT
      DO 30 I=1,NB
        U = Y
        Y = U*B2(I) + X1(I)
        X1(I) = U*B1(I) - Y*C1(I) + X2(I)
        X2(I) = U*B0(I) - Y*C0(I)
  30  CONTINUE
      RESP = Y
      RETURN
      END
C
C-----------------------------------------------------------------------
C SUBROUTINE:   CODE3
C STORE PAIRING AND ORDERING
C
      SUBROUTINE CODE3(L, J, N)
C
      DIMENSION J(16), L(16)
      DO 10 I=1,N
        J(I) = L(I)
  10  CONTINUE
      RETURN
      END
C
C-----------------------------------------------------------------------
C SUBROUTINE:   CODE4
C RESTORE PAIRING AND ORDERING
C-----------------------------------------------------------------------
C
      SUBROUTINE CODE4(L, J, N)
C
      DIMENSION J(16), L(16)
      DO 10 I=1,N
        L(I) = J(I)
  10  CONTINUE
      RETURN
      END
C
C-----------------------------------------------------------------------
C SUBROUTINE:   CODE5
C SEARCH FOR A GIVEN CODE ARRAY IN LL
C RESTORE PAIRING AND ORDERING TO L
C-----------------------------------------------------------------------
C
      SUBROUTINE CODE5(L, J, N, LL, M)
C
      DIMENSION L(16), J(16), LL(16)
      M = 2
      DO 30 I=1,N
        IQ = J(I)
        DO 10 II=1,N
          IF (IQ.EQ.LL(II)) GO TO 20
  10    CONTINUE
        RETURN
  20    L(I) = IQ
  30  CONTINUE
      M = 1
      RETURN
      END
C
C-----------------------------------------------------------------------
C SUBROUTINE:   QUAN
C CHECK THE ACTUAL COEFFICIENT QUANTISATION
C-----------------------------------------------------------------------
C
      SUBROUTINE QUAN(ACO, ACQ)
C
      COMMON /COPTSP/ IB, JSEQN(16), JSEQD(16), AMAX, SCA, ALSBI
C
      Q = ABS(ACO)
      IF (Q.NE.AINT(Q)) GO TO 10
      ACQ = 1.
      RETURN
  10  Q = Q*ALSBI
      ACQ = 0.5/ALSBI
      GO TO 30
  20  Q = Q*0.5
      ACQ = ACQ*2.
  30  IF (Q.EQ.AINT(Q)) GO TO 20
      RETURN
      END
C
C-----------------------------------------------------------------------
C FUNCTION:     ACOEF
C RETRANSFORM ONE COEFFICIENT
C-----------------------------------------------------------------------
C
      FUNCTION ACOEF(AS, IE, ID, IR)
C
      Q = AS*2.**(IR-IE)
      ACOEF = Q
      IF (ID.LE.(-100)) RETURN
      AD = 2.**ID
      ACOEF = Q - SIGN(AD,AS)
      RETURN
      END
C
C-----------------------------------------------------------------------
C FUNCTION:     ACOEFS
C TRANSFORM ONE COEFFICIENT
C-----------------------------------------------------------------------
C
      FUNCTION ACOEFS(AC, IE, ID, IR)
C
      AD = 0.
      IF (ID.GT.(-100)) AD = 2.**ID
      ACOEFS = (AC-SIGN(AD,AC))*2.**(IE-IR)
      RETURN
      END
C
C-----------------------------------------------------------------------
C SUBROUTINE:   COPY01
C COPY COEFFICIENT FIELD /FILT/ IN TO THE FIELD /SFILT/
C-----------------------------------------------------------------------
C
      SUBROUTINE COPY01
```

```
C
      COMMON /FILT/ NB, FACT, B2(16), B1(16), B0(16), C1(16), C0(16)
      COMMON /SFILT/ NBS, SFACT, SB2(16), SB1(16), SB0(16), SC1(16),
     *    SC0(16)
C
      DIMENSION ACO(16,5), SACO(16,5)
      EQUIVALENCE (ACO(1,1),B2(1)), (SACO(1,1),SB2(1))
C
      NBS = NB
      SFACT = FACT
      DO 20 J=1,5
        DO 10 I=1,NB
          SACO(I,J) = ACO(I,J)
  10    CONTINUE
  20  CONTINUE
      RETURN
      END
C
C-----------------------------------------------------------------------
C SUBROUTINE:   COPY02
C COPY COEFFICIENT FIELD /SFILT/ INTO THE FIELD /FILT/
C-----------------------------------------------------------------------
C
      SUBROUTINE COPY02
C
      COMMON /FILT/ NB, FACT, B2(16), B1(16), B0(16), C1(16), C0(16)
      COMMON /SFILT/ NBS, SFACT, SB2(16), SB1(16), SB0(16), SC1(16),
     *    SC0(16)
C
      DIMENSION ACO(16,5), SACO(16,5)
      EQUIVALENCE (ACO(1,1),B2(1)), (SACO(1,1),SB2(1))
C
      NB = NBS
      FACT = SFACT
      DO 20 J=1,5
        DO 10 I=1,NB
          ACO(I,J) = SACO(I,J)
  10    CONTINUE
  20  CONTINUE
      RETURN
      END
C
C-----------------------------------------------------------------------
C SUBROUTINE:   COPY03
C COPY POLE LOCATIONS
C-----------------------------------------------------------------------
C
      SUBROUTINE COPY03
C
      COMPLEX ZP, ZPS
C
      COMMON /FILT/ NB, FACT, B2(16), B1(16), B0(16), C1(16), C0(16)
      COMMON /CPOL/ ZP(16,2), ZPS(16,2)
C
      DO 20 I=1,NB
        DO 10 J=1,2
          ZP(I,J) = ZPS(I,J)
  10    CONTINUE
  20  CONTINUE
      RETURN
      END
C
C-----------------------------------------------------------------------
C FUNCTION:     DELLK
C CALCULATE COMPLETE ELLIPTIC INTEGRAL OF FIRST KIND
C-----------------------------------------------------------------------
C
      DOUBLE PRECISION FUNCTION DELLK(DK)
C
      DOUBLE PRECISION DPI, DOMI
      DOUBLE PRECISION DE, DGEO, DK, DRI, DARI, DTEST
C
      COMMON /CONST/ PI, FLMA, FLMI, FLER
      COMMON /CONST2/ DPI, DOMI
C
      DATA DE /1.D00/
C
      DGEO = DE - DK*DK
      IF (DGEO) 10, 10, 20
  10  DELLK = DBLE(FLMA)
      RETURN
C
  20  DGEO = DSQRT(DGEO)
      DRI = DE
  30  DARI = DRI
      DTEST = DARI*DOMI
      DRI = DGEO + DRI
C
      IF (DARI-DGEO-DTEST) 50, 50, 40
  40  DGEO = DSQRT(DARI*DGEO)
      DRI = 0.5D00*DRI
      GO TO 30
  50  DELLK = DPI/DRI
      RETURN
      END
C
C-----------------------------------------------------------------------
C FUNCTION:     DSN2
C CALCULATION OF THE JAKOBI'S ELLIPTIC FUNCTION SN(U,K)
C
C EXTERNAL CALCULATION OF THE PARAMETER NECESSARY
C DK = K($K)
C DQ = EXP(-PI*K'/K) ... (JACOBI'S NOME)
C-----------------------------------------------------------------------
C
      DOUBLE PRECISION FUNCTION DSN2(DU, DK, DQ)
C
      DOUBLE PRECISION DPI, DOMI
      DOUBLE PRECISION DE, DZ, DPI2, DQ, DM, DU, DK, DC, DQQ, DH, DQ1,
     *    DQ2
C
      COMMON /CONST2/ DPI, DOMI
C
      DATA DE, DZ /1.D00,2.D00/
C
      DPI2 = DPI/DZ
      IF (DABS(DQ).GE.DE) GO TO 30
C
      DM = DPI2*DU/DK
      DC = DZ*DM
      DC = DCOS(DC)
C
      DM = DSIN(DM)*DK/DPI2
      DQQ = DQ*DQ
      DQ1 = DQ
      DQ2 = DQQ
C
      DO 10 I=1,100
        DH = (DE-DQ1)/(DE-DQ2)
        DH = DH*DH
```

```
      DH = DH*(DE-DZ*DQ2*DC+DQ2*DQ2)
      DH = DH/(DE-DZ*DQ1*DC+DQ1*DQ1)
      DM = DM*DH
C
      DH = DABS(DE-DH)
      IF (DH.LT.DOMI) GO TO 20
C
      DQ1 = DQ1*DQQ
      DQ2 = DQ2*DQQ
   10 CONTINUE
C
      GO TO 30
C
   20 DSN2 = DM
      RETURN
C
   30 DSN2 = 0.D00
      RETURN
      END
C
C-----------------------------------------------------------------------
C SUBROUTINE:   DELI1
C ELLIPTIC FUNCTION
C-----------------------------------------------------------------------
C
      SUBROUTINE DELI1(RES, X, CK)
C
      DOUBLE PRECISION RES, X, CK, ANGLE, GEO, ARI, PIM, SQGEO, AARI,
     *    TEST, DPI
      DOUBLE PRECISION DOMI
C
      COMMON /CONST2/ DPI, DOMI
      IF (X) 20, 10, 20
   10 RES = 0.D0
      RETURN
C
   20 IF (CK) 40, 30, 40
   30 RES = DLOG(DABS(X)+DSQRT(1.D0+X*X))
      GO TO 130
C
   40 ANGLE = DABS(1.D0/X)
      GEO = DABS(CK)
      ARI = 1.D0
      PIM = 0.D0
   50 SQGEO = ARI*GEO
      AARI = ARI
      ARI = GEO + ARI
      ANGLE = -SQGEO/ANGLE + ANGLE
      SQGEO = DSQRT(SQGEO)
      IF (ANGLE) 70, 60, 70
C
C REPLACE 0 BY A SMALL VALUE
C
   60 ANGLE = SQGEO*DOMI
   70 TEST = AARI*DOMI*1.D+05
      IF (DABS(AARI-GEO)-TEST) 100, 100, 80
   80 GEO = SQGEO + SQGEO
      PIM = PIM + PIM
      IF (ANGLE) 90, 50, 50
   90 PIM = PIM + DPI
      GO TO 50
  100 IF (ANGLE) 110, 120, 120
  110 PIM = PIM + DPI
  120 RES = (DATAN(ARI/ANGLE)+PIM)/ARI
  130 IF (X) 140, 150, 150
  140 RES = -RES
  150 RETURN
      END
C
C-----------------------------------------------------------------------
C FUNCTION:     SINH
C-----------------------------------------------------------------------
C
      FUNCTION SINH(X)
C
      SINH = (EXP(X)-EXP(-X))/2.
      RETURN
      END
C
C-----------------------------------------------------------------------
C FUNCTION:     COSH
C-----------------------------------------------------------------------
C
      FUNCTION COSH(X)
C
      COSH = (EXP(X)+EXP(-X))/2.
      RETURN
      END
C
C-----------------------------------------------------------------------
C FUNCTION:     ARSINH
C-----------------------------------------------------------------------
C
      FUNCTION ARSINH(X)
C
      ARSINH = ALOG(X+SQRT(X*X+1.))
      RETURN
      END
C
C-----------------------------------------------------------------------
C FUNCTION:     ARCOSH
C-----------------------------------------------------------------------
C
      FUNCTION ARCOSH(X)
C
      IF (X.LT.1.) GO TO 10
      ARCOSH = ALOG(X+SQRT(X*X-1.))
      RETURN
   10 ARCOSH = 0.
      RETURN
      END
C
C-----------------------------------------------------------------------
C SUBROUTINE:   DSQRTC
C COMPUTATION OF THE COMPLEX SQUARE ROOT
C IN DOUBLE PRECISION
C DU + J*DV = SQRT ( DX + J*DY )
C-----------------------------------------------------------------------
C
      SUBROUTINE DSQRTC(DX, DY, DU, DV)
C
      DOUBLE PRECISION DPI, DOMI, D1MACH
      DOUBLE PRECISION DX, DU, DY, DV, DQ, DP
      COMMON /CONST2/ DPI, DOMI
C
      DQ = DX
      DP = DY
C
      DV = 0.5D00*DQ
      DU = DQ*DQ + DP*DP
```

```
      DU = DSQRT(DU)
      DU = 0.5D00*DU
      DV = DU - DV
      DU = DV + DQ
      IF (DABS(DU).LE.3.D0*D1MACH(3)) DU = 0.D0
      DU = DSQRT(DU)
      IF (DABS(DV).LE.3.D0*D1MACH(3)) DV = 0.D0
      DV = DSQRT(DV)
      IF (DP.LT.(-DOMI)) DU = -DU
      RETURN
      END
C
C-----------------------------------------------------------------------
C SUBROUTINE:   ERROR
C OUTPUT OF AN ERROR MESSAGE
C-------------------------------------------------------------------
C
      SUBROUTINE ERROR(IERCOD)
C
      COMMON /CANPAR/ KA1, KA2, KA3, KA4, KA5, LINE
C
      WRITE (KA2,9999) IERCOD
9999  FORMAT (22H *** ERROR *** NUMBER , I3)
      IF (IERCOD.GE.20) STOP
      RETURN
      END
```

6.2

Program for Minimum −p Synthesis of Recursive Digital Filters

A. G. Deczky

Institut de Microtechnique
Universite de Neuchatel
Rue de la Maladière 71
2000 Neuchatel, Switzerland

1. Purpose

This program performs the synthesis of recursive digital filters using a minimum p error criterion and the method of Fletcher and Powell [2] for function minimization.

2. Method

The method is outlined in detail in [1]. The Fletcher-Powell subroutine is taken from [3], but has been modified extensively such that the version included here should be used.

3. Program Description

3.1 Usage

The program is self contained with all the necessary subroutines for input and output. The input section is also documented with a short description of each of the input parameters.

The program expects the primary input parameters on unit IND=I1MACH(1) and outputs the results on unit IOUTD=I1MACH(2). The final filter coefficients are also output on unit IOUTP=I1MACH(3). This could be used for later input, or for producing punched cards, etc. Finally in the case of the group delay equalizer, the coefficients of the filter to be equalized are read in from unit INP which is set to I1MACH(1). The format expected here is:

```
E14.7,I4 = CONST, N
5E14.7   = A0(1), A1(1), A2(1), B1(1), B2(1)
   :              :
5E14.7   = A0(N), A1(N), A2(N), B1(N), B2(N)
```

This same format is used for outputting the final filter coefficients on unit IOUTP=I1MACH(3).

The program minimizes one of the following three functions:

$$\text{F1} = \sum_{J=1}^{NP} \text{WH(J)*\{X(N)*H(X,J) - FS(J)\}**IP} \tag{1}$$

$$\text{F2} = \sum_{J=1}^{NP} \text{WG(J)*\{DG(X,J) - DGS(J) - X(N)\}**IQ} \tag{2}$$

$$\text{F3} = \text{ALFA} \sum_{J=1}^{NP} \text{WH(J)*\{X(N-1)*H(X,J) - FS(J)\}**IP}$$

$$+ (1 - \text{ALFA}) \sum_{J=1}^{NP} \text{WG(J)*\{DG(X,J) - DGS(J) - X(N)\}**IQ} \tag{3}$$

where

F1 is used for magnitude approximation

F2	is used for group delay approximation or group delay equalization
F3	is used for combined magnitude and group delay approximation
X	is the argument vector and contains the zeros and poles in polar coordinates
J	corresponds to a frequency, as discussed below
FS	is the desired magnitude function at the frequency point corresponding to J
DGS	is the desired group delay function at the frequency point corresponding to J
ALFA	is a weighting coefficient between 0 and 1 for the magnitude in the case of combined approximation.
WH(J)	is a weighting function for the magnitude at the frequency point corresponding to J
WG(J)	is a weighting function for the group delay at the frequency point corresponding to J
H(X,J)	is the magnitude of the filter which is a function of the argument vector X and the frequency point J
DG(X,J)	is the group delay of the filter which is a function of the argument vector X and the frequency point J.
IP,IQ	are indices for the magnitude and group delay error respectively. They should be positive and even.

The above functions are computed on a number NK of intervals. These intervals are specified by their end points FIK(I) and FIK(I+1) and their number I, I = 1, NK. The number of points on each of these intervals is designated by NPK(I), I = 1, NK. The frequency points J are assigned to each interval depending on the parameter WEIGHT(I) as follows:

WEIGHT(I)	J corresponds to the frequency	
1	(FIK(I+1)–FIK(I))*(J–JO–1)/NPK(I)+FIK(I)	equispaced
2	(FIK(I+1)–FIK(I))*SIN{(J–JO–1)*(PI/2)/NPK(I)}+FIK(I)	sine half cycle
3	(FIK(I)–FIK(I+1))*COS{(J–JO–1)*(PI/2)/NPK(I)}+FIK(I+1)	cosine half cycle
4	0.5*(FIK(I)–FIK(I+1))*COS{(J–JO–1)*PI/NPK(I))}+ 0.5*(FIK(I)+FIK(I+1))	cosine full cycle

where

$$JO = \sum_{K=1}^{I-1} NPK(K)$$

If NPK(I) = 0, no frequency points are assigned to that interval. This could be used to designate a "don't care" interval.

The real weight functions WH(J) and WG(J) are constant over any given interval I and are specified by the functions WHK(I) and WGK(I), i.e.,

$$\begin{aligned} WH(J+JO) &= WHK(I) \quad \text{for } J = 1,NPK(I) \\ WG(J+JO) &= WGK(I) \quad \text{for } J = 1,NPK(I) \end{aligned}$$

The indices IP, IQ are specified by the input parameters IP(II), IQ(II), II = 1,IPK. They should be even. If a large index is to be used a sequence such as 2, 4, 10 is recommended. The largest index used should be about 10.

The argument vector X contains the zeros and poles in polar coordinates. On input it specifies the initial position of the zeros and poles, while on output it gives their position at the attained minimum of F. The zeros are partitioned into real ones, ones on the unit circle and complex ones, while the poles are partitioned into real and complex ones. Real zeros and poles are specified by their radii. Complex zeros and poles are specified by their angle expressed as a frequency (i.e., angle in radians/(2π) $\times$ sampling frequency) followed by their radii, in pairs. Finally zeros on the unit circle are specified by their angle. The number of real zeros, zeros on the unit circle, complex zeros, real poles and complex poles are specified by NZR, NZU, NZC, NPR, NPC in that order. (Note that only zeros (poles) with positive angles are specified, their complex conjugate pairs being understood.) The last parameter X(N) is a multiplying constant KO for the magnitude for magnitude approximation and an added constant TAUO for the group delay for group delay synthesis or equalization. For the case of combined magnitude and group delay equalization X(N−1) is KO and X(N) is TAUO.

The required magnitude and group delay functions FS and DGS have to be provided by separate subroutines. A typical subroutine structure is given with the examples below. In the case of group delay equalization, the array DGS(J) is set equal to minus the group delay of the filter to be equalized, the latter being calculated on the basis of the coefficients read in on unit INP=I1MACH(1).

3.2 Types of Output

The program essentially computes an argument vector X and a gradient vector G corresponding to the minimum of the function F1, F2, or F3 above. Based on X, the zeros and poles as well as the coefficients of the cascade realization of the filter are computed. Optionally the frequency response is also given.

Generally two types of output may be obtained. The first type is printed in 72 column format for output on a teletype or printer, and consists of either a short output summarizing the results of the minimization (LONG=.FALSE.) or a longer output also giving the progress of the minimization (LONG=.TRUE.). Examples of both are given below.

The second type of output is strictly for debugging purposes, and produces extensive intermediate results in 72 column format to monitor the progress of the entire program. This should only be used as a debugging aid by persons thoroughly familiar with the program. This output is suppressed by setting both DEBFP and DEB to .FALSE. in the input sequence.

3.3 Description of Parameters

This is given in the input section of the program and is reproduced here for convenience.

MODE Type of approximation

=1 magnitude only
=2 group delay only
=3 group delay equalization
=4 magnitude and group delay

NZR Number of real zeros

NZU Number of zeros on the unit circle

NZC Number of complex zeros

NPR Number of real poles

NPC Number of complex poles

X Initial parameter vector. Zeros and poles specified in polar coordinates as follows. Real zeros first followed by zeros on the unit circle, complex zeros, real poles, and finally complex poles. Real zeros (poles) are specified as radius1,radius2,...; zeros on the unit circle are specified as angle1,angle2,...; complex zeros (poles) are specified as angle1,radius1,angle2,radius2,..., where the angles are specificed in Hz as phi/(2π)*fsample and only zeros (poles) with positive angles are specified, their complex conjugate pairs being understood. The last parameter(s) is (are):

KO if mode = 1
TAUO if mode =2,3
KO,TAUO if mode =4

NK Number of intervals

NPK(I) Number of points on the interval FIK(I)−FIK(I+1)

WEIGHT(I) Spacing of the points NPK(I) on the Ith interval

= 1 points equispaced
= 2 points spaced on a sine abscissa half cycle
= 3 points spaced on a cosine abscissa, half cycle
= 4 points spaced on a cosine abscissa, full cycle

WFK(I) Relative weighting of the Ith interval for the magnitude

WGK(I) Relative weighting of the Ith interval for the group delay

FIK(I) Starting point of the Ith approximation interval

IPK Number of successive indices of approximation

IP(II) Successive indices of approximation for the magnitude

IQ(II) Successive indices of approximation for the group delay

INV Number of total cycles if no convergence

F1 Lower frequency limit for the frequency response

F2 Upper frequency limit for the frequency response

FSAMPL Sampling frequency

IZ Number of points for the frequency response (IF IZ $\leqslant 0$ no response is calculated)

ALFA Relative weighting of the magnitude
Relative weighting of group delay is 1.−ALFA
This parameter is used only if MODE=4

LONG Logical parameter for type of output

= .TRUE. for LONG output
= .FALSE. for SHORT output

LIMIT Number of iterations per cycle before restarting with steepest descent

EPS Error parameter for FMFP subroutine — typically 10^{-3}

DEB For debugging printout set = .TRUE.

DEBFP For debugging printout from FMFP set = .TRUE.

This completes the description of the input parameters.

IER Used on output to indicate:

= −1 repeated failure of iteration
= 0 convergence
= 1 failure of convergence in the specified number of iterations
= 2 failure of linear search to find a minimum along the direction vector

If IER = 0 and the results are not satisfactory, the program could be restarted with the final argument vector as the initial condition. If INV is greater than 1 when IER = 0, this will happen automatically.

3.4 Dimension Restrictions

The number of parameters in X is restricted to 40. The number of filter sections is restricted to 20. The total number of points over which F is computed is restricted to 201. The total number of points for the frequency response is also restricted to 201. The number of successive indices IP, IQ is restricted to 10. The number of intervals FIK is restricted to 10. To change any of these, modify the appropriate DIMENSION or COMMON statement(s).

4. Examples

Five examples will be given to illustrate how the program works. For each example we give the input sequence, the functions FS and DGS (or the input coefficients of the filter to be equalized in example 2), followed by the complete output as specified by the input sequence.

Example 1

This example illustrates the synthesis of a recursive low pass filter of order 4. The magnitude of the filter is to approximate 1 in the range 0 to 3 and 0 in the range 3.4 to 4. There are thus three intervals, namely from 0 to 3, from 3 to 3.4 and from 3.4 to 4, with the second one being a "don't care" interval -- hence NPK(2) = 0. The initial argument vector is given as two zeros on the unit circle specified by the parameters 3.6 and 3.8, plus 2 complex poles specified by the parameters 1.6, 0.8 and 2.4, 0.8. (Note that only the zeros and poles with positive angles are specified -- the complex conjugate ones are understood.) The program output is specified as LONG. The input parameters, the magnitude specification function FS and the program output are given in Tables 1, 2 and 3 respectively.

Example 2

This example illustrates the group delay equalization of a recursive low pass filter. The coefficients of the filter to be equalized are used as input on unit INP=I1MACH(1). They are reproduced in Table 4, together with the input parameters. LONG output is again specified and is given in Table 5.

Example 3

This example illustrates the approximation of the ideal differentiator FS(FI) = FI on the interval 0 to 1 (with a sampling frequency of 2). Note that now 3 indices IP(II) are specified in the sequence 2, 4, 10 and the program should restart a maximum of 4 times if no convergence is attained. The input parameters for this example are given in Table 6, the magnitude specification function FS is given in Table 7 and the (specified) LONG output is given in Table 8.

Example 4

This example illustrates the approximation of a rising group delay function DGS(FI) = 15.0*FI using an all pass filter of order 8. The input parameters for this example are given in Table 9, the group delay specification function DGS is given in Table 10 and the (LONG) output in Table 11.

Example 5

This example illustrates the approximation of a recursive low pass filter with both specified magnitude and a group delay. The magnitude is to be 1 from 0 to 0.3 and 0 from 0.6 to 1.0 while the group delay is to be 4 from 0 to 0.5. (The sampling frequency is 2.0.) Thus now there are four intervals; 0-0.3, 0.3-0.5, 0.5-0.6 and 0.6-1.0. The second and third intervals are "don't care" intervals for the magnitude with WHK(2) = WHK(3) = 0, while the last two are "don't care" intervals for the group delay with WGK(3) = WGK(4) = 0. The input parameters for this example are given in Table 12, the magnitude and group delay specification functions FS and DGS are given in Table 13, and the (LONG) output in Table 14.

References

1. A. G. Deczky, "Synthesis of Recursive Digital Filters Using the Minimum-p Error Criterion", *IEEE Trans. on Audio and Electroacoustics,* Vol. AU-20, No. 4, pp. 257-263, Oct. 1972.

2. R. Fletcher and M. J. D. Powell, "A rapidly convergent descent method for minimization", *Comput. J.*, Vol. 6, No. 2, pp. 163-168, July 1963.

3. IBM, "System/360 Scientific Subroutine Package Version III", Program No. 360A-CM-03X.

Table 1

```
      1
      0
      2
      0
      0
      2
    3.6
    3.8
    1.6
    0.8
    2.4
    0.8
    1.0
      3
     41
      0
     41
      2
      1
      3
     10.
      0.
      1.
      0.
      0.
      0.
    0.0
    3.0
    3.4
    4.0
      1
      2
      0
      2
    0.0
    4.0
    8.0
     41
    1.0
T
F
F
```

Table 2

```
C
C-----------------------------------------------------------------------
C FUNCTION:  FS
C FUNCTION FOR SPECIFYING THE REQUIRED MAGNITUDE
C-----------------------------------------------------------------------
C
      FUNCTION FS(FI)
      COMMON /CM40/ FIK(10), NPK(10), NK
      COMMON /UNIT/ IND, INP, IOUTD, IOUTP, PI, SPM
C
      DO 60 K=2,NK
        K1 = K - 1
C
        IF (FI.LT.FIK(K1) .OR. FI.GE.FIK(K)) GO TO 60
C
        GO TO (10, 20, 30, 40, 50), K1
 10     FS = 1.0
        GO TO 60
 20     FS = 0.0
        GO TO 60
 30     FS = 0.0
        GO TO 60
 40     FS = 0.0
        GO TO 60
 50     FS = 0.0
 60   CONTINUE
      RETURN
      END
```

Table 3

```
KO = 0.92072797E-01

F= 0.20112E 01      IER= 0      NO OF ITERATIONS=  0

NO OF FUNCT EVALUATIONS=   2


INITIAL ARG. VECTOR AND GRADIENT

         X( 1)= 0.90000000E 00         G( 1)=-0.48376090E 00
         X( 2)= 0.95000001E 00         G( 2)=-0.21596835E 00
         X( 3)= 0.40000000E 00         G( 3)=-0.14348531E 01
         X( 4)= 0.80000000E 00         G( 4)= 0.44140776E 01
         X( 5)= 0.60000000E 00         G( 5)=-0.41605789E 01
         X( 6)= 0.80000000E 00         G( 6)=-0.15939708E 01
         X( 7)= 0.92072797E-01         G( 7)= 0.85156908E-06
OLDF =  0.2011E 01          ITERATION NO.  1
OLDF =  0.1843E 01          ITERATION NO.  2
OLDF =  0.9440E 00          ITERATION NO.  3
OLDF =  0.9437E 00          ITERATION NO.  4
OLDF =  0.2332E 00          ITERATION NO.  5
OLDF =  0.1010E 00          ITERATION NO.  6
OLDF =  0.7646E-01          ITERATION NO.  7
OLDF =  0.4707E-01          ITERATION NO.  8
OLDF =  0.3230E-01          ITERATION NO.  9
OLDF =  0.2245E-01          ITERATION NO. 10
OLDF =  0.2088E-01          ITERATION NO. 11
OLDF =  0.1844E-01          ITERATION NO. 12
OLDF =  0.1840E-01          ITERATION NO. 13
OLDF =  0.1840E-01          ITERATION NO. 14
OLDF =  0.1299E-01          ITERATION NO. 15
OLDF =  0.1226E-01          ITERATION NO. 16
OLDF =  0.8329E-02          ITERATION NO. 17
OLDF =  0.3661E-02          ITERATION NO. 18
OLDF =  0.3357E-02          ITERATION NO. 19
OLDF =  0.3193E-02          ITERATION NO. 20
OLDF =  0.2968E-02          ITERATION NO. 21
OLDF =  0.2671E-02          ITERATION NO. 22
OLDF =  0.2317E-02          ITERATION NO. 23
OLDF =  0.2179E-02          ITERATION NO. 24
OLDF =  0.1915E-02          ITERATION NO. 25
OLDF =  0.1560E-02          ITERATION NO. 26
OLDF =  0.1333E-02          ITERATION NO. 27
OLDF =  0.1188E-02          ITERATION NO. 28

F= 0.11195E-02      IER= 1      NO OF ITERATIONS= 28

NO OF FUNCT EVALUATIONS= 135

         X( 1)= 0.94482371E 00         G( 1)= 0.86673560E-02
         X( 2)= 0.97314056E 00         G( 2)= 0.36474567E-02
         X( 3)= 0.72813510E 00         G( 3)= 0.85005024E-02
         X( 4)= 0.43455667E 00         G( 4)= 0.71129587E-02
         X( 5)= 0.76811282E 00         G( 5)=-0.50341815E-02
         X( 6)= 0.87526403E 00         G( 6)= 0.14805895E-01
         X( 7)= 0.33795486E 00         G( 7)=-0.11955116E-01
OLDF =  0.1120E-02          ITERATION NO.  1
OLDF =  0.1116E-02          ITERATION NO.  2
OLDF =  0.1114E-02          ITERATION NO.  3
OLDF =  0.1040E-02          ITERATION NO.  4
OLDF =  0.1023E-02          ITERATION NO.  5
OLDF =  0.9362E-03          ITERATION NO.  6
OLDF =  0.8319E-03          ITERATION NO.  7
OLDF =  0.7426E-03          ITERATION NO.  8
OLDF =  0.6654E-03          ITERATION NO.  9
OLDF =  0.6060E-03          ITERATION NO. 10
OLDF =  0.4856E-03          ITERATION NO. 11
OLDF =  0.3706E-03          ITERATION NO. 12
OLDF =  0.3171E-03          ITERATION NO. 13
OLDF =  0.2556E-03          ITERATION NO. 14
OLDF =  0.2345E-03          ITERATION NO. 15
OLDF =  0.2041E-03          ITERATION NO. 16
OLDF =  0.1728E-03          ITERATION NO. 17
OLDF =  0.1621E-03          ITERATION NO. 18
```

Table 3
(Continued)

```
OLDF =  0.1555E-03          ITERATION NO. 19
OLDF =  0.1466E-03          ITERATION NO. 20
OLDF =  0.1423E-03          ITERATION NO. 21
OLDF =  0.1392E-03          ITERATION NO. 22
OLDF =  0.1375E-03          ITERATION NO. 23
OLDF =  0.1369E-03          ITERATION NO. 24
OLDF =  0.1367E-03          ITERATION NO. 25
OLDF =  0.1366E-03          ITERATION NO. 26
OLDF =  0.1366E-03          ITERATION NO. 27
OLDF =  0.1366E-03          ITERATION NO. 28

F= 0.13661E-03     IER= 0     NO OF ITERATIONS= 28

NO OF FUNCT EVALUATIONS= 246

         X( 1)= 0.85934433E 00          G( 1)=-0.12731826E-04
         X( 2)= 0.93765157E 00          G( 2)= 0.46204799E-04
         X( 3)= 0.80925703E 00          G( 3)= 0.62992023E-04
         X( 4)= 0.51230380E 00          G( 4)= 0.12864991E-04
         X( 5)= 0.78067586E 00          G( 5)=-0.11647338E-04
         X( 6)= 0.88864766E 00          G( 6)=-0.43512307E-04
         X( 7)= 0.44070618E 00          G( 7)=-0.42195573E-04


SYNTHESIS OF DIGITAL FILTER WITH SPEC. MAGNITUDE AND/OR GROUP DELAY

THE ERROR CRITERION USED IS MINIMUM 2 FOR THE MAGNITUDE AND

                            MINIMUM 0 FOR THE GROUP DELAY

F= 0.13661E-03     IER= 0     NO OF ITERATIONS= 56

NO OF FUNCT EVALUATIONS= 246


COMPUTED ARG. VECTOR AND GRADIENT

         X( 1)= 0.85934433E 00          G( 1)=-0.12731826E-04
         X( 2)= 0.93765157E 00          G( 2)= 0.46204799E-04
         X( 3)= 0.80925703E 00          G( 3)= 0.62992023E-04
         X( 4)= 0.51230380E 00          G( 4)= 0.12864991E-04
         X( 5)= 0.78067586E 00          G( 5)=-0.11647338E-04
         X( 6)= 0.88864766E 00          G( 6)=-0.43512307E-04
         X( 7)= 0.44070618E 00          G( 7)=-0.42195573E-04
                          Z PLANE ZEROS

           ANGLE/2PI*FS                     RADIUS

          0.34373773E 01            0.10000000E 01
          0.37506063E 01            0.10000000E 01
                          Z PLANE POLES

           ANGLE/2PI*FS                     RADIUS

          0.32370281E 01            0.51230380E 00
          0.31227034E 01            0.88864766E 00

CONST= 0.44070618E 00


                          NUMERATOR COEFFICIENTS


            A0                      A1                      A2

      0.10000000E 01          0.18078962E 01          0.10000000E 01
      0.10000000E 01          0.19617561E 01          0.10000000E 01


                          DENOMINATOR COEFFICIENTS
```

Table 3
(Continued)

B0	B1	B2
0.10000000E 01	0.84608650E 00	0.26245519E 00
0.10000000E 01	0.13718319E 01	0.78969466E 00

FREQUENCY RESPONSE

FREQUENCY	MAGNITUDE	LOSS IN DB	PHASE	GROUP DELAY
0.	0.99734E 00	0.23141E-01	0.	0.41631E 00
0.1000E 00	0.99736E 00	0.22927E-01	-0.32716E-01	0.41705E 00
0.2000E 00	0.99744E 00	0.22285E-01	-0.65548E-01	0.41927E 00
0.3000E 00	0.99756E 00	0.21217E-01	-0.98614E-01	0.42301E 00
0.4000E 00	0.99773E 00	0.19725E-01	-0.13204E 00	0.42832E 00
0.5000E 00	0.99795E 00	0.17812E-01	-0.16594E 00	0.43528E 00
0.6000E 00	0.99822E 00	0.15486E-01	-0.20046E 00	0.44399E 00
0.7000E 00	0.99853E 00	0.12756E-01	-0.23573E 00	0.45460E 00
0.8000E 00	0.99889E 00	0.96374E-02	-0.27192E 00	0.46726E 00
0.9000E 00	0.99929E 00	0.61515E-02	-0.30919E 00	0.48219E 00
0.1000E 01	0.99973E 00	0.23305E-02	-0.34772E 00	0.49964E 00
0.1100E 01	0.10002E 01	-0.17812E-02	-0.38774E 00	0.51992E 00
0.1200E 01	0.10007E 01	-0.61210E-02	-0.42948E 00	0.54342E 00
0.1300E 01	0.10012E 01	-0.10604E-01	-0.47320E 00	0.57060E 00
0.1400E 01	0.10017E 01	-0.15111E-01	-0.51922E 00	0.60204E 00
0.1500E 01	0.10022E 01	-0.19485E-01	-0.56790E 00	0.63843E 00
0.1600E 01	0.10027E 01	-0.23515E-01	-0.61966E 00	0.68065E 00
0.1700E 01	0.10031E 01	-0.26921E-01	-0.67499E 00	0.72979E 00
0.1800E 01	0.10034E 01	-0.29345E-01	-0.73451E 00	0.78722E 00
0.1900E 01	0.10035E 01	-0.30329E-01	-0.79891E 00	0.85469E 00
0.2000E 01	0.10034E 01	-0.29314E-01	-0.86908E 00	0.93449E 00
0.2100E 01	0.10030E 01	-0.25655E-01	-0.94610E 00	0.10296E 01
0.2200E 01	0.10022E 01	-0.18691E-01	-0.10313E 01	0.11443E 01
0.2300E 01	0.10009E 01	-0.79214E-02	-0.11265E 01	0.12844E 01
0.2400E 01	0.99924E 00	0.65813E-02	-0.12340E 01	0.14591E 01
0.2500E 01	0.99732E 00	0.23335E-01	-0.13570E 01	0.16834E 01
0.2600E 01	0.99565E 00	0.37861E-01	-0.15004E 01	0.19846E 01
0.2700E 01	0.99538E 00	0.40248E-01	0.14694E 01	0.24188E 01
0.2800E 01	0.99836E 00	0.14274E-01	0.12546E 01	0.31137E 01
0.2900E 01	0.10048E 01	-0.41574E-01	0.96552E 00	0.43889E 01
0.3000E 01	0.99246E 00	0.65750E-01	0.53160E 00	0.69448E 01
0.3100E 01	0.82696E 00	0.16503E 01	-0.15353E 00	0.10246E 02
0.3200E 01	0.42377E 00	0.74575E 01	-0.93579E 00	0.87367E 01
0.3300E 01	0.13648E 00	0.17299E 02	-0.14902E 01	0.56188E 01
0.3400E 01	0.19414E-01	0.34238E 02	0.12814E 01	0.40069E 01
0.3500E 01	0.16017E-01	0.35909E 02	0.10007E 01	0.32261E 01
0.3600E 01	0.17455E-01	0.35162E 02	0.76557E 00	0.28006E 01
0.3700E 01	0.65969E-02	0.43613E 02	0.55644E 00	0.25453E 01
0.3800E 01	0.60683E-02	0.44339E 02	0.36320E 00	0.23889E 01
0.3900E 01	0.15293E-01	0.36310E 02	0.17938E 00	0.23024E 01
0.4000E 01	0.18610E-01	0.34605E 02	0.45134E-08	0.22747E 01

```
 COEFFICIENTS OF DIGITAL FILTER WITH SPEC. AMPLITUDE AND/OR GROUP DELAY
N= 2
 0.4407062E 00    2
 0.1000000E 01 0.1807896E 01 0.1000000E 01 0.8460865E 00 0.2624552E 00
 0.1000000E 01 0.1961756E 01 0.1000000E 01 0.1371832E 01 0.7896947E 00
```

Table 4

```
     3
     0
     0
     0
     0
     5
0.5
0.8
1.0
0.8
1.5
0.8
2.0
0.8
2.5
0.8
0.0
     1
    41
     2
0.0
1.0
0.0
3.1
     1
     0
     2
     2
0.0
4.0
8.0
    41
0.0
T
F
F
 0.4054801E+00   2
 0.1000000E+01 0.1808052E+01 0.1000000E+01 0.7221175E+00 0.2360188E+00
 0.1000000E+01 0.1963544E+01 0.1000000E+01 0.1344155E+01 0.8002729E+00
```

Table 5

```
TO = 0.23540250E 02

F= 0.46328E 01      IER= 0      NO OF ITERATIONS=  0

NO OF FUNCT EVALUATIONS=   2


INITIAL ARG. VECTOR AND GRADIENT

         X( 1)= 0.12500000E 00          G( 1)= 0.65278458E 02
         X( 2)= 0.80000000E 00          G( 2)= 0.41243368E 01
         X( 3)= 0.25000000E 00          G( 3)= 0.22710607E 02
         X( 4)= 0.80000000E 00          G( 4)= 0.12409674E 02
         X( 5)= 0.37500000E 00          G( 5)= 0.55761530E 01
         X( 6)= 0.80000000E 00          G( 6)= 0.13062503E 02
         X( 7)= 0.50000000E 00          G( 7)=-0.14036452E 02
         X( 8)= 0.80000000E 00          G( 8)= 0.15493840E 02
         X( 9)= 0.62500000E 00          G( 9)=-0.68205060E 02
         X(10)= 0.80000000E 00          G(10)= 0.45851613E 01
         X(11)= 0.23540250E 02          G(11)= 0.10770746E-05
OLDF =  0.4633E 01         ITERATION NO.  1
OLDF =  0.1453E 01         ITERATION NO.  2
OLDF =  0.1027E 01         ITERATION NO.  3
OLDF =  0.9522E 00         ITERATION NO.  4
OLDF =  0.9090E 00         ITERATION NO.  5
OLDF =  0.8530E 00         ITERATION NO.  6
OLDF =  0.8498E 00         ITERATION NO.  7
OLDF =  0.8458E 00         ITERATION NO.  8
OLDF =  0.8437E 00         ITERATION NO.  9
OLDF =  0.8253E 00         ITERATION NO. 10
OLDF =  0.7907E 00         ITERATION NO. 11
```

Table 5
(Continued)

```
OLDF =  0.7892E 00            ITERATION NO. 12
OLDF =  0.7771E 00            ITERATION NO. 13
OLDF =  0.7758E 00            ITERATION NO. 14
OLDF =  0.7757E 00            ITERATION NO. 15

F= 0.77572E 00      IER= 0      NO OF ITERATIONS= 15

NO OF FUNCT EVALUATIONS=  60

         X( 1)= 0.69931745E-01          G( 1)=-0.10378439E-02
         X( 2)= 0.74191473E 00          G( 2)=-0.57872886E-03
         X( 3)= 0.21490648E 00          G( 3)= 0.18699270E-02
         X( 4)= 0.74735180E 00          G( 4)=-0.32778834E-03
         X( 5)= 0.35960912E 00          G( 5)=-0.15411069E-02
         X( 6)= 0.75370811E 00          G( 6)= 0.70288699E-03
         X( 7)= 0.50457408E 00          G( 7)=-0.82738208E-03
         X( 8)= 0.76478518E 00          G( 8)= 0.10225077E-02
         X( 9)= 0.65060432E 00          G( 9)= 0.92509451E-03
         X(10)= 0.78936825E 00          G(10)=-0.72606624E-03
         X(11)= 0.23742240E 02          G(11)= 0.63875690E-04


SYNTHESIS OF DIGITAL FILTER WITH SPEC. MAGNITUDE AND/OR GROUP DELAY

THE ERROR CRITERION USED IS MINIMUM 0 FOR THE MAGNITUDE AND

                            MINIMUM 2 FOR THE GROUP DELAY

F= 0.77572E 00      IER= 0      NO OF ITERATIONS= 15

NO OF FUNCT EVALUATIONS=  60


COMPUTED ARG. VECTOR AND GRADIENT

         X( 1)= 0.69931745E-01          G( 1)=-0.10378439E-02
         X( 2)= 0.74191473E 00          G( 2)=-0.57872886E-03
         X( 3)= 0.21490648E 00          G( 3)= 0.18699270E-02
         X( 4)= 0.74735180E 00          G( 4)=-0.32778834E-03
         X( 5)= 0.35960912E 00          G( 5)=-0.15411069E-02
         X( 6)= 0.75370811E 00          G( 6)= 0.70288699E-03
         X( 7)= 0.50457408E 00          G( 7)=-0.82738208E-03
         X( 8)= 0.76478518E 00          G( 8)= 0.10225077E-02
         X( 9)= 0.65060432E 00          G( 9)= 0.92509451E-03
         X(10)= 0.78936825E 00          G(10)=-0.72606624E-03
         X(11)= 0.23742240E 02          G(11)= 0.63875690E-04

                        Z PLANE ZEROS

           ANGLE/2PI*FS                 RADIUS

          0.27972698E 00           0.13478638E 01
          0.85962593E 00           0.13380579E 01
          0.14384365E 01           0.13267736E 01
          0.20182963E 01           0.13075567E 01
          0.26024173E 01           0.12668359E 01

                        Z PLANE POLES

           ANGLE/2PI*FS                 RADIUS

          0.27972698E 00           0.74191473E 00
          0.85962593E 00           0.74735180E 00
          0.14384365E 01           0.75370811E 00
          0.20182963E 01           0.76478518E 00
          0.26024173E 01           0.78936825E 00

CONST= 0.25809068E-01


                        NUMERATOR COEFFICIENTS
```

Table 5
(Continued)

A0	A1	A2
0.10000000E 01	-0.26309318E 01	0.18167368E 01
0.10000000E 01	-0.20890137E 01	0.17903990E 01
0.10000000E 01	-0.11327730E 01	0.17603281E 01
0.10000000E 01	0.37577558E-01	0.17097046E 01
0.10000000E 01	0.11545468E 01	0.16048731E 01
0.10000000E 01	0.18080520E 01	0.10000000E 01
0.10000000E 01	0.19635440E 01	0.10000000E 01

DENOMINATOR COEFFICIENTS

B0	B1	B2
0.10000000E 01	-0.14481634E 01	0.55043747E 00
0.10000000E 01	-0.11667867E 01	0.55853470E 00
0.10000000E 01	-0.64350106E 00	0.56807592E 00
0.10000000E 01	0.21978977E-01	0.58489637E 00
0.10000000E 01	0.71940067E 00	0.62310224E 00
0.10000000E 01	0.72211750E 00	0.23601880E 00
0.10000000E 01	0.13441550E 01	0.80027290E 00

FREQUENCY RESPONSE

FREQUENCY	MAGNITUDE	LOSS IN DB	PHASE	GROUP DELAY
0.	0.99397E 00	0.52569E-01	0.	0.13272E 02
0.1000E 00	0.99402E 00	0.52095E-01	-0.10481E 01	0.13483E 02
0.2000E 00	0.99418E 00	0.50674E-01	0.10184E 01	0.13887E 02
0.3000E 00	0.99445E 00	0.48308E-01	-0.79897E-01	0.13997E 02
0.4000E 00	0.99483E 00	0.45000E-01	-0.11673E 01	0.13634E 02
0.5000E 00	0.99532E 00	0.40758E-01	0.92243E 00	0.13182E 02
0.6000E 00	0.99591E 00	0.35593E-01	-0.10632E 00	0.13098E 02
0.7000E 00	0.99661E 00	0.29525E-01	-0.11475E 01	0.13474E 02
0.8000E 00	0.99740E 00	0.22582E-01	0.91600E 00	0.13948E 02
0.9000E 00	0.99830E 00	0.14809E-01	-0.18501E 00	0.13989E 02
0.1000E 01	0.99928E 00	0.62713E-02	-0.12679E 01	0.13537E 02
0.1100E 01	0.10003E 01	-0.29378E-02	0.82968E 00	0.13100E 02
0.1200E 01	0.10015E 01	-0.12684E-01	-0.19654E 00	0.13126E 02
0.1300E 01	0.10026E 01	-0.22783E-01	-0.12446E 01	0.13612E 02
0.1400E 01	0.10038E 01	-0.32979E-01	0.80726E 00	0.14078E 02
0.1500E 01	0.10050E 01	-0.42924E-01	-0.29869E 00	0.13972E 02
0.1600E 01	0.10060E 01	-0.52152E-01	-0.13748E 01	0.13401E 02
0.1700E 01	0.10069E 01	-0.60041E-01	0.73341E 00	0.12992E 02
0.1800E 01	0.10076E 01	-0.65785E-01	-0.28978E 00	0.13169E 02
0.1900E 01	0.10079E 01	-0.68356E-01	-0.13479E 01	0.13811E 02
0.2000E 01	0.10077E 01	-0.66493E-01	0.68727E 00	0.14265E 02
0.2100E 01	0.10068E 01	-0.58735E-01	-0.42571E 00	0.13945E 02
0.2200E 01	0.10050E 01	-0.43571E-01	-0.14914E 01	0.13190E 02
0.2300E 01	0.10023E 01	-0.19834E-01	0.63383E 00	0.12807E 02
0.2400E 01	0.99857E 00	0.12424E-01	-0.38315E 00	0.13227E 02
0.2500E 01	0.99424E 00	0.50145E-01	-0.14591E 01	0.14197E 02
0.2600E 01	0.99038E 00	0.83919E-01	0.54186E 00	0.14653E 02
0.2700E 01	0.98939E 00	0.92664E-01	-0.58274E 00	0.13789E 02
0.2800E 01	0.99536E 00	0.40426E-01	0.15298E 01	0.12469E 02
0.2900E 01	0.10098E 01	-0.84676E-01	0.57264E 00	0.12194E 02
0.3000E 01	0.98849E 00	0.10056E 00	-0.44639E 00	0.14160E 02
0.3100E 01	0.72063E 00	0.28458E 01	0.15006E 01	0.15321E 02
0.3200E 01	0.31031E 00	0.10164E 02	0.45139E 00	0.11112E 02
0.3300E 01	0.95332E-01	0.20415E 02	-0.28369E 00	0.79653E 01
0.3400E 01	0.13550E-01	0.37361E 02	-0.84098E 00	0.63978E 01
0.3500E 01	0.11106E-01	0.39089E 02	-0.13067E 01	0.55386E 01
0.3600E 01	0.12283E-01	0.38214E 02	0.14220E 01	0.50138E 01
0.3700E 01	0.49781E-02	0.46059E 02	0.10425E 01	0.46763E 01
0.3800E 01	0.35983E-02	0.48878E 02	0.68428E 00	0.44624E 01
0.3900E 01	0.98545E-02	0.40127E 02	0.33906E 00	0.43428E 01
0.4000E 01	0.12105E-01	0.38341E 02	0.85404E-08	0.43042E 01

Table 5
(Continued)

```
 COEFFICIENTS OF DIGITAL FILTER WITH SPEC. AMPLITUDE AND/OR GROUP DELAY
N= 7
 0.2580907E-01    7
 0.1000000E 01-0.2630932E 01 0.1816737E 01-0.1448163E 01 0.5504375E 00
 0.1000000E 01-0.2089014E 01 0.1790399E 01-0.1166787E 01 0.5585347E 00
 0.1000000E 01-0.1132773E 01 0.1760328E 01-0.6435011E 00 0.5680759E 00
 0.1000000E 01 0.3757756E-01 0.1709705E 01 0.2197898E-01 0.5848964E 00
 0.1000000E 01 0.1154547E 01 0.1604873E 01 0.7194007E 00 0.6231022E 00
 0.1000000E 01 0.1808052E 01 0.1000000E 01 0.7221175E 00 0.2360188E 00
 0.1000000E 01 0.1963544E 01 0.1000000E 01 0.1344155E 01 0.8002729E 00
```

Table 6

```
   1
   2
   0
   0
   2
   0
1.0
0.0
0.4
-0.4
1.0
   1
  51
   2
1.0
0.0
0.0
1.0
   3
   2
   4
  10
   0
   0
   0
   4
0.0
1.0
2.0
  51
1.0
T
F
F
```

Table 7

```
C
C-----------------------------------------------------------------------
C FUNCTION:  FS
C FUNCTION FOR SPECIFYING THE REQUIRED MAGNITUDE
C-----------------------------------------------------------------------
C
      FUNCTION FS(FI)
      COMMON /CM40/ FIK(10), NPK(10), NK
      COMMON /UNIT/ IND, INP, IOUTD, IOUTP, PI, SPM
C
      DO 60 K=2,NK
        K1 = K - 1
C
        IF (FI.LT.FIK(K1) .OR. FI.GE.FIK(K)) GO TO 60
C
        GO TO (10, 20, 30, 40, 50), K1
  10    FS = FI
        GO TO 60
  20    FS = 0.0
        GO TO 60
```

Table 7
(Continued)

```
30      FS = 0.0
        GO TO 60
40      FS = 0.0
        GO TO 60
50      FS = 0.0
60    CONTINUE
      RETURN
      END
```

Table 8

```
K0 = 0.40636079E 00

F= 0.73584E-03      IER= 0      NO OF ITERATIONS=  0

NO OF FUNCT EVALUATIONS=   2


INITIAL ARG. VECTOR AND GRADIENT

         X( 1)= 0.10000000E 01           G( 1)=-0.89594565E-08
         X( 2)= 0.                       G( 2)=-0.83907518E-02
         X( 3)= 0.40000000E 00           G( 3)= 0.71825904E-02
         X( 4)=-0.40000000E 00           G( 4)= 0.93726298E-02
         X( 5)= 0.40636079E 00           G( 5)=-0.44460990E-07
OLDF =  0.7358E-03           ITERATION NO.  1
OLDF =  0.6962E-03           ITERATION NO.  2
OLDF =  0.5559E-03           ITERATION NO.  3
OLDF =  0.3074E-03           ITERATION NO.  4
OLDF =  0.2654E-03           ITERATION NO.  5
OLDF =  0.2616E-03           ITERATION NO.  6
OLDF =  0.2614E-03           ITERATION NO.  7
OLDF =  0.2613E-03           ITERATION NO.  8
OLDF =  0.2612E-03           ITERATION NO.  9
OLDF =  0.2276E-03           ITERATION NO. 10
OLDF =  0.2251E-03           ITERATION NO. 11
OLDF =  0.2057E-03           ITERATION NO. 12
OLDF =  0.2056E-03           ITERATION NO. 13
OLDF =  0.2029E-03           ITERATION NO. 14
OLDF =  0.1928E-03           ITERATION NO. 15
OLDF =  0.1618E-03           ITERATION NO. 16
OLDF =  0.1610E-03           ITERATION NO. 17
OLDF =  0.1605E-03           ITERATION NO. 18
OLDF =  0.1341E-03           ITERATION NO. 19
OLDF =  0.1180E-03           ITERATION NO. 20

F= 0.99983E-04      IER= 1      NO OF ITERATIONS= 20

NO OF FUNCT EVALUATIONS=  87

         X( 1)= 0.93417000E 00           G( 1)=-0.11521710E-02
         X( 2)=-0.37288283E 00           G( 2)=-0.22680329E-02
         X( 3)=-0.32908125E-01           G( 3)= 0.16926230E-02
         X( 4)=-0.50697450E 00           G( 4)= 0.32463408E-02
         X( 5)= 0.38296990E 00           G( 5)=-0.35211345E-02
OLDF =  0.9998E-04           ITERATION NO.  1
OLDF =  0.9836E-04           ITERATION NO.  2
OLDF =  0.9728E-04           ITERATION NO.  3
OLDF =  0.7447E-04           ITERATION NO.  4
OLDF =  0.7266E-04           ITERATION NO.  5
OLDF =  0.7207E-04           ITERATION NO.  6
OLDF =  0.6405E-04           ITERATION NO.  7
OLDF =  0.6369E-04           ITERATION NO.  8
OLDF =  0.6308E-04           ITERATION NO.  9
OLDF =  0.5820E-04           ITERATION NO. 10
OLDF =  0.3910E-04           ITERATION NO. 11
OLDF =  0.2897E-04           ITERATION NO. 12
OLDF =  0.2036E-04           ITERATION NO. 13
OLDF =  0.1656E-04           ITERATION NO. 14
OLDF =  0.1597E-04           ITERATION NO. 15
OLDF =  0.1592E-04           ITERATION NO. 16
OLDF =  0.1591E-04           ITERATION NO. 17
```

Table 8
(Continued)

```
 F= 0.15910E-04       IER= 0       NO OF ITERATIONS= 17

 NO OF FUNCT EVALUATIONS= 151

          X( 1)= 0.10000424E 01          G( 1)= 0.30941774E-06
          X( 2)=-0.67870385E 00          G( 2)= 0.67787406E-05
          X( 3)=-0.14765522E 00          G( 3)=-0.22261859E-05
          X( 4)=-0.72136589E 00          G( 4)=-0.75488852E-05
          X( 5)= 0.36585334E 00          G( 5)= 0.96522615E-06
 OLDF =  0.6192E-09          ITERATION NO.  1
 OLDF =  0.6140E-09          ITERATION NO.  2
 OLDF =  0.6140E-09          ITERATION NO.  3

F= 0.61399E-09       IER=-1       NO OF ITERATIONS=  3

NO OF FUNCT EVALUATIONS= 159

         X( 1)= 0.10000470E 01          G( 1)= 0.39802203E-08
         X( 2)=-0.67867164E 00          G( 2)=-0.21001670E-08
         X( 3)=-0.14766660E 00          G( 3)=-0.58512773E-08
         X( 4)=-0.72140586E 00          G( 4)= 0.14038922E-07
         X( 5)= 0.36587823E 00          G( 5)= 0.21759798E-07
OLDF =  0.6140E-09          ITERATION NO.  1
OLDF =  0.6140E-09          ITERATION NO.  2
OLDF =  0.6140E-09          ITERATION NO.  3
OLDF =  0.6140E-09          ITERATION NO.  4

F= 0.61399E-09       IER=-1       NO OF ITERATIONS=  4

NO OF FUNCT EVALUATIONS= 169

         X( 1)= 0.10000470E 01          G( 1)= 0.39722658E-08
         X( 2)=-0.67867164E 00          G( 2)=-0.21268438E-08
         X( 3)=-0.14766659E 00          G( 3)=-0.58366352E-08
         X( 4)=-0.72140592E 00          G( 4)= 0.14067840E-07
         X( 5)= 0.36587815E 00          G( 5)= 0.21716322E-07
OLDF =  0.6140E-09          ITERATION NO.  1
OLDF =  0.6140E-09          ITERATION NO.  2
OLDF =  0.6140E-09          ITERATION NO.  3
OLDF =  0.6140E-09          ITERATION NO.  4

F= 0.61399E-09       IER=-1       NO OF ITERATIONS=  4

NO OF FUNCT EVALUATIONS= 180

         X( 1)= 0.10000470E 01          G( 1)= 0.39662328E-08
         X( 2)=-0.67867164E 00          G( 2)=-0.21494927E-08
         X( 3)=-0.14766657E 00          G( 3)=-0.58250289E-08
         X( 4)=-0.72140595E 00          G( 4)= 0.14092620E-07
         X( 5)= 0.36587810E 00          G( 5)= 0.21683346E-07


SYNTHESIS OF DIGITAL FILTER WITH SPEC. MAGNITUDE AND/OR GROUP DELAY

THE ERROR CRITERION USED IS MINIMUM 4 FOR THE MAGNITUDE AND

                            MINIMUM 0 FOR THE GROUP DELAY

F= 0.61399E-09       IER=-1       NO OF ITERATIONS= 48

NO OF FUNCT EVALUATIONS= 180


COMPUTED ARG. VECTOR AND GRADIENT

         X( 1)= 0.10000470E 01          G( 1)= 0.39662328E-08
         X( 2)=-0.67867164E 00          G( 2)=-0.21494927E-08
         X( 3)=-0.14766657E 00          G( 3)=-0.58250289E-08
         X( 4)=-0.72140595E 00          G( 4)= 0.14092620E-07
         X( 5)= 0.36587810E 00          G( 5)= 0.21683346E-07

                         Z PLANE ZEROS
```

Table 8
(Continued)

ANGLE/2PI*FS	RADIUS
0.	0.10000470E 01
0.	-0.67867164E 00

Z PLANE POLES

ANGLE/2PI*FS	RADIUS
0.	-0.14766657E 00
0.	-0.72140595E 00

CONST= 0.36587810E 00

NUMERATOR COEFFICIENTS

A0	A1	A2
0.10000000E 01	-0.10000470E 01	0.
0.10000000E 01	0.67867164E 00	0.

DENOMINATOR COEFFICIENTS

B0	B1	B2
0.10000000E 01	0.14766657E 00	0.
0.10000000E 01	0.72140595E 00	0.

FREQUENCY RESPONSE

FREQUENCY	MAGNITUDE	LOSS IN DB	PHASE	GROUP DELAY
0.	0.14602E-04	0.96712E 02	0.	0.85654E 00
0.2000E-01	0.19535E-01	0.34184E 02	0.15491E 01	0.36860E 00
0.4000E-01	0.39078E-01	0.28161E 02	0.15263E 01	0.36013E 00
0.6000E-01	0.58635E-01	0.24637E 02	0.15038E 01	0.35923E 00
0.8000E-01	0.78213E-01	0.22134E 02	0.14812E 01	0.35971E 00
0.1000E 00	0.97820E-01	0.20191E 02	0.14585E 01	0.36081E 00
0.1200E 00	0.11746E 00	0.18602E 02	0.14358E 01	0.36233E 00
0.1400E 00	0.13715E 00	0.17256E 02	0.14130E 01	0.36422E 00
0.1600E 00	0.15688E 00	0.16089E 02	0.13900E 01	0.36645E 00
0.1800E 00	0.17666E 00	0.15057E 02	0.13669E 01	0.36900E 00
0.2000E 00	0.19650E 00	0.14133E 02	0.13437E 01	0.37188E 00
0.2200E 00	0.21641E 00	0.13295E 02	0.13202E 01	0.37508E 00
0.2400E 00	0.23638E 00	0.12528E 02	0.12965E 01	0.37859E 00
0.2600E 00	0.25642E 00	0.11821E 02	0.12726E 01	0.38242E 00
0.2800E 00	0.27653E 00	0.11165E 02	0.12485E 01	0.38657E 00
0.3000E 00	0.29671E 00	0.10553E 02	0.12240E 01	0.39103E 00
0.3200E 00	0.31697E 00	0.99797E 01	0.11993E 01	0.39580E 00
0.3400E 00	0.33729E 00	0.94399E 01	0.11743E 01	0.40088E 00
0.3600E 00	0.35768E 00	0.89300E 01	0.11489E 01	0.40627E 00
0.3800E 00	0.37814E 00	0.84469E 01	0.11232E 01	0.41196E 00
0.4000E 00	0.39865E 00	0.79881E 01	0.10972E 01	0.41794E 00
0.4200E 00	0.41921E 00	0.75513E 01	0.10707E 01	0.42420E 00
0.4400E 00	0.43982E 00	0.71346E 01	0.10438E 01	0.43073E 00
0.4600E 00	0.46044E 00	0.67365E 01	0.10166E 01	0.43751E 00
0.4800E 00	0.48108E 00	0.63556E 01	0.98887E 00	0.44452E 00
0.5000E 00	0.50172E 00	0.59908E 01	0.96071E 00	0.45174E 00
0.5200E 00	0.52233E 00	0.56411E 01	0.93209E 00	0.45914E 00
0.5400E 00	0.54290E 00	0.53056E 01	0.90301E 00	0.46668E 00
0.5600E 00	0.56340E 00	0.49837E 01	0.87345E 00	0.47432E 00
0.5800E 00	0.58381E 00	0.46746E 01	0.84340E 00	0.48202E 00
0.6000E 00	0.60410E 00	0.43779E 01	0.81288E 00	0.48973E 00
0.6200E 00	0.62424E 00	0.40930E 01	0.78186E 00	0.49738E 00
0.6400E 00	0.64421E 00	0.38194E 01	0.75037E 00	0.50493E 00
0.6600E 00	0.66399E 00	0.35568E 01	0.71842E 00	0.51230E 00
0.6800E 00	0.68355E 00	0.33046E 01	0.68600E 00	0.51945E 00

Table 8
(Continued)

```
0.7000E 00      0.70288E 00     0.30623E 01     0.65315E 00     0.52634E 00
0.7200E 00      0.72199E 00     0.28294E 01     0.61987E 00     0.53294E 00
0.7400E 00      0.74089E 00     0.26050E 01     0.58618E 00     0.53932E 00
0.7600E 00      0.75961E 00     0.23882E 01     0.55209E 00     0.54563E 00
0.7800E 00      0.77823E 00     0.21778E 01     0.51761E 00     0.55220E 00
0.8000E 00      0.79687E 00     0.19722E 01     0.48268E 00     0.55967E 00
0.8200E 00      0.81570E 00     0.17694E 01     0.44723E 00     0.56919E 00
0.8400E 00      0.83496E 00     0.15667E 01     0.41107E 00     0.58272E 00
0.8600E 00      0.85494E 00     0.13613E 01     0.37386E 00     0.60349E 00
0.8800E 00      0.87600E 00     0.11499E 01     0.33498E 00     0.63643E 00
0.9000E 00      0.89838E 00     0.93078E 00     0.29348E 00     0.68835E 00
0.9200E 00      0.92202E 00     0.70517E 00     0.24793E 00     0.76656E 00
0.9400E 00      0.94603E 00     0.48194E 00     0.19653E 00     0.87389E 00
0.9600E 00      0.96807E 00     0.28187E 00     0.13773E 00     0.99885E 00
0.9800E 00      0.98422E 00     0.13812E 00     0.71360E-01     0.11068E 01
0.1000E 01      0.99025E 00     0.85114E-01     0.22831E-08     0.11506E 01
COEFFICIENTS OF DIGITAL FILTER WITH SPEC. AMPLITUDE AND/OR GROUP DELAY
N= 2
 0.3658781E 00    2
 0.1000000E 01-0.1000047E 01 0.              0.1476666E 00 0.
 0.1000000E 01 0.6786716E 00 0.              0.7214059E 00 0.
```

Table 9

```
     2
     0
     0
     0
     0
     4
0.28
0.83
0.49
0.86
0.68
0.89
0.78
0.93
0.0
     1
    51
     2
0.0
1.0
0.1
0.9
     2
     0
     0
     2
     4
     2
0.0
1.0
2.0
    51
0.
T
F
F
```

Table 10

```
C
C-----------------------------------------------------------------------
C FUNCTION:  DGS
C FUNCTION FOR SPECIFYING THE REQUIRED GROUP DELAY
C-----------------------------------------------------------------------
C
      FUNCTION DGS(FI)
      COMMON /CM40/ FIK(10), NPK(10), NK
      COMMON /UNIT/ IND, INP, IOUTD, IOUTP, PI, SPM
C
      DO 60 K=2,NK
        K1 = K - 1
C
        IF (FI.LT.FIK(K1) .OR. FI.GE.FIK(K)) GO TO 60
C
        GO TO (10, 20, 30, 40, 50), K1
  10    DGS = 15.0*FI
        GO TO 60
  20    DGS = 4.0
        GO TO 60
  30    DGS = 0.0
        GO TO 60
  40    DGS = 0.0
        GO TO 60
  50    DGS = 0.0
  60  CONTINUE
      RETURN
      END
```

Table 11

```
T0 = 0.79225496E 01

F= 0.50424E 02      IER= 0      NO OF ITERATIONS=   0

NO OF FUNCT EVALUATIONS=   2


INITIAL ARG. VECTOR AND GRADIENT

         X( 1)= 0.28000000E 00          G( 1)=-0.36053095E 01
         X( 2)= 0.83000000E 00          G( 2)= 0.39729306E 02
         X( 3)= 0.49000000E 00          G( 3)=-0.27526696E 02
         X( 4)= 0.86000000E 00          G( 4)= 0.52982008E 02
         X( 5)= 0.68000000E 00          G( 5)= 0.87300517E 02
         X( 6)= 0.89000000E 00          G( 6)= 0.11213860E 03
         X( 7)= 0.78000000E 00          G( 7)=-0.34945496E 03
         X( 8)= 0.93000000E 00          G( 8)= 0.54749140E 03
         X( 9)= 0.79225496E 01          G( 9)= 0.12665987E-06
OLDF =  0.5042E 02          ITERATION NO.  1
OLDF =  0.1331E 02          ITERATION NO.  2
OLDF =  0.6873E 01          ITERATION NO.  3
OLDF =  0.1406E 01          ITERATION NO.  4
OLDF =  0.5441E 00          ITERATION NO.  5
OLDF =  0.2941E 00          ITERATION NO.  6
OLDF =  0.2792E 00          ITERATION NO.  7
OLDF =  0.1421E 00          ITERATION NO.  8
OLDF =  0.8885E-01          ITERATION NO.  9
OLDF =  0.8829E-01          ITERATION NO. 10
OLDF =  0.7818E-01          ITERATION NO. 11
OLDF =  0.6680E-01          ITERATION NO. 12
OLDF =  0.6096E-01          ITERATION NO. 13
OLDF =  0.5107E-01          ITERATION NO. 14
OLDF =  0.4558E-01          ITERATION NO. 15
OLDF =  0.4314E-01          ITERATION NO. 16
OLDF =  0.3764E-01          ITERATION NO. 17
OLDF =  0.3505E-01          ITERATION NO. 18
OLDF =  0.3377E-01          ITERATION NO. 19
OLDF =  0.3310E-01          ITERATION NO. 20
OLDF =  0.3286E-01          ITERATION NO. 21
OLDF =  0.3268E-01          ITERATION NO. 22
OLDF =  0.3263E-01          ITERATION NO. 23
OLDF =  0.3260E-01          ITERATION NO. 24
```

Table 11
(Continued)

```
OLDF =  0.3244E-01          ITERATION NO. 25
OLDF =  0.3244E-01          ITERATION NO. 26
OLDF =  0.3230E-01          ITERATION NO. 27
OLDF =  0.3196E-01          ITERATION NO. 28
OLDF =  0.3077E-01          ITERATION NO. 29
OLDF =  0.3020E-01          ITERATION NO. 30
OLDF =  0.2964E-01          ITERATION NO. 31
OLDF =  0.2942E-01          ITERATION NO. 32
OLDF =  0.2914E-01          ITERATION NO. 33
OLDF =  0.2865E-01          ITERATION NO. 34
OLDF =  0.2803E-01          ITERATION NO. 35
OLDF =  0.2752E-01          ITERATION NO. 36

F= 0.27454E-01      IER= 1      NO OF ITERATIONS= 36

NO OF FUNCT EVALUATIONS= 136

         X( 1)= 0.41772590E 00          G( 1)=-0.30741183E-02
         X( 2)= 0.57610147E 00          G( 2)=-0.12525613E-02
         X( 3)= 0.62142190E 00          G( 3)= 0.49062657E-02
         X( 4)= 0.63898080E 00          G( 4)=-0.27420288E-02
         X( 5)= 0.10983226E 01          G( 5)=-0.21828591E-01
         X( 6)= 0.75217267E 00          G( 6)= 0.47436717E-02
         X( 7)= 0.77566455E 00          G( 7)= 0.84268712E-02
         X( 8)= 0.68576224E 00          G( 8)=-0.63354241E-02
         X( 9)= 0.86191942E 01          G( 9)= 0.17568368E-02
OLDF =  0.2745E-01          ITERATION NO.  1
OLDF =  0.2745E-01          ITERATION NO.  2
OLDF =  0.2745E-01          ITERATION NO.  3
OLDF =  0.2745E-01          ITERATION NO.  4
OLDF =  0.2745E-01          ITERATION NO.  5
OLDF =  0.2745E-01          ITERATION NO.  6
OLDF =  0.2745E-01          ITERATION NO.  7
OLDF =  0.2745E-01          ITERATION NO.  8
OLDF =  0.2745E-01          ITERATION NO.  9
OLDF =  0.2745E-01          ITERATION NO. 10

F= 0.27449E-01      IER= 0      NO OF ITERATIONS= 10

NO OF FUNCT EVALUATIONS= 176

         X( 1)= 0.41735033E 00          G( 1)= 0.23355838E-05
         X( 2)= 0.57608889E 00          G( 2)=-0.63836675E-06
         X( 3)= 0.62094770E 00          G( 3)= 0.41995328E-05
         X( 4)= 0.63927549E 00          G( 4)=-0.48311127E-07
         X( 5)= 0.10985726E 01          G( 5)= 0.11442379E-04
         X( 6)= 0.75270749E 00          G( 6)= 0.16899898E-05
         X( 7)= 0.77532603E 00          G( 7)= 0.43452516E-05
         X( 8)= 0.68637722E 00          G( 8)= 0.31803929E-05
         X( 9)= 0.86208143E 01          G( 9)=-0.64447522E-06
OLDF =  0.5880E-02          ITERATION NO.  1
OLDF =  0.5097E-02          ITERATION NO.  2
OLDF =  0.4842E-02          ITERATION NO.  3
OLDF =  0.4310E-02          ITERATION NO.  4
OLDF =  0.4123E-02          ITERATION NO.  5
OLDF =  0.3709E-02          ITERATION NO.  6
OLDF =  0.3623E-02          ITERATION NO.  7
OLDF =  0.3590E-02          ITERATION NO.  8
OLDF =  0.3527E-02          ITERATION NO.  9
OLDF =  0.3213E-02          ITERATION NO. 10
OLDF =  0.3060E-02          ITERATION NO. 11
OLDF =  0.2971E-02          ITERATION NO. 12
OLDF =  0.2940E-02          ITERATION NO. 13
OLDF =  0.2932E-02          ITERATION NO. 14
OLDF =  0.2931E-02          ITERATION NO. 15

F= 0.29311E-02      IER= 0      NO OF ITERATIONS= 15

NO OF FUNCT EVALUATIONS= 254

         X( 1)= 0.40242405E 00          G( 1)= 0.34703542E-04
         X( 2)= 0.59617415E 00          G( 2)= 0.57860917E-04
         X( 3)= 0.61030855E 00          G( 3)=-0.37734724E-04
```

Table 11
(Continued)

```
X( 4)= 0.66355933E 00          G( 4)= 0.35411865E-04
X( 5)= 0.11052927E 01          G( 5)= 0.78650220E-04
X( 6)= 0.77201111E 00          G( 6)=-0.12829921E-03
X( 7)= 0.76772848E 00          G( 7)=-0.15171362E-03
X( 8)= 0.71009994E 00          G( 8)=-0.34973891E-04
X( 9)= 0.87403825E 01          G( 9)= 0.19682339E-05


SYNTHESIS OF DIGITAL FILTER WITH SPEC. MAGNITUDE AND/OR GROUP DELAY

THE ERROR CRITERION USED IS MINIMUM 0 FOR THE MAGNITUDE AND

                            MINIMUM 4 FOR THE GROUP DELAY

F= 0.29311E-02      IER= 0      NO OF ITERATIONS= 61

NO OF FUNCT EVALUATIONS= 254


COMPUTED ARG. VECTOR AND GRADIENT

X( 1)= 0.40242405E 00          G( 1)= 0.34703542E-04
X( 2)= 0.59617415E 00          G( 2)= 0.57860917E-04
X( 3)= 0.61030855E 00          G( 3)=-0.37734724E-04
X( 4)= 0.66355933E 00          G( 4)= 0.35411865E-04
X( 5)= 0.11052927E 01          G( 5)= 0.78650220E-04
X( 6)= 0.77201111E 00          G( 6)=-0.12829921E-03
X( 7)= 0.76772848E 00          G( 7)=-0.15171362E-03
X( 8)= 0.71009994E 00          G( 8)=-0.34973891E-04
X( 9)= 0.87403825E 01          G( 9)= 0.19682339E-05

                    Z PLANE ZEROS

ANGLE/2PI*FS                   RADIUS

0.40242405E 00             0.16773622E 01
0.61030855E 00             0.15070242E 01
0.89470732E 00             0.12953181E 01
0.76772848E 00             0.14082525E 01

                    Z PLANE POLES

ANGLE/2PI*FS                   RADIUS

0.40242405E 00             0.59617415E 00
0.61030855E 00             0.66355933E 00
0.89470732E 00             0.77201111E 00
0.76772848E 00             0.71009994E 00

CONST= 0.47031828E-01


                    NUMERATOR COEFFICIENTS


      A0                     A1                     A2

0.10000000E 01        -0.10123400E 01        0.28135440E 01
0.10000000E 01         0.10237209E 01        0.22711221E 01
0.10000000E 01         0.24501904E 01        0.16778491E 01
0.10000000E 01         0.20993461E 01        0.19831751E 01


                    DENOMINATOR COEFFICIENTS


      B0                     B1                     B2

0.10000000E 01        -0.35980954E 00        0.35542362E 00
0.10000000E 01         0.45075555E 00        0.44031099E 00
0.10000000E 01         0.14603163E 01        0.59600115E 00
0.10000000E 01         0.10585783E 01        0.50424192E 00
```

Table 11
(Continued)

FREQUENCY RESPONSE

FREQUENCY	MAGNITUDE	LOSS IN DB	PHASE	GROUP DELAY
0.	0.10000E 01	-0.12943E-06	0.	0.25380E 01
0.2000E-01	0.10000E 01	-0.12943E-06	-0.15963E 00	0.25459E 01
0.4000E-01	0.10000E 01	-0.12943E-06	-0.32026E 00	0.25696E 01
0.6000E-01	0.10000E 01	0.12943E-06	-0.48288E 00	0.26098E 01
0.8000E-01	0.10000E 01	-0.12943E-06	-0.64857E 00	0.26673E 01
0.1000E 00	0.10000E 01	0.12943E-06	-0.81845E 00	0.27434E 01
0.1200E 00	0.10000E 01	-0.12943E-06	-0.99375E 00	0.28400E 01
0.1400E 00	0.10000E 01	0.	-0.11758E 01	0.29592E 01
0.1600E 00	0.10000E 01	0.	-0.13661E 01	0.31036E 01
0.1800E 00	0.10000E 01	0.12943E-06	-0.15664E 01	0.32764E 01
0.2000E 00	0.10000E 01	-0.12943E-06	0.13631E 01	0.34810E 01
0.2200E 00	0.10000E 01	0.	0.11370E 01	0.37209E 01
0.2400E 00	0.10000E 01	0.	0.89467E 00	0.39993E 01
0.2600E 00	0.10000E 01	0.19414E-06	0.63358E 00	0.43182E 01
0.2800E 00	0.10000E 01	0.19414E-06	0.35118E 00	0.46772E 01
0.3000E 00	0.10000E 01	0.	0.45080E-01	0.50716E 01
0.3200E 00	0.10000E 01	0.19414E-06	-0.28666E 00	0.54908E 01
0.3400E 00	0.10000E 01	0.12943E-06	-0.64506E 00	0.59166E 01
0.3600E 00	0.10000E 01	-0.12943E-06	-0.10298E 01	0.63246E 01
0.3800E 00	0.10000E 01	0.19414E-06	-0.14389E 01	0.66885E 01
0.4000E 00	0.10000E 01	0.12943E-06	0.12726E 01	0.69887E 01
0.4200E 00	0.10000E 01	0.	0.82590E 00	0.72209E 01
0.4400E 00	0.10000E 01	0.12943E-06	0.36633E 00	0.74012E 01
0.4600E 00	0.10000E 01	0.12943E-06	-0.10375E 00	0.75620E 01
0.4800E 00	0.10000E 01	0.12943E-06	-0.58437E 00	0.77434E 01
0.5000E 00	0.10000E 01	0.19414E-06	-0.10780E 01	0.79817E 01
0.5200E 00	0.10000E 01	0.19414E-06	0.15525E 01	0.82988E 01
0.5400E 00	0.10000E 01	0.38829E-06	0.10190E 01	0.86945E 01
0.5600E 00	0.10000E 01	0.32357E-06	0.45890E 00	0.91398E 01
0.5800E 00	0.10000E 01	0.19414E-06	-0.12938E 00	0.95794E 01
0.6000E 00	0.10000E 01	0.19414E-06	-0.74340E 00	0.99489E 01
0.6200E 00	0.10000E 01	0.32357E-06	-0.13773E 01	0.10209E 02
0.6400E 00	0.10000E 01	0.19414E-06	0.11174E 01	0.10374E 02
0.6600E 00	0.10000E 01	0.12943E-06	0.46125E 00	0.10513E 02
0.6800E 00	0.10000E 01	0.32357E-06	-0.20509E 00	0.10714E 02
0.7000E 00	0.10000E 01	0.19414E-06	-0.88765E 00	0.11035E 02
0.7200E 00	0.10000E 01	0.12943E-06	0.15473E 01	0.11473E 02
0.7400E 00	0.10000E 01	0.12943E-06	0.81139E 00	0.11948E 02
0.7600E 00	0.10000E 01	0.12943E-06	0.47784E-01	0.12334E 02
0.7800E 00	0.10000E 01	0.19414E-06	-0.73502E 00	0.12555E 02
0.8000E 00	0.10000E 01	0.19414E-06	-0.15275E 01	0.12664E 02
0.8200E 00	0.10000E 01	0.12943E-06	0.81401E 00	0.12827E 02
0.8400E 00	0.10000E 01	0.12943E-06	-0.21088E-02	0.13190E 02
0.8600E 00	0.10000E 01	0.12943E-06	-0.84736E 00	0.13728E 02
0.8800E 00	0.10000E 01	-0.12943E-06	0.14170E 01	0.14138E 02
0.9000E 00	0.10000E 01	-0.12943E-06	0.52951E 00	0.13991E 02
0.9200E 00	0.10000E 01	-0.51772E-06	-0.32665E 00	0.13167E 02
0.9400E 00	0.10000E 01	-0.38829E-06	-0.11181E 01	0.12012E 02
0.9600E 00	0.10000E 01	-0.64715E-06	0.13029E 01	0.10971E 02
0.9800E 00	0.10000E 01	-0.38829E-06	0.63709E 00	0.10293E 02
0.1000E 01	0.10000E 01	-0.51772E-06	0.19966E-07	0.10063E 02

```
COEFFICIENTS OF DIGITAL FILTER WITH SPEC. AMPLITUDE AND/OR GROUP DELAY
N= 4
 0.4703183E-01    4
 0.1000000E 01-0.1012340E 01 0.2813544E 01-0.3598095E 00 0.3554236E 00
 0.1000000E 01 0.1023721E 01 0.2271122E 01 0.4507555E 00 0.4403110E 00
 0.1000000E 01 0.2450190E 01 0.1677849E 01 0.1460316E 01 0.5960012E 00
 0.1000000E 01 0.2099346E 01 0.1983175E 01 0.1058578E 01 0.5042419E 00
```

Table 12

```
   4
   1
   2
   1
   0
   3
0.42
0.632
0.851
0.238
0.507
0.101
0.619
0.274
0.629
0.458
0.732
1.0
0.0
   4
  21
  21
   0
  21
   2
   2
   1
   3
1.0
0.0
0.0
1.0
1.0
1.0
0.0
0.0
0.0
0.3
0.5
0.6
1.0
   1
   2
   2
   1
0.0
1.0
2.0
  51
0.9
T
F
F
```

Table 13

```
C
C-----------------------------------------------------------------------
C FUNCTION:  FS
C FUNCTION FOR SPECIFYING THE REQUIRED MAGNITUDE
C-----------------------------------------------------------------------
C
      FUNCTION FS(FI)
      COMMON /CM40/ FIK(10), NPK(10), NK
      COMMON /UNIT/ IND, INP, IOUTD, IOUTP, PI, SPM
C
      DO 60 K=2,NK
        K1 = K - 1
C
        IF (FI.LT.FIK(K1) .OR. FI.GE.FIK(K)) GO TO 60
C
        GO TO (10, 20, 30, 40, 50), K1
  10    FS = 1.0
```

Table 13
(Continued)

```
        GO TO 60
   20   FS = 0.0
        GO TO 60
   30   FS = 0.0
        GO TO 60
   40   FS = 0.0
        GO TO 60
   50   FS = 0.0
   60 CONTINUE
      RETURN
      END
C
C-----------------------------------------------------------------------
C FUNCTION:  DGS
C FUNCTION FOR SPECIFYING THE REQUIRED GROUP DELAY
C-----------------------------------------------------------------------
C
      FUNCTION DGS(FI)
      COMMON /CM40/ FIK(10), NPK(10), NK
      COMMON /UNIT/ IND, INP, IOUTD, IOUTP, PI, SPM
C
      DO 60 K=2,NK
        K1 = K - 1
C
        IF (FI.LT.FIK(K1) .OR. FI.GE.FIK(K)) GO TO 60
C
        GO TO (10, 20, 30, 40, 50), K1
   10   DGS = 4.0
        GO TO 60
   20   DGS = 4.0
        GO TO 60
   30   DGS = 0.0
        GO TO 60
   40   DGS = 0.0
        GO TO 60
   50   DGS = 0.0
   60 CONTINUE
      RETURN
      END
```

Table 14

```
 K0 = 0.18284121E-01                 T0 = 0.35229725E 01

 F= 0.34480E-02       IER= 0         NO OF ITERATIONS=  0

 NO OF FUNCT EVALUATIONS=    2


 INITIAL ARG. VECTOR AND GRADIENT

           X( 1)= 0.42000000E 00            G( 1)= 0.21118077E-02
           X( 2)= 0.63200000E 00            G( 2)= 0.28182383E-02
           X( 3)= 0.85100000E 00            G( 3)= 0.63655715E-03
           X( 4)= 0.23800000E 00            G( 4)=-0.73575538E-02
           X( 5)= 0.50700000E 00            G( 5)= 0.39115127E-02
           X( 6)= 0.10100000E 00            G( 6)= 0.20516445E-01
           X( 7)= 0.61900000E 00            G( 7)=-0.42940497E-02
           X( 8)= 0.27400000E 00            G( 8)= 0.17985253E-01
           X( 9)= 0.62900000E 00            G( 9)=-0.18634771E-01
           X(10)= 0.45800000E 00            G(10)=-0.13497053E 00
           X(11)= 0.73200000E 00            G(11)= 0.83689370E-01
           X(12)= 0.18284121E-01            G(12)= 0.18122631E-06
           X(13)= 0.35229725E 01            G(13)= 0.84692147E-08
 OLDF =  0.3448E-02             ITERATION NO.  1
 OLDF =  0.2780E-02             ITERATION NO.  2
 OLDF =  0.2675E-02             ITERATION NO.  3
 OLDF =  0.2158E-02             ITERATION NO.  4
 OLDF =  0.1901E-02             ITERATION NO.  5
 OLDF =  0.1798E-02             ITERATION NO.  6
 OLDF =  0.1665E-02             ITERATION NO.  7
 OLDF =  0.1493E-02             ITERATION NO.  8
```

Table 14
(Continued)

```
OLDF =  0.1283E-02          ITERATION NO.  9
OLDF =  0.1191E-02          ITERATION NO. 10
OLDF =  0.1023E-02          ITERATION NO. 11
OLDF =  0.7016E-03          ITERATION NO. 12
OLDF =  0.6314E-03          ITERATION NO. 13
OLDF =  0.5193E-03          ITERATION NO. 14
OLDF =  0.4512E-03          ITERATION NO. 15
OLDF =  0.4389E-03          ITERATION NO. 16
OLDF =  0.4170E-03          ITERATION NO. 17
OLDF =  0.4020E-03          ITERATION NO. 18
OLDF =  0.3969E-03          ITERATION NO. 19
OLDF =  0.3826E-03          ITERATION NO. 20
OLDF =  0.3754E-03          ITERATION NO. 21
OLDF =  0.3713E-03          ITERATION NO. 22
OLDF =  0.3663E-03          ITERATION NO. 23
OLDF =  0.3616E-03          ITERATION NO. 24
OLDF =  0.3543E-03          ITERATION NO. 25
OLDF =  0.3468E-03          ITERATION NO. 26
OLDF =  0.3433E-03          ITERATION NO. 27
OLDF =  0.3215E-03          ITERATION NO. 28
OLDF =  0.3211E-03          ITERATION NO. 29
OLDF =  0.3184E-03          ITERATION NO. 30
OLDF =  0.3156E-03          ITERATION NO. 31
OLDF =  0.3127E-03          ITERATION NO. 32
OLDF =  0.3087E-03          ITERATION NO. 33
OLDF =  0.3052E-03          ITERATION NO. 34
OLDF =  0.3019E-03          ITERATION NO. 35
OLDF =  0.2990E-03          ITERATION NO. 36
OLDF =  0.2911E-03          ITERATION NO. 37
OLDF =  0.2890E-03          ITERATION NO. 38
OLDF =  0.2853E-03          ITERATION NO. 39
OLDF =  0.2834E-03          ITERATION NO. 40
OLDF =  0.2831E-03          ITERATION NO. 41
OLDF =  0.2823E-03          ITERATION NO. 42
OLDF =  0.2815E-03          ITERATION NO. 43
OLDF =  0.2799E-03          ITERATION NO. 44
OLDF =  0.2792E-03          ITERATION NO. 45
OLDF =  0.2784E-03          ITERATION NO. 46
OLDF =  0.2777E-03          ITERATION NO. 47
OLDF =  0.2771E-03          ITERATION NO. 48
OLDF =  0.2761E-03          ITERATION NO. 49
OLDF =  0.2753E-03          ITERATION NO. 50
OLDF =  0.2750E-03          ITERATION NO. 51
OLDF =  0.2746E-03          ITERATION NO. 52

F= 0.27332E-03      IER= 1      NO OF ITERATIONS= 52

NO OF FUNCT EVALUATIONS= 191

          X( 1)= 0.24982625E 00         G( 1)= 0.34937362E-02
          X( 2)= 0.62820711E 00         G( 2)=-0.13012964E-02
          X( 3)= 0.84752128E 00         G( 3)=-0.31967661E-03
          X( 4)= 0.15254241E 00         G( 4)=-0.89132080E-03
          X( 5)= 0.29627715E 00         G( 5)= 0.57309923E-02
          X( 6)= 0.10178863E 00         G( 6)=-0.81774487E-04
          X( 7)= 0.54562625E 00         G( 7)=-0.49900725E-03
          X( 8)= 0.31160252E 00         G( 8)= 0.13535585E-02
          X( 9)= 0.56832202E 00         G( 9)=-0.65854154E-03
          X(10)= 0.50460494E 00         G(10)= 0.21895326E-02
          X(11)= 0.64320938E 00         G(11)=-0.33977294E-03
          X(12)= 0.29779218E-02         G(12)=-0.20130444E 00
          X(13)= 0.30811412E 01         G(13)= 0.11298486E-03


SYNTHESIS OF DIGITAL FILTER WITH SPEC. MAGNITUDE AND/OR GROUP DELAY

THE ERROR CRITERION USED IS MINIMUM 2 FOR THE MAGNITUDE AND

                            MINIMUM 2 FOR THE GROUP DELAY

F= 0.27332E-03      IER= 1      NO OF ITERATIONS= 52

NO OF FUNCT EVALUATIONS= 191
```

Table 14
(Continued)

```
COMPUTED ARG. VECTOR AND GRADIENT

        X( 1)= 0.24982625E 00            G( 1)= 0.34937362E-02
        X( 2)= 0.62820711E 00            G( 2)=-0.13012964E-02
        X( 3)= 0.84752128E 00            G( 3)=-0.31967661E-03
        X( 4)= 0.15254241E 00            G( 4)=-0.89132080E-03
        X( 5)= 0.29627715E 00            G( 5)= 0.57309923E-02
        X( 6)= 0.10178863E 00            G( 6)=-0.81774487E-04
        X( 7)= 0.54562625E 00            G( 7)=-0.49900725E-03
        X( 8)= 0.31160252E 00            G( 8)= 0.13535585E-02
        X( 9)= 0.56832202E 00            G( 9)=-0.65854154E-03
        X(10)= 0.50460494E 00            G(10)= 0.21895326E-02
        X(11)= 0.64320938E 00            G(11)=-0.33977294E-03
        X(12)= 0.29779218E-02            G(12)=-0.20130444E 00
        X(13)= 0.30811412E 01            G(13)= 0.11298486E-03

                       Z PLANE ZEROS

         ANGLE/2PI*FS              RADIUS

        0.                         0.24982625E 00
        0.                         0.40027819E 01
        0.62820711E 00             0.10000000E 01
        0.84752128E 00             0.10000000E 01
        0.15254241E 00             0.29627715E 00
        0.15254241E 00             0.33752181E 01

                       Z PLANE POLES

         ANGLE/2PI*FS              RADIUS

        0.10178863E 00             0.54562625E 00
        0.31160252E 00             0.56832202E 00
        0.50460494E 00             0.64320938E 00

CONST= 0.29779218E-02


                       NUMERATOR COEFFICIENTS


          A0                  A1                   A2

      0.10000000E 01     -0.42526081E 01      0.10000000E 01
      0.10000000E 01      0.78394464E 00      0.10000000E 01
      0.10000000E 01      0.17748885E 01      0.10000000E 01
      0.10000000E 01     -0.52580426E 00      0.87780149E-01
      0.10000000E 01     -0.59900133E 01      0.11392097E 02


                      DENOMINATOR COEFFICIENTS


          B0                  B1                   B2

      0.10000000E 01     -0.10359316E 01      0.29770800E 00
      0.10000000E 01     -0.63414777E 00      0.32298992E 00
      0.10000000E 01      0.18609760E-01      0.41371830E 00
      0.10000000E 01      0.                  0.
      0.10000000E 01      0.                  0.


FREQUENCY RESPONSE

FREQUENCY      MAGNITUDE     LOSS IN DB       PHASE         GROUP DELAY

0.             0.98200E 00   0.15780E 00    0.              0.60749E 01
0.2000E-01     0.98309E 00   0.14814E 00   -0.38178E 00     0.60788E 01
0.4000E-01     0.98617E 00   0.12095E 00   -0.76402E 00     0.60889E 01
0.6000E-01     0.99071E 00   0.81041E-01   -0.11470E 01     0.61007E 01
0.8000E-01     0.99600E 00   0.34840E-01   -0.15306E 01     0.61087E 01
0.1000E 00     0.10013E 01  -0.11414E-01    0.12272E 01     0.61085E 01
0.1200E 00     0.10061E 01  -0.53237E-01    0.84362E 00     0.60983E 01
```

Table 14
(Continued)

0.1400E 00	0.10102E 01	-0.88524E-01	0.46099E 00	0.60802E 01
0.1600E 00	0.10136E 01	-0.11694E 00	0.79619E-01	0.60594E 01
0.1800E 00	0.10160E 01	-0.13830E 00	-0.30054E 00	0.60425E 01
0.2000E 00	0.10175E 01	-0.15047E 00	-0.67991E 00	0.60351E 01
0.2200E 00	0.10171E 01	-0.14766E 00	-0.10592E 01	0.60397E 01
0.2400E 00	0.10139E 01	-0.11948E 00	-0.14391E 01	0.60553E 01
0.2600E 00	0.10059E 01	-0.51280E-01	0.13214E 01	0.60765E 01
0.2800E 00	0.99148E 00	0.74290E-01	0.93892E 00	0.60959E 01
0.3000E 00	0.96885E 00	0.27488E 00	0.55552E 00	0.61062E 01
0.3200E 00	0.93692E 00	0.56594E 00	0.17188E 00	0.61032E 01
0.3400E 00	0.89543E 00	0.95935E 00	-0.21119E 00	0.60886E 01
0.3600E 00	0.84490E 00	0.14639E 01	-0.59313E 00	0.60690E 01
0.3800E 00	0.78632E 00	0.20880E 01	-0.97394E 00	0.60544E 01
0.4000E 00	0.72079E 00	0.28438E 01	-0.13542E 01	0.60536E 01
0.4200E 00	0.64928E 00	0.37513E 01	0.14066E 01	0.60698E 01
0.4400E 00	0.57264E 00	0.48423E 01	0.10244E 01	0.60967E 01
0.4600E 00	0.49196E 00	0.61614E 01	0.64057E 00	0.61169E 01
0.4800E 00	0.40905E 00	0.77644E 01	0.25639E 00	0.61040E 01
0.5000E 00	0.32679E 00	0.97147E 01	-0.12523E 00	0.60316E 01
0.5200E 00	0.24884E 00	0.12081E 02	-0.49999E 00	0.58849E 01
0.5400E 00	0.17886E 00	0.14950E 02	-0.86327E 00	0.56684E 01
0.5600E 00	0.11943E 00	0.18458E 02	-0.12113E 01	0.54035E 01
0.5800E 00	0.71615E-01	0.22900E 02	-0.15419E 01	0.51179E 01
0.6000E 00	0.35037E-01	0.29109E 02	0.12871E 01	0.48360E 01
0.6200E 00	0.84207E-02	0.41493E 02	0.99160E 00	0.45738E 01
0.6400E 00	0.98863E-02	0.40099E 02	0.71175E 00	0.43391E 01
0.6600E 00	0.21544E-01	0.33334E 02	0.44571E 00	0.41340E 01
0.6800E 00	0.28034E-01	0.31046E 02	0.19166E 00	0.39574E 01
0.7000E 00	0.30599E-01	0.30286E 02	-0.52120E-01	0.38064E 01
0.7200E 00	0.30248E-01	0.30386E 02	-0.28713E 00	0.36778E 01
0.7400E 00	0.27784E-01	0.31124E 02	-0.51469E 00	0.35685E 01
0.7600E 00	0.23845E-01	0.32452E 02	-0.73591E 00	0.34756E 01
0.7800E 00	0.18939E-01	0.34453E 02	-0.95174E 00	0.33967E 01
0.8000E 00	0.13470E-01	0.37413E 02	-0.11630E 01	0.33297E 01
0.8200E 00	0.77640E-02	0.42198E 02	-0.13704E 01	0.32730E 01
0.8400E 00	0.20841E-02	0.53622E 02	0.15671E 01	0.32252E 01
0.8600E 00	0.33559E-02	0.49484E 02	0.13658E 01	0.31852E 01
0.8800E 00	0.83817E-02	0.41533E 02	0.11667E 01	0.31520E 01
0.9000E 00	0.12852E-01	0.37820E 02	0.96953E 00	0.31250E 01
0.9200E 00	0.16655E-01	0.35569E 02	0.77388E 00	0.31037E 01
0.9400E 00	0.19701E-01	0.34110E 02	0.57941E 00	0.30874E 01
0.9600E 00	0.21922E-01	0.33182E 02	0.38580E 00	0.30761E 01
0.9800E 00	0.23274E-01	0.32663E 02	0.19276E 00	0.30693E 01
0.1000E 01	0.23727E-01	0.32495E 02	0.60857E-08	0.30671E 01

```
 COEFFICIENTS OF DIGITAL FILTER WITH SPEC. AMPLITUDE AND/OR GROUP DELAY
N= 5
 0.2977922E-02    5
 0.1000000E 01-0.4252608E 01 0.1000000E 01-0.1035932E 01 0.2977080E 00
 0.1000000E 01 0.7839446E 00 0.1000000E 01-0.6341478E 00 0.3229899E 00
 0.1000000E 01 0.1774889E 01 0.1000000E 01 0.1860976E-01 0.4137183E 00
 0.1000000E 01-0.5258043E 00 0.8778015E-01 0.             0.
 0.1000000E 01-0.5990013E 01 0.1139210E 02 0.             0.
```

Appendix

```
C
C-----------------------------------------------------------------------
C  MAIN PROGRAM: COMBINED AMPLITUDE AND GROUP DELAY SYNTHESIS USING A
C                MINIMUM P ERROR CRITERION AND THE FLETCHER POWELL
C                SUBROUTINE FOR FUNCTION MINIMIZATION
C
C  AUTHOR:       A. G. DECZKY
C                INSTITUT DE MICROTECHNIQUE, UNIVERSITE DE NEUCHATEL
C                RUE DE LA MALADIERE 71,  2000 NEUCHATEL, SWITZERLAND
C
C  REFERENCE:    'SYNTHESIS OF RECURSIVE DIGITAL FILTERS USING THE
C                MINIMUM P ERROR CRITERION', A. G. DECZKY, IEEE TRANS.
C                AUDIO ELECTROACOUST., VOL. AU-20, NO. 5, PP. 257-263,
C                OCTOBER 1972.
C-----------------------------------------------------------------------
C
      LOGICAL DEB, DEBFP, LONG
      INTEGER WEIGHT
      COMMON /CM05/ AS(201), DS(201), WF(201), WG(201)
      COMMON /CM10/ NP, NPD
      COMMON /CM15/ DEB
      COMMON /CM20/ C0(20), C1(20), C2(20), D1(20), D2(20)
      COMMON /CM25/ DEBFP, LONG
      COMMON /CM30/ FV(100), XIT(100), KOUNT
      COMMON /CM35/ FD(201), FDN(201)
      COMMON /CM40/ FIK(10), NPK(10), NK
      COMMON /CM45/ WEIGHT(10)
      COMMON /CM50/ N1, N2, N3, N4, N5, MODE
      COMMON /CM55/ ALFA, BETA, NIT
      COMMON /CM60/ WFK(10), WGK(10)
      COMMON /UNIT/ IND, INP, IOUTD, IOUTP, PI, SPM
      DIMENSION X(40), G(40), IP(10), IQ(10)
C
C INITIALIZE MACHINE DEPENDENT CONSTANTS
C
      IND = I1MACH(1)
      INP = I1MACH(1)
      IOUTD = I1MACH(2)
      IOUTP = I1MACH(3)
      PI = 4.0*ATAN(1.0)
      R1 = R1MACH(1)
      SPM = SQRT(R1)
C
C INPUT DATA
C
C MODE = TYPE OF APPROXIMATION
C = 1 MAGNITUDE ONLY
C = 2 GROUP DELAY ONLY
C = 3 GROUP DELAY EQUALISATION
C = 4 MAGNITUDE AND GROUP DELAY
C
      READ (IND,9999) MODE
C
C NZR = NO OF REAL ZEROS
C NZU = NO OF ZEROS ON THE UNIT CIRCLE
C NZC = NO OF OF COMPLEX ZEROS
C NPR = NO OF REAL POLES
C NPC = NO OF OF COMPLEX POLES
C
      READ (IND,9999) NZR, NZU, NZC, NPR, NPC
      N1 = NZR
      N2 = N1 + NZU
      N3 = N2 + NZC*2
      N4 = N3 + NPR
      N5 = N4 + NPC*2 - 1
      N = N5 + 2
      IF (MODE.EQ.4) N = N + 1
C
C X = INITIAL PARAMETER VECTOR
C ZEROES AND POLES SPECIFIED IN POLAR COORDINATES AS FOLLOWS
C REAL ZEROS COME FIRST, THEN ZEROS ON THE UNIT CIRCLE, THEN COMPLEX
C ZEROS, THEN REAL POLES AND FINALLY COMPLEX POLES
C REAL ZEROES (POLES) ARE SPECIFIED AS RADIUS1,RADIUS2,...
C ZEROES ON THE UNIT CIRCLE ARE SPECIFIED AS ANGLE1,ANGLE2,...
C COMPLEX ZEROES (POLES) ARE SPECIFIED AS ANGLE1,RADIUS1,ANGLE2,
C RADIUS2,...
C WHERE THE ANGLES ARE SPECIFIED IN HZ AS PHI/(2 PI)*FSAMPLE
C AND ONLY ZEROES (POLES) WITH POSITIVE ANGLES ARE SPECIFIED
C THEIR COMPLEX CONJUGATE PAIRS BEING UNDERSTOOD
C THE LAST PARAMETER(S) IS (ARE): K0 IF MODE=1, TAU0 IF MODE=2,3,
C K0,TAU0 IF MODE=4
C
      READ (IND,9998) (X(I),I=1,N)
C
C NK = NO OF INTERVALS
C
      READ (IND,9999) NK
C
C NPK(I) = NO. OF POINTS ON THE INTERVAL FIK(I) - FIK(I+1)
C
      READ (IND,9999) (NPK(I),I=1,NK)
C
C WEIGHT(I) = SPACING OF THE POINTS NPK(I) ON THE I TH INTERVAL
C = 1   POINTS EQUISPACED
C = 2   POINTS SPACED ON A SINE ABSCISSA, HALF CYCLE
C = 3   POINTS SPACED ON A COS  ABSCISSA, HALF CYCLE
C = 4   POINTS SPACED ON A COS  ABSCISSA, FULL CYCLE
C
      READ (IND,9999) (WEIGHT(I),I=1,NK)
C
C WFK(I) = RELATIVE WEIGHTING OF THE I TH INTERVAL FOR THE MAGNITUDE
C
      READ (IND,9998) (WFK(I),I=1,NK)
C WGK(I) = RELATIVE WEIGHTING OF THE I TH INTERVAL FOR THE GROUP DELAY
C
      READ (IND,9998) (WGK(I),I=1,NK)
C
      NK = NK + 1
C
C FIK(I) = STARTING POINT OF THE  I  TH APPROX. INTERVAL
C
      READ (IND,9998) (FIK(I),I=1,NK)
C
C IPK = NO OF SUCCESSIVE INDICES OF APPROXIMATION
C
      READ (IND,9999) IPK
C
C IP(II) = SUCCESSIVE INDICES OF APPROXIMATION FOR THE MAGNITUDE
C
      READ (IND,9999) (IP(II),II=1,IPK)
C
C IQ(II) =  SUCCESSIVE INDICES OF APPROXIMATION FOR THE GROUP DELAY
C
      READ (IND,9999) (IQ(II),II=1,IPK)
C
C INV = NUMBER OF TOTAL CYCLES IF NO CONVERGENCE
C
      READ (IND,9999) INV
C
C F1 = LOWER FREQUENCY LIMIT FOR THE FREQUENCY RESPONSE
```

```
C F2 = UPPER FREQUENCY LIMIT FOR THE FREQUENCY RESPONSE
C FSAMPL = SAMPLING FREQUENCY
C IZ  = NUMBER OF POINTS FOR THE FREQUENCY RESPONSE
C (IF IZ LE 0 NO RESPONSE CALCULATED )
C
      READ (IND,9998) F1, F2, FSAMPL
      READ (IND,9999) IZ
      WS = FSAMPL/2.0
C
C ALFA= RELATIVE WEIGHTING OF THE MAGNITUDE
C RELATIVE WEIGHTING OF THE GROUP DELAY = 1.0 - ALFA
C THIS PARAMETER IS ONLY USED IF MODE = 4
C
      READ (IND,9998) ALFA
      BETA = 1.0 - ALFA
C
C LONG = LOGICAL PARAMETER FOR TYPE OF OUTPUT
C = .TRUE. FOR LONG OUTPUT
C = .FALSE. FOR SHORT OUTPUT
C
C
      READ (IND,9997) LONG
C
C LIMIT = NUMBER OF ITERATIONS PER CYCLE BEFORE RESTARTING WITH
C STEEPEST DESCENT
C
      LIMIT = 4*N
C
C EPS = ERROR PARAMETER FOR FMFP SUBROUTINE
C
      EPS = 1.E-3
C
C DEBUG  = .TRUE . FOR DEBUGGING PRINTOUT
C
      DEB = .FALSE.
C
C DEBFP  = .TRUE . FOR DEBUGGING PRINTOUT FROM FMFP  SUBROUTINE
C
      DEBFP = .FALSE.
      READ (IND,9997) DEB, DEBFP
C
9999  FORMAT (I4)
9998  FORMAT (F16.9)
9997  FORMAT (L1)
      CALL COSYFP(X, G, IZ, F1, F2, WS, N, EPS, IP, IQ, LIMIT, INV, IPK)
      STOP
      END
C
C-----------------------------------------------------------------------
C SUBROUTINE:  COSYFP
C MAIN SUBROUTINE FOR AMPLITUDE AND/OR G.D. SYNTHESIS USING FP
C-----------------------------------------------------------------------
C
      SUBROUTINE COSYFP(X, G, IZ, F1, F2, WS, N, EPS, IP, IQ, LIMIT,
     *    INV, IPK)
      LOGICAL DEB, DEBFP, LONG
      INTEGER WEIGHT
      COMMON /CM05/ AS(201), DS(201), WF(201), WG(201)
      COMMON /CM10/ NP, NPD
      COMMON /CM15/ DEB
      COMMON /CM20/ C0(20), C1(20), C2(20), D1(20), D2(20)
      COMMON /CM25/ DEBFP, LONG
      COMMON /CM30/ FV(100), XIT(100), KOUNT
      COMMON /CM35/ FD(201), FDN(201)
      COMMON /CM40/ FIK(10), NPK(10), NK
      COMMON /CM45/ WEIGHT(10)
      COMMON /CM50/ N1, N2, N3, N4, N5, MODE
      COMMON /CM55/ ALFA, BETA, NIT
      COMMON /CM60/ WFK(10), WGK(10)
      COMMON /CM65/ IPX, IQX
      COMMON /CM70/ A(201), B(201)
      COMMON /UNIT/ IND, INP, IOUTD, IOUTP, PI, SPM
      DIMENSION GDH(201)
      DIMENSION X(1), G(1), H(201), U0(20), V0(20), U(20), V(20),
     *    A0(20), A1(20), A2(20), B0(20), B1(20), B2(20), AMPL(201),
     *    PHI(201), DB(201), GD(201), IP(10), IQ(10)
C
C INITIALIZE CONSTANTS
C
      NPD = 0
      NPA = 0
      NP = 0
      KP = 1
      KOUNT = 0
      NIT = 0
      IER = 0
      NIT = 0
      NKOUNT = 0
      IQX = IQ(1)
      IPX = IP(1)
      NN = N/4
      CONST = 1.0
      DO 10 I=1,20
        A0(I) = 1.0
        B0(I) = 1.0
        A1(I) = 0.0
        A2(I) = 0.0
        B1(I) = 0.0
        B2(I) = 0.0
 10   CONTINUE
C
C INITIALIZE X(N) AND X(N-1) (IF NECESSARY)
C
      GO TO (20, 30, 30, 40), MODE
 20   X(N) = 1.0
      GO TO 50
 30   X(N) = 0.0
      GO TO 50
 40   X(N-1) = 1.0
      X(N) = 0.0
 50   WSPI = PI/WS
C
C COMPUTE FREQUENCY POINTS ON THE VARIOUS INTERVALS
C
      DO 110 K=2,NK
        NPX = NPK(K-1)
        NPX1 = NPX - 1
        FNPX = FLOAT(NPX)
        FI2 = FIK(K)
        FI1 = FIK(K-1)
        IWG = WEIGHT(K-1)
        WGK1 = WGK(K-1)
        WFK1 = WFK(K-1)
C
        IF (NPX.EQ.0) GO TO 110
C
        DO 100 J=1,NPX
          JK = J + NP
          WG(JK) = WGK1
          WF(JK) = WFK1
```

```
C
          FLJNP = FLOAT(J-1)/FNPX
          GO TO (60, 70, 80, 90), IWG
   60     FDN(JK) = FLJNP*(FI2-FI1) + FI1
          GO TO 100
   70     FDN(JK) = SIN(FLJNP*PI/2.0)*(FI2-FI1) + FI1
          GO TO 100
   80     FDN(JK) = COS(FLJNP*PI/2.0)*(FI1-FI2) + FI2
          GO TO 100
   90     FDN(JK) = 0.5*(FI1+FI2) + COS(FLJNP*PI)*(FI1-FI2)*0.5
  100   CONTINUE
C
        IF (WFK1.NE.0.) NPA = NPA + NPX
        IF (WGK1.NE.0.) NPD = NPD + NPX
        NP = NP + NPX
  110 CONTINUE
C
      DO 120 J=1,NP
        FD(J) = FDN(J)*WSPI
  120 CONTINUE
C
      N1P = N1 + 1
      N2P = N2 + 1
      N4P = N4 + 1
C
C NORMALISE THE ANGLE OF THE POLES AND ZEROES
C
      IF (N1P.GT.N2) GO TO 140
      DO 130 J=N1P,N2
        X(J) = X(J)/WS
  130 CONTINUE
  140 IF (N2P.GT.N3) GO TO 160
      DO 150 J=N2P,N3,2
        X(J) = X(J)/WS
  150 CONTINUE
  160 IF (N4P.GT.N5) GO TO 180
      DO 170 J=N4P,N5,2
        X(J) = X(J)/WS
  170 CONTINUE
C
C READ IN FILTER COEFFICIENTS
C
  180 IF (MODE.NE.3) GO TO 190
      READ (INP,9999) CONST0, M
 9999 FORMAT (E14.7, I4)
      CALL INCOCA(M, CONST0, C0, C1, C2, D1, D2)
      CALL DGSPEC(GDH, M)
C
C SET UP ARRAYS OF DESIRED FUNCTIONS FOR MAGN. AND GD
C
  190 DO 240 J=1,NP
        FJ = FDN(J)
        GO TO (200, 210, 220, 230), MODE
  200   AS(J) = FS(FJ)
        GO TO 240
  210   DS(J) = DGS(FJ)
        GO TO 240
  220   DS(J) = -GDH(J)
        GO TO 240
  230   AS(J) = FS(FJ)
        DS(J) = DGS(FJ)
  240 CONTINUE
C
      CALL FUNCT(N, X, F, G)
C
C COMPUTE INITIAL APPROX. TO X(N) AND X(N-1)
C
      CALL XNXM1(X, N, NP)
C
      GO TO (250, 260, 260, 270), MODE
  250 IF (LONG) WRITE (IOUTD,9983) X(N)
      GO TO 280
  260 IF (LONG) WRITE (IOUTD,9982) X(N)
      GO TO 280
  270 IF (LONG) WRITE (IOUTD,9981) X(N-1), X(N)
  280 CALL FUNCT(N, X, F, G)
      IF (LONG) WRITE (IOUTD,9998) F, IER, KOUNT, NIT
      IF (LONG) WRITE (IOUTD,9994)
      IF (LONG) WRITE (IOUTD,9995) (J,X(J),J,G(J),J=1,N)
C
C MAIN LOOP CALLING THE F.P. SUBROUTINE
C
  290 CALL FMFP(N, X, F, G, EPS, LIMIT, IER, H)
C
      NKOUNT = NKOUNT + KOUNT
      IF (LONG) WRITE (IOUTD,9998) F, IER, KOUNT, NIT
      IF (LONG) WRITE (IOUTD,9995) (J,X(J),J,G(J),J=1,N)
C
      IF (IER.EQ.0) GO TO 310
      IF (INV-1) 310, 310, 300
  300 INV = INV - 1
      GO TO 290
  310 IF (IER.NE.0 .OR. KP.GE.IPK) GO TO 320
      KP = KP + 1
      IQX = IQ(KP)
      IPX = IP(KP)
      GO TO 290
C
C COMPUTE QUADRATIC FACTORS, POLES AND ZEROS FROM ARG. VECTOR
C
  320 CALL FACTOR(X, A1, A2, B1, B2, U0, V0, U, V, CONST, JA, JB, JZ,
     *    JP, WS)
C
      NN = MAX0(JA-1,JB-1)
      JZ1 = JZ - 1
      JP1 = JP - 1
C
      IF (MODE.NE.3) GO TO 350
      JA = JA + M
      JB = JB + M
      NNP = NN + 1
      NM = NN + M
C
      IF (NNP.GT.NM) GO TO 340
      DO 330 J=NNP,NM
        JM = J - NN
        A1(J) = C1(JM)
        A2(J) = C2(JM)
        B1(J) = D1(JM)
        B2(J) = D2(JM)
  330 CONTINUE
  340 NN = NM
C
  350 IF (DEB) WRITE (IOUTD,9980) JA, JB
      IF (DEB) WRITE (IOUTD,9979) (A1(J),A2(J),B1(J),B2(J),J=1,NN)
C
      GO TO (360, 390, 370, 380), MODE
  360 CONST = X(N)
      GO TO 390
  370 CONST = CONST*CONST0
```

```
      GO TO 390
 380  CONST = X(N-1)
C
C OUTPUT SECTION
C
 390  WRITE (IOUTD,9996)
      WRITE (IOUTD,9986) IPX, IQX
      WRITE (IOUTD,9998) F, IER, NKOUNT, NIT
      WRITE (IOUTD,9993)
      WRITE (IOUTD,9995) (J,X(J),J,G(J),J=1,N)
      IF (JZ1.GT.0) WRITE (IOUTD,9989) (U0(J),V0(J),J=1,JZ1)
      IF (JP1.GT.0) WRITE (IOUTD,9988) (U(J),V(J),J=1,JP1)
      WRITE (IOUTD,9987) CONST
      WRITE (IOUTD,9985) (A0(J),A1(J),A2(J),J=1,NN)
      WRITE (IOUTD,9984) (B0(J),B1(J),B2(J),J=1,NN)
      IF (IZ.LE.0) GO TO 400
      CALL FREDIC(NN, CONST, A0, A1, A2, B1, B2, F1, F2, WS, IZ, AMPL,
     *    PHI, DB, GD, FD)
      WRITE (IOUTD,9997) (FD(J),AMPL(J),DB(J),PHI(J),GD(J),J=1,IZ)
C
 400  WRITE (IOUTP,9992) NN
      WRITE (IOUTP,9991) CONST, NN
      WRITE (IOUTP,9990) (A0(J),A1(J),A2(J),B1(J),B2(J),J=1,NN)
      RETURN
C
C FORMATS
C
9998  FORMAT (/3H F=, E12.5, 5X, 4HIER=, I2, 5X, 17HNO OF ITERATIONS=,
     *    I3//1X, 24HNO OF FUNCT EVALUATIONS=, I4)
9997  FORMAT (1H1//19H FREQUENCY RESPONSE//1X, 9HFREQUENCY, 6X,
     *    9HMAGNITUDE, 5X, 10HLOSS IN DB, 8X, 5HPHASE, 7X, 9HGROUP DEL,
     *    2HAY//(E11.4, 4E15.5))
9996  FORMAT (1H1//49H SYNTHESIS OF DIGITAL FILTER WITH SPEC. MAGNITUDE,
     *    20H AND/OR GROUP DELAY /)
9995  FORMAT (/(10X, 2HX(, I2, 2H)=, E15.8, 10X, 2HG(, I2, 2H)=, E15.8))
9994  FORMAT (//33H INITIAL ARG. VECTOR AND GRADIENT)
9993  FORMAT (//34H COMPUTED ARG. VECTOR AND GRADIENT)
9992  FORMAT (52H COEFFICIENTS OF DIGITAL FILTER WITH SPEC. AMPLITUDE,
     *    19H AND/OR GROUP DELAY, / 3H N=, I2)
9991  FORMAT (E14.7, I4)
9990  FORMAT (5E14.7)
9989  FORMAT (/24X, 13HZ PLANE ZEROS//12X, 12HANGLE/2PI*FS, 16X,
     *    6HRADIUS//(2E25.8))
9988  FORMAT (/24X, 13HZ PLANE POLES//12X, 12HANGLE/2PI*FS, 16X,
     *    6HRADIUS//(2E25.8))
9987  FORMAT (/1X, 6HCONST=, E15.8)
9986  FORMAT (/36H THE ERROR CRITERION USED IS MINIMUM, I2, 9H FOR THE ,
     *    13HMAGNITUDE AND//29X, 7HMINIMUM, I2, 14H FOR THE GROUP,
     *    6H DELAY)
9985  FORMAT (//24X, 22HNUMERATOR COEFFICIENTS///14X, 2HA0, 20X, 2HA1,
     *    20X, 2HA2//(3E22.8))
9984  FORMAT (//23X, 24HDENOMINATOR COEFFICIENTS///14X, 2HB0, 20X, 2HB1,
     *    20X, 2HB2//(3E22.8))
9983  FORMAT (1H1//5H K0 =, E15.8)
9982  FORMAT (1H1//5H T0 =, E15.8)
9981  FORMAT (1H1//5H K0 =, E15.8, 10X, 5H T0 =, E15.8)
9980  FORMAT (1H1//2I10//)
9979  FORMAT (//(4E18.5))
C
      END
C
C-----------------------------------------------------------------------
C SUBROUTINE:  XNXM1
C SUBROUTINE TO CALCULATE INITIAL APPROX. TO X(N),X(N-1)
C-----------------------------------------------------------------------
C
      SUBROUTINE XNXM1(X, N, NP)
      DIMENSION X(N)
      COMMON /CM05/ AS(201), DS(201), WF(201), WG(201)
      COMMON /CM50/ N1, N2, N3, N4, N5, MODE
      COMMON /CM70/ A(201), B(201)
      COMMON /UNIT/ IND, INP, IOUTD, IOUTP, PI, SPM
C
      AGD = 0.0
      AGN = 0.0
      AD = 0.0
      AN = 0.0
      DO 40 J=1,NP
        GO TO (10, 20, 20, 30), MODE
  10    WFJ = WF(J)
        IF (WFJ.LT.SPM) WFJ = SPM
        HJ = A(J)/WFJ + AS(J)
        WHJ = HJ*WFJ
        AN = AN + WHJ*AS(J)
        AD = AD + HJ*WHJ
        GO TO 40
  20    AGN = AGN + B(J)
        AGD = AGD + WG(J)
        GO TO 40
  30    WFJ = WF(J)
        IF (WFJ.LT.SPM) WFJ = SPM
        HJ = A(J)/WFJ + AS(J)
        WHJ = HJ*WFJ
        AN = AN + WHJ*AS(J)
        AD = AD + HJ*WHJ
        AGN = AGN + B(J)
        AGD = AGD + WG(J)
  40  CONTINUE
C
      GO TO (50, 60, 60, 70), MODE
  50  X(N) = AN/AD
      GO TO 80
  60  X(N) = AGN/AGD
      GO TO 80
  70  X(N) = AGN/AGD
      X(N-1) = AN/AD
  80  RETURN
      END
C
C-----------------------------------------------------------------------
C FUNCTION:  RANGE
C FUNCTION TO LIMIT THE RANGE OF ANGLES (X) TO 0 TO PI
C-----------------------------------------------------------------------
C
      FUNCTION RANGE(X)
C
      X = ABS(X)
      IX = IFIX(X)
      MX = 2*(IX/2)
      RANGE = X - FLOAT(MX)
      IF (RANGE.GT.1.0) RANGE = 2.0 - RANGE
      RETURN
      END
C
C-----------------------------------------------------------------------
C SUBROUTINE:  DGSPEC
C TO CALCULATE THE GROUP DELAY OF THE NON-EQUALISED FILTER
C-----------------------------------------------------------------------
C
      SUBROUTINE DGSPEC(GDH, M)
```

```
      DIMENSION GDH(201)
      COMMON /CM10/ NP, NPD
      COMMON /CM20/ C0(20), C1(20), C2(20), D1(20), D2(20)
      COMMON /CM35/ FD(201), FDN(201)
      COMMON /UNIT/ IND, INP, IOUTD, IOUTP, PI, SPM
      COMPLEX NUMC, NUMD, DENC, DEND
C
      DO 20 J=1,NP
        FJ = FD(J)
        GDH(J) = 0.
        CS = COS(FJ)
        SN = SIN(FJ)
        CS2 = CS + CS
        SN2 = SN + SN
C
        DO 10 I=1,M
          C11 = C0(I) + C2(I)
          C12 = C1(I)
          C22 = C0(I) - C2(I)
          D11 = 1. + D2(I)
          D12 = D1(I)
          D22 = 1. - D2(I)
          NUMC = CMPLX(CS2+C12,SN2)
          NUMD = CMPLX(CS2+D12,SN2)
          DENC = CMPLX(C11*CS+C12,C22*SN)
          DEND = CMPLX(D11*CS+D12,D22*SN)
          GDH(J) = GDH(J) + REAL(NUMD/DEND) - REAL(NUMC/DENC)
 10     CONTINUE
C
 20   CONTINUE
      RETURN
      END
C
C-----------------------------------------------------------------------
C SUBROUTINE:  INCOCA
C READ COEFFICIENTS FOR ONE DIGITAL CASCADE FILTER
C E14.7 FORMAT, FILE ON UNIT INP
C-----------------------------------------------------------------------
C
      SUBROUTINE INCOCA(NTERM, CONST, A0, A1, A2, B1, B2)
      COMMON /UNIT/ IND, INP, IOUTD, IOUTP, PI, SPM
      REAL A0(NTERM), A1(NTERM), A2(NTERM), B1(NTERM), B2(NTERM)
      DO 10 N=1,NTERM
        READ (INP,9999) A0(N), A1(N), A2(N), B1(N), B2(N)
 10   CONTINUE
9999  FORMAT (5E14.7)
      RETURN
      END
C
C-----------------------------------------------------------------------
C SUBROUTINE:  FREDIC
C SUBROUTINE TO COMPUTE THE FREQUENCY RESPONSE OF A DIGITAL FILTER IN
C CASCADE FORM
C-----------------------------------------------------------------------
C
      SUBROUTINE FREDIC(NTERM, CONST, A0, A1, A2, B1, B2, FU, FO, WS,
     *    IZ, AMPL, PHI, DB, GD, FD)
      COMMON /UNIT/ IND, INP, IOUTD, IOUTP, PI, SPM
      COMPLEX N, D, GWS, NN, DD, N1, N2
      DIMENSION A0(NTERM), A1(NTERM), A2(NTERM), B1(NTERM), B2(NTERM)
      DIMENSION AMPL(IZ), PHI(IZ), DB(IZ), FD(IZ), GD(IZ)
      DIMENSION AA(40), AB(40), BA(40), BB(40)
C
      WSXPI = PI/WS
      WU = FU
      Z = (FO-FU)/FLOAT(IZ-1)
C
      DO 10 I=1,NTERM
        AA(I) = A0(I) + A2(I)
        AB(I) = A0(I) - A2(I)
        BA(I) = 1.0 + B2(I)
        BB(I) = 1.0 - B2(I)
 10   CONTINUE
C
      DO 40 J=1,IZ
        WT = WU + Z*FLOAT(J-1)
        FD(J) = WT
        WT = WT*WSXPI
        CWT = COS(WT)
        SWT = SIN(WT)
        D = CMPLX(1.0,0.)
        N = D
        QL = 0.0
        QQ = QL
C
        DO 20 I=1,NTERM
          NN = CMPLX(AA(I)*CWT+A1(I),AB(I)*SWT)
          DD = CMPLX(BA(I)*CWT+B1(I),BB(I)*SWT)
          N2 = CMPLX(2.0*CWT+B1(I),2.0*SWT)
          N1 = CMPLX(2.0*A0(I)*CWT+A1(I),2.0*A0(I)*SWT)
          N = N*NN
          D = D*DD
C
          IF (CABS(DD).LT.SPM) DD = CMPLX(SPM,SPM)
          Q2 = REAL(N2/DD)
C
          IF (CABS(NN).LT.SPM) NN = CMPLX(SPM,SPM)
          Q1 = REAL(N1/NN)
          IF (ABS(Q1-QL).GT.50.) Q1 = 1.0
          QL = Q1
          QQ = QQ + Q2 - Q1
 20     CONTINUE
C
        GD(J) = QQ
C
        IF (CABS(D).LT.SPM) D = CMPLX(SPM,SPM)
        GWS = N/D
C
        AMPL(J) = CONST*CABS(GWS)
        GABS = AMPL(J)
        IF (GABS.NE.0.) GO TO 30
C
        PHI(J) = 0.0
        DB(J) = 1000.
        GO TO 40
C
 30     GWSI = AIMAG(GWS)
        GWSR = REAL(GWS)
        PHI(J) = ATAN(GWSI/GWSR)
        DB(J) = -ALOG10(GABS)*20.0
 40   CONTINUE
C
      RETURN
      END
C
C-----------------------------------------------------------------------
C SUBROUTINE:  FACTOR
C SUBROUTINE TO RECALCULATE QUADRATIC FACTORS
C AND POLES AND ZEROS FROM PARAMETER VECTOR X
C-----------------------------------------------------------------------
```

```
C
      SUBROUTINE FACTOR(X, A1, A2, B1, B2, U0, V0, U, V, CONST, JA, JB,
     *    JZ, JP, WS)
      COMMON /CM50/ N1, N2, N3, N4, N5, MODE
      COMMON /UNIT/ IND, INP, IOUTD, IOUTP, PI, SPM
      DIMENSION X(1), A1(1), A2(1), B1(1), B2(1), U0(1), V0(1)
      DIMENSION U(1), V(1)
C
      J = 0
      JA = 1
      JB = 1
      JZ = 1
      CONST = 1.0
   10 IF (J.GE.N1) GO TO 40
      J = J + 1
      R = X(J)
      GO TO (20, 30, 30, 30), MODE
   20 A2(JA) = 0.0
      A1(JA) = -R
      U0(JZ) = 0.
      V0(JZ) = R
      JA = JA + 1
      JZ = JZ + 1
      GO TO 80
   30 A1(JA) = -R - 1.0/R
      A2(JA) = 1.0
      U0(JZ) = 0.0
      V0(JZ) = R
      U0(JZ+1) = 0.0
      V0(JZ+1) = 1.0/R
      JA = JA + 1
      JZ = JZ + 2
      GO TO 80
C
   40 IF (J.GE.N2) GO TO 50
      J = J + 1
      XJ = X(J)
      XJ = RANGE(XJ)
      A1(JA) = -2.0*COS(XJ*PI)
      A2(JA) = 1.0
      V0(JZ) = 1.0
      U0(JZ) = XJ*WS
      JA = JA + 1
      JZ = JZ + 1
      GO TO 80
C
   50 IF (J.GE.N3) GO TO 90
      J = J + 2
      XJM = X(J-1)
      XJM = RANGE(XJM)
      CSP = COS(XJM*PI)
      R = X(J)
      GO TO (60, 70, 70, 70), MODE
   60 A2(JA) = R*R
      A1(JA) = -(R+R)*CSP
      U0(JZ) = XJM*WS
      V0(JZ) = R
      JA = JA + 1
      JZ = JZ + 1
      GO TO 80
   70 A2(JA) = R*R
      A1(JA) = -(R+R)*CSP
      A2(JA+1) = 1.0/A2(JA)
      A1(JA+1) = -2.0/R*CSP
      V0(JZ) = R
      V0(JZ+1) = 1.0/R
      U0(JZ) = XJM*WS
      U0(JZ+1) = XJM*WS
      JA = JA + 2
      JZ = JZ + 2
   80 IF (J.LT.N3) GO TO 10
C
   90 NN = MAX0(N4,N5)
      JP = 1
  100 IF (J.GE.N4) GO TO 140
      J = J + 1
      R = X(J)
      GO TO (110, 120, 120, 110), MODE
  110 B2(JB) = 0.0
      B1(JB) = -R
      U(JP) = 0.0
      V(JP) = R
      GO TO 130
  120 B2(JB) = 0.0
      B1(JB) = -R
      A1(JA) = -1.0/R
      A2(JA) = 0.0
      CONST = CONST*R
      U(JP) = 0.0
      V(JP) = R
      U0(JZ) = 0.0
      V0(JZ) = R
      JA = JA + 1
      JZ = JZ + 1
  130 JB = JB + 1
      JP = JP + 1
      GO TO 180
C
  140 IF (J.GT.N5) GO TO 190
      J = J + 2
      R = X(J)
      XJM = X(J-1)
      XJM = RANGE(XJM)
      GO TO (150, 160, 160, 150), MODE
  150 B2(JB) = R*R
      B1(JB) = -(R+R)*COS(XJM*PI)
      U(JP) = XJM*WS
      V(JP) = R
      GO TO 170
  160 B2(JB) = R*R
      B1(JB) = -(R+R)*COS(XJM*PI)
      A1(JA) = B1(JB)/B2(JB)
      A2(JA) = 1.0/B2(JB)
      CONST = CONST*B2(JB)
      U(JP) = XJM*WS
      V(JP) = R
      U0(JZ) = XJM*WS
      V0(JZ) = 1.0/R
      JA = JA + 1
      JZ = JZ + 1
  170 JB = JB + 1
      JP = JP + 1
  180 IF (J.LT.NN) GO TO 100
C
  190 RETURN
      END
C
C-----------------------------------------------------------------------
C SUBROUTINE:  FUNCT
C SUBROUTINE TO CALCULATE THE FUNCTION TO BE MINIMIZED
```

```
C FROM THE PARAMETER VECTOR X AS A FUNCTION OF FREQUENCY
C-----------------------------------------------------------------------
C
      SUBROUTINE FUNCT(N, X, F, G)
      REAL NUM, NUMM, NUMP
      LOGICAL DEB
      COMMON /CM05/ AS(201), DS(201), WF(201), WG(201)
      COMMON /CM10/ NP, NPD
      COMMON /CM15/ DEB
      COMMON /CM35/ FD(201), FDN(201)
      COMMON /CM50/ N1, N2, N3, N4, N5, MODE
      COMMON /CM55/ ALFA, BETA, NIT
      COMMON /CM65/ IP, IQ
      COMMON /CM70/ A(201), B(201)
      COMMON /UNIT/ IND, INP, IOUTD, IOUTP, PI, SPM
      DIMENSION X(N), G(N), GZR(20), GZU(20), GPR(20), GZCP(20),
     *    GZCR(20), GPCP(20), GPCR(20), GPRD(20), GPCPD(20), GPCRD(20)
C
      RSPM = SQRT(SPM)
      F = 0.0
C
      DO 10 J=1,N
        G(J) = 0.0
   10 CONTINUE
      XN1 = X(N-1)
      XN = X(N)
      NIT = NIT + 1
      IF (DEB) WRITE (IOUTD,9999) NIT
 9999 FORMAT (//27H NO. OF FUNCT. EVALUATIONS=, I4//)
C
      DO 400 I=1,NP
C
        FI = FD(I)
        CS = COS(FI)
        CS2 = 2.0*CS
        H = XN
        IF (MODE.EQ.4) H = XN1
        HD = 1.0
        HN = 1.0
        DG = 0.0
C
        NN = MAX0(N4,N5)
C
        J = 0
   20   J = J + 1
        IF (J.GT.N1) GO TO 50
        R = X(J)
        GO TO (30, 190, 190, 40), MODE
   30   NUM = 1. + R*R - R*CS2
        IF (NUM.LT.SPM) NUM = SPM
        HN = HN*NUM
        GZR(J) = (R-CS)/NUM
        GO TO 190
   40   NUM = 1.0/R + R - CS2
        IF (NUM.LT.SPM) NUM = SPM
        HN = HN*NUM**2
        GZR(J) = (1.0-1.0/R**2)/NUM
        GO TO 190
C
   50   IF (J.GT.N2) GO TO 70
        GO TO (60, 190, 190, 60), MODE
   60   P = X(J)*PI
        NUMP = 2.0*(1.0-COS(FI+P))
        NUMM = 2.0*(1.0-COS(FI-P))
        IF (NUMM.LT.SPM) NUMM = SPM
        IF (NUMP.LT.SPM) NUMP = SPM
        HN = HN*NUMM*NUMP
        GZU(J) = -SIN(FI-P)/NUMM + SIN(FI+P)/NUMP
        GO TO 190
C
   70   IF (J.GE.N3) GO TO 100
        R = X(J+1)
        P = X(J)*PI
        GO TO (80, 190, 190, 90), MODE
   80   RP = 1.0 + R*R
        R2 = R + R
        CSNM = COS(FI-P)
        CSNP = COS(FI+P)
        NUMM = RP - R2*CSNM
        NUMP = RP - R2*CSNP
        IF (NUMP.LT.SPM) NUMP = SPM
        IF (NUMM.LT.SPM) NUMM = SPM
        HN = HN*NUMM*NUMP
        GZCP(J) = R*(-SIN(FI-P)/NUMM+SIN(FI+P)/NUMP)
        GZCR(J) = (R-CSNM)/NUMM + (R-CSNP)/NUMP
        J = J + 1
        GO TO 190
   90   CSNP = 2.0*COS(FI+P)
        CSNM = 2.0*COS(FI-P)
        R2 = R + 1.0/R
        NUMM = R2 - CSNM
        NUMP = R2 - CSNP
        IF (NUMP.LT.RSPM) NUMP = RSPM
        IF (NUMM.LT.RSPM) NUMM = RSPM
        HN = HN*(NUMM*NUMP)**2
        GZCP(J) = 2.0*(-SIN(FI-P)/NUMM+SIN(FI+P)/NUMP)
        GZCR(J) = (1.0-1.0/R**2)*(1.0/NUMP+1.0/NUMM)
        J = J + 1
        GO TO 190
C
  100   IF (J.GT.N4) GO TO 140
        R = X(J)
        RSQ = R**2
        R2 = 2.0*R
        DEN = 1.0 + RSQ - R2*CS
        IF (DEN.LT.RSPM) DEN = RSPM
        GO TO (110, 120, 120, 130), MODE
  110   HD = HD*DEN
        GPR(J) = (R-CS)/DEN
        GO TO 190
  120   DG = DG + 2.0*(1.0-R*CS)/DEN
        GPRD(J) = 2.0*((1.0+RSQ)*CS-R2)/DEN**2
        GO TO 190
  130   HD = HD*DEN
        GPR(J) = (R-CS)/DEN
        DG = DG + (1.0-R*CS)/DEN
        GPRD(J) = ((1.0+RSQ)*CS-R2)/DEN**2
        GO TO 190
C
  140   P = X(J)*PI
        R = X(J+1)
        R2 = R + R
        RP = 1.0 + R*R
        CSNP = COS(FI+P)
        CSNM = COS(FI-P)
        SNDP = SIN(FI+P)
        SNDM = SIN(FI-P)
        DENP = RP - R2*CSNP
        DENM = RP - R2*CSNM
        IF (ABS(DENP).LT.RSPM) DENP = RSPM
```

```
      IF (ABS(DENM).LT.RSPM) DENM = RSPM
      GO TO (150, 160, 160, 170), MODE
150   HD = HD*DENM*DENP
      GPCR(J) = (R-CSNM)/DENM + (R-CSNP)/DENP
      GPCP(J) = R*(-SNDM/DENM+SNDP/DENP)
      GO TO 180
160   DENP2 = DENP**2
      DENM2 = DENM**2
      DG = DG + 2.0*((1.0-R*CSNM)/DENM+(1.0-R*CSNP)/DENP)
      GPCPD(J) = R2*(2.0-RP)*(SNDM/DENM2-SNDP/DENP2)
      GPCRD(J) = 2.0*((RP*CSNM-R2)/DENM2+(RP*CSNP-R2)/DENP2)
      GO TO 180
170   HD = HD*DENM*DENP
      GPCR(J) = (R-CSNM)/DENM + (R-CSNP)/DENP
      GPCP(J) = R*(-SNDM/DENM+SNDP/DENP)
      DENP2 = DENP**2
      DENM2 = DENM**2
      DG = DG + (1.0-R*CSNM)/DENM + (1.0-R*CSNP)/DENP
      GPCPD(J) = R*(2.0-RP)*(SNDM/DENM2-SNDP/DENP2)
      GPCRD(J) = (RP*CSNM-R2)/DENM2 + (RP*CSNP-R2)/DENP2
180   J = J + 1
C
190   IF (J.LT.NN) GO TO 20
      XNP = FLOAT(NP)
      XIP = FLOAT(IP)
      XNPD = FLOAT(NPD)
      XIQ = FLOAT(IQ)
      GO TO (200, 210, 210, 220), MODE
200   H = H*SQRT(HN/HD)
      E = H - AS(I)
      IF (ABS(E).LT.SPM) E = SPM
      A(I) = E*WF(I)
      EP = E**IP/XNP
      E2H = XIP*(EP/E)*H*WF(I)
      E2P = E2H*PI
      GO TO 230
C
210   ED = DG - DS(I) - XN
      IF (ABS(ED).LT.SPM) ED = SPM
      B(I) = ED*WG(I)
      EDP = ED**IQ/XNPD
      ED2H = XIQ*(EDP/ED)*WG(I)
      ED2P = ED2H*PI
      GO TO 230
C
220   H = H*SQRT(HN/HD)
      E = H - AS(I)
      IF (ABS(E).LT.SPM) E = SPM
      A(I) = E*WF(I)
      EP = E**IP/XNP
      E2H = XIP*(EP/E)*H*WF(I)*ALFA
      E2P = E2H*PI
      ED = DG - DS(I) - XN
      IF (ABS(ED).LT.SPM) ED = SPM
      B(I) = ED*WG(I)
      EDP = ED**IQ/XNPD
      ED2H = XIQ*(EDP/ED)*WG(I)*BETA
      ED2P = ED2H*PI
C
230   J = 0
240   J = J + 1
      IF (J.GT.N1) GO TO 250
      G(J) = G(J) + E2H*GZR(J)
      GO TO 360
C
250   IF (J.GT.N2) GO TO 260
      G(J) = G(J) + E2P*GZU(J)
      GO TO 360
C
260   IF (J.GE.N3) GO TO 270
      JJ = J + 1
      G(J) = G(J) + E2P*GZCP(J)
      G(JJ) = G(JJ) + E2H*GZCR(J)
      J = JJ
      GO TO 360
C
270   IF (J.GT.N4) GO TO 310
      GO TO (280, 290, 290, 300), MODE
280   G(J) = G(J) - E2H*GPR(J)
      GO TO 360
290   G(J) = G(J) + ED2H*GPRD(J)
      GO TO 360
300   G(J) = G(J) - E2H*GPR(J) + ED2H*GPRD(J)
      GO TO 360
C
310   JJ = J + 1
      GO TO (320, 330, 330, 340), MODE
320   G(J) = G(J) - E2P*GPCP(J)
      G(JJ) = G(JJ) - E2H*GPCR(J)
      GO TO 350
330   G(J) = G(J) + ED2P*GPCPD(J)
      G(JJ) = G(JJ) + ED2H*GPCRD(J)
      GO TO 350
340   G(J) = G(J) - E2P*GPCP(J) + ED2P*GPCPD(J)
      G(JJ) = G(JJ) - E2H*GPCR(J) + ED2H*GPCRD(J)
350   J = JJ
360   IF (J.LT.NN) GO TO 240
C
      GO TO (370, 380, 380, 390), MODE
370   F = F + EP*WF(I)
      G(N) = G(N) + E2H/XN
      GO TO 400
380   F = F + EDP*WG(I)
      G(N) = G(N) - ED2H
      GO TO 400
390   F = F + EP*WF(I)*ALFA + EDP*WG(I)*BETA
      G(N-1) = G(N-1) + E2H/XN1
      G(N) = G(N) - ED2H
C
400   CONTINUE
      RETURN
      END
C
C-----------------------------------------------------------------------
C SUBROUTINE:  FMFP
C TO FIND THE LOCAL MINIMUM OF A FUNCTION OF SEVERAL VARIABLES
C BY THE METHOD OF FLETCHER AND POWELL
C-----------------------------------------------------------------------
C
      SUBROUTINE FMFP(N, X, F, G, EPS, LIMIT, IER, H)
      COMMON /CM25/ DEB, LONG
      COMMON /CM30/ FV(100), XIT(100), KOUNT
      COMMON /CM50/ M1, M2, M3, M4, M5, MODE
      COMMON /UNIT/ IND, INP, IOUTD, IOUTP, PI, SPM
      LOGICAL DEB, LONG
      DIMENSION H(201), X(N), G(N), Y(40)
C
C COMPUTE FUNCTION VALUE AND GRADIENT VECTOR FOR INITIAL ARGUME
C
      CALL FUNCT(N, X, F, G)
C
```

```
C
C RESET ITERATION COUNTER AND GENERATE IDENTITY MATRIX
C
      KOUNT = 0
      IER = 0
      N2 = N + N
      N3 = N2 + N
      N31 = N3 + 1
   10 K = N31
      IF (DEB) WRITE (IOUTD,9982)
      DO 30 J=1,N
        NJ = N - J
        H(K) = 1.0
        IF (NJ.LE.0) GO TO 40
        DO 20 L=1,NJ
          KL = K + L
          H(KL) = 0.0
   20   CONTINUE
        K = KL + 1
   30 CONTINUE
C
C START ITERATION LOOP, UPDATE COUNTER AND SAVE FUNCTION VALUE
C
   40 KOUNT = KOUNT + 1
      FV(KOUNT) = F
      XIT(KOUNT) = KOUNT
C
C SAVE FUNCTION VALUE, ARGUMENT VECTOR AND GRADIENT VECTOR
C
      IF (DEB .OR. LONG) WRITE (IOUTD,9999) F, KOUNT
      IF (DEB) WRITE (IOUTD,9997)
      IF (DEB) WRITE (IOUTD,9998) (X(J),G(J),J=1,N)
      OLDF = F
C
C DETERMINE DIRECTION VECTOR H
C
      DO 70 J=1,N
        K = N + J
        H(K) = G(J)
        K = K + N
        H(K) = X(J)
        T = 0.0
        K = J + N3
        DO 60 L=1,N
          T = T - G(L)*H(K)
          IF (L.GE.J) GO TO 50
          K = K + N - L
          GO TO 60
   50     K = K + 1
   60   CONTINUE
        H(J) = T
   70 CONTINUE
      IF (DEB) WRITE (IOUTD,9981)
      IF (DEB) WRITE (IOUTD,9996) (J,H(J),J=1,N)
C
C CHECK WHETHER FUNCTION WILL DECREASE STEPPING ALONG H
C
      GNRM = 0.0
      HNRM = 0.0
      DY = 0.0
C
C CALCULATE DIRECTIONAL DERIVATIVE AND TEST VALUES FOR DIRECTION
C VECTOR H AND GRADIENT VECTOR G.
C
      DO 80 J=1,N
        HNRM = HNRM + ABS(H(J))
        GNRM = GNRM + ABS(G(J))
        DY = DY + H(J)*G(J)
   80 CONTINUE
      IF (DEB) WRITE (IOUTD,9995) HNRM, GNRM
      IF (DEB) WRITE (IOUTD,9992) DY, F
C
C REPEAT SEARCH IN DIRECTION OF STEEPEST DESCENT IF DIRECTIONAL
C DERIVATIVE APPEARS TO BE POSITIVE OR ZERO OR IF H IS
C
C
      IF (DY.GE.0.0 .OR. (HNRM/GNRM).LE.EPS) GO TO 500
C
C SEARCH MINIMUM ALONG DIRECTION H
C SEARCH ALONG H FOR POSITIVE DIRECTIONAL DERIVATIVE
C
      FY = F
      HN = 1.0
      IF (HNRM.GT.1.0) HN = 1.0/HNRM
      EPSC = 2.*R1MACH(3)
      UC = 1.0 - EPSC
C
C FIND MAX. STEP SIZE TO ENSURE STABILITY
C
      IP1 = M3 + 1
      IP2 = M4
      IP3 = M4 + 2
      IP4 = M5 + 1
C
C CONSIDER REAL POLES
C
      IF (M3.EQ.M4) GO TO 100
      DO 90 J=IP1,IP2
        IF (ABS(X(J)+H(J)*HN).LT.UC) GO TO 90
        IF (H(J).LT.0.0) HN = (-UC-X(J))/H(J)
        IF (H(J).GT.0.0) HN = (UC-X(J))/H(J)
   90 CONTINUE
C
C CONSIDER COMPLEX POLES
C
  100 IF (M4.EQ.M5+1) GO TO 120
      DO 110 J=IP3,IP4,2
        R = X(J) + H(J)*HN
        IF (R.GT.0. .AND. R.LT.UC) GO TO 110
        IF (H(J).GT.0.0) HN = (UC-X(J))/H(J)
        IF (H(J).LT.0.0) HN = -X(J)/H(J)
  110 CONTINUE
  120 IF (DEB) WRITE (IOUTD,9994) HN
      DO 130 J=1,N
        Y(J) = X(J) + HN*H(J)
  130 CONTINUE
      AMBDA = HN
      CALL FUNCT(N, Y, F, G)
      DYN = DY*HN
      DYY = 0.0
      DO 140 J=1,N
        DYY = DYY + H(J)*G(J)
  140 CONTINUE
      IF (DEB) WRITE (IOUTD,9991) DYY, F
      IF (DYY.NE.0.0) GO TO 160
      DO 150 J=1,N
        X(J) = Y(J)
  150 CONTINUE
      GO TO 390
  160 CONTINUE
```

```
      DEN = F - FY - DYN
      IF (DEN.EQ.0.0) GO TO 170
      ALFA = -0.5*DYN/DEN*HN
      IF (DEB) WRITE (IOUTD,9990) ALFA, HN
C
C USE ESTIMATED STEP SIZE IF POSITIVE OTHERWISE USE HN
C
      IF (ALFA.LE.0.0) GO TO 170
      IF (F.GT.FY .AND. DYY.LT.0.) GO TO 170
      IF (DYY*(ALFA-HN).GE.0.0) GO TO 170
      AMBDA = ALFA
 170  ALFA = 0.0
C
C SAVE FUNCTION AND DERIVATIVE VALUES FOR OLD ARGUMENT
C
 180  DX = DY
      FX = FY
C
C STEP ARGUMENT ALONG H
C FIND MAX. STEP SIZE TO ENSURE STABILITY
C CONSIDER REAL POLES FIRST
C
      IF (M3.EQ.M4) GO TO 200
      DO 190 J=IP1,IP2
        IF (ABS(X(J)+H(J)*AMBDA).LT.UC) GO TO 190
        IF (H(J).LT.0.0) AMBDA = (-UC-X(J))/H(J)
        IF (H(J).GT.0.0) AMBDA = (UC-X(J))/H(J)
 190  CONTINUE
C
C CONSIDER COMPLEX POLES
C
 200  IF (M4.EQ.M5+1) GO TO 220
      DO 210 J=IP3,IP4,2
        R = X(J) + H(J)*AMBDA
        IF (R.GT.0. .AND. R.LT.UC) GO TO 210
        IF (H(J).GT.0.0) AMBDA = (UC-X(J))/H(J)
        IF (H(J).LT.0.0) AMBDA = -X(J)/H(J)
 210  CONTINUE
 220  DO 230 I=1,N
        X(I) = X(I) + AMBDA*H(I)
 230  CONTINUE
      IF (DEB) WRITE (IOUTD,9994) AMBDA
      IF (DEB) WRITE (IOUTD,9993) (X(I),I=1,N)
C
C COMPUTE FUNCTION VALUE AND GRADIENT FOR NEW ARGUMENT
C
      CALL FUNCT(N, X, F, G)
      FY = F
C
C COMPUTE DIRECTIONAL DERIVATIVE DY FOR NEW ARGUMENT. TERMINATE
C SEARCH IF DY IS POSITIVE. IF DY IS ZERO THE MINIMUM IS FOUND
C
      DY = 0.0
      DO 240 I=1,N
        DY = DY + H(I)*G(I)
 240  CONTINUE
      IF (DEB) WRITE (IOUTD,9989) DY, F
      DYMN = AMIN1(-DX,DY)
      OF = F
      IF (DY) 250, 400, 260
C
C TERMINATE SEARCH ALSO IF THE FUNCTION VALUE INDICATES THAT
C A MINIMUM HAS BEEN PASSED
C
 250  IF (FY.GE.FX) GO TO 260
      IF (ABS(DX-DY).LE.EPSC) GO TO 520
C
C REPEAT SEARCH AND DOUBLE STEP SIZE FOR FURTHER SEARCHES
C
      AMBDA = AMBDA + ALFA
      ALFA = AMBDA
C
C END OF SEARCH LOOP
C TERMINATE IF THE CHANGE IN ARGUMENT GETS VERY LARGE
C
      IF (HNRM*AMBDA.LE.1.0) GO TO 180
C
C LINEAR SEARCH TECHNIQUE INDICATES THAT NO MINIMUM EXISTS
C
      IER = 2
      RETURN
C
C INTERPOLATE CUBICALLY IN THE INTERVAL DEFINED BY THE SEARCH
C ABOVE AND COMPUTE THE ARGUMENT X FOR WHICH THE INTERPOLATION
C POLYNOMIAL IS MINIMIZED
C
 260  T = 0.0
 270  CONTINUE
      IF (DEB) WRITE (IOUTD,9987) FX, FY
      IF (DEB) WRITE (IOUTD,9986) DX, DY
      IF (AMBDA.EQ.0.0) GO TO 390
      Z = 3.0*(FX-FY)/AMBDA + DX + DY
      DALFA = Z*Z - DX*DY
      IF (DALFA.LT.0.0) GO TO 500
      W = SQRT(DALFA)
      ALFA = DY - DX + W + W
      IF (ALFA.EQ.0.0) GO TO 280
      ALFA = (DY-Z+W)/ALFA
      GO TO 290
 280  ALFA = (Z+DY-W)/(Z+DX+Z+DY)
 290  ALFA = ALFA*AMBDA
      DO 300 I=1,N
        X(I) = X(I) + (T-ALFA)*H(I)
 300  CONTINUE
C
C TERMINATE IF THE VALUE OF THE ACTUAL FUNCTION AT X IS LESS
C THAN THE FUNCTION VALUES AT THE INTERVAL ENDS. OTHERWISE REDUCE
C THE INTERVAL BY CHOOSING ONE END POINT EQUAL TO X AND REPEAT
C THE INTERPOLATION. WHICH END POINT IS CHOSEN DEPENDS ON THE
C VALUE OF THE FUNCTION AND ITS GRADIENT AT X.
C
      CALL FUNCT(N, X, F, G)
      DALFA = 0.0
      DO 310 I=1,N
        DALFA = DALFA + H(I)*G(I)
 310  CONTINUE
      IF (DEB) WRITE (IOUTD,9988) DALFA, F
      FYE = 1.0 - FY/F - EPS
      FXE = 1.0 - FX/F - EPS
      IF (DEB) WRITE (IOUTD,9985) FXE, FYE
      IF (FXE.GT.0.0) GO TO 320
      IF (FYE.LE.0.0) GO TO 390
 320  DALFA = 0.0
      DO 330 I=1,N
        DALFA = DALFA + H(I)*G(I)
 330  CONTINUE
      IF (DALFA.GE.0.0) GO TO 360
      IF (FXE) 350, 340, 360
 340  IF (DX.EQ.DALFA) GO TO 390
 350  FX = F
```

```
      AMBDA = ALFA
      T = AMBDA
      DX = DALFA
      GO TO 270
 360  IF (FYE) 380, 370, 380
 370  IF (DY.EQ.DALFA) GO TO 390
 380  AMBDA = AMBDA - ALFA
      DY = DALFA
      FY = F
      GO TO 260
C
C IF DIR. DER. AT X NOT << DIR. DER. AT INTERVAL ENDS, REPEAT INTERP
C
 390  DER = ABS(DALFA/DYMN)
      IF (DER.LE.1.0E-2 .OR. ABS(1.0-OF/F).LE.EPS) GO TO 400
      OF = F
      IF (DEB) WRITE (IOUTD,9983) DER, OLDF
      IF (DALFA) 350, 350, 380
 400  CONTINUE
C
C TERMINATE IF FUNCTION HAS NOT BEEN REDUCED DURING LAST ITERATION
C
      IF (OLDF+EPS.LT.F) GO TO 500
C
C COMPUTE DIFFERENCE VECTORS OF ARGUMENT AND GRADIENT FROM
C TWO CONSECUTIVE ITERATIONS
C
      DO 410 J=1,N
        K = N + J
        H(K) = G(J) - H(K)
        K = N + K
        H(K) = X(J) - H(K)
 410  CONTINUE
      IF (DEB) WRITE (IOUTD,9988) DALFA, F
C
C TEST LENGTH OF ARGUMENT DIFFERENCE VECTOR AND DIRECTION VECTOR
C IF AT LEAST N ITERATIONS HAVE BEEN EXECUTED. TERMINATE IF
C BOTH ARE LESS THAN EPS
C
      IER = 0
      IF (KOUNT.LT.N) GO TO 430
      T = 0.0
      DO 420 J=1,N
        K = N2 + J
        T = T + ABS(H(K))
 420  CONTINUE
      IF (HNRM.GT.EPS) GO TO 430
      IF (T.LE.EPS) GO TO 540
C
C TERMINATE IF NUMBER OF ITERATIONS WOULD EXCEED LIMIT
C
 430  IF (KOUNT.GE.LIMIT) GO TO 490
C
C PREPARE UPDATING OF MATRIX H
C
      Z = 0.0
      ALFA = Z
      DO 460 J=1,N
        W = 0.0
        K = J + N3
        DO 450 L=1,N
          KL = N + L
          W = W + H(KL)*H(K)
          IF (L.GE.J) GO TO 440
          K = K + N - L
          GO TO 450
 440      K = K + 1
 450    CONTINUE
        K = N + J
        KN = K + N
        ALFA = ALFA + W*H(K)
        Z = Z + H(K)*H(KN)
        H(J) = W
 460  CONTINUE
      IF (DEB) WRITE (IOUTD,9984) Z, ALFA
C
C REPEAT SEARCH IN DIRECTION OF STEEPEST DESCENT IF RESULTS
C ARE NOT SATISFACTORY
C
      IF (Z*ALFA.EQ.0.0) GO TO 10
C
C UPDATE MATRIX H
C
      K = N31
      DO 480 L=1,N
        KL = N2 + L
        DO 470 J=L,N
          NJ = N2 + J
          H(K) = H(K) + H(KL)*H(NJ)/Z - H(L)*H(J)/ALFA
          K = K + 1
 470    CONTINUE
 480  CONTINUE
      GO TO 40
C
C END OF ITERATION LOOP
C NO CONVERGENCE AFTER LIMIT ITERATIONS
C
 490  IER = 1
      RETURN
C
C RESTORE OLD VALUES OF FUNCTION AND ARGUMENTS
C
 500  DO 510 J=1,N
        K = N2 + J
        X(J) = H(K)
 510  CONTINUE
      CALL FUNCT(N, X, F, G)
C
C REPEAT SEARCH IN DIRECTION OF STEEPEST DESCENT IN DERIVATIVE
C FAILS TO BE SUFFICIENTLY SMALL
C
      IF (GNRM.LE.EPS) GO TO 530
C
C TEST FOR REPEATED FAILURE OF ITERATION
C
 520  IF (IER.LT.0) GO TO 540
      IER = -1
      GO TO 10
 530  IER = 0
 540  RETURN
C
C FORMATS
C
9999  FORMAT (8H OLDF = , E11.4, 10X, 13HITERATION NO., I3)
9998  FORMAT (//(2E30.5))
9997  FORMAT (//31H FUNCTION AND GRADIENT VECTORS-)
9996  FORMAT (3(5H   H(, I3, 4H) = , E13.6))
9995  FORMAT (//9H HNRM   =, E12.5, 10X, 7HGNRM  =, E12.5)
9994  FORMAT (//11H STEP SIZE=, E12.5)
9993  FORMAT (16H NEW ARG. VECTOR/(4E15.7))
```

```
9992  FORMAT (//9H DY(XO) =, E12.5, 10X, 7HF(XO) =, E12.5)
9991  FORMAT (//10H DY(XO+H)=, E12.5, 10X, 8HF(XO+H)=, E12.5)
9990  FORMAT (//6H ALFA=, E12.5, 10X, 3HHN=, E12.5)
9989  FORMAT (//12H DY(XO+A H)=, E12.5, 10X, 10HF(XO+A H)=, E12.5)
9988  FORMAT (//7H DYMIN=, E14.7, 10X, 4HFMN=, E14.7)
9987  FORMAT (//4H FX=, E14.7, 10X, 3HFY=, E14.7)
9986  FORMAT (//4H DX=, E14.7, 10X, 3HDY=, E14.7)
9985  FORMAT (//5H FXE=, E14.7, 10X, 4HFYE=, E14.7)
9984  FORMAT (//3H Z=, E14.7, 10X, 5HALFA=, E14.7)
9983  FORMAT (/5H DER=, E14.7, 10X, 6H OLDF=, E14.7)
9982  FORMAT (//38H ************** START STEEPEST DESCENT)
9981  FORMAT (//30X, 20H DIRECTION VECTOR H //)
      END
```

6.3

An Optimization Program for the Design of Digital Filter Transfer Functions

M. T. Dolan and *J. F. Kaiser*

Digital Systems Research Dept.
Bell Laboratories
Murray Hill, NJ 07974

1. Purpose

Given the cascade realization of a digital filter transfer function, this program varies the coefficents until the performance meets arbitrary frequency-domain specifications on the magnitude. The response is calculated as insertion loss in dB.

2. Method

The method solves the design problem by working directly with the coefficients of the transfer function

$$H(z^{-1};\boldsymbol{\alpha}) = \alpha_0 \prod_{i=1}^{n} \frac{\alpha_{1i}z^{-2} + \alpha_{2i}z^{-1} + 1}{\alpha_{3i}z^{-2} + \alpha_{4i}z^{-1} + 1} \tag{1}$$

where $\boldsymbol{\alpha} = [\alpha_0, \alpha_{11}, \alpha_{21}, \alpha_{31}, \alpha_{41}, \ldots, \alpha_{4n}]'$, $z^{-1} = e^{-sT}$ and T is the sampling period. Note that $N{=}4n$ is the total number of coefficients in n second-order sections. For a given sampling rate, the system is completely defined if n and $\boldsymbol{\alpha}$ are known. This initial design is usually obtained by standard approximation techniques [1].

Let $L(\omega;\boldsymbol{\alpha})$ be a function of the filter on which insertion loss specifications are imposed and $l_1(\omega)$, $l_2(\omega)$ be two given real-valued functions of ω defined, respectively, on Ω_1 and Ω_2, two not-necessarily-disjoint intervals on the frequency axis. Then, all the common filter or equalizer specifications can be expressed by requiring the following inequalities to be satisfied:

$$f_1(\omega;\boldsymbol{\alpha}) \equiv L(\omega;\boldsymbol{\alpha}) - l_1(\omega) \geqslant 0, \quad \omega \subset \Omega_1 \tag{2}$$

$$f_2(\omega;\boldsymbol{\alpha}) \equiv l_2(\omega) - L(\omega;\boldsymbol{\alpha}) \geqslant 0, \quad \omega \subset \Omega_2$$

A vector $\boldsymbol{\alpha}$ of a given dimension which satisfies the inequalities belongs to the *specification set S*,

$$S \equiv \{\boldsymbol{\alpha} |\ f_1(\omega;\boldsymbol{\alpha}) \geqslant 0,\ \omega \subset \Omega_1;\ f_2(\omega;\boldsymbol{\alpha}) \geqslant 0,\ \omega \subset \Omega_2\} \tag{3}$$

A stable digital filter for which $\boldsymbol{\alpha} \subset S$ will be called *acceptable,* the philosophy being that a solution (not necessarily a global optimum) satisfies the specifications. There are no constraints on stability. Since we are dealing with insertion loss only, if a final design is unstable, it is made stable by inversion of the poles which lie inside the unit circle in the z^{-1} plane.

The specification set S is defined in (3) through an infinite number of constraints. Because of the discrete nature of the computation, the user must select a finite number of discrete frequencies ω_k in Ω_1 and Ω_2. The program works with the discretized specification set

$$S_d \equiv \{\ \boldsymbol{\alpha} | f_1(\omega_k;\boldsymbol{\alpha}) \equiv L(\omega_k;\boldsymbol{\alpha}) - l_1(\omega_k) \geqslant 0,\ \omega_k \subset \Omega_1,\ k{=}1,2,\ldots,m_1; \tag{4}$$

$$f_2(\omega_k;\boldsymbol{\alpha}) \equiv l_2(\omega_k) - L(\omega_k;\boldsymbol{\alpha}) \geqslant 0,\ \omega_k \subset \Omega_2,\ k{=}m_1{+}1,\ldots,m\}$$

Figure 1 is an example of a lowpass filter $H(z^{-1};\boldsymbol{\alpha})$ with frequency response $L(\omega_k;\boldsymbol{\alpha})$. The filled circles mark the selected frequencies ω_k. The vector $\boldsymbol{\alpha}$ belongs to the discretized specification set S_d of (4) which in this case is a subset of S since the specifications are not violated between the selected frequencies. Note that S_d is, of course, not necessarily a subset of S because there may exist $\boldsymbol{\alpha} \subset S_d$ and $\omega \subset \Omega_1 \cup \Omega_2$ which do not satisfy the original specifications (2). This makes the selection of the

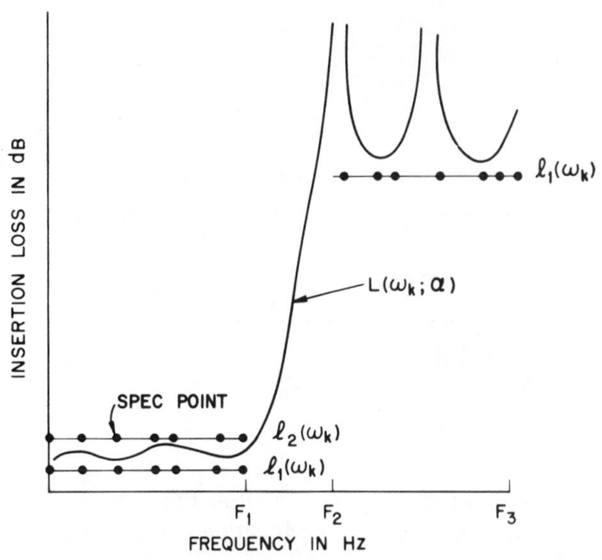

Fig. 1 Lowpass filter response meeting discretized specifications
• — selected frequencies ω_k
$\Omega_1: \omega_k,\ 0 \leqslant \omega_k \leqslant F_1,\ k = 1,2,...,m_0$ and $F_2 \leqslant \omega_k \leqslant F_3,\ k = m_0+1,...,m_1$
$\Omega_2: \omega_k,\ 0 \leqslant \omega_k \leqslant F_1,\ k = m_1+1,...,m$

frequencies ω_k somewhat of an art because one would like to keep their number, m, as low as possible (in order to economize computation) but, at the same time, consider enough constraints to have a reasonable assurance that the solution α^* of an optimization problem based on S_d will also belong to S.

The optimality criterion adopted is of the following min-max character: If a filter of a given order is acceptable, then maximize the minimum amount by which it exceeds the specifications; if it is not acceptable, then minimize the maximum amount by which it fails.

The problem is expressed mathematically by making use of an additional variable α_{N+1} (excursion from specifications) and introducing non-negative weights w_{1k} and w_{2k} multiplying α_{N+1} in each constraint. The problem becomes:
minimize

$$\alpha_{N+1}$$

subject to

$$C_{1k}(\tilde{\alpha}) \equiv w_{1k}\alpha_{N+1} + f_1(\omega_k;\alpha) \geqslant 0,\ k=1,2,...,m_1 \tag{5}$$

$$C_{2k}(\tilde{\alpha}) \equiv w_{2k}\alpha_{N+1} + f_2(\omega_k;\alpha) \geqslant 0,\ k=m_1+1,m_1+2,...,m$$

The optimization is performed in the space of the augmented vector $\tilde{\alpha} = [\alpha|\alpha_{N+1}]'$. Clearly, if α^*_{N+1}, the final value of α_{N+1}, is nonpositive, then the filter meets the discretized specifications. If $\alpha^*_{N+1} > 0$, then the final filter is not acceptable and α^*_{N+1} represents the minimum of the maximum amount by which the constraints of S_d in (4) cannot be satisfied. Similar formulations have been used in approximation methods [2], in the parameter optimization of networks in the time [3] or frequency domain [4,5,6], and in the design of antennas [7].

The values of the input weights may be adjusted by the user so that certain regions of Ω_1 or Ω_2 require greater attention than others.

The solution of the nonlinear program of (5) by a sequence of unconstrained minimizations is accomplished through the use of an auxiliary or "penalty" function $A(\tilde{\alpha};r_\nu)$, where r_ν is a positive parameter. The auxiliary function $A(\tilde{\alpha};r_\nu)$ consists of the objective function plus a penalty term which is a monotonic function of the constraints multiplied by r_ν and so formulated that it reduces the value of $A(\tilde{\alpha};r_\nu)$ when $\tilde{\alpha}$ belongs to $\tilde{S}_d$ and/or "penalizes" the minimization of $A(\tilde{\alpha};r_\nu)$ when $\tilde{\alpha}$ is on the boundary or the exterior of the specification set $\tilde{S}_d$. $A(\tilde{\alpha};r_\nu)$ is minimized sequentially over all $\tilde{\alpha} \subset R^{N+1}$ for a sequence of r_ν, $\nu = 1,2,...$, giving minima $\tilde{\alpha}^*(r_\nu)$. If $A(\tilde{\alpha};r_\nu)$ and the sequence $<r_\nu>$ are properly constructed, then the sequence $<\tilde{\alpha}^*(r_\nu)>$ converges to the solution $\tilde{\alpha}^*$ of the original nonlinear program [8]. In all cases, as the auxiliary function is minimized for a sequence of values r_ν, the penalty term tends to zero so that the successive minima of the auxiliary function converge to the minimum of the objective function.

Two auxiliary functions used in the program are the following:

$$A_1(\tilde{\alpha};r_\nu) = \alpha_{N+1} + r_\nu \sum_{k=1}^{m_1} \frac{1}{C_{1k}(\tilde{\alpha})} + r_\nu \sum_{k=m_1+1}^{m} \frac{1}{C_{2k}(\tilde{\alpha})} \tag{6}$$

$$A_2(\tilde{\alpha};r_\nu) = \alpha_{N+1} - r_\nu \sum_{k=1}^{m_1} \ln(C_{1k}(\tilde{\alpha})) - r_\nu \sum_{k=m_1+1}^{m} \ln(C_{2k}(\tilde{\alpha})) \tag{7}$$

When either of these functions is minimized sequentially, for a monotonically decreasing sequence of positive r_ν (which tend to 0), then the sequence of the successive global minima $\tilde{\alpha}^*(r_\nu)$ converges to the solution of (5). A detailed treatment of the properties of these techniques is given in chapter 3 of reference [8].

Functions (6) and (7) are often called "interior" because they have the property that, if the initial value of $\tilde{\alpha}$ belongs to $\tilde{S}_d$, then all the subsequently generated solutions $\tilde{\alpha}^*(r_\nu)$ are in $\tilde{S}_d$ as well. Both (6) and (7) approach infinity as any constraint approaches zero; therefore, any minimization procedure will tend to avoid the boundary of the constraint set. In interior methods, each time r_ν is reduced, the weighting of the penalty term is decreased while the weighting of the objective function is increased. As for the initial value of $\tilde{\alpha}$ belonging to $\tilde{S}_d$, one can always find such an initial point by taking the initial value of α_{N+1} large enough. When an interior penalty function is selected, the program automatically determines this initial value for the user.

Auxiliary functions which can work with points outside of $\tilde{S}_d$ are called "exterior". The program incorporates one such function as an option:

$$A_3(\tilde{\alpha};r_\nu) = \alpha_{N+1} + r_\nu \sum_{k=1}^{m_1} [\hat{C}_{1k}(\tilde{\alpha})]^2 + r_\nu \sum_{k=m_1+1}^{m} [\hat{C}_{2k}(\tilde{\alpha})]^2 \tag{8}$$

where

$$\begin{aligned} \hat{C}_{1k}(\tilde{\alpha}) &= 0 && \text{if } C_{1k}(\tilde{\alpha}) \geqslant 0 \\ &= C_{1k}(\tilde{\alpha}) && \text{if } C_{1k}(\tilde{\alpha}) < 0 \end{aligned}$$

and similarly for $\hat{C}_{2k}(\tilde{\alpha})$. This function is minimized sequentially for a monotonically increasing sequence of positive r_ν (which tend to ∞) and has analogous convergence properties with those of (6) or (7). See [9] or Chapter 4 of [8]. In exterior methods, the penalty term prevents the points from straying too far from the feasible region. Only the constraints (5) which are violated enter the computation at any time. It is especially true in exterior methods that the sign of α_{N+1} is not significant before convergence. Convergence implies that subsequent values of r_ν will reduce α_{N+1} by amounts too small to justify the expenditure. In a wide range of problems, four values of r_ν, i.e., four major iterations have been sufficient. Typically, in exterior methods, the initial r_ν is 1 and is multiplied by 10 in each succeeding major iteration. In interior methods, the initial r_ν is usually less than 1 and is divided by 10 in succeeding major iterations.

The test program of the Appendix provides an option for automatic selection of many of the optimization parameters for both types of auxiliary functions.

The unconstrained minimizations of the above auxiliary functions begin with an initial point $\tilde{\boldsymbol{\alpha}}_0$ and generate a sequence of points $<\tilde{\boldsymbol{\alpha}}_i>$, such that

$$A(\tilde{\alpha}_{i+1};r_\nu) \leqslant A(\tilde{\alpha}_i\,;r_\nu) \tag{9}$$

by taking a step σ_i along a direction vector $\boldsymbol{d}_i \subset R^{N+1}$. Thus,

$$\tilde{\alpha}_{i+1} = \tilde{\alpha}_i + \sigma_i \boldsymbol{d}_i \tag{10}$$

A class of unconstrained minimization methods, known as quasi-Newton methods, compute the direction vectors as

$$d_i = -Q_i g_i \tag{11}$$

where $g_i = g(\tilde{\alpha}_i)$ is the gradient vector of $A(\tilde{\alpha};r_\nu)$ with respect to $\tilde{\alpha}$ and Q_i is a positive-definite $(N+1)\times(N+1)$ matrix which is updated at each iteration. Broyden [10] suggested the following general updating expressions

$$Q_{i+1} = Q_i - Q_i\gamma_i p_i^t + \beta_i q_i^t \tag{12}$$

where

$$\gamma_i \equiv g_{i+1} - g_i \tag{13}$$

$$\beta_i \equiv \tilde{\alpha}_{i+1} - \tilde{\alpha}_i \tag{14}$$

and the vectors p_i and q_i are arbitrary except for the normalizing condition

$$p_i^t\gamma_i = q_i^t\gamma_i = 1 \tag{15}$$

The specialization of (12) included in the program is the well-known method of Davidon-Fletcher-Powell [11].

$$p_i = (\gamma_i^t Q_i \gamma_i)^{-1} Q_i^t \gamma_i \tag{16a}$$

$$q_i = (\beta_i^t \gamma_i)^{-1} \beta_i \tag{16b}$$

The length of the step σ_i is determined at each iteration by minimizing $A(\tilde{\alpha};r_\nu)$ along d_i by means of a quadratic approximation of the function along d_i.

Finally, there is a choice of algorithms for "stop" criteria. Note that convergence to a negative α_{N+1} indicates that the frequency response satisfies the specifications at the frequencies chosen by the user. Users should check the resulting transfer function by generating a frequency response using the subroutine CASC of the Appendix. Comments in the subroutine include notes for its use outside the package. If specifications are violated between the frequencies selected, additional frequencies should be included. The optimization is greatly facilitated by a good initial design and a judicious choice of frequencies at which the specifications must be met.

3. Program Description

3.1 Usage

The package consists of a test program and nine subroutines - ZR, FLPWL, STEP, QUAD, FUNC, FG, CASC, RINV, FINISH. The test program and the subroutines are in the Appendix.

The test program reads in the data required to describe the digital filter transfer function and the frequency-domain specifications calculated as insertion loss in dB. Various optimization parameters must also be read in. The coefficients are varied by the appropriate subroutines until the "stop" criterion is satisfied.

For the sake of portability, all data is read in as formatted input. In some situations such as in the case of linear frequencies or analytic specifications, users may want to modify the test program to generate such data or read it from files.

ZR calculates the initial α_{N+1} (ALPHA) and sets the initial value r_ν (R) for the chosen penalty function (6), (7), or (8). Users have the option to set the initial r_ν.

FLPWL computes the direction vector (11) for each iteration.

STEP sets the initial value of σ_i of (10).

QUAD performs on the function to be minimized quadratic interpolations for successive minima.

FUNC evaluates the penalty function.

FG evaluates the penalty function and its gradient.

CASC calculates the frequency response and its gradient. The response is calculated as insertion loss in dB.

RINV checks that the modified transfer function printed after each major iteration is stable. Since the coefficients are in the z^{-1} plane, poles are inverted if necessary so that they lie outside the unit circle.

FINISH checks whether or not the stop criterion has been met and takes the appropriate action.

3.2 Description of Required Data

The notation below is not quite consistent with that in the earlier general discussion; it is instead consistent with the actual code and printout.

NORD (ORDER) The order of the filter must be an even number, NORD$\leqslant$ 20 (NORD$=2n$ where n is the number of second-order sections). For odd order filters, use the next higher even order, set appropriate coefficients to zero and request that they remain constant. For example, if the numerator and denominator are both to be seventh order, set $A2_4$ and $B2_4$ equal to zero in the array of coefficients described below.

T (SAMPLING TIME IN SECONDS) T = 1/sampling-frequency.

XK (GAIN FACTOR) On input, XK contains the gain factor associated with the initial design. On output, it is the gain factor associated with the final design.

X (ARRAY of COEFFICIENTS) On input, X contains the coefficients of the transfer function of the initial design. For example, let

$$H(z^{-1};x) = g\times\prod_{i=1}^{n}\frac{A2_iz^{-2}+A1_iz^{-1}+1}{B2_iz^{-2}+B1_iz^{-1}+1}$$

The system expects the number of coefficients to be twice the order of the filter and in the following sequence:

$$x = [A2_1,A1_1,B2_1,B1_1,A2_2,A1_2,B2_2,B1_2,\ldots,A2_n,A1_n,B2_n,B1_n]'$$

On output, X contains the coefficients of the optimized transfer function.

NC (NUMBER OF COEFFICIENTS TO REMAIN CONSTANT) Any of the $4n$ coefficients may be held fixed during the optimization. NC = total count of those coefficients that must not change.

NL (ARRAY OF INDICES OF FIXED COEFFICIENTS) NL contains the indices of coefficients to be held fixed. In the coefficient vector, x, index 1 refers to $A2_1$, 2 refers to $A1_1$, etc.

NG (GAIN OPTION) If NG = 1, the optimizer changes the initial gain factor in addition to the coefficients free to change. If NG = 0, the gain remains unchanged. In most cases, it is advisable that the gain factor be free to vary.

NF (NUMBER OF FREQUENCIES FOR OPTIMIZATION) NF = count of the frequencies available to the optimizer - limit is 42.

FREQ (ARRAY OF FREQUENCIES IN HZ) The goal of the optimizer is to satisfy the specifications at these selected frequencies.

N1 (INDEX FOR TOP RANGE (1 - N1) OF LOWER SPECIFICATIONS) If, for example, the lower specifications apply to the 1st through 10th frequencies, the user would input 10 for N1. The 1 at the beginning is fixed. This is not a serious restriction since frequencies do not have to be in any special order. The first N1 can be chosen from anywhere in the range. Of course, the desired order must be set up in the frequency array. Note that the goal of the optimizer is to keep the loss above (greater than) the lower specifications at the given frequencies.

SPEC1 (ARRAY OF LOWER SPECIFICATION DATA) Contains lower specifications calculated as loss in dB.

TH1 (ARRAY OF WEIGHTS FOR LOWER SPECIFICATIONS) Contains non-negative weights associated with the lower specifications.

N2 (LOWER INDEX FOR RANGE (N2 - N3) OF UPPER SPECIFICATIONS) If, for example, the upper specifications apply at the 10th through 20th frequencies, the user inputs N2 equal to 10.

N3 (UPPER INDEX FOR RANGE (N2 - N3) OF UPPER SPECIFICATIONS) When the upper specifications apply at the 10th through 20th frequencies, e.g., the user inputs N3 equal to 20. Note that the optimizer works to keep the loss below (less than) the upper specifications at the given frequencies.

SPEC2 (ARRAY OF UPPER SPECIFICATION DATA) Contains the upper specification data calculated as loss in dB.

TH2 (ARRAY OF WEIGHTS FOR UPPER SPECIFICATIONS) Contains non-negative weights associated with the upper specifications.

NGID (OPTIMIZATION GUIDE OPTION) Set NGID=1 and no additional input is necessary. The remaining parameters will be set by the test program. Set NGID=0 to guide the optimization, i.e., to choose values for the remaining parameters.

NOUT (OUTPUT OPTION) NOUT=0 for final results of major iterations, NOUT=1 for detailed output of minor iterations. Along with the initial coefficients and the R value, final results include the final coefficients for each major iteration, the gain factor, maximum excursion from specifications (ALPHA), penalty function value, modified coefficients (poles in z^{-1} plane made to lie outside unit circle), corresponding modified gain factor, the gradient vector, the number of function evaluations, and the number of gradient evaluations. Detailed output includes, of course, the results described above and in addition, for each minor iteration, the coefficient vector, gain factor, maximum excursion from specifications (ALPHA), penalty function value, initial step, and the number of times the constraints were violated during quadratic fit. Note that the number of minor iterations corresponds to the number of gradient evaluations. On rare occasions, it may be of interest to follow the optimization, but the user is warned that NOUT=1 often produces a wallpaper effect.

NN (INITIAL STEP OPTION - NN=1 FOR STEP BASED ON MAX. COEF. EXPECTED; 2 FOR STEP BASED ON APPROX. FUNC. MIN.; 3 FOR OPTION 1 ONCE THEN DIVISION BY 2) For option 1 user is expected to provide an estimate of the maximum coefficient expected (VBND).

$$\text{STEP}=\min\left(2,\ \frac{\text{VBND}}{\textit{max. abs. component of direction vector}}\right)$$

For option 2 user is expected to provide an estimate of the function minimum (VBND).

$$\text{STEP}=\min\left(2,\ \frac{-2\times(\textit{func.}-\text{VBND})}{\textit{gradient transpose}\times\textit{direction vector}}\right)$$

Since option 3 uses option 1 once, user is expected to provide VBND. Subsequently, the current step will be divided by 2 for a new step.

VBND (PARAMETER FOR INITIAL STEP OPTION) For step options 1 or 3, VBND is the maximum coefficient expected. If user inputs 0., default option sets VBND equal to .05 × (*initial maximum absolute coefficient*). For step option 2, VBND is the approximate function minimum. There is no default option.

NPEN (PENALTY FUNCTION - NPEN=1 FOR RECIPROCAL INTERIOR; 2 FOR LOGARITHMIC INTERIOR; 3 FOR PARAMETRIC EXTERIOR) Penalty Function:

$$P = \alpha + r_\nu\,\phi[c(x)]$$

where α = the maximum excursion from specifications and r_ν is the positive parameter of (6), (7), and (8) of the general discussion.

Option 1 sets

$$\phi[c(x)] = \sum_{k=1}^{\text{N1}} \frac{1}{c_{1k}} + \sum_{k=\text{N2}}^{\text{N3}} \frac{1}{c_{2k}}$$

Option 2 sets

$$\phi[c(x)] = -\left[\sum_{k=1}^{N1} ln(c_{1k}) + \sum_{k=N2}^{N3} ln(c_{2k})\right]$$

Option 3 sets

$$\phi[c(x)] = \sum_{k=1}^{N1} \hat{c}_{1k}^2 + \sum_{k=N2}^{N3} \hat{c}_{2k}^2$$

$$\text{where } \hat{c}_{1k} = \begin{cases} 0 & \text{if } c_{1k} \geqslant 0 \\ c_{1k} & \text{if } c_{1k} < 0 \end{cases}$$

and similarly for $\hat{c}_{2k}$.

The constraint set $c(x)$ is defined as follows

$$c_{1k} = L(f_k) + w_{1k}\alpha - \hat{l}_{1k} \geqslant 0 \qquad k = 1,2,3,\ldots,N1$$

$$c_{2k} = -L(f_k) + w_{2k}\alpha + \hat{l}_{2k} \geqslant 0 \qquad k = N2,N2+1,\ldots,N3$$

where

$L(f_k)$ is the filter loss in dB at the kth frequency.

w_{1k} is the weight associated with the lower specification at the kth frequency.

$\hat{l}_{1k}$ is the lower specification at the kth frequency.

w_{2k} is the weight associated with the upper specification at the kth frequency.

$\hat{l}_{2k}$ is the upper specification at the kth frequency.

α is the maximum excursion from the specifications.

KP (STOP CRITERION FOR MINOR ITERATIONS. KP=1 FOR STOP BASED ON GRADIENT CHANGES; 2 FOR STOP BASED ON FUNCTION CHANGES; 3 FOR OPTION 2 SATISFIED A GIVEN NO. OF TIMES; 4 FOR STOP BASED ON COEFFICIENT CHANGES; 5 FOR STOP AFTER FIXED NO. OF ITERATIONS) Option 1 - stop when gradient transpose times gradient is less than some value given by user. Option 2 - stop when present function value minus previous function value is less than .0001 times present value. Option 3 - stop when option 2 is satisfied a number of times set by the user. Option 4 - stop when maximum absolute coefficient minus previous maximum absolute coefficient is less than a given value set by the user. Option 5 - stop after a given number of iterations set by user.

EPS (EPS - PARAMETER FOR MINOR STOP OPTION) This parameter is used in stop options 1 and 4. The default value for option 1 is gradient-transpose times gradient divided by 1.E+06. Default value for option 4 is .0005 times present maximum absolute coefficient. Input 0. for default.

KTIME (PARAMETER FOR MINOR STOP OPTIONS) This parameter is used in stop options 3 and 5. For option 3 the default number is 4.

NFINR (STOP CRITERION FOR MAJOR ITERATIONS. NFINR=1 FOR STOP BASED ON ALPHA CHANGES OR SIZE OF CONSTRAINTS-SQUARED; 2 FOR STOP AFTER FIXED NO. OF R-VALUES) Option 1 - for interior reciprocal and logarithmic penalty functions - stop when present ALPHA value minus previous ALPHA value is less than a given percentage of the present value. User provides percentage. Option 1 - for parametric exterior penalty function - Program stops when sum of constraints-squared is less than a given value. User provides the value. Option 2 - may be chosen for any penalty function - User provides the number of major iterations desired.

PERR (PERCENTAGE) This parameter is used for stop option 1 with interior penalty functions.

TAZ (SUM OF CONSTRAINTS-SQUARED) Parameter used only with stop option 1 with exterior penalty function.

NR (NUMBER OF R-VALUES) User is expected to provide this number for stop option 2.

INIR (R GUIDE OPTION - INPUT 0 FOR AN AUTOMATIC INITIAL R-VALUE AND CHANGE FACTOR, OTHERWISE INPUT 1)
If the user inputs 0, a good starting r-value will be calculated based on the assumption that the two terms in the penalty function be of the same order of magnitude. Also the change factor is set to 10.

RINI (INITIAL R-VALUE) Used only if the user set INIR=1.

RCHNG (R CHANGE FACTOR) Used only if user set INIR=1.

4. Test Problem

Design a lowpass filter with cutoff frequency 4000 Hz for a sampling frequency of 32000 Hz. The following limits are imposed on the magnitude-frequency characteristic:

LOWER SPECIFICATIONS

In the passband, the insertion loss is required to be greater than

$$-0.1 + 20.\times\log_{10}\left(\frac{\sin(\pi f/8000)}{\pi f/8000}\right)^2 \quad dB$$

In the stopband, the insertion loss is required to be greater than

$$30. + 20.\times\log_{10}\left(\frac{\sin(\pi f/8000)}{\pi f/8000}\right)^2 \quad dB$$

UPPER SPECIFICATIONS

In the passband, the insertion loss is required to be less than

$$+0.1 + 20.\times\log_{10}\left(\frac{\sin(\pi f/8000)}{\pi f/8000}\right)^2 \quad dB$$

For an initial design, the choice is a lowpass sixth-order elliptic filter with a minimum stopband loss of 30 dB and a ripple factor of 0.2 dB.

The main program (Appendix) reads in the parameters described in Section 3.2 and for each major iteration prints, among other things, the transfer function, the gain factor, and the maximum excursion from the specifications, ALPHA. Convergence to a negative ALPHA means that the filter meets the specifications at the frequencies selected.

Table 1 contains the input data and Table 2 the output. Figure 2 shows the frequency response for the

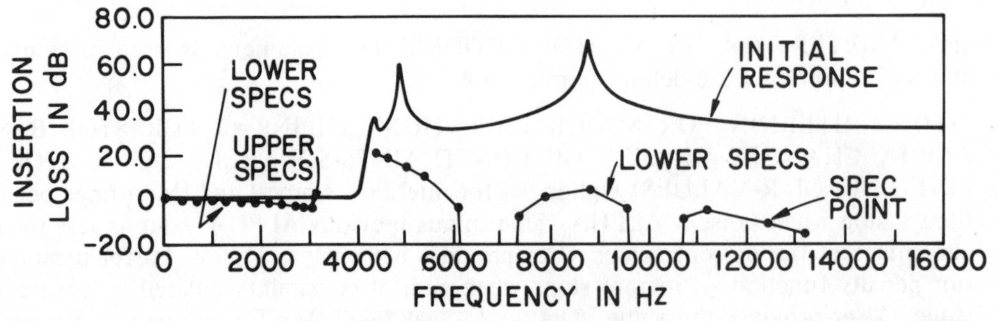

Fig. 2 Example - initial design

initial design. Filled circles mark the discrete frequencies ω_k of (4). Note that the specifications are satisfied in the stopband but violated in the passband. Figure 3 is a blowup of the passband region of

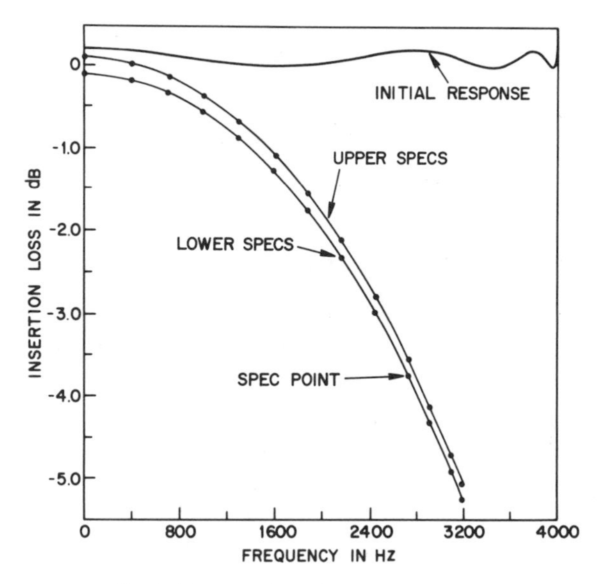

Fig. 3 Blowup of passband - initial design

the initial design showing that the specifications are violated by approximately 5 dB.

Figure 4 is the optimized design. The stopband underwent little

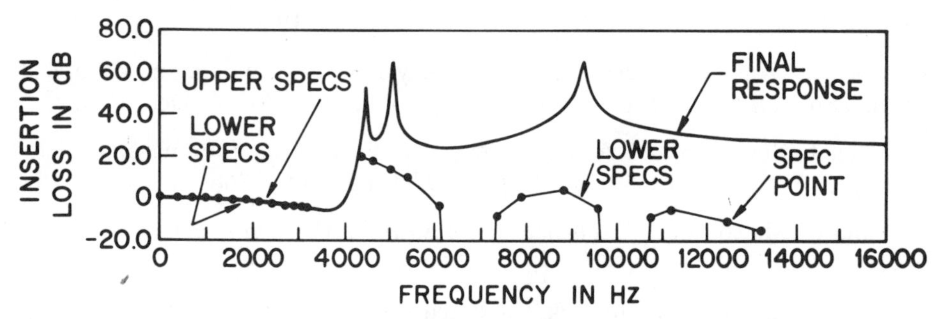

Fig. 4 Example - final design

change during the optimization. Figure 5 is a blowup of the passband region of the optimized

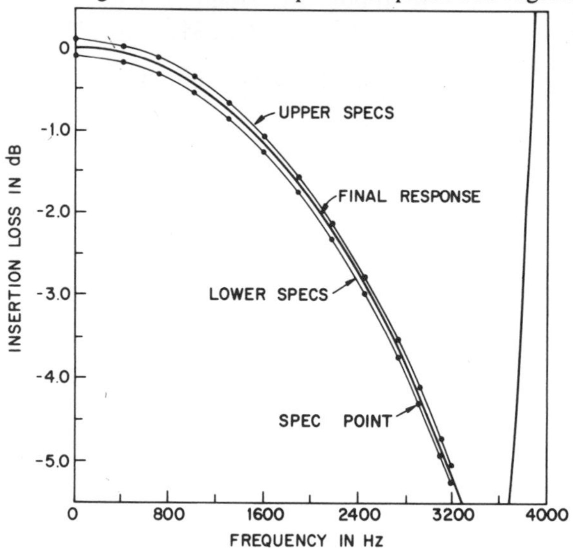

Fig. 5 Blowup of passband - final design

design where most of the change took place. The specifications are clearly satisfied.

5. Acknowledgment

The authors are grateful to Dr. J. A. Athanassopoulos, formerly of Bell Laboratories, for his many valuable contributions to the formulation of the frequency-domain design problem for digital filters as a nonlinear program.

References

1. J. F. Kaiser, "Digital Filters", in *Systems Analysis by Digital Computer,* F. F. Kuo and J. F. Kaiser, eds., John Wiley & Sons, Inc., New York, pp. 218-285, 1966.
2. K. Steiglitz, "The Equivalence of Digital and Analog Signal Processing", *Information and Control,* Vol. 8, No. 5, pp. 455-467, Oct. 1965.
3. See, for example, section 3 of: P. Rabinowitz, "Applications of Linear Programming to Numerical Analysis", *SIAM Review,* Vol. 10, No. 2, pp. 121-159, April 1968.
4. J. A. Athanassopoulos, J. D. Schoeffler and A. D. Waren, "Time-Domain Synthesis by Nonlinear Programming", *Proceedings of the Fourth Annual Allerton Conference on Circuit and System Theory,* pp. 766-775, Oct. 1966.
5. L. S. Lasdon and A. D. Waren, "Optimal Design of Filters with Bounded, Lossy Elements", *IEEE Trans. on Circuit Theory,* Vol. CT-13, No. 2, pp. 175-187, June 1966.
6. Y. Ishizaki and H. Watanabe, "An Iterative Chebyshev Approximation Method for Network Design", *IEEE Trans. on Circuit Theory,* Vol. CT-15, No. 4, pp. 326-336, Dec. 1968.
7. L. S. Lasdon, D. F. Suchman and A. D. Waren, "Nonlinear programming applied to linear array design", *J. Acoust. Soc. Am.,* Vol. 40, No. 5, pp. 1197-1200, Nov. 1966.
8. A. V. Fiacco and G. McCormick, "*Sequential Unconstrained Minimization Techniques*", John Wiley & Sons, Inc., New York, 1968.
9. W. I. Zangwill, "Nonlinear Programming via Penalty Functions", *Management Science,* Series A, Vol. 13, No. 5, pp. 344-358, Jan. 1967.
10. C. G. Broyden, "Quasi-Newton Methods and their Application to Function Minimisation", *Mathematics of Computation,* Vol. 21, No. 99, pp. 368-381, July 1967.
11. R. Fletcher and M. J. D. Powell, "A rapidly convergent descent method for minimization", *The Computer Journal,* Vol. 6, No. 2, pp. 163-168, June 1963.

Table 1

```
6  0.31250000E-04  0.48525318E-01
0.10000000E+01 -0.13200034E+01  0.95917120E+00 -0.13700062E+01
0.10000000E+01 -0.11366976E+01  0.79029629E+00 -0.13095079E+01
0.10000000E+01  0.30500371E+00  0.43064909E+00 -0.11934469E+01
3
1  5  9
1 28
0.10000000E-02  0.39979457E+03  0.69890112E+03  0.99680567E+03
0.12930097E+04  0.15870320E+04  0.18784126E+04  0.21667174E+04
0.24515404E+04  0.27325081E+04  0.29175082E+04  0.31005501E+04
0.31913099E+04  0.44034865E+04  0.46480381E+04  0.50430803E+04
0.54222495E+04  0.61329350E+04  0.67815157E+04  0.73711900E+04
0.79061884E+04  0.88310266E+04  0.95931836E+04  0.10225487E+05
0.10754695E+05  0.11201857E+05  0.12446131E+05  0.13191388E+05
28
-0.10000000E+00 -0.17149741E+00 -0.31933432E+00 -0.54880904E+00
-0.86123592E+00 -0.12584475E+01 -0.17428471E+01 -0.23174821E+01
-0.29861407E+01 -0.37534815E+01 -0.43226320E+01 -0.49401417E+01
-0.52677316E+01  0.18688911E+02  0.16903418E+02  0.13416951E+02
 0.90909537E+01 -0.42527329E+01 -0.27946299E+03 -0.86185069E+01
 0.21700557E+00  0.30703544E+01 -0.56281188E+01 -0.27946299E+03
-0.99938936E+01 -0.58035957E+01 -0.11648717E+02 -0.16014496E+02
 0.10000000E+01  0.10000000E+01  0.10000000E+01  0.10000000E+01
 0.10000000E+01  0.10000000E+01  0.10000000E+01  0.10000000E+01
 0.10000000E+01  0.10000000E+01  0.10000000E+01  0.10000000E+01
 0.10000000E+01  0.10000000E+01  0.10000000E+01  0.10000000E+01
 0.10000000E+01  0.10000000E+01  0.10000000E+01  0.10000000E+01
 0.10000000E+01  0.10000000E+01  0.10000000E+01  0.10000000E+01
 0.10000000E+01  0.10000000E+01  0.10000000E+01  0.10000000E+01
1 13
 0.10000000E+00  0.28502590E-01 -0.11933433E+00 -0.34880904E+00
-0.66123592E+00 -0.10584475E+01 -0.15428471E+01 -0.21174821E+01
-0.27861407E+01 -0.35534815E+01 -0.41226320E+01 -0.47401417E+01
-0.50677316E+01
 0.10000000E+01  0.10000000E+01  0.10000000E+01  0.10000000E+01
 0.10000000E+01  0.10000000E+01  0.10000000E+01  0.10000000E+01
 0.10000000E+01  0.10000000E+01  0.10000000E+01  0.10000000E+01
 0.10000000E+01
 0
```

Table 2

```
INITIAL COEFFICIENTS FOR R=  0.26582312E-03
       A2               A1              B2              B1
 0.10000000E 01 -0.13200034E 01  0.95917120E 00 -0.13700062E 01
 0.10000000E 01 -0.11366976E 01  0.79029629E 00 -0.13095079E 01
 0.10000000E 01  0.30500371E 00  0.43064909E 00 -0.11934469E 01
GAIN= 0.48525318E-01 ALPHA= 0.51585356E 01 FUNCTION= 0.10317071E 02

FINAL COEFFICIENTS FOR R=  0.26582312E-03
       A2               A1              B2              B1
 0.10000000E 01 -0.11005577E 01  0.86160736E 00 -0.13928926E 01
 0.10000000E 01 -0.12672789E 01  0.74878288E 00 -0.12533824E 01
 0.10000000E 01  0.44910363E 00  0.42624463E 00 -0.10688564E 01
GAIN= 0.51341804E-01 ALPHA=-0.15522700E-01 FUNCTION= 0.67633536E-01

MODIFIED COEFFICIENTS
 0.10000000E 01 -0.11005577E 01  0.86160736E 00 -0.13928926E 01
 0.10000000E 01 -0.12672789E 01  0.74878288E 00 -0.12533824E 01
 0.10000000E 01  0.44910363E 00  0.42624463E 00 -0.10688564E 01
MODIFIED GAIN= 0.51341804E-01

GRADIENTS
 0.             -0.15056353E 00  0.27614075E 00  0.33629320E 00
 0.             -0.19561886E 00  0.22825324E 00  0.28943125E 00
 0.             -0.48712567E-01  0.15260250E 00  0.21883522E 00
  377 FUNCTION EVALUATIONS          51 GRADIENT EVALUATIONS

XXXXXXXXXXXXXXXXXXXXXXXXXXXXXXXXXXXXXXXXXXXXXXXXXXXXXXXXXXXXXXXXXXXXXX
```

Table 2
(Continued)

```
INITIAL COEFFICIENTS FOR R=  0.26582312E-04
       A2                A1                B2                B1
 0.10000000E 01 -0.11005577E 01  0.86160736E 00 -0.13928926E 01
 0.10000000E 01 -0.12672789E 01  0.74878288E 00 -0.12533824E 01
 0.10000000E 01  0.44910363E 00  0.42624463E 00 -0.10688564E 01
GAIN= 0.51341804E-01 ALPHA=-0.15522700E-01 FUNCTION=-0.72070768E-02

FINAL COEFFICIENTS FOR R=  0.26582312E-04
       A2                A1                B2                B1
 0.10000000E 01 -0.10786655E 01  0.87831046E 00 -0.13160157E 01
 0.10000000E 01 -0.12562839E 01  0.78747962E 00 -0.12743440E 01
 0.10000000E 01  0.48262715E 00  0.44985502E 00 -0.10219863E 01
GAIN= 0.72560707E-01 ALPHA=-0.73472910E-01 FUNCTION=-0.47247530E-01

MODIFIED COEFFICIENTS
 0.10000000E 01 -0.10786655E 01  0.87831046E 00 -0.13160157E 01
 0.10000000E 01 -0.12562839E 01  0.78747962E 00 -0.12743440E 01
 0.10000000E 01  0.48262715E 00  0.44985502E 00 -0.10219863E 01
MODIFIED GAIN= 0.72560707E-01

GRADIENTS
 0.             -0.35273176E 00  0.53998295E 00  0.72532947E 00
 0.             -0.48105144E 00  0.44776873E 00  0.69719212E 00
 0.             -0.10700078E 00  0.21568865E 00  0.44558926E 00
  198 FUNCTION EVALUATIONS          32 GRADIENT EVALUATIONS

XXXXXXXXXXXXXXXXXXXXXXXXXXXXXXXXXXXXXXXXXXXXXXXXXXXXXXXXXXXXXXXXXXXXXXX


INITIAL COEFFICIENTS FOR R=  0.26582312E-05
       A2                A1                B2                B1
 0.10000000E 01 -0.10786655E 01  0.87831046E 00 -0.13160157E 01
 0.10000000E 01 -0.12562839E 01  0.78747962E 00 -0.12743440E 01
 0.10000000E 01  0.48262715E 00  0.44985502E 00 -0.10219863E 01
GAIN= 0.72560707E-01 ALPHA=-0.73472910E-01 FUNCTION=-0.70850371E-01

FINAL COEFFICIENTS FOR R=  0.26582312E-05
       A2                A1                B2                B1
 0.10000000E 01 -0.10733848E 01  0.87859791E 00 -0.13189261E 01
 0.10000000E 01 -0.12641738E 01  0.78693668E 00 -0.12579032E 01
 0.10000000E 01  0.48517703E 00  0.46097852E 00 -0.10228557E 01
GAIN= 0.76552721E-01 ALPHA=-0.91377584E-01 FUNCTION=-0.83240161E-01

MODIFIED COEFFICIENTS
 0.10000000E 01 -0.10733848E 01  0.87859791E 00 -0.13189261E 01
 0.10000000E 01 -0.12641738E 01  0.78693668E 00 -0.12579032E 01
 0.10000000E 01  0.48517703E 00  0.46097852E 00 -0.10228557E 01
MODIFIED GAIN= 0.76552721E-01

GRADIENTS
 0.             -0.68659899E-01  0.80960995E-01  0.21131991E 00
 0.             -0.12261476E 00 -0.66190683E-02  0.14751522E 00
 0.             -0.11043458E-01 -0.15977386E 00 -0.45411769E-01
   89 FUNCTION EVALUATIONS          12 GRADIENT EVALUATIONS

XXXXXXXXXXXXXXXXXXXXXXXXXXXXXXXXXXXXXXXXXXXXXXXXXXXXXXXXXXXXXXXXXXXXXXX


INITIAL COEFFICIENTS FOR R=  0.26582312E-06
       A2                A1                B2                B1
 0.10000000E 01 -0.10733848E 01  0.87859791E 00 -0.13189261E 01
 0.10000000E 01 -0.12641738E 01  0.78693668E 00 -0.12579032E 01
 0.10000000E 01  0.48517703E 00  0.46097852E 00 -0.10228557E 01
GAIN= 0.76552721E-01 ALPHA=-0.91377584E-01 FUNCTION=-0.90563842E-01

FINAL COEFFICIENTS FOR R=  0.26582312E-06
       A2                A1                B2                B1
 0.10000000E 01 -0.10733774E 01  0.87862835E 00 -0.13189427E 01
 0.10000000E 01 -0.12641572E 01  0.78697857E 00 -0.12578720E 01
 0.10000000E 01  0.48517863E 00  0.46098580E 00 -0.10228711E 01
GAIN= 0.76564977E-01 ALPHA=-0.96739767E-01 FUNCTION=-0.94341949E-01
```

Table 2
(Continued)

```
MODIFIED COEFFICIENTS
 0.10000000E 01 -0.10733774E 01  0.87862835E 00 -0.13189427E 01
 0.10000000E 01 -0.12641572E 01  0.78697857E 00 -0.12578720E 01
 0.10000000E 01  0.48517863E 00  0.46098580E 00 -0.10228711E 01
MODIFIED GAIN= 0.76564977E-01

GRADIENTS
 0.              0.16562078E 01 -0.25811159E 01 -0.35206183E 01
 0.              0.23699104E 01 -0.20665732E 01 -0.32827339E 01
 0.              0.48901627E 00 -0.71702110E 00 -0.19263913E 01
   46 FUNCTION EVALUATIONS          4 GRADIENT EVALUATIONS

XXXXXXXXXXXXXXXXXXXXXXXXXXXXXXXXXXXXXXXXXXXXXXXXXXXXXXXXXXXXXXXXXXXXXXXX
```

Appendix

```
C
C-----------------------------------------------------------------------
C MAIN PROGRAM: OPTIMIZATION PROGRAM FOR THE DESIGN OF
C               DIGITAL FILTER TRANSFER FUNCTIONS
C AUTHORS:      M. T. DOLAN AND J. F. KAISER
C               BELL LABORATORIES, MURRAY HILL, NEW JERSEY 07974
C
C INPUT:        NORD  IS THE ORDER, MUST BE EVEN, .LE. 20
C               T     IS THE SAMPLING TIME
C               XK    IS THE GAIN FACTOR
C               X     IS THE COEFFICIENT VECTOR
C               NC    IS THE NUMBER OF COEFS TO REMAIN CONSTANT
C               NL    IS THE ARRAY OF INDICES OF FIXED COEFS
C                       IF NC=0, DO NOT INPUT NL ARRAY
C               NG    IS THE GAIN OPTION
C               NF    IS THE NUMBER OF FREQUENCIES FOR OPTIMIZATION
C               FREQ  IS THE ARRAY OF FREQUENCIES
C               N1    IS THE INDEX FOR TOP RANGE OF LOWER SPECS
C               SPEC1 IS THE ARRAY OF LOWER SPECS
C               TH1   IS THE ARRAY OF WEIGHTS FOR LOWER SPECS
C               N2    IS THE LOWER INDEX FOR RANGE OF UPPER SPECS
C               N3    IS THE UPPER INDEX FOR RANGE OF UPPER SPECS
C               SPEC2 IS THE ARRAY OF UPPER SPECS
C               TH2   IS THE ARRAY OF WEIGHTS FOR UPPER SPECS
C               NGID  IS THE OPTIMIZATION GUIDE OPTION
C               NOUT  IS THE OUTPUT OPTION
C               NN    IS THE INITIAL STEP OPTION
C               VBND  IS THE PARAMETER USED BY INITIAL STEP OPTIONS
C               NPEN  IS THE PENALTY FUNCTION CHOICE
C               KP    DETERMINES STOP CRITERION FOR MINOR ITERATIONS
C               EPS   IS A PARAMETER USED BY 2 MINOR STOP OPTIONS
C               KTIME IS A PARAMETER USED BY 2 MINOR STOP OPTIONS
C               NFINR DETERMINES STOP CRITERION FOR MAJOR ITERATIONS
C               PERR  IS THE PERCENTAGE VALUE USED FOR MAJOR STOP
C                       OPTION 1 WITH INTERIOR PENALTY FUNCTIONS
C               TAZ   IS THE SUM OF CONSTRAINTS-SQUARED USED FOR MAJOR
C                       STOP OPTION 1 WITH EXTERIOR PENALTY FUNCTION
C               NR    IS THE NUMBER OF R VALUES
C               INIR  IS THE R GUIDE OPTION - 1 TO GUIDE
C                                           - 0 FOR AUTOMATIC
C               RINI  IS THE INITIAL R VALUE SET BY ROUTINE ZR IF INIR=0
C               RCHNG IS THE R CHANGE FACTOR SET BY MAIN IF INIR=0
C-----------------------------------------------------------------------
C
      DIMENSION X(42), S(42), XB(42), H(42,42), G(42), Y(42), GB(42)
      DIMENSION SPEC1(42), SPEC2(42), TH1(42), TH2(42), FREQ(42), W(42)
      COMMON X, F, DEL, VBND, EPS, KTIME, NIT, NFE, NGE, NQ, K
      COMMON /SUBC/ S, XB, GB, FB, H, G, IPANIC
      COMMON /FGVARY/ T, XK, FREQ, W, SPEC1, SPEC2, TH1, TH2, R, NX, N,
     *    NF, N1, N2, N3
      COMMON /GRAD/ TAZ, NC, NL(42), NOUT, NVI, NPEN, NFINR
      COMMON /RVALUE/ RINI, INIR
      COMMON /WRITE/ IW
C
C IR AND IW ARE MACHINE DEPENDENT READ AND WRITE DEVICE NUMBERS.
C THE WRITE DEVICE NUMBER, IW, IS PASSED TO THE SUBROUTINES
C WHICH NEED IT THROUGH LABELLED COMMON.
C
      IR = I1MACH(1)
      IW = I1MACH(2)
C
C READ ORDER, SAMPLING TIME, AND GAIN FACTOR
C
      READ (IR,9999) NORD, T, XK
9999  FORMAT (I3, 2E16.8)
C
      N = NORD
      NW = 2*N
C
C READ THE COEFFICIENT VECTOR
C
      READ (IR,9998) (X(I),I=1,NW)
9998  FORMAT (4E16.8)
C
C READ NUMBER OF COEFS TO REMAIN CONSTANT
C
      READ (IR,9997) NC
9997  FORMAT (10I3)
      IF (NC.EQ.0) GO TO 10
C
C READ ARRAY OF INDICES OF FIXED COEFS IF NC .NE. 0
C
      READ (IR,9997) (NL(I),I=1,NC)
C
C READ GAIN OPTION AND NUMBER OF FREQS FOR OPTIMIZATION
C
  10  READ (IR,9997) NG, NF
C
C READ ARRAY OF FREQUENCIES
C
      READ (IR,9998) (FREQ(I),I=1,NF)
C
C READ INDEX FOR TOP RANGE OF LOWER SPECS
C
      READ (IR,9997) N1
C
C READ ARRAY OF LOWER SPECS
C
      READ (IR,9998) (SPEC1(I),I=1,N1)
C
C READ ARRAY OF WEIGHTS FOR LOWER SPECS
C
      READ (IR,9998) (TH1(I),I=1,N1)
C
C READ LOWER AND UPPER INDICES FOR RANGE OF UPPER SPECS
C
      READ (IR,9997) N2, N3
C
      N23 = N3 - N2 + 1
C
C READ ARRAY OF UPPER SPECS
C
      READ (IR,9998) (SPEC2(I),I=1,N23)
C
C READ ARRAY OF WEIGHTS FOR UPPER SPECS
C
      READ (IR,9998) (TH2(I),I=1,N23)
C
C NGID IS THE OPTIMIZATION GUIDE OPTION.  INPUT 0
C FOR AUTOMATIC SELECTION OF PARAMETERS AND NO FURTHER
C INPUT CARDS ARE NECESSARY
C READ OPTIMIZATION GUIDE OPTION
C
      READ (IR,9997) NGID
      IF (NGID.EQ.0) GO TO 20
C
C READ ADDITIONAL CARDS ONLY FOR NGID=1
C TO SIMPLIFY MATTERS, ALL DATA IS ENTERED.
C USERS SHOULD INPUT 0 FOR PARAMETERS THAT WILL NOT BE USED.
C
```

```
C IF NGID=1, READ OUTPUT OPTION, INITIAL STEP OPTION,
C PARAMETER FOR STEP OPTION, PENALTY FUNCTION CHOICE,
C STOP CRITERION FOR MINOR ITERATIONS, TWO PARAMETERS
C FOR STOP OPTION, STOP CRITERION FOR MAJOR ITERATIONS,
C THREE PARAMETERS FOR STOP OPTION, R-GUIDE OPTION,
C INITIAL R VALUE, AND THE R CHANGE FACTOR.
C
      READ (IR,9996) NOUT, NN, VBND, NPEN, KP, EPS, KTIME
      READ (IR,9995) NFINR, PERR, TAZ, NR, INIR, RINI, RCHNG
      GO TO 30
9996  FORMAT (2I3, E12.4, 2I3, E12.4, I3)
9995  FORMAT (I3, 2E12.4, 2I3, 2E12.4)
C
C AUTOMATIC SELECTION FOR NGID=0
C FOR CONSISTENCY, PARAMETERS NOT USED WILL BE SET TO 0
C
   20 NOUT = 0
      NN = 2
      VBND = 0.
      NPEN = 1
      KP = 2
C
C EPS AND KTIME WILL NOT BE USED
C
      EPS = 0.
      KTIME = 0
      NFINR = 2
C
C PERR AND TAZ WILL NOT BE USED
C
      PERR = 0.
      TAZ = 0.
      NR = 4
      INIR = 0
C
C RINI AND RCHNG WILL BE SET BY THE PROGRAM
C
      RINI = 0.
      RCHNG = 0.
C
   30 NX = NW + 2
      NVI = 0
      IF (INIR.EQ.0) RCHNG = 10.
      NXM1 = NX - 1
      X(NXM1) = SQRT(ABS(XK))
      IF (NG.NE.0) GO TO 40
      NC = NC + 1
      NL(NC) = NXM1
   40 TPI = 8.*ATAN(1.)
      NQ = NX
      DO 50 I=1,NF
        W(I) = TPI*FREQ(I)
   50 CONTINUE
      XK = X(NXM1)**2
      CALL ZR(X)
   60 NIT = 0
      NFE = 0
      NGE = 0
      K = KP
      OLDXNX = X(NX)
   70 CALL FLPWL
      IF (K) 100, 100, 80
   80 CALL STEP(NN)
      CALL QUAD
      IF (K) 100, 100, 90
   90 CALL FINISH
  100 CONTINUE
      IF (K) 110, 110, 70
  110 WRITE (IW,9994) R
      WRITE (IW,9993)
9994  FORMAT (26H0FINAL COEFFICIENTS FOR R=, E16.8)
9993  FORMAT (54H          A2                A1                B2
     *     4H  B1)
      WRITE (IW,9998) (X(I),I=1,NW)
      XK = X(NXM1)**2
      WRITE (IW,9992) XK, X(NX), F
9992  FORMAT (6H GAIN=, E15.8, 7H ALPHA=, E15.8, 10H FUNCTION=, E15.8)
      CALL RINV(X, Y, NX)
      WRITE (IW,9991)
9991  FORMAT (22H0MODIFIED COEFFICIENTS)
      WRITE (IW,9998) (Y(I),I=1,NW)
      DXK = Y(NXM1)**2
      WRITE (IW,9990) DXK
9990  FORMAT (15H MODIFIED GAIN=, E15.8)
      WRITE (IW,9989)
9989  FORMAT (10H0GRADIENTS)
      WRITE (IW,9998) (G(I),I=1,NW)
      WRITE (IW,9986) NFE, NGE
      WRITE (IW,9988)
      WRITE (IW,9987)
9988  FORMAT (54H0XXXXXXXXXXXXXXXXXXXXXXXXXXXXXXXXXXXXXXXXXXXXXXXXXXXXXXX,
     *     18HXXXXXXXXXXXXXXXXXX)
9987  FORMAT (1H0)
9986  FORMAT (1H , I5, 21H FUNCTION EVALUATIONS, I10, 14H GRADIENT EVAL,
     *     7HUATIONS)
      IF (NFINR.EQ.2) GO TO 120
      IF (ABS(X(NX)-OLDXNX)-ABS(PERR*.01*X(NX))) 160, 160, 130
  120 NR = NR - 1
      IF (NR.EQ.0) STOP
  130 GO TO (140, 140, 150), NPEN
  140 R = R/RCHNG
      GO TO 60
  150 R = R*RCHNG
      GO TO 60
  160 STOP
      END
C
C-----------------------------------------------------------------------
C SUBROUTINE:  ZR
C CALCULATES X(NX)=ALPHA AND INITIAL R
C-----------------------------------------------------------------------
C
      SUBROUTINE ZR(X)
C
      DIMENSION X(42), G(42), FF(42), BOUND1(42), BOUND2(42),
     *     DF(42,42), HM(42), HG(42,42)
      DIMENSION SPEC1(42), SPEC2(42), TH1(42), TH2(42), FREQ(42), W(42)
      COMMON /FGVARY/ T, XK, FREQ, W, SPEC1, SPEC2, TH1, TH2, R, NX, N,
     *     NF, N1, N2, N3
      COMMON /GRAD/ TAZ, NC, NL(42), NOUT, NVI, NPEN, NFINR
      COMMON /RVALUE/ RINI, INIR
      COMMON /WRITE/ IW
C
      NXM1 = NX - 1
      XK = X(NXM1)**2
      CALL CASC(N, X, W, NF, T, XK, HM, FF, 0, HG, DF)
      DO 10 J=1,N1
        IF (TH1(J).EQ.0.) GO TO 10
        BOUND1(J) = (FF(J)-SPEC1(J))/TH1(J)
   10 CONTINUE
```

```
      JJ = 0
      DO 20 J=N2,N3
        JJ = JJ + 1
        IF (TH2(JJ).EQ.0.) GO TO 20
        BOUND2(JJ) = (-FF(J)+SPEC2(JJ))/TH2(JJ)
   20 CONTINUE
      DO 30 J=1,N1
        IF (BOUND1(J).LT.0.) GO TO 100
   30 CONTINUE
      JJ = 0
      DO 40 J=N2,N3
        JJ = JJ + 1
        IF (BOUND2(JJ).LT.0.) GO TO 100
   40 CONTINUE
C
C ALL POSITIVE.  FIND MINIMUM HIT
C
      DO 50 J=1,N1
        IF (BOUND1(J).EQ.0.) GO TO 50
        X(NX) = BOUND1(J)
        GO TO 70
   50 CONTINUE
      JJ = 0
      DO 60 J=N2,N3
        JJ = JJ + 1
        IF (BOUND2(JJ).EQ.0.) GO TO 60
        X(NX) = BOUND2(JJ)
        GO TO 70
   60 CONTINUE
   70 DO 80 J=1,N1
        IF (BOUND1(J).EQ.0.) GO TO 80
        IF (X(NX).LE.BOUND1(J)) GO TO 80
        X(NX) = BOUND1(J)
   80 CONTINUE
      JJ = 0
      DO 90 J=N2,N3
        JJ = JJ + 1
        IF (BOUND2(JJ).EQ.0.) GO TO 90
        IF (X(NX).LE.BOUND2(JJ)) GO TO 90
        X(NX) = BOUND2(JJ)
   90 CONTINUE
      X(NX) = -X(NX)*.9999
      ZETA = X(NX)
      GO TO 150
C
C AT LEAST ONE BOUND IS NEGATIVE.  FIND THE MAXIMUM MISS.
C
  100 X(NX) = 0.
      DO 120 J=1,N1
        IF (BOUND1(J)) 110, 110, 120
  110   TXNX = ABS(BOUND1(J))
        IF (TXNX.LT.X(NX)) GO TO 120
        X(NX) = TXNX
  120 CONTINUE
      JJ = 0
      DO 140 J=N2,N3
        JJ = JJ + 1
        IF (BOUND2(JJ)) 130, 130, 140
  130   TXNX = ABS(BOUND2(JJ))
        IF (TXNX.LT.X(NX)) GO TO 140
        X(NX) = TXNX
  140 CONTINUE
      X(NX) = X(NX)*1.00001
      ZETA = X(NX)
  150 AZ = 0.
      DO 220 J=1,N1
        BOUND1(J) = FF(J) + ZETA*TH1(J) - SPEC1(J)
        IF (NPEN.EQ.3) GO TO 190
        IF (BOUND1(J)) 320, 320, 160
  160   GO TO (170, 180), NPEN
  170   BOUND1(J) = 1./BOUND1(J)
        AZ = AZ + BOUND1(J)
        GO TO 220
  180   AZ = AZ - ALOG(BOUND1(J))
        GO TO 220
  190   IF (BOUND1(J)) 200, 220, 210
  200   AZ = AZ + BOUND1(J)**2
        GO TO 220
  210   BOUND1(J) = 0.
  220 CONTINUE
      JJ = 0
      DO 290 J=N2,N3
        JJ = JJ + 1
        BOUND2(JJ) = -FF(J) + ZETA*TH2(JJ) + SPEC2(JJ)
        IF (NPEN.EQ.3) GO TO 260
        IF (BOUND2(JJ)) 320, 320, 230
  230   GO TO (240, 250), NPEN
  240   BOUND2(JJ) = 1./BOUND2(JJ)
        AZ = AZ + BOUND2(JJ)
        GO TO 290
  250   AZ = AZ - ALOG(BOUND2(JJ))
        GO TO 290
  260   IF (BOUND2(JJ)) 270, 290, 280
  270   AZ = AZ + BOUND2(JJ)**2
        GO TO 290
  280   BOUND2(JJ) = 0.
  290 CONTINUE
      IF (INIR.EQ.1) GO TO 310
      IF (AZ.EQ.0.) GO TO 300
      R = ABS(ZETA/AZ)
      IF (R.EQ.0.) R = 1.
      RETURN
  300 R = 1.
      RETURN
  310 R = RINI
      RETURN
  320 WRITE (IW,9999)
 9999 FORMAT (54H0TROUBLE - CHECK THAT SPECS ARE MET AT FREQUENCIES WIT,
     *    11HH 0 WEIGHTS)
      STOP
      END
C
C-----------------------------------------------------------------------
C SUBROUTINE:  FLPWL
C COMPUTES DIRECTION VECTOR
C-----------------------------------------------------------------------
C
      SUBROUTINE FLPWL
C
      DIMENSION X(42), S(42), XB(42), H(42,42), G(42)
      DIMENSION SIGMA(42), GB(42), Y(42), YTH(42), HY(42)
      COMMON X, F, DEL, VBND, EPS, KTIME, NIT, NFE, NGE, NX, K
      COMMON /SUBC/ S, XB, GB, FB, H, G, IPANIC
      COMMON /WRITE/ IW
C
C FIND GRADIENT USING NEWEST X-ARRAY
C
      CALL FG
      NGE = NGE + 1
      IF (IPANIC) 20, 20, 10
```

```
   10  WRITE (IW,9999)
 9999  FORMAT (24H0  FATAL IPANIC IN FLPWL)
       STOP
C
C IF THIS IS THE FIRST ITERATION, SET H = UNIT MATRIX
C
   20  IF (NIT) 30, 30, 90
   30  DO 70 I=1,NX
         DO 60 J=1,NX
           IF (I-J) 40, 50, 40
   40      H(I,J) = 0.
           GO TO 60
   50      H(I,J) = 1.
   60    CONTINUE
   70  CONTINUE
C
C IN THIS CASE, SLOPE = -GRADIENT
C G NOW BECOMES THE OLD G        STORE IT IN GB
C
       DO 80 I=1,NX
         GB(I) = G(I)
         S(I) = -G(I)
   80  CONTINUE
       GO TO 210
C
C IF THIS IS NOT THE FIRST ITERATION, CONTROL REACHES HERE.
C PREPARE TO UPDATE H
C
   90  STY = 0.
       DO 100 I=1,NX
C
C SIGMA VECTOR REPRESENTS CHANGE IN THE X-ARRAY
C
         SIGMA(I) = X(I) - XB(I)
C
C Y-VECTOR REPRESENTS CHANGE IN THE GRADIENT
C
         Y(I) = G(I) - GB(I)
C
C ONCE THE NEW G-VECTOR IS USED TO FIND Y, IT BECOMES THE OLD G
C
         GB(I) = G(I)
         STY = SIGMA(I)*Y(I) + STY
  100  CONTINUE
       IF (STY) 110, 180, 110
C
C A-MATRIX = (SIGMA-VECTOR TIMES SIGMA-TRANSPOSE) DIVIDED BY
C (SIGMA-TRANSPOSE TIMES Y-VECTOR)
C B-MATRIX = -(H-MATRIX TIMES Y-VECTOR) TIMES
C (Y-TRANSPOSE TIMES H-MATRIX) ALL DIVIDED BY
C (Y-TRANSPOSE TIMES H-MATRIX TIMES Y-VECTOR)
C
  110  DO 130 I=1,NX
         HY(I) = 0.
         YTH(I) = 0.
         DO 120 J=1,NX
           HY(I) = H(I,J)*Y(J) + HY(I)
           YTH(I) = H(J,I)*Y(J) + YTH(I)
  120    CONTINUE
  130  CONTINUE
C
       YTHY = 0.
       DO 140 I=1,NX
         YTHY = YTH(I)*Y(I) + YTHY
  140  CONTINUE
       IF (YTHY) 150, 180, 150
C
C UPDATE H
C H-MATRIX = H-MATRIX + A-MATRIX + B-MATRIX
C
  150  DO 170 I=1,NX
         DO 160 J=1,NX
           H(I,J) = -HY(I)*HY(J)/YTHY + H(I,J) + (SIGMA(I)*SIGMA(J))/STY
  160    CONTINUE
  170  CONTINUE
C
C CALCULATE THE SLOPE VECTOR  S-VECTOR=-(H-MATRIX TIMES G-VECTOR)
C
  180  DO 200 I=1,NX
         S(I) = 0.
         DO 190 J=1,NX
           S(I) = -H(J,I)*G(J) + S(I)
  190    CONTINUE
  200  CONTINUE
C
  210  RETURN
       END
C
C-----------------------------------------------------------------------
C SUBROUTINE:  STEP
C SETS INITIAL STEP
C-----------------------------------------------------------------------
C
       SUBROUTINE STEP(N)
C
       DIMENSION X(42), S(42), XB(42), H(42,42), G(42), GB(42)
       COMMON X, F, DEL, VBND, EPS, KTIME, NIT, NFE, NGE, NX, K
       COMMON /SUBC/ S, XB, GB, FB, H, G, IPANIC
       COMMON /GRAD/ TAZ, NC, NL(42), NOUT, NVI, NPEN, NFINR
       COMMON /WRITE/ IW
C
       IF (N-2) 10, 120, 160
C
C OPTION 1
C USER PROVIDES VMAX OR VMAX=.05*(MAX ABS X)
C STEP=MIN(2.,(VMAX/(MAX ABS S))
C
   10  VMAX = VBND
       IF (VMAX) 60, 20, 60
   20  XMAX = ABS(X(1))
       DO 40 I=2,NX
         IF (ABS(X(I))-XMAX) 40, 40, 30
   30    XMAX = ABS(X(I))
   40  CONTINUE
       VMAX = .05*XMAX
       IF (VMAX) 60, 50, 60
   50  VMAX = .01
   60  SMAX = ABS(S(1))
       DO 80 I=2,NX
         IF (ABS(S(I))-SMAX) 80, 80, 70
   70    SMAX = ABS(S(I))
   80  CONTINUE
       IF (SMAX) 90, 90, 110
   90  DEL = 2.
  100  IF (NOUT.EQ.0) RETURN
       WRITE (IW,9999) DEL
 9999  FORMAT (14H0INITIAL STEP=, E16.8)
       RETURN
  110  FAC2 = VMAX/SMAX
       DEL = AMIN1(2.,FAC2)
```

```
      GO TO 100
C
C OPTION 2
C USER PROVIDES ESTIMATE OF FUNCTION MINIMUM CALLED VMIN.
C STEP=MIN(2.,-2.*(F-VMIN))/(G-TRANSPOSE*S-VECTOR)
C
 120  VMIN = VBND
      DEN = 0.
      DO 130 I=1,NX
        DEN = G(I)*S(I) + DEN
 130  CONTINUE
      IF (DEN) 140, 90, 140
 140  FAC2 = -2.*(F-VMIN)/DEN
      IF (FAC2) 90, 90, 150
 150  DEL = AMIN1(2.,FAC2)
      GO TO 100
C
C OPTION 3  -  IF THIS IS THE FIRST TIME THROUGH, USE OPTION 1.
C IF THIS IS NOT THE FIRST TIME, DIVIDE PRESENT STEP BY 2.
C
C OPTION 4 -  DO NOTHING         USER PROVIDED STEP
C
 160  IF (N-4) 170, 190, 190
 170  IF (NIT) 180, 10, 180
 180  DEL = DEL/2.
      GO TO 100
 190  RETURN
      END
C
C-----------------------------------------------------------------------
C SUBROUTINE:  QUAD
C PERFORMS QUADRATIC INTERPOLATION
C-----------------------------------------------------------------------
C
      SUBROUTINE QUAD
C
      DIMENSION X(42), S(42), XB(42), H(42,42), G(42), GB(42)
      DIMENSION X0(42), X1(42), X2(42)
      COMMON X, F, DEL, VBND, EPS, KTIME, NIT, NFE, NGE, NX, K
      COMMON /SUBC/ S, XB, GB, FB, H, G, IPANIC
      COMMON /WRITE/ IW
C
C EVALUATE THE FUNCTION WITH THE GIVEN X-ARRAY AND CALL IT F0.
C AT THE SAME TIME, SAVE THIS ARRAY IN XB AS THE OLD X-ARRAY
C AND SAVE ITS FTN VALUE IN FB AS THE OLD FTN VALUE.  LATER
C WE WILL STORE THE NEW ARRAY IN X AND ITS FTN VALUE IN F.
C
      CALL FUNC(X, F)
      NFE = NFE + 1
      DO 10 I=1,NX
        XB(I) = X(I)
        X0(I) = X(I)
  10  CONTINUE
      FB = F
      F0 = F
C
C GENERATE X1-ARRAY AND F1
C
  20  DO 30 I=1,NX
        X1(I) = X0(I) + DEL*S(I)
  30  CONTINUE
      CALL FUNC(X1, F1)
      NFE = NFE + 1
      IF (IPANIC) 50, 50, 40
  40  DEL = DEL/5.
      GO TO 20
C
C CHECK IF F1 IS SMALLER THAN F0
C
  50  IF (F1-F0) 60, 60, 120
  60  DO 70 I=1,NX
        X2(I) = X1(I) + DEL*S(I)
  70  CONTINUE
      CALL FUNC(X2, F2)
      NFE = NFE + 1
      IF (IPANIC) 90, 90, 80
  80  DEL = DEL/5.
      GO TO 60
  90  IF (F2-F1) 100, 170, 170
C
C IF F2 IS SMALLER THAN F1, DISCARD F0, SHIFT ALL RESULTS LEFT,
C DOUBLE THE INCREMENT, AND GENERATE A NEW X2-ARRAY AND F2
C
 100  DO 110 I=1,NX
        X0(I) = X1(I)
        X1(I) = X2(I)
 110  CONTINUE
      F0 = F1
      F1 = F2
      DEL = 2.*DEL
      GO TO 60
C
C IF CONTROL REACHES HERE, F1 WAS LARGER AND F0.
C CHECK IF IT WAS MORE THAN 2.*F0
C
 120  TF0 = 2.*F0
      IF (F1-TF0) 130, 160, 160
 130  DO 140 I=1,NX
        X2(I) = X1(I)
 140  CONTINUE
      F2 = F1
      DEL = DEL/2.
      DO 150 I=1,NX
        X1(I) = X0(I) + DEL*S(I)
 150  CONTINUE
      CALL FUNC(X1, F1)
      NFE = NFE + 1
      IF (F1-F0) 170, 170, 130
C
C F1 WAS MORE THAN 2.*F0, DIVIDE INCREMENT BY 7 AND
C GENERATE NEW X1-ARRAY AND F2
C
 160  DEL = DEL/7.
      GO TO 20
C
C INTERPOLATE
 170  DO 190 I=1,NX
        IF (S(I)) 180, 190, 180
 180    B = (X1(I)-X0(I))/S(I)
        GO TO 200
 190  CONTINUE
      GO TO 270
 200  C = B + DEL
      B2 = B**2
      C2 = C**2
      FAC1 = .5*(F0*(C2-B2)-C2*F1+B2*F2)
      IF (FAC1) 210, 260, 210
 210  FAC2 = F0*(C-B) - C*F1 + B*F2
      IF (FAC2) 220, 260, 220
 220  AL = .5*(F0*(C2-B2)-C2*F1+B2*F2)/(F0*(C-B)-C*F1+B*F2)
```

```
      DO 230 I=1,NX
        X0(I) = X0(I) + AL*S(I)
  230 CONTINUE
      CALL FUNC(X0, F0)
      NFE = NFE + 1
C
C COMPARE RESULT WITH F1 AND RETURN THE BETTER ONE
C
      IF (F1-F0) 270, 240, 240
  240 DO 250 I=1,NX
        X(I) = X0(I)
  250 CONTINUE
      F = F0
      GO TO 290
  260 K = 0
      WRITE (IW,9999)
 9999 FORMAT (31H0UNDERFLOW DURING INTERPOLATION)
  270 DO 280 I=1,NX
        X(I) = X1(I)
  280 CONTINUE
      F = F1
  290 RETURN
      END
C
C-----------------------------------------------------------------------
C SUBROUTINE:  FUNC
C EVALUATES THE PENALTY FUNCTION
C-----------------------------------------------------------------------
C
      SUBROUTINE FUNC(X, F)
C
      DIMENSION SPEC1(42), SPEC2(42), TH1(42), TH2(42), FREQ(42), W(42)
      DIMENSION HG(42,42), DF(42,42), S(42), XB(42), H(42,42), GB(42)
      DIMENSION X(42), G(42), FF(42), BOUND1(42), BOUND2(42), HM(42)
      COMMON /FGVARY/ T, XK, FREQ, W, SPEC1, SPEC2, TH1, TH2, R, NX, N,
     *    NF, N1, N2, N3
      COMMON /SUBC/ S, XB, GB, FB, H, G, IPANIC
      COMMON /GRAD/ TAZ, NC, NL(42), NOUT, NVI, NPEN, NFINR
C
      NXM1 = NX - 1
      XK = X(NXM1)**2
      CALL CASC(N, X, W, NF, T, XK, HM, FF, 0, HG, DF)
      AZ = 0.
      IPANIC = 0
      ZETA = X(NX)
      DO 70 J=1,N1
        BOUND1(J) = FF(J) + ZETA*TH1(J) - SPEC1(J)
        IF (NPEN.EQ.3) GO TO 40
        IF (BOUND1(J)) 150, 150, 10
   10   GO TO (20, 30), NPEN
   20   BOUND1(J) = 1./BOUND1(J)
        AZ = AZ + BOUND1(J)
        GO TO 70
   30   AZ = AZ - ALOG(BOUND1(J))
        GO TO 70
   40   IF (BOUND1(J)) 50, 70, 60
   50   AZ = AZ + BOUND1(J)**2
        GO TO 70
   60   BOUND1(J) = 0.
   70 CONTINUE
      JJ = 0
      DO 140 J=N2,N3
        JJ = JJ + 1
        BOUND2(JJ) = -FF(J) + ZETA*TH2(JJ) + SPEC2(JJ)
        IF (NPEN.EQ.3) GO TO 110
        IF (BOUND2(JJ)) 150, 150, 80
   80   GO TO (90, 100), NPEN
   90   BOUND2(JJ) = 1./BOUND2(JJ)
        AZ = AZ + BOUND2(JJ)
        GO TO 140
  100   AZ = AZ - ALOG(BOUND2(JJ))
        GO TO 140
  110   IF (BOUND2(JJ)) 120, 140, 130
  120   AZ = AZ + BOUND2(JJ)**2
        GO TO 140
  130   BOUND2(JJ) = 0.
  140 CONTINUE
      F = ZETA + R*AZ
      GO TO 160
  150 IPANIC = 1
      NVI = NVI + 1
  160 RETURN
      END
C
C-----------------------------------------------------------------------
C SUBROUTINE:  FG
C EVALUATES BOTH THE PENALTY FUNCTION AND ITS GRADIENT
C-----------------------------------------------------------------------
C
      SUBROUTINE FG
C
C EVALUATES THE PENALTY FUNCTION AND ITS GRADIENT FOR A FUNCTION
C WITH VALUES FF(J) AND DERIVATIVE VALUES DF(K,J) AT POINTS J FOR
C ELEMENT NUMBER K. THE ELEMENT VALUES ARE STORED AS X(K), THE
C PENALTY FUNCTION AS F, AND THE GRADIENT AS G(K).
C
      DIMENSION X(42), G(42), FF(42), BOUND1(42), BOUND2(42),
     *    DF(42,42), HM(42), HG(42,42)
      DIMENSION SPEC1(42), SPEC2(42), TH1(42), TH2(42), FREQ(42), W(42)
      DIMENSION S(42), XB(42), H(42,42), GG(42), GB(42)
      COMMON /FGVARY/ T, XK, FREQ, W, SPEC1, SPEC2, TH1, TH2, R, NX, N,
     *    NF, N1, N2, N3
      COMMON /GRAD/ TAZ, NC, NL(42), NOUT, NVI, NPEN, NFINR
      COMMON /SUBC/ S, XB, GB, FB, H, G, IPANIC
      COMMON X, F, DEL, VBND, EPS, KTIME, NIT, NFE, NGE, NQ, KDUM
      COMMON /WRITE/ IW
C
      CONS = -20./ALOG(10.)
      NXM1 = NX - 1
      NXM2 = NX - 2
      XK = X(NXM1)**2
      CALL CASC(N, X, W, NF, T, XK, HM, FF, 1, HG, DF)
      AZ = 0.
      ZETA = X(NX)
      IPANIC = 0
      DO 70 J=1,N1
        BOUND1(J) = FF(J) + ZETA*TH1(J) - SPEC1(J)
        IF (NPEN.EQ.3) GO TO 40
        IF (BOUND1(J)) 340, 340, 10
   10   GO TO (20, 30), NPEN
   20   BOUND1(J) = 1./BOUND1(J)
        AZ = AZ + BOUND1(J)
        GO TO 70
   30   AZ = AZ - ALOG(BOUND1(J))
        GO TO 70
   40   IF (BOUND1(J)) 50, 70, 60
   50   AZ = AZ + BOUND1(J)**2
        GO TO 70
   60   BOUND1(J) = 0.
   70 CONTINUE
```

```
      JJ = 0
      DO 140 J=N2,N3
        JJ = JJ + 1
        BOUND2(JJ) = -FF(J) + ZETA*TH2(JJ) + SPEC2(JJ)
        IF (NPEN.EQ.3) GO TO 110
        IF (BOUND2(JJ)) 340, 340, 80
 80     GO TO (90, 100), NPEN
 90     BOUND2(JJ) = 1./BOUND2(JJ)
        AZ = AZ + BOUND2(JJ)
        GO TO 140
 100    AZ = AZ - ALOG(BOUND2(JJ))
        GO TO 140
 110    IF (BOUND2(JJ)) 120, 140, 130
 120    AZ = AZ + BOUND2(JJ)**2
        GO TO 140
 130    BOUND2(JJ) = 0.
 140  CONTINUE
      F = ZETA + R*AZ
      IF ((NPEN.EQ.3) .AND. (NFINR.EQ.1)) GO TO 150
      GO TO 160
 150  IF (AZ.GT.TAZ) GO TO 160
      KDUM = 0
C
C THIS NEXT SECTION DETERMINES THE GRADIENT OF THE PENALTY
C FUNCTION WITH RESPECT TO THE VARIABLES X(K), BUT NOT INCLUDING
C IS X(NX).  NOR DOES IT INCLUDE XK WHICH IS X(NX-1)
C
 160  NXM2 = NX - 2
      GO TO (170, 200, 230), NPEN
 170  DO 180 J=1,N1
        BOUND1(J) = BOUND1(J)*BOUND1(J)
 180  CONTINUE
      N4 = N3 - N2 + 1
      DO 190 J=1,N4
        BOUND2(J) = BOUND2(J)*BOUND2(J)
 190  CONTINUE
      GO TO 260
 200  DO 210 J=1,N1
        BOUND1(J) = 1./BOUND1(J)
 210  CONTINUE
      N4 = N3 - N2 + 1
      DO 220 J=1,N4
        BOUND2(J) = 1./BOUND2(J)
 220  CONTINUE
      GO TO 260
C
C SIGN CHANGES BELOW SO CODING AT 2 CAN BE USED
C
 230  DO 240 J=1,N1
        BOUND1(J) = -2.*BOUND1(J)
 240  CONTINUE
      N4 = N3 - N2 + 1
      DO 250 J=1,N4
        BOUND2(J) = -2.*BOUND2(J)
 250  CONTINUE
 260  DO 290 K=1,NXM2
        AZ = 0.0
        DO 270 J=1,N1
          AZ = AZ - DF(K,J)*BOUND1(J)
 270    CONTINUE
        JJ = 0
        DO 280 J=N2,N3
          JJ = JJ + 1
          AZ = AZ + DF(K,J)*BOUND2(JJ)
 280    CONTINUE
        G(K) = R*AZ
 290  CONTINUE
C
C THE PARTIAL OF THE FTN WRT XK - SAME FOR RECIP LOG & EXT
C BECAUSE OF WHAT IS CURRENTLY STORED IN BOUNDS FOR EACH
C
      AZ = 0.
      DO 300 J=1,N1
        AZ = AZ - BOUND1(J)
 300  CONTINUE
      JJ = 0
      DO 310 J=N2,N3
        JJ = JJ + 1
        AZ = AZ + BOUND2(JJ)
 310  CONTINUE
      G(NXM1) = R*AZ*CONS*2./X(NXM1)
C
C THE GRADIENT OF THE PENALTY FUNCTION WITH RESPECT TO ZETA
C SAME CODING FOR ALL PENALTY FTNS
C
      AZ = 0.0
      DO 320 J=1,N1
        AZ = AZ + TH1(J)*BOUND1(J)
 320  CONTINUE
      JJ = 0
      DO 330 J=N2,N3
        JJ = JJ + 1
        AZ = AZ + TH2(JJ)*BOUND2(JJ)
 330  CONTINUE
      G(NX) = 1. - R*AZ
      GO TO 350
 340  IPANIC = 1
      WRITE (IW,9999)
9999  FORMAT (29H0  CONSTRAINTS VIOLATED IN FG)
 350  IF (NIT.EQ.0) GO TO 370
      IF (NOUT.EQ.0) GO TO 380
      WRITE (IW,9998) NVI
9998  FORMAT (23H0  CONSTRAINTS VIOLATED, I5, 22H TIMES DURING QUADRATI,
     *    5HC FIT)
      NVI = 0
      WRITE (IW,9997)
9997  FORMAT (13H0COEFFICIENTS)
 360  WRITE (IW,9996)
9996  FORMAT (54H         A2                A1                B2          ,
     *    4H  B1)
      WRITE (IW,9995) (X(I),I=1,NXM2)
9995  FORMAT (4E16.8)
      XK = X(NXM1)**2
      WRITE (IW,9994) XK, X(NX), F
9994  FORMAT (6H GAIN=, E15.8, 7H ALPHA=, E15.8, 10H FUNCTION=, E15.8)
      GO TO 380
 370  WRITE (IW,9993) R
9993  FORMAT (28H0INITIAL COEFFICIENTS FOR R=, E16.8)
      GO TO 360
 380  IF (NC.EQ.0) GO TO 400
      DO 390 I=1,NC
        IFIX = NL(I)
        G(IFIX) = 0.
 390  CONTINUE
 400  RETURN
      END
C
C-----------------------------------------------------------------------
C SUBROUTINE:  CASC
C CALCULATES THE FREQUENCY RESPONSE AND ITS GRADIENT
```

```
C-----------------------------------------------------------------------
C
      SUBROUTINE CASC(N, AL, W, NF, T, XK, HM, HL, JJ, HG, HGL)
C
C NOTES FOR USE OUTSIDE THE PACKAGE:
C CASC CAN BE USED TO CALCULATE THE FREQUENCY RESPONSE AND ITS
C GRADIENT FOR ANY NUMBER OF FREQUENCY VALUES, NF.
C THE 42 OF THE CURRENT DIMENSION STATEMENT LIMITS THE TRANSFER
C FUNCTION TO 10 SECTIONS EACH HAVING 4 COEFFICIENTS.
C THE EXTRA 2 LOCATIONS ARE USED BY THE PACKAGE AND ARE NOT
C NECESSARY IF CASC IS USED INDEPENDENTLY.  TO USE CASC FOR
C A TRANSFER FUNCTION WITH M SECTIONS WHERE M IS GREATER
C THAN 10, CHANGE THE 42 TO 4*M AND CHANGE THE SIZE OF AH TO M.
C THERE ARE 4 COEFFICIENTS PER SECTION BECAUSE THE CONSTANTS MUST
C BE 1.
C          N    - INTEGER INPUT - ORDER OF THE FILTER .LE. 20
C          AL   - REAL INPUT - ARRAY OF COEFFICIENTS  OF THE
C                 TRANSFER FUNCTION.  SEE "DESCRIPTION OF
C                 REQUIRED DATA".
C          W    - REAL INPUT - ARRAY OF FREQUENCY VALUES MULTIPLIED
C                 BY 2 PI
C          NF   - INTEGER INPUT - NUMBER OF FREQUENCY VALUES
C          T    - REAL INPUT - SAMPLING TIME
C          XK   - REAL INPUT - GAIN FACTOR
C          HM   - REAL OUTPUT - ARRAY OF MAGNITUDE VALUES
C          HL   - REAL OUTPUT - ARRAY OF INSERTION LOSS VALUES
C          JJ   - INTEGER INPUT OPTION FOR CALCULATION OF THE
C                 GRADIENT  JJ=0 FOR NO GRADIENT
C                           JJ=1 FOR GRADIENT
C          HG   - REAL OUTPUT MATRIX - GRADIENT OF THE MAGNITUDE
C                 WHERE HG(I,J) IS THE DERIVATIVE AT FREQUENCY J
C                 FOR COEFFICIENT I
C          HGL  - REAL OUTPUT MATRIX - GRADIENT OF THE INSERTION LOSS
C                 WHERE HGL(I,J) IS THE DERIVATIVE AT FREQUENCY J
C                 FOR COEFFICIENT I
C
      DIMENSION AL(42), W(NF), HM(NF), HL(NF), HG(42,NF), HGL(42,NF),
     *    AH(10)
      CONS = -20./ALOG(10.)
      NU = N/2
C
      DO 100 I=1,NF
C
C CALCULATE QUANTITIES CONSTANT WITH FREQUENCY
C
        WT = W(I)*T
        CWT = COS(WT)
        TWT = 2.*WT
        CTWT = COS(TWT)
        SWT = SIN(WT)
        STWT = SIN(TWT)
C
C INITIALIZE MAGNITUDE OF H
C
        HM(I) = 1.
C
        DO 30 K=1,NU
C
C CALCULATE QUANTITIES CONSTANT WITH ALPHA(K)
C
          K4 = 4*K
          K4M3 = K4 - 3
          K4M2 = K4 - 2
          K4M1 = K4 - 1
          A = AL(K4M3)
          B = AL(K4M2)
          C = AL(K4M1)
          D = AL(K4)
          XM1 = A*CTWT + B*CWT + 1.
          XM2 = C*CTWT + D*CWT + 1.
          XMU1 = A*STWT + B*SWT
          XMU2 = C*STWT + D*SWT
          XM12 = XM1*XM2
          XMU12 = XMU1*XMU2
          XM1MU2 = XM1*XMU2
          XM2MU1 = XM2*XMU1
          DW = XM2**2 + XMU2**2
C
C SAVE MAGNITUDE OF H(K)
          RE = (XM12+XMU12)/DW
          XIM = (XM1MU2-XM2MU1)/DW
          AH(K) = SQRT(RE**2+XIM**2)
          IF (JJ) 20, 20, 10
C
C IF NECESSARY, MAKE PARTIAL CALCULATION OF GRADIENT
C
  10      DW2 = DW**2
          CST = XM2*CTWT + XMU2*STWT
          CS = XM2*CWT + XMU2*SWT
          XXR = XM12 + XMU12
          XXI = XM1MU2 - XM2MU1
          PRA = CST/DW
          PRB = CS/DW
          PRC = (XM1*CTWT+XMU1*STWT)/DW - 2.*XXR*CST/DW2
          PRD = (XM1*CWT+XMU1*SWT)/DW - 2.*XXR*CS/DW2
          PIA = (XMU2*CTWT-XM2*STWT)/DW
          PIB = (XMU2*CWT-XM2*SWT)/DW
          PIC = (XM1*STWT-XMU1*CTWT)/DW - 2.*XXI*CST/DW2
          PID = (XM1*SWT-XMU1*CWT)/DW - 2.*XXI*CS/DW2
          HG(K4M3,I) = (RE*PRA+XIM*PIA)/AH(K)
          HG(K4M2,I) = (RE*PRB+XIM*PIB)/AH(K)
          HG(K4M1,I) = (RE*PRC+XIM*PIC)/AH(K)
          HG(K4,I) = (RE*PRD+XIM*PID)/AH(K)
C
C MULTIPLY MAGNITUDE OF H BY LATEST MAGNITUDE OF H(K)
C
  20      HM(I) = HM(I)*AH(K)
  30    CONTINUE
C
        HM(I) = HM(I)*XK
C
C FIND INSERTION LOSS
C
        HL(I) = -20.*ALOG10(HM(I))
C
        IF (JJ) 100, 100, 40
C
C IF NECESSARY, COMPLETE CALCULATION OF GRADIENT
C
  40    DO 70 K=1,NU
          K4 = 4*K
          DO 60 L=1,NU
            IF (L-K) 50, 60, 50
  50        HG(K4-3,I) = HG(K4-3,I)*AH(L)
            HG(K4-2,I) = HG(K4-2,I)*AH(L)
            HG(K4-1,I) = HG(K4-1,I)*AH(L)
            HG(K4,I) = HG(K4,I)*AH(L)
  60      CONTINUE
  70    CONTINUE
        DO 80 K=1,NU
```

```
          K4 = 4*K
          HG(K4-3,I) = HG(K4-3,I)*XK
          HG(K4-2,I) = HG(K4-2,I)*XK
          HG(K4-1,I) = HG(K4-1,I)*XK
          HG(K4,I) = HG(K4,I)*XK
   80   CONTINUE
C
C FIND GRADIENT OF INSERTION LOSS
C
        CH = CONS/HM(I)
        DO 90 K=1,NU
          K4 = 4*K
          HGL(K4-3,I) = CH*HG(K4-3,I)
          HGL(K4-2,I) = CH*HG(K4-2,I)
          HGL(K4-1,I) = CH*HG(K4-1,I)
          HGL(K4,I) = CH*HG(K4,I)
   90   CONTINUE
C
  100 CONTINUE
C
      RETURN
      END
C
C-----------------------------------------------------------------------
C SUBROUTINE:  RINV
C INVERTS POLES OUTSIDE UNIT CIRCLE
C-----------------------------------------------------------------------
C
      SUBROUTINE RINV(X, Y, NX)
C
      DIMENSION X(42), Y(42)
C
C SQRT OF CORRECTION ON GAIN IS USED BECAUSE GAIN
C IS SQUARED BY CALLING PROGRAM. OPTIMIZER WORKS
C WITH SQRT OF GAIN TO KEEP IT POSITIVE
      NW = NX - 2
      NXM1 = NX - 1
      DO 10 I=1,NX
        Y(I) = X(I)
   10 CONTINUE
      DO 100 KK=3,NW,4
        IF (Y(KK).NE.0.) GO TO 20
        GO TO 40
   20   DISC = Y(KK+1)**2 - 4.*Y(KK)
        IF (DISC.GE.0.) GO TO 50
        IF (Y(KK).GT.1.) GO TO 30
        GO TO 100
   30   SAVE = Y(KK)
        Y(KK) = 1./Y(KK)
        Y(KK+1) = Y(KK+1)/SAVE
        Y(NXM1) = Y(NXM1)/SQRT(ABS(SAVE))
        GO TO 100
   40   IF (ABS(Y(KK+1)).LE.1.) GO TO 100
C
C FOR REAL ROOTS MULT GAIN BY ROOT BEFORE IT'S FLIPPED
C
        SAVE = Y(KK+1)
        Y(KK+1) = 1./Y(KK+1)
        Y(NXM1) = Y(NXM1)/SQRT(ABS(SAVE))
        GO TO 100
   50   R1 = (-Y(KK+1)+SQRT(DISC))/(2.*Y(KK))
        R2 = (-Y(KK+1)-SQRT(DISC))/(2.*Y(KK))
        IF (ABS(R1).LT.1.) GO TO 60
        GO TO 70
   60   Y(NXM1) = Y(NXM1)*SQRT(ABS(R1))
        R1 = 1./R1
   70   IF (ABS(R2).LT.1.) GO TO 80
        GO TO 90
   80   Y(NXM1) = Y(NXM1)*SQRT(ABS(R2))
        R2 = 1./R2
C
C RECONSTRUCT THE QUADRATIC
C
   90   Y(KK) = 1./(R1*R2)
        Y(KK+1) = -(R1+R2)/(R1*R2)
  100 CONTINUE
      RETURN
      END
C
C-----------------------------------------------------------------------
C SUBROUTINE:  FINISH
C CHECKS THE STOP CRITERION
C-----------------------------------------------------------------------
C
      SUBROUTINE FINISH
C
      COMMON X, F, DEL, VBND, EPS, KTIME, NIT, NFE, NGE, NX, N
      COMMON /SUBC/ S, XB, GB, FB, H, G, IPANIC
      DIMENSION X(42), S(42), XB(42), H(42,42), G(42), XX(42), GB(42)
C
C OPTIONS 1,2,4,5 INITIALIZE THE COUNTER KT IN OPTION 3 IN
C CASE THERE IS SWITCHING TO AND FROM OPTION 3
C
C IF N=0 ON RETURN FROM THIS SUBROUTINE, THEN THE STOP
C CRITERION HAS BEEN SATISFIED.     NIT, THE NUMBER OF
C ITERATIONS IS RESET TO ZERO.
C
C N NEGATIVE ON RETURN IS A WARNING THAT OVER 1000 ITERATIONS
C HAVE OCCURRED.
C
C N POSITIVE ON RETURN IS NORMAL AND INDICATES THAT THE
C ITERATIONS ARE TO CONTINUE.
C
      GO TO (10, 50, 60, 110, 200), N
C
C OPTION 1
C STOP WHEN GRADIENT-TRANSPOSE TIMES GRADIENT (GTG) IS LESS
C THAN SOME NUMBER GIVEN BY USER OR SET TO GTG/1.E+06
C
   10 KT = 0
      GTG = 0.
      DO 20 I=1,NX
        GTG = G(I)*G(I) + GTG
   20 CONTINUE
      IF (EPS) 30, 30, 40
   30 EPS = GTG/1.E+06
   40 IF (GTG-EPS) 210, 220, 220
C
C OPTION 2
C STOP WHEN PRESENT FUNCTION VALUE MINUS PREVIOUS FUNCTION
C VALUE IS LESS THAN .0001 TIMES PRESENT VALUE
C
   50 KT = 0
      IF (ABS(F-FB)-ABS(.0001*F)) 210, 210, 220
C
C OPTION 3
C STOP WHEN OPTION 2 IS SATISFIED A GIVEN NUMBER OF TIMES.
C IF NUMBER IS NOT SPECIFIED IT IS SET TO 4.
C
   60 IF (KTIME) 70, 70, 80
```

```
   70   KTIME = 4
   80   IF (ABS(F-FB)-ABS(.0001*F)) 100, 100, 90
   90   KT = 0
        GO TO 220
  100   KT = KT + 1
        IF (KT-KTIME) 220, 210, 210
C
C OPTION 4
C STOP WHEN MAX ABS (X-NEW MINUS X-OLD) IS LESS THAN A GIVEN
C NUMBER.  IF NOT SPECIFIED, THE NUMBER IS SET TO .0005 MAX X
C
  110   KT = 0
        DELTA = EPS
        IF (DELTA) 120, 120, 160
  120   XMAX = ABS(X(1))
        DO 140 I=2,NX
          IF (ABS(X(I))-XMAX) 140, 140, 130
  130     XMAX = ABS(X(I))
  140   CONTINUE
        DELTA = .0005*XMAX
        IF (DELTA) 150, 150, 160
  150   DELTA = .0005
  160   DO 170 I=1,NX
          XX(I) = X(I) - XB(I)
  170   CONTINUE
        DMAX = ABS(XX(1))
        DO 190 I=2,NX
          IF (ABS(XX(I))-DMAX) 190, 190, 180
  180     DMAX = ABS(XX(I))
  190   CONTINUE
        IF (DMAX-DELTA) 210, 210, 220
C
C OPTION 5
C STOP AFTER A GIVEN NUMBER OF ITERATIONS
C
  200   KT = 0
        IF (NIT-KTIME) 220, 210, 210
C
  210   N = 0
        NIT = 0
        RETURN
  220   NIT = NIT + 1
        IF (NIT-1000) 240, 230, 230
  230   N = -1
  240   RETURN
        END
```

6.4

A Program for Designing Finite Word-Length IIR Digital Filters

Kenneth Steiglitz and *Bruce D. Ladendorf*

Department of Electrical Engineering and Computer Science
Princeton University
Princeton, NJ 08540

1. Purpose

This program performs the design of finite word-length IIR digital filters to meet magnitude specifications in the frequency domain.

2. Method

This program is a somewhat generalized implementation of the algorithm described in [1]. The reader is referred to that paper for a detailed problem statement and description of the method. The program begins with a set of high precision coefficients for an IIR digital filter in cascade form. This filter has been designed by other means, and its magnitude transfer function meets tolerance requirements in pass- and stopbands in the frequency domain. The problem is to design a new filter which still meets these requirements but has coefficients with no more than NBIT1 bits after the decimal point. The program obtains a starting point by rounding off the high precision coefficients to NBIT1 bits, and then uses a randomized version of the Hooke and Jeeves search algorithm [2] to optimize the coefficients with an initial search increment of NBIT2 bits (NBIT2 $\leqslant$ NBIT1). If the final function value is no more than 1, the desired tolerances have been met. Because the search algorithm is randomized, different runs with different settings of the random number initialization, NSET, will in general give different answers, so that the best of several runs may be used. The input data is read in by SUBROUTINE INDATA; the heading of that subroutine describes the arrangement of the input data. The random number generator used is UNI (see Standards Section).

3. Program Description

The program is organized in the following subprograms:

3.1 MAIN Program

This is the executive program, which calls the others. The subprograms communicate largely through COMMON blocks ABC, RAW, and STEER. ABC has information about the coefficients X and other input parameters, such as the desired accuracy. RAW has information about the frequency domain specifications — a grid of frequency points is supplied, together with the corresponding desired values of the transfer function and tolerances. COMMON block STEER has three logical variables, PRINT, TWOPT, and FIXCOF, which determine whether the final results are printed by FUNCT, whether two-opt is performed, and whether the second numerator coefficients are fixed at 1, respectively. After optimization SUBROUTINE FUNCT is called with PRINT = .TRUE., which evaluates the transfer function on a finer grid and prints out the final results.

3.2 SUBROUTINE INDATA

This subroutine reads in all the problem data; the details are described in its heading. It shares this data through COMMON with the other programs. Note that the user should set NTWOPT = 1 to use two-opt, and 0 otherwise.

3.3 SUBROUTINE FUNCT

This computes the value of the maximum weighted error function on the coarse grid, and at the end, on a finer grid. The constant multiplier of the filter, A, is computed to high precision and not quantized. This calculation assumes that the desired transfer function values are 0 or 1; and that the weights in all the passbands are equal; otherwise this subroutine must be modified (see Section 4).

3.4 SUBROUTINE HANDJ

This is a randomized version of the Hooke and Jeeves Pattern Search. A block diagram is given in [1].

3.5 SUBROUTINE EXPLOR

This is a local exploration program called by HANDJ; a block diagram is also given in [1]. The local exploration is controlled by the logical variables TWOPT and FIXCOF. If TWOPT is .TRUE. then the denser two-opt search is performed. If FIXCOF is .TRUE. then the second numerator coefficient in each section is frozen at the value 1.

3.6 SUBROUTINE SET

This is a program which simply sets one vector equal to another.

3.7 SUBROUTINE SHUFF

This program randomly orders the list of coordinates, LIST. It is used by EXPLOR.

4. Calculations of the Gain Constant A

The gain constant A is kept to high precision and not optimized; it is recalculated on the basis of the coarse grid each time FUNCT is called, and at the end on the basis of a finer grid. It is important, therefore, to have an efficient way of obtaining A. In this implementation the following heuristic is used, which somewhat restricts the class of filters to which the program is applicable. The minimum and maximum values of the magnitude of the unnormalized transfer function N/D, Y_1 and Y_2, are calculated in the passbands only, and A is calculated by

$$A = 2/(Y_1+Y_2)$$

This centers the magnitude characteristic in the passbands; assuming that the desired transfer function values are 1 in all the passbands (and 0 in all the stopbands), and that the tolerances in the passbands are all equal. It is heuristic even with these assumptions because it ignores the error in the stopbands. In most cases, however, there is little to be gained by de-centering the magnitude characteristic in the passband to reduce the error in the stopband.

The exact calculation of A requires the solution of a linear program; at each point we write the constraints

$$AY - Y_d \leqslant de$$
$$-(AY-Y_d) \leqslant de$$

where Y is the magnitude of N/D, Y_d is the desired magnitude characteristic, and d is the tolerance. The linear programming problem is to find A and e such that e is a minimum.

5. Comments

A. If NCOEF is specified as 4, 4 coefficients per stage are read in, and all of them optimized. If NCOEFF is specified as 3, it is assumed that the second numerator coefficient of each section is fixed at 1. This is often a reasonable assumption when the filter has stopbands, and has the advantages of a simpler filter implementation and an optimization problem with 3*NSECT instead of 4*NSECT parameters.

B. The check mentioned in [1] that the first numerator coefficient be less than 2 in magnitude is omitted in FUNCT. Thus it is conceivable that the numerator coefficients become large in magnitude, requiring more bits for their integer parts.

C. The grid is not generated automatically, but must be read in point by point. Experience has shown that it can have as few as 2.5*N points, where N is the number of adjustable parameters in the problem, provided that the points are chosen carefully. Since the running time of FUNCT is roughly proportional to the number of grid points, it is important to keep this number as small as possible. It has been found best to choose the points in clusters near the transition regions, and less densely elsewhere.

D. The dominant computational burden is the DO 50 loop in FUNCT. This uses a formula close to that in [3] and requires 10 multiplication/additions per section per grid point, so the total number of such operations in this loop is 10*NSECT*M*KOUNT, where NSECT = number of sections, M = number of grid points, and KOUNT = number of function evaluations. Thus, in the test problem, NSECT = 4, M = 28, and (for NSET = 100) KOUNT = 3181; so that there are 3,562,720 multiplications/additions in the DO 50 loop. The total execution time for this problem (excluding compilation) was 4.26 seconds, using FORTRAN H on an IBM 360/91; 10.23 seconds using FORTRAN G on the same machine; and 38.8 seconds on the Honeywell 6080N.

6. Test Problem

A listing of the program is given in the Appendix. We use example 1 from [1], the bandpass example of Avenhaus and Schuessler [4], as a test example. The filter has the passband [0.411111, 0.466666]; and stopbands [0.0, 0.383333] and [0.494444, 1.]. The test data is listed in Table 1, and the corresponding program output in Table 2.

The first card (in FORMAT (5I3)) sets

NCOEF = 3
NSET = 100
NTWOPT = 1
NBIT1 = 6
NBIT2 = 5

The next 28 cards establish the grid and weighting, the next card says there are 4 sections, and the remaining 12 cards give the high precision coefficients. (See the heading of SUBROUTINE INDATA for detailed formatting information.)

With this value of NSET = 100, the local optimum is found after 3181 function calls, and corresponds to a value of F = 0.7681 on the fine grid.

Acknowledgments

This work was supported by NSF Grant GK-42048, and by the U. S. Army Research Office, Durham, under Grant DAAG29-75-G-0192.

References

1. K. Steiglitz, "Designing Short-Word Recursive Digital Filters", *Proc. 9th Annual Allerton Conf. on Circuit and System Theory,* pp. 778-788, October 1971. Reprinted in *Digital Signal Processing II,* Edited by the Digital Signal Processing Committee of the IEEE Group on ASSP, IEEE Press, N. Y., 1976.

2. R. Hooke and T. A. Jeeves, "`Direct Search' Solution of Numerical and Statistical Problems", *Jour. Assoc. Comput. Mach.,* Vol. 8, No. 2, pp. 212-229, April 1961.

3. C. Charalambous and M. J. Best, "Optimization of Recursive Digital Filters with Finite Word Lengths", *IEEE Trans. Acoust., Speech, and Signal Processing,* Vol. ASSP-22, No. 6, pp. 424-431, December 1974. Reprinted in *Digital Signal Processing II,* Edited by the Digital Signal Processing Committee of the IEEE ASSP Society, IEEE Press, N. Y., 1976. See Eq. 6.

4. E. Avenhaus and H. W. Schuessler, "On the Approximation Problem in the Design of Digital Filters with Limited Word-Length", *Archiv der Elecktrischen Uebertragung,* Vol. 24, No. 12, pp. 571-572, December 1970.

Table 1

```
   3100  1  6  5
0.          0.          .01
0.36        0.          .01
0.3658      0.          .01
0.383333    0.          .01
0.411111    1.          .03
0.412       1.          .03
0.4129      1.          .03
0.414       1.          .03
0.4157      1.          .03
0.4185      1.          .03
0.423       1.          .03
0.4275      1.          .03
0.433       1.          .03
0.4386      1.          .03
0.4443      1.          .03
0.45        1.          .03
0.4545      1.          .03
0.459       1.          .03
0.461916    1.          .03
0.4636      1.          .03
0.464833    1.          .03
0.465749    1.          .03
0.466666    1.          .03
0.494444    0.          .01
0.51        0.          .01
0.51448     0.          .01
0.52        0.          .01
1.          0.          .01
   4
  0.4512591755000000
 -0.2855823838000000
  0.9116860616000000
 -1.0986945504000000
 -0.4457342317000000
  0.9130786730000000
 -0.0099665157000000
 -0.1969016650000000
  0.9695353354000000
 -0.7304169706000000
 -0.5515388641000000
  0.9705899009000000
```

Table 2

```
 ***** INPUT DATA *****
NUMBER OF COEFFICIENTS PER STAGE IS   3
RANDOM INITIALIZATION IS 100
TWOPT IS  1
FINAL ( AND INITIAL ROUNDING ) PRECISION IS  6 BITS
INITIAL SEARCH DELTA IS  5 BITS
```

GRID SPECIFICATIONS

POINT NO.	FREQUENCY	DESIRED Y	TOLERANCE
1	0.	0.	0.1000000000D-01
2	0.3600000000D 00	0.	0.1000000000D-01
3	0.3658000000D 00	0.	0.1000000000D-01
4	0.3833330016D 00	0.	0.1000000000D-01
5	0.4111110016D 00	0.1000000000D 01	0.3000000000D-01
6	0.4120000000D 00	0.1000000000D 01	0.3000000000D-01
7	0.4129000000D 00	0.1000000000D 01	0.3000000000D-01
8	0.4140000000D 00	0.1000000000D 01	0.3000000000D-01
9	0.4157000000D 00	0.1000000000D 01	0.3000000000D-01
10	0.4185000000D 00	0.1000000000D 01	0.3000000000D-01
11	0.4230000000D 00	0.1000000000D 01	0.3000000000D-01
12	0.4275000000D 00	0.1000000000D 01	0.3000000000D-01
13	0.4330000000D 00	0.1000000000D 01	0.3000000000D-01
14	0.4386000000D 00	0.1000000000D 01	0.3000000000D-01
15	0.4443000000D 00	0.1000000000D 01	0.3000000000D-01
16	0.4500000000D 00	0.1000000000D 01	0.3000000000D-01
17	0.4545000000D 00	0.1000000000D 01	0.3000000000D-01
18	0.4590000000D 00	0.1000000000D 01	0.3000000000D-01
19	0.4619160000D 00	0.1000000000D 01	0.3000000000D-01
20	0.4636000000D 00	0.1000000000D 01	0.3000000000D-01

Table 2
(Continued)

```
       21     0.4648329984D 00     0.1000000000D 01     0.3000000000D-01
       22     0.4657489984D 00     0.1000000000D 01     0.3000000000D-01
       23     0.4666660032D 00     0.1000000000D 01     0.3000000000D-01
       24     0.4944440000D 00     0.                   0.1000000000D-01
       25     0.5100000000D 00     0.                   0.1000000000D-01
       26     0.5144800000D 00     0.                   0.1000000000D-01
       27     0.5200000064D 00     0.                   0.1000000000D-01
       28     0.1000000000D 01     0.                   0.1000000000D-01

INITIAL HIGH PRECISION COEFFICIENTS FOR  4 SECTIONS ARE
SECTION  1
    0.4512591744D 00      0.1000000000D 01
   -0.2855823808D 00      0.9116860672D 00
SECTION  2
   -0.1098694544D 01      0.1000000000D 01
   -0.4457342336D 00      0.9130786688D 00
SECTION  3
   -0.9966515712D-02      0.1000000000D 01
   -0.1969016640D 00      0.9695353344D 00
SECTION  4
   -0.7304169664D 00      0.1000000000D 01
   -0.5515388608D 00      0.9705899008D 00

THE INITIAL COEFFICIENTS, ROUNDED TO   6 BITS, ARE
SECTION  1
    0.4531249984D 00      0.1000000000D 01
   -0.2812500000D 00      0.9062499968D 00
SECTION  2
   -0.1093750000D 01      0.1000000000D 01
   -0.4531249984D 00      0.9062499968D 00
SECTION  3
   -0.1562500000D-01      0.1000000000D 01
   -0.2031250000D 00      0.9687500032D 00
SECTION  4
   -0.7343750016D 00      0.1000000000D 01
   -0.5468750016D 00      0.9687500032D 00

NEXT FOLLOWS A REPORT FROM THE SEARCH ALGORITHM
NO. CALLS PAT. SIZE           DELTA          OLD VALUE          NEW VALUE
      264           1   0.31250000D-01   0.25259811D 01   0.17236617D 01
      805           1   0.31250000D-01   0.17236617D 01   0.17160663D 01
     1602           1   0.15625000D-01   0.17160663D 01   0.94060213D 00
     2126           1   0.15625000D-01   0.94060213D 00   0.73309429D 00
     2650           1   0.15625000D-01   0.73309429D 00   0.71935856D 00

FINAL VALUES- NO. OF FUNCT CALLS=  3181 FUNCT=     0.7193585600D 00

THE FINAL COEFFICIENTS, HAVING  6 BITS, ARE
SECTION  1
    0.4375000000D 00      0.1000000000D 01
   -0.2812500000D 00      0.9062499968D 00
SECTION  2
   -0.1093750000D 01      0.1000000000D 01
   -0.4531249984D 00      0.9062499968D 00
SECTION  3
   -0.1562500000D-01      0.1000000000D 01
   -0.1875000000D 00      0.9687500032D 00
SECTION  4
   -0.7343750016D 00      0.1000000000D 01
   -0.5625000000D 00      0.9687500032D 00

***** FINAL RESULTS *****
THE CONSTANT A IS      0.7493357696D-02
FUNCTION VALUE ON COARSE GRID IS      0.7193585600D 00
          FREQUENCY          DESIRED Y            ACTUAL Y              ERROR
  0.                  0.                  0.70283624D-02   0.70283624D 00
  0.36000000D 00      0.                  0.64137194D-02   0.64137194D 00
  0.36580000D 00      0.                  0.67483316D-02   0.67483316D 00
  0.38333300D 00      0.                  0.52589518D-02   0.52589518D 00
  0.41111100D 00      0.10000000D 01      0.10215808D 01   0.71935856D 00
  0.41200000D 00      0.10000000D 01      0.10204763D 01   0.68254401D 00
  0.41290000D 00      0.10000000D 01      0.10148222D 01   0.49407221D 00
  0.41400000D 00      0.10000000D 01      0.10066453D 01   0.22150962D 00
```

Table 2
(Continued)

```
  0.41570000D 00  0.10000000D 01  0.99765911D 00  0.78030010D-01
  0.41850000D 00  0.10000000D 01  0.99730467D 00  0.89844010D-01
  0.42300000D 00  0.10000000D 01  0.10136004D 01  0.45334529D 00
  0.42750000D 00  0.10000000D 01  0.10161582D 01  0.53860584D 00
  0.43300000D 00  0.10000000D 01  0.99671587D 00  0.10947098D 00
  0.43860000D 00  0.10000000D 01  0.98587546D 00  0.47081791D 00
  0.44430000D 00  0.10000000D 01  0.10004271D 01  0.14235485D-01
  0.45000000D 00  0.10000000D 01  0.10154313D 01  0.51437758D 00
  0.45450000D 00  0.10000000D 01  0.10021237D 01  0.70790976D-01
  0.45900000D 00  0.10000000D 01  0.97841925D 00  0.71935856D 00
  0.46191600D 00  0.10000000D 01  0.97883618D 00  0.70546071D 00
  0.46360000D 00  0.10000000D 01  0.98927915D 00  0.35736149D 00
  0.46483300D 00  0.10000000D 01  0.99955200D 00  0.14933567D-01
  0.46574900D 00  0.10000000D 01  0.10052380D 01  0.17460133D 00
  0.46666600D 00  0.10000000D 01  0.10045265D 01  0.15088357D 00
  0.49444400D 00  0.              0.53608885D-02  0.53608885D 00
  0.51000000D 00  0.              0.70073326D-02  0.70073326D 00
  0.51448000D 00  0.              0.71485085D-02  0.71485085D 00
  0.52000000D 00  0.              0.67435740D-02  0.67435740D 00
  0.10000000D 01  0.              0.70871288D-02  0.70871288D 00

FINAL RESULTS ON FINE GRID- 10 POINTS PER GRID POINT
       FREQUENCY        DESIRED Y         ACTUAL Y             ERROR
***** BAND FROM GRID POINT     1 TO     2 *****
  0.              0.              0.70355889D-02  0.70355889D 00
  0.36000000D-01  0.              0.70067577D-02  0.70067577D 00
  0.72000000D-01  0.              0.69162904D-02  0.69162904D 00
  0.10800000D 00  0.              0.67511103D-02  0.67511103D 00
  0.14400000D 00  0.              0.64849851D-02  0.64849851D 00
  0.18000000D 00  0.              0.60693616D-02  0.60693616D 00
  0.21600000D 00  0.              0.54133609D-02  0.54133609D 00
  0.25200000D 00  0.              0.43383098D-02  0.43383098D 00
  0.28800000D 00  0.              0.24698003D-02  0.24698003D 00
  0.32400000D 00  0.              0.98742474D-03  0.98742474D-01
***** BAND FROM GRID POINT     2 TO     3 *****
  0.36000000D 00  0.              0.64203139D-02  0.64203139D 00
  0.36058000D 00  0.              0.64798328D-02  0.64798328D 00
  0.36116000D 00  0.              0.65348297D-02  0.65348297D 00
  0.36174000D 00  0.              0.65848927D-02  0.65848927D 00
  0.36232000D 00  0.              0.66295783D-02  0.66295783D 00
  0.36290000D 00  0.              0.66684091D-02  0.66684091D 00
  0.36348000D 00  0.              0.67008707D-02  0.67008707D 00
  0.36406000D 00  0.              0.67264089D-02  0.67264089D 00
  0.36464000D 00  0.              0.67444262D-02  0.67444262D 00
  0.36522000D 00  0.              0.67542784D-02  0.67542784D 00
***** BAND FROM GRID POINT     3 TO     4 *****
  0.36580000D 00  0.              0.67552702D-02  0.67552702D 00
  0.36755330D 00  0.              0.66964453D-02  0.66964453D 00
  0.36930660D 00  0.              0.65250871D-02  0.65250871D 00
  0.37105990D 00  0.              0.62103614D-02  0.62103614D 00
  0.37281320D 00  0.              0.57130142D-02  0.57130142D 00
  0.37456650D 00  0.              0.49827854D-02  0.49827854D 00
  0.37631980D 00  0.              0.39549038D-02  0.39549038D 00
  0.37807310D 00  0.              0.25452870D-02  0.25452870D 00
  0.37982640D 00  0.              0.64388579D-03  0.64388579D-01
  0.38157970D 00  0.              0.18946638D-02  0.18946638D 00
  0.38333300D 00  0.              0.52643590D-02  0.52643590D 00
***** BAND FROM GRID POINT     5 TO     6 *****
  0.41111100D 00  0.10000000D 01  0.10226311D 01  0.75437147D 00
  0.41119990D 00  0.10000000D 01  0.10228557D 01  0.76185765D 00
  0.41128880D 00  0.10000000D 01  0.10229910D 01  0.76636676D 00
  0.41137770D 00  0.10000000D 01  0.10230429D 01  0.76809520D 00
  0.41146660D 00  0.10000000D 01  0.10230170D 01  0.76723329D 00
  0.41155550D 00  0.10000000D 01  0.10229190D 01  0.76396509D 00
  0.41164440D 00  0.10000000D 01  0.10227540D 01  0.75846824D 00
  0.41173330D 00  0.10000000D 01  0.10225274D 01  0.75091390D 00
  0.41182220D 00  0.10000000D 01  0.10222440D 01  0.74146667D 00
  0.41191110D 00  0.10000000D 01  0.10219085D 01  0.73028457D 00
***** BAND FROM GRID POINT     6 TO     7 *****
  0.41200000D 00  0.10000000D 01  0.10215256D 01  0.71751907D 00
  0.41209000D 00  0.10000000D 01  0.10210939D 01  0.70313092D 00
  0.41218000D 00  0.10000000D 01  0.10206224D 01  0.68741216D 00
  0.41227000D 00  0.10000000D 01  0.10201150D 01  0.67049975D 00
```

Table 2
(Continued)

```
  0.41236000D 00  0.10000000D 01  0.10195757D 01  0.65252410D 00
  0.41245000D 00  0.10000000D 01  0.10190083D 01  0.63360920D 00
  0.41254000D 00  0.10000000D 01  0.10184162D 01  0.61387271D 00
  0.41263000D 00  0.10000000D 01  0.10178028D 01  0.59342618D 00
  0.41272000D 00  0.10000000D 01  0.10171713D 01  0.57237521D 00
  0.41281000D 00  0.10000000D 01  0.10165246D 01  0.55081958D 00
***** BAND FROM GRID POINT     7 TO     8 *****
  0.41290000D 00  0.10000000D 01  0.10158656D 01  0.52885347D 00
  0.41301000D 00  0.10000000D 01  0.10150473D 01  0.50157765D 00
  0.41312000D 00  0.10000000D 01  0.10142192D 01  0.47397244D 00
  0.41323000D 00  0.10000000D 01  0.10133853D 01  0.44617820D 00
  0.41334000D 00  0.10000000D 01  0.10125497D 01  0.41832464D 00
  0.41345000D 00  0.10000000D 01  0.10117159D 01  0.39053136D 00
  0.41356000D 00  0.10000000D 01  0.10108873D 01  0.36290838D 00
  0.41367000D 00  0.10000000D 01  0.10100667D 01  0.33555656D 00
  0.41378000D 00  0.10000000D 01  0.10092570D 01  0.30856820D 00
  0.41389000D 00  0.10000000D 01  0.10084608D 01  0.28202742D 00
***** BAND FROM GRID POINT     8 TO     9 *****
  0.41400000D 00  0.10000000D 01  0.10076803D 01  0.25601064D 00
  0.41417000D 00  0.10000000D 01  0.10065098D 01  0.21699227D 00
  0.41434000D 00  0.10000000D 01  0.10053884D 01  0.17961268D 00
  0.41451000D 00  0.10000000D 01  0.10043218D 01  0.14406032D 00
  0.41468000D 00  0.10000000D 01  0.10033147D 01  0.11049151D 00
  0.41485000D 00  0.10000000D 01  0.10023710D 01  0.79033543D-01
  0.41502000D 00  0.10000000D 01  0.10014936D 01  0.49787442D-01
  0.41519000D 00  0.10000000D 01  0.10006849D 01  0.22830594D-01
  0.41536000D 00  0.10000000D 01  0.99994657D 00  0.17808973D-02
  0.41553000D 00  0.10000000D 01  0.99927970D 00  0.24010040D-01
***** BAND FROM GRID POINT     9 TO    10 *****
  0.41570000D 00  0.10000000D 01  0.99868488D 00  0.43836976D-01
  0.41598000D 00  0.10000000D 01  0.99786248D 00  0.71250407D-01
  0.41626000D 00  0.10000000D 01  0.99723406D 00  0.92198169D-01
  0.41654000D 00  0.10000000D 01  0.99679534D 00  0.10682197D 00
  0.41682000D 00  0.10000000D 01  0.99654003D 00  0.11533237D 00
  0.41710000D 00  0.10000000D 01  0.99646012D 00  0.11799615D 00
  0.41738000D 00  0.10000000D 01  0.99654624D 00  0.11512560D 00
  0.41766000D 00  0.10000000D 01  0.99678792D 00  0.10706939D 00
  0.41794000D 00  0.10000000D 01  0.99717386D 00  0.94204776D-01
  0.41822000D 00  0.10000000D 01  0.99769208D 00  0.76930922D-01
***** BAND FROM GRID POINT    10 TO    11 *****
  0.41850000D 00  0.10000000D 01  0.99833011D 00  0.55663123D-01
  0.41895000D 00  0.10000000D 01  0.99957423D 00  0.14192199D-01
  0.41940000D 00  0.10000000D 01  0.10010407D 01  0.34690503D-01
  0.41985000D 00  0.10000000D 01  0.10026754D 01  0.89178679D-01
  0.42030000D 00  0.10000000D 01  0.10044248D 01  0.14749376D 00
  0.42075000D 00  0.10000000D 01  0.10062374D 01  0.20791344D 00
  0.42120000D 00  0.10000000D 01  0.10080639D 01  0.26879637D 00
  0.42165000D 00  0.10000000D 01  0.10098581D 01  0.32860352D 00
  0.42210000D 00  0.10000000D 01  0.10115775D 01  0.38591673D 00
  0.42255000D 00  0.10000000D 01  0.10131836D 01  0.43945445D 00
***** BAND FROM GRID POINT    11 TO    12 *****
  0.42300000D 00  0.10000000D 01  0.10146425D 01  0.48808469D 00
  0.42345000D 00  0.10000000D 01  0.10159251D 01  0.53083512D 00
  0.42390000D 00  0.10000000D 01  0.10170070D 01  0.56690020D 00
  0.42435000D 00  0.10000000D 01  0.10178694D 01  0.59564508D 00
  0.42480000D 00  0.10000000D 01  0.10184982D 01  0.61660646D 00
  0.42525000D 00  0.10000000D 01  0.10188847D 01  0.62949036D 00
  0.42570000D 00  0.10000000D 01  0.10190250D 01  0.63416685D 00
  0.42615000D 00  0.10000000D 01  0.10189199D 01  0.63066217D 00
  0.42660000D 00  0.10000000D 01  0.10185744D 01  0.61914826D 00
  0.42705000D 00  0.10000000D 01  0.10179979D 01  0.59993034D 00
***** BAND FROM GRID POINT    12 TO    13 *****
  0.42750000D 00  0.10000000D 01  0.10172030D 01  0.57343289D 00
  0.42805000D 00  0.10000000D 01  0.10159582D 01  0.53193950D 00
  0.42860000D 00  0.10000000D 01  0.10144452D 01  0.48150652D 00
  0.42915000D 00  0.10000000D 01  0.10127028D 01  0.42342644D 00
  0.42970000D 00  0.10000000D 01  0.10107730D 01  0.35910105D 00
  0.43025000D 00  0.10000000D 01  0.10087000D 01  0.29000047D 00
  0.43080000D 00  0.10000000D 01  0.10065288D 01  0.21762531D 00
  0.43135000D 00  0.10000000D 01  0.10043042D 01  0.14347290D 00
  0.43190000D 00  0.10000000D 01  0.10020702D 01  0.69007842D-01
  0.43245000D 00  0.10000000D 01  0.99986912D 00  0.43628569D-02
***** BAND FROM GRID POINT    13 TO    14 *****
```

Table 2
(Continued)

```
 0.43300000D 00  0.10000000D 01  0.99774069D 00  0.75310269D-01
 0.43356000D 00  0.10000000D 01  0.99568654D 00  0.14378228D 00
 0.43412000D 00  0.10000000D 01  0.99378179D 00  0.20727361D 00
 0.43468000D 00  0.10000000D 01  0.99205839D 00  0.26472054D 00
 0.43524000D 00  0.10000000D 01  0.99054424D 00  0.31519196D 00
 0.43580000D 00  0.10000000D 01  0.98926318D 00  0.35789432D 00
 0.43636000D 00  0.10000000D 01  0.98823476D 00  0.39217477D 00
 0.43692000D 00  0.10000000D 01  0.98747431D 00  0.41752302D 00
 0.43748000D 00  0.10000000D 01  0.98699283D 00  0.43357239D 00
 0.43804000D 00  0.10000000D 01  0.98679699D 00  0.44010034D 00
***** BAND FROM GRID POINT   14 TO   15 *****
 0.43860000D 00  0.10000000D 01  0.98688913D 00  0.43702874D 00
 0.43917000D 00  0.10000000D 01  0.98727659D 00  0.42411345D 00
 0.43974000D 00  0.10000000D 01  0.98795361D 00  0.40154633D 00
 0.44031000D 00  0.10000000D 01  0.98890878D 00  0.36970726D 00
 0.44088000D 00  0.10000000D 01  0.99012605D 00  0.32913181D 00
 0.44145000D 00  0.10000000D 01  0.99158467D 00  0.28051073D 00
 0.44202000D 00  0.10000000D 01  0.99325934D 00  0.22468875D 00
 0.44259000D 00  0.10000000D 01  0.99512014D 00  0.16266201D 00
 0.44316000D 00  0.10000000D 01  0.99713278D 00  0.95573713D-01
 0.44373000D 00  0.10000000D 01  0.99925879D 00  0.24707332D-01
***** BAND FROM GRID POINT   15 TO   16 *****
 0.44430000D 00  0.10000000D 01  0.10014557D 01  0.48523386D-01
 0.44487000D 00  0.10000000D 01  0.10036776D 01  0.12258821D 00
 0.44544000D 00  0.10000000D 01  0.10058758D 01  0.19585872D 00
 0.44601000D 00  0.10000000D 01  0.10079990D 01  0.26663173D 00
 0.44658000D 00  0.10000000D 01  0.10099947D 01  0.33315810D 00
 0.44715000D 00  0.10000000D 01  0.10118103D 01  0.39367734D 00
 0.44772000D 00  0.10000000D 01  0.10133937D 01  0.44645757D 00
 0.44829000D 00  0.10000000D 01  0.10146952D 01  0.48984066D 00
 0.44886000D 00  0.10000000D 01  0.10156687D 01  0.52229130D 00
 0.44943000D 00  0.10000000D 01  0.10162735D 01  0.54244880D 00
***** BAND FROM GRID POINT   16 TO   17 *****
 0.45000000D 00  0.10000000D 01  0.10164754D 01  0.54917973D 00
 0.45045000D 00  0.10000000D 01  0.10163331D 01  0.54443719D 00
 0.45090000D 00  0.10000000D 01  0.10159152D 01  0.53050826D 00
 0.45135000D 00  0.10000000D 01  0.10152176D 01  0.50725415D 00
 0.45180000D 00  0.10000000D 01  0.10142406D 01  0.47468723D 00
 0.45225000D 00  0.10000000D 01  0.10129894D 01  0.43298007D 00
 0.45270000D 00  0.10000000D 01  0.10114742D 01  0.38247160D 00
 0.45315000D 00  0.10000000D 01  0.10097101D 01  0.32367061D 00
 0.45360000D 00  0.10000000D 01  0.10077177D 01  0.25725621D 00
 0.45405000D 00  0.10000000D 01  0.10055223D 01  0.18407528D 00
***** BAND FROM GRID POINT   17 TO   18 *****
 0.45450000D 00  0.10000000D 01  0.10031541D 01  0.10513703D 00
 0.45495000D 00  0.10000000D 01  0.10006481D 01  0.21604847D-01
 0.45540000D 00  0.10000000D 01  0.99804357D 00  0.65214538D-01
 0.45585000D 00  0.10000000D 01  0.99538348D 00  0.15388379D 00
 0.45630000D 00  0.10000000D 01  0.99271450D 00  0.24284985D 00
 0.45675000D 00  0.10000000D 01  0.99008619D 00  0.33046045D 00
 0.45720000D 00  0.10000000D 01  0.98755054D 00  0.41498217D 00
 0.45765000D 00  0.10000000D 01  0.98516140D 00  0.49462008D 00
 0.45810000D 00  0.10000000D 01  0.98297383D 00  0.56753911D 00
 0.45855000D 00  0.10000000D 01  0.98104336D 00  0.63188776D 00
***** BAND FROM GRID POINT   18 TO   19 *****
 0.45900000D 00  0.10000000D 01  0.97942525D 00  0.68582494D 00
 0.45929160D 00  0.10000000D 01  0.97856902D 00  0.71436620D 00
 0.45958320D 00  0.10000000D 01  0.97788085D 00  0.73730509D 00
 0.45987480D 00  0.10000000D 01  0.97737438D 00  0.75418718D 00
 0.46016640D 00  0.10000000D 01  0.97706251D 00  0.76458304D 00
 0.46045800D 00  0.10000000D 01  0.97695714D 00  0.76809520D 00
 0.46074960D 00  0.10000000D 01  0.97706901D 00  0.76436645D 00
 0.46104120D 00  0.10000000D 01  0.97740730D 00  0.75308967D 00
 0.46133280D 00  0.10000000D 01  0.97797941D 00  0.73401947D 00
 0.46162440D 00  0.10000000D 01  0.97879041D 00  0.70698625D 00
***** BAND FROM GRID POINT   19 TO   20 *****
 0.46191600D 00  0.10000000D 01  0.97984262D 00  0.67191280D 00
 0.46208440D 00  0.10000000D 01  0.98055988D 00  0.64800384D 00
 0.46225280D 00  0.10000000D 01  0.98135653D 00  0.62144912D 00
 0.46242120D 00  0.10000000D 01  0.98223124D 00  0.59229199D 00
 0.46258960D 00  0.10000000D 01  0.98318216D 00  0.56059447D 00
 0.46275800D 00  0.10000000D 01  0.98420682D 00  0.52643932D 00
 0.46292640D 00  0.10000000D 01  0.98530202D 00  0.48993239D 00
```

Table 2
(Continued)

0.46309480D 00	0.10000000D 01	0.98646385D 00	0.45120503D 00
0.46326320D 00	0.10000000D 01	0.98768750D 00	0.41041687D 00
0.46343160D 00	0.10000000D 01	0.98896724D 00	0.36775870D 00
***** BAND FROM GRID POINT 20 TO 21 *****			
0.46360000D 00	0.10000000D 01	0.99029633D 00	0.32345566D 00
0.46372330D 00	0.10000000D 01	0.99129625D 00	0.29012479D 00
0.46384660D 00	0.10000000D 01	0.99231493D 00	0.25616929D 00
0.46396990D 00	0.10000000D 01	0.99334844D 00	0.22171894D 00
0.46409320D 00	0.10000000D 01	0.99439249D 00	0.18691706D 00
0.46421650D 00	0.10000000D 01	0.99544236D 00	0.15192135D 00
0.46433980D 00	0.10000000D 01	0.99649286D 00	0.11690478D 00
0.46446310D 00	0.10000000D 01	0.99753830D 00	0.82056549D-01
0.46458640D 00	0.10000000D 01	0.99857251D 00	0.47582919D-01
0.46470970D 00	0.10000000D 01	0.99958876D 00	0.13708201D-01
***** BAND FROM GRID POINT 21 TO 22 *****			
0.46483300D 00	0.10000000D 01	0.10005797D 01	0.19324343D-01
0.46492460D 00	0.10000000D 01	0.10012949D 01	0.43164152D-01
0.46501620D 00	0.10000000D 01	0.10019884D 01	0.66279129D-01
0.46510780D 00	0.10000000D 01	0.10026564D 01	0.88547527D-01
0.46519940D 00	0.10000000D 01	0.10032952D 01	0.10984092D 00
0.46529100D 00	0.10000000D 01	0.10039007D 01	0.13002403D 00
0.46538260D 00	0.10000000D 01	0.10044686D 01	0.14895458D 00
0.46547420D 00	0.10000000D 01	0.10049945D 01	0.16648317D 00
0.46556580D 00	0.10000000D 01	0.10054736D 01	0.18245317D 00
0.46565740D 00	0.10000000D 01	0.10059010D 01	0.19670060D 00
***** BAND FROM GRID POINT 22 TO 23 *****			
0.46574900D 00	0.10000000D 01	0.10062716D 01	0.20905412D 00
0.46584070D 00	0.10000000D 01	0.10065804D 01	0.21934500D 00
0.46593240D 00	0.10000000D 01	0.10068212D 01	0.22737191D 00
0.46602410D 00	0.10000000D 01	0.10069882D 01	0.23294069D 00
0.46611580D 00	0.10000000D 01	0.10070755D 01	0.23584995D 00
0.46620750D 00	0.10000000D 01	0.10070767D 01	0.23589117D 00
0.46629920D 00	0.10000000D 01	0.10069855D 01	0.23284891D 00
0.46639090D 00	0.10000000D 01	0.10067950D 01	0.22650112D 00
0.46648260D 00	0.10000000D 01	0.10064986D 01	0.21661944D 00
0.46657430D 00	0.10000000D 01	0.10060891D 01	0.20296961D 00
0.46666600D 00	0.10000000D 01	0.10055594D 01	0.18531197D 00
***** BAND FROM GRID POINT 24 TO 25 *****			
0.49444400D 00	0.	0.53664005D-02	0.53664005D 00
0.49599960D 00	0.	0.23290146D-02	0.23290146D 00
0.49755520D 00	0.	0.57134124D-04	0.57134124D-02
0.49911080D 00	0.	0.19262916D-02	0.19262916D 00
0.50066640D 00	0.	0.33826998D-02	0.33826998D 00
0.50222200D 00	0.	0.45079746D-02	0.45079746D 00
0.50377760D 00	0.	0.53664970D-02	0.53664970D 00
0.50533320D 00	0.	0.60093980D-02	0.60093980D 00
0.50688880D 00	0.	0.64775363D-02	0.64775363D 00
0.50844440D 00	0.	0.68037441D-02	0.68037441D 00
***** BAND FROM GRID POINT 25 TO 26 *****			
0.51000000D 00	0.	0.70145375D-02	0.70145375D 00
0.51044800D 00	0.	0.70569896D-02	0.70569896D 00
0.51089600D 00	0.	0.70921265D-02	0.70921265D 00
0.51134400D 00	0.	0.71203702D-02	0.71203702D 00
0.51179200D 00	0.	0.71421190D-02	0.71421190D 00
0.51224000D 00	0.	0.71577488D-02	0.71577488D 00
0.51268800D 00	0.	0.71676144D-02	0.71676144D 00
0.51313600D 00	0.	0.71720504D-02	0.71720504D 00
0.51358400D 00	0.	0.71713735D-02	0.71713735D 00
0.51403200D 00	0.	0.71658821D-02	0.71658821D 00
***** BAND FROM GRID POINT 26 TO 27 *****			
0.51448000D 00	0.	0.71558586D-02	0.71558586D 00
0.51503200D 00	0.	0.71376699D-02	0.71376699D 00
0.51558400D 00	0.	0.71134734D-02	0.71134734D 00
0.51613600D 00	0.	0.70837068D-02	0.70837068D 00
0.51668800D 00	0.	0.70487784D-02	0.70487784D 00
0.51724000D 00	0.	0.70090696D-02	0.70090696D 00
0.51779200D 00	0.	0.69649370D-02	0.69649370D 00
0.51834400D 00	0.	0.69167132D-02	0.69167132D 00
0.51889600D 00	0.	0.68647096D-02	0.68647096D 00
0.51944800D 00	0.	0.68092172D-02	0.68092172D 00
***** BAND FROM GRID POINT 27 TO 28 *****			
0.52000000D 00	0.	0.67505077D-02	0.67505077D 00
0.56800000D 00	0.	0.22088895D-03	0.22088895D-01

Table 2
(Continued)

```
 0.61600000D 00  0.                 0.31906682D-02  0.31906682D 00
 0.66400000D 00  0.                 0.48591221D-02  0.48591221D 00
 0.71200000D 00  0.                 0.57711975D-02  0.57711975D 00
 0.76000000D 00  0.                 0.63127353D-02  0.63127353D 00
 0.80800000D 00  0.                 0.66508561D-02  0.66508561D 00
 0.85600000D 00  0.                 0.68657264D-02  0.68657264D 00
 0.90400000D 00  0.                 0.69986081D-02  0.69986081D 00
 0.95200000D 00  0.                 0.70712714D-02  0.70712714D 00
 0.10000000D 01  0.                 0.70944158D-02  0.70944158D 00

THE VALUE OF A FROM THE FINE GRID IS     0.7501062336D-02

THE FINAL VALUE OF THE ERROR ON THE FINE GRID IS     0.7680951936D 00
```

Appendix

```
C
C-----------------------------------------------------------------------
C  MAIN PROGRAM:  FWIIR
C  AUTHORS:       KENNETH STEIGLITZ AND BRUCE D. LADENDORF
C                 PRINCETON UNIVERSITY, PRINCETON, NJ 08540
C                 VERSION OCTOBER 15, 1978
C
C  INPUT:         A GRID OF FREQUENCY POINTS AND A SET OF
C                 (ESSENTIALLY) INFINITE PRECISION COEFFICIENTS.
C                 SEE SUBROUTINE INDATA FOR DETAILED INPUT INFORMATION
C
C  THIS PROGRAM DESIGNS FINITE WORD-LENGTH IIR DIGITAL FILTERS.
C  THE METHOD USES RANDOMIZED PATTERN SEARCH OF HOOKE AND JEEVES,
C  AND IS DESCRIBED IN "DESIGNING SHORT-WORD RECURSIVE DIGITAL FILTERS,"
C  BY KENNETH STEIGLITZ, IN PROC. 9TH ANNUAL ALLERTON CONF. ON CIRCUIT
C  AND SYSTEM TH., PP. 778-788; OCT. 1971. REPRINTED IN DIGITAL SIGNAL
C  PROCESSING II, EDITED BY THE DIGITAL SIGNAL PROCESSING COMMITTEE
C  OF THE GROUP ON ASSP, IEEE PRESS, N. Y., 1976.
C-----------------------------------------------------------------------
C
      DOUBLE PRECISION W(100), Y(100), WEIGHT(100), X(36)
      DOUBLE PRECISION SDELTA, DELTAZ, DUMMY, RHO, EST, F
      LOGICAL PRINT, TWOPT, FIXCOF
      COMMON /ABC/ X, NSET, NTWOPT, NBIT1, NBIT2, NSECT, N
      COMMON /RAW/ W, Y, WEIGHT, M, KOUNT
      COMMON /STEER/ PRINT, TWOPT, FIXCOF
      COMMON /MACH/ IND, IOUTD
C
C SET UP MACHINE CONSTANTS
C
      IND = I1MACH(1)
      IOUTD = I1MACH(2)
C
C READ IN PROBLEM DATA
C
      CALL INDATA
C
C SET SMALL AND LARGE DELTA
C
      SDELTA = (.5D0)**NBIT1
      DELTAZ = (.5D0)**NBIT2
C
C SET RANDOM NUMBER GENERATOR BY CALLING IT NSET TIMES
C
      DO 10 J=1,NSET
        DUMMY = UNI(0.)
  10  CONTINUE
C
C TURN OFF THE PRINTING OPTION FOR FUNCT
C
      PRINT = .FALSE.
C
C ROUND OFF THE COEFFICIENTS TO NBIT1 BITS
C
      DO 20 J=1,N
        X(J) = DSIGN(SDELTA*FLOAT(IDINT(DABS(X(J))/SDELTA+.5D0)),X(J))
  20  CONTINUE
      WRITE (IOUTD,9999) NBIT1
9999  FORMAT (38H1THE INITIAL COEFFICIENTS, ROUNDED TO , I3, 8H BITS, A,
     *    2HRE)
      DO 30 J=1,NSECT
        WRITE (IOUTD,9998) J, X(4*J-3), X(4*J-2), X(4*J-1), X(4*J)
  30  CONTINUE
9998  FORMAT (8H SECTION, I3/1H , 2D20.10/1H , 2D20.10)
C
C SET PARAMETER VALUES FOR H-AND-J; THESE MAY BE CHANGED IF DESIRED
C H-AND-J MULTIPLIES THE STEP SIZE BY RHO; TAKE RHO = .5
C
      RHO = .5D0
C
C H-AND-J STOPS IF THE FUNCTION VALUE FALLS BELOW EST; TAKE EST = 0.
C
      EST = 0.
C
C H-AND-J STOPS IF WE EXCEED "LIMIT" FUNCT CALLS; TAKE LIMIT = 10,000
C
      LIMIT = 10000
      WRITE (IOUTD,9997)
9997  FORMAT (48H0NEXT FOLLOWS A REPORT FROM THE SEARCH ALGORITHM)
      CALL HANDJ(N, DELTAZ, SDELTA, RHO, EST, LIMIT, X)
      WRITE (IOUTD,9996) NBIT1
9996  FORMAT (31H0THE FINAL COEFFICIENTS, HAVING, I3, 10H BITS, ARE)
      DO 40 J=1,NSECT
        WRITE (IOUTD,9998) J, X(4*J-3), X(4*J-2), X(4*J-1), X(4*J)
  40  CONTINUE
C
C PRINT OUT THE FINAL FINE GRID AND QUIT
C
      PRINT = .TRUE.
      CALL FUNCT(N, X, F)
      STOP
      END
C
C-----------------------------------------------------------------------
C SUBROUTINE:    INDATA
C THIS SUBROUTINE READS IN THE PROBLEM DATA:
C CARD 1 HAS THE NUMBER OF COEFFICIENTS PER STAGE ( 3 OR 4), THE
C INITIAL SETTING OF THE RANDOM NUMBER GENERATOR, WHETHER TWO-OPT
C SHOULD BE USED ( 0 IS NO, 1 IS YES), THE DESIRED BIT LENGTH OF THE
C COEFFICIENTS, AND THE INITIAL SEARCH DELTA ( IN BITS) FOR THE
C HOOKE AND JEEVES ALGORITHM.
C
C THE NEXT SET OF CARDS SPECIFIES THE GRID, ONE GRID POINT PER CARD:
C EACH CARD HAS FIRST THE FREQUENCY IN FRACTIONS OF THE NYQUIST
C FREQUENCY, NEXT THE DESIRED TRANSFER FUNCTION MAGNITUDE, AND THIRD
C THE TOLERANCE. TO INDICATE THE END OF THE GRID POINTS,
C SPECIFY A FREQUENCY OF 1.  THE M CARDS ARE COUNTED, M.LE.100 .
C THE PRESENT SUBROUTINE FUNCT ASSUMES FOR THE A CALCULATION THAT THE
C DESIRED TRANSFER FUNCTION TAKES ON ONLY THE VALUES 0 OR 1. THAT
C IS, THE FILTER HAS ONLY PASS AND STOP BANDS, AND EVERY PASS BAND
C HAS DESIRED VALUE 1, AND THE SAME TOLERANCE.
C
C THE NEXT CARD HAS THE NUMBER OF SECOND-ORDER SECTIONS, NSECT.
C
C THE NEXT CARDS HAVE THE "INFINITE PRECISION" COEFFICIENTS, ONE PER
C CARD. THE COEFFICIENTS FOR SECTION 1 ARE GIVEN FIRST, NEXT
C SECTION 2, ETC. WITHIN EACH SECTION THEY ARE IN THE ORDER: NUMERATOR
C COEFFICIENT OF Z**(-1), NUMERATOR COEFFICIENT OF Z**(-2) ( IF 4
C PER SECTION), DENOMINATOR COEFFICIENT OF Z**(-1), DENOMINATOR
C COEFFICIENT OF Z**(-2). THERE ARE 3*NSECT OR 4*NSECT COEFFICIENTS
C READ IN, DEPENDING ON WHETHER THERE ARE 3 OR 4 COEFFICIENTS PER
C SECTION, NSECT.LE.9 . IN THE CASE OF 3 COEFFICIENTS PER SECTION,
C THE SECOND NUMERATOR COEFFICIENT IS SET TO 1 AND FROZEN.
C-----------------------------------------------------------------------
C
      SUBROUTINE INDATA
      DOUBLE PRECISION W(100), Y(100), WEIGHT(100), X(36)
      LOGICAL PRINT, TWOPT, FIXCOF
      COMMON /ABC/ X, NSET, NTWOPT, NBIT1, NBIT2, NSECT, N
      COMMON /RAW/ W, Y, WEIGHT, M, KOUNT
```

```
      COMMON /STEER/ PRINT, TWOPT, FIXCOF
      COMMON /MACH/ IND, IOUTD
C
C INITIALIZE KOUNT, THE NUMBER OF FUNCTION EVALUATIONS
C
      KOUNT = 0
C
C READ AND WRITE CARD 1 PARAMETERS
C
      READ (IND,9999) NCOEF, NSET, NTWOPT, NBIT1, NBIT2
9999  FORMAT (5I3)
      WRITE (IOUTD,9998) NCOEF, NSET, NTWOPT, NBIT1, NBIT2
9998  FORMAT (24H1 ***** INPUT DATA *****/26H NUMBER OF COEFFICIENTS PE,
     *    10HR STAGE IS, I3/26H RANDOM INITIALIZATION IS , I3/7H TWOPT ,
     *    2HIS, I3/44H FINAL ( AND INITIAL ROUNDING ) PRECISION IS, I3,
     *    5H BITS/24H INITIAL SEARCH DELTA IS, I3, 5H BITS)
C
C SET LOGICAL VARIABLE TWOPT IF TWO-OPT IS CALLED FOR
C
      TWOPT = .FALSE.
      IF (NTWOPT.EQ.1) TWOPT = .TRUE.
C
C READ AND WRITE THE GRID
C
      WRITE (IOUTD,9997)
9997  FORMAT (20H0GRID SPECIFICATIONS/30H POINT NO.          FREQUENCY,
     *    40H           DESIRED Y            TOLERANCE)
      M = 0
  10  M = M + 1
      READ (IND,9996) W(M), Y(M), WEIGHT(M)
9996  FORMAT (3F10.5)
      WRITE (IOUTD,9995) M, W(M), Y(M), WEIGHT(M)
9995  FORMAT (1H , I9, 3D20.10)
      WEIGHT(M) = 1.D0/WEIGHT(M)
      IF (W(M).LT.1.D0) GO TO 10
C
C READ AND WRITE THE COEFFICIENTS
C
      READ (IND,9999) NSECT
      WRITE (IOUTD,9994) NSECT
9994  FORMAT (40H0INITIAL HIGH PRECISION COEFFICIENTS FOR, I3, 6H SECTI,
     *    7HONS ARE)
      N = 4*NSECT
      IF (NCOEF.EQ.3) GO TO 20
      FIXCOF = .FALSE.
      READ (IND,9993) (X(J),J=1,N)
9993  FORMAT (F20.16)
      GO TO 40
  20  FIXCOF = .TRUE.
      READ (IND,9993) (X(4*J-3),X(4*J-1),X(4*J),J=1,NSECT)
      DO 30 J=1,NSECT
        X(4*J-2) = 1.D0
  30  CONTINUE
  40  CONTINUE
      DO 50 J=1,NSECT
        WRITE (IOUTD,9992) J, X(4*J-3), X(4*J-2), X(4*J-1), X(4*J)
  50  CONTINUE
9992  FORMAT (8H SECTION, I3/1H , 2D20.10/1H , 2D20.10)
      RETURN
      END
C
C-----------------------------------------------------------------------
C SUBROUTINE:    FUNCT
C THIS COMPUTES THE VALUE OF THE ERROR FUNCTION, USING THE
C GRID. IF PRINT.EQ. .TRUE., IT COMPUTES THE ERROR FUNCTION
C AT A FINER GRID ( 10 POINTS FOR EACH GRID POINT), AND PRINTS
C THE RESULTS; THIS IS USED TO PRINT OUT THE FINAL TRANSFER FUNCTION.
C
C NOTE THAT THE CALCULATION OF THE CONSTANT MULTIPLIER A ASSUMES
C THAT ALL THE PASSBANDS HAVE THE SAME DESIRED VALUE OF 1. THIS
C MUST BE CHANGED TO A BAND-BY-BAND CALCULATION IF THERE IS MORE
C THAN ONE PASSBAND DESIRED VALUE.
C-----------------------------------------------------------------------
C
      SUBROUTINE FUNCT(N, X, F)
      DOUBLE PRECISION W(100), Y(100), WEIGHT(100)
      DOUBLE PRECISION COST2(100), COS2T4(100)
      DOUBLE PRECISION SUB1(100), SUB2(100), SUB3(100)
      DOUBLE PRECISION SUB6(100), SUB7(100), SUB8(100)
      DOUBLE PRECISION X(36), YHT(100), E(100)
      DOUBLE PRECISION YHTHT(100,11)
      DOUBLE PRECISION A, F, PI, NUM, DEN, Y1, Y2, FREQ, CCC, SSS, A1,
     *    ERROR
      LOGICAL PRINT, TWOPT, FIXCOF
      COMMON /RAW/ W, Y, WEIGHT, M, KOUNT
      COMMON /STEER/ PRINT, TWOPT, FIXCOF
      COMMON /MACH/ IND, IOUTD
C
C CHECK THAT THE POLES OF THE FILTER ARE INSIDE OR ON THE
C UNIT CIRCLE; NO CHECKS ARE MADE ON NUMERATOR COEFFICIENTS
C
      K = N/4
      DO 10 J=1,K
        J4 = J*4
        IF (X(J4).LE.1.D0.AND.1.D0+X(J4).GE.DABS(X(J4-1))) GO TO 10
        F = 1.D12
        RETURN
  10  CONTINUE
      PI = 4.D0*DATAN(1.D0)
      IF (KOUNT.NE.0) GO TO 30
C
C COMPUTE AND SAVE THE TRIG FUNCTIONS AT THE GRID POINTS
C THE FIRST TIME THROUGH
C
      DO 20 I=1,M
        COST2(I) = 2.D0*DCOS(PI*W(I))
        COS2T4(I) = COST2(I)**2
  20  CONTINUE
C
C MOVE THE CALCULATION OF CONSTANTS OUTSIDE THE INNER LOOP
C
  30  DO 40 J=1,K
        J4 = J*4
        SUB1(J) = X(J4-3)**2 + (X(J4-2)-1.D0)**2
        SUB2(J) = X(J4-3)*(X(J4-2)+1.D0)
        SUB3(J) = X(J4-2)
        SUB6(J) = X(J4-1)**2 + (X(J4)-1.D0)**2
        SUB7(J) = X(J4-1)*(X(J4)+1.D0)
        SUB8(J) = X(J4)
  40  CONTINUE
C
C EVALUATE THE MAGNITUDE OF THE TRANSFER FUNCTION, YHT, AT EACH
C OF THE M GRID POINTS
C
      DO 60 I=1,M
        NUM = 1.D0
        DEN = 1.D0
        DO 50 J=1,K
          NUM = NUM*(SUB1(J)+SUB2(J)*COST2(I)+SUB3(J)*COS2T4(I))
          DEN = DEN*(SUB6(J)+SUB7(J)*COST2(I)+SUB8(J)*COS2T4(I))
```

```
   50    CONTINUE
         YHT(I) = DSQRT(NUM/DEN)
   60 CONTINUE
C
C FIND THE LARGEST AND SMALLEST VALUES OF YHT IN THE PASSBANDS
C
      Y1 = 1.D12
      Y2 = 0.D0
      DO 70 I=1,M
        IF (Y(I).EQ.0.D0) GO TO 70
        Y1 = DMIN1(Y1,YHT(I))
        Y2 = DMAX1(Y2,YHT(I))
   70 CONTINUE
C
C DEFINE THE CONSTANT MULTIPLIER A TO BE THE RECIPROCAL OF THE AVERAGE
C OF THE SMALLEST AND LARGEST YHT"S IN THE PASSBANDS. THIS MAKES SENSE
C IF ALL THE PASSBANDS HAVE THE SAME DESIRED VALUE OF 1 AND THE SAME
C TOLERANCE; OTHERWISE THIS CALCULATION OF A MUST BE MODIFIED.
C
      A = 2.D0/(Y1+Y2)
C
C CALCULATE THE WEIGHTED ERROR AT THE GRID POINTS
C
      F = 0.D0
      DO 80 I=1,M
        YHT(I) = A*YHT(I)
        E(I) = WEIGHT(I)*DABS(YHT(I)-Y(I))
        F = DMAX1(F,E(I))
   80 CONTINUE
      KOUNT = KOUNT + 1
      IF (.NOT.PRINT) RETURN
C
C IF REQUIRED, PRINT OUT THE RESULTS ON THE COARSE GRID
C
      WRITE (IOUTD,9999) A, F
 9999 FORMAT (26H1***** FINAL RESULTS *****/18H THE CONSTANT A IS,
     *    D20.10/33H FUNCTION VALUE ON COARSE GRID IS, D20.10)
      WRITE (IOUTD,9998)
 9998 FORMAT (1H , 7X, 9HFREQUENCY, 7X, 9HDESIRED Y, 8X, 8HACTUAL Y,
     *    11X, 5HERROR)
      DO 90 I=1,M
        WRITE (IOUTD,9997) W(I), Y(I), YHT(I), E(I)
   90 CONTINUE
 9997 FORMAT (1H , 4D16.8)
C
C TO RE-CALCULATE A, CALCULATE N/D=YHTHT ON THE FINE GRID
C THE SAME RESERVATION ABOUT THIS CALCULATION OF A APPLIES AS ABOVE
C
      Y1 = 1.D12
      Y2 = 0.D0
      MM = M - 1
      DO 130 I=1,MM
C
C SKIP TRANSITION BANDS, DEFINED AS THOSE WHERE Y(I) CHANGES
C
        IF (Y(I).NE.Y(I+1)) GO TO 130
        DO 120 II=1,11
          IF (II.NE.11) GO TO 100
C
C FIND CASES WHERE THE LAST POINT OF THE BAND WILL NOT BE COMPUTED LATER
C
          IF (I.EQ.MM) GO TO 100
          IF (Y(I+1).NE.Y(I+2)) GO TO 100
          GO TO 120
  100     FREQ = W(I) + (W(I+1)-W(I))*.1D0*FLOAT(II-1)
          CCC = 2.D0*DCOS(PI*FREQ)
          SSS = CCC**2
          NUM = 1.D0
          DEN = 1.D0
          DO 110 J=1,K
            J4 = J*4
            NUM = NUM*(X(J4-3)**2+(X(J4-2)-1.D0)**2+X(J4-3)*(X(J4-2)
     *            +1.D0)*CCC+X(J4-2)*SSS)
            DEN = DEN*(X(J4-1)**2+(X(J4)-1.D0)**2+X(J4-1)*(X(J4)+1.D0)*
     *            CCC+X(J4)*SSS)
  110     CONTINUE
          YHTHT(I,II) = DSQRT(NUM/DEN)
          IF (Y(I).EQ.0.D0) GO TO 120
          Y1 = DMIN1(Y1,YHTHT(I,II))
          Y2 = DMAX1(Y2,YHTHT(I,II))
  120   CONTINUE
  130 CONTINUE
      A = 2.D0/(Y1+Y2)
C
C NOW REPEAT THE SAME LOOPS, FINDING THE ERROR AND PRINTING THE RESULTS
C
      WRITE (IOUTD,9996)
 9996 FORMAT (53H1FINAL RESULTS ON FINE GRID- 10 POINTS PER GRID POINT)
      WRITE (IOUTD,9998)
      F = 0.D0
      DO 160 I=1,MM
        IF (Y(I).NE.Y(I+1)) GO TO 160
        IP = I + 1
        WRITE (IOUTD,9995) I, IP
 9995   FORMAT (27H ***** BAND FROM GRID POINT, I5, 3H TO, I5, 6H *****)
        DO 150 II=1,11
          IF (II.NE.11) GO TO 140
          IF (I.EQ.MM) GO TO 140
          IF (Y(I+1).NE.Y(I+2)) GO TO 140
          GO TO 150
  140     FREQ = W(I) + (W(I+1)-W(I))*.1D0*FLOAT(II-1)
          A1 = A*YHTHT(I,II)
          ERROR = WEIGHT(I)*DABS(A1-Y(I))
          F = DMAX1(F,ERROR)
          WRITE (IOUTD,9997) FREQ, Y(I), A1, ERROR
  150   CONTINUE
  160 CONTINUE
      WRITE (IOUTD,9994) A, F
 9994 FORMAT (37H0THE VALUE OF A FROM THE FINE GRID IS, D20.10/6H0THE F,
     *    43HINAL VALUE OF THE ERROR ON THE FINE GRID IS, D20.10)
      RETURN
      END
C
C-----------------------------------------------------------------------
C SUBROUTINE:    HANDJ
C THIS IS AN IMPLEMENTATION OF A RANDOMIZED HOOKE AND JEEVES
C SEARCH ALGORITHM. IT FINDS LOCAL OPTIMA; THE COORDINATES ARE
C RANDOMLY ORDERED EACH TIME EXPLOR IS INVOKED. FLOW CHARTS FOR
C THIS SUBROUTINE AND THE NEXT ARE INCLUDED IN THE REFERENCED PAPER.
C-----------------------------------------------------------------------
C
      SUBROUTINE HANDJ(N, DELTAZ, SDELTA, RHO, FMIN, LIMIT, PSI)
      DOUBLE PRECISION PSI(36), THETA(36), PHI(36)
      DOUBLE PRECISION W(100), Y(100), WEIGHT(100)
      DOUBLE PRECISION FPSI, FPHI, DELTA, RHO, DELTAZ, SDELTA, FMIN
      COMMON /RAW/ W, Y, WEIGHT, M, KOUNT
      COMMON /MACH/ IND, IOUTD
      CALL FUNCT(N, PSI, FPSI)
      DELTA = DELTAZ
      WRITE (IOUTD,9999)
```

```
9999  FORMAT (10H NO. CALLS, 10H PAT. SIZE, 11X, 5HDELTA, 7X, 7HOLD VAL,
     *    2HUE, 7X, 9HNEW VALUE)
  10  IF (DELTA.LT.SDELTA .OR. FPSI.LT.FMIN .OR. KOUNT.GT.LIMIT) GO TO
     *    50
      CALL SET(PHI, PSI, N)
      FPHI = FPSI
      CALL EXPLOR(PHI, FPHI, N, DELTA)
      I = 0
  20  IF (FPHI.GE.FPSI .OR. FPSI.LT.FMIN .OR. KOUNT.GT.LIMIT) GO TO 40
      I = I + 1
      WRITE (IOUTD,9998) KOUNT, I, DELTA, FPSI, FPHI
9998  FORMAT (4X, I6, 4X, I6, 3D16.8)
      CALL SET(THETA, PSI, N)
      CALL SET(PSI, PHI, N)
      FPSI = FPHI
      DO 30 J=1,N
        PHI(J) = 2.*PHI(J) - THETA(J)
  30  CONTINUE
      CALL FUNCT(N, PHI, FPHI)
      CALL EXPLOR(PHI, FPHI, N, DELTA)
      GO TO 20
  40  IF (I.GT.0) GO TO 10
      DELTA = RHO*DELTA
      GO TO 10
  50  WRITE (IOUTD,9997) KOUNT, FPSI
9997  FORMAT (34H0FINAL VALUES- NO. OF FUNCT CALLS=, I6, 7H FUNCT=,
     *    D20.10)
      RETURN
      END
C
C-----------------------------------------------------------------------
C SUBROUTINE:    EXPLOR
C THIS IS A LOCAL EXPLORATION SUBROUTINE USED BY H-AND-J
C-----------------------------------------------------------------------
C
      SUBROUTINE EXPLOR(PHI, FPHI, N, DELTA)
      DOUBLE PRECISION PHI(36), FPHI, DELTA, SAVE, FNEW, SAVEI, SAVEJ
      INTEGER LIST(36)
      COMMON /STEER/ PRINT, TWOPT, FIXCOF
      LOGICAL PRINT, TWOPT, FIXCOF
C
C RANDOMIZE THE ORDER OF THE COORDINATES
C
      CALL SHUFF(N, LIST)
      ICOUNT = 1
  10  IF (ICOUNT.GT.N) RETURN
      I = LIST(ICOUNT)
C
C IF ONLY THREE COEFFICIENTS WERE READ IN PER STAGE
C THEN DO NOT CHANGE THE SECOND NUMERATOR COEFFICIENT
C
      IF (FIXCOF .AND. I-(I/4)*4.EQ.2) GO TO 100
      SAVE = PHI(I)
      PHI(I) = PHI(I) + DELTA
      CALL FUNCT(N, PHI, FNEW)
      IF (FNEW.GE.FPHI) GO TO 20
      FPHI = FNEW
      GO TO 100
  20  PHI(I) = PHI(I) - 2.*DELTA
      CALL FUNCT(N, PHI, FNEW)
      IF (FNEW.GE.FPHI) GO TO 30
      FPHI = FNEW
      GO TO 100
  30  PHI(I) = SAVE
      IF (.NOT.TWOPT) GO TO 100
C
C BEGINNING OF 2-OPT
C
      JCOUNT = ICOUNT + 1
  40  IF (JCOUNT.GT.N) GO TO 100
      J = LIST(JCOUNT)
C
C SIMILARLY FOR TWOPT, IF THREE COEFFICIENTS PER STAGE WERE
C READ IN, DO NOT CHANGE THE SECOND NUMERATOR COEFFICIENT
C
      IF (FIXCOF .AND. J-(J/4)*4.EQ.2) GO TO 90
      SAVEI = PHI(I)
      SAVEJ = PHI(J)
      PHI(I) = PHI(I) + DELTA
      PHI(J) = PHI(J) + DELTA
      CALL FUNCT(N, PHI, FNEW)
      IF (FNEW.GE.FPHI) GO TO 50
      FPHI = FNEW
      GO TO 90
  50  PHI(I) = PHI(I) - 2.*DELTA
      CALL FUNCT(N, PHI, FNEW)
      IF (FNEW.GE.FPHI) GO TO 60
      FPHI = FNEW
      GO TO 90
  60  PHI(J) = PHI(J) - 2.*DELTA
      CALL FUNCT(N, PHI, FNEW)
      IF (FNEW.GE.FPHI) GO TO 70
      FPHI = FNEW
      GO TO 90
  70  PHI(I) = PHI(I) + 2.*DELTA
      CALL FUNCT(N, PHI, FNEW)
      IF (FNEW.GE.FPHI) GO TO 80
      FPHI = FNEW
      GO TO 90
  80  PHI(I) = SAVEI
      PHI(J) = SAVEJ
  90  JCOUNT = JCOUNT + 1
      GO TO 40
C
C END OF 2-OPT
C
 100  ICOUNT = ICOUNT + 1
      GO TO 10
      END
C
C-----------------------------------------------------------------------
C SUBROUTINE:    SET
C THIS SETS VECTOR A EQUAL TO VECTOR B
C-----------------------------------------------------------------------
C
      SUBROUTINE SET(A, B, N)
      DOUBLE PRECISION A(36), B(36)
      DO 10 I=1,N
        A(I) = B(I)
  10  CONTINUE
      RETURN
      END
C
C-----------------------------------------------------------------------
C SUBROUTINE:    SHUFF
C THIS RANDOMLY ORDERS LIST
C-----------------------------------------------------------------------
C
      SUBROUTINE SHUFF(N, LIST)
      INTEGER LIST(36)
```

```
      DO 10 I=1,N
        LIST(I) = I
 10   CONTINUE
      DO 20 LL=1,N
        L = N - LL + 1
        J = INT(FLOAT(L)*UNI(0.)) + 1
        K = LIST(L)
        LIST(L) = LIST(J)
        LIST(J) = K
 20   CONTINUE
      RETURN
      END
```

CHAPTER 7

Cepstral Analysis

A. V. Oppenheim

Introduction

This section consists of two programs related to the computation of the cepstrum and the generation of the minimum phase equivalent of a nonminimum phase signal. The cepstrum method is a nonlinear signal analysis technique which has found application to a variety of problem areas including echo detection [1], speech analysis [2], geophysical data processing [3] and a variety of others. Computation of the cepstrum is also an intermediate step in homomorphic filtering for deconvolution [4].

The cepstrum is defined as the inverse Fourier transform of the logarithm of the Fourier transform of the input sequence and, in general, this requires the evaluation of the phase as a continuous function of frequency. This in turn requires computation of an "unwrapped" phase, i.e., a phase curve for which the discontinuities associated with computation of the phase modulo 2π are removed. The cepstrum obtained in this manner is referred to as the *complex cepstrum* since it requires the use of the complex logarithm.

An alternative form of the cepstrum is based on the use of the logarithm of the *magnitude* rather than the complex logarithm of the Fourier transform. This form of the cepstrum is often referred to as the *real cepstrum*. For a minimum phase signal the complex cepstrum is easily obtained from this form of the cepstrum by simple windowing in the cepstral domain and the need for phase unwrapping is thereby eliminated. Computation of the cepstrum using only the magnitude of the Fourier transform followed by appropriate windowing in the cepstral domain and the inverse cepstral transformation also provides a means of transforming a signal to its minimum-phase equivalent. A discussion of the theoretical details can be found in Oppenheim and Schafer [4].

This section contains two programs related to cepstral analysis. The first program is directed at the computation of the complex cepstrum and contains as a major component the program for phase unwrapping. It should be noted that the phase unwrapping program included in Section 7.1 has many other applications in addition to its importance in cepstral analysis. For example, phase unwrapping arises in the context of tracking propagation times in the ocean using sinusoidal sources. The second program computes the real cepstrum and implements the appropriate windowing and inverse transformation to obtain the minimum-phase reconstruction of the signal. In applying these programs and cepstral analysis in general, it is often important to avoid data which has been oversampled since this will introduce a frequency band with small amplitudes and low signal-to-noise ratio which will lead to unreliable phase unwrapping and will tend to dominate the cepstrum because of the logarithmic transformation. Thus the use of modulation and decimation/interpolation filtering may be required prior to cepstral analysis. This and a variety of other computational issues and concerns are discussed in detail in Tribolet [3].

References

1. B. P. Bogert, M. J. R. Healey, J. W. Tukey, "The Quefrency Alanysis of Time Series", *Proc. Symp. Time Series Analysis,* M. Rosenblatt, Ed., New York, John Wiley & Sons, pp. 209-243, 1963.

2. A. V. Oppenheim and R. W. Schafer, "Homomorphic Analysis of Speech", *IEEE Trans. Audio Electroacoust.,* Vol. AU-16, No. 2, pp. 221-226, June 1968.

3. J. Tribolet, *Application of Homomorphic Filtering to Seismic Signal Processing,* Prentice-Hall Inc., Englewood Cliffs, New Jersey, 1979.

4. A. V. Oppenheim and R. W. Schafer, *Digital Signal Processing,* Chapter 10, Prentice-Hall Inc., Englewood Cliffs, New Jersey, 1975.

7.1

Computation of the Complex Cepstrum

J. M. Tribolet

Instituto Superior Tecnico
Lisbon, Portugal

and

T. F. Quatieri

Massachusetts Institute of Technology
Cambridge, MA 02139

1. Purpose

The program is designed to compute the complex cepstrum $\hat{x}(n)$ of a real sequence $x(n)$. The complex cepstrum is defined as the inverse Fourier transform of the complex logarithm of the Fourier transform of the input sequence.

2. Method

The complex cepstrum $\hat{x}[n]$ of a real sequence $x[n]$ is evaluated by means of the DFT computational realization as discussed by Oppenheim and Schafer [1]

$$\hat{x}[n] = \mathrm{IDFT}(\log[\mathrm{DFT}(x[n])]) \tag{1}$$

where the DFT length is sufficient to avoid cepstral aliasing. The complex logarithmic operation involved in this computation makes it necessary to evaluate the phase of $x[n]$ as a continuous function in frequency. This evaluation, called phase unwrapping, is accomplished by means of the adaptive numerical integration algorithm proposed by Tribolet [2]. This technique uses both the principal value of the phase and the phase derivative to compute the unwrapped phase at each frequency. The evaluation of the phase derivative requires the knowledge of the DFTs of both $x(n)$ and $nx(n)$.

The basic idea of the phase unwrapping algorithm is the following. Let Ω_1 be an arbitrary frequency value, $ARG[X(e^{j\Omega_1})]$ be the principle value of the phase, $arg[X(e^{j\Omega_1})]$ be the unwrapped phase, and $arg'[X(e^{j\Omega_1})]$ the phase derivative at Ω_1. The set of permissible phase values at Ω_1 is given by

$$\{ARG[X(e^{j\Omega_1})] + 2\pi l, \quad l \text{ integer}\} \tag{2}$$

The phase unwrapping problem amounts to determining the correct integer value $l_c(\Omega_1)$ such that

$$arg[X(e^{j\Omega_1})] = ARG[X(e^{j\Omega_1})] + 2\pi l_c(\Omega_1)\ . \tag{3}$$

This is done through the use of numerical integration of the phase derivative. We adopt here the trapezoidal integration rule. Assuming the unwrapped phase to be known at a frequency $\Omega_0 < \Omega_1$, we define a phase estimate at Ω_1, $\tilde{arg}[X(e^{j\Omega_1})|\Omega_0]$ by

$$\tilde{arg}[X(e^{j\Omega_1})|\Omega_0] = arg[X(e^{j\Omega_0})] + \frac{\Omega_1 - \Omega_0}{2}\,[arg'[X(e^{j\Omega_0})] + arg'[X(e^{j\Omega_1})]]\ . \tag{4}$$

Clearly, this estimate improves as the step interval $\Delta\Omega = \Omega_1 - \Omega_0$ becomes smaller. We define the phase estimate at Ω_1 to be consistent if it lies within a predefined distance of one of the permissible phase values at Ω_1, that is, if there exists an $l_c(\Omega_1)$ such that

$$|\tilde{arg}[X(e^{j\Omega_1})|\Omega_0] - ARG[X(e^{j\Omega_1})] + 2\pi l_c(\Omega_1)| < \text{THLCON} < \pi \ . \tag{5}$$

The basic idea of this algorithm is thus to adapt the step size $\Delta\Omega$ until a consistent phase estimate is found. The resultant $l_c(\Omega_1)$ in Eq. (5) is used in Eq. (3) to form the unwrapped phase at Ω_1. This unwrapped phase is then used to form $\tilde{arg}[X(e^{j\Omega_2})|\Omega_1]$, $\Omega_2 > \Omega_1$, and so on.

For this algorithm to be practical, one must take full advantage of the FFT algorithm and reduce the number of extra discrete Fourier transform (DFT) computations to a reasonably small number. Let us denote by

$$\{\omega_K = (2\pi/N)k, \quad k=0,1,...,N-1\} \tag{6}$$

the set of uniformly spaced frequencies with interval $2\pi/N$ where $N = 2^M$ (or, in general, any highly composite number). The phase derivative and the principal value of the phase at these frequencies may then be computed using FFT's to evaluate the DFT's of $x(n)$ and $nx(n)$. At each ω_K, a phase estimate is initially formed by one-step trapezoidal integration, starting at ω_{K-1}. If the resultant estimate is not consistent, the adaptive integration scheme is applied within the interval $[\omega_{K-1}, \omega_K]$. The step size adaptation was carefully designed to minimize the number of extra DFT's required. The search for consistency is done by consecutively splitting the step interval in half. As the required phase derivatives and principal values are computed, they are stored in a stack fashion. As soon as a consistent estimate is found, the corresponding data are moved out of the stack to a register that holds the most recent consistent estimate of the phase at some frequency within $[\omega_{k-1}, \omega_k]$. New estimates are always formed by integrating from the most recent estimate to the frequency corresponding to the top of the stack.

3. Program Description

The desired sequence $x(n)$ is generated within the main program CCMAIN through subroutine COEFF. A diagram of this mainline program is given in Fig. 1. The FFT length and threshold levels THLCON (of Eq. (5)) and THLINC (see Section 5 on computational accuracy for a description of this threshold) are initialized. The computation of the complex cepstrum is accomplished by calling subroutine CCEPS and proceeds as follows. First, the FFTs of both $x[n]$ and $nx[n]$ are computed. Then, the log spectral magnitude, the principal value of the phase, and the phase derivative of $x[n]$ on the FFT frequency grid are evaluated. Next, adaptive phase unwrapping is successfully performed between these equispaced frequencies, through the use of function PHAUNW. A block diagram of this program is given in Fig. 2. After the successful completion of the unwrapping, the linear phase component is removed and an IFFT taken, to yield the complex cepstrum $\hat{x}[n]$. A block diagram of CCEPS is given in Fig. 3. Within CCMAIN the first and last 32 values of the complex cepstrum are then printed, along with its corresponding sign and linear phase component.

The code for these routines and additional secondary routines is given in the Appendix where dimension requirements are specified.

4. Test Problem

In order to facilitate testing, as well as help the user getting acquainted with the potentialities and limitations of this program, the subroutine COEFF is provided for generation of test signals of finite length, with prescribed z-transforms.

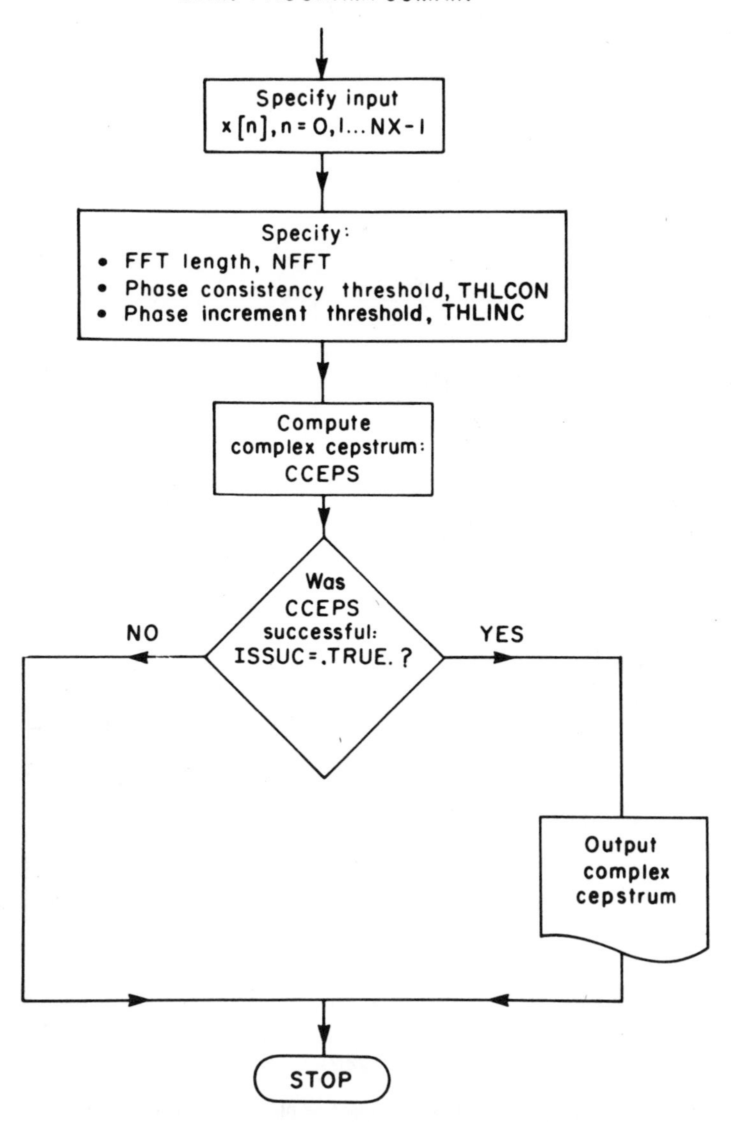

Fig. 1 Block diagram of main program: CCMAIN.

The specifications included in CCMAIN will generate a sequence $x[n]$ whose z-transform $X(z)$ has 6 zeros at $z_1, z_1^*, z_2, z_2^*, z_3, z_3^*$, where:

$$z_1 = (0.9)\, e^{(j\pi/4)}$$
$$z_2 = (1.1)\, e^{[j(\frac{\pi}{4} + \frac{\pi}{8192})]}$$
$$z_3 = (0.9)\, e^{[j(\frac{\pi}{4} + \frac{2\pi}{8192})]}$$

The output of CCMAIN is given in Table 1. The results match those evaluated analytically, using the equation

$$\hat{x}[n] = \begin{cases} -\left(\dfrac{z_1^n + z_1^{*n} + z_3^n + z_3^{*n}}{n}\right) & n > 0 \\[2ex] \left(\dfrac{z_2^n + z_2^{*n}}{n}\right) & n < 0 \end{cases} \tag{7}$$

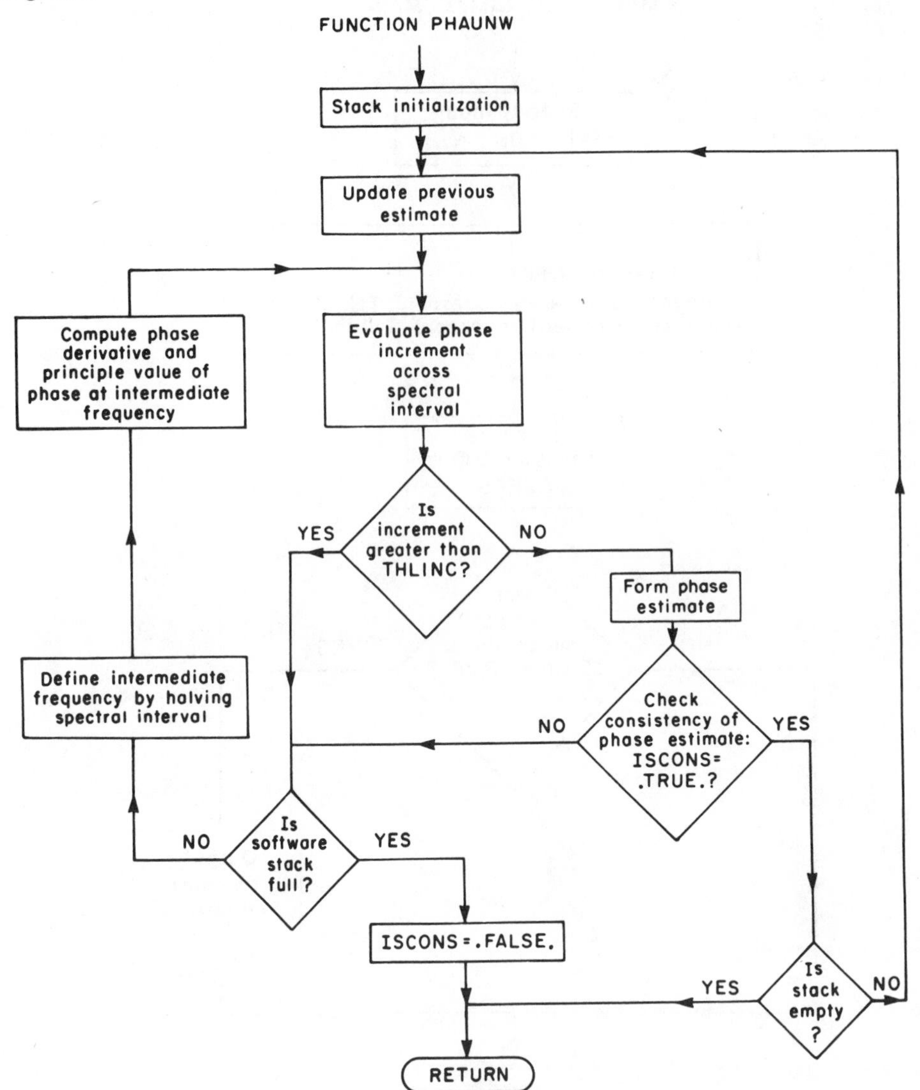

Fig. 2 Block diagram of Tribolet's adaptive phase unwrapping algorithm.

5. Computational Accuracy

It is important to realize that when the input data $x(n)$ has a number of very sharp zeros near the unit circle (although not on it), the computational noise generated, for example, in the FFTs, may lead to erroneous values for the principal value of the phase and for the phase derivative near such zeros.

Subroutine PHAUNW is designed to detect such cases and interrupt the unwrapping process, typing the message "Phase Estimation Failed". This detection is controlled by the values THLINC and THLCON.

THLCON is defined by Eq. (5) of Section 2. and, for most seismic and speech signals of our interest, is set at 0.5. THLINC is used to detect the possibility of an erroneous $2\pi l_c(\Omega_1)$ of Eq. (3) in a region of a sharp zero; that is, a poor integration step within Eq. (4) may lead to a consistent estimate but in error by a 2π multiple. Setting THLINC to 1.5 will avoid any such phenomenon, by requiring further step adaption, when the unwrapped phase increment is greater than 1.5. However, in certain circumstances the stack memory may not be sufficient for the given thresholds.

For example, if the magnitudes of z_1,z_2 and z_3 in the example above are 0.9999, 1.0001 and 0.9999, respectively, then the unwrapping fails.

The authors tested a version of this program on an ECLIPSE minicomputer, using precisely the code published in this book, except in that its compilation (including the FFTs) was done using the Eclipse-Fortran 5 Double Precision Compiler Option. This has the effect of forcing all real variables and constants to become double precision. Furthermore, all single precision Fortran library functions

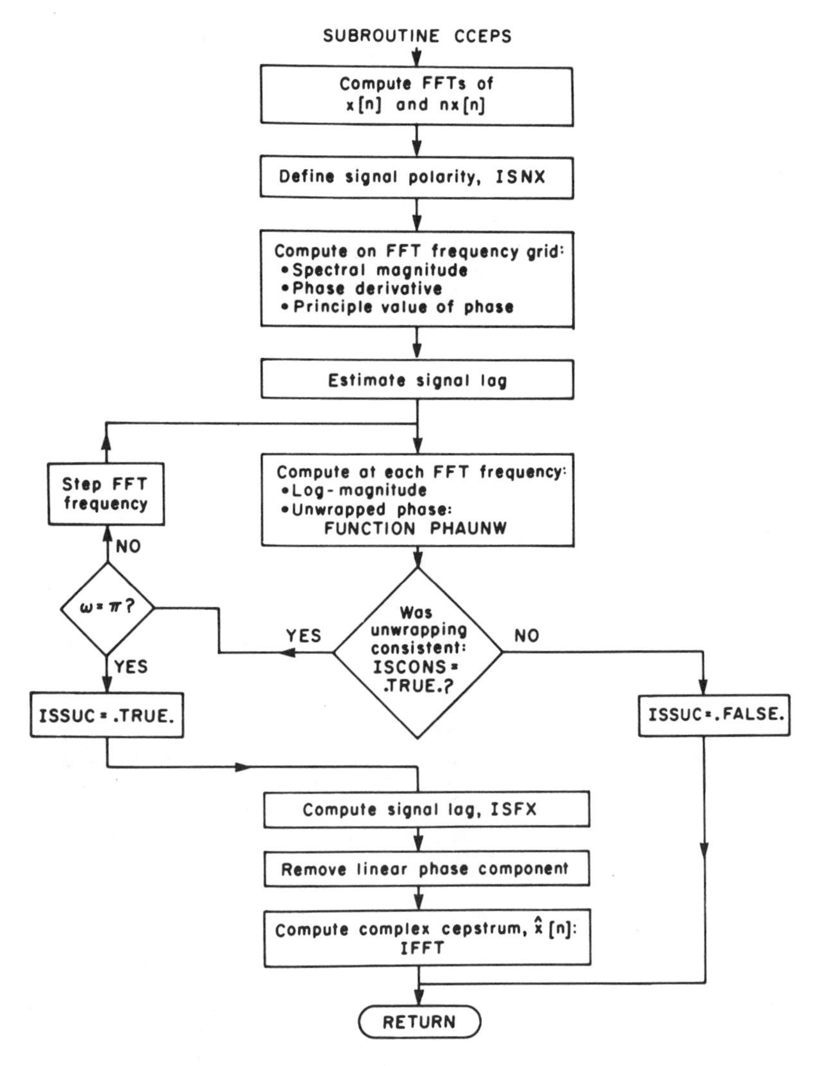

Fig. 3 Block diagram of program for computation of complex cepstrum.

having double precision counterparts are replaced by them.

Such double precision version of this code was capable of correctly unwrapping the data of the example above. An important issue in the usage of this program is therefore the selection of threshold levels THLINC and THLCON.

Small values of these thresholds are associated with high probability of error detection, and thus with greater confidence in the program's output. Unfortunately, the use of small threshold values is also associated with longer computation times.

The user is thus strongly encouraged to run this program under a variety of test signals, so as to determine the accuracy requirements and maximum threshold levels that might be appropriate for his particular application.

References

1. A. V. Oppenheim and R. W. Schafer, *Digital Signal Processing,* Chapter 10, Prentice-Hall, Inc., Englewood Cliffs, New Jersey, 1975.

2. J. M. Tribolet, "A New Phase Unwrapping Algorithm", *IEEE Trans. Acoust., Speech, and Signal Processing,* Vol. ASSP-25, No. 2, pp. 170-177, April 1977.

Table 1 Output of main program: CCMAIN.

```
INPUT SIGNAL-LENGTH (IN SAMPLES) =    7

 X(    1)=   1.0000  X(    2)= -4.0996  X(    3)=  8.4057  X(    4)=-10.1765
 X(    5)=   7.7801  X(    6)= -3.5142  X(    7)=  0.7939  X(    8)=  0.

SIGN= 1

LINEAR PHASE=  -2

COMPLEX CEPSTRUM


CX( 993)= -0.0030 CX( 994)= -0.0024 CX( 995)= -0.0000 CX( 996)=  0.0030
CX( 997)=  0.0050 CX( 998)=  0.0040 CX( 999)=  0.0001 CX(1000)= -0.0052
CX(1001)= -0.0085 CX(1002)= -0.0069 CX(1003)= -0.0001 CX(1004)=  0.0090
CX(1005)=  0.0149 CX(1006)=  0.0123 CX(1007)=  0.0001 CX(1008)= -0.0164
CX(1009)= -0.0272 CX(1010)= -0.0227 CX(1011)= -0.0002 CX(1012)=  0.0314
CX(1013)=  0.0531 CX(1014)=  0.0453 CX(1015)=  0.0003 CX(1016)= -0.0664
CX(1017)= -0.1166 CX(1018)= -0.1040 CX(1019)= -0.0004 CX(1020)=  0.1753
CX(1021)=  0.3415 CX(1022)=  0.3546 CX(1023)=  0.0006 CX(1024)= -1.2852

CX(   1)=  0.1906 CX(   2)= -2.5446 CX(   3)=  0.0012 CX(   4)=  0.6881
CX(   5)=  0.6561 CX(   6)=  0.3334 CX(   7)= -0.0008 CX(   8)= -0.1938
CX(   9)= -0.2152 CX(  10)= -0.1213 CX(  11)=  0.0005 CX(  12)=  0.0810
CX(  13)=  0.0941 CX(  14)=  0.0550 CX(  15)= -0.0004 CX(  16)= -0.0390
CX(  17)= -0.0463 CX(  18)= -0.0276 CX(  19)=  0.0002 CX(  20)=  0.0203
CX(  21)=  0.0243 CX(  22)=  0.0146 CX(  23)= -0.0002 CX(  24)= -0.0110
CX(  25)= -0.0133 CX(  26)= -0.0080 CX(  27)=  0.0001 CX(  28)=  0.0062
CX(  29)=  0.0075 CX(  30)=  0.0045 CX(  31)= -0.0001 CX(  32)= -0.0035
```

Appendix

```
C
C-----------------------------------------------------------------------
C MAIN PROGRAM: MAIN LINE PROGRAM TO COMPUTE THE COMPLEX CEPSTRUM
C               OF A REAL SEQUENCE X(N)
C AUTHORS:      JOSE M. TRIBOLET
C               INSTITUTO SUPERIOR TECNICO, LISBON, PORTUGAL
C               THOMAS F. QUATIERI
C               M.I.T., CAMBRIDGE, MASS. 02139
C-----------------------------------------------------------------------
C
      DIMENSION X(1024), CX(1026), AUX(1026), SPECM(12), SPECP(12)
      COMMON PI, TWOPI, THLINC, THLCON, NFFT, NPTS, N, L, H, H1, DVTMN2
      LOGICAL ISSUC
C
C DIMENSION REQUIREMENTS:
C
C (SEE SUBROUTINE CCEPS)
C
C DESCRIPTION OF ARRAYS:
C
C      X = ARRAY CONTAINING THE SEQUENCE X(N)
C     CX = ARRAY CONTAINING THE COMPLEX CEPSTRUM OF X(N)
C    AUX = AUXILIARY ARRAY
C SPECM = MAGNITUDES OF SPECTRAL ZEROS (SEE COEFF)
C SPECP = PHASE OF SPECTRAL ZEROS (SEE COEFF)
C
C DESCRIPTION OF VARIABLES:
C
C  IOUTD = OUTPUT DEVICE NUMBER
C     NX = LENGTH OF THE SEQUENCE X(N)
C   NFFT = LENGTH OF THE FFT
C THLINC = PHASE INCREMENT THRESHOLD(USED TO OBTAIN MORE
C          CONFIDENT PHASE ESTIMATE NEAR SHARP ZEROS)
C THLCON = PHASE CONSISTENCY THRESHOLD
C  ISSUC = .TRUE. IF COMPLEX CEPSTRUM SUCCESSFULLY COMPUTED;
C          .FALSE. OTHERWISE(PHASE ESTIMATION INCOMPLETE)
C
C SUBROUTINES CALLED:
C
C COEFF = SUBROUTINE TO COMPUTE A SEQUENCE X(N),BY
C         PRESCRIBING ITS SPECTRAL ZEROS
C CCEPS = SUBROUTINE TO COMPUTE THE COMPLEX CEPSTRUM
C         OF A REAL SEQUENCE X(N)
C
C
C
C
C SET LENGTH OF FFT,THRESHOLD LEVELS,AND OTHER CONSTANTS
C
      IOUTD = I1MACH(2)
      NFFT = 1024
      THLINC = 1.5
      THLCON = .5
      PI = 4.*ATAN(1.)
      TWOPI = 2.*PI
C
C GENERATE THE TEST SIGNAL WITH ZEROS AT:
C
C !0.9!EXP(+/-JPI/4)
C !1.1!EXP(+/-J(PI/4+PI/8192))
C !0.9!EXP(+/-J(PI/4+2*PI/8192))
C
C **************************************
C *                                    *
C *       P L E A S E   N O T E        *
C *                                    *
C *    USER IS STRONGLY ENCOURAGED TO   *
C *    RUN TEST PROGRAM WITH A VARIETY  *
C *    OF TEST SIGNALS TO BECOME FULLY  *
C *    ACQUAINTED WITH THIS PROGRAM'S   *
C *    CAPABILITIES.PHASE UNWRAPPING    *
C *    MAY OFTEN BE NOT POSSIBLE DUE    *
C *    EITHER TO THEORETICAL REASONS    *
C *     (ZEROS ON THE UNIT CIRCLE )     *
C *    OR TO COMPUTATIONAL REASONS      *
C *     (ZEROS TOO CLOSE TO UNIT CIRCLE) *
C *                                    *
C **************************************
C
      NX = 6
      SPECM(1) = .9
      SPECP(1) = (PI/4.)
      SPECM(2) = 1.1
      SPECP(2) = (PI/4.+PI/8192.)
      SPECM(3) = .9
      SPECP(3) = (PI/4.+TWOPI/8192.)
      CALL COEFF(NX, X, SPECM, SPECP)
C
C WRITE INPUT DATA SPECIFICATIONS
C
      WRITE (IOUTD,9999)
9999  FORMAT (1X, 10X, 35H*** CCMAIN- TEST PROGRAM OUTPUT ***//)
      WRITE (IOUTD,9998) NX
9998  FORMAT (1X, 34HINPUT SIGNAL-LENGTH (IN SAMPLES) =, I4/)
      I = NX/4
      J = 4*I
      IF (NX.EQ.J) GO TO 20
      J = J + 4
      K = NX + 1
      DO 10 I=K,J
        X(I) = 0.
  10  CONTINUE
  20  WRITE (IOUTD,9997) (I,X(I),I=1,J)
9997  FORMAT (1X/4(4H  X(, I4, 2H)=, F8.4))
C
C COMPUTE THE COMPLEX CEPSTRUM
C
      CALL CCEPS(NX, X, ISNX, ISFX, ISSUC, CX, AUX)
C
C CHECK WHETHER PHASE UNWRAPPING SUCCESSFUL; IF SO
C WRITE OUT DATA;OTHERWISE STOP!
C
      IF (ISSUC) GO TO 30
      WRITE (IOUTD,9996)
9996  FORMAT (1X///1X, 23HPHASE ESTIMATION FAILED)
      STOP
C
C WRITE SIGN,LINEAR PHASE,AND LAST AND FIRST 32 VALUES OF
C THE COMPLEX CEPSTRUM
C
  30  WRITE (IOUTD,9995) ISNX, ISFX
9995  FORMAT (1X/1X, 5HSIGN=, I2//1X, 13HLINEAR PHASE=, I4//)
      INITL = NFFT - 31
      WRITE (IOUTD,9994)
9994  FORMAT (1X, 17HCOMPLEX CEPSTRUM //)
      WRITE (IOUTD,9993) (I,CX(I),I=INITL,NFFT)
      WRITE (IOUTD,9993) (I,CX(I),I=1,32)
9993  FORMAT (1X/4(4H CX(, I4, 2H)=, F8.4))
      STOP
      END
C
```

```
C-----------------------------------------------------------------------
C SUBROUTINE: COEFF
C COMPUTES A SEQUENCE C(N) WITH NC SAMPLES BY PRESCRIBING
C THE MAGNITUDES AND PHASES OF ITS SPECTRAL ZEROS.
C-----------------------------------------------------------------------
C
      SUBROUTINE COEFF(NC, C, SM, SP)
      DOUBLE PRECISION S(2,31), Z1, Z2, Z3, Z4
      REAL M(12), P(12), C(12), SM(1), SP(1)
C
C DESCRIPTION OF ARGUMENTS
C
C NC    -ON INPUT EQUALS THE TOTAL NUMBER OF ZEROS OF C(Z)
C       -ON OUTPUT EQUALS THE SEQUENCE LENGTH
C C     -DSEQUENCE WITH PRESCRIBED Z-TRANSFORM
C M     -ARRAY OF SPECTRAL MAGNITUDES.ONLY ONE ENTRY
C        PER EACH COMPLEX CONJUGATE ZERO PAIR MUST BE SPECIFIED
C P     -ARRAY OF SPECTRAL PHASES IN RADIANS.
C ONLY ONE ENTRY PER EACH COMPLEX CONJUGATE ZERO PAIR MUST
C BE SPECIFIED.
C
      N = NC
      I = 1
      IND = 1
  10  CONTINUE
      M(I) = SM(IND)
      P(I) = SP(IND)
      IND = IND + 1
      IF (P(I).EQ.0.) GO TO 20
      M(I+1) = M(I)
      P(I+1) = -P(I)
      I = I + 1
  20  I = I + 1
      IF (I.LT.N) GO TO 10
      Y = P(1)
      S(1,1) = -DBLE(M(1)*COS(Y))
      S(2,1) = -DBLE(M(1)*SIN(Y))
      S(1,2) = 1.
      S(2,2) = 0.
      IF (N.EQ.1) GO TO 50
      DO 40 J=2,N
        S(1,J+1) = S(1,J)
        S(2,J+1) = S(2,J)
        Y = P(J)
        Z3 = -DBLE(M(J)*COS(Y))
        Z4 = -DBLE(M(J)*SIN(Y))
        DO 30 K=1,J
          M0 = J - K
          M1 = M0 + 1
          Z1 = S(1,M1)
          Z2 = S(2,M1)
          IF (M0.EQ.0) GO TO 30
          S(1,M1) = S(1,M0) + Z1*Z3 - Z2*Z4
          S(2,M1) = S(2,M0) + Z1*Z4 + Z2*Z3
  30    CONTINUE
        S(1,1) = Z1*Z3 - Z2*Z4
        S(2,1) = Z1*Z4 + Z2*Z3
  40  CONTINUE
  50  CONTINUE
      NC = N + 1
      DO 60 I=1,NC
        J = NC - I + 1
        C(I) = S(1,J)
  60  CONTINUE
      C(1) = 1.
      RETURN
      END
C
C-----------------------------------------------------------------------
C SUBROUTINE: CCEPS
C SUBROUTINE TO COMPUTE THE COMPLEX CEPSTRUM OF A SEQUENCE X(N)
C-----------------------------------------------------------------------
C
      SUBROUTINE CCEPS(NX, X, ISNX, ISFX, ISSUC, CX, AUX)
C
C DESCRIPTION OF ARGUMENTS:
C
C    NX = LENGTH OF THE SEQUENCE X(N)
C     X = ARRAY CONTAINING THE SEQUENCE X(N)
C  ISNX = INTEGER VALUE,EITHER +1 OR  -1 DEPENDING
C         ON THE SIGN REVERSAL OF X(N)
C  ISFX = INTEGER VALUE INDICATING THE AMOUNT OF SHIFT
C         ON X(N) DUE TO LINEAR PHASE REMOVAL
C ISSUC = .TRUE. IF PHASE ESTIMATION COMPLETE;
C         .FALSE. OTHERWISE
C    CX = ARRAY CONTAINING THE COMPLEX CEPSTRUM
C   AUX = AUXILIARY ARRAY
C
      DIMENSION X(1), CX(1), AUX(1)
C
C DIMENSION REQUIREMENTS:
C
C NX        .LE. NFFT
C DIM(X)    .LE. NFFT
C DIM(CX)   .GE. NFFT+2
C DIM(AUX)  .GE. NFFT+2
C
      COMMON PI, TWOPI, THLINC, THLCON, NFFT, NPTS, N, L, H, H1, DVTMN2
      LOGICAL ISSUC
C
C DESCRIPTION OF VARIABLES:
C
C THLINC = PHASE INCREMENT THRESHOLD(USED TO OBTAIN MORE
C          CONFIDENT ESTIMATE NEAR SHARP ZEROS)
C THLCON = PHASE CONSISTENCY THRESHOLD
C   NFFT = LENGTH OF FFT
C   NPTS = HALF THE LENGTH OF THE FFT
C DVTMN2 = TWICE THE MEAN OF THE PHASE DERIVATIVE
C
C SUBROUTINES CALLED:
C
C FFA,FFS-SUBROUTINES TO COMPUTE FFT AND IFFT (RADIX 2)
C
C FUNCTIONS CALLED:
C
C AMODSQ = FUNCTION TO COMPUTE THE MODULUS
C          SQUARED OF A COMPLEX NUMBER
C PHADVT = FUNCTION TO COMPUTE THE PHASE DERIVATIVE OF
C          A SPECTRAL VALUE
C PPVPHA = FUNCTION TO COMPUTE THE PRINCIPLE VALUE OF THE
C          PHASE OF A SPECTRAL VALUE
C PHAUNW = FUNCTION TO COMPUTE THE UNWRAPPED PHASE OF A
C          SPECTRAL VALUE
C
C INITIALIZATION
C
      NPTS = NFFT/2
      N = 12
      L = 2**N
      H = FLOAT(L)*FLOAT(NFFT)
```

```
      H1 = PI/H
      ISSUC = .TRUE.
      ISNX = 1
C
C TRANSFORM X(N) AND NX(N):FFT
C
      DO 10 I=1,NX
        CX(I) = X(I)
        AUX(I) = FLOAT(I-1)*X(I)
  10  CONTINUE
      INITL = NX + 1
      IEND = NFFT + 2
      DO 20 I=INITL,IEND
        CX(I) = 0.0
        AUX(I) = 0.0
  20  CONTINUE
C
C USE RADIX 2 FFT
C
      CALL FFA(CX, NFFT)
      CALL FFA(AUX, NFFT)
C
C CHECK IF SIGN REVERSAL IS REQUIRED
C
      IF (CX(1).LT.0.0) ISNX = -1
C
C COMPUTE MAGNITUDE OF SPECTRUM:STORE IN ODD-INDEXED
C VALUES OF AUX
C COMPUTE PHASE DERIVATIVE OF SPECTRUM:STORE IN EVEN-INDEXED
C VALUES OF AUX
C COMPUTE LINEAR PHASE ESTIMATE(MEAN OF THE PHASE DERIVATIVE):
C STORE TWICE THE ESTIMATE IN DVTMN2
C
      IO = -1
      DVTMN2 = 0.0
      IEND = NPTS + 1
      DO 30 I=1,IEND
        IO = IO + 2
        IE = IO + 1
        AMAGSQ = AMODSQ(CX(IO),CX(IE))
        PDVT = PHADVT(CX(IO),CX(IE),AUX(IO),AUX(IE),AMAGSQ)
        AUX(IO) = AMAGSQ
        AUX(IE) = PDVT
        DVTMN2 = DVTMN2 + PDVT
  30  CONTINUE
      DVTMN2 = (2.*DVTMN2-AUX(2)-PDVT)/FLOAT(NPTS)
C
C COMPUTE LOGMAGNITUDE:STORE IN ODD-INDEXED
C VALUES OF CX
C COMPUTE UNWRAPPED PHASE:STORE IN EVEN-INDEXED
C VALUES OF CX
C
      PPDVT = AUX(2)
      PPHASE = 0.0
      PPV = PPVPHA(CX(1),CX(2),ISNX)
      CX(1) = .5*ALOG(AUX(1))
      CX(2) = 0.0
      IO = 1
      DO 50 I=2,IEND
        IO = IO + 2
        IE = IO + 1
        PDVT = AUX(IE)
        PPV = PPVPHA(CX(IO),CX(IE),ISNX)
        PHASE = PHAUNW(X,NX,ISNX,I,PPHASE,PPDVT,PPV,PDVT,ISSUC)
C
C IF PHASE ESTIMATION SUCCESSFUL,CONTINUE;OTHERWISE RETURN
C
        IF (ISSUC) GO TO 40
        ISSUC = .FALSE.
        RETURN
  40    PPDVT = PDVT
        PPHASE = PHASE
        CX(IO) = .5*ALOG(AUX(IO))
        CX(IE) = PHASE
  50  CONTINUE
C
C REMOVE LINEAR PHASE COMPONENT
C
      ISFX = (ABS(PHASE/PI)+.1)
      IF (PHASE.LT.0.0) ISFX = -ISFX
      H = PHASE/FLOAT(NPTS)
      IE = 0
      DO 60 I=1,IEND
        IE = IE + 2
        CX(IE) = CX(IE) - H*FLOAT(I-1)
  60  CONTINUE
C
C COMPUTE THE COMPLEX CEPSTRUM:IFFT
C
      CALL FFS(CX, NFFT)
      RETURN
      END
C
C-----------------------------------------------------------------------
C SUBROUTINE: SPCVAL
C SUBROUTINE TO COMPUTE A SPECTRAL VALUE AT FREQUENCY
C FREQ(RADIANS) FOR SEQUENCES X(N) AND N*X(N)
C-----------------------------------------------------------------------
C
      SUBROUTINE SPCVAL(NX, X, FREQ, XR, XI, YR, YI)
      DIMENSION X(1)
      DOUBLE PRECISION U0, U1, U2, W0, W1, W2, A, B, C, D, A1, A2, SA0,
     *    CA0
C
C DESCRIPTION OF ARGUMENTS:
C
C   NX = # VALUES IN SEQUENCE X(N)
C    X = ARRAY CONTAINING INPUT SEQUENCE X(N)
C FREQ = FREQUENCY(RADIANS)
C   XR = REAL PART OF THE SPECTRAL VALUE OF X(N)
C   XI = IMAGINARY PART OF THE SPECTRAL VALUE OF X(N)
C   YR = REAL PART OF THE SPECTRAL VALUE OF NX(N)
C   YI = IMAGINARY PART OF THE SPECTRAL VALUE OF NX(N)
C
C METHOD:
C--MODIFIED GOERTZEL ALGORITHM AS PROPOSED BY BONZANIGO
C
C (IEEE TRAS. ASSP,VOL. 26,NO. 1,FEB 78)
C
C INITIALIZATION
C
      CA0 = DBLE(COS(FREQ))
      SA0 = DBLE(SIN(FREQ))
      A1 = 2.D+0*CA0
      U1 = 0.D+0
      U2 = U1
      W1 = U1
      W2 = U1
C
C MAIN LOOP (GOERTZEL ALGORITHM)
```

```
C
      DO 10 J=1,NX
        XJ = DBLE(X(J))
        U0 = XJ + A1*U1 - U2
        W0 = DBLE(FLOAT(J-1))*XJ + A1*W1 - W2
        U2 = U1
        U1 = U0
        W2 = W1
        W1 = W0
   10 CONTINUE
C
C BONZANIGO'S PHASE CORRECTION
C
      A = U1 - U2*CA0
      B = U2*SA0
      C = W1 - W2*CA0
      D = W2*SA0
      A2 = DBLE(FREQ*FLOAT(NX-1))
      U1 = DCOS(A2)
      U2 = -DSIN(A2)
      XR = SNGL(U1*A-U2*B)
      XI = SNGL(U2*A+U1*B)
      YR = SNGL(U1*C-U2*D)
      YI = SNGL(U2*C+U1*D)
      RETURN
      END
C
C-----------------------------------------------------------------------
C FUNCTION: PHAUNW
C PHASE UNWRAPPING BASED ON TRIBOLET'S ADAPTIVE INTEGRATION SCHEME.
C THE UNWRAPPED PHASE ESTIMATE IS RETURNED IN PHAUNW.
C-----------------------------------------------------------------------
C
      FUNCTION PHAUNW(X, NX, ISNX, I, PPHASE, PPDVT, PPV, PDVT, ISCONS)
C
C DESCRIPTION OF ARGUMENTS:
C
C       X = ARRAY CONTAINING SEQUENCE X(N)
C      NX = # POINTS IN SEQUENCE X(N)
C    ISNX = INTEGER VALUE,EITHER +1 OR -1 DEPENDING
C           ON THE SIGN REVERSAL OF X(N)
C       I = INDEX OF PHASE ESTIMATE ON EQUALLY SPACED FFT
C           FREQUENCY GRID
C  PPHASE = PHASE ESTIMATE AT INDEX I-1
C   PPDVT = PHASE DERIVATIVE AT INDEX I-1
C     PPV = PHASE PRINCIPLE VALUE AT INDEX I
C    PDVT = PHASE DERIVATIVE AT INDEX I
C  ISCONS = .FALSE. IF PHASE ESTIMATION UNSUCCESSFUL
C           (STACK DIMENSION EXCEEDED BEFORE CONSISTENT ESTIMATE
C           FOUND);.TRUE. OTHERWISE
C
C
C SUBROUTINES CALLED:
C
C  SPCVAL = SUBROUTINE TO COMPUTE SPECTRAL VALUE
C  PHCHCK = SUBROUTINE TO CHECK PHASE CONSISTENCY
C
C FUNCTIONS CALLED:
C
C  PPVPHA = FUNCTION TO COMPUTE PRINCIPLE VALUE OF PHASE
C  PHADVT = FUNCTION TO COMPUTE PHASE DERIVATIVE
C  AMODSQ = FUNCTION TO COMPUTE MODULUS SQUARED OF A
C           COMPLEX NUMBER
C
C
      DIMENSION SDVT(17), SPPV(17), X(1)
      INTEGER SINDEX(17), PINDEX, SP
      LOGICAL ISCONS, FIRST
      COMMON PI, TWOPI, THLINC, THLCON, NFFT, NPTS, N, L, H, H1, DVTMN2
C
C DESCRIPTION OF ARRAYS AND VARIABLES:
C
C SINDEX = INDEX STACK
C   SDVT = PHASE DERIVATIVE STACK
C   SPPV = PHASE PRINCIPLE VALUE STACK
C     SP = STACK POINTER
C
C
C
C
C
C INITIALIZATION
C
      FIRST = .TRUE.
      PINDEX = 1
      SP = 1
      SPPV(SP) = PPV
      SDVT(SP) = PDVT
      SINDEX(SP) = L + 1
C
C ENTER MAJOR LOOP
C
      GO TO 40
C
C UPDATE PREVIOUS ESTIMATE
C
   10 PINDEX = SINDEX(SP)
      PPHASE = PHASE
      PPDVT = SDVT(SP)
      SP = SP - 1
      GO TO 40
C
C IF SOFTWARE STACK DIMENSION DOES NOT ALLOW FURTHER STEP
C REDUCTION, RETURN.
C
   20 IF ((SINDEX(SP)-PINDEX).GT.1) GO TO 30
      ISCONS = .FALSE.
      PHAUNW = 0.
      RETURN
C
C DEFINE INTERMEDIATE FREQUENCY(I.F.):
C W = (TWOPI/NFFT)*(I-2+(K-1)/L)
C
   30 K = (SINDEX(SP)+PINDEX)/2
C
C CALCULATE I.F.
C
      FREQ = TWOPI*(FLOAT(I-2)*FLOAT(L)+FLOAT(K-1))/H
      CALL SPCVAL(NX, X, FREQ, XR, XI, YR, YI)
C
C COMPUTE PHASE DERIVATIVE AND PRINCIPLE VALUE OF THE PHASE
C AT I.F.;UPDATE STACK
C
      SP = SP + 1
      SINDEX(SP) = K
      SPPV(SP) = PPVPHA(XR,XI,ISNX)
      XMAG = AMODSQ(XR,XI)
      SDVT(SP) = PHADVT(XR,XI,YR,YI,XMAG)
C
C EVALUATE THE PHASE INCREMENT ACROSS SPECTRAL INTERVAL
```

```
C
  40  DELTA = H1*FLOAT(SINDEX(SP)-PINDEX)
      PHAINC = DELTA*(PPDVT+SDVT(SP))
C
C IF PHASE INCREMENT,REDUCED BY EXPECTED LINEAR PHASE INCREMENT,
C IS GREATER THAN SPECIFIED THRESHOLD,ADAPT STEP SIZE
C
      IF (ABS(PHAINC-DELTA*DVTMN2).GT.THLINC) GO TO 20
C
C FORM PHASE ESTIMATE;CHECK CONSISTENCY
C
      PHASE = PPHASE + PHAINC
      CALL PHCHCK(PHASE, SPPV(SP), ISCONS)
      IF (.NOT.ISCONS) GO TO 20
C
C IF RESULTING PHASE INCREMENT IS GREATER THAN PI,ADAPT STEP SIZE
C FOR MORE CONFIDENT ESTIMATE;OTHERWISE UPDATE PREVIOUS ESTIMATE
C IF STACK IS NOT EMPTY
C
      IF (ABS(PHASE-PPHASE).GT.PI) GO TO 20
C
C WHEN STACK IS EMPTY,THE UNWRAPPED PHASE AT
C W = TWOPI*(I-1)/NFFT IS HELD IN PHASE
C
      IF (SP.NE.1) GO TO 10
      PHAUNW = PHASE
      RETURN
      END
C
C-----------------------------------------------------------------------
C FUNCTION: PPVPHA
C COMPUTE THE PRINCIPLE VALUE OF THE PHASE OF A SPECTRAL VALUE
C-----------------------------------------------------------------------
C
      FUNCTION PPVPHA(XR, XI, ISNX)
C
C DESCRIPTION OF ARGUMENTS:
C
C   XR = REAL PART OF THE SPECTRAL VALUE
C   XI = IMAGINARY PART OF THE SPECTRAL VALUE
C ISNX = SIGN OF THE SPECTRAL VALUE AT ZERO FREQUENCY
C
      IF (ISNX.EQ.1) PPVPHA = (ATAN2((XI),(XR)))
      IF (ISNX.EQ.(-1)) PPVPHA = (ATAN2(-(XI),-(XR)))
      RETURN
      END
C
C-----------------------------------------------------------------------
C FUNCTION: PHADVT
C COMPUTE THE PHASE DERIVATIVE OF A SPECTRAL VALUE OF A SEQUENCE X(N)
C A SPECTRAL VALUE OF A SEQUENCE X(N)
C-----------------------------------------------------------------------
C
      FUNCTION PHADVT(XR, XI, YR, YI, XMAG)
C
C DESCRIPTION OF ARGUMENTS:
C
C   XR = REAL PART OF THE SPECTRAL VALUE OF X(N)
C   XI = IMAGINARY PART OF THE SPECTRAL VALUE OF X(N)
C   YR = REAL PART OF THE SPECTRAL VALUE OF NX(N)
C   YI = IMAGINARY PART OF THE SPECTRAL VALUE OF NX(N)
C XMAG = MAGNITUDE SQUARED OF THE SPECTRAL VALUE OF X(N)
C
      PHADVT = -SNGL((DBLE(XR)*DBLE(YR)+DBLE(XI)*DBLE(YI))/DBLE(XMAG))
      RETURN
      END
C
C-----------------------------------------------------------------------
C FUNCTION: AMODSQ
C COMPUTE THE SQUARE OF THE MODULUS OF A COMPLEX NUMBER
C-----------------------------------------------------------------------
C
      FUNCTION AMODSQ(ZR, ZI)
C
C DESCRIPTION OF ARGUMENTS:
C
C ZR = REAL PART OF THE COMPLEX NUMBER
C ZI = IMAGINARY PART OF THE COMPLEX NUMBER
C
      AMODSQ = SNGL(DBLE(ZR)*DBLE(ZR)+DBLE(ZI)*DBLE(ZI))
      RETURN
      END
C
C-----------------------------------------------------------------------
C SUBROUTINE: PHCHCK
C SUBROUTINE TO CHECK CONSISTENCY OF A PHASE ESTIMATE
C-----------------------------------------------------------------------
C
      SUBROUTINE PHCHCK(PH, PV, ISCONS)
C
C DESCRIPTION OF ARGUMENTS:
C
C     PH = PHASE ESTIMATE
C     PV = PRINCIPLE VALUE OF PHASE AT FREQUENCY
C          OF PHASE ESTIMATE
C ISCONS = .FALSE. IF PHASE ESTIMATE NOT CONSISTENT;
C          .TRUE. IF PHASE ESTIMATE CONSISTENT,
C          PHASE RETURNED IN PH
C
      COMMON PI, TWOPI, THLINC, THLCON, NFFT, NPTS, N, L, H, H1, DVTMN2
      LOGICAL ISCONS
C
C FIND THE TWO ADMISSIBLE PHASE VALUES CLOSEST TO PH
C
      A0 = (PH-PV)/TWOPI
      A1 = FLOAT(IFIX(A0))*TWOPI + PV
      A2 = A1 + SIGN(TWOPI,A0)
      A3 = ABS(A1-PH)
      A4 = ABS(A2-PH)
C
C CHECK CONSISTENCY
C
      ISCONS = .FALSE.
      IF (A3.GT.THLCON .AND. A4.GT.THLCON) RETURN
      ISCONS = .TRUE.
C
C FIND THE CLOSEST UNWRAPPED PHASE ESTIMATE
C
      PH = A1
      IF (A3.GT.A4) PH = A2
      RETURN
      END
```

7.2

Computation of the Real Cepstrum and Minimum-Phase Reconstruction

T. F. Quatieri

Massachusetts Institute of Technology
Cambridge, MA 02139

and

J. M. Tribolet

Instituto Superior Tecnico
Lisbon, Portugal

1. Purpose

The program is designed to compute the real cepstrum and the minimum-phase reconstruction of an arbitrary phase, real sequence $x(n)$. The real cepstrum is defined as the inverse Fourier transform of the log magnitude of the Fourier transform of the input sequence.

2. Method

The real cepstrum $x(n)$ of a real sequence $\hat{x}(n)$ is evaluated by means of the DFT computational realization as discussed by Oppenheim and Schafer [1].

$$\hat{x}(n) = \text{IDFT}\ (\log|\text{DFT}(x(n))|) \tag{1}$$

where the DFT length is sufficient to avoid cepstral aliasing.

For the minimum-phase reconstruction of a sequence $x[n]$, the real cepstrum $\hat{x}[n]$ is windowed by $w[n]$:

$$w[n] = \begin{cases} 0 & n < 0 \\ 1 & n = 0 \\ 2 & n > 0 \end{cases} \tag{2}$$

With $x(n)$ having a rational z-transform, this windowing operation is equivalent to mapping maximum-phase poles and zeros to their conjugate symmetric counterparts within the unit circle of the z-plane. Finally, the minimum-phase reconstructed sequence $y[n]$ is computed by inverse complex cepstral mapping $\hat{y}[n] = \hat{x}[n]w[n]$. This mapping is evaluated as follows:

$$y[n] = \text{IDFT}(\exp[\text{DFT}(y[n])]) \tag{3}$$

The inverse DFT length must likewise be sufficient to avoid aliasing in the time domain.

The inverse mapping may also be used after windowing the real cepstrum or complex cepstrum to accomplish separation of two convolutionally combined sequences whose cepstra are additively combined. For example, a "low-time" rectangular cepstral window is often applied in speech processing to retrieve the cepstrum of the vocal tract impulse response, and a "high-time" window is used in seismic processing to recover the earth's reflector series (Oppenheim and Schafer [1]).

3. Program Description

The three steps involved in the computation of the forward mapping are respectively, an FFT, determination of the log magnitude spectrum, and an IFFT. This task is performed by subroutine

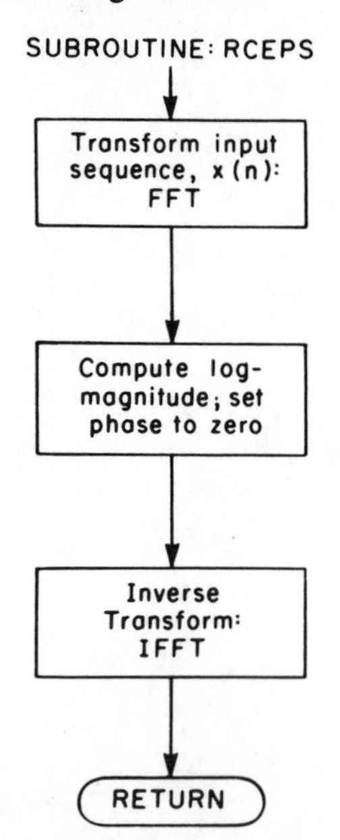

Fig. 1 Block diagram of subroutine RCEPS.

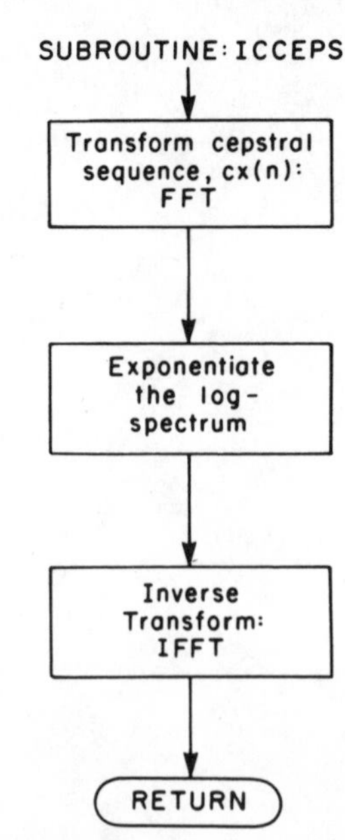

Fig. 2 Block diagram of subroutine ICCEPS.

RCEPS, whose flow diagram is depicted in Fig. 1.

The three steps involved in the inverse mapping are respectively, an FFT, complex exponentiation, and an IFFT. This task is performed by subroutine ICCEPS, whose flow diagram is depicted in Fig. 2.

The main program RCPICP first computes the real cepstrum of a specified sequence and FFT length by calling RCEPS. The first and last 32 values of this sequence are printed. Next, the minimum-phase reconstruction is determined, after appropriate windowing, by ICCEPS, and the first and last 32 values are again printed.

The FFT lengths of the forward and inverse computations within RCPICP are equal, but in general these lengths can be tailored to the particular aliasing problems. A flow diagram of RCPICP is given in Fig. 3.

The code for these routines and additional secondary routines are given in the Appendix along with array dimension requirements.

4. Test Problem

The input $x[n]$ is a sequence whose z-transform $X(z)$ contains one zero outside the unit circle:

$$X(z) = (1.+(1./0.99)z^{-1}) \tag{4}$$

The output of RCPICP is given in Table 1. These results are consistent with the analytical evaluation for both the real cepstrum and the minimum-phase reconstruction

$$\hat{x}[0] = \log(1./0.99)$$
$$\hat{x}[n] = -(-0.495)^n \quad |n| > 0 \tag{5}$$
$$X_{\min}[n] = \begin{cases} 1.0101 & n = 0 \\ 1.0000 & n = 1 \\ 0.0000 & \text{elsewhere} \end{cases}$$

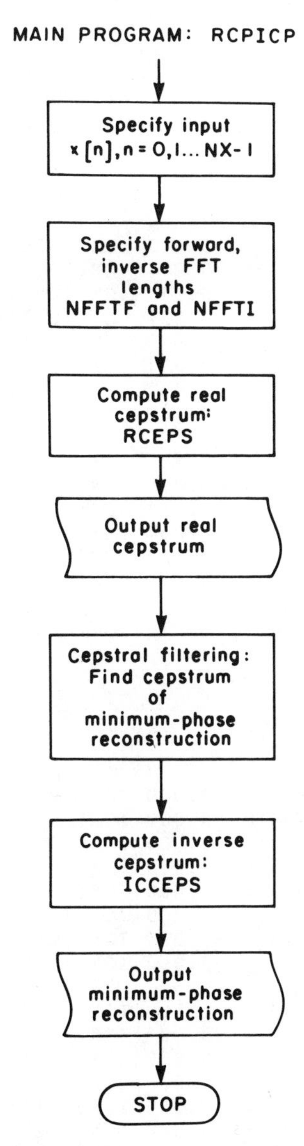

Fig. 3 Block diagram of main program RCPICP.

References

1. A. V. Oppenheim and R. W. Schafer, *Digital Signal Processing,* Chapter 10, Prentice-Hall Inc., Englewood Cliffs, New Jersey, 1975.

Table 1

```
*** RCPCIP- TEST PROGRAM OUTPUT ***

REAL CESTRUM


CX( 993)= -0.011  CX( 994)=  0.012  CX( 995)= -0.012  CX( 996)=  0.013
CX( 997)= -0.013  CX( 998)=  0.014  CX( 999)= -0.015  CX(1000)=  0.016
CX(1001)= -0.016  CX(1002)=  0.017  CX(1003)= -0.018  CX(1004)=  0.019
CX(1005)= -0.020  CX(1006)=  0.022  CX(1007)= -0.023  CX(1008)=  0.025
CX(1009)= -0.027  CX(1010)=  0.029  CX(1011)= -0.031  CX(1012)=  0.034
CX(1013)= -0.037  CX(1014)=  0.041  CX(1015)= -0.045  CX(1016)=  0.051
CX(1017)= -0.058  CX(1018)=  0.067  CX(1019)= -0.078  CX(1020)=  0.095
CX(1021)= -0.120  CX(1022)=  0.162  CX(1023)= -0.245  CX(1024)=  0.495

CX(   1)=  0.010  CX(   2)=  0.495  CX(   3)= -0.245  CX(   4)=  0.162
CX(   5)= -0.120  CX(   6)=  0.095  CX(   7)= -0.078  CX(   8)=  0.067
CX(   9)= -0.058  CX(  10)=  0.051  CX(  11)= -0.045  CX(  12)=  0.041
CX(  13)= -0.037  CX(  14)=  0.034  CX(  15)= -0.031  CX(  16)=  0.029
CX(  17)= -0.027  CX(  18)=  0.025  CX(  19)= -0.023  CX(  20)=  0.022
CX(  21)= -0.020  CX(  22)=  0.019  CX(  23)= -0.018  CX(  24)=  0.017
CX(  25)= -0.016  CX(  26)=  0.016  CX(  27)= -0.015  CX(  28)=  0.014
CX(  29)= -0.013  CX(  30)=  0.013  CX(  31)= -0.012  CX(  32)=  0.012


MINIMUM-PHASE RECONSTRUCTION


ICX( 993)=  0.000 ICX( 994)= -0.000 ICX( 995)=  0.000 ICX( 996)=  0.000
ICX( 997)=  0.000 ICX( 998)= -0.000 ICX( 999)= -0.000 ICX(1000)= -0.000
ICX(1001)= -0.000 ICX(1002)= -0.000 ICX(1003)= -0.000 ICX(1004)= -0.000
ICX(1005)= -0.000 ICX(1006)=  0.000 ICX(1007)= -0.000 ICX(1008)=  0.000
ICX(1009)=  0.000 ICX(1010)=  0.000 ICX(1011)= -0.000 ICX(1012)=  0.000
ICX(1013)=  0.000 ICX(1014)= -0.000 ICX(1015)= -0.000 ICX(1016)=  0.000
ICX(1017)=  0.000 ICX(1018)= -0.000 ICX(1019)= -0.000 ICX(1020)=  0.000
ICX(1021)= -0.000 ICX(1022)=  0.000 ICX(1023)=  0.000 ICX(1024)= -0.000

ICX(   1)=  1.010 ICX(   2)=  1.000 ICX(   3)=  0.000 ICX(   4)=  0.000
ICX(   5)=  0.000 ICX(   6)=  0.000 ICX(   7)=  0.000 ICX(   8)= -0.000
ICX(   9)= -0.000 ICX(  10)= -0.000 ICX(  11)= -0.000 ICX(  12)= -0.000
ICX(  13)= -0.000 ICX(  14)= -0.000 ICX(  15)= -0.000 ICX(  16)=  0.000
ICX(  17)= -0.000 ICX(  18)=  0.000 ICX(  19)= -0.000 ICX(  20)=  0.000
ICX(  21)=  0.000 ICX(  22)=  0.000 ICX(  23)= -0.000 ICX(  24)=  0.000
ICX(  25)=  0.000 ICX(  26)=  0.000 ICX(  27)= -0.000 ICX(  28)=  0.000
ICX(  29)= -0.000 ICX(  30)=  0.000 ICX(  31)= -0.000 ICX(  32)=  0.000
```

Appendix

```
C
C-------------------------------------------------------------------
C MAIN PROGRAM: MAIN LINE PROGRAM TO COMPUTE THE REAL CEPSTRUM
C               AND THE MINIMUM-PHASE RECONSTRUCTION OF AN
C               ARBITRARY-PHASE, REAL SEQUENCE X(N)
C AUTHORS:      THOMAS F. QUATIERI
C               M.I.T., CAMBRIDGE MASS. 02139
C               JOSE M. TRIBOLET
C               INSTITUTO SUPERIOR TECNICO, LISBON, PORTUGAL
C-------------------------------------------------------------------
C
      DIMENSION X(512), CX(1026)
      REAL ICX(1026)
C
C DIMENSION REQUIREMENTS:
C
C (SEE SUBROUTINES RCEPS AND ICEPS)
C
C
C DESCRIPTION OF ARRAYS:
C
C    X = ARRAY CONTAINING THE SEQUENCE X(N)
C   CX = ARRAY CONTAINING THE ZERO-PHASE(REAL)
C        OR MINIMUM-PHASE CEPSTRUM OF X(N)
C  ICX = ARRAY CONTAINING THE INVERSE CEPSTRUM
C         (MINIMUM-PHASE RECONSTRUCTION OF X(N))
C
C
C DESCRIPTION OF VARIABLES:
C
C    NX = LENGTH OF THE SEQUENCE X(N)
C NFFTF = LENGTH OF THE FFTS IN FORWARD CEPSTRUM
C NFFTI = LENGTH OF THE FFTS IN INVERSE CEPSTRUM
C          (INVERSE LENGTH NEED NOT EQUAL FORWARD LENGTH)
C
C
C SUBROUTINES CALLED:
C
C RCEPS = SUBROUTINE TO COMPUTE THE REAL CEPSTRUM OF A
C          REAL SEQUENCE
C
C ICCEPS = SUBROUTINE TO COMPUTE THE INVERSE COMPLEX CEPSTRUM
C           (NOTE THAT "COMPLEX" REFERS TO AN ARBITRARY,BUT
C           REAL-VALUED SEQUENCE)
C
C
C SET FFT LENGTHS AND DEFINE OUTPUT DEVICE IODEV
C
      NFFTF = 1024
      NFFTI = 1024
      IODEV = I1MACH(2)
C
C GENERATE THE TEST SIGNAL WITH Z-TRANSFORM:
C
C X(Z)=1.0+(1.0/.99)*Z**(-1) ---ONE ZERO AT 1.01010101
C
      NX = 2
      X(1) = 1.0
      X(2) = 1.01010101
C
C PAD X(N) WITH ZEROS:STORE IN CX
C
      DO 10 I=1,NX
        CX(I) = X(I)
  10  CONTINUE
      INITL = NX + 1
      IEND = NFFTF + 2
      DO 20 I=INITL,IEND
        CX(I) = 0.0
  20  CONTINUE
C
C COMPUTE THE REAL CEPSTRUM
C
      CALL RCEPS(NFFTF, CX)
C
C WRITE LAST AND FIRST 32 VALUES OF THE REAL CEPSTRUM
C
      WRITE (IODEV,9999)
9999  FORMAT (1X, 10X, 35H*** RCPCIP- TEST PROGRAM OUTPUT ***//)
      INITL = NFFTF - 31
      WRITE (IODEV,9998)
9998  FORMAT (1X, 15H REAL CEPSTRUM //)
      WRITE (IODEV,9997) (I,CX(I),I=INITL,NFFTF)
      WRITE (IODEV,9997) (I,CX(I),I=1,32)
9997  FORMAT (1X/4(5H  CX(, I4, 2H)=, F7.3))
C
C FIND THE MINIMUM-PHASE COUNTERPART OF THE ZERO-PHASE(REAL)
C CEPSTRUM AND PAD WITH ZEROS:STORE IN ICX (NOTE THAT THE FORWARD
C FFT LENGTH EQUALS THE INVERSE FFT LENGTH IN THIS EXAMPLE.
C IN GENERAL,THE FOLLOWING PROCEDURE SHOULD BE TAILORED TO THE
C PARTICULAR FFT LENGTHS).
C
      ICX(1) = CX(1)
      IEND = NFFTF/2 + 1
      DO 30 I=2,IEND
        ICX(I) = 2.0*CX(I)
  30  CONTINUE
      INITL = IEND + 1
      IEND = NFFTI + 2
      DO 40 I=INITL,IEND
        ICX(I) = 0.0
  40  CONTINUE
C
C COMPUTE THE INVERSE CEPSTRUM
C
      CALL ICCEPS(NFFTI, ICX)
C
C WRITE LAST AND FIRST 32 VALUES OF THE MINIMUM-PHASE RECONSTRUCTION
C
      INITL = NFFTI - 31
      WRITE (IODEV,9996)
9996  FORMAT (///29H MINIMUM-PHASE RECONSTRUCTION//)
      WRITE (IODEV,9995) (I,ICX(I),I=INITL,NFFTI)
      WRITE (IODEV,9995) (I,ICX(I),I=1,32)
9995  FORMAT (1X/4(5H ICX(, I4, 2H)=, F7.3))
      STOP
      END
C
C-------------------------------------------------------------------
C SUBROUTINE: RCEPS
C COMPUTE THE REAL CEPSTRUM OF A REAL SEQUENCE X(N)
C-------------------------------------------------------------------
C
      SUBROUTINE RCEPS(NFFT, CX)
C
C DESCRIPTION OF ARGUMENTS:
C
C NFFT = LENGTH OF THE FFT
C   CX = ARRAY CONTAINING THE REAL CEPSTRUM
C         (ALSO USED AS AN AUXILIARY ARRAY)
```

```
C
      DIMENSION CX(1)
C
C DIMENSION REQUIREMENTS:
C
C DIM(CX) .GE. NFFT+2
C
C
C SUBROUTINES CALLED:
C
C FFA,FFS-SUBROUTINES TO COMPUTE THE FFT AND IFFT (RADIX 2)
C
C FUNCTIONS CALLED:
C
C AMODSQ = FUNCTION TO COMPUTE THE MODULUS SQUARED
C          OF A COMPLEX NUMBER
C
C
C
C
C TRANSFORM X(N):FFT
C
      CALL FFA(CX, NFFT)
C
C COMPUTE THE LOGMAGNITUDE OF THE SPECTRUM:STORE IN THE
C ODD-INDEXED VALUES OF CX;CLEAR THE EVEN-INDEXED VALUES OF CX
C
      IEND = NFFT/2 + 1
      IO = -1
      DO 10 I=1,IEND
        IO = IO + 2
        IE = IO + 1
        AMAGSQ = AMODSQ(CX(IO),CX(IE))
        CX(IO) = .5*ALOG(AMAGSQ)
        CX(IE) = 0.0
  10  CONTINUE
C
C COMPUTE THE REAL CEPSTRUM:IFFT
C
      CALL FFS(CX, NFFT)
      RETURN
      END
C
C-------------------------------------------------------------------
C SUBROUTINE: ICCEPS
C COMPUTE THE INVERSE COMPLEX CEPSTRUM (NOTE THAT "COMPLEX" REFERS
C TO AN ARBITRARY, BUT REAL-VALUED CEPSTRAL SEQUENCE)
C-------------------------------------------------------------------
C
      SUBROUTINE ICCEPS(NFFT, ICX)
C
C DESCRIPTION OF ARGUMENTS:
C
C NFFT = LENGTH OF THE FFT
C  ICX = ARRAY CONTAINING THE INVERSE CEPSTRUM
C        (ALSO USED AS AN AUXILIARY ARRAY)
C
      REAL ICX(1)
C
C DIMENSION REQUIREMENTS:
C
C DIM(ICX) .GE. NFFT+2
C
C
C SUBROUTINES CALLED:
C
C FFA,FFS-SUBROUTINES TO COMPUTE THE FFT AND IFFT (RADIX 2)
C
C
C
C TRANSFORM CX(N):FFT
C
      CALL FFA(ICX, NFFT)
C
C EXPONENTIATE:STORE THE REAL AND IMAGINARY VALUES IN THE
C ODD-INDEXED AND EVEN-INDEXED LOCATIONS OF ICX, RESPECTIVELY
C
      IEND = NFFT/2 + 1
      IO = -1
      DO 10 I=1,IEND
        IO = IO + 2
        IE = IO + 1
        RMAG = EXP(ICX(IO))
        ICX(IO) = RMAG*COS(ICX(IE))
        ICX(IE) = RMAG*SIN(ICX(IE))
  10  CONTINUE
C
C INVERSE TRANSFORM:IFFT
C
      CALL FFS(ICX, NFFT)
      RETURN
      END
C
C-------------------------------------------------------------------
C FUNCTION: AMODSQ
C COMPUTE THE SQUARE OF THE MODULUS OF A COMPLEX NUMBER
C-------------------------------------------------------------------
C
      FUNCTION AMODSQ(ZR, ZI)
C
C DESCRIPTION OF ARGUMENTS:
C
C ZR = REAL PART OF THE COMPLEX NUMBER
C ZI = IMAGINARY PART OF THE COMPLEX NUMBER
C
      AMODSQ = SNGL(DBLE(ZR)*DBLE(ZR)+DBLE(ZI)*DBLE(ZI))
      RETURN
      END
```

CHAPTER 8

Interpolation and Decimation

R. E. Crochiere

Introduction

The processes of interpolation and decimation are fundamental operations in digital signal processing. They are used whenever it is necessary to change from one sampling rate to another in a digital system. For example, in speech processing, estimates of speech parameters are often computed at a low sampling rate for low bit-rate storage or transmission. This operation can often be conveniently defined in terms of concepts of decimation. For reconstructing a synthetic speech signal or a decoded replica of the original speech from the low bit-rate representation, the speech parameters are normally required at much higher sampling rates. In such cases, the sampling rate must be increased by a digital interpolation process.

The techniques of interpolation and decimation are also useful in defining highly efficient ways for implementing narrow band FIR filters and modulators. By careful design these techniques can be used for bandpass and highpass filters as well as the traditional lowpass filter. Computational efficiencies comparable to those of recursive elliptic filters are possible while still retaining the desirable properties of linear phase response and limit-cycle free operation. Yet another application of decimation and interpolation concepts can be found in digital phase shifting and in implementations of fractional sample delays.

Two types of computational issues generally arise in the design and implementation of interpolation and decimation systems. The first issue involves the design of appropriate filters around which the decimation or interpolation is based. The second issue involves the actual implementation of the decimation or interpolation processing. Three programs are provided in this section, in addition to the programs in other sections, which are useful in accomplishing these tasks.

The design of interpolation or decimation filters involves the use of lowpass (or bandpass) digital filters. Such filters can be designed in a variety of ways, e.g. window designs, equiripple designs, etc. For many applications this design can be achieved by the use of the filter design programs in Chapter 5. However, for interpolation designs with mean square minimization of interpolation error and preservation of original signal values, specialized procedures exist.

Program 8.1 is one such example in which this mean square error criterion is used. The program applies to the design of interpolation filters in which the sampling rate is increased by an integer factor of R. The resulting filter design allows the original input samples to pass through the interpolator unchanged and it interpolates $R-1$ sample values between each pair of original samples in such a way that the mean square error between these samples and their theoretically ideal values is minimized. The design is achieved through the formulation of an $L \times L$ matrix inversion where $2L$ is the number of original sample values used to interpolate each new sample. The resulting interpolation filter is a $2RL + 1$ tap, linear phase, FIR, digital filter. The size of L is limited to be 10 or less and the program requires the use of a matrix inversion program which is provided.

The implementation of interpolators or decimators involves the implementation of digital filters in which the input and output sampling rates are different. Because of this difference in sampling rates a straight-forward implementation of FIR digital filter structures is not practical. Instead, special considerations must be taken in interfacing the different sampling rates and in efficiently implementing the digital filter. The next two programs in this section provide alternative approaches to this problem based on whether the ratio between the two sampling rates is of the order of unity (say for example 5/8 or 16/15) and can be expressed as a ratio of integers, or whether the ratio is a very large integer value which can be factored into smaller integer values (for a cascade implementation).

Program 8.2 covers the first category in which the ratio of output to input sampling rates can be expressed as L/M where L and M are arbitrary integers. In the case where $M = 1$ an integer interpolation by a factor of L is performed and when $L = 1$ an integer decimation by a factor of M is performed. The program requires a coefficient array of a symmetric FIR decimating or interpolating filter which must be supplied by the user (any of the above designs can be used). After an initialization call, data can be processed in a consecutive block-by-block manner by repeated calls to the program. Data is received and returned in blocks whose sizes are related by the ratio L/M to accomodate the difference in input and output sampling rates.

Program 8.3 covers the second category of decimation and interpolation implementations in which the ratio of output to input sampling rates is a large integer (in the range 15 and up) which can be factored into smaller integer values. Generally it is more efficient in this case to perform the interpolation (or decimation) by a series of cascaded stages of interpolators (or decimators). In this way the filter requirements and computation at each stage can often be substantially reduced over that of a single stage implementation. For example a 1:100 interpolator might be more efficiently implemented by a three stage cascade of interpolators with conversion rates of 1:2, 1:5, and 1:10 than by a single stage conversion. For such a case, speed improvements of up to 50 can be obtained. The filter designs for each of the stages must be supplied by the user. As in Program 8.2, after an initialization call, data is processed in a consecutive block by block manner with unequal input and output block sizes to accomodate the different input and output sampling rates. The program can implement 1, 2 or 3 stage decimator or interpolator designs. It also has the capability of implementing a cascade of a (1, 2, or 3 stage) decimator and a (1, 2, or 3 stage) interpolator for the efficient implementation of narrowband FIR filters.

In addition to the programs and methods presented here, a number of other techniques and design methods for decimation and interpolation have been proposed in the literature. Bellanger et al. [1,2], for example, have proposed design techniques based on single stage or cascaded stages of 2:1 decimators or interpolators with "half band" digital filters. In this case advantage can be taken of the fact that half of the coefficients in the digital filter are exactly zero. Goodman and Carey [3] have proposed a number of specialized, low order, filter designs for cascaded stages of decimators or interpolators. Parks and Kolba [4] have recently proposed a new minimax filter design for interpolation to complement the least square designs of Program 8.1.

References

1. M. G. Bellanger, J. L. Daguet, and G. P. Lepagnol, "Interpolation, Extrapolation, and Reduction of Computation Speed in Digital Filters", *IEEE Trans. Acoust., Speech, Signal Processing,* Vol. ASSP-22, No. 4, pp. 231-235, Aug. 1974.

2. M. G. Bellanger, "Computation Rate and Storage Estimation in Multirate Digital Filtering with Half-Band Filters", *IEEE Trans. Acoust., Speech, Signal Processing,* Vol. ASSP-25, No. 4, pp. 344-346, Aug. 1977.

3. D. J. Goodman and M. J. Carey, "Nine Digital Filters for Decimation and Interpolation", *IEEE Trans. Acoust., Speech, Signal Processing,* Vol. ASSP-25, No. 2, pp. 121-126, April 1977.

4. T. W. Parks and D. P. Kolba, "Interpolation Minimizing Maximum Normalized Error for Band-Limited Signals", *IEEE Trans. Acoust., Speech, Signal Processing,* Vol. ASSP-26, No. 4, pp. 381-384, Aug. 1978.

8.1

A Computer Program for Digital Interpolator Design

G. Oetken

Institut für Nachrichtentechnik
Universitat Erlangen-Nürnberg
Cauerstrasse 7
8520 Erlangen, Germany

T. W. Parks

Department of Electrical Eng.
Rice University
Houston, TX 77001

H. W. Schüssler

Institut für Nachrichtentechnik
Universitat Erlangen-Nürnberg
Cauerstrasse 7
8520 Erlangen, Germany

1. Purpose

This program [1] designs a length $2rL + 1$ FIR interpolating filter with unit pulse response $h(n)$, $n = -rL,...,rL$, which, given every r-th sample of a signal $x(n)$, interpolates the remaining samples.

2. Method

The unit pulse response $h(n)$ is designed so that

$$\|(x \cdot \sqcup\!\sqcup_r) * h - x\|^2$$

is minimized for bandlimited signals $x(n)$. The symbol * represents convolution and · represents multiplication. The symbol $\sqcup\!\sqcup_r$ represents sampling,

$$\sqcup\!\sqcup_r(n) = \begin{cases} 1 & n=0 \bmod r \\ 0 & \text{otherwise} \end{cases} .$$

The signal $x(n)$ is assumed to be bandlimited with

$$X(\omega) = 0 \quad \text{for} \quad |\omega| \geqslant \alpha\,\pi/r \,.$$

The squared norm of the error is

$$\|e\|^2 = \sum_{n=-\infty}^{\infty} e(n)^2 \,.$$

The method divides $h(n)$ into r subsequences $h_\tau(l) = h(rl+\tau)$, $\tau = 0,1,...,r-1$ and solves the following set of linear equations for each τ

$$\sum_{l=-L}^{L-1} h(rl+\tau)\phi(r(l-m)) = \phi(rm+\tau) \quad m = -L,...,L-1$$

where $\phi(k) = \int_{-\alpha\pi/r}^{\alpha\pi/r} |X(\omega)|^2 e^{-jk\omega} d\omega$ is the autocorrelation of x. In terms of matrices these equations become

$$\Phi h_\tau = \Phi_\tau \,.$$

Rather than invert the $2L \times 2L$ matrix Φ, the algorithm, using symmetries of Φ, inverts 2 $L \times L$ matrices. These inverse matrices are then used to calculate only the first half of the pulse response $h(-rL),...,h(o)$ since $h(n)$ is an even function.

3. Program Description

3.1 Usage

The subroutine DODIF requires subroutine SMINVD and a user defined function DPH1.

3.2 Subroutines and Functions

SUBROUTINE DODIF (L,R,ALPHA,DIF) is called from a main program and designs the optimal digital interpolating filter.

SUBROUTINE SMINVD (A,N,EPS,IER) inverts positive definite symmetric matrix using double precision.

DOUBLE PRECISION FUNCTION DPH1 (X,ALPHA) calculates the autocorrelation of the input data.

3.3 Description of Parameters

R – Wanted increase in sampling rate, i.e., the filter calculates R−1 samples between each pair of input samples. Range for R is any integer >1.

L – Integer, L, which determines the degree of the filter. The degree is $2 \cdot L \cdot$R. Range for L is 1 to 10, may be increased at the risk of numerical instability (change LMAX and array bounds of AP,AM,D for $L > 10$).

ALPHA – Cutoff frequency (radians) of the filter input data divided by π/R. Range for ALPHA is $0 <$ ALPHA $\leqslant 1.00$ (double precision).

DIF – Double precision array in which the first half ($L \cdot$R+1 values) of the unit sample response is returned. (The optimal filter has linear phase.)

A – Double precision upper triangular part of given positive definite symmetric $N \times N$ matrix stored columnwise. On return A contains the upper triangular part of the resulting inverse matrix in double precision.

N – The number of rows (columns) in the given matrix.

EPS – Relative tolerance for test on loss of significance.

IER – Resulting error parameter IER $= -1$ indicates error in inverting matrix possibly due to loss of significance.

X – Point for which the autocorrelation is desired.

3.4 User Requirements

The user must supply the main program and the autocorrelation function DPH1. In the main program the dimension of DIF must be fixed at a value greater than or equal to $L \cdot$R+1.

Big values of L and small values of ALPHA increase the possibility of problems with matrix inversion (depending upon the wordlength of the computer). A very good possibility to control the quality of the matrix inversion is to check the magnitude of the unit sample response DIF(K) at K $= M \cdot$R + 1, $M = 0$ to $L-1$, which ideally have to be zero.

4. Test Problem

A test program was run to design an interpolating filter for data with a flat spectrum, i.e.,

$$|X(\omega)|^2 = 1 \text{ for } |\omega| < \alpha\pi/r .$$

The parameters were R = 4, L = 4, and ALPHA = 0.500.

References

1. G. Oetken, T. W. Parks, and H. W. Schüssler, "New Results in the Design of Digital Interpolators", *IEEE Trans. Acoust., Speech, and Signal Processing,* Vol. ASSP-23, No. 3, pp. 301-309, June 1975.

Table 1

```
OPTIMAL INTERPOLATING FILTER DATA .


L=   4         R=   4         ALPHA= 0.50


THERE ARE    17 VALUES IN HALF OF THE FILTER


H(-16)=    0.
H(-15)=  -0.4559319E-02
H(-14)=  -0.6777514E-02
H(-13)=  -0.5177760E-02
H(-12)=  -0.8881784E-15
H(-11)=   0.2578440E-01
H(-10)=   0.3945777E-01
H( -9)=   0.3118661E-01
H( -8)=  -0.8881784E-15
H( -7)=  -0.8770110E-01
H( -6)=  -0.1426581E 00
H( -5)=  -0.1220465E 00
H( -4)=  -0.3552714E-14
H( -3)=   0.2910058E 00
H( -2)=   0.6098364E 00
H( -1)=   0.8713054E 00
H(  0)=   0.1000000E 01
```

Appendix

```
C
C-----------------------------------------------------------------------
C  MAIN PROGRAM: TEST PROGRAM FOR OPTIMAL DIGITAL INTERPOLATING FILTER
C  AUTHORS:      GEERD OETKEN, H. W. SCHUSSLER AND T. W. PARKS
C                T. W. PARKS
C                DEPT. OF ELECTRICAL ENGINEERING
C                RICE UNIV.
C                HOUSTON, TEXAS   77001
C
C                G. OETKEN AND H. W. SCHUSSLER
C                INSTITUT FUER NACHRICHTENTECHNIK
C                UNIVERSITAET ERLANGEN-NUERNBERG
C                8520 ERLAN
C                CAUERSTRASSE 7
C                WEST GERMANY
C
C  INPUT:        NUM IS THE NUMBER OF TIMES TO LOOP CALLING DODIF
C                L DETERMINES THE DEGREE OF THE FILTER
C                    THE DEGREE IS 2*L*R
C                ALPHA IS THE CUTOFF FREQUENCY (RADIANS) OF THE FILTER
C                    INPUT DATA.  RANGE: 0.D0 < ALPHA <= 1.D0
C                R IS THE WANTED INCREASE IN THE SAMPLING RATE
C                    RANGE: AN INTEGER > 1
C-----------------------------------------------------------------------
C
      DOUBLE PRECISION ALPHA, DIF(100)
      INTEGER R
C
C DEFINE I/O DEVICE CODES
C
      IN = I1MACH(1)
      IOUT = I1MACH(2)
C
      READ (IN,9999) NUM
9999  FORMAT (I5)
      DO 20 II=1,NUM
        READ (IN,9998) L, ALPHA, R
9998    FORMAT (I5, F5.3, I5)
        LR = L*R + 1
        WRITE (IOUT,9997)
9997    FORMAT (1H0, 35H  OPTIMAL INTERPOLATING FILTER DATA)
        WRITE (IOUT,9996)
9996    FORMAT (1H0)
        WRITE (IOUT,9995) L, R, ALPHA
9995    FORMAT (1H0, 5H  L= , I2, 10H        R= , I2, 14H        ALPHA= ,
     *    F4.2)
        WRITE (IOUT,9996)
        CALL DODIF(L, R, ALPHA, DIF)
        WRITE (IOUT,9994) LR
9994    FORMAT (1H0, 10H THERE ARE, I5, 28H VALUES IN HALF OF THE FILTE,
     *    1HR)
        WRITE (IOUT,9996)
        J = -LR
        DO 10 I=1,LR
          J = J + 1
          WRITE (IOUT,9993) J, DIF(I)
  10    CONTINUE
9993    FORMAT (1H0, 4H  H(, I3, 2H)=, E16.7)
  20  CONTINUE
      STOP
      END
C
C-----------------------------------------------------------------------
C SUBROUTINE: DODIF
C SUBROUTINE TO CALCULATE THE OPTIMAL DIGITAL INTERPOLATING
C FILTER
C-----------------------------------------------------------------------
C
      SUBROUTINE DODIF(L, R, ALPHA, DIF)
      DOUBLE PRECISION AP(100), AM(100), D(200)
      DOUBLE PRECISION DBLE, DFLOAT, S1, S2, DPHI, ALPHA, H, X, Y,
     *    DIF(1)
      INTEGER R
C
C INPUT:
C     R      -WANTED INCREASE IN SAMPLING RATE, I.E. THE FILTER
C                  CALCULATES R-1 SAMPLES BETWEEN EACH PAIR OF
C                  INPUT SAMPLES. RANGE FOR R/ ANY INTEGER > 1
C     L      -INTEGER WHICH DETERMINES THE DEGREE OF THE FILTER.
C                  THE DEGREE IS 2*L*R.
C                  RANGE FOR L/ 1 TO 10, MAY BE INCREASED AT THE
C                  RISK OF NUMERICAL INSTABILITY (CHANGE LMAX AND
C                  ARRAY BOUNDS OF AP,AM,D FOR L > 10).
C     ALPHA  -CUTOFF FREQUENCY (RADIANS) OF THE FILTER INPUT DATA
C                  DIVIDED BY PI/R.
C                  RANGE FOR ALPHA/ 0.D0< ALPHA <= 1.D0 (DOUBLE
C                  PRECISION).
C     DIF    -DOUBLE PRECISION ARRAY IN WHICH THE FIRST HALF (L*R+1
C                  VALUES) OF THE UNIT SAMPLE RESPONSE IS RETURNED.
C                  (THE OPTIMAL FILTER HAS LINEAR PHASE).
C
C FORTRAN 4 USAGE/
C     CALL DODIF(L,R,ALPHA,DIF)
C
C REMARKS/
C     BIG VALUES OF L AND SMALL VALUES OF ALPHA INCREASE THE
C     POSSIBILITY OF PROBLEMS WITH MATRIX INVERSION (DEPENDING UPON
C     THE WORDLENGTH OF THE COMPUTER). A VERY GOOD POSSIBILITY TO
C     CONTROL THE QUALITY OF THE MATRIX INVERSION IS TO CHECK THE
C     MAGNITUDE OF THE UNIT SAMPLE RESPONSE DIF(K) AT K=M*R+1,
C     M=0 TO L-1, WHICH IDEALLY HAVE TO BE ZERO.
C
C SUBROUTINES AND FUNCTIONS REQUIRED/
C     DPHI, SMINVD
C
      DATA LMAX /10/
C
C (MAXIMUM VALUE OF L SPECIFIED BY ARRAY BOUNDS)
C
      DATA EPS /1.E-6/
C
C (RELATIVE TOLERANCE FOR LOSS OF SIGNIFICANCE TEST IN MATRIX)
C
      DFLOAT(IF) = DBLE(FLOAT(IF))
      IODEV = I1MACH(2)
C
C (DEVICE NO. ON WHICH ERROR MESSAGES ARE TO HAPPEN)
C
C
C TEST IF L WITHIN CURRENT BOUNDS
C
      IF (L.GT.LMAX) GO TO 100
C
C CALCULATE THE TWO L*L MATRICES
C
      LR = 0
      DO 20 M=1,L
        DO 10 K=1,M
          K1 = K + LR
          S1 = K - M
```

```
            S1 = DABS(S1)
            S2 = DFLOAT(2*L+1-M-K)
            S1 = DPHI(S1,ALPHA)
            S2 = DPHI(S2,ALPHA)
            AP(K1) = S1 + S2
            AM(K1) = S1 - S2
 10       CONTINUE
          LR = LR + M
 20     CONTINUE
C
C INVERT BOTH MATRICES USING IBM/SSP SUBROUTINES
C
        CALL SMINVD(AP, L, EPS, IER1)
        CALL SMINVD(AM, L, EPS, IER2)
        IF (IER1.EQ.(-1) .OR. IER2.EQ.(-1)) GO TO 110
C
C CALCULATE THE L*2L MATRIX D
C
        DO 40 M=1,L
          M1 = (M-1)*L
          DO 30 K=1,L
            K1 = 2*(M1+K) - 1
            K2 = (K-1)*K/2 + M
            IF (M.GT.K) K2 = (M-1)*M/2 + K
            S1 = AP(K2) + AM(K2)
            S2 = AP(K2) - AM(K2)
            D(K1) = S1
            D(K1+1) = S2
 30       CONTINUE
 40     CONTINUE
C
C CALCULATE UNIT SAMPLE RESPONSE FROM D AND DPHI
C
        J = 0
        DO 90 I=1,L
          I1 = (I-1)*2*L
          DO 80 M=1,R
            M1 = M - 1
            GO TO 60
 50         M1 = R
 60         X = DFLOAT(M1)/DFLOAT(R)
            H = 0.D0
            DO 70 K=1,L
              Y = DFLOAT(L-(K-1)) - X
              IT = I1 + 2*K - 1
              H = H + D(IT)*DPHI(Y,ALPHA)
              Y = DFLOAT(L-K) + X
              IT = I1 + 2*K
              H = H + D(IT)*DPHI(Y,ALPHA)
 70         CONTINUE
            J = J + 1
C
C FIRST HALF OF UNIT SAMPLE RESPONSE IS STORED IN DIF(J)
C
            DIF(J) = .5D0*H
            IF (I.EQ.L .AND. M1.EQ.(R-1)) GO TO 50
 80       CONTINUE
 90     CONTINUE
        RETURN
C
C ERROR MESSAGES
C
 100    WRITE (IODEV,9999) LMAX
9999    FORMAT (18H L IS GREATER THAN, I4)
        GO TO 120
 110    WRITE (IODEV,9998)
9998    FORMAT (54H ERROR DURING MATRIX INVERSION, ILL CONDITIONED SYSTEM)
 120    RETURN
        END
C
C-----------------------------------------------------------------------
C FUNCTION: DPHI
C FUNCTION REQUIRED BY DODIF
C SPECIAL VERSION FOR LOWPASS INTERPOLATORS
C-----------------------------------------------------------------------
C
        DOUBLE PRECISION FUNCTION DPHI(X, ALPHA)
        DOUBLE PRECISION X, ALPHA, PI
        PI = 4.D0*DATAN(1.D0)
        IF (DABS(X).LT.1.D-10) GO TO 10
        X = ALPHA*PI*X
        DPHI = DSIN(X)/X
        RETURN
 10     CONTINUE
        DPHI = 1.D0
        RETURN
        END
C
C-----------------------------------------------------------------------
C SUBROUTINE: SMINVD
C AUTHOR: DEAN KOLBA
C         DEPT. OF ELEC. ENG.
C         RICE UNIVERSITY
C         HOUSTON, TX. 77001
C
C SUBROUTINE TO INVERT A POSITIVE-DEFINITE SYMMETRIC MATRIX
C-----------------------------------------------------------------------
C
        SUBROUTINE SMINVD(A, N, EPS, IER)
        DOUBLE PRECISION A(1), SUM, AD
C
C CALCULATE THE CHOLESKY DECOMPOSITION
C
        IER = 0
        ID = 0
        DO 70 K=1,N
          ID = ID + K
          IV = ID
          KM1 = K - 1
          TEST = EPS*SNGL(A(ID))
          DO 60 I=K,N
            SUM = 0.D0
            IF (KM1.EQ.0) GO TO 20
            DO 10 J=1,KM1
              J1 = ID - J
              J2 = IV - J
              SUM = SUM + A(J1)*A(J2)
 10         CONTINUE
 20         SUM = A(IV) - SUM
            IF (I.GT.K) GO TO 40
            IF (SNGL(SUM).GT.ABS(TEST)) GO TO 30
            IF (SUM.LE.0.D0) GO TO 150
            IF (IER.GT.0) GO TO 30
            IER = KM1
 30         A(ID) = DSQRT(SUM)
            GO TO 50
 40         A(IV) = SUM/A(ID)
 50         IV = IV + I
 60       CONTINUE
 70     CONTINUE
```

```
C
C INVERT UPPER TRIANGULAR MATRIX RESULTING FROM CHOLESKY DECOMPOSITION
C
      NV = N*(N+1)/2
      ID = NV
      DO 110 K=1,N
        A(ID) = 1.D0/A(ID)
        JH = N
        KM1 = K - 1
        IF (KM1.EQ.0) GO TO 100
        JL = N - KM1
        IV = NV - K + 1
        DO 90 I=1,KM1
          SUM = 0.D0
          JH = JH - 1
          J1 = ID
          J2 = IV
          DO 80 J=JL,JH
            J1 = J1 + J
            J2 = J2 + 1
            SUM = SUM + A(J1)*A(J2)
 80       CONTINUE
          A(IV) = -SUM*A(ID)
          IV = IV - JH
 90     CONTINUE
100     ID = ID - JH
110   CONTINUE
C
C CALCULATE THE INVERSE OF A USING THE INVERTED TRIANGULAR MATRIX
C
      ID = 0
      DO 140 K=1,N
        ID = ID + K
        IV = ID
        DO 130 I=K,N
          SUM = 0.D0
          J1 = IV
          JL = I - K
          DO 120 J=I,N
            J2 = J1 + JL
            SUM = SUM + A(J1)*A(J2)
            J1 = J1 + J
120       CONTINUE
          A(IV) = SUM
          IV = IV + I
130     CONTINUE
140   CONTINUE
      RETURN
150   IER = -1
      RETURN
C
C IER=-1    INVERSION FAILED
C IER=0     GOOD INVERSION
C IER=I     LOSS OF SIGNIFICANCE AT THE I+1 STAGE
C
      END
```

8.2

A General Program to Perform Sampling Rate Conversion of Data by Rational Ratios

R. E. Crochiere

Bell Laboratories
Murray Hill, NJ 07974

1. Purpose

This program can be used to convert the sampling rate of data by factors of L/M where L and M are arbitrary positive integers. Input data is provided through an input buffer in blocks of length $N_D M$ and output data is received through an output buffer in blocks of $N_D L$ where N_D is any positive integer.

2. Method

The program is based on the one stage interpolator/decimator method of R. E. Crochiere and L. R. Rabiner [1,2]. Conceptually the method is depicted in Fig. 1.

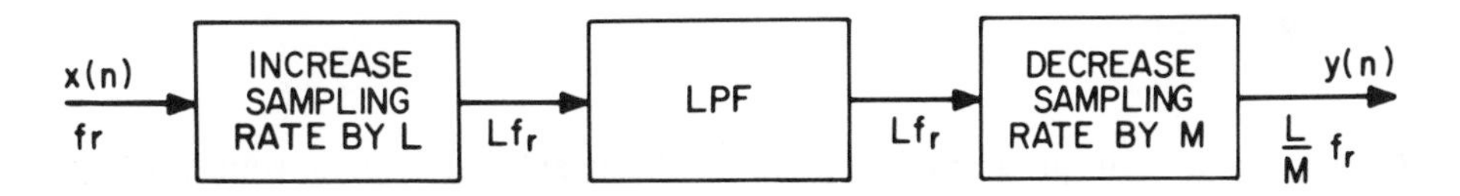

Fig. 1. Sampling rate converter

The input sampling rate f_r is increased by a factor L by inserting $L-1$ zero valued samples between each pair of input samples. This signal is then filtered with an FIR lowpass filter whose stopband cutoff frequency is $f_r/2$ or $Lf_r/(2M)$, whichever is smaller. The output signal of the filter is then reduced by a factor of M by keeping only one out of every M samples. By the elimination of multiplications by zero and elimination of computing unnecessary output samples it has been shown in [1] that the output $y(n)$ can be computed by the relation

$$y(n) = \sum_{k=0}^{Q-1} h(kL + (nM) \oplus L)\, x\left(\left[\frac{nM}{L}\right] - k\right) \tag{1}$$

where $(\ )\oplus L$ implies the quantity in parentheses modulo L and $[\ ]$ corresponds to the integer less than or equal to the number in brackets. The sequence $h(n)$, $n = 0, 1, \ldots, N-1$ represents the coefficients of the FIR filter and N is the number of taps in the filter such that

$$N \leqslant QL \tag{2}$$

and Q is an arbitrary positive integer.

Figure 2 shows the flow diagram of the program which implements relation (1). Input data $x(n)$ is supplied through BUFM and output data $y(n)$ is received through BUFL. QBUF stores the necessary internal state variables and COFS stores the coefficients $h(n)$ in a prescribed "scrambled" order such that they can be sequentially accessed (see [1]). ICTR is a control memory which is generated by the initialization program and is used to control the indexing of data and coefficients in the program.

The group delay of the sampling rate converter at the output sampling rate is

$$\tau_g = \frac{1}{M}\left(\frac{N-1}{2} - (M-1)\right) \quad (samples) \tag{3}$$

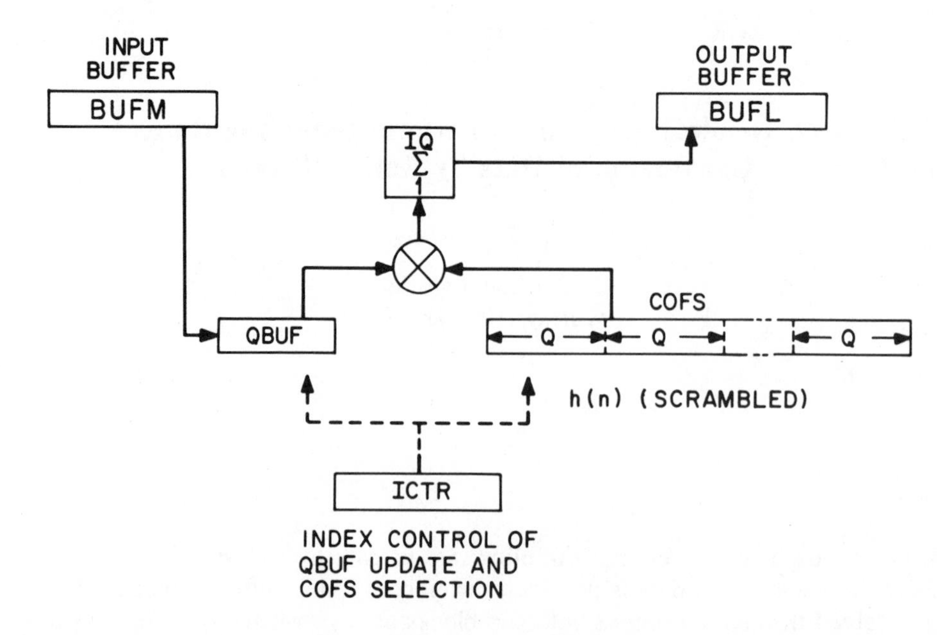

Fig. 2. Flow diagram of program

where it is assumed that the FIR filter is a linear phase, i.e., symmetric, filter. If a zero delay conversion is required, the filter length N can be selected to make τ_g an integer and the first τ_g output samples can be discarded.

3. Program Description

3.1 Usage

The program consists of a main program (supplied by the user) and two subroutines, one for initialization (SRINIT) and one for processing (SRCONV). On initialization the user supplies values for the parameters M, L, NQ, COEF, N, NC and NI. For processing the user supplies values for BUFM, BUFL, and ND. Once the program is initialized SRCONV can be used as often as necessary for processing contiguous blocks of data.

3.2 Description of Parameters

M	Decimation ratio (see Fig. 1), M.
L	Interpolation ratio (see Fig. 1), L.
QBUF	Internal buffer for storage of state variables of filter.
NQ	Size of QBUF, equal to or greater than $2*Q$, where Q is the next greatest integer of N/L.
COEF	Buffer containing FIR symmetric filter coefficients. Because of symmetry the buffer should contain only half of the filter coefficients beginning at the middle impulse response coefficient.
N	No. of taps in FIR filter, N.
COFS	Buffer for storing scrambled coefficients.
NC	Size of buffer COFS, equal to or greater than $L*Q$ where Q is the next greatest integer of N/L.
ICTR	Buffer for storing control data generated by SRINIT and used by SRCONV.
NI	Size of ICTR, greater than or equal to $2*L$.
IERR	Error code for debugging = 0 No errors found in initialization = 1 QBUF (NQ) too small

= 2 COFS (NC) too small
= 3 ICTR (NI) too small

BUFM Input data buffer for processing, size N_D*M.

BUFL Output data buffer for processing, size N_D*L.

ND Any positive integer, N_D.

3.3 Dimension Requirements

The DIMENSION statement in the main program should be modified according to the requirements of each particular problem. The required minimum dimensions of the arrays are

QBUF must be of dimension equal to or greater than 2* (the next greatest integer of N/L).

COEF must be of dimension $[(N+1)/2]$ where $[x]$ is the integer part of x.

COFS must be of dimension $Q*L$ where Q is equal to or greater than the next largest integer of N/L.

ICTR must be of dimension $2*L$.

BUFM must be of dimension N_D*M.

BUFL must be of dimension N_D*L.

3.4 Summary of User Requirements

(1) Specify M, L, NQ, N, NC, NI

(2) Fill in coefficient buffer COEF

(3) Adjust dimensions for QBUF, COFS, ICTR, COEF, BUFM, BUFL

(4) CALL SRINIT once for initialization

(5) Supply BUFM with input data

(6) CALL SRCONV and obtain output samples in BUFL

(7) Repeat steps (5) and (6) as often as desired

4. Test Problem

Compute the unit sample response of a sampling rate converter which converts data from a sampling rate of 10 kHz to 16 kHz.

The ratio of conversion is 16/10 or preferably 8/5. Therefore, $L = 8$ and $M = 5$. The input sampling rate f_r, is 10 kHz and the output sampling rate Lf_r/M is 16 kHz. As seen from Fig. 1 the intermediate sampling rate Lf_r is 80 kHz which corresponds to the effective sampling rate of the lowpass filter. Since 10 kHz is the lowest sampling rate involved, the lowpass filter should have a stopband cutoff of one half of this rate or 5 kHz. The filter was designed using the McClellan, et al. program [3].

The test program reads in the filter coefficients from the standard input unit and outputs them to the standard output unit. It initializes the conversion routine by calling SRINIT and then generates a unit sample in the BUFM array by zeroing it out and setting the first sample value to one. It converts the sampling rate of this array by the factor 8/5 by calling SRCONV. BUFL then contains the interpolated impulse response of the sampling rate converter at the 16 kHz sampling rate. Figure 3a shows a plot of this response.

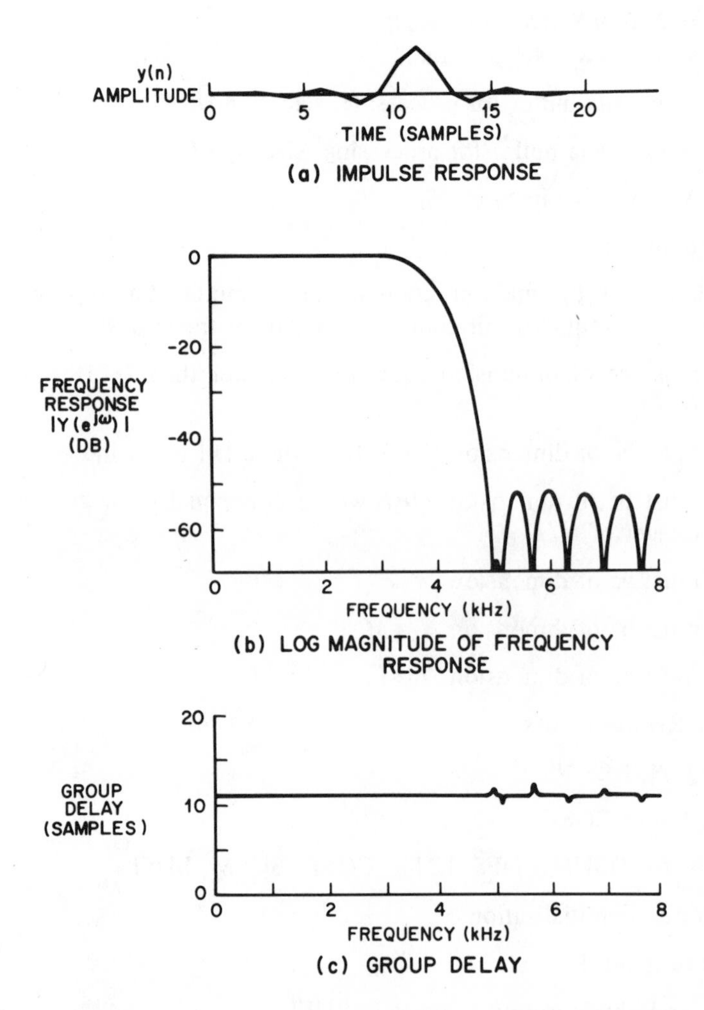

Fig. 3. Impulse response, frequency response, and group delay of test example.

Figure 3b shows the log magnitude of the frequency response obtained by taking the Fourier transform of Fig. 3a. The filter characteristics are clearly evident. Figure 3c shows the group delay of the sampling rate converter. Note that the delay is 11 samples as predicted by Eq. (3) and is exactly flat with frequency because of the linear phase (symmetric) FIR filter.

Table 1 shows the output of the test program for the case $L = 8$, $M = 5$, and $N = 119$; the coefficients of the 119 tap FIR filter input to test program are echoed back in the output of the test example. Also the unit sample response of Fig. 3a is given in Table 1.

References

1. R. E. Crochiere and L. R. Rabiner, "Optimum FIR Digital Filter Implementation for Decimation, Interpolation, and Narrow Band Filtering", *IEEE Trans. Acoust., Speech, Signal Processing,* Vol ASSP-23, No. 5, pp. 444-456, October 1975.

2. R. E. Crochiere, L. R. Rabiner, and R. R. Shively, "A Novel Implementation of Digital Phase Shifters", *Bell Syst. Tech. Jour.,* Vol. 54, No. 8, pp. 1497-1502, October 1975.

3. J. H. McClellan, T. W. Parks and L. R. Rabiner, "A Computer Program for Designing Optimum FIR Linear Phase Digital Filters", *IEEE Trans. Audio Electroacoust.,* Vol. AU-21, No. 6, pp. 506-526, December 1973; see program 5.1 in this book.

Table 1

```
COEFFICIENTS
   .098344    .096741    .092027    .084489    .074579
   .062882    .050073    .036869    .023975    .012040
   .001611   -.006898   -.013228   -.017283   -.019127
  -.018966   -.017123   -.014002   -.010055   -.005738
  -.001477    .002359    .005488    .007720    .008970
   .009248    .008652    .007342    .005525    .003429
   .001279   -.000722   -.002410   -.003668   -.004435
  -.004700   -.004502   -.003916   -.003043   -.001998
  -.000896    .000157    .001076    .001799    .002292
   .002545    .002576    .002416    .002112    .001715
   .001277    .000844    .000452    .000129   -.000112
  -.000268   -.000345   -.000357   -.000319   -.000533

IERR=  0

UNIT SAMPLE RESPONSE
  -.002144    .010217    .020362   -.007169   -.037602
   .010230    .073986   -.011818   -.151726    .012884
   .503055    .786755    .503055    .012884   -.151726
  -.011818    .073986    .010230   -.037602   -.007169
   .020362    .010217   -.002144    .000000
```

Appendix

```
C
C-----------------------------------------------------------------------
C MAIN PROGRAM: TEST PROGRAM TO PERFORM SAMPLING RATE CONVERSION
C               BY A RATIONAL RATIO
C AUTHOR:       R E CROCHIERE
C               BELL LABORATORIES, MURRAY HILL, NEW JERSEY 07974
C INPUT:        COEF = ARRAY OF COEFFICIENTS FOR FIR INTERPOLATING FILTER
C-----------------------------------------------------------------------
C
      COMMON /SRCOM/ IQ, JQ, IL
      DIMENSION COEF(60), COFS(120), QBUF(30), ICTR(16), BUFL(24),
     *    BUFM(15)
      IQ = 0
      JQ = 0
      IL = 0
C
C THIS PROGRAM CONVERTS THE SAMPLING RATE OF A SIGNAL BY A RATIO OF L/M
C THE PROGRAM CALLS SRINIT TO INITIALIZE AND THEN CALLS SRCONV SUPPLYING
C INPUT DATA THROUGH BUFM AND TAKING OUTPUT DATA FROM BUFL
C
C INPUT: COEFFICIENTS ARE READ FROM THE STANDARD INPUT UNIT (IND)
C
C OUTPUT: ALL OUTPUT IS WRITTEN ON THE STANDARD OUTPUT UNIT (LPT)
C
      IND = I1MACH(1)
      LPT = I1MACH(2)
      NN = 60
      READ (IND,9999) (COEF(K),K=1,NN)
      WRITE (LPT,9998)
      WRITE (LPT,9997) (COEF(K),K=1,NN)
      WRITE (LPT,9996)
C
C INITIALIZE CONVERSION ROUTINE
C
      N = 119
      NC = 120
      NI = 16
      L = 8
      M = 5
      NQ = 30
      CALL SRINIT(M, L, QBUF, NQ, COEF, N, COFS, NC, ICTR, NI, IERR)
      WRITE (LPT,9995) IERR
      WRITE (LPT,9996)
C
C GENERATE UNIT SAMPLE
C
      DO 10 I=1,15
        BUFM(I) = 0.
  10  CONTINUE
      BUFM(1) = 1.
C
C PROCESS DATA
C
      ND = 3
      CALL SRCONV(BUFM, BUFL, ND, QBUF, COFS, ICTR)
      WRITE (LPT,9994)
      NN = 24
      WRITE (LPT,9997) (BUFL(K),K=1,NN)
9999  FORMAT (7F11.0)
9998  FORMAT (13H COEFFICIENTS)
9997  FORMAT (5F10.6)
9996  FORMAT (1H )
9995  FORMAT (6H IERR=, I3)
9994  FORMAT (21H UNIT SAMPLE RESPONSE)
      STOP
      END
C
C-----------------------------------------------------------------------
C SUBROUTINE: SRINIT
C INITIALIZATION FOR SRCONV WHICH CONVERTS THE SAMPLING RATE
C OF A SIGNAL BY THE RATIO OF L/M
C-----------------------------------------------------------------------
C
      SUBROUTINE SRINIT(M, L, QBUF, NQ, COEF, N, COFS, NC, ICTR, NI,
     *    IERR)
      COMMON /SRCOM/ IQ, JQ, IL
      DIMENSION QBUF(1), COEF(1), COFS(1), ICTR(1)
C
C     M = DECIMATION RATIO
C     L = INTERPOLATION RATIO
C  QBUF = STATE VARIABLE BUFFER
C    NQ = SIZE OF QBUF, GREATER OR EQUAL TO 2*(THE NEXT
C         GREATEST INTEGER OF N/L)
C  COEF = ARRAY OF COEFFICIENTS FOR FIR INTERPOLATING FILTER
C     N = NO. OF TAPS IN THE FIR INTERPOLATING FILTER
C  COFS = SCRAMBLED COEFFICIENT VECTOR GENERATED BY SRINIT
C    NC = SIZE OF COFS VECTOR, EQUAL TO OR GREATER THAN
C         L*(THE NEXT GREATEST INTEGER OF N/L)
C  ICTR = CONTROL ARRAY GENERATED BY SRINIT AND USED BY SRCONV
C    NI = SIZE OF ICTR VECTOR EQUAL OR GREATER THAN 2*L
C  IERR = ERROR CODE FOR DEBUGGING
C       = 0  NO ERRORS FOUND IN INITIALIZATION
C       = 1  QBUF (NQ) TOO SMALL
C       = 2  COFS (NC) TOO SMALL
C       = 3  ICTR (NI) TOO SMALL
C
      IERR = 0
      IL = L
C
C COMPUTE IQ
C
      IQ = N/L
      IF (N.NE.(IQ*L)) IQ = IQ + 1
      NP = IQ*L
      IF (NQ.LT.(2*IQ)) IERR = 1
      IF (NC.LT.NP) IERR = 2
      NCF = (N+1)/2
      FL = L
C
C ZERO OUT QBUF
C
      DO 10 I=1,NQ
        QBUF(I) = 0
  10  CONTINUE
C
C SCRAMBLE COEFFICIENTS
C
      I = 1
      DO 30 ML=1,L
        DO 20 MQ=1,IQ
          MX = (ML-1) + (MQ-1)*L
          IF (MX.LT.NCF) MM = NCF - MX
          IF (MX.GE.NCF) MM = MX - (N-NCF-1)
          IF (MM.LE.NCF) COFS(I) = COEF(MM)*FL
          IF (MM.GT.NCF) COFS(I) = 0.
          I = I + 1
  20    CONTINUE
  30  CONTINUE
C
C SETUP OF MOVING ADDRESS POINTER
```

```
C
      JQ = IQ
C
C GENERATE CONTROL ARRAY ICTR
C
      LM = L*M
      IF (NI.LT.(2*L)) IERR = 3
      LC = 0
      MC = 0
      INCR = 0
      K = 1
      DO 50 I=1,LM
        IF (LC.EQ.0) INCR = INCR + 1
        IF (MC.LT.(M-1)) GO TO 40
C
C NO OF SAMPLES TO UPDATE QBUF
C
        ICTR(K) = INCR
        INCR = 0
        K = K + 1
C
C STARTING LOCATION IN COFS VECTOR
C
        ICTR(K) = LC*IQ
        INCR = 0
        MC = -1
        K = K + 1
  40    LC = LC + 1
        IF (LC.GE.L) LC = 0
        MC = MC + 1
  50  CONTINUE
      RETURN
      END
C
C-----------------------------------------------------------------------
C SUBROUTINE: SRCONV
C CONVERTS THE SAMPLING RATE OF A SIGNAL BY THE RATIO L/M.
C SRINIT MUST BE CALLED PRIOR TO CALLING THIS ROUTINE.
C-----------------------------------------------------------------------
C
      SUBROUTINE SRCONV(BUFM, BUFL, ND, QBUF, COFS, ICTR)
      COMMON /SRCOM/ IQ, JQ, IL
      DIMENSION BUFM(1), BUFL(1), QBUF(1), COFS(1), ICTR(1)
C
C BUFM = INPUT DATA BUFFER OF SIZE ND*M
C BUFL = OUTPUT DATA BUFFER OF SIZE ND*L
C   ND = ANY POSITIVE INTEGER
C QBUF = STATE VARIABLE BUFFER
C COFS = SCRAMBLED COEFFICIENT VECTOR GENERATED BY SRINIT
C ICTR = CONTROL ARRAY GENERATED BY SRINIT AND USED BY SRCONV
C
      MB = 1
      LB = 1
      L = IL
      DO 50 I=1,ND
C
C MB = INDEX ON BUFM
C COMPUTE L OUTPUT SAMPLES
C
        K = 1
        DO 40 J=1,L
          JD = ICTR(K)
          IC = ICTR(K+1)
          K = K + 2
C
C UPDATE QBUF
C
  10      IF (JD.EQ.0) GO TO 20
          QBUF(JQ) = BUFM(MB)
          JQ1 = JQ + IQ
          QBUF(JQ1) = BUFM(MB)
          MB = MB + 1
          JQ = JQ - 1
          IF (JQ.EQ.0) JQ = IQ
          JD = JD - 1
          GO TO 10
C
C COMPUTE 1 SAMPLE OF OUTPUT DATA AND STORE IN BUFL
C
  20      SUM = 0.
          DO 30 KQ=1,IQ
            ICOF = KQ + IC
            IQB = KQ + JQ
            SUM = SUM + QBUF(IQB)*COFS(ICOF)
  30      CONTINUE
          BUFL(LB) = SUM
          LB = LB + 1
  40    CONTINUE
  50  CONTINUE
      RETURN
      END
```

8.3

A Program for Multistage Decimation, Interpolation, and Narrow Band Filtering

R. E. Crochiere and *L. R. Rabiner*

Acoustics Research Dept.
Bell Laboratories
Murray Hill, NJ 07974

1. Purpose

This program decimates, interpolates, or narrow band filters an input signal using a multistage realization with up to 3 stages for both decimation and interpolation. The output sequence is written directly over the input sequence.

2. Method

The program is based on the method of R. E. Crochiere and L. R. Rabiner [1,2,3]. The user can decimate and interpolate (by an integer ratio) or narrow band filter (by a combination of integer decimation and interpolation) a signal by performing these operations in a series of stages. Up to three stages can be used for decimation and interpolation. For each stage a decimation ratio is required as well as a set of filter coefficients (of a direct form linear phase FIR digital filter) for the filtering to be performed in that stage. A block of samples of the input sequence is supplied and a block of samples of the output sequence is generated by the program. Thus to process a large sequence of samples the input sequence is partitioned into blocks and the program is called once for each block of samples.

A flow diagram for the implementation of a three stage decimator in cascade with a three stage interpolator is shown in Fig. 1(a) and its corresponding control sequence is given in Fig. 1(b). Together the cascade results in the implementation of a narrow band lowpass filter. To realize only a decimator or interpolator, appropriate parts of this structure are partitioned off (in the program) from the main structure.

The decimator has three data buffers S1D, S2D, and S3D for storage of internal data for its three stages. These storage buffers are of durations N'_1, N'_2, and N'_3 words respectively. These lengths are derived according to the relationship

$$N'_i = Q_i D_i \geqslant N_i \, , \quad i = 1,2,3 \, , \tag{1}$$

where Q_i is the integer greater than or equal to N_i/D_i, N_i is the filter length for the ith stage and D_i is its decimation ratio. Each data buffer is partitioned into Q_i blocks of data of length D_i. Three additional buffers hold the coefficients for the filters in each of the three stages. The interpolator has six data buffers associated with it, three (S1I, S2I, and S3I) of lengths Q_1, Q_2, and Q_3 respectively and three (T1I, T2I, and T3I) of lengths D_1, D_2, and D_3 respectively. In addition, each stage i has a buffer for the "scrambled" coefficients which are partitioned into D_i blocks of Q_i samples each. The operation of this structure is depicted by the control sequence in Fig. 1(b). The process begins by reading D_1 samples from the main I/O buffer into S1D. One output sample is then computed from stage one of the decimator and stored in S2D. This process is repeated D_2 times until D_2 samples have been computed and stored in S2D. One output is then computed for stage 2 of the decimator and stored in S3D. The above process is repeated D_3 times until D_3 samples have been stored in S3D at which point one output sample is calculated from stage 3 of the decimator. This completes one cycle of the decimator; $D = D_1 D_2 D_3$ samples have been read from the main I/O buffer and one output sample has been computed. A similar computation cycle can now proceed for the interpolator. The output sample from the

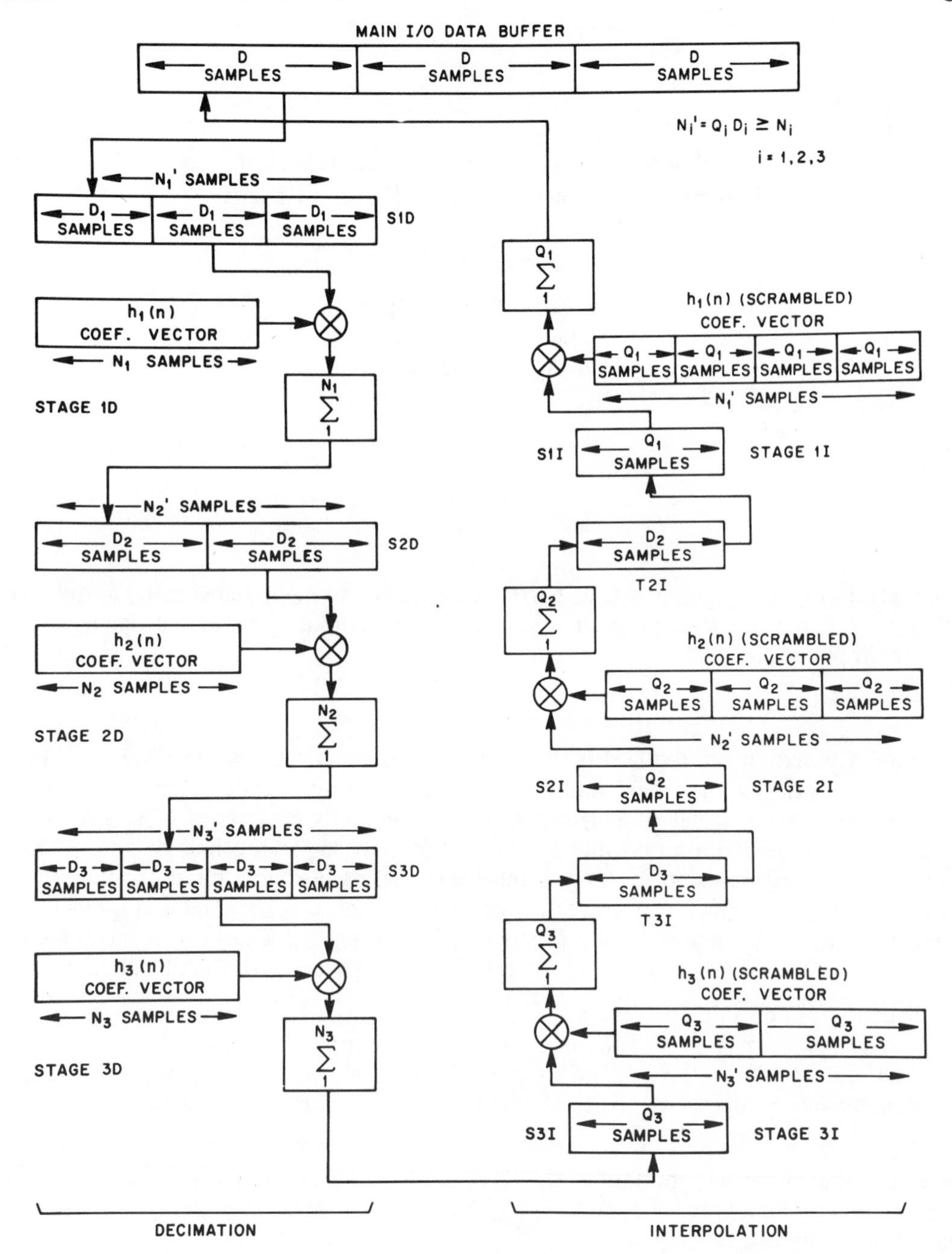

Fig. 1(a) Flowchart for a 3 stage decimator followed by a 3 stage interpolator.

decimator is stored into S3I. D_3 output samples are computed from stage 3 of the interpolator and temporarily stored in T3I. One sample from T3I is then stored in S2I and D_2 output samples are computed from stage 2 and stored in T2I. S1I is then updated by one sample from T2I and D_1 outputs are computed and stored in the main I/O buffer. This is repeated until all D_2 samples in T2I are removed. T2I is then refilled by updating S2I with one sample from T3I and computing D_2 more samples. Upon completion of the interpolator cycle one input was transferred into the interpolator and D output samples are computed and stored in the main I/O data buffer. The process can then be repeated on the next block of D samples.

Because of the delay associated with the linear phase filters the output (decimated, interpolated, or filtered) waveform will have a delay associated with it with respect to the input. This delay can be determined as follows: let

$$\tau_{N1} = \frac{N_1 + 1}{2} \quad \text{(samples)}$$

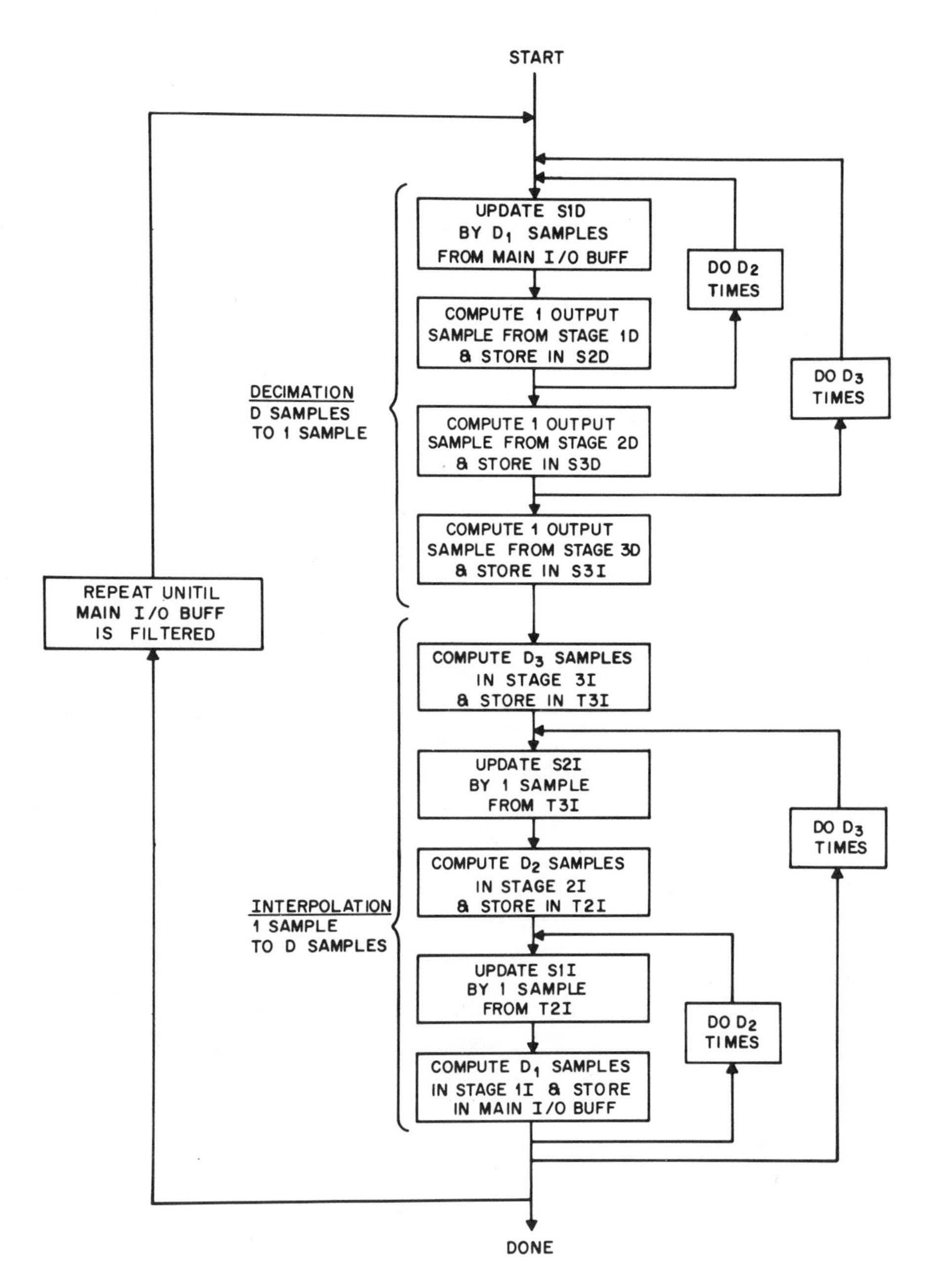

Fig. 1(b) The control sequence for Fig. 1(a).

$$\tau_{N2} = \frac{N_2 + 1}{2} \qquad \text{(samples)}$$

and

$$\tau_{N3} = \frac{N_3 + 1}{2} \qquad \text{(samples)}$$

be the delays (group delay) of the filters for stages 1, 2, and 3. If only two stages are used let $\tau_{N3} = 0$ and if only one stage is used let $\tau_{N2} = \tau_{N3} = 0$. The group delay, τ_L, of the multistage decimator at the output (the low sampling rate) is then

$$\tau_L = \frac{1}{D_1 D_2 D_3} \{\tau_{N1} + D_1 \tau_{N2} + D_1 D_2 \tau_{N3} - (D_1 D_2 D_3 - 1)\} \qquad \text{(samples)}$$

(For decimator output)

where $D_3 = 1$ for a two stage implementation and $D_2 = D_3 = 1$ for a one stage implementation. The group delay for the interpolator output (at the high sampling rate) is

$$\tau_H = \tau_{N1} + D_1\tau_{N2} + D_1D_2\tau_{N3} \quad \text{(samples)}$$

(For interpolator output)

and for the filter (at the high sampling rate)

$$\tau_H = 2(\tau_{N1}+D_1\tau_{N2}+D_1D_2\tau_{N3}) - (D_1D_2D_3-1) \quad \text{(samples)}$$

(For the filter output)

For the decimator and interpolator the group delay at the high sampling rate τ_H is related to the group delay at the low sampling rate, τ_L, by the expression $\tau_H = D_1D_2D_3\tau_L$. The duration of the overall impulse response is approximately twice the group delay.

3. Program Description

3.1 Usage

The program consists of a main program (supplied by the user) and two subroutines, one for initialization (DIINIT), and one for performing the required computation (DIFILT). The user is required to specify the number of stages, the decimation (interpolation) ratio for each stage, and a set of filter coefficients (of a direct form FIR linear phase filter) for each stage. In addition, the user supplies an input-output buffer to hold both the input and output sequences, and a scratch buffer of sufficient size to store all the internal buffers required in the computation. An error return is used to indicate any difficulty in initializing the package. Otherwise no output (other than the output array) is provided.

3.2 Subroutines Required

SUBROUTINE DIINIT (KD, ID1D, ID2D, ID3D, N1D, N2D, N3D, COEF1, COEF2, COEF3, ITYPED, BUFF, IDJD, SBUFF, ISBD, IERR), called from main program, initializes the decimation-interpolation package. SUBROUTINE DIFILT (COEF1, COEF2, COEF3, BUFF, SBUFF) is called from the main program, once for each buffer of input samples, performs the required computation (i.e., decimation, interpolation, or narrow band filtering) and returns the output sequence in SBUFF. Assume BUFF is of size LD where L is an integer and $D = D_1D_2D_3$. For a D:1 decimator fill BUFF with LD input samples and receive as output the first L samples in BUFF. For an interpolator supply only L samples to the first L locations in BUFF and receive as output LD samples. For a filter supply LD samples for input and receive LD samples as output.

3.3 Description of Parameters

KD Number of stages of decimation (and/or interpolation)

ID1D Decimation ratio for stage 1 (i.e., D_1)

ID2D Decimation ratio for stage 2 (i.e., D_2)

ID3D Decimation ratio for stage 3 (i.e., D_3)
(ID3D = 1 if KD = 2 and ID2D = ID3D = 1 if KD = 1)

N1D Filter duration (in samples) of linear phase FIR digital filter for stage 1

N2D Filter duration (in samples) of linear phase FIR digital filter for stage 2

N3D Filter duration (in samples) of linear phase FIR digital filter for stage 3
(N3D = 0 if KD = 2 and N2D = N3D = 0 if KD = 1)

COEF1 Buffer containing filter coefficients for stage 1

COEF2 Buffer containing filter coefficients for stage 2

COEF3 Buffer containing filter coefficients for stage 3.
If the filter duration is odd, the program automatically changes COEFn(1) to COEFn(1)/2

for $n = 1,2,3$. Also COEFn contains half of the filter impulse response beginning at the middle impulse response coefficient.

ITYPED = 1 for a decimator alone
= 2 for an interpolator alone
= 3 for a narrow band filter (decimator followed by interpolator)

BUFF Storage buffer for input and output sequence (Required dimension given below)

IDJD Dimension of BUFF. (See below.)

SBUFF Storage buffer for internal variables, and scrambled coefficients for interpolation stages). (Required dimension given below.)

ISBD Dimension of SBUFF (see below).

IERR Error Return for initialization (useful for debugging purposes).

IERR = 0 No errors
= 1 Means KD is out of range, i.e., KD > 3 or KD < 1
= 2 Means dimension for BUFF is too small or not an integer multiple of $(D_1 * D_2 * D_3)$
= 3 Means dimension for SBUFF is too small.

3.4 Dimension Requirements

The DIMENSION statement in the main program should be modified according to the requirements of each particular problem. The required dimensions of the arrays are

COEF1 must be of dimension [(N1D+1)/2] where $[x]$ is the integer part of x

COEF2 must be of dimension [(N2D+1)/2]

COEF3 must be of dimension [(N3D+1)/2]

BUFF must be of dimension IDJD which is computed by the rule

$$\text{IDJD} = \text{L}*(\text{ID1D}*\text{ID2D}*\text{ID3D})$$

where L is any nonzero integer, i.e., IDJD is an integer multiple of the total decimation (interpolation) ratio of the system.

SBUFF must be of dimension ISBD which is computed by the rule:

(a) For decimator only (ITYPED = 1)

$$\text{ISBD} \geqslant 2*(\text{N1P}+\text{N2P}+\text{N3P})$$

where

$$\text{N1P} = \left(\left[\frac{\text{N1D}}{\text{ID1D}}\right] + 1\right)*\text{ID1D}$$

$$\text{N2P} = \left(\left[\frac{\text{N2D}}{\text{ID2D}}\right] + 1\right)*\text{ID2D}$$

$$\text{N3P} = \left(\left[\frac{\text{N3D}}{\text{ID3D}}\right] + 1\right)*\text{ID3D}$$

and $[x]$ corresponds to the integer part of x (i.e., N1P, N2P, N3P correspond to N'_i, $i = 1,2,3$ in Eq. (1)).

(b) For interpolator only (ITYPED = 2)

$$\text{ISBD} \geqslant (\text{N1P}+\text{N2P}+\text{N3P}) + (\text{ID2D}+\text{ID3D}) + 2*\left(\frac{\text{N1P}}{\text{ID1D}} + \frac{\text{N2P}}{\text{ID2D}} + \frac{\text{N3P}}{\text{ID3D}}\right)$$

where N1P, N2P, and N3P are defined above in (a).

(c) For narrow band filter (ITYPED = 3)

$$\text{ISBD} \geqslant \text{ISBD}\Big|_{\substack{\text{for}\\ \text{decimator}\\ (a)}} + \text{ISBD}\Big|_{\substack{\text{for}\\ \text{interpolator}\\ (b)}}$$

3.5 Summary of User Requirements

(1) Specify KD, ID1D, ID2D, ID3D

(2) Choose filter lengths N1D, N2D, N3D and fill in filter coefficient buffers COEF1, COEF2, COEF3

(3) Adjust dimension for BUFF and SBUFF

(4) CALL DIINIT once for initialization

(5) Supply BUFF with input data

(6) CALL DIFILT and obtain output samples back in BUFF

(7) Repeat steps (5) and (6) as often as necessary for continuous processing on blocks of IDJD samples.

4. Test Problems

1. Decimate the allpass signal

$$x(n) = \begin{cases} 0 & n < 0 \\ -a & n = 0 \\ (1-a^2)a^{n-1} & n \geqslant 1 \end{cases}$$

for $a = 0.9$, by a factor of 10 to 1 using a 2 stage decimator with decimation ratios of 5 to 1 (for stage 1D - Fig. 1) and 2 to 1 (for stage 2D). The filter impulse response durations are 25 samples for stage 1D, and 28 samples for stage 2D.

The main program will print out the filter coefficients for each stage, and will print out both the input sequence, and the decimated output sequence. The reader should note that the input sequence is decimated using 3 calls to the decimation subroutine in which consecutive blocks of 20 samples are decimated to give consecutive blocks of 2 samples.

The parameters for this example are: KD = 2, ID1D = 5, ID2D = 2, ID3D = 1, N1D = 25, N2D = 28, N3D = 0, ITYPED = 1, IDJD = 20, ISBD = 106.

2. Interpolate the allpass signal of example 1 using a 2 stage interpolator with interpolation ratios of 1 to 2 (for stage 2I†) and 1 to 5 (for stage 1I).

The filter impulse responses are the same as used in example 1. The main program will print out both the input sequence, and the interpolated output sequence. The reader should again note that the input sequence is interpolated using 3 calls to the interpolation subroutine in which consecutive blocks of 2 samples are interpolated to give consecutive blocks of 20 samples.

The parameters for this example are: KD = 2, ID1D = 5, ID2D = 2, ID3D = 1, N1D = 25, N2D = 28, N3D = 0, ITYPED = 2, IDJD = 20, ISBD = 106.

3. Filter coefficients for a one stage decimator (or interpolator) with a decimation ratio of 10 to 1 are printed out to test out the one-stage implementation.

Filter coefficients for a three stage decimator with decimation ratios of 10 to 1, 5 to 1 and 2 to 1 (for stages 1D, 2D and 3D respectively) are printed out to test the three-stage implementation.

The output for these three examples including coefficient values, are contained in Table 1.

---

† The reader will recall that interpolation stages are numbered in reverse order because of their duality with decimation stages.

References

1. R. E. Crochiere and L. R. Rabiner, "Optimum FIR Digital Filter Implementation for Decimation, Interpolation, and Narrow Band Filtering", *IEEE Trans. Acoustics, Speech, and Signal Processing,* Vol. ASSP-23, No. 5, pp. 444-456, October 1975.

2. L. R. Rabiner and R. E. Crochiere, "A Novel Implementation for Narrow-Band FIR Digital Filters", *IEEE Trans. Acoustics, Speech, and Signal Processing,* Vol. ASSP-23, No. 5, pp. 457-464, October 1975.

3. R. E. Crochiere and L. R. Rabiner, "Further Considerations in the Design of Decimators and Interpolators", *IEEE Trans. Acoustics, Speech, and Signal Processing,* Vol. ASSP-24, No. 4, pp. 296-311, August 1976.

Table 1

```
        OUTPUT OF TEST PROGRAM
COEFFICIENTS FOR STAGE 1
  .168728   .158866   .131836   .094397   .055181   .021904  -.000775
 -.012024  -.014156  -.011033  -.006346  -.002682  -.000174
COEFFICIENTS FOR STAGE 2
  .352610   .202371   .022946  -.064816  -.045444   .007674   .027651
  .011124  -.007792  -.009267  -.001169   .003120   .001807  -.000303
INPUT SEQUENCE
 -.9000  .1900  .1710  .1539  .1385  .1247  .1122  .1010  .0909  .0818
  .0736  .0662  .0596  .0537  .0483  .0435  .0391  .0352  .0317  .0285
DECIMATED OUTPUT SEQUENCE
  .0000 -.0003
INPUT SEQUENCE
  .0257  .0231  .0208  .0187  .0168  .0152  .0136  .0123  .0110  .0099
  .0089  .0081  .0072  .0065  .0059  .0053  .0048  .0043  .0039  .0035
DECIMATED OUTPUT SEQUENCE
  .0011 -.0017
INPUT SEQUENCE
  .0031  .0028  .0025  .0023  .0020  .0018  .0017  .0015  .0013  .0012
  .0011  .0010  .0009  .0008  .0007  .0006  .0006  .0005  .0005  .0004
DECIMATED OUTPUT SEQUENCE
 -.0004  .0079
INPUT SEQUENCE
 -.9000  .1900
INTERPOLATED OUTPUT SEQUENCE
 -.0000 -.0000 -.0000 -.0000 -.0000 -.0000  .0000  .0002  .0003  .0005
  .0006  .0005  .0003 -.0001 -.0008 -.0015 -.0024 -.0034 -.0043 -.0051
INPUT SEQUENCE
  .1710  .1539
INTERPOLATED OUTPUT SEQUENCE
 -.0055 -.0057 -.0055 -.0048 -.0035 -.0017  .0006  .0033  .0064  .0096
  .0126  .0154  .0177  .0193  .0198  .0189  .0168  .0130  .0078  .0014
INPUT SEQUENCE
  .1385  .1247
INTERPOLATED OUTPUT SEQUENCE
 -.0063 -.0146 -.0234 -.0321 -.0397 -.0457 -.0499 -.0514 -.0501 -.0453
 -.0366 -.0247 -.0091  .0091  .0291  .0500  .0712  .0910  .1084  .1208
COEFFICIENTS FOR A 1-STAGE DESIGN
  .072497   .071163   .068543   .064731   .059859   .054098   .047646
  .040717   .033539   .026336   .019326   .012707   .006652   .001303
 -.003232  -.006889  -.009636  -.011483  -.012473  -.012680  -.012202
 -.011157  -.009672  -.007882  -.005918  -.003903  -.001948  -.000148
  .001423   .002713   .003690   .004343   .004679   .004723   .004512
  .004091   .003513   .002830   .002094   .001355   .000653   .000022
 -.000513  -.000939  -.001247  -.001440  -.001524  -.001515  -.001428
 -.001284  -.001101  -.000898  -.000692  -.000497  -.000324  -.000179
 -.000069   .000007   .000049   .000506
COEFFICIENTS FOR A 3-STAGE DESIGN
STAGE 1
  .089335   .086183   .080139   .071694   .061511   .050351   .038992
  .028149   .018410   .010192   .003719  -.000967  -.003987  -.005579
 -.006050  -.005730  -.004931  -.003918  -.002890  -.001977  -.001247
 -.000715  -.000364  -.000156  -.000052
```

Table 1
(Continued)

STAGE 2

.182858	.160221	.120688	.073819	.029920	-.003025	-.021307
-.025777	-.020599	-.011188	-.002219	.003589	.005662	.004955
.003002	.001103	-.000076	-.000482	-.000407	-.000202	

STAGE 3

.430137	.168113	-.068326	-.080180	.026939	.054430	-.010516
-.041032	.001691	.032172	.003666	-.025510	-.007056	.020130
.009159	-.015612	-.010347	.011750	.010849	-.008433	-.010828
.005600	.010404	-.003210	-.009681	.001235	.008742	.000352
-.007666	-.001579	.006515	.002478	-.005350	-.003079	.004213
.003422	-.003148	-.003542	.002179	.003484	-.001332	-.003281
.000614	.002976	-.000036	-.002601	-.000410	.002192	.000727
-.001770	-.000931	.001366	.001033	-.000990	-.001054	.000662
.001008	-.000381	-.000918	.000160	.000794	.000009	-.000658
-.000123	.000517	.000196	-.000387	-.000226	.000273	.000233
-.000193	-.000229	.000174	.000422	.000276		

Appendix

```
C
C-----------------------------------------------------------------------
C MAIN PROGRAM: TEST PROGRAM FOR DECIMATION,INTERPOLATION AND FILTERING
C AUTHOR:       R E CROCHIERE, L R RABINER
C               BELL LABORATORIES, MURRAY HILL, NEW JERSEY, 07974
C INPUT:        COEF1 = ARRAY OF FILTER COEFFICIENTS FOR FIRST STAGE
C                       FOR TEST PROBLEM 1.
C               COEF2 = ARRAY OF FILTER COEFFICIENTS FOR SECOND STAGE
C                       FOR TEST PROBLEM 1.
C               COEF11 = ARRAY OF FILTER COEFFICIENTS FOR FIRST STAGE
C                        FOR 1-STAGE EXAMPLE IN TEST PROBLEM 3.
C               COEF1A = ARRAY OF FILTER COEFFICIENTS FOR FIRST STAGE
C                        FOR 3-STAGE EXAMPLE IN TEST PROBLEM 3.
C               COEF2A = ARRAY OF FILTER COEFFICIENTS FOR SECOND STAGE
C                        FOR 3-STAGE EXAMPLE IN TEST PROBLEM 3.
C               COEF3A = ARRAY OF FILTER COEFFICIENTS FOR THIRD STAGE
C                        FOR 3-STAGE EXAMPLE IN TEST PROBLEM 3.
C-----------------------------------------------------------------------
C
      COMMON /DIFC1/ K, ID1, ID2, ID3, N1, N2, N3, ITYPE, ID, ISB, N1P,
     *    N2P, N3P
      COMMON /DIFC2/ IDD, J1, J2, J3, J4, J5, J6, J7, J8, J9, J10, J11,
     *    J12, J1S, J2S, J3S, NCF1, NCF2, NCF3, NF1, NF2, NF3, IQ1,
     *    IQ2, IQ3, K1S, K2S, K3S
      DIMENSION X(60), COEF1(13), COEF2(14), SBUFF(106), BUFF(20),
     *    COEF3(1)
      DIMENSION COEF11(60)
      DIMENSION COEF1A(25), COEF2A(20), COEF3A(75)
      INTEGER TTI, TTO
C
C DEFINE I/O DEVICE CODES
C INPUT: INPUT TO THIS PROGRAM IS USER INTERACTIVE
C        THAT IS - A QUESTION IS WRITTEN ON THE USER
C        TERMINAL (TTO) AND THE USER TYPES IN THE ANSWER.
C
C OUTPUT: ALL OUTPUT IS WRITTEN ON THE STANDARD
C         OUTPUT UNIT (LPT).
C
      TTI = I1MACH(1)
      TTO = I1MACH(4)
      LPT = I1MACH(2)
C
C TEST EXAMPLE 1
C
      A = 0.9
      X(1) = -A
      X(2) = 1. - A*A
      DO 10 I=3,60
        X(I) = X(I-1)*A
  10  CONTINUE
      READ (TTI,9999) (COEF1(I),I=1,13)
      READ (TTI,9999) (COEF2(I),I=1,14)
      WRITE (TTO,9998)
      WRITE (TTO,9999) (COEF1(I),I=1,13)
      WRITE (TTO,9997)
      WRITE (TTO,9999) (COEF2(I),I=1,14)
9999  FORMAT (7F10.6)
9998  FORMAT (25H COEFFICIENTS FOR STAGE 1)
9997  FORMAT (25H COEFFICIENTS FOR STAGE 2)
      CALL DIINIT(2, 5, 2, 1, 25, 28, 0, COEF1, COEF2, COEF3, 1, BUFF,
     *    20, SBUFF, 106, IER)
      IF (IER.NE.0) STOP
      DO 30 I=1,3
        DO 20 J=1,20
          JX = (I-1)*20 + J
          BUFF(J) = X(JX)
  20    CONTINUE
        WRITE (TTO,9996)
9996    FORMAT (15H INPUT SEQUENCE)
        WRITE (TTO,9995) (BUFF(J),J=1,20)
9995    FORMAT (1H , 10F7.4)
        CALL DIFILT(COEF1, COEF2, COEF3, BUFF, SBUFF)
        WRITE (TTO,9994)
9994    FORMAT (26H DECIMATED OUTPUT SEQUENCE)
        WRITE (TTO,9995) (BUFF(J),J=1,2)
  30  CONTINUE
C
C TEST EXAMPLE 2
C RESTORE COEF1(1)
C
      COEF1(1) = COEF1(1)*2.
      CALL DIINIT(2, 5, 2, 1, 25, 28, 0, COEF1, COEF2, COEF3, 2, BUFF,
     *    20, SBUFF, 106, IER)
      IF (IER.NE.0) STOP
      WRITE (TTO,9993)
9993  FORMAT (1H /1H )
      DO 50 I=1,3
        DO 40 J=1,2
          JX = (I-1)*2 + J
          BUFF(J) = X(JX)
  40    CONTINUE
        WRITE (TTO,9996)
        WRITE (TTO,9995) (BUFF(J),J=1,2)
        CALL DIFILT(COEF1, COEF2, COEF3, BUFF, SBUFF)
        WRITE (TTO,9992)
9992    FORMAT (29H INTERPOLATED OUTPUT SEQUENCE)
        WRITE (TTO,9995) (BUFF(J),J=1,20)
  50  CONTINUE
C
C TEST EXAMPLE 3
C
      WRITE (TTO,9993)
      READ (TTI,9999) (COEF11(I),I=1,60)
      WRITE (TTO,9991)
9991  FORMAT (34H COEFFICIENTS FOR A 1-STAGE DESIGN)
      WRITE (TTO,9999) (COEF11(I),I=1,60)
      WRITE (TTO,9993)
      READ (TTI,9999) (COEF1A(I),I=1,25)
      READ (TTI,9999) (COEF2A(I),I=1,20)
      READ (TTI,9999) (COEF3A(I),I=1,75)
      WRITE (TTO,9990)
9990  FORMAT (34H COEFFICIENTS FOR A 3-STAGE DESIGN)
      WRITE (TTO,9989)
      WRITE (TTO,9999) (COEF1A(I),I=1,25)
      WRITE (TTO,9988)
      WRITE (TTO,9999) (COEF2A(I),I=1,20)
      WRITE (TTO,9987)
      WRITE (TTO,9999) (COEF3A(I),I=1,75)
9989  FORMAT (8H STAGE 1)
9988  FORMAT (8H STAGE 2)
9987  FORMAT (8H STAGE 3)
      STOP
      END
C
C-----------------------------------------------------------------------
C SUBROUTINE: DIINIT
C INITIALIZATION FOR DIFILT WHICH DECIMATES, INTERPOLATES OR
C FILTERS A SIGNAL
C-----------------------------------------------------------------------
C
```

```
      SUBROUTINE DIINIT(KD, ID1D, ID2D, ID3D, N1D, N2D, N3D, COEF1,
     *    COEF2, COEF3, ITYPED, BUFF, IDJD, SBUFF, ISBD, IERR)
      COMMON /DIFC1/ K, ID1, ID2, ID3, N1, N2, N3, ITYPE, ID, ISB, N1P,
     *    N2P, N3P
      COMMON /DIFC2/ IDD, J1, J2, J3, J4, J5, J6, J7, J8, J9, J10, J11,
     *    J12, J1S, J2S, J3S, NCF1, NCF2, NCF3, NF1, NF2, NF3, IQ1,
     *    IQ2, IQ3, K1S, K2S, K3S
      DIMENSION COEF1(1), COEF2(1), COEF3(1), SBUFF(1), BUFF(1)
C
C     KD = NO. OF STAGES OF DECIMATION AND /OR INTERPOLATION
C   ID1D = DECIMATION RATIO FOR STAGE 1
C   ID2D = DECIMATION RATIO FOR STAGE 2 = 1 IF KD=1
C   ID3D = DECIMATION RATIO FOR STAGE 3 = 1 IF KD=1 OR 2
C    N1D = FILTER LENGTH FOR STAGE 1
C    N2D = FILTER LENGTH FOR STAGE 2 = 0 IF KD = 1
C    N3D = FILTER LENGTH FOR STAGE 3 = 0 IF KD = 1 OR 2
C  COEF1 = COEF. ARRAY FOR STAGE 1      (SIZE = [(N1D+1)/2] )[X]=INTEGER
C  COEF2 = COEF. ARRAY FOR STAGE 2      (SIZE = [(N2D+1)/2] )    PART OF
C  COEF3 = COEF. ARRAY FOR STAGE 3      (SIZE = [(N3D+1)/2] )
C ITYPED = 1 FOR DECIMATOR
C        = 2 FOR INTERPOLATOR
C        = 3 FOR FILTER
C   BUFF = STORAGE BUFFER FOR INPUT/OUTPUT
C   IDJD = SIZE OF BUFF = INTEGER*ID1D*ID2D*ID3D
C  SBUFF = SCRATCH STORAGE BUFFER FOR INTERNAL VARIABLES
C   ISBD = SIZE OF SBUFF
C        = 2*(N1P+N2P+N3P)                  FOR DECIMATOR
C        = (N1P+N2P+N3P)+ID1D+ID2D          FOR INTERPOLATOR
C        = 3*(N1P+N2P+N3P)+ID1D+ID2D        FOR FILTER
C WHERE  N1P = ([N1D/ID1D]+1)*ID1D       [X]=INTEGER PART OF X
C        N2P = ([N2D/ID2D]+1)*ID2D           OR X-1 IF X=INTEGER
C        N3P = ([N3D/ID3D]+1)*ID3D
C   IERR = ERROR CODE FOR DEBUGGING PURPOSES
C        = 0     NO ERRORS ENCOUNTERED IN INITIALIZATION
C        = 1     KD NOT EQUAL TO 1, 2, OR 3
C        = 2     IDJD NOT AN INTEGER MULTIPLE OF ID1D*ID2D*ID3D
C
C INITIALIZATION AND SETUP
C TRANSFER OF DUMMY ARGUMENTS TO COMMON
C
      K = KD
      ID1 = ID1D
      ID2 = ID2D
      ID3 = ID3D
      N1 = N1D
      N2 = N2D
      N3 = N3D
      ITYPE = ITYPED
      ID = IDJD
      ISB = ISBD
      IERR = 0
C
C PRELIMINARY CHECKS
C
      IF ((K.GT.3) .OR. (K.LT.1)) IERR = 1
      M = 1
      IF (K.EQ.3) GO TO 10
      N3 = 0
      ID3 = 0
      IQ3 = 0
      IF (K.EQ.2) GO TO 20
      N2 = 0
      ID2 = 0
      IQ2 = 0
      GO TO 30
  10  M = M*ID3
      IQ3 = N3/ID3
      IF (N3.NE.(IQ3*ID3)) IQ3 = IQ3 + 1
  20  M = M*ID2
      IQ2 = N2/ID2
      IF (N2.NE.(IQ2*ID2)) IQ2 = IQ2 + 1
  30  M = M*ID1
      IQ1 = N1/ID1
      IF (N1.NE.(IQ1*ID1)) IQ1 = IQ1 + 1
      IDD = ID/M
      IF (ID.NE.(M*IDD)) IERR = 2
      N1P = IQ1*ID1
      N2P = IQ2*ID2
      N3P = IQ3*ID3
C
C SETUP OF ADDRESS LOCATIONS IN SBUFF FOR INTERNAL STORAGE BUFFERS
C AND ZERO OUT SBUFF FOR INITIALIZATION
C
      J1 = 0
      J2 = J1 + 2*N1P
      J3 = J2 + 2*N2P
      J4 = J3 + 2*N3P
      IF (ITYPE.EQ.2) J4 = 0
      J5 = J4 + IQ1*ID1
      J6 = J5 + IQ2*ID2
      J7 = J6 + IQ3*ID3
      J8 = J7 + 2*IQ1
      J9 = J8 + 2*IQ2
      J10 = J9 + 2*IQ3
      J11 = J10 + ID2
      J12 = J11 + ID3
      IF (ITYPE.EQ.1) J12 = J4
      IF (ISB.LT.J12) IERR = 3
      DO 40 M=1,J12
        SBUFF(M) = 0.0
  40  CONTINUE
C
C SETUP OF SCRAMBLED COEFFICIENT SETS FOR INTERPOLATION,
C INITIALIZATION OF MOVING ADDRESS POINTERS AND
C HALVING OF FIRST COEFFICIENT IN SETS COEF1 TO COEF3
C
C STAGE 1
C
      IDX = J4 + 1
      NCF1 = (N1+1)/2
      SUM = ID1
      IF (ITYPE.EQ.1) GO TO 70
      DO 60 MD=1,ID1
        DO 50 MQ=1,IQ1
          M = (MD-1) + (MQ-1)*ID1
          IF (M.LT.NCF1) MM = NCF1 - M
          IF (M.GE.NCF1) MM = M - (N1-NCF1-1)
          IF (MM.LE.NCF1) SBUFF(IDX) = COEF1(MM)*SUM
          IF (MM.GT.NCF1) SBUFF(IDX) = 0.
          IDX = IDX + 1
  50    CONTINUE
  60  CONTINUE
  70  IF (N1.EQ.(2*NCF1-1)) COEF1(1) = COEF1(1)/2.0
      NF1 = (N1+2)/2
      J1S = J1 + N1P
      K1S = J7 + IQ1
      IF (K.LT.2) GO TO 140
C
C STAGE 2
C
```

```
      IDX = J5 + 1
      NCF2 = (N2+1)/2
      SUM = ID2
      IF (ITYPE.EQ.1) GO TO 100
      DO 90 MD=1,ID2
        DO 80 MQ=1,IQ2
          M = (MD-1) + (MQ-1)*ID2
          IF (M.LT.NCF2) MM = NCF2 - M
          IF (M.GE.NCF2) MM = M - (N2-NCF2-1)
          IF (MM.LE.NCF2) SBUFF(IDX) = COEF2(MM)*SUM
          IF (MM.GT.NCF2) SBUFF(IDX) = 0.
          IDX = IDX + 1
 80     CONTINUE
 90   CONTINUE
100   IF (N2.EQ.(2*NCF2-1)) COEF2(1) = COEF2(1)/2.0
      NF2 = (N2+2)/2
      J2S = J2 + N2P
      K2S = J8 + IQ2
      IF (K.LT.3) GO TO 140
C
C STAGE 3
C
      IDX = J6 + 1
      NCF3 = (N3+1)/2
      SUM = ID3
      IF (ITYPE.EQ.1) GO TO 130
      DO 120 MD=1,ID3
        DO 110 MQ=1,IQ3
          M = (MD-1) + (MQ-1)*ID3
          IF (M.LT.NCF3) MM = NCF3 - M
          IF (M.GE.NCF3) MM = M - (N3-NCF3-1)
          IF (MM.LE.NCF3) SBUFF(IDX) = COEF3(MM)*SUM
          IF (MM.GT.NCF3) SBUFF(IDX) = 0.
          IDX = IDX + 1
110     CONTINUE
120   CONTINUE
130   IF (N3.EQ.(2*NCF3-1)) COEF3(1) = COEF3(1)/2.0
      NF3 = (N3+2)/2
      J3S = J3 + N3P
      K3S = J9 + IQ3
140   RETURN
      END
C
C-----------------------------------------------------------------------
C SUBROUTINE: DIFILT
C DIFILT DECIMATES, INTERPOLATES OR FILTERS A SIGNAL,
C DIINIT MUST BE CALLED PRIOR TO CALLING THIS ROUTINE
C-----------------------------------------------------------------------
C
C FOR DECIMATOR, FILL BUFF WITH IDJD SAMPLES, CALL DIFILT, AND
C     RECEIVE (IDJD/ID1D*ID2D*ID3D) SAMPLES
C FOR INTERPOLATOR, SUPPLY (IDJD/ID1D*ID2D*ID3D) INPUT SAMPLES
C     TO BUFF, CALL DIFILT, AND RECEIVE IDJD OUTPUT SAMPLES
C FOR FILTER, SUPPLY AND RECEIVE IDJD SAMPLES FROM BUFF
C
      SUBROUTINE DIFILT(COEF1, COEF2, COEF3, BUFF, SBUFF)
      COMMON /DIFC1/ K, ID1, ID2, ID3, N1, N2, N3, ITYPE, ID, ISB, N1P,
     *    N2P, N3P
      COMMON /DIFC2/ IDD, J1, J2, J3, J4, J5, J6, J7, J8, J9, J10, J11,
     *    J12, J1S, J2S, J3S, NCF1, NCF2, NCF3, NF1, NF2, NF3, IQ1,
     *    IQ2, IQ3, K1S, K2S, K3S
      DIMENSION COEF1(1), COEF2(1), COEF3(1), SBUFF(1), BUFF(1)
C
C  COEF1 = COEF. ARRAY FOR STAGE 1      (SIZE = [(N1D+1)/2] )[X]=INTEGER
C  COEF2 = COEF. ARRAY FOR STAGE 2      (SIZE = [(N2D+1)/2] )     PART OF
C  COEF3 = COEF. ARRAY FOR STAGE 3      (SIZE = [(N3D+1)/2] )
C   BUFF = STORAGE BUFFER FOR INPUT/OUTPUT
C  SBUFF = SCRATCH STORAGE BUFFER FOR INTERNAL VARIABLES
C
      MM = 1
      NN = 1
      IF (ITYPE.NE.2) GO TO 20
      MM = ID - IDD + 1
      DO 10 M=1,IDD
        BUFF(MM) = BUFF(M)
        MM = MM + 1
 10   CONTINUE
      MM = ID - IDD + 1
 20   IF (K.EQ.2) GO TO 100
      IF (K.EQ.3) GO TO 220
C
C ONE STAGE DECIMATION AND/OR INTERPOLATION
C THIS SECTION OF CODE IS REMOVABLE IF 1 STAGE (KD=1) IMPLEMENTATION
C ARE NOT USED         (SUBSTITUTE       200 GO TO 2000)
C ONE STAGE DECIMATION
C READ ID1 SAMPLES FROM BUFF INTO S1D BUFFER
C
C
 30   IF (ITYPE.NE.2) GO TO 40
      SUM = BUFF(MM)
      MM = MM + 1
      GO TO 70
 40   M1 = J1S - ID1
      MD = J1S + N1P
 50   SBUFF(J1S) = BUFF(MM)
      SBUFF(MD) = BUFF(MM)
      MM = MM + 1
      J1S = J1S - 1
      MD = MD - 1
      IF (M1.LT.J1S) GO TO 50
C
C COMPUTE ONE FILTER OUTPUT FOR STAGE 1D
C
      JD = J1S + NCF1
      JU = J1S + NF1
      SUM = 0.0
      M1 = 0
 60   M1 = M1 + 1
      SUM = SUM + COEF1(M1)*(SBUFF(JU)+SBUFF(JD))
      JD = JD - 1
      JU = JU + 1
      IF (M1.LT.NCF1) GO TO 60
      IF (J1S.LE.J1) J1S = J1 + N1P
      IF (ITYPE.NE.1) GO TO 70
      BUFF(NN) = SUM
      NN = NN + 1
      IF (MM.LE.ID) GO TO 40
      GO TO 390
C
C ONE STAGE INTERPOLATION
C STORE DATA INTO S1I
C
 70   SBUFF(K1S) = SUM
      MD = K1S + IQ1
      SBUFF(MD) = SUM
      MQ = J4
C
C COMPUTE ID1 SAMPLES FROM STAGE 1I
C
 80   M1 = K1S
```

```
      SUM = 0.0
   90 MQ = MQ + 1
      SUM = SUM + SBUFF(MQ)*SBUFF(M1)
      M1 = M1 + 1
      IF (M1.LT.MD) GO TO 90
C
C STORE OUTPUT INTO BUFF
C
      BUFF(NN) = SUM
      NN = NN + 1
      IF (MQ.LT.J5) GO TO 80
      K1S = K1S - 1
      IF (K1S.LE.J7) K1S = J7 + IQ1
      IF (MM.LE.ID) GO TO 30
      GO TO 390
C
C TWO STAGE DECIMATION AND/OR INTERPOLATION:
C THIS SECTION OF CODE IS REMOVABLE IF 2 STAGE (KD=2) IMPLEMENTATION
C ARE NOT USED        (SUBSTITUTE        300 GO TO 2000)
C TWO STAGE DECIMATION
C
  100 IF (ITYPE.NE.2) GO TO 110
      SUM = BUFF(MM)
      MM = MM + 1
      GO TO 160
  110 M2 = J2S - ID2
      MQ = J2S + N2P
C
C READ ID1 SAMPLES FROM BUFF INTO S1D BUFFER
C
  120 M1 = J1S - ID1
      MD = J1S + N1P
  130 SBUFF(J1S) = BUFF(MM)
      SBUFF(MD) = BUFF(MM)
      MM = MM + 1
      J1S = J1S - 1
      MD = MD - 1
      IF (M1.LT.J1S) GO TO 130
C
C COMPUTE ONE FILTER OUTPUT FOR STAGE 1D
C
      JD = J1S + NCF1
      JU = J1S + NF1
      SUM = 0.0
      M1 = 0
  140 M1 = M1 + 1
      SUM = SUM + COEF1(M1)*(SBUFF(JD)+SBUFF(JU))
      JD = JD - 1
      JU = JU + 1
      IF (M1.LT.NCF1) GO TO 140
      IF (J1S.LE.J1) J1S = J1 + N1P
C
C STORE DATA INTO S2D
C
      SBUFF(J2S) = SUM
      SBUFF(MQ) = SUM
      J2S = J2S - 1
      MQ = MQ - 1
      IF (M2.LT.J2S) GO TO 120
C
C COMPUTE ONE FILTER OUTPUT FOR STAGE 2D
C
      JD = J2S + NCF2
      JU = J2S + NF2
      SUM = 0.0
      M2 = 0
  150 M2 = M2 + 1
      SUM = SUM + COEF2(M2)*(SBUFF(JD)+SBUFF(JU))
      JD = JD - 1
      JU = JU + 1
      IF (M2.LT.NCF2) GO TO 150
      IF (J2S.LE.J2) J2S = J2 + N2P
      IF (ITYPE.NE.1) GO TO 160
      BUFF(NN) = SUM
      NN = NN + 1
      IF (MM.LE.ID) GO TO 110
      GO TO 390
C
C TWO STAGE INTERPOLATION
C STORE DATA INTO S2I
C
  160 SBUFF(K2S) = SUM
      MD = K2S + IQ2
      SBUFF(MD) = SUM
      IDX = J10 + 1
      MQ = J5
C
C COMPUTE OUTPUT FOR STAGE 2I
C
  170 M2 = K2S
      SUM = 0.0
  180 MQ = MQ + 1
      SUM = SUM + SBUFF(MQ)*SBUFF(M2)
      M2 = M2 + 1
      IF (M2.LT.MD) GO TO 180
C
C STORE OUTPUT IN T2I
C
      SBUFF(IDX) = SUM
      IDX = IDX + 1
      IF (IDX.LE.J11) GO TO 170
      K2S = K2S - 1
      IF (K2S.LE.J8) K2S = J8 + IQ2
      IDX = J10 + 1
C
C STORE DATA INTO S1I
C
  190 SBUFF(K1S) = SBUFF(IDX)
      MD = K1S + IQ1
      SBUFF(MD) = SBUFF(IDX)
      IDX = IDX + 1
      MQ = J4
  200 M1 = K1S
      SUM = 0.0
  210 MQ = MQ + 1
      SUM = SUM + SBUFF(MQ)*SBUFF(M1)
      M1 = M1 + 1
      IF (M1.LT.MD) GO TO 210
      BUFF(NN) = SUM
      NN = NN + 1
      IF (MQ.LT.J5) GO TO 200
      K1S = K1S - 1
      IF (K1S.LE.J7) K1S = J7 + IQ1
      IF (IDX.LE.J11) GO TO 190
      IF (MM.LE.ID) GO TO 100
      GO TO 390
C
C THREE STAGE DECIMATION AND/OR INTERPOLATION
C THIS SECTION OF CODE IS REMOVABLE IF 3 STAGE (KD=3) IMPLEMENTATION
C THREE STAGE DECIMATION
```

```
C
 220  IF (ITYPE.NE.2) GO TO 230
      SUM = BUFF(MM)
      MM = MM + 1
      GO TO 300
 230  M3 = J3S - ID3
      M = J3S + N3P
 240  M2 = J2S - ID2
      MQ = J2S + N2P
C
C READ ID1 SAMPLES FROM BUFF INTO S1D BUFFER
C
 250  M1 = J1S - ID1
      MD = J1S + N1P
 260  SBUFF(J1S) = BUFF(MM)
      SBUFF(MD) = BUFF(MM)
      MM = MM + 1
      J1S = J1S - 1
      MD = MD - 1
      IF (M1.LT.J1S) GO TO 260
C
C COMPUTE ONE FILTER OUTPUT FOR STAGE 1D
C
      JD = J1S + NCF1
      JU = J1S + NF1
      SUM = 0.0
      M1 = 0
 270  M1 = M1 + 1
      SUM = SUM + COEF1(M1)*(SBUFF(JD)+SBUFF(JU))
      JD = JD - 1
      JU = JU + 1
      IF (M1.LT.NCF1) GO TO 270
      IF (J1S.LE.J1) J1S = J1 + N1P
C
C STORE DATA INTO S2D
C
      SBUFF(J2S) = SUM
      SBUFF(MQ) = SUM
      J2S = J2S - 1
      MQ = MQ - 1
      IF (M2.LT.J2S) GO TO 250
C
C COMPUTE ONE FILTER OUTPUT FOR STAGE 2D
C
      JD = J2S + NCF2
      JU = J2S + NF2
      SUM = 0.0
      M2 = 0
 280  M2 = M2 + 1
      SUM = SUM + COEF2(M2)*(SBUFF(JD)+SBUFF(JU))
      JD = JD - 1
      JU = JU + 1
      IF (M2.LT.NCF2) GO TO 280
      IF (J2S.LE.J2) J2S = J2 + N2P
C
C STORE DATA INTO S3D
C
      SBUFF(J3S) = SUM
      SBUFF(M) = SUM
      J3S = J3S - 1
      M = M - 1
      IF (M3.LT.J3S) GO TO 240
C
C COMPUTE ONE FILTER OUTPUT FOR STAGE 3D
C
      JD = J3S + NCF3
      JU = J3S + NF3
      SUM = 0.0
      M3 = 0
 290  M3 = M3 + 1
      SUM = SUM + COEF3(M3)*(SBUFF(JD)+SBUFF(JU))
      JD = JD - 1
      JU = JU + 1
      IF (M3.LT.NCF3) GO TO 290
      IF (J3S.LE.J3) J3S = J3 + N3P
      IF (ITYPE.NE.1) GO TO 300
      BUFF(NN) = SUM
      NN = NN + 1
      IF (MM.LE.ID) GO TO 230
      GO TO 390
C
C THREE STAGE INTERPOLATION
C STORE DATA INTO S3I
C
 300  SBUFF(K3S) = SUM
      MD = K3S + IQ3
      SBUFF(MD) = SUM
      M = J11 + 1
      MQ = J6
C
C COMPUTE OUTPUT FOR STAGE 3I
C
 310  M3 = K3S
      SUM = 0.0
 320  MQ = MQ + 1
      SUM = SUM + SBUFF(MQ)*SBUFF(M3)
      M3 = M3 + 1
      IF (M3.LT.MD) GO TO 320
C
C STORE OUTPUT IN T3I
C
      SBUFF(M) = SUM
      M = M + 1
      IF (M.LE.J12) GO TO 310
      K3S = K3S - 1
      IF (K3S.LE.J9) K3S = J9 + IQ3
      M = J11 + 1
C
C STORE DATA INTO S2I
C
 330  SBUFF(K2S) = SBUFF(M)
      MD = K2S + IQ2
      SBUFF(MD) = SBUFF(M)
      M = M + 1
      IDX = J10 + 1
      MQ = J5
C
C COMPUTE OUTPUT FOR STAGE 2I
C
 340  M2 = K2S
      SUM = 0.0
 350  MQ = MQ + 1
      SUM = SUM + SBUFF(MQ)*SBUFF(M2)
      M2 = M2 + 1
      IF (M2.LT.MD) GO TO 350
C
C STORE OUTPUT IN T2I
C
      SBUFF(IDX) = SUM
      IDX = IDX + 1
```

```
      IF (IDX.LE.J11) GO TO 340
      K2S = K2S - 1
      IF (K2S.LE.J8) K2S = J8 + IQ2
      IDX = J10 + 1
C
C STORE DATA INTO S1I
C
 360  SBUFF(K1S) = SBUFF(IDX)
      MD = K1S + IQ1
      SBUFF(MD) = SBUFF(IDX)
      IDX = IDX + 1
      MQ = J4
C
C COMPUTE OUTPUT FOR STAGE 1I
C
 370  M1 = K1S
      SUM = 0.0
 380  MQ = MQ + 1
      SUM = SUM + SBUFF(MQ)*SBUFF(M1)
      M1 = M1 + 1
      IF (M1.LT.MD) GO TO 380
C
C STORE OUTPUT IN BUFF
C
      BUFF(NN) = SUM
      NN = NN + 1
      IF (MQ.LT.J5) GO TO 370
      K1S = K1S - 1
      IF (K1S.LE.J7) K1S = J7 + IQ1
      IF (IDX.LE.J11) GO TO 360
      IF (M.LE.J12) GO TO 330
      IF (MM.LE.ID) GO TO 220
 390  RETURN
      END
```

Reviewers

The successful completion of this program book represents the culmination of a joint effort on the part of many contributors. The purpose of this section is to acknowledge the extensive efforts expended by the reviewers. Their comprehensive testing contributed to the high quality of the programs included in the book.

The following is a complete list of program titles, authors and their affiliations, reviewers and their affiliations, as well as computers, operating systems, and FORTRAN compilers used by the reviewers.

1.1 FOUREA - A Short Demonstration Version of the FFT
C. M. Rader
M.I.T. Lincoln Laboratory, Lexington, MA 02173
Reviewed by M. T. Dolan
Bell Laboratories, Murray Hill, NJ 07974
Honeywell 6080N computer
GCOS
FORTRAN Y compiler

1.2 Fast Fourier Transform Algorithms
G. D. Bergland and M. T. Dolan
Bell Laboratories, Murray Hill, NJ 07974
Reviewed by L. R. Morris
Carleton University, Ottawa, Canada
PDP-11/55, 11/60 computers
RSX-11M
FORTRAN IV PLUS compiler
RT-11
FORTRAN IV compiler

1.3 FFT Subroutines for Sequences With Special Properties
L. R. Rabiner
Bell Laboratories, Murray Hill, NJ 07974
Reviewed by M. T. Dolan
Bell Laboratories, Murray Hill, NJ 07974
Honeywell 6080N computer
GCOS
FORTRAN Y compiler

1.4 Mixed Radix Fast Fourier Transforms
R. C. Singleton
SRI International, Menlo Park, CA 94025
Reviewed by M. T. Dolan
Bell Laboratories, Murray Hill, NJ 07974
Honeywell 6080N computer
GCOS
FORTRAN Y compiler

1.5 Optimized Mass Storage FFT Program
D. Fraser
CSIRO, Canberra City, Australia
Reviewed by N. M. Brenner
IBM, Yorktown Heights, NY 10598
IBM 370/168 computer
VM/370
FORTRAN IV compiler
Reviewed by C. A. McGonegal, L. R. Rabiner, and M. T. Dolan
Bell Laboratories, Murray Hill, NJ 07974
Data General Eclipse S/230 computer
RDOS
FORTRAN V compiler
Honeywell 6080N computer
GCOS
FORTRAN Y compiler

1.6 Chirp z-Transform Algorithm Program
L. R. Rabiner
Bell Laboratories, Murray Hill, NJ 07974
Reviewed by R. M. Mersereau
Georgia Institute of Technology, Atlanta, GA 30332
Data General Eclipse S/230 computer
RDOS
FORTRAN 5 compiler (Rev. 6)

1.7 Complex General-N Winograd Fourier Transform Algorithm (WFTA)
J. H. McClellan and H. Nawab
M.I.T., Cambridge, MA 02139
Reviewed by D. Kolba
Rice University, Houston, TX 02881
IBM 370/155 computer
TSO
FORTRAN G1 compiler

1.8 Time-Efficient Radix-4 Fast Fourier Transform
L. R. Morris
Carleton University, Ottawa, Canada
Reviewed by C. M. Rader
M.I.T. Lincoln Laboratory, Lexington, MA 02173
IBM 370/168 computer
CMS
FORTRAN IV compiler

1.9 Two-Dimensional Mixed Radix Mass Storage Fourier Transform
R. C. Singleton
SRI International, Menlo Park, CA 94025
Reviewed by M. T. Dolan
Bell Laboratories, Murray Hill, NJ 07974
Honeywell 6080N computer
GCOS
FORTRAN Y compiler

2.1 Periodogram Method for Power Spectrum Estimation
L. R. Rabiner
Bell Laboratories, Murray Hill, NJ 07974
R. W. Schafer and D. Dlugos
Georgia Institute of Technology, Atlanta, GA 30332
Reviewed by D. E. Dudgeon
M.I.T. Lincoln Laboratory, Lexington, MA 02173
PDP11/34 computer
RT-11 version 3
FORTRAN IV compiler (version 2.04)

2.2 Correlation Method for Power Spectrum Estimation
L. R. Rabiner
Bell Laboratories, Murray Hill, NJ 07974
R. W. Schafer and D. Dlugos
Georgia Institute of Technology, Atlanta, GA 30332
Reviewed by D. E. Dudgeon
M.I.T. Lincoln Laboratory, Lexington, MA 02173
PDP11/34 computer
RT-11 version 3
FORTRAN IV compiler (version 2.04)

2.3 A Coherence and Cross Spectral Estimation Program
C. Clifford Carter and James F. Ferrie
Naval Underwater Systems Center, New London, CT 06320
Reviewed by Jont B. Allen
Bell Laboratories, Murray Hill, NJ 07974
Data General Eclipse S/230 computer
RDOS
FORTRAN V compiler

3.1 FASTFILT - An FFT Based Filtering Program
Jont B. Allen
Bell Laboratories, Murray Hill, NJ 07974
Reviewed by A. E. Filip
M.I.T. Lincoln Laboratory, Lexington, MA 02173
IBM 370/168 computer
CMS
FORTRAN IV compiler

4.1 Linear Prediction Analysis Programs (AUTO-COVAR)
A. H. Gray Jr.
University of California, Santa Barbara, CA 93106
J. D. Markel
Signal Technology Inc., Santa Barbara, CA 93101
Reviewed by Victor W. Zue
M.I.T., Cambridge, MA 02139
DEC System 20 computer
TOPS-20
Standard FORTRAN compiler

4.2 Efficient Lattice Methods for Linear Prediction
R. Viswanathan and J. Makhoul
Bolt, Beranek and Newman, Inc., Cambridge, MA 02138
Reviewed by Lloyd J. Griffiths
University of Colorado, Boulder, CO 80309
CDC 6400 computer
KRONOS
RUN, FTN, and MNF FORTRAN IV compilers

4.3 Linear Predictor Coefficient Transformations Subroutine LPTRN
A. H. Gray Jr.
University of California, Santa Barbara, CA 93106
J. D. Markel
Signal Technology Inc., Santa Barbara, CA 93101
Reviewed by Steven F. Boll and George Randall
University of Utah, Salt Lake City, Utah 84112
PDP11/45 computer
RSX11M
FORTRAN V3 compiler

5.1 FIR Linear Phase Filter Design Program
J. H. McClellan
M.I.T., Cambridge, MA 02139
T. W. Parks
Rice University, Houston, TX 77001
L. R. Rabiner
Bell Laboratories, Murray Hill, NJ 07974
Reviewed by Michael T. McCallig
Sperry Research Center, Sudbury, MA 01776
UNIVAC 1106 computer
EXEC8 version 33/R3
FORTRAN V compiler

5.2 FIR Windowed Filter Design Program - WINDOW
L. R. Rabiner and C. A. McGonegal
Bell Laboratories, Murray Hill, NJ 07974
D. B. Paul
M.I.T. Lincoln Laboratory, Lexington, MA 02173
Reviewed by D. B. Paul
M.I.T. Lincoln Laboratory, Lexington, MA 02173
PDP11/45 computer
UNIX
Princeton FORTRAN compiler

5.3 Design Subroutine (MXFLAT) for Symmetric FIR Low Pass Digital Filters with Maximally-Flat Pass and Stop Bands
J. F. Kaiser
Bell Laboratories, Murray Hill, NJ 07974
Reviewed by D. B. Paul
M.I.T. Lincoln Laboratory, Lexington, MA 02173
PDP11/45 computer
UNIX
Princeton FORTRAN compiler

5.4 A Subroutine for Finite Wordlength FIR Filter Design
Ulrich Heute

Institut fuer Nachrichtentechnik, Universitaet Erlangen-Nuernberg
Cauerstrasse 7, D-8520 Erlangen, Germany
Reviewed by Steven L. Wood
University of Rhode Island, Kingston, RI 02881
IBM 370 computer
ITEL 5
FORTRAN Level G compiler
Reviewed by J. F. Kaiser and M. T. Dolan
Bell Laboratories, Murray Hill, NJ 07974
Honeywell 6080N computer
GCOS
FORTRAN Y compiler

6.1 Program for the Design of Recursive Digital Filters
Gunter F. Dehner
Institut fuer Nachrichtentechnik, Universitaet Erlangen-Nuernberg
Cauerstrasse 7, D-8520 Erlangen, Germany
Reviewed by R. M. Mersereau
Georgia Institute of Technology, Atlanta, GA 30332
Data General Nova 830 computer
RDOS
FORTRAN 5 compiler
Reviewed by J. W. Woods and J. F. McDonald
Rensselaer Polytechnic Institute, Troy, NY 12181
PRIME 500 computer
PRIMOS
FORTRAN IV compiler
Reviewed by J. F. Kaiser and M. T. Dolan
Bell Laboratories, Murray Hill, NJ 07974
Honeywell 6080N computer
GCOS
FORTRAN Y compiler
Reviewed by T. W. Parks
Rice University, Houston, TX 77001
IBM 370/155 computer
TSO
FORTRAN G1 compiler
Reviewed by Horacio G. Martinez
Instituto de Ingenieria, Mexico 20, D.F., Mexico
IBM 370/155 computer at Rice University
TSO
FORTRAN G1 compiler

6.2 Program for Minimum-p Synthesis of Recursive Digital Filters
A. G. Deczky
Institut de Microtechnique, Universite de Neuchatel
Rue Pierre-a-Mazel 7, 2000 Neuchatel, Switzerland
Reviewed by Kenneth Steiglitz
Princeton University, Princeton, NJ 08540
IBM 360/91 computer
OS/360
FORTRAN G compiler

6.3 An Optimization Program for the Design of
Digital Filter Transfer Functions
M. T. Dolan and J. F. Kaiser
Bell Laboratories, Murray Hill, NJ 07974
Reviewed by Horacio G. Martinez
Instituto de Ingenieria, Mexico 20, D.F., Mexico
Burroughs B6700 computer
MCP
B6700/B7700 FORTRAN compiler

6.4 A Program for Designing Finite Word-Length IIR Digital Filters
Kenneth Steiglitz and Bruce D. Ladendorf
Princeton University, Princeton, NJ 08540
Reviewed by John W. Woods
Rensselaer Polytechnic Institute, Troy, NY 12108
IBM 370 Model 3033 computer
MTD
G and H FORTRAN compilers

7.1 Computation of the Complex Cepstrum
J. M. Tribolet
Instituto Superior Tecnico, Lisbon, Portugal
T. F. Quatieri
M.I.T., Cambridge, MA 02139
Reviewed by T. Ulrych
University of British Columbia, Vancouver, Canada
IBM 370/168 computer
MTS
IBM FORTRAN G compiler

7.2 Computation of the Real Cepstrum and Minimum-Phase Reconstruction
T. F. Quatieri
M.I.T., Cambridge, MA 02139
J. M. Tribolet
Instituto Superior Tecnico, Lisbon, Portugal
Reviewed by T. Ulrych
University of British Columbia, Vancouver, Canada
IBM 370/168 computer
MTS
IBM FORTRAN G compiler

8.1 A Computer Program for Digital Interpolator Design
T. W. Parks
Rice University, Houston, TX 77001
G. Oetken and H. W. Schuessler
Institut fuer Nachrichtentechnik, Universitaet Erlangen-Nuernberg
Cauerstrasse 7, D-8520 Erlangen, Germany
Reviewed by J. M. Tribolet
Instituto Superior Tecnico, Lisbon, Portugal
Data General Eclipse S/230 computer
RDOS
FORTRAN V compiler

8.2 A General Program to Perform Sampling Rate Conversion of Data by Rational Ratios
R. E. Crochiere
Bell Laboratories, Murray Hill, NJ 07974
Reviewed by Don H. Johnson
Rice University, Houston, TX 77001
IBM 370/155 computer
TSO
FORTRAN G1 compiler

8.3 A Program for Multistage Decimation, Interpolation, and Narrow Band Filtering
R. E. Crochiere and L. R. Rabiner
Bell Laboratories, Murray Hill, NJ 07974
Reviewed by Maurice G. Bellanger
Telecommunications, Radioelectriques et Telephoniques
5 Ave. Reaumur, 92350 - Plesis, Robinson, France
Philips P880 computer
ANSI FORTRAN IV compiler